DISEASES OF TREES AND SHRUBS

Second Edition

DISEASES OF TREES AND SHRUBS

SECOND EDITION

WAYNE A. SINCLAIR and
HOWARD H. LYON

COMSTOCK PUBLISHING ASSOCIATES, a division of
CORNELL UNIVERSITY PRESS | ITHACA AND LONDON

First published 2005 by Cornell University Press

Printed in China

Library of Congress Cataloging-in-Publication Data

Sinclair, Wayne A., 1936-
  Diseases of trees and shrubs / Wayne A. Sinclair and Howard H. Lyon.–2nd ed.
    p. cm.
  Includes bibliographical references and index.
  ISBN-13: 978-0-8014-4371-8 (cloth : alk. paper)
  ISBN-10: 0-8014-4371-7 (cloth : alk. paper)
  1. Trees–Diseases and pests–United States. 2. Shrubs–Diseases and pests–United States. 3. Trees–Diseases and pests–Canada. 4. Shrubs–Diseases and pests–Canada. 5. Trees–Wounds and injuries–United States. 6. Shrubs–Wounds and injuries–United States. 7. Trees–Wounds and injuries–Canada. 8. Shrubs–Wounds and injuries–Canada. I. Lyon, Howard H. II. Title.
  SB762.S56 2005
  634.9′6–dc22
                                                                          2005012282

Cornell University Press strives to use environmentally responsible suppliers and materials to the fullest extent possible in the publishing of its books. Such materials include vegetable-based, low-VOC inks and acid-free papers that are recycled, totally chlorine-free, or partly composed of nonwood fibers. For further information, visit our website at www.cornellpress.cornell.edu.

Cloth printing   10 9 8 7 6 5 4 3 2 1

Arrangement of subjects

Contents

vii

Diseases of Trees and Shrubs, 2nd ed., is primarily a survey of diseases of forest and shade trees and woody ornamental plants of the temperate zone in the USA and Canada. Illustrations and textual information have nearly equal emphasis. All major and many lesser categories of infectious and noninfectious diseases of trees and shrubs are discussed. Disorders caused by more than 470 pathogens and 50 chemical, physical, or unknown agents to plants of more than 300 species are portrayed, and hundreds of other diseases and pathogens related to those shown are noted. The book is designed to serve as a diagnostic aid, as an authoritative reference to the diseases and pathogens that are illustrated, and as a guide to further information; however, it is neither encyclopedic nor designed to serve as a stand-alone diagnostic reference. Many diseases of trees and shrubs are omitted, and the symptom descriptions and host lists for many diseases and pathogens, respectively, are synoptic. Diseases and injuries caused by insects and mites are discussed only as they relate to diseases caused by other agents.

The descriptions and discussions are written to accommodate readers with general knowledge of biology as well as those with training in plant pathology and related fields. The discussions emphasize biology and ecology. Scientific terms other than Latin names are used only as necessary for precise communication. Information about disease control (biological and cultural, but usually not chemical) and resistant plants, when available, is included if space is sufficient. A glossary is provided.

A great deal of information relevant to pathology of woody plants has been published since the first edition was compiled, and several new diseases have appeared. This edition updates information previously presented and adds numerous new topics and images. Major advances have been made in taxonomy and nomenclature of plant-pathogenic organisms, and the text reflects them. Latin names of plants are used and common names are given parenthetically, once per species per page, where space permits. Color plates are numbered serially. Halftone photos and line drawings illustrating microscopic features of pathogens are called figures and have the same numbers as the plates they face. All of the plates and figures have been constructed from digital files and edited electronically.

Many more fungal diseases of trees and shrubs are known than diseases caused by other biotic agents, and the book reflects this imbalance: there are 174 plates for diseases caused by fungi and 53 for all other biotic agents. Coverage of disorders caused by environmental or unknown factors is much less extensive (34 plates) and does not reflect their frequency or importance.

The plates show symptoms and signs visible with the unaided eye or with a hand lens. Microscopic details are described in the text only as necessary for practical understanding and diagnostic assistance. For each image, the place (state or province) and month or season of photography—or, for specimens photographed after collection, the place and month of collection—are indicated. Postal codes, listed on page x, are used for province and state abbreviations. Places and months are indicated both to reveal where observations and photos were made and because some woody plants and disease symptoms and signs vary in appearance from one region and season to another.

Many colleagues generously contributed color or black-and-white images. Contributors are identified by name below the appropriate plate captions or in the figure captions and more fully in the Acknowledgments. Color images not credited are owned by the authors or by Cornell University. Together they represent observations in more than 50 states and provinces. Most of the line drawings are copied or adapted from published sources that are credited with the figures.

Names and classification of organisms and viruses. Scientific names are the only ones available for most microorganisms. Names of fungi used herein are those currently accepted in *Index Fungorum* (ref. 597) and/or the USDA-ARS Systematic Botany and Mycology Databases (ref. 4108) or are adopted from recent research publications. Synonyms (abbreviated *syn.*) are provided for many pathogen names. Authorities for names are available in databases

and the references. Relationships among pathogens are revealed by indicating the order and family in which each is classified. The classification of fungi is that used in *Index Fungorum*, which updates the classification in *Ainsworth & Bisby's Dictionary of the Fungi*, ninth edition (ref. 2077). Bacterial names and classification used herein are as sanctioned in *Approved Lists of Bacterial Names* (ref. 3683) or in the *International Journal of Systematic and Evolutionary Microbiology*. Virus names, abbreviations, and classification used herein are as presented in *Plant Viruses Online: Descriptions and Lists from the VIDE Database* (ref. 555) or in recent research papers. Formal and vernacular names of viruses are used as specified in *Virus Taxonomy: Classification and Nomenclature of Viruses* (ref. 3247). Names of other microorganisms are adopted from recent research publications.

Plant names are those adopted in the USDA-NCRS PLANTS database (ref. 4110), the USDA-ARS *GRIN Taxonomy* database (ref. 4109), and/or the Missouri Botanical Garden's VAST (VAScular Tropicos) database (ref. 2676). Common names of plants are as given in the USDA databases noted above or in *Scientific and Common Names of 7000 Vascular Plants in the United States* (ref. 475). A two-word common name or part of one is hyphenated if the plant is not of the genus or species most commonly associated with the noun in the name (e.g., European mountain-ash) or if the name could be misconstrued without a hyphen (e.g., lead-tree). Common names are listed and cross-referenced to Latin names in the index. Most plant cultivar names are as listed in Dirr's *Manual of Woody Landscape Plants* (ref. 965). Cultivar names are set in single quotation marks unless they follow the word *cultivar* or the abbreviation *cv.*

Binomial Latin names are spelled out when they are first used in a text article. Thereafter, a genus name is usually shortened to its first letter until another genus name beginning with the same letter appears. *Plasmopara viburni*, for example, would be abbreviated *P. viburni* after its first appearance, but the name would again be spelled in full after the appearance of a genus name such as *Peronospora*.

Plant lists. Host ranges of pathogens and lists of plants that resist pathogens or tolerate environmental stresses are given as space permits. This information is incomplete because we lacked the resources and space to provide comprehensive lists. In lists of plant genera, if only one species of a genus has the characteristic under discussion, then a common name of the species rather than of the genus is often provided parenthetically. Pathogen-host distribution records from Canada and the USA were readily available to us. Information from Mexico is fragmentary but was included as encountered. Space sufficient for a list of affected plants was unavailable for some worthy disease topics. We addressed the missing-list problem in a few instances by providing index entries under appropriate plant genera or species as if the lists were present in the text. Readers who need more extensive host-pathogen data are directed to references listed for each disease; to the indices by Farr et al., Conners, and Ginns (refs. 1099, 779, and 1298, respectively); and to the fungal databases maintained by the USDA-ARS Systematic Botany and Mycology Laboratory (ref. 4108).

References. All are numbered serially in the bibliography near the end of the book. Appropriate numbers follow each discussion. Reference lists corresponding to the numbers after each discussion are on the disc provided with the book. The references contain source and additional information, illustrations of diseases and injuries, descriptions and microscopic illustrations of pathogens, and information about other diseases related to those discussed. For many topics, only a small fraction of the relevant references are listed. See How To Use This Book, page x, for information about listings of leaflets and book chapters.

Index. Information is arranged primarily under the Latin names of the plants and pathogens. Common names are cross-indexed to Latin names and are given parenthetically with Latin names. If multiple species of a plant genus would have identical entries for a particular disease or pathogen, then the subject is often indexed only for the plant genus. Illustrations related to listed subjects are indicated by boldface page numbers.

Wayne A. Sinclair
Howard H. Lyon

Ithaca, New York

As a Diagnostic Aid

Enter the index at the genus name (e.g., *Abies*) and also at the species name (e.g., *Abies balsamea*) of the affected plant, and there look for names of symptoms, diseases, or pathogens that may be relevant. It is necessary to search under both genus and species names because index entries for diseases are not duplicated for every susceptible plant species within a genus. If a plant species is not indexed, relevant information about its diseases may nonetheless be found under the genus name. All genus names are printed in **boldface** at first appearance.

If the Latin name of a plant genus or species is not known to the user, then enter the index at the common name of the genus or species. For the example above, "fir" is cross-indexed to *Abies*, and "fir, balsam" to *Abies balsamea*. Usually only one common name is indexed per species. Additional common names can be found in horticultural references, in *GRIN Taxonomy* (ref. 4109, searchable using either common or Latin names), and also in *Scientific and Common Names of 7000 Vascular Plants in the United States* (ref. 475).

Continuing the example, if a disorder of *A. balsamea* is characterized by cankers and dieback, then both symptoms may be indexed and the user will be led to diagnostic possibilities such as Cytospora canker and Phomopsis canker, as well as to names of pathogens. Commonly used names of many diseases are indexed. If multiple pathogens in the same genus affect a plant, then the index may mention only the pathogen genus even though two or more species are discussed in the text. Page numbers in boldface type indicate the presence of illustrations.

References that are listed after each discussion contain additional information and illustrations. Chapters in books are usually not listed separately. Some reference list entries refer to large sets of leaflets, notably *IMI Descriptions of Pathogenic Fungi and Bacteria* (ref. 1859), rather than to individual publications. Individual publications in such sets can be found via their indices or in electronic databases.

As an Instructional Aid

Topics are arranged according to type of pathogen and secondarily according to plant parts affected and types of symptoms. An overview of each major disease type or pathogen group is provided, and relevant concepts are included. Many disease cycles are discussed explicitly.

To locate information about disease or pathogen groups, enter the Arrangement and Contents and look first under Arrangement of Subjects. After locating the category of interest, proceed to the Contents, which is an expanded version of Arrangement of Subjects.

Specific subjects can be located via the index. Items indexed as instructional aids include illustrated examples of various fungal structures; mycological and plant-pathological terms; key words for concepts; and references to chemicals, structures, and processes that are important in pathogenesis and resistance.

Geographic Abbreviations

AB, Alberta; AK, Alaska; AL, Alabama; AR, Arkansas; AZ, Arizona; BC, British Columbia; CA, California; CO, Colorado; CT, Connecticut; DC, District of Columbia; DE, Delaware; FL, Florida; GA, Georgia; IA, Iowa; ID, Idaho; IL, Illinois; IN, Indiana; KS, Kansas; KY, Kentucky; LA, Louisiana; MA, Massachusetts; MB, Manitoba; MD, Maryland; ME, Maine; MI, Michigan; MN, Minnesota; MO, Missouri; MS, Mississippi; MT, Montana; NA, North America; NB, New Brunswick; NC, North Carolina; ND, North Dakota; NE, Nebraska; NH, New Hampshire; NJ, New Jersey; NM, New Mexico; NS, Nova Scotia; NV, Nevada; NY, New York; OH, Ohio; OK, Oklahoma; ON, Ontario; OR, Oregon; PA, Pennsylvania; PR, Puerto Rico; QC, Quebec; RI, Rhode Island; SC, South Carolina; SD, South Dakota; SK, Saskatchewan; TN, Tennessee; TX, Texas; UK, United Kingdom; USA, United States of America; UT, Utah; VA, Virginia; VT, Vermont; WA, Washington; WI, Wisconsin; WV, West Virginia; WY, Wyoming.

This book is dedicated to the memory of our colleague, Prof. Warren T. Johnson, who conceived the project that resulted in the first edition and who authored the companion volume, *Insects That Feed on Trees and Shrubs*. Prof. Johnson made many of the photographs that appear in this book. The book is further dedicated to the thousands of plant pathologists, mycologists, and other scientists whose observations, research, scholarship, and insights are represented herein or provided essential background for our interpretations.

Preparation of both editions of this book was facilitated by the New York State College of Agriculture and Life Sciences at Cornell University through its Department of Plant Pathology. We gratefully acknowledge the support and academic environment that made the project feasible. The resources of the Albert R. Mann Library were invaluable. Photo Specialist Kent Loeffler of the Department of Plant Pathology provided advice and major help for the second edition by making many studio photos and transferring many hundreds of images from negatives and transparencies to digital format.

Special acknowledgment goes to arborist George Callaway, whose financial support and unflagging interest in the project were unique stimuli.

The following colleagues contributed time, knowledge, and materials that aided us in fieldwork and photography for the first edition. The locations and organizational affiliations indicated were in effect when assistance was given.
Canada: H. L. Gross, D. T. Myren, and R. D. Whitney of the Canadian Forest Service, Sault Ste. Marie, ON; G. B. Ouellette of the Canadian Forest Service, St. Foy, QC; S. Navratil of Lakehead University, Thunder Bay, ON.
Arizona: S. M. Alcorn, H. E. Bloss, and R. L. Gilbertson of the University of Arizona, Tucson; F. H. Mahr of the University of Arizona, Mesa.
California: F. W. Cobb Jr., J. W. Kloepper, and R. D. Raabe of the University of California, Berkeley; G. Nyland of the University of California, Davis; J. Ryder, apple grower, of Watsonville; R. Sanborn of Cooperative Extension, Contra Costa County.
Colorado: F. G. Hawksworth, T. E. Hinds, and J. M. Staley of the USDA Forest Service, Fort Collins; J. G. Laut of the Colorado Forest Service, Fort Collins.
Connecticut and New York: D. R. Houston and P. M. Wargo of the USDA Forest Service, Hamden, CT.
District of Columbia: R. Hammerschlag and H. V. Wester of the USDA National Park Service.
Florida: S. A. Alfieri Jr., J. C. Denmark, and R. P. Esser of the Department of Agriculture and Consumer Services, Division of Plant Industry, Gainesville; E. W. Barnard of the Division of Forestry, Gainesville; G. M. Blakeslee and R. E. McCoy of the University of Florida, Gainesville and Fort Lauderdale, respectively.
Georgia: W. C. Bryan and W. A. Campbell, formerly of the USDA Forest Service, Athens; F. F. Hendrix Jr. and W. Wynn of the University of Georgia, Athens.
Idaho: A. D. Partridge of the University of Idaho, Moscow.
Illinois: E. B. Himelick of the Illinois Natural History Survey, Urbana; D. Stenger of the Sinnissippi Forest, Oregon.
Louisiana: G. E. Holcomb of Louisiana State University, Baton Rouge; A. L. Welden of Tulane University, New Orleans.
Michigan: J. H. Hart of Michigan State University, East Lansing.
Minnesota: R. A. Blanchette and D. W. French of the University of Minnesota, St. Paul.
Mississippi: D. J. Blasingame of Mississippi State University, State College; T. H. Filer Jr., R. Lewis Jr., and F. I. McCracken of the USDA Forest Service, Stoneville; A. G. Kais of the USDA Forest Service, Gulfport.
Missouri: V. Dropkin of the University of Missouri, Columbia; A. Foudin of the USDA Animal and Plant Health Inspection Service, Columbia.
Nebraska: G. W. Peterson and J. W. Riffle of the USDA Forest Service, Lincoln.
New Hampshire: A. L. Shigo of the USDA Forest Service, Durham.
New York: G. Callaway, arborist, of Argyle; J. D. Castello, P. D. Manion, and R. A. Zabel of the State University College of Environmental Science and Forestry, Syracuse; G. W. Hudler of Cornell University, Ithaca; C. F. Scheer of Suffolk County Cooperative Extension.
North Carolina: R. I. Bruck of North Carolina State University, Raleigh.
North Dakota and northeastern states: R. W. Stack of North Dakota State University, Fargo; formerly of Cornell University, Ithaca, NY.
Oklahoma: G. Simmons of the Samuel Roberts Noble Foundation, Ardmore.
Oregon: H. R. Cameron, E. M. Hansen, and L. F. Roth of Oregon State University, Corvallis.

Pennsylvania: W. Merrill and F. A. Wood of The Pennsylvania State University, University Park; B. Towers of the Department of Environmental Resources, Middletown.
South Carolina: O. W. Barnett of Clemson University, Clemson.
Tennessee: C. Hadden of the University of Tennessee, Knoxville.
Texas: J. Amador of Texas A&M University, Weslaco; C. W. Horne and D. Rosburgh of Texas A&M University, College Station.
Virginia: S. A. Alexander, R. J. Stipes, and R. L. Wick of Virginia Polytechnic Institute and State University, Blacksburg.
Washington: K. W. Russell of the Department of Natural Resources, Olympia; L. F. Roth of Oregon State University, Corvallis; J. M. Staley of the USDA Forest Service, Fort Collins, CO.

The present edition includes images from many sources. Their makers or contributors are gratefully acknowledged. Ownership of contributed images, other than those in public domain, remains with the contributors. The organizational affiliations indicated below were those in effect when photos or drawings were made or assistance was given. Numbers refer to plates unless preceded by "Fig."
E. A. Allen, Canadian Forest Service, Victoria, BC, for 29E, 106G, 117G, 166G, 168E
R. L. Anderson, USDA Forest Service, Asheville, NC, for 9C
P. A. Angwin, USDA Forest Service, Redding, CA, for 170K
C. L. Ash, University of Minnesota St. Paul, for 121A–D
E. L. Barnard, Florida Division of Forestry, Gainesville, for 109G–J, 123I, 127A–E, 144F, G, 150E, 176G–J
O. W. Barnett, North Carolina State University, Raleigh, for 203K, L, 206A, 207F, G, 213F
D. M. Benson, North Carolina State University, Raleigh, for 34G
D. R. Bergdahl, University of Vermont, Burlington, for 57I, 149H
A. L. Bishop, Yoder Brothers, Inc., Alva, FL, for 183C, D, G, H, 185I, J
G. M. Blakeslee, University of Florida, Gainesville, for 167C, D
M. Blua, University of California, Riverside, for 192H–J
E. J. Braun, Cornell University, Ithaca, NY, for Fig. 194
M. F. Brown, University of Missouri, Columbia, for Fig. 2
M. Brundrett, University of Western Australia, Crawley, W. Australia, for 259L
C. L. Burlock, Canadian Forest Service, Fredericton, NB, for 107A
B. E. Callan, Canadian Forest Service, Victoria, BC, for Figs. 44A–D, 168A, B
K. Camilli, Texas Forest Service, Austin, and D. N. Appel, Texas A&M University, College Station, for 226D
Canadian Forest Service Great Lakes Forestry Centre, Sault Ste. Marie, ON, for 24H, I, 32K, 105C, 112A, C, 143G, 168A
Canadian Forest Service Laurentian Forestry Centre, St. Foy, QC, for 30A, B
Canadian Forest Service Pacific Forestry Centre, Victoria, BC, for 149D
J. D. Castello, State University College of Environmental Science and Forestry, Syracuse, NY, for 202A, 207H
A. R. Chase, Chase Research Gardens Inc., Mt. Aukum, CA, for 184I, J, 185A, B
M. N. Cline, Monsanto Co., St. Louis, MO, for Fig. 36
C. E. Cordell, USDA Forest Service, for 108C
D. L. Coyier, Oregon State University, Corvallis, for 209A, B
M. L. Daughtrey, Cornell University, Riverhead, NY, for 52A, C–F, 120A, B, I, K
J. M. Davidson, University of California, Davis, for 178K, M
D. R. Deal, Cornell University, Ithaca, NY, for 259M
S. H. De Boer, Agriculture Canada, Charlottetown, PEI, for 184C
P. Dmytrasz, Urban Forestry Services, Toronto, ON, for 171B
C. E. Dorworth, Canadian Forest Service, Victoria, BC, for 219D, E
L. D. Dwinell, USDA Forest Service, Athens, GA, for 147F, H
A. W. Engelhard, University of Florida Gulf Coast Research & Education Center, Bradenton, for 8I
L. Englander, University of Rhode Island, Kingston, for Fig. 179B
H. G. Eno, USDA Forest Service, Hamden, CT, for 162H
A. H. Epstein, Iowa State University, Ames, for 208E–H
P. Fenn, University of Arkansas, Fayetteville, for 99B, C, 195C, D
S. J. Frankel, USDA Forest Service, Vallejo, CA, for 178A
D. W. French, University of Minnesota, St. Paul, for 118I, J
M. Garbelotto, University of California, Berkeley, for 178 F, J
R. Garbutt, Canadian Forest Service, Victoria, BC, for 112B
D. J. Goheen, USDA Forest Service, Central Point, OR, for 180C–F
E. Goheen, USDA Forest Service, Central Point, OR, for 114B
T. R. Gordon, University of California, Davis, for 93C
T. R. Gottwald, USDA Agricultural Research Service, Orlando, FL, for 185D, E, G, H
J. H. Graham, University of Florida, Lake Alfred, for 179F
R. J. Green, Purdue University, West Lafayette, IN, for 181G, 195H

M. Gryzenhout, University of Pretoria, South Africa, for Fig. 80
A. K. Hagan, Auburn University, Auburn, AL, for 186A, B
K. J. Harrison, Canadian Forest Service, Fredericton, NB, for 58D, I, J, 107C
J. R. Hartman, University of Kentucky, Lexington, for 3B, 127F
E. Hecht-Poinar, University of California, Berkeley, for 60G
C. R. Hibben, Brooklyn Botanic Garden, Ossining, NY, for 52J, K, 196D
K. D. Hickey, Pennsylvania State University, Biglerville, PA, for 201A, C, H
J. Hopkins, Canadian Forest Service, Victoria, BC, for 25H, 66G, 106F, H
D. R. Houston, formerly of USDA Forest Service, Hamden, CT, for 91H
G. W. Hudler, Cornell University, Ithaca, NY, for 32A–D, 117A, 119B, 129E, 139G, 154H, 177A–G, 234F, 241B, C, 257E; also Figs. 32A, B, and 178B–E
J. D. Janse, Plant Protection Service, Wageningen, Netherlands, for 189I
S. N. Jeffers, Cornell University, Ithaca, NY, for 179G, H
A. L. Jones, Michigan State University, East Lansing, for 199I, M
R. K. Jones, North Carolina State University, Raleigh, for 34I, 35F
K. Julin, Marin County (CA) Fire Department, for 178D, E
W. B. Kendrick, University of Waterloo, ON, for part of Fig. 43B
K. J. Kessler, USDA Forest Service, Carbondale, IL, for 9H–J, 51C
K. S. Kim, University of Arkansas, Fayetteville, for 211D, 213A–C
J. T. Kliejunas, USDA Forest Service, Vallejo, CA, for 108A
E. S. Kondo, Canadian Forest Service, Sault Ste. Marie, ON, for 168G
J. C. Krug, University of Toronto, ON, for 171A
G. Laflamme, Canadian Forest Service Laurentian Forestry Centre, St. Foy, QC, for 30E–G
J. A. Laurence, Boyce Thompson Institute, Ithaca, NY, for 241G
T. D. Leininger, USDA Forest Service, Stoneville, MS, for 63H
S. E. Lindow, University of California, Berkeley, for 186G
M. J. Linit, University of Missouri, Columbia, for 216E, F
K. E. Loeffler, Cornell University, Ithaca, NY, for 178C, I
R. F. Lucas, Cornell University, Ithaca, NY, for 259C, N
P. T. Marshall, Indiana Department of Natural Resources, Vallonia, for 9G
E. Masters and USDA Forest Service for 236D
J. A. Matteoni, Cornell University, Ithaca, NY, for 51E, 125E 157C
W. Merrill, Pennsylvania State University, University Park, for 23B, 24C, 26G, H, 27F, 28F, G, 71C–G, 139B
J.-M. Moncalvo, University of Toronto, ON, for 173E, F
D. J. Morrison, Canadian Forest Service, Victoria, BC, for 163H, 166A–C
J. M. Mullen, Auburn University, Auburn, AL, for 63B, 184E, F
G. P. Munkvold, University of Illinois, Champaign-Urbana, for 21A–F and Fig. 21A
T. R. Nag Raj, University of Waterloo, ON, for permission to adapt his published illustrations for Figures 38A, 94A, B, 95A–D
M. E. Ostry, USDA Forest Service, St. Paul, MN, for 11A–E, 40C, 76D; also Fig. 46A and part of B
J. N. Pinkerton, USDA Agricultural Research Service, Corvallis, OR, for 78A
B. C. Raju, University of California, Davis, for 194C, H
D. C. Ramsdell, Michigan State University, East Lansing, for 201B, E
S. C. Redlin of USDA Animal and Plant Health Inspection Sevice, Raleigh, NC, for 53D, E, 77H; also Figs. 48A–E, 52A–D, 53A–D, 77F, G
R. Reich, Canadian Forest Service, Victoria, BC, for 168D
D. F. Ritchie, North Carolina State University, Raleigh, for 186C, E
D. M. Rizzo, University of California, Davis, for 178G, H, L; also Fig. 179A
D. A. Rosenberger, Cornell University, Highland, NY, for 129H–L, 199J, L
C. L. Schoulties, Florida Division of Agriculture & Consumer Services, Gainesville, for 112D–G
H.-J. Schroers, Centraalbureau voor Schimmelcultures, Utrecht, Netherlands, for Fig. 122B–D
E. C. Setliff, Lakehead University, Thunder Bay, ON, for 159C–H
A. L. Shigo, USDA Forest Service, Durham, NH, for 107J
J. M. Skelly, Pennsylvania State University, University Park, for 92A–E, 239A, D–F, 240F, G, 241F
C. D. Smart, University of California, Davis, for 194A, B, E, 199H
J. D. Solomon, USDA Forest Service, Stoneville, MS, for 70G
J. M. Staley, USDA Forest Service, Parsons, WV, and Fort Collins, CO, for 22F, 165H, I
G. R. Stanosz, University of Wisconsin, Madison, for 39F–H, 54B
D. A. Stark, Maine Forest Service, Augusta, for 107B
T. J. Stephens, Cornell University, Ithaca, NY, for 259J, K
R. J. Stipes, Virginia Polytechnic Institute & State University, for 123G, H
P. Svihra, University of California Cooperative Extension, Novato, for 178B
B. L. Teviotdale, University of California, Kearny Agricultural Center, Parlier, for 115F, G, 186H–J, 188G, H
W. G. Thies, USDA Forest Service, Corvallis, OR, for 109D–F, 114C–E, 170B, F, I
N. A. Tisserat, Kansas State University, Manhattan, for 57C, G, H, 60H, I, 92F–I, 95B, E, 122H, 181F, 212I; also Fig. 60B
W. Uddin, Pennsylvania State University, University Park, for 69C, D, 184D, H

USDA Forest Service for 40E, 108C, 125D
USDA Forest Service Southern Hardwoods Laboratory, Stoneville, MS, for 70B–D, H, 92L
B. C. van Dam, Institute for Forestry and Nature Research, Wageningen, Netherlands, for 189I
T. J. Volk, University of Wisconsin, La Crosse, for 164G
J. A. Walla, North Dakota State University, Fargo, for 9F
P. M. Wargo, USDA Forest Service, Hamden, CT, for 162I
H. E. Waterworth, USDA, Beltsville, MD, for 204G, 213I
N. Wenner and W. Merrill, Pennsylvania State University, for 27D–F, 28C
J. J. Worrall, USDA Forest Service, Gunnison, CO, for 24D–G
E. I. Zehr, Clemson University, Clemson, SC, for 186D, F

Numerous publishers granted permission for our use of copyright-protected illustrations. We gratefully acknowledge the following:
A.D.A.C.–Cryptogamie (Association des Amis des Cryptogames), Paris, for part of Figure 43A
American Journal of Botany for Figure 33C
American Phytopathological Society, St. Paul, MN, for Figures 3B, 11C, 17B, 37A, 83A–C, 89B, C, 96B, C, 109A–D, 113A–F, 176A–C, 178A, 179C–G, 181A–E; and Plates 68D, 203K, 206A
Arnold Arboretum of Harvard University for Figure 89A, copyrighted by the President and Fellows of Harvard College, Harvard University, Cambridge, MA
Blackwell Publishing, Oxford, UK, for Figures 19 and 82C, D
Brill Academic Publishers, Leiden, Netherlands, for Figure 216F–K
CAB International for Figures 9, 11A, B, D, 14A, 17D, 23A, B, 27A, 31, 32C, 39A, 41, 49B, 53E, 54B, C, 58A–C, 59A, B, 60A, C, 63A, B, 64A, B, 66A–C, 72, 76A, B, 77D, E, 82A, B, 90, 92, 121D, E, 128A, 134E
Cambridge University Press, Cambridge, UK, for Figures 17A, 46C, 67E–G, 71A–C
Elsevier Science Publishers, Oxford, UK, for Figures 35A, B, 110A, B, 122C, 123A–D
Elsevier GmbH, Gustav Fischer, Stuttgart, Germany, for Figures 3A, C–E
Ferdinand Berger & Söhne Gesellschaft m.b.H, Horn, Austria, for Figures 13 and 16B
Gebrüder Borntraeger Verlagsbuchhandlung, of E. Schweizerbart'sche Verlagsbuchhandlung, Stuttgart, Germany (http://www.schweizerbart.de), for Figures 7C, 12, 17C, 18B, 30, 39B, parts of 43A and 43B, 44D, 49A, 54A, 55, 56A, B, 61A, C, 62A, B, 68C, D, 75A, B, 87A–C, 115A–D, 121A–C, 125B, 127A, C
International Academic Publishers, Beijing, China, for Figure 16A, D
Kluwer Academic/Plenum Publishers, New York, for Figure 37B
Masson, Paris, France, for Figure 18A
Mycological Society of America for Figures 7D, 8B, C, 15, 21B, C, 46B, 52A–D, 57A, 74A–C, 77F, G, 81C, 93A–C, 99A–C, 100B–D, 122A, B, 125A, C
Mycological Society of Japan for Figure 67B–D
Mycotaxon, Ltd., for Figures 48A–E, 59C, D, 100E
Natural Resources Canada, Canadian Forest Service, Victoria, BC, for Plate 66G; Figure 44A–D, 168A, E
Natural Resources Canada, Canadian Forest Service, Sault Ste. Marie, ON, for Figure 168A
North Carolina Academy of Sciences for Figure 217
NRC Research Press, Ottawa, ON, for Figures 7B, 46D–F, 51A, 61B; Plate 184C
Public Works & Government Services Canada for Figures 6A–C, 16C, 40A–C, 75C, 134D
Royal Botanic Gardens, Kew, UK, for Figure 127B, D, E
Universitetsforlaget, Oslo, Norway, for Figures 103B, 104
John Wiley & Sons, Inc., Hoboken, NJ, for Figures 47B, 175
All illustrations and photographs other than those acknowledged here and on text pages are from the collections of the authors or the Department of Plant Pathology, Cornell University, and are reproduced by permission.

We are deeply indebted to many colleagues who reviewed the manuscript and provided information, corrections, and helpful suggestions. Those who saw the entire manuscript were:
Robert A. Blanchette, University of Minnesota, St. Paul
John N. Gibbs, formerly of the Forestry Commission, UK
Everett M. Hansen, Oregon State University, Corvallis
John R. Hartmann, University of Kentucky, Lexington
William L. MacDonald, West Virginia University, Morgantown
Ned A. Tisserat, Colorado State University, Ft. Collins
Michael J. Wingfield, University of Pretoria, Pretoria, South Africa

Those who reviewed particular sections were:

David N. Appel, Texas A&M University, College Station
Bonnie L. Appleton, Virginia Polytechnic Institute & State University, Virginia Beach
Cynthia L. Ash, USDA Forest Service, Durham, NH
Yilmaz Balcì, West Virginia University, Morgantown
Edward L. Barnard, Florida Division of Forestry, Gainesville
Steven V. Beer, Cornell University, Ithaca, NY
D. Michael Benson, North Carolina State University, Raleigh
Dale R. Bergdahl, University of Vermont, Burlington
Alan R. Biggs, West Virginia University, Kearneysville
George M. Blakeslee, University of Florida, Gainesville
Pierluigi Bonello, Ohio State University, Columbus
Kerry O. Britton, USDA Forest Service, Arlington, VA,
Thomas J. Burr, Cornell University, Geneva, NY
Frank L. Caruso, University of Massachusetts, East Wareham
John D. Castello, State University College of Environmental Science & Forestry, Syracuse, NY
Ann R. Chase, Chase Research Gardens, Inc., Mount Aukum, CA
Gary A. Chastagner, Washington State University, Puyallup
James C. Correll, University of Arkansas, Fayetteville
Pedro W. Crous, Centraalbureau voor Schimmelcultures, Utrecht, Netherlands
Margery L. Daughtrey, Cornell University, Riverhead, NY
Donald D. Davis, Pennsylvania State University, University Park
L. David Dwinell, USDA Forest Service, Athens, GA
Scott A. Enebak, Auburn University, Auburn, AL
David Gadoury, Cornell University, Geneva, NY
Matteo Garbelotto, University of California, Berkeley
Mark L. Gleason, Iowa State University, Ames
Donald J. Goheen, USDA Forest Service, Central Point, OR
George L. Good, Cornell University, Ithaca, NY
Thomas R. Gordon, University of California, Davis
Stewart M. Gray, USDA Agricultural Research Service, Ithaca, NY
Shari Halik, University of Vermont, Burlington
Thomas C. Harrington, Iowa State University, Ames
Kenneth J. Harrison, Canadian Forest Service, Fredericton, NB
Nigel A. Harrison, University of Florida, Fort Lauderdale
Yasuyuki Hiratsuka, Canadian Forest Service, Edmonton, AB
Charles S. Hodges, formerly of USDA Forest Service, Raleigh, NC
Donald L. Hopkins, University of Florida, Apopka
R. Kenneth Horst, Cornell University, Ithaca, NY
David R. Houston, formerly of USDA Forest Service, Hamden, CT
George W. Hudler, Cornell University, Ithaca, NY
Richard S. Hunt, Canadian Forest Service, Victoria, BC
William R. Jacobi, Colorado State University, Fort Collins
Karel A. Jacobs, Morton Arboretum, Lisle, IL
Steven N. Jeffers, Clemson University, Clemson, SC
Jennifer Juzwik, USDA Forest Service, St. Paul, MN

Gaston Laflamme, Canadian Forest Service, Ste. Foy, QC
Theodor D. Leininger, USDA Forest Service, Stoneville, MS
William H. Livingston, University of Maine, Orono
Paul D. Manion, formerly of State University College of Environmental Science & Forestry, Syracuse, NY
Jeanne D. Mihail, University of Missouri, Columbia
Charles W. Mims, University of Georgia, Athens
Jean-Marc Moncalvo, Royal Ontario Museum, Toronto
Kenneth W. Mudge, Cornell University, Ithaca, NY
George Newcombe, University of Idaho, Moscow
Daniel L. Nickrent, Southern Illinois University, Carbondale
Steven W. Oak, USDA Forest Service, Asheville, NC
Michael E. Ostry, USDA Forest Service, St. Paul, MN
John N. Pinkerton, USDA Agricultural Research Service, Corvallis, OR
Donald R. Reynolds, Natural History Museum of Los Angeles County, CA
Scott C. Redlin, USDA Animal & Plant Health Inspection Service, Raleigh, NC
David M. Rizzo, University of California, Davis
Timothy S. Schubert, Florida Division of Plant Industry, Gainesville
Edson C. Setliff, Lakehead University, Thunder Bay, ON
James L. Sherald, National Park Service, Washington, DC
Walter C. Shortle, USDA Forest Service, Durham, NH
R. Jay Stipes, Virginia Polytechnic Institute & State University, Blacksburg
David M. Sylvia, Pennsylvania State University, University Park
Beth L. Teviotdale, University of California Kearney Agricultural Center, Parlier
Walter G. Thies, USDA Forest Service, Corvallis, OR
James A. Walla, North Dakota State University, Fargo
Philip M. Wargo, formerly of USDA Forest Service, Hamden, CT
Stephen Woodward, University of Aberdeen, Aberdeen, Scotland
James J. Worrall, USDA Forest Service, Gunnison, CO
Paul J. Zambino, USDA Forest Service, Moscow, ID

We also acknowledge with gratitude the compilers of the computer databases that were used to discover relevant references published since the mid-1980s and to verify or correct taxonomic and nomenclatural information. CAB Abstracts and BIOSIS Previews were the principal sources of references. Databases consulted for taxonomic and nomenclatural information are indicated on page ix. We are grateful also to the creators of the versatile computer programs Adobe® Photoshop®, Streamline®, and PageMaker®, with which images were edited and the plates and figures constructed.

Special thanks go also to Candace Akins and Scott Levine, of Cornell University Press, and to Melinda N. Conner, for editorial and production services.

W.A.S.
H.H.L.

DISEASES OF TREES AND SHRUBS

Second Edition

Diseases Caused by Fungi (Plates 1–174)
Overview of Fungal Diseases of Woody Plants

Fungi and woody plants are linked in health, disease, and the cycling of carbon and minerals. Parasitic fungi derive sustenance from living tissues or kill them and utilize the remains. Endophytes colonize various plant parts and live benignly until host stress or senescence triggers further growth and reproduction. Symbiotic fungi associate with roots. Saprobes decompose what dies. Many fungi have dual or alternating roles as parasites and saprobes. This unit presents selected characteristics of fungi that cause disease in woody plants along with some relevant terms and concepts. Some of the mycological terms used in this book are explained here, others in the Glossary. Readers seeking an introduction to the fungi should consult the listed references and Internet sites. A brief introduction to the Oomycota, which were formerly considered to be fungi, appears on page 352.

Plant parasitism has evolved in numerous groups within Kingdom Fungi. At the beginning of the 21st century, pathogens of woody plants were arrayed in 18 of the 55 orders of Phylum Ascomycota (ascomycetes) and 8 of the 33 orders of Basidiomycota (basidiomycetes). Nearly all fungi grow as filaments, have an absorptive mode of nutrition, and reproduce by nonmotile spores. Their energy sources are absorbed organic compounds. The principal structural polymers of their cell walls are chitin and β-glucans. Their main energy storage compounds are carbohydrates and lipids. Most fungi reproduce both sexually and asexually. They are polymorphic, presenting different forms during colony growth, asexual reproduction, sexual reproduction, and quiescence (during conditions unfavorable for growth). Sexual reproductive forms are called *teleomorphs*. Asexual reproductive forms are called *anamorphs*. Some fungi produce more than one anamorph. Most anamorphs are ascomycetes. The term *holomorph* refers to all forms and developmental phases of a fungus. The name of a teleomorph applies to all forms of a fungus, but anamorph names are also necessary because they are linked to descriptions that are needed for fungal identification.

Fungal forms and functions. A person who would diagnose plant diseases or have detailed knowledge about them must become familiar with various types of fungal structures and their functions. Illustrations of the structures mentioned below are presented in the plates and/or in figures in the text. Examples can be located via the Index.

Fungal filaments (*hyphae*) are microscopic in diameter, branched, tubular, and compartmented. They elongate at their tips, ramify in substrates, and absorb nutrients which are then translocated to sites of growth and reproduction. Multiple hyphae in a network or loose mass constitute *mycelium*. A fungus can grow indefinitely as mycelium. Hyphae may grow individually or collectively, forming strands, rhizomorphs, stromata, fruit bodies, and survival structures. A *rhizomorph* is a rootlike organ in which the outer layers of hyphae differentiate into a protective rind and the inner hyphae are specialized for translocation. A *stroma* is a pad- or cushionlike hyphal matrix. Fruit bodies and other spore-producing structures of ascomycetes often form on or in stromata.

Typical hyphae are haploid and may have one or more nuclei per cell. If the nuclei are of one genotype, a hypha is *homokaryotic*. If nuclei of two genotypes are present as a result of fusion of two fungal individuals, the hypha is *heterokaryotic*, and the phenotype of the hypha reflects both genotypes. Diploids form within hyphae or in specialized cells prior to meiosis. In most fungi, meiosis occurs in cells called *asci* or *basidia,* for which the Ascomycota and Basidiomycota, respectively, are named. Asci and basidia produce haploid spores—*ascospores* and *basidiospores,* respectively—that contain nuclei derived by meiosis. These spores are typically produced in or on fruiting bodies—*ascocarps* and *basidiocarps,* respectively.

A fungus may reproduce by forming spores or multicellular resting structures or by fragmentation of hyphae so that individual cells become spores. Fungal spores are walled propagules capable of being dispersed and of initiating new colonies. They are diverse in origin, form, size, color, and number of cells. Nearly all are microscopic. Spores other than those derived by hyphal fragmentation form on or in specialized generative cells produced by hyphae. Spores produced asexually—that is, whose nuclei are products of mitosis—have various names that reflect different modes of formation and different functions. Two of these names are used broadly in this book: *conidium* for any asexually produced spore that becomes detached when mature, and *chlamydospore* for any thick-walled spore specialized for survival. Many ascomycetes produce conidia in or on tiny fruiting bodies that are collectively called *conidiomata*. A cup- or flasklike conidioma is a *pycnidium;* a blisterlike conidioma is an *acervulus*. Anamorphs that produce these types of conidiomata are herein called *coelomycetes*. Anamorphs that produce conidia on the exterior of simple or complex hyphal structures are herein called *hyphomycetes*. These terms refer to forms, not taxonomic categories. Some conidia can function either as propagules or as gametes.

Many ascomycetes and some basidiomycetes produce *spermatia,* gametes that function as male cells in sexual reproduction. They are dispersed by water and insects and are referred to as spores, but they do not germinate. Spermatia of some ascomycetes and rust fungi are produced in tiny *spermagonia* that resemble pycnidia. Spermagonia are sometimes the only fungal structures on diseased specimens received by diagnosticians.

Plant-pathogenic ascomycetes produce ascocarps that vary in form from microscopic spheres or flasks (*cleistothecia, perithecia*) to stromata with fertile cavities (*pseudothecia*) to cuplike or linear organs up to several centimeters in diameter or length that open widely when mature (*apothecia*). Plant-pathogenic basidiomycetes other than rust fungi produce basidiocarps that vary from flat sheets of fertile tissue to large, elaborate bracket forms (*conks*) and mushrooms. The rust fungi (basidiomycetes of the Uredinales) have unique fruiting structures, spore types, and terminology that are discussed in the section on rust diseases.

Resting structures are crucial for fungal continuity. Plant-parasitic fungi in living plant parts are protected from desiccation and from competition or antagonism by other organisms during pathogenesis, but most pathogens are no match for saprobes in killed tissue. Their survival apart from living hosts depends on being able to produce microbe-resistant, desiccation-tolerant, energy-storing resting structures or barriers that enclose and protect active hyphae. Most resting or survival structures are composed of, or have a rind or multicellular wall made of, thick-walled darkly pigmented (melanized) cells. Examples of such structures include spores of some types, specialized hyphae, microsclerotia, sclerotia, rhizomorphs, perennial stromata or fruiting bodies, ascocarps of powdery mildew fungi, and telia of some rust fungi. Many wood-decaying fungi produce melanized sheets of tightly packed hyphae (*pseudosclerotial plates*) that enclose and protect active mycelium.

Dispersal. Plant-pathogenic fungi are dispersed as spores and other propagules and also as mycelium in plants and other substrates. Spores can travel in air or water or on (sometimes within) vectors. Vectors, in plant pathology, are agents that transmit pathogens. Most vectors are insects, but humans play a large role in the dispersal of many fungi. Tree pathogens introduced to the environments of highly susceptible species through commerce in plants and plant products have caused famous destructive epidemics. Examples include butternut canker, canker-stain of *Platanus,* chestnut blight, Dothistroma blight of pines, dogwood anthracnose, Dutch elm disease, pitch canker of pines, Seiridium canker of cypress, sudden oak death, and white pine blister rust. Severe damage has also developed where trees have been grown in monocultures in conditions that promote buildup of populations of indigenous pathogens. Armillaria root disease, fusiform rust of pines, and brown spot needle blight of pines are examples.

Nutritional habits. Plant-pathogenic fungi vary from strictly *biotrophic* to *necrotrophic*. Biotrophs (obligate parasites) enter intact physiologically active leaves, fruit, and young stems, either by sending specialized hyphae into stomata or by penetrating the cuticle and epidermis. Most biotrophs obtain nutrients by penetrating living cells and producing, between the wall and the cell membrane, an absorptive structure called a *haustorium*. The cell membranes of haustorium and host are in broad contact, facilitating signal exchange and nutrient withdrawal. The largest groups of biotrophic pathogens of woody plants are powdery mildews (ascomycetes of the Erysiphales) and rust fungi.

Necrotrophs (facultative parasites) kill host tissues and utilize them as both food and physical substrates. Necrotrophs enter all plant parts, either through wounds or by penetrating intact surfaces. They reproduce only on or in dead tissue. They vary in attack habits: some enter intact tissues, others prefer or require wounds; some attack growing organs, others prefer mature or dormant tissues. Many necrotrophs that infect bark and sapwood can enter whenever temperature and moisture permit their growth, and they tend to be aggressive (or to become active after endophytic residence) in plants stressed by disease, injury, or environmental factors such as drought or freezing.

Host selection. It has been said that fungi are good taxonomists, a reference to the fact that host specialization, when expressed by a fungal species, is usually at the plant genus or family level. Biotrophs tend to be quite specialized with respect to both host range and mode of attack, while opportunistic necrotrophs tend to be nonspecialized in both respects.

Attack factors. Fungal nutrition and pathogenesis are linked. Fungal pathogens produce extracellular enzymes, some wall-bound and others secreted to the hyphal environment, that digest organic polymers in plant cuticle, cell walls, membranes, and intracellular structures. The resulting small molecules can be assimilated by hyphae. Some pathogens also secrete toxins that disrupt plant cell membranes and/or organelles and thus cause cell and tissue death and nutrient release. Many pathogens secrete or possess on their surfaces molecules—mainly proteins (including enzymes) and polysaccharides—that are sensed by plant cells and that elicit cell and tissue death. If the elicited response is rapid, it contributes to plant resistance. If the response is slow, it exacerbates disease. Some attack factors are resistance elicitors. Many attack factors are produced constitutively at low levels; the main arsenal is induced as a pathogenic fungus senses its chemical and physical environment. Some fungal pathogens also secrete plant hormones or induce hormonal imbalances that lead to aberrant forms such as galls and witches'-brooms.

Evading or defeating host defenses. Fungi that parasitize plants are able to evade or defeat constitutive and inducible defenses that render a given plant species immune to the majority of pathogenic fungi and effectively resistant to many others. Plants' constitutive defenses are both physical (e.g., corky bark, barrier zones in wood) and chemical (e.g., phenolic or cyanogenic compounds that are released from damaged cells). Induced defenses are initially biochemical, such as production of phytoalexins, toxic free radicals, and enzymes that can degrade fungal cell walls. Self-destruction of cells and tissues, such that an invader is denied sustenance from living cells, is also an important mode of defense. After an infection is halted, further host responses produce barriers between damaged and healthy tissues (discussed with Plates 254–257). Plants have less resistance to necrotrophic pathogens during dormancy than they do during the growing season. Therefore, disease-induced dieback of twigs and branches is more common (although less conspicuous) during host dormancy than during the growing season.

Relationship of fungal life cycles to disease cycles. Life cycles of fungal pathogens are diverse and often lack regular alternation of haploid and diploid phases. Some fungi reproduce only sexually or only asexually. Many fungi grow vegetatively and/or reproduce asexually for long periods. The generalized life cycle of an ascomycete or basidiomycete can be represented as beginning with an ascospore or basidiospore that germinates to produce a haploid hypha (*germ tube*) that gives rise to mycelium. The mycelium grows for a time (days to decades) and may produce asexual spores or resting structures, each capable of starting another colony like the first. Mutations and introduction of "foreign" DNA slowly increase variability in such a population. Eventually the fungus begins a heterokaryotic phase as the result of either hyphal fusion (*anastomosis*) between itself and another individual of different genotype or fusion between a receptive hyphal structure and a gamete from a different individual of the same species. Somatic anastomosis is enabled or prevented by *vegetative-compatibility* genes. Sexual reproduction is governed by *mating type* genes. Sexual reproduction takes precedence over somatic incompatibility.

A heterokaryon may grow indefinitely and/or reproduce asexually or sexually. Sexual reproduction in most ascomycetes and basidiomycetes occurs in or on a fruit body. It involves formation of a diploid by fusion of haploid nuclei, then meiosis and formation of ascospores or basidiospores. Some basidiomycetes have an extended diploid phase in hyphae. A mycelium that produces either asexual or sexual reproductive structures may remain capable of indefinite growth and further reproduction.

The concept of *disease cycle* integrates the behavior of pathogen and host as conditioned by environment through time. A disease cycle is typically represented as beginning with coincident seasonal onset or resumption of plant growth and pathogen activity (growth, sporulation, dispersal) and continuing with *inoculation* (contact between pathogen and susceptible plant organ), *incubation period* (encompassing events such as spore germination and host penetration that precede the onset of symptoms), *infection* (active host-pathogen interaction), further colonization (by necrotrophic pathogens), and pathogen reproduction resulting in new *inoculum* (spores or resting structures capable of starting another cycle). In regions with distinct seasons, a *primary cycle* is the first of the year, and all subsequent cycles initiated during the same plant growth period are called secondary. Secondary cycles occur in many foliar diseases, but most other diseases of trees and shrubs have only annual or multiyear cycles.

Names of fungi. Most known fungi have only Latin names. Many fungi have different names for their teleomorphs and anamorphs because different forms require different labels. Names of fungi sometimes change as new information necessitates reclassification, as duplicate or improperly used names are reduced to synonymy, and as early validly published names are discovered. The earliest valid name for a teleomorph or anamorph must be adopted under rules of the *International Code of Botanical Nomenclature,* which governs the naming of plants, fungi, and oomycetes. Names of famous fungi or those with complex nomenclatural histories are sometimes officially "conserved" to prevent confusion that could be caused by strict adherence to nomenclatural rules. Diagnosticians usually encounter plant-pathogenic fungi in their asexual forms, so many anamorph names have enduring usefulness. An anamorph genus name is usually linked to just one ascomycete or basidiomycete genus, but exceptions occur because different teleomorphs may have highly similar anamorphs. Some ascomycete and basidiomycete genera include multiple anamorphs and therefore are linked to multiple anamorph names.

Disease types as related to fungal groups. Most pathogens of trees and shrubs "nibble" on expendable parts—leaves, twigs, small branches, rootlets—that would eventually die and be shed even if they were not diseased. Relatively few leaf and twig pathogens remove so much foliage or so many branches that they interfere with plant growth and longevity. Most leaf and twig diseases are caused by ascomycetes or by rust fungi. The majority of these ascomycetes are in the orders Helotiales, Mycosphaerellales, Pleosporales, and Rhytismatales.

Most canker and dieback diseases are caused by ascomycetes in the Dothideales, Helotiales, Hypocreales, Diaporthales, Diatrypales, or Xylariales; or by basidiomycetes in the Hymenochaetales and Polyporales. Pathogens in the Helotiales and Hypocreales colonize bark and cambium primarily. Pathogens in the Dothideales, Diaporthales, and Diatrypales colonize bark, and many species also colonize sapwood. Pathogens in the Xylariales and in the basidiomycete orders named above colonize wood primarily and then kill bark. Many fungi that colonize bark and/or sapwood of stems or roots can grow well at relatively low water potentials that impair the hosts' defense responses. Accordingly, many canker and dieback diseases and some root diseases caused by fungi are exacerbated by drought. The pathogens in such diseases are viewed as opportunists.

Wood decay is caused mainly by basidiomycetes of the Agaricales, Hymenochaetales, Polyporales, and Russulales; and ascomycetes of the Xylariales. The majority of wood decomposers in trees are wound invaders that follow other organisms and do not themselves kill tree tissues. Wood-decay fungi that kill and decay sapwood and bark cause important root and butt rots and trunk rots.

Fungi that invade xylem systemically and cause wilt diseases are ascomycetes of the Microascales and Ophiostomatales and their anamorphs, and species of *Fusarium, Nalanthamala,* and *Verticillium* that have no known sexual states. Nearly all are able to sporulate in xylem sap, which facilitates systemic spread in their hosts.

Major groups of root-infecting fungi include ascomycetes of the Hypocreales and Ophiostomatales, assorted anamorphs that have broad host ranges, basidiomycetes of the Ceratobasidiales, and the wood-decaying groups mentioned above.

References: 40, 123, 159, 218, 649, 1056, 1081, 1099, 1280, 1298, 1369, 1476, 1853, 1927, 1928, 2037, 2077, 2297, 2703, 2704, 2713, 3344, 3380, 3534, 3669, 3722, 3913, 4117, 4265

Diseases Caused by Ascomycota (Plates 1–123)
Taphrina Diseases (Plates 1, 2)

Taphrina species (Taphrinomycetes, Taphrinales) cause localized overgrowth and distortion of leaves, flowers, or fruit; some also cause witches'-brooms. Blisterlike distortions on leaves, exemplified by peach leaf curl (Plate 1A, B) and oak leaf blister (Plate 1E–H), are the most common symptoms. The approximately 95 species in *Taphrina* are considered to be primitive ascomycetes. They have no multicellular fruiting structures; their asci are produced on the surfaces of diseased plant parts (Fig. 2). Some 44 species are reported to occur on woody plants in eight families in the USA and Canada. Diseases caused by the fungi listed for each host genus below are leaf blisters or curls unless otherwise noted. Geographic distributions are identified by region or with postal abbreviations (NA = North America).

Acer: Taphrina aceris, UT; *T. carveri,* central and southern USA, ON; *T. darkeri,* northwestern USA, western Canada; *T. dearnessii,* central and eastern NA; *T. letifera,* eastern and north-central USA, NS; *T. sacchari,* central and eastern USA, QC
Aesculus californica: T. aesculi, CA, TX
Alnus: T. alni (catkin hypertrophy), AK, GA, south-central Canada, Europe, Japan; *T. japonica,* northwestern NA, Japan; *T. occidentalis* (catkin hypertrophy), northwestern NA, QC; *T. robinsoniana* (catkin hypertrophy), central and eastern NA; *T. tosquinetii* (catkin hypertrophy), NH, NS, Europe
Amelanchier alnifolia and *Amelanchier* sp.: *T. amelanchieris* (witches'-broom) and *T. japonica,* respectively, CA
Betula: T. americana (witches'-broom), NA; *T. bacteriosperma,* Northern Hemisphere; *T. boycei,* NV, central and western Canada; *T. carnea,* northeastern USA, eastern and western Canada, Europe; *T. flava,* northeastern USA, eastern Canada; *T. nana,* WY, eastern and western Canada, Europe; *T. robinsoniana,* NF
Carpinus caroliniana: T. australis, eastern USA, ON
Castanopsis: T. castanopsidis, CA, OR
Corylus: T. coryli, USA, Japan
Malus: T. bullata, WA, Europe
Ostrya virginiana: T. virginica, central and eastern USA, eastern Canada
Populus: T. johansonii (catkin hypertrophy), USA, eastern Canada, Europe, Japan; *T. populi-salicis,* northwestern NA; *T. populina,* Northern Hemisphere; *T. rhizophora,* NY, WI, Europe
Potentilla: T. potentillae, USA, Europe
Prunus: T. armeniacae, USA; *T. communis* (plum fruit hypertrophy), NA; *T. confusa* (fruit hypertrophy and witches'-broom of *P. virginiana*), USA, eastern Canada; *T. deformans,* cosmopolitan; *T. farlowii* (cherry leaf curl, fruit and shoot hypertrophy), central and eastern USA; *T. flavorubra* (cherry and plum fruit and shoot hypertrophy), USA, central and eastern Canada; *T. flectans* (cherry witches'-broom), western USA, BC; *T. jenkinsoniana,* NV; *T. pruni* (plum pocket), Northern Hemisphere; *T. pruni-subcordatae* (plum witches'-broom), western USA; *T. thomasii* (witches'-broom on *P. ilicifolia*), CA; *T. wiesneri* (cherry witches'-broom), cosmopolitan
Pyrus: T. bullata, WA, BC, Europe
Quercus: T. caerulescens, NA, Europe, northern Africa
Rhus: T. purpurascens, central and eastern USA, Europe
Salix laevigata: T. populi-salicis, CA
×*Sorbopyrus auricularis: T. bullata,* BC
Ulmus americana: T. ulmi, central and eastern USA, QC, Europe

Taphrina species on leaves often cause unthickened spots that turn brown and go unnoticed. Blisterlike deformities and curled leaves, on the other hand, attract attention. These result from abnormal cell division (hyperplasia) and enlargement (hypertrophy) that the parasites induce in part by secreting phytohormones. Several species, including *T. deformans* and *T. wiesneri,* are known to produce cytokinins (cell-division factors) and auxins (promoters of cell enlargement). Bulging spots on leaves must be examined closely to distinguish between Taphrina leaf blisters and symptoms caused by eriophyid mites (see *Insects that Feed on Trees and Shrubs*).

Disease cycles. Taphrina species infect growing leaves, flower parts, or fruit annually in spring and produce spores in late spring or early summer. They survive during the rest of the year by growing as saprobes on plant surfaces during moist weather at moderate temperatures and by persisting as quiescent spores on twigs and among bud scales during dry periods and in winter. They assume different forms in their parasitic and saprobic phases. The parasitic phase consists of intercellular or subcuticular mycelium that may

stimulate host overgrowth and pigment formation and that produces a single layer of asci on the host surface. Asci of *Taphrina* are microscopic sacs in each of which many spores develop. When fresh the layer of asci appears as a granular, sometimes glistening deposit, usually white or translucent. The spores are discharged in late spring to midsummer, depending on the species and locality, and the diseased plant tissue dies soon afterward.

Spores from asci give rise to the saprobic phase, in which the fungus produces successive generations of spores on plant surfaces by budding like a yeast during warm, moist weather. Some of these spores (termed *blastospores,* or *bud-conidia*) survive winter on twigs or among bud scales and are thus on hand to cause new infections or to produce more spores that in turn infect young plant tissue in spring. *Taphrina* species can be isolated from noninfected plant surfaces during most of the year. They resemble yeasts when growing on laboratory media. Species with perennial mycelium in twigs have the same type of annual cycle outlined above. In addition, however, occasional infections progress into stem tissue and become perennial, causing distortion or broom formation.

Peach leaf curl. *Taphrina deformans* causes leaf curl and deformity of succulent shoots of *Prunus persica* (peach and nectarine) worldwide. *P. dulcis* (almond) is a reported host in British Columbia, Europe, Morocco, and New Zealand; and *P. armeniaca* (apricot) in Florida and New Zealand. Symptoms appear as leaves unfold. Leaves become puckered and curled during the development of thickened bulging areas that change from green to yellow and then become reddish. Succulent shoots may also be deformed. Within a few weeks a white bloom appears on the deformed part of the leaf, usually on the upper surface. This is the layer of asci, similar in magnified view to that of *T. wiesneri* (Plate 2F). Infected parts of leaves soon degenerate, and the leaves fall prematurely. Severe infection can cause such defoliation that a second flush of shoot growth occurs in early summer. Blossoms sometimes become infected and then shrivel and fall. Photosynthesis is greatly suppressed in diseased leaves, and respiration and transpiration are elevated.

Only undifferentiated tissues are susceptible. Thus most infections occur on leaves during or just after emergence from buds. Mycelium of the fungus grows intercellularly in and beneath either the upper or lower epidermis and develops extensively beneath the cuticle, giving rise there to the layer of asci. Spores from asci on deformed leaves splash about on the plant during rains and multiply by budding. The fungus is thereby distributed all over the plant and keeps pace with the growth of new shoots but seldom infects them. The spores can tolerate hot, dry summers and freezing winters. Spores lodged in winter buds start primary infections annually. Secondary cycles possibly occur but are never important. Periods of cool, wet weather during leaf emergence favor leaf curl. *T. deformans* in its saprobic phase grows most rapidly at temperatures near 20°C.

Peach and nectarine cultivars vary in resistance to leaf curl. Cultivars grown commercially in the USA are all at least moderately susceptible. Several cultivars from Germany and elsewhere in Eurasia appeared more resistant than any from the USA in evaluations performed in West Virginia.

Plum pocket, or bladder plum. Plum pocket (Plate 1C, D) occurs around the Northern Hemisphere. In North America it is usually caused by *T. communis,* occasionally by *T. pruni.* In Eurasia,

---

A, B. Peach leaf curl, caused by *Taphrina deformans.* A. Diseased leaves are puckered and develop red pigment (OR, Jul). B. A superficial layer of asci (white area) produces spores (CA, Jun).

C, D. Bladder plum, or plum pocket, caused by *T. communis,* on *Prunus nigra* (Canadian plum). Enlarged fruits that contain multiple hollow chambers darken and fall after spore production. Dark spots on green fruits (C) are injuries made by an insect, the plum curculio (NH, Jun).

E–H. Oak leaf blister, caused by *T. caerulescens.* E, F. On *Quercus velutina* (black oak) (NY, Jul). G. On *Q. phellos* (willow oak) (MS, May). H. Asci on the lower surface of a *Q. rubra* (northern red oak) leaf. These areas turn brown after spore production (NY, Jun).

T. pruni is common and *T. communis* absent. Swollen fruits infected by these fungi appear most often on native species, especially *Prunus americana* (American plum) and *P. nigra* (Canadian plum). These diseases are less common on *P. domestica* (domestic or European plum) and are uncommon on *P. salicina* (Japanese plum). Other susceptible species include *P. angustifolia, P. besseyi, P. hortulana, P. maritima,* and *P. munsoniana.*

Infection by *T. communis* or *T. pruni* apparently begins soon after blossoms open and is first evident on fruits 6–12 mm in diameter as elevated, spongy, greenish white spots. These usually enlarge until an entire fruit is involved. It grows quickly to three or four times normal size, becomes hollow with pocketlike invaginations, usually lacks a seed, and develops a leathery surface. A white layer of asci develops on the epidermis and breaks through the cuticle. Sporulation occurs in late March to early July, depending on climate. After spore dispersal the affected fruit darkens, shrivels, and falls. *T. communis* and *T. pruni* also cause thickening and deformity of succulent stems, and *T. pruni* produces perennial mycelium in them. In other respects the disease cycles resemble that of peach leaf curl.

Oak leaf blister. *Taphrina caerulescens,* which causes oak leaf blister (Plate 1E–H), occurs around the globe on more than 50 *Quercus* species. In the USA the disease is more common and severe in southeastern states than in the North, perhaps because the pathogen population remains high on plant surfaces throughout the winter in southern areas. Leaf blister is favored by mild, moist conditions during the early phases of leaf growth.

The symptoms are bulges, or depressions as viewed from the reverse side, usually 3–20 mm across on either leaf surface. These distortions may in turn cause severe curling on narrow-leaved species such as *Q. phellos* (willow oak). When numerous, blisters sometimes coalesce and involve entire leaves. A layer of asci forms on the blister, usually on the concave surface. This layer is colorless at first; it darkens after spores have been released (Plate 1H). Severely diseased leaves may drop prematurely. *Q. nigra* (water oak) with severe leaf blister, for example, has been observed to lose half of its leaves by the end of May in Mississippi.

The primary cycle of oak leaf blister is as outlined for Taphrina diseases in general. Leaves become infected as the buds open; expanded leaves are not susceptible. The mycelium develops subcuticularly and intercellularly in the epidermis. Asci mature and disperse their spores between late March and mid-August, depending on latitude and elevation. Secondary cycles occur occasionally as bud-conidia infect young leaves from buds that open unseasonably in summer.

T. caerulescens, the only named *Taphrina* species on oaks, may be a group of biologically distinct organisms. Its asci vary in form and in relationship to epidermal cells on different oaks, and there is circumstantial evidence of pathogenic specialization. More than one observer has noted a severely diseased oak of one species growing amid healthy oaks of a second species known to be susceptible.

Leaf curl and witches'-broom of cherry. *Taphrina wiesneri* (Plate 2A–F) causes leaf curl and witches'-broom on wild and domestic cherries across North America and around the world in both temperate zones. Hosts include *Prunus armeniaca, P. avium, P. cerasus, P. emarginata, P. fruticosa, P. pensylvanica, P. pseudo-cerasus, P. serrulata,* and *P. ×yedoensis.* The disease achieved minor notoriety in the 1920s when it occurred in *P. ×yedoensis* (Tokyo or Yoshina cherry) plantings in Washington, DC, from which it was quickly eradicated by pruning. It is seen yearly in cherry orchards of the Pacific Northwest and on wild *P. pensylvanica* (pin cherry) in eastern states and provinces.

Leaf curl precedes broom formation and may occur independent of brooms. The leaf-curl phase of the disease apparently has an annual cycle like that of peach leaf curl. Diseased parts of leaves become slightly thickened, discolored yellowish to reddish or reddish brown, and curled, crinkled, or droopy. Asci develop on the lower leaf surface, rarely on both surfaces, about a month after leaves begin growth. Spores are dispersed from a given leaf during a 3–4-day period, both night and day. Diseased parts of leaves darken and die soon after sporulation.

Brooms are induced by perennial intercellular mycelium in twigs. Whether the fungus grows through petioles to stems or infects them directly is unclear. Once in a stem, however, *T. wiesneri* causes localized swelling and twig proliferation that may continue for several years. Symptoms develop on nearly all leaves produced on a broom, an indication that the fungus is systemically distributed within its twigs. This distribution is probably due to colonization of buds and ability of the fungus to keep pace with twig growth. Brooms range in length from a few centimeters to more than a meter. They produce foliage earlier than normal and bear no flowers or fruit. Leaves in brooms are heavily infected early in the season and soon die or fall. New normal leaves then develop, making brooms conspicuous during much of the growing season. The knotlike swellings at the bases of old brooms sometimes exude gum. Twigs in brooms tend to die back in winter.

Yellow blisters on poplar leaves. *Taphrina populina* (Plate 2G–I) is globally distributed and causes yellow leaf blister of various *Populus* (poplar) species coast to coast in North America as far south as South Carolina and Louisiana. *T. populi-salicis* causes yellow leaf blister in California and the Pacific Northwest on *P. balsamifera* subsp. *trichocarpa* (black cottonwood), *P. fremontii* (Fremont cottonwood), and *Salix laevigata* (red willow). Infected areas vary from round depressions or mounds a few millimeters across to large, irregularly shaped, thickened areas on distorted leaves. The upper surfaces turn light green or greenish yellow and the lower ones bright yellow when asci develop. The yellow color is imparted by pigment in oil droplets within asci during the period of sporulation. The annual cycle is as generalized for the genus *Taphrina.*

References: 29, 126, 609, 622, 779, 1099, 1214, 1298, 1609, 1859, 1956, 2047, 2686, 2867, 2896, 3206, 3207, 3466, 3467, 3517, 3769, 3965, 3976, 4015

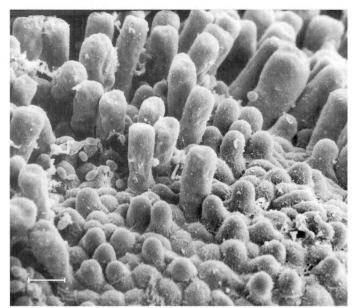

Figure 2. Asci of *Taphrina deformans* on a *Prunus persica* (peach) leaf as seen in a scanning electron micrograph. Scale bar = 10 μm. Courtesy of M. F. Brown.

A–F. Witches'-broom and leaf curl of *Prunus pensylvanica* (pin cherry), caused by *Taphrina wiesneri.* A. A broom several years old at the beginning of the growing season. Winter-killed twigs are evident (NY, May). B, C. Diseased foliage on a small broom and on a nonbroomed branch (NY, Jun). D. A swollen knot identifies the initial site of perennial infection at the base of a broom (NY, Jun). E. A diseased leaf. The bleached area on the lower surface is a layer of asci (NY, Jun). F. Magnified view of asci. The midvein of the leaf is at left (NY, Jun).

G–I. Yellow leaf blister of *Populus balsamifera* subsp. *trichocarpa* (black cottonwood) caused by *Taphrina populi-salicis.* G. Small blisters on the upper surface of a leaf (ID, Jul). H, I. Upper and lower surfaces of distorted leaves. Yellow patches on lower surfaces are large groups of asci (ID, Jul).

Powdery Mildews (Plates 3–6)

Powdery mildew diseases are characterized by superficial growth and sporulation of the causal fungi on plant surfaces. The fungal colonies often appear powdery. The term *powdery mildew* may refer either to one of the pathogens or to a disease. The pathogens are obligately biotrophic ascomycetes of the order Erysiphales, family Erysiphaceae. Most of them penetrate and parasitize only the epidermal cells of their hosts, producing therein absorbing structures (haustoria) that extract water and nutrients to support growth of the superficial hyphae and later of reproductive structures. Hyphae of some species grow both superficially and within leaves, producing haustoria in both epidermal and internal (mesophyll and palisade) cells or only within internal cells. The superficial hyphae form white or light-colored wefts or mats that become brownish with age in some species. Asexual reproduction occurs by formation of conidiophores and conidia on the superficial mats or, in *Leveillula*, on conidiophores that emerge from stomata.

Symptoms and impact. Powdery mildews depress growth and cause dwarfing and distortion of organs infected when young, blemished or aborted fruits, and low yields. Leaves may develop chlorosis and senesce and die prematurely. Photosynthesis is greatly reduced in parasitized leaves, even before they discolor, and transpiration is usually reduced. A few powdery mildews induce witches'-brooms. Powdery mildews have significant economic impact in horticultural situations but are unimportant in North American forests.

The pathogens and their hosts. Experts recognize more than 500 powdery mildew species in 13 genera on more than 7000 plant species worldwide. All hosts are angiosperms. Most of the powdery mildew species are arrayed in 3 genera: *Erysiphe*, *Phyllactinia*, and *Podosphaera*. The asexual states (anamorphs) of powdery mildew fungi are assigned to 4 genera, discussed on page 10. Recent progress in taxonomy of powdery mildews has resulted in the merging of some well-known genera. *Erysiphe* now includes nearly all fungi formerly classified in *Microsphaera* and *Uncinula*, while *Podosphaera* includes fungi formerly classified in *Sphaerotheca*. Several species that affect woody plants have been transferred from *Erysiphe* to a new genus, *Golovinomyces*. At the same time, some old broadly construed species have been divided into numerous species linked to particular hosts. Powdery mildew genera and species are differentiated on morphological characters of both asexual and sexual states. Conidia and conidiophores of *Podosphaera*, *Cystotheca*, and *Sawadaea* contain conspicuous light-refractive particles (*fibrosin bodies*). Host specialization is an aid in practical diagnosis. The extent to which powdery mildew species are biologically isolated has not been tested in most cases. Genetic research is needed to support or correct the classification.

At least 85 currently recognized powdery mildew species, representing 11 genera, occur on woody hosts in temperate North America. The list below shows mildew-host combinations that are verified or probable. The list was compiled by examining distribution records for contemporary species together with host records for former species complexes. Abbreviations in parentheses indicate former classification: *E.* for *Erysiphe*, *M.* for *Microsphaera*, *S.* for *Sphaerotheca*, *U.* for *Uncinula* or *Uncinuliella*. Geographic distributions are identified with postal abbreviations of states and provinces or more broadly as follows: c, central; e, east; n, north(ern); s, south; w, west; Can, Canada; Eur, Europe; hemis, hemisphere; NA, North America; SA, South America. The list omits many associations that are inconsistent with knowledge of contemporary species or cannot be translated reliably from old records.

Brasiliomyces trina on *Lithocarpus, Quercus*: sw USA
Cystotheca lanestris on *Lithocarpus, Quercus*: c, s, w USA; Mexico; Japan
Erysiphe (*M.*) *abbreviata* on *Quercus*: NA
E. (*U.*) *adunca* on *Populus, Salix*: n hemis. This is a complex within which new species will soon be described.
E. castaneae (*M. americana*) on *Castanea, Castanopsis*: e USA, Canada
E. aggregata on *Alnus*: NA
E. (*M.*) *penicillata* on *Alnus*: n hemis
E. (*M.*) *alphitoides* on *Quercus* in NA, additional genera elsewhere: global
E. (*U.*) *australiana* on *Lagerstroemia*: global
E. (*M.*) *azaleae* on *Rhododendron periclymenoides, R.* sp.: USA, Can
E. (*M.*) *calocladophora* on *Quercus*: se USA

E. (*M.*) *caryae* on *Carya*: c, e USA
E. (*M.*) *ceanothi* on *Ceanothus americanus*: USA
E. cinnamomicola (*M. cinnamomi*) on *Cinnamomum camphora*: LA
E. (*U.*) *circinata* on *Acer*: c, e USA; c, e, w Can
E. (*U.*) *clintonii* on *Tilia americana*: c, e USA; e Can
E. (*M.*) *diffusa* on *Bauhinia, Cassia, Robinia, Sophora*: n hemis, SA
E. (*M.*) *elevata* on *Catalpa*: c, e USA; e Can
E. (*M.*) *ellisii* on *Carpinus, Corylus, Ostrya*: c, e USA; e Can
E. (*M.*) *erineophila* on *Fagus*: e USA, e Can
E. (*M.*) *euonymi* on *Euonymus*: c USA
E. (*M.*) *euonymi-japonici* on *Euonymus japonicus*: global
E. (*M.*) *extensa* on *Quercus*: c, e USA; e Can
E. (*U.*) *flexuosa* on *Aesculus*: c, e USA
E. (*U.*) *geniculata* on *Morus rubra*: c, e USA
E. (*M.*) *grossulariae* on *Ribes, Sambucus*: NA
E. (*M.*) *hommae* on *Corylus*: n hemis
E. (*M.*) *juglandis-nigrae* on *Juglans nigra*: USA
E. liriodendri on *Liriodendron*: e USA
E. (*M.*) *lonicerae* on *Lonicera*: n hemis
E. (*U.*) *macrospora* on *Celtis, Ulmus*: c, e USA; e Can
E. (*M.*) *magnifica* on *Magnolia*: e USA, Japan
E. (*U.*) *necator* on *Ampelopsis, Cissus, Parthenocissus, Vitis*: global
E. (*M.*) *neglecta* on *Ulmus*: c USA
E. (*M.*) *nemopanthi* on *Ilex, Nemopanthus*: e USA, Can
E. (*M.*) *neomexicana* on *Forestiera pubescens*, AZ

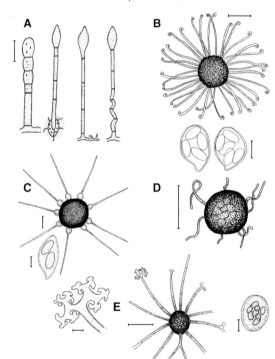

Figure 3. Powdery mildew fungi. A. Modes of conidial production in the four anamorph genera: left to right, *Oidium*, *Oidiopsis*, *Ovulariopsis*, and *Streptopodium*. B. Cleistothecium and asci of *Erysiphe adunca*. C. Cleistothecium and ascus of *Phyllactinia guttata*. D. Cleistothecium of *Podosphaera pannosa*. E. Cleistothecium, ascus, and detail of an appendage tip of *Podosphaera clandestina*. Scale bars = 100 μm for cleistothecia, 20 μm for other structures. A, C–E adapted from reference 494 by permission of Elsevier GmbH; B adapted from reference 290 by permission of the American Phytopathological Society.

A. *Erysiphe syringae* on *Syringa vulgaris* (common lilac) (NY, Sep).
B. Leaf curl, epinasty, reddish purple discoloration, and marginal necrosis caused by *Erysiphe pulchra* on *Cornus florida* (flowering dogwood) (KY, Sep).
C. Colonies of *Brasiliomyces trina* on *Quercus agrifolia* (California live oak) (CA, Mar).
D. *Erysiphe euonymi-japonici* on lower and upper surfaces of *Euonymus japonicus* leaves (CA, Apr).
E. *Phyllactinia guttata* on the undersurface of a leaf of *Q. kelloggii* (California black oak). Black dots are cleistothecia (CA, Sep).
F. Cleistothecia of *P. guttata* in various stages of development on a sparse mycelial mat on the lower surface of a *Calycanthus occidentalis* (western sweetshrub) leaf (CA, Nov).

E. (M.) ornata on *Betula*: NA, Asia
E. (M.) palczewskii on *Caragana arborescens*, nw USA, Eurasia
E. (U.) parvula on *Celtis*: c, e, nw USA
E. poeltii on *Hydrangea*: USA
E. (M.) platani on *Platanus*: NA, Eur, Australia
E. pseudopusilla (*M. pusilla*) on *Euonymus*: c, e USA
E. (M.) pulchra on *Cornus*: c, e NA; e Asia
E. (M.) ravenelii on *Gleditsia triacanthos*: c, e USA; Can
E. (M.) semitosta on *Cephalanthus occidentalis*: NA
E. sepulta on *Chrysothamnus nauseosus*: w USA
E. (M.) sydowiana on *Ceanothus velutinus*: WA
E. (M.) symphoricarpi on *Symphoricarpos*: NA
E. (M.) syringae on *Forestiera, Fraxinus, Ligustrum, Syringa*: n hemis, Australia
E. (M.) tortilis on *Cornus*: MO
E. (M.) vaccinii on *Andromeda, Epigaea, Gaultheria, Gaylussacia, Kalmia, Ledum, Lyonia, Menziesia, Rhododendron, Vaccinium*: NA
E. (M.) vanbruntiana var. *vanbruntiana* on *Sambucus*: n hemis
E. viburni (*M. sparsa*) on *Viburnum*: n hemis, New Zealand
Golovinomyces (*E.*) *californicus* on *Eriogonum*: w USA
G. (*E.*) *cichoracearum* on *Artemisia, Baccharis, Chrysothamnus*, and herbaceous Asteraceae: global
G. (*E.*) *orontii* on *Erica, Eucalyptus, Hibiscus, Hydrangea*, others in various families: global
G. (*E.*) *pseudosepultus* on *Chrysothamnus*: w USA
Leveillula taurica on *Hibiscus, Prosopis* in USA, more woody plants elsewhere: global
Oidium caricae-papayae on *Carica*: s USA, other subtropical and tropical regions
O. hortensiae on *Hydrangea macrophylla*: global
O. tingitaninum on *Citrus*: NA, SA, Africa, Asia
Ovulariopsis farinosa on *Zanthoxylum*: FL, Mesoamerica
Phyllactinia angulata on *Castanea, Fagus, Quercus, Ulmus*: NA
P. elaeagni on *Elaeagnus commutata*: nw USA
P. fraxini on *Fraxinus*: n hemis
P. guttata on *Acer, Aesculus, Alnus, Amelanchier, Aralia, Berberis, Betula, Calycanthus, Carpinus, Carya, Castanea, Catalpa, Celastrus, Cephalanthus, Chionanthus, Cladrastis, Cornus, Corylus, Crataegus, Cydonia, Elaeagnus, Fagus, Gleditsia, Hamamelis, Holodiscus, Ilex, Juglans, Lagerstroemia, Liriodendron, Magnolia, Malus, Melia, Morus, Ostrya, Philadelphus, Physocarpus, Platanus, Populus, Prunus, Quercus, Rhus, Ribes, Robinia, Rosa, Rubus, Salix, Sambucus, Sassafras, Shepherdia, Sorbus, Tilia, Ulmus, Vaccinium, Zanthoxylum*, and other woody plants outside NA: global
P. mali on *Crataegus* in NA, and on other woody rosaceous plants elsewhere: n hemis
Pleochaeta polychaeta on *Celtis*: se USA, SA
P. prosopidis on *Prosopis*: TX, SA
Podosphaera aphanis var. *aphanis* on *Eucalyptus, Potentilla, Rubus*: global
P. aphanis var. *physocarpi* on *Physocarpus*: e, c USA; Can
P. biuncinata on *Hamamelis*: e USA, e Can
P. (*S.*) *caricae-papayae* on *Carica papaya*: global
P. clandestina on *Amelanchier, Crataegus, Cydonia, Diospyros, Holodiscus, Malus, Prunus, Pyracantha, Pyrus, Spiraea, Symphoricarpos*: n hemis, Australia, New Zealand
P. leucotricha on *Cydonia, Malus, Photinia, Prunus, Pyrus*: global
P. (*S.*) *mors-uvae* on *Ribes*: n hemis
P. myrtillina on *Vaccinium*: n hemis
P. (*S.*) *pannosa* on *Lycium, Photinia, Prunus, Rosa*: global
P. (*S.*) *phytoptophila* on *Celtis*: c, ne USA; ON
P. (*S.*) *pruinosa* on *Rhus*: NA
P. (*S.*) *shepherdiae* on *Shepherdia*: w USA, w Can
P. (*S.*) *spiraeae* on *Spiraea*: n hemis
P. tridactyla on *Prunus, Sorbus, Spiraea*: global except Africa
Sawadaea bicornis on *Acer*: nw USA, BC; widespread in Eurasia
Typhulochaeta couchii on *Quercus arizonica*: AZ

Host specialization of powdery mildew species on woody plants is usually at the plant genus or family level, although some mildew species have only one known host. *Erysiphe liriodendri* on *Liriodendron tulipifera* (tuliptree) and *E. euonymi-japonici* on *Euonymus japonicus* (evergreen euonymus) are examples. Mildew species such as *Leveillula taurica* and *Phyllactinia guttata* that have hosts in diverse families are believed to be assemblages of biologically separate but similar host-specialized populations. Further host specialization occurs by natural selection when pathogens are confronted by resistant cultivars of susceptible plant species. A resistant cultivar functions as a selective medium for growth of any strain that can overcome or evade the resistance.

Mildew strains that can overcome or evade host resistance are usually called races.

Signs. The characteristic fruiting structure of powdery mildew fungi is a closed ascocarp, or cleistothecium, a tiny sphere usually 0.1–0.2 mm in diameter that forms on or in a mycelial mat on the plant surface (Plate 3F). When mature and hydrated, it ruptures to liberate ascospores from microscopic sacs (asci) within the sphere. Cleistothecia are first colorless, then yellow, brown, and finally black in most species. They may appear partially collapsed when dry. Their functions, in addition to sexual reproduction, are survival and dispersal. They have appendages of various forms that diagnosticians consider together with internal features (characteristics of wall cells, single versus multiple asci, ascus size) for identification. The appendages provide anchorage, initially to the mycelial colony, and later, after dispersal, to surfaces of survival sites such as bark of a host stem. Cleistothecia of a few species are so distinctive as to be identifiable with just a hand lens. Those of *Brasiliomyces trina* on *Quercus* (oaks) in California, for example, are yellow-brown, relatively transparent, and consistently abundant. Those of *Pleochaeta polychaeta* on *Celtis laevigata* (sugarberry) are two to three times as large as most others, reaching more than 0.3 mm in diameter and easily visible to the unaided eye. Young powdery mildew colonies lack cleistothecia, and some species produce them rarely or not at all.

Diagnosis of powdery mildew anamorphs (asexual states) is based on microscopic features of conidiophores and conidia and identity of host plants. Species of *Cystotheca, Erysiphe, Golovinomyces*, and *Podosphaera* have anamorphs in *Oidium* (mycelium external and conidia produced in chains). *Phyllactinia* species have anamorphs in *Ovulariopsis* (mycelium both external and internal, the former producing conidia singly atop the conidiophores). *Pleochaeta* species have anamorphs in *Streptopodium* (mycelium both external and internal, the former producing conidia singly atop twisted conidiophores). *Leveillula* species have anamorphs in *Oidiopsis* (external and internal mycelium with conidiophores emerging from stomata and producing conidia singly; Fig. 3). The conidia of powdery mildews are single-celled, colorless, and relatively large; those of most species are in the range 20–50 × 12–20 μm.

Disease cycles. Powdery mildew colonies are initiated by ascospores and conidia. Both types of spores germinate by producing germ tubes that differentiate into branched hyphae. Appressoria (swollen hyphal outgrowths that adhere to the host surface) form on germ tubes and on superficial hyphae and produce narrow penetrating hyphae (penetration tubes) that initiate infection and give rise to haustoria. Hyphae of *Leveillula* species may enter hosts through stomata. A powdery mildew cycle fits one of three patterns. (1) The fungus may produce mycelium and conidia throughout the year, a behavior common in warm regions and in greenhouses. The conidia initiate multiple overlapping cycles of disease. *Erysiphe euonymi-japonici* on *Euonymus japonicus* (Plate 3D) persists this way. (2) The fungus may survive winter or hot, dry summer as cleistothecia, then start new (primary) infections by means of ascospores dispersed in air to susceptible organs. Conidia produced on the growing colonies then initiate secondary cycles, and cleistothecia form near the end of the host's growth period. *Erysiphe syringae* on *Syringa* (lilac; Plate 3A) and *Brasiliomyces trina*

A, B. *Podosphaera clandestina* on fruits of *Symphoricarpos albus* (snowberry). Berries that were attacked while young ceased growth. Black specks are cleistothecia (CA, Sep).

C. *P. clandestina* on *Prunus ilicifolia* (holly-leaf cherry). Infection causes distortion and reddening of leaf tissue beneath the white mat of mildew (CA, Jul).

D–F. *Podosphaera pannosa* var. *rosae* on *Rosa* sp. Uniformly mildewed leaves (D) grew from an infected bud. Discrete colonies on leaves (E) represent secondary infections. Powdery mildew on canes (F) typically appears as white, feltlike patches. Cleistothecia are most common on canes and thorns (CA; Aug, Sep, & Jul, respectively).

G. *Erysiphe* sp., probably *E. poeltii*, causing premature senescence and browning of *Hydrangea* leaves (CA, Nov).

on *Quercus* (oak; Plate 3C) have this pattern. Cleistothecia do not always function for survival, however. Those of *Podosphaera pannosa* on *Rosa* and *P. clandestina* on *Crataegus* (hawthorn), for example, degenerate during winter in some areas. (3) The fungus may colonize primordial organs within buds and spend the winter there. Shoots that develop from diseased buds are stunted, distorted, and covered by mildew, from which conidia are dispersed and initiate new colonies. Severe disease on some young shoots adjacent to shoots with no mildew indicates that the pathogen overwintered in buds. Powdery mildew fungi that overwinter in colonized buds include *Cystotheca lanestris* on *Quercus agrifolia* (California live oak; Plate 5A, B), *Erysiphe alphitoides* on *Quercus*, *E. australiana* on *Lagerstroemia indica* (crape-myrtle; Plate 5F, G), *Podosphaera clandestina* on *Crataegus*, *P. leucotricha* on *Malus* (apple; Plate 5C, D), *P. mors-uvae* on *Ribes* (currant), and *P. pannosa* on *Rosa* (Plate 4D–F) and *Prunus persica* (peach). Some powdery mildews, such as *E. syringae* on *Syringa* (lilac) and *Uncinula necator* on *Vitis* (grapevine), have more than one mode of survival, but in a given situation, one predominates.

Epidemiology. Spore germination and penetration of the plant surface usually occur in 6 hours or less except in cool weather. Many powdery mildews can produce a new generation of spores in 4–6 days under favorable conditions with temperatures in the range 15–28°C. A mildew colony may continue producing spores for several weeks. Conidia are released in greatest numbers during daytime in response to changes in humidity or wetting of the plant surface. They are dispersed primarily by air, but also by water. Unlike spores of most other fungi, powdery mildew spores have high water content, do not need to imbibe water before germination, and can germinate on dry surfaces. Many powdery mildews are inhibited by free water. Conidia of some species (*Leveillula taurica*, for example) can germinate when humidity is low, but colonies of all these fungi grow best in humid air. They increase markedly on shaded plant parts during dry weather with warm days and cool, humid nights.

Many plants are most susceptible while new shoots are developing and leaves expanding; mature leaves and fruits are resistant. Cultural measures, such as trimming, fertilizing, and cultivating, that stimulate or prolong succulent plant growth encourage powdery mildew on such plants. In other plant-mildew interactions, mature foliage is highly susceptible, and disease builds up markedly in late summer.

References: 7, 29, 60, 290, 489, 494, 497, 498, 683, 785, 1099, 1298, 1676, 1859, 2727, 2999, 3310, 3379, 3673, 3774, 4162, 4459, 4460, 4474, 4476

Representative diseases and pathogens. Powdery mildew of lilac (Plate 3A) is caused by *Erysiphe syringae*, which infects various plants in the Oleaceae. Colonies become noticeable on leaves in summer and increase during the remainder of the growing season. *Syringa vulgaris* (common lilac) cultivars are more severely affected than those of most other lilacs. In California *E. syringae* may overwinter in buds, as indicated by mildew on scattered new shoots among predominantly healthy ones in spring. This pattern is not observed in regions with cold winters.

E. pulchra (Plate 3B) infects *Cornus* (dogwood) in eastern North America and eastern Asia. *C. florida* (flowering dogwood) is severely affected. Mycelium develops on green leaves and causes reddish or yellowish blotches, reddish purple discoloration, leaf curl, epinasty (downward bending of petioles), marginal necrosis, and premature leaf drop. Some cultivars are resistant.

E. euonymi-japonici (Plate 3D), although found only on *Euonymus japonicus*, occurs wherever its host plant grows. It forms a thick felt with abundant conidia on leaves, causing some yellowing and defoliation. Its cleistothecia have been found in Europe, but elsewhere only its anamorph, *Oidium euonymi-japonici*, is known.

Phyllactinia guttata (Plate 3E, F) is probably a complex of morphologically similar biological species. Host-specialized subgroups include *P. guttata* formae speciales *alni, betulae, carpini, coryli,* and *fagi* on *Alnus* (alder), *Betula* (birch), *Carpinus* (hornbeam), *Corylus* (hazelnut), and *Fagus* (beech), respectively. *P. guttata* sometimes causes distortion of succulent late-season shoots on deciduous plants and premature yellowing of year-old leaves on broadleaved evergreens. It probably overwinters as cleistothecia in temperate areas.

Podosphaera clandestina attacks primarily members of the Rosaceae, including the genera listed on page 10 plus *Mespilus* and *Sorbus* in Europe. Nonrosaceous hosts include *Diospyros virginiana* (common persimmon) and *Symphoricarpos* (snowberry; Plate 4A, B). In North America *P. clandestina* is most important on *Prunus* (stone-fruit trees). It overwinters as cleistothecia on *Prunus* (cherry) leaves and bark in interior northwestern USA, but as mycelium in buds of *Crataegus* in England.

Podosphaera pannosa (Plate 4D–F) is the common powdery mildew of roses. Its variety *rosae* attacks *Rosa* and *Photinia*; the variety *persicae* attacks *Prunus armeniaca* (apricot), *P. persica,* and *P. domestica* (common plum). It survives winter in buds or, in warm areas and greenhouses, reproduces continually on succulent parts. Its cleistothecia, uncommon in warm regions, form in the mycelium around thorns and flower receptacles. In England the ascospores in cleistothecia degenerate during winter. Rose species and cultivars vary from highly susceptible to apparently immune to *S. pannosa*. Races of the pathogen vary in ability to attack resistant cultivars.

Erysiphe poeltii, a pathogen of *Hydrangea* (Plate 4G), is among the many species that were collectively called *E. polygoni*. It is one of several powdery mildews, including *Golovinomyces orontii* and *Oidium hortensiae*, that affect hydrangea. The latter two are also widespread.

Brasiliomyces trina (Plate 3C) and *Cystotheca lanestris* (Plate 5A, B) infect *Quercus* species in California. The latter fungus often causes shoots growing from diseased buds to develop into witches'-brooms with dwarfed, pale leaves covered with mycelium. Brooms are most common on *Q. agrifolia* and occur also on *Q. chrysolepis, Q. ilex, Q. kelloggii, Q. laurifolia, Q. nigra,* and *Q. phellos* (canyon, holly, California black, laurel, water, and willow oaks) and *Lithocarpus densiflorus* (tanoak). In southern California, *C. lanestris* colonies survive winter on *Q. agrifolia* leaves. Farther north, the fungus overwinters in buds. The witches'-brooms often die during winter. Those that survive may produce either more infected shoots or normal ones the next spring. Two other powdery mildews, *Podosphaera phytoptophila* on *Celtis* (hackberry and sugarberry; Plate 6A–E) and *P. aphanis* var. *physocarpi* on *Physocarpus* (ninebark; Plate 5E), are also believed to cause witches'-brooms.

Podosphaera leucotricha (Plate 5C, D) infects several members of the rose family but is most important on *Malus* (apple, crabapple). Severe infection can kill shoots and blossoms. The fungus overwinters in buds. Cleistothecia develop in some areas but are apparently unimportant for the pathogen's survival. Diseased buds and shoots are abnormally susceptible to death by freezing in winter. Some mildew-resistant crabapples are listed opposite Plate 42.

Erysiphe australiana causes powdery mildew of *Lagerstroemia indica* (crape-myrtle; Plate 5F, G). The fungus overwinters in buds and causes stunting and floral abortion on young shoots. Diseased leaves and buds usually drop within a few weeks, but the tips of diseased shoots often outgrow infection. The disease subsides during hot summer weather and develops again in autumn. Cleistothecia are rare.

References: 7, 60, 304, 489, 782, 813, 1054, 1099, 1221, 1239, 1298, 1629, 1642, 1859, 2056, 2057, 2091, 2525, 2526, 2896, 2897, 3148, 3219, 3743, 3774, 3805, 4162, 4299, 4389, 4474

A, B. Witches'-brooms caused by *Cystotheca lanestris* on *Quercus agrifolia* (California live oak). Shoots from diseased buds developed into brooms covered with white mycelium and spores. Year-old leaves are unaffected (CA, Jul).

C, D. *Podosphaera leucotricha* on a *Malus pumila* (apple) shoot that grew from a diseased bud. Mildew at leaf axils enters buds such as the one that was exposed (arrow in D) by removing a leaf near the base of the shoot (NY, Jun).

E. Witches'-broom believed to be caused by *P. aphanis* var. *physocarpi* on *Physocarpus opulifolius* (ninebark); the dark color is caused by large numbers of black cleistothecia on leaves (NY, Oct).

F, G. *Erysiphe australiana* on *Lagerstroemia indica* (crape-myrtle). F. A shoot from a diseased bud is white with powdery mildew; green leaves at its base are from a different bud (FL, Apr). G. Powdery mildew-induced distortion and death of leaves and flowers (CA, Oct).

13

Witches'-broom of Hackberry and Black Witches'-broom of Serviceberry (Plate 6)

Witches'-broom of Hackberry

Brooms attributed to the joint action of a powdery mildew fungus and an eriophyid mite disfigure *Celtis occidentalis* (hackberry) throughout much of its range. *C. laevigata* (sugarberry) is also affected. The powdery mildew fungus is *Podosphaera phytoptophila* (syn. *Sphaerotheca phytoptophila*) and the mite is *Aceria celtis* (syn. *Eriophyes celtis*). The two agents are readily found on buds and shoots of brooms, but neither has been tested for ability to induce broom formation. In Illinois, a survey of 39 trees with brooms revealed mites in the buds of brooms on 37; cleistothecia of the powdery mildew fungus were found in the buds on only 15 trees, and then only in buds infested with mites. Powdery mildew fungi other than *S. phytoptophila* are sometimes found associated with the mites. One interpretation of these observations is that the mites induce brooms and the mildew fungus exploits the habitat thus created. The severity of attack varies greatly among trees in close proximity to one another, but reasons for this are unknown. Trees growing in the open are more frequently and severely affected than those in woodlands. Large trees sometimes bear hundreds of brooms without obvious loss of vitality.

The brooms arise initially from single mite-infested buds, each of which produces shoots with more infested buds. A loose broom, often with subsidiary tight clusters of twigs along its axis, may form on a vigorously growing branch (Plate 6B). Slowly growing branches lose the apically dominant habit and develop tight brooms. Each broom consists of numerous short twigs that arise close together, often at a conspicuous swelling or knot on the branch. Many twigs in the broom die back during the dormant season. Buds on surviving twigs are abnormally numerous, unusually grayish, and broader with looser scales than those on normal twigs. Loose bud scales are the result of swollen internal parts.

Mites in all stages of development can be found throughout the year beneath bud scales and on the primordial shoots within buds (Plate 6D, E). They are most numerous in late summer, when populations as large as 2000 per bud have been reported. In the central USA, females begin to produce eggs in May, and new generations develop throughout the growing season. An eriophyid mite similar to those associated with *Celtis* is shown magnified in Plate 208.

During spring and early summer, a sparse mat of mycelium and conidiophores of the powdery mildew fungus grows on green stems, petioles, buds, and sometimes the lower surfaces of leaves of the brooms. Fruit bodies (cleistothecia) 60–75 µm in diameter form on this mat and may be found throughout the year, most numerous on and between bud scales (Plate 6C). Ascospores mature beginning in autumn and are presumed to be dispersed in spring. Whether primary infections in spring are initiated by ascospores or conidia is unknown. Conidia could have this function if they are produced on diseased shoots growing from buds in which the fungus overwintered as mycelium. Conidia are presumed to cause secondary infections.

References: 67, 498, 2029, 2031, 3747

Black Witches'-broom of Serviceberry

Various species of *Amelanchier* (serviceberry) are affected by black witches'-broom from coast to coast in Canada and as far south as the Carolinas and New Mexico. The disease is common in woodlands, especially on moist sites. Landscape specimens are seldom affected. The pathogen, *Apiosporina collinsii* (Pleosporales, Venturiaceae), produces perennial mycelium in branches and dark mycelium, conidiophores, and pseudothecia on leaf surfaces (Fig. 6).

Symptoms and signs. The brooms consist of clusters of twigs on short, swollen branches. Brooms growing on shaded branches remain compact, and the diseased branch bends toward the ground at the point of initial infection. Diseased branches on trees in the open develop loose brooms with more regular branching. If a broom maintains a distinct axis, this may become S-shaped as twig tips turn upward. Many twigs in brooms die back during winter. Leaves on a broom may be dwarfed and yellowish, depending on the plant species and on age and vigor of the broom. Where summers are dry, as in the Northwest, the first crop of diseased leaves may be cast and

replaced at midsummer by a second crop more sickly than the first and likely to be killed by autumn frost. Branches with brooms lose vigor as years pass, but the disease has little effect on the general health of a plant unless brooms are very numerous. Some brooms snap off under the weight of snow.

Signs are present throughout the year. Soon after leaves unfold from diseased buds, hyphae emerge from stomata or from between epidermal cells and form a mycelial mat on which conidiophores and conidia develop. This anamorph has been reported as a *Cladosporium*. A different generic assignment is needed for it, however, because *Cladosporium* species are anamorphs of *Mycosphaerella*, unrelated to the Venturiaceae. The mycelial mat is initially thin and olive colored, but it thickens, expands, and turns dark brown to black, eventually covering the entire undersurface of the leaf except for the midrib and major veins. Pseudothecia begin developing in early summer. When mature, they are black spheres 150–250 µm across, each with an irregular pore through which spores are released in spring. Pseudothecia may cover the entire lower leaf surface. Leaves die soon after pseudothecia form, but many remain attached for at least a year after death.

Disease cycle. Both ascospores and conidia are disseminated by wind during an extended period in spring and early summer, and both are believed capable of initiating primary infections. The spores germinate in tiny bark fissures or axils of leaves or buds, and hyphae growing from these sites invade twigs, growing primarily in the cortex and phelloderm toward the twig apex. They cause hypertrophy and hyperplasia, which account for the swelling of diseased twigs. Buds, petioles, leaf blades, and flowers are also colonized. Broom formation begins when shoots grow from colonized buds. Conidia and immature pseudothecia are produced on current-season leaves. Ascospores are released from pseudothecia on dead year-old leaves. Whether secondary cycles occur is unknown.

Pruning is feasible for control of black witches'-boom on ornamental *Amelanchier* specimens because the pathogen does not grow far toward mainstems. Host defenses, expressed as cork formation in healthy cortex adjacent to the diseased tissue, are believed to restrict proximal growth of the pathogen.

References: 804, 1614, 1859, 2041, 3806

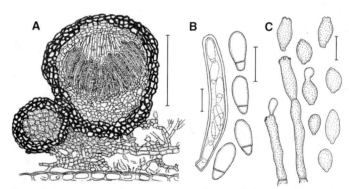

Figure 6. Apiosporina collinsii. A. Section of a pseudothecium on a leaf. B. Ascus and ascospores. C. Conidiophores and conidia. Scale bars: A = 100 µm; B, C = 10 µm. Adapted from reference 804 by permission of Public Works & Government Services Canada.

A–E. Witches'-brooms on *Celtis occidentalis* (hackberry). A, B. Brooms on a mature tree (NY, May). C. A diseased bud with black cleistothecia of the powdery mildew *Podosphaera phytoptophila* (NY, Sep). D, E. A bud dissected to detect eriophyid mites. Mites, about 200 µm long, are visible in magnified view (arrows in E) (NY, Sep).

F–H. Black witches'-broom of *Amelanchier* (serviceberry), caused by *Apiosporina collinsii*. F. Witches'-broom on *Amelanchier canadensis* (Canadian serviceberry) (NY, Jul). G. Shoots of *Amelanchier* sp. in the first year of infection. Lesions or entire leaves have a dark mycelial mat on their undersides except on veins (NY, Jul). H. Pseudothecia of *Apiosporina collinsii* on the lower surface of a dead leaf (NY, Aug).

Dark Fungi on Plants: Sooty Molds, Black Mildews, and Others (Plate 7)

Many and diverse fungi with brown or black cell walls darken the surfaces of plants. Broad groups of these fungi are popularly called sooty molds and black mildews. Although conspicuous, most of them do not cause much damage.

Sooty molds and generalized saprobes. Sooty molds, as popularly understood, are saprobes that grow on plant surfaces and derive nourishment from insect and plant secretions. Some of these fungi are specialized for life on living plants; others are generalized opportunists that grow wherever nutrients and moisture are adequate. Mycologists often apply the term *sooty mold* only to ascomycetes and related anamorphs in the orders Capnodiales and Chaetothyriales whose habitats are the surfaces of living plants. More than 40 genera of these "true" sooty molds are recognized, arrayed in the Antennulariellaceae, Capnodiaceae, Chaetothyriaceae, Euantennariaceae, and allied families. Genera with numerous or well-known species include *Capnodium*, *Chaetothyrium*, *Euantennaria*, *Metacapnodium*, and *Scorias*. Most of these fungi have no known host preferences. Multiple species may intermingle on a leaf or stem.

Sooty mold colonies vary in gross appearance from thin dark patches or coverings to irregular masses. Sooty mold mats on leaves sometimes resemble black tissue paper that tears after aging and drying, as shown on a *Citrus* leaf (Plate 7F). Some sooty mold fungi produce spongelike masses of thick-walled hyphae on leaves, twigs, and branches. *Scorias spongiosa*, for example (Fig. 7A; Plate 7C), occurs on *Alnus* (alder), *Betula* (birch), *Fagus* (beech), *Pinus* (pine), and other trees in eastern North America. The hyphal masses, called stromata, are soft when moist, brittle when dry. Pycnidia and later ascocarps develop on the stromata.

Sooty-mold fungi reproduce by ascospores and/or conidia, and some species form spores by fragmentation of hyphae. The species that have been studied grow readily in pure cultures on artificial media. Sooty-mold hyphae are hygroscopic, so they can absorb water from humid air as well as from wet surfaces. Their principal nutritional substrate is honeydew, a liquid secretion released from the anus of aphids, soft scales, mealybugs, psyllids, and some species of leafhoppers. Honeydews are complex mixtures of sugars, amino acids, proteins, other organic substances, and minerals derived from the insects' host plants. Droplets of honeydew often fall from insect-infested leaves or branches and stick to other plant parts or to other objects. Honeydew and spores of sooty-mold fungi are dispersed in water during rain. Thus, dark fungal deposits often occur on plant parts and other surfaces not infested by insects. Some sooty molds grow on organic substances secreted by normal leaves and green twigs. For example, sooty molds commonly associate with the glandular trichomes on *Catalpa* leaves in eastern USA.

Plant surfaces where organic nutrients accumulate are also colonized by common dark hyphomycetes such as *Aureobasidium* and *Cladosporium* species, which are ubiquitous opportunists that exploit diverse habitats. *A. pullulans* is notorious for its ability to discolor wood and painted surfaces. The generalized saprobes and "true" sooty molds are not usually found together, however. In one study of dark fungi on *Citrus* leaves, "true" sooty molds dominated samples from Florida, while generalized opportunists were found in similar circumstances in California.

Sooty molds in the broad sense are economically important on some tree crops that grow in warm, humid regions—notably *Citrus*, *Coffea* (coffee), and *Mangifera indica* (mango)—because they diminish the value of blemished fruit. Heavy accumulation of sooty mold blocks light and can suppress net photosynthesis, as demonstrated in *Carya illinoinensis* (pecan), but there are no reliable reports of growth or yield reduction due to sooty mold.

Black mildews and similar fungi. The term *black mildew* also has both broad and narrow connotations. The broad connotation includes both parasitic and saprobic dark fungi. Black mildews as considered here, however, are biotrophs of the order Meliolales that grow on the surfaces of leaves or green twigs and derive nourishment by penetrating epidermal cells and producing haustoria in them. Black mildews occur around the world in tropical and humid warm-temperate regions, appearing as dark patches or mats, often with prominent setae (sterile hairs). They reproduce only by means of ascospores. Nearly 2000 species have been described, based in part on the unsupported presumption that each species is

host specialized. The principal genera are *Amazonia*, *Asteridiella*, *Irenopsis*, and *Meliola*. *M. palmicola*, which infects *Serenoa repens* (scrub palmetto; Plate 7G, H), is typical of the group. It appears as circular black patches that on close examination are seen to consist of branched, radiating hyphae that have many short, lateral, microscopic appressoria (formerly called hyphopodia). These provide anchorage and points from which the parasite sends haustoria into epidermal cells.

Other fungi sometimes called black mildews include species of *Asterina* and *Lembosina* of the Asterinaceae (a family related to Capnodiales) that form dark patches on tree leaves. *Lembosina quercina*, as an example, forms black patches of hyphae that are initially superficial on *Quercus* (oak) leaves in southeastern USA. By late summer, however, the fungus penetrates the cuticle and develops between the cuticle and epidermis, causing browning and disorganization of epidermal cells.

Other dark fungi on plants. Many fungi in addition to those discussed above parasitize leaves and produce dark hyphae and spores on the surfaces. Some are related to the saprobic sooty molds and are classified in the Capnodiales and Chaetothyriales. For example, *Rasutoria abietis* (syn. *Epipolaeum abietis*), a member of the Euantennariaceae, forms dark patches on the lower surfaces of *Abies* (fir) needles (Plate 7B) in the Northwest. Colonized needles senesce prematurely, so *R. abietis* is presumed to be pathogenic. Other noteworthy groups of pathogens that cause premature senescence or localized necrosis and produce dark superficial hyphae and fruiting structures include *Alternaria* (Plate 41), *Apiosporina* (Plate 6), *Herpotrichia* (Plate 31), and *Venturia* (Plates 42–45).

References: 226, 231, 901, 1099, 1214, 1298, 1490, 1499, 1500, 1809, 2392, 2396, 2745, 2746, 2957, 3267–3271, 3669, 4424

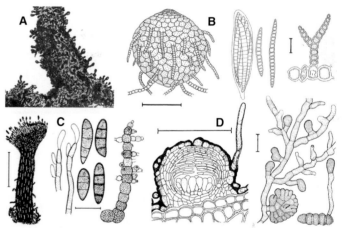

Figure 7. A–D. Sooty molds. A. *Scorias spongiosa*: part of a stroma with stalked pseudothecia extending from its surface. B. *Ophiocapnocoma phloiophilia*: perithecium, ascus, ascospores, and a branching hypha. C. Anamorphs of *Metacapnodium spongiosum*: a synemma with conidiophores and conidia, and a large hypha with short phialides. D. The black mildew *Meliola floridensis*: section of a young perithecium, hyphae with appressoria, and a germinating ascospore. Scale bars: synemma = 200 μm, perithecium = 100 μm, other structures = 25 μm. A adapted from reference 3267 by permission of the author; B from reference 226 by permission of NRC Research Press; C from reference 3669 by permission of Gebrüder Borntraeger Verlagsbuchhandlung; D from reference 2396 by permission of the Mycological Society of America.

A. Unidentified sooty mold on year-old needles of *Pinus ponderosa* (ponderosa pine) (OR, Jul).

B. *Rasutoria abietis* on lower surfaces of *Abies grandis* (grand fir) needles, associated with premature senescence (ID, Jul).

C. *Scorias spongiosa* on a *Fagus grandifolia* (American beech) twig. The fungus is growing on remains of an aphid colony (NY, Aug).

D–F. Unidentified sooty molds on *Gardenia jasminoides* (common gardenia) (FL, Apr), *Ixora coccinea* (scarlet ixora) (PR, Feb), and *Citrus* sp. (FL, Nov).

G, H. *Meliola palmicola*, a black mildew, on *Serenoa repens* (scrub palmetto). Close view shows branched, radiating dark hyphae (FL, Mar).

Spot Anthracnoses and Scabs Caused by *Elsinoë* and *Sphaceloma* (Plate 8)

More than 50 host-specialized species of *Elsinoë* (Myriangiales, Elsinoaceae) and the related anamorph genus *Sphaceloma* cause small lesions and scabby eruptions on leaves, fruits, green shoots, and other succulent parts of woody plants. *Elsinoë* species produce minute (<100 μm) black fruit bodies (ascostromata) in killed tissue on living plant parts (Fig. 8A). The ascostromata are preceded by acervuli of the *Sphaceloma* state, in which conidia are produced. The acervuli are near-microscopic cuticular blisters (Fig. 8B) and often are the only fruiting structures found. If the ascostromal state is known, we use the name *Elsinoë*; if not, *Sphaceloma*.

The term *spot anthracnose* refers to tiny, sunken lesions caused by *Elsinoë* or *Sphaceloma*. *Scab*, on the other hand, may mean a disease caused by one of these fungi or by a *Venturia* species (Plates 42–45). Scab diseases caused by *Elsinoë* or *Sphaceloma* are characterized by localized scabby overgrowth of host tissues. The overgrowth results in part from the action of hormones secreted by the pathogens.

Spot anthracnose of snowberry. *Sphaceloma symphoricarpi* causes this disease, which disfigures leaves and fruits of *Symphoricarpos albus* (snowberry) and *S. orbiculatus* (coralberry) across North America. Dark purple to black lesions appear on leaves in spring and enlarge slowly, their centers becoming grayish. The lesions sometimes coalesce into large, irregular blotches of dead, cracked tissue. Multiple lesions on expanding leaves cause deformity, and severe disease partially defoliates the plants. Lesions also form on flower and leaf buds and green shoots. On fruits, circular, somewhat sunken, pink-brown lesions, most common at the calyx end, may cause lopsided growth. Diseased berries are often invaded by secondary fungi, particularly *Alternaria* species, that cause berry rot. Killed berries shrivel and remain in place for a time as brown mummies.

Euonymus scab. *Elsinoë euonymi-japonici* induces lesions that disfigure *Euonymus japonicus* (Japanese euonymus). Discovered first in Japan and later in the USA, the causal fungus may have come to North America with its host plant. Lesions are up to 2 mm across, grayish white with a raised orange-cinnamon, waxy-appearing margin and, in the larger spots, a raised, dark center on which minute white dots are visible. Lesions are more frequent on upper than lower leaf surfaces and may coalesce when numerous. The centers of large spots sometimes fall out. Lesions on stems are usually darker with raised, wrinkled, or fissured surfaces.

Spot anthracnose of dogwood. *Elsinoë corni* causes spot anthracnose of *Cornus asperifolia* (small rough-leaf cornel), *C. florida* (flowering dogwood), *C. kousa* (Korean dogwood), and *C. mas* (cornelian-cherry). This disease, not to be confused with anthracnose caused by *Discula destructiva* (Plate 52), occurs from Florida into northeastern USA and also in California. It is significant only on *C. florida*. Bracts, leaves, petioles, fruit, peduncles, and green stems are affected, resulting in malformed leaves and bracts and dead spots on stems and fruits. Lesions are 2 mm or less in diameter and reddish purple with tan centers. Spots on young leaves usually have yellow borders. Grayish tissues in foliar lesions tend to fall out, giving severely diseased leaves a ragged appearance.

Acervuli form in lesions until late in the growing season, but they are inconspicuous and often sparse. Primary infection in spring may be caused by conidia from overwintered lesions on twigs. Conidia from acervuli are presumed responsible for secondary cycles and some increase of the disease within a season. Black ascostromata mature in leaf and twig lesions in August in the mid-Appalachian region and are postmature by September. The role of ascospores in spot anthracnose is unknown. The disease is favored by wet weather. If uncontrolled for several years under conducive conditions, it can weaken trees so that flowers and foliage become sparse.

C. florida cultivars with pink bracts are less often disfigured than those with white bracts. The cultivars Cherokee Princess, First Lady, Fragrant Cloud, Purple Glory, Springtime, and Plena have shown apparent resistance in Tennessee.

Poinsettia scab. *Sphaceloma poinsettiae* causes scab on *Euphorbia pulcherrima* (poinsettia) and at least one annual species of *Euphorbia*. The fungus occurs in subtropical and tropical parts of the Americas and the South Pacific region. Severe infection causes wrinkling and distortion of leaves, premature leaf drop, and dieback of young stems. The most striking symptom, however, is a dramatic

stem elongation due to a growth-regulating chemical produced by the fungus that causes internodes to lengthen and shoots to extend 15 cm or more above those on healthy plants. Lesions on stems are raised, pale buff, up to 1 cm long, often surrounded by a reddish purple margin, and usually numerous. Lesions on petioles and leaf midribs are similar to those on stems. On leaf blades, however, the fungus causes brown spots, concave on the upper surface and up to 3 mm across or sometimes larger, that may cause a puckered appearance. Old stem lesions become grayish brown and covered with a velvety layer of conidiophores and conidia. Conidia, dispersed by splashing water, cause new infections. Symptoms appear within 7 days, and a new generation of spores is produced 14–30 days after infection.

Other *Elsinoë* and *Sphaceloma* pathogens on woody plants in the USA and Canada and their woody host genera are listed below. Plant species are listed if they are the only reported hosts in their genera.

Elsinoë cinnamomi on *Cinnamomum camphora*; *E. diospyri* on *Diospyros virginiana*; *E. fawcettii* on *Citrus, Clausena, Fortunella, Poncirus*; *E. ilicis* on *Ilex cornuta*; *E. jasmini* on *Jasminum*; *E. ledi* on *Gaultheria, Ledum, Leucothoe*; *E. lepagei* on *Manilkara, Pouteria*; *E. leucospila* on *Camellia, Ternstroemia*; *E. magnoliae* on *Magnolia grandiflora*; *E. mangiferae* on *Mangifera indica*; *E. mattirolianum* on *Arbutus*; *E. mulleri* on *Polyscias guilfoylei*; *E. parthenocissi* on *Parthenocissus*; *E. piri* on *Cydonia, Malus, Pyrus*; *E. quercicola* on *Quercus laurifolia*; *E. quercus-falcatae* on *Quercus falcata*; *E. randii* on *Carya illinoinensis*; *E. rosarum* on *Rosa*; *E. sesbaniae* on *Sesbania exaltata*; *E. spondiadis* on *Spondias purpurea*; *E. tiliae* on *Tilia*; *E. vaccinii* on *Vaccinium reticulatum*; *E. veneta* on *Rubus*; *E. ampelina* on *Vitis*.

Sphaceloma araliae on *Aralia, Schefflera*; *S. bucidae* on *Bucida buceras*; *S. cercocarpi* on *Cercocarpus betuloides*; *S. choisyae* on *Choisya dumosa*; *S. crataegi* on *Crataegus*; *S. hederae* on *Hedera helix*; *S. morindae* on *Morinda royoc*; *S. murrayae* on *Salix*; *S. oleandri* on *Nerium oleander*; *S. perseae* on *Persea americana*; *S. psidii* on *Acca, Psidium*; *S. punicae* on *Punica granatum*; *S. ribis* on *Ribes*; *S. viburni* on *Viburnum*.

Undetermined or undescribed species of *Elsinoë* or *Sphaceloma* have been reported on *Bignonia capreolata, Camellia, Catalpa speciosa, Conocarpus erectus, Forestiera segregata, Leucospermum, Myrica cerifera, Prunus ilicifolia, Rhododendron macrophyllum*, and *Sambucus canadensis*.
References: 245, 573, 1099, 1153, 1759, 1920, 1922, 1923, 1925, 3366, 3795, 3919, 3930

Figure 8. A. *Elsinoë veneta*: section of an ascostroma with asci and ascospores. B, C. *Sphaceloma crataegi*. B. Section of an acervulus. C. Conidiophores and conidia. Scale bars = 50 μm for A, B; 10 μm for C. A adapted from reference 573; B, C adapted from reference 3919 by permission of the Mycological Society of America.

A, B. Spot anthracnose on *Symphoricarpos albus* (snowberry) leaves, caused by *Sphaceloma symphoricarpi*. Fruit rot in A was caused by a different fungus, perhaps secondary to *S. symphoricarpi* (NY, Sep).

C, D. Scab of *Euonymus japonicus* (evergreen euonymus) caused by *S. euonymi-japonici*. Foliar lesions may coalesce and be enlarged by secondary fungi (FL, Apr).

E–H. Spot anthracnose of *Cornus florida* (flowering dogwood), caused by *Elsinoë corni*. E–G. Foliar lesions and associated distortion (MS, May; NY, Jul). H. Lesions on bracts (FL, Mar).

I, J. Scab caused by *S. poinsettiae* on a leaf and stem of *Euphorbia pulcherrima* (poinsettia) (FL, Apr).

Photo credit: I—A. W. Engelhard

Diseases Caused by *Mycosphaerella* Species and Related Anamorphs (Plates 9–17)

Overview

Mycosphaerella (Mycosphaerellales, Mycosphaerellaceae) is a large, taxonomically difficult genus associated with a wide array of plant diseases. The number of *Mycosphaerella* species that occur on woody plants in North America is unknown but probably exceeds 120. Many species are known only as anamorphs (asexual forms). Widely distributed woody genera such as *Acer*, *Alnus*, *Eucalyptus*, *Populus*, and *Quercus* are each host to many species. This account presents a sample of *Mycosphaerella* species and of the diversity of symptoms and signs.

Mycosphaerella species produce small black pseudothecia with two-celled, usually colorless, ascospores (Fig. 9). The pseudothecia lack pseudoparaphyses (hyphae among asci), which distinguishes *Mycosphaerella* from morphologically similar genera. Pseudothecia typically form in dead fallen leaves of deciduous species but sometimes in lesions on living leaves of evergreens and occasionally in dead stems or lesions on living stems. Usually only anamorphs of these fungi are found associated with disease symptoms. Anamorphs of *Mycosphaerella* are diverse. Some are coelomycetes, producing conidia in acervuli or pycnidia. Others are hyphomycetes, producing conidiophores and conidia on the surface of diseased tissue, either scattered or in minute clumps. Conidial shape, septation, and color and conidiophore form and pigmentation vary widely among these fungi. The following anamorph genera occur on woody plants and are currently accepted and linked to *Mycosphaerella*. Coelomycetous forms include *Colletogloeopsis*, *Colletogloeum*, *Dothistroma* (Plate 13), *Lecanosticta* (Plate 12), *Phaeophleospora*, *Phloeospora*, *Septoria* (Plate 11), and *Sonderhenia* (Plate 17). Hyphomycetous forms include *Batcheloromyces*, *Cercospora* (Plates 15, 16), *Cercosporella*, *Miuraea*, *Passalora* (Plate 15), *Phaeoisariopsis*, *Pseudocercospora* (Plates 15, 16), *Pseudocercosporella*, *Ramularia*, *Ramulispora*, *Stenella*, *Stigmina* (Plate 14), *Thedgonia*, *Uwebraunia*, and *Xenostigmina*. Spermagonial forms are assignable to *Asteromella* and *Asteroma*. Some genus names no longer appropriately linked to *Mycosphaerella* or reduced to synonymy under other genera include *Ascochyta*, *Cercosporidium*, *Cercostigmina*, *Cladosporium*, *Cylindrosporium*, *Marssonina*, *Phoma*, *Phyllosticta*, and several that are less well known in dendropathology. In the list and discussion that follow, names enclosed in quotation marks are misapplied, but no other name is available.

Many *Mycosphaerella* species have spermatial states in the genera *Asteromella* or *Asteroma*. These forms consist of tiny (<100 μm in diameter) flasklike spermagonia in which spermatia, sometimes called microconidia, are produced. Spermatia are sporelike cells on the order of 3–4 × 1 μm that function as gametes and are noninfectious. Spermatial states of a number of *Mycosphaerella* species were formerly classified in *Phyllosticta* or *Phoma*.

Relatively few of the *Mycosphaerella* diseases of woody plants in North America have been studied beyond the point of describing and naming the associated fungi. Some of the pathogens, their conidial (c) and spermagonial (s) states if reported, and their hosts are listed below or are discussed in relation to illustrations in Plates 9–11. Many more pathogens of woody plants are treated in the listed references.

Mycosphaerella caryigena: c *Pseudocercosporella caryigena*, downy spot of *Carya illinoinensis* (pecan)
M. cerasella: c *Pseudocercospora circumscissa*, s *Asteromella*, leaf spot of *Prunus* (cherry)
M. colorata: s *Asteromella*, leaf spot of *Kalmia latifolia* (mountain-laurel)
M. dendroides: leaf blotch of *Carya illinoinensis*
M. laricina: c *Pseudocercospora* sp., needle blight of *Larix* (larch)
M. liriodendri: c "*Phyllosticta*" *liriodendrica* on leaves of *Liriodendron tulipifera* (tuliptree)
M. milleri: c *Passalora* (syn. *Cercosporidium*) *magnoliae*, s *Asteromella*, leaf spot of *Magnolia*
M. mori: c *Phloeospora maculans*, leaf spot of *Morus* (mulberry)
M. pomi: c *Pseudocercosporella pomi*, Brooks fruit spot of *Chaenomeles* (flowering quince) and *Cydonia* (quince) and *Malus* (apple)
M. pruni-persicae: c *Miuraea persicae*, frosty mildew of *Prunus persica* (peach)
M. psilospora: c *Septoria querceti*, leaf spot of *Quercus* (oak)
M. pyri: c *Septoria pyricola*, leaf fleck of *Malus*, *Pyrus* (pear)
M. ribis: c *Septoria ribis*, s *Asteromella*, leaf spot of *Ribes* (currant)

M. tulipiferae: c *Phaeoisariopsis liriodendri*, s *Asteromella*, leaf spot of *Liriodendron*
References: 70, 490, 493, 495, 779, 806, 807, 847, 849, 850, 853, 1086, 1099, 1295, 1298, 1415, 1702, 1921, 2978, 3522, 3669, 3672, 3858, 3915, 3921, 4132

Leaf spots of ash. *Mycosphaerella effigurata* and *M. fraxinicola* cause necrotic spots on *Fraxinus* (ash) leaves. The latter fungus, which causes the disease illustrated in Plate 9A–E, occurs across North America on *F. americana* (white ash), *F. latifolia* (Oregon ash), *F. nigra* (black ash), *F. pennsylvanica* (green ash), and *F. quadrangulata* (blue ash). The brown lesions, usually 5–15 mm in diameter with yellow borders, become conspicuous in late summer. When numerous, they coalesce to form large brown blotches. Severely diseased leaflets may drop 4–6 weeks early, but no significant harm is done to the trees. In fact, premature defoliation of ash seedlings by leaf-spotting fungi in early autumn is welcomed in some nurseries because it accelerates the development of dormancy and thus allows early lifting for winter storage.

M. fraxinicola has three forms. Its pseudothecia mature in fallen leaves during host dormancy. Ascospores are discharged from pseudothecia from late spring to early summer and presumably initiate the first infections each year. In summer the anamorph *Pseudocercosporella fraxini* (syn. *Cylindrosporium fraxini*) develops on upper (and some lower) surfaces of lesions and produces masses of colorless, cylindrical, multiseptate conidia (Plate 9D). These spores presumably cause secondary cycles of disease. A spermatial state of the *Asteromella* type develops in late summer as tiny, black, flasklike structures in the lower surface that impart a dark hue to the lesions (Plate 9E).

M. effigurata also infects leaves of *Fraxinus* across the continent and can sometimes be found on leaves colonized by *M. fraxinicola*. Lesions caused by *M. effigurata* are usually numerous and less than 3 mm across (Plate 9F, G). Its pseudothecia develop in fallen

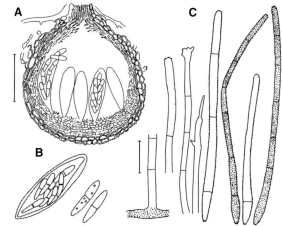

Figure 9. Mycosphaerella citri. A. Pseudothecium. B. Ascus and ascospores. C. Conidiophores and elongate multicelled conidia of the *Stenella* anamorph. Scale bars = 25 mm for A, 10 mm for B, C. Adapted from reference 3672 by permission of CAB International.

A–E. Leaf spots caused by *Mycosphaerella fraxinicola* on *Fraxinus* (ash). A, B, D, E. On *F. latifolia* (Oregon ash). A, B. Lesions with yellow margins. D. Magnified view of acervuli with masses of colorless conidia on the upper surface of a lesion. E. Magnified view of spermagonia on the lower surface of a lesion (OR, Sep). C. Lesions on *Fraxinus pennsylvanica* (green ash).

F, G. Leaf spots caused by *M. effigurata* on *Fraxinus* seedlings. F. On *F. pennsylvanica* (ND, Oct). G. On *F. americana* (white ash) (IN, summer).

H–J. Leaf spot caused by *M. juglandis* on *Juglans nigra* (black walnut) (H, I) and on a hybrid, *J. sieboldiana × cinerea* (J) (discussion, p. 22).

Photo credits: C—R. L. Anderson; F—J. A. Walla; G—P. T. Marshall; H–J—K. J. Kessler

overwintered leaves and produce ascospores that cause the first infections in spring. In summer the fungus produces colorless acervuli with two-celled colorless conidia on the lower surfaces of lesions. This life stage is *Septoria fraxinicola* (syn. *Marssonina fraxini*). It is followed by black spermagonia of an *Asteromella* state, also on the lower leaf surface.
References: 492, 495, 1614, 3290, 4412, 4415

Leaf spot of walnut. A leaf spot caused by *M. juglandis* (Plate 9H–J) induces premature yellowing and defoliation of *Juglans nigra* (black walnut) in midwestern USA and has also been found in North Carolina. The disease also affects *J. ailanthifolia* (Japanese walnut), *J. cinerea* (butternut), *J. regia* (English walnut), and hybrids. Lesions, initiated by ascospores from pseudothecia in overwintered leaves, develop shortly after leaves reach full size in late spring. The lesions are angular and up to 4 mm wide. Dark acervuli 50–100 μm in diameter develop mainly on the lower surfaces of lesions and produce colorless, elongate, multiseptate conidia. This form, "*Cylindrosporium*" *juglandis*, resembles *Septoria* species and induces secondary cycles of disease during summer. The conidia are dispersed by splashing and dripping water. Lesions often coalesce along veins and sometimes at leaflet tips. Severely diseased leaflets turn yellow and drop prematurely.
References: 2052, 2053

Greasy spot of Citrus. *Mycosphaerella citri* and some closely related species that attack *Citrus* cause diseases called greasy spot. *M. citri* occurs in the Americas and on Okinawa. *M. horii* and an unnamed species occur in Japan and Australia, respectively. The disease caused by *M. citri* disfigures fruit, on which it is called greasy spot rind blotch, and causes premature leaf drop beginning in autumn and continuing into spring. Leaf infections occur on all commercial *Citrus* and *Fortunella* (kumquat) cultivars, *Poncirus* (citrumelo and trifoliate orange), *Murraya paniculata* (orange-jasmine), and *Aeglopsis chevalieri*. Leaf loss, if early and severe, reduces fruit yield the next season.

Symptoms and signs. Tiny, raised, blisterlike eruptions appear on lower surfaces of leaves, and inconspicuous yellow mottling develops on upper surfaces. The blisters darken through shades of orange and brown to black, and both leaf surfaces appear as though spattered with a black greasy substance (Plate 10A—C). The greasy spots consist of groups of mesophyll cells that become swollen, necrotic, and infused with gum. Severely affected leaves turn yellow except where chlorophyll is retained close to the lesions. On fruit, pinpoint black specks appear between the oil glands, and development of yellow or orange color is delayed in the vicinity of the specks.

The conidial state of *M. citri*, *Stenella citri-grisea* (Fig. 9), develops superficially on living leaves, producing conidiophores and conidia on sparse hyphae without obvious relationship to lesions. If infected tissues are damaged by a secondary agent such as frost, however, the fungus produces clumps of conidiophores with conidia on necrotic parts of living leaves. Pseudothecia form in fallen diseased leaves and open through both leaf surfaces.

Disease cycle. Although infectious conidia are available all year, most infections are started by ascospores, which are discharged from pseudothecia in response to rain or sprinkler irrigation. The peak period of ascospore maturation and dispersal and new infection occurs in early summer. Ascospore dispersal may continue throughout summer, depending on weather and locality. The fungus enters leaves or fruit through stomata. Lesions visible to the unaided eye arise through the mass action of many individual infections close together. A period of 1–8 months, depending on weather and host, elapses between infection and the appearance of symptoms.

Although ascospore release requires leaf wetting, liquid water is not essential for germination and penetration. Relative humidity near 100% and temperature near 30°C allow spores to germinate and produce a sparse, superficial mycelium. The amount of infection that follows is determined in part by organic nutrients on the plant surface. Greasy spot often becomes severe, for example, if leaves are coated with honeydew. This insect-secreted sugary coating promotes growth of superficial mycelium and increases the frequency of hyphal penetration of the plant surface. Citrus rust mite infestation also promotes greasy spot severity.
References: 3147, 3672, 4031, 4332, 4334–4336, 4338

Leaf spot of rose. *Mycosphaerella rosicola* occurs wherever roses are grown in North America and elsewhere, causing round lesions, usually 2–4 mm but occasionally up to 10 mm across, that may coalesce into irregular dead areas. The lesions are light brown to tan or gray with a narrow, dark border or sometimes with a diffuse purple halo (Plate 10D–F). When severe, the disease causes premature defoliation. Microscopic stromata of the conidial state, *Passalora rosicola* (syn. *Cercospora rosicola*), with brown conidiophores and elongate olivaceous conidia, develop on the lesions. The conidia measure 35–100 × 3–6 μm and have one to six (mostly three or four) transverse septa. Successive generations of conidia, dispersed by air and splashing water, cause disease increase during summer. The spermatial state, of the *Asteromella* type, supplants the conidial state in late summer. The spermagonia are 50–70 μm in diameter, and the spermatia are rod-shaped cells measuring 4 × 1 μm. Pseudothecia form in fallen leaves and discharge ascospores in spring. These spores are presumed to cause the first infections each year. *M. rosicola* has been studied by plant chemists because it synthesizes the plant hormone abscisic acid, which is presumed to play a role in defoliation induced by the fungus.

Leaf spot of mountain-ash. A fungus named "*Phyllosticta*" *sorbi* (Plate 10G–I) is found in brown lesions on leaves of *Sorbus americana* (American mountain-ash) and *S. aucuparia* (European mountain-ash) from Maine to Illinois and Texas and in Europe. Severe infection induces premature yellowing and casting of leaflets, but the disease is usually inconsequential. *P. sorbi*, not a true *Phyllosticta*, has been suggested to be a spermatial state of *Mycosphaerella aucupariae*, but this is still a matter for research. In the Pacific Northwest, a true *Phyllosticta*, *P. globigera*, also spots *Sorbus* leaves.

Leaf spot of sycamore. The disease of *Platanus occidentalis* (sycamore) illustrated in Plate 10J–L, although undiagnosed, seems linked to *Mycosphaerella*, possibly to *M. stigmina-platani*. The mature lesions are round to irregular, up to 1 cm in diameter, tan to grayish in the center, have indefinite reddish brown margins, and often coalesce. When numerous, they induce generalized yellowing and necrosis. A microconidial fungus consistent with *Asteromella* (spermatial state of *Mycosphaerella*) occurs in these lesions. The spermagonia break through the lower leaf surface and appear as dark brown specks barely visible with a hand lens. They are subspherical, up to 90 μm in diameter, and contain spores measuring about 3 × 1 μm. Microconidial forms similar to this one have been reported from eastern and central North America as "*Phyllosticta*" *platani* and have been linked to *M. stigmina-platani*. Full diagnosis cannot be made based on microconidial forms alone, however. *M. stigmina-platani* is reported to cause necrotic spots and blotches on *P. occidentalis* and to produce spermagonia. It also produces, during summer, microscopic stromata with two conidial states (*Pseudocercospora* and *Xenostigmina*) not seen on the leaves shown here. *Platanus* leaves in the USA are also colonized by *M. albocrustata* and *M. platani*, but diseases associated with these fungi have not been characterized.
References: 431, 851, 900, 1758, 3963

A–C. Greasy spot, caused by *Mycosphaerella citri*, on leaves of *Citrus ×paradisi* (grapefruit) (A) and *C. sinensis* 'Valencia' (sweet orange) (B, C). In magnified view (C), lesions appear as black flecks or small, black, blisterlike eruptions (FL, Mar–Apr).

D–F. Leaf spot of *Rosa* sp., caused by *Mycosphaerella rosicola*. Minute spermagonia of the *Asteromella* state (arrows in F) are visible with a hand lens in the centers of the lesions (NH, Sep).

G–I. Leaf spot of *Sorbus americana* (American mountain-ash) with pycnidia of the microconidial fungus "*Phyllosticta*" *sorbi* (NH, Sep).

J–L. Leaf spot of *Platanus occidentalis* (sycamore). Tiny pycnidia of an *Asteromella* formerly called "*Phyllosticta*" *platani* appear as dark brown dots in L. They are probably the spermagonial state of a *Mycosphaerella* (MD, Sep).

Septoria diseases of poplar. *Mycosphaerella populicola* and *M. populorum* in their conidial states, *Septoria populicola* and *S. musiva*, respectively (coelomycetes), cause leaf spots on *Populus* (aspen and poplar) across North America. The latter fungus also causes cankers. The diseases are innocuous in natural stands, but *M. populorum* can devastate young plantings of highly susceptible genotypes. Diagnosis is based on identification of the conidial state of each pathogen. Canker diagnosis often requires isolation and identification of *M. populorum* in pure culture because secondary fungi such as *Cytospora chrysosperma* (Plate 85) may prevent sporulation of the *Septoria* in cankers.

All North American *Populus* species are at least somewhat susceptible to *M. populorum*. Hybrids and Eurasian species sustain the most damage. *M. populicola*, common in the Pacific Northwest, affects fewer species and is less virulent than *M. populorum*, usually causing only leaf spots. Hosts of *M. populicola* include *P. angustifolia* (narrowleaf cottonwood), *P. balsamifera* (balsam poplar) and its subspecies *trichocarpa* (black cottonwood), and *P. deltoides* (eastern cottonwood).

Symptoms and signs. The appearance of Septoria leaf spots varies within and between poplar species. Usually they first appear as sunken black flecks that enlarge to more or less round spots 1–15 (mostly 2–5) mm across, coalescing and forming dead blotches where they are numerous. Dead tissue fades to shades of brown, tan, or white with a brown or black margin, as shown on *P. balsamifera* subsp. *trichocarpa* (Plate 11F). Lesions on plants that are somewhat resistant usually remain small (1–2 mm) and may appear silvery.

Cankers develop only on trees with leaf infections. Most cankers occur on young trees, usually within 1.5 m of the ground. Infection begins at wounds, lenticels, stipules, and leaf bases. Cankers on side shoots of young plants commonly extend to the mainstem during the second year of infection. Diseased bark is initially black and often becomes tan in the center of a lesion as pycnidia develop. Pycnidia are common in young cankers but difficult to find in older ones. Cankers may girdle stems during the first season or may be halted by host defenses. Water stress may diminish these defenses. Woundwood ridges at the margins of arrested lesions eventually grow together, restoring stem symmetry. Trunks of highly susceptible trees thus attacked may be girdled within 1–2 additional years. Stem breakage at cankers is common in highly susceptible genotypes.

Disease cycle. Pseudothecia of *M. populicola* and *M. populorum* mature in fallen overwintered leaves, and pseudothecia of the latter fungus occasionally form in cankers. The first leaf infections annually are caused by airborne ascospores, which are discharged from pseudothecia for many weeks beginning in spring. Infection by *M. populorum* may also be started by conidia from overwintered pycnidia in cankers. Ascospores are discharged in greatest numbers at 22–26°C during moist weather. Leaves become infected soon after they unfold, and lesions develop 1–2 weeks later, most numerous on the foliage of low branches. Pycnidia develop 3–4 weeks after infection and appear as black specks on one or both surfaces of a lesion. Pink masses or tendrils of conidia extrude from pycnidia under moist conditions. These spores, dispersed by water, infect leaves and stems that have not yet formed corky bark. Leaf lesions increase rapidly in size and number under favorable conditions.

Poplars vary greatly in susceptibility to *Mycosphaerella* species, so diseases caused by these fungi can be controlled through selection and breeding of resistant trees. The hybrid cultivar Assiniboine, developed for the northern prairie region, is reportedly resistant, as are hybrids in which *P. deltoides* genes predominate.
References: 361, 609, 2159, 2312, 2352, 2382, 2517, 2836, 2839, 2924, 3492, 3670, 4017, 4236, 4503

Other Septoria leaf spots. Several dozen species of *Septoria* parasitize leaves of woody plants in North America. Most of these fungi cause brown spots and sometimes premature leaf drop but are otherwise little known. They produce inconspicuous black pycnidia in lesions on green leaves (Plate 11H). The conidia of most species are colorless, long and narrow, multicelled with rounded ends, and straight or curved. Although *Septoria* species are anamorphic *Mycosphaerella*, the sexual states of some species are unknown.

S. exotica causes necrotic leaf spots on *Hebe* species and on plants in the related herbaceous genus *Veronica* around the world. The spots are round, mostly 1–3 mm across, and brown at first, becoming grayish brown and finally gray to white with dark borders. When numerous, they cause premature yellowing and abscission. Green color remains for a time around spots on yellowing leaves.

S. azaleae causes brown, usually angular leaf spots with yellow halos on *Rhododendron* species (evergreen azaleas) in areas of mild, humid climate and occasionally in greenhouses. The disease is called angular leaf spot or scorch, the latter name referring to the browning of leaf tips and margins that may occur when lesions are numerous. The lesions, usually 1.5–4 mm across, are yellow to reddish yellow at first and become rusty brown with age. The beaks of pycnidia protrude from both surfaces. The disease usually appears late in the growing season and intensifies slowly during cool weather. Conidia are dispersed by rain and overhead irrigation. Temperatures of 16–28°C favor spore germination and growth of the fungus. Up to 2 months may elapse between infection and symptom appearance, presumably because host defenses restrict the pathogen in young leaves. *S. azaleae* occasionally causes defoliation that makes nursery plants unsalable and that, if repeated, leads to bud death. Significant outbreaks have been noted on many azalea cultivars in Japan, Europe, and North America. Hiryu (*R. obtusum*), macranthum (*R. indicum*), and snow (*R. mucronatum*) azaleas are notably susceptible.

Several *Septoria* species affect *Cornus* (dogwood) leaves. The lesions are more or less angular, often limited by veins, at first brown and later tan to grayish or nearly white in the center with a purplish brown margin. *S. floridae* causes colorful but usually inconsequential leaf spots on *Cornus florida* (flowering dogwood). The spots are 1–6 mm across and may have a two-toned halo, reddish next to the lesion and yellow beyond. They appear in early July in the South but later in northern areas. *S. cornicola* and *S. cornina* each infect several *Cornus* species other than *C. florida*. *S. cornimaris* infects *C. mas* (cornelian-cherry). These diseases intensify during the latter part of the growing season. The pathogens can be distinguished reliably only by microscopic examination.
References: 1097, 1153, 1599, 1614, 1925, 2814, 3388, 3980, 4450

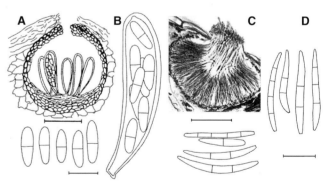

Figure 11. Mycosphaerella populorum. A. Section of pseudothecium. B. Ascus and ascospores. C. Section of pycnidium. D. Conidia. Scale bars = 50 μm for fruit bodies, 10 μm for other structures. A, B, D adapted from reference 3670 by permission of CAB International; C from reference 4017 by permission of the American Phytopathological Society.

A–F. Septoria cankers and leaf spots of *Populus* (poplars) caused by *Mycosphaerella populorum* (anamorph *Septoria musiva*). A, B. Lesions on year-old stems of hybrid poplars (MN, summer). C. An inactive, sunken lesion on an older hybrid (MN, May). D–F. Leaf spots on *Populus* hybrids (D, E) and *P. balsamifera* (balsam poplar, F) (WI, MN, & ON, respectively, Jul).

G, H. Spots caused by *S. floridae* on leaves of *Cornus florida* (flowering dogwood). G. Affected leaves (GA, Aug). H. Magnified view of a lesion with pycnidia. Minute buff-colored specks with black edges are masses of conidia atop the protruding beaks of black pycnidia. The short white spines are leaf hairs (PA, Aug).

I. Leaf spot caused by *Septoria exotica* on *Hebe speciosa* (New Zealand hebe) (CA, Mar).

J. Angular leaf spots and associated yellowing caused by *S. azaleae* on *Rhododendron indicum* (macranthum azalea) (FL, Apr).

Photo credits: A–E—M. E. Ostry

Brown Spot Needle Blight of Pines

Brown spot needle blight kills foliage and retards growth of *Pinus* species (pines). Long known for causing defoliation and growth loss in *P. palustris* (longleaf pine) in southeastern USA, the disease also causes damage to landscape and Christmas tree plantings of *P. ponderosa* (ponderosa pine) and *P. sylvestris* (Scots pine) in the central plains and Great Lakes regions. It occurs from Central America northward to Oregon, Manitoba, Ontario, and Vermont, also in Cuba, South America, Europe, and eastern Asia. The pathogen was probably carried to China with *P. elliottii* (slash pine) from the USA.

The pathogen is *Mycosphaerella dearnessii* (syn. *Eruptio acicola, Scirrhia acicola*). Its reported hosts, in addition to those mentioned above, include *P. attenuata, P. ayacahuite, P. banksiana, P. caribaea, P. cembroides, P. clausa, P. contorta, P. cubensis, P. echinata, P. glabra, P. halepensis, P. longaeva, P. maestrensis, P. massoniana, P. maximinoi, P. monticola, P. mugo, P. nigra, P. oocarpa, P. patula, P. pinaster, P. pinea, P. pityusa, P. radiata, P. resinosa, P. rigida, P. serotina, P. strobus, P. taeda, P. tecumumanii, P. thunbergii, P. uncinata, P. virginiana,* and various hybrids.

Damage to *P. palustris* is linked to that pine's peculiar growth habit. Young seedlings grow only slightly in height until about the age of 5 years and then rise rapidly. Before the onset of height growth, seedlings are said to be in the grass stage. They are subject to severe infection because moisture conditions and abundant inoculum at ground level favor disease buildup. Severe blight in 3 successive years can kill seedlings. Less severe infection may delay the onset of height growth 2–10 years.

Symptoms and signs. Brown spot needle blight is most damaging on low branches and small trees. Lesions begin as spots and often enlarge to bands that encircle needles and cause death of distal parts. In southern USA, spots appear at any time of year but mostly from May to October. They appear in June–August in the central plains and Great Lakes regions. The lesions are of two kinds, each about 3 mm long. The more common type is initially straw yellow, becoming light brown with a dark border. Lesions of this type on *P. palustris* eventually appear raised because unaffected tissue between lesions shrinks as the needles die. The second type of lesion is a brown spot on an amber-yellow band. The yellow tissue is infiltrated with resin. Diseased needles often have dead tips, central zones with several spots in green tissue, and green bases. Killed needles turn reddish brown, then fade to drab hues before dropping.

After successive years of severe infection, *P. palustris* seedlings beyond the grass stage may have a needle-free stem surmounted by a tuft of diseased foliage on the terminal shoot. In central and northern parts of the disease range, killed needles drop during autumn, leaving bare branches if infection was severe. Buds on defoliated branches usually remain alive and, during the next year, produce foliage that in turn becomes infected.

In southern USA, *M. dearnessii* produces both asexual and sexual fruiting structures. They arise as small dark masses of mycelium (stromata) within dead tissue. Fertile stromata produce spores in a central cavity. Stromata of the asexual state, *Lecanosticta acicola,* develop throughout the year, rupture the epidermis, and liberate multicellular, crooked or curved, olive-colored conidia about 15–35 × 3–4 μm in size. Some stromata differentiate into sexual fruit bodies (pseudothecia) 6–8 weeks after conidial formation. Pseudothecia can be found throughout the year on necrotic tips of *P. palustris* needles. In central and northern areas, the conidial state develops in mid to late summer and is the only fruiting state.

Brown spot can be confused with Dothistroma needle blight, caused by *M. pini* (Plate 13). Both pathogens produce conidia in cavities in erumpent black stromata. Conidia of *Dothistroma septosporum,* the anamorph of *M. pini,* are similar to those of *L. acicola* in size, shape, and septation but are colorless.

Disease cycle. M. dearnessii overwinters as stromata in lesions or dead needles on trees and in fallen needles. Conidia, extruded from these structures in gelatinous masses, are dispersed by splashing rain, wind-driven water, and contaminated equipment. After spores germinate on needles, the fungus enters through stomata. Conidia initiate infection from spring into summer, and symptoms appear 1

to several months later, depending on pine species and the age of needles infected. Succulent needle tissue is more susceptible than mature tissue. Symptoms are thought to be induced or exacerbated by fungal toxin(s). A toxin to which pines and other plants are sensitive has been purified from cultures of *M. dearnessii,* but its detection in pine tissue has not been reported. Stromata appear 2 weeks or more after lesions form, and conidia from them cause repeating infections in areas of warm climate. In the North, one cohort of stromata forms annually in summer and releases conidia beginning in early autumn and again in spring. Most infections in the North occur in spring. Generally 2–3 years with conducive weather are required for brown spot needle blight to reach epidemic status in a susceptible plantation.

In southern USA, ascospore production increases from early spring to a peak in late summer. The ascospores, discharged in response to rain, dew, or fog, are windblown and account for long-distance dispersal. Ascospore showers are blamed for occasional rapid development of brown spot needle blight in nursery beds and plantations previously free from the disease.

Warm, wet weather favors this disease. Fruit bodies produce and expel spores while moist or wet. Spores germinate and penetrate needles only when they are wet. Although infection can occur over a wide temperature range, it is most rapid in *P. palustris* if day and night temperatures are about 30°C and 21°C, respectively.

Susceptibility to brown spot needle blight varies within and among *Pinus* species. Some trees in severely affected natural *P. palustris* populations possess heritable resistance. They remain free from brown spot and grow more rapidly than diseased neighbors. In *P. sylvestris,* short-needle varieties from southern France and Spain are especially susceptible, and long-needle varieties from Germany and Austria are resistant. The Chinese species *P. fenzeliana, P. massoniana,* and *P. taiwanensis* are highly resistant.

Damage by brown spot needle blight in plantations is minimized by planting seedlings derived from resistant trees or populations, planting disease-free seedlings on sites free from *M. dearnessii,* leaving space for ventilation between plants, and avoiding cultural operations that spread spores while foliage is wet. Control of the disease in highly susceptible species in Christmas tree plantations requires fungicides. In *P. palustris* stands, prescribed ground fires suppress brown spot because young trees in the grass stage can

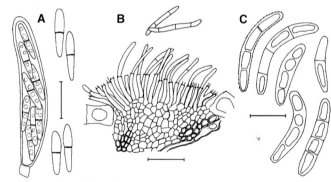

Figure 12. Mycosphaerella dearnessii. A. An ascus and ascospores. B. The *Lecanosticta* anamorph. A stroma has broken through the epidermis of a pine needle and is flanked by two epidermal cells. C. Conidia. Scale bars = 25 μm for stroma, 10 μm for other structures. Adapted from reference 3669 by permission of Gebrüder Borntraeger Verlagsbuchhandlung.

Brown spot needle blight, caused by *Mycosphaerella dearnessii.*
A. Needle browning on *Pinus elliottii* var. *elliottii* (slash pine) (MS, Apr).
B, D. Symptoms on *P. palustris* (longleaf pine) seedlings in the grass stage. Year-old needles are brown, and elongating green needles of the current season have dead tips that bear conidial stromata (arrows in D) (MS, May).
C, E, F. Symptoms on *P. mugo* (mugo pine). C. First-year needles have blighted tips, and many second-year needles are missing (NY, Dec). E, F. Lesions and conidial stromata (gray to black spots) on first-year needles, typical of brown spot needle blight on many pine species (NY, Sep).

survive low-intensity fires that destroy dead, pathogen-infested needles.

References: 236, 800, 853, 1086, 1214, 1790, 1859, 1934, 1936, 1998, 1999, 2213, 2320, 3038, 3290, 3629, 3669, 3684, 3750, 3952, 4414, 4481

Dothistroma Needle Blight of Pines

Dothistroma needle blight, also called red-band needle blight, is notorious for retarding growth and sometimes killing *Pinus* species in plantations around the world. In New Zealand, where outbreaks occur in *P. radiata* (Monterey pine) plantations, Dothistroma blight was the first forest tree disease to be controlled on a practical basis by application of fungicides from aircraft. The causal fungus, *Mycosphaerella pini* (syn. *Eruptio pini*, *Scirrhia pini*), was probably distributed widely with pine planting stock. It is usually encountered in its conidial state, *Dothistroma septosporum*. *M. pini* is widespread in natural forests in western North America but causes significant damage mainly in parks, shelterbelts, and plantations where pines are planted outside their natural circumstances. An exception occurs in northwest British Columbia where *M. pini* has severely damaged *P. contorta* (lodgepole pine) plantations and natural stands, in some instances killing 55-year-old trees. In the East, the fungus occurs in Newfoundland and in scattered locations from Ontario and Vermont to Virginia and westward to the Great Plains. It has not been reported from the western Great Plains or central Rocky Mountains. Young trees are likely to be damaged more than older ones. Growth loss in *P. radiata* becomes significant if more than 25% of a tree's foliage is killed, and growth is virtually halted if defoliation reaches 75%. Severely diseased seedlings may die within a year.

More than 35 pine species and hybrids are subject to infection. *P. nigra* (Austrian pine), *P. ponderosa* (ponderosa pine), and *P. radiata* sustain severe damage. Other susceptible pines include *P. albicaulis*, *P. attenuata*, *P. ayacahuite*, *P. brutia*, *P. bungeana*, *P. canariensis*, *P. caribaea*, *P. cembra*, *P. contorta*, *P. culminicola*, *P. densiflora*, *P. echinata*, *P. elliottii*, *P. flexilis*, *P. halepensis*, *P. jeffreyi*, *P. koraiensis*, *P. longaeva*, *P. maximinoi*, *P. michoacana*, *P. monticola*, *P. mugo*, *P. pinaster*, *P. pinea*, *P. resinosa*, *P. tabuliformis*, *P. taeda*, *P. tecumumanii*, *P. thunbergii*, *P. wallichiana*, and others. Pines with considerable resistance include *P. muricata*, *P. patula*, *P. resinosa*, and *P. sylvestris*. Other conifers growing adjacent to severely diseased pines sometimes develop mild symptoms. *M. pini* has been found on *Abies vejarii*, *Larix decidua*, *Picea abies*, *P. omorika*, *P. sitchensis*, and *Pseudotsuga menziesii*.

Symptoms and signs. Lesions begin as water-soaked or translucent spots or bands that appear from September onward in the Great Plains but may develop at any time of year in areas of moderate climate near the Pacific Coast. Lesions become tan to brown or reddish brown with abrupt transition from discolored to normal green tissue. Needles with necrotic bands die back to the bands. When disease is severe, dying needle tips become prominent 2–3 weeks after spots appear. Needle bases sometimes remain green. Damage is most severe within 2 m of the ground and in most areas is slight on foliage more than 6 m above ground. Diseased needles drop prematurely, the older ones first. *Pinus ponderosa* needles infected during their first growing season usually have no symptoms until autumn and often persist until late summer of the following year. Newly infected second-year needles may drop during autumn or as late as summer of the next year. The red color in lesions is due to accumulation of dothistromin, a toxin that *M. pini* produces. Dothistromin kills leaf tissue in advance of the fungus. Strong light enhances symptoms caused by the toxin, and shade suppresses them.

Within a few weeks after lesions form, or after winter in inland areas, small black stromata of the *Dothistroma* anamorph break through the epidermis singly or in groups. Fertile stromata produce conidia in one or two cavities (Fig. 13), and massed conidia exude during wet weather. Several varieties of *D. septospora* were formerly recognized on the basis of conidial size and shape, but critical studies revealed a continuum of variation that does not sustain varietal distinctions. Pseudothecia of *M. pini* have been collected in scattered localities throughout the world. In North

America they have been found in California and the Pacific Northwest. They differentiate from stromata in needles beginning in autumn of the year of infection and continuing the next year. They may rupture the epidermis in autumn, but ascospores do not develop in them until spring. Observations to detect or assess Dothistroma blight are best made in spring and summer except near the Pacific Coast, where symptoms and signs are visible much of the year.

Disease cycle. In the Great Plains, observations of *P. nigra* and *P. ponderosa* have indicated one cycle of disease per year with infection occurring over a period of months. Second-year and older needles become infected from May into October, and first-year needles are attacked beginning in midsummer. After the first year of disease occurrence, lesions in all stages of symptom development and sporulation can be detected throughout the growing season. In mild Pacific areas, spore production and infection occur throughout the year as moisture permits. The interval from infection until appearance of symptoms varies from 4 weeks to 6 months, depending on environment and host factors. It is shortest in spring to early summer and longest through winter. Secondary cycles occur in areas of mild, moist climate. The pathogen survives winter or dry summer mainly in foliage on the tree.

Most infections are caused by conidia, which are dispersed by splashing or wind-driven rain or mist. Cloud mist was found to carry spores from diseased pine plantations at high elevations in East Africa. In California, conidia have also been trapped from air apparently devoid of water droplets. At least 3 days of cool, rainy, or overcast weather after a period of conidial dispersal are required for spore germination and infection. For most strains of *M. pini*, temperatures of 5–25°C permit conidial germination and mycelial growth. These processes are most rapid at 17–20°C. Ascospores are dispersed mainly in late spring, although they have been trapped all year in California. The role of ascospores in development of epidemics is unknown.

Epidemics of Dothistroma needle blight develop most quickly in areas of mild climate with high rainfall or frequent fog or mist. The disease typically increases exponentially for several years in young plantations. Highly susceptible trees may be killed in 5 years or less. They can be protected by fungicidal sprays timed to prevent infection.

Resistance to *M. pini* varies widely within and among pine species. Proposed mechanisms of resistance, all supported by research, include hypersensitivity to dothistromin, fungal inhibition

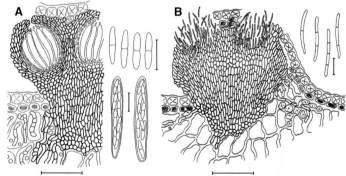

Figure 13. Mycosphaerella pini. A. An ascostroma, asci, and ascospores. B. The *Dothistroma* anamorph with conidia. Scale bars = 50 µm for fruit bodies, 10 µm for other structures. Adapted from reference 584 by permission of Ferdinand Berger & Söhne Gesellschaft.

Dothistroma needle blight of *Pinus nigra* (Austrian pine).

A. Foliage browned by disease on low branches of a mature landscape tree (NE, Jun).

B. Blighted tips on second-year needles. First-year needles become susceptible shortly after the stage of development shown here (IL, Jun).

C–E. Successively closer views of diseased needles. Reddish pigment is sometimes prominent in tan or brown spots and bands. Conidial stromata are visible in lesions on second-year needles in E (IL, Jun).

by fatty acids and resin acid derivatives on needle surfaces, phytoalexin synthesis, and lignification of tissues at lesion margins. Benzoic acid, produced in *P. radiata* needles in response to dothistromin, apparently functions as a phytoalexin. Heritable blight resistance that increases with age has been detected in *P. radiata*. In the Great Plains, *P. ponderosa* from certain sources in Arizona, New Mexico, and Nebraska and *P. nigra* from the Tara Plateau in Yugoslavia have shown high resistance.
References: 18, 19, 236, 474, 584, 643, 730, 800, 1086, 1174, 1214, 1269, 1270, 1859, 1878, 2213, 3035, 3036, 3552, 3873, 4430

Stigmina Leaf Spot of Palms

This disease is caused by *Stigmina palmivora*, a hyphomycete related to *Mycosphaerella*. It is common in humid tropical regions around the world and also occurs on palms grown under glass in temperate areas. In the USA it occurs from Florida to Texas and is found most often in nurseries.

Palms reported to be affected in the USA include *Acoelorraphe wrightii* (Everglades palm), *Adonidia merrillii* (Christmas palm), *Bismarckia nobilis* (Bismarck palm), *Borassus aethiopum* (African fan palm), *Brahea armata* (Mexican blue palm), *Butia capitata* (South American jelly palm), *Caryota mitis* (Burmese fishtail palm), *C. urens* (wine palm), *Chamaerops humilis* (European fan palm), *Cocos nucifera* (coconut palm), *Dypsis lutescens* (Madagascar or areca palm), *Howea forsteriana* (kentia palm), *Livistona chinensis* (Chinese fan palm), *Phoenix canariensis* (Canary Island date palm), *P. dactylifera* (date palm), *P. loureiri*, *P. reclinata* (Senegal date palm), *P. roebelenii* (pygmy date palm), *P. rupicola* (cliff date palm), *Rhapis excelsa* (lady palm), *Roystonea elata* (Florida royal palm), *Sabal palmetto* (cabbage palmetto), *Serenoa repens* (saw palmetto), *Syagrus romanzoffiana* (queen palm), *Thrinax morrisii* (key palm), and *Washingtonia robusta* (desert, thread, or Washington palm).

Prominent brown spots or dead tips on leaflets draw attention to the disease. These lesions arise as tiny, circular, translucent tan spots that become depressed and often somewhat irregular or elongate with a dark brown to black spot in the center, an indefinite brown margin, and a diffuse yellow halo. Mature lesions are elongate, 2–3 × 1–4mm or larger, and may coalesce to form irregular dead blotches that in time turn grayish brown. These often cause or contribute to dieback of leaflet tips. Lesions also occur on frond rachises.

The pathogen produces minute stromata that fill substomatal cavities and expand through the stomata on both surfaces of lesions. A stroma then becomes visible as a minute, dark brown to black dome about 150 μm wide on which dark-pigmented conidiophores and ovoid, usually four-celled brown conidia are produced (Fig. 14A). Conidia measure 60–120 (mostly 64–89) × 7–10 (mostly 4–7) μm. They are dispersed by air and water and cause new infections on both surfaces of wet leaves.

Epidemiological aspects of Stigmina leaf spot of palms have not been reported beyond the observation that the disease is common on plants grown under glass or with insufficient light.
References: 46, 704, 1052, 1099, 1153, 1806, 1859, 3918

Mycosphaerella Leaf Spot of *Yucca*

A *Mycosphaerella* species causes brown lesions with tan centers, usually 15 mm or less in extent, on the leaves of *Yucca filamentosa* (Adam's needle), *Y. glauca* (soapweed yucca), and *Y. gloriosa* (moundlily yucca). The disease is noticed mainly in warm, humid regions. If they are numerous, the spots cause premature yellowing and death of leaves. Foliage on an affected plant may slowly become unsightly as the number of lesions increases. In spring the fungus produces a conidial state that fits the description of *Pseudocercospora* (syn. *Cercostigmina*) *concentrica*. A similar fungus is illustrated in Figure 14B. *P. concentrica* appears as tiny dark eruptions that create starlike patterns as they break through the cuticle on both surfaces of a lesion. Each projection consists of a stroma bearing a dense tuft of olive-brown conidiophores and conidia. In late spring, spermagonia differentiate in stromata that may still be producing conidia. After conidial production ceases, pseudothecia develop. These are brown at first and black at maturity.

The disease cycle has not been studied in detail. From observations, we infer that inoculum in the form of conidia or

ascospores is available at least during spring and early summer. Conidia of the *Pseudocercospora* state can be dispersed by either wind or water, and ascospores are dispersed by wind. The random distribution of lesions on leaves indicates that leaves are susceptible during the entire period of elongation and perhaps much longer.

The appropriate name of the *Mycosphaerella* state is uncertain. *M. yuccae* is recorded as widely associated with leaf spots on *Yucca* species in the USA. *M. deightonii* is recorded as the teleomorph of *P. concentrica*. Comparison of collections bearing these *Mycosphaerella* names is required to ascertain whether *M. deightonii* and *M. yuccae* are distinct or synonymous.
References: 490, 806, 928, 3635

Blights and Leaf Spots Caused by Fungi of the *Cercospora* Complex

The *Cercospora* complex comprises more than a dozen genera of mostly plant-parasitic hyphomycetes including hundreds of species that attack leaves or in some cases green shoots of trees and shrubs. Most of the genera are anamorphs of *Mycosphaerella*. More than 250 fungi of this complex have been recorded on woody plants in the USA and Canada. They cause foliar symptoms ranging from chlorotic or necrotic lesions to diffuse infected areas that remain alive and only slightly discolored while the pathogen sporulates on them. These diseases occur in all temperate habitats and abound in warm, humid regions.

Genera of the *Cercospora* complex (cercosporoid fungi) are similar to, or were segregated from, *Cercospora*, formerly a heterogeneous group with more than 3000 species names. Many fungi once in *Cercospora* are now arrayed in *Asperisporium*, *Camptomeris*, *Cercosporella*, *Cladosporium*, *Passalora* (syn. *Cercosporidium*, *Mycovellosiella*, *Phaeoramularia*), *Phaeoisariopsis*, *Pseudocercospora* (syn. *Cercostigmina*), *Ramularia*, *Stenella*, *Stigmina*, and other genera. These groups are differentiated on the basis of conidial ontogeny; shape, surface texture, wall thickness, and pigmentation (colorless, olivaceous, brown) of conidiophores

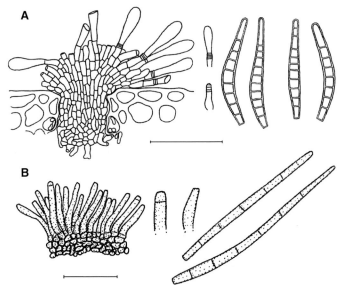

Figure 14. A. *Stigmina palmivora*: habit on palm leaflet, conidiophores, and conidia. B. *Pseudocercospora* (syn. *Cercostigmina*) sp.: habit on leaf, conidiophores, and conidia. Scale bars = 50 μm in A, 20 μm in B. A adapted from reference 1806 by permission of CAB International; B adapted from reference 495.

A–C. Stigmina leaf spot of *Dypsis lutescens* (Madagascar or areca palm). A. Dieback of leaflets, the result of severe infection and coalesced lesions. B, C. Successively closer views of killed portions of a leaflet, showing sporulation of *Stigmina palmivora*. Dark brown spots in the centers of lesions are characteristic (FL, Apr).

D–G. Leaf spot caused by *Mycosphaerella* sp. on *Yucca gloriosa* (moundlily yucca). In spring, pseudothecia occur in dark brown to black erumpent stromata such as those in F and at right in G. While pseudothecia are immature, the conidial (*Cercostigmina*) state of the pathogen may be found on the same stromata. Tan stromata, as at left in G, bear only the conidial state (MS, Apr).

and conidia; arrangement, prominence, and pigmentation of scars on conidiophores where conidia were released; clustered versus solitary conidiophores; location of stromata on which conidiophores develop (superficial or beneath host surface); and presence or absence of superficial mycelium. Conidia are needlelike or wider and vary in number and arrangement of septa. Species are differentiated according to the degree of their expression of particular generic characters; dimensions and colors of stromata, conidiophores, and conidia; host plants; and host symptoms. Most of the species seem to be specialized for attack of one or a few closely related plant taxa, but this matter has not been investigated by means of extensive cross inoculations.

References: 490, 491, 495, 496, 734, 800, 847, 850, 853, 929, 930, 1415, 1416, 1859, 3858

"Cercospora" Blights of Cupressaceae. The most destructive tree diseases caused by fungi in the *Cercospora* complex in North America are the needle blights of evergreens in the Cupressaceae (cypress family) caused by *Passalora sequoiae* (syn. *Asperisporium sequoiae, Cercospora sequoiae*) and *Pseudocercospora juniperi* (syn. *C. sequoiae* var. *juniperi*) (Plate 15A–C). Neither fungus has a known sexual state. We retain the disease name Cercospora blight because it is familiar to many readers. *Passalora sequoiae* is distributed throughout the southeastern USA and occurs as far north and west as Pennsylvania and Nebraska. This fungus also occurs in South America and Japan. It has caused severe damage to Christmas tree plantations of *Cupressus arizonica* (Arizona cypress) and ornamental plantings of *Platycladus orientalis* (Oriental arborvitae) and *Juniperus chinensis* (Chinese juniper). Plants grown in nurseries are sometimes affected, especially those held for several years. Other hosts include *Cryptomeria japonica* (Japanese cedar) and its variety *elegans*, ×*Cupressocyparis leylandii* (Leland cypress), *Cupressus lusitanica* (Mexican cypress), *C. macrocarpa* (Monterey cypress), *C. sempervirens* (Italian cypress), *Juniperus communis* (common juniper) and its variety *depressa* (low juniper), *J. virginiana* (eastern red-cedar), *Sequoiadendron giganteum* (giant sequoia), *S. sempervirens* (redwood), *Taxodium distichum* (bald cypress), and *Thuja plicata* (western red cedar).

Pseudocercospora juniperi occurs in British Columbia and eastern, central, and north-central USA and adjacent Canada on *C. arizonica, J. communis, J. scopulorum,* and *J. virginiana,* sometimes causing severe damage in windbreaks and other plantings in the Great Plains. This fungus has also been recorded in Japan on *Cryptomeria japonica.* A third fungus, *Asperisporium juniperinum,* sometimes confused with *P. juniperi,* causes a needle blight of *J. communis* in north-central USA, Ontario, and Romania.

Symptoms and signs. Passalora sequoiae and *Pseudocercospora juniperi* both cause progressive browning and loss of foliage beginning on low branches close to the mainstem and spreading upward and outward from year to year until the plant is dead or until green shoots remain only on the topmost branches. Cankers that sometimes girdle young stems or deform older ones develop in *C. japonica* plantations in Japan. Symptoms develop during summer in temperate regions and, except for cankers just noted, are similar on the various hosts. Leaves (scale-leaves or needles) on small shoots near branch bases become bronze to tan or light brown and eventually grayish. Often all leaves on a shoot are affected. As the inner, lower part of the crown turns brown and later bare, juvenile foliage often grows from dormant or adventitious buds on denuded low branches. Highly susceptible plants may be killed within 1–3 years.

Soon after diseased leaves die, dark, cushionlike stromata of the causal fungi break through the epidermis (Plate 15C). Brownish conidiophores form on the surface of a stroma and produce dry, elongate, multicelled, brownish conidia. The stromata of *Passalora sequoiae* are up to 50 μm across; those of the *Pseudocercospora juniperi* are much larger (to 140 μm). *P. juniperi* also has narrower conidia (30–60 × 5–5.5 μm versus 42–57 × 2.5–3 μm) and shorter conidiophores.

Green branch tips and juvenile foliage on low branches, where it ordinarily does not occur, serve to distinguish Cercospora blights from Phomopsis, Kabatina, and Sclerophoma blights of *Juniperus*

and other members of the Cupressaceae (Plate 72). The latter diseases are characterized by tip blight.

Disease cycles and epidemiology. The pathogens overwinter in diseased leaves on living plants. Given sufficient moisture, their conidia are produced throughout the growing season in temperate regions, most abundantly during wet weather in late spring and summer. Conidia are dispersed by splashing or wind-driven water. They germinate on wet foliage, and the fungi enter new or year-old leaves through stomata or by direct penetration of the cuticle. Infections occur throughout the summer. Symptoms appear beginning 2–3 weeks after infection from midsummer to late autumn.

Wet weather and moderate temperature favor Cercospora blights. Growth of the pathogens in culture is most rapid at about 24°C. At this temperature, conidial germination begins within 6 hours after wetting and is nearly complete within 16 hours. In one test, spores of *P. juniperi* germinated at 18–28°C but failed to germinate at 8 or 32°C. Shipment of diseased plants accounts for long-distance movement of these pathogens.

Resistance to Cercospora blights varies within and between species. In the Great Plains, the disease caused by *P. juniperi* develops more rapidly and to greater intensity on *Juniperus scopulorum* than on *J. virginiana,* while *J. chinensis* is apparently highly resistant. *J. monosperma* was more severely damaged than *J. virginiana* in one comparison.

References: 183, 800, 850, 1700, 1859, 1874, 2035, 3033, 3035, 3091, 3916, 4043

Some "Cercospora" leaf spots. The fungus associated with the leaf spot of *Eucalyptus cinerea* (spiral eucalyptus) shown in Plate 15E is *Pseudocercospora eucalyptorum.* The lesions develop on either leaf surface, vary in size from pinpoints to more than 1 cm across, often coalesce, and may then involve entire leaves. The centers of large lesions, initially chestnut brown, become grayish tan. The disease occurs around the world on several *Eucalyptus* species. For many years it was erroneously attributed to *Phaeophleospora epicoccoides* (syn. *Cercospora epicoccoides*). A similar disease affects *E. globulus* (Tasmanian blue gum) in California.

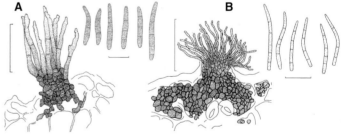

Figure 15. A. *Passalora sequoiae.* B. *Pseudocercospora juniperi.* Scale bars = 50 μm for stromata, 20 μm for conidia. Adapted from reference 3916 by permission of the Mycological Society of America.

A. "Cercospora" blight of *Cupressus sempervirens* (Italian cypress), caused by *Passalora sequoiae.* In contrast to this example, symptoms usually start on low branches and progress upward (FL, Apr).

B, C. "Cercospora" blight of *Juniperus virginiana* (eastern red-cedar) caused by *Pseudocercospora juniperi* (NE, Jun). B. A diseased branch with only the terminal shoots remaining green. C. Dark brown stromata on killed leaves.

D. Leaf spot of *Ligustrum japonicum* (wax-leaf privet) caused by *Pseudocercospora ligustri.* These lesions, atypically small, were either immature or arrested in development. Yellow halos around spots are typical (FL, Apr) (discussion, p. 34).

E. Leaf spot of *Eucalyptus cinerea* (spiral eucalyptus) caused by *Pseudocercospora eucalyptorum* (FL, Apr).

F. Leaf spot caused by *Pseudocercospora liquidambaris* on *Liquidambar styraciflua* (sweetgum) (GA, Aug) (discussion, p. 34).

G. Diffuse leaf spot of *Nandina domestica* (sacred-bamboo) caused by *Pseudocercospora nandinae* (FL, Apr) (discussion, p. 34).

33

Ligustrum species in southeastern USA are attacked by several species of *Cercospora* and related fungi. These include *C. adusta, Pseudocercospora lilacis,* and *P. ligustri.* The last-named fungus occurs on *L. amurense* (Amur privet), *L. japonicum* (wax-leaf privet) (Plate 15D), *L. lucidum* (Chinese or glossy privet), *L. ovalifolium* (California privet), and *L. vulgare* (common privet). It also occurs in Kansas and in France and Japan. A similar fungus spots *Ligustrum* leaves in California. *P. ligustri* causes circular lesions that appear as chlorotic spots 10–12 days after inoculation. Lesions enlarge to 5–15 mm across and become depressed and tan to brown in the center with wide reddish purple margins. A chlorotic halo surrounds most lesions. Small brown stromata (to about 40 μm across) arise on the upper surfaces of lesions 2–3 weeks after infection. In pathogenicity tests, *C. adusta, P. lilacis,* and *P. ligustri* have caused distinct symptoms on *L. japonicum* and *L. lucidum*; nevertheless, precise diagnosis of these leaf spots on privets requires mycological expertise.

Pseudocercospora liquidambaris (Fig. 16A) spots the leaves of *Liquidambar styraciflua* (sweetgum; Plate 15F) and *L. formosana.* The fungus is widespread in southeastern USA as far north as Delaware and Maryland. It also occurs in China. The lesions are angular to nearly round, 2–10 mm across, and dark brown with a purplish black border and a diffuse purplish halo. The pathogen sporulates on both surfaces of lesions, producing conidia on dark brown stromata 20–40 μm in diameter. No biological studies have been reported.

Pseudocercospora nandinae causes irregular reddish to rust-colored blotches, some with much darker centers, on the upper surfaces of leaves of *Nandina domestica* (sacred-bamboo; Plate 15G). Stromata seldom form, but the lower surface of each lesion is slightly darkened, sometimes appearing grayish, by olivaceous conidiophores and conidia. The fungus, which apparently came to North America with its host from Japan, is widely distributed in southeastern USA and westward to Oklahoma. It occurs also in China and New Zealand.

Pseudocercospora kalmiae (Plate 16A–C; Fig. 16B) causes irregular to circular necrotic spots on the leaves of *Kalmia latifolia* (mountain-laurel) in eastern USA and in Japan. Severe infection seems to retard plant growth and suppress flowering, particularly on plants in moist, shady places. Mature lesions are usually 5–20 mm across, at first medium to dark brown on both surfaces but fading to grayish brown in the center of the upper surface. The margin remains dark brown to purplish brown. Black stromata 50–150 μm in diameter dot the upper surface. Most infections begin on newly expanded leaves shortly before and during bloom, and lesions appear in summer. The fungus grows and sporulates most rapidly at temperatures near 25°C. Most leaves, even if severely infected, remain on the plants until the next spring. Thus lesions on year-old leaves seem to be the source of inoculum for new infections. Tests in Japan indicated *P. kalmiae* from *Kalmia* to be host specialized; other ericaceous plants did not become infected.

Rhododendron species (rhododendrons and azaleas) of several types are attacked by *Pseudocercospora handelii* (syn. *Cercospora handelii*), the anamorph of *Mycosphaerella handelii,* which causes irregular to circular lesions on both sides of the leaves (Plate 16D) and occasionally on petioles. The pathogen (Fig. 16C) is widespread in North America and occurs also in Europe, Asia, New Zealand, and South Africa. Host species include *R. catawbiense* (catawba rhododendron), *R. indicum* (macranthum azalea), *R. obtusum* (Hiryu azalea), *R. ponticum* (Pontian rhododendron), *R. simsii* (Sims azalea), and in Florida *Actinidia arguta* (tara vine). The lesions are various shades of brown or, after weathering, grayish brown above and brown below, and measure 2–10 mm across with a narrow black margin and an orange halo. Under moist conditions, a thin, olive-colored mat of hyphae with conidiophores and conidia may develop on the lesions. Dark brown stromata 15–70 μm in diameter form on both surfaces but chiefly on the upper one. The multiseptate conidia measure 90–160 × 2–2.5 μm. Immature pseudothecia of an unidentified sexual state sometimes develop on the lower surface. Severe disease on azaleas causes chlorosis and premature defoliation.

At least six cercosporoid fungi infect the leaves of *Diospyros virginiana* (common persimmon). Criteria for distinguishing these fungi include the type of lesion (clearly demarcated versus indistinct), effuse sporulation versus conidial production on tufts of conidiophores, and the color and dimensions of conidiophores and conidia. *Pseudocercospora fuliginosa* causes leaf spots (Plate 16E, F) in most places where *D. virginiana* grows naturally. The fungus also occurs in China on *D. kaki.* The lesions develop on both leaf surfaces and are purple-black with reddish brown centers, usually angular, 0.4–4 mm across, and may have indistinct chlorotic halos. The pathogen usually sporulates on the lower surface. Stromata, formed inconsistently, are up to 50 μm in diameter. Severely spotted leaves turn yellow and drop prematurely.

Pseudocercospora pittospori (Fig. 16D) causes angular leaf spot of *Pittosporum tobira* (Japanese pittosporum; Plate 16G). Chlorotic to yellowish brown or dull brown angular spots 1–5 mm (occasionally up to 12 mm) in diameter form on both variegated and green varieties, appearing first on the upper leaf surface. Infection occurs during the warm season. The spots are pale green initially, but during a period of months they pass through shades of yellow to brown and remain bounded by veinlets of the leaves. The causal fungus grows out through stomata and, in humid air, produces an olive-colored felt of hyphae, conidiophores, and conidia on the lower surfaces of old spots. Many diseased leaves remain attached; thus symptoms are visible all year. Severe infection of young leaves may cause distortion or premature leaf drop. The pathogen occurs in southeastern USA, Hawaii, and China.

References: 491, 496, 734, 852, 856, 857, 930, 1290, 1752, 1415, 1416, 2529, 3040, 3083, 3089, 3753, 4518

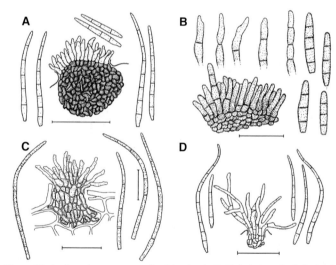

Figure 16. A. *Pseudocercospora liquidambaris.* B. *P. kalmiae.* C. *P. handelii.* D. *P. pittospori.* Scale bars = 40 μm, except 20 μm for *P. handelii* conidia. A, D adapted from reference 1415 by permission of International Academic Publishers; B from reference 496 by permission of Ferdinand Berger & Söhne Gesellschaft m.b.H.; C from reference 3083 by permission of Public Works & Government Services Canada.

A–C. Leaf spot of *Kalmia latifolia* (mountain-laurel) caused by *Pseudocercospora kalmiae.* A. Lesions on year-old leaves. B. Lower (left) and upper surfaces of leaves with typical lesions. C. Close view of the upper surface of a lesion with stromata of *P. kalmiae* (NY, Jun–Jul).

D. Necrotic spots caused by *P. handelii* on leaves of *Rhododendron* sp. (CA, Oct).

E, F. Leaf spots caused by *P. fuliginosa* on *Diospyros virginiana* (common persimmon). E. Severely spotted leaves are prematurely senescent. F. Close view of typical black lesions (GA, Aug).

G. Angular leaf spot of *Pittosporum tobira* (Japanese pittosporum), caused by *Pseudocercospora pittospori.* Diseased spots remain chlorotic for months (FL, Apr).

Linden Leaf Blotch, and Leaf Spots Caused by *Coniothyrium, Microsphaeropsis, Sonderhenia,* and *"Hendersonia"* (Plate 17)

Linden Leaf Blotch

A leaf blotch of *Tilia* species (linden, lime), caused by the ascomycete *Didymosphaeria petrakiana* (Dothideales, Didymosphaeriaceae), occurs in the USA from Wisconsin eastward and in Europe. The disease affects at least three species: *Tilia americana* (basswood, American linden), *T. cordata* (littleleaf linden), and *T. platyphyllos* (bigleaf linden). Blotches develop after midsummer, the upper surface appearing dark brown with feathery margins. In close view, the margins consist of dendritic branches of dark brown leaf tissue extending from the central part of the lesion. These extensions develop beneath branched bundles of subcuticular hyphae. The lower surface of a lesion is tan with an indistinct margin. Wide chlorotic halos develop around the blotches as leaves become prematurely senescent. Diseased leaves shrivel and drop. A severely affected tree may be nearly defoliated in early September.

Although records of linden leaf blotch extend back to the early 1800s, the pathogen remained incompletely identified until the 1990s. This situation occurred because the only fruit bodies or spores associated with the leaf blotch symptoms are of a presumed spermatial state, known as *Asteromella tiliae* (syn. *Asteroma tiliae*), that has minute pale brown pycnidia with microconidia. These structures develop in autumn, usually in fallen leaves. The spermagonia open to the lower leaf surface. They are 60–120 µm in diameter and produce colorless microconidia 4–6 × 3–5 µm. *D. petrakiana* produces its pseudothecia and ascospores in overwintered fallen leaves.

The disease cycle is inferred from knowledge of the symptoms and life stages of the pathogen. The disease is apparently monocyclic, with infection initiated in spring by airborne ascospores from pseudothecia in fallen overwintered leaves. Extensive colonization and blotch formation are delayed until mid to late summer, presumably because of transient host resistance.
References: 586, 3913, 4108

Leaf Spots Caused by *Coniothyrium, Microsphaeropsis, Sonderhenia,* and *"Hendersonia"*

Fungi in these genera, all coelomycetes, inhabit leaves and the cortical and woody tissues of stems. Some species cause leaf spots, twig dieback, or cankers; others are secondary invaders of dying or dead plant parts; and some species have both roles. Relatively few of the diseases caused by these fungi have been studied. We enclose *"Hendersonia"* in quotation marks because this name has been rejected for nomenclatural reasons. It still appears in plant disease literature and fungal host indices because many of the species have not yet been transferred to appropriate genera.

Diseases caused by *Coniothyrium* and *Microsphaeropsis.* Fungi in these genera were once all classified in *Coniothyrium.* This genus, as it is currently understood, comprises at least 25 species that produce globose, brown to black, thin-walled pycnidia with brown, thick-walled, one- or two-celled conidia on annelides (conidiophores that repeatedly produce spores by enlargement at the apex, leaving a series of scars). Some *Coniothyrium* species are conidial states of ascomycetes in the genera *Kalmusia, Leptosphaeria,* and *Paraphaeosphaeria.*

Microsphaeropsis species are similar to *Coniothyrium,* but their brown conidia are unicellular and are produced by phialides (open-ended spore-producing cells of conidiophores that extrude conidia). Several fungi still classified in *Coniothyrium,* including the well-known species *C. fuckelii,* are expected to be transferred to *Microsphaeropsis* or to an older genus with nomenclatural priority when these groups next receive systematic study.

C. fuckelii, the conidial state of *Leptosphaeria coniothyrium* (syn. *Kalmusia coniothyrium;* Pleosporales, Leptosphaeriaceae), occurs worldwide and causes diverse diseases, including shoot blight of *Juniperus* (juniper) and *Thuja* (arborvitae); leaf spot of *Magnolia;* stem cankers of *Malus* (apple), *Parthenocissus* (Virginia creeper), and *Rosa* (rose); cane blight of *Ribes* (currant) and *Rubus* (brambles); and leaf blight of *Quercus glauca* (ring-cup oak) in Japan. This species is also a common inhabitant of conifer leaves, tree bark, and soil. Lesions caused by *C. fuckelii* on leaves of *Magnolia grandiflora* (southern magnolia; Plate 17D–F) are up to 1 cm across, at first brown with a nearly black outer ring, and sometimes with a diffuse brown border that merges gradually with surrounding green tissue. The center of the lesion becomes light grayish brown, and numerous pycnidia develop beneath its upper surface. Many other species of

Coniothyrium are associated with symptoms on leaves or stems of woody plants in North America, but details and confirmation of pathogenicity are lacking.

Microsphaeropsis includes few confirmed pathogens. *M. concentrica* causes leaf spots of *Agave* and *Yucca. M. olivacea,* although often found associated with plant diseases, is a saprobe or secondary invader of stressed tissues. *M. olivacea* and *C. fuckelii* possibly suppress such common foliar and stem pathogens as *Lophodermium* species on pines and *Leucostoma* species on stone-fruit trees.
References: 235, 1099, 1758, 1783, 1811, 1815, 1816, 2077, 2137, 2679, 2717, 2726, 3502, 3669, 3754, 3913, 4108

Diseases associated with *Sonderhenia* and other fungi of the *"Hendersonia"* group. Fungi of this group have black, globose pycnidia that are at first immersed in bark or leaf tissue and may later become erumpent. The conidia are elongate, usually brown, and multicellular. Species from this group have been reclassified in various genera, listed here with the corresponding teleomorph genera, if known, in parentheses: *Camarosporium* (*Cucurbitaria*), *Macrodiplodiopsis* (*Splanchnonema*), *Phaeophleospora, Sclerostagonospora* (*Montagnula*), *Seimatosporium* (*Discostroma*), *Sonderhenia* (*Mycosphaerella*), and *Stagonospora* (*Didymella*). We use *"Hendersonia"* for species not yet reclassified.

Fungi in the groups noted above are associated with the following disorders, among others. *Sonderhenia eucalypticola* on *Eucalyptus globulus* (blue gum) in California causes small, brown to tan lesions with dark purplish borders, visible on both sides of the leaf (Plate 17G–I). Apparently only the juvenile foliage is susceptible. *Seimatosporium lichenicola* causes leaf blight of *Prunus dulcis* (almond) in California. *Macrodiplodiopsis desmazieri* is associated with a minor twig dieback and canker of *Platanus* (plane, sycamore) species from New Jersey to California. *"Hendersonia"* pinicola apparently causes a needle blight of *Pinus contorta* (lodgepole pine) in northwestern North America, often in conjunction with *Lophodermella concolor. H. acicola* in Europe invades pine needles already infected by the needle-cast fungus *Lophodermella sulcigena* and suppresses fruiting of the latter.

More than 30 fungi remaining in *"Hendersonia"* are reported to occur on woody plants in the USA and Canada. They are associated with leaf spots or leaf blight on *Aralia, Carya* (hickory), *Crataegus* (hawthorn), *Magnolia, Malus, Pyrus* (pear), *Rhododendron,* and *Viburnum;* and with twig cankers on *Acer negundo* (box-elder) and *Elaeagnus angustifolia* (Russian-olive), among other hosts.
References: 852, 1099, 1214, 1810, 2137, 2681, 2896, 3819, 3913, 3929, 4108

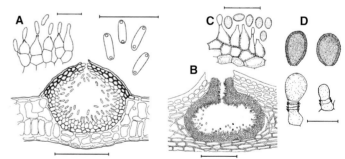

Figure 17. A. *Asteromella tiliae,* the spermatial state of *Didymosphaeria petrakiana.* B. *Coniothyrium wernsdorffiae,* cause of brand canker on rose canes. C, D. The manner of conidial production in *Coniothyrium.* Scale bars: pycnidia = 50 µm, other structures = 10 µm. A, B, C, D adapted from references 586, 1758, 3669, and 3913 by permission of Cambridge University Press, the American Phytopathological Society, Gebrüder Borntraeger Verlagsbuchhandlung, and CAB International, respectively.

A–C. Linden leaf blotch, caused by *Didymosphaeria petrakiana,* on *Tilia americana* (basswood or American linden).

D–F. Leaf spot caused by *Leptosphaeria coniothyrium* on *Magnolia grandiflora.* Tiny pycnidia dot the center of a lesion as viewed with a hand lens (MS, Aug).

G–I. Leaf spot caused by *Sonderhenia eucalypticola* on *Eucalyptus globulus* (blue gum) (CA, Apr).

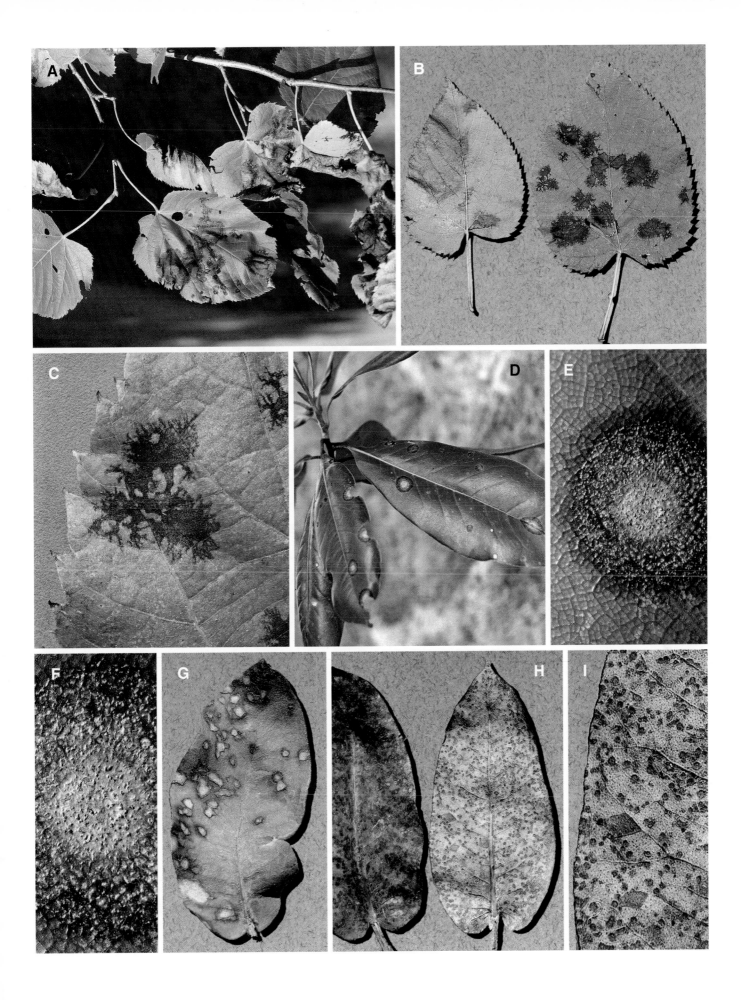

Ascochyta Blight of Lilac (Plate 18)

Little known and seldom noticed, Ascochyta blight is occasionally destructive to *Syringa josikaea* (Hungarian lilac) and *S. vulgaris* (common lilac). The pathogen, *Ascochyta syringae* (syn. *Phyllosticta syringae*), a coelomycete, was described in the late 1800s in Europe, where it is widespread. It also occurs in Asia and Australia. In the Western Hemisphere it has been recorded in Chile and in scattered locations across northern USA and Canada. In eastern USA it occurs from North Carolina northward. In Europe *A. syringae* also occurs on living leaves and dry fruits of *Fraxinus angustifolia* (narrowleaf ash) and *F. excelsior* (European ash).

Ascochyta blight on *S. vulgaris* has two phases, one on shoots in spring and the other on leaves later in the season. Shoot blight develops as elongate lesions girdle soft green twigs and flower stalks, causing the parts beyond to droop, shrivel, and turn brown. Lesions are usually restricted to tissues of the current season but may spread into apical parts of year-old twig segments, which then die back to the next branch. Lesions caused by *A. syringae* are prominent, even on brown dead shoots, because of their distinct lower margins, tan to grayish color, and shriveled appearance. During wet weather or when specimens are incubated in high humidity, white mycelium and dark gray pycnidia of the causal fungus break through the surface, giving the lesion a speckled gray appearance (Plate 18C). This disease is readily distinguished from bacterial blight (Plate 183) or Phytophthora dieback in that the grayish lesions remain distinct on dead shoots.

The foliar phase develops in late summer and autumn as leaf spots or as spreading, coalescing necrotic blotches. The lesions are initially olive green and water-soaked, but they soon turn tan to light brown with indefinite margins. Lesions centered on major veins tend to become elongate, but most are more or less round.

No sexual state of *A. syringae* is known. It produces pycnidia prolifically on the brownish parts of foliar lesions, mainly on the upper surface. Young pycnidia lack pigment and can best be perceived by viewing lesions under magnification with a bright light behind the leaf. Thus illuminated, the pycnidia appear as minute light spots. They darken with age, however, becoming grayish brown and readily visible (Plate 18E). The two-celled conidia are colorless with blunt ends and measure 8–15 × 3.0–3.5 μm.

The disease cycle is not known from research, but its main elements are apparent from observations of symptoms and signs. Pycnidia of *A. syringae* were observed on bud scales during winter in Italy, and shoot blight developed soon after buds opened. Thus the pathogen overwinters in buds, and shoot blight probably results either from latent infections in buds or, more likely, from spores dispersed from infected bud scales during wet weather in early spring. Rain splash is the principal means of local dispersal of conidia of *Ascochyta* and allied fungi. By this means conidia from lesions on shoots are dispersed to leaves, and foliar infection develops in summer to early autumn. Whether buds or bud scales are infected simultaneously with leaves or by spores dispersed from leaves in autumn is unknown.

A. syringae can readily be isolated from diseased tissues on laboratory media. Its pathogenicity has been proven.

References: 122, 504, 873, 1099, 1298, 2235, 2589, 4108

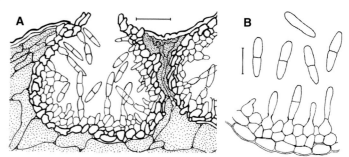

Figure 18. Generic characteristics of *Ascochyta* (illustration of *A. syringae* unavailable). A. Pycnidia immersed in host tissue. B. Conidia are mostly two-celled, colorless, and produced by short phialides. Scale bars = 10 μm. A and B adapted from references 2235 and 122, respectively, the latter by permission of Gebrüder Borntraeger Verlagsbuchhandlung.

A, B. Shoot blight of *Syringa vulgaris* (common lilac) caused by *Ascochyta syringae*. A. Lesions in soft parts of green twigs cause shoots to droop and shrivel. An affected stem then dies back to a node with living leaves (NY, Jun).

C. Enlarged view of a stem lesion, showing abrupt lower margin, tan surface, and fruiting of *A. syringae* (NY, Jun).

D, E. Foliar phase of Ascochyta blight on *S. vulgaris*. D. Upper and lower surfaces of leaves with spreading lesions. E. Part of a lesion, magnified to show gray-brown pycnidia of *A. syringae* (NH, Sep).

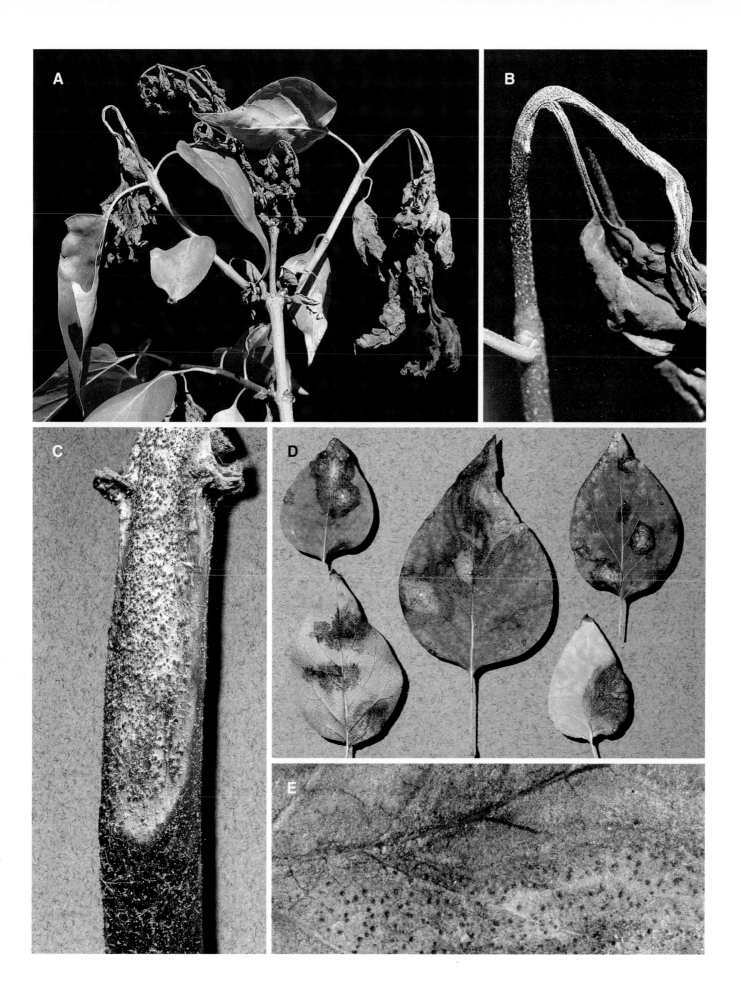

Leaf Spots and Blotches Caused by *Guignardia* and *Phyllosticta* (Plates 19, 20)

Diseases Caused by *Guignardia* Species

Guignardia species (Dothideales, Botryosphaeriaceae) cause leaf spots, leaf blotches, and fruit rots of broadleaved woody plants, and needle blights or shoot blights of some conifers. *G. aesculi* causes leaf blotch of *Aesculus* (horse-chestnut and buckeye) in the eastern half of North America and in Europe. *G. bidwellii*, the cause of black rot of grape and leaf spot of *Ampelopsis* (monk's-hood vine) and *Parthenocissus* (Virginia creeper and Boston ivy), occurs around the globe. *G. citricarpa*, the cause of black spot of *Citrus*, is globally distributed in citrus-growing regions except California. *G. gaultheriae* (anamorph *Phyllosticta gaultheriae*) causes leaf spots on *Gaultheria shallon* in northwestern USA. *G. mangiferae* (anamorph *P. capitalensis*) is a cosmopolitan endophyte in many woody plants. An undetermined *Guignardia* species with a *Phyllosticta* anamorph causes leaf spots on *Cornus florida* in southeastern USA.

Guignardia species are usually encountered as anamorphs of the *Phyllosticta* type, characterized by black pycnidia with globose to ovoid, single-celled, colorless conidia, each with a gelatinous sheath that extends as an appendage at the apical end (Fig. 19A). The name *Phyllosticta* denotes the typical development of these pycnidia in discrete, usually small, necrotic foliar lesions. Lesions of this sort (leaf spots) represent a functional balance between host and parasite. The host leaf, initially susceptible, soon mobilizes defenses and restricts infection to a small area. Thus, infections even if numerous usually cause no great loss of photosynthetic surface unless they trigger abscission. On the other hand, the lesion is a sufficient substrate for reproduction of the parasite. Given conducive conditions—wet weather in spring and early summer in successive years—populations of some phyllostictas may build up enough to kill a significant proportion of leaf area or to cause premature leaf shedding. The usual situation, however, is inconsequential spotting of leaves on low branches. *Guignardia* species typically overwinter as immature pseudothecia in dead fallen leaves or other organs that were first colonized parasitically. They liberate ascospores in spring in temperate zones. Some species also overwinter as pycnidia.

Guignardia blotch of horse-chestnut and buckeye. This disease disfigures the foliage of *Aesculus hippocastanum* and *A. ×carnea* (common and red horse-chestnuts), *A. turbinata* (Japanese horse-chestnut), and buckeyes: *A. californica* (California buckeye), *A. glabra* (Ohio buckeye), *A. octandra* (yellow buckeye), *A. pavia* (red buckeye), and some lesser-known species. Observations in Illinois revealed probable resistance in *A. glabra* varieties *arguta, monticola,* and *sargentii* and in *A. parviflora* (bottlebrush buckeye) and its variety *serotina*. *A. parviflora* seems resistant in Italy.

Symptoms and signs. Lesions first appear as water-soaked irregular areas that enlarge rapidly. Within a few days they turn reddish brown to brown, often bordered by a yellow band that merges gradually with green tissue. Small lesions may be limited by veins. Large lesions frequently coalesce and cause distortion and partial shriveling of leaflets. Severe disease causes premature defoliation. Petioles and immature fruits become infected occasionally. Lesions on these parts are small reddish brown spots, somewhat elongate when on petioles.

Black pycnidia of the conidial anamorph, *Phyllosticta sphaeropsoidea*, appear soon after lesions form. The pycnidia are nearly globose, 80–165 µm in diameter, and usually open through the upper leaf surface (Fig. 19A). When mature and moist, a pycnidium extrudes a mass of colorless, one-celled conidia that are dispersed by water. They measure 10–20 × 9–13 µm, and each has an apical appendage 6–7.5 µm long when fresh. In late summer black spermagonia and immature pseudothecial stromata appear on both surfaces of lesions. The spermagonia (of the *Leptodothiorella* type) are 40–110 µm in diameter (Fig. 19B). The pseudothecial stromata appear similar to pycnidia but produce no spores until after winter.

Disease cycle. *G. aesculi* overwinters as pycnidia and immature pseudothecia in fallen leaves. Pseudothecia mature in early spring and liberate ascospores to the air during wet weather while leaves are developing. The ascospores initiate infections if deposited on leaves that remain wet for several hours. Conidia, splash-dispersed from pycnidia in overwintered leaves, also have this capability. Blotches appear after 10–20 days or longer. New pycnidia develop beginning in late spring, and conidia from them initiate secondary cycles throughout the summer as permitted by wet weather.

Guignardia blotch typically becomes severe in plantings where tree crowns are close together, because dense foliage retards drying of leaves after rainfall. The disease tends to develop after most of the annual growth of host plants is complete. Thus it does not greatly influence growth.

References: 1, 154, 1099, 1804, 1859, 2779, 2787, 2803, 2808, 3468, 3669, 3861

Guignardia leaf spot of Boston ivy and black rot of grapevine. *Guignardia bidwellii*, the cause of these diseases, was once confined to North America, but grape-infecting strains were apparently carried with propagating material to every vegetated continent except Australia. Black rot is a threat to grape crops in eastern North America. *G. bidwellii* comprises host-specialized subgroups called formae speciales. *G. bidwellii* f.sp. *parthenocissi* infects *Ampelopsis acontifolia* (monk's-hood vine), *Parthenocissus quinquefolia* (Virginia creeper), and *P. tricuspidata* (Boston ivy). *G. bidwellii* f.sp. *euvitis* infects *V. labrusca* (fox grape) and *V. vinifera* (European grape). *G. bidwellii* f.sp. *rotundifolia* infects *V. rotundifolia* (muscadine grape) and *V. vinifera*.

Symptoms and signs. Lesions on *Parthenocissus* leaves are small (up to 6–8 mm across), tan to brown, and often angular where limited by leaf veins. They have dark brown margins with an abrupt transition to green tissue. The dead tissue eventually breaks away, leaving ragged holes. If numerous spots develop before leaf growth is complete, the leaves become distorted as green tissues continue expanding around the dead spots. As soon as lesions attain full size, black pycnidia protrude through the upper surfaces, either scattered or arranged in a ring. These fruit bodies are 60–200 µm across and extrude colorless, one-celled conidia when moistened. Conidia measure 5–12 × 4–7 µm and have a slime layer and an apical appendage nearly the length of the spore. White masses of these spores sometimes adhere to the tips of pycnidia. The pycnidial anamorph of *G. bidwellii* is *Phyllosticta ampelicida*. Spermagonia of the *Leptodothiorella* type and immature pseudothecia develop in the same lesions in late summer.

Leaf spots caused by *G. bidwellii* on *Vitis* are similar to those on *Parthenocissus* but are much larger on some varieties of *V. vinifera*.

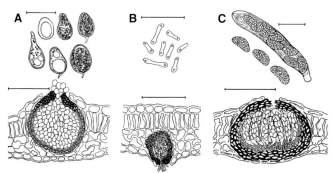

Figure 19. Spore-producing states of *Guignardia aesculi* in leaves of *Aesculus hippocastanum* (horse-chestnut). A–C, respectively: pycnidium with conidia, spermagonium with spermatia, and pseudothecium with asci and ascospores. Bars = 100 µm for fruit bodies, 10 µm for other structures. Adapted from reference 3468 by permission of Blackwell Publishing.

A–D. Leaf blotch of *Aesculus hippocastanum* (horse-chestnut) caused by *Guignardia aesculi*. A. Foliage distorted by severe infection. B. Typical lesions. C, D. Pycnidia in the upper surface of a lesion, magnified. The light dot in the center of each pycnidium is the pore through which conidia emerge when the leaf is moist (NY, Aug–Sep).

E–H. Leaf spot of *Parthenocissus tricuspidata* (Boston ivy) caused by *G. bidwellii* f.sp. *parthenocissi*. E, F. Lesions viewed on upper and lower leaf surfaces (NY, Sep). G, H. Pycnidia on upper surfaces of lesions (NY, Jun).

I. Leaf spot of *Cornus kousa* (Korean or kousa dogwood) attributed to *Phyllosticta* sp. (NJ, Jun) (discussion, p. 42).

J. Leaf spot of *Corylus* sp. (hazelnut), attributed to *Phyllosticta* sp. but more likely caused by *Phoma* sp. (NJ, Jun) (discussion, p. 42).

Mature leaves of *Vitis* are resistant but may support latent infections. The fungus also causes small lesions on shoots, fruit stalks, and tendrils. Black rot of the fruit typically appears when berries are about half grown; more mature berries are resistant. Fruit lesions begin as pale spots within dark rings, but the entire fruit turns black, shriveled, and hard (mummified) within 7–10 days. Pycnidia develop in the mummified fruit and on stems, tendrils, and fallen leaves. Pseudothecia form in fallen leaves.

Disease cycle. The pathogen overwinters as immature pseudothecia and (on *Vitis*) as pycnidia in dead fruits, stems, and tendrils. Infection in spring is initiated by ascospores from pseudothecia in fallen leaves and conidia from pycnidia in dead tissues on vines or on the ground. The annual peak of ascospore dispersal occurs in late spring. Conidia are liberated throughout the season but in greatest numbers in summer after pycnidia develop in new foliar lesions. Conidia can initiate secondary cycles of disease on late-developing leaves. Rains lasting 1–3 hours are optimal for release of both types of spores. Release begins within an hour after the onset of rain and may continue for several hours. Infection occurs within 6–24 hours, depending on temperature, if plant surfaces remain wet. Infection by spores of either type is most rapid at 26–27°C. The incubation period (interval from infection to appearance of symptoms) is 1–3 weeks, depending on environmental conditions. Pycnidia usually appear within an additional 3–6 days. Spores of both types are susceptible to desiccation, surviving 48 hours or less on dry leaves. *References:* 278, 1123, 1124, 1859, 2183–2185, 2393, 2999, 3237, 3560, 3801

Phyllosticta Leaf Spots and Blotches

Many *Phyllosticta* species have not been linked to named *Guignardia* species. Those that truly lack teleomorphs are presumed to overwinter as pycnidia. Species that perpetuate themselves only by conidial production and infect deciduous plants are assumed to survive winter in fallen leaves and, in spring, to produce conidia that are dispersed to new foliage and start another cycle of disease. Alternative overwintering sites for some species are lesions on twigs, from which conidia are dispersed in spring, as was shown for the anamorph of *Guignardia bidwellii* (Plate 19E–H) on grapevine. *Phyllosticta* conidia are adapted for dispersal by water, which in part accounts for leaf spot outbreaks in wet growing seasons. The sheath, or extracellular matrix, on the spores permits them to adhere to poorly wettable surfaces such as leaves. Adhesion is necessary for germination and host penetration. Some phyllostictas live for a time as latent endophytic colonists of green leaves.

In earlier classifications, more than 2000 names were proposed in the genus *Phyllosticta* for fungi that produce black pycnidia in leaves. These fungi, although similar in gross form and in producing colorless, unicellular spores, were heterogeneous in microscopic characteristics and affiliations with diverse sexual states. Some species were described and named more than once. Many fungi formerly classified in *Phyllosticta* have been more appropriately placed in *Phoma*, *Phomopsis*, and *Ascochyta*. Others that produce tiny spores on the order of 1 × 3–4 µm are microconidial (spermatial) states of ascomycetes and would properly be classified in *Asteromella* or *Leptodothiorella* (for spermatial states of *Mycosphaerella* or *Guignardia*, respectively). The tiny spores are noninfectious and are believed to function as gametes. Many fungi named *Phyllosticta* were not described sufficiently to either confirm their classification or reassign them to more appropriate genera.

Leaf spots of dogwood. *P. globifera* and perhaps additional phyllostictas infect *Cornus* in the USA. *P. globifera* has been recorded in the Appalachian region and Iowa. Its pycnidia are 100–140 µm in diameter; the conidia measure 6.2–8.8 × 5.4–7.1 µm. The identities of other true phyllostictas on *Cornus* are uncertain. Many collections identified as *P. cornicola* have been made on various *Cornus* species in western and eastern USA, also in Europe and Australia. Unfortunately, the fungus first described as *P. cornicola* was really a *Septoria* (page 24), so the appropriate names for true phyllostictas among fungi called *P. cornicola* are unknown. Similarly, fungi once named *P. corni* and *P. starbaeckii* that infect *Cornus* actually belong to *Phoma* and *Asteromella*, respectively. Lesions caused by *Phyllosticta* (Plate 19I) are round to irregular spots 1–6 mm across with tan to gray centers and brown or reddish brown borders. Pycnidia arise in the tan portion, which becomes fragile

and tends to break out of mature lesions. Phyllosticta leaf spots on *C. florida* are similar to lesions caused by *Ascochyta cornicola*. The most obvious difference between *Phyllosticta* and *Ascochyta* is that the latter has two-celled conidia (Fig. 18). Severe disease caused by *A. cornicola* can result in blackening and shriveling of leaves on low branches in mid-June. Phyllostictas on *Cornus* have not been associated with such severe blight.

Leaf spot of hazelnut. One or more fungi until recently known as *P. coryli* occur across the USA in leaf spots on *Corylus* (filbert, hazelnut): *C. americana* (American filbert), *C. avellana* (European filbert), and *C. cornuta* (beaked filbert). The lesions (Plate 19J) are round to irregular with dark margins, light brown centers, and sometimes yellow halos. Coalescing lesions give rise to irregular dead patches at the edges of leaves. Pycnidia break through the upper surface of lesions. Unfortunately, neither of the fungi originally described under the name *P. coryli* is a *Phyllosticta*. One is *Phoma exigua*; the other is *Asteroma coryli*. Whether a true *Phyllosticta* occurs on *Corylus* is unknown.

Leaf blight of witch-hazel. A true *Phyllosticta*, *P. hamamelidis* (Plate 20A–E) causes necrotic spots and sometimes kills leaves of *Hamamelis virginiana* (witch-hazel). The disease is widespread east of the Great Plains. When severe, it disfigures entire plants. The spots are irregular in outline, brown, and darker on the upper surface than the lower. Each has a narrow, dark purple margin. Lesions often coalesce to involve large areas or whole leaves. Black pycnidia about 100 µm in diameter break through both surfaces of the lesions. The ovoid conidia measure 9–10 × 6–10 µm and have a thick slime layer but no apical appendage.

Phyllosticta spots of maples. *P. minima* (Plate 20G–L) causes eye spot, or purple-bordered leaf spot, of maples: *Acer campestre* (hedge maple), *A. ginnala* (Amur maple), *A. glabrum* (Rocky Mountain maple), *A. mono* (mono maple), *A. negundo* (box-elder), *A. nigrum* (black maple), *A. palmatum* (Japanese maple), *A. platanoides* (Norway maple), *A. pseudoplatanus* (sycamore maple), *A. rubrum* (red maple), *A. saccharinum* (silver maple), *A. saccharum* (sugar maple), *A. spicatum* (mountain maple), and *A. tataricum* (Tatarian maple). It is common from the Great Plains eastward but is seldom reported from the West. Severe infections have caused partial defoliation of *A. rubrum*. The spots are irregularly round and usually less than 5 mm across, at first brown and later with a tan central portion and dark border. The border is often reddish or purplish. Lesions sometimes coalesce into large, irregular dead areas. Pycnidia, up to 150 or rarely 200 µm in diameter, are visible on both sides of a lesion, but most of them open to the upper surface. Often they are arranged in a circle. The conidia of this fungus measure 7.5–12 × 4.5–8 µm; each has a slimy coating and an apical appendage when fresh.

A fungus long known as *Phyllosticta negundinis* but perhaps more properly classified in *Phoma* affects *Acer negundo* (box-elder) throughout central and eastern North America, also in Europe and China. The lesions (Plate 20F) are up to 8 mm across, yellowish brown at the edges, and pale yellow in the center, where they become thin and translucent. Each is bounded by a narrow ridge. The fragile central tissue often breaks out, leaving a ragged hole. Severe outbreaks may impart yellowish color and sparseness to foliage in tree crowns viewed from a distance. Pycnidia are usually numerous in lesions, more so on the upper side than on the lower, scattered or arranged in a circle. Their small size (100–125 µm diameter) makes them barely visible to the unaided eye. The conidia measure 6–7 × 3–4 µm.

References: 1, 2, 383, 1099, 1116, 1153, 1925, 2779, 2787, 3522, 3669, 3913

A–E. Leaf spot and leaf blight of *Hamamelis virginiana* (witch-hazel) caused by *Phyllosticta hamamelidis* (PA, Jun). A–C. Symptoms on leaves and shoots. D. Close view of lesions with pycnidia. E. Magnified view of pycnidia.

F. Leaf spot caused by *P. negundinis* on *Acer negundo* (box-elder) (NY, Aug).

G–L. Leaf spots caused by *P. minima* on maples. G–I. On *A. saccharinum* (silver maple) (NY, Jul), *A. saccharum* (sugar maple) (NY, Jun), and *A. palmatum* (Japanese maple) (NY, Jun), respectively. J–L. Single lesions with pycnidia on leaves of the same respective collections.

Tubakia Leaf Spot of Oaks and Other Trees, and Fly-speck Leaf Spot of *Vaccinium* (Plate 21)

Tubakia Leaf Spot

Leaf spots caused by *Tubakia dryina* (a coelomycete) occur on diverse trees and shrubs. The fungus is distributed throughout the eastern half of the USA and has been recorded in Oregon and New Brunswick. It is widespread in Europe and occurs also in New Zealand. In Italy it has been linked to a sexual state, the ascomycete *Dicarpella dryina* (Diaporthales, Melanconidaceae), which forms in fallen overwintered leaves. *Quercus* species (oaks) are often affected by Tubakia leaf spot. Additional host genera include *Acer* (maple), *Carya* (hickory, pecan), *Castanea* (chestnut, chinkapin), *Cercis* (redbud), *Eucalyptus*, *Fraxinus* (ash), *Malus* (apple), *Nyssa* (blackgum, tupelo), *Oxydendrum* (sourwood), *Toxicodendron* (poison-ivy), *Rosa* (rose), *Sassafras*, and *Ulmus* (elm), although some of these are rarely affected.

Lesions usually appear in midsummer and become more prominent as the season advances. They are initially water-soaked and later turn tan or reddish brown to dark brown. Mature lesions are 1–15 mm in diameter and often angular, their size depending on host susceptibility and environmental conditions. They tend to become large on leaves stressed by other factors such as nutrient shortage, as on the chlorotic *Q. palustris* (pin oak) featured in Plate 21. Lesions on some plants—*Castanea*, for example—may have chlorotic halos and concentric patterns in the necrotic zones. Numerous and confluent lesions on oak leaves can kill large areas of leaf tissue and cause prominent foliar browning and defoliation.

Fruit bodies of *T. dryina* form on upper, lower, or both surfaces of lesions. They are superficial dark brown to black disclike structures mostly 60–110 μm in diameter and 20–40 μm high, called pycnothyria. The disc has thick hyphae radiating from its center toward the edges (Fig. 21) and is supported by a short central column. Conidia form in a sticky matrix beneath the disc, and with continued production are pushed out beyond the edges. The conidia are elliptic, initially colorless, single celled, and mostly 8–14 × 6–10 μm. They turn light brown with age. *T. dryina* often fruits on lesions caused by other organisms or abiotic stress. This habit probably reflects the presence of initially latent colonies in leaves and twigs, because *T. dryina* is a common endophyte of *Quercus* in both Europe and the USA. In Europe *T. dryina* has been found producing pycnidia on twigs of *Q. rubra* (red oak) while producing pycnothyria on leaves.

The disease cycle has not been described in detail. The first infections each year are believed caused by conidia from overwintered pycnothyria on fallen leaves. Ascospores may have the same function in areas where the *Dicarpella* state forms. Secondary cycles of infection have not been documented, but inoculations of mature oak leaves in August have indicated full susceptibility of the leaves at that time. Late-season increase in symptoms could be due to infections initiated by conidia from the first lesions, delayed appearance of lesions resulting from latent infections, or both.

Conidia of *T. dryina* adhere to leaf surfaces and, given moist conditions and moderate temperature, produce germ tubes and appressoria. Germ tubes enter leaves via stomata or produce appressoria on epidermal cells and over stomata. After penetration of mature leaves, the fungus produces inter- and intracellular hyphae that ramify in the tissue. Host cells soon die and collapse, and a lesion appears. Under the most favorable conditions, *T. dryina* begins to differentiate fruiting bodies within 3 days after conidial germination.

Symptoms caused by *T. dryina* are presumed due in part to its secretion of nonspecific phytotoxins, four of which have been isolated from liquid cultures: isosclerone, 3-hydroxyisosclerone, 6-hydroxyisosclerone, and 6-hydroxymellein. Each compound caused necrosis when applied to leaves of *Q. alba* (white oak), *Q. rubra*, and various herbaceous plants.

Some strains of *T. dryina* exhibit host selectivity. Isolates from several *Quercus* species and *Castanea crenata* (Japanese chestnut) caused lesions on *Quercus* leaves but not on leaves of *Acer saccharum*, *Nyssa sylvatica*, *Cercis canadensis*, *Carya ovata*, *Fraxinus americana*, or *Liquidambar styraciflua*, all of which are recorded hosts of the fungus.

References: 293, 428, 1099, 1153, 1254, 1302, 1719, 1965, 2319, 2761, 2762, 3912, 3975, 3977, 4108, 4147, 4514

Fly-Speck Leaf Spot of *Vaccinium*

This inconsequential disease, which affects species of *Gaylusaccia* and *Vaccinium* (blueberry, farkleberry, huckleberry), is caused by *Ophiodothella vaccinii* (Phyllachorales, Phyllachoraceae) and characterized by distinctive colorful lesions. Hosts include *V. arboreum*, *V. candicans*, *V. corymbosum*, *V. stamineum* var. *melanocarpum*, *V. ashei*, *V. tenellum*, and *V. virgatum*. The pathogen occurs widely in southeastern USA and also in Illinois and California. Symptoms appear in early summer as nearly round yellow spots 1–5 mm across on upper leaf surfaces. The spots enlarge to 10–15 mm or more and sometimes turn orange to red, often with a reddish purple border or a yellow halo. They become visible on the lower surface as yellowish brown areas with indistinct margins. Hyphae of the parasite grow inter- and intracellularly throughout the lesion. From early summer until early autumn, pigmented hyphae proliferate in groups of epidermal cells on both sides of a lesion, giving rise to numerous black stromata, the "fly specks." Fly specks on *V. arboreum* remain separate, but those on other hosts often become so crowded that the entire lesion appears black. Colorless, linear, flexuous conidia, 18–42 × 1 μm, are produced in acervuli, mostly 500–600 μm broad, that arise beneath the blackened epidermis during summer and autumn. The conidial state is *Acerviclypeatus poriformans*. During wet weather conidia extrude in a gelatinous matrix through a central pore in the cover of each acervulus. The conidia are apparently dispersed by water, but their function has not been studied. After leaves drop, the lesions turn black. Perithecia begin to develop in the lesions during early autumn and mature in fallen leaves in late spring. Some perithecia are unusual in opening simultaneously to both sides of a leaf.

More than 25 other species of *Ophiodothella* cause leaf spots of various plants in tropical and warm-temperate regions around the world. In North America, *O. fici* and *O. floridana* cause tar spot, also known as black leaf spot, and premature leaf drop of *Ficus aurea* (strangler fig) and several other *Ficus* species in Florida and the Caribbean region. The spots consist of raised, uneven, shiny black stromata measuring 1–10 mm across, solitary or in groups, visible on both leaf surfaces, and often surrounded on the lower surface by a yellow halo. *O. ferruginea* occurs on *Lyonia ferruginea* (staggerbush) in Florida, and *O. leucospila* occurs on *Platanus occidentalis* (sycamore) in Georgia.

References: 467, 1153, 1475, 1477

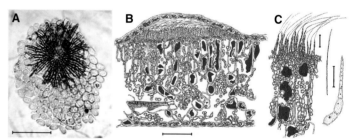

Figure 21. A. *Tubakia dryina:* a pycnothyrium with conidia that have been forced from beneath the shield. B, C. *Ophiodothella vaccinii* in a *Vaccinium arboreum* (farkleberry) leaf. B. Section of a subepidermal acervulus subtended by hyphae in and between leaf cells. C. Enlarged views of part of the spore-producing zone, a linear conidium, and a conidiophore. Scale bars = 50 μm in A, 100 μm in B, 10 μm in C. A courtesy of G. P. Munkvold, B and C adapted from reference 467 by permission of the Mycological Society of America.

A–F. Tubakia leaf spot of oaks, caused by *Tubakia dryina*. A, B. Foliar browning due to severe disease on *Quercus palustris* (pin oak) (IL, Sep). C. Black pycnothyria forming on a young lesion (IL, Jun). D–F. Mature lesions on *Q. palustris* and *Q. velutina* (black oak), with numerous pycnothyria visible in F (IL, Aug–Oct).

G–J. Fly speck leaf spot of *Vaccinium arboreum* (farkleberry). G, H. Spotted leaves. I, J. Close views of upper and lower leaf surfaces bearing black acervuli of *Ophiodothella vaccinii* (GA, Aug).

Photo credits: A–F—G. P. Munkvold.

Lophodermium Needle Casts (Plate 22)

More than 60 species of *Lophodermium* (Rhytismatales, Rhytismataceae) colonize the needles of coniferous trees and shrubs around the world. *Lophodermium* species are ubiquitous on *Pinus* (pines), although most species are endophytes, minor pathogens, or saprobes. Only one species, *L. seditiosum*, is a major pathogen in North America. Several others invade needles 1 or more years old and cause premature shedding. Additional lophodermia are minor pathogens or saprobic colonists of broadleaved plants. Fungi in several other genera of Rhytismataceae (e.g., *Davisomycella, Elytroderma, Isthmiella, Lophodermella, Ploioderma*, and others) also cause needle-cast diseases. Lophodermia of conifers in North America, their hosts, and attack habits are listed below. Reported distributions are identified as follows: c, central; e, east; n, north; s, south; w, west; NA, North America.

On *Pinus:*

L. australe on two- and three-needle pines; saprobic; warm-temperate to tropical regions around the world

L. baculiferum on *P. jeffreyi* (Jeffrey pine) and *P. ponderosa* (ponderosa pine); pathogen that infects year-old needles, which drop 1 year later; nw USA

L. canberrianum on *P. ponderosa;* pathogen; Australia, nw USA

L. conigenum on two-, three-, and some five-needle pines; an endophyte that causes no known damage; e Canada, USA, Europe, New Zealand

L. durilabrum on five-needle pines; weak pathogen associated with needle browning; c, nw NA, Korea

L. molitoris on two- and three-needle pines; associated with needle cast; c Canada, se USA

L. nitens on five-needle pines, but on *P. densiflora* (Japanese red pine) in Japan; endophyte and weak pathogen that causes spots, yellowing, and casting of old needles; NA, Asia

L. pinastri on two-, three-, and possibly five-needle pines, reported also on *Larix* (larch); endophyte and weak pathogen that causes yellowing and casting of old needles; NA, Eurasia, Australia

L. pini-excelsae on two-, three-, and five-needle pines; associated with needle cast; Ontario, nw USA, Europe, Asia

L. ravenelii on *Pinus* sp.; habit unreported; s USA

L. seditiosum on two-, three-, and some five-needle pines; pathogen that infects new needles and kills them before the second growing season; NA, Eurasia

L. staleyi on *P. elliottii* (slash pine), *P. ponderosa, P. sylvestris* (Scots pine); pathogen that causes yellowing of 2-year-old needles in late summer; Oregon, Asia

On *Abies* (fir) and *Picea* (spruce):

L. consociatum on *Abies;* associated with needle cast; w Canada, nw USA

L. crassum on *P. breweriana* (Brewer spruce); associated with needle cast; California

L. decorum on *Abies;* associated with needle cast; w Canada, nw USA

L. lacerum on *Abies;* associated with needle cast; e, w NA

L. nanakii on *Picea;* saprobe; NA, Eurasia

L. piceae on *Picea*, occasionally *Abies;* common endophyte of *Picea* needles; NA, Eurasia

L. uncinatum on *Abies;* associated with needle cast; w Canada, nw USA

On *Calocedrus* (incense cedar), *Chamaecyparis* (false-cypress), *Juniperus* (juniper), *Thuja* (arborvitae):

L. chamaecyparidis on *J. communis* (common juniper); British Columbia, Asia

L. juniperinum on *Calocedrus decurrens* (incense cedar), *Chamaecyparis, Juniperus;* saprobe or weak pathogen; NA, Europe

L. thujae on *Thuja;* c, e, w NA

Symptoms and signs. Pathogenic *Lophodermium* species on conifers induce chlorotic to necrotic spots and bands and dieback on needles, and/or premature needle senescence and casting. Symptoms appear in late summer to early spring, depending on tree species, pathogen, and local environment. Vigor of affected branches or entire small trees diminishes if disease repeatedly removes first-year or year-old age classes of needles. The causal fungi fruit on dead needles, producing inconspicuous spermagonia and then elongate apothecia called hysterothecia. The spermagonia, of the *Leptostroma* type, appear as tiny, elongate brown marks on a tan background (Plate 22G). The spores produced in spermagonia are not infectious and are presumed to be male gametes. Hysterothecia are gray-brown to black and develop in or just beneath the epidermis, causing the needle surface to bulge (Plate 22E). When mature, they open widely by a medial longitudinal split. *Lophodermium* species grow readily in culture, and spermagonia sometimes develop on culture media.

Life cycles. Lophodermium species complete their life cycles in not less than 1 year. Species such as *L. piceae* that grow as endophytes in green needles probably have a cycle of variable length, with infection occurring on needles of two or more age classes that live 1 or more years after infection. Infection is initiated by ascospores that are linear, colorless, and have sticky sheaths that serve for adhesion. They germinate on wet needles or scale-leaves, and the germ tubes produce appressoria from which host penetration occurs. Hyphal growth or residency in green needles lasts at least several months before symptoms are induced. The time of fruit body maturation and ascospore release varies among species.

Needle cast caused by *L. seditiosum*. This disease, which was erroneously attributed to *L. pinastri* until the 1970s, is economically important in Europe and North America. Epidemics have occurred in nurseries, Christmas tree plantations, and young planted forest stands. *Pinus nigra, P. resinosa*, and *P. sylvestris* (Austrian, red, and Scots pines, respectively) are damaged most in North America. Other hosts include *Pinus canariensis* (Canary Island pine), *P. cembra* (Swiss stone pine), *P. halepensis* (Aleppo pine), *P. leucodermis* (Bosnian pine), *P. montezumae* (rough-bark Mexican pine), *P. mugo* (mugo pine), *P. pinaster* (cluster pine), *P. taiwanensis* (Taiwan pine), and *P. virginiana* (Virginia pine). Resistance to *L. seditiosum* varies within and among pine species.

L. seditiosum attacks the current season's foliage and kills it before the next growing season. Lesions appear as yellow spots in late autumn to early spring and become brown with yellow margins, merging as they enlarge. Browning is most conspicuous in April and May, after which new foliage partially hides dead needles and bare twigs. Dead needles gradually turn from reddish brown to straw colored and either droop for a time or are cast, usually in June–July. Some diseased needles remain partly green with dead spots and tips. Twigs that bear only diseased needles often wither during late winter or early spring, and buds that survive produce stunted shoots and leaves. The disease is most severe on low branches, but entire small trees may be affected.

L. seditiosum fruits on dead year-old needles on the ground or lodged among branches, on needles still attached to twigs, and even on cones. Its apothecia are 0.8–1.5 mm long and lie entirely beneath the epidermis (Fig. 23A, p. 48), which causes them to appear gray unless they are wet. They mature in late summer and discharge ascospores in response to rainfall during August–November. Most infections occur during August and September.

Requirements for destructive outbreaks of needle cast, given a susceptible host, are nearby sources of abundant ascospores and cool, moist conditions. Cultural measures that prevent severe disease include using clean planting stock, separating new and old plantations, using non-host species for windbreaks near susceptible plants, and controlling weeds in young plantations. Weed control promotes air circulation, thus tending to shorten periods of foliar wetness conducive to infection.

References: 643, 886, 887, 924, 1099, 1214, 1490, 1832, 1859, 2213, 2598, 2600, 2603, 2662–2667, 2913, 2914, 3038, 3623, 3851

A–D. Needle cast caused by *Lophodermium seditiosum* on *Pinus sylvestris* (Scots pine). A, C. Christmas trees with severe symptoms. Year-old needles are dead and have begun to drop. New shoots are stunted, and many buds failed to open. Small tree size is the result of repeated loss of year-old needles (WA, Jun). B, D. Less severe symptoms on year-old needles: brown spots with yellow borders, dead tips. Gaps along the twig in D are where severely affected needles were cast (IL & WA, Jun).

E, F. Mature apothecia of *L. seditiosum* on *P. sylvestris*, closed when dry (E), and fully open for ascospore discharge after wetting (NY & WA, Jun).

G. Apothecia (elliptic structures) and spermagonia (small brown marks) of *L. pinastri* on a needle of *Pinus thunbergii* (Japanese black pine) (NY, Sep).

H, I. Apothecia of *L. juniperi* on *Juniperus* foliage that dried following mechanical damage in winter (NY, May).

Photo credit: F—J. M. Staley

Ploioderma Needle Casts of Pines (Plate 23)

Hard pines, those with two or three needles per fascicle, are subject to needle browning and defoliation caused by several species of *Ploioderma* (Rhytismatales, Rhytismataceae). Members of this small genus (perhaps nine species) have been reported from North America and Asia.

Generalized symptoms and signs. Yellow spots develop on mature first-year needles in autumn to winter, then enlarge and coalesce into brown bands that in turn cause death of all tissues distal to the lesions. Severe infection causes general browning of tree crowns and then openness due to premature needle drop. *Ploioderma* is similar to several other genera of needle-cast fungi in producing inconspicuous spermagonia, followed by dark brown to black apothecia (also called hysterothecia) embedded in dead needle tissues. The hysterothecia of most species are more or less elliptic in outline and open at maturity by a medial longitudinal slit, exposing the asci. The colorless, unicellular ascospores are more or less rod shaped, varying among species from about three to nine times as long as wide, and have a gelatinous sheath. *Ploioderma* is distinguished from related genera by a combination of microscopic characteristics including ascospore shape.

Generalized disease cycle. *Ploioderma* diseases have a 1-year cycle. The pathogens produce spermagonia and apothecia sequentially in autumn to spring, depending on species and latitude, in dead needles or lesions in partly green needles. Ascospores are presumed responsible for all infections. They are dispersed by air in spring and infect elongating needles. Symptoms develop 6 months or more after infection, typically during autumn and winter. The pathogens overwinter as mycelium in diseased needles on the tree or fallen needles on the ground.

Needle cast caused by *Ploioderma lethale*. This fungus is the best known and most widely distributed *Ploioderma* in North America. It occurs from Maine to Florida and across southeastern USA to Texas and Missouri. It infects native and introduced pines in natural and planted stands, including Christmas tree plantations, and in nurseries and landscapes. Hosts include *Pinus caribaea, P. clausa, P. echinata, P. elliottii, P. glabra, P. mugo, P. nigra, P. pungens, P. resinosa, P. rigida, P. serotina, P. taeda, P. thunbergii,* and *P. virginiana.*

P. lethale infects newly formed needles in summer, and symptoms develop in winter and early spring. Yellow spots progress to brown bands that give year-old needles a mottled appearance. In late spring the bands turn straw color to grayish brown, and the intervening tissue and needle tips turn brown, needle bases may remain green. Needles with tip necrosis tend to fall prematurely, or dead parts break off. Needle drop during spring and summer may leave just the new foliage on a severely diseased tree. This foliage grows in tufts at the branch tips because shoots from defoliated twigs do not attain normal length, causing needle bundles to remain close together. Damage is usually most severe on low branches or those near the base of a tree's crown, but entire trees of any size can be severely affected.

Colorless spermagonia develop in early spring, embedded in the lesions, and appear as minute longitudinal wrinkles in the epidermis. Apothecia develop in May in the South and June in the Northeast. They appear as black lines 0.4–1.4 mm long, covered by the epidermis (Fig. 23B), on all surfaces of straw-colored needles or in dead bands on green needles. Sometimes they aggregate into continuous lines up to 5 mm long. Apothecia open, and ascospores are expelled, during wet weather in late spring to early summer, the time depending on locality. Their gelatinous sheath allows the spores to stick to foliage and germinate to start new infections.

Resistance to *P. lethale* is commonly expressed by individual pines of susceptible species. For example, healthy individuals of *Pinus taeda* (loblolly pine) and *P. elliottii* (slash pine) may occur among severely diseased trees during epidemics. Resistance in these two species is strongly heritable and not correlated with growth rates of the trees. *P. palustris* (longleaf pine) in the South and *P. resinosa* (red pine) in the North are generally unaffected by *P. lethale*. The resistance of *P. resinosa* was verified by inoculation of seedlings with ascospores. The seedlings remained healthy, while those of similarly inoculated southern pines developed needle cast.

Other *Ploioderma* species. Three additional American fungi are known: *P. hedgcockii, P. lowei,* and *P. pedatum. P. hedgcockii* occurs in southeastern USA from Virginia to Texas and in Mexico. It attacks *Pinus caribaea, P. clausa, P. echinata, P. hartwegii, P. palustris, P. pseudostrobus, P. rigida, P. taeda,* and *P. virginiana. Ploioderma lowei* affects *Pinus elliottii* in Louisiana and Mississippi, and *Ploioderma pedatum* occurs on *Pinus radiata* (Monterey pine) in California.

Ploioderma hedgcockii and *P. lowei* furnish exceptions to the generalized description and cycle. The former species is unusual in producing pycnidia (spermagonia?) and apothecia in green as well as dead tissues and in having two-celled conidia (spermatia?), roughly 20 × 5 μm, that may have a propagative function. In *P. lowei* the apothecia are nearly circular in outline and mature in autumn, and ascospores are dispersed before midwinter. Thus, they probably initiate infection in mature needles.

Outbreaks of Ploioderma needle casts in southeastern USA are often accompanied by other needle diseases, notably needle cast caused by *Lophodermella cerina*.

References: 462, 643, 876, 886, 887, 1832, 1859, 1935, 1937, 1938, 2146, 2603, 2663, 2664, 3038

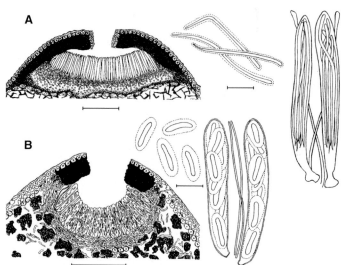

Figure 23. A. *Lophodermium seditiosum:* apothecium in cross section, ascospores, and asci. B. *Ploioderma lethale:* apothecium in cross section, ascospores, and asci. Hyphae among asci are paraphyses, thought to serve as spacers in a hymenium. Both fungi produce apothecia entirely beneath the epidermis. A black roof of fungal tissue covers the apothecium of *L. seditiosum*, which therefore appears elliptic when viewed from above. Black tissue in *P. lethale* is not as wide as the apothecium, which therefore appears as a dark line on the needle surface. Scale bars = 200 μm for apothecia, 25 μm for other structures. Adapted from references 2664 and 2665 by permission of CAB International.

Needle cast caused by *Ploioderma lethale*.
A. On *Pinus elliottii* var. *elliottii* (slash pine). Resistant trees in the background contrast with the severely affected tree (GA, May).
B–G. On *Pinus nigra* (Austrian pine). B. One-sided defoliation of a young tree (PA, Apr). C. Early symptoms: yellow to brown spots and bands (NY, Feb). D. Fully developed symptoms in spring. Green basal parts of needles grew after infection occurred the previous year (NY, Apr). E–G. Symptomatic needles with apothecia forming in some lesions (arrow in E). Tip necrosis, a secondary symptom, results from withering of green tips beyond fungus-killed bands (NY, Jun).
H, I. Mature apothecia of *Ploioderma lethale* on needles of *Pinus nigra* before wetting (H) and after opening in response to moisture (NY, Jun).
Photo credit: B—W. Merrill

Canavirgella and Davisomycella Needle Casts of Pines (Plate 24)

Canavirgella Needle Cast of White Pines

Scattered trees or groups of related individuals of *Pinus strobus* (eastern white pine) throughout much of its natural range are affected by a needle cast caused by *Canavirgella banfieldii* (Rhytismatales, Rhytismataceae). The disease has been known for nearly a century, but symptoms now thought to be caused by *C. banfieldii* were formerly attributed to other fungi or to abiotic agents. Uncertainty remains about mimicking symptoms, however, as noted below. Most individuals of *P. strobus* escape damage and are presumed to be resistant. *P. peuce* (Macedonian pine) growing adjacent to diseased *P. strobus* was found affected in Vermont.

Symptoms and signs. Lesions appear in June and July as pinkish tan spots on the stomatal (adaxial) surfaces of elongating needles and develop into pinkish tan to yellowish tan bands with indistinct margins. Lesions expand primarily toward needle tips and kill them. Needle bases usually remain green. Dead parts turn reddish brown by early autumn, then fade to grayish tan. Severe disease can result in browning of nearly all first-year needles on a highly susceptible tree growing in the open, but in most years such trees have infection only on scattered needles. Diseased needles are retained until the end of the second growing season. Many dead needle tips break off before that time, however.

Inconspicuous spermagonia of *C. banfieldii* develop in late summer to early autumn beneath the abaxial surface of the recently killed part of a needle. They appear as elliptic areas 0.5 × 0.2 mm on average, slightly darker than the surrounding surface. Hysterothecia (apothecia that open by splitting longitudinally) begin to form during autumn, also beneath the abaxial surface, and mature in spring. Each one appears initially as a dull gray stripe 15–50 mm long that becomes gray and black when mature. The ascospores are colorless, single celled, narrow-elliptic, and measure 19–36 × 4–6.5 μm (average near 25 × 5 μm). They have a sticky sheath that permits adhesion to host needles. Hyphae of *C. banfieldii* grow only transiently from either ascospores or recently infected pine needles on artificial media. Needles inhabited by this fungus usually become colonized by secondary fungi, notably "*Hendersonia*" *pinicola* (see p. 36), which can cause diagnostic confusion.

Disease cycle. The 1-year cycle begins with infection by ascospores of immature needles in late spring to early summer, the time depending on latitude and elevation. Hyphae associate with walls of mesophyll cells, which collapse as lesions form. Lesions appear while needles are still growing. The pathogen overwinters as developing hysterothecia in diseased needles on the tree. Hysterothecia mature in late spring to early summer, then open and discharge ascospores in response to wetting, completing the cycle.

The syndrome attributed to *C. banfieldii* is similar to "semimature-tissue needle blight" (SNB) of *P. strobus* (Plate 238), which has been attributed hypothetically to transient severe water stress in succulent needle tissue and also to ozone injury. Further research on biology of *C. banfieldii* and its association with needle blight is needed to clarify the matter.
References: 821, 1859, 2336, 2338, 2339, 2340, 2603, 2605, 4291

Davisomycella Needle Casts

Two- and three-needle pines that grow in northern and western North America often undergo needle browning and defoliation caused by *Davisomycella* species (Rhytismatales, Rhytismataceae). Two the seven recognized species in North America are featured here.

Davisomycella ponderosae, endemic in the Rocky Mountain region from Colorado to Montana, causes sporadic prominent needle browning and defoliation of *Pinus ponderosa* (ponderosa pine). Needles 1–5 years old die or develop necrotic bands in spring, the symptoms often appearing on needles of several age classes in the same year. One to all needles in a fascicle may be affected. Fascicles with more than one needle diseased are shed prematurely.

Inconspicuous spermagonia mature in dead light brown needles 2–5 years old that are still attached to twigs or on dead parts of living needles. Hysterothecia develop in the same needles in spring and mature in summer. They form beneath the epidermis, usually on the abaxial needle face (outward with respect to axis of needle bundle).

They are ovoid to linear, frequently sinuous, and measure 1–45 × 0.5–1.1 mm when open. The colorless ascospores are linear and measure 120–170 × 3–4.5 μm, each with a gelatinous sheath about 7 μm thick when hydrated. Ascospores are dispersed in summer in response to rainfall. Some hysterothecia remain capable of dispersing spores until mid-October. *D. ponderosae* produces reddish pigment in pure cultures.

The disease cycle has not been elucidated. A likely scenario involves infection of first-year needles by ascospores in summer, then 1 to several years of latent infection before symptoms appear. Fruiting bodies develop soon after lesions appear or needles die, but do not occur on all faded needles before they drop. Ascospore dispersal and infection in summer complete the cycle. Epidemics are initiated in response to wet weather in summer, and symptoms are possibly exacerbated by drought in subsequent years. *D. ponderosae* and drought were associated with severe needle browning in southern Colorado in 1999–2001. In theory, severe water shortage during the latent phase of infection could interfere with host defenses and cause infected needles of several age classes to fade and die in the same year.

Davisomycella ampla is distributed across North America and also occurs in Brazil and New Zealand. It causes a disease called tar-spot needle cast of *Pinus banksiana* (jack pine) and *P. contorta* (lodgepole and shore pines), and it has been recorded on *P. pinaster*, *P. radiata*, and *P. strobus*. It infects current-year needles, and lesions are apparent by the following spring. Mature lesions girdle the needles and cause distal parts to die. Severe infection leads to dramatic browning and premature casting of year-old foliage, which may leave only current-season needles on a tree by late summer. Damage to planted *P. banksiana* has been observed to vary with seed source.

Hysterothecia of *D. ampla* form about a year after infection in bleached (buff-colored) tissue beneath the epidermis of either surface on needles that usually are still partly green. The bleached areas are typically bordered by narrow orange-brown bands, beyond which narrow chlorotic zones separate dead from green tissue. The hysterothecia are black, elliptic in outline, 0.7–1.35 mm long, and often confluent. When mature and wet, they open and release club-shaped colorless ascospores 60–130 × 8 μm, each with a gelatinous sheath 3–7 μm thick. Inconspicuous spermagonia, embedded in the needles, precede the apothecia. Secondary fungi, notably *Sarcotrochila macrospora* in British Columbia, sometimes suppress fruiting of *D. ampla*. Diseased needles drop in summer after spore release is complete.
References: 635, 886, 887, 1005, 1214, 1818, 1832, 1859, 2663, 2776, 3820, 4441, 4527

A–C. Needle cast of *Pinus strobus* (eastern white pine) attributed to *Canavirgella banfieldii*. A. Foliage in the lower two-thirds of the crown of a pole-size tree discolored from infection (NY, Sep). B. Typical yellow to brown lesions, some having spread to needle tips (NY, Sep). C. Magnified views of needles with signs of the causal fungus: at top an elliptic spermagonium with two dark pores for release of spermatia (PA, Sep); at center, part of an immature apothecium, appearing as a gray-and-black stripe (VT, Mar); at bottom, part of a mature apothecium, widely open after wetting, exposing the hymenium of asci (ME, Jun).

D–G. Needle cast of *P. ponderosa* (ponderosa pine) caused by *Davisomycella ponderosae*. D. Faded foliage on severely diseased trees contrasts with predominantly green crowns of less affected trees (CO, May). E. Sparse foliage and stunted twigs on trees defoliated 2 years earlier (CO, May). F. Portions of maturing, unopened hysterothecia on a killed needle (CO, May). G. Mature, open hysterothecia (CO, Oct).

H, I. Needle cast of *Pinus banksiana* (jack pine) caused by *Davisomycella ampla*. H. Faded and brown year-old needles, soon to be cast, contrast with newly grown current year's foliage (ON, summer). I. Diseased needles with mature hysterothecia (ON, early summer).

Photo credits: C—W. Merrill; D–G—J. J. Worrall; H, I—Canadian Forest Service Great Lakes Forestry Centre

Elytroderma Needle Casts of Pines (Plate 25)

Two species of *Elytroderma* (Rhytismatales, Rhytismataceae) are known, one in North America and one in southern Europe. Both cause needle casts of two- and three-needle *Pinus* species (pines). They are unusual in that they may perennate in buds, twigs, and branches and cause witches'-brooms on some hosts. The North American fungus is *E. deformans;* the species in Europe is *E. torresjuanii* (syn. *E. lusitanicum*). Disease caused by the former is featured here.

E. deformans induces needle cast and witches'-brooms on *P. attenuata, P. contorta, P. coulteri, P. jeffreyi,* and *P. ponderosa* (knobcone, lodgepole, Coulter, Jeffrey, and ponderosa pines, respectively), but causes only needle cast on its other known hosts: *P. banksiana, P. cembroides, P. echinata,* and *P. edulis* (jack, Mexican pinyon, shortleaf, and pinyon pines, respectively). This fungus occurs from Arizona and New Mexico northward into British Columbia and Northwest Territories, and eastward to South Dakota and Ontario. It has also been collected in New Brunswick on *P. banksiana* and in Georgia on *P. echinata*. Severe outbreaks of disease caused by *E. deformans*, although uncommon, can cause great damage that may be exacerbated by bark beetle attacks. One outbreak in *P. ponderosa* in Oregon resulted in premature harvest to salvage more than 100 million board-feet of sawtimber. Repeated outbreaks on *P. jeffreyi* have been noted in the Sierra Nevada. The information below is from accounts of this disease in *P. ponderosa* and *P. jeffreyi*.

Symptoms and signs. Foliar symptoms appear in spring. Lesions on needles infected by spores are reddish brown spots that expand or coalesce to involve entire needles except for a few millimeters at the base. When needles are infected by mycelium growing out from a twig, most or all year-old needles on the twig discolor simultaneously, turning red-brown except for the basal 6–12 mm. Inconspicuous spermagonia of *E. deformans*, up to 1.2 mm long and the same color as the leaf surface, appear as minute blisters in newly reddened year-old needles in spring. Translucent beads or tendrils of spores emerge from these in moist weather. In May–June, dark lines between stomatal rows signal the development of apothecia. These arise beneath the epidermis of all surfaces of dead parts of year-old needles and become mature from midsummer until spring. They are 2–10 mm long and 0.35–0.45 mm wide and open by a longitudinal fissure between the lines of stomata. The ascospores are two celled, cylindric, colorless, measure 90–118 × 6–8 μm, and have a gelatinous sheath 6–10 μm thick.

Twigs with red-brown foliage, sometimes called "flags," represent infections that began in an earlier year and became systemic in the twigs. Flags gradually fade and become partly obscured by new foliage, and by late summer are inconspicuous among branches with normally senescing old needles. Most diseased needles fall during autumn and winter, but a few remain attached, take on a weathered gray appearance, and slowly disintegrate. Twigs that display flags usually do so each spring for many years, slowly lose vigor, and die. Initially vigorous branches with flags may grow slowly for some years and curve upward. Twigs flagged in two or more consecutive years lack old needles. This characteristic and lost vigor may make tree crowns appear thin.

In bark *E. deformans* often causes small, elongate, resinous lesions that appear as reddish brown streaks in longitudinal view (Plate 25E) or as a ring of dots in living bark near the cambium in cross sections. The lesions, sometimes called resin cysts, are visible in branch segments 2–12 years old. Infection of vigorous branches often results in broom formation. Small trees in dense stands are sometimes completely converted into loose, flat-topped brooms.

Trees of all ages and sizes may be affected. Severe disease in 2–3 consecutive years can cause nearly complete defoliation and is sometimes lethal, especially to saplings in dense stands. This situation is infrequent, however, because the natural habitats of *P. ponderosa* seldom have successive years with weather suitable for severe infection. Mainstems of seedlings and saplings sometimes become infected by hyphal growth from needles. Seedlings in their first few years of growth are seldom killed but may become permanently stunted. The disease commonly results in growth loss and increased susceptibility of older trees to killing by bark beetles and root-infecting fungi.

Disease cycle. The cycle is incompletely known. Apothecia liberate two-celled, sticky-sheathed ascospores to the air during wet weather beginning in spring in some years, but usually in summer and early autumn. These spores are believed to cause all infections, most of which apparently begin in early autumn. New needles, as well as those a year or more old, have become infected when ascospores were used as inoculum in experiments conducted in spring and summer. Needles inoculated in spring remained asymptomatic until autumn, and those inoculated in summer or autumn showed symptoms the next spring. The 1-year version of the disease cycle is completed when apothecia produce ascospores in needles infected the previous year. Longer versions of the cycle involve perennial infection of stems and buds.

The interval between needle infection and display of flags is unknown and perhaps variable. During this interval the fungus grows in phloem of needles and then twigs, and thence into buds, new shoots, and new needles. Perennial infection of twigs perpetuates needle disease in the absence of new infections started by spores. Hyphae of *E. deformans* colonize buds as much as a year before symptoms develop in needles grown from those buds. The fungus probably does not grow far in the proximal direction from the point of entry in a twig.

Epidemiology. Incidence of Elytroderma needle cast is low in most areas within the pathogen's geographic range, but local epidemics develop occasionally and last for several years because of perennating infections. Major waves of infection began in some northwestern *P. ponderosa* forests early in the 20th century and again during the 1930s and 1940s. The most destructive epidemics have occurred in localities near 1500 m elevation in Oregon and 1000 m in northern Washington. Outbreaks start in sheltered places such as narrow valley bottoms, edges of meadows, thickets of saplings, and on low branches on north and northwest sides of pole-size and larger trees. Outbreaks tend to begin almost simultaneously hundreds of miles apart, indicating regional occurrence of conditions favorable for new infections by spores, but the specific conditions have not been described. The infrequency of destructive epidemics indicates that conditions favoring ascospore production and discharge, germination, and penetration of needles seldom all occur in a single year. Hot, dry weather in spring, while allowing development of foliar symptoms, suppresses spermatial production and is followed by suppressed apothecial development or failure of spore production in apothecia. This relationship is a basis for the unproven presumption that the spores herein called spermatia do have a sexual function in the life cycle of the fungus.

In areas where susceptible pines are subject to severe infection, economic damage can be minimized by thinning or selective logging to eliminate severely diseased trees and those with infections near the top of the crown. This practice also reduces the impact of secondary pests and pathogens. Individual *P. ponderosa* trees practically free from symptoms have been noted in areas of severe disease, an indication of possible resistance.

References: 643, 724, 886, 887, 1099, 1214, 1298, 1832, 1859, 2213, 2603, 2663, 3038, 3350, 3631, 4240

Symptoms and signs of needle cast caused by *Elytroderma deformans* on *Pinus ponderosa*.

A. Brown foliage on systemically infected branches. These "flags" are reddish brown in spring and fade through shades of brown during the summer (ID, Jul).

B. Twig and branch dieback and witches'-brooms. Diseased twigs turn upward (ID, Jul).

C. Brooms, usually less than 60 cm but occasionally up to 2 m broad, are tight clusters of twigs on which only needles of the current season are green. Dead needles remain attached until autumn or winter (ID, Jul).

D, F. Perennially infected branches have stunted twigs and needles, dieback, and needle blight. Usually all needles in a fascicle die simultaneously (CO, Jun; ID, Jul).

E. Brown lesions in inner bark of diseased twigs are diagnostic but not always present (CO, Jun).

G, H. Linear apothecia in year-old needles from infected twigs. G. Some diseased needles retain green bases (ID, Jul). H. Open apothecia (BC, Aug).

Photo credit: H—J. Hopkins

Lophodermella and Cyclaneusma Needle Casts of Pines (Plate 26)

Lophodermella Needle Casts

Six species of *Lophodermella* (Rhytismatales, Rhytismataceae) in North America and at least three others in Eurasia kill and cause premature casting of pine needles. *Lophodermella* species, their hosts, and synoptic information about their disease cycles, if reported, are listed below. Geographic distributions are identified by postal codes and NA for North America; e, east; n, north; s, south; w, west.

L. arcuata on *Pinus albicaulis, P. flexilis, P. lambertiana, P. monticola*: spores dispersed in July–August, lesions appear in spring, apothecia mature in early summer, needles drop in autumn; w NA

L. cerina on *P. contorta, P. elliottii, P. ponderosa, P. sylvestris, P. taeda*: spores dispersed in spring, lesions appear in autumn, apothecia mature in spring, needles drop 1–2 years after infection; se, sw USA, CO

L. concolor on *P. albicaulis, P. banksiana, P. contorta, P. ponderosa, P. sylvestris*: spores dispersed in June–July, lesions appear in spring, apothecia mature in late spring, needles drop 12–14 months after infection; ON, QC, w NA

L. conjuncta on *P. mugo, P. nigra, P. sylvestris*: spores dispersed all year with a summer peak in mild climates, lesions appear after 12–18 months, apothecia mature after 12–24 months, needles drop after 2 years; a major pathogen in Europe

L. maueri on *P. ayacahuite*: spore dispersal in June–July; Mexico

L. montivaga on *P. contorta, P. flexilis, P. monticola, P. ponderosa, P. radiata, P. sylvestris*: spores dispersed in July–September, lesions appear after 1 month, apothecia mature during the dormant season, needles drop in summer; causes dramatic needle browning on *P. contorta* in western Canada in some years; w NA

L. morbida on *P. attenuata, P. ponderosa*: spores dispersed in June–August, lesions appear after 1 month, apothecia mature during the dormant season, needles drop 13–15 months after infection; has caused severe growth loss in young *P. ponderosa* plantations on moist, high-elevation sites normally occupied by other tree species on the western slope of the Cascade Mountains but is inconsequential on sites typically colonized by *P. ponderosa*; CA, OR, WA

L. orientalis on *P. kesiya*: Asia

L. sulcigena on *P. mugo, P. nigra, P. sylvestris*: spores dispersed in June–August, lesions appear after 1 month, apothecia mature during the dormant season, needles drop 15 months after infection; Europe

All of these fungi are native to the continents where they occur. The North American species typically cause sporadic needle-cast epidemics in natural forests in single years. If conditions permit severe infections in successive years, needles of all but the youngest age class are removed and tree growth slows. Only *L. cerina* and *L. morbida* are reliably known to produce spermagonia, which appear with or in advance of the apothecia.

A representative Lophodermella needle cast. L. concolor causes a typical pine needle cast with a 1-year-long cycle that is initiated in first-year needles by ascospores in late spring to early summer. Diseased needles remain green in the year of infection, and the fungus overwinters in them. By spring these needles are dead and reddish brown, and by July they are straw colored (Plate 26B). Apothecia (also called hysterothecia) mature in them in late spring and early summer. Apothecia of *L. concolor*, unusual in being colorless, appear as shallow, elliptic depressions, 0.4–0.8 mm long, in all surfaces of a needle (Plate 26D). The apothecia open by a longitudinal fissure during wet weather and expel elongate, colorless, club-shaped, unicellular spores that have sticky, gelatinous sheaths. Spores that germinate on young needles produce appressoria from which the needles are penetrated, renewing the cycle. Second-year needles begin to fall as apothecia mature, and casting is complete before autumn. During outbreaks of this disease, nearly all of the second-year foliage in the lower third or half of the crown may die and fall. In the West conditions favoring infection by *L. concolor* are common in mountain valleys and in some coastal areas where fog provides moisture for infection.
References: 643, 877, 940, 1214, 1490, 1543, 1703, 1831, 1832, 1859, 2635, 2663, 2680, 3038, 3563, 3822, 4448, 4484, 4527

Cyclaneusma Needle Cast

Cyclaneusma minus (Rhytismatales, Rhytismataceae) causes needle cast of many pine species around the world. It is significant mainly in young plantations. Recorded hosts include *Pinus brutia, P. caribaea, P. contorta, P. flexilis, P. halepensis, P. jeffreyi, P. mugo, P. nigra, P. pinaster, P. ponderosa, P. radiata, P. sabiniana, P. stankewiczii, P. strobus, P. sylvestris, P. uncinata,* and *P. virginiana.* In North America damage by *C. minus* to *P. sylvestris* grown for Christmas trees is noteworthy because the premature yellowing and casting of 1- and 2-year-old needles reduces the value of the trees. *C. niveum,* the only other known species of *Cyclaneusma,* is a widely distributed but inconsequential colonist of senescent and dead needles of some of the hosts listed above.

Symptoms and signs. C. minus infects and resides as an endophyte in green first- and second-year needles and causes yellowing and casting of second-year needles. Yellowing begins in late summer and casting in autumn, but some diseased needles remain green until spring. The lesions appear as spots that are first light green, then yellowish, enlarging to yellow bands. Affected needles eventually become entirely yellow. Transverse brown bars appear on yellow needles of *P. sylvestris* (Plate 26H), but this symptom is uncommon or indistinct on other hosts. On *P. radiata,* whole needles or bands on them may turn reddish brown. Severely diseased trees of *P. sylvestris* appear distinctly yellow before needles fall (Plate 26E). After several years of severe disease, trees retain only the current season's needles.

Apothecia of *C. minus* develop in the brown bars, if these are present, and then beneath the entire surface of the needle, appearing at first as elliptic swellings 0.2–0.6 mm long, the same color as the rest of the needle. When mature they open by a longitudinal fissure (Plate 26J) and expand when wet, pushing flaps of host epidermis upward like tiny covers with hinges at their sides. The dull cream-white to yellow hymenium (spore-producing layer) is thus exposed. In some regions the apothecia are preceded and accompanied by inconspicuous pycnidia, also colorless, that are deeply embedded in the needles. The spores produced in these are not infectious. In Pennsylvania, where Cyclaneusma needle cast has been studied intensively, pycnidia have not been observed in nature, although they form readily on laboratory media. *C. minus* expels ascospores into the air when mature apothecia are moistened by rainfall or dew at temperatures from near freezing to near 30°C (optimum 22°C). Peak concentrations of airborne spores occur 4–6 hours after the onset of rain.

Disease cycles. C. minus liberates ascospores throughout the year, resulting in overlapping generations. In Pennsylvania most infections begin in mid-April to late August, and a secondary wave of infection begins in late autumn. Needles of all ages are susceptible. The fungus enters them through stomata. Diseased needles turn yellow after 10–15 months, and apothecia form before or after needles drop. The length of the incubation period, before needles turn yellow, may depend in part on host vigor. The pathogen's cycle on *P. radiata* is just 1 year long, which may reflect the fact that this tree species grows in climates milder than that of Pennsylvania.
References: 583, 643, 901, 961, 1176, 1220, 1859, 2086, 2601, 2603, 2604, 2663, 3038, 3196

A–D. Needle cast of *Pinus contorta* var. *contorta* (shore pine) caused by *Lophodermella concolor*. A. Diseased year-old needles in the lower half of the crown are brown to straw colored. B. Twigs with healthy developing needles, dead second-year needles, spaces where some diseased needles were cast previously, and green third-year needles that escaped infection while young. C. Diseased needles turn brown, then straw colored. D. Apothecia become visible as small depressions in needle surfaces (all WA, Jun).

E–J. Cyclaneusma needle cast of *Pinus sylvestris* (Scots pine). E. Discolored second-year needles on a sheared tree (MI, Aug). F. Close view of diseased needles (MI, Aug). G. Winter view of a tree that sustained severe needle cast the previous year (PA, Feb). H. Brown bars on yellow needles are useful for diagnosis; normal senescent needles are shown for comparison (PA, Oct). I. Apothecia form in lesions and are the same color as their substrate (MI, Aug). J. Mature apothecia open by median fissures (NY, Oct). *Photo credits*: G, H—W. Merrill

55

Rhizosphaera and Isthmiella Needle Casts (Plate 27)

Rhizosphaera Needle Casts

Six species of *Rhizosphaera* (coelomycetes) cause needle cast or needle blight of conifers. *R. kalkhoffii* and *R. pini* are the best-known members of this group. They occur as facultative parasites (organisms that have both saprobic and parasitic capability) on diverse conifers around the world. *R. kalkhoffii* is the more important species, causing partial defoliation of *Picea* species (spruces) growing outside their native ranges in nurseries, plantations, and landscapes. It is inconsequential in natural forests. *P. engelmannii* (Engelmann spruce) and *P. pungens* (Colorado blue spruce) are highly susceptible; the latter species has been severely damaged in some Christmas tree plantations in eastern USA. *P. glauca* (white spruce) is intermediate in susceptibility, and *P. abies* (Norway spruce) is relatively resistant. Other reported spruce hosts are *P. mariana*, *P. omorika*, *P. orientalis*, *P. rubens*, *P. schrenkiana*, and *P. sitchensis*. Hosts in other genera include *Abies alba*, *A. grandis*, *A. procera*, *A. sibirica*, *Cedrus deodara*, *Pinus cembroides*, *P. densiflora*, *P. jeffreyi*, *P. mugo*, *P. nigra*, *P. radiata*, *P. strobus*, *P. thunbergii*, *P. wallichiana*, *Pseudotsuga menziesii*, *Sequoia sempervirens*, and *Tsuga heterophylla*.

Symptoms and signs. This description applies primarily to *Picea*. Symptoms often appear in mid to late summer as yellow mottling of first-year or older needles, some of which later turn brown or, in *P. pungens*, purplish brown. Given conducive conditions in successive years, symptoms spread upward and around the tree. Branches die if year-old needles are removed each year for 3–4 years. Trees of any size may be infected, but serious defoliation leading to branch death is confined mainly to trees less than 7 m tall.

Rhizosphaera species produce pycnidia that are rooted in stomata and plug the pores. Thus pycnidia of *R. kalkhoffii* are found in rows on all faces of spruce needles, usually on discolored needles in spring, but often at other times and on needles that are still green. Mature pycnidia appear as black spheres less than 0.1 mm in diameter, each capped by a bit of white wax that was pushed out of the stoma by hyphae that gave rise to the fruit body. They produce colorless, single-celled conidia 4.5–8.6 × 2.5–4.6 μm.

Disease cycle. The pathogen overwinters as mycelium in, and pycnidia on, living and recently dead needles. Conidia are dispersed by water beginning in spring and continuing into autumn. They germinate relatively slowly. The fungus requires about 48 hours to infect wet spruce needles at temperatures near 25°C. The spores can survive for many weeks on *Picea* needles under conditions that do not permit germination. First-year needles on low branches are commonly infected, but needles of any age may be colonized. Diseased needles typically turn brown and pycnidia develop on them during late winter and early spring. They drop during summer and autumn, 12–15 months after the onset of infection. Pycnidia sometimes appear on discolored *P. pungens* needles in autumn of their first year.

Christmas tree culture favors *R. kalkhoffii* because its conidia may be splashed or mechanically transferred among closely spaced wet trees, and also because dense foliage on sheared trees retains moisture longer than the foliage of unsheared trees. *R. kalkhoffii* is both a primary pathogen and a colonist of senescent needles and those stressed by other pathogens or environmental insults. It apparently contributes to damage in freeze-injured *Picea rubens* (red spruce) needles, and it colonizes *Picea* foliage stressed by ozone or sulfur dioxide more readily than it does healthy foliage. It is soon replaced by saprobes in dead needles, however. In Japan development of needle blight caused by *R. kalkhoffii* on *Pinus densiflora* (Japanese red pine) was accelerated in foliage previously stressed by exposure to sulfur dioxide. *R. kalkhoffii* grows most rapidly at 18–25°C on laboratory media, producing mycelium, yeastlike bud cells, and pycnidia.

R. pini is similar to *R. kalkhoffii* but has larger conidia, 15–30 × 7–12.5 μm. It causes needle yellowing and browning of year-old and older needles of firs: *Abies balsamea*, *A. cephalonica*, *A. fraseri*, *A. grandis*, *A. nordmanniana*, and *A. veitchii*. Its development in Christmas tree plantations seems to be favored by moist conditions associated with dense foliage. In forests it causes partial defoliation of low branches of *A. balsamea* and occasionally kills seedlings up to 6 years old. *R. pini* has also been recorded on *Picea mariana*, *Pinus* sp., and *Tsuga diversifolia*.

References: 588, 643, 722, 954, 956, 994, 1099, 1214, 1298, 1340, 1341, 1988, 2182, 2213, 2417, 2460, 2599, 2603, 2846, 4235

Isthmiella Needle Casts

Four species of *Isthmiella* (Rhytismataceae, Rhytismatales) cause needle cast of *Abies* (fir) and *Picea* (spruce) in North America. *I. abietis* and *I. quadrispora* occur on *Abies lasiocarpa*; *I. faullii* on *A. balsamea* and *Picea mariana*; and *I. crepidiformis* on *A. balsamea*, *P. abies*, *P. engelmannii*, *P. glauca*, *P. mariana*, and *P. rubens*. These fungi are found mainly in natural forests but sometimes disfigure ornamental trees. *I. crepidiformis* is common in moist spruce-fir forests. Its spermagonia and apothecia (hysterothecia) are always found on dead needles 2 or more years old. It does no economic damage, although it often causes conspicuous foliar browning of *P. mariana* (black spruce). Spruce needles killed by *I. crepidiformis* remain attached until they weather off after several years. Such needles sometimes persist longer than needles that senesce normally, and so can be seen on otherwise bare branch segments.

Apothecia of *I. crepidiformis* (Plate 27I) are black, up to 3 mm long, and 0.3–0.6 mm wide. When on *Abies*, they open on the lower needle surface. They produce elongate ascospores that at maturity are constricted in the middle, ensheathed in a mucilaginous substance, and are shot into the air in response to needle wetting.

The life cycle of *I. crepidiformis* is incompletely known but seems to be similar to those of the better known *I. faullii* and *I. quadrispora*. They have 2-year cycles characterized by ascospore dispersal and infection of first-year needles in summer, death of diseased needles the next spring, production of spermagonia 1 year after infection, and maturation of apothecia after an additional year.

References: 886, 887, 1214, 1298, 1683, 1683a, 1832, 3038, 3823, 4525

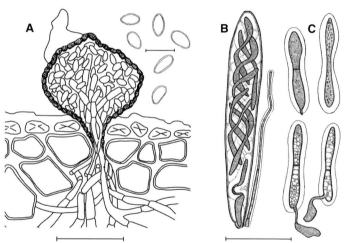

Figure 27. A. *Rhizosphaera kalkhoffii*: a pycnidium and conidia on a *Picea* (spruce) needle. The pycnidium has a wax cap. Scale bars = 50 μm for pycnidium, 10 μm for conidia. B, C. An ascus and ascospores of *Isthmiella crepidiformis*. Two of the spores have germinated. Scale bar = 50 μm. A adapted from reference 956 by permission of CAB International; B, C adapted from reference 886.

A–C. Rhizosphaera needle cast of spruce. A, B. Foliar yellowing and browning caused by *Rhizosphaera kalkhoffii* on (A) a landscape specimen of *Picea glauca* (white spruce) (NY, Jul) and (B) young *P. pungens* (Colorado blue spruce) in a plantation (PA, Oct). C. Magnified view of a *P. pungens* needle with pycnidia of *R. kalkhoffii* (NY, Sep).

D–F. Rhizosphaera needle blight of fir. D. Needles on a low branch of *Abies balsamea* (balsam fir) affected by *R. pini* are brown except near twig tips (NH, Jun). E. Foliar yellowing attributed to *R. pini* on *Abies fraseri* (NC, Feb). F. Pycnidia of *R. pini* on the lower (stomatal) surface of an *A. balsamea* needle (NH, Jul).

G–I. Isthmiella needle cast of *Picea mariana* (black spruce). G. A branch with third-year and older needles browned by *Isthmiella crepidiformis* (ON, Jun). H. A third-year segment of a *P. mariana* twig with needles killed by *I. crepidiformis* (ON, Aug). I. Apothecia of *I. crepidiformis* (ON, Aug).

Photo credits: D–F—N. Wenner and W. Merrill

Rhabdocline and Swiss Needle Casts of Douglas-fir (Plate 28)

Needle casts of *Pseudotsuga menziesii* (Douglas-fir), caused by fungi native to the forests of western North America, came to prominence first in Europe and later in North America. The pathogens, *Rhabdocline* species (Helotiales, Hemiphacidiaceae) and *Phaeocryptopus gaeumannii* (soon to be renamed and reclassified as noted below), are restricted to *Pseudotsuga* and have been transported around the world with their host.

Rhabdocline Needle Cast

Needle browning and defoliation by Rhabdocline needle cast leads to suppressed growth and value loss in Christmas trees and ornamental nursery stock. Seedlings that become infected in nurseries can become foci of subsequent damage in plantations. The disease sometimes disfigures landscape trees.

Both *Pseudotsuga menziesii* and *P. macrocarpa* (bigcone Douglas-fir) are affected. Similar symptoms are caused by five *Rhabdocline* taxa: *R. pseudotsugae*; comprising subspecies *epiphylla* and *pseudotsugae*; and *R. weirii*, comprising subspecies *oblonga*, *obovata*, and *weirii*. These taxa may soon be recognized as distinct species. *R. weirii* subsp. *oblonga* and *R. pseudotsugae* subsp. *pseudotsugae* are the common forms in the Great Lakes region and eastern North America. Various subspecies intermingle throughout the range of the hosts in the West. The common form in Europe is *R. pseudotsugae* subsp. *pseudotsugae*. A third species, *R. parkeri*, is an innocuous needle endophyte. *Meria laricis*, a foliar pathogen of *Larix* (Plate 29), is an anamorphic *Rhabdocline*.

Symptoms and signs. Chlorotic lesions appear in late summer or early autumn on one or both surfaces of first-year and occasionally older needles. Lesions darken to purplish brown spots and bands by late autumn to early spring. Often they coalesce and involve entire needles except for a short basal portion. Such needles begin casting during winter. Those less severely affected are retained until the next summer. Discolored foliage is prominent in early spring, but the only conspicuous symptom during summer may be missing foliage. Severe symptoms are often limited to branches within 6–8 m of the ground.

Apothecia of *Rhabdocline* species develop beneath the epidermis of needles attached to twigs. They mature in May–June, appearing as swellings up to several millimeters long, the same color as the lesion. They open by splitting the needle surface, usually near the midrib, and pushing a flap of epidermis aside. Apothecia of all *Rhabdocline* taxa except *R. pseudotsugae* subsp. *epiphylla* open along the stomatal lines on the lower needle surface. Anamorphs of *Rhabdocline* are unknown except in *R. weirii* subsp. *weirii* and *R. parkeri*. The former fungus produces acervuli of the *Rhabdogloeum* type that precede apothecia. Whether the *Rhabdogloeum* spores are infectious is unknown. The anamorph of *R. parkeri* is *Meria parkeri*.

Disease cycle. Pathogenic *Rhabdocline* species overwinter as mycelium in lesions in needles attached to twigs. Ascospores mature in May–June, depending on local climate, and are dispersed by air. Infection occurs on new needles during the period of shoot elongation. The fungus penetrates directly through the cuticle. Colonization and cellular disruption proceed slowly during summer but accelerate during autumn as symptoms become apparent. The cycle is completed as apothecia and asci mature the next spring. Whether lesions on needles more than a year old indicate infection of mature needles or fungal development lasting more than 1 year is unknown.

Epidemiology and management. Rhabdocline needle cast is favored by cool, moist environments. In British Columbia *R. pseudotsugae* subsp. *pseudotsugae* requires about 3 days of cool (e.g., 13°C), wet conditions for maximum infection from ascospores. Severe disease occurs in young plantations where weed growth, close spacing of trees, or dense foliage induced by shearing impedes air circulation and prolongs foliar wetness. *Rhabdocline* species do not complete their life cycles in fallen needles or in foliage on cut branches. Therefore, sanitation by cutting diseased low branches and severely affected trees can contribute to disease suppression in plantations.

P. menziesii strains from seed sources in New Mexico and the northwest coastal region are least affected by *R. pseudotsugae* subsp. *pseudotsugae*. The relation of seed origin to susceptibility to other *Rhabdocline* forms is less clear. Some trees within highly susceptible populations remain free from the disease under epidemic conditions, indicating resistance.

References: 476, 1214, 1261, 1262, 1704, 1859, 2555, 2959, 2960, 3849, 3869, 3882

Swiss Needle Cast

This disease, also called Phaeocryptopus needle cast, can be destructive in regions where the climate is moist. It has caused significant damage in Christmas tree plantations across North America and in young forest stands of *Pseudotsuga* in Oregon and Europe. It develops insidiously because *Phaeocryptopus gaeumannii* can produce pseudothecia and spores inconspicuously on green needles. *P. gaeumannii* is known from DNA sequence comparisons to be closely related to *Mycosphaerella* and not closely related to other *Phaeocryptopus* species; its reclassification was pending as this was written.

Symptoms and signs. Much green foliage is already diseased when symptoms first become noticeable. Needles turn yellow-green mottled with brown or entirely brown before dropping prematurely. Defoliation begins with the oldest needles. In severe attacks all but the youngest needles may fall and tree growth is retarded. Cut Christmas trees with green, infected needles dry out and lose needles more rapidly than healthy trees do. The signs are black pseudothecia 40–100 μm in diameter that grow from the stomata in lines on the lower needle surface. Pseudothecia arise throughout the year on year-old or older needles and are sometimes visible on first-year needles as early as August. Diseased needles may produce pseudothecia for up to three seasons before falling.

Disease cycle. The pathogen overwinters as mycelium in, and pseudothecia on, needles on the tree. Ascospores are dispersed in response to moisture from April through September, with peak dispersal in late spring and early summer in western USA. Rainy summer weather is conducive to infection, which usually occurs on first-year needles during the time of shoot elongation. The fungus does not kill needle tissues rapidly, but it does occlude stomata. Pseudothecia begin to appear within a year after infection. Needles senesce and drop after 1–3 years, when about half of their stomata are occluded. The pathogen has overlapping generations and thus causes overlapping cycles of disease. No anamorph of *P. gaeumannii* is known.

Heritable variation in resistance of *Pseudotsuga* to *P. gaeumannii* has been noted. The average level of resistance is positively correlated with the amount of summer rainfall in localities of seed collection. This correlation has been interpreted as evidence that *P. gaeumannii* and *Pseudotsuga* have coevolved in western North America.

Several "true" species of *Phaeocryptopus* are known, all on gymnosperm foliage, but they are not significant pathogens. *P. nudus* occurs on *Abies* (fir) and *Tsuga* (hemlock) around the Northern Hemisphere. *P. pinastri* occurs on *Pinus* (pine) in the USA, and *P. abietis* on *Abies* in China. Three additional species occur on native gymnosperms in South America.

References: 459, 582, 708, 709, 1214, 1493, 1738, 2213, 2422, 2459, 2551, 2616, 2617, 2817, 3869, 4401

A–F. Rhabdocline needle cast. A. Dead low branches and purplish brown year-old needles mark a highly susceptible tree in spring after several years of severe disease (NY, Apr). B. Purplish brown year-old needles on low branches in spring (NY, May). C. Close view of lesions in early spring (PA, Apr). D. Twigs with severely diseased year-old needles. Those produced during a second growth flush in midsummer escaped infection (NY, Mar). E. Apothecia of *Rhabdocline weirii* subsp. *oblonga* opening on lower surfaces of needles (NY, May). F. Moist, fully open apothecia with mature asci at the time of ascospore discharge (PA, Jun).

G–K. Swiss needle cast, caused by *Phaeocryptopus gaeumannii*. G. Defoliated low branches (PA, May). H. An affected branch from which most second- and third-year needles have fallen. The current season's shoots are stunted because of previous needle loss on the branch (NH, Jul). I. Mottled, chlorotic 1- and 2-year-old needles. Deformity of first-year needles was caused by an insect, *Adelges cooleyi* (Cooley spruce gall adelgid) (NH, Jul). J, K. Pseudothecia of *P. gaeumannii* on lines of stomata on lower surfaces of needles. Fruiting on the needle in K is prolific in a lesion caused by *R. weirii* (NY, Nov).

Photo credits: C—N. Wenner and W. Merrill; F, G—W. Merrill

59

Meria Needle Blight of Larch and Didymascella Leaf Blight of Cedar (Plate 29)

Meria Needle Blight

Meria laricis, a hyphomycete, causes browning of *Larix* (larch) needles. The disease affects trees of all sizes but is most damaging to small ones. Loss of foliage slows growth, diminishes resistance to opportunistic pathogens, and sometimes kills seedlings. *M. laricis* has caused major losses in forest nurseries in western North America, Great Britain, and Europe. It is noticed occasionally in the Great Lakes region, and it was apparently carried with its hosts to New Zealand. Suscepts include *Larix decidua* (European larch), *L. gmelinii* (Dahurian larch), *L. kaempferi* (Japanese larch), *L. lyallii* (Lyall larch), *L. occidentalis* (western larch), and *L. sibirica* (Siberian larch). Asian species are less susceptible than European and North American ones. Hybrids between European and Japanese larches (e.g., Dunkeld larch) display intermediate susceptibility. *L. occidentalis*, which is native in interior northwestern North America, cannot be grown on the moist west side of the Cascade Mountains without risk of death from repeated defoliation by *M. laricis*. The sexual state of *M. laricis*, if any, is unknown. Results of DNA analysis indicate that its closest relatives are species of *Rhabdocline*, which are parasites of *Pseudotsuga* (Douglas-fir) needles.

Symptoms and signs. Symptoms appear rather suddenly during or after wet weather while shoots are still growing. Yellow to brown spots arise on any needle surface and enlarge to involve the whole needle. If moist conditions continue, the pathogen produces colorless conidia in tiny clusters along the lines of stomata, mainly on the lower needle surface. The spores develop on conidiophores that emerge from stomata. Viewed with a hand lens, the spore clusters appear as white dots more prominent than the wax deposits normally associated with stomata. Brown needles wither and soon drop.

Disease cycle. M. laricis overwinters in fallen needles and in dead terminal tufts of needles that adhere to twigs. The latter site is particularly important on seedlings in nurseries. Spores splashed from these sources and possibly other sites on larger trees initiate the first infections in spring. The usual mode of penetration into larch needles has not been reported, but the fungus is known to colonize sites of feeding injury by adelgids. It soon sporulates on diseased needles, providing inoculum for secondary cycles. Although *M. laricis* is capable of producing many generations of spores per year, infection apparently occurs only during the early part of the annual growth period while foliage is succulent. Needles that escape infection during their succulent phase remain green. The mode of dispersal of the fungus through the crowns of trees in the forest has not been studied but may involve transport of conidia with mist and cloud droplets as well as by splashing.

Dry weather suppresses the disease. If dry conditions prevail during the first part of the growth period, and wet weather follows, the first-formed foliage may remain green while that toward the shoot apex becomes blighted. Experiments in Great Britain indicated that temperatures of 10–25°C favor infection. Slight infection occurred at 0–5° and none at 30°C. Optimum temperature for both hyphal growth and sporulation was 17–20°C. Inoculation of various *Larix* species with isolates from the same or other species revealed no host specialization. Larch performance trials in Europe and western North America have indicated that selection for rapid growth within *L. decidua* and *L. occidentalis* yields the correlated benefit of partial resistance to *M. laricis*.

At times the symptoms of Meria needle blight arise so quickly as to suggest frost damage, but the two disorders can be distinguished. Lesions caused by the fungus arise as discrete, enlarging spots, whereas frost quickly kills entire needles. Also, *M. laricis* kills only needles, whereas frost kills succulent stem tissue as well. Needles infected by *M. laricis* usually fall after becoming totally brown, whereas frost-killed needles tend to hang on for a few weeks.

Gross symptoms of Meria needle blight can also be confused with needle cast caused by *Hypodermella laricis* (Rhytismatales, Rhytismataceae) in the Rocky Mountain region. Needles on dwarf shoots become infected by *H. laricis* as they emerge in April, and they turn brown around June 1. Needles on terminal shoots usually escape browning because the terminals elongate after the pathogen's spores have dispersed. Killed needles tend to remain on the short shoots. Fruit bodies (hysterothecia) develop in some of these needles and produce ascospores the next spring. Hypodermella needle cast and Meria needle blight sometimes occur together.

References: 253, 761, 790, 800, 1214, 1234, 1261, 2988, 2989, 3248, 3249

Didymascella Leaf Blight

Didymascella thujina (Helotiales, Hemiphacidiaceae) causes a significant leaf blight, often called Keithia blight, of *Thuja plicata* (western red cedar) and *Platycladus orientalis* (Oriental arborvitae) and is an inconsequential pathogen of *T. occidentalis* (eastern arborvitae) and *Chamaecyparis lawsoniana* (Lawson-cypress). It is widespread in western North America, is collected occasionally from the Great Lakes region eastward, and has also been reported from Texas and Virginia. It has caused major losses of 2-year-old *T. plicata* seedlings in forest nurseries in western North America and in Great Britain and Europe, where it was introduced with that host. Seedling mortality rates up to 97% have been recorded in forest localities. The disease is most common and severe on moist sites and in damp places such as ravines where snow lingers and on shaded foliage in dense stands.

Symptoms and signs. Lesions appear on year-old foliage in spring as small bleached areas, one or more per leaf, usually on the upper surface. These become confluent, turn brown, and involve the entire leaf. On seedlings, foliage of the juvenile type is most commonly infected. On older plants, foliage of all ages seems equally susceptible. Severe disease retards growth and may kill seedlings less than 4 years old or make low branches of larger trees appear scorched. Foliage on upper branches may also become generally infected, but never as intensely as that on lower branches.

Fruit bodies (apothecia) become conspicuous by late June. Young apothecia appear as slightly swollen spots, usually on the upper leaf surface, red-brown to olive-brown against the dull brown background of the killed tissue. Usually one apothecium, but occasionally as many as three, form per leaf and become nearly black at maturity. They are circular to elliptic and measure 0.3–0.7 × 0.2–0.5 mm. Under moist conditions they swell until the epidermis splits, exposing the olive-colored spore-producing layer. This fungus produces two ellipsoid ascospores per ascus. The spores measure 20–25 × 14–16 µm, and they become spiny, unequally two celled, and brown with age. Twigs with severely diseased scale-leaves fall during autumn. Killed leaves that remain on the tree weather to a gray color and often contain dark cavities where spent apothecia shriveled or fell out.

Disease cycle. The cycle is 1 year long. Ascospores are expelled from apothecia in June–October, with a peak in late summer and early autumn. They can be dispersed many kilometers in air without loss of viability. They are thick walled and have a sticky sheath by which they may adhere to foliage for a long time before germination. No other spore type is known. Although some infections may begin in autumn, most start in spring when ascospores that have overwintered on green foliage germinate and the fungus penetrates leaves. It sometimes spreads from one scale-leaf to others at the same position on a twig, but it does not move distally or proximally. Severe damage to seedlings is a consequence of pathogen establishment on the plants in one year and population buildup the next year.

T. plicata varies in susceptibility, resistant individuals having been detected in the midst of populations affected by natural epidemics. Resistance to *D. thujina* appears to be simply inherited. In Europe *T. plicata* and *T. standishii* (Japanese arborvitae) have been hybridized with the aim of combining the size and rapid growth of the American species with the resistance of the Asian one.

Four additional species of *Didymascella* are known, all inconsequential colonists of conifer needles. *D. chamaecyparidis* infects *Chamaecyparis thyoides* (Atlantic white cedar) in New Jersey, and *D. tetramicrospora* is known from collections on *Juniperus* in Colorado. *D. oxycedri* and *D. tetraspora* occur on *Juniperus* in Europe.

References: 1214, 1859, 2127, 2952, 2984, 4088, 4277

A–D. Meria needle blight of *Larix occidentalis* (western larch) (ID, Jul). A, B. Blighted foliage extending high into tree crowns. C. A branch with blighted foliage and twig dieback. Cause of the dieback is unknown. D. Close view of diseased leaves on dwarf shoots.

E, F. Didymascella needle blight of *Thuja plicata* (western red cedar). E. Affected branches (BC, Aug). F. Brown to black apothecia in lesions that involve entire scale-leaves (OR, Jun).

Photo credit: E—E. A. Allen

Snow Blights (Plate 30)

Various fungi, collectively called snow molds, kill the foliage of conifers beneath snow. Those producing annual mycelium that disappears from plant surfaces soon after snowmelt cause snow blights, while fungi with brown mycelium that grows perennially on plant surfaces cause brown felt blights (Plate 31). Snow blights are widespread in alpine and boreal regions and may cause severe damage to small trees in natural and planted forests and nurseries where deep snow persists. Hundreds of thousands of seedlings may be killed in an outbreak. Small trees beyond the seedling stage, although often severely defoliated, usually retain some green foliage and survive. Mycelium of snow blight fungi grows from one needle to another, and it ceases growth when the snow melts.

For many years nearly all snow blight was attributed to one fungus, *Phacidium infestans* (syn. *P. pini-cembrae*; Helotiales, Phacidiaceae), but research in Europe and North America showed that several fungi, somewhat specialized in host preferences, are involved. Two pathogens account for most of the snow blight records in North America. *Phacidium abietis* is the common snow mold on *Abies amabilis* (Pacific silver fir), *A. balsamea* (balsam fir), *A. concolor* (white fir), *A. grandis* (grand fir), *A. lasiocarpa* (subalpine fir), and *Pseudotsuga menziesii* (Douglas-fir). It is pathogenic also to *Picea abies* (Norway spruce), *P. glauca* (white spruce), *P. pungens* (blue spruce), and *P. rubens* (red spruce)—although not to *P. mariana* (black spruce)—and to *Pinus strobus* (eastern white pine) and *Tsuga canadensis* (eastern hemlock). It produces fruit bodies (apothecia) only on *Abies*. *Lophophacidium hyperboreum* (Helotiales, Phacidiaceae), which occurs in North America, Europe, and China, attacks mainly spruces: *Picea abies*, *P. engelmannii* (Engelmann spruce), *P. glauca*, *P. mariana*, *P. pungens*, and *P. rubens*. It is also pathogenic to *Abies balsamea* and *Pinus strobus* but does not sporulate on the former. *Phacidium infestans*, important in Europe on various conifers, has been reported to occur in North America, but the reports predate modern taxonomic treatments of *Phacidium* and are suspect or simply erroneous.

Various additional fungi related to these first three are listed as snow molds or are blamed for snow blight, but most of them require more study before their roles can be known with certainty. *Phacidium dearnessii* on *Taxus brevifolia* (Pacific yew) and *P. sherwoodiae* on *Thuja plicata* (western red cedar) are examples. *Sarcotrochila piniperda* (Helotiales, Hemiphacidiaceae) is associated with snow blight on *Picea abies*, *P. glauca*, *P. mariana*, *P. rubens*, and *P. sitchensis* (Sitka spruce), but its pathogenicity is unproved. *S. balsameae* and *Nothophacidium phyllophilum* (Helotiales, Dermateaceae) on *Abies*, and *S. macrospora* and *Hemiphacidium longisporum* on *Pinus*, colonize foliage killed by needle-cast fungi and inhibit their fruiting. *H. planum* fruits on *Pinus albicaulis* (whitebark pine), *P. flexilis* (limber pine), *P. edulis* (Colorado pinyon), *P. monticola* (western white pine), *P. ponderosa* (ponderosa pine), and *P. strobus*. This fungus apparently follows a needle-cast fungus on *P. ponderosa*, but this possible sequence has not been studied for other hosts. *H. convexum* on *Pinus rigida* (pitch pine) is rare and its pathogenicity unproved. (Note: *Hemiphacidium* may soon be synonymized under *Sarcotrochila*.) *Korfia tsugae* on *Tsuga canadensis* (eastern hemlock) is little known and also of unproved pathogenicity. *Phacidium taxicola* on *Taxus canadensis* (Canadian yew) is a proven pathogen but not a typical snow mold. Apparently it infects wounded twigs and then grows into needles, on which it eventually sporulates. Typical snow molds do not infect stems. In Japan, the sterile soilborne fungus *Racodium therryanum*, which is not related to phacidiaceous fungi, attacks coniferous seeds and seedlings, causing a disease that is also called snow blight.

Symptoms and signs. Snow blight, regardless of the causal fungus, can be recognized by death and discoloration of patches of foliage up to 60–70 cm in diameter in spring at about the time that snow melts away from nursery beds or from low branches of trees. Blighted foliage appears ashy or glaucous (waxy) brown (Plate 30C). During the ensuing summer it becomes somewhat bleached, and it slowly weathers away during the next 1–2 years.

During the first summer after blight, small brown to black apothecia develop, usually on the lower sides of needles if on *Abies*. The apothecia of *Phacidium abietis* are oval to round, 0.3–0.5 mm wide, and arise in rows on each side of the midrib. Apothecia of *Lophophacidium hyperboreum* are elongate, appearing as a dark line that commonly extends half the length of the needle and

sometimes the entire length. *P. abietis* and *L. hyperboreum* grow readily on laboratory media.

Disease cycles. *P. abietis* and *L. hyperboreum* spend the summer as mycelium and developing apothecia in blighted foliage. The apothecia mature in autumn, open by irregular fissures, and discharge ascospores into the air during wet weather from late September until early November. These spores infect first-year foliage after snow covers it. The next spring during snowmelt, mycelium emerges from stomata of diseased needles and grows to adjacent healthy ones on the same or neighboring plants, including plants of other species. Therefore, green needles are infrequent within a patch of snow blight.

Some snow blight fungi have been reported to produce minute black structures, termed microsclerotia, on recently killed needles during summer. The potential role of these in survival of the fungi or initiation of infection is unknown. Their identity as structures of snow molds requires verification.

The Eurasian snow blight fungus, *P. infestans*, has been studied more than any other. Some authorities view it as a broad species that encompasses Eurasian and North American forms (*P. infestans* varieties *infestans* and *abietis*, respectively) and host-specialized, geographically separate populations with preferences for *Pinus* versus *Abies* or *Picea*. This interpretation requires testing through genetic research. *P. infestans* in the narrow sense grows on laboratory media at temperatures from −3 to above 20°C, growing most rapidly at about 15°C. It is favored by moist autumn weather, which presumably enhances ascospore production and dispersal. Such weather presages severe snow blight, while dry weather before snowfall leads to less disease. In Finland snow blight is most severe on sites deficient in potassium, and a dense ground cover of ericaceous plants inhibits spread of the pathogen. Such relationships have not been studied in North America.

References: 123, 384, 800, 901, 960, 1103, 1214, 1262, 1859, 2190, 2217, 3252, 3325, 3697, 3698, 3702, 4527

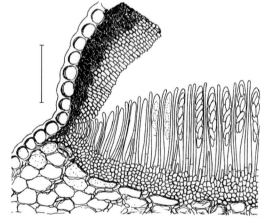

Figure 30. *Phacidium infestans:* part of an apothecium with asci and ascospores in a conifer needle, viewed in cross section. The apothecium developed beneath the epidermis and ruptured it when mature. Scale bar = 100 μm. Adapted from reference 123 by permission of Gebrüder Borntraeger Verlagsbuchhandlung.

A. Snow blight of *Abies balsamea* (balsam fir) caused by *Phacidium abietis*. The green top of this sapling remained above snow during the previous winter (QC, Jun).

B. *Sarcotrochila balsameae*, a putative snow blight pathogen, fruiting on *A. balsamea* needles (QC, autumn).

C. Snow blight of *Abies bifolia* (Rocky Mountain subalpine fir). The mottled grayish brown color is typical of symptoms caused by *Phacidium abietis*, which would have produced apothecia several weeks after the photo was made (CO, Jun).

D. Snow blight of *Picea engelmannii* (Engelmann spruce) (CO, Jun).

E–G. Snow blight of *Picea glauca* (white spruce) seedlings caused by *Lophophacidium hyperboreum* in nurseries. E. Seedlings killed during winter in a transplant bed. F. Mycelium and brown needles revealed as snow melts away from a seedling. G. Typical symptoms and webs of mycelium visible shortly after disappearance of snow (QC, spring).

Photo credits: A, B—Canadian Forest Service Laurentian Forestry Centre; E–G—G. Laflamme

Brown Felt Blights (Plate 31)

Several fungi, collectively called snow molds, kill the foliage of conifers beneath snow. Those producing annual mycelium that disappears from plants soon after snow melts cause snow blights (Plate 30). Those with brown mycelium that grows perennially on plant surfaces are sometimes called brown or black snow molds and cause brown felt blights. Brown felt fungi produce dense mats of dark brown mycelium that overgrows needles and twigs, binding them together and killing them. The felts develop while branches are beneath snow. They become visible and cease growth as the snow melts in spring and early summer. On sites exposed to harsh weather, brown felts are most common on the leeward sides of trees. Felt-covered twigs and branches often die, but they may survive and resume growth if their apical buds are not covered by the felt.

Two fungi commonly cause brown felt blights. *Herpotrichia juniperi* (Pleosporales, Lophiostomataceae) attacks gymnosperms, usually *Abies* (fir) and *Picea* (spruce). *Neopeckia coulteri* (a fungus of uncertain placement in the Dothideales) usually attacks *Pinus* (pine). In North America these fungi are common only in the mountains of the West, where they may interfere with conifer regeneration at high elevations but do not cause economic damage. In Europe, however, both occasionally do cause damage, killing *Picea abies* (Norway spruce) and other conifers in northern nurseries and on mountains where seedlings have been planted for avalanche control.

Hosts of *H. juniperi* include *Abies alba* (silver fir), *A. amabilis* (Pacific silver fir), *A. bifolia* (Rocky Mountain subalpine fir), *A. concolor* (white fir), *A. grandis* (grand fir), *A. lasiocarpa* (subalpine fir), *A. magnifica* (red fir), *A. procera* (noble fir), *Calocedrus decurrens* (incense cedar), *Chamaecyparis nootkatensis* (Alaska or yellow cedar), *Juniperus communis* (common juniper), *J. deppeana* (alligator juniper), *J. horizontalis* (creeping juniper), *J. scopulorum* (Rocky Mountain juniper), *Phyllodoce empetriformis* (mountain-heather), *Picea abies* (Norway spruce), *P. engelmannii* (Engelmann spruce), *P. sitchensis* (Sitka spruce), *Pinus albicaulis* (whitebark pine), *P. contorta* (lodgepole pine), *P. monticola* (western white pine), *P. mugo* (mugo pine), *Pseudotsuga menziesii* (Douglas-fir), *Taxus brevifolia* (Pacific yew), *Thuja plicata* (western red cedar), *Tsuga heterophylla* (western hemlock), and *T. mertensiana* (mountain hemlock).

Hosts of *N. coulteri* include *Picea engelmannii*, *P. glauca* (white spruce), *Pinus albicaulis*, *P. balfouriana* (foxtail pine), *P. contorta*, *P. flexilis* (limber pine), *P. jeffreyi* (Jeffrey pine), *P. lambertiana* (sugar pine), *P. longaeva* (Great Basin bristlecone pine), *P. monticola*, *P. mugo*, *P. nigra* (Austrian pine), *P. ponderosa* (ponderosa pine), *P. strobus* (eastern white pine), and *Taxus brevifolia*.

Two additional species of *Herpotrichia* occur on gymnosperms in Europe and North America, although not as brown felts. *H. macrotricha*, a widely distributed saprobe on various plants, has been mistaken for *H. juniperi* on *Picea mariana* (black spruce) and *Abies balsamea* (balsam fir) in eastern Canada. *H. parasitica* is a pathogen of conifer foliage in Europe.

Signs and symptoms. *H. juniperi* and *N. coulteri* are seen in late spring and summer as thick brown mycelial mats and decomposing mat fragments on dead foliage previously covered by snow. As matted foliage is released from snow, the mycelium at first appears dull grayish and fuzzy. Within days it turns dark and often somewhat shiny as the hyphae become thick walled and brown. Felts in this condition can withstand sunlight and drying without rapid loss of viability. Close inspection of felts during summer may reveal nearly globose pseudothecia, 0.25–0.5 mm in diameter, embedded in the mycelium with their tips protruding slightly (Plate 31H). These become conspicuous after some of the felt weathers away. They produce ascospores during summer and early autumn. The role of these spores in disease cycles is unclear. During felt development, both fungi produce compact masses of cells, called either stromata or microsclerotia, that plug stomata. The two pathogens can be distinguished on the basis of host preferences, temperature response, and form. *H. juniperi* grows well over a broader temperature range, and studies in Europe have shown it to occur at a broader range of elevations than does *N. coulteri*. The ascocarps of *H. juniperi* are globose with a flattened base, while those of *N. coulteri* are somewhat pear shaped. Mature ascospores of *H. juniperi* are four to five celled and brown, while those of *N. coulteri* are two celled and brown. Both fungi have asexual states

in the genus *Pyrenochaeta*, which develop in culture. They are characterized by tiny pycnidia with colorless single-celled conidia. Whether these spores are infectious or serve as gametes in the sexual cycle has not been reported.

Disease cycles. The life cycles and related pathogenic behavior of *H. juniperi* and *N. coulteri* are incompletely known. Both fungi are quiescent in summer, during which their ascocarps mature. The life cycle of *H. juniperi* requires 2 years. Needles and twigs are killed by mycelial felts during the first winter after felt formation, and the fungus completes its sexual development, producing pseudothecia and ascospores, during the second summer. *N. coulteri* survives summer as microsclerotia and as mycelium at the bases of needles and among bud scales. Apparently both fungi can be dispersed as ascospores and as felt fragments that fall with dead twigs and needles from diseased trees. Litter infested with mycelium and ascospores falls onto snow and is deposited on branches below as the snow melts. This mechanism provides for eventual infection. Another possible means of infection is contact between healthy foliage and litter on the ground as branches are forced down by the weight of snow.

Under snow in early spring, hyphae grow from the survival structures. *N. coulteri* penetrates the cuticles of living needles, producing haustoria in cells beneath. *H. juniperi*, on the other hand, seems to remain superficial while the needles are alive, but it penetrates both needles and twigs after they die. The mycelial plugs in stomata have been credited with a "smothering" action, but the mechanisms by which brown felts actually kill needles have not been studied.

N. coulteri and *H. juniperi* grow most rapidly at 10 and 15–18°C, respectively, in pure cultures, but can grow at 0 to –3°C. Temperatures in this range may occur near ground level in deep snow even when the air above is much colder. Most felt development occurs at temperatures slightly above 0°C in early spring when melting snow provides both free water and a water-saturated atmosphere around buried branches. Felt growth both in and above snow has been noted in dense young spruce stands in the Baltic region. The mycelium grows from needle to needle and also from one plant to another, even among different kinds of plants. Growth from coniferous hosts may explain the reports of brown felt blight on *Arceuthobium abietinum* (fir dwarf mistletoe) and *Phyllodoce empetriformis* (mountain-heather), a tiny evergreen plant in the Ericaceae.

References: 230, 231, 271, 1099, 1190, 1213, 1214, 1298, 1501, 1859, 3637, 3669, 3671

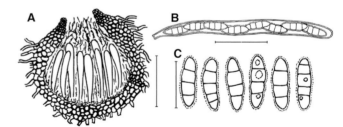

Figure 31. Herpotrichia juniperi. A. A pseudothecium. B. An ascus containing ascospores. C. Ascospores. Scale bars = 170, 50, and 25 μm, respectively. Adapted from reference 3671 by permission of CAB International.

A–C, G–I. *Herpotrichia juniperi* on *Picea engelmannii* (Engelmann spruce) (CO, Jun). A, B. Young trees shortly after snowmelt. New felts are nearly black; old, weathered ones are gray-brown. C. Green tips on a felted twig are capable of growth. G. A year-old felt with pseudothecia on a dead twig. H. Pseudothecia embedded in a felt, visible as pimplelike protrusions. I. Pseudothecia exposed after brown mycelium has weathered away.

D. *H. juniperi* on *Abies lasiocarpa* (subalpine fir). A branch affected for several years has new felts (dark), grayish tan felts formed the previous year, and bare twigs where old felts with dead needles have weathered off (WA, Jun).

E. New felt on *A. bifolia* (Rocky Mountain subalpine fir) just after release from snow (CO, Jun).

F. Brown felt blight on *Pinus albicaulis* (whitebark pine). This felt, in its first year, lacked fruit bodies and spores necessary for identification of the fungus. Some needles are detached from the twig but are bound by mycelium (OR, Sep).

Tar Spots (Plate 32)

Most tar spots on trees are caused by *Rhytisma* species (Rhytismatales, Rhytismataceae), of which at least 17 occur in temperate North America. All but 2 occur on woody plants. Three species occur on *Acer* (maples) and are featured here: *R. acerinum*, *R. americanum*, and *R. punctatum*. Other tar-spot fungi on woody plants in the USA and Canada are listed below. Geographic distributions are identified as follows: c, central; e, east; hemis, hemisphere; n, north(ern); s, south; w, west; NA, North America.

On ericaceous plants: *R. andromedae* on *Andromeda, Kalmia, Leucothoe, Lyonia,* and *Pieris* (n hemis); *R. arbuti* on *Arbutus menziesii, Menziesia,* and *Rhododendron* (NA); *R. decolorans* on *Leucothoe* and *Lyonia; R. vaccinii* on *Arctostaphylos, Gaylusaccia, Lyonia, Oxydendrum,* and *Vaccinium* (NA).

On *Ilex* and related plants: *R. concavum* (c, e USA), *R. ilicincola* (c, se USA), and *R. velatum* (se USA) on *Ilex; R. prini* on *Ilex* and *Nemopanthus* (NA).

On other plants: *R. liriodendri* on *Liriodendron tulipifera* (se USA, Europe), *R. salicinum* on *Alnus* and *Salix* (n hemis), *R. tostum* on *Quercus* (se USA), *R. ulmi* on *Ulmus* (e, nc USA), *R. vitis* on *Vitis* (c, e USA). The disease of *Salix* (willows) caused by *R. salicinum* (Plate 32K) has not been studied beyond the point of describing the pathogen and identifying some of its hosts.

A few tar-spot diseases of woody plants and several such diseases of herbaceous plants are caused by fungi unrelated to *Rhytisma*. For example, *Phyllachora* species (Phyllachorales, Phyllachoraceae) cause tar spots on *Acacia* and *Cocos* (coconut palm).

Maple tar spots. The most common of the maple tar-spot pathogens in North America was misidentified for decades as *R. acerinum* but is now known as *R. americanum*. It causes prominent raised tar spots 5–15 mm across, usually with a furrowed surface (Plate 32E, G). It occurs wherever maples grow in moist environments in North America. A second species, *R. punctatum*, occurs across in North America from the latitude of North Carolina northward and also in Europe. It causes clusters of small tar spots, each about 1 mm in diameter. The disease is sometimes called speckled tar spot (Plate 32J). A third fungus, *R. acerinum*, has become common in northeastern USA on European maple species, notably *Acer platanoides* (Norway maple). Its stromata begin as separate tiny black spots that coalesce into discs up to 3 cm in diameter, raised only slightly above the leaf surface. This species, originally described in Europe, was found in Ohio in 1940 but remained scarcely known in North America until outbreaks began in New York State during the 1980s. Most records of *R. acerinum* in North America actually refer to *R. americanum*, so we have translated American records accordingly in the discussion below. European records of *R. acerinum* are believed valid.

Maples reported susceptible to both *R. americanum* and *R. punctatum* are *Acer glabrum* (Rocky Mountain maple), *A. macrophyllum* (bigleaf maple), *A. negundo* (box-elder), *A. pseudoplatanus* (sycamore maple), *A. rubrum* (red maple), *A. saccharinum* (silver maple), *A. saccharum* (sugar maple), and *A. spicatum* (mountain maple). *R. punctatum* is the only form recorded on *A. circinatum* (vine maple) and *A. pensylvanicum* (striped maple). *R. acerinum* is the only form recorded on *A. campestre* (hedge maple), *A. ginnala* (Amur maple), and *A. platanoides*. It also occurs on *A. pseudoplatanus*.

Symptoms and signs. Spots arise in late spring and early summer after leaves attain full size. At first the diseased tissue turns light green or yellowish green. Then, during mid to late summer, black stromata develop on yellow spots on the upper leaf surfaces. They arise within the upper epidermis and rupture the vertical walls of epidermal cells. The lower surface of a leaf beneath a large tar spot turns brown, but the surface beneath speckled tar spots remains yellow. Leaves with multiple tar-spot lesions may wither and drop prematurely, but not so early or in such numbers as to threaten the general health of the tree.

Rhytisma species produce spermagonia of the *Melasmia* type during summer. Spermagonia of *R. americanum* and *R. punctatum* appear briefly as black dots on yellow leaf spots where black stromata soon develop. In *R. acerinum*, clusters of black spermagonia develop on yellow lesions and produce spermatia until late summer, by which time they are embedded in large black stromata. *R. acerinum* also produces germinable asexual spores in laboratory cultures, but they are not infectious.

Disease cycle. During autumn and early spring, apothecia develop in stromata on fallen leaves that remain moist. They ripen in spring, split open the stromatal surface, and eject sticky, needle-shaped ascospores into the air. Spores that alight on young leaves may start a new annual cycle. Symptoms appear 1–2 months after infection. Conspicuous outbreaks of tar spots caused by *R. americanum* and *R. punctatum* are infrequent except in moist, sheltered locations favorable for winter survival of the causal fungi. *R. acerinum* apparently has less exacting requirements and can cause severe disease in *A. platanoides* on well-ventilated sites.

Strains of *R. acerinum* differ in host preference. In experiments, ascospores from apothecia on *Acer campestre, A. platanoides,* and *A. pseudoplatanus* caused abundant infections on the respective source species but few or no spots on other maples. Cuticular wax from leaves of each of three maple species, mixed into a germination medium, stimulated conidial germination of *Rhytisma* strains specialized for parasitism of the respective species. For example, more spores of an isolate from Norway maple germinated on a medium containing cuticular wax of that species than on media with wax from other maples.

References: 637, 638, 1099, 1214, 1298, 1797, 1800, 1968, 2288, 3506, 4421

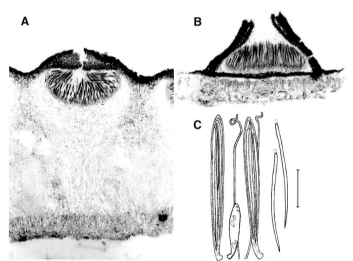

Figure 32. A, B. Apothecia of *Rhytisma* species as seen in sections through stromata on dead maple leaves collected from the ground at the end of winter. Each apothecium, less than 0.3 mm wide, appears as a cavity with a thick black cover that splits to allow discharge of ascospores. Vertically oriented elements within the cavity are asci. A. Part of a thick stroma of *R. americanum* on *Acer rubrum.* The dark band across the bottom of the photo is leaf tissue; the lighter tissue above is composed of hyphae. B. The thin stroma of *R. acerinum* rests directly on dead leaf tissue of *A. platanoides* (NY, May). C. Asci, paraphyses, and ascospores of *R. acerinum.* Scale bar = 25 µm. A, B. Courtesy of G. W. Hudler; C adapted from reference 637 by permission of CAB International.

A–D. Tar spot caused by *Rhytisma acerinum* on *Acer platanoides* (Norway maple) (NY, Aug). A. Brown foliage, soon to be prematurely shed, caused by severe infection during an epidemic. B. Typical lesions with developing stromata. C. Contrast between symptoms on shaded (left) and sunlit leaves. D. Close view of a lesion with developing stromata.

E, F. Tar spot caused by *R. americanum* on *Acer rubrum* (red maple). E. Typical lesions with immature apothecial stromata (NY, Aug). F. A severely affected leaf with many small lesions on which atypically tiny stromata will form (NY, Jul).

G. Close view of developing stromata of *Rhytisma* sp. (similar to *R. americanum*) on *A. pseudoplatanus* (sycamore maple). The convoluted upper surface of a stroma shows the arrangement of numerous immature apothecia (UK, Sep).

H, I. Small tar spot, or speckled leaf spot, caused by *R. punctatum* on *A. macrophyllum* (bigleaf maple). Chlorophyll is often retained around such lesions (the green-island effect) on senescing leaves (OR, Sep).

J. Developing stromata of *R. punctatum* on *A. pensylvanicum* (striped maple) (NY, Sep).

K. Tar spot caused by *R. salicinum* on *Salix* sp. (willow) (ON, summer).

Photo credits: A–D—G. W. Hudler; K—Canadian Forest Service Great Lakes Forestry Centre

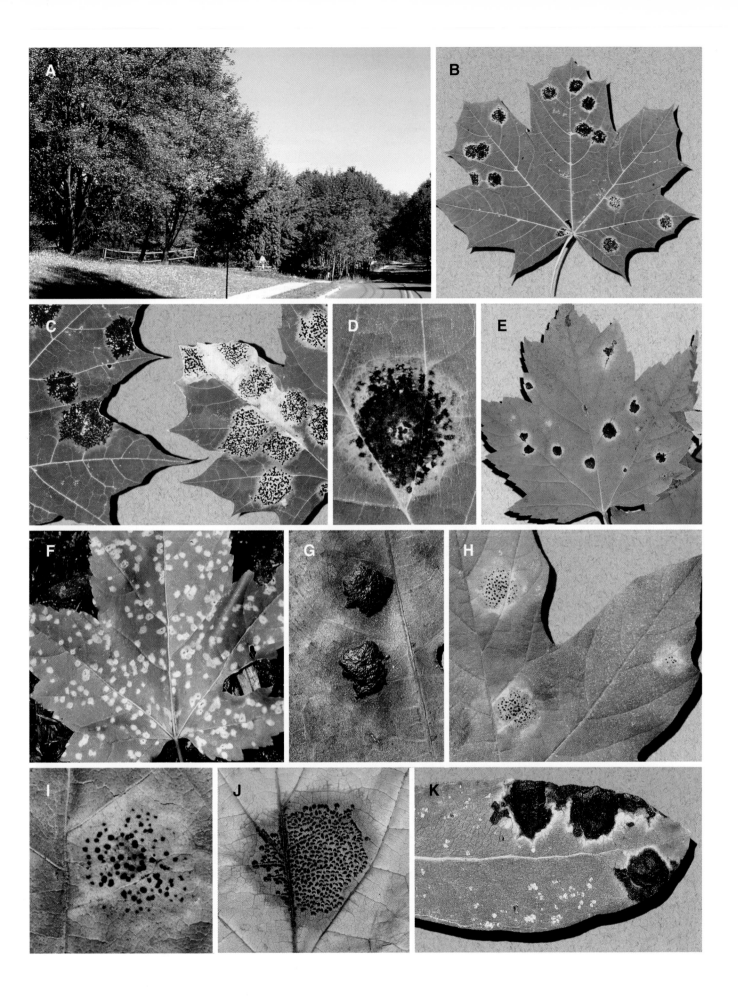

67

Ink Spot Leaf Blight of Aspen and Poplar (Plate 33)

Populus tremuloides (trembling aspen), which grows all across North America, is accompanied in northern areas and in the central Rocky Mountains by *Ciborinia whetzelii* (Helotiales, Sclerotiniaceae), one of two or possibly three fungi that cause ink spot leaf blight. These fungi kill leaves in late spring and quickly produce in them prominent dark brown to black sclerotia, or ink spots. The diseases are sporadic in young stands of aspen, killing 25–100% of the foliage in localized outbreaks but not causing long-term or consistent suppression of tree growth. Saplings in dense thickets can be killed by severe, repeated defoliation, but this mortality is properly viewed as a natural thinning required for normal development of the survivors.

C. whetzelii occasionally infects, but never causes severe leaf blight on, other *Populus* species, including *P. grandidentata* (big-tooth aspen), *P. deltoides* (eastern cottonwood), *P. balsamifera* (balsam poplar), *P. nigra* (black poplar), and hybrids. *P. tremuloides* and *P. grandidentata* sometimes hybridize naturally, and hybrid families commonly display the susceptibility of the trembling aspen parent.

Symptoms and signs. Reddish brown blotches develop on young full-grown leaves and usually expand until the entire leaf blade is dead. During expansion the blotches sometimes display alternating concentric zones of light and dark color. Dead leaves soon become tan. One to four discrete, dark, circular to oval, waferlike sclerotia form in the dead tissue of each diseased leaf. They measure 2–4 × 3–6 mm, are about 0.5 mm thick, and include all leaf tissues at the site of formation (Plate 33D). Initially light brown, they slowly darken to nearly black. Dead leaf tissue breaks at the edges of fully formed sclerotia, allowing them to fall out during summer. The dead leaves, conspicuously perforated, hang on for a while.

The sclerotia, which function as survival structures, are rich in storage compounds that sustain the fungus through the host's dormant period and support reproduction the next spring. The dark (melanized) outer layer, or rind, of a sclerotium is relatively impervious to microbial attack and accounts for the ability of sclerotia to survive in soil. The main storage compounds are carbohydrates in the cell walls and intercellular matrix of internal tissue, and protein bodies in the cytoplasm.

Disease cycle. Annual events begin in spring as host leaves are expanding. At that time sclerotia that have wintered in contact with the soil produce stalked, cuplike ascocarps (apothecia; Fig. 33), one to three per sclerotium. The apothecia, seldom noticed except by collectors of fungi, are light tan to brown, usually 4–6 mm across, with stalks up to 1.5 cm long or occasionally longer. During wet weather, asci in these structures shoot ascospores into the air. Spores that are deposited on wet developing leaves germinate immediately and initiate infection. Symptoms appear after 2–3 weeks, and formation of sclerotia in blighted leaves is apparent after a further 2–4 weeks. Sclerotia are released to the ground during July or August—the time varying with locality—and ripen during autumn and early spring.

Severe outbreaks of ink spot leaf blight are favored by dense stands of young trees, low temperature and high humidity during the period of spore dispersal and leaf expansion, and survival of a large number of sclerotia on the ground throughout winter. These conditions occur erratically, and thus outbreaks are unpredictable.

C. whetzelii was one of many endophytes and latent pathogens isolated during a study of fungal inhabitants of *P. tremuloides* bark and wood in western Canada. Whether this detection represented an anomalous occurrence of the fungus or an as yet undescribed aspect of the pathogen's life cycle is unknown.

In western North America, ink spot leaf blight was formerly attributed to *C. seaveri*, but it has not been clearly demonstrated that this fungus is distinct from *C. whetzelii*. On the other hand, *C. pseudobifrons*, a distinct although less common species, occurs on *P. tremuloides* from Alberta to Quebec and New York.

Several additional species of *Ciborinia* are tree pathogens. *C. foliicola* and *C. wisconsinensis* cause "black rib" of *Salix* (willow), a disease named for the formation of dark sclerotia along major veins in dead and dying leaves. *C. foliicola*, widespread in Canada, occurs in Colorado and has been reported from Great Britain. *C. wisconsinensis* has been identified only from collections in Wisconsin. *C. candolleana* causes necrotic blotches on leaves of *Quercus rubra* (red oak) in the USA and on several oaks in Europe. It has been isolated from leaves, twigs, and bud scales of *Q. rubra. C. gracilipes*, parasitic on petals of *Magnolia virginiana* (sweetbay), is known from New Jersey. *C. camelliae*, widespread in the USA and Japan, causes flower blight of *Camellia* species (Plate 34).

Ciborinia species grow readily and produce sclerotia and spermatia in laboratory cultures. They do not produce infectious asexual spores.

References: 195, 254, 779, 1099, 1214, 1298, 1398, 1730, 1840, 2116, 2117, 3115, 3917

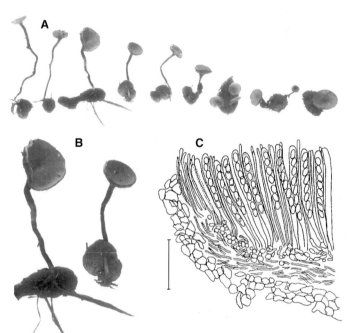

Figure 33. A, B. Apothecia of *Ciborinia whetzelii*, approximately natural size and enlarged about twofold, respectively. C. Partial section of an apothecium of *Ciborinia gracilipes*, a pathogen of *Magnolia*, showing the arrangement of asci with ascospores in members of the genus. Scale bar = 50 µm. C adapted from reference 254 by permission of the American Journal of Botany.

A–C. Symptoms and signs of ink spot leaf blight on *Populus tremuloides*. Usually the entire leaf but not the petiole is blighted. Brown stromata form in blighted leaves, turn nearly black, and drop out, leaving holes with smooth edges (NY & ON, Jul).

D. Magnified view of a young sclerotium in a blighted leaf of hybrid aspen (*P. grandidentata* × *P. tremuloides*) (ON, Jul).

69

Ovulinia Petal Blight of *Rhododendron* and Ciborinia Flower Blight of *Camellia* (Plate 34)

Disfiguring flower blights occur on *Rhododendron* species (azalea and rhododendron) and *Camellia* species throughout regions of relatively mild climate in the USA. Ovulinia petal blight is caused by *Ovulinia azaleae*. Camellia flowers are blighted by *Ciborinia camelliae*. Both fungi belong to the Sclerotiniaceae, order Helotiales, and infect only the flowers of their hosts. They may have been introduced to North America from the Orient, as they were unknown in the USA until the 1930s.

Ovulinia Petal Blight of *Rhododendron*

The disease occurs in North America and Hawaii, Europe and Great Britain, Japan, Australia, and New Zealand. Many kinds of native and introduced *Rhododendron* as well as *Kalmia latifolia* (mountain-laurel) are affected in the USA from southern New England southward to Florida and west to California. Petal blight is most destructive in the humid South. It may occur on greenhouse-grown azaleas, but this is usually prevented by rigorous exclusionary measures. Cultivars of *R. indicum* (Indian azalea) are notably susceptible; those of Kurume hybrids are somewhat less so. Petal blight can be confused with frost damage on the latter types. *R. carolinianum* and *R. catawbiense* (Carolina and catawba rhododendrons) and related hybrids as well as *K. latifolia* are most likely to become infected when growing near azaleas. Flowers of three species of *Vaccinium* (blueberry, huckleberry) and *Gaylussacia baccata* (black huckleberry) proved susceptible when inoculated experimentally.

Symptoms and signs. Petal blight begins as spots about 1 mm across, whitish on colored petals or brownish on white petals. Within 24 hours the spots are 5–10 mm across, and if numerous, they coalesce and cause a petal or an entire flower to collapse. Symptoms ordinarily progress from first spots to full blight in about 3 days, but development can be delayed by dry, cold, or very hot weather. In a locality where the population of *O. azaleae* is large, nearly all the azalea blossoms may seem to collapse simultaneously when mild, moist weather favors the disease. Limp, moist petal tissue, degraded by fungal enzymes, disintegrates to mush if touched lightly. Many blighted flowers remain attached or collapse against nearby leaves, dry out, and remain hanging on the plants. *O. azaleae* reproduces asexually by conidia formed on short conidiophores within moist diseased blossoms. The conidia are large (60×35 µm), colorless, and very unequally two celled, with an ovoid apical cell and a tiny basal cell. Conidia en masse rupture a petal's surface and form a superficial layer from which they are dispersed. If moist conditions continue, sclerotia (fungal survival structures) form in blighted petals and mature 6–8 weeks after infection. The sclerotia are variously shaped, flattened black objects 3–12 mm long and up to 8 mm broad. They eventually fall out or are carried with falling flower parts to the ground. Symptoms of Ovulinia petal blight can be confused with those of Botrytis blight (Plate 35), but large sclerotia do not form in lesions caused by *Botrytis*.

Disease cycle. Multiple cycles of azalea petal blight develop each year, beginning when azaleas start to bloom—January–February near the Gulf Coast but much later in the North. Sclerotia of *O. azaleae*, having wintered on or just below the soil surface beneath host branches, produce tiny, stalked fruiting structures (apothecia) shaped like wineglasses, similar to those shown in Figure 33. They are 1.5–2.5 mm across, dull reddish brown, with light brown to buff-colored stalks 3–10 mm long. One (usually) to three apothecia may grow from one sclerotium. Ascospores are shot from apothecia into the air during moist weather. Considered collectively, apothecia in a locality develop and discharge spores during a period of 4–5 weeks, but an individual apothecium requires only 5–8 days to complete these functions. Ascospores that alight on moist petals germinate, and the fungus penetrates immediately. If mild, moist weather continues, new cycles of disease initiated by successive crops of conidia may occur at intervals as short as 3–4 days. Conidia are dispersed by wind, splashing rain, and pollinating insects. Bumblebees and carpenter bees are the most important vectors.

Weather conditions that favor infection are fog, dew, or rain and continual high humidity at temperatures of 10–22°C (optimum near 18°C). Losses in greenhouses have been greatest at 10–15°C, probably because this temperature range promotes condensation or retention of water on plants. Symptoms and sclerotial development may be arrested by a change to hot, dry weather.

Sclerotia, rich in storage compounds, sustain the fungus through the host's dormant period and support reproduction the next spring.

The main storage compounds are carbohydrates in the cell walls and intercellular matrix of internal tissues, and protein bodies in the cytoplasm. The dark (melanized) outer layer of a sclerotium is relatively impervious to microbial attack. Sclerotia of *O. azaleae* can survive in soil for 2 years or longer, which explains the failure of attempts to control flower blight by removing all flower buds from azalea plants for a year.

In southern areas, early-blooming azaleas are most likely to be disfigured, and late ones may escape because of diminished conidial production by the causal fungus. In the North, however, the pattern is reversed because ascospores are not produced until after the blooming period of early varieties.

References: 825, 1153, 1730, 1967, 2116, 4283

Ciborinia Flower Blight of *Camellia*

This disease is similar to Ovulinia petal blight of azalea in most respects. The important differences are that *Ciborinia* species produce only ascospores and have one generation per year, so there are no secondary cycles and no significant vector activities of insects. *C. camelliae* occurs where camellias grow outdoors in North America from Virginia south to Florida and the Gulf Coast and west to California and Oregon. It also occurs in Europe and New Zealand. It attacks all varieties of *Camellia japonica* (common camellia) and *C. sasanqua* (sasanqua camellia) but is unknown on other plants. The fungus causes rapidly developing flower blight and produces sclerotia in the bases of killed flowers. Sclerotia overwinter on or just beneath the soil surface, and apothecia develop in late winter or spring as camellias come into bloom. Ascospores infect the blossoms. The disease is favored by wet conditions and temperatures of 15–21°C during the period of bloom.

Infection occurs any time after the tips of the petals emerge from buds and begin to show color. Sometimes stamens or sepals are also attacked. Brownish specks appear on expanding petals as soon as 24 hours after spore deposition. Lesions enlarge until an entire flower is dead and dull brown. Killed tissues are dry or leathery except during wet weather. In early stages of disease, veins in infected petals are darker than surrounding tissue. The resulting netted appearance and the scattered distribution of affected flowers on a plant distinguish flower blight from injury by frost or wind. Blighted flowers tend to retain their shape and soon drop to the ground. A ring of fluffy white mycelium around the petal bases on fallen flowers is diagnostic. Under moist conditions dark, shiny streaks or masses of spermatia form on blighted flowers. Sclerotia usually form within petal bases and frequently unite into flattened compound structures up to 25 mm or more across, mimicking the original arrangement of the sepals. They are initially buff-olive and become nearly black with age. In spring some sclerotia produce apothecia, while others remain dormant for a year or more. Some sclerotia produce apothecia annually for several years. The apothecia (Plate 34I) are buff-olive and shaped like wineglasses, similar to those of *C. whetzelii* (Fig. 33). They are 5–20 mm in diameter with stalks 3–40 mm long; stalk length is related to depth of sclerotia in the soil.

Most outbreaks of camellia flower blight can be traced to shipments of plants in containers that presumably carried sclerotia of *C. camelliae*. Once established in a locality, the pathogen cannot be eradicated by practical means, but the chance of introduction to noninfested areas can be minimized by shipping only bare-rooted plants on which flower buds show no color.

References: 1419, 1498, 1730, 1967, 2116, 3972, 4164

A–F. Petal blight of *Rhododendron* species caused by *Ovulinia azaleae*. A, B. Early stages of symptoms on two azalea cultivars. Small lesions enlarge quickly (FL, Mar–Apr). C–F. Late stages of symptoms and associated sclerotial development on blighted rhododendron blossoms (PA, Jun).

G, H. Ciborinia flower blight on two *Camellia* cultivars. G. Blighted flowers that dropped from plants. White ring of mycelium at petal bases is diagnostic (NC, Mar). H. Darkened veins are also diagnostic (CA, Feb).

I. Apothecia of *Ciborinia camelliae* among leaves and pine needles on the ground where diseased flowers dropped in an earlier year (NC, Mar).

Photo credits: G—D. M. Benson; I—R. K. Jones

Botrytis Blight (Plate 35)

Botrytis cinerea, a hyphomycete, causes flower, leaf, and shoot blights; fruit rots; and sometimes cankers in succulent stems of hundreds of woody and herbaceous plant species worldwide. The diseases are often called gray mold blight or gray mold rot. The fungus also causes damping-off of coniferous seedlings, storage mold and rot of horticultural plants and tree seedlings, and snow mold of *Cryptomeria japonica* (Japanese cedar) and *Pseudotsuga menziesii* (Douglas-fir). It is a colonist of dead and dying vegetable matter and living tissues predisposed by inadequate or unbalanced nutrition, low light intensity, high or low temperature, prolonged succulence, senescence, or toxic chemicals, including air pollutants. It often colonizes injured or dead plant parts first and then spreads into uninjured tissue.

B. cinerea is considered by some authorities to be a set of morphologically indistinguishable strains that differ in host specificity; however, host specificity has not been demonstrated among strains that attack woody plants. Strains from *Malus* (apple), *Vitis* (grape), and some other hosts are anamorphic *Botryotinia fuckeliana* (Helotiales, Sclerotiniaceae). These strains could be called *B. fuckeliana*, but most plant pathologists use the familiar name of the anamorph, *Botrytis*.

Woody hosts of *B. cinerea* include species of *Abies, Actinidia, Alnus, Amelanchier, Berberis, Calluna, Camellia, Catharanthus, Chamaecyparis, Citrus, Cornus, Corylus, Crataegus, Cryptomeria, Cupressus, Cydonia, Cytisus, Diospyros, Eucalyptus, Euonymus, Euphorbia, Fagus, Ficus, Fuchsia, Gardenia, Hibiscus, Hydrangea, Ilex, Juniperus, Larix, Lavandula, Lonicera, Macadamia, Maclura, Magnolia, Malus, Mangifera, Nerium, Osmanthus, Philadelphus, Picea, Pinus, Pistacia, Prunus, Pseudotsuga, Punica, Pyracantha, Pyrus, Rhododendron, Ribes, Rosa, Rubus, Sambucus, Sequoia, Sequoiadendron, Stewartia, Syringa, Thuja, Tsuga, Vaccinium, Viburnum,* and *Vitis*. Latent infections—those not immediately causing visible symptoms—are common in many plants, including flowers of *Hydrangea, Pyrus, Ribes,* and *Rosa*; and flowers and fruit of *Vitis*.

Symptoms and signs. Symptoms on many plants are similar but are not diagnostic. Expanding tan to brown lesions that eventually turn grayish are common. Lesions in delicate leaves or blossoms become wrinkled as they dry. Those in thick leaves such as those of *Rhododendron* or *Ficus* (fig) may have concentric zones of lighter and darker color. Succulent shoots collapse and wither soon after infection begins. Leaf blight of *Cornus florida* (flowering dogwood) occurs when blighted bracts fall on developing leaves. In nursery storages *B. cinerea* may cause a mushy brown decay of bark and cambium of plants packed too tightly or stored too long or warmed during refrigeration failures.

In humid air *B. cinerea* produces a sparse web of gray-brown mycelium on diseased parts. Branched conidiophores that bear colorless to pale brown conidia on swollen heads (Fig. 35), readily visible with a hand lens, arise on the web (Plate 35D, H). The spores are dry, and clouds of them rise when disturbed. Many strains of *B. cinerea* produce small black sclerotia in moist blighted tissue. These vary in form from disc shaped to hemispheric or loaf shaped and are 2–7 mm in length and width, 1–3 mm thick. The sclerotia are survival structures.

Disease cycle. *B. cinerea* can reproduce indefinitely by colonizing dead and dying plant materials; thus it is likely to accompany plants or plant debris in humid places. The fungus overwinters as sclerotia and as mycelium in decaying vegetation. The mycelium produces conidiophores and conidia as environmental conditions allow. Sclerotia germinate in spring to produce either conidia or stalked apothecia of the sexual state, from which ascospores are discharged to the air. Ascospores may cause some infections but are considered unimportant in the epidemiology of Botrytis blight.

B. cinerea enters plants by direct penetration of healthy blossoms, leaves, succulent stems, and fruits. It also enters through wounds and dead parts. Infection may be initiated by conidia, ascospores, or hyphae. Conidia are released to the air during periods of rapidly rising or falling humidity. They also are dispersed during rain as shock waves of air from impacting drops dislodge dry spores. In addition, conidia may also be carried 1–2 m on the surfaces of splashing drops. Various insects, notably fruit flies, fungus gnats, shore flies, and thrips, act as conidial vectors. Conidia may go for long periods without germinating, held in check by lack of moisture or nutrients or by microbial antagonism.

The usual mode of penetration of intact surfaces involves germination of a conidium, adhesion of the germ tube to the host surface, and differentiation of a holdfast structure, the appressorium, at the tip of the germ tube. A penetration hypha from the appressorium (or from a germ tube on some flowers) then breaches the cuticle and epidermal wall. Penetration by hyphae is the rule where superficial mold has already developed, as occurs in cold storage. Apparently *B. cinerea* kills plant tissues by a combination of enzymatic and toxic actions. It produces cutinolytic, pectolytic, cellulolytic, and lipid-degrading enzymes with which it can penetrate the cuticle and cell walls, disrupt cell membranes, and macerate soft tissues. It also produces several types of phytotoxic compounds and is able to degrade phytoalexins (defensive chemicals) of many plants.

B. cinerea flourishes where air is moist and stagnant. It is common on low leaves of plants in dense stands, notably on conifer seedlings in nurseries. Weather events likely to promote blight are a warm period in early spring as plant growth begins, then cool, humid weather, which prolongs succulence in developing leaves and shoots. Frost or freeze damage, whether it kills or only weakens plant tissues, also sets the stage for attack, especially if followed by damp or wet weather. Sublethal frost injury causes abnormal nutrient leakage on plant surfaces. The nutrients stimulate germination of *B. cinerea* spores. Volatile substances from leaves of several conifers can apparently also stimulate spore germination, even at low temperatures such as occur in storage.

Given high humidity, *B. cinerea* grows at temperatures from 0 to above 25°C. The optimum for growth and conidial production is 20–22°C, but both processes can occur, albeit slowly, at 0–5°C. The minimum time required for spore germination and infection is about 20 hours when the air is saturated with water and the temperature is near 20°C. Conidia can germinate on wet or dry plant surfaces if the ambient relative humidity is more than 94%. A few days of warm, dry weather will prevent or check the disease.

Although plants vary in susceptibility to *B. cinerea*, no major genes for resistance are known in either woody or herbaceous plants. Losses to Botrytis blight can be minimized by fertility management that curtails unnecessary succulent growth and by measures that prevent plant stress or promote air circulation and thus keep plant surfaces dry. Various organisms have given partial protection in biological control trials.

References: 29, 769, 770, 775, 874, 992, 1047, 1048, 1099, 1119, 1162, 1606, 1683a, 1731, 1907, 1917, 2371, 2683, 3041, 3412, 3426, 4130, 4151, 4370, 4513

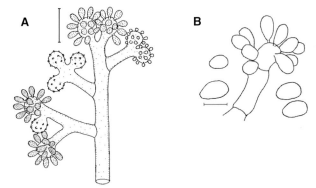

Figure 35. A. The apical part of a conidiophore of *Botrytis cinerea*. B. Conidia. Scale bars = 25 and 10 μm, respectively. Adapted from reference 29 by permission of Elsevier.

A. Early stage of Botrytis blight, caused by *Botrytis cinerea*, on blossoms of *Magnolia stellata* (star magnolia) (CA, Apr).

B. Bracts and a flower of *Cornus florida* (flowering dogwood) killed by *B. cinerea* (NY, Jun).

C. Botrytis blight of a *Cercis canadensis* (redbud) leaf. This sprout leaf was growing in humid air in deep shade (NY, Aug).

D, E. Botrytis blight of rose. Flowers collapse, and the causal fungus grows as a gray mold on dead parts (NY, Jul).

F. Magnified view of sclerotia of *B. cinerea* on killed stems of young *Rhododendron* (azalea) seedlings. These sclerotia began growth within the stems (NC, Mar).

G. A spreading lesion on a *Rhododendron* leaf. *B. cinerea* is often a secondary invader of lesions like this one that are caused by other agents (NY, Jul).

H. Sporulation of *B. cinerea* on a dead blossom of *Bauhinia purpurea* (orchid-tree) (FL, Mar).

Photo credit: F—R. K. Jones

73

Cristulariella Leaf Spots (Plate 36)

Cristulariella leaf spots, also known as target, zonate, or white leaf spots, are caused by *Cristulariella depraedens*, *C. moricola*, and *C. pruni*, all hyphomycetes. The latter two fungi are anamorphs of fungi that produce stalked apothecia, *Grovesinia pyramidalis* and *G. pruni* (Helotiales, Sclerotiniaceae), respectively. No sexual state of *C. depraedens* has been reported. *G. pruni*, which affects *Prunus* in Japan, is not considered further here.

C. moricola, the most important of the three pathogens, attacks diverse herbaceous and woody plants throughout the eastern half of the USA and in eastern Asia. It causes economically significant defoliation of *Carya illinoinensis* (pecan) and *Juglans nigra* (black walnut). Other woody hosts include species of *Acer, Ailanthus, Alnus, Amelanchier, Ampelopsis, Ardisia, Asimina, Bauhinia, Broussonetia, Bucida, Camellia, Carya, Catalpa, Celtis, Chionanthus, Cinnamomum, Cornus, Ficus, Fraxinus, Halesia, Hibiscus, Juglans, Lagerstroemia, Lindera, Liriodendron, Magnolia, Malus, Melia, Millettia, Morus, Oxydendrum, Parthenocissus, Pentas, Persea, Platanus, Prunus, Punica, Rhus, Sassafras, Tilia, Ulmus, Vaccinium, Vernicia, Viburnum, Vitex, Vitis,* and *Ziziphus.* Pathogenicity tests with isolates of this fungus from various plants have given no indication of host specialization and have shown its ability to parasitize many species not reported to be affected in nature.

Zonate leaf spot caused by *C. moricola* is occasionally severe in pecan orchards in southeastern USA. Major commercial pecan cultivars ranked in decreasing order of susceptibility are Moneymaker, Schley, Stuart, and Desirable. Cultivars Cape Fear, Elliott, Sumner, and Sioux are moderately susceptible.

C. depraedens occurs in eastern and northwestern North America, Brazil, Great Britain, and Europe. It has been reported most often on *Acer* species, and it occurs also on *Aruncus dioicus* (goatsbeard) and *Bucida buceras* (black-olive) in North America. In Europe it occurs also on *Carpinus, Cornus, Potentilla, Quercus, Rubus, Salix, Sorbus, Tilia,* and assorted herbaceous plants.

Symptoms and signs. *C. depraedens* and *C. moricola* cause similar symptoms on *Acer* species (maples). Lesions begin as water-soaked greenish gray spots and when fully developed are of all sizes up to about 3 cm, round or irregular, grayish brown with dark margins. Some lesions are pointed along leaf veins. When weather is cool and moist, lesions may coalesce quickly and give a scalded appearance to sections of leaf blades. Dead tissue soon begins to weather away, leaving holes in leaves; severely damaged leaves drop.

C. moricola on other hosts causes lesions that vary from pinpoints to irregular or target-shaped, grayish brown blotches, usually on leaves but sometimes on other succulent parts such as the pods of *Bauhinia* (Plate 36F). On most hosts, spots that remain small have light centers and dark margins. Target spots have light centers and alternating light and dark concentric rings. Severe disease causes premature leaf senescence and defoliation. White crystalline deposits often appear in concentric rings (Plate 36B) or as a film on a lesion (Plate 36H).

Beginning a few days after lesion formation, distinctive reproductive structures (called either conidia or propagules) may develop. They are found mostly on lower surfaces of lesions and sometimes on green tissue at their edges and are visible with a hand lens. The propagules of *C. depraedens* are scattered, colorless, multicellular spheres, about 150 µm across, on slender stalks. *C. moricola* produces conical structures that average about 400 µm long and 150 µm broad at the base. These also are stalked, multicellular, and colorless (Plate 36I; Fig. 36). Sometimes they are branched. The propagules function as infectious spores. Their abundance varies with host and weather conditions. They eventually become detached from the stalks and turn light tan late in the season.

Black resting bodies (sclerotia) form along the veins in large lesions on some hosts, having been noted on *Acer* species, *Fraxinus americana* (white ash), and *Magnolia fraseri* (Fraser magnolia) in the USA. Sclerotia sometimes begin to develop while leaves are attached, but they are usually found on fallen leaves near the end of the growing season. They form also in laboratory cultures after 2–3 weeks of mycelial growth. Those in culture are round in outline, 2–4 mm wide, 1–2 mm thick, and arrayed in concentric rings or clusters. *C. moricola* also produces distinctive eight-sided crystals in culture media.

Disease cycle. Sclerotia presumably play a role in winter survival and development of the apothecial states of *Cristulariella* species,

but facts are not yet available. Apothecia have developed from sclerotia of *C. moricola* in culture but have not been reported from field collections. The form and source of inoculum for initiation of Cristulariella leaf spots in spring are unknown.

Propagules of *Cristulariella* are dispersed by splashing water and can cause infection at 6–27°C. Under conditions highly conducive to infection (wet leaves at 21°C), propagules of *C. moricola* germinate within about 6 hours by multiple germ tubes. These penetrate the host surface, and internal colonization begins within 10–24 hours. Propagules remain in place for a time after germination; thus a propagule can often be found in the center of a young small lesion. Weather permitting, several generations of propagules form during summer and cause secondary cycles of disease. A new generation of *C. moricola* propagules may form as few as 3 days after inoculation of *Vitis* (grapevine) leaves, and new propagules of *C. depraedens* may appear as soon as 5 days after inoculation of *Acer* leaves.

C. moricola apparently kills plant tissues by a combination of toxic and enzymatic activities. It produces toxic amounts of oxalic acid in colonized tissue, and it degrades cell walls by means of pectic enzymes that are synergized by the acid.

Epidemics of Cristulariella leaf spots are favored by cool, wet weather in midsummer. In temperate regions the disease becomes noticeable on low branches by the middle of July and can completely defoliate susceptible plants within a month. Normally, however, it is retarded by weather fluctuations. Lesions stop expanding and no propagules form when relative humidity is less than about 94%. The pathogen remains alive in lesions, however, and may produce propagules again if, after wetting, lesions are kept moist for 4 days or longer.

References: 457, 748, 749, 1099, 1214, 1351, 1503, 2136, 2188, 2231, 2246, 2247, 2855, 3240, 32258, 4087, 4239

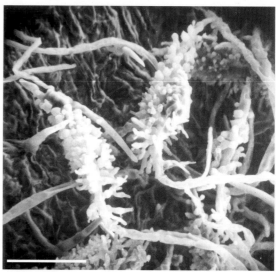

Figure 36. Propagules of *Cristulariella moricola* on a leaf as viewed with a scanning electron microscope. Scale bar = 100 µm. Courtesy of M. N. Cline.

Leaf and fruit diseases caused by *Cristulariella moricola.*

A, B. Variation in lesions on leaves of *Asimina triloba* (pawpaw), including large lesions with concentric markings (target spots) (NY, Sep).

C. An irregular, expanding necrotic blotch on a leaflet of *Carya ovata* (shagbark hickory) (NY, Aug).

D, E. Lesions on *Fraxinus americana* (white ash). Purple borders are typical of small lesions on this host (NY, Aug).

F. Lesions on a pod of *Bauhinia variegata* (mountain-ebony) (FL, Mar).

G. Spots, blotches, and shriveling of *Acer negundo* (box-elder) leaflets; such leaves soon drop (NY, Aug).

H, I. Propagules of *C. moricola*. H. The lower surface of a lesion on *Acer saccharinum* (silver maple) with abundant propagules and a white crystalline deposit (WV, Aug). I. Magnified view showing conical form of a propagule on an *A. negundo* leaflet (NY, Aug).

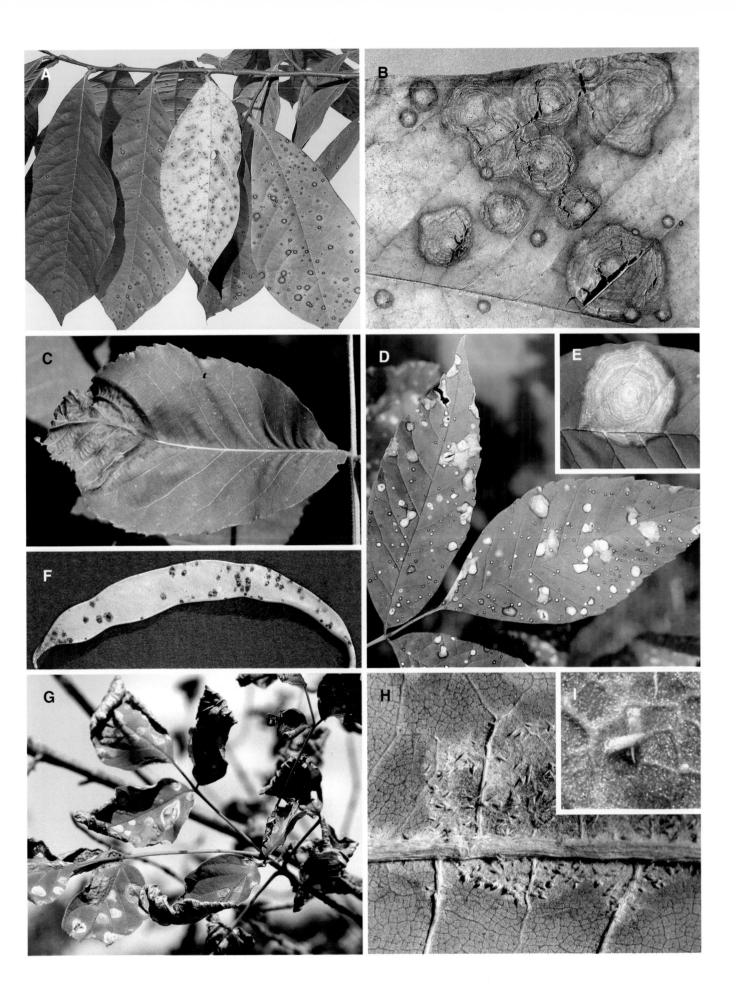

Brown Rot of Stone Fruits (Plate 37)

Stone-fruit trees around the world are subject to brown rot diseases caused by *Monilinia* species (Helotiales, Sclerotiniaceae). These diseases are characterized by blossom and shoot blight, twig cankers, and fruit rot. Common hosts are *Prunus armeniaca* (apricot), *P. avium* (sweet cherry), *P. cerasus* (sour cherry), *P. domestica* (plum and prune), *P. dulcis* (almond), and *P. persica* (nectarine and peach). Ornamental and wild *Prunus* species are also affected. Brown rot of stone-fruit species in North America is usually caused by either *M. fructicola* or *M. laxa*. The former is native to North America and the latter to Eurasia. *M. fructicola* predominates in the East, and *M. laxa* occurs there in scattered northern localities. *M. laxa* predominates in California, where *M. fructicola* is sporadically important. Both species are important in the Pacific Northwest.

Symptoms and signs. The two fungi may cause nearly identical symptoms. Both cause blossom blight. *M. laxa* is the more common cause of twig cankers and dieback, while *M. fructicola* predominates in fruit rot. The first symptom annually is sudden collapse and browning of blossoms. All parts of blossoms may be penetrated directly after germination of *Monilinia* conidia or ascospores. Blight appears 3–6 days after infection. Primary infection is often restricted to flower parts, but shoot wilt and twig dieback may follow during the next 3–4 weeks as the fungus grows from flowers into spurs and twigs. On plum and sometimes cherry, brown rot fungi may infect young leaves and then spread into shoots, resulting in the symptom called wither-tip.

Nongirdling cankers on twigs become apparent as small darkly discolored areas from which gum often exudes. A dead twig or fruit spur typically extends from the center of each lesion. Cankers usually remain inconspicuous except for gummosis, but some enlarge during more than one season. This enlargement is usually the result of invasion by secondary fungi, especially *Leucostoma* species (Plate 84). Gum production at cankers varies with host and size of infected branch. Small twigs may be girdled and die back without gummosis except perhaps at the junction with the parent branch. Apricot trees produce gum most profusely.

Fruit rot is usually initiated by conidia from blossoms or peduncles. Green fruits may have only latent infections, but brown lesions expand rapidly in ripe or ripening fruits and spread from one fruit into another if they are in contact. Some decaying fruits fall, but many remain attached and shrivel. Shriveled fruits, their flesh having been digested and mostly replaced by fungal tissue, are called mummies or sclerotized fruit.

Diagnostic signs are powdery buff to gray tufts (sporodochia) of conidiophores and conidia that erupt from the surfaces of diseased organs. These anamorphs are of the *Monilia* type. The conidia are colorless, unicellular, ovoid to lemon shaped, and borne dry in chains. Those of *M. fructicola* measure 14–16 × 9–11.5 μm; those of *M. laxa* are more elongate, mostly 12–18 × 9–10 μm.

Disease cycle. Brown rot fungi persist in blighted blossoms and peduncles, twig cankers, and mummified fruit on the tree, producing conidia during the growing season and again in late winter and early spring as temperature and moisture permit. These spores and/or ascospores (discussed below) cause blossom blight and thus start the annual cycle. Where winters are cold, brown rot fungi survive poorly on trees but readily persist in fallen sclerotized fruits. Sclerotia are rich in storage compounds that sustain a fungus through the host's dormant period and support reproduction the next year. These structures become sites of apothecium formation and sexual reproduction if they are in or on moist soil for more than 2 months at low temperature in winter.

In spring *Monilinia* species that overwintered as sclerotia produce conidia on the sclerotized fruit or produce apothecia from which ascospores are ejected into the air during host bloom. Apothecia of *M. fructicola* are light brown and shaped somewhat like wineglasses. Most are 10–15 mm across, with stems rising from buried or partly buried sclerotia. The interior surface becomes light buff and at maturity consists of a layer of asci. One sclerotium typically produces several apothecia. Apothecia of *M. laxa*, rarely found, are smaller than those of *M. fructicola*.

Conidia from blighted flowers may infect additional flowers and occasionally leaves and thus contribute to intensification of disease. More important, blighted flowers and peduncles are sources of conidia that infect fruit. Wild hosts such as *P. angustifolia* (chickasaw plum) are also sources of spores that infect domesticated

species. Conidia are dispersed by splashing water, air, and insects. Although brown rot fungi can penetrate leaves and fruits by way of stomata and hair sockets, the usual routes of entry are cracks and other tiny wounds in the cuticle. The most important insect vectors are those that make or visit wounds in fruit. These include *Conotrachelus nenuphar* (plum curculio), *Grapholita molesta* (Oriental fruit moth), beetles of the Nitidulidae (sap or dried-fruit beetles), and *Drosophila* species (fruit flies). Given free moisture and moderate temperature, most conidia germinate within 2–4 hours; ascospores require at least 6 hours. Temperatures most favorable for spore release, germination, infection, and disease development are in the range 15–25°C.

M. fructicola and *M. laxa* differ in several habits. *M. laxa* kills many twigs and produces buff-gray sporodochia on them. *M. fructicola*, by contrast, causes much fruit rot but relatively little twig blight and produces dusty buff to fawn-colored sporodochia. In pure cultures, *M. fructicola* forms colonies with regular margins; *M. laxa* produces lobed colonies. Both fungi occasionally infect rosaceous plants other than stone fruits. *M. fructicola* causes fruit rot of *Malus* (apple), and *M. laxa* causes blossom wilt and rarely fruit rot of *Pyrus* (pear). Both species infect *Chaenomeles* (flowering quince) and *Cydonia* (quince).

About 20 other *Monilinia* species cause diseases similar to the two described here. Pathogens of stone-fruit trees include *M. demissa* on *Prunus virginiana* (chokecherry), *M. padi* on *P. cerasus*, and *M. seaveri* on *P. serotina* (black cherry) and *P. cerasus*. *Amelanchier* (serviceberry) is affected by *M. amelanchieris*, and *Crataegus* (hawthorn) by *M. johnsonii*. In Europe *M. fructigena* and *M. laxa* forma *mali* affect *Chaenomeles*, *Cydonia*, *Malus*, *Mespilus* (medlar), and *Pyrus*. *M. fructigena* also occurs on *Pyrus* in at least one North American location. *M. azaleae* causes blossom and shoot blight of *Rhododendron* (azalea), and *M. vaccinii-corymbosi* causes mummyberry disease of *Vaccinium* (cultivated blueberries). The anamorph *Monilia polystroma* causes brown rot of pome fruits in Japan.

References: 214, 255, 256, 371, 375, 595, 665, 862, 1178, 1729, 1735–1737, 1943, 2228, 2279, 2896, 2897, 3215, 3450, 3992, 4174, 4252

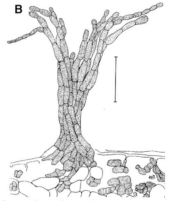

Figure 37. The *Monilia* anamorph of *Monilinia fructicola*. A. Chains of conidia on a sporodochium. B. Conidiophores with conidia in an early stage of sporodochium formation. Scale bars = 25 μm. A adapted from reference 2897 by permission of the American Phytopathological Society; B adapted from reference 214 by permission of Kluwer Academic/Plenum Publishers.

A, B. Shoot blight, branch dieback, and gummosis on *Prunus armeniaca* (apricot) caused by *Monilinia laxa* (CA, Jun).

C, D. Shoot blight, twig dieback, and fruit rot caused by *M. fructicola* on *Prunus domestica* (common plum) (NY, summer).

E. Twig dieback caused by *M. laxa* on *P. persica* (peach). The twig was girdled by an infection two nodes above the fruit (OR, Jul).

F, G. *M. laxa* on *P. serrulata* (Japanese flowering cherry). F. Shoot and flower clusters that wilted and died when the parent twig was girdled by a brown-rot canker. G. Sporulation on calyces of blighted blossoms (WA, Jun).

H. *M. laxa* causing fruit rot of *P. avium* (sweet cherry). Powdery gray sporodochia cover the lesions (OR, Jul).

I. *M. fructicola* sporulating on a fallen fruit of *P. domestica* (plum) (NY, Sep).

J. Apothecia of *M. fructicola* (NY, May).

Entomosporium Leaf Spot of Maloideae (Plate 38)

Diplocarpon mespili (Helotiales, Dermateaceae) in its conidial state, *Entomosporium mespili* (a coelomycete), causes necrotic spots on leaves, fruits, and sometimes succulent stems of more than 60 plant species around the world, all in the subfamily Maloideae of the Rosaceae. Diseases attributed to it are called Entomosporium leaf spot, Diplocarpon or Fabraea scald, leaf spot, leaf blight, and fruit spot. Susceptible plants occur in the genera *Amelanchier, Aronia, Chaenomeles, Cotoneaster, Crataegus, Cydonia, Eriobotrya, Heteromeles, Malus, Mespilus, Photinia, Pyracantha, Pyrus, Rhaphiolepis,* and *Sorbus. Prunus persica* (peach) has sometimes been listed as susceptible, but this may be an error.

Damage by *D. mespili* is both economic and aesthetic. Leaf and fruit spot of susceptible *Pyrus* (pear) caused by this fungus can result in trees being defoliated and fruit spotted, deformed, and cracked by July. Yield is reduced accordingly. Defoliation of *Cydonia* (quince) understocks can cause them to harden prematurely and thus be unsuitable for grafting with pear scions. Young seedlings of *Eriobotrya japonica* (loquat) have been killed by multiple lesions on succulent stems. Highly susceptible species of *Amelanchier* (saskatoon or serviceberry) develop twig and branch dieback if they are defoliated in several successive years. Highly susceptible *Crataegus* (hawthorn) can be defoliated before midsummer.

Symptoms and signs. Lesions arise as minute dots and when fully developed appear as brown to gray irregular spots 2–5 mm in diameter. Those on leaf veins or succulent stems are elongate. Dead spots up to 1 cm in diameter develop on highly susceptible plants such as *Rhaphiolepis umbellata* (Japanese- or yedda-hawthorn). Lesions may be scattered or so numerous that they coalesce to form large dead blotches. Mature lesions in *Amelanchier* leaves and in plants with thick leaves, such as *Eriobotrya, Photinia,* and *Rhaphiolepis,* commonly have reddish to dark brown borders or halos (Plate 38E–G). Leaves that turn yellow in response to infection, as on *Crataegus, Cydonia,* and *Malus* (crabapple), often retain green color around the lesions. Yellowing presages premature leaf cast. If plants defoliate before midsummer, new leaves may develop, but these soon also become infected.

Tiny, blisterlike, irregularly shaped fruiting structures (acervuli) form beneath the cuticle in the centers of lesions soon after death of the plant tissue (Plate 38F–H). Acervuli vary in size up to 1 mm or more in diameter, depending on host and environment, and they often coalesce. Black at first, they and the surrounding leaf surface become gray with the accumulation of colorless conidia that appear white in mass. The conidia are five celled, with four cells normally visible and seemingly arranged in a cross. Each cell has a thin appendage (Fig. 38). These spores remind some people of insects, as the name *Entomosporium* suggests.

Disease cycle. D. mespili overwinters in fallen leaves and in lesions on persistent leaves and young green twigs (and in fruit of *Amelanchier*). Conidia can survive winter in acervuli on dead leaves, and some new conidia are produced there in spring. Primary infection begins in spring when conidia from these sources are dispersed by splashing water to developing leaves. Each conidial cell is capable of germinating. Germ tubes produce appressoria from which the host surface is penetrated. Studies of the disease in *Photinia* revealed that leaf infection occurs most frequently on the abaxial (lower) surface. The fungus usually penetrates directly through the cuticle and cell wall but sometimes enters through stomata, especially on mature leaves. In early stages of infection, hyphae grow intercellularly and send absorptive structures (haustoria) into host cells, especially cells with chloroplasts. Lesions soon become visible on the adaxial surface, and acervuli form there. Secondary cycles of disease start thereafter as successive generations of conidia cause new lesions. Secondary cycles begin in early summer in northern areas but can continue all year on evergreen hosts such as *Eriobotrya, Heteromeles, Photinia,* and *Rhaphiolepis* in regions with a mild winter climate. Young leaves of these plants are more susceptible than mature ones.

In autumn in northern areas, the fungus shifts from conidial production to development of its sexual phase. Structures that look like abnormally small acervuli appear, especially on lower surfaces of lesions, and produce tiny, noninfectious one-celled spores that may function as spermatia. Apothecia then begin to form in fallen leaves. They break through the leaf surface and complete their development in spring. At maturity they are black, cup shaped, and only about 150 μm wide and 100 μm deep. They produce colorless,

two-celled ascospores that at maturity are ejected into the air. The role of ascospores in initiating disease in spring is unknown. They are generally considered to be unimportant in warm regions because of continual conidial production and the apparent absence of apothecia there. During prolonged wet periods in spring, conidia sometimes begin to form in place of asci in apothecia on fallen leaves.

Mild, wet summer weather promotes epidemics. Conditions that favor severe disease are temperatures of 15–25°C and wet plants for at least 9–12 hours. Conidia germinating on leaves can survive temporary drying (up to 12 hours tested) and remain infectious if not exposed to sun.

Some strains of *D. mespili* may be specialized to attack particular plant types. In one study isolates from *Pyrus* and *Cydonia* were highly virulent on *Pyrus* and *Eriobotrya japonica* but not on *Photinia glabra* or *P. serrulata*, while isolates from the two *Photinia* species were virulent on their original hosts and also on *Pyrus* and *E. japonica. D. mespili* will grow and produce conidia on laboratory media, but cultures lose virulence rapidly.

Hosts of *D. mespili* vary widely in susceptibility. *Malus* and *Chaenomeles* (flowering quince) usually develop only small lesions and usually do not defoliate prematurely. Among pears, *Pyrus amygdaliformis, P. betulifolia, P. calleryana, P. pyrifolia, P. ussuriensis,* and other East Asian species are resistant, as are hybrids between these species and *P. communis.* Among hawthorns, *Crataegus phaenopyrum* (Washington thorn) has remained only lightly spotted while nearby *C. laevigata* 'Paulii' was completely defoliated. *C. crus-galli* (cockspur thorn) is also resistant. *Rhaphiolepis delacourii* 'Eskimo' is resistant, as are *R. umbellata* 'Indian Princess' and 'Olivia' in Alabama. *Photinia serratifolia* was less susceptible than *P. glabra* or hybrids between these species in detached-leaf assays. *Amelanchier alnifolia* 'Success' is resistant.

Cultural methods have been suggested but not critically tested for suppression of Entomosporium leaf spot. Removal of fallen leaves might be helpful for plants on which the fungus does not overwinter on twigs or persistent leaves. Avoidance of sprinkler irrigation may help, but in the humid eastern part of North America, rainfall is usually adequate for severe disease on susceptible plants. Summer pruning should be avoided because it tends to promote a flush of highly susceptible succulent new growth.

References: 258, 259, 300, 1431, 1753, 1897, 2659, 2660, 2779, 3327, 3844, 3885, 3913

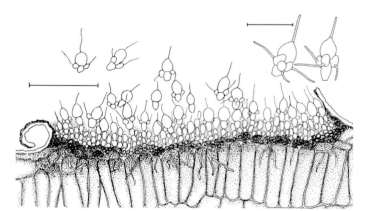

Figure 38. An acervulus with insectlike conidia of the *Entomosporium* state of *Diplocarpon maculatum;* two conidia enlarged at right. Scale bars = 50 and 20 μm, respectively. Adapted from reference 2779 by permission of T. R. Nag Raj.

Entomosporium leaf spot.

A–C. Symptoms on *Crataegus laevigata* 'Paulii' (Paul's scarlet hawthorn) (NY, Jul). A. A tree defoliated by the disease. B, C. Lesions viewed on upper and lower leaf surfaces and on a stem.

D. Lesions on leaves and fruit of *Cydonia oblonga* (quince) (NY, Sep).

E. Lesions on year-old leaves of *Heteromeles salicifolia* (toyon) (CA, May).

F. Lesions and associated reddening on *Photinia* leaves (MS, May).

G, H. Lesions on leaves of *Rhaphiolepis indica* (Yedda-hawthorn). Dark acervuli form in the centers of lesions on upper and lower leaf surfaces (CA, Nov).

Black Spot of Rose and Blumeriella Leaf Spot of Cherry and Plum (Plate 39)

Black Spot of Rose

Diplocarpon rosae (Helotiales, Dermateaceae) causes black spot, which occurs wherever roses are grown and is their most important disease. The fungus is limited to *Rosa*. Yellow- or gold-flowering roses tend to be more susceptible than red or pink ones. Austrian Brier roses, Pernetians, Polyanthas, Tea roses, and Hybrid Teas are notably susceptible.

Symptoms and signs. Black spots or blotches up to 15 mm across and sometimes larger develop on leaves throughout the season. They usually arise on the upper surface and have fimbriate (fringed) margins. Subcuticular hyphae, organized into strands radiating from the point of infection, account for the fringed appearance. Minute black blisters (acervuli) of the conidial state, *Marssonina rosae*, arise beneath the cuticle in the centers of spots. The acervuli (Fig. 39A) are 0.1–0.4 mm in diameter, glisten when wet, and produce colorless, two-celled conidia. Leaves with black-spot lesions turn yellow and drop. Petioles, fruit, and flower parts except petals develop symptoms similar to those on leaves. Petal infections, relatively unimportant, appear as red dots with some distortion of surrounding tissues. Rose canes are susceptible during their first season of growth. Lesions on canes develop as purplish red, raised, irregular blotches that usually remain small and become black and dotted with acervuli but lack fimbriate margins. Repeated severe defoliation weakens plants, so they produce fewer and poorer blossoms, are difficult to graft, are abnormally sensitive to winter cold, survive poorly in storage, and fail to thrive after planting.

Disease cycle. Multiple cycles of black spot develop each year. The first, or primary, cycle is usually initiated by conidia that are dispersed with splashing water from overwintered fruit bodies (modified acervuli) on fallen leaves or diseased canes. The first infections give rise to successive generations of conidia and spots. Infection occurs through either leaf surface as spores germinate and produce appressoria from which penetration pegs pierce the cuticle. Hyphae then form a radiating network of subcuticular and intercellular mycelium from which branches enter epidermal and palisade cells and produce absorptive structures (haustoria) in them. The infected cells soon die and darken.

D. rosae grows as a saprobe in fallen rose leaves and in spring disperses either conidia from acervuli or ascospores from apothecia. The apothecia, found in some localities in northern USA, Canada, and Europe, are black, globose to disc shaped, and less than 300 μm in diameter. The ascospores, although infectious, are considered to be less important than conidia as inoculum.

Environmental conditions most conducive to black spot are wet plants, high humidity, and temperatures of 20–24°C for several days. Under these conditions symptoms appear 4–5 days after inoculation, and acervuli with a new generation of conidia form within 10–11 days. Severity of the disease and susceptibility of the plants both seem to reach annual peaks in late summer. Development of black spot on susceptible cultivars is retarded somewhat in mixed plantings of susceptible and resistant roses. Maintaining space between plants retards buildup of black spot somewhat by promoting ventilation and drying of plants after rain.

D. rosae has numerous races and pathotypes that vary in virulence. The mixture of races in the pathogen population varies geographically and over time. Thus, roses resistant in one locality may become spotted if grown in another area inhabited by different races.

Several rose species and many cultivars are sufficiently resistant to *D. rosae* to be useful either for ornament or as breeding stock. *Rosa banksiae*, *R. bracteata*, *R. carolina*, *R. laevigata*, *R. laxa*, *R. multiflora*, *R. roxburghii*, *R. pimpinellifolia*, *R. rugosa*, *R. virginiana*, and *R. wichuraiana* are among the species utilized as sources of resistance, winter hardiness, and desired growth habits. Some resistant cultivars secrete substances that inhibit conidial germination or cuticular penetration of leaves. Others exhibit hypersensitivity or limit colony growth after penetration, and some delay symptom development and sporulation, which retards disease increase.

References: 181, 671, 784, 923, 978, 998, 1758, 1859, 2433, 2434, 3239, 3663, 3743, 3776, 3913, 3924, 4289, 4359, 4464, 4486

Blumeriella Leaf Spot of Cherry and Plum

The most common fungal leaf spot of *Prunus* is caused by *Blumeriella jaapii* (Helotiales, Dermateaceae), which occurs across North America and Europe on cherries and plums and rarely on *Oemleria cerasiformis*, the Indian-plum or osoberry. The anamorph *Phloeosporella padi* (Fig. 39B) is the state found sporulating on leaves. The disease is economically important in cherry orchards, especially in *Prunus cerasus* (sour cherry), because severe infection causes defoliation and reduces yield and winter hardiness. Defoliation of *P. serotina* (black cherry) seedlings by cherry leaf spot in forest stands and nurseries suppresses growth and contributes to low survival. Other plants affected include *P. americana*, *P. angustifolia*, *P. armeniaca*, *P. avium*, *P. caroliniana*, *P. domestica*, *P. dulcis*, *P. emarginata*, *P. fruticosa*, *P. laurocerasus*, *P. mahaleb*, *P. nigra*, *P. pensylvanica*, *P. persica*, *P. pumila*, *P. salicina*, *P. serrulata*, *P. tenella*, and *P. virginiana*. Strains of *B. jaapii* that infect cherries are generally unable to attack plums, and vice versa. Leaf-spot resistance occurs in *Prunus avium*, *P. besseyi*, *P. maakii*, *P. padus*, *P. pseudocerasus*, and *P. tomentosa*.

Leaf spot lesions usually appear about the time leaves reach full size. Initially dark purple, they turn reddish brown and attain diameters of 0.5–3 mm. New lesions continue to appear until late summer, either scattered or aggregated on one part of a leaf blade. Many lesions close together cause large sections of the leaf to turn brown. During wet weather, white masses of conidia extrude from acervuli, usually on the lower leaf surface. Leaves with numerous lesions turn yellow and drop. If diseased leaves remain attached, the lesions become isolated by cork barriers and may drop out, leaving shot holes. Shot holes alone are not diagnostic, however, because various agents induce them in *Prunus* leaves. Lesions sometimes develop on fruits and pedicels but are not important there.

B. jaapii overwinters in fallen leaves and produces inconspicuous brownish apothecia about 0.2 mm in diameter in spring. The first infections annually are caused by airborne ascospores from this source. The fungus enters leaves via stomata, grows intercellularly, and sends haustoria into living cells, which soon die. Symptoms appear as soon as 5–8 days, and acervuli with conidia as soon as 6 days, after infection. Water-disseminated conidia cause repeating cycles of disease. Conidial germination and infection occur only on wet foliage and are most rapid at 20–28°C. Mild, wet summer weather promotes cherry leaf spot. Photosynthesis is severely suppressed in diseased leaves.

References: 123, 1043, 1099, 1235, 1648, 2030, 2856, 2896, 3830

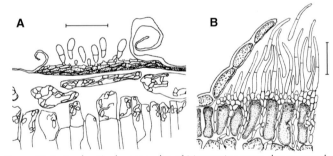

Figure 39. A. A subcuticular acervulus of *Marssonina rosae*, the anamorph of *Diplocarpon rosae*. B. A subepidermal acervulus of *Phloeosporella padi*, the anamorph of *Blumeriella jaapii*. Scale bars = 25 μm. A adapted from reference 3913 by permission of CAB International; B adapted from reference 123 by permission of Gebrüder Borntraeger Verlagsbuchhandlung.

A–C. Black spot symptoms and signs on rose leaflets. Lesions in B and C have typical fimbriate, or fringed, appearance. Leaflets showing yellow will soon drop (NY, Jun–Aug).

D. Magnified view of part of a black spot lesion with acervuli, which appear as minute black blisters, some located by arrows (CA, Jul).

E. Leaf spot caused by *Blumeriella jaapii* on *Prunus cerasus* (sour cherry). Leaves showing yellow will soon drop (NY, Jul).

F–H. Leaf spot caused by *B. jaapii* on *P. serotina* (black cherry). F. Severely affected leaves on a first-year seedling. G. Angular lesions, premature senescence, and foliar browning due to lesion coalescence. H. Magnified view of fresh acervuli of the *Phloeosporella* state of *B. jaapii* on brown lesions on the lower surface of a dying leaf (PA, Aug).

Photo credits: F–H—G. R. Stanosz.

81

Marssonina Spots and Blights (Plate 40)

Leaf spots and blights of aspen and poplar. Marssonina leaf spots and blights of *Populus* (poplar and aspen) occur worldwide and cause defoliation, dieback, and lost growth of trees that are used in intensive forestry programs. Three fungi are involved primarily: *Drepanopeziza populi-albae*, *D. populorum*, and *D. tremulae* (syn. *D. punctiformis*) (Helotiales, Dermateaceae). Their conidial states are the coelomycetes *Marssonina castagnei*, *M. populi*, and *M. brunnea*, respectively.

Symptoms and signs. Dark brown spots or blotches appear on leaves, and pustules develop on green twigs. Acervuli up to 400 μm in diameter with unequally two-celled, colorless conidia form in the epidermal cell layer of the lesions. Severe outbreaks may cause foliar browning in midsummer and nearly complete defoliation by early August. Regrowth occurs in late summer and early autumn, and twig dieback may follow in winter because late-season shoots lack normal cold hardiness. Seedlings are sometimes killed.

D. tremulae is the most important of these pathogens. Its anamorph, *Marssonina brunnea*, has several host-specialized formae speciales, of which f.sp. *brunnea* and f.sp. *trepidae* are noteworthy. The former typically infects *Populus balsamifera* (balsam poplar), *P.* ×*canadensis* (Carolina poplar), *P. deltoides* (eastern cottonwood), *P. fremontii* (Fremont cottonwood), *P. nigra* (black and Lombardy poplars), and various Euroamerican poplar hybrids. The latter infects *P. tremula* (European aspen) and *P. tremuloides* (trembling aspen) but not *P. deltoides* or Euroamerican hybrids. Strains of *D. tremulae*, *D. populi-albi*, and *D. populorum* also vary in host specialization.

D. tremulae causes dark brown, discrete spots about 1 mm in diameter (Plate 40A) that appear soon after leaves expand or later in the season. On green twigs this fungus causes tiny cankers that appear as pustules. An acervulus in the center of each lesion appears at first faintly orange, then grayish white as conidia are exposed. The conidia are mostly 13–18 × 4.5–6 μm. Leaves with numerous lesions develop general yellowing or browning, especially along the margins, and drop prematurely. *D. tremulae* also infests poplar seed and may infect seedlings grown from infested seed.

D. populi-albi typically attacks *P. alba* (white poplar) and *P. canescens* (gray poplar) and sometimes *P. grandidentata* (bigtooth aspen). It causes dark brown to reddish brown lesions, often with lighter centers, mostly about 3 mm in diameter, that often coalesce into irregular blotches. Acervuli develop primarily on upper surfaces, and the conidia are mostly 16–21 × 6–7.5 μm.

D. populorum typically attacks *P. balsamifera* and its subspecies *trichocarpa* (black cottonwood), *P. deltoides*, and *P. tremuloides*. It causes irregular bronze to chestnut brown spots 2–5 mm in diameter that usually coalesce into vein-limited blotches. These may involve nearly the entire leaf. The upper surfaces of lesions are dotted with grayish white acervuli. The conidia are mostly 18–25 × 8.5–10.5 μm. This fungus occasionally causes spectacular outbreaks of leaf blight and defoliation of *P. tremuloides* in the central Rocky Mountain region.

A fourth species, *Marssonina balsamiferae*, spots the leaves of *P. balsamifera* in Manitoba and Ontario. Its importance and ascigerous state are unknown.

Disease cycles. Annual cycles of the three pathogens are similar. They survive winter as tiny stromata in fallen leaves and in twig lesions. Water-dispersed conidia from acervuli on the stromata initiate primary infection soon after leaves emerge in spring. Airborne ascospores also cause primary infections in areas where the ascigerous states occur. Ascospores are liberated from brown apothecia 150–350 μm across on overwintered fallen leaves. Symptoms commonly appear 14–16 days after inoculation. Conidia from acervuli in primary lesions on leaves and twigs initiate repeating cycles of disease on leaves. Disease increases as wet weather permits until leaves fall. Symptoms intensify and seem to ascend trees as the season advances. In late summer and early autumn the pathogens produce microconidia (<6 μm long) in the same lesions. Microconidia are noninfectious and may function as spermatia in the sexual process that culminates in ascospore production in apothecia in spring.

Aspen and poplar clones vary widely in susceptibility to particular strains of *Marssonina*. For example, natural clones of *P. tremuloides* in the Rocky Mountain region sometimes differentiate into a mosaic of green (resistant), brown (highly susceptible), and intermediate

foliar hues during outbreaks of leaf blight caused by *M. populi*. Hybrid poplar clones also vary in responses to *Marssonina* species.

References: 609, 1214, 1473, 1506, 1682, 1958, 2001, 2625, 2831, 3079, 3080, 3084–3086, 3779–3781, 3783, 3784, 3787, 3798, 3799

Anthracnose of birch. *Marssonina betulae* causes a striking but little-noticed anthracnose of *Betula papyrifera* (paper birch) and *B. pendula* (European white birch) in eastern Canada and adjacent USA and Europe. The lesions are large, dark reddish brown spots with radiate margins. Subcuticular acervuli form on the upper surfaces. The acervuli are flat, dark brown, and difficult to discern. They produce colorless oblong spores that at maturity are two celled and measure 17–22 × 5–10 μm. Leaves with multiple lesions turn yellow and drop prematurely. The ascigerous state of *M. betulae*, if any, is unknown, and the disease cycle has not been studied. For more birch anthracnoses, see Plate 53.

References: 780, 1298, 1397, 3065

Anthracnose of bittersweet. *Celastrus scandens* is occasionally attacked by *Marssonina celastri*, which infects leaves, green stems, and fruit. Lesions on leaves are dark, 1–3 mm across, and are sometimes surrounded by a yellow zone. Lesions on fruit are shallow, yellowish to light tan cankers about 1 mm across with abrupt dark edges. Lesions on green stems are elliptic, up to 3 mm long, and flat to slightly raised. Where numerous, lesions may coalesce and kill stems, fruit, or leaves. The surfaces of young lesions are covered by raised, light yellowish acervuli that contain two-celled conidia 22–33 × 8–14 μm. The apical cell is usually wider than the basal cell. The spores are colorless individually but appear yellow in mass. *M. celastri* occurs also on *C. orbiculatus* (Oriental bittersweet) in Korea. No information is available on the disease cycle or the life history of the pathogen. Its distribution and history are confused with those of *Septogloeum thomasianum* (syn. *M. thomasiana*), which is reported to occur on *Celastrus* and *Euonymus atropurpurea* (burning-bush) in the USA and Europe.

References: 445, 1098, 1924, 3605, 3913

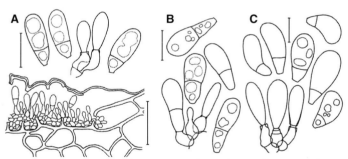

Figure 40. A. *Marssonina brunnea*: conidia and part of an acervulus in vertical section. B, C. Conidia of *M. castagnei* and *M. populi*, respectively. Scale bars = 30 μm for acervulus, 10 μm for conidia. Adapted from references 3084–3086 by permission of Public Works & Government Services Canada.

A, B. Leaf spots caused by *Marssonina brunnea* on *Populus deltoides* (eastern cottonwood). A. Typical punctate lesions. B. Magnified view of the upper surface of a leaf bearing tiny dark brown lesions with grayish centers where massed conidia in acervuli raise the cuticle (NY, Aug).

C. Lesions caused by *M. brunnea* on young twigs of *Populus* sp. (WI, summer).

D. Leaf spots caused by *M. castagnei* on *Populus alba* (white poplar) (NY, Sep).

E. Leaf blotch caused by *M. populi* on *P. tremuloides* (UT, summer).

F, G. Leaf blotch of *Betula pendula* (European white birch) caused by *M. betulae*. F. Diseased leaves, one prematurely yellow. G. Close view of the upper surface of a lesion, showing the typical radiate margin (NY, Aug).

H–K. Anthracnose of *Celastrus scandens* (American bittersweet) caused by *M. celastri*. H. Leaf blight and shoot blight resulting from coalesced lesions. I. Lesions on fruit. J. Lesions on stems. K. Close view of lesions on leaves, each covered by a large acervulus with conidia that are yellowish in mass (NY, Aug).

Photo credits: C—M. E. Ostry; E—USDA Forest Service

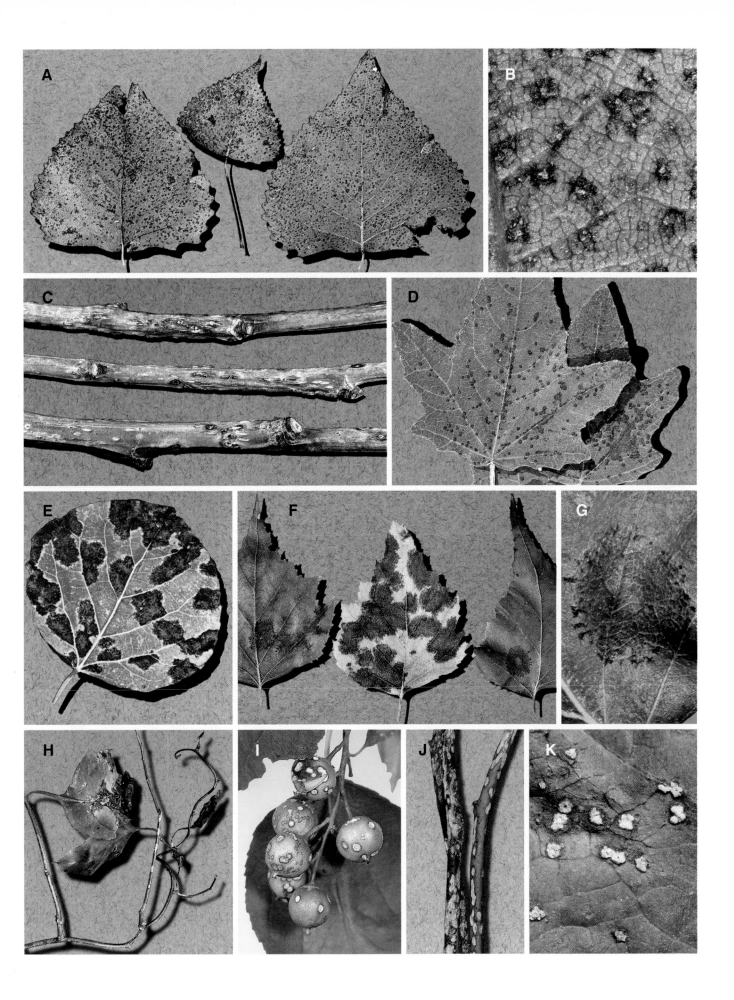

Alternaria Blights and Leaf Spots (Plate 41)

Alternaria is a ubiquitous genus of dark-spored hyphomycetes that includes both plant pathogens and saprogens. The pathogens inflict slight damage on most woody plants, usually causing necrotic spots or blotches on leaves and decay of fruit. Both pathogens and saprogens are often secondary invaders of foliage, fruit, and woody parts killed or weakened by other agents. *Alternaria* species sporulate on dead plant surfaces, producing brown conidiophores and conidia singly or in chains. These structures may be visible as a brownish or black mold. The conidia are large (e.g., 30–300 μm long), club or pear shaped, dry, thick walled, and multicellular, often with both transverse and longitudinal or oblique septa and often with a long, tapered apical cell (beak). The species are described, but often not clearly differentiated, in terms of presence or absence of conidial chains, conidial chain length, size and shape of conidia, and characteristics of conidial beaks. Even within a species the spores vary in size, shape, septation, and surface ornamentation. Some species are anamorphs of ascomycetes in the genus *Lewia* (Pleosporales, Pleosporaceae).

Many species names in *Alternaria* are associated with diseases of trees and shrubs, but relatively few species are distinguished by diagnosticians. Some of those reported in the USA and Canada are listed below. Plants in many of the listed genera are hosts of both identified and undetermined *Alternaria* species. We list a host genus for undetermined *Alternaria* only if the genus is not listed with a named *Alternaria*. This practice avoids duplication but understates the diversity of disease records.

Alternaria alternata: leaf spot or leaf blight of *Atriplex, Cytisus, Dypsis, Hedera, Hibiscus, Ilex, Jasminum, Juglans, Ligustrum, Magnolia, Pieris, Pistacia, Pittosporum, Ptelea, Rhus, Robinia, Rosa, Syagrus, Syringa, Vaccinium, Yucca*
  – damping-off and wood stain of *Pinus*
  – decline of *Mangifera* (accompanying other pathogens)
  – flower spot of *Rhododendron*
  – fruit spot or fruit rot of *Carica, Citrus, Malus, Pistacia, Prunus, Pyracantha, Vaccinium, Vitis*
  – seedling tip necrosis of Citrus
A. araliae: leaf spot of *Polyscias, Schefflera*
A. brassicae: leaf spot of *Rosa, Syringa*
A. brassicicola: leaf spot of *Rosa*
A. camelliae: leaf spot of *Camellia*
A. catalpae: leaf spot of *Catalpa*
A. citri: leaf spot of *Citrus, Fortunella, Murraya, Phoenix*
  – fruit rot of *Citrus, Prunus, Psidium*
  – seedling necrosis of *Poncirus*
A. euphorbiicola: leaf blight of *Euphorbia*
A. gomphrenae: leaf spot of *Vitis*
A. limicola: leaf spot of *Citrus*
A. mali: leaf spot of *Malus*
  – fruit rot of *Cydonia, Malus*
A. panax: leaf spot of *Aralia, Eleutherococcus, Fatsia, Polyscias, Populus, Robinia, Schefflera, Tupidanthus*
A. raphani: leaf blight of *Populus*
A. tenuissima: leaf spot of *Pittosporum, Pistacia, Prunus, Vaccinium*
Undetermined *Alternaria*: leaf spot or leaf blight of *Acacia, Acer, Aucuba, Coccoloba, Cycas, Euonymus, Ficus, Forsythia, Gardenia, Ixora, Juniperus, Nerium, Pandanus, Prunus, Psidium, Pyrus, Ribes, Sorbus, Swietenia, Torreya, Triadica, Ulmus, Wisteria*
  – fruit rot of *Diospyros, Fortunella, Mangifera, Persea, Pyrus, Symphoricarpos*
  – twig dieback of *Thuja, Vaccinium*
  – flower blight of *Rosa*
  – damping-off of *Rhododendron*

Plant-pathogenic *Alternaria* species survive in plant debris and in or on bark between growing seasons. Their conidia are dispersed by wind, or locally by water, and have no special requirements for germination. These fungi enter plants by direct penetration from appressoria or by hyphal growth through stomata or into wounds. Given favorable temperature and moisture, they sporulate as soon as they colonize a food base of dead tissue. They produce various toxins that act as virulence factors and (or) host range determinants by weakening or killing tissues that are then exploited for nutrition.

The list above indicates that some *Alternaria* species are generalists while others are host specialized. Some host-specialized

taxa, notably *A. citri* on *Citrus* and *A. mali* on *Malus*, are morphologically inseparable from *A. alternata*. Accordingly, some authorities regard these taxa as formae speciales or pathotypes of *A. alternata* (Fig. 41), the species most commonly identified on or from plants. In practice, diagnosticians apply this name to various fungi that produce branching chains of short-beaked conidia in the size range 40–70 × 15–20 μm. When found on necrotic spots or blight lesions, *A. alternata* is often considered to be a secondary invader, although it can act as a primary pathogen. It causes a petal blight of *Rosa* with symptoms similar to those of Botrytis blight (Plate 35). The rose disease has been noted in southern USA on field-grown Floribunda and Hybrid Tea roses during wet summer weather. Spores produced on dead blossoms or rose hips are responsible for infection, *A. alternata* also infects seeds or seedlings of various plants, causing, for example, diminished germinability and albinism in *Citrus* and damping-off of *Pinus elliottii*. This fungus penetrates healthy fruit of some plants (e.g., *Diospyros, Malus, Mangifera, Vaccinium*), establishes tiny colonies, and remains quiescent until the fruit becomes senescent during ripening. Then the colonies resume growth and cause decay. *A. alternata* and another generalist, *A. tenuissima*, have also been detected in skin of humans with weakened immune systems.

A. panax, which causes leaf blight of plants in the Araliaceae, is one of the host-specialized species that attack woody plants. It occurs in nurseries and landscapes in southern USA and occasionally in greenhouses elsewhere. It causes discrete necrotic spots or spreading dark lesions, often surrounded by yellow zones, in leaves of *Eleutherococcus, Polyscias, Schefflera, Tupidanthus,* and herbaceous species. Lesions either remain discrete or expand into large blotches. The fungus soon sporulates on lesions in humid air. Severe disease causes defoliation. Dispersal of the conidia by air or splashing rain leads to secondary cycles of disease, with new symptoms appearing 4–5 days after inoculation. Severe disease is favored by temperatures of 24–27°C, shade, relatively low nutrient levels, and intermittent wetting of foliage. The wetness is necessary for spore germination and penetration. Isolates of *A. panax* have shown no host preference within the Araliaceae.
References: 46, 139, 140, 297, 370, 706, 712, 944, 1055, 1099, 1153, 1789, 2192, 2263, 2948, 3158, 3347, 3636, 3757, 3758, 4032, 4111

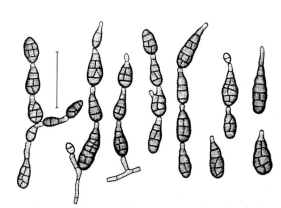

Figure 41. Alternaria alternata. This species produces its dark conidia in branched chains. Bar = 50 μm. Adapted from reference 1055 by permission of CAB International.

A–C. Necrotic spots and blotches on *Schefflera actinophylla* leaves caused by *Alternaria panax*. Dark mold on a lesion in C is a sporulating colony of the pathogen (FL, Mar–Apr).

D. Necrotic spots and blotches on leaves of *Polyscias* sp. (aralia) caused by *A. panax* (FL, Mar).

E, F. Foliar lesions apparently caused by *Alternaria* sp. on *Hibiscus* sp. Black dots in the lesion in F are clumps of *Alternaria* conidiophores and conidia (FL, Mar).

G, H. Leaf blight of hybrid poplar. *Alternaria* sp. that sporulates on such lesions (magnified in H) under moist conditions is presumed to cause them (MI, Aug).

Scab Diseases Caused by *Venturia* Species (Plates 42–45)

Plant pathogens of the genus *Venturia* species (Pleosporales, Venturiaceae) and related anamorphs infect succulent aboveground parts of their hosts, causing localized swelling and distorted necrotic spots on leaves, or blight of leaves and shoots. They produce olive or olive-brown mats of conidiophores and conidia on infected or recently killed parts. The pathogens overwinter as mycelium or immature pseudothecia in dead, previously diseased leaves or in twigs. Primary infections are initiated by ascospores, conidia, or both, depending on the disease. If weather is conducive, conidia cause secondary cycles of disease as long as susceptible host tissue is available. These pathogens display considerable host specificity; each fungal species is restricted to one plant genus or a few closely related genera. The generalized life cycle of these fungi involves an alternation of parasitic anamorphs (asexual states) and saprobic teleomorphs (ascigerous states). The anamorphs are in the genera *Fusicladium* (syn. *Pollaccia, Spilocaea*) and *Fuscicladosporium*. One species, *Venturia rhamni* on *Frangula californica* (California buckthorn) in California, is exceptional in that it completes its life cycle in living leaves during the growing season. Some anamorphic scab fungi have not been associated with teleomorphs.
References: 227, 232, 1859, 2787, 3494a, 3668, 3669

Apple Scab

The most famous and economically important scab disease of trees is apple scab caused by *Venturia inaequalis* (anamorph *Fusicladium pomi*). Hosts include all commercial cultivars of *Malus pumila* (apple) except those bred for immunity, most crabapples, and other *Malus* species. The list includes but is not restricted to cultivars of *M. angustifolia, M. ×arnoldiana, M. baccata, M. brevipes, M. coronaria, M. florentina, M. glaucesens, M. ioensis, M. micromalus, M. ×platycarpa, M. ×purpurea, M. ×scheideckeri, M. sieboldii,* and *M. sylvestris.* Other hosts include several species of *Sorbus* (mountain-ash)—*S. aria, S. aucuparia, S. domestica,* and *S. torminalis*—as well as *Cotoneaster affinis, C. integerrimus, Pyracantha* (firethorn), and *Pyrus communis* (common pear). Strains of *V. inaequalis* from *Malus* do not infect plants in other genera, nor do strains from those plants infect *Malus.*

Symptoms and signs. Scab lesions on leaves and fruit arise in spring as olive-green spots, becoming mostly 4–10 mm in diameter, that darken but remain superficial and more or less circular with radiate margins. They become velvety as masses of olivaceous brown conidiophores and conidia develop on the epidermis and break through the cuticle (described and illustrated with Plate 43). Chlorosis and death of tissues beneath the colonies follow during the next several weeks. Lesions on expanded leaves are often diffuse and tend to be associated with veins. Tissues that become infected while enlarging are at first stimulated to overgrowth that results in raised lesions (scabs) on fruit, bumps on succulent twigs, and curled or puckered leaves. Diseased tissues later cease growth prematurely, resulting in further deformity as surrounding healthy tissues continue growth. Lesions on fruit and twigs become delimited by a cork layer, and the surfaces become cracked and rough. Mature tissues resist new infections but may continue to support previously established colonies of the parasite. Photosynthesis is suppressed in diseased leaves, and severely affected leaves and fruit fall prematurely. Symptoms on *Sorbus* and other hosts are similar to those on apple, with emphasis on foliar lesions and premature leaf fall.

Disease cycle and epidemiology. The apple scab fungus overwinters as immature pseudothecia in fallen, previously diseased leaves and as mycelium and conidia in apple buds. It can also overwinter as mycelium in twig lesions on highly susceptible trees in areas of mild climate. Pseudothecia produce ascospores over a period of 1–3 months, depending on local climate, beginning with the onset of host growth in spring. Ascospores are shot into the air during rain and (rarely) after dew. Ascospores typically initiate the first (primary) cycle of disease annually. Conidia from diseased buds and twigs may also start primary infections in regions where winters are relatively warm and spring weather is wet.

Leaves (both surfaces), flower parts, fruit, and succulent twigs become infected. Germ tubes from either ascospores or conidia produce appressoria from which penetration pegs puncture the cuticle and differentiate into infection hyphae. Colonies expand radially between the cuticle and epidermis and draw nutrients from living tissues beneath, but do not enter those tissues. Colonies several days old begin to produce conidia, and massed

conidiophores and conidia rupture the cuticle. Conidia cause secondary infections on leaves. After diseased leaves fall, *V. inaequalis* colonizes tissues beneath former scab lesions and produces pseudothecia during the dormant season, thus completing its life cycle. Twig lesions become inactive during winter or during the second growing season, when periderm formation separates them from living tissues.

Apple scab is favored by mild, rainy weather, and scab infection periods can be predicted based on weather forecasts. Rain induces expulsion of ascospores from pseudothecia and disperses conidia. Given wet plants and available spores, scab infections may occur within 6 hours if temperature is near the optimum of 20°C. Longer wet periods lead to more abundant infections. Symptoms appear and conidial production begins usually 8–18 days after infection, but the incubation period is longer under cool conditions or if dry weather intervenes during incubation.

V. inaequalis includes subpopulations specialized for parasitism of plants in particular genera of Maloideae or for particular *Malus* genotypes. Formae speciales specialized for attack of *Cotoneaster, Malus, Pyracantha,* and *Sorbus* are known. Strains from *Malus* and *Pyracantha,* formerly thought to represent different species, have been shown by DNA analyses and mating tests to belong to the same species. Races of *V. inaequalis* cause scab on species and cultivars that are resistant to the general population of the fungus. Race 6 is noteworthy for its ability to overcome resistance conferred by the *Vf* gene from *M. floribunda,* which has been bred into many modern apple cultivars. Scab has been noticed on *M. floribunda* in recent years.

Natural controls on apple scab, aside from resistance or dry weather, are few. In moist areas, earthworms and litter-decomposing microorganisms degrade fallen leaves and reduce the overwintering population of the scab fungus. Removal of fallen leaves in autumn also reduces the population. Decomposition of collected leaves can be hastened by composting them and applying nitrogenous fertilizer to the compost pile. Dolomitic lime applied to the ground suppresses maturation of the pathogen in overwintering leaves and thus suppresses ascospore production in spring.

Resistant cultivars. Many *Malus* species and cultivars are resistant—and some are apparently immune—to *V. inaequalis.* Resistance is conferred by several unlinked genes and is expressed as hypersensitivity and/or phytoalexin accumulation at sites of challenge by the pathogen. Some crabapple types resistant to apple scab and also to powdery mildew and cedar-apple rust include *Malus* cultivars Adams, Baskatong, Bob White, David, Henry Kohankie, Mary Potter, Molten Lava, Ormiston Roy, Professor Sprenger, Red Jewel, Sentinel, and Sugartyme; also *M. baccata* cv. Jackii, *M. floribunda, M. halliana* cv. Parkmanii, *M. sargentii, M. sargentii* cv. Tina, and *M. zumi* var. *calocarpa.* Additional cultivars with scab resistance include Adirondack, Christmas Holly, Evereste, Golden Raindrops, Jewelberry, Jewelcole, Midwest, Molazam, Prairie Maid, Prairifire, Strawberry Parfait, Purple Prince, Silver Moon, Tom's Pink, White Angel, Winter Gem, *M. halliana* var. *spontanea,* and *M. zumi* cv. Zumi Wooster. Overwintered leaves of resistant *M. pumila* cultivars produce less ascospore inoculum and produce it later in spring than do leaves of highly susceptible cultivars.

A second apple scab fungus, *V. asperata,* described in New Zealand, has been found on 'Almey' crabapple in Ontario. Its North American distribution and host range are unknown.
References: 152, 163, 164, 277, 302, 371, 617, 646, 711, 1663, 1785, 1859, 1898, 2181, 2410, 2411, 2521, 2592, 2611, 2645, 2741, 2787, 2849, 2850, 2896, 3465, 3494a, 3626, 3668, 3743, 3804, 3848

A–F. Apple scab. A. Foliar lesions and premature yellowing on *Malus* sp. (crabapple). Chlorophyll is retained (green-island effect) beneath lesions. B. Crabapple shoots defoliated by scab. C. Chlorosis and distortion of a diseased leaf. D–F. Scab on leaves and fruit of *Malus pumila* (apple). E. Close view of olive-brown colonies of *Venturia inaequalis* on a leaf; radiate margins of colonies are characteristic. F. Old lesions on mature fruit (NY, Jun–Aug).

G. Scab caused by *V. inaequalis* on *Sorbus aucuparia* (European mountain-ash). Lesions, often concentrated along midveins of leaflets, cause premature yellowing and death or casting (NY, Aug).

Scabs of Firethorn, Loquat, and Toyon

Scab is perhaps the most important disease of *Pyracantha* (firethorn). It is caused by *Venturia inaequalis* f.sp. *pyracanthae*, formerly known as *Spilocaea pyracanthae*. This fungus is morphologically indistinguishable from *Fusicladium pomi*, the anamorph of the apple scab pathogen, but it is incapable of infecting *Malus* (apple, crabapple). Conversely, *V. inaequalis* f.sp. *pomi* causes apple scab but is incapable of infecting *Pyracantha*. The firethorn pathogen also infects *Eriobotrya japonica* (loquat), *Kageneckia oblonga*, and *Heteromeles salicifolia* (toyon). It occurs across the USA and Canada and is widespread in Europe, from where it is believed to have been imported to North America.

Symptoms and signs. Scab on the hosts noted above appears as olive-green to black velvety spots on leaves, fruit, and young twigs. These are colonies of dark septate hyphae that grow between the cuticle and epidermis. The centers of colonies become thickened, and the cuticle ruptures as conidiophores and conidia form. Conidia of *V. inaequalis* f.sp. *pyracanthae* are colorless or nearly so, one celled, somewhat club shaped, and measure 12–22 × 6–9 µm. They are borne solitary on the tips of colorless, one-celled conidiophores that measure 50–60 × 6–8 µm. Scab on *Pyracantha* is more common on fruit than on leaves. However, severe infection of highly susceptible firethorns such as *P. coccinea* cv. Lalandei may cause nearly all the leaves and fruit to drop, thus greatly reducing the ornamental value of the shrub. Lesions on *Eriobotrya* leaves may coalesce to form large darkened areas, and lesions on fruit may cause deformity. Affected leaves may turn yellow or red around lesions. Severe infection of this plant also leads to premature casting of leaves and fruit. Severe scab on young leaves and shoots of *Heteromeles* may kill those parts (Plate 43E), even though the pathogen is confined to surface cells. Infection of fully expanded leaves results in discrete spots, and the leaves gradually turn red to brown. Scab causes green *Heteromeles* fruit to shrivel, and it can cause severe defoliation or death of *K. oblonga*.

Disease cycle. As the host plants of *V. inaequalis* f.sp. *pyracanthae* are evergreen, the pathogen can survive winter on diseased leaves as well as twigs. Conidia dispersed from year-old leaves and twigs by water and perhaps also by air cause the first infections in spring. The incubation period after conidial inoculation in California was about 2 weeks at 16°C and 3–5 weeks under natural conditions at 6–17°C. The pathogen reproduces only asexually in nature, so far as is known.

Pyracantha species and cultivars vary greatly in susceptibility. Those affected to various degrees include *P. angustifolia, P. atalantioides, P. coccinea, P. crenulata, P. fortuneana,* and *P. koidzumii*. When exposed to natural inoculation in Washington State, the following cultivars seemed resistant: *P. coccinea* Government Red and Prostrata; *P. koidzumii* Bella, Duval, and Santa Cruz Prostrata; *P. crenulata* var. *rogersiana* Flava; and the hybrid *Pyracantha* Shawnee. In eastern trials, *Pyracantha* Fiery Cascade and *P. coccinea* Rutgers were resistant. Several hybrid cultivars, resistant to scab plus other diseases, notably fire blight, have been released in the USA and Europe. These include Apache, Cadange Saphyr Orange, Cadrou Saphyr Rouge, Navaho, Pueblo, and Teton. *References:* 121, 227, 598, 617, 1038–1040, 2536, 2787, 3194, 3743, 4143, 4157

Pear Scab

Two fungi that are nearly alike in form cause pear scab. *Venturia pyrina* is the pathogen in the Western Hemisphere and Europe. Its host plants include most cultivars of *Pyrus communis* (common pear), *P. pyrifolia* (Asian pear), *P. syriaca*, and in England *Eriobotrya japonica* (loquat). Pear scab in Japan and eastern Asia is caused by *V. nashicola*, which affects primarily the Asian species *P. bretschneideri* (Chinese white pear), *P. pyrifolia*, and *P. ussuriensis* (sand pear), and causes no damage on *P. communis*. Most cultivated Asian pears are *P. pyrifolia*. *V. nashicola* is not considered further here.

Pear scab, which is economically significant in the Pacific Northwest, is similar to apple scab (Plate 42) in its modes of infection, symptoms, annual cycle, and epidemiology. Primary infection is initiated in spring by ascospores liberated from pseudothecia in overwintered fallen leaves, and by conidia from lesions on twigs (see below). Subcuticular, more or less circular,

olivaceous lesions develop on leaf blades, petioles, flower peduncles, and fruit. The causal fungus produces olivaceous mounds of conidiophores and conidia that break through the cuticle, permitting conidial dispersal by water or air. Diseased fruit may become malformed as fruit growth under scab lesions is suppressed. Conidia cause secondary cycles of disease as conditions permit. The most significant difference between apple scab and pear scab is that twig infections by the pear scab fungus are more numerous and persistent than are those of apple scab. Twig infections cause roughening and distinct lesions, initially visible as pustules 1–2 mm in size, on young green stems. Pustules soon break open, exposing conidiophores and conidia. The fungus remains active and produces conidia in twig lesions through as many as three seasons.

V. pyrina differs from *V. inaequalis* in several features in addition to host range. Its two-celled ascospores are longer than those of *V. inaequalis*, and the apical cell is longer than the basal cell (Fig. 43). Its conidial state is *Fusicladium pyrorum*. Its conidia are light yellowish green, at first nonseptate but becoming equally two celled with age, pointed at both ends, and 28–30 × 7–9 µm in size. They are produced on short conidiophores and leave prominent raised scars when they disperse.

Host-specialized races and biotypes occur in *V. pyrina*. Five races were differentiated on varieties of *P. communis* and *P. syriaca* in Israel. Four biotypes were distinguished on *P. communis* cultivars in England, and numerous biotypes differing in virulence on various pear cultivars were detected in the USA. Most genotypes of Asian *Pyrus* species, such as *P. amygdaliformis, P. calleryana, P. pyrifolia,* and *P. ussuriensis*, are resistant to *V. pyrina*, as are hybrids between *P. communis* and these species. The cultivar Orcas, a hybrid clone of unknown parentage, is also resistant, as is Pitoma Slanopadja, a hybrid of *P. amygdaliformis* and *P. communis*, selected in Europe. *References:* 3, 227, 649, 1859, 1868, 1869, 2233, 2896, 3494a, 3547, 3668, 3669, 3802

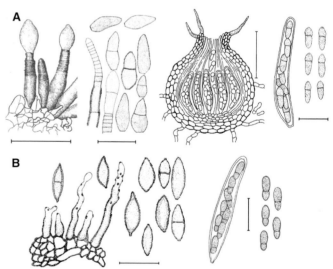

Figure 43. A. *Venturia inaequalis:* conidiophores, conidia, pseudothecium, ascus, and ascospores. B. *V. pyrina:* conidiophores, conidia, ascus, and ascospores. Scale bars = 50 µm for pseudothecium, 20 µm for other structures. A adapted from references 123, 3669, and 4157 by permission of Gebrüder Borntraeger Verlagsbuchhandlung and A.D.A.C.—Cryptogamie; B adapted from references 649 and 3669 by permission of W. B. Kendrick and Gebrüder Borntraeger Verlagsbuchhandlung.

A–F. Scab caused by *Venturia inaequalis* f.sp. *pyracanthae*. A–C. Scab on fruit of *Pyracantha coccinea* 'Lalandei' (NY & MD, Sep–Oct). D. Scab on a yellow-fruited *Pyracantha* (CA, Sep). E. Discoloration and death of leaves and shoots of *Heteromeles salicifolia* (toyon) resulting from infection of immature tissues (CA, Sep). F. Velvety scab lesions on an *Eriobotrya japonica* (loquat) leaf (CA, Jun).

G. *Pyrus communis* 'Bartlett' (pear) fruit with scab caused by *Venturia pyrina*. A large primary lesion is surrounded by small lesions resulting from secondary infection (OR, Jul).

89

Venturia Leaf and Shoot Blights of Aspen and Poplar

At least 10 species of *Venturia* cause leaf and shoot blight diseases of *Populus* (aspen and poplar). Five named species are known in North America. *V. moreletii* and *V. populina* are significant pathogens, reported to have transcontinental distributions, that cause damage mainly in young forest stands. *V. moreletii* (syn. *V. macularis*, *V. tremulae* var. *grandidentatae*; anamorph *Fusicladium radiosum* var. *lethiferum* (syn. *Pollaccia radiosa*) usually attacks species in *Populus* section *Populus*: *P. alba* (white poplar), *P. grandidentata* (bigtooth aspen), *P. tremuloides* (trembling aspen), and hybrids. *V. populina* (anamorph *F. elegans*, syn. *Pollaccia elegans*) usually attacks species in *Populus* sections *Aigeiros* and *Tacamahaca* (poplars and balsam poplars): *P. balsamifera* (balsam poplar), *P. balsamifera* subsp. *trichocarpa* (black cottonwood), *P. nigra* (black poplar), and hybrids. Both fungi occasionally infect additional *Populus* species. Both fungi also occur in Europe, and *V. populina* is in India as well. A recently discovered third pathogen, *V. inopina*, causes leaf-and-shoot blight of *P. balsamifera* subsp. *trichocarpa* and hybrids between it and *P. deltoides* in the Pacific Northwest. In that region, moreover, *V. populina* is found only on the introduced *P. nigra*, not on *P. balsamifera* subsp. *trichocarpa*. This finding has called into question the host and geographic distributions of *V. populina* in North America as reported above. A fourth species, *V. borealis* (anamorph *F. borealis*), causes an inconsequential purple-brown leaf spot of *Populus tremuloides* in northwestern Canada. The fifth, *V. orbicularis* is reported to cause leaf spot of aspens (*Populus* section *Populus*) in eastern North America but is not well documented.

In Europe *V. populi-albae*, *V. tremulae*, and *V. populina* are the main pathogens. The first of these causes necrotic spots on leaves of *P. alba*, *P. tremula* (European aspen), and *P. tremuloides*. *V. tremulae* causes leaf spots and blight of *P. alba*, *P. canescens* (gray poplar), *P. tremula*, and hybrids. *V. maculosa* causes leaf spots on various species. In China *V. mandschurica* causes gray leaf spot of poplars.

Symptoms and signs. Only *V. moreletii* and *V. populina* are considered further here. Both species infect succulent leaves and shoots, but they differ in the patterns and seasonal development of symptoms induced. The first infections by *V. moreletii* on aspen occur in spring on leaf blades or petioles and on succulent stems. Dark brown to black lesions expand rapidly, causing leaves and shoots to droop, wither, and become brittle. Within a few days after a lesion forms, the blackened surface turns olive-green as a layer of conidiophores and conidia develops (Plate 44H). The conidia are ellipsoid to clavate, one–two-septate, and measure 15–42 × 6–11 μm. In wet years this pathogen may kill nearly all terminal shoots in aspen stands that are regenerating by sprouts after the harvest of mature trees. This damage reduces height growth and results in crooked stems where lateral shoots become dominant after terminals are killed (Plate 44F, G). Plants less than 3 m tall are at greatest risk; damage becomes negligible as trees grow taller than 5 m. Most infections occur on terminal shoots because these grow throughout the season. Lateral shoots cease elongation early in the season and become resistant. Thus, most of the foliage of diseased trees remains intact.

V. populina attacks plants of all ages and sizes and may kill leaves and shoots throughout the crown. Diseased tissues turn black, and killed shoots droop and become brittle. The conidial state appears as an olive to olive-brown mat or sometimes tufts of conidiophores and conidia on recently killed parts. The conidia are elliptic to broadly fusoid, three celled with the central cell largest, and measure 26–35 × 9–13 μm. Mature parts of the current year's shoots and all parts formed in previous years escape infection. *P. balsamifera* severely damaged by *V. populina* can be recognized at any time of year. During the dormant season, damaged branches appear to be crowded, stunted, and decadent. The crowding results from development of adventitious shoots and multiple leaders after death of terminals. Canker-causing fungi enter through blighted terminals and contribute to the decadence.

V. moreletii and *V. populina* are readily distinguishable. The former has smaller conidia as noted, smaller pseudothecia (40–180 μm versus 160–220 μm in diameter), and smaller ascospores that are septate just above the middle rather than in the lower third like those of *V. populina*.

Disease cycles. *V. moreletii* overwinters as developing pseudothecia in fallen, previously blighted leaves and as mycelium in the stubs of blighted shoots. Primary infection can be caused by airborne ascospores discharged from pseudothecia or by conidia from blighted twig stubs. Secondary cycles, promoted by wet weather, are initiated throughout the period of shoot elongation by conidia from newly blighted shoots. New shoots frequently grow adjacent to blighted ones and are killed in turn, but lesions do not extend into woody twigs.

V. populina overwinters as developing pseudothecia in blighted shoots. In spring its ascospores cause primary infections on the tips of leaves emerging from buds adjacent to the dead shoots. Germinating ascospores produce appressoria from which epidermal cells are penetrated directly, causing conspicuous lesions within 4 days. Conidial production begins within an additional 7 days. The killing of leaf tips results in distortion, but the leaf bases typically stay alive. Conidia, dispersed from diseased leaves to growing shoots, initiate secondary cycles characterized by leaf spot and shoot blight. In eastern Ontario the primary cycle typically begins near the end of May, and secondary cycles early in June. During midsummer *V. populina* ceases to sporulate on twigs that became infected in June. This cessation apparently signals the onset of sexual reproduction, for immature pseudothecia can be found in blighted shoots beginning in August. They remain immersed in cortical tissue and mature in early spring. Given wet surfaces, ascospores and conidia of both pathogens germinate over a wide range of temperatures. Germination is most rapid at 15–25°C, depending on the isolate.

Resistance to *V. populina* has been detected in hybrids between *P. balsamifera* subsp. *trichocarpa* and *P. deltoides*, *P. nigra*, and *P. maximowiczii* in western Canada and in Europe. Resistance of *P. nigra* was indicated in two studies by development of lesions on which the pathogen failed to sporulate. In Europe, however, *V. populina* can defoliate *P. nigra*. *P. tremula* and *P. ×berolinensis* (Berlin poplar) are also susceptible. Resistance of aspens and their hybrids to *V. moreletii* has not been reported.

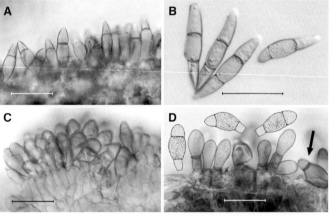

Figure 44. A, B. *Venturia moreletii:* conidia of the anamorph, *Fusicladium radiosum* var. *lethiferum*, on an aspen leaf and viewed singly. C, D. *V. inopina:* a stroma of the *Fusicladium* anamorph and developing conidia on a leaf of *Populus balsamifera* subsp. *trichocarpa*. Drawings of typical two-septate conidia of *V. populina* are provided for comparison with the one-septate conidia of *V. inopina* (arrow). Scale bars = 25 μm. Photos courtesy of B. E. Callan, Canadian Forest Service. Drawings of conidia in D adapted from reference 3669 by permission of Gebrüder Borntraeger Verlagsbuchhandlung.

A, B. Leaf-and-shoot blight of *Populus balsamifera* (balsam poplar) believed caused by *Venturia populina*. A. Severe blight on saplings (AB, Jun). B. A blighted shoot and foliar lesions (QC, Aug).

C, D. Blight of *P. balsamifera* subsp. *trichocarpa* (black cottonwood) sprout leaves, probably caused by *V. inopina*. C. Diseased leaves. D. Tufts of conidiophores and conidia on a lesion (WA, Jun).

E–H. Leaf-and-shoot blight of *P. tremuloides* (trembling aspen) caused by *V. moreletii*. E. A diseased branch (AB, Jun). F. Deformed tops of saplings in a young forest stand (ON, Jun). G. Foliar lesions, apical crook on a dead shoot, and a bend in the twig where a lateral shoot assumed dominance after death of the terminal (ON, Jul). H. A young foliar lesion; conidia are produced on the olive-green area (VT, May).

91

References: 227, 232, 609, 882, 883, 1099, 1214, 1215, 1298, 1393, 1859, 2721, 2722, 2831, 2835, 2838, 3370, 3494a, 3668, 3669

Scab and Black Canker of Willow

Two diseases of *Salix* (willow), scab and black canker, are characterized by rapid blighting of leaves and shoots. Scab is caused by *Venturia saliciperda* and black canker by *Glomerella miyabeana*. The latter fungus and its relatives comprise a family, Glomerellaceae, which in 2004 was awaiting placement in the higher classification of ascomycetes. The two pathogens often occur together and can kill highly susceptible trees by repeated defoliation and destruction of shoots. The name willow blight is often used for these simultaneous infections. Both pathogens apparently came to North America from Europe. Willow blight appeared in the Maritime Provinces in the 1920s and soon became widespread in eastern Canada and northeastern USA, greatly reducing populations of highly susceptible species and varieties. Subsequent reports extended its known range into the Appalachian and Great Lakes regions and the Pacific Northwest.

Most published information about the relative susceptibilities of various willows to blight applies to trees naturally attacked by both pathogens. Species most often damaged in North America are *Salix alba* (white willow) and its subspecies *vitellina* (golden willow), *S.* ×*pendulina* (Niobe willow), *S. capraea* (goat willow), *S. cordata* (heartleaf willow), and *S. nigra* (black willow). Others affected include *S. amygdaloides* (peachleaf willow), *S. aurita* (eared willow) *S. bebbiana* (Bebb willow), *S. discolor* (pussy willow), *S. fragilis* (crack willow), *S. lucida* (shining willow), *S.* ×*mollissima* (sharpstipule willow), *S. myrsinifolia* (dark-leaf willow), and *S. sericea* (silky willow). Plants considered to be more or less resistant are *S. alba* var. *tristis*, *S. babylonica* (weeping willow), *S. pentandra* (bay-leaved willow), *S. purpurea* (purple willow), *S. triandra* (almond-leaf willow), and *S. viminalis* (basket or osier willow). *S. alba* subsp. *caerulea* (cricket-bat willow) is reportedly immune. Willows with partial resistance to black canker develop leaf spots and drop diseased leaves early but do not sustain shoot blight and cankers.

Willow scab. The name scab indicates the relationship of *V. saliciperda* to the well-known pathogens of apple and pear (Plates 42, 43). Attacks begin in spring on the youngest leaves, which are often killed as they emerge from buds. Dead leaves shrivel and gradually drop. The fungus grows through petioles into twigs, causing girdling cankers and shoot blight. Olive-brown velvety masses of conidiophores and two- or three-celled conidia of the anamorph *Fusicladium* (syn. *Pollaccia*) *saliciperdum* develop within a few days as tiny mounds or as a continuous layer on diseased leaves and shoots (Plate 45C, D). The conidia measure 16–23 × 6–9 μm and have a rounded apex and truncate base. This sporulation is most common along major veins on the lower sides of leaves.

Conidia cause repeating cycles of scab during wet weather, leading not only to death of the first complement of leaves and shoots but also to infection of adventitious shoots that arise after defoliation. Only the current season's shoots are attacked, however, and both leaves and stems become less susceptible as the growing season advances. The fungus overwinters in killed twigs and in spring produces conidia that cause primary infections. Wet seasons promote epidemics, and dry weather halts them. The ascigerous state of *V. saliciperda* has been produced in laboratory cultures but has not been reported to occur in nature. Thus, ascospores probably have no role in the disease cycle.

Three other species of *Venturia* occur on willows in North America, but diagnostic confusion is unlikely. *V. chlorospora* causes inconsequential leaf spots. Its conidial state, a *Fusicladosporium*, has developed in laboratory cultures but has not been found in nature. *V. minuta* and *V. subcutanea* have no known conidial states and are not recorded pathogens.

Black canker. This disease and its cause, *G. miyabeana*, were described first in Japan and were soon found in Europe and later in North America and New Zealand. Conidia or ascospores, both from overwintering sites on twigs killed the previous year, cause primary infections. The first symptoms are blackened areas on leaf blades, extending quickly toward the leaf base. Lesions may remain discrete, but usually entire leaves die, droop, and shrivel (Plate 45E, F). Lesions on twigs most often arise at nodes where the pathogen has grown down petioles. Cankers either encircle twigs or remain

more or less elliptic and usually 2–3 cm long but sometimes extend 5–8 cm. They become depressed, and the dead bark splits as twigs increase in girth. *G. miyabeana* tends to attack leaves and twigs later in the season and to cause cankers on larger woody twigs than does the scab pathogen.

Acervuli of the conidial anamorph of *G. miyabeana*, a *Colletotrichum*, develop beginning early in summer on cankers and killed twigs, rarely on blighted leaves or shoot tips. The conidia are light pinkish in mass (Plate 45G). Individual conidia are elliptic, unicellular, nearly colorless, and measure 12–24 × 3–6 μm. Dispersed by water, they germinate on wet leaves or succulent stems and produce appressoria from which the fungus penetrates a host. This process initiates secondary cycles of disease. Conidial germination and mycelial growth are most rapid at temperatures near 25°C. Lesions have appeared within 40 hours after wet leaves were inoculated with conidia and incubated at 25°C. In autumn, perithecia follow or accompany acervuli in the lesions. The perithecia are nearly globose and 140–200 μm in diameter at maturity. They remain immersed in plant tissues. They liberate one-celled, colorless ascospores in spring.

The conidial state of *G. miyabeana* is morphologically indistinguishable from *Colletotrichum gloeosporioides*, the anamorph of *G. cingulata* (Plate 56), which causes many plant diseases. Even so, analyses of ribosomal DNA of New Zealand isolates indicated that *G. miyabeana* is more closely related to *C. acutatum* than to *C. gloeosporioides*.

In recent years, although damage by scab and black canker persist in wild willow populations, these diseases have become uncommon in managed landscapes. Perhaps this scarcity indicates depleted populations of highly susceptible genotypes in areas where climatic conditions favor these diseases.

References: 227, 232, 751, 1206, 1522, 1859, 1957, 2787, 2788, 2878, 3494a, 3668, 3669, 3796

Venturia Leaf Blotch of Maple

Venturia acerina causes a leaf blotch of *Acer* (maples) in the Great Lakes and Appalachian regions and eastward to Newfoundland. The pathogen's conidial state is *Fusicladosporium humile* (syn. *Cladosporium humile*, *Fusicladium humile*), a hyphomycete that has also been reported in Great Britain. The disease affects *Acer negundo* (box-elder), *A. nigrum* (black maple), *A. rubrum* (red maple), *A. spicatum* (mountain maple), *A. saccharinum* (silver maple), and *A. saccharum* (sugar maple). It is best known on *A. rubrum*, on which necrotic lesions up to 2 cm in diameter develop in midsummer. The lesions are more or less round except where bounded by major veins or a leaf edge. On the upper surface they have deep reddish brown centers and dark brown edges with diffuse margins. The lower surface is grayish green to grayish tan. Coalescing lesions kill large areas of leaf blades. Severe infection of *A. rubrum* leads to premature leaf reddening and casting.

The conidial state develops under humid conditions on a network of light brown mycelium on both surfaces of lesions. The olive-brown, one- or two-celled conidia form in chains on short conidiophores that occur singly or in clusters.

V. acerina overwinters as mycelium and developing pseudothecia in fallen leaves. Its ascospores mature in spring and are presumed responsible for primary infection. Germinable conidia have been found on overwintered leaves, but their role in the disease is unknown.

An anamorph reported as *Cladosporium humile* causes a leaf spot of *Populus* in nurseries in India. Comparisons of Indian collections with authentic anamorphic *V. acerina* have not been reported.

References: 227, 288, 1099, 1298, 2973a, 3090, 3494a, 3668, 3669

A–D. Willow scab, caused by *Venturia saliciperda*. A, B. *Salix* sp. defoliated by the disease (NY, Jul). C, D. Masses of olive-brown conidia on a killed stem tip and on leaves of *S. purpurea* (NY, Jun).

E–G. Black canker caused by *Glomerella miyabeana*. E, F. Shoot blight, leaf blight, and twig canker on *Salix* sp. (WA, May). G. Acervuli with pinkish masses of conidia on killed buds in a twig canker on *Salix* sp. (NY, Sep).

H, I. Leaf blotch caused by *Venturia acerina* on *Acer rubrum* (red maple). H. Dark brown blotch lesions on green leaves. The small lighter brown lesions are Phyllosticta leaf spots. I. Magnified view of part of a lesion where the pathogen is sporulating (arrows) (NY, Aug).

93

Bronze Leaf Disease of Aspen and Poplar and Linospora Leaf Blight of Balsam Poplars (Plate 46)

Bronze Leaf Disease

Several species of *Populus* (aspen, poplar) in section *Populus* (syn. *Leuce*) of that genus are subject to bronze leaf disease, caused by *Apioplagiostoma populi* (Diaporthales, Gnomoniaceae). The disease is known only in North America in the region extending from Iowa and Manitoba to Pennsylvania, Massachusetts, and Quebec. *P.* ×*canescens* (gray poplar), *P. grandidentata* (bigtooth aspen), *P. tremula* (European aspen), *P. tremuloides* (trembling aspen), and hybrids involving these species or *P. alba* (white poplar) are susceptible.

Symptoms and signs. Foliar symptoms arise in spring and summer. Undersized chlorotic leaves develop in spring from buds on branches that were diseased the previous year. Drab green lesions expand between the veins of these leaves and turn bronze or orange-brown to reddish brown, sometimes chocolate brown, during summer. Lesions often have diffuse yellow to golden margins in late summer. The major veins remain green for a time, but veins in and at the edges of bronzed areas often turn black. Bronzed areas remain moist initially; then the reddish brown parts dry and the edges curl upward. Most of the leaves on affected branches usually have symptoms, and these typically appear first on low branches. Both lateral twigs and branch terminals are affected. Killed leaves adhere to some diseased twigs throughout winter. Lesions as described above also develop during summer in scattered leaves that were infected by airborne ascospores in spring.

Xylem of diseased stems becomes pigmented in streaks and diffuse zones, the color varying from pale yellow or gold to bronze or sometimes bright reddish brown. The pigment persists during the first year after death of a branch. Diseased branches either have symptoms in successive years or die prematurely. *Populus* clones vary in susceptibility as indicated by incidence and severity of symptoms.

In summer minute brown acervuli develop beneath the cuticle on the upper surfaces of reddish brown leaves (Plate 46F; Fig. 46A). They produce unicellular, colorless conidia that average 7×3.3 µm. These spores have not germinated for researchers. Perhaps they function as spermatia. Perithecia begin to develop in brown leaves in early autumn and mature in spring in both fallen leaves and those remaining on the tree. The perithecia are flask shaped and nearly black when mature. They lie parallel to the leaf surface, and each has a beak that extends through the lower leaf surface at a right angle to the axis of the fruit body. The ascospores are colorless and unequally two celled, $10–13 \times 3.5–6$ µm.

Disease cycle. *A. populi* overwinters as immature perithecia, as noted above, and also as perennial mycelium in diseased twigs and branches. The former mode leads to infection of scattered leaves when ascospores are discharged from perithecia in spring (from late April into June in Minnesota). Ascospore discharge and germination and foliar infection are favored by temperatures of 15–20°C, and in one study were inhibited at 23°C. Mycelium that overwinters in twigs (possibly also in buds) grows into green shoots in spring. This ability has been verified by grafting experiments in which symptoms spread from diseased scions into initially healthy recipient plants. Proximal spread from diseased twigs has not been studied. *A. populi* is fastidious, as all attempts to obtain pure cultures from spores or diseased tissues have failed. Its pathogenicity has been demonstrated by controlled inoculations with ascospores.

The ubiquitous endophyte *Epicoccum nigrum* has often been isolated from *Populus* leaves and stems affected by bronze leaf disease. This fungus produces bronze pigment in culture but did not cause symptoms when inoculated to *Populus* clones known to be susceptible to bronze leaf disease. Its relationship to *A. populi* and to bronze leaf symptoms is unclear.

A. populi is the only known North American species in its genus. In Europe *A. carpinicolum* occurs on dead leaves of *Carpinus betulus* (European hornbeam), and *A. aceriferum* is found on leaves of *Acer* species.
References: 229, 233, 669, 881, 1198, 2699, 2740, 2874, 3723

Linospora Leaf Blight

Linospora tetraspora (Diaporthales, Gnomoniaceae) causes a leaf blight of *Populus balsamifera* (balsam poplar), its subspecies *trichocarpa* (black cottonwood), and hybrids involving these taxa.

The disease occurs from Quebec and New England to the Pacific Northwest. Severe outbreaks are confined to small localities and usually to pure stands of susceptible trees, on which 80–100% of the foliage may be blighted.

Symptoms and signs. Lesions vary in shape and size and often involve whole leaves. The upper surfaces are dark brown near the margins, becoming grayish brown to ashen in older parts. The lower surface is reddish brown. Enlarging lesions follow leaf veins and thus have irregular outlines. Infections usually spread into petioles, causing any remaining green leaf tissue to die and shrivel.

Black stromata about 0.5 mm across, consisting of both fungal and leaf tissue, develop in the lesions in midsummer. In late summer these structures produce acervuli with microconidia, which are apparently noninfectious, on the upper surface. Perithecia develop in the stromata in fallen leaves and mature in late spring. Elongate multicellular ascospores dispersed from these fruit bodies are presumed responsible for starting the single annual cycle of infection. The perithecia are unusual in that their bodies are aligned in the plane of the leaf blade while their long necks bend upward at right angles and project above the surface of the stroma.

L. tetraspora is a proven pathogen; inoculations with asci plus ascospores dissected out of perithecia have resulted in leaf blight. The fungus grows slowly in pure cultures on laboratory media.

One other *Linospora* species, *L. capraea*, is recorded in North America. It attacks *Salix* in the Pacific Northwest. In Europe several *Linospora* species, including *L. capraea*, are found on *Salix*, and *L. ceuthocarpa* causes leaf blight on *Populus tremula*.
References: 233, 234, 609, 635, 1298, 2118, 2641, 2699, 2834, 3078, 4016

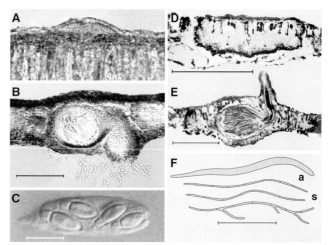

Figure 46. A–C. *Apioplagiostoma populi.* A. A blisterlike acervulus between the epidermis and cuticle of an upper leaf surface, not yet releasing conidia. B. A perithecium in an overwintered leaf. Asci from a separate specimen are posed with the perithecium. C. An ascus and ascospores. D–F. *Linospora tetraspora.* D. Section of a stroma in a leaf. E. A perithecium in an overwintered leaf. F. An ascus (a) and ascospores (s), one shown germinating. Scale bars = 100 µm, except 10 µm in C. Image A and asci in B courtesy of M. E. Ostry; section of perithecium in B adapted from reference 669 by permission of the Mycological Society of America; C adapted from reference 1198 by permission of Cambridge University Press; D–F adapted from reference 4016 by permission of NRC Research Press.

A–E. Bronze leaf disease, caused by *Apioplagiostoma populi.* A, C. Reddish brown blighted leaves of *Populus tremuloides* (trembling aspen) in contrast with normal foliage (NY, Sep). B, D. Leaf blight and twig dieback of hybrid aspen (*P. grandidentata* × *P. alba*). E. Bronze to reddish brown stain in xylem of a hybrid aspen branch, darkest at the base of a diseased lateral twig. F. Brown acervuli of the anamorph of *A. populi* on the upper surface of a lesion in a hybrid aspen leaf (NY, Aug).

G–I. Linospora leaf blight of *Populus balsamifera* (balsam poplar). G. Typical lesions with a dark brown, irregular margin and an ash gray center with stromata of *Linospora tetraspora.* H. Leaves distorted by lesions. I. Close view of stromata on an upper leaf surface (VT, Aug).

Anthracnose Overview and Black Spot of Elm (Plate 47)

Overview

Anthracnoses are diseases caused by fungi that produce conidia in blisterlike acervuli (Figures 47–56) in necrotic lesions on leaves, stems, flower parts, and fruit. The pathogens are anamorphic ascomycetes in several orders and families: mainly Gnomoniaceae and Valsaceae of the Diaporthales, Dermateaceae of the Helotiales, and Glomerellaceae, an unclassified family. The anamorphs are classified in genera such as *Asteroma, Aureobasidium, Colletotrichum, Cryptocline, Diplodina, Discella, Discosia, Discula,* and *Monostichella.* Anthracnose symptoms range from leaf spots through leaf and shoot blights to cankers and dieback of twigs and branches. Given wet weather at moderate temperatures, the pathogens cause several cycles of disease annually. Conidia, dispersed by water, start each cycle, and more conidia are produced in acervuli that form in the lesions. In warm regions on evergreen hosts, this pattern is interrupted only by dry weather or periods when plants are between growth flushes and resistant. In temperate regions, however, the pathogens overwinter as mycelium in lesions, most often on twigs, or as immature perithecia or apothecia in twig cankers or fallen leaves. New symptoms in spring may result from expansion of twig lesions initiated the previous season or from primary infections initiated by conidia or ascospores. Some anthracnose pathogens enter leaves and twigs and persist as endophytes, causing no symptoms until the plant part becomes senescent or is damaged by a second agent such as thrips or a gall-inducing insect. *Diplodina acerina* and *Aureobasidium apocryptum* on *Acer* (maple), and *Discula* species on *Fagus* (beech) and *Quercus* (oak) are examples.

References: 46, 122, 585, 635, 643, 677, 1099, 1298, 1452, 2785

Black Spot of Elm

Black spot, also called anthracnose, was a well-known disease of *Ulmus* (elms) in North America before the advent of Dutch elm disease. It is caused by the native fungus *Stegophora ulmea* (Diaporthales, Gnomoniaceae) and is responsible for severe defoliation of highly susceptible trees in wet years wherever elms grow from the Great Plains eastward. It occurs also in California and in India, China, and the Russian far east. Many *Ulmus* species, including all of those native to the USA and Canada, and also *Zelkova serrata* (Japanese zelkova), are susceptible to varying degrees.

Symptoms and signs. The usual symptoms are black spots and the shedding of diseased leaves before the normal time, but severe infection early in the season can cause leaf-and-shoot blight or complete defoliation by early August. Lesions form on petioles and succulent stems as well as leaf blades. Symptoms progress from low branches to higher ones. Green fruits of some elm species develop a crumpled appearance as the result of infection.

Foliar lesions at first appear as yellow spots about 1 mm across on upper surfaces beginning while leaves are unfolding from buds. After several days, a subcuticular acervulus forms in the center of a lesion and a dark stroma forms beneath the acervulus (Fig. 47). The stroma is visible as a black dot about 0.5 mm in diameter. Usually several acervuli and stromata develop close together and may coalesce to an irregular black mass up to 5 mm wide, surrounded by a narrow band of whitish dead tissue. Sometimes the stroma covers an entire lesion. The cuticle over each acervulus splits irregularly, exposing a white mass of conidia.

Two anamorphs develop successively in the acervuli. The first produces colorless, unicellular conidia 8–10 × 3–3.5 μm in size during spring and early summer. This state, responsible for secondary cycles and long known as *Gloeosporium ulmicolum,* lacks a proper contemporary name. In midsummer a transition to a putative spermatial form (*Asteroma*) occurs. The spermatia typically measure 4–6 × 1–2 μm. Perithecia begin to develop while spermatia are present. The beaks of immature perithecia protrude through the lower surfaces of lesions during late summer and autumn. When mature in early spring in fallen leaves, perithecia are flask shaped with bodies 200–385 μm wide and 150–230 μm deep, and beaks 80–100 μm long.

Disease cycle. In northern areas *S. ulmea* overwinters as developing perithecia and as acervuli in buds. In parts of California where some elms are evergreen, the fungus overwinters as mycelium and acervuli in foliar lesions. Perithecia expel colorless, unequally two-celled ascospores into the air in spring under conditions of alternating wetting and drying after several days with temperatures of at least 7°C. Ascospore dispersal is usually synchronized with the onset of elm leaf growth and causes infections mainly on low branches. Infections in early spring are probably also caused by conidia from diseased bud scales, accounting for some of the lesions on the youngest leaves. New acervuli with conidia mature 10–20 days after infection. Conidia, dispersed by rain and sprinkler irrigation, cause secondary cycles of disease during spring and early summer. After leaf fall, developing perithecia in fallen leaves and persistent acervuli in buds bring the cycle to completion. Foliar disease continues all year on persistent leaves of *U. parvifolia* and *U. pumila* in parts of California.

S. ulmea is adapted to activity during both cool spring weather and warm conditions of early summer. The optimum temperature for germination of its ascospores is about 8°C. In experiments, germination was reduced at temperatures above 16°C and did not occur at 24°C. Conidia, by contrast, germinated at 4–28°C. The optimum temperature for induction of disease by conidial inoculation was 16°C.

Black spot is chronically severe in areas where cool, moist weather is prolonged and (or) fog is common in spring and early summer. The disease is arrested during dry summers, and even severely blighted parts then seem to recover as the result of growth from buds that would normally remain dormant until the next spring.

Resistance to *S. ulmea* varies greatly within and between elm species and is strongly heritable. One expression of resistance is nonsporulating lesions. In Wisconsin, elms ranked in order of increasing susceptibility as follows: *U. thomasii* (cork or rock elm), *U. laciniata* (Manchurian elm), *U. parvifolia* (Chinese elm), *U. pumila* (Siberian elm), *U. minor* (smooth-leaf elm), *U. japonica* (Japanese elm), *U. americana* (American elm), *U. glabra* (Scotch elm), and *U. laevis* (European white elm). *U. parvifolia* has a dominant gene for resistance to Wisconsin strains of *S. ulmea* that apparently confers resistance to interspecific hybrids. In coastal California, however, *U. parvifolia* is considered highly susceptible.

References: 122, 229, 307, 1099, 1298, 2559, 2560, 2631, 2699, 3114, 3913, 4316

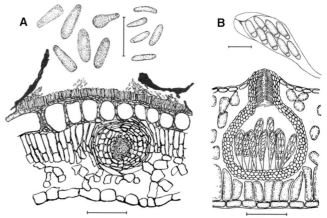

Figure 47. Stegophora ulmea. A. An acervulus after rupture of the cuticle; macro- and microconidia enlarged above. B. A perithecium opening to the abaxial surface of an overwintered leaf, and a mature ascus. Scale bars = 50 μm for fruit bodies, 10 μm for conidia and ascus. A adapted from reference 2631; B adapted from reference 40 (3rd ed.) by permission of John Wiley & Sons, Inc.

A, B. Leaf-and-shoot blight of *Ulmus parvifolia* (Chinese elm) caused by *Stegophora ulmea.* A. Overview. B. Lesions with black stromata are raised above normal leaf surface (CA, Jun).

C, F, G. Symptoms and signs of black spot, caused by *S. ulmea,* on *U. laevis* (European white elm). C. Leaf spots accompanied by general necrosis where lesions are most numerous. F. Magnified view of raised lesions, each with a small but prominent black stroma. G. Stromata with whitish borders remain prominent on killed parts of leaves (NY, Aug).

D, E. Black spot on *U. americana* (American elm). D. A senescent leaf with unusually large black stromata. E. Magnified view of stromata on the leaf shown in D (NY, Sep).

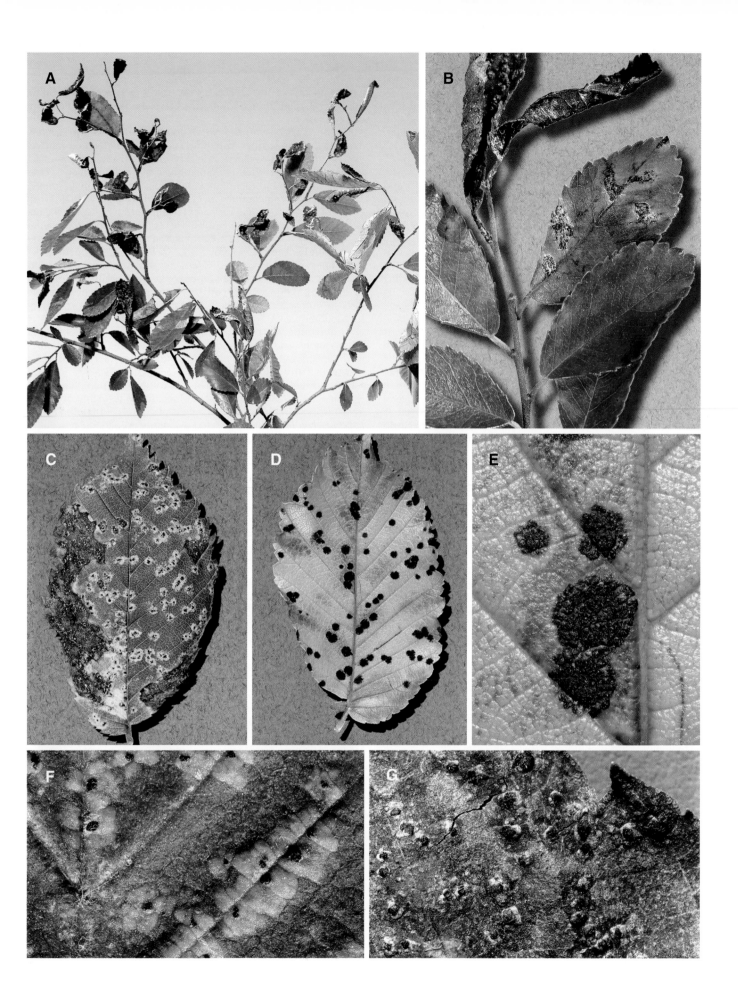

Ash Anthracnose (Plate 48)

Ash anthracnose is common across Canada and the USA wherever *Fraxinus* species (ash) grow in relatively cool climates. It is seen occasionally in warmer areas, having been noted in Arizona and Louisiana. The pathogen is *Gnomoniella fraxini* (Diaporthales, Gnomoniaceae). Its conidial state is *Discula fraxinea*, which was formerly included in the *Discula umbrinella* complex. Hosts of *G. fraxini* include *Fraxinus americana* (white ash), *F. nigra* (black ash), *F. pennsylvanica* (green or red ash), *F. quadrangulata* (blue ash), and *F. velutina* (velvet ash). *F. velutina* var. *glabra* (Arizona or Modesto ash) is highly susceptible; anthracnose is its most important disease. *F. pennsylvanica* is sometimes severely affected in the northern Great Plains and northeastern USA, although it appears relatively resistant in California. *Chionanthus retusis* (Chinese fringe tree) was found with leaf blight caused by *G. fraxini* in Delaware.

Symptoms and signs. Lesions appear in spring on expanding shoots and leaves as water-soaked spots, enlarging and coalescing rapidly and becoming greenish brown. Infection at this growth stage of the tree often kills young leaves and shoots. Lesions that girdle leaf rachises or the bases of young shoots cause the distal parts to droop and shrivel. In years of prolonged wet spring weather with moderate temperatures, the leaf-blight and shoot-blight phases of anthracnose may kill almost the entire first flush of shoots on highly susceptible trees (Plate 48A). This damage is similar to frost injury in timing and appearance. Severe defoliation by anthracnose in several successive years may lead to dieback, as has been noted on *F. velutina* var. *glabra* in California and *F. pennsylvanica* in North Dakota. Usually, however, the defoliation is restricted to low branches, and these produce new leaves by midsummer.

As leaves attain full size, they become somewhat resistant to infection, lesion expansion is arrested, and lesions bleach to tan. Mature diseased leaves typically have tan necrotic blotches on distorted leaflets. The distortion is caused by mechanical stress in growing healthy tissue surrounding lesions. Lesions on expanded leaves are often associated with insect feeding injuries, notably those made by *Tropidosteptes* species (ash plant bugs), an indication that growth into damaged tissue permits the pathogen to evade the resistance presented by intact leaves. This idea is consistent with results of laboratory assays in which lesions formed around heat-damaged but not intact inoculation sites on leaf discs. Ash stems also become resistant as they become woody, and lesions on nongirdled twigs remain as tiny elliptical cankers. Ash samaras (winged fruit) also become infected, but symptoms on samaras are not often noticed.

Acervuli form in lesions a few days after infection and produce masses of conidia. The acervuli, elliptic when on veins, petioles, and twigs, but disc-shaped when on interveinal leaf tissue, are initially the same color as the substrate. They are generally 50–250 μm in diameter and can be discerned with a magnifying lens (Plate 48F). With age they become somewhat darker and more prominent and resemble those of *Discula quercina* (Plate 49G, H). Acervuli with viable spores can be found on dead petioles for up to a year after infection. The conidia are one celled, colorless, narrow-elliptic, 5–12 × 2.1–5.6 μm, and dull white to faintly pinkish in mass. In the northern Great Plains, perithecia mature in formerly diseased leaves on the ground in spring. Information about perithecia in other regions is lacking. The perithecia are shaped like long-necked flasks, with their globular bodies embedded in leaf tissue and the necks protruding (Fig. 48).

Disease cycle. The anthracnose pathogen overwinters in twigs, petioles, leaf veins, and samaras, and in spring liberates conidia from acervuli on these structures. Petioles colonized by *G. fraxini* and bearing acervuli often persist through winter on diseased *F. pennsylvanica* in the northern Great Plains and on *F. velutina* in California. These petioles provide a major source of inoculum close to the buds from which susceptible shoots emerge in spring. Conidia, dispersed by water, cause primary infections on succulent parts. Conidia from fallen plant parts cause infections on low foliage. Ascospores, dispersed in air from perithecia in leaves on the ground, presumably also cause primary infections, but their importance in the disease cycle is unknown. Conidia cause repeating cycles of disease if cool, wet weather retards tree growth and prolongs the period of susceptibility. New infections and

expansion of existing lesions are arrested when hot, dry weather follows initial infection. Prolonged wet weather favors development of large lesions.

Female *F. pennsylvanica* were observed to be more severely affected than male trees in North Dakota. This difference, not yet explained through research, could be due to overwintering of *G. fraxinea* in colonized fruit hanging on the trees, which would provide a source of primary inoculum not present on male trees.

Studies of temperature effects on conidial germination of Californian isolates of *G. fraxini* and leaf infection by North Dakota isolates have given similar results. The conidia germinated well at 15–27°C but poorly at 9°C. Germination began 8–12 hours after wetting and was most rapid at 24–27°C. Lesions developed at 15–25°C but not at 10° or 30°C. Thus, the relatively cool conditions that favor anthracnose do not promote the most rapid development of the causal fungus. Instead, cool temperatures may promote disease primarily by retarding host development and prolonging susceptibility.

Fraxinus species differ in their susceptibility to anthracnose. Observations in Illinois indicated the following susceptibilities in diminishing order: *F. chinensis* (Chinese ash) > *F. angustifolia* (narrowleaf ash) > *F. pennsylvanica* > *F. mandshurica* (Manchurian ash) = *F. excelsior* (European ash) > *F. tomentosa* (pumpkin ash) = *F. americana*. *F. quadrangulata* (blue ash) was highly resistant. Intraspecific variability in resistance is also likely. In North Dakota *F. pennsylvanica* severely affected by anthracnose tends to occur in groups (disease foci), and the individuals most and least damaged in any year are again the most and least damaged in later years. Differences in disease severity among *F. pennsylvanica* cultivars have been observed in Illinois.

References: 122, 677, 1099, 1298, 1364, 1893, 2603, 2895, 3244, 3913

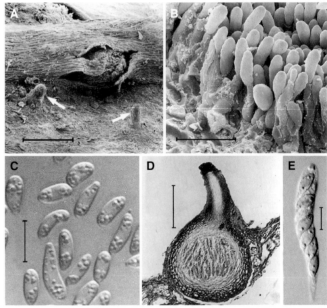

Figure 48. Gnomoniella fraxini. A, B. Scanning electron micrographs. A. An acervulus of the anamorph, *Discula fraxinea*, on a vein on the abaxial surface of an overwintered leaf of *Fraxinus pennsylvanica* (green ash). Beaks of two perithecia (arrows) are protruding from the surface in the foreground. B. Conidia in an acervulus. C–E. Light micrographs. C. Conidia. D. Section of a perithecium. E. Ascospores in a mature ascus. Scale bars in A–E = 200, 10, 10, 100, and 10 μm, respectively. All from reference 3244 courtesy of S. C. Redlin with permission of Mycotaxon, Ltd.

A–C. Anthracnose on *Fraxinus velutina* var. *glabra* (Modesto ash). A. Defoliation from an early-season attack. B. Pinpoint lesions and spreading necrosis. C. Mature lesions (CA, Apr).

D–F. Anthracnose on *F. americana* (white ash). D, E. Leaf deformity associated with lesions (NY, May–Jun). F. Inconspicuous acervuli (arrows) with masses of conidia on the lower surface of a young lesion (NY, May).

99

Oak Anthracnose (Plate 49)

Anthracnose diseases of *Quercus* species (oaks) occur throughout their range in the eastern half of North America, in Pacific Coast states and British Columbia, and in Europe. More than 20 oak species in North America are reported hosts of anthracnose fungi, and this number probably underestimates the true host range. Anthracnose is innocuous in most years, but during outbreaks on highly susceptible trees it may kill nearly all the foliage and many twigs.

The disease in North America and southern Europe is caused most often by *Apiognomonia quercina* (Diaporthales, Gnomoniaceae), a fungus that colonizes primarily *Quercus*, in which it is a ubiquitous endophyte. Its conidial state is *Discula quercina* (syn. *D. umbrinella*). Additional fungi reported associated with oak anthracnoses include *A. errabunda*, *A. veneta* (Plate 50), and various *Discula* strains. Isolates of *D. quercina* from oaks have, in limited tests, infected several oak species, and particular isolates infected *Juglans nigra* (black walnut) and *Platanus occidentalis* (sycamore) as well.

Symptoms and signs. Oak species, growth stage at the time of infection, and weather influence symptoms. *Q. alba*, a highly susceptible species, exhibits three types of symptoms: leaf-and-shoot blight characterized by death and shriveling of young leaves (Plate 49A), irregular dead areas on distorted leaves that otherwise remain green (Plate 49C), and necrotic spots on mature leaves. Enlarging lesions on leaves tend to follow the veins or midrib and often kill tissue on only one side of a rib or vein, thus causing distortion if the leaf is still growing. After drying, lesions have a papery texture and turn tan to nearly white before weathering to grayish white. The edges of lesions are marked by an abrupt transition from brown to green tissue. Leaves approaching full size become resistant, which accounts for the small lesion size on mature leaves. Large irregular lesions and blight are infrequent on most other oaks. The usual foliar symptoms on *Q. velutina* (black oak) are brown spots up to 2 cm in diameter. Leaf tips on *Q. virginiana* (live oak) are frequently killed.

Twig infections and dieback associated with anthracnose fungi are recorded for several oak species in eastern USA and California. Twig blight occurring before buds open in spring is inconspicuous. Midsummer twig blight has occurred in California on *Q. agrifolia*, *Q. lobata*, and *Q. wislizenii* (California live, valley, and interior live oaks, respectively).

In Mississippi a fungus similar to the *Discula* state of *A. quercina* has caused epidemics of necrotic spots on leaves of several oak species. The spots vary from pinpoint size to 2–3 mm in diameter. They are dark brown to black on *Q. phellos* (willow oak) and *Q. laurifolia* (laurel oak) and brown with yellow halos on *Q. pagoda* (cherrybark oak). When very numerous the spots cause leaf tips to wither and become ragged.

Acervuli of *D. quercina* form on lower surfaces of foliar lesions and on killed twigs (Plate 49E–H; Fig. 49). Those on leaves appear as raised brown flecks, circular to elliptic in outline, often about 250 μm in diameter, and are more common on or adjacent to major veins than in interveinal areas. Acervuli on twigs appear as orange-brown pustules that later darken and may resemble pycnidia. The colorless, unicellular conidia are mostly 9–15 × 4–7 μm. When growing in culture, the fungus also produces microconidia that average about half this size. Colonies appear pale yellow when viewed through the bottom of a culture plate.

A microconidial (spermatial) state of *A. quercina*, named *Gloeosporidina moravica*, forms in fallen leaves during autumn and winter in some areas and is succeeded by the perithecial state. Perithecia mature in spring. They are black, 130–440 μm in diameter, and have beaks at least as long as the perithecium is wide. The ascospores are colorless and unequally two celled, 13–15 × 3–4.5 μm in size.

Disease cycle. The *Discula* state of *A. quercina* is a year-round inhabitant of oak buds, twigs, and leaves. Foliar infections in spring are presumed to be initiated mainly by conidia from acervuli on dead twigs, but overwintered mycelium in buds may also play a role. In addition, ascospores, shot into the air from perithecia in fallen leaves, are infectious. Infection and colonization are believed to proceed as described for the conidial state of *Apiognomonia errabunda* on *Fagus sylvatica* (European beech). Conidia adhere to leaves and produce germ tubes or appressoria from which the epidermis is penetrated by fine hyphae. Many infections become localized beneath the cuticle or epidermis and remain latent, but if colonization proceeds, hyphae grow inter- and intracellularly and cause lesions. Acervuli in lesions on oak leaves and young stems produce conidia capable of causing secondary infections. Under optimum weather conditions, conidial production is abundant within 2 weeks after infection. The mode of fungal entry into stems has not been studied. Direct penetration, wound colonization, and hyphal growth down petioles are possibilities.

Acervular fungi identified as *D. quercina* or *D. umbrinella* are widespread in asymptomatic leaves, twigs, and acorns of various oaks in North America and Europe. These fungi are considered by some to be harmless endophytes, but the weight of evidence indicates they have pathogenic capability that is expressed in immature tissues or those altered by insect feeding or stressed by some other types of injuries. For example, a *Discula* that is common in healthy-appearing leaves kills certain insect-induced leaf galls on *Q. garryana* (Oregon oak) and *Q. agrifolia* (California live oak) in western USA. In the Northeast, anthracnose outbreaks on oaks infested by *Asterolecanium variolosum*, the golden oak scale, seem to trigger more dieback than would be caused by either agent alone.

Anthracnose outbreaks are promoted by rainy weather in spring and subside before midsummer, but succulent shoots such as those developing on trees defoliated by insects may be attacked any time a rainy period occurs during the growing season. Infection occurs over a wide temperature range if plant surfaces remain wet for several hours. Symptoms develop rapidly at 20–28°C, but lower temperatures (16–20°C) promote the greatest eventual severity of symptoms, presumably by prolonging the host's period of high susceptibility.

Oak species and individual trees vary in susceptibility to anthracnose, but systematic observations or selections for resistance have not been reported. In Italy inoculated seedlings of *Q. cerris* (European turkey oak) proved more susceptible than *Q. pubescens* (pubescent oak) or *Q. robur* (English oak).

References: 122, 229, 234, 587, 600, 677, 1099, 1298, 1430, 1584, 2213, 2699, 2811, 2970, 3199–3203, 3498, 3913, 4068, 4379, 4380

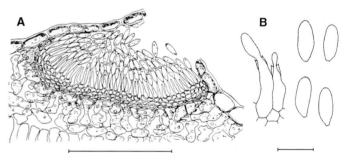

Figure 49. Discula umbrinella, which is indistinguishable from *D. quercina*, the conidial state of *Apiognomonia quercina*. A. An acervulus. B. Conidiophores with phialides and conidia. Scale bars = 100 μm and 10 μm, respectively. A adapted from reference 122 by permission of Gebrüder Borntraeger Verlagsbuchhandlung; B from reference 3913 by permission of CAB International.

A–C. Anthracnose on *Quercus alba* (white oak). A. Leaf blight; surviving parts of affected leaves are distorted. B. Dieback and tufted sprout growth in the crown of a tree perennially damaged by anthracnose and *Asterolecanium variolosum* (golden oak scale). C. Foliar lesions ranging from spots a few millimeters across to large irregular dead areas (NY, Jun).

D–F. Anthracnose on *Q. rubra* (northern red oak). D. Leaves deformed by lesions that developed during leaf expansion (WV, Jul). E. Twig cankers at nodes, with pimplelike bulges where acervuli of the *Discula* state of *A. quercina* are forming (NY, Apr). F. Close view of young acervuli on a killed twig (NY, May).

G, H. Successively magnified views of acervuli (brown dots) of the *Discula* state of *Apiognomonia quercina* on and beside leaf veins on the lower surface of an anthracnose lesion on *Q. alba* (NY, Jun).

Sycamore Anthracnose (Plate 50)

Anthracnose of *Platanus* species (planetree, sycamore) is characterized by leaf-and-shoot blight, twig cankers and dieback, and branch deformity. It occurs in North and South America, Europe, South Africa, Australia, and New Zealand. Trees that repeatedly sustain severe damage by anthracnose are weakened, as evidenced by loss of vigor, dieback, and apparent increased susceptibility to borers. They also become severely deformed due to repeated loss of buds and twigs. The pathogen, *Apiognomonia veneta* (Diaporthales, Gnomoniaceae), closely resembles the oak anthracnose pathogen (Fig. 49), but in cross-inoculation tests it has shown preference for *Platanus*. It attacks *P.* ×*acerifolia* (London plane, syn. *P. hybrida*), *P. occidentalis* (eastern sycamore), *P. orientalis* (Oriental plane), *P. racemosa* (California sycamore), and *P. wrightii* (Arizona sycamore).

Symptoms and signs. Sycamore anthracnose has three phases corresponding to the plant parts attacked. Listed in order of appearance in the season they are canker formation (including bud and twig mortality), shoot blight, and leaf blight. Year-to-year weather variation influences the occurrence and severity of each phase relative to the others. Cankers form and buds and twigs die during the dormant period. Many twig lesions are restricted to the immediate vicinity of buds, but others girdle twigs. Damage increases as trees age. Cankers at twig bases may girdle stems 2 or more years old. In years of severe bud and twig disease more than 95% of the buds may be killed, and trees so affected stand leafless until summer except for tufts of foliage at branch tips. Foliation begins by midsummer, however, as shoots grow from buds that would have remained dormant. Leaves on these late-growing shoots usually escape significant damage. Most anthracnose cankers enlarge during only one dormant season, but some expand during two or more seasons. In these cases the tree produces callus and woundwood at lesion edges during the growing season. During the ensuing dormant season the pathogen kills these tissues plus some surrounding bark and cambium. Woundwood eventually covers lesions that do not girdle stems. Elongate anthracnose cankers form occasionally on the mainstems of young, highly susceptible trees, especially those stressed by drought or untimely pruning.

Shoot blight involves sudden death of expanding shoots and leaves and resembles injury by late spring frost. It occurs when the pathogen either kills twigs on which shoot growth has begun or enters succulent new shoots soon after they emerge from buds. Shoot blight tends to develop during or immediately after a period of cold spring weather. Leaf blight, initiated by spores, is most severe on low branches and intensifies in wet seasons, causing premature leaf drop. Foliar lesions characteristically extend along the veins, although irregular marginal lesions develop occasionally.

Fruit bodies of *A. veneta* develop on twigs and leaves. The conidial state, *Discula platani*, varies in form and appearance according to the substrate and time of year. In early spring, black pycnidia develop in the bark of new cankers and newly killed twigs. The pycnidia are cup-shaped or nearly covered, walled structures 500–900 µm in diameter, easily visible to the unaided eye, that arise beneath the cork layer of the outer bark. They are preceded by and develop beneath small masses of fungal tissue (pressure cushions) that rupture the bark surface. Conidiomata similar to those on twigs are found on overwintered leaves on the ground. Acervuli, similar to those of *Discula quercina* (Fig. 49), form during the growing season on the lower surfaces of leaf lesions, especially along veins. The acervuli are intraepidermal, brownish, and 100–300 µm in diameter. All types of conidiomata produce colorless, unicellular conidia, mostly 8–15 × 4–6 µm. Gray to black spermagonia with spermatia (the *Gloeosporidina* state of *A. veneta*) form in fallen leaves during the dormant season and are succeeded there by perithecia. Perithecia are mostly 150–300 µm in diameter with beaks protruding 50–100 µm from the leaves. They produce unequally two-celled, colorless ascospores measuring usually 13–15 × 3–4.5 µm.

Disease cycle. *A. veneta* overwinters as mycelium and fruit bodies in lesions at nodes on twigs, in cankers, and in fallen leaves. Perithecia mature during late winter and expel ascospores into the air in spring. These spores are presumed to cause infections on young leaves and shoots, but the principal damage is due to activity of the fungus in overwintered lesions. Prolonged cold weather in spring delays tree growth while allowing *A. veneta* to further colonize and kill dormant bark and buds. Conidia from pycnidia in cankers and killed twigs are dispersed by water beginning about the time sycamore buds open, and they start most leaf and shoot infections. Acervuli soon form in the new lesions and produce more conidia that cause secondary infections. Germinating conidia produce appressoria from which young leaves, and presumably tender young stems, are penetrated directly. The fungus may grow beneath the cuticle for a time or remain quiescent before invading tissues beneath. Mature leaves are also penetrated, but unless wounded they resist colonization until they become senescent. *A. veneta* enters twigs via hyphal growth from petioles during the growing season and then remains more or less quiescent until host dormancy, restrained by host defenses in bark during the growing season. Buds are often killed because their location immediately above leaf scars is very close to the pathogen's overwintering sites.

Environmental determinants of anthracnose severity are known from field observations and experiments. The canker phase, including bud and twig mortality, and shoot blight are governed primarily by temperature. Mild weather during host dormancy promotes fungal activity and lesion formation in woody parts. Temperatures averaging below 12–13°C for the first 2 weeks after buds open prolong twig susceptibility and are conducive to severe shoot blight. In contrast, temperatures above 15–16°C favor quick shoot growth and cessation of twig killing. Wet conditions are important mainly for leaf blight. Given young wet leaves, spore germination and foliar infection are highly favored at 16–23°C. Conidia may germinate within as short a time as 6 hours at temperatures near 20°C. Wounds made by insects with piercing-sucking mouthparts seem to promote foliar infection. For example, anthracnose is exacerbated on trees infested by *Corythucha ciliata* (sycamore lace bug) .

Platanus species and cultivars vary in susceptibility. Severe disease develops on *P. occidentalis* and *P. racemosa*, while *P. orientalis* is resistant. Trees identified as *P.* ×*acerifolia* (= *P. occidentalis* × *P. orientalis*) vary in susceptibility because many of them are actually complex hybrids that contain differing proportions of genes from the two species. *P. orientalis* gives significant resistance to its progeny when crossed with susceptible species, but unfortunately the hybrids are readily pollinated by susceptible sycamores, and resistance in seedling offspring is diluted. Resistance is maintained in clonal lines, however. Examples are the 'Bloodgood' clone of London plane and the hybrid clones 'Columbia' and 'Liberty.'

References: 64, 122, 229, 677, 1033, 1268, 1664, 2699, 2802, 2809, 2810, 3398–3400, 3498, 3533, 3905, 3925

Anthracnose on *Platanus occidentalis* (eastern sycamore) and *P.* ×*acerifolia* (London plane).

A. A large sycamore with severe twig blight, nearly devoid of foliage at the time leaves would normally be approaching full size (NY, Jun).

B. A sycamore branch, crooked because of repeated twig blight (NY, Jun).

C, D. Twigs killed by anthracnose, one before and the other after shoot growth; also scorch-type leaf blight. Dark brown dots on twig in D are old acervuli of the *Discula* state of *Apiognomonia veneta* that discharged spores at the time nearby buds opened (NY, Aug).

E. Twig blight in close view: brown tissue in a girdling canker at a node and young shoots killed by a separate infection at a lower node (NY, Jun).

F. A 2-year-old canker on a small branch. Healthy woundwood beginning to cover the lesion indicates that the pathogen is no longer active (NY, Aug).

G. An anthracnose canker on the main stem of a young *P.* ×*acerifolia* (London planetree). Bark cracks at canker edges are caused by woundwood formation (NY, Aug).

H. Shoot blight on *P.* ×*acerifolia* (NY, Jun).

I. Vein-associated leaf blight lesions on *P. occidentalis* (NY, Jul).

103

Walnut Anthracnose and Gnomonia Leaf Spots of Hickory and Pecan (Plate 51)

Walnut Anthracnose

Walnut anthracnose occurs worldwide where *Juglans* species grow in humid climates. It is common throughout eastern USA and adjacent Canada and also in the Pacific Northwest. Its main effects are defoliation and diminished nut quality. Anthracnose builds up during wet seasons, causing highly susceptible trees to be nearly leafless by late July or early August. Nuts from severely defoliated trees may have dark, somewhat shriveled kernels. Anthracnose has little effect on tree growth, however, unless trees are defoliated repeatedly, because annual growth is nearly complete by the time diseased leaves senesce. The pathogen is *Gnomonia leptostyla* (Diaporthales, Gnomoniaceae). Its conidial anamorph has been called *Marssoniella juglandis* (syn. *Marssonina juglandis*). Both of these names are inappropriate, and the anamorph awaits a suitable name.

Trees affected in North America include *J. cinerea* (butternut), *J. nigra* (black walnut), *J. californica* (southern California walnut), *J. hindsii* (Hinds or northern California walnut), *J. regia* (English or Persian walnut), and assorted hybrids. *J. nigra*, although not the most susceptible species in comparative tests, is the most severely affected under natural conditions. *J. major* (Arizona walnut) and *J. hindsii* are more susceptible than *J. nigra*, while *J. ailanthifolia* (heartnut, Japanese walnut), *J. cinerea*, *J. microcarpa* (little walnut), and *J. regia* are less susceptible. Cultivars of *J. regia* range from highly susceptible to quite resistant. In California, the principal area of commercial walnut production in North America, *J. regia* escapes significant infection because of hot, dry summer weather.

Symptoms and signs. Symptoms on leaves are more or less circular brown lesions that appear first on the undersurfaces about the time leaves reach full size. Lesions soon become equally apparent on both sides. They range in size from pinpoints to about 5 mm in diameter and are typically surrounded by a yellow zone. When numerous they cause yellowing, marginal browning, curling, and casting of leaflets. Defoliation, although a valid measure of anthracnose severity, is not closely correlated with the number of lesions per leaflet. Lesions also form on petioles, rachises, fruit, and occasionally on green twigs. The lesions on fruit are tiny dark depressions that if numerous cause premature fruit drop. Lesions on twigs are circular to oval, gray-brown spots up to 2 mm across, with dark margins.

Brown acervuli of the conidial anamorph form in lesions, particularly along leaf veins. Viewed when moist, the acervuli are tiny blisters 100–200 μm in diameter. They form beneath the epidermis and produce crescent-shaped, colorless, two-celled conidia 14–30 × 3–4 μm in size. Conidia are dispersed mainly by splashing and windblown water. Beginning in late summer as night temperatures drop below about 10°C, microconidia may be found in some acervuli. These cells are rod shaped, 6–12 × 1 μm in size, and noninfectious. They function as spermatia in the sexual cycle of the fungus. Black flask-shaped perithecia form in fallen leaflets, rachises, and fruit husks during host dormancy. They are mostly 200–350 μm in diameter with beaks (necks) 200–400 μm long. They produce two-celled, colorless, cylindric ascospores measuring 17–25 × 2.5–4.5 μm.

Disease cycle. The pathogen overwinters as developing perithecia and, in some regions, as mycelium and acervuli in twig lesions. Perithecia mature in spring and expel ascospores in response to rain during a period of several weeks. Airborne ascospores cause most primary infections. Water-dispersed conidia from twig lesions may also initiate primary infections.

Wet plant surfaces are required for spore germination and infection. After germination the parasite produces appressoria from which penetration pegs breach intact leaf or fruit surfaces. Increases in lesion numbers can be perceived 15–17 days after periods when leaves are wet for 12 hours or longer, and acervuli develop beginning 3 days after lesions appear. Thereafter, successive cycles of infection are caused by conidia.

Environmental conditions that favor anthracnose are relatively well understood. Perithecial formation is most favored at temperatures of 7–10°C, and ascospore production at about 10°C. Expulsion of ascospores begins about 1 hour after the onset of rain and reaches a peak after 5–6 hours. Conidia are dispersed whenever precipitation is sufficient for water to run or splash on plant surfaces. Temperatures of 18–26°C are most favorable for production of acervuli and conidia, and both ascospores and conidia germinate most rapidly at 24–26°C. Susceptibility of *J. nigra* to anthracnose is influenced by tree nutrition. Application of nitrogenous fertilizers to the root zones of trees has resulted in more growth and less severe disease than occurred in untreated trees.

References: 122, 229, 343, 385, 386, 750, 1099, 1108, 2507, 2699, 2804, 3329, 4053

Gnomonia Leaf Spots of Hickory and Pecan

Hickories throughout eastern USA and southern Ontario are host to *Gnomonia caryae* (Diaporthales, Gnomoniaceae), which causes brown lesions with yellowish indefinite edges. The fungus has also been reported from Mexico. Susceptible species include *Carya cordiformis* (bitternut hickory), *C. glabra* (pignut hickory), *C. illinoinensis* (pecan), *C. ovata* (shagbark hickory), and *C. tomentosa* (mockernut hickory). Individual lesions are nearly circular and 5–20 mm in diameter. Where confluent they cause browning and curling of large areas of leaflets, premature yellowing, and leaf cast. A conidial (spermatial?) state, *Asteroma caryae*, becomes prominent on the lower surface of each lesion. Acervuli 70–150 μm in diameter, at first pale pink and later brown, develop in abundance. The conidia are unicellular, colorless, and 7–10 × 1–2 μm in size. No test of their pathogenicity has been reported.

Perithecia of *G. caryae* form in fallen leaves during winter. Ascospores presumably function as primary inoculum, but this disease cycle has not been studied. Symptoms are invariably delayed until late summer and early autumn. One possible explanation is latent infection. Young leaves perhaps become infected in spring but resist colonization and fail to show symptoms until late summer.

C. illinoinensis in the Gulf Coast states is subject to two or three additional Gnomonia leaf spots, depending on the authority one follows. *G. pecanae* causes brown lesions bounded by major lateral veins in leaflets (liver spot), and *G. nerviseda* causes elongate lesions along the veins (vein spot). The latter fungus also occurs on pecan in Illinois. The conidial states of both fungi are similar to that of *G. caryae* and appear late in the season. *G. dispora* causes necrotic spots on pecan leaves in Florida and Georgia. It is unusual, inasmuch as its perithecia mature in leaves on the tree during late summer and have only two ascospores per ascus. This fungus has no known conidial state.

References: 229, 677, 765, 766, 932, 1099, 2699, 3392, 3913, 4410

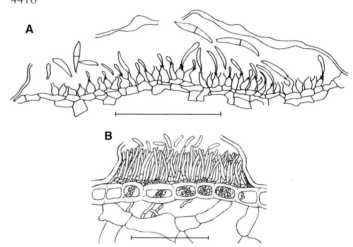

Figure 51. Subcuticular acervuli of *Gnomonia leptostyla* (A) and *G. caryae* (B) on leaves of *Juglans* and *Carya*, respectively. Scale bars = 50 μm. A adapted from reference 3329 by permission of NRC Research Press; B adapted from reference 4410.

A–E. Walnut anthracnose, caused by *Gnomonia leptostyla*, on *Juglans nigra* (black walnut). A. Yellowing, browning, and curling of diseased leaflets (NY, Sep). B. Close view of a leaflet with numerous small lesions that have caused marginal browning (NY, Sep). C. Leaflets with larger lesions and disease-induced yellowing (KY, Aug). D. Close view of one lesion; dark dots are acervuli (NY, Sep). E. Magnified view of blisterlike acervuli (IL, Jul).

F, G. Leaf spot of *Carya ovata* (shagbark hickory) caused by *Gnomonia caryae*. F. Typical lesions. G. Magnified view of brown acervuli of the conidial state of the pathogen on the lower surface of a lesion (NY, Sep).

Photo credits: C—K. J. Kessler; E—J. A. Matteoni

Dogwood Anthracnose (Plate 52)

Dogwood anthracnose, unknown before the late 1970s, has devastated populations of *Cornus florida* (flowering dogwood) in parts of eastern USA and caused prominent damage to *C. nuttallii* (Pacific dogwood) in the Pacific Northwest. It is caused by *Discula destructiva*, an acervular coelomycete. DNA analyses of this fungus have revealed that it is related to *Gnomonia* (Diaporthales, Gnomoniaceae) and has little genetic diversity, the latter fact indicating that a small founding population was probably introduced from abroad. The origin and the sexual state of the fungus, if any, are unknown.

In the East the malady was first noticed near New York City, but within 20 years it had spread as far as Alabama, Kansas, Michigan, and Maine. In the Northwest it apparently spread from western Washington to Oregon and northern California, Idaho, and southern British Columbia. Disease incidence and severity increased rapidly in areas with cool, moist conditions in spring and early summer. Mortality approached 100% on many sites in southern New England and New York and in the Appalachian region, but dogwoods growing in drier, warmer habitats sustained less damage and little mortality.

C. florida and *C. nuttallii* are the only species severely damaged under natural conditions. *C. kousa* (Korean or kousa dogwood) usually sustains only minor foliar lesions. Species found slightly susceptible when tested under conditions highly conducive to infection include *C. alba* (Tatarian dogwood), *C. controversa* (giant dogwood), *C. kousa* var. *chinensis* (Chinese dogwood), and *C. sericea* (red-osier dogwood).

Symptoms and signs. Anthracnose affects all aboveground parts of *C. florida* and *C. nuttallii*. In typical order, leaves and tender stems (also bracts and fruit), twigs and small branches, epicormic sprouts, and finally scaffold limbs and the trunk may be attacked. Leaves of trees on moist, shady sites develop necrotic spots with chlorotic halos or necrotic blotches at leaf tips and edges, or whole leaves die. Leaves exposed to sun typically develop purple-rimmed lesions a few millimeters across, some later becoming shot holes. Living diseased leaves sometimes drop prematurely. Leaves killed rapidly often shrivel and remain attached until the next spring. Diseased bracts have reddish purple spots or brown blotches.

Twig dieback on dogwoods in the open or at woodland edges tends to progress upward from low branches. Woodland understory trees tend to be uniformly affected. Twig lesions result either from direct penetration of tender stems in spring or from growth of the pathogen through petioles into stems later in the year. The former process leads to small lesions that are limited by host defenses. The latter process leads to twig death, which occurs mainly during the dormant season. Twig dieback stimulates bud release along branches and trunks of *C. florida*, resulting in growth of epicormic shoots that also become diseased. Elliptic annual cankers involving bark and cambium form on branches and trunks around the bases of killed twigs and epicormic sprouts. Cankers may become noticeable during the next growing season if sufficient callus and woundwood form around them to cause the dead bark to appear sunken. When numerous, cankers coalesce and girdle stems. Eventually the dead bark cracks and weathers away, exposing discolored wood. Severely affected trees may die within a few years, but those growing in locations less favorable for the disease survive with symptoms that vary in severity from year to year. Surviving trees eventually enclose cankers beneath woundwood.

D. destructiva produces subcuticular acervuli on leaves (mainly on undersurfaces) and twigs. Those on leaves develop as swellings beneath trichomes (a diagnostic feature; Fig. 52). They are round in outline and up to 135 µm wide, while those on twigs are elliptic and up to 340 µm wide. Single-celled conidia in a mucilaginous matrix rupture the cuticle and exude in a pallid to pale pinkish mass. The conidia are narrow-elliptic with one slightly blunt end, and measure 10–18 × 2.5–4 µm. The fungus grows readily but slowly in culture. Colonies are initially white, later gray, and finally black as viewed through the bottom of the culture dish.

Disease cycle. *D. destructiva* overwinters as mycelium and acervuli in twigs and adhering blighted leaves. Conidia from these sources, dispersed by rain and wind, initiate infection on succulent parts in spring. Acervuli with a new generation of conidia soon form in the lesions, and these conidia cause secondary infections on leaves as conditions permit. Conidial germ tubes penetrate leaves

and tender stems directly. Further growth and host colonization seem to depend on hyphal secretion of toxins and/or enzymes, because necrosis of palisade and spongy parenchyma cells precedes hyphal advance. Phytotoxic phenols as well as cellulase, pectinases, and hemicellulases are produced by the fungus in culture.

Conidial germination and hyphal growth of *D. destructiva* are most rapid at 21–24°C. In one study the fungus failed to grow at 27°C. Somewhat lower temperatures, 18–21°C, have been correlated with severe disease outbreaks in the field.

Overland dispersal of *D. destructiva* occurs in several ways. Conidia are transported by insects and possibly birds and on wind-carried water droplets during storms. Many arthropods have been found carrying the conidia, and the ability of arthropods to act as vectors has been demonstrated using *Hippodamia convergens* (convergent lady beetle). Long-distance spread beyond the Appalachian region has in some instances been linked to shipment of nursery stock.

Shade and prior drought predispose dogwoods to more severe anthracnose than develops on open-grown, well-watered trees. Disease severity is inversely related to light intensity and the potential for leaf drying as related to sky exposure. Topography, amount of light, and moisture relations of dogwood habitats also influence carbohydrate reserves and thus disease impact.

C. florida and *C. nuttallii* can be grown in the presence of the anthracnose fungus provided moist, shady sites are avoided. When irrigation is needed, water can be applied at ground level rather than by sprinkler. Anthracnose-resistant dogwoods may be useful in landscapes where the disease has been severe. *C. mas* is highly resistant or immune. Resistance is also available in *C. kousa* 'Steeple' and 'Milky Way'; *C. kousa* × *florida* hybrid clones such as 'Celestial,' 'Stardust,' and 'Stellar Pink'; and *C. florida* 'Appalachian Spring.'

References: 515, 535, 599, 600, 643, 645, 677, 713, 889, 890, 1073, 1732, 3219, 3241, 3242, 3582, 3739, 3743, 4080, 4512

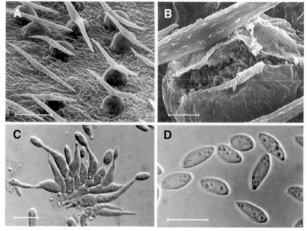

Figure 52. Discula destructiva. A, B. Acervuli that arise beneath trichomes, with conidia visible in B. C. Conidiophores with young conidia. D. Mature conidia. Scale bars = 100, 20, 10, and 10 µm, respectively. From references 3241 and 3242 courtesy of S. C. Redlin, with permission of the Mycological Society of America.

Dogwood anthracnose on *Cornus florida* (flowering dogwood).

A, B. Disease-induced dieback of low branches of trees at a forest edge (NY, Jun) and on one tree in closer view. Dying branches on the tree in B have yellowish red foliage (NY, Aug).

C, D. Foliar lesions (NY, Jun).

E. Cankers on a young main stem at the bases of twigs killed by anthracnose (NY, Apr).

F. Withered foliage clinging to twigs that were diseased in the previous year (NY, May).

G, H. Epicormic sprouts already diseased or dead on low branches of trees with disease-induced dieback on higher limbs (NY, Aug).

I. Dead leaves clinging to diseased branches in late autumn (NY, Nov).

J, K. Magnified views of acervuli of *Discula destructiva*, some with conidial cirrhi (spore tendrils), on killed twigs (NY, Jun).

Photo credits: A, C–F—M. L. Daughtrey; J, K—C. R. Hibben

Anthracnoses of Birch, Filbert, and Redbud (Plate 53)

Birch Anthracnoses

Several anthracnose fungi affect *Betula* (birch) in North America and Europe. Foliar diseases caused by the coelomycetes *Discula betulina* and *Cryptocline betularum* are described here. *D. betulina* is reported to occur on *B. nigra* (river birch), *B. papyrifera* (canoe or paper birch), and *B. populifolia* (gray birch) in eastern states and provinces; on *B. occidentalis* (mountain or water birch) in Utah; and on *B. pendula* in Great Britain. Landscape specimens as well as forest trees are affected. This fungus was first described as causing spots 5–10 mm across, but the prevailing symptoms are large brown blotches that have indefinite margins and become surrounded by yellow tissue. Affected leaves often fall while part of the blade is still green. The disease is most prominent on low branches and may progress up the tree. Brown acervuli form on the undersides of lesions, both along veins and scattered on interveinal areas (Plate 53B, C). The colorless, ovoid, unicellular conidia measure 7–17 × 2.0–4.5 µm.

D. betulina has not been associated with a sexual state, and the disease cycle is undescribed. A cycle may be hypothesized based on observed symptoms and knowledge of fungi similar to *D. betulina* that are anamorphs of ascomycetes in the Gnomoniaceae. Given the apparent absence of twig lesions, it may be presumed that the fungus overwinters primarily in fallen leaves in which a sexual state develops during the dormant season. Ascospores from this source, possibly augmented by conidia from an overwintering site on the tree, cause primary infections in spring. Conidia from acervuli in foliar lesions cause secondary cycles.

The most commonly collected birch anthracnose fungus in North America is *C. betularum*, which causes leaf spots on *B. lenta* (black or sweet birch) and *B. nigra* throughout central and eastern regions. Disease caused by this fungus induces premature leaf loss on low branches of both wild and ornamental trees. Epidemics have occurred in wet years.

Lesions caused by *C. betularum* occur only on leaves and are pale brown, up to 3 mm in diameter, and surrounded by a dark brown ring and sometimes also a chlorotic border around the brown zone. Lesions have been noted as early as late May to early June. A single lesion per leaf is often enough to induce abscission, and partial defoliation by mid-August is common. The causal fungus produces brown acervuli 200–650 µm in diameter on both leaf surfaces. They are covered initially by discolored host cuticle. The unicellular, ellipsoid conidia measure 8–15 × 5–8 µm.

The cycle of anthracnose caused by *C. betularum* is known, although some aspects have not been confirmed experimentally. The pathogen overwinters in fallen leaves and there produces perithecia of an as yet nameless species of *Apiognomonia*. The ascospores are mature at about the time birches begin growth in spring and are presumed to be dispersed by air. Ascospores from overwintered leaves and conidia from lesions in living leaves each caused typical symptoms on young leaves in pathogenicity tests. Acervuli with conidia developed within 14 days after inoculation. Conidia apparently initiate secondary cycles of disease during the growing season. However, conidia probably do not start primary infections, because overwintered acervuli on fallen leaves were found devoid of spores.

Four or five additional fungi are associated with anthracnose diseases on birches, but most are known only from herbarium records and mycological descriptions. An anthracnose caused by *Marssonina betulae* is shown in Plate 40. An anthracnose caused by *Pyrenopeziza betulicola* (anamorph *Cylindrosporium* sp.) was recently described in Finland.

References: 643, 779, 1099, 1298, 1743, 2939, 3065

Filbert Anthracnose

Monostichella coryli (syn. *Piggotia coryli*), a coelomycete, causes brown spots on leaves of *Corylus* species (filbert, hazelnut) across North America and in Europe and Japan. The spots range from small with rather abrupt edges (Plate 53F, G) to large with indefinite edges. Hosts include *C. americana* (American filbert), *C. avellana* (European filbert), *C. cornuta* (beaked filbert) and its variety *californica* (California hazelnut), and *C. heterophylla* (Siberian hazelnut). Severe infection leads to marginal scorch, premature yellowing, and leaf cast, but significant defoliation before late

summer has not been reported. The causal fungus forms subcuticular acervuli, 60–150 µm in diameter, that are more numerous on the upper than the lower surfaces of lesions. The overlying cuticle is stained dark brown. The conidia are single celled, pale brown, have a rounded apex and truncate base, and measure 9–16 × 4–7 µm. The sexual state of this fungus, if any, is unknown, and the disease cycle has not been reported. In Great Britain and Europe *M. coryli* or a similar fungus attacks catkins, leaves, and fruits of *C. avellana* and has also been associated with bud rot and twig canker of this tree.

References: 122, 1099, 1298, 1384, 2709, 2936, 3913

Redbud Anthracnose

Cercis canadensis (redbud) has no important foliar diseases. The anthracnose shown in Plate 53 is uncommon and apparently undescribed. It is characterized by necrotic lesions that at first follow veins and then expand to become large, irregular interveinal blotches with abrupt transition between dead and green or chlorotic tissue. A species of *Aureobasidium* (syn. *Kabatiella*) forms acervuli on the undersides of the lesions, especially along veins, and produces colorless, unicellular conidia that appear white in mass (Plate 53K).

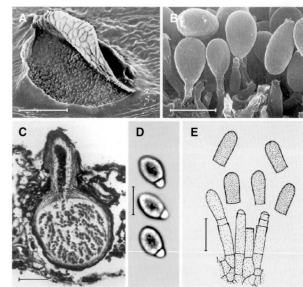

Figure 53. A–D. A birch anthracnose fungus, *Apiognomonia* sp., on *Betula nigra*. A, B. An acervulus, conidiogenous cells, and conidia of the anamorph, *Cryptocline betularum*. C. A mature perithecium, its neck extending through the abaxial surface of an overwintered leaf. D. Ascospores. E. The filbert anthracnose fungus, *Monostichella coryli*. Scale bars = 100 µm in A and C, others 10 µm. A–D courtesy of S. C. Redlin; E adapted from reference 3913 by permission of CAB International.

A–C. Anthracnose of *Betula papyrifera* (paper birch) caused by *Discula betulina*. A. Brown blotches associated with yellowing and leaf cast (NY, Jul). B, C. Magnified views of dark brown acervuli along veins (B) and in interveinal areas of undersurfaces of lesions (C) (NY & QC, Jul).

D, E. Anthracnose of *B. nigra* (river birch) caused by *Cryptocline betularum*. D. Lesions viewed on upper and lower leaf surfaces, and a severely affected leaf that was about to be cast. E. Enlarged view of typical lesions with yellow halos (MD, Aug).

F–I. Leaf spots of *Corylus* caused by *Monostichella coryli*. F. Lesions on *C. avellana* (European filbert) (OR, Aug). G. Numerous lesions associated with minor veins on *C. americana* (American filbert), causing yellowing and marginal scorch (WA, Jun). H, I. Successively magnified views of the undersurface of a lesion, showing tiny dark brown acervuli (OR, Aug).

J, K. An undescribed anthracnose of *Cercis canadensis* (redbud) (NY, Jun). J. Lesions that form initially along veins expand to involve interveinal and marginal areas. K. Close view of the undersurface of a lesion. White masses of conidia of the putative causal fungus, *Aureobasidium* sp., are associated with veins.

Photo credits: D, E—S. C. Redlin

Anthracnoses and Didymosporina Leaf Spot of Maples (Plate 54)

Maple Anthracnoses

Several fungi cause anthracnoses of *Acer* species (maples). The best known of these pathogens in North America are *Discula campestris*, *D. umbrinella*, *Aureobasidium apocryptum* (syn. *Kabatiella apocrypta*), and the omnivorous *Colletotrichum gloeosporioides* (discussed with Plate 56). All are coelomycetes. These fungi induce necrotic spots, vein-associated lesions, and/or irregular, spreading necrotic blotches. Young leaves may become blackened and shriveled, but those infected after reaching mature size retain nearly normal form. Young succulent shoots are sometimes killed. New shoots growing on trees previously defoliated by insects have been severely damaged. Each of the first three pathogens named above has been recorded as causing severe foliar browning on maples in scattered years and localities. *D. umbrinella* (Fig. 49) causes leaf spots and twig cankers on *Fagus* (beech), *Fraxinus* (ash), and *Quercus* (oak), as well as on *Acer*.

D. campestris, which is similar to *D. umbrinella*, was originally described in Europe on *Acer campestre* (hedge maple). It is widespread in north-central and eastern North America, affecting *A. saccharum* (sugar maple) and occasionally *A. rubrum* (red maple). It causes damage primarily to seedlings and to succulent shoots of larger plants, especially to parts damaged by *Thrips* or gall-inducing insects. This fungus was found to be one of the factors regulating survival of *A. saccharum* seedlings in Pennsylvania forests. Lesions on mature leaves are often associated with insect damage along veins, and then are elongate, brown above and tan below. Leaf sectors bounded by dead veins turn yellow or red, then brown. These symptoms become prominent in late summer. The causal fungus produces inconspicuous light brown acervuli on the lower surfaces of lesions, mainly on and immediately adjacent to the veins.

A second, unidentified *Discula* species causes necrotic spots that often but not invariably center on veins of *A. saccharum* (Plate 54D, E). The mature lesions are reddish brown with tan centers. Brown acervuli become prominent on the upper surfaces, especially along veins.

An unidentified *Discula* causes vein-associated lesions on *A. platanoides* (Norway maple; Plate 54F, G). This fungus was involved in a defoliation episode attributed primarily to the tar-spot fungus *Rhytisma acerinum* (Plate 32) in New York State in the 1990s.

A. apocryptum occurs on many *Acer* species across North America and in Europe, causing necrotic spots or scorchlike necrosis on leaves and occasional blight of succulent sprouts. This fungus is also an endophyte of *Quercus petraea* (sessile oak) and *Q. robur* (English oak) in Europe. Symptoms on maples develop during or shortly after wet weather from late spring until late summer as discrete lesions that merge and kill large areas of a leaf. Severe infection leads to premature defoliation. Inconspicuous light brown acervuli form on the lower surfaces of lesions, especially along veins. This fungus is distinguished from others on maples by its formation of one-celled, colorless conidia in groups of four to eight on the tips of short, broad conidiophores (Fig. 54A). In culture *A. apocryptum* does not form fruit bodies but instead produces conidia on hyphal cells and lacks distinct conidiophores. These habits are characteristic of its genus.

Disease cycles of maple anthracnoses have not been studied in detail. Also, the ascigerous states, if any, of most anthracnose fungi on maples have not been identified. (One exception is *Cryptodiaporthe hystrix*, anamorph *Diplodina acerina*, the perithecia of which occur on dead twigs.) The nature and sources of inoculum for primary infections are mostly unknown. It is likely that the pathogens overwinter in killed buds and twigs and perhaps also in petioles and veins of fallen leaves, and there produce spores that initiate infection in spring.
References: 46, 122, 635, 788, 1099, 1298, 1452, 1584, 1619, 2518, 2785, 3831, 3832, 4131

Didymosporina Leaf Spot of Maples

Didymosporina aceris, another coelomycete, causes large necrotic spots on the leaves and samaras of *Acer campestre* (hedge maple), *A. macrophyllum* (bigleaf maple), *A. negundo* (box-elder), and *A. platanoides* (Norway maple). The disease occurs in northeastern and northwestern USA and in Europe. It appears in late summer. The spots on *A. platanoides* are distinctly two toned with orange-brown centers, olive-gray outer portions, and an abrupt transition to normal green tissue. They coalesce when numerous, making leaves appear blighted. Dark acervuli about 100 μm in diameter develop between the epidermis and cuticle of the lower surfaces of lesions and produce two-celled, yellow-brown conidia in masses that rupture the cuticle. The conidia measure 8–10 × 4–5 μm and have broadly rounded ends. The disease cycle has not been studied, and no ascigerous state of the pathogen is known.
References: 1099, 1711, 3913, 4108

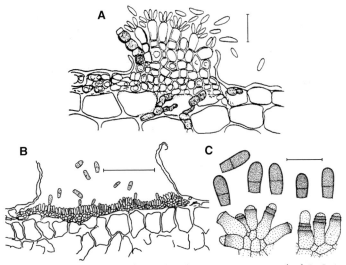

Figure 54. A. An acervulus of *Aureobasidium apocryptum* on a leaf. B, C. An acervulus, conidiophores, and conidia of *Didymosporina aceris*. Scale bars = 20 μm in A, 40 μm in B, 10 μm in C. A is from reference 122 by permission of Gebrüder Borntraeger Verlagsbuchhandlung; B and C are adapted from reference 3913 by permission of CAB International.

A–C. Anthracnose of *Acer saccharum* (sugar maple) caused by *Discula campestris*. A, C. Vein-associated lesions (NH, Sep). In magnified view, the lower surface of a lesion has acervuli visible as brown flecks on the vein. B. Marginal and vein-associated lesions on a first-year seedling (PA, Jun).

D, E. Anthracnose of *A. saccharum* caused by *Discula* sp. Dark brown acervuli are prominent on the upper surfaces of foliar lesions (NY, Aug).

F, G. Anthracnose caused by *Discula* sp. on *A. platanoides* (Norway maple) (NY, Jul). F. Vein-associated lesions and necrosis of large interveinal areas. Chlorotic lesions with black centers were caused by the tar-spot pathogen *Rhytisma acerinum* (Plate 32). G. Magnified view of lesions with acervuli of *Discula* sp. visible as dark dots on tan portions of veins. Black spots on interveinal tissue are young stromata of *R. acerinum*.

H. Anthracnose caused by *Aureobasidium apocryptum* on *A. palmatum* 'Atropurpureum' (NY, Jun).

I, J. Didymosporina leaf spot of *A. platanoides* (NY, Sep). I. Olive-gray lesions, some with orange-brown centers. J. Magnified view of the lower surface of a lesion with minute acervuli visible as dark brown stippling.

Photo credit: B—G. R. Stanosz

111

Hornbeam Anthracnose (Plate 55)

Apiosporopsis carpinea (syn. *Sphaerognomonia carpinea*; Diaporthales, Gnomoniaceae) in its conidial state, *Monostichella robergei*, causes anthracnose of *Carpinus caroliniana* (American hornbeam) and *Ostrya virginiana* (hop-hornbeam) in landscapes and forests in much of eastern North America. The reported limits of its distribution are in Quebec, Georgia, Oklahoma, and Wisconsin. The same disease is widespread on *Carpinus betulus* (European hornbeam) in Europe and affects *C. japonica* (Japanese hornbeam) in Japan. Infection of the latter two species in North America has not been reported. *M. robergei* has been reported to occur on *Eucalyptus* species in New Zealand, but the identification needs corroboration. Pathogenicity of isolates from *Eucalyptus* has not been tested.

Symptoms and signs. The usual symptoms on hornbeam leaves are irregular necrotic spots ranging in size from pinpoints to 5–6 mm in diameter. Lesions are reddish brown on *O. virginiana* and brown to gray-brown on *C. caroliniana*. They increase in number and become confluent during the season, causing marginal and apical browning, curling, and leaf cast. Subcuticular acervuli develop on both surfaces of lesions, mainly the upper one. These acervuli are visible with a magnifying lens as dark gray to nearly black dots. They are 60–120 µm in diameter and produce colorless, unicellular conidia that measure 12–16 × 5–7 µm (Fig. 55). Diseased leaves fall prematurely, and low branches of *O. virginiana* are defoliated by August in some years.

A twig canker phase of the disease occurs on *O. virginiana*. Reddish brown cankers, dramatic in contrast to greenish healthy bark, are centered on leaf scars. Axillary buds die (Plate 55B, C), and the cankers often encircle 1-year-old twigs, causing dieback. These events usually occur during the dormant season and thus are inconspicuous. Occasionally the canker phase becomes severe, killing both 1- and 2-year-old twigs not only during the dormant season but also after leaves expand. Pathogenicity tests have confirmed that *A. carpinea* can cause cankers during the period of host growth in spring. Canker expansion during the growing season results in prominent small "flags" consisting of wilted and dead shoots with curled, somewhat bleached leaves. A twig canker epidemic in western New York in the mid-1970s was associated with summer drought, which was presumed to have impaired the ability of the trees to restrict the pathogen in twigs.

Since the twigs of *O. virginiana* are delicate, those killed in the dormant season tend to dry, shrivel, and fail to support fruiting of the pathogen. Acervuli do form, however, in cankers and on some of the largest twigs killed by the fungus. These fruit bodies arise in spring and summer in cortical tissue, break through the periderm (Plate 55C), and become much larger and more conspicuous than those in leaves. They are black, persistent (Plate 55D), and contain conidia for up to a year. Cankers do not enlarge after the first season, but their locations may become conspicuous as twigs swell when callus and woundwood develop around the lesions.

The ascigerous state of *A. carpinea* develops in fallen leaves and matures in spring. The black perithecia appear as somewhat flattened spheres 110–250 × 80–170 µm in size. The ascospores are colorless and unicellular, and most measure 10–16 × 3–6.5 µm. Perithecia of *A. carpinea* have been collected from fallen leaves of *Acer, Alnus, Betula, Castanea, Corylus, Fraxinus,* and *Quercus* as well as *Ostrya* and *Carpinus*. However, *M. robergei* has been associated with only the latter two genera. This disparity raises two

questions for research. Does *M. robergei* colonize living leaves of plants other than those in *Carpinus* and *Ostrya*, and if so, does it cause disease in them? The reported connection between *A. carpinea* and *M. robergei* is real, because cultures of the latter were obtained from ascospores of *A. carpinea* in Germany.

Disease cycle. From observations of the seasonal development of symptoms and signs and knowledge of the pathogen's sexual state, we suggest that the annual cycle of hornbeam anthracnose is as follows. The pathogen overwinters as mycelium and acervuli in twig cankers and killed twigs of *O. virginiana* and perhaps in twigs of *C. caroliniana*. It also overwinters as developing perithecia in fallen leaves. In early spring additional acervuli develop in twig lesions. Water-dispersed conidia from twigs, augmented by airborne ascospores from fallen leaves, initiate the primary cycle of disease, which occurs on leaves. Conidia from acervuli in foliar lesions start secondary cycles. Twig infections occur either by conidial infection of fresh leaf scars or by mycelial growth of the fungus into twigs via petioles.

A. carpinea grows readily in pure culture, producing acervuli if pieces of host tissue are present, and conidia without organized fruit bodies elsewhere on the medium. It also causes amber-brown crystals to form in the medium, a characteristic useful for diagnosis. *References:* 122, 229, 635, 1099, 2699, 3126, 3253, 3470, 3649

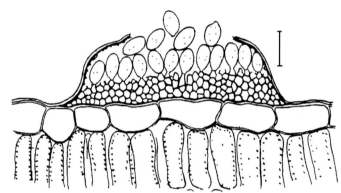

Figure 55. An acervulus of *Monostichella robergei*, the anamorph of *Apiosporopsis carpinea*, on a leaf. Scale bar = 20 µm. From reference 122 by permission of Gebrüder Borntraeger Verlagsbuchhandlung.

Anthracnose caused by *Apiosporopsis carpinea* on *Ostrya virginiana* (hop-hornbeam) and *Carpinus caroliniana* (American hornbeam).

A–D. Twig blight and cankers on *O. virginiana*. A. Dead shoots with bleached leaves identify twigs girdled during the growing season. B. Twig dieback and cankers centered on leaf scars. C. Enlarged view of a canker with acervuli of the pathogen on a year-old twig. D. Enlarged view of year-old cankers in 2-year-old twigs. Swelling is caused by woundwood formation at canker margins. Black acervuli are still prominent in the canker at right (NY, Jul).

E. Foliar symptoms on *O. virginiana* (upper leaves) and *C. caroliniana* (NY, Aug).

F, G. Magnified views of acervuli, some identified by arrows, on foliar lesions. F. Lower surface of *O. virginiana* leaf. G. Upper surface of *C. caroliniana* leaf (NY, Aug).

113

Anthracnoses and Diebacks Caused by *Glomerella* and *Colletotrichum* (Plate 56)

Glomerella species and their anamorphs in the genus *Colletotrichum* (coelomycetes) are common anthracnose pathogens worldwide, attacking annual and perennial plants in hundreds of genera. These fungi comprise a family of perithecial ascomycetes, the Glomerellaceae. They cause various symptoms, including damping-off, seedling blight, leaf spots, leaf-and-shoot blights, blossom blight, fruit rot, cankers, dieback, and death of small plants. Two species, *G. cingulata* (anamorph *C. gloeosporioides*) and *G. acutata* (anamorph *C. acutatum*), account for most of the records on woody plants. Perithecia of *G. acutata* have been seen only in cultures.

Colletotrichum acutatum was distinguished from *C. gloeosporioides* in disease records beginning late in the 20th century. Both species are heterogeneous, each including groups of strains that seem to be biologically distinct taxa but cannot be reliably distinguished on the basis of form or cultural characteristics. Woody plants attacked by one or both species occur in more than 115 genera, mostly of dicots but also of palms and gymnosperms. The listed references include numerous descriptions of diseases of woody plants caused by these fungi in North America. Two other *Colletotrichum* species associated with woody plants in recent records are *C. dematium* and *C. neriicolum*. The former is widespread and nonspecialized; the latter was described recently from *Nerium oleander* (oleander) in Turkey.

Diagnostic signs. Colletotrichum species produce subepidermal or subcuticular acervuli in recently killed tissue on any aboveground organ. Acervuli vary in appearance from pinkish eruptions to small dark blisters up to 0.5 mm in diameter. The pinkish color, when present, is that of conidia in a mucilaginous matrix. Acervuli of *C. gloeosporioides* often contain dark setae. Conidia of typical forms of *C. gloeosporioides* are cylindric with rounded ends, unicellular, individually colorless, and measure 12–17 × 3.5–6 µm. They become two celled shortly before germination. Conidia of *C. acutatum* are spindle shaped, abruptly tapered at each end, and measure 8.5–16.5 × 2.5–4 µm. These fungi also differ in color of pure cultures, *C. gloeosporioides* appearing dark gray and *C. acutatum* pink to carmine as viewed from the undersides of culture dishes. Perithecia of *G. cingulata* develop singly or in clusters and, on woody plants, break through the bark surface of cankers and dead twigs in spring. They are globose to somewhat pear shaped, 85–300 µm in diameter. The ascospores are colorless and unicellular, average approximately 12 × 4 µm in size, and are released in response to rainfall. They occasionally develop a single transverse septum and become faintly brown before germination.

Disease cycles. These pathogens perennate on and in leaves and stems of woody hosts, producing acervuli (and perithecia of *G. cingulata*) on killed parts. They also survive for a time in plant debris on the ground. In warm regions on evergreen plants such as *Citrus*, conidia may be present all year on foliage, their numbers sustained by new arrivals from acervuli during rains. Infection is initiated by conidia or ascospores, the former dispersed by water and insects and the latter primarily by air. Both spore types germinate under wet conditions and form dark appressoria from which plant surfaces are penetrated directly. Penetration may occur soon after germination or after a delay of weeks, with the appressoria acting as resting structures in the interim. If penetration is immediate, lesions may enlarge quickly, or the fungus may establish itself in microlesions and remain quiescent for a period before resuming host colonization. Wounds are also common sites of infection. Although capable of causing lesions in vigorous young tissues, these fungi most often become aggressive in plant parts weakened by environmental stress, improper nutrition, or natural senescence. Penetration activity of *Colletotrichum* strains that rot fruit is stimulated by volatile substances, notably ethylene, that emanate from ripening fruit.

G. cingulata and *C. acutatum* have several formae speciales, among which two on woody plants are well known. *G. cingulata* f.sp. *camelliae* causes cankers and dieback of *Camellia* species, and *C. acutatum* f.sp. *pineum* causes a shoot blight of *Pinus* seedlings, called terminal crook disease, in the Southern Hemisphere.
References: 16, 23, 168, 213, 339, 503, 512, 534, 714, 1099, 1121, 1148, 1163, 1164, 1179, 1214, 1376, 1403, 1404, 1539, 1573, 1873, 1957, 2394, 2425, 2467, 2685, 3157, 3738, 3811, 3843, 3887, 3922, 3987, 3988, 3992, 4028, 4163, 4254, 4339, 4531

Glomerella canker and dieback of camellia. This disease affects *Camellia* species in most areas where they grow in the USA and also in Great Britain, New Zealand, and Taiwan. Susceptible species include *C. hiemalis*, *C. japonica* (common camellia), *C. oleifera* (tea-oil plant), *C. reticulata*, *C. saluensis*, *C. sasanqua* (sasanqua camellia), and *C. sinensis* (tea). *C. oleifera* and *C. sasanqua* are more susceptible than *C. japonica*. Cultivars of affected species vary in susceptibility. The *C. japonica* cultivars Governor Mouton, Professor Sargent, Rose Emery, and Woodville Red are somewhat resistant. So also are the *C. sasanqua* cultivars Daydream and Setsugekka. The understock *C. hiemalis* 'Kanjiro' is reported to be resistant.

Dieback begins in early spring. Young shoots wilt and turn dull green and then brown. A band of darkly discolored bark and wood, usually 20–25 mm long, may be found at the junction of dead and living stem tissue. Death of tissues in this zone, from which *G. cingulata* f.sp. *camelliae* can be isolated readily, causes the shoots beyond to die from lack of water and nutrients. Woody twigs and small branches also may be girdled. Sometimes the fungus causes necrotic spots on young leaves, and the affected leaves drop. Lesions that do not encircle twigs persist as elliptic cankers. Usually less than 25 mm long, they are at first slightly sunken and eventually are surrounded by a ridge of woundwood. The fungus remains active in cankers, slowly killing sapwood until symptoms of stress (sparse foliage and dieback) develop in the parts beyond the canker.

Acervuli form in cankers and fallen leaves and branches. Conidial production by the pathogen peaks annually in early spring at about the time old leaves drop. The spores are dispersed by water and germinate rapidly under warm (18–27°C), wet conditions. Infection occurs within 16 hours if spores are deposited in fresh leaf scars or wounds. Freeze injuries, graft unions, and wounds are also common sites of infection. Dieback may develop within 7 days on new shoots or after months to years as cankers affect branches and mainstems.
References: 268–270, 958, 2844

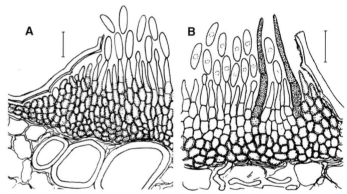

Figure 56. Partial sections of acervuli of the *Colletotrichum* state of *Glomerella cingulata* on different substrates. The acervulus in B contains stout, dark, erect hyphae called setae. Scale bars = 20 µm. Adapted from reference 122 by permission of Gebrüder Borntraeger Verlagsbuchhandlung.

A–D. Canker and dieback of *Camellia* caused by *Glomerella cingulata* f.sp. *camelliae*. A. Dried leaves cling to a killed twig of *C. sasanqua* (sasanqua camellia). B. Dieback on *C. japonica* (common camellia). C, D. Cankers that originated at twig bases on *C. japonica* and *C. sasanqua*, respectively (FL, Apr).

E, F. Anthracnose of *Ficus elastica* (Indian rubber-tree). E. Lesions on leaves. F. A canker with pinkish masses of *Colletotrichum* conidia on a stem (FL, Apr).

G, H. Anthracnose on a *Citrus limon* (lemon) leaf. *Colletotrichum* acervuli are arrayed concentrically (FL, Mar).

I, J. Dieback of *Aucuba japonica* (Japanese aucuba) caused by *C. acutatum*. I. Leaves distal to a lethal stem infection droop and turn brown. J. Magnified view of acervuli above the leaf scar in I (TN, Apr).

K–M. Anthracnose of *Acacia saligna* (beach acacia or orange wattle). K, L. Lesions on a twig and a leaf. M. Foliar lesions with minute dark acervuli of *Colletotrichum* (FL, Apr).

Butternut Canker (Plate 57)

Juglans cinerea (butternut), throughout most of its range in North America, is subject to cankers and dieback that shorten its life span, suppress reproduction, and greatly devalue its wood for lumber. Most of the damage is due to one disease called butternut canker or Sirococcus canker, which occurs from New Brunswick to Minnesota in the north and from the Carolinas to Missouri in the South. It has nearly eliminated the tree in the Carolinas and threatens its survival elsewhere. The disease was first noticed in Wisconsin in the 1960s, and it seemed to spread rapidly. The pathogen, *Sirococcus clavigignenti-juglandacearum* (in this discussion simply called *Sirococcus*), is a coelomycete with no known sexual state. Its origin is unknown, but it is thought to have been introduced to North America, because it displays limited genetic variation, as would be expected if the population had grown rapidly from a small founding stock.

Symptoms and signs. Cankers develop on stems of all sizes and on buttress roots, and affected trees usually have multiple cankers. The cankers are elliptic to lens shaped, expanding more rapidly along the axis of a stem than around it. Newly infected bark—and soon the wood beneath—turns dark brown to nearly black. If bark is shaved from stems with multiple lesions, the lesions are seen to be connected by dark streaks of dead sapwood. Single cankers may girdle twigs or small branches, but large limbs and trunks succumb only after girdling by multiple lesions. Coalescing cankers on trunks and buttress roots kill entire trees.

During lesion formation the inner bark is degraded to dark brown amorphous matter. Liquid oozes through small openings to the bark surface and spreads out as dark spots. After drying, the spots remain as brownish black to sooty patches on smooth gray bark. The bark in cankers later splits and shreds as the unaffected part of the stem continues growth. Sprouts commonly grow from the edges of cankers on trunks or limbs, and these sprouts are soon killed by new infections or by enlargement of nearby lesions. Cankers on limbs and trunks are perennial. Their expansion alternates with formation of callus and woundwood at the margins. Concentric ridges of woundwood become visible after the bark sloughs off.

Within a few weeks after a lesion forms, *Sirococcus* begins to produce a thin black stroma in the outer bark. Hyphae of the stroma differentiate into tiny pillars, called hyphal pegs, about 0.5 mm thick and 1.5–1.9 mm tall, that lift and rupture the papery outermost bark layer. Irregularly shaped, globose to flattened black pycnidia, 90–320 × 200–490 μm, develop on the stroma or sometimes at the bases or in the tips of hyphal pegs. Conidia are individually colorless, spindle shaped with pointed ends (rounded after imbibing water), and have a medial septum. They measure 9–17 × 1–1.5 μm.

Disease cycle. The causal fungus overwinters in cankers. During wet weather from early spring until mid-autumn, light tan to cream-colored tendrils (cirrhi) of conidia extrude from pycnidia and are dispersed by water. Sporulation is greatest on small branches in tree crowns. Infection on young twigs occurs via leaf scars, lenticels, and insect feeding wounds. Older stems are entered via adventitious buds, natural bark cracks, and wounds of all kinds. Infection in spring often leads to canker formation within 3 weeks, but the fungus may remain quiescent at the point of infection for several weeks before beginning to colonize and kill bark rapidly. If infection occurs in autumn through leaf scars, symptoms usually appear the next spring. Hyphae of the parasite grow via the medullary rays from bark into sapwood and then up and down the stem in xylem vessels and parenchyma cells. Hyphae in sapwood grow beyond the limits of cankers in the bark, and they cause new cankers where they grow outward and kill the cambium. The pathogen can be isolated readily from diseased bark or wood.

Conidial dispersal occurs in water that runs down wet bark and in splashing and windblown droplets. Conidia have been trapped from air at distances up to 40 m from diseased trees, but most are deposited close to their sources. Thus Sirococcus canker intensifies most rapidly in low branches subject to rain splash from lesions above or in adjacent trees. Sporulation on limbs or trunks continues for up to 20 months after a tree dies. Modes of long-distance dispersal of *Sirococcus* are mostly unknown, but transmission by seed to young seedlings of *J. cinerea* and *J. nigra* (black walnut) is known to occur, and transmission by insects is suspected. Beetle species representing eight families, collected from diseased or dead

butternut, have been found contaminated with conidia of the pathogen. Their possible roles as vectors were being evaluated as this was written.

Wet weather is required conidial germination and infection. Conidia perish quickly if exposed to sunlight and drying, but they remain germinable for many days on insects and can pass unharmed through the gut of some insects. Fungal growth and lesion development are most rapid at 24–28°C.

Although *Sirococcus* has thus far caused significant damage only to butternut, inoculation tests have revealed that most species and hybrids of walnut and also *Carya ovata* (shagbark hickory) are susceptible to some degree. *J. nigra* seedlings are highly susceptible if inoculated under suitable conditions, but few diseased trees of this species have been found in nature, and all were growing near severely affected *J. cinerea*. There is one report of natural infection of *J. ailanthifolia* var. *cordiformis* (heartnut), which is the least susceptible taxon yet tested.

For many years the disease now known as Sirococcus canker was called Melanconis dieback or Melanconis canker because the fungus *Melanconis juglandis* (Plate 67) was a prominent colonist of lesions and dead twigs and branches. *M. juglandis* quickly invades twigs or the bark of branches and trunks of *J. cinerea* weakened or newly killed by other agents. This fungus is a proven pathogen, but only to predisposed tissues. It is also known to be carried by insects. *References:* 1216, 1358, 1444, 2021, 2055, 2213, 2781, 2782, 2922, 3860, 4040–4042

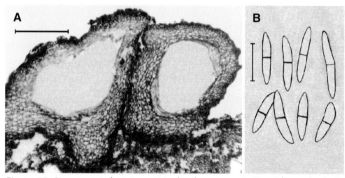

Figure 57. Sirococcus clavigignenti-juglandacearum. A. Pycnidia on bark. B. Conidia in water. Scale bars = 100 μm and 10 μm, respectively. A adapted from reference 2782 by permission of the Mycological Society of America; B drawn from a photo by S. Halik.

Siroccocus canker of *Juglans cinerea* (butternut), caused by *Sirococcus clavigignenti-juglandacearum.*

A. Twig and branch dieback on a tree declining because of multiple cankers on limbs and trunk (NY, Aug).

B, C. Perennial cankers on trunks and buttress roots (NY, May; WI, Jun).

D. Stained cracking bark indicates locations of cankers (NY, May).

E. A canker with dark brown wood beneath, exposed by removal of bark. Coalescing cankers have girdled the stem above the light-colored sapwood. Sprouts grew from adventitious buds that formed below cankers (MN, Aug).

F. The mainstem of a sapling killed by Sirococcus canker. Cracked, shredding bark, dark brown stain on the bark, and sprout growth below cankers are typical. Tiny black lumps on the bark are spore masses of the secondary fungus, *Melanconis juglandis* (MN, Aug).

G. Dead inner bark and sapwood, indicating internal spread of the pathogen, are revealed where intact outer bark was removed from the trunk of a sapling. The fungus entered the trunk from the diseased branch at upper left (WI, Nov).

H. Rows of hyphal pegs of *S. clavigignenti-juglandacearum* exposed where periderm fell away from a recently killed young stem (WI, May).

I. Hyphal pegs and pycnidia on a twig where the periderm is rolled back. White dots beside and atop pegs are masses of conidia extruded from pycnidia (VT, May).

J. A twig with its bark surface broken by hyphal pegs and pycnidia of the pathogen (MN, Aug).

Photo credit: C, G, H—N. A. Tisserat; I—D. R. Bergdahl

Sirococcus Blight of Conifers (Plate 58)

Sirococcus conigenus (syn. *S. strobilinus*), a coelomycete, causes shoot blight and kills seedlings of conifers. Outbreaks of Sirococcus blight occur in localities with mild, moist weather, such as the coastal fog belt in northern California, coastal Nova Scotia, and elsewhere during cool, wet summers. The causal fungus, described in the mid-1800s in Europe, occurs in nurseries, plantations, natural forests, and occasional landscape plantings coast to coast in Canada and the northern half of the USA, as well as in Europe and North Africa.

Hosts include the following (asterisk indicates species most often damaged): *Abies concolor* (white fir), *Larix decidua* (European larch), *L. laricina* (eastern larch or tamarack), *Picea abies* (Norway spruce), *P. engelmannii* (Engelmann spruce), *P. glauca* (white spruce), *P. mariana* (black spruce), *P. pungens* (blue spruce), *P. rubens* (red spruce), *P. sitchensis* (Sitka spruce), *Pinus albicaulis* (whitebark pine), *P. banksiana* (jack pine), *P. cembra* (Swiss stone pine), *P. contorta* (lodgepole pine), *P. coulteri* (Coulter pine), *P. halepensis* (Aleppo pine), *P. jeffreyi* (Jeffrey pine), *P. lambertiana* (sugar pine), *P. mugo* (mugo or Swiss mountain pine), *P. ponderosa* (ponderosa pine), *P. resinosa* (red pine), *P. sylvestris* (Scots pine), *Pseudotsuga menziesii* (Douglas-fir), and *Tsuga heterophylla* (western hemlock). The fungus is probably an endophyte in some species, as it has been detected in healthy-appearing needles of *L. decidua, Picea abies,* and *Pinus cembra* in Europe.

Symptoms and signs. Infection begins in succulent tissues. Seedlings, young shoots, and often year-old twigs are killed. Lesion formation on elongating shoots or needles causes them to droop. Shoot tips wither and turn brown within several days to a few weeks. Shoot death and needle droop occur from June into August, depending on plant species and region. The spread of infection into year-old twigs is common in *Picea pungens, Pinus resinosa,* and *T. heterophylla.*

In forests near the Pacific Coast, *S. conigenus* causes foliar blight and twig dieback of *P. contorta* and *P. ponderosa, T. heterophylla,* and, rarely, *Pseudotsuga menziesii.* In the Midwest and East, similar damage has occurred on *Pinus resinosa* in natural forests and *Picea pungens* in Christmas tree plantations. Infection on pole-size conifers is usually confined to low branches and has little effect on general health. Branch mortality in pole-size *Pinus resinosa,* however, may spread upward in years with disease-conducive weather. Susceptible trees in the understory beneath larger diseased trees are at risk of severe damage. Small *P. resinosa, Picea sitchensis,* and *T. heterophylla* in this circumstance have been grossly disfigured or killed and reproduction of these species temporarily curtailed by the disease in some localities.

Sirococcus blight in nurseries usually appears in patches that enlarge as the result of rain-splash dispersal of conidia. In *Pinus,* the disease often appears adjacent to windbreaks that contain diseased trees. In *Picea,* symptoms show up randomly in seedlings to which *S. conigenus* was transmitted in infested seed. First-year seedlings are often killed; older seedlings may become deformed and unfit for planting. Infested seed and planting stock contribute to long-distance dispersal. Outbreaks involving up to 30% of spruce seedlings in seedbeds have been traced to seed lots in which only 0.1–3% of seeds were infested. Seed infestation has in turn been traced to the colonization of spruce cones by *S. conigenus* in forests of the interior West. Seed infestation also occurs in *Picea abies* in Europe.

Pycnidia appear on killed stems and the bases of killed needles in late summer to early autumn of the year of infection or the next spring, depending on the climatic zone and the time when infection began. On *Pinus resinosa,* pycnidia develop abundantly beneath the basal sheaths of drooping needles. On *Picea* and *Tsuga,* pycnidia are most numerous on the stem. They also occur on cones and bud scales of *Picea.* Pycnidia are brown initially and blacken with age. They are irregularly round, average near 100 μm in diameter (often much larger on *P. resinosa* needles), and open by a wide irregular pore. The conidia are colorless, two celled, irregularly spindle shaped, and most measure 11–15 × 2.5–3 μm. Some pycnidia contain a small proportion of three- and four-celled conidia that are proportionately longer than the two-celled spores. *S. conigenus* can be isolated from infected tissues readily and forms pycnidia within 2 weeks in culture. Strains that attack *T. heterophylla* differ in DNA sequence, cultural characteristics, and average conidial size and shape from strains found on other conifers. The strains from *Tsuga* will probably be reclassified as a different species.

Disease cycle. The cycle is completed in 1 year, although spore dispersal from killed parts may continue for an additional 10 months. The fungus overwinters in killed needles, shoots, and cone scales. Conidia are dispersed by splashing and dripping water during spring and summer, with a peak of dispersal during the time of host shoot growth. Infection occurs on or adjacent to needle bases on new shoots and, presumably, young cones. Conditions favoring infection are succulent plant parts, a wet period lasting at least 24 hours, and temperatures of 16–20°C. Longer wet periods favor increasingly severe disease. Infection can occur between 10° and 25°C. Most new infections occur within a few meters of spore sources. Symptoms appear about 2 weeks after infection of highly susceptible plants, and lesion expansion on a given shoot may be complete after 4–6 weeks. Under optimum circumstances, pycnidia may form within an additional 2 weeks. Often, however, pycnidial formation is delayed until the next spring.

Factors that promote Sirococcus blight, in addition to moist weather, include low light intensity, as beneath a forest canopy, and magnesium deficiency. Fertilization of *Picea abies* with magnesite compounds, which supply magnesium and calcium, diminished the severity of Sirococcus shoot blight in experiments in Germany.

Tree species resistant or immune to *S. conigenus* include *Abies balsamea* (balsam fir), *Pinus nigra* (Austrian pine), and *P. strobus* (eastern white pine), which have remained healthy when exposed to spores from diseased *P. resinosa. P. contorta* seedling populations, either tested by inoculation or naturally infected, have shown differences in disease incidence related to seed source. Disease incidence in inoculated seedlings was lowest among those from sources in the Sierra Nevada.

References: 166, 800, 1212–1214, 1464, 1851, 1911, 2213, 2888, 3471, 3548, 3711, 3735, 3908, 3913, 4195, 4354

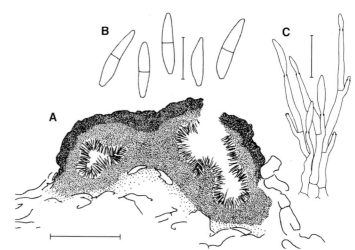

Figure 58. Sirococcus conigenus. A. Section of a pycnidial stroma. B. Conidia. C. Conidiophores with young conidia. Scale bars = 200, 10, and 10 μm, respectively. Adapted from reference 3913 by permission of CAB International.

A–C. Sirococcus shoot blight of *Pinus resinosa* (red pine). A, B. Dead branch tips with brown needles on current-season and year-old twigs. Needles infected before reaching maturity droop, die, and adhere for a year or longer. *P. strobus* (eastern white pine), its branches intermingled with diseased *P. resinosa* in A, remains unaffected (NY, Jul). C. Pycnidia of *Sirococcus conigenus* on the base of a dead needle 1 year after infection (NY, Jun).

D. Shoot blight caused by *S. conigenus* on *Larix laricina* (eastern larch) (NB, Jun).

E–H. Sirococcus shoot blight of *Picea pungens* (blue spruce) (NH, Jun). E. A 2-year-old branch with most twig tips killed. Diseased twigs slowly die back to the parent branch. F. A twig tip killed during growth the previous year has a characteristic crook. G. Some needles are discolored and others are missing at a lesion midway between the base and the terminal bud of a year-old shoot. H. Pycnidia of *S. conigenus* on a twig tip that became infected 11 months earlier.

I, J. Pycnidia of *S. conigenus* on a killed twig and a dying first-year seedling of *Picea* sp. (NS, Sep).

Photo credits: D, I, J—K. J. Harrison, Canadian Forest Service

Botryosphaeria Cankers and Diebacks, Plates 59–63

Botryosphaeria species (Dothideales, Botryosphaeriaceae) are opportunistic fungi that cause cankers, dieback, and fruit rots of many plants. We discuss *B. dothidea* and *B. ribis* (Plates 59, 60), *B. stevensii* (Plate 60), *B. quercuum* (Plate 61), *B. obtusa* (Plate 62), and *B. rhodina* (Plate 63). These fungi are usually encountered as anamorphs in the genera *Dothiorella*, *Fusicoccum*, *Diplodia*, *Lasiodiplodia*, and *Sphaeropsis*. Some authors synonymize the first two anamorph names as *Fusicoccum* and the latter three names as *Diplodia*. Contemporary reports of *Botryosphaeria* on woody plants include the species discussed on these pages as well as several that can only be mentioned. *B. australis* occurs on several native plants and on *Sequoiadendron* in Australia; *B. eucalyptorum* on *Eucalyptus* in South Africa; *B. laricina* on *Larix* (larch) in Japan; *B. lutea* on various plants in Portugal and New Zealand; *B. parva* on various plants in Australasia, Portugal, South Africa, and Hawaii; and *B. populi* on *Populus* (poplar) in Portugal. All of these species have *Fusicoccum* anamorphs.

Cankers, Dieback, and Leaf Blight Caused by *Botryosphaeria dothidea* and *B. ribis*

Several fungi with similar morphological characteristics and broad host ranges compose a complex to which the names *Botryosphaeria dothidea* and *B. ribis* have most often been applied. The anamorphs of *B. dothidea* and *B. ribis* are *Fusicoccum aesculi* and *F. ribis*, respectively. Fungi of the *B. dothidea–B. ribis* complex attack primarily woody plants in at least 170 genera of dicots, monocots, and gymnosperms, including most of the genera important in North American horticulture. These fungi occur globally in temperate and tropical regions, causing cankers, dieback, and fruit rots and colonizing twigs and branches that are dying or recently dead from other causes. They cause severe diseases on plants stressed by wounds, drought, freezing, defoliation, or planting outside their native ranges. They also occur as endophytes in leaves, fruit, and bark. Pathogenicity tests of isolates from particular plant species have usually not indicated host specialization. DNA sequence analyses and critical morphological studies of conidial states have confirmed that *B. dothidea* and *B. ribis* are distinct species. Unfortunately, most mycological and plant pathological records do not reliably distinguish between them or their anamorphs. Therefore, we refer specifically to *B. dothidea* or *B. ribis* only to relate information about reliably identified strains.

Symptoms and signs. Symptoms vary with the host, the nature and size of the organ attacked, and the extent of predisposition. Lesions in stems vary from small spots in outer bark to spreading, usually elongate lesions without marginal swelling, or sunken cankers delimited by woundwood. Diseased twigs usually die, but trunks and large branches usually have sufficient resistance to restrict the pathogens to discrete cankers, often centered on twig or branch stubs. Killed bark becomes depressed as surrounding tissues grow, and it becomes roughened and darkened where fungal stromata break the surface. Lesions may expand during and/or between growing seasons. Leaves on diseased branches may wilt, or the buds fail to open in spring. Wood beneath cankers is discolored brown, and this discoloration extends several centimeters above and below a canker. Lesions in young stems of some smooth-barked plants, notably *Malus* (apple, crabapple), appear as blisters due to separation of the thin periderm from deeper bark layers. Diseased tissues in stems of gum-producing plants such as *Liquidambar styraciflua* (sweetgum) and *Prunus* (peach and others) become infiltrated with gum, which may exude as blobs on the bark surface—a symptom called gummosis. Leaf blight occurs on plants with thick long-lived leaves, such as *Rhododendron* and *Macadamia*. Fruit rots occur in diverse plants. *B. dothidea* has been reported to kill drought-stressed pines after entering roots, but how it reached the roots is unknown.

The asexual states of fungi in the *B. dothidea–B. ribis* complex are characterized by black pycnidial stromata that differentiate within killed bark and break the surface when mature. Stromata vary in shape and are 1–4 mm in the longest dimension. Each stroma contains one to several locules (cavities) 150–250 μm in diameter with colorless contents that appear white when sliced (Plates 59E, 60F). When diseased bark is moist or wet, colorless one-celled conidia, somewhat narrowed at the ends, extrude in a mass from the pore at the top of each pycnidium. Conidial dimensions of *B.*

dothidea are typically 20–35 × 5–7.5 μm (length width ratio >3.5), while those of *B. ribis* are typically 15–24 × 5–11 μm (length width ratio usually <3.0).

Pycnidial stromata develop throughout the year as temperature permits, usually beginning within days to weeks after diseased tissue dies. This development depends on sufficient moisture in the tissue and on the kind of plant. Cankers caused by *B. dothidea* on *Metasequoia*, for example, may lack fruit bodies, although isolates from these cankers readily produce pycnidial stromata in culture or after inoculation into bark of other plants.

Pseudothecia, preceded by spermagonia that contain nongerminable microconidia, may develop in the same stromata that previously produced conidia or in new stromata, usually on bark dead for several months to a year or more. Pseudothecia are not found on the current season's twigs or on recently killed bark in cankers. One to several pseudothecia may develop in a single stroma. The ascospores of both species are colorless, unicellular, and ovoid, 17–23 × 7–10 μm in size. Sometimes both the conidial and ascigerous states can be found together, either in separate stromata or in the same one, and sometimes the conidial state is apparently not followed by pseudothecia.

Disease cycle. *B. dothidea* and *B. ribis* overwinter as pycnidial or pseudothecial stromata in dead bark and as small endophytic colonies in bark and perennial leaves. Endophytic colonies are positioned for rapid exploitation of the substrate when it senesces or comes under environmental stress. Probably most infections are initiated by conidia, which are dispersed by dripping or splashing water during much of the year and in greatest numbers in late spring and early summer. Where pseudothecia develop, ascospores are dispersed by air and water during much of the growing season and, like conidia, are most abundant during late spring to early summer. These pathogens may also be transmitted by pruning tools.

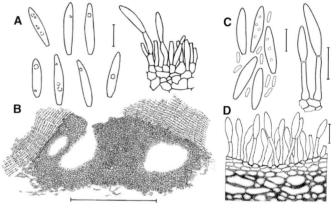

Figure 59. Anamorphs of *Botryosphaeria dothidea* and *B. ribis*. A, B. Conidia, conidiophores, and a section through a conidial stroma of *B. dothidea*. C, D. Conidia, microconidia, conidiophores, and part of the interior of a conidial stroma of *B. ribis*. Scale bars = 500 μm for the stroma in bark, 10 μm for other structures. A, B adapted from reference 3913 by permission of CAB International; C, D adapted from reference 2725 by permission of Mycotaxon, Ltd.

Dieback and cankers caused by fungi of the *Botryosphaeria dothidea–B. ribis* complex.

A–C, E. Dieback and cankers on *Cercis canadensis* (redbud). A. Dieback. B. An inconspicuous canker with cracked bark at its edges. C. Old cankers with prominent woundwood ridges at margins. E. Magnified view of a bark surface with pycnidia visible as groups of tiny black objects in cracks (upper arrow). White contents of pycnidia are revealed by shaving off the bark surface (lower arrow) (OH, Jun).

D. A canker girdling a branch of *Cornus sericea* (red-osier dogwood) (NY, Jun).

F–I. Cankers on small branches of *Fraxinus quadrangulata* (blue ash). F. Multiple lesions on a slowly growing branch. G. Close view of a canker that developed at a wound; black dots are pycnidia. H. Magnified view of pycnidial stromata on a twig above a leaf scar. I. Old inactive lesions covered by woundwood (OH, May).

121

B. dothidea and *B. ribis* commonly infect wounds, but lenticels and growth cracks in bark are also avenues of entry. Leaves of some plants are invaded via stomata. Lesions in stems may develop within days to weeks after inoculation, but a long incubation period sometimes precedes symptoms. Incubation periods of 3–13 months have been recorded for infections in *Prunus persica* (peach) twigs and 4–10 months for infections in *Ulmus americana* (American elm) twigs. Temperature permitting, sporulation usually begins within a few weeks after lesion formation, completing the cycle.

B. dothidea and *B. ribis* colonize both bark and sapwood during parasitic attack and when growing as saprobes. They produce macerating enzymes that probably contribute to cellular death. Isolates vary greatly in aggressiveness, but an isolate from one host is normally capable of causing significant disease in an array of unrelated plants. Given suitable moisture, spores of both fungi germinate rapidly at 24–32°C. Colonies grow rapidly at 25–35°C, which suggests that these fungi are often favored by temperatures above the optimum for plant growth.

Plants that are intensively selected for horticultural characteristics (e.g., *Malus* species [apple, crabapple], *P. persica*, and *Vaccinium* species [blueberries]) are susceptible to attack by *Botryosphaeria* species under a wide array of circumstances, but resistance is the normal condition for most host plants. Environmental stress predisposes them to infection. *Botryosphaeria* cankers and dieback of *Cercis canadensis* (redbud) and *Liquidambar styraciflua* (sweetgum) are more common and severe on plants growing in shade than on those in full light. Water shortage has been shown to predispose *Pistacia chinensis* (pistachio) to attack by *B. dothidea*. In experiments, drought stress insufficient to cause wilting permitted infection of *Betula pendula* (European white birch), *Cornus sericea* (red-osier dogwood), *L. styraciflua*, and *Sorbus aucuparia* (European mountain-ash) by a putative *B. dothidea* strain. The plants regained resistance within days after being resupplied with water. Rapid freezing, simulating temperature drops after mild weather in winter, caused temporarily elevated susceptibility of the test species named above in the absence of visible damage. It follows that prevention of disease caused by these fungi depends to a large extent on avoiding or minimizing stresses that induce susceptibility. References: 125, 228, 231, 371, 372, 513, 526, 538, 539, 854, 934, 1065, 1099, 1259, 1698, 1859, 1896, 2395, 2401, 2402, 2577, 2618, 2634, 2725, 2801, 2901, 3063, 3064, 3123, 3182–3185, 3479, 3480, 3689, 3712, 3717, 3719, 3720, 3913, 3920, 4248, 4259, 4303, 4416, 4438, 4519, 4520

Dieback and Cankers Caused by *Botryosphaeria stevensii* and *Diplodia quercina*

Botryosphaeria stevensii (anamorph *Diplodia mutila*) is widespread in North America and Europe, causing twig and branch cankers, dieback, and sometimes fruit rot on diverse plants. It also occurs in parts of South America, Africa, Asia, Australia, and New Zealand. This fungus, in common with several other *Botryosphaeria* species, is an opportunistic pathogen. It is associated with and apparently exacerbates drought-induced oak decline in southern Europe, but it is also capable of causing cankers and twig dieback in nonstressed plants. Recorded woody host genera of *B. stevensii* or *D. mutila* include *Abies, Acacia, Acer, Actinidia, Cedrus, Celtis, Ceratonia, Chamaecyparis, Corylus, Cupressus, Eriobotrya, Ficus, Fraxinus, Juglans, Juniperus, Laburnum, Malus, Phoenix, Picea, Pinus, Pittosporum, Platanus, Populus, Prosopis, Protea, Prunus, Pyrus, Quercus, Rhus, Rosa, Syringa, Tecomanthe, Thuja, Ulmus,* and *Vitis*. Isolates from *Quercus* have been shown to be pathogenic to *Quercus* and *Malus*.

The history and taxonomy of *B. stevensii* are entangled with those of related fungi. This species is similar to *B. quercuum* (Plate 61) and *B. obtusa* (Plate 62), but its conidial characteristics permit diagnostic differentiation from those fungi. Its pycnidial stromata, often 0.5–1 mm broad, are immersed in bark singly or in groups and have one fertile locule. The conidia are colorless when released, and the majority measure 25–31 × 10–13 μm with a length/width ratio near 2.3. Some conidia turn brown and two celled with age.

Diplodia quercina has caused outbreaks of twig and branch cankers and dieback of drought-stressed *Quercus agrifolia* (California live oak) in California (Plate 60G). Symptoms vary from scattered twig dieback to death of limbs and entire trees. Brown staining of sapwood beneath bark lesions and roughening of bark by pycnidial stromata occur more or less as shown for oaks affected by *B. quercuum* (Plate 61). *D. quercina* is similar to *D. mutila* and has genetic characters in common with isolates identified as that fungus as well as with isolates identified as *B. quercuum*. Its conidia differ from those of the latter fungus in being mostly longer than 30 μm with a length/width ratio near 2.0. They are initially single celled and colorless but eventually turn brown and equally two celled, occasionally three celled.

A fungus identified as *B. stevensii* causes cankers and dieback in *Juniperus* species in central and eastern USA. Elliptic lesions, sometimes appearing resinous and often coalescing, form on twigs, branches, and mainstems, commonly at branch crotches. The foliage on parts girdled by lesions turns yellow and then brown. The disease occurs on *J. sabina* (savin juniper) and several cultivars of *J. scopulorum* (Rocky Mountain juniper), and the causal fungus has been found associated with dieback on *J. virginiana* (eastern redcedar). Its pathogenicity was proven on these species and also on *J. chinensis* (Chinese juniper) and *Cupressus sempervirens* (Italian cypress). Juniper strains, however, were not pathogenic to apple fruit or twigs, and a pathogenic apple isolate of *B. stevensii* did not cause lesions when inoculated to juniper.

Black stromata with pycnidia 250–500 μm in diameter form in the cankers and on killed branches. If pycnidia are sliced open, the contents appear white or nearly so. Conidia of *B. stevensii* on

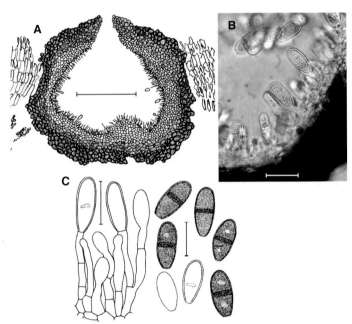

Figure 60. Diplodia mutila, the anamorph of *Botryosphaeria stevensii.* A. Section of a pycnidium in bark. B. Microscopic view of part of a pycnidium with young conidia. C. Conidiophores and young and mature conidia. Scale bars = 200 μm in A, 20 μm in B and C. A and C adapted from reference 3913 by permission of CAB International; B courtesy of N. A. Tisserat.

A–F. Diseases caused by fungi of the *Botryosphaeria dothidea–B. ribis* complex. A. Dieback of *Cornus florida* (flowering dogwood) (NY, Jun). B–F. Leaf blight and dieback of *Rhododendron* 'Nova Zembla' (NY, Aug). B. Leaf blight. C. Wilting due to a canker on a mainstem. D. Bark has been removed to show the lesion (NY, Aug). E. Magnified view of pycnidial stromata breaking through the surface of a twig (NY, Aug). F. Magnified portion of a blighted rhododendron leaf showing inconspicuous pimplelike pycnidia, a cluster of which at center has been sliced at the level of the leaf surface to reveal white contents (NY, Aug).

G. Dieback of *Quercus agrifolia* (California live oak) caused by *Diplodia quercina* (CA, Nov).

H, I. Dieback and a branch canker (sunken area partially in shadow) on *Juniperus scopulorum* (Rocky Mountain juniper) caused by *B. stevensii* (KS, Apr).

Photo credits: G—E. Hecht-Poinar; H, I—N. A. Tisserat

Juniperus are broadly elliptic in outline and while young are colorless and unicellular with smooth thick walls (Fig. 60). The majority measure 23–32 × 12–15 μm. These spores eventually turn tan to brown and become equally two celled. Pseudothecia develop later in the same or similar stromata on killed parts. Microconidia, nongerminable and presumably functioning as spermatia, precede asci in the locules of these stromata. Similar or perhaps identical fungi have been associated with dieback on *J. sabina* in Ontario and on *Cupressus arizonica* (Arizona cypress) in Georgia, USA.

Most new infections of *Juniperus* probably occur in late spring and summer. Spore trapping near diseased junipers in Kansas revealed an annual peak in conidial release in late spring. Given moist conditions, conidia germinated in 4–6 hours at 25–30°C. Studies with four host species revealed high susceptibility of stems wounded and inoculated in May and August but not in December.

The *Diplodia* state of *B. stevensii* on *Juniperus* in the USA is similar to *Diplodia pinea* f.sp. *cupressi*, which causes a serious canker and dieback syndrome of *Cupressus sempervirens* (Italian cypress) in the Mediterranean region and occurs also in South Africa. Fungal strains representing USA and Mediterranean populations were found to have many similar morphological characteristics and multiple genetic markers in common. *B. stevensii* and *D. pinea* f.sp. *cupressi* also have in common the production of nonspecific phytotoxins (sphaeropsidins and other compounds) that cause lesions and dieback when injected into *C. sempervirens* and other plants. Accordingly, these fungi may be united under one name. However, it is not clear that this name will continue to be *B. stevensii*, because the pathogens of cypress and juniper differ in a number of respects from strains of *B. stevensii* that parasitize *Malus*, *Quercus*, and other angiosperms.

References: 934, 1089, 1090, 1156, 1195, 2324, 2380, 2381, 2389, 2723, 3153, 3198, 3204, 3205, 3669, 3759, 3835, 3838, 3913, 3931, 4045, 4127, 4520

Cankers and Dieback Caused by *Botryosphaeria quercuum* and Related Fungi on Oak

Several *Botryosphaeria* species and related asexual fungi cause cankers and dieback of *Quercus* (oak). The group includes at least *B. dothidea* and *B. ribis* (Plates 59, 60), *B. stevensii* (Plate 60), *B. quercuum* (Plate 61), *B. obtusa* (Plate 62), *B. rhodina* (Plate 63), and *B. melanops*. The conidial states of these fungi can be distinguished on the basis of spore form and size, color, septation, and developmental changes in these characters. However, some of the ascigerous states are morphologically indistinguishable. In addition, pathological records include asexual strains identified as *Botryodiplodia*, *Diplodia*, *Dothiorella*, *Fusicoccum*, or *Sphaeropsis* that do not align neatly with the species named above. Relationships among these fungi are not understood.

B. quercuum (anamorph *Botryodiplodia* or *Dothiorella*) is probably the most common member of its genus on oaks in eastern North America, occurring at least from Nova Scotia west to Ontario and Minnesota, and south to Texas and Florida. It also occurs in California and British Columbia. Several authors have referred to the pycnidial form of this fungus as either *Botryodiplodia gallae* or *Dothiorella quercina*, and we include records of these forms with those of *Botryosphaeria quercuum*. This fungus occurs on *Carya* sp. (hickory), *Fagus grandifolia* (American beech), many *Quercus* species, *Vitis* sp. (grapevine), and possibly many other plants. It is a proven pathogen of *Q. alba* (white oak) and *Q. ellipsoidalis* (northern pin oak). It has been reported to occur on other plants in more than 35 genera in North America and Eurasia, but records not accompanied by detailed diagnostic data should be considered with caution because the taxonomic history of *B. quercuum* is entangled with the histories of *B. obtusa* and *B. stevensii*, which have broad geographic and host ranges.

Symptoms and signs. Lesions caused by *B. quercuum* develop in summer and, by killing segments of twigs or small branches, cause distal leaves to wilt and turn brown. The foliar symptoms are indistinguishable from those caused by certain insects, notably *Elaphidionoides* species (twig pruners), *Agrilus* and *Oncideres* species (twig girdlers), and members of the Cicadidae (cicadas), but distinct lesions on the stems indicate fungal disease. Diseased bark turns dark brown to almost black, shrivels, and eventually cracks at the junction with living tissue. The outer sapwood beneath a lesion in bark is discolored brown to black, and brown streaks in the sapwood extend up and down the twig or branch for 2–25 cm. In transverse view streaks appear as spots or tangential bands. Most lesions remain localized in twigs and small branches, but occasionally the fungus advances downward perennially, causing progressive dieback or, on large stems, sunken strips of dead bark bordered by woundwood ridges. Sprouts sometimes develop along trunks or branches of trees that have sustained dieback. *B. quercuum* has been blamed for the death of entire trees, but it seems unlikely that it alone causes such severe damage.

Black pycnidial stromata of *B. quercuum* develop in bark lesions beginning soon after leaves wilt. Maturing stromata break through the bark surface. They measure 0.6–1.3 mm across and protrude up to 0.5 mm above the contour of the twig. Each stroma contains one to several cavities in which conidia form. If the tops of mature stromata are shaved off, pycnidial cavities appear as white spots on a dark background, as illustrated for *B. dothidea* in Plate 59. Pycnidia also form on dead oak galls and acorns. The conidia have smooth walls 1.5–2 μm thick and measure 18–25 × 12–17 μm, with the length/width ratio near 1.5. They are at first colorless and unicellular. Some conidia later turn light brown and become two celled. Pseudothecia differentiate and produce ascospores in overwintered stromata in bark. The pathogen can be isolated from discolored bark or wood, and it forms pycnidia with characteristic conidia in culture.

Spore dispersal and natural infection by *B. quercuum* have not been studied. Twig cankers and flagging caused by this fungus tend to occur in outbreaks that subside after a year or two. Outbreaks have been associated with infestations of *Asterolecanium variolosum* (golden oak scale) on *Quercus alba* and *Q. prinus* (chestnut oak) and with attack by *Agrilus* sp. (twig girdler) on various oaks.

Other oak pathogens. In southern and southeastern USA, twig canker and dieback of various oaks are sometimes caused by *D. longispora*. Similar symptoms on drought-stressed *Quercus agrifolia* (California live oak) caused by *Diplodia quercina* were noted with the discussion of *B. stevensii*. Relationships among these three fungi and *Botryosphaeria melanops* require clarification. The latter in its *Fusicoccum* (syn. *Dothiorella*) state produces colorless conidia that are mostly 47–50 × 10–11 μm. It is better known in southern Europe than in North America.

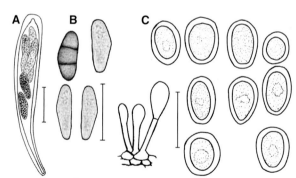

Figure 61. Botryosphaeria quercuum. A. An ascus. B. Ascospores. C. Conidiophores and conidia. Scale bars = 25 μm. A and C adapted from reference 3669 by permission of Gebrüder Borntraeger Verlagsbuchhandlung; B derived from reference 3606 by permission of NRC Research Press.

Twig canker and dieback of *Quercus rubra* (northern red oak) caused by *Botryosphaeria quercuum* (NY, Aug).

A, B. Wilting or browned foliage identifies diseased twigs.

C. Bark removed to show the location of a lesion.

D. Bark in a lesion on a year-old twig is nearly black in contrast to normal dark green; necrotic streaks in xylem extend several centimeters beyond the lesion.

E, F. Bark in twig lesions is roughened by swellings where pycnidia are developing, especially beneath lenticels. Arrow in E indicates lesion margin. The tattered membrane in F is the original epidermis.

G, H. Mature pycnidial stromata of *B. quercuum*. The magnified view in H reveals clustered stromata.

125

References: 46, 125, 461, 659, 843, 1099, 1298, 1614, 1854, 3456, 3606, 3669, 3855, 4128, 4520

Cankers and Dieback Caused by *Botryosphaeria obtusa*

Botryosphaeria obtusa, known primarily as a pathogen of fruit trees, is a nonspecialized, opportunistic fungus that colonizes bark of many kinds of plants as a saprobe and attacks living parts if they are predisposed by environmental stress, wounds, or intrinsic high susceptibility. This fungus occurs in temperate zones around the world on woody and some nonwoody plants in at least 55 families. In North America *B. obtusa* is best known for causing dieback of *Vitis* (grapevine); canker, gummosis, and dieback of *Prunus persica* (peach); canker and dieback of *Quercus* (oaks); fruit rot of *Citrus*; and three disorders of *Malus* (apples and crabapples): frogeye leaf spot, black rot of fruit, and black rot canker.

Additional genera of woody hosts include the following:

Acer, Aesculus, Ailanthus, Amelanchier, Alnus, Amorpha, Ampelopsis, Aronia, Artemisia, Baccharis, Betula, Bignonia, Broussonetia, Callicarpa, Calycanthus, Carissa, Carpinus, Carya, Castanea, Catalpa, Ceanothus, Celastrus, Celtis, Cephalanthus, Cercis, Chaenomeles, Chamaecyparis, Chamaedaphne, Clematis, Cocculus, Cornus, Corylus, Cotinus, Cotoneaster, Crataegus, Cupressus, Cydonia, Cytisus, Diospyros, Elaeagnus, Eriobotrya, Euonymus, Ficus, Fraxinus, Gelsemium, Genista, Gleditsia, Hamamelis, Hedera, Hibiscus, Ilex, Jasminum, Juglans, Juniperus, Koelreuteria, Lagerstroemia, Lantana, Larix, Leucaena, Ligustrum, Lindera, Liquidambar, Liriodendron, Lonicera, Lyonia, Maclura, Magnolia, Melia, Morus, Myrica, Nannorrhops, Nerium, Osteomeles, Ostrya, Oxydendrum, Parthenocissus, Paulownia, Persea, Phyllanthus, Physocarpus, Pinus, Pistacia, Platanus, Poncirus, Populus, Pouteria, Prosopis, Pyracantha, Pyrus, Rhamnus, Rhododendron, Rhus, Ribes, Robinia, Rosa, Rubus, Salix, Sambucus, Sassafras, Smilax, Sorbus, Spiraea, Styphnolobium, Symphoricarpos, Syringa, Tabebuia, Tectona, Thuja, Tilia, Ulmus, Vaccinium, Viburnum, Wisteria, and *Yucca.*

Symptoms and signs. Symptoms caused by *B. obtusa,* except for frogeye leaf spot on *Malus,* are similar to those caused by several other fungi, notably *B. dothidea* and *B. rhodina.* Frogeye leaf spot first appears on young leaves as circular spots with indefinite purple margins. By early summer the lesions are brown, 2–10 mm (average 4 mm) in diameter, and have abrupt margins. Some lesions undergo secondary enlargement and become lobed. The name frogeye leaf spot refers to lobed spots. Severe leaf infection causes premature yellowing and defoliation. Lesions in green fruit appear as tiny purple pimples or black spots. Ripe or ripening fruit rots rapidly and either drops or shrivels and remains on the tree. Dense masses of black pycnidia form on decayed fruit.

Lesions on stems of many hosts develop at wounds, around the bases of dead twigs, and on upper or southwest-facing sides of limbs following freezing injury or sunburn. Some lesions remain confined in the outer phloem; others extend to the wood and either cease enlargement during the first year or develop perennially, attaining lengths of 30 cm or more. They enlarge during the growing season, most rapidly during times of water stress. Cracked bark at the margins of a canker indicates that the lesion has at least temporarily ceased enlarging. Cankers that kill limbs during the growing season cause either wilting or premature yellowing and casting of leaves. The fungus penetrates the wood only slightly except in twigs, where both bark and wood are extensively invaded. Dead bark remains attached for at least a year, then slowly cracks and falls away.

B. obtusa causes cankers and gummosis on *Prunus persica,* but gum production is a nonspecific response. *B. dothidea* and *B. rhodina* also cause these symptoms and may be found together with *B. obtusa* in peach orchards in southeastern USA. Spores of *B. obtusa* often lodge in peach buds, resulting in lethal bud infections in spring.

Diagnosis of disease caused by *B. obtusa* usually depends on recognition of the anamorph, *Sphaeropsis* sp., which produces black pycnidia up to 0.5 mm in diameter in the bark of dead twigs and branches and in cankers. Single-celled brown conidia, most measuring 20–26 × 9–12 µm, form in the pycnidia. Fruiting in cankers that develop late in a growing season is sometimes delayed until the next spring. Pycnidia may be followed by black stromata up to 3 mm wide, in which pseudothecia form. Pycnidia sometimes form in foliar lesions on *Malus.*

Disease cycle. B. obtusa overwinters as immature and mature pycnidia and pseudothecia in bark. It also survives in buds of peach and in mummified fruit of some apple cultivars, such as Cortland. Under moist conditions during the growing season, tendrils of brown conidia ooze from the pycnidia and are dispersed by splashing water and insects. *Hippodamia convergens,* the convergent ladybird beetle, was found carrying conidia of *B. obtusa* in New Hampshire. Pseudothecia develop in the second and subsequent seasons after infection and produce ascospores that are dispersed by water and air.

Both conidia and ascospores are released throughout the season but most abundantly in spring. Spore release and infection are highly favored by wet weather with moderate temperatures (20–28°C) for 12–24 hours or more. Infection occurs in wounds of all kinds and in apparently intact thin bark stressed by heat or freezing. Leaves are penetrated through stomata. Leaf spots on *Malus* may be observed 2–3 weeks after the first large release of conidia in spring. *B. obtusa* secretes macerating enzymes and phytotoxins (mellein and related compounds) that probably play a role in pathogenesis, because host cells die slightly in advance of hyphal progress in lesions.

There seems to be just one cycle of disease per year on most hosts, but it can start any time during the growing season. Thus, pathogen generations overlap. Secondary cycles may occur on *Malus* when pycnidia form in foliar lesions and produce conidia that infect fruit or woody parts.

B. obtusa populations build up in neglected or environmentally stressed plantings, the fungus reproducing on pruned or broken twigs or those killed by other diseases. In *Malus,* *Botryosphaeria* species often colonize bark killed by fire blight (Plate 187). Control depends in part on avoiding or minimizing stress caused by water shortage or temperature extremes. Sanitary pruning eliminates twigs and branches on which the fungus sporulates. Lists of disease-free or resistant crabapples are available, but direct tests of their susceptibility to *B. obtusa* have not been made.

A second black rot fungus, *B. stevensii,* less common than *B. obtusa* in North America, occurs on *Malus, Quercus, Vitis,* and other plants and is discussed with Plate 60.

References: 104, 105, 371, 514, 539, 934, 1099, 1195, 1622, 1728, 1859, 1896, 2230, 3182, 3184, 3572, 3606, 3669, 3743, 3856, 3857, 3920, 4148, 4520

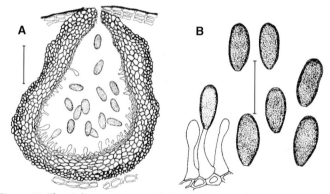

Figure 62. The *Sphaeropsis* state of *Botryosphaeria ribis.* A. Section of a pycnidium in bark. B. Conidiophores and conidia. Scale bars = 200 µm for pycnidium, 25 µm for conidia. Adapted from reference 3669 by permission of Gebrüder Borntraeger Verlagsbuchhandlung.

A–E. Dieback (A) and black rot cankers caused by *Botryosphaeria obtusa* on *Malus* 'Rosybloom' (crabapple). B. Large pruning wounds, apparently made for disease control, were sites of infection that killed one branch (right arrow) and has spread into the limb at left. The branch at center has a linear canker (arrow) that developed after removal of nearby branches exposed the bark to bright sun. C, D. Killed bark turns tan, then dark and rough where pycnidial stromata of *B. obtusa* break the surface. Cracks in bark (D) identify the margin of a canker that has temporarily ceased expansion. E. Magnified view of pycnidial stromata breaking the bark surface (ND, Jul).

F. Gummosis and an elongate flat area on a *Prunus persica* (peach) limb identify the location of a Botryosphaeria canker (GA, May).

G. Frogeye leaf spot, caused by *B. obtusa,* on *Malus* 'Rosybloom' (ND, Jul).

127

Cankers and Diebacks Caused by *Botryosphaeria rhodina*

Botryosphaeria rhodina is a versatile, opportunistic fungus that attacks woody and herbaceous plants in at least 280 genera. It occurs around the world, mainly between latitudes 40°N and 40°S. It is likely to attack trees weakened by heat, drought, freezing, or other pathogens. *B. rhodina* is usually found in its pycnidial state, *Lasiodiplodia theobromae* (syn. *Diplodia gossypina, D. natalensis, Botryodiplodia theobromae*), which develops on killed parts of leaves, stems, or fruits. The fungus causes canker, dieback, twig blight, fruit rot, seed rot, leaf blight, gummosis, root rot, collar rot, and black stain of wood. In the USA it causes, for example, cankers on *Citrus*, cankers and dieback of *Cornus florida* (flowering dogwood) in southern USA and *Elaeagnus angustifolia* (Russian-olive) in the Great Plains, and shoot blight of *Pistacia chinensis* (pistachio) in California. It contributes to decline of *Platanus occidentalis* (sycamore) in southern states, often attacking trees affected also by *Phomopsis scabra* (Plate 70) or *Xylella fastidiosa,* the cause of bacterial scorch (Plate 193). In Texas it sometimes follows or accompanies the oak wilt pathogen in *Quercus virginiana* and *Q. fusiformis* (live oaks) (Plate 118). *B. rhodina* is one of three *Botryosphaeria* species that cause cankers, twig dieback, and gummosis of *Prunus persica* (peach) trees.

Other genera of woody hosts of *B. rhodina* include the following: *Acacia, Acca, Acer, Ailanthus, Albizia, Anacardium, Araucaria, Artocarpus, Aucuba, Baccharis, Brachychiton, Bucida, Buddleja, Camellia, Carissa, Carya, Catharanthus, Cassia, Castanea, Catalpa, Chamaecyparis, Chamaerops, Cinnamomum, Cocos, Cotinus, Cotoneaster, Diospyros, Eranthemum, Eriobotrya, Eucalyptus, Eugenia, Euphorbia, Ficus, Fortunella, Gelsemium, Gleditsia, Grevillea, Hevea, Hibiscus, Ilex, Illicium, Juniperus, Ligustrum, Liquidambar, Liriodendron, Macadamia, Maclura, Magnolia, Malus, Mangifera, Melia, Metrosideros, Morus, Myrica, Nerium, Osmanthus, Pandanus, Parkinsonia, Persea, Phoenix, Photinia, Pinus, Pittosporum, Polyscias, Poncirus, Pouteria, Prosopis, Prunus, Psidium, Psychotria, Pyrus, Quercus, Rhododendron, Rhus, Rosa, Rubus, Salix, Sambucus, Sassafras, Terminalia, Theobroma, Thuja, Tilia, Torreya, Vaccinium, Vernicia, Viburnum, Vitis, Yucca, Zanthoxylum,* and *Ziziphus.*

Symptoms and signs. The cankers are usually elongate, as the fungus grows along a stem much more rapidly than around it. On *Elaeagnus angustifolia,* for example, *B. rhodina* kills long strips of bark. Plants capable of producing gum or resin (e.g., *Araucaria, Elaeagnus, Prunus*) do so at sites of infection. The gum or resin permeates diseased tissue and may exude at the edges of lesions. Lesions may otherwise be indistinct, because dead bark tends to persist and woundwood formation may be suppressed on stressed stems. Dieback caused by *B. rhodina* on *Platanus occidentalis* begins on broken or wounded twigs and advances into larger branches. The fungus grows in the sapwood, killing and staining it darkly 30–40 cm in advance of progress in the bark.

Fruit bodies of *B. rhodina* develop in killed bark and fruit lesions. Pycnidia form in black stromata 2–4 mm in diameter, single or grouped, that break through the bark surface and open by an irregular rupture. Conidia exude in a mass that soon appears black. The conidia are thick walled, elliptic in outline, initially unicellular and colorless, and measure 20–30 × 10–15 μm. They soon become equally two celled and brown with distinctive longitudinal striations. Pseudothecia, also black, are less common than pycnidia. They usually appear in crowded groups. If pseudothecia or young pycnidia are cut through with a sharp blade and examined with a hand lens, they exhibit white contents and black walls. This feature is illustrated for other *Botryosphaeria* species and related fungi in Plates 59 and 60.

Disease cycle. B. rhodina usually enters stems through wounds and broken twigs. Conidia start most new infections. They are discharged during or after rainfall, are dispersed singly or in clumps by running or splashing water and also by air, and are available as inoculum throughout the growing season. Mature conidia tolerate desiccation; they are capable of surviving from one growing season to the next on the bark of citrus trees. The importance of ascospores as inoculum is unknown. *B. rhodina* can also survive in soil as chlamydospores derived from hyphae, an attribute that may partially explain root infections. It is seedborne in *Eriobotrya, Pinus,* and presumably other plants on which it infects fruits. Pycnidia typically develop in the first autumn after infection on stems in temperate regions and may be found through the year. Pseudothecia form in spring of the year after the onset of infection.

B. rhodina causes black stain in the wood of various trees in tropical forests. The stain is associated with wounds made by boring insects. Insect vectors of the fungus are suspected to occur but have not been documented for tree diseases.

Strains of *B. rhodina* vary in virulence. Some strains can cause lesions on apparently nonstressed trees, but damage by all strains is enhanced by host stress, particularly that caused by heat and associated water shortage. *B. rhodina* grows at temperatures up to nearly 40°C and grows most rapidly at 30–34°C, several degrees above the optimum temperature for growth of most plants. Cankers develop most rapidly during hot weather and faster in drought-stressed trees than in nonstressed ones. A few examples can be cited from many available. An outbreak of dieback caused by *B. rhodina* in *Platanus occidentalis* in the Mississippi Delta in the 1950s was apparently triggered by drought, and it subsided when normal precipitation returned to that region. Water stress induced in *P. occidentalis* by *Xylella fastidiosa* (Plates 192, 193) is suspected to predispose the trees to damage by *B. rhodina.* An outbreak of cankers that led to wilt and dieback on *Cornus florida* in Alabama was associated with drought. Freeze damage or nutrient deficiency may also predispose trees to attack. Trunk cankers on *Citrus* in Texas occurred after freeze damage, and severity of stem infections on *Eucalyptus* in Brazil was linked to boron deficiency. Most strains show little host specificity, although exceptions are known. For example, a strain that occurs in Brazil is pathogenic to *Cocos nucifera* (coconut palm) but not to *Theobroma cacao* (cacao). *References:* 46, 128, 539, 540, 735, 908, 934, 1058, 1059, 1099, 1130, 1168, 1896, 2237, 2286, 2304, 2305, 2619, 2756, 3032, 3163, 3184, 3213, 3290, 3633, 3854, 3913, 4171, 4445

Figure 63. Lasiodiplodia theobromae, the anamorph of *Botryosphaeria rhodina.* A. Section through a pycnidium. B. Conidiophores, young conidia, and mature striate conidia. Scale bars = 250 μm for pycnidium, 10 μm for other structures. Adapted from reference 3163 by permission of CAB International.

A. Dieback of *Elaeagnus angustifolia* (Russian-olive) (NE, Jun).
B. An elongate canker caused by *Botryosphaeria rhodina* on the mainstem of a young *Cornus florida* (flowering dogwood). Bark was shaved to reveal the otherwise inconspicuous lesion (AL, Jul).
C–H. Dieback and cankers in *Platanus occidentalis* (sycamore). C. Dieback (FL, May). D, E. An elongate perennial canker that originated at a trunk wound. An arrowhead in D and shaved bark in E locate the edge of the lesion near its upper limit (MS, Apr). Both *Botryosphaeria rhodina* and *Ceratocystis fimbriata* f. *platani* (Plate 116) can cause such cankers. F. The mainstem of a young top-killed sycamore with bark shaved to display the edge of the lesion. *Phomopsis scabra* and *B. rhodina* were both associated with dieback in this tree (MS, Apr). G. Woundwood ridges around a canker indicate that *B. rhodina* was halted after moving into the mainstem from a branch (MS, Apr). H. Close view of sycamore bark with numerous black conidial stromata of *B. rhodina* (MS, Jun).

Photo credits: B—J. Mullen; H—T. D. Leininger

Diplodia (Sphaeropsis) Blight of Pines and Other Conifers (Plate 64)

Diplodia blight is a sometimes devastating disease of *Pinus*, especially two- and three-needle species. Although common in other conifers, it seldom causes much damage to them. The causal fungus, *Diplodia pinea* (syn. *Sphaeropsis sapinea*, a coelomycete related to *Botryosphaeria*), is globally distributed in temperate and tropical zones. It is infamous for opportunistic behavior because it causes severe damage to pines stressed by water shortage, heat, compacted soil, shade, insect damage, or frost. In a given region, Diplodia blight may be severe in pine species introduced from elsewhere, but it usually does not cause much damage to species planted within their natural ranges. It is seldom noticed in natural forests. Although *D. pinea* contaminates some seed lots, it causes losses in nurseries only where its spores are dispersed from diseased pines near seedbeds.

Damage to conifers other than pines is usually associated with environmental stress and a large supply of spores from pines. The host list includes species of *Abies* (fir), *Araucaria*, *Cedrus* (cedar), *Chamaecyparis* (false-cypress), *Cupressus* (cypress), *Larix* (larch), *Picea* (spruce), *Pseudotsuga* (Douglas-fir), *Thuja* (arborvitae), and many *Pinus* species. Those most often damaged in North America are *Pinus nigra* (Austrian pine), *P. ponderosa* (ponderosa pine), *P. resinosa* (red pine), and *P. sylvestris* (Scots pine).

A fungus known as *Diplodia pinea* (*Sphaeropsis sapinea*) f.sp. *cupressi*, formerly thought to be a host-specialized variant of *D. pinea*, causes cankers and dieback in trees of the Cupressaceae (cypress family) in Europe, the Near East, and Africa. Several lines of evidence indicate that this fungus is a different species.

Symptoms and signs. *D. pinea* causes tip blight, resinous cankers on mainstems and branches, dieback and misshapen tops, death of cones, seedling blight, gray to black stain of sapwood, and sometimes death of entire trees. The most common symptoms in North America are tip blight and death of low branches. Stunted dead shoots with short needles glued in place by resin are diagnostic markers. Lesions are initiated on growing buds, shoots, needles, and cones, causing them to cease growth. Dying shoots soon fade to straw color. Lesions are resin soaked, dark reddish brown, and often exude resin. In time the resin crystallizes, making a dead shoot hard and brittle. When disease is severe, clusters of shoots are blighted and branches are deformed.

D. pinea usually kills only current-season buds and shoots and second-year cones, unless trees have been stressed. Given host predisposition by environmental stress, the fungus progresses into older twigs and branches, first in the pith and then into sapwood and bark. Cankers, often marked by exuding resin and persistent dead needles, arise as either nongirdling shoot infections or lesions centered on wounds. Dead tops, cankers on mainstems, or scattered dead branches in otherwise green crowns indicate wound infection. Wounds made by insects, hail, or other agents are suitable entry points. Severe damage attributed to infection at insect wounds has occurred in *Pinus nigra* and *P. sylvestris* attacked by *Aphrophora parallela* (pine spittlebug), *P. nigra* infested by *Gastrodes grossipes* (a cone bug), *Pinus sabiniana* (digger pine) infested by *Petrova sabiniana* (a pitch nodule moth), and *Pinus elliottii* (slash pine) attacked by *Pissodes nemorensis* (deodar weevil). Dieback and death of whole trees result from invasion of the wood and bark of trunks and, rarely, of the major roots of severely stressed individuals.

Pycnidia, typically near 200–250 μm in diameter, break through the surfaces of killed needles, fascicle sheaths, cone scales, and the bark of twigs or branches beginning in late summer of the year of infection and continuing the next spring. On pines with scaly bark, pycnidia also form beneath superficial bark scales, appearing as small, flattened black pads accompanied by grayish mycelium. Pycnidia on needle bases may be exposed by pulling off the fascicle sheath. Short, straw-colored to gray needles adjacent to those held in place by resin are most suitable for this test. The conidia are elliptic to club shaped, dark brown, and measure 30–45 × 10–16 μm. Most are unicellular, but a few are two celled. Three morphological types of *D. pinea*, designated A, B, and C, have been recognized. They differ in conidial size, texture of conidial walls (smooth versus pitted), cultural characteristics, and aggressiveness. All three occur in North America. Type B is now considered to be a separate species, *D. scrobiculata*.

Disease cycle. *D. pinea* overwinters in dead needles, stems, and cones on trees and on the ground. Its conidia, extruded from pycnidia and dispersed throughout the year as temperature and moisture permit, cause infections on growing or wounded stems and cones. Infections leading to tip blight begin in spring on buds, succulent parts of elongating shoots, and sometimes on immature needles. The fungus can penetrate intact epidermis of succulent stems; it enters needles through stomata. If it does not immediately begin killing the substrate, it may persist as a latent colonist, or endophyte, in living asymptomatic leaves, stems, and the root-collar region of pine seedlings. Latent colonies become active and cause damage when the plants become stressed, as by drought.

Cones become infected while growing rapidly in the spring of their second year. Cone infections are inconsequential for general health of the trees but are important for their contribution to epidemics, because cones are sources of inoculum for infection of shoots and wounds. Severe tip blight on pines of less than cone-bearing age is uncommon in North America.

Given predisposed trees, epidemics of Diplodia blight are promoted by wet spring weather, which favors spore production, dispersal, and infection. Conidia are capable of germinating within a few hours at 12–36°C. A wet period of 12 hours suffices for germination and infection. In warm weather, symptoms appear 3–4 days later. The pathogen grows most rapidly at about 28°C.

The most important cultural measures for avoidance or mitigation of damage by *D. pinea* are to provide and conserve water during drought. Trees with Diplodia blight may be pruned to improve their appearance, but this practice does not affect the likelihood of new infections because conidia are released from diseased cones on green branches. Nitrogen fertilization, while not necessarily stressing pines, may predispose them to *Diplodia* infection. Severe disease has been positively correlated with soil and plant nitrogen levels, perhaps because nitrogen promotes growth of succulent, susceptible shoots.

References: 157, 408–410, 525, 728, 800, 901, 934, 1089, 1426, 1953, 2469, 3034, 3035, 3165, 3687, 3718, 3719, 3833, 3834, 3837–3839, 3913, 3931, 3932, 4234, 4310, 4311, 4395, 4519, 4535

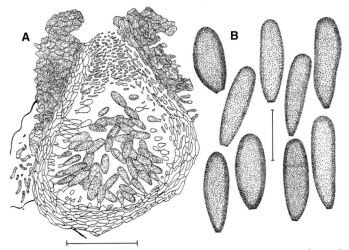

Figure 64. *Diplodia pinea.* A. Section through a pycnidium. B. Conidia. Scale bars = 100 μm for pycnidium, 25 μm for spores. Adapted from reference 3165 by permission of CAB International.

A. *Pinus nigra* (Austrian pine) with branch tips blighted and low branches killed by *Diplodia pinea* (NY, Sep).

B. A shoot tip of *P. mugo* (mugo pine) killed by *D. pinea* the previous summer (NY, May).

C, D. Basal infection of partly grown needles of *P. nigra* halts their growth and causes them to fade to straw color (NY, Jun)

E, F. Successively closer views of the basal parts of *P. nigra* needles killed by *D. pinea*. The fascicle sheath has been removed, revealing a bleached basal zone where pycnidia have ruptured the epidermis (NY, May).

G, H. Cones of *P. nigra*, killed by *D. pinea* during their second season of development and observed the next spring, have tightly closed scales with profuse pycnidia (NY, May).

Sphaeropsis Canker and Dieback of Elm (Plate 65)

Ulmus species (elms) are affected by cankers and dieback caused by various fungi. *Sphaeropsis hypodermia* (syn. *S. ulmicola*, *Botryodiplodia hypodermia*), a coelomycete with no known teleomorph, is commonly associated with this damage. Our previous treatment of this subject included the concept of two similar elm-colonizing species, *B. hypodermia* and *S. ulmicola*, for which North American records were tangled. Mycological opinion now favors the concept of one species with characteristics that encompass those described for both. If the fungus is truly one species and is classified in *Sphaeropsis*, then the appropriate name for it is *S. hypodermia*. Taxonomy and nomenclature of fungi related to *S. hypodermia* were in flux as this was written, and a suggestion had been made that *Sphaeropsis* be considered a synonym of *Diplodia*. If the suggestion is fully implemented, then the elm pathogen will be renamed yet again.

S. hypodermia in the inclusive sense is widespread in North America, its reported range in the North extending from Montana and Saskatchewan to Nova Scotia, and in the South from Oklahoma to Virginia. Its recorded history is dichotomous, as follows. The fungus later called *B. hypodermia* was first described from *Ulmus minor* (European field elm, smooth-leaf elm) in Germany in 1870. European reports of it in international publications are scarce, however. It was found on *U. minor* in Massachusetts in the late 1920s and was shown to be capable of causing cankers in small elm stems. Strains identified as *B. hypodermia* began to attract attention in the USA in the 1970s as plant pathologists investigated shelterbelt deterioration in the Great Plains. Surveys in that region pointed to *B. hypodermia* as the most common and important of the fungi associated with cankers and dieback in *U. pumila* (Siberian elm). Pathogenicity of the fungus was confirmed.

S. ulmicola was first described from southern Ontario in 1891 and was recognized as a significant pathogen of elms beginning in 1920. At that time it was associated with severe twig and branch dieback of *U. americana* (American elm) shade trees and nursery stock in Wisconsin. It was shown capable of causing lesions in detached elm twigs, but its pathogenicity was not tested further. We merge the accounts for *B. hypodermia* and *S. ulmicola* under the name *S. hypodermia* at this point.

Severe damage to elms by *S. hypodermia* appears linked to stress that reduces trees' ability to resist infection. For example, although *U. pumila* has been widely planted in central USA because of its reputation for hardiness and drought tolerance, it is often stressed there by heat, drought, freeze-thaw cycles, and perhaps herbicides. Thus predisposed, it is attacked by opportunistic fungi, including *S. hypodermia*, that are usually inconsequential on elms in eastern North America. Similarly, the severe damage shown in Plate 65 on *U. parvifolia* (Chinese elm) occurred in a too-dense planting on a site prone to water shortage. *U. glabra* (Scotch or wych elm) on stressful sites is also attacked by *S. hypodermia*. The damage reported for *U. americana* in Wisconsin in 1920 followed a drought. Sphaeropsis dieback is usually of minor importance on *U. americana*, even in north-central USA.

Symptoms and signs. Cankers develop on twigs, branches, or trunks and may girdle small limbs within one season. Newly infected bark is water-soaked, soft, and reddish brown to brownish black. The transition from diseased to normal tissue is abrupt. The sapwood beneath diseased bark and cambium also turns reddish brown and later dark brown. During the year of infection discolored wood does not extend much beyond the killed bark. In subsequent years discolored wood extending from cankers may indicate colonization by secondary organisms. Lesions caused by *S. hypodermia* on *U. pumila* typically kill branches up to 10 cm in diameter, but cumulative damage sometimes leads to death of larger limbs or whole trees. Foliage above girdling cankers on *U. parvifolia* and *U. pumila* wilts and dies, sometimes turning yellow first. Diseased twigs and branches of *U. americana* display bright yellow leaves in mid to late summer; these are shed prematurely but do not wilt. Adventitious sprouts often develop below cankers on large stems and create a bushy appearance. On *U. parvifolia* the bark sloughs off old cankers, leaving conspicuous scars.

Pycnidia, mostly 200–330 μm in diameter, develop in killed bark within 2 months after summer infection, causing prominent roughening of the surface (Plate 65F). Pycnidia that are cut through by shaving the surface of colonized bark appear black walled with white contents (Plate 65G, H). The conidia are thick walled and initially single celled and colorless, but they eventually become brown, and some become two celled. Conidia of strains identified as *B. hypodermia* measure mostly 20–36 × 13–18 μm; those of strains identified as *S. ulmicola* measure 20–30 × 11–15 μm.

Disease cycle. Mature conidia are present all year and exude in masses from pycnidia after wetting. They are dispersed by splashing or running water. Other possible modes of transmission are unexplored. The fungus infects wounds and other injuries to bark. It also colonizes bark at the ends of cut or broken branches and may thus be present in trees without cankers or dieback. Winter-injured twigs and branches of *U. americana* are commonly infected. In experiments, bark of intact twigs was not penetrated. The fungus causes most damage if infection occurs during the growing season while the trees are stressed by insufficient water. In Nebraska, lesions up to 72 cm long developed on mainstems of *U. pumila* during the first season after inoculation during July–September, and many of the lesions enlarged and caused dieback and death of trees during the second season. Inoculation at other times led to small cankers that became inactive after one season and later were covered by woundwood. In a separate study in North Dakota *U. americana* was most susceptible during April–June.

Lesions in stressed trees expand rapidly at 15–30°C. *S. hypodermia* grows into sapwood beneath bark lesions and has been observed near the pith of diseased twigs. Perhaps this fungus also colonizes normal bark and persists as an endophyte until host defenses are impaired, as related fungi do in various hosts. Some observers have speculated that herbicides predispose *U. pumila* to infection, because in central USA this tree commonly exhibits herbicide-induced growth abnormalities together with symptoms caused by *S. hypodermia*. No direct test of the influence of herbicides on severity of damage by this fungus has been reported, however.

Strains of *S. hypodermia* vary in aggressiveness to *U. pumila*, as reflected by differences in lesion size and incidence of dieback on inoculated branches. On the other side of the interaction, clones of *U. pumila* have been shown to vary in susceptibility as indicated by the same criteria. These facts indicate that elm lines resistant to *S. hypodermia* could be selected or bred.

References: 564, 854, 934, 1099, 1298, 1794, 1795, 2157, 2158, 2160, 2931, 3054, 3286, 3290, 3816, 3871, 4519

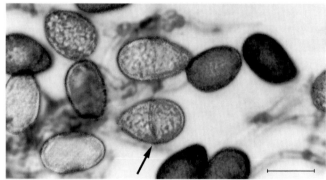

Figure 65. Conidia of *Sphaeropsis hypodermia* with differing degrees of pigmentation. A septum (arrow) was forming in one spore when it was photographed. Scale bar = 20 μm. Adapted from reference 3290.

Dieback and cankers caused by *Sphaeropsis hypodermia*.

A. Dieback and mortality of *Ulmus pumila* (Siberian elm) in a windbreak (ND, Jul).

B–E. Cankers and dieback on *U. parvifolia* (Chinese elm). B, C. Cankers and adventitious shoots, some shoots wilting as the result of canker extension. Bark is sloughing from cankers. D. Adhering bark almost masks a year-old trunk canker, its presence indicated by the slightly raised and cracked bark at its edges. E. A lesion similar to that in D but with bark removed (IL, Jun)

F. Pycnidia developing in *U. pumila* bark, causing numerous minute swellings (ND, Oct).

G, H. Successively magnified views of pycnidia exposed by shaving the surface of dead bark. Pycnidial walls are dark and the contents white to gray depending on frequency of dark-pigmented spores in them (ND, Oct).

Sphaeropsis Knot and Diplodia Gall (Plate 66)

Sphaeropsis tumefaciens, a coelomycete, causes galls and brooms on woody plants in Florida, Mexico, the Caribbean region, and Hawaii. Sphaeropsis knot, as the disease is called, is best known for its damage to *Citrus*, especially in Jamaica. It was once common in citrus groves in Florida but is now confined mainly to ornamental and wild plants there.

The known host list of *S. tumefaciens* includes nearly 50 woody angiosperm species in more than 30 genera, as follows:

Albizia lebbek, *Bauhinia galpinii* (red bauhinia), *B. variegata* (mountain-ebony), *Callistemon citrinus* (crimson bottlebrush), *C. viminalis* (weeping bottlebrush), *Carissa macrocarpa* (Natal-plum), *Cinnamomum camphora* (camphor tree), *Citrus aurantifolia* (lime), *C. aurantium* (sour orange), *C. jambhiri* (rough lemon), *C. madurensis* (calamondin), *C. maxima* (pummelo), *C. medica* (citron), *C. ×paradisi* (grapefruit), *C. sinensis* (sweet orange), *Crataegus* species (hawthorn), *Eucalyptus cinerea*, *Eugenia uniflora* (Surinam-cherry), *Ficus benjamina* (weeping fig), *Ilex ×attenuata* (topal holly), *I. cassine* (dahoon), *I. cornuta* (Chinese holly), *I. crenata* (Japanese holly), *I. opaca* (American holly). *I. vomitoria* (yaupon), *Ixora* sp., *Jatropha* sp., *Lagerstroemia indica* (crape-myrtle), *Ligustrum japonicum* (Japanese privet), *L. sinense* (Chinese privet), *Melaleuca decora*, *Morella cerifera* (wax-myrtle), *Nerium oleander* (oleander), *Parkinsonia aculeata* (Jerusalem thorn), *Persea americana* (avocado), *Philadelphus inodorus* (scentless mock-orange), *Pittosporum tobira* (Japanese cheesewood), *Poncirus trifoliata* (trifoliate orange), *Portlandia grandiflora*, *Prunus* species (stone-fruit trees), *Pyracantha coccinea* (firethorn), *Rhus* species (sumac), *Schinus terebinthifolius* (Brazilian pepper-tree), *Ulmus parvifolia* (Chinese elm), *U. pumila* (Siberian elm), *Vitis rotundifolia* (muscadine grape), and *Wisteria sinensis* (Chinese wisteria).

Cupressus sempervirens (Italian cypress) is also a listed host, but that record requires corroboration. *S. tumefaciens* is nonspecialized and is likely to be found on additional species.

Symptoms and signs. Symptoms on all hosts include localized swellings (hypertrophy) and dieback of twigs and branches. Broom formation by growth of multiple shoots from hypertrophic tissue occurs on most hosts. The galls, also called knots, are woody; they are nearly the color of normal bark while young and become rough, fissured, and dark with age. Eventually the bark sloughs off dead galls. On *Ilex* symptoms range from swelling of elongate segments of young twigs to irregular knobby galls on mature branches. Numerous stunted, sometimes leafless, twigs arise close together, forming loose brooms (Plate 66A, C). Foliage on diseased branches turns yellow and drops, and dieback ensues. Typical symptoms on *Callistemon viminalis* are globose galls 1–9 cm in diameter on twigs and small branches (Plate 66F), sometimes in a row along a branch. The galls become darker than normal bark and have knobby surfaces and deep indentations. Galls at nodes often give rise to multiple buds that produce brooms consisting of somewhat stunted shoots. Galls on internodes frequently remain devoid of shoots. Occasionally cankers develop instead of galls; hypertrophy and roughening occur at canker margins. Bark within cankers separates from the wood at an early stage. In time, diseased branches die back to below the galls or cankers. *Carissa macrocarpa* produces either greatly roughened, often elongate galls (Plate 66E) or sunken cankers in response to *S. tumefaciens*.

Sapwood within galls or beneath cankers often becomes darkly stained as a result of colonization by dark hyphae of the parasite. The staining sometimes extends several centimeters in each direction from the original locus of infection, and galls that arise in a series on a branch may be connected by hyphae in the wood. Internal spread and development of secondary knots up to 90 cm from the point of infection in citrus branches have been documented.

Black pycnidia of *S. tumefaciens*, 180–220 μm in diameter, arise singly or in small groups in dead outer bark of cankers, galls, and dead twigs. They produce large (20–34 × 6–10 μm), colorless to faintly yellow, mostly unicellular conidia. Pycnidia on *Persea americana* have been found containing microconidia (see below). Usually, however, pycnidia are hard to find. Firm diagnosis may then depend on isolation and identification of the pathogen in pure culture. It grows readily from gall or canker tissue onto laboratory media, producing black mycelium, pycnidia, and conidia as above and also microconidia measuring 3–5 × 1.5 μm. Two-celled conidia also are sometimes produced in the pycnidia that form in pure cultures. Growth in culture is most rapid near 30°C, a higher optimum than most plant pathogens have. Isolates from one host

plant readily infect wounds on other reported hosts. No sexual state of this fungus is known.

Disease cycle. The mode of transmission of *S. tumefaciens* in nature is unknown. Wounds are common infection courts, but unwounded succulent stem tissues or buds may also be susceptible. These matters and the duration of the disease cycle need study. The parasite can persist several years in colonized tissues, first in a living stem and then for a time in dead bark. Inoculation of various hosts has resulted in galls or cankers that become noticeable after 6–15 weeks in most cases. Galls may enlarge for several months before attaining final size.

Plants affected by relatively few knots or cankers can be cured by removing the affected parts, provided that each cut is made at least 15 cm below a swelling or gall. The disease was controlled in commercial citrus through hygiene in propagation activities and by sanitary pruning and destruction of galls in groves.

In the North, *Diplodia tumefaciens*, a fungus similar to *S. tumefaciens* in appearance and cultural characteristics, causes galls and rough bark on various *Populus* species (aspen and poplar) from Ontario westward through Canada and northern USA to British Columbia (Plate 66G). *D. tumefaciens* has not been reported to invade sapwood or to induce brooms, however. No comparisons of *D. tumefaciens* with *S. tumefaciens* have been reported.

Taxonomy and nomenclature of fungi related to *S. tumefaciens* and *D. tumefaciens* were in flux as this was written, and a suggestion had been made that *Sphaeropsis* be considered a synonym of *Diplodia*. If the suggestion is fully implemented, then the pathogens mentioned here must be compared and at least one renamed.

References: 46, 609, 934, 1099, 1153, 1203, 1213, 1582, 1683, 1720, 2471, 2472, 3284, 4495, 4496, 4499

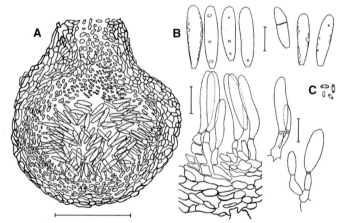

Figure 66. Sphaeropsis tumefaciens. A. Section through a pycnidium. B. Conidia and conidiophores. C. Microconidia. Scale bars = 100 μm for pycnidium, 10 μm for other structures. Adapted from reference 1720 by permission of CAB International.

A–D. Brooms and dieback on *Ilex cassine* (dahoon holly), caused by *Sphaeropsis tumefaciens*. A. Overview of twig and branch dieback. B. A diseased branch tip with upturned, dark brown leafless shoots arising from the infected portion and prematurely senescent leaves beyond the point of infection. C. A diseased branch tip infected at two points: on a lateral twig, causing dieback, and at the branch tip, causing broom formation. D. A branch tip killed by *S. tumefaciens*, showing swelling of the once-infected portion of the branch axis and blackened upright twigs of a broom (FL, Apr).

E. Stem galls caused by *S. tumefaciens* on *Carissa macrocarpa* (Natal-plum). Numerous small galls cause enlarged and greatly roughened stems (FL, Apr).

F. Sphaeropsis knot of *Callistemon viminalis* (bottlebrush). Multiple shoots growing from the knot are typical of the disease on this plant (FL, Apr).

G. Galls caused by *Diplodia tumefaciens* on a branch of *Populus* sp. (poplar) (BC, Jun).

Photo credit: G—J. Hopkins, Canadian Forest Service

135

Melanconis Diebacks and Sooty Canker (Plate 67)

Melanconis Diebacks

Several species of *Melanconis* (Diaporthales, Melanconidaceae) cause cankers, dieback of stressed trees, or both. Their anamorphs (conidial states) are coelomycetes in the genus *Melanconium* that develop in recently killed bark. Some fungi in this group also grow as endophytes in living twigs. Known pathogens in North America, with names of their anamorphs in brackets, include: *Melanconis alni* [*Melanconium apiocarpum*] on *Alnus* (alder); *Melanconis juglandis* [*Melanconium oblongum*] on *Juglans* (butternut, walnut); *Melanconis stilbostoma* [*Melanconium bicolor*] on *Betula* (birch); and *Melanconis marginalis* [*Melanconium marginale*] on *Alnus*.

Melanconis stilbostoma colonizes bark of *Betula* species when the trees are stressed by insect attack (e.g., by *Agrilus anxius*, the bronze birch borer), drought, or other environmental conditions. This fungus occurs across North America and in Europe and Japan. Colonized bark of young stems, normally dark, turns orange (Plate 67F). Fungal stromata that develop into acervuli cause the bark to rise in circular to elliptic bulges, which eventually break open and release conidia. The stromata are 1–2 mm across and short-conical with white tips. Brown, single-celled, obovoid conidia, most measuring 9–14 × 5.5–7 µm, form in cavities surrounding the center of each stroma (Fig. 67B, C) and are eventually released en masse as a black cirrhus or blob. Conidia are dispersed by water. Perithecia may form in the same stromata later.

Melanconis juglandis invades twigs or the bark of branches and trunks of *Juglans cinerea* (butternut) and other *Juglans* species weakened or newly killed by other agents. It is common in eastern North America and occurs in Korea and Japan. In pathogenicity tests it has caused lesions in stems of *J. ailanthifolia* (Japanese walnut), *J. cinerea*, *J. nigra* (black walnut), and *J. regia* (Persian walnut). *M. juglandis* var. *caryae* occurs on *Carya glabra* and *C. tomentosa* (pignut and mockernut hickories, respectively). *M. juglandis* var. *tiliae* occurs on *Tilia* (linden). Acervuli of *M. juglandis* form in the bark and produce unicellular, brown, ovoid conidia, mostly 19–25 × 9–12 µm in size. Massed conidia are extruded as black tendrils or rounded black masses on the bark surface and are dispersed by water. Perithecia follow the acervuli but are less conspicuous. Their bodies are embedded in the bark, and their necks protrude. A similar fungus, *M. carthusiana* (anamorph *Melanconium juglandinum*), causes twig and branch dieback of *Juglans regia* (Persian walnut) and other walnuts in Europe.

References: 294, 298, 1099, 1298, 1358, 2105, 2107, 3624, 4271

Sooty Canker

This disease, also known as branch wilt, limb wilt, and Nattrassia dieback, occurs around the world in diverse trees that grow where summer heat and water stress become intense. The pathogen is an unnamed *Botryosphaeria*. Its anamorphs are the coelomycete *Fusicoccum mangiferum* (syn. *Nattrassia mangiferae, Hendersonula toruloidea*) and the thallospore-forming *Scytalidium dimidiatum*. It occurs in western North America as far north as British Columbia. Plants damaged by it in North America include *Arbutus menziesii* (madrone), *Castanea sativa* (European chestnut), *Citrus* species, *Eucalyptus camaldulensis* (Murray red gum), *Euonymus, Ficus carica* (fig), *Juglans hindsii* (Hinds walnut), *J. regia* (Persian walnut), *Morus alba* (white mulberry), *Populus fremontii* (Fremont cottonwood), *Prunus armeniaca* (apricot), *P. dulcis* (almond), *P. persica* (peach), and *Sequoiadendron giganteum* (giant sequoia). Numerous additional species are affected elsewhere.

Symptoms and signs. Cankers and/or dieback develop on scattered branches. Foliage on these either shrivels and clings to the twigs or turns yellow and drops prematurely. Lesions appear sooty because a dry black mass of spores develops beneath the periderm and causes it to flake off. Bark beneath the spore mass is brown to nearly black, and the sapwood is stained gray to black. Sapwood may be killed beyond the margins of lesions in bark. Citrus and stone-fruit trees often exude gum at edges of lesions. Cankers become sunken on branches that survive infection.

The sooty spore mass arises as a dense mat of branched, septate brown hyphae, the cells of which separate into one- and two-celled spores (thallospores). Pycnidial stromata up to 0.5 mm in diameter may also form beneath the periderm and rupture it. The stromata

contain single or multiple cavities in which conidia form. Conidia in black blobs or cirrhi exude during wet weather. Mature conidia are three celled, usually 10–14 × 4–6 µm, and narrowed at both ends. The central cell is darker than the others (Fig. 67F). The teleomorph is not seen in nature.

Disease cycle. The pathogen persists as mycelium and thallospores in and on killed stems. Thallospores are dispersed by air or water, can survive hot, dry weather for long periods, germinate readily in fresh wounds, and cause new infections. The conidia are dispersed primarily by water. Most infections are thought to begin in cracks or wounds during mild, wet weather in winter and to remain latent or restricted until host defenses are impaired in summer. Branches with sunburn become highly susceptible, even if the injury is mild. Water shortage, waterlogged soil, freezing, and predisposition by other pathogens have also been linked to susceptibility.

References: 604, 912, 1065, 1859, 2495, 2675, 2789, 2896, 2985, 3690, 3762. 3914, 4095, 4382

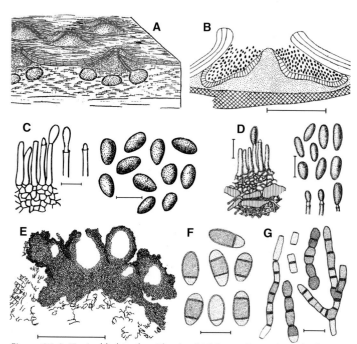

Figure 67. A. Typical habit of perithecia of *Melanconis* species in bark. B, C. The anamorph *Melanconium bicolor* on *Betula*. B. An acervulus with a sterile center. C. Conidiophores and conidia. D. *Melanconium oblongum* on *Juglans*. E–G. Anamorphs of the *Botryosphaeria* species that causes sooty canker. E, F. *Fusicoccum mangiferum*: a pycnidial stroma and conidia. G. *Scytalidium dimidiatum*: arthric hyphae that at maturity break up into dark thallospores. Scale bars = 500 µm in B and E; 10 µm in C, F, and G; 25 µm in D. The sketch in A is not scaled. A adapted from reference 4271; B–D adapted from reference 2105 by permission of the Mycological Society of Japan; E–G adapted from reference 3914 by permission of Cambridge University Press.

A–F. Melanconis dieback of *Betula pendula* (European white birch). A. Branch dieback on a stressed tree. B, E. Part of a trunk with an elongate canker associated with borer galleries. Blisters in bark at left in E are sites of developing fruiting bodies of *Melanconis bicolor*. The canker margin is revealed where the periderm was shaved off. C, D. Successively closer views of black masses of conidia on dead bark. F. Close view of acervuli of *M. bicolor* with black masses of conidia still beneath the periderm of a twig (all NY, May).

G. Part of a *Juglans cinerea* (butternut) stem killed by Sirococcus canker (Plate 57), colonized by *Melanconis juglandis*, and bearing black masses of conidia of this fungus (MN, Aug).

H. A twig of *J. microcarpa* (little walnut) with rounded black masses of conidia of *M. juglandis* (IL, Jun).

I–K. Sooty canker of *Morus alba* (white mulberry). I. Dieback. J, K. Diseased branches with periderm cracked and peeling, exposing sooty masses of spores (AZ, Apr).

137

Diaporthe and Phomopsis Cankers and Diebacks, Plates 68–72

Overview

Numerous *Diaporthe* species (Diaporthales, Diaporthaceae) cause or are found associated with leaf spots, cankers, and dieback of woody plants. These fungi range from aggressive parasites to saprobes that colonize only dead plant parts. Many are opportunistic invaders of bark and sapwood predisposed or injured by wounds, transplanting shock, drought, freezing, or other pests and pathogens.

Diaporthe species have black perithecia shaped more or less like flasks with spherical bodies and cylindrical necks. The bodies, often somewhat flattened, are embedded in bark or the outermost wood, sometimes in a distinct stroma that may be black or light at the surface. Perithecial necks protrude (Fig. 68; Plate 69I). Two-celled, colorless ascospores are produced in the perithecia. These fungi also produce pycnidial anamorphs (conidial states) in the genus *Phomopsis*. Pycnidia (Figs. 68, 71, 72) precede perithecia in killed tissues, often in the same stroma. In a typical *Phomopsis*, each pycnidium produces two kinds of colorless, unicellular conidia— ellipsoid to spindle-shaped cells that germinate readily and curved, linear cells that do not germinate (α and β conidia, respectively). The α conidia usually contain two oil droplets, called guttules, one near each end of the spore (Fig. 68D). Unfortunately for diagnosticians, the majority of *Phomopsis* species do not have a known *Diaporthe* state and do not produce β spores readily.

There are hundreds of *Diaporthe* and *Phomopsis* species names, the majority reflecting names of plants colonized by these fungi, but the number of species that can be distinguished by morphological and cultural criteria is much smaller. Some species are host specialized; others such as *D. eres* (a species complex) colonize diverse plants. Some associate with their hosts as endophytes and later as opportunistic pathogens, finally producing fruit bodies in dead tissues. The endophytes apparently establish latent colonies that are held in check by plant defenses until the host organ dies or is weakened by senescence or stress. *P. occulta* on many plants and *P. salicina* on *Salix* (willow) are common endophytes of leaves and/ or twigs. One basis for distinguishing between aggressive parasites and opportunistic colonists of debilitated tissue is the time of year when they cause lesions. Aggressive parasites such as *P. arnoldiae* (Plate 68), *P. amygdali* (Plate 69), and *P. juniperivora* (Plate 72) can invade and kill previously healthy tissue during the growing season. Most of the fungi mentioned with Plates 68–72, however, are opportunists that invade stressed or dying tissues. Even so, they may cause severe damage.

Ascospores and/or conidia of *Diaporthe* and *Phomopsis* are dispersed in splashing and windblown water droplets. When a pycnidial substrate is moist, conidia in a mucilaginous matrix extrude from the fruit body. Under wet conditions the spore mass appears as a blob that spreads out when diluted with water. If the substrate is somewhat drier, the spore mass emerges as a tendril, also called a cirrhus (Plates 69G, 70F). Ascospores in mixture with partially dissolved asci extrude similarly from perithecia. Conidia or ascospores with their sticky matrix are also well suited for dispersal by vagrant insects, but the importance of this mode of dispersal is unknown for most fungi in this group. Long-distance dispersal with diseased or infested plants is common but poorly documented.

Diaporthe and *Phomopsis* survive winter as mycelium and often as fruit bodies in dead bark and sometimes in leaves or fruit. The mycelium also colonizes outer sapwood. Infection may be caused by either conidia or ascospores. Fresh wounds, including natural ones such as leaf scars, are common sites of infection. Some species, however, can penetrate intact young leaves, shoots, or fruits; *D. citri* on *Citrus* and *D. alleghaniensis* on *Betula alleghaniensis* (yellow birch) are examples. Secondary cycles of disease are generally unimportant or absent because most plants are susceptible during only one part of the year, either the dormant period or the first part of the growth period.

References: 124, 233, 677, 1099, 1298, 1859, 3202, 3250, 3913, 4115, 4270

Phomopsis Canker of Russian-olive

The Asian tree *Elaeagnus angustifolia* (Russian-olive, or oleaster) has been planted widely in North America because of its winter hardiness, tolerance of drought and salt spray, and attractive silvery foliage. Initially it seemed relatively free from pests and diseases, but three serious canker diseases were eventually noticed in midwestern USA. Phomopsis canker, caused by *Phomopsis arnoldiae*, is one of them. The others are caused by *Botryosphaeria rhodina* (Plate 63) and *Tubercularia ulmea* (Plate 88).

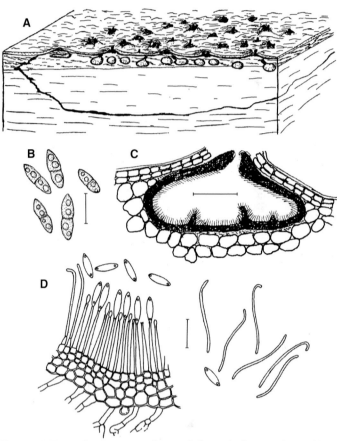

Figure 68. Diaporthe eres. A. Its fruiting habit in bark. Pycnidia and later perithecia develop in a stroma composed of fungal hyphae and plant tissue, delineated by a pseudosclerotial plate, the heavy black line beneath fruit bodies. B. Ascospores. C. A pycnidium shown diagrammatically. D. Conidiophores and α and β conidia. Scale bars = 100 μm for pycnidium, 10 μm for spores. The sketch in A is not scaled. A and B from reference 4270; C and D adapted from reference 124 by permission of Gebrüder Borntraeger Verlagsbuchhandlung.

A–F, I. Phomopsis canker of *Elaeagnus angustifolia* (Russian-olive).
  A. A young tree with its top killed by a girdling canker on the trunk. B. Lesions in smooth bark of young stems are reddish brown near the margin and dark brown to nearly black in the center. C. Gum, amber-brown and translucent when fresh, exudes from the margins of some expanding cankers; on drying it persists as a smooth, hard, black deposit. D. The bark surface in the center of a canker is soon roughened by developing fruit bodies of the pathogen. E. Magnified view of young, grayish tan pycnidial stromata of *Phomopsis arnoldiae* that have recently broken through the bark surface. F. During the year after infection, bark in nongirdling cankers splits. I. Old pycnidial stromata appear as clusters of tiny black cushions on the bark of a year-old canker (all NY, Sep).

G, H. Lesions apparently caused by *Phomopsis* sp. on year-old leaves of a *Rhododendron* hybrid. G. Young lesions are brown; older ones develop grayish white centers. H. Close view of a lesion with numerous pycnidia (NY, May).

J, K. Surface and transverse views of a canker on a *Liriodendron tulipifera* (tuliptree) sapling. J. Numerous pycnidia of *Phomopsis* sp. roughen the surface of the elongate lesion. K. Transverse view reveals that the lesion formed between periods of host growth and that sapwood beneath the canker is dead or dying (NY, Sep).

Within 15 years after Phomopsis canker was first described in 1967, it was found in several midwestern and eastern states, Ontario, Quebec, and British Columbia. In the USA, infected nursery stock was planted in widely scattered landscapes where wilting, dieback, and cankers later attracted attention. In Canada the disease was noted only in nurseries, usually on plants imported from Europe. In some instances stock that appeared healthy on arrival developed cankers during the next year.

Symptoms and signs. P. arnoldiae kills seedlings and saplings and causes dieback and cankers on larger plants. Shriveled, faded foliage clinging to killed branches or to entire small trees from midsummer through autumn calls attention to the disease. Seedlings, twigs on larger plants, and occasionally saplings up to 2 m tall may wilt and die without developing distinct cankers. Usually, however, a definite lesion 10–30 cm long is visible at the base of the dead or dying part.

Young cankers in smooth-barked branches or stems are reddish brown to nearly black, and the underlying sapwood is reddish brown. Gum often exudes, forming amber-brown blobs. These spread out somewhat during wet weather and in time darken and dry to a nearly black crust. Cankers on trunks and scaffold limbs appear as dark depressed areas where the bark splits as the result of drying in the lesion and growth of the stem around the edges of the lesion. Whether cankers enlarge during more than one season has not been reported.

Within 1–4 weeks after infection, pimplelike eruptions appear on the surface of a lesion. The bark soon breaks at these points, exposing clusters of pycnidial stromata. These structures, initially grayish tan, darken with age and eventually become nearly black. Usually they are abundant by the time the disease is detected, and they remain prominent for at least a year. Pycnidia have even been found on roots of diseased seedlings. Individual stromata are nearly 1 mm in diameter and contain one to several fertile cavities with two types of colorless, unicellular spores: spindle-shaped α conidia measuring $5–11 \times 1.5–2.5\ \mu m$ and curved, filamentous, nongerminating β conidia $15–20 \times 0.7–1\ \mu m$ in size. *P. arnoldiae* can be isolated readily. It commonly forms pycnidia in pure cultures within 3 weeks at 24–30°C. Development is slower at lower temperatures.

Disease cycle. P. arnoldiae probably overwinters as pycnidial stromata in cankers and releases conidia when temperature and moisture permit. Its transmission in nature has not been studied. Fungi similar to this species are dispersed as spores in splashing or wind-driven rain and may be transmitted by contact when plants are wet. Inoculation studies have shown that fresh wounds in stems or roots of *E. angustifolia* are readily infected by *P. arnoldiae* at any time during the growing season. Lesions become apparent within 1–2 weeks, and small stems wilt within 3–4 weeks. Severe outbreaks of Phomopsis canker on saplings, as are seen occasionally in nurseries, could result from transmission to pruning wounds. This is a matter for research.

P. arnoldiae has also been reported to cause dieback of *Juglans nigra* (black walnut) seedlings in nurseries. Lesions form at terminal buds or leaf scars, kill only a portion of the shoot, and expand during only one season. Most seedlings survive but develop multiple stems unless pruned back to leave only one or two buds. Whether the fungus from *J. nigra* can infect *Elaeagnus* or vice versa is unknown.

P. arnoldiae may be the anamorph of *Diaporthe elaeagni*, which is widely distributed in Europe, but this connection has not been proved. In North America *D. elaeagni* has been found on dead branches of *Elaeagnus commutata* (silverberry) and *E. multiflora* (cherry elaeagnus, cherry silverberry) but has not been associated with Phomopsis canker of *E. angustifolia* or *J. nigra*.
References: 110, 111, 664, 1361, 2718, 3913

Diseases of *Rhododendron*

A *Phomopsis* species develops in more or less circular brown lesions with a narrow purple-brown line at the junction of healthy and diseased tissue on year-old leaves of rhododendrons (Plate 68G, H). In time the centers of the lesions whiten. Pycnidia break through the surface in both the white and the brown areas. This fungus is but one of several *Phomopsis* and *Diaporthe* species—including *D.*

vaccinii and the ubiquitous *D. eres*—associated with leaf spots and twig dieback of rhododendrons and azaleas. *P. ericaceana* occurs on leaves of *R. macrophyllum* (Pacific rhododendron) in California. In southeastern USA a *Phomopsis* isolated from azalea was inconsistently pathogenic when inoculated into azalea stems. It was described as mildly pathogenic to stems of two other species of *Rhododendron*, *Kalmia latifolia* (mountain-laurel), *Leucothoe axillaris* (coastal doghobble, swamp dog-laurel), and *Vaccinium arboreum* (farkleberry), all in the Ericaceae.
References: 1101, 2650, 4108

Phomopsis Canker of Almond and Peach

Phomopsis amygdali (syn. *Fusicoccum amygdali*) causes cankers and dieback of *Prunus dulcis* (almond), *P. persica* (peach), and *P. triloba* (flowering almond, flowering plum). This fungus occurs in North and South America, Europe, northern Africa, and Japan. The disease on *P. persica*, the plant usually affected in the USA, is called constriction canker, Fusicoccum canker, or Phomopsis dieback. The principal symptoms are elongate cankers on twigs and wilting and death of leaves and shoots beyond the cankers. Strains of the fungus in New Jersey also cause large brown leaf spots on peach, but foliar lesions are usually not observed in southeastern states. Lesions develop at nodes on twigs in spring and early summer, reflecting the role of leaf scars and buds as sites of infection. Leaf scars become infected in autumn, buds in either autumn or spring, and fresh wounds any time during the growing season if wet weather favoring spore dispersal and infection occurs at the time of wounding. Gum infiltrates tissues in lesions and the sapwood beneath them, but exudation of gum from lesions is usually slight. Pycnidia form in both foliar and twig lesions and usually produce only fusiform (spindle-shaped) α conidia $(5–10 \times 2.3–3.8\ \mu m)$, but occasionally produce both α and β conidia. Wilting is caused by girdling lesions and by uncontrolled transpiration from diseased shoots, induced by the phytotoxin fusicoccin, which is secreted by the pathogen and translocated in xylem. This toxin causes permanent stomatal opening.

When Phomopsis dieback first became common in peach-growing areas in southeastern USA, the pathogen there was thought to differ from *P. amygdali*. Comparisons of strains in culture and of their nuclear DNA sequences, however, revealed that the strains from southeastern USA were highly similar to strains of *P. amygdali* from Spain and Portugal.
References: 22, 187, 516, 680, 1100, 2219, 2896, 2897, 3705, 4103, 4114

Cankers and Diebacks Associated with the *Diaporthe eres* Complex

Fungi in the *D. eres* complex occur around the world and are associated with cankers and dieback of hundreds of plant species in more than 60 genera. The complex includes an unknown number of fungi that vary in host preference and in minor characters of the

---

A–D. Phomopsis canker and dieback of *Prunus triloba* (flowering almond) and *P. persica* (peach) caused by *Phomopsis amygdali*. A, B. *Prunus triloba*: dieback and close view of the base of a canker that apparently originated at a bud (NY, Sep). C, D. *P. persica*: lesions on a year-old twig in spring and on a current-year twig in autumn (GA, Apr & Oct).

E–G. Dieback of *Rosa* sp. apparently caused by *Phomopsis* sp. E. The lower part of a girdling canker shows color transition from brown to green. The brown area is speckled with pycnidia. F, G. Magnified views of the canker surface with masses of conidia extruded as tendrils from pycnidia (NY, June).

H–J. Canker and dieback of *Philadelphus coronarius* (mock-orange) caused by *Diaporthe eres*. H. The lower margin of a canker in which bark and sapwood colonized by the fungus are discolored brown. I. Magnified view of the bark surface in a canker, with black necks of perithecia protruding from a crack. J. Magnified view of killed bark, shaved to show embedded bodies of perithecia (NY, Aug).
Photo credits: C, D—W. Uddin

141

Phomopsis state but are not reliably distinguishable in the perithecial state. Most records of *D. eres* or the associated anamorph *Phomopsis oblonga* do not distinguish between saprobic and pathogenic occurrence. Other fungi of the *D. eres* complex include *D. ambigua* (anamorph *P. ambigua*) and *D. perniciosa* (anamorph *P. prunorum*) on angiosperms, and *D. conorum* (anamorph *P. conorum*) and *D. occulta* (anamorph *P. occulta*) on gymnosperms (Plate 71).

D. eres in the broad sense is the presumptive cause of the symptoms and signs shown on *Liriodendron tulipifera* (tuliptree) in Plate 68 and on *Rosa* sp. and *Philadelphus coronarius* (mock-orange) in Plate 69. The disease of *L. tulipifera* occurred in a nursery where numerous saplings had elongate cankers in which a *Phomopsis* was fruiting. When isolated from canker margins and inoculated into healthy trees, however, the fungus did not cause disease. Therefore, either the fungus had invaded bark killed by another agent or, more likely, it had killed tissues in which defense responses were impaired by cold injury or transplanting stress. When diseased stems were severed for diagnostic examination, sapwood beneath cankers was found to be dead or dying in a wedge-shaped sector extending toward the pith (Plate 68K), a common fate of sapwood beneath lesions that involve the vascular cambium. Rose canes weakened by cold injury or other stress may be colonized by *D. eres*. The *Phomopsis* shown in Plate 69E–G colonized an overwintered rose cane. *Philadelphus coronarius* stems colonized by *D. eres* (Plate 69H–J) develop girdling lesions with diffuse margins; these cause dieback of distal parts during winter or wilting and browning of affected branches during the growing season. Other diseases attributed to *D. eres*, in addition to those caused by other named fungi in the *D. eres* complex (see list below), include canker and dieback of *Ilex aquifolium* (English holly), *Paulownia*, and *Ulmus* (elm); and dieback and sapwood stain of *Acer* (maple).
References: 677, 1099, 1213, 1294, 1298, 2110, 3190, 3282, 3705, 3706

Phomopsis Dieback of Poplar

Phomopsis macrospora (Plate 70A–F) causes elliptic or diamond-shaped cankers and dieback on *Populus* species in nurseries and young plantations. Its ascigerous state, unreported in North America, is *Diaporthe santonensis*. Susceptible plants include *Populus ×canadensis* (Carolina poplar), *P. deltoides* (eastern cottonwood), *P. maximowiczii*, *P. nigra*, and numerous hybrid clones. Outbreaks have been reported from Minnesota to Mississippi and in Japan. Loss of bare-root trees in winter storage has occurred also. Infection begins in wounds and perhaps leaf scars. Fresh wounds are suitable for infection at any time of year in warm areas, but trees are more susceptible during dormancy than at other times. Warm (20–30°C), wet weather in November has favored the disease in Mississippi. Trees stressed by water shortage are especially vulnerable. Cankers on thin stems cause dieback; those on larger stems usually do not girdle, but stems may break at lesions. Stem growth around nongirdling cankers makes them appear greatly sunken within a year.
References: 1129, 1183, 2106, 2108

Diaporthe Canker and Dieback of Sycamore

Platanus occidentalis (eastern sycamore) in eastern and southern USA is subject to cankers and dieback caused by one or more species of *Phomopsis*. *P. scabra* and the unnamed *Phomopsis* state of *Diaporthe arctii* have been associated with the symptoms. *D. arctii* is a common endophyte of sycamore, having been detected in 5–34% of twig and leaf samples from plantations analyzed in Mississippi. When young sycamores are stressed, the fungus can cause cankers that girdle twigs (Plate 70G, H), branches, or even mainstems, causing dieback. These cankers range in length from several centimeters to more than a meter. Pycnidia develop in the killed bark. Small trees severely affected by cankers that form during autumn and winter may produce small, off-color leaves in spring and may die within a few weeks.
References: 65, 2286

Phomopsis Dieback of Weeping Fig

Ficus benjamina (weeping fig), which grows naturally and is planted as a shade tree in tropical and subtropical areas, is also a popular long-lived indoor ornamental plant. When stressed by inadequate light, water, or nutrition, it becomes susceptible to *Diaporthe cinerascens*, a member of the *D. eres* complex. The fungus causes branch cankers and dieback. The disease is usually noticed when leaf drop and twig dieback begin on one or more branches. The cankers are diffuse and initially undetectable except by shaving the surface to reveal brown discoloration of bark and sapwood. In time the bark surface becomes rough as the tops of black pycnidia of the anamorph, *Phomopsis cinerascens* (Plate 70I, J), break through the surface in profusion. The parasite probably becomes established in pruning wounds or other minor injuries while the plants are being grown for sale, but it causes no damage as long as the plants are vigorous. *D. cinerascens* is better known as a pathogen of *F. carica* (common fig) trees in orchards, where it causes cankers and dieback after trees have been stressed by drought or winter injury.
References: 1469, 2896

Other Cankers and Diebacks Caused by *Diaporthe* and *Phomopsis*

In addition to the pathogens and diseases illustrated or discussed above, the following are noteworthy (anamorphs indicated parenthetically):

Diaporthe alleghaniensis (anamorph *Phomopsis* sp.): leaf-and-shoot blight and canker of *Betula alleghaniensis* (yellow birch); associated with birch dieback (Plate 230); occasionally destructive in seedling stands; in the Great Lakes region and eastward, also in Japan.

D. ambigua and *D. perniciosa* (*P. ambigua* and *P. prunorum*, respectively): canker and dieback of *Acer* (maple), *Malus* (apple), *Prunus* (cherry and peach), *Pyrus* (pear), *Sorbus* (mountain-ash); fruit rot of *Malus* and *Cydonia* (quince); twig dieback of *Aesculus carnea* (red horse-chestnut) and *Juglans regia* (English walnut); widespread in North America and Europe, also occur in South Africa and China. These fungi are possibly one species and are in the *D. eres* complex.

D. citri (*Phomopsis citri*): dieback and melanose of *Citrus*. Melanose disfigures fresh fruit and occurs in most citrus-growing regions.

D. dubia (*Phomopsis* sp.): canker and dieback of *Acer saccharum* and other maples in eastern North America and Japan.

D. japonica (*P. japonica*): twig blight of *Kerria japonica* (Japanese-rose) in eastern and southern USA and Portugal. This fungus is in the *D. eres* complex.

D. kalmiae (*P. kalmiae*): leaf and twig blight of *Kalmia latifolia* (mountain-laurel) in eastern USA.

D. oncostoma (*P. oncostoma*): canker and dieback of *Robinia pseudoacacia* (black locust) in eastern USA and Eurasia; reported also on *Elaeagnus umbellata* (autumn-olive) in Japan.

D. pruni (*P. pruni*): twig blight and dieback of *Prunus* (cherry) in the eastern half of North America.

A–F. Cankers and dieback of *Populus deltoides* (eastern cottonwood) caused by *Phomopsis macrospora*. A. Dieback on a young stem, the surface of the dead part roughened by pycnidia (MS, Apr). B. An elliptic twig canker that began at a bud (MS, Jan); C. A lenticular canker with pycnidia on a young stem (LA, May). D. An inactive canker appearing sunken as the result of stem growth around it (MS, May). E, F. Magnified views of the surface of a young stem killed by *P. macrospora*, with ostioles of pycnidia protruding (E) and off-white, translucent masses of conidia extruding (F) (MS, Apr).

G, H. Girdling cankers caused by *Phomopsis scabra* on *Platanus occidentalis* (sycamore) twigs. Infection occurred at nodes. Pycnidia are prominent in the lesions (MS, May & Mar, respectively)

I, J. Fruiting of *Phomopsis cinerascens*, the anamorph of *Diaporthe cinerascens*, on bark of *Ficus benjamina* (weeping fig). The bark surface, broken by the ostioles of numerous pycnidia, is shown in successively magnified views. Lens-shaped transverse protrusions are lenticels (indoor plant; NY, Oct).

Photo credits: B–D, H—USDA Forest Service Southern Hardwoods Laboratory; G—J. D. Solomon

D. vaccinii (*P. vaccinii*): canker and dieback of *Vaccinium* (blueberry) and other ericaceous plants; widespread in USA.

Phomopsis diospyri: twig dieback of *Diospyros virginiana* (common persimmon) in southeastern USA, Europe.

Phomopsis sp.: canker and dieback of *Prunus domestica* (common plum) in eastern USA.

References: 109, 516, 665, 680, 920, 1099, 1101, 1298, 1354, 2620, 2787, 3282, 3705, 4108, 4129, 4270, 4276, 4338

Diaporthe and Phomopsis Cankers of Conifers

Several species of *Diaporthe* and *Phomopsis* cause twig and branch cankers and dieback. These fungi are opportunistic pathogens likely to cause damage mainly to trees predisposed by other agents. *D. conorum*, *D. lokoyae*, and *D. occulta* are most commonly encountered, usually as the anamorphs (asexual states) *Phomopsis conorum*, *P. lokoyae*, and *P. occulta*, respectively. A fourth fungus, *Phomopsis porteri* (no ascigerous state known) causes twig and branch cankers of *Pseudotsuga menziesii* (Douglas-fir) in western Canada. The information and illustrations with Plate 68 provide general information about *Diaporthe* and *Phomopsis*.

D. conorum is widespread on many conifers around the Northern Hemisphere. In morphological characteristics it is aligned with the *D. eres* complex (Plates 68, 69). *D. conorum* has often been considered to be saprobic, having been found, for example, on snow-blighted seedlings of *Picea engelmannii* (Engelmann spruce) as well as on dead twigs, branches, and tops of *Abies* (fir), *Picea*, and *Pinus* (pine). In Austria the *Phomopsis* state of this fungus was associated with twig and crown dieback of *Picea abies* (Norway spruce) in young plantations, particularly where seedlings were planted in elevation zones above where their parents grew. In pathogenicity tests this fungus has either failed to cause disease or has done so only if given a start in tissues killed by some other agent. Even so, it deserves mention because of its association with cankers such as the one illustrated here.

D. lokoyae has been associated with sporadic dieback of young *Pseudotsuga menziesii* and (rarely) *Metasequoia* (dawn-redwood), *Thuja plicata* (western red cedar), and *Tsuga heterophylla* (western hemlock) in forests and nurseries near the Pacific Coast. In Pennsylvania it has been associated with cankers and dieback of young *Abies fraseri* (Fraser fir) and *Picea pungens* (blue spruce). Symptoms on *Pseudotsuga menziesii* begin as elliptic annual cankers, mainly at or on branch bases, that develop while the plants are dormant. Pycnidia form in the cankers during spring and summer and are often followed by perithecia. When conditions have favored unusually severe disease, *D. lokoyae* has been associated with lesions nearly 2 m long and 2.5–5 cm wide, extending down mainstems from killed tops. Predisposition of the trees by aberrant weather was indicated in outbreaks in California in 1930 and on Vancouver Island in 1966. Severe drought preceded the California outbreak, and damage was greatest on sites unfavorable for growth. On Vancouver Island undetermined factors greatly suppressed tree growth in the year the cankers formed. In Pennsylvania *D. lokoyae* has been isolated from nursery seedlings of *A. fraseri* affected by top and branch cankers (Plate 71F, G) and from *Picea pungens* similarly affected in a young plantation. The spruces were chlorotic and growing on a soil of high clay content and low fertility.

D. occulta, yet another member of the *D. eres* complex, occurs around the Northern Hemisphere. It causes canker and dieback of tops and branches of *P. abies* in Europe and of *P. pungens* and *Pseudotsuga menziesii* in North America, most often in nurseries. Other hosts include *Cryptomeria japonica* (Japanese cedar) and species of *Cupressus* (cypress), *Juniperus* (juniper), *Larix* (larch), *Sequoia* (redwood), *Sequoiadendron* (giant sequoia), *Taxodium* (bald cypress), *Taxus* (yew), *Thuja* (arborvitae), *Thujopsis* (hiba-arborvitae), and *Tsuga* (hemlock). This fungus is capable of causing lesions that expand on young shoots during the growing season. It has caused cankers, tip blight, and dieback of *Picea pungens* in Michigan and Wisconsin nurseries. The symptoms include chlorotic flecks on the basal halves of needles near the shoot tips, followed by small cankers on stems, from which resin often exudes. Distal parts of girdled stems shrivel, and the needles turn brown. Pathogenicity of *D. occulta* isolated from *P. pungens* was proven by inoculation to four *Picea* species, among which *P. pungens* sustained the most damage. Either mycelium or conidia caused lesions when inocu-

lated to wounded seedlings, and conidia caused infection of intact seedlings. The importance of predisposing stress was confirmed by an experiment in which root-pruned and nonpruned seedlings were inoculated and the resulting lesions counted and measured. Lesions formed on 90% of root-pruned seedlings versus 50% of nonpruned controls, and the lesions were much larger on the root-pruned seedlings. *D. occulta* is also a common endophyte of *Abies alba* (silver fir), *Acer macrophyllum* (bigleaf maple), *P. abies*, *Pinus nigra* (Austrian pine), and *Sequoia sempervirens* (coast redwood), among other species.

Phomopsis strains on conifers are often identified only to genus because few *Phomopsis* species have distinctive morphological features. Therefore, many reports of disease associations remain incomplete. Such is the case for an unidentified *Phomopsis* associated with twig necrosis of *Picea rubens* (red spruce) in forest stands in West Virginia. The fungus apparently kills twigs stressed by maturation of galls induced by *Adelges abietis*, the eastern spruce gall adelgid, at twig bases. These galls normally die while the supporting twigs remain alive after the adelgids depart. Pathogenicity of the fungus was proven by inoculation of *P. rubens* seedlings.

References: 145, 285, 458, 800, 985, 1099, 1209, 1213, 1298, 1434–1436, 1846, 2104, 2606, 3396, 3622, 4115, 4270

Pathogens confused with *Phomopsis*. Two other fungi that cause twig and branch cankers of conifers deserve mention here because their anamorphs have been confused with *Phomopsis* and were once classified in that genus. *Phacidium coniferarum* (Helotiales, Phacidiaceae), anamorph *Allantophomopsis pseudotsugae* (syn. *Phomopsis pseudotsugae*, *P. strobi*), occurs in eastern and western North America, Eurasia, and New Zealand. It colonizes many conifers including pines. The other fungus, *Potebniamyces balsamicola* (Rhytismatales, Rhytismataceae), anamorph *Phacidiopycnis balsamicola* (syn. *Phomopsis boycei*), also occurs in both eastern and western North America but is unknown elsewhere. It infects only *Abies* species (firs). These fungi are opportunists that cause damage mainly on young plants stressed by factors such as frost or wounds.

References: 960, 1210, 1211, 1213, 1434, 1435, 1859, 3696, 3701, 4120, 4351

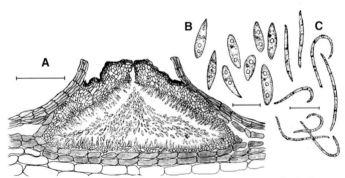

Figure 71. Phomopsis conorum. A. A pycnidium in conifer bark. B. α conidia. C. β conidia. Scale bars = 100 μm for pycnidium, 10 μm for spores. Adapted from reference 1434 by permission of Cambridge University Press.

A, B. Top dieback of *Abies balsamea* (balsam fir). A. Branches and foliage above a girdling canker have dried and the foliage has turned reddish brown. *Phomopsis* sp. was fruiting in the canker. B. The lower edge of the canker (NY, Apr).

C–E. Cankers putatively caused by *Phomopsis* sp. on *Picea abies* (Norway spruce). C. A split portion of the mainstem near the top of a mature tree that had sustained top dieback because of a girdling canker. Dead, sunken, resin-soaked bark is revealed. Arrows show the lower limit of cambial necrosis. D. A branch canker outlined by cracked bark. E. Magnified view of the beaks of pycnidia (arrows) extending above the surface of a canker (PA, Feb).

F, G. Canker and dieback of *Abies fraseri* (Fraser fir) seedlings caused by *Diaporthe lokoyae*. F. Dead branches on a 3-year-old seedling. G. A girdling canker on the leader (apical twig) of a seedling (PA, Apr).

Photo credits: C–G—W. Merrill

Phomopsis, Kabatina, and Sclerophoma Blights of Juniper and Other Gymnosperms (Plate 72)

Phomopsis Blight

Many members of the Cupressaceae (cypress family) are hosts of *Phomopsis juniperivora*, a coelomycete that causes shoot blight, twig cankers, and dieback collectively called Phomopsis blight. This is one of the major diseases of *Juniperus* (juniper) in North America. It causes disfigurement and occasional failure of landscape plantings, kills seedlings or young grafted stock in nurseries, and impairs survival of nursery-grown seedlings after outplanting. Junipers and other hosts in natural stands do not sustain significant damage.

Phomopsis blight occurs throughout eastern North America, in the Pacific Northwest, and also in Europe, Africa, China, Australia, and New Zealand. In North America it appears most often on *Cupressus arizonica* (Arizona cypress), *Juniperus horizontalis* (creeping juniper), *J. sabina* (savin juniper), *J. scopulorum* (Rocky Mountain juniper), and *J. virginiana* (eastern red-cedar). More than 25 additional species in *Chamaecyparis* (false-cypress), *Cupressus*, *Juniperus*, and *Thuja* (arborvitae) are susceptible to various degrees, and the fungus is sometimes found on other gymnosperms: *Cedrus* (cedar), *Cephalotaxus* (plum-yew), *Larix* (larch), *Pinus* (pine), *Pseudotsuga* (Douglas-fir), *Sequoiadendron* (giant sequoia), and *Taxus* (yew).

Symptoms and signs. The usual symptom is tip blight. This begins with infection of immature scale-leaves or needles. Mature leaves are resistant. The lesions first appear as tiny yellowish spots. Soon they involve xylem, and diseased shoots fade to light green, then turn reddish brown. A grayish band, which marks the site of infection, appears at the base of the dead part. Killed shoots usually remain on a plant for several months and turn gray. When lateral shoots of highly susceptible plants become infected, the parasite grows into woody stems and causes cankers that may extend several centimeters. Stems smaller than about 1 cm in diameter may be girdled. *J. virginiana* seedlings in nurseries have been killed in this manner.

Pycnidia develop in the gray band at the base of the killed shoot or in the bark of cankers and killed stems. They appear pale orange to gray or black, depending on location, and range in diameter up to about 400 μm. They break through the surface when maturing. During wet weather, pale yellowish to cream-colored masses of spores in a mucilaginous matrix extrude from pycnidia, either as tiny slimy blobs or as hairlike tendrils. Plates 69 and 70 show spore tendrils of other *Phomopsis* species.

P. juniperivora produces two types of single-celled, colorless spores: ellipsoid α conidia measuring about 7.5–10×2.2–2.8 μm, with a prominent oil droplet in each end of the spore; and filamentous, slightly curved β conidia measuring 20–27×1 μm. Only the ellipsoid cells germinate. The oil droplets in the α conidia distinguish *P. juniperivora* from other microfungi on *Juniperus* and related plants. Pycnidia in plant material often contain only the ellipsoid spores, which complicates diagnosis, but both types of spores are usually abundant in pycnidia that form in cultures. *P. juniperivora* grows readily in culture, and most isolates produce a deep yellow pigment accompanied by orange-red crystals consisting primarily of a tetrahydroquinone. The pigment and crystals are useful diagnostic characters.

Disease cycle. Infection occurs whenever succulent young foliage is available. Infection in spring is caused by conidia from shoots or twigs killed the previous year. Late-season infection may be caused by conidia from shoots killed in spring or the previous year, because *P. juniperivora* persists as a saprobe and sporulates on killed tissues for as long as 2 years. Conidia are dispersed by splashing or wind-driven rain. They can tolerate temporary drying yet remain germinable. During warm, wet weather the first symptoms may appear 3–5 days after infection, but much longer incubation periods have also been reported. Pycnidia appear on shoots beginning 3–4 weeks after infection.

Infection can begin within as few as 7 hours on continuously wet plant surfaces at 20–24°C. Longer wet periods promote increasingly severe disease. Temperatures as low as 8–12°C delay but do not prevent spore germination and infection. Severe symptoms are promoted by relatively high temperatures (e.g., 26–32°C). Late-season infections are likely if plant growth is prolonged by shearing or by high nutrient levels and plenty of water.

Few if any junipers are immune to *P. juniperivora*. Varieties and cultivars considered resistant on the basis of field performance are listed with similar information for Kabatina blight and juniper rusts on page 262.

References: 800, 1433, 1436, 1439, 1859, 3019, 3031, 3035, 3038, 3164, 3290, 3476, 3743

Kabatina and Sclerophoma Blights

Two additional fungi, both coelomycetes, cause symptoms similar to those of Phomopsis blight. *Kabatina juniperi* kills year-old twigs in spring, and *Sclerophoma pythiophila* attacks year-old and sometimes older parts in spring. These fungi, apparently incapable of penetrating intact healthy foliage, enter through wounds made by insects or injuries caused by ice or snow. Both fungi are widely distributed in North America and Europe. In some areas—Kentucky and Ontario, for example—*K. juniperi* causes more damage than does *P. juniperivora*.

Symptoms of Kabatina blight and Sclerophoma blight resemble those of Phomopsis blight in that grayish lesions appear at the bases of dead shoots. Microscopic examination is necessary to differentiate the three fungi. The fruit bodies of *K. juniperi* are erumpent acervuli, numerous in spring and diminishing in number during the remainder of the season. Pycnidia of *S. pythiophila* (not shown) closely resemble those of *Phomopsis* but never contain two types of spores. The conidia of both *K. juniperi* and *S. pythiophila* are similar to α conidia of *P. juniperivora*—colorless, single-celled, and ellipsoid—but lack oil droplets. In pure cultures *Kabatina* species and *S. pythiophila* form slimy colonies and produce conidia without forming fruit bodies. The colonies are similar to those of saprobic *Hormonema* species and the cosmopolitan *Aureobasidium pullulans*, which can cause diagnostic confusion. For tentative differentiation of juniper tip blights, *Kabatina* is usually found on tips that turn brown in spring as new growth begins; *Phomopsis* is more often associated with tip dieback in late spring and summer.

K. juniperi was for a time thought to be a variety of *K. thujae*, a fungus known mainly in Europe and New Zealand. The two fungi are distinguished in part by host preference: *Juniperus* versus other genera of Cupressaceae. Both species infect *Calocedrus* (incense cedar), *Chamaecyparis*, *Juniperus*, and *Thuja*.

S. pythiophila is common on diverse conifers including species of *Abies* (fir), *Calocedrus*, *Juniperus*, *Larix*, *Pinus*, *Picea* (spruce), *Pseudotsuga*, *Thuja*, and *Tsuga* (hemlock). Although sometimes detected in living leaves, it is primarily a colonist of moribund leaves and stems. Only *S. pythiophila* among the fungi considered with Plate 72 has a sexual state. It is *Sydowia polyspora* (Dothideales, Dothioraceae), which is occasionally found on dead conifer twigs.

References: 501, 643, 1213, 1214, 1706, 1741, 1788, 1859, 2917, 3022, 3035, 3290, 3469, 3743, 3913, 4043, 4511

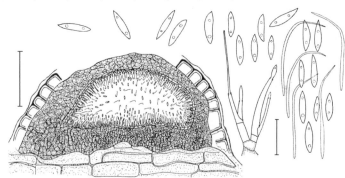

Figure 72. Phomopsis juniperivora: a pycnidium, conidiophores, and α and β conidia. Scale bars = 100 μm for pycnidium, 10 μm for other structures. Adapted from reference 3164 by permission of CAB International.

A–D. Phomopsis blight of juniper, caused by *Phomopsis juniperivora*. A. New infections, indicated by fading foliage, are confined to the current season's growth (NY, Jun). B, C. A twig killed by a girdling lesion (grayish band) at the base of the brown part. The pathogen is fruiting on the lesion (NY, Jun). D. Magnified view of a diseased twig with pycnidia of *P. juniperivora* (NY, Aug).
E. Juniper dieback attributed primarily to cankers caused by *P. juniperivora* (NY, Jun).
F, G. Kabatina blight of juniper. F. Symptoms in spring (TN, Apr). G. Erumpent acervuli of *Kabatina juniperi* on a dead year-old twig of *Juniperus virginiana* (eastern red-cedar) (NE, Jun).

146

Phomopsis Galls (Plate 73)

In the course of early studies of crown gall (Plate 190) and other plant tumors during the 1920s and 1930s, many gall and tumor diseases were discovered that could not be attributed to known causal agents among bacteria, fungi, and insects. Many galls, being economically insignificant, were never satisfactorily diagnosed (see Plate 253). Galls caused or putatively caused by *Phomopsis* species (coelomycetes) are considered with Plates 73 and 74.

Phomopsis species were first recognized as possible gall inducers in the 1930s. Phomopsis gall and canker of *Gardenia* (Plate 74) was described in California, and isolates of unnamed *Phomopsis* were obtained from galls on various woody plants in eastern and central USA. Affected plants included *Acer rubrum* and *A. saccharum* (red and sugar maples), *Carya* species (hickories), *Jasminum nudiflorum* (winter jasmine), *Ligustrum vulgare* (common privet), *Quercus* species (oaks), *Ulmus americana* (American elm), *Vaccinium corymbosum* (highbush blueberry), and *Viburnum opulus* (cranberry-bush). Phomopsis galls, as they came to be called, are found on *Acer*, *Carya*, and *Quercus* throughout central and eastern USA, more commonly in the North than in the South. Affected *Quercus* species include *Q. alba* (white oak), *Q. macrocarpa* (bur oak), *Q. prinus* (chestnut oak), *Q. rubra* (northern red oak), *Q. stellata* (post oak), and *Q. velutina* (black oak). Galls on *V. corymbosum* occur in several northern states across the continent.

Phomopsis galls range from pea size up to 25 cm or more in diameter, depending on plant species and age and gall location on the plant. Most are more or less spherical, but the largest ones appear as hemispheric protrusions from the lower parts of tree trunks. Affected plants typically occur singly or in small clusters and bear many galls while neighbor plants of the same species have none. Galls often develop for several years and then die. Multiple galls seem to cause general loss of vigor of affected plants, and gall formation on twigs and small branches apparently causes dieback.

Some Phomopsis galls appear as clusters of nodules pressed tightly together (Plate 73B, F, H). Beneath the roughened bark, each nodule consists of hard wood that is anatomically disorganized in comparison with normal wood. Intercellular mycelium has been observed in young, enlarging galls. The galls on *Acer* differ in form from those on other hosts. They begin as smooth swellings on which the bark eventually cracks and roughens, and they lack the nodular construction. Internally, however, they also are composed of hard wood.

The gall-inducing *Phomopsis* was originally detected when it grew from bits of internal gall tissue onto laboratory media, where it produced pycnidia with colorless, single-celled conidia typical of the genus *Phoma*. When cultures were refrigerated for several months, however, pycnidia with the two kinds of conidia typical of *Phomopsis* (Fig. 68) formed. Similarly, fresh cultures from chilled galls formed pycnidia with two kinds of conidia. *Phomopsis* pycnidia have been observed on galls on *Viburnum* and *Ligustrum*, but they are uncommon or perhaps do not form on living galls on *Acer*, *Carya*, *Quercus*, or *Ulmus*.

The *Phomopsis* isolated from galls on *Carya* and *Quercus* was reported to cause galls when inoculated into wounded twigs of *Viburnum*, *Jasminum*, and *Ligustrum*, in addition to the original hosts. An isolate from *Quercus* also caused galls on *Vaccinium*. *Phomopsis* isolates from galls on *Ulmus* were reported to cause small, slow-growing galls on *Jasminum*, *Ligustrum*, and *Ulmus*, and an isolate from *Acer* also caused galls on *Ligustrum*. Thus it seemed possible that one fungus lacking significant host specificity was responsible for galls on these and other hosts. These early reports were never corroborated, however. Later investigators of galls on *Acer* failed to detect an associated *Phomopsis*. This subject deserves fresh study.

The disease cycle has not been studied. From observations it seems that infection may start only on young twigs and that most galls found on major limbs and trunks were initiated on lateral twig bases. Modes of transmission of the pathogen(s) are unknown except as may be inferred from knowledge of other *Phomopsis* species (Plates 68–72). No sexual state of a gall-inducing *Phomopsis* has been found on the plants discussed with this plate.

References: 541, 542, 895

A, B. Phomopsis galls on *Carya cordiformis* (bitternut hickory). A. Affected trees typically bear many galls that arise at twig bases. B. Close view of active (brown) and apparently dead (brownish black) galls (NY, May).

C–E. Galls attributed to *Phomopsis* on *Rhododendron* sp. (azalea). C. Galls of various sizes and ages, some associated with twig dieback. Branches with galls on this plant were less vigorous than gall-free branches. D. Close view of galls on a 6-year-old branch segment. E. Close view of small galls on a 3-year-old branch segment (NY, May).

F–H. Galls attributed to *Phomopsis* on *Forsythia*. The galls are nodular, each with multiple centers of growth, and are usually associated with twig dieback (NY, Apr–Jun).

Phomopsis Canker of *Gardenia* and Nectriella Gall of Ornamental Plants (Plate 74)

Phomopsis Canker of *Gardenia*

This major disease of *Gardenia jasminoides* (Cape jasmine or gardenia) causes damage in nurseries and landscape plantings in Florida and California and in greenhouses in cooler parts of North America. It occurs also in Europe, India, South America, and Japan. Shipment of diseased plants accounts for its wide distribution. The pathogen, *Phomopsis gardeniae*, causes cankers and cankerous galls on branches and mainstems, commonly near the soil line. The cankers are perennial, enlarge slowly, and eventually girdle stems. Diseased branches lose vigor and wilt or drop leaves prematurely, and parts distal to cankers die. Slow growth and reversible wilting may call attention to cankers on mainstems.

Lesions first appear as small brown spots, usually centered on wounds, around which callus and woundwood soon begin to form. As a lesion expands and woundwood forms beyond the margin, a dark brown, elliptic canker with a greatly roughened surface and furrowed edges develops. Bark in the canker tends to break away and leave the wood exposed. If the surface of the most recently formed bark at the edge of a canker is cut away, living inner tissues there may be bright yellow or yellow-orange, a diagnostic character. Normal inner bark is greenish white.

Cankerous galls often grow to several times the diameter of affected branches. They have one flat or concave face with an irregular outline. Growth of callus and woundwood around a lesion is the main basis of gall enlargement. Yellowish orange inner bark may extend a few centimeters beyond a gall but not without swelling and superficial roughening that indicates infection. Roughened scaly bark and enlarged nodal areas (Plate 74D) occur occasionally in the absence of cankers. Leaves and roots are also susceptible to *P. gardeniae* if the parasite is introduced into wounds. Brown necrotic blotches form on leaves, and cankers develop on roots. Normally, however, only stem infections are significant.

Black pycnidia 300–500 μm in diameter develop just beneath the surface of dead bark in cankers and may also develop on leaves on moist soil. During wet weather conidia in mucilaginous masses or tendrils extrude from pycnidia, as illustrated in Plates 69 and 70. Conidia (of the two types discussed with Plate 68) are dispersed by running and splashing water and seem suited to dispersal by vagrant insects as well. The α conidia measure 7–12 × 2.4–4 μm; the β conidia, not always present, measure 25 × 1 μm. Perithecia of a *Diaporthe* have occasionally been found on dead galls in Florida and have developed in laboratory cultures. The two-celled ascospores were found capable of causing infection, although most infections are caused by conidia. The name *Diaporthe gardeniae* has been used for this fungus, but no such species has been validly described and named.

P. gardeniae is a wound invader that causes problems primarily in gardenias propagated by cuttings. Swellings and cankers develop near the bases of cuttings, above or below the soil line, where leaves were removed before the cuttings were stuck in rooting medium.

Cultivars of *G. jasminoides* apparently vary in susceptibility to *P. gardeniae*, but this subject has not been studied enough to provide a basis for recommending resistant or tolerant cultivars. Sanitary propagation procedures reduce the chance of outbreaks.
References: 891, 1153, 1495, 1496, 2566, 2637

Nectriella Gall

Nectriella pironii (Hypocreales, Bionectriaceae) causes galls or cankers on many woody and herbaceous plants in Florida. Susceptible woody species occur in at least 48 genera: *Alnus, Albizia, Aphelandra, Barleria, Bauhinia, Berberis, Callistemon, Catalpa, Cercis, Clerodendrum, Codiaeum, Conocarpus, Cuphea, Cyphomandra, Dombeya, Dracaena, Eupatorium, Euphorbia, Fatsia, Ficus, Forsythia, Gelsemium, Hedera, Hibiscus, Hydrangea, Jasminum, Lansium, Lantana, Leea, Leucophyllum, Ligustrum, Lonicera, Lysiloma, Melia, Nandina, Osmanthus, Parkinsonia, Parmentiera, Pittosporum, Polyscias, Psychotria, Rosa, Salix, Sambucus, Schefflera, Trevesia, Triadica,* and *Ulmus.*

The disease was first noted in 1955 in *Codiaeum variegatum* (garden croton), which develops irregular, corky, greatly roughened proliferations of callus up to 5 cm in diameter along the stems and on petioles and midveins of leaves. Galls are light tan while expanding but turn brown when mature. Numerous galls cause loss of vigor and then dieback of affected stems. Old galls are frequently covered with algae and support large populations of saprophagous mites. *N. pironii* causes similar galls on its other host plants apart from *Clerodendrum bungei* (glory-bower), on the stems of which it causes deep, roughened cankers.

The fungus produces two fruiting states in corky superficial gall tissues, often on the same gall. White tufts (sporodochia) of the anamorph *Kutilakesa pironii* arise first. Yellow to nearly white perithecia follow. The sporodochia, 250–750 μm in the longest dimension, produce unicellular, ellipsoid, pale orange conidia that measure 6–9 × 2–2.5 μm and pale yellow setae (hairs) 50–150 μm long (Fig. 74B, C). The pear-shaped perithecia are immersed to various degrees in gall tissue. Some have only the tips exposed; others are nearly superficial. They develop singly or in groups of two to six and measure 165–235 μm high and 115–185 μm wide. The ascospores are pale orange, spindle shaped, and equally two celled, and most measure 12.5–16 × 2.5–4.5 μm (Fig. 74A). Superficial gall tissues containing sporodochia and perithecia tend to flake off easily, so these structures are sometimes difficult to find.

N. pironii is a wound parasite, apparently unable to penetrate intact surfaces. Fresh leaf scars are common sites of infection. The modes of transmission in nature are unknown. Fungi of this sort are adapted for transmission by rain splash and by vagrant insects or mites. Two species of saprophagous mites have been found contaminated with *N. pironii*, but no report of vector activity has appeared.
References: 44–47, 1050, 3345

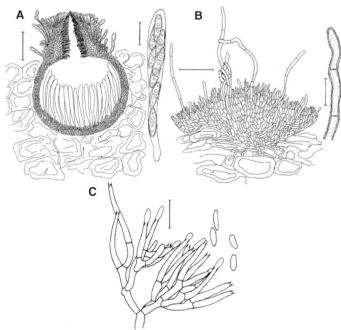

Figure 74. Nectriella pironii. A. A perithecium and one mature ascus with ascospores. B, C. The *Kutilakesa* anamorph. B. A sporodochium; one of the stout hairs that protrude from such sporodochia is shown enlarged. C. Conidiophores and conidia. Scale bars = 50 μm for perithecium and sporodochium, 10 μm for other structures. Adapted from reference 47 by permission of the Mycological Society of America.

A–D. Phomopsis canker and gall of *Gardenia jasminoides*. A. Leaf cast and dieback as secondary symptoms on small branches bearing galls. B. A cankerous gall on a small branch. C. Enlarged view of the same gall shown in B, the face cut away to reveal diagnostic orange-yellow color of infected inner bark. D. Cutaway view of one end of a roughened infected area, showing contrast between swollen orange-yellow bark in the diseased region and normal greenish white inner bark beyond (FL, Mar).

E–G. Nectriella gall of *Codiaeum variegatum* (croton). E. Galls and dieback on stems in a landscape planting. Green appearance of old galls is due to algae. F, G. Successively closer views of typical rough-textured galls. New gall tissue is light in color. Dead parts turn dark brown and eventually weather to grayish or to grayish green if colonized by algae (FL, Mar).

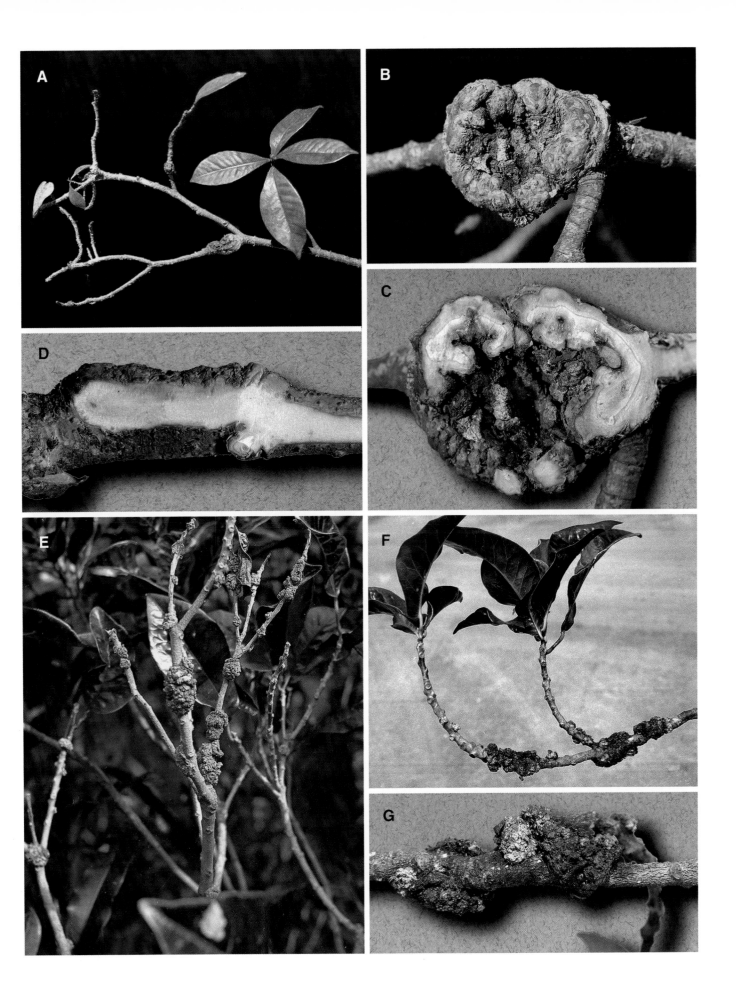

151

Black Knot of *Prunus* (Plate 75)

Black knot affects at least 25 *Prunus* species, mainly cherries and plums. The pathogen is *Apiosporina morbosa* (syn. *Dibotryon morbosum*; Pleosporales, Venturiaceae), a North American fungus that occurs throughout Canada and the USA where hosts grow in regions with moist climate. It causes rough, more or less elongate galls (knots) on twigs and branches of all sizes. Infection of limbs or trunks leads to the formation of cankerous galls and dieback.

Outbreaks of black knot are common on *P. domestica* (European plum) in abandoned orchards and on *P. americana* (American or wild plum) and *P. virginiana* (chokecherry). Severely diseased fruit trees become worthless because of dieback. Large galls or cankerous knots as long as 60 cm or more may form on trunks of *P. serotina* (black cherry), reducing the trees' value for lumber. Other plants affected to varying degrees include the following:

P. angustifolia (chickasaw plum), *P. armeniaca* (apricot), *P. avium* (sweet cherry), *P. cerasifera* (myrobalan plum), *P. cerasus* (sour cherry), *P. domestica* var. *insititia* (bullace or Damson plum), *P. emarginata* (bitter cherry), *P. mahaleb* (mahaleb cherry), *P. maritima* (beach plum), *P. munsoniana* (wildgoose plum), *P. nigra* (Canadian plum), *P. padus* (European bird cherry), *P. pensylvanica* (pin cherry), *P. persica* (peach), *P. pumila* (sand cherry) and its variety *besseyi* (western sand cherry), *P. salicina* (Japanese plum), *P. spinosa* (blackthorn), *P. subcordata* (Klamath or sierra plum), *P. tomentosa* (Nanking cherry), *P. triloba* (flowering almond), and *P. umbellata* (flatwoods or hog plum).

Additional *Prunus* species would probably be found susceptible if tested.

Symptoms and signs. Knots first appear in autumn as swellings on twigs of the current season that became infected in spring. Enlargement is interrupted by winter dormancy but resumes in spring. Often the diseased twig bends sharply at a knot because of one-sided overgrowth. Superficial bark on the knot splits during growth, revealing an olive-green fungal stroma that is corky in texture when young. The outer portion is composed entirely of fungal tissue; the inner part is a mixture of host and fungal cells. The olive-green surface in early summer is covered with conidiophores and light brown, single-celled or occasionally two-celled conidia measuring $4-9 \times 3.5-5.5$ μm. The conidial state is a hyphomycete in the genus *Fusicladium*. The stroma darkens during the growing season and turns hard and black during winter. Black pseudothecia differentiate on the stromatal surface during autumn and winter. They are more or less globose, 150–300 μm in diameter, and produce unequally two-celled, club-shaped, olivaceous ascospores measuring $13-18 \times 4.5-7.5$ μm.

The black part of a knot dies after ascospores are liberated. Often the twig beyond the knot either fails to leaf out or wilts and dies in early summer. If the twig remains alive, the bark-covered parts of the knot continue growth and stromal expansion. Thus the knot becomes perennial and elongates in annual increments toward the base of the twig. On reaching the subtending branch, the parasite slowly spreads along and around it. The branch is likely to be girdled and die; if it survives, a swollen, rough, often gum-encrusted canker with a sunken center eventually forms. Cankerous galls may also develop after wound infection on limbs and trunks. Black-knot galls on limbs or trunks of *P. serotina* sometimes retain normal bark from which discrete stromata of the pathogen emerge as small black cushions.

Disease cycle. A. morbosa overwinters as stromata on living knots. Its pseudothecia are mature in early spring, and ascospores are ejected into the air throughout the spring in response to rain. When ascospores are present on new shoots, a wet period as brief as 6 hours in the optimum temperature range 21–24°C is sufficient for infection. Some infections begin in wounds, but the great majority occur on intact green shoots. After spore germination the parasite penetrates the cuticle and grows intercellularly in the cortex, cambial region, and xylem, causing no symptoms during the first several months. It secretes indoleacetic acid and other growth-regulating chemicals that eventually stimulate cambial activity and swelling of the infected area. The amount of new xylem and phloem produced, the proportion of parenchyma in the xylem, and the size of parenchyma cells are all abnormally large.

The duration of the disease cycle is normally 2 years, but pseudothecial maturation 1 year after infection has been noted occasionally in southern Ontario and may be common in regions with a long growing season. On a perennial knot, the year-old portion bears pseudothecia in spring and the newly exposed stroma produces conidia in summer. After the first year of knot growth, pseudothecia in various stages of formation and degeneration can be seen throughout the year.

Most infections are considered to be initiated by ascospores. Conidia are also infectious, but their importance is unknown. Plants are less susceptible during the time of conidial abundance in summer than during the period of ascospore abundance in spring. *A. morbosa* apparently also grows saprobically on twig and branch surfaces and produces dark chlamydospores there. Their role in the disease cycle is unknown.

A. morbosa apparently has host-specialized subpopulations. Early experiments and field observations indicated that spores from black knot on *P. virginiana* would infect this plant but not *P. americana*. Susceptible domestic plums and cherries near thickets of diseased wild species sometimes remained free from disease. *P. serotina* was severely damaged by black knot in some areas but remained free from it in other places where the disease was severe on other species. This subject deserves further study.

Many plum cultivars resistant to *A. morbosa* have been released. Black knot in plum orchards is suppressed by planting resistant cultivars or, for susceptible cultivars, by sanitation and fungicidal sprays. Effective sanitation involves removing knots, cutting at least 10 cm beyond the swelling, before spore dispersal begins in early spring. Pruned material must be destroyed or removed from the site.

A. morbosa is among the fungal pathogens that have been tested for biological control of forest weeds. It was shown capable of suppressing young unwanted *P. pensylvanica* (pin cherry) in plots on reforestation sites in eastern Canada, where disease incidence was enhanced by importing diseased trees to the plots prior to ascospore dispersal.

References: 123, 232, 805, 1362, 1859, 2113, 2114, 2558, 2787, 2873, 2876, 3669, 3710, 4189, 4191

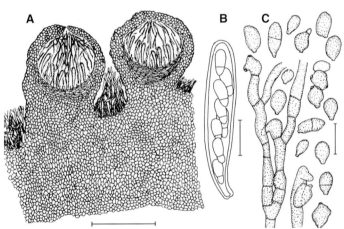

Figure 75. Apiosporina morbosa. A. Part of a stroma with two pseudothecia in section. B. A mature ascus with ascospores. C. Conidiophores and conidia. Scale bars = 200 μm for stroma, 10 μm for other structures. A and B adapted from reference 123 by permission of Gebrüder Borntraeger Verlagsbuchhandlung; C adapted from reference 805 by permission of Public Works & Government Services Canada.

A. Black knot on twigs and branches of *Prunus* sp., a wild plum (NY, Feb).

B–D. Black knot on *P. serotina* (black cherry). B. Knots developing into cankerous galls and causing dieback on a young tree. C. A gall about 40 cm in diameter on a limb. D. Close view of the surface of a large gall similar to that in C, with scattered black stromata of *Apiosporina morbosa*, the larger one about 1 cm long (NY, Jun).

E–G. Symptoms and signs of black knot on *P. virginiana* (chokecherry). E. A 2-year-old knot that caused a sharp bend in the twig because of one-sided overgrowth in the first year. The twig tip and host tissues in the black part of the knot have died. F. Close view of the surface of a living portion of a knot. The outer bark has split, exposing olive-green fungal tissue that darkens slowly as pseudothecia mature. G. Magnified view of the surface of a knot covered with mature pseudothecia of *A. morbosua* (NY, Jun).

153

Cryptodiaporthe Cankers (Plates 76, 77)

Cryptodiaporthe Canker of Poplar

Cryptodiaporthe canker, also known as Discosporium canker or, formerly, Dothichiza canker, is one of the major diseases of *Populus* (poplars) around the Northern Hemisphere and in South America. The pathogen is *Cryptodiaporthe populea* (Diaporthales, Gnomoniaceae), which is usually encountered in its conidial state, *Discosporium populeum*. Known in North America since 1915, it is widespread from eastern Canada to Florida and the Southwest, and it occurs also in the Pacific Northwest. It is common and destructive in trees weakened by transplanting, drought, freezing, waterlogged or infertile soil, or wounding of stems or roots. Damage is most severe in landscape plantings of *P. nigra* (black poplar), especially the cultivar Italica (Lombardy poplar), which is often killed at a young age. Poplars in North American forest plantations sustain relatively little damage from this disease, and those in natural forests are virtually unaffected. In southern Europe and in China, on the other hand, damage is widespread in nurseries and plantations of many species and cultivars. Most information about the disease comes from European research.

Susceptibility of poplar and aspen species varies widely, with damage generally restricted to those in *Populus* sections *Aigeiros* and *Tacamahaca*. These sections include most cottonwoods, balsam poplars, and many hybrid cultivars. Species commonly affected include *P. alba*, *P. balsamifera*, *P. nigra*, and *P. simonii* (white, balsam, black, and Simon poplars); also *P. ×acuminata*, *P. cathayana*, *P. laurifolia*, *P. ×petrowskiana*, and *P. tristis*. Plants with known resistance or no record of severe damage include *P. balsamifera* subsp. *trichocarpa* (black cottonwood), *P. ×berolinensis*, *P. ×canadensis* (Carolina poplar), *P. canescens* (gray poplar), *P. deltoides* (eastern cottonwood) and its subspecies *monilifera* (plains cottonwood), *P. fremontii* (Fremont cottonwood), *P. maximowiczii*, and assorted hybrids. Species in *Populus* section *Populus*, such as *P. grandidentata*, *P. tremula*, and *P. tremuloides* (bigtooth, European, and trembling aspens), are resistant.

Symptoms and signs. Cankers develop in stems of all sizes and ages, most often in twigs and small branches, especially at branch axils. Girdling or coalescing lesions cause dieback. The cankers range from tiny brown spots in bark just beneath the periderm to spreading lesions that involve both bark and cambium. The bark surface may be only slightly or not at all discolored, but the tissue beneath appears brown, grayish brown, or black, depending on tree species and time since death. The sapwood beneath lesions in bark is invaded and discolored pale brown. Adventitious sprouts commonly develop below large cankers and soon also become diseased. Leaves on stems stressed by multiple lesions may turn yellow and drop prematurely.

Many infections are halted by host defenses, even in highly susceptible poplars. Defense processes include production of a barrier of lignified parenchyma, then cork, that isolates a lesion. Callus and woundwood develop at the margin of most cankers that involve the cambium. Many small cankers are arrested and then overgrown by callus and woundwood during the first growing season after infection. Woundwood formation and drying of killed bark lead to cracking and eventual sloughing of bark, which exposes the wood.

From early spring until late autumn, scattered black pycnidia of the *Discosporium* state of *C. populea* develop just beneath the periderm in lesions and on killed twigs distal to lesions and break through the bark surface (Plate 76G, H). Pycnidia are 700–800 μm in diameter and open widely at maturity. While bark is moist, greenish to amber masses or tendrils of conidia issue from pycnidia and dry to a brown color or are dispersed by water. The single-celled conidia appear colorless and measure 9–11 × 6–8 μm. These spores can survive for several months.

Perithecia develop after pycnidia but are apparently uncommon, having been recorded in England and Italy. They are long necked, 500–600 μm in diameter, and tend to form in groups, pushing up the periderm in tiny mounds. Dying and dead twigs and branches beyond cankers are quickly colonized by other fungi, especially *Valsa* species (Plate 85).

Disease cycle. The pathogen survives winter as mycelium in cankers and as conidia in late-developing, unopened pycnidia. Conidia cause most infections, which occur via buds and bud scale scars in spring, via leaf scars in autumn, and via lenticels and wounds at various times of year. Ascospores are presumed to be infectious also, but this attribute has not been studied. *C. populea* may remain latent in bark up to 2 years after infection, but it normally causes lesions during the first dormant period. These may elongate as much as 30 cm before the next growing season. Some lesions in large stems enlarge during 2 or more years. A toxin produced by *C. populea* is implicated in lesion formation. Pycnidia form within lesions and on killed distal parts of twigs within several weeks after death of the bark. The cycle is completed with production of a new generation of conidia.

Temperature and moisture relationships of poplar bark are correlated with seasonal trends in susceptibility. Bark is less well hydrated and more susceptible during the dormant period than during the growing season. In certain poplars, resistance reactions, including cork formation, were found to be weak or absent and lesions developed at 12°C or less, but resistance was strong and lesions did not develop at 16°C or higher. Similarly, lesions expanded more rapidly in bark stressed by water shortage than in well-hydrated tissues.

Strains of *C. populea* differ in virulence and in behavior in culture. On average, cultures grow most rapidly near 20°C, and spore germination is most rapid and reliable at 16–20°C. Conidia can germinate at 5–35°C and have been reported capable of germinating without free water if relative humidity is above 90%.

Severe outbreaks of Cryptodiaporthe canker require either predisposing stress on the trees or extremely susceptible poplar genotypes. Outbreaks in North America have occurred on hybrid poplars stressed by drought and on *P. nigra* 'Italica' regardless of known predisposition. An interesting disease interaction reported in Europe is the infection by *C. populea* of leaf scars that result from defoliation by a *Melampsora* rust (Plates 141, 142).

References: 91, 233, 444, 581, 1366, 1791, 3065, 3080, 3290, 3913, 4238, 4293

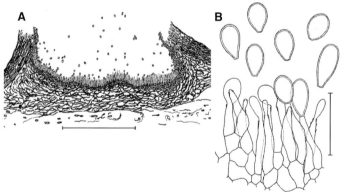

Figure 76. Discosporium populeum, the coelomycete anamorph of *Cryptodiaporthe populea*. A. Section through a widely open pycnidium in bark. B. Conidiophores and conidia. Scale bars = 400 μm and 20 μm, respectively. A adapted from reference 3913, B from reference 444; both by permission of CAB International.

Cryptodiaporthe canker, caused by *Cryptodiaporthe populea*.

A–C, E–H. Symptoms and signs on *Populus nigra* cv. Italica. A. Dieback in mature trees (NY, Oct). B. Dead twigs and foliage, an early stage of damage to trees in a windbreak (NY, Aug). C. A small branch with bark shaved to reveal a girdling canker at center and numerous small cankers below, the latter having been invisible while covered by periderm (NY, Aug). E, F. Old cankers on a branch and a trunk. Swelling and cracks in bark result from formation of woundwood at canker margins (NY, Mar). G. Pycnidia of *C. populea* in dead bark near a year-old canker from which bark has sloughed (NY, Aug). H. Magnified view of killed bark on a young stem with the surface raised or broken by pycnidia (NY, Jul).

D. Close view of a girdling canker with black pycnidia producing spore tendrils on a young branch of a *Populus* hybrid. White objects on bark are lenticels.

Photo credit: D—M. E. Ostry

Cryptodiaporthe Canker of Willow

Cryptodiaporthe canker, one of several diseases that cause dieback of *Salix* (willow) twigs and branches, is usually of minor importance. The pathogen, *Cryptodiaporthe salicella* (Diaporthales, Gnomoniaceae), is an opportunist that occurs across North America and is common also in Europe. It is usually found in its conidial state, *Diplodina microsperma*, a coelomycete. It causes damage to willows stressed by environment, injuries, or other diseases. Drought, frost, and nitrogen shortage are known predisposing factors. Severe twig dieback caused by *C. salicella* has been recorded on landscape specimens of *S. discolor* (pussy willow) and *S. caprea* (goat willow) and on other *Salix* species in plantations. Plantations of *S. viminalis* (basket willow or osier) have been affected in Sweden, where this plant is grown to provide fuel for power generation. *C. salicella* occasionally kills stems up to 6–7 m tall (3 years old) in these plantations.

Additional recorded hosts of *C. salicella* include *S. alba*, *S. babylonica*, *S. bebbiana*, *S. cordata*, *S. exigua*, *S. fragilis*, *S. hookeriana*, *S. nigra*, and *S. scouleriana*. The fungus also occurs as an endophyte in living willow twigs and is common as a saprobe on dead twigs, sometimes including those of *Acer* (maple), *Alnus* (alder), and *Populus* (poplar).

Symptoms and signs. Cankers develop during host dormancy at lenticels, nodes, and wounds made by insects and other agents. Twigs and small branches that are girdled die back to the next living side branch. Lesions are yellow to brown, in contrast to the normal greenish to yellow surface of young willow bark. Bark remains on killed parts but appears shriveled with cracks at the junction between dead and living tissue. In spring, black disc-shaped pycnidia form beneath the periderm. Pycnidia measure 0.5–1 mm in diameter. Their tips break the surface and release conidia in amorphous masses during wet weather or in tendrils under somewhat drier conditions. Conidia are equally two celled, colorless, and spindle shaped, and measure 13–18 × 3.5–5 µm. They are dispersed by water. Perithecia form later, singly or in small groups, typically where pycnidia appeared first. Perithecia are long necked, black, mostly 0.3–0.5 mm in diameter, and produce colorless, two-celled ascospores.

The cycle of Cryptodiaporthe canker of willow has not been studied but is probably similar to that of the corresponding disease of poplars (Plate 76). Lesions expand during host dormancy, and callus and woundwood form at lesion edges during the growing season. Lesions on highly susceptible trees may expand during the growing season or in more than one year, but the more common situation is cessation of lesion growth after one dormant season. Impairment of defense reactions in bark during dormancy is linked to temporary partial drying of twig or branch bark. Research with *S. hookeriana* and *S. scouleriana* revealed that dormant twig cuttings were resistant when well hydrated but became susceptible as their water content diminished to less than about 80% of capacity.

C. salicella is mentioned in some references as one of two similar fungi on willows, the second being *C. salicina*. Asci and ascospores of the latter are reported to be larger than those of *C. salicella*. Anamorph names associated with both teleomorph names include *Discella carbonacea*, *D. salicis*, and *Diplodina salicis*. Most contemporary authorities recognize only *C. salicella* with the anamorph *D. microsperma*.
References: 138, 229, 233, 363, 677, 1099, 1614, 3065, 3913, 4270

Golden Canker of Alternate-Leaf Dogwood

Cryptodiaporthe corni (Diaporthales) causes the most common branch canker and dieback of *Cornus alternifolia* (alternate-leaf or pagoda dogwood). The disease, called golden canker, occurs throughout nearly the entire range of the tree, from Mississippi, Missouri, and Minnesota to the Atlantic seaboard in the USA and Canada. The fungus is believed to occur only on *C. alternifolia*; reports of it on other plants are suspect. It is likely to be reclassified in *Cryphonectria* or *Endothia*, based on DNA sequence similarities.

Symptoms and signs of the disease are conspicuous and distinctive because killed bark of small branches and twigs turns yellowish and then tan in contrast to the greenish purple to reddish purple of healthy smooth bark. Diseased bark near lesion edges develops numerous small yellowish blisters that coalesce into continuous yellowish areas. Orange spots, dull to bright in color intensity and less than 1 mm across, appear as fungal stromata form

just beneath the bark surface. Pigment in hyphae imparts the orange color. Pycnidia of the asexual state, *Myxosporium nitidum* (syn. *Zythia aurantiaca*), a coelomycete, differentiate in these stromata, raise and break through the surface at lenticels, and release conidia. The orange pycnidia are subspheric to conic, 215–340 µm in diameter. Conidia are single celled, colorless, and ellipsoid, and most measure 6–9 × 2.5–3 µm. Perithecia also develop in stromata, usually in clusters of two to five in the bark of branches and mainstems, seldom on twigs. Perithecia are black, globose or nearly so, 275–390 µm in diameter, with beaks that break through the bark surface. Orange pigment may surround the beaks. The ascospores are colorless, unequally two celled with narrowed ends, and most measure 11–18 × 5–7 µm. The fungus is readily culturable on various media, and orange pigment appears in the cultures, associated with irregularly shaped granules in the hyphae.

The disease cycle has not been studied. Both ascospores and conidia germinate readily in culture, so both types of spores probably function as inoculum. The ubiquitous association of *Cryptodiaporthe corni* with *Cornus alternifolia* indicates that the fungus is an endophyte and opportunistic pathogen of this tree. In this role it may hasten the death of shaded branches and cause dieback of exposed branches predisposed by environmental stress. Most of the dieback occurs during host dormancy, as may be inferred from the usual absence of shriveled leaves on dead branches.
References: 229, 233, 677, 779, 1099, 1298, 3243, 3245, 3913, 4270

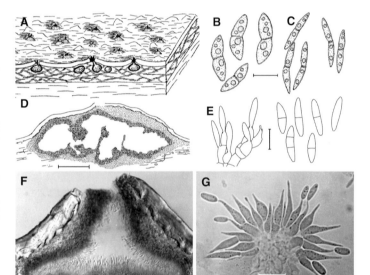

Figure 77. A. *Cryptodiaporthe salicella:* habit of perithecia in bark. B, C. Ascospores identified as those of *C. salicina* and *C. salicella*, respectively. D, E. A pycnidium, conidiophores, and conidia of *Diplodina microsperma*, the anamorph of *C. salicella*. F, G. A pycnidium, conidiophores, and conidia of *Myxosporium nitidum*, the anamorph of *Cryptodiaporthe corni*. Scale bars = 200 µm in D, 100 µm in F; all others = 10 µm. The sketch in A is unscaled. A–C adapted from reference 4270; D, E adapted from reference 3913 by permission of CAB International; F, G courtesy of S. C. Redlin with permission of the Mycological Society of America.

A–D. Cryptodiaporthe canker and dieback of *Salix caprea* (goat willow). A. Dieback progressing down a small branch. B. The basal part of a twig canker with cracked, peeling periderm and black pycnidia of the *Diplodina* state of *Cryptodiaporthe salicella*. C, D. Magnified views of pycnidia, dry (C) and producing masses of conidia while bark is moist (D) (NY, May).

E–I. Golden canker of *Cornus alternifolia* (alternate-leaf dogwood). E. An affected tree at a forest edge. Wilted leaves reveal that lesions in some branches have extended during the growing season. F, G. Lesions spreading toward the bases of small branches. Tan bark around the twig crotch in F is the site where a lesion spread from the twig into the branch. Recently colonized bark has a blistered appearance. H, I. Magnified views of bark in which pycnidial stromata have developed. Orange color encircles the ostioles of pycnidia in H. Orange color is more diffuse in I, and blobs of conidia have extruded from pycnidia (arrows) (E–G, I; NY, summer.

Photo credit: H—S. C. Redlin

157

Eastern Filbert Blight (Plate 78)

Corylus species (filbert, hazelnut) are subject to eastern filbert blight, a destructive canker disease caused by *Anisogramma anomala* (Diaporthales, Valsaceae). The fungus is apparently restricted to North America and to *Corylus*. Affected species include *C. americana* (American filbert), *C. avellana* (European filbert), *C. colurna* (Turkish hazelnut), *C. cornuta* (beaked filbert), and *C. maxima* (giant filbert). *A. anomala* occurs in the region extending from Manitoba and Illinois to Nova Scotia and North Carolina, and in Oregon, Washington, and British Columbia. It was first descri-bed in northeastern USA in the late 1800s and was discovered in the Pacific Northwest in 1968. Economic damage in *C. avellana* orchards there has been severe, and many affected orchards have been removed because of the disease. In New York State it has caused severe dieback of *C. americana* in arboreta and other landscape plantings and of young *C. colurna* grown as shade trees.

Symptoms and signs. Dead branches, sometimes with shriveled leaves adhering, call attention to elongate, girdling cankers that begin on twigs. The lesions enlarge perennially, attaining lengths greater than 1 m on highly susceptible plants and killing progres-sively larger branches. Highly susceptible trees, such as the *C. avellana* cultivars Daviana, Ennis, and Tonda Romana, may be killed within 5–12 years if diseased parts are not removed. Cankers on *C. avellana* typically elongate 30 cm or more per year. Lateral expansion is much slower. Lesions remain covered with bark and become sunken as surrounding tissues grow. Cankers on resistant trees remain small or enlarge slowly.

A. anomala produces distinctive stromata in the cankers. They develop in spring to early summer within bark in rows parallel to the stem axis. They raise and rupture the bark surface, appearing as elliptic mounds 1.5–3 × 2–10 mm and 1–2 mm high closely adjacent to green tissue. The exposed stromatal surface is lenticular and white initially; it darkens as black ostioles of perithecia appear. Some stromata extend to the wood. A year-old twig lesion may contain only a single row of stromata, but older, larger lesions have double or multiple rows. Stromata also appear in crevices on branches with rough bark.

Perithecia begin to form within the stromata in spring and are mature by late summer. Mature perithecia are ovate to pear shaped with long necks, 0.25–0.8 mm in diameter and 1–2 mm high. Ascospores form during late summer and autumn and are discharged after a maturation period of 2–4 months. These spores are colorless, unequally two celled, and measure 8–12 × 4–5 μm, with the smaller cell appearing as a barely discernible cap on one end of the larger cell. The larger cell contains a refractive body resembling an oil drop near each end. Perithecial stromata die after spore discharge but persist for a time, crumbly and with bleached surfaces, on dead ends of diseased branches.

Disease cycle. The cycle of eastern filbert blight requires 2 years and may extend a third year in some resistant trees. It was elucidated in western Oregon, where winters are moist and mild. Events and processes that occur during winter there are probably postponed until spring in central and eastern North America. *A. anomala* is apparently an obligate parasite under natural conditions. It overwinters as perithecial stromata in which ascospores are already mature. Spore expulsion from perithecia begins during autumn rains and continues intermittently in response to rain until late spring. The spores are dispersed within and between tree canopies by splashing and running water and over longer distances by air. They are short-lived, and only those deposited in spring on susceptible parts of growing filbert plants have a role in the disease. Susceptible parts are young, succulent internodes near the shoot apex and unfolding leaves. Ascospores adhere there, germinate within 24 hours, and produce hyphae that penetrate epidermal cells. Infection occurs at 8–25°C and is facilitated by a 24–72-hour period of high humi-dity, but free water is not required. Hyphae apparently grow intracellularly from the epidermis into vascular tissues and after a few months can be found in the phloem, cambium, and outer xylem. They produce "haustoria-like" cells with a presumed nutrient-absorbing function within living parenchyma in phloem and xylem rays. Host response to this invasion is not apparent for several months, but in spring the colonized phloem and cambium die and turn brown. Lesions expand during March–October in

Oregon. Hyphae in the outer 0.5–1 mm of xylem at the edge of a lesion are believed responsible for the perennial expansion of cankers. These hyphae grow beneath the cork barrier that formed during the previous season and colonize dormant tissues beyond the lesion.

Perithecial stromata begin to form in spring a year (occasionally 2 years) after infection. An extended cold period is required to initiate stromatal formation. Ascospores are dispersed from the perithecia during the next winter and spring, completing the cycle.

A. anomala has no known saprobic phase and does not reproduce asexually. It does not grow on common media. Slow-growing mycelial colonies can be induced to form on special media, however.

Control of eastern filbert blight depends ultimately on high-level host resistance because the slow accumulation of damage on moderately resistant plants gradually reduces their productivity or ornamental value. Diseased branches can be pruned, but this practice halts the extension of cankers only if cuts are made in wood not yet colonized by the pathogen. Small cankers and asymptomatic first-year infections are missed. The productivity of moderately susceptible cultivars in Oregon has been maintained by annual pruning complemented by fungicide applications.

Resistance to *A. anomala* strains that occur in Oregon has been found in certain selections or cultivars of *C. americana*, *C. avellana*, *C. colurna*, *C. cornuta* var. *cornuta* (beaked hazelnut), *C. cornuta* var. *californica* (California hazelnut or western hazel), *C. hetero-phylla* (Siberian hazelnut), and *C. sieboldiana* (Japanese hazelnut). The hybrid cultivars Lewis and Clark, released in Oregon in 2000–2001, are resistant. Qualitative resistance, controlled by single dominant genes, occurs in several genotypes of *C. avellana* that have shown no visible response when inoculated with spores of the pathogen. Quantitative resistance, apparently conferred by multiple genes, has also been detected. Plants with quantitative resistance, such as the *C. avellana* cultivars Tonda di Giffoni and Gem, develop fewer and smaller lesions than occur on fully susceptible plants. Information about resistance should be interpreted cautiously until putatively resistant plants have been challenged with both eastern and western strains of the pathogen.

A disease similar to eastern filbert blight affects *Betula* species (birches). *Anisogramma virgultorum* is associated with cankers and dieback of birch twigs in northeastern USA and Europe. The birch disease is rarely noticed.

References: 233, 779, 827, 1099, 1199, 1330, 1954, 1955, 2581, 2582, 2916, 3071–3075, 3992, 4270, 4408

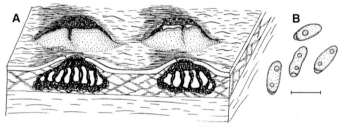

Figure 78. Anisogramma anomala. A. Perithecial stromata in bark. B. ascospores. Scale bar = 10 μm. The sketch in A is unscaled. Adapted from reference 4270.

A. Severe damage from eastern filbert blight on *Corylus avellana* (European filbert) (OR, Aug).

B–I. Eastern filbert blight on *C. americana* (American filbert). B. Branch dieback. C. A canker with a double row of perithecial stromata. The bark below the canker has numerous small bulges where young stromata are developing. D, E. Perithecial stromata in cankers on small branches; the canker margin in D is delimited by a green ridge. In close view (F), stromata adjacent to the ridge have white surfaces broken by the black tips of perithecia. G. Close view of a stromatal surface. H, I. Longitudinal dissection of a perithecial stroma and surrounding tissues, revealing black bodies of perithecia deep in the stroma and brown necrotic phloem, cambium, and outer sapwood adjacent to it (NY, Jul).

Photo credit: A—J. N. Pinkerton

Chestnut Blight (Plate 79)

Chestnut blight became internationally famous when it nearly eliminated *Castanea dentata* (American chestnut) in eastern North America during the first half of the 20th century. The causal fungus, *Cryphonectria parasitica* (Diaporthales, Valsaceae), was introduced from the Orient, probably with chestnut seeds or seedlings from Japan. It spread throughout eastern USA and into southern Ontario and also appeared in chestnut plantings in the Pacific Coast states and British Columbia. Nearly all large American chestnut trees were killed. The species now survives mainly as scattered sprout stems in the forest understory. Chestnut blight also became widespread in Europe, affecting primarily *C. sativa* (European or Spanish chestnut). It occurs in Asian *Castanea* species in China, Japan, and northern India.

Cryphonectria parasitica colonizes many woody plants, mainly of the Fagaceae, causing cankers and dieback or invading moribund bark and then growing as a saprobe. *Castanea dentata* is the most susceptible species. *C. sativa* is less so. *C. crenata* (Japanese chestnut) and *C. mollissima* (Chinese chestnut) and hybrids derived from them are moderately to highly resistant. Other plants affected are *C. pumila* (Allegheny chinkapin) and its variety *ozarkensis* (Ozark chinkapin) and several oaks, notably *Quercus alba*, *Q. coccinea*, *Q. ilex*, *Q. pubescens*, *Q. stellata*, and *Q. virginiana* (white, scarlet, holly, pubescent, red, post, and live oaks, respectively). Various other plants, including *Eucalyptus*, have proven susceptible in tests and may occasionally be naturally infected but do not sustain significant damage.

Symptoms and signs. Cankers develop on stems of all sizes. Lesions on highly susceptible trees usually involve bark, cambium, and sapwood. Water conduction is blocked, and parts beyond girdling cankers die. Typical cankers on smooth-barked stems are shades of brown in contrast to normal greenish gray or greenish brown bark. Cankers on *C. dentata* expand laterally 14–25 cm per year, the rate depending on temperature. Infection often spreads from small stems into larger ones; thus many cankers are found at branch axils. Cankers on stems with thick bark are inconspicuous unless the bark begins to swell or crack. Lesions that induce swelling may appear as spindle-shaped areas on small branches or as raised elongate areas with longitudinal cracks in thick bark. Dead bark eventually falls from cankers that involve the cambium. Adventitious sprouts commonly develop below cankers, often from the root collar of *C. dentata*. Sprouting has thus far prevented extirpation of this species. *C. mollissima* and other resistant chestnuts affected by chestnut blight may develop twig and branch dieback or targetlike perennial cankers on limbs and trunks.

Numerous orange to yellowish brown pycnidial stromata break through the bark surface in lesions or develop in crevices if the bark is thick and fissured. Stromata are up to 3 mm long or wide and about 0.5 mm deep, with multiple fertile chambers where conidia are produced (Fig. 79). Solitary pycnidia, about 250 μm in diameter, may develop on wood or on the inner surface of loosening bark. The conidia are unicellular, rod shaped, colorless, and measure 3–5 × 1–1.5 μm.

Perithecia form later (several weeks to a year) in the same stromata that gave rise to pycnidia. Both types of fruit bodies may be present simultaneously. Perithecial stromata are usually reddish brown. Perithecia are flask shaped with long necks extending to the surface. The ascospores are colorless, 7–12 × 3.5–5 μm in size, equally two celled with rounded ends, and slightly constricted in the middle.

Disease cycle. The pathogen overwinters in lesions and other colonized bark. Spore dispersal and infection occur throughout the year, given mild weather. Yellow to buff tendrils of conidia extrude from pycnidia when bark is moist. Perithecia, when moistened, eject ascospores into the air from early spring until late autumn. Airborne ascospores and conidia carried by birds and insects that visit contaminated bark account for overland spread of the pathogen.

Most infections on young stems begin when spores germinate in fresh wounds in bark. The susceptibility of wounded tissues diminishes after several days. Hyphae grow into and between cortical cells and also into the outer sapwood, killing tissues as they advance. Hyphae in the bark organize into flat, spreading, buff to orange-buff plaques (mycelial fans) that disrupt living bark and cambium by combined chemical and mechanical actions. Death of tissues is attributed to direct and indirect effects of acidification by

oxalic and gallic acids and to the action of macerating enzymes. A phytotoxin identified as a galactan polymer (a carbohydrate) may be involved as well. It is produced by the fungus in culture but has not been sought in infected bark. Water stress in the host interferes with resistance and thus promotes lesion expansion. Pycnidial stromata usually develop within several weeks after infection and produce conidia. When a limb or tree dies, the fungus grows into bark beyond the lesion and sporulates there as a saprobe for 2 years or longer.

Chestnut trees resist *Cryphonectria parasitica* in several ways. Their bark produces chitinases and enzyme inhibitors that suppress the parasite, and they form a zone of lignified cells and subsequently a cork barrier around a lesion. These responses proceed faster and more effectively in resistant than in highly susceptible trees. Mycelial fans of *C. parasitica* break through incomplete cork barriers or grow beneath them. In resistant plants *C. parasitica* causes most damage to those predisposed by frost, drought, or other stressing factors.

Scattered American chestnuts have survived the blight although disfigured by cankers. Some of these trees are somewhat resistant, but their survival in most instances represents either escape or infection by hypovirulent strains of the pathogen. In Europe, however, numerous *C. sativa* trees with blight resistance have been identified and cloned.

Chestnut breeders in the USA are developing trees with the vigor, form, and fruit characteristics of *C. dentata* and blight resistance from *C. mollissima*. Resistance in *C. mollissima* is apparently controlled by two unlinked genes. Breeders first obtained resistant hybrids, then began transferring resistance genes into the genetic background of *C. dentata* by producing successive backcross generations in which resistant individuals were crossed with pure *C. dentata*. The program is expected to produce seeds adapted to various geographic regions by 2015. Transformation of *C. dentata* by genetic engineering with resistance genes also appears feasible but had not advanced as far as the breeding program when this was written.

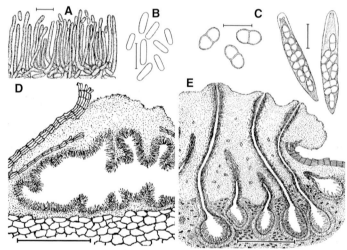

Figure 79. Cryphonectria parasitica. A, B. Conidiophores and conidia. C. Ascospores and asci. D, E. Sections of pycnidial and perithecial stromata, respectively. Scale bars = 10 μm for A and C. 5 μm for B, and 500 μm for C and D. A and C adapted from reference 84; D and E from reference 1571.

A–G. Chestnut blight in *Castanea dentata* (American chestnut). A. A young sprout recently killed by *Cryphonectria parasitica* (NY, Jul). B. A sapling stem girdled by a blight lesion (NY, Jul). C. Swollen, cracked bark in a blight canker on a small stem (PA, Aug). D, E. Distant and close views of typical cankers with perithecial stromata on a stem about 15 cm in diameter (NY, May). F, G. Close views of the advancing margin of a canker with pycnidial stromata. A buff-colored mycelial fan of *C. parasitica* (arrow in G) is visible where the outer bark was shaved (NY, Jul).

H. A perennial canker caused by *C. parasitica* on *Castanea mollissima* (Chinese chestnut). Infection began at a pruning wound (NY, Jul).

I. An old perennial canker caused by *Cryphonectria parasitica* on *Quercus virginiana* (live oak). Woundwood and cracked bark mark the margin of the canker (VA, May).

In parts of Europe and in a few American localities, blight severity has moderated due to infection of the fungal pathogen by viruses, primarily of the *Hypovirus* group, that reduce fungal virulence. They are transmitted by hyphal fusion and by conidia. Hypovirulent fungal strains grow less rapidly in bark, have less macerating enzyme activity, and sporulate less prolifically than wild-type strains. They are resisted more or less effectively by host defenses, and lesions containing them tend to be confined to swollen superficial bark. The discovery of infectious hypovirulence raised hopes for practical biological control of the blight, but most hypovirulent strains of the fungus in North America do not transmit virus to many other strains because of vegetative-compatibility barriers and do not compete effectively with virulent strains for chestnut substrates. *Vegetative compatibility* refers to ability of different fungal strains of the same species to accomplish viable hyphal fusion. Strains that differ at genetic loci for vegetative compatibility are unlikely to accomplish hyphal fusion. This barrier to transmission can be overcome by genetically engineering hypovirulent strains with hypoviral DNA sequences in their nuclear DNA. Such strains constitutively express hypovirulence and can hybridize with wild-type strains, passing hypovirulence to their progeny. In 2004 such strains were about to be tested for ability to provide biological control of chestnut blight.

Damage by *C. parasitica* to most trees other than *Castanea* is slight, but the fungus does cause cankers on *Quercus coccinea* (scarlet oak), *Q. rubra* (red oak), and *Q. stellata* (post oak) in the Appalachian region and on *Q. virginiana* (live oak) from Virginia to Florida and Mississippi. Perennial, slowly expanding, elongate cankers form on *Q. virginiana* trunks and limbs. The crowns of severely affected trees decline slowly, displaying sparse foliage and dead and dying branches. Drought stress promotes damage to oaks by *C. parasitica*.

References: 69, 84, 85, 167, 330, 360, 576, 577, 812, 814, 815, 915, 1091, 1207, 1228, 1229, 1372, 1373, 1551, 1569, 1571, 1577, 1588, 1589, 1709, 1859, 1963, 2161, 2213, 2316, 2348, 2408, 2409, 2534, 2573, 2614, 2615, 2632, 2633, 2902, 3061, 3308, 3434, 3571, 3721, 4067, 4137

Chrysoporthe Canker of *Eucalyptus* (Plate 80)

Chrysoporthe cubensis (syn. *Cryphonectria cubensis*; Diaporthales, Valsaceae) causes an important canker disease of *Eucalyptus* in tropical and subtropical regions around the world. *E. grandis* (rose gum) and *E. camaldulensis* (Murray red gum) have been affected in southern Florida. Butt logs of diseased trees are devalued by stain and decay that begin where cankers involve the cambium, and severely affected trees may be girdled and killed. Sprouts growing from stumps of diseased trees often become infected and die, which jeopardizes stand reproduction by coppice. Many eucalypts are susceptible to some extent, and several have been ranked as follows: *Corymbia maculata* and *E. saligna*, highly susceptible; *E. grandis*, *E. propinqua*, and *E. tereticornis*, moderately susceptible; *E. microcorys*, *E. paniculata*, *E. pellita*, and *E. robusta*, moderately resistant; *C. citriodora*, *C. torelliana*, and *E. urophylla*, highly resistant. *Chrysoporthe cubensis* has also been found in cankers on *Quercus virginiana* (live oak) in Florida and on *Syzygium aromaticum* (clove) and other species abroad.

Symptoms and signs. Perennial lesions form in outer bark (old secondary phloem) on the basal parts of *Eucalyptus* trunks beginning when they are less than 2 years old. Multiple infections occur on some trees. Lesions slowly expand and deepen, and some eventually encircle the butt and may extend nearly 2 m above the ground. Occasional cankers occur higher, often centered on branch stubs. The only external symptom in early stages may be a slightly depressed area. Later the outer bark cracks longitudinally and becomes rough. The canker margin is abrupt but barely apparent unless the bark surface is shaved to reveal it. Strips of diseased outer bark fall away from some trees. On others, lesions deepen and involve the cambium and then may either girdle the tree or become delimited by callus and woundwood. Where the cambium is killed, sectors of discolored wood extend inward to the pith. In Florida many cankers cease development by the time trees are 6–8 years old.

Pycnidia form on killed bark, especially in crevices and on the inner surface of loosened bark, and are often followed by perithecia.

Sometimes both types of fruit bodies are found together. Pycnidia are superficial or nearly so, conic to pear shaped, and solitary or fused at their bases in groups of two to five. Reddish brown at first, they darken to nearly black except at the tip, which remains brown. They measure 0.4–1.2 mm tall and 0.2–0.8 mm in diameter. The conidia, extruded in yellowish tendrils under moist conditions, are individually colorless, unicellular, ovoid with one end narrowed, and measure 2.5–4 × 1.8–2.2 μm.

The perithecia are black, partly immersed in bark, solitary or in small groups, and sometimes aligned in a crevice. Their necks vary in length up to about 0.5 mm in humid locations near ground level but are much shorter where the bark remains relatively dry. Ascospores are colorless, equally two celled, elliptic, and 5.8–8.2 × 2.2–3.0 μm in size.

Disease cycle. Details of the cycle have not been published. Infection probably occurs during wet weather in wounds that become contaminated with airborne ascospores or waterborne (or insect-borne) conidia. Sprouts on diseased stumps are invaded by hyphae growing from the stump. Pycnidia are probably produced within the first year of lesion development, so a new cycle may start a year or less after the first infection. Surveys in Florida and India revealed increases in disease incidence from 15% to 57% over 4 years and from 1% to 27% in 3 years, respectively.

Chrysoporthe canker occurs primarily in regions with abundant rainfall. The pathogen grows most rapidly at 28–32°C, and lesions enlarge more rapidly in well-watered plants than in those under drought stress. Southern Florida is somewhat unfavorable for the disease because winters are relatively cool by tropical standards, and summers, although hot, are somewhat dry.

Hypovirulent strains of *C. cubensis* infected with double-stranded RNA viruses have been found in South America. The viruses are not expected to exert significant control of the fungus, however, because it has vegetative-incompatibility factors that restrict virus spread. Vegetative compatibility permits hyphae of different fungal colonies to fuse, which is necessary for virus transmission.

Clones and seedling families of *E. grandis* from different geographic sources vary significantly in susceptibility. Therefore, selection for resistance offers promise for control of Chrysoporthe canker. Phenolic compounds that occur in eucalyptus bark have been shown to inhibit growth and spore germination of *C. cubensis* in culture, but their possible role in field resistance is undetermined.

Two pathogens formerly confused with *C. cubensis*—*C. austroafricana* and *Cryphonectria eucalypti*—cause cankers and dieback of *Eucalyptus* in South Africa, and the latter fungus also occurs in Australia.

References: 43, 210, 233, 781, 1400, 1401, 1699, 1701, 1859, 2614, 2615, 2772–2775, 2903, 3357, 4149, 4397, 4536, 4537

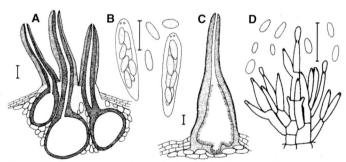

Figure 80. Chrysoporthe cubensis. A. Perithecia. B. Asci and ascospores. C. Pycnidium. D. Conidiophores and conidia. Scale bars: A and C = 100 μm, others = 10 μm. All courtesy of M. Gryzenhout.

Chrysoporthe canker of *Eucalyptus grandis* (rose gum) (FL, Apr).

A. A 12-year-old plantation affected by the disease in southern Florida.

B, C. A basal canker extending nearly 2 m up a trunk. Roughened bark extending to the upper margin (C) is typical.

D. The base of a healthy tree. The bark surface is mostly smooth, as old bark naturally falls away in strips.

E. The abrupt junction of healthy and diseased bark at the top of a basal canker is revealed where bark was shaved.

F, G. Successively closer views of pycnidia of *Chrysoporthe cubensis* on the inner surface of a dead strip of bark.

163

Endothia Canker (Plate 81)

Endothia gyrosa (syn. *Cryphonectria gyrosa*; Diaporthales, Valsaceae) causes twig and branch dieback and perennial cankers on branches, trunks, and exposed roots of various woody angiosperms. The fungus is widespread in eastern USA and causes noticeable damage to *Quercus* species (oaks) in the Southeast from Virginia to Texas. It also occurs in California. Records from abroad come from Europe, Asia, South Africa, Philippines, Australia, and New Zealand. In most circumstances the fungus causes localized, slowly expanding cankers, but it may girdle and kill branches and trunks of trees under environmental stress. The disease on *Quercus* is sometimes called orange hobnail canker. In coastal Virginia, cankers caused by *E. gyrosa* have contributed to a dieback syndrome of *Quercus palustris* (pin oak) called pin oak blight.

Recorded oak hosts include *Quercus agrifolia* (California live oak), *Q. alba* (white oak), *Q. bicolor* (swamp white oak), *Q. cerris* (European turkey oak), *Q. coccinea* (scarlet oak), *Q. falcata* (southern red oak), *Q. georgiana* (Georgia oak), *Q. ilicifolia* (bear oak), *Q. imbricaria* (shingle oak), *Q. lyrata* (overcup oak), *Q. macrocarpa* (bur oak), *Q. marilandica* (blackjack oak), *Q. nigra* (water oak), *Q. palustris*, *Q. phellos* (willow oak), *Q. prinus* (chestnut oak), *Q. robur* (English oak), *Q. rubra* (northern red oak), *Q. suber* (cork oak), *Q. velutina* (black oak), and *Q. virginiana* (southern live oak). Other hosts include *Acer saccharinum* (silver maple), *Castanea dentata* (American chestnut), *Corylus avellana* (European hazelnut), *Fagus grandifolia* (American beech), *F. sylvatica* (European beech), *Ilex opaca* (American holly), *Liquidambar formosana* (Formosan gum), *L. styraciflua* (sweetgum), *Prunus laurocerasus* (cherry-laurel), *Ulmus americana* (American elm), and *Vitis labrusca* (fox grape). Records of *E. gyrosa* in Australia and South Africa are now properly linked to the fungus *Cryphonectria eucalypti*, which causes cankers and dieback on *Eucalyptus*.

Symptoms and signs. Cankers on *Quercus* and *Liquidambar* limbs or trunks are usually depressed or sunken. On *Q. palustris* predisposed to severe disease, girdling cankers on trunks and branches result in scorch and browning of leaves, premature defoliation, dieback, and general decline. The fungus has been reported to cause a yellow-orange decay of the wood beneath cankers on oaks. On *Liquidambar*, purple staining and gum exudation at canker margins are common. Cankers on slowly growing *Fagus* limbs cause little change in bark contours until several years have elapsed. The diseased bark sometimes turns orange-pink; later it dries and cracks (Plate 81D) but remains in place for several years. Woundwood ridges at canker margins vary from absent to prominent.

E. gyrosa fruits on killed bark, commonly near the edges of wounds, producing conspicuous persistent stromata that are orange initially and darken to brick red or cinnamon brown. Their internal tissue remains bright orange. Stromata measure 1–2 mm in diameter and 1.5–2 mm high. Often they are arranged in parallel rows (Plate 81C). An unusual diagnostic feature of the stroma is the instant release of beet red pigment where it is touched by dilute potassium hydroxide solution. Young stromata contain irregular chambers where conidia form. When stromata are moist, amber blobs of conidia ooze out (Plate 81H). The conidia are individually colorless, rod shaped, and 3–4 × 0.75–1 µm in size.

At a later stage of development (in autumn or winter in the USA), long-necked perithecia form in the same stromata, and stromatal surfaces then have dark, nipplelike projections that are the tips of pycnidia and (or) tips of perithecial necks. The perithecia are globose, 250–300 µm in diameter, with necks to 500 µm long. The ascospores are unicellular, colorless, curved with tapering ends, and measure 6–10 × 2–2.5 µm.

Disease cycle. *E. gyrosa* infects broken branches and wounds on stems and exposed roots. Infection can occur at any time of year in a warm climate. Cankers develop most rapidly during summer. Dispersal of the pathogen has not been studied but is presumed to be similar to that of related fungi such as *Cryphonectria* species. The conidia may be dispersed by rain splash, insects, and possibly pruning tools. Ascospores are probably airborne. The interval from infection to formation of pycnidial stromata and dispersal of the next generation of conidia may be as short as 4 months, based on

observations in southeastern USA. Therefore, although the cankers are perennial, the disease cycle may be completed in less than a year. In cool-temperate regions, however, host dormancy probably interrupts the cycle.

Environmental predisposition of hosts seems to be a prerequisite for severe damage. Early reports characterized *E. gyrosa* as a weak parasite because it colonized experimentally inoculated branches or roots slowly. Later research showed that it causes rapidly expanding lesions in the bark of *Quercus* species stressed by water shortage. Small *Fagus* or *Quercus* trees that become infected during the period of stress after transplanting are at risk of girdling by the fungus. Rates of canker elongation up to 9 cm per month in *Q. palustris* occurred in experiments. The formation of cankers 1.0–1.6 m long and 8–12 cm wide in a single year was recorded for *Quercus* species in Texas, but such observations may represent coalescence of multiple cankers.

E. gyrosa grows readily in culture, producing orange pigment in most media. Such pigment production is characteristic of several *Endothia* and *Cryphonectria* species.

References: 98, 99, 114, 229, 233, 1053, 1400, 1833, 1859, 2614, 2615, 2772, 2901, 3308, 3309, 3571, 3748, 3872, 4149, 4150, 4218, 4280, 4491

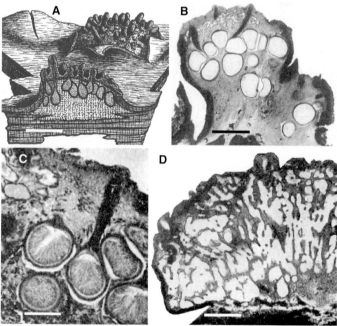

Figure 81. Endothia gyrosa. A. Habit of perithecial stromata in bark. B, C. Sections of perithecial stromata. The large voids in B are empty perithecial bodies. Small voids near the top of the stroma were sites of conidial production. Dark bands extending to the top of the stroma are perithecial necks. C. Dark perithecial walls and the entire neck of one perithecium are apparent. D. A conidial stroma with labyrinthine chambers where conidia were produced. Scale bars = 500 µm. A adapted from reference 1053; B and D adapted from reference 3571; C adapted from reference 3309 by permission of the Mycological Society of America.

A. Dieback caused in part by *Endothia gyrosa* on *Quercus palustris* (pin oak). This canker-dieback syndrome has been called pin oak blight (VA, May).

B, C. Cankers caused by *E. gyrosa* at pruning wounds on branches of *Fagus sylvatica* (European beech) (NY, Aug).

D, E. Close views of cankers on *F. sylvatica*, with persistent dead, cracked bark and numerous orange-brown stromata (PA, Jun).

F, G. Brick red mature stromata (F) and young orange ones (magnified) on a root buttress of *Quercus rubra* (northern red oak) (NY, Apr).

H. Magnified view of stromata on bark of *F. sylvatica*. Amber blobs of conidia (arrows) are visible (PA, Jun).

Leucostoma and Valsa Cankers (Plates 82–86)

Overview

Leucostoma and *Valsa* species (Diaporthales, Valsaceae) and their anamorphs (conidial states) in the genus *Cytospora* cause cankers and dieback on woody plants in hundreds of species. The pathogens are opportunistic invaders of bark on twigs, branches, or mainstems that have been weakened by drought, de-icing salt, freezing, heat, mechanical injuries, poor nutrition, or other diseases. These fungi are also ubiquitous colonists of dying bark and therefore are commonly found on dead twigs and branches. Sometimes they colonize cankers caused by other agents (e.g., see Plate 86). Parasitic species show limited host specialization, usually occurring on plants in one genus or several related genera, but sometimes many unrelated genera. In one case recorded in British Columbia, *C. chrysosperma*, which is best known as a pathogen of *Populus* (aspen, poplar) and *Salix* (willow), colonized scorched but living shrubs and trees in eight families after a ground fire.

Leucostoma and *Valsa* species produce conidia and later perithecia and ascospores in stromata shaped like short cones 1–4 mm across at the base, with flattened tops, that develop beneath the periderm or in the cambial region of thin bark. The tops (discs) break through the bark surface and appear as black to gray or brown, tan, or even white dots. Each conidial stroma contains one or more irregularly lobed fertile chambers that open through pores (ostioles) in the disc. Perithecia, 5–30 per stroma, are arranged in a group or a circle with their necks converging at the disc (Plate 82E; see also Figs. 82, 83, and Plates 85, 86). *Leucostoma* species have perithecial stromata with a black basal layer adjacent to the underlying bark; *Valsa* species lack this feature. Young stromata with conidia are closely associated with disease symptoms.

Conidia produced by these fungi are tiny (e.g., 4–6 × 1–2 μm), individually colorless, single celled, and allantoid (curved with rounded ends). When a mature conidial stroma is moistened, a yellow to orange mass of conidia in a gelatinous matrix extrudes as a curling tendril, or spore horn, from the opening in the top of the stroma. When the conidial mass is wet, the spores disperse over the bark in a film of water and then may be transported by dripping and splashing rain.

Ascospores are released in response to moisture. Studies of *V. ceratosperma* and *L. persoonii* (Plate 84) revealed two mechanisms of ascospore release. If a stroma remains moist for a long time, a mass of asci containing ascospores extrudes from each perithecium in a manner resembling the release of conidia. If dry bark with stromata becomes wet, however, ascospores may be forcibly expelled into the air. Ascospores are colorless, unicellular, allantoid, and 5–30 μm long, depending on species.

References: 229, 921, 926, 1099, 1213, 1298, 2042, 2045, 2046, 3381, 3777, 3913

Valsa Cankers and Diebacks of Conifers

Many *Valsa* and *Cytospora* species have been described from coniferous trees or shrubs around the world, but most of the species are poorly delimited, and relatively few are proven pathogens. Identification is difficult because of similarities among species and variability within them. Diagnosticians often find only the *Cytospora* state of a fungus in this group and usually identify it only to the generic level. Three species common in North America that are proven pathogens are noted below with some of their hosts. Asterisks identify species or genera in which pathogenicity has been demonstrated by inoculation.

Valsa abietis (conidial state *C. abietis*) occurs around the Northern Hemisphere on *Abies* species (fir), *Cedrus deodara* (deodar cedar), *Chamaecyparis nootkatensis* (Alaska cedar), *C. obtusa* (Hinoki-cypress), *Cryptomeria japonica* (Japanese cedar), *Juniperus communis* (common juniper), *Larix* species (larch), *Picea* species (spruce), *Pinus koraiensis* (Korean pine), *P. ponderosa* (ponderosa pine), *Platycladus orientalis* (Oriental arborvitae), *Pseudotsuga menziesii* (Douglas-fir), *Thuja occidentalis* (arborvitae), *T. plicata* (western red cedar), *Tsuga canadensis* (eastern hemlock), and *T. heterophylla* (western hemlock). The perithecial state of *V. abietis* is reportedly rare in western North America but is common on several tree species in Quebec. *V. abietis* causes branch mortality and top dieback and sometimes kills seedlings and saplings of *Abies amabilis* (subalpine fir), *A. grandis* (grand fir), *A. magnifica* (red fir),

and *A. procera* (noble fir). Damage is common in forests on the east slopes of the Cascades and Sierra Nevada, where trees are often subject to drought stress or winter drying. Cankers often develop on branches infected by dwarf mistletoe. Symptoms, the progression of disease, and the influences of predisposing factors are similar to those described for Leucostoma canker of spruce (Plate 83) except that sunken cankers are relatively more common on firs. The pathogen grows in culture at 5–30°C and grows most rapidly at 25–30°C.

Valsa friesii (conidial state *Cytospora pinastri*) also occurs around the Northern Hemisphere. Its hosts include several species each of *Abies*, *Juniperus*, *Larix*, *Picea*, and *Pinus*, as well as *Sequoia sempervirens* (coast redwood) and *Tsuga canadensis*. *V. friesii* also colonizes dying needles of *Abies* and *Pinus* species and has been credited with causing needle blight on these trees.

Valsa pini (conidial state *Cytospora pini*) occurs in North America and Europe. This fungus causes cankers and branch or twig dieback on *Abies balsamea*, *A. grandis*, *Pinus* species, and *Picea* species. *Leucostoma kunzei*, the most important pathogen of the Valsaceae on conifers, is discussed with Plate 83.

References: 1099, 1135, 1213, 1298, 3226, 3251, 3437, 3700, 4442

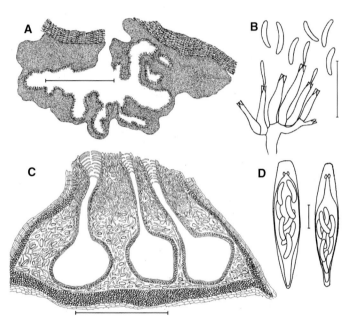

Figure 82. A, B. *Cytospora chrysosperma* (discussed with Plate 85). A. Section of a pycnidium with a labyrinthiform chamber where conidia form. B. Conidiophores and conidia. C, D. *Leucostoma persoonii* (discussed with Plate 84). C. Section of a perithecial stroma. D. Asci with ascospores. Scale bars = 500 μm for fruit bodies, 10 μm for other structures. A and B adapted from reference 3913 by permission of CAB International; C and D adapted from reference 2046 by permission of Blackwell Publishing.

A–E. Cytospora canker of *Tsuga canadensis* (eastern hemlock). A. A canker with a dark, resin-coated surface at the base of a dead branch on a young tree. B, C. A canker similar to that shown in A, with bark shaved in C to show location and length of the lesion. The dark brown line at the edge of the lesion is a zone where the host's defenses have halted the parasite. D. Magnified view of the bark surface, raised where pycnidial stromata are developing beneath the periderm. E. Pycnidial stromata of *Cytospora* sp., viewed after slicing off their tops. Fertile chambers with multiple lobes are revealed (NY, Sep).

F, G. Cytospora canker of *Abies balsamea* (balsam fir). F. A sapling girdled by a canker at a branch whorl. G. A closer view showing erumpent pycnidia near the upper margin of the original canker. The bark above died because of the girdling lesion (NY, Jun).

H–J. Perithecia of *Valsa* sp. in bark of *Pinus resinosa* (red pine). This fungus was associated with dieback of drought-stressed trees. H. Groups of perithecial necks breaking through the bark surface on a branch approximately 2 cm in diameter. I, J. Successively closer views of clusters of perithecial necks (MN, Aug).

Leucostoma Canker of Spruce and Other Conifers

Leucostoma canker of conifers, known also as Cytospora canker or spruce canker, occurs around the Northern Hemisphere. It is the most common and damaging bark disease of *Picea* species in the American Midwest and East. It affects ornamental trees and those in windbreaks, forest plantations, and sometimes natural stands, especially where trees are stressed by drought, frost, or poor site conditions. The pathogen is *Leucostoma kunzei* (conidial state *Cytospora kunzei*). Three varieties of it have been recognized: *L. kunzei* var. *piceae* on spruces, var. *superficialis* on pines, and var. *kunzei* on other conifers. These designations, however, are not widely used. Host plants include *Abies balsamea* and *A. fraseri* (balsam and Fraser firs); *Larix decidua*, *L. kaempferi*, and *L. laricina* (eastern, European, and Japanese larches); *Picea abies*, *P. engelmannii*, *P. glauca*, *P. mariana*, *P. orientalis*, *P. pungens*, and *P. rubens* (Norway, Engelmann, white, black, Oriental, Colorado blue, and red spruces); *Pinus banksiana*, *P. resinosa*, *P. strobus*, and *P. wallichiana* (jack, red, eastern white, and Himalayan white pines); *Pseudotsuga menziesii* (Douglas-fir); *Thuja plicata* (western red cedar); and *Tsuga canadensis* (eastern hemlock).

Picea pungens sustains the greatest damage, being attacked wherever it is grown east of its natural range in the Rocky Mountains. Within that range, however, Leucostoma canker is rarely noticed. Pathogenicity of *Leucostoma kunzei* to *Picea* and *Pseudotsuga* has been proved, but isolates from different hosts have varied widely in virulence when inoculated to *Picea pungens*.

Symptoms and signs. The disease seldom kills trees, but it disfigures them by killing branches and causing resin exudation from perennial lesions on branches or trunks. Infections on *P. orientalis* and *P. pungens* are usually confined to branches, although both trunks and branches of other spruces are attacked. Old branches sustain more damage than young ones. The following description applies to *P. pungens* in particular. In spring and early summer the foliage on one or more branches fades and turns brown, an indication that a branch or mainstem has been girdled. Brown needles persist during much of the growing season and drop off during winter, leaving bare twigs and branches. This process recurs more or less annually, moving from low branches to higher ones and destroying the symmetry of ornamental trees. Dead twigs and branches persist for many years. The disease occasionally starts in a high branch and, exceptionally, may kill the top of an otherwise green tree. Damage usually does not begin until trees are at least 10–15 years old. In landscape nurseries, however, small branches of young *P. pungens* or occasionally *P. glauca* may be killed.

Lesions often start at the bases of small twigs and develop into elliptic or sometimes diamond-shaped cankers. Lesions that originate on branches near the mainstem may spread into it. Bark killed by *L. kunzei* has brown to reddish brown internal color and is infiltrated with resin. The underlying sapwood, although killed and colonized by the pathogen, is scarcely discolored. Amber resin exudes copiously from the edges of cankers, runs down the bark, or drips onto lower branches or the ground, and hardens in a white crust. Except for the resin, canker locations on spruces remain obscure for several years because diseased bark is glued in place by resin, and woundwood formation at canker margins is slight or absent. Cankers on trunks eventually appear sunken because of the growth of surrounding tissues. Several years or sometimes decades may pass before a trunk or large limb is girdled, if the process is completed at all. Symptoms on other conifers are similar to those on spruces except that resin exudation is usually less prominent. Inconspicuous branch cankers are the rule on pines.

Stromata of the *Cytospora* state of *L. kunzei* form annually in recently killed bark within cankers and, more abundantly, beyond girdling cankers. Pycnidial stromata are shaped like short cones, 1–2 mm in diameter, with fertile chambers radiating from the center and opening through a common pore at the top. When moistened they produce yellow tendrils of conidia. An individual stroma does this only once, however. Conidia are unicellular, allantoid (sausage shaped), and 4–6 × 0.5–1 μm in size. Perithecial stromata, which mature in spring, are also short-conic and 1–2 mm in diameter. The interior tissue is pale yellow to grayish brown with 5–30 black perithecia embedded in it. Perithecia are 200–600 μm in diameter, and their necks converge at the disclike top of the stroma. The disc is usually 0.2–1.0 mm in diameter and gray to black on the surface.

The ascospores are colorless, unicellular, curved, and measure 5–8 × 1–2 μm.

Disease cycle. Infection begins in recent wounds, such as those made by tools or insects, or in cracks resulting from the burden of snow or ice. Most infections are believed to start in early spring, although the possibility of spring renewal of latent infections initiated in previous seasons deserves study. *L. kunzei* can be found in outer bark of apparently healthy branches, so latent infection may occur long before lesions develop. Both conidia and ascospores are infectious. Conidia are released during wet weather, most abundantly in spring but also throughout summer and autumn. Conidia can withstand freezing, and they germinate at 20–33°C. The optimum temperature for both conidial germination and initial growth of the fungus is near 27°C. Ascospores are released in spring and early summer. Dispersal of both types of spores by running and splashing water accounts for the year-to-year intensification and upward progression of symptoms. Both spore types can also be trapped from air in the vicinity of diseased trees. Conidia are more numerous than ascospores in both water and air. Just how conidia become airborne is uncertain; perhaps they are initially carried on tiny rain-splash droplets and become free in air as droplets evaporate. Airborne spores and insects account for spread from tree to tree.

After a branch or stem has been girdled the pathogen rapidly colonizes large areas of bark beyond the girdle, forming numerous pycnidial and later perithecial stromata. The duration of the disease cycle is probably 1 year, because pycnidial stromata form within the first year of lesion enlargement.

Of the factors that predispose conifers to damage by *L. kunzei*, drought is most commonly named. In experiments with *P. pungens*, the incidence and rate of development of cankers were greater in drought-stressed than in nonstressed inoculated trees. Damage by Leucostoma canker to *P. glauca* in Ontario was preceded by years of low rainfall. Frost injury and drought have both been linked to damage by *L. kunzei* (sometimes called pitch girdle) in plantations of *Pseudotsuga menziesii*. Other reported predisposing factors include hail damage and injury to roots by plant-parasitic nematodes.

References: 229, 1099, 1298, 1371, 1974, 2003, 2045, 2252, 3154, 3155, 3251, 3481, 4237, 4443

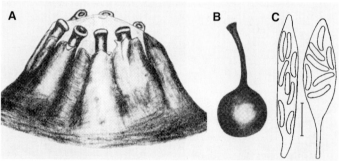

Figure 83. Valsa sordida (discussed with Plate 85). A. A perithecial stroma, drawn as if bark had been removed, with necks of perithecia arranged in a circle around the edge of the disc (the apex of a former pycnidial stroma). B. An isolated perithecium. C. Asci with ascospores. The ascus at left is nearing maturity; that at right is mature. Scale bar = 10 μm. A and B are unscaled. All adapted from reference 731 by permission of the American Phytopathological Society.

Leucostoma canker of *Picea pungens* (Colorado blue spruce).
A, B. Leafless twigs and faded or brown foliage on branches girdled by cankers (NY, Jun).
C. Twig dieback on a young tree in a nursery (NY, Sep).
D. Crystallized resin exuded on diseased and dead bark. Arrow indicates a location, beyond the canker, where fruit bodies of the pathogen would be found most readily (NY, Mar).
E. Magnified view of bark, similar to that identified by arrow in D, beyond a girdling canker. After removal of loose bark scales, the tops of pycnidial or perithecial stromata are apparent (NY, May).
F. Magnified view of moist bark at the margin of a canker. The white deposit is crystallized resin. The amber blobs (arrows) are masses of conidia that have oozed up from pycnidia in the bark (NY, Oct).

169

Leucostoma Cankers of *Prunus*

Leucostoma cincta and *L. persoonii* cause dieback and perennial cankers on *Prunus* species (stone-fruit trees) in orchards, landscapes, and forests. The diseases have several names, including Cytospora canker, Leucostoma canker, Valsa canker, and perennial canker. When severe in orchard trees, Leucostoma cankers cause marked decline in productivity. The causal fungi are most often found in their conidial states, *Cytospora cincta* and *C. leucostoma*, respectively. Both fungi occur across North America and also in Europe, Turkey, and Japan. *L. persoonii* is more commonly isolated than *L. cincta* in eastern USA and California, while the latter fungus is the more common in Ontario, northwestern fruit-growing areas, and Europe.

Leucostoma species are opportunistic colonists of wounded or dying twigs or bark, which serve as beachheads for invasion of healthy bark. These fungi are aggressive in plants weakened by environmental insults. A given strain is usually able to infect several *Prunus* species and some other rosaceous plants. Strains found on *Prunus* vary from strictly saprobic to virulently pathogenic. Low virulence in some strains is associated with infection by mycoviruses, but these viruses are not known to spread readily in the fungal population.

The host ranges of *L. cincta* and *L. persoonii* are similar: *Prunus* and other rosaceous plants and a few records from other woody angiosperms. Hosts include the following (the initials Lc or Lp appear if only one of the pathogens has been reported): *Amelanchier alnifolia* (western serviceberry or saskatoon, Lp), *Amelanchier* sp. (Lc), *Cotoneaster salicifolius* (Lc), *Crataegus* sp. (hawthorn, Lc), *Cydonia oblonga* (quince, Lp), *Liriodendron tulipifera* (tuliptree, Lp), *Malus baccata* (Siberian crabapple, Lp), *M. coronaria* (American crabapple, Lp), *M. pumila* (apple), *M. ioensis* (prairie crabapple, Lp), *M. sylvestris* (European crabapple), *Populus tremuloides* (trembling aspen, Lp), *Prunus americana* (wild plum, Lp), *P. angustifolia* (chickasaw plum, Lp), *P. armeniaca* (apricot), *P. avium* (sweet cherry), *P. caroliniana* (Carolina laurel cherry, Lp), *P. cerasus* (sour cherry), *P. davidiana* (Chinese wild peach, Lp), *P. domestica* (common plum), *P. dulcis* (almond), *P. munsoniana* (wild-goose plum, Lp), *P. padus* (European bird cherry), *P. pensylvanica* (pin cherry), *P. persica* (peach), *P. serotina* (black cherry, Lp), *P. serrulata* (Japanese flowering cherry, Lp), *P. spinosa* (blackthorn, Lc), *P. virginiana* (chokecherry), *Pyrus communis* (pear, Lp), *Robinia pseudoacacia* (black locust, Lc), and *Sorbus sitchensis* (Pacific or Sitka mountain-ash, Lp).

Symptoms and signs. This description applies to highly susceptible *Prunus* species. Affected trees usually have multiple cankers. Symptoms include dead buds, twigs, and branches; lesions on stems of all sizes; gum exudation at sites of active infection; dried gum on old cankers; and yellowing or wilting and death of leaves on diseased branches. Dead twigs or spurs often extend from the centers of branch cankers. Inner bark in lesions turns reddish brown and shrinks, causing the canker surface to appear depressed. Swelling due to woundwood formation at lesion edges results in torn or broken bark on old cankers. Typical old cankers on limbs or trunks are elongate, sunken, and black, with concentric ridges of roughened woundwood. Eventually they girdle limbs.

Gum accumulates in cavities in diseased inner bark and oozes out through cracks and ruptured lenticels. When wet, the gum swells, dissolves partially, and spreads over nearby bark. Debris and dark-colored saprobic fungi accumulate in the gum, which eventually becomes a blackened crust. Gum also infiltrates sapwood beneath and beyond cankers, causing the wood to cease sap conduction. This cessation accounts for the foliar symptoms. Concentrations of several nutrient elements, especially calcium, are subnormal in foliage beyond cankers.

Pycnidial stromata form in cankers and on girdled twigs and branches distal to cankers. Stromata are superficially black, with gray to grayish brown internal tissue, and are underlain by a black layer. Stromal development is first evident as pimplelike swellings at the tips of which the bark ruptures to expose the top, or disc, of a stroma. The disc of *L. persoonii* is white or frosted in appearance; that of *L. cincta* is gray to brownish gray. When pycnidia are mature and bark is moist, a flesh-colored to orange tendril of conidia in a gelatinous matrix extrudes from each stroma. Conidia of both *L. cincta* and *L. persoonii* are individually colorless, allantoid (sausage shaped), and measure 5–10 × 1–2 μm.

Perithecia form much later, sometimes 2–3 years after branch death, also within stromata. Ascospores are released from perithecia during spring and early summer. Asci and ascospores of *L. cincta* are larger than those of *L. persoonii*: asci are 45–80 versus 35–55 μm long, and the spores are 15–30 × 4–8 versus 10–18 × 2–5 μm.

Disease cycle. The pathogens overwinter in bark as mycelium, fruiting bodies, and conidia. Germinable conidia are present throughout the year, are most abundant in spring, and cause the majority of infections. Infection can occur whenever temperature and moisture permit, but late winter to early spring is the most common time. Lesions enlarge more rapidly in spring than during summer and autumn, when host resistance is more effective. During host growth, cork barriers and ridges of callus and woundwood often form at canker margins. At other times of year the parasites breach these barriers and kill bark on and beyond the woundwood. This sequence recurs annually at canker margins.

Both conidia and ascospores are dispersed by dripping or wind-blown water, and ascospores of *L. persoonii* are also dispersed by air during and after rain. Air dispersal of *L. cincta* has not been studied. Infection resulting from contaminated pruning tools and insects has also been reported.

Sites of infection include leaf scars, fruit scars, wounds made during pruning and other cultural operations, blossom spurs, fruit racemes (on *P. pensylvanica* and *P. serotina*), freeze-injured buds or twigs, lesions caused by brown rot fungi (Plate 37), and injuries caused by various insects. The pathogens invade bark, wood, and pith of twigs and outer sapwood beneath and beyond cankers. In one study *Leucostoma* species were isolated from wood 2 cm beyond canker margins on twigs. Host cells die before they are colonized by hyhae, indicating the involvement of toxin(s). *L. persoonii* secretes oxalic acid, which may promote cell death by acidifying tissue and sequestering calcium in oxalate crystals. The fungus also secretes polypeptide phytotoxins in culture; these have not been sought in diseased bark. Bark colonization may also be aided by cellulase, as this enzyme was produced by an isolate from *P. armeniaca*. Pycnidial stromata form beginning several weeks after death of bark. Conidia from spring-initiated lesions perhaps start a secondary cycle of disease, but this activity has not been documented. The spores of both fungi germinate over a wide temperature range. *L. persoonii* grows at higher temperatures (up to about 35°C) than does *L. cincta*.

Prunus cerasus, *P. dulcis*, and various stone-fruit tree and breeding lines have resistance to *Leucostoma* species. In addition, host vigor influences disease damage. Factors that impair vigor and predispose trees to infection under field conditions include water shortage, potassium deficiency, and freeze damage. Peach rootstocks that enhance the winter hardiness of stems also enhance resistance to Leucostoma canker.

References: 6, 193, 347–349, 364, 367, 368, 371, 688, 689, 947, 1099, 1390, 1470, 1598, 1656, 1859, 2045, 2046, 2896, 3186, 3501, 3502, 3749, 3926, 3982, 4075, 4210, 4292, 4403

A–F. Leucostoma canker of *Prunus serrulata* 'Kwanzan' (flowering cherry). A. Bud failure and arrested shoot growth on small branches indicate the presence of a canker on the supporting branch. B, C. Intact bark hides the canker, but bark necrosis is revealed by shaving off the periderm. D. Pycnidial stromata raising the periderm several weeks after canker formation (NY, May). E, F. Yellow and brown leaves call attention to a branch recently girdled by a canker. The canker's lower margin is below the dead shoot (arrow in E and lower arrow in F). One year earlier the lesion was localized around a dead twig (upper arrow in F) (NY, May–Jun).

G–I. Leucostoma canker of *Prunus persica* (peach). G. Gum exuding from a lesion at a twig base. H. Bark shaved to reveal the lesion. Reddish discoloration of sapwood and brown interior of the bud on the left side of the stem are symptoms of freeze damage, which may have predisposed the stem to infection (NY, May). I. Pycnidial stromata of *Leucostoma persoonii* on a dead peach twig (GA, May).

Leucostoma and Valsa Cankers of Poplar and Willow

Populus (aspen, poplar) and *Salix* (willow) are commonly affected by cankers and dieback caused by *Valsa sordida* and *Leucostoma niveum* (syn. *Valsa nivea*), which are usually encountered as the anamorphs *Cytospora chrysosperma* and *C. nivea*, respectively. These pathogens are opportunists, most likely to cause damage to weakened or predisposed plants. *V. sordida*, the more common and aggressive species, is also capable of causing severe damage to vigorous plants. It causes losses in nurseries, storages, newly established plantations, landscape and shelterbelt plantings, and sprout stands that grow after forest harvests. This fungus often inhabits apparently healthy bark and buds and is thus in a position to colonize parts that become weakened. In addition, both fungi colonize dying or recently dead bark. They occur around the temperate Northern Hemisphere, and *V. sordida* accompanied its host plants to Australia. Illustrations and general information about *Valsa*, *Leucostoma*, and their anamorphs accompany Plates 82 and 83.

L. niveum occurs on at least 6 *Populus* and 10 *Salix* species. Records for *V. sordida* or its anamorph include at least 16 *Populus* species and numerous hybrids as well as 21 *Salix* species and assorted trees and shrubs in 10 other genera. The poplar and willow records doubtless understate the true host ranges of these fungi in the Salicaceae. Records of *V. sordida* on other plants are, in some references, mixed with records for the ubiquitous *V. ambiens* (Plate 86) because the conidial states of these species cannot be distinguished reliably. A modern assessment of differences between these species is needed. Numerous reports also exist of other or unnamed *Valsa* or *Cytospora* species on *Populus* and *Salix*.

Symptoms and signs. V. sordida and *L. niveum* cause indistinguishable symptoms: spreading necrosis of weakened bark or, on stems of intermediate susceptibility, localized annual cankers or slowly expanding perennial cankers. Symptoms on *Populus* are considered here. Cankers are usually irregular in outline and more or less elongate. They form on trunks, limbs, and small branches, but twigs are killed without the formation of discrete lesions. Branch cankers often form at the bases of dead twigs. New lesions in smooth bark appear as sunken brownish patches. The inner bark turns nearly black and emits a foul salty odor. The sapwood appears reddish brown and water-soaked. The surface of *P. tremuloides* (trembling aspen) bark recently killed by *V. sordida* may appear yellow-orange. Cankers on old stems with thick bark may at first be imperceptible, but dead bark eventually sloughs off. Cankers a meter or more long may form on highly susceptible trees. Severely affected trees often die branch by branch.

Blackstem disease of *P. deltoides* (eastern and plains cotton-woods) seedlings and cuttings, attributed primarily to *V. sordida*, causes severe losses in storage and sometimes in nursery beds. Symptoms arise in autumn as tiny lesions at the ends of cuttings or at leaf scars and lenticels, usually on stems but sometimes on roots. Lesions enlarge during winter, becoming dark brown to black and water-soaked with sharply defined margins.

Fruiting structures of *V. sordida* and *L. niveum* are stromata in which pycnidial cavities and eventually perithecia form. They arise just beneath the periderm beginning within a few weeks after the bark dies in and distal to cankers. The exposed part, or disc, of a pycnidial stroma is typically gray-brown to black in *V. sordida* and white to gray-brown (usually white) in *L. niveum*. Each stroma at maturity is 1–2 mm in diameter and gray-brown internally and contains a fertile chamber with several irregular lobes that radiate downward and outward from an opening in the disc (Fig. 82A). Conidia of both species are single-celled, allantoid (sausage shaped), and colorless, and most measure 4–6 × 0.8–1.3 μm. If the bark is moist but not wet, massed conidia exude as cirrhi, or spore horns, up to several millimeters long (Plate 85H, I). These vary in color from reddish orange through yellow and eventually bleach nearly white if they are not first dissolved by water. If diseased bark is thick and fissured, cirrhi appear in the fissures.

Perithecia of *V. sordida* and *L. niveum* are less common than pycnidia. They form in the same stromata with pycnidia or in new stromata beginning in autumn and winter after pycnidial formation. Perithecial bodies, approximately 0.5 mm in diameter, are black and are arranged several in a group or ring in the lower, outer part of the stroma. Their necks converge at the surface of the disc (Figs. 82C, 83A). Sometimes a pycnidial cavity persists in the center of a stroma. Ascospores are liberated as described with Plate 82.

Disease cycle. Most infections are presumed to be initiated by conidia because they are more abundant than ascospores and are liberated throughout the year as temperature and moisture permit. Buds, nodes, lenticels, and wounds of various kinds are sites of infection. In experiments, wounds in *P. tremuloides* bark remained susceptible for at least 10 days in water-stressed trees but became resistant after as few as 2 days in the absence of environmental stress. Latent infections may occur, as *V. sordida* has been detected in symptomless buds and stems of *P. deltoides* and limbs of *P. tremuloides*. Cankers usually develop during the dormant season and enlarge at rates related to temperature in the range 2–30°C. Rates of canker enlargement up to 40 mm per day have been recorded. The pathogens grow most rapidly in culture at temperatures near 25°C.

The rate of canker development and the annual or perennial nature of a canker depend also on effectiveness of host defenses. The blackstem symptom in *P. deltoides* indicates a near absence of defense reactions in bark. The cortex is rapidly colonized and cellular contents digested. In a slowly enlarging canker, in contrast, the pathogen is continually inhibited by tannins and other phenolic compounds (and probably other defenses) in and among cells at the margin of the lesion. These reactions are somewhat weaker in sapwood than in bark; thus the fungus is able to advance in the cambium and sapwood beyond the externally visible edge of the lesion. If fungal advance is slowed sufficiently or halted, a periderm differentiates in bark around the lesion (process described with Plate 257), and the killed sapwood becomes compartmentalized (Plates 254, 255). A canker may resume enlargement if the surrounding periderm is breached by new wounds or by wedges of aggregated hyphae in the vicinity of phloem fiber bundles, as has been observed in lesions caused by *V. sordida*.

Damage to aspens and poplars by *V. sordida* has been linked to unfavorable sites for growth, water shortage, summer heat, winter drying, freeze-thaw cycles, defoliation, nitrogen deficiency, damage by browsing animals, miscellaneous wounds, and infection by other pathogens. Bark below large wounds is usually attacked because it no longer receives enough nutrients and water for effective defense. Blackstem symptoms invariably develop in cottonwood cuttings weakened by prolonged or improper storage. However, epidemics of Valsa canker in the absence of known stresses have also been recorded. Strains of *V. sordida* are known to vary in aggressiveness. This fungus also invades bark weakened or killed by other pathogens, such as *Septoria musiva* (Plate 11), and may obscure the primary pathogen.

Tree breeders have searched for *Populus* species and clones resistant to *V. sordida*, and some have reported finding heritable resistance. To date, however, heritable resistance has not been distinguished from heritable tolerance of predisposing factors at test sites. Unfortunately, the adaptability of planting stock to specific planting sites is seldom known.

References: 374, 414, 418, 575, 609, 731, 1129, 1418, 1672, 1792, 1859, 1948, 2042, 2045, 2046, 2561, 3475, 3488, 3778, 4199

A–C. Dieback and cankers caused mainly by *Leucostoma niveum* in a fastigiate cultivar of *Populus alba* (white poplar). A. Severe dieback. B. New and year-old cankers on small branches. The canker at right has numerous raised white discs of stromata of *L. niveum*, shown magnified in C. A circle of perithecia within a stroma is revealed where its top was sliced off (NY, Jul).

D. An elongate Cytospora canker that arose at a pruning wound on a mature *P. alba* (NY, Jun).

E, F. Cankers caused by *Valsa sordida* on *Salix capraea* (goat willow) (NY, Jun).

G. Blackstem symptoms on cuttings of *P. deltoides* (eastern cottonwood) (ND, May).

H, I. Cirrhi (spore horns) of *Cytospora chrysosperma*, the conidial state of *V. sordida*, on *P. tremuloides* (trembling aspen) and *Salix* sp., respectively. Dark color of bark around the base of the cirrhus in I indicates the location and size of the stroma beneath the periderm. Young stromata are developing nearby (NY, May–Jun).

173

Valsa Cankers of Maples

Valsa ambiens and *V. ceratosperma* are associated with branch cankers and dieback of *Acer* species (maples) in forests, landscapes, and nurseries in eastern North America. These fungi occur on numerous woody angiosperms across North America and in Europe. *V. ambiens* comprises two subspecies, *ambiens* and *leucostomoides* (syn. *V. leucostomoides*), with anamorphs (pycnidial states) *Cytospora leucosperma* and *C. annulata*, respectively. Both subspecies are opportunistic pathogens and saprobic colonists of woody plants weakened or killed by diseases or injuries. The recorded host range of *V.a.* subsp. *ambiens* includes representatives of more than 40 genera, while that of *V.a.* subsp. *leucostomoides* includes only *Acer* and occasional records for *Fraxinus* (ash), *Malus* (apple, crabapple), and *Pyrus* (pear) in North America. The host range of *V. ceratosperma* (anamorph *C. sacculus*) includes species in more than 50 genera.

Valsa cankers of maples, causing dieback of scattered branches, are usually noticed on either *A. platanoides* (Norway maple) or *A. saccharum* (sugar maple). Other maple hosts include *A. negundo, A. rubrum,* and *A. saccharinum* (box-elder, red maple, and silver maple, respectively). Outbreaks on the first two species occur in occasional years, sometimes during or following drought. Diseased branches may die at any time of year. Those dying in spring through midsummer become conspicuous when foliage turns yellow prematurely or wilts and turns brown. Distinct girdling cankers sometimes cause the dieback, but often no clearly defined lesion is apparent. The branch dies back to its junction with a larger limb. A dark olive-green to greenish black line typically separates healthy sapwood from dead wood at that location (Plate 86B). *V.a.* subsp. *leucostomoides* also invades and induces stain in freshly wounded maple sapwood. It is commonly associated, for example, with columns of dark olive to greenish black stain that start from tap holes drilled in *A. saccharum* trunks for sap extraction. Pycnidial stromata, most often those of *C. annulata*, develop in recently killed bark and may be found at any time of year. Perithecia of *V.a.* subsp. *leucostomoides* develop in the same or new stromata in spring to early summer, following the pycnidia.

The forms and life histories of *Valsa* species on maple are similar to those described for related fungi on other plants (Plates 82–85). Young stromata with conidia (the *Cytospora* state) are closely associated with disease symptoms. This state of *V.a.* subsp. *ambiens* has a grayish tan to brown or black disc up to 1.1 mm across with one–eight ostioles (openings for spore release). The conidia are single celled, colorless, sausage shaped, and mostly 3–8 × 0.8–2.0 μm in size. The *Cytospora* state of *V.a.* subsp. *leucostomoides* has a white to gray disc up to 0.7 mm across, sometimes with a dark border, one to four ostioles, and conidia 3–6 × 0.75–1 μm. Conidia are extruded and dispersed as described and illustrated with Plates 82 and 85.

Strains of both subspecies of *V. ambiens* isolated from maples and tested for pathogenicity caused small cankers in inoculated stems of *A. platanoides, A. rubrum,* and *A. saccharum,* but neither subspecies was an aggressive colonist of healthy bark. *V.a.* subsp. *ambiens* is suspected to be pathogenic to additional plants including *Amelanchier* (serviceberry), *Malus* (apple), *Rosa* (rose), and *Salix* (willow).

V. ceratosperma may be only a saprobe or a colonist of dying bark on maples, as several strains of it failed to cause lesions when inoculated to maple branches. This fungus causes a significant canker disease of *Malus* in Japan, however, so it must be considered a potential pathogen of other woody plants.
References: 1099, 1298, 1859, 3704, 3777, 3778, 3807

Cryptosporiopsis Canker of Red Maple (Plate 86)

Cryptosporiopsis canker of *Acer rubrum* (red maple) caused economic damage in nurseries and appeared in young trees in some landscapes in northeastern USA in the late 1970s but has attracted no attention since then. The most severely affected cultivars were October Glory and Autumn Flame. The pathogen, an unidentified species of *Cryptosporiopsis* (a coelomycete), infects oviposition wounds made by *Oecanthus niveus* (snowy or narrow-winged tree cricket) in trunks and low branches. The cankers are annual, usually less than 20 cm long, and are often only several millimeters wide, tapering toward the ends. They reach their greatest length and width in the cambial region, and each is centered on a tiny hole from which dark brown fluid bleeds. Streaks of discolored sapwood extend as far as 50 cm above cankers. Small trees may be girdled by these lesions, and larger stems may be disfigured by multiple cankers.

Unfortunately, no description or illustration of the pathogen has been published. Generic characteristics of *Cryptosporiopsis* include fruit bodies that vary in form from acervular to stromatic. Those in bark are dark stromata with a spore-producing locule of variable shape and an irregular opening. The conidia are colorless, single celled, ovoid, and relatively large (e.g., 10 × 25 μm) in most species. *Cryptosporiopsis* species are considered to be anamorphs of *Pezicula* and *Neofabraea* (Helotiales, Dermateaceae).

Infection of maples occurs at the time of cricket oviposition in autumn, cankers develop mainly during the dormant season, and bleeding is conspicuous in spring. Vigorous trees produce prominent woundwood at canker margins by late summer and may enclose the smallest cankers beneath woundwood within a year. Often, however, opportunistic secondary fungi, especially *Valsa ambiens*, colonize the outer bark of cankers and perhaps cause canker expansion. The relationship between insect and fungus in this disease is not fully understood, but inoculation is believed to occur as the female tree cricket scrapes fragments of superficial bark infested with the fungus into the oviposition hole to form a plug.

The activity of the *Cryptosporiopsis* on *A. rubrum* is consistent with information about related fungi that cause cankers, leaf spots, and fruit rot on various trees. *Pezicula acericola* (anamorph *Cryptosporiopsis* sp.) occurs on twigs and branches of *Acer* species, including *A. rubrum,* in northeastern North America. *P. cinnamomea* is an endophyte and causes a canker of *Quercus rubra* (red oak) in Europe. *P. sporulosa* (anamorph *C. abietina*) is an endophyte of *Acer macrophyllum* (bigleaf maple) twigs and gymnosperm leaves and twigs. *Neofabraea malicorticis* (syn. *Pezicula malicorticis*, anamorph *C. curvispora*) and *N. perennans* (syn. *P. perennans*, anamorph *C. perennans*) cause cankers on *Malus* and other rosaceous trees. *N. alba* (syn. *P. alba*, anamorph *Phlyctema vagabunda*) causes small circular cankers (coin canker) on *Fraxinus* species and cankers on *Malus* and other woody plants.
References: 1959, 1973, 2028, 3343, 3913, 3973, 3974, 4152

A–G. Cankers on *Acer* species caused by *Valsa ambiens* subsp. *leucostomoides*. A. Branches of *A. platanoides* (Norway maple) with faded and brown leaves (NY, Jun). B. The base of a diseased *A. saccharum* (sugar maple) branch with a characteristic dark line between healthy and dead tissues (NY, Sep). C. Magnified view of *A. saccharum* bark where rows of stromata of the anamorph *Cytospora annulata* are raising the surface (NY, Sep). D. Rows of white dots, the tops of pycnidial and perithecial stromata, nearly natural size, on *A. platanoides* bark (NY, Jul). E. Magnified view of *A. platanoides* bark with the surface shaved off to reveal dark gray-brown pycnidial stromata 2–3 mm in diameter (NY, Sep). F. Magnified view of pycnidia breaking the bark surface at the edge of a canker on *A. rubrum* (red maple); healthy bark is at upper right. G. Magnified view of fresh perithecial stromata in *A. rubrum* bark. Black objects rising above the bark surface (arrow at top) are the clustered tops of perithecia that have convergent necks. Where the topmost part of a stroma was cut off (left arrow), multiple necks of perithecia appear as black dots on a light gray background. Gelatinous contents of perithecia are visible where deeper parts of stromata were cut away (NY, Apr).

H–K. Cryptosporiopsis canker on *Acer rubrum* cv. October Glory (NY, Jun). H, I. A stem with two small bleeding cankers before (H) and after (I) removing surface bark of one canker to reveal a dark yellowish brown zone in the cambial region. J. A canker that developed during the dormant season. The wound at the center of the canker was presumed to be the site of infection. The bark surface in this canker is roughened by pycnidia of *Cytospora annulata*, a secondary invader. K. A year-old inactive canker, sunken as the result of woundwood formation at margins.

Nectria Cankers and Diebacks (Plates 87–91)

Many species of *Nectria* and related genera (Hypocreales, Nectriaceae) colonize trees and shrubs, and some cause cankers and dieback. The genera *Nectria* and *Neonectria* comprise species that produce perithecia (often brightly colored) on more or less superficial stromata, mostly on woody substrates. Probably many of the species have pathogenic capability, but only a few that cause significant diseases have been much studied. Those considered here are *Nectria cinnabarina* (Plates 87, 88), *N. austro-americana* (Plate 89), *Neonectria galligena* (Plate 90), and *Neonectria coccinea* var. *faginata* (Plate 91).

Coral-spot Nectria Canker

The disease is caused by *Nectria cinnabarina*, known as the coral-spot fungus because it produces erumpent sporodochia that are coral pink when young. It is an opportunistic pathogen and cosmopolitan saprobe on many plant species, mostly in the temperate zones, worldwide. Its pathogenicity has been proven by inoculations to many plant species. It colonizes bark, cambium, and outer sapwood of stems weakened by freezing, water shortage, mechanical injuries, or other diseases. It may cause significant damage to recently transplanted trees or shrubs on landscape or forest sites.

N. cinnabarina occurs on plants in more than 90 genera, primarily woody angiosperms but also a few gymnosperms, palms, and herbaceous plants. Woody host genera of the fungus or its anamorph, *Tubercularia vulgaris*, include the following:
Acer (maple), *Aesculus* (buckeye, horse-chestnut), *Agathis* (kauri), *Ailanthus* (tree-of-heaven), *Albizia* (mimosa), *Alnus* (alder), *Amelanchier* (serviceberry), *Ampelopsis, Aralia, Asimina* (pawpaw), *Berberis* (barberry), *Betula* (birch), *Bougainvillea, Broussonetia* (paper-mulberry), *Buxus* (boxwood), *Caesalpinia* (nicker), *Callicarpa* (beautyberry), *Calycanthus* (sweetshrub), *Caragana* (pea-tree), *Carya* (hickory), *Celastrus* (bittersweet), *Catalpa, Cercis* (redbud), *Celtis* (hackberry), *Chaenomeles* (flowering quince), *Choisya* (Mexican-orange), *Citrus, Cornus* (dogwood), *Cotoneaster, Cydonia* (quince), *Cytisus* (broom), *Daphne, Elaeagnus* (autumn-olive, Russian-olive), *Eriobotrya* (loquat), *Erythrina, Euonymus, Fagus* (beech), *Ficus* (fig), *Firmiana* (parasol tree), *Frangula* (buckthorn), *Fraxinus* (ash), *Gleditsia* (honeylocust), *Hibiscus* (mallow), *Holodiscus, Hydrangea, Juglans* (walnut), *Kerria* (Japanese-rose), *Koelreuteria* (golden-rain tree), *Laburnum* (bean-tree), *Larix* (larch), *Ligustrum* (privet), *Liquidambar* (sweetgum), *Lonicera* (honeysuckle), *Maclura* (osage-orange), *Magnolia, Malus* (apple, crabapple), *Melia* (chinaberry), *Morus* (mulberry), *Philadelphus* (mock-orange), *Phoradendron* (mistletoe), *Picea* (spruce), *Pinus* (pine), *Populus* (aspen, poplar), *Prunus* (stone-fruit trees), *Pseudotsuga* (Douglas-fir), *Pyracantha* (firethorn), *Pyrus* (pear), *Quercus* (oak), *Rhodotypos* (jet-bead), *Rhus* (sumac), *Ribes* (currant), *Robinia* (locust), *Rosa* (rose), *Roystonea* (royal palm), *Rubus* (brambles), *Salix* (willow), *Sambucus* (elder), *Schinus* (pepper-tree), *Sorbus* (mountain-ash), *Spiraea, Styphnolobium* (pagoda tree), *Symphoricarpos* (snowberry), *Tilia* (linden), *Ulmus* (elm), *Umbellularia* (California bay), *Vaccinium* (blueberry), *Vernicia* (tung-oil tree), *Viburnum, Vitis* (grapevine), and *Wisteria.*

Symptoms and signs. Lesions usually form during host dormancy. Twigs or branches with girdling lesions fail to produce leaves in spring, or foliage on them may suddenly wilt and shrivel after growth begins. Discrete cankers may form on branches or stems of small plants weakened by freezing. These cankers often appear as slightly sunken areas associated with wounds. The killed bark is discolored brown to dark brown. If the outer bark is corky, the dead area remains imperceptible until the fungus begins to produce sporodochia or until growth of surrounding tissues leads to cracks at canker margins. Cankers usually expand during host dormancy in only one year and then become sunken as callus and woundwood form around the edges. In severely stressed trees, however, cankers may develop during the growing season and may enlarge during more than one year.

Lesions caused by *N. cinnabarina* are seldom noticed before the fungus begins to form cushionlike sporodochia. These usually appear in spring and early summer, beginning with eruption of stromata (mycelial cushions) through the bark surface at points of natural weakness such as lenticels in young bark or cracks in corky bark. Sporodochia in bark cracks, often aligned in rows, range from less than 0.5 mm to about 1.5 mm in diameter and height. They vary in color with host and stage of maturity from creamy or coral pink to pink-orange or light purplish red when young, and to tan, brown,

or nearly black in aging specimens, especially after exposure to frost. Sporodochia swell and have a viscid surface when wet. Conidia, individually colorless, single celled, and measuring 5–7 × 2–3 μm, are produced in a gelatinous mass on the cushion. The surface becomes crusty when dry.

In summer and autumn, globose orange-red perithecia form singly and in clusters of up to 15, either among the sporodochia or replacing them on the same stromata. The perithecia are mostly 375–450 μm in diameter, have rough surfaces, and turn dark reddish brown with age. They may persist through winter on dead bark. The ascospores are colorless, equally two celled with a slight constriction at the septum, narrowed at the ends, and mostly 12–20 × 4–7 μm in size.

Disease cycle. The disease is considered to be monocyclic (one cycle per year) with multiple starting times. *N. cinnabarina* invades pruning wounds, injured buds, cracks caused by the weight of ice or snow, and bark damaged by frost, hail, insects, and mammals. Injuries sustained in autumn and winter are most important as infection sites because hosts' wound-isolation responses are minimal during the dormant season. The fungus overwinters as mycelium in bark and as sporodochia and perithecia on bark.

Both conidia and ascospores of *N. cinnabarina* are infectious, and infection may occur throughout the year when temperature and moisture permit. Conidia are dispersed by splashing and dripping water and perhaps on pruning tools. During wet weather, massed

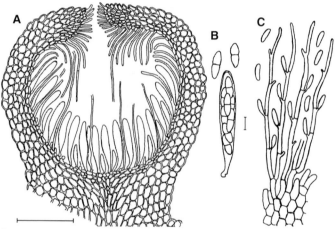

Figure 87. Nectria cinnabarina. A. Section of a perithecium. B. An ascus and ascospores. C. Part of the surface of a sporodochium with conidiophores and conidia. Scale bars = 100 μm for perithecium, 10 μm for other structures. Adapted from reference 123 by permission of Gebrüder Borntraeger Verlagsbuchhandlung.

Coral-spot dieback and cankers and signs of the causal fungus, *Nectria cinnabarina.*

A. Coral-spot dieback of *Rosa* (rose). A rough wound near the top of the pruned cane was the site of infection. Pink sporodochia are visible (NY, Nov).

B–E. Coral-spot Nectria canker of *Acer saccharum* (sugar maple). B, C. Two views of the stem of a sapling transplanted the previous year, nearly girdled by cankers initiated at pruning wounds (NY, Sep). D. Close view of a canker with pink sporodochia and, in the older part of the canker, red perithecia (NY, Jun). E. Partial cross section of the stem shown in B. A very narrow growth ring (upper arrows) indicates that severe stress occurred during the year of transplanting when infection began. Lower arrow identifies the beginning of the previous growth ring. Cracks in the wood, unrelated to disease, formed as the specimen dried.

F–I. Magnified views of *N. cinnabarina* on *A. saccharum* bark. F. Young sporodochia (NY, Sep). G. Sporodochia and developing perithecia, the latter with scaly surfaces. H. Mature perithecia. I. Sporodochia and young perithecia on wet bark (NY, Jun).

J. Magnified view of overwintered sporodochia on bark of *Gleditsia triacanthos* (honeylocust) bark (NY, Jun).

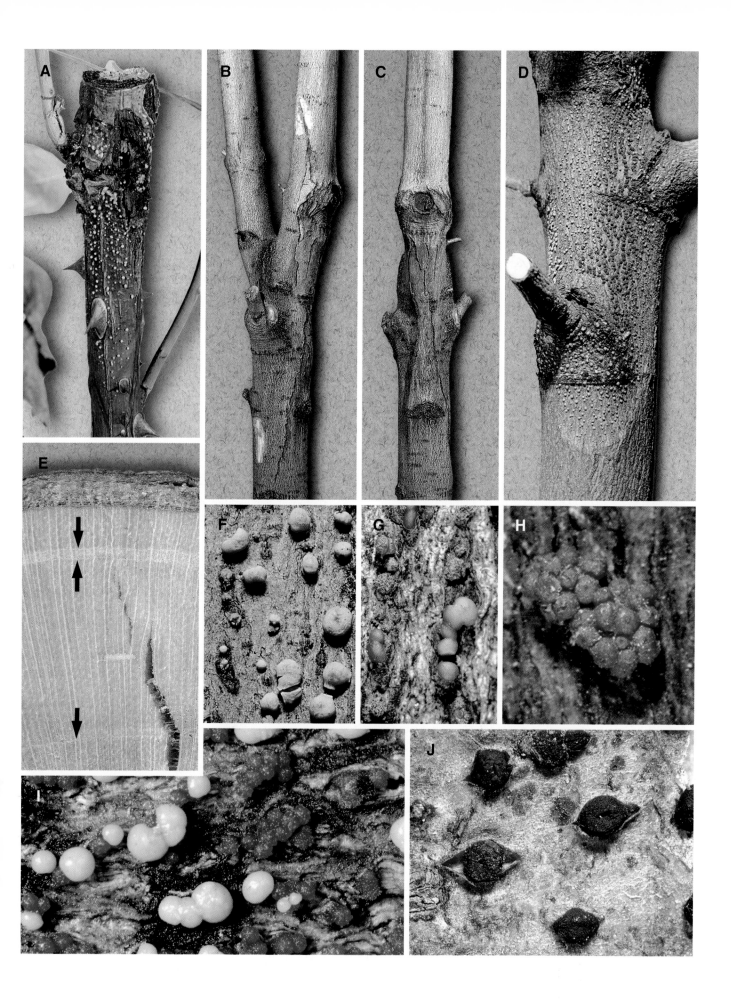

ascospores extrude in a blob from a pore in the top of each perithecium and are dispersed mainly by water. Ascospore dispersal continues even in winter unless interrupted by freezing weather. Conidia and ascospores germinate most rapidly at 21–26°C, but the latter can germinate, albeit slowly, at temperatures as low as 0°C. The fungus colonizes bark, cambium, and the outer sapwood. In *Ribes* and probably in other hosts, the advance of mycelium in outer sapwood precedes that in the cambium and phloem. The interval from infection to spore production varies with the season. Isolates of *N. cinnabarina* vary widely in aggressiveness, but host specialization has not been reported.

N. cinnabarina commonly infects plants predisposed by cultural treatment or weather that compromises cold hardiness. Untimely pruning or fertilization (e.g., in late summer or early autumn), root pruning, and transplanting predispose plants to infection. In Minnesota, treatments of *Gleditsia triacanthos* (honeylocust) that increased vigor predisposed trees to more severe damage by *N. cinnabarina* than occurred in comparable less vigorous trees, apparently because cold acclimation in autumn was delayed in the more vigorous trees. Rapid freezing after warm days in late autumn and late winter, when woody plants are not maximally cold hardy, also predisposes them to attack because frost insufficient to kill bark may impair host defenses. In experiments with *Euonymus alatus* (winged enonymus), the coral-spot fungus caused cankers or dieback on cold-treated stems that lacked evidence of direct frost injury, while the same isolates caused no significant damage to comparable untreated stems. Predisposition due to sublethal cold shock is temporary, however. Normal resistance to pathogens returns after several days. When *N. cinnabarina* invades stems injured by freezing, it increases the amount of dieback. Damage by *N. cinnabarina* can be avoided or diminished by planting trees or shrubs known to be adapted to the climate of the planting site, by minimizing water stress, and by avoiding untimely pruning or other wounding.

In north-central USA and Canada *Tubercularia ulmea* causes diseases similar to those caused by *N. cinnabarina*. *T. ulmea* is similar to *T. vulgaris*, the anamorph of *N. cinnabarina*, and some authorities suspect that the former fungus is a variant of the latter. Genetic studies are needed to resolve the matter.

References: 123, 280–282, 441, 607, 1099, 1859, 3342, 3345, 3479, 3480, 3482, 3617

Tubercularia Canker and Dieback

Tubercularia ulmea, a hyphomycete, causes annual cankers and dieback of several species of woody plants in the northern prairie region from Illinois (where it was originally described) to Manitoba and Alberta. The fungus causes damage mainly to trees or shrubs weakened by environmental stresses such as sudden freezing and is therefore found most commonly on dead twigs and branches. Reported hosts include *Acer ginnala* (Amur maple), *A. negundo* (box-elder), *Caragana* sp. (pea-shrub, pea-tree), *Elaeagnus angustifolia* (Russian-olive), *Euonymus alatus* (winged euonymus), *Frangula alnus* (European alder-buckthorn), *Fraxinus pennsylvanica* (green ash), *Gleditsia triacanthos* (honeylocust), *Picea pungens* (Colorado blue spruce), *Sambucus racemosa* var. *rubens* (catberry), *Ulmus americana* (American elm), and *U. pumila* (Siberian elm). The fungus has been reported from all across Canada on *U. pumila*. Additional hosts are probable because those already known are diverse.

Symptoms and signs. Lesions caused by *T. ulmea* are usually not noticed until they girdle and kill twigs and branches. The resulting dieback becomes apparent when buds fail to open or when foliage wilts in late spring and early summer. Nongirdling cankers are oval to elongate, often taking the form of longitudinal strips. New lesions in smooth bark of young *U. pumila* stems appear reddish brown and slightly shriveled, but lesions in older stems with corky outer bark are imperceptible until cracks form at margins. Cankers on *Elaeagnus* sometimes exude gum at the margins. Woundwood formation around the edges of nongirdling cankers causes bark to crack there, and bark in a canker eventually shreds and sloughs. Small cankers on vigorous stems may be enclosed beneath woundwood within a year, but usually several years are required.

Vigorous sprouts may develop below girdling cankers. Cankers on *U. pumila* are most common on small branches and twigs in the interior of the canopy. *Frangula alnus* 'Tallhedge' (columnar buckthorn) is subject to girdling cankers near stem bases.

Cushionlike fruiting structures (sporodochia) of *T. ulmea* begin to break through the surface of killed bark during spring and early summer. They are cream colored at first and slowly darken through shades of rust-pink to brown and finally black. At maturity they are up to 1.5 mm in diameter and 0.9 mm high. Unicellular, ovoid, colorless conidia, mostly 4.6–6.2 × 1.5–2.5 μm, form on their surfaces. Mature sporodochia may be found in dead bark throughout the summer and autumn.

Beginning in autumn and continuing the next spring, sporodochia are sometimes followed by pale pinkish stromata on which bright red perithecia form. This sexual state cannot be distinguished from *Nectria cinnabarina* (Plate 87). Therefore *T. ulmea* may be only a variant of *T. vulgaris*, the anamorph of *N. cinnabarina*.

Disease cycle. A 1-year cycle is probable, with infection occurring during the dormant season. Cankers form in spring, enlarge during a period of several weeks, and then cease enlargement permanently. Sporulation may begin within a few weeks after infection and continue throughout the year as temperature and moisture conditions permit, or it may be delayed for up to a few months. Conidia of *T. ulmea* are dispersed by water. They germinate within 24–40 hours on laboratory media and produce rapidly growing white colonies. In experiments with *U. pumila*, cankers caused by *T. ulmea* began to form within 10 days after inoculation in April, and sporodochia began to emerge within an additional 10 days. When stems of *Gleditsia triacanthos* were inoculated in October, formation of sporodochia was delayed until the next spring.

Rapid drops in temperature to below −20°C, as may happen after relatively mild weather in late autumn or late winter in the northern prairie region, commonly lead to infection by *T. ulmea*. The predisposing effect of sudden freezing has been demonstrated experimentally with *F. alnus* 'Tallhedge.' Basal cankers were caused in part by the direct effects of freezing and in part by *T. ulmea*. Tubercularia dieback has also been observed on heavily fertilized 'Tallhedge' plants and on those growing near lights that artificially extended the photoperiod during winter. These conditions, by delaying or preventing normal cold hardiness, predispose plants to infection. Isolates of *T. ulmea* vary significantly in aggressiveness (amount of damage they are capable of causing in a given host), but host specialization has not been detected.

References: 661, 2054, 3290, 3477, 3479, 3871, 4200

Thyronectria Canker of Honeylocust

Nectria austroamericana (syn. *Thyronectria austroamericana*) and *N. cinnabarina* cause cankers and dieback of *Gleditsia triacanthos* (honeylocust). The former species, considered here, has a narrow host range; the latter (Plate 87) has hundreds of hosts. The name Thyronectria canker is still used to distinguish this disease from

---

A–E. Tubercularia canker and dieback of *Elaeagnus angustifolia* (Russian-olive). A. A tree with two mainstems girdled by cankers. Wilted foliage adheres to twigs on the stem that died early in the growing season. Epicormic sprouts are developing below the cankers (ND, Jul). B. Gum exuding from longitudinal cracks at the margin of a Tubercularia canker. Gum exudation may be the only external indicator of disease on a stem with thick bark (ND, Jul). C. A diseased branch with bark shaved off to show the extent of a lesion (CO, Jul). D. A branch canker with the bark surface about to be broken by developing sporodochia (CO, Jul). E. Young sporodochia of *Tubercularia ulmea*, rust-pink in color, breaking through the bark surface in a canker (ND, Jul).

F–H. Magnified views of *T. ulmea* and its perithecial state, indistinguishable from *Nectria cinnabarina*, on the bark of *Ulmus pumila* (Siberian elm). F. Sporodochia of various ages, the oldest nearly black. G. Old sporodochia (black cushion-shaped structures) and young perithecial stromata (pink structures). H. Clusters of perithecia (ND, May).

those caused by the closely related *N. cinnabarina*. Thyronectria canker occurs from Colorado and Oklahoma eastward to Georgia and Massachusetts. It is common in central USA and is cited as a major disorder of *Gleditsia triacanthos* planted as shade trees in Colorado and Illinois and as windbreak trees in Oklahoma. It is uncommon in northeastern and Atlantic Coast states, and it causes only branch cankers and minor dieback on *G. triacanthos* in natural wooded areas. The causal fungus also occurs as an apparent saprobe, producing pycnidial stromata and perithecia on dead trunks and branches, in areas where the canker disease has not been reported.

N. austroamericana also occurs occasionally on *G. japonica* (Japanese honeylocust) and *Albizia julibrissin* (mimosa), causing wilt and death of the former species. On *Albizia* it is restricted to senescent branches or to injured trees or those affected by mimosa wilt (Plate 123). It occurs on *Acacia* in South America, where it was first described, and it has been found on *Fraxinus pennsylvanica* (green ash), presumably a rarely colonized host, in Nebraska. All known hosts except the last named are in the Fabaceae.

Symptoms and signs. *N. austroamericana* has been characterized as causing two diseases: canker of *G. triacanthos* and wilt of *G. japonica*. Symptoms on *G. triacanthos* include annual and perennial lesions on trunks and branches of all sizes. Trunk cankers are often associated with crotches, pruning wounds, and sunburned bark. Stems girdled by cankers develop yellow or wilted foliage and dieback and do not usually produce epicormic sprouts. Typical cankers are elongate and slightly depressed when young and become sunken if woundwood ridges develop at the edges. Woundwood development varies from none to prominent, according to the host's vigor. The surface of killed bark on young trunks or limbs is often orange-brown initially. It bleaches somewhat with time, often to bright yellow-orange, and the periderm cracks and begins to peel after the emergence of fruit bodies. In southern USA a gummy substance sometimes exudes from the lesions. Thick bark on old stems obscures cankers. Reddish brown discoloration develops in sapwood beneath and near the cankers and may extend to the heartwood. In Colorado, Thyronectria canker is often found on trees affected by a collar rot that occurs on poorly drained or overirrigated sites. The cause of the latter disease has not been identified.

In *G. japonica* the parasite invades the current season's sapwood systemically, causing reddish orange streaks that correspond to clusters of xylem vessels in which it moves. The disease may kill an entire tree within one season. The foliage may wilt suddenly in spring or turn yellow and drop from diseased parts later in the season. If the tree survives until the next year, excessive fruit production may presage death.

Numerous clusters of stromata bearing the pycnidial state of the pathogen, *Gyrostroma austroamericanum*, break through the bark surface, mainly through lenticels, several weeks after the bark dies. Stromata are pinkish at first and gradually darken through shades of tan and brown to almost black. Conidia are produced on long, branched conidiophores in irregular fertile chambers in stromata, and pinkish orange to cream-colored masses of conidia are released after stromata become wet. Conidial masses sometimes take the form of cirrhi (spore horns). The conidia are individually colorless, one celled, and 2.6–3.8 × 1.0–1.8 µm in size.

Perithecia develop beginning in autumn in clusters of 10–100 or more on the same stromata previously occupied by pycnidia and are mature the next spring. The perithecia, 200–450 µm in diameter, have yellow-brown bodies with black tops. The ascospores are yellowish, elliptic in outline, divided irregularly into many cells, and measure 8–16 × 4.5–9 µm.

Disease cycle. An annual cycle with variable starting time is probable. Conidia are dispersed by water. Ascospores are dispersed in air. They are shot up to 2 mm into the air when mature perithecia become wet. Both conidia and ascospores have caused infection when introduced at various times during the growing season into wounds less than about 3 weeks old. Wounds become progressively less susceptible with time. Lesions expand most rapidly at temperatures near 28°C or perhaps higher, and may elongate as much as 30 cm per year. More and larger lesions form on trees that become

infected in summer and autumn than on those inoculated in other seasons. Some cankers are perennial, but most cease expansion within a year. In *Gleditsia triacanthos* the fungus is usually restricted to killed bark and to the sapwood beneath and immediately adjacent to it. New stromata form in killed bark and produce conidia and usually ascospores within the first year after infection, completing the cycle. Wood chip mulch from diseased trees is also a potential source of inoculum because the pathogen can survive 2 years or longer in the chips.

Conidia within stromata of *N. austroamericana* remain germinable for many weeks. These spores are resistant to desiccation and heat and can germinate in humid air (above 75% relative humidity) without free water. The fungus grows most rapidly at 28–32°C, and its conidia germinate at 15–40°C. Germination is maximal at 25–35°C. In culture the fungus produces pink-orange mycelium and dark pycnidial stromata. Isolates vary in aggressiveness.

Honeylocust cultivars vary in susceptibility to *N. austroamericana*. Tests in Kentucky resulted in smaller lesions on cultivars Skyline and Trueshade than on Rubylace, Sunburst, and Shademaster, while lesion size on Imperial was variable. When tested in Colorado, Skyline and Imperial were damaged less than Sunburst, and an unnamed selection of *G. triacanthos* var. *inermis* (thornless honeylocust) was more resistant than any of the cultivars. In an Illinois test, canker incidence on inoculated stems was least on cultivars Holka, Imperial, and Shademaster; intermediate on Moraine and Skyline; and greatest on Sunburst.

References: 282, 448, 858, 1099, 1802, 1885–1887, 1890, 2132, 2813, 3129, 3289, 3290, 3342, 3525, 3526

Perennial Nectria Canker

Neonectria galligena (syn. *Nectria galligena*) causes perennial cankers and dieback, often simply called Nectria canker, on various angiosperm trees in temperate regions worldwide except Australia. In eastern North America, Nectria canker is among the most important stem diseases of *Betula* species (birches) and *Juglans nigra* (black walnut), reducing or destroying the value of the butt logs. The disease affects *Malus* (apple) and *Pyrus* (pear) trees in orchards, and on these hosts is often called European canker. *N. galligena* may

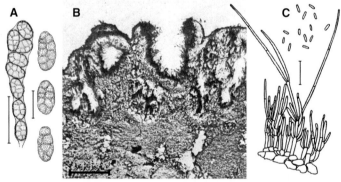

Figure 89. Nectria austroamericana. A. An ascus and ascospores. B. Section through part of a conidial stroma; conidia are produced in the cavities. C. Conidiophores and conidia. Scale bars = 200 µm for stroma, 25 µm for ascus, 10 µm for others. A adapted from reference 3525 by permission of the Arnold Arboretum of Harvard University; B and C adapted from reference 448 by permission of the American Phytopathological Society.

A–C. Dieback and canker caused by *Nectria austroamericana* on *Gleditsia triacanthos* var. *inermis* (thornless honeylocust). A. Dieback. B. A canker with prominent conidial stromata. C. Close view of conidial stromata near a canker edge (MN, Aug).

D–F. Thyronectria canker and associated reddish stain of sapwood on a *G. triacanthos* limb about 8 cm in diameter. D. A canker with typical yellowish bark roughened by conidial stromata. E, F. Successively closer views of conidial stromata (CO, Jul).

G. Magnified view of wet conidial stromata producing cream-colored masses of conidia (NY, Aug).

H. Magnified view of clusters of perithecia on dead bark (NY, Jun).

also rot apple fruit in storage. The same fungus is involved in beech bark disease (Plate 91).

Host plants of *N. galligena* represent at least 38 genera:

Acer (maple), *Aesculus* (horse-chestnut), *Alnus* (alder), *Amelanchier* (serviceberry), *Betula*, *Carpinus* (hornbeam), *Carya* (hickory, pecan), *Cercis* (redbud), *Cornus* (dogwood), *Corylus* (filbert), *Crataegus* (hawthorn), *Cydonia* (quince), *Fagus* (beech), *Fraxinus* (ash), *Ilex* (holly), *Juglans* (walnut, butternut), *Liquidambar* (sweetgum), *Magnolia* (magnolia), *Malus* (apple), *Morus* (mulberry), *Morella* (candleberry-myrtle), *Nyssa* (tupelo), *Ostrya* (hop-hornbeam), *Oxydendrum* (sourwood), *Persea* (avocado, red bay), *Populus* (aspen, poplar), *Prunus* (cherry), *Pyrus* (pear), *Quercus* (oak), *Rhus* (sumac), *Salix* (willow), *Sassafras*, *Sorbus* (mountain-ash), *Swietenia* (mahogany), *Tilia* (basswood), *Ulmus* (elm), and *Umbellularia* (California laurel).

Symptoms and signs. Lesions first appear as small, dark, depressed areas on young smooth-barked stems, and they often kill twigs by girdling. Most cankers are centered on small branch stubs or their remains, and a diseased tree usually has more than one canker. Cankers on many hosts become perennial and target shaped. They expand slowly, usually during host dormancy, often less than 1 cm tangentially per year. During the growing season callus and woundwood form around a lesion. The woundwood ridge is invaded and killed—wholly or in part—during the next dormant period. This alternation may continue for decades. If woundwood formation is slight, as in *Tilia americana* (basswood), cankers become craterlike. Target cankers seldom girdle stems larger than twigs, but compound cankers develop where lesions coalesce. Some cankers or parts of their margins become inactive and then are slowly enclosed by woundwood.

In western North America, Nectria cankers on *Acer macrophyllum* (bigleaf maple) and *Malus pumila* (apple) sometimes expand not only during host dormancy but also during the summer dry season and thus prevent formation of marginal callus and woundwoood. Elongate cankers that form in this manner on *A. macrophyllum* sometimes girdle small trunks or branches, causing conspicuous flags as foliage wilts or turns yellow prematurely and dies.

Bright red or reddish orange, lemon-shaped perithecia 250–425 µm in diameter develop in autumn–spring on killed bark. They form singly or in groups on young cankers still covered with bark or on woundwood ridges and in bark crevices at the margins of old cankers. New perithecia may form up to 30 months after death of the bark; they darken with age. The ascospores are colorless, equally two celled, oval to somewhat spindle shaped, and slightly constricted at the septum. They measure 14–22 × 6–9 µm.

N. galligena has a conidial state, *Cylindrocarpon heteronema*, that forms during moist weather before perithecia appear. Inconspicuous, creamy white, cushionlike sporodochia of this state form in cracks or protrude from lenticels. The conidia are colorless and of two types: nearly cylindric macroconidia with rounded ends, mostly 50–62 × 4–7 µm, with two to five or more cells; and ellipsoid microconidia 4–7 × 1–2 µm.

Disease cycle. N. galligena overwinters as perithecia and as mycelium in dead bark. Ascospores are expelled from perithecia and dispersed by wind or water any time of year during rain or moist periods when the temperature is above freezing, but primarily in spring and autumn. Conidia are dispersed by rain splash throughout the year in areas of warm climate and during the growing season in colder areas. Infection by either ascospores or conidia apparently can occur at any time of year when temperature and moisture permit, although most infections are thought to occur in autumn. Fresh wounds such as leaf scars, points of fruit detachment, cracks in twig axils, and sunscald lesions, as well as senescing low branches, are sites of infection. Given water, conidia and ascospores can germinate at 0–30°C and do so most rapidly at 20–25°C.

N. galligena kills bark, cambium, and the outermost sapwood in advance of its hyphae. The fungus secretes protease and pectic enzymes, which are presumed partly responsible for tissue necrosis and the soft decay of diseased bark, but it does not decay wood. It also produces indoleacetic acid, which may stimulate formation of the prominent woundwood around canker margins. Wood decay beneath Nectria cankers on most hosts is uncommon. Nectria cankers on birches, however, are sometimes infection sites for the canker-rot fungus *Inonotus obliquus* (Plate 156).

N. galligena displays limited host specificity. When a given isolate is inoculated to several known host species, the typical result is canker formation on the host of origin plus several others but not on all species tested. Two formae speciales, *N. galligena* f.sp. *mali* and f.sp. *fraxini*, were proposed on the basis of differing host specificity in Ireland, but this scheme is not supported by North American observations. Strains of *N. galligena* vary widely in aggressiveness, which is controlled by multiple genes. The fungus is homothallic (self-fertile) but displays considerable genetic variability, indicating that genetically different strains often mate.

Perennial Nectria canker is most common in cool, humid climates. Where it occurs at relatively high elevations, it is often found on exposed slopes with shallow or infertile soils. At low elevations the disease is common in cold pockets and on poorly drained soils.

Several additional fungi related to *N. galligena* and possessing *Cylindrocarpon* anamorphs are associated with cankers on angiosperm trees in North America. *Neonectria coccinea* and *N. ditissima*, proven pathogens, are widespread on various trees but are uncommonly collected. Both fungi cause significant diseases of *Fagus* in Europe. The little-known *Nectria magnoliae* (probably assignable to *Neonectria*) causes cankers on members of the Magnoliaceae, such as *Magnolia fraseri* (Fraser magnolia) and *Liriodendron tulipifera* (tuliptree), in southeastern USA.

References: 46, 129, 209, 441, 442, 461, 1004, 1099, 1147, 1298, 1357, 2213, 2358, 2369, 2370, 2452, 2896, 3096, 3097, 3345, 4002, 4285, 4458, 4498

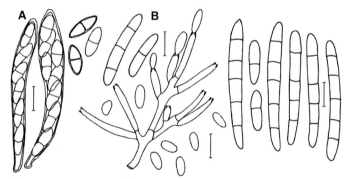

Figure 90. Neonectria galligena. A. Asci and ascospores. B. The *Cylindrocarpon* anamorph: a conidiophore and micro- and macroconidia. Scale bars = 10 µm. Adapted from reference 442 by permission of CAB International.

A. Old confluent Nectria cankers on *Betula lenta* (sweet birch). The large ridge of woundwood at left indicates an inactive portion of the canker margin (NY, May).

B. Craterlike Nectria cankers on *Tilia americana* (basswood) (NY, Apr).

C–E, H. Nectria canker of *Acer macrophyllum* (bigleaf maple). C. Dead and dying branches girdled by cankers (WA, Aug). D, E. A girdling canker on the trunk of a young tree. Cracked bark at upper arrows marks the locations of woundwood ridges formed in a previous year. Dark red perithecia of *Neonectria galligena*, visible as tiny dots several centimeters above the canker margins (lower arrows), are shown in closer view in H (OR, Sep).

F. A "target" canker caused by *N. galligena* on *Juglans nigra* (black walnut) (NY, Apr).

G. Magnified view of young perithecia of *N. galligena* on bark of *Acer platanoides* (Norway maple) (NY, Sep).

Beech Bark Disease (Plate 91)

Bark necrosis of *Fagus grandifolia* (American beech) and *F. sylvatica* (European beech) that results from sequential attack by scale insects and certain fungi in the genus *Neonectria* (Hypocreales, Nectriaceae) is called beech bark disease. It is widespread in eastern North America and Europe. The principal insect involved is *Cryptococcus fagisuga*, the beech scale. A second scale insect, *Xylococculus betulae*, contributes to the problem in North America. Bark predisposed by severe beech scale infestation is colonized and killed by *N. coccinea* in Europe and by *N. coccinea* var. *faginata* or *N. galligena* in North America. *N.c.* var. *faginata* is known only on beech. Information about *N. galligena* is presented primarily with Plate 90.

Beech bark disease was noticed in Europe beginning in the mid-1800s. It appeared in *F. grandifolia* in Nova Scotia near 1920 and then spread slowly west and south into the St. Lawrence Valley and the Appalachian region. By 2004 it occurred as far west and south as Michigan and western North Carolina. Large numbers of beech trees, some with trunks roughened or deformed by numerous small cankers, survive in areas where the killing front of the disease has passed. Thickets of beech sprouts and seedlings grow in openings where large beech were killed or salvaged. Sprouts and seedlings are not susceptible to beech bark disease until they grow large enough to provide habitat for *C. fagisuga*. Landscape specimens of *F. sylvatica* have generally escaped severe damage.

The beech scale was apparently introduced to Nova Scotia with planting stock of *F. sylvatica* from Europe. The origin of *N.c.* var. *faginata* is unknown, but genetic analyses have revealed its close relationship to European strains of *N. coccinea*. The indigenous *N. galligena* is often the first fungus to attack scale-infested beech, but *N.c.* var. *faginata* replaces it after several years.

Signs and symptoms. Colonies of the beech scale become prominent on trunks and large limbs before symptoms develop. Small colonies appear as scattered woolly white tufts in tiny bark crevices. This appearance is due to wax filaments that cover the insects. The scale population increases for several years, often appearing as vertical lines of colonies in tiny crevices and later as broad strips of white wax felt, especially on the north side of a trunk. The adult insect is soft bodied, yellow, elliptic in outline, and 0.5–1.0 mm long. It reproduces parthenogenetically, one generation per year. Males are unknown. Eggs are laid during mid to late summer, and the adults then die. First-stage nymphs (crawlers) soon emerge. They are about 0.3 mm long and are the only mobile stage. They move about briefly before settling on a permanent feeding site or falling from the tree. Airborne crawlers, lofted above the forest canopy by wind, are responsible for overland spread of the population. After settling and beginning to feed, nymphs secrete the woolly wax cover and pass the winter. In spring they molt to adulthood.

Xylococculus betulae, apparently indigenous in eastern North America, infests *Betula* (birch) species as well as *Fagus*. Adult females are soft bodied, reddish orange, and up to 4 mm long. They are found embedded, except for the posterior, in cystlike cavities in bark beneath or along the sides of fissures. A long wax tube for honeydew conduction extends from the insect's anus to the bark surface or beyond. Feeding by this insect induces bark swelling and fissures that are in turn colonized by the next generation. Young scales become embedded as the bark grows around them. Males are winged and do not feed.

Some bark necrosis is caused by the scales, but most is due to *Neonectria* species. As bark dies, beech scale colonies discolor and disappear. Slimy red-brown fluid often bleeds from tiny breaks in diseased bark. Lesions may be diffuse with imperceptible margins or may be clearly delimited patches or strips with the edges raised and cracked due to formation of cork, callus, and woundwood. Branches above large areas of dead bark either fail to produce leaves or may produce sparse chlorotic foliage. Dead bark begins to fall away from large lesions after 2–3 years. Secondary insects and many wood-decaying fungi invade dying and dead parts of trees. Beech trees with living crowns and extensive trunk decay may break and fall without warning.

N. coccinea and *N. galligena* produce pale cream to white sporodochia with conidia on newly killed bark in summer and autumn. The conidial state of *N.c.* var. *faginata* is *Cylindrocarpon faginatum*. The conidia are colorless and range from oval and single-celled microconidia, 11–14 × 3 μm, to cylindric or slightly curved, two–eight-celled macroconidia with rounded ends. Most macroconidia measure 40–110 × 5–7 μm. Red lemon-shaped perithecia of *N.c.* var. *faginata*, 200–300 μm in diameter and 250–400 μm high, develop in groups and mature in autumn. During wet weather they expel masses of colorless ascospores that, after drying, appear as white dots on the tips of the perithecia. The ascospores are two celled, 10.5–12.7 × 4.8–6.2 μm, and have broadly pointed ends. *Neonectria* species found on beech can be distinguished from one another by the sizes of their ascospores and conidia. *N. coccinea* and *N. galligena* have larger ascospores and smaller conidia than those of *N.c.* var. *faginata*.

Disease cycle. A cycle of indefinite length begins when *C. fagisuga* colonizes a tree. After the scale population weakens bark defenses, *Neonectria* conidia or ascospores infect it. Likely sites of infection are tiny wounds made by scales' stylets and cracks resulting from drying of infested bark. Latent presence of *N. coccinea* in healthy *F. sylvatica* bark has been demonstrated, which could expedite fungal invasion after scale infestation. The fungi involved in beech bark disease kill bark, cambium, and the outermost sapwood. They overwinter as perithecia and as mycelium in dead bark. Most infections and most canker expansion occur in autumn, although severely stressed bark can be colonized at any time if temperature permits. *Neonectria* conidia are dispersed mainly by water, and the ascospores by wind and water. Ascospore dispersal of *N.c.* var. *faginata* continues throughout winter and spring when air temperature is above approximately 0°C. Sporodochia and perithecia are produced sequentially on killed bark within the first year after infection, leading to intensification and spread of the disease. Susceptibility of *F. sylvatica*, and presumably also of *F. grandifolia*, to colonization by *Neonectria* is increased by water shortage and nutritional deficiency.

The physiology of beech predisposition and the mechanisms of attack by *Neonectria* species are not known in detail. The beech scale secretes pectinases in its saliva that may play a role by degrading parenchyma cell walls and middle lamellae. The saliva is also suspected to contain chemicals that induce slight swelling and cracking of bark. *N. coccinea* secretes pectic and cellulolytic enzymes capable of degrading cell walls and, theoretically, of killing cells.

Most natural controls on the beech scale and associated *Neonectria* species are ineffectual. The mycoparasite *Nematogonum ferrugineum*, for example, suppresses sporulation of *N.c. faginata*, but only on a small proportion of diseased trees. In both European and North American forests, on the other hand, scattered trees resist the beech scale and therefore do not become predisposed to infection by *Neonectria*. Resistant *F. grandifolia* have less nitrogen in their bark than do susceptible trees. In *F. sylvatica*, scale resistance is associated with a high proportion of lignified cells (stone cells) in outer bark.

References: 441, 488, 1042, 1764, 1771, 1773, 1775–1777, 1857, 1859, 1861, 2358, 2364, 2423, 3020, 3097, 3345, 4188, 4222

A. Chlorosis and dieback in the crown of a *Fagus grandifolia* (American beech) dying of beech bark disease (NY, Jun).

B–D. *Cryptococcus fagisuga* (beech scale) on *F. grandifolia*. B. Severe infestation on a mature tree. C. Close view of bark in an early stage of infestation, with wax filaments protecting scales in tiny bark crevices. D. Magnified view of yellowish mature insects exposed by removing wax filaments (NY, Sep–Oct).

E, F, I–K. *Neonectria coccinea* var. *faginata* on *F. grandifolia*. E. Coalescing cankers are marked by irregular rings of red perithecia (NY, Oct). F. Cankers on a somewhat resistant tree. Many of the lesions are confined to outer bark and isolated from healthy tissues by a cork layer (NY, Jun). I. Magnified view of sporodochia of *Cylindrocarpon faginatum*, the conidial state of *N.c.* var. *faginata* (NY, Oct). J. Young perithecia (bright red) adjacent to older, darker ones. K. Magnified view of young perithecia (NY, Oct).

G, H. *Xylococculus betulae* on *F. grandifolia*. G. Bark in a crevice similar to those in B, dissected to reveal a mature female, anterior still embedded, with wax tubes extending from the posterior (NY, Oct). H. Wax tubes with honeydew droplets, extending beyond the bark surface (VT, Aug).

L. Buff-brown mycelium and spores of a mycoparasite, *Nematogonum ferrugineum*, on bark killed by *N.c.* var. *faginata*. The former fungus parasitizes the latter but has little or no influence on the fate of a diseased tree (NY, Jun).

Photo credit: H—D. R. Houston

Fusarium Cankers (Plates 92, 93)

Many species of *Fusarium* (hyphomycetes) cause cankers and dieback of trees and shrubs. Fungi of this genus produce colorless, elongate, multicelled conidia that usually have curved and pointed (beaked) ends and thus are canoe shaped in outline. The species are distinguished on the basis of characteristics in culture: size and shape of conidiophores; size, shape, and septation of conidia; formation of chlamydospores; and pigment formation. Sporodochia, formed by some species on plants and in culture, are useful for tentative field diagnoses. All fusaria are anamorphs of ascomycetes in the Hypocreales. As examples: *F. circinatum* (Plate 93) and *F. lateritium* are the conidial states of *Gibberella circinata* and *G. baccata*, respectively, and some strains of the *F. solani* complex are anamorphic *Haematonectria haematococca*, all in the Nectriaceae. However, many species are not linked to known teleomorphs. *Fusarium* species that cause plant diseases also include strains that are secondary invaders of weakened or dying plants, so diagnosis of *Fusarium*-associated tree diseases often requires pathogenicity tests.

Taxonomy and species concepts in *Fusarium* were changing as this was written. Species have traditionally been distinguished entirely on morphological criteria. Some of the species names, notably *F. solani* and *F. lateritium*, refer to complexes of morphologically inseparable but biologically distinct species. Analyses of base sequences in conserved genes (genes in which mutations are rare) have supported this concept. Unfortunately, natural relationships among fusaria deduced from DNA sequences are not in close agreement with older classifications based on morphology. Therefore, taxonomists are revising the classification of *Fusarium* to reflect natural relationships and link DNA sequence data to appropriate morphological data. For the present, however, most records of tree diseases caused by fusaria are still based on traditional morphological diagnoses.

F. solani and *F. lateritium* are the species most often associated with cankers and dieback. *F. compactum*, *F. decemcellulare*, *F. pallidoroseum* (syn. *F. semitectum*), *F. sporotrichioides*, and *F. tricinctum* are identified less often, and other species rarely. Both pathogenic and saprobic fusaria are common inhabitants of soil and plants, including the bark of healthy and diseased trees. They are dispersed mainly as conidia by water, air, and insects. Trunk and branch cankers caused by fusaria are usually initiated at wounds or other injuries. They form primarily during host dormancy and later become enclosed beneath woundwood, leaving T-shaped defects as viewed in cross-section. The top of a T indicates the width of a lesion and the location of the vascular cambium when the lesion formed. The leg of the T, pointing toward the surface of the stem, is the mark left where woundwood ridges met and enclosed the lesion.

Cankers Caused by *Fusarium solani* or *F. lateritium*

Fusarium solani in the broad sense is believed to comprise more than 50 biological species, strains of which have been associated with cankers or twig dieback on *Acer*, *Broussonetia*, *Citrus*, *Coccoloba*, *Codiaeum*, *Cornus*, *Euphorbia*, *Fraxinus*, *Ilex*, *Juglans*, *Lagerstroemia*, *Liquidambar*, *Liriodendron*, *Nyssa*, *Populus*, *Prunus*, *Quercus*, *Rhododendron*, *Robinia*, and *Tectona*, among other woody plants. Fungi in the *F. solani* group also stain the wood of some trees and cause root rot and damping-off of hundreds of plant species. *F. solani* sometimes produces sporodochia at lenticels but often presents no macroscopic signs. It is usually detected by isolation from diseased tissues. Some canker-inducing strains are self-fertile and produce perithecia (*Haematonectria haematococca*) in culture.

F. lateritium is also a species complex. It includes several host-specialized formae speciales. *F. lateritium* f.sp. *mori* (also designated as variety *mori*), for example, is pathogenic to *Robinia pseudoacacia* (black locust), *Albizia julibrissin* (mimosa), and *Morus* (mulberry) but not to assorted other plants that have been tested. In North America, *F. lateritium* in the broad sense has been associated with twig dieback or cankers on *Ailanthus*, *Albizia*, *Araucaria*, *Carpinus*, *Citrus*, *Cotoneaster*, *Euonymus*, *Ficus*, *Fraxinus*, *Hibiscus*, *Juglans*, *Laburnum*, *Malus*, *Melia*, *Morus*, *Populus*, *Prunus*, *Pyrus*, *Robinia*, *Salix*, *Skimmia*, *Sorbus*, *Styphnolobium*, and *Thuja*, among other woody plants. Some *F. lateritium* strains produce peach-colored sporodochia at lenticels on recently colonized bark. Perithecia of

the sexual state, *Gibberella baccata*, form on some woody hosts but are not closely associated with disease symptoms.

Damage by canker-inducing fusaria is usually fostered by environmental stress. In Mississippi, for example, cankers caused by *F. solani* have developed on the trunks of *Populus deltoides* (eastern cottonwood), *Quercus* species (oaks), and *Nyssa aquatica* (water tupelo) following prolonged flooding of lowland forests. In Ohio, cankers were caused by *F. solani* in *Liriodendron tulipifera* (tuliptree) following drought. In Minnesota and Wisconsin, *F. solani* and *F. sporotrichioides* attacked freeze-damaged *Juglans nigra* (black walnut). *F. lateritium* becomes aggressive in *Populus balsamifera* subsp. *trichocarpa* (black cottonwood) cuttings, and *F. solani* in *Morus alba* (white mulberry) seedlings, when plants are stressed by water shortage.

F. solani and *F. lateritium* are associates of certain ambrosia beetles that normally bore into weakened or dying trees. The former fungus has caused girdling cankers around holes bored by *Xyleborus sayi* and *Xylosandrus germanus* in young *L. tulipifera*. Sequential attack of young *Juglans nigra* by *X. germanus* and *F. lateritium* has occasionally caused cankers and some mortality in midwestern plantations.

References: 88, 108, 296, 326, 363, 443, 469, 864, 865, 972, 974, 1099, 1129, 1567, 1587, 2050, 2051, 2413, 2491, 2701, 2819,

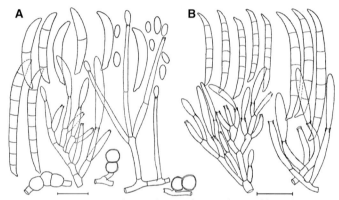

Figure 92. A. *Fusarium solani*, a fungus that produces chlamydospores as well as macro- and microconidia. B. *F. lateritium*. Scale bars = 20 μm. Adapted from reference 443 by permission of CAB International.

A–E. Cankers caused by *Fusarium solani* on *Acer saccharum* (sugar maple). A. Sunken cankers 1–2 years old. Woundwood growth at canker edges has cracked the bark; dead bark is already loose. B. Multiple lesions, probably less than a year old, their locations marked by cracks and exudation of dark liquid on a small trunk. C. Old cankers becoming enclosed by woundwood. Basal sprouts grew in response to partial girdling of the stem by cankers. D, E. T-shaped defects in the wood where lesions were enclosed by woundwood (PA, Sep–Dec).

F–H. Cankers caused by *F. solani* on *Juglans nigra* (black walnut). F. A young lesion developing at the former location of a twig (KS, Sep). G. A year-old canker with bark partially removed, revealing a woundwood ridge; recent necrosis beyond the woundwood is evident near the branch stub at lower center (KS, Sep). H. An old canker from which bark has sloughed, slowly being enclosed by woundwood (WI, Apr).

I. *Rhus glabra* (smooth sumac) with an elongate canker in which *F. solani* was found (KS, Sep).

J, K. *F. lateritium* sporulating on a branch canker on *Laburnum anagyroides* (golden-chain tree). J. The mature lesion is tan with a purple border; sporodochia have erupted from the bark at lenticels. K. Magnified view of sporodochia (NY, Jul).

L. The trunk of a young *Populus deltoides* (eastern cottonwood), about 15 cm in diameter, with bark removed to reveal multiple lesions caused by *F. solani*, all apparently formed in the same year (MS, Oct).

Photo credits: A–E—J. M. Skelly; F–I—N. A. Tisserat; L—USDA Forest Service Southern Hardwoods Laboratory, Stoneville, MS.

187

2890, 3039, 3345, 3682, 3899, 4039, 4060, 4062, 4173, 4273, 4427

Pitch Canker of Pines

Pitch canker, named for the copious exudation of resin from lesions, is caused by *Gibberella circinata* (Hypocreales, Nectriaceae) in its conidial state *Fusarium circinatum* (syn. *F. subglutinans* f.sp. *pini*). The disease was first described in the Appalachian region in the mid-1940s, but no great attention was paid to it until a widespread epidemic began in the mid-1970s. Pitch canker occurs in southeastern USA from Virginia to Florida and Texas, in California, and also in Haiti, Mexico, Chile, Spain, South Africa, Japan, and South Korea. It kills branches and thus deforms trees, suppresses height and volume growth, reduces seed crops, and kills highly susceptible trees. Seedlings in nurseries and young plantations are also at risk. *Pinus elliottii* (slash pine) and *P. echinata* (shortleaf pine) are severely affected in southeastern USA, and *P. radiata* (Monterey pine) in California, where the damage is aggravated by insects, mainly bark beetles, that make wounds and act as vectors. Several quarantines and import regulations intended to prevent interstate and international movement of *F. circinatum* are in force.

Hosts in addition those named above include *P. arizonica, P. attenuata, P. canariensis, P. clausa, P. douglasiana, P. estevezii, P. glabra, P. halepensis, P. leiophylla, P. luchuensis, P. muricata, P. palustris, P. patula, P. pinea, P. pungens, P. rigida, P. strobus, P. taeda,* and *P. virginiana.* Additional pine species not found affected in nature have proven susceptible when tested. *Pseudotsuga menziesii* (Douglas-fir) was found diseased where growing among diseased pines in California.

Symptoms and signs. The primary symptoms are resin-soaked lesions that encircle twigs and small branches or develop as perennial cankers on limbs, trunks, and exposed roots. Diseased bark turns dark brown, and the underlying sapwood becomes yellowish brown where infiltrated with resin. Resin-soaked wood may extend to the pith. A perennial canker may eventually girdle the affected part. Bark remains on the canker. Resin usually exudes and may run down the bark or drip on foliage. Female flowers, growing cones, and seeds may be infected or killed. Dieback is prominent on *P. attenuata, P. clausa, P. echinata, P. elliottii, P. muricata, P. radiata, P. taeda,* and *P. virginiana.* Trunk cankers are common on *P. palustris, P. radiata, P. strobus,* and *P. virginiana.*

Most shoot and branch dieback develops during autumn through spring. Girdling lesions or resin-soaked wood in branches cause distal shoots to droop and wither, sometimes taking the form of shepherd's crooks. Needles on girdled twigs and branches turn yellow, then brown. Dead needles, glued in place by crystallized resin on or below lesions, often persist for more than a year, slowly weathering to dull grayish brown. Trees usually recover from outbreaks of shoot dieback. Remission in *Pinus radiata* has been attributed to systemic induced resistance.

F. circinatum is transmitted into nurseries with seed. Diseased seedlings often occur in clusters where the pathogen has spread from an initially diseased individual. Seedlings may wilt, or their needles may turn yellow and then brown due to resinous lesions on the lower stem, root collar, or taproot.

F. circinatum produces multicelled macroconidia on cushion-like, salmon-orange sporodochia on killed shoots and twigs. Sporodochia are 0.2–2 mm broad and up to 1 mm high. They are most common at scars where needle fascicles have fallen. In warm areas sporodochia may be visible at any time of year. They are most abundant in autumn and winter on branches that still have reddish brown needles. Apparently they are rare, or at least are rarely noticed, on trunk and limb cankers.

The fungus grows readily in pure cultures, most rapidly near 24°C, producing sporodochia after 2–3 weeks. Its conidia are colorless and of two types. Single-celled, ovoid microconidia, mostly 8.5–11 × 2.8–3.6 μm, are borne in droplets (false heads) atop branched conidiophores. Macroconidia have one–four cells, are usually 33–42 × 3.4–3.7 μm in size, and are curved with pointed ends. Coiled sterile hyphae form in certain media. Isolates vary in virulence, but isolates from one pine species can readily infect others. Perithecia of *G. circinata* have not been found associated with pitch canker. The association between conidial and perithecial states is known from cultural studies in which fertile perithecia formed after compatible isolates were paired.

Disease cycle. Pitch canker seems to be monocyclic (one cycle or less per year) in most circumstances, but with variable timing and duration of elements in the cycle. This variation is associated with weather, host growth stage, and seasonal activity of insects that make wounds suitable for infection. Conidia from sporodochia are dispersed by air, water, and insects. Dispersal occurs throughout the year in areas where the climate is mild and moist. *F. circinatum* is also transported with pine seeds. Infection occurs in wounds caused by insects and other agents at any time of year. Injuries remain susceptible up to several days, and lesion formation begins soon after inoculation. Most infections probably occur in late summer and autumn in southeastern USA.

In eastern USA, wounds suitable for infection include those made on current-year shoots by *Monochamus* (pine sawyers) and *Pissodes nemorensis* (deodar weevil). Injuries from hail, wind, and cone harvesting also become infected. Limbs and trunks of trees in seed orchards sustain infections at injuries made by cone-harvesting equipment. In California, insects involved either as vectors or as wounding agents include the cone and twig feeders *Conophthorus radiatae, Ernobius punctulatus,* and *Pityophthorus* species; also three species of *Ips* (bark beetles). Wounds made on shoots by the spittlebug *Aphrophora canadensis* and by ice or wind damage also become infected. Fertilization at high rates, especially with nitrogen, enhances pines' susceptibility to pitch canker.

Pine species vary in susceptibility to pitch canker. *P. echinata, P. radiata,* and *P. virginiana* are highly susceptible, and *P. elliottii* is somewhat less so. *P. strobus* and *P. taeda* are somewhat resistant, as are *P. canariensis, P. halepensis,* and *P. pinea. P. thunbergii* is resistant. Heritable variation in resistance occurs within susceptible species, but resistant seed sources are likely to be developed only for *P. radiata* in the near future.

References: 243, 244, 390, 436, 517, 518, 810, 811, 942, 1017, 1020, 1165, 1321–1323, 1617, 1746, 2168, 2213, 2819, 2866, 2891, 3756, 3883, 3899, 4160

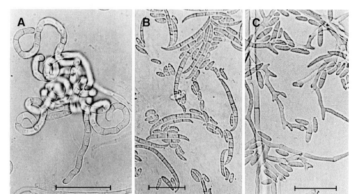

Figure 93. Fusarium circinatum. A. Coiled hyphae typical of the fungus in culture. B. Macro- and microconidia from a sporodochium. C. Conidiophores and conidia in culture. Scale bars = 25 μm. Adapted from reference 2866 by permission of the Mycological Society of America.

Pitch canker of pines, caused by *Fusarium circinatum.*

A, B. Top killing and deformity of young *Pinus elliottii* var. *elliottii* (slash pine) caused by pitch canker. In B, the top of a diseased tree shows evidence of repeated leader death. The original leader at center died the previous year and was replaced as leader by the lateral branch at left, which in turn was killed and is being replaced by the branch at right (FL, Apr).

C. Severe damage on *P. radiata* (Monterey pine) (CA, Jul).

D. Resin dripping from a trunk canker on *P. strobus* (eastern white pine) (TN, Apr).

E. Close view of pitch canker on a *P. elliottii* twig, with dead needles glued in place by crystallized resin (FL, Apr).

F. Magnified view of part of a killed twig of *Pinus elliottii* with pink sporodochia of *Fusarium circinatum* on needle fascicle scars (FL, Apr).

Photo credit: C—T. R. Gordon

Pestalotiopsis Spots, Blights, and Diebacks (Plate 94)

Pestalotiopsis species and similar fungi (coelomycetes) are often associated with dead or dying spots, blotches, tips, or margins of leaves; twig dieback; cankers; and fruit rot. These fungi have many hosts and may be found fruiting on dead tissues, or they may grow in cultures from diseased or apparently healthy plant parts. They often colonize tissues made susceptible by senescence, damage by other pathogens or insects, freezing, sunscald, or other injuries. These fungi have often been dismissed as saprobes or inconsequential pathogens, but this judgment is inappropriate until tested for a given host-pathogen combination.

Most species of *Pestalotiopsis* were formerly classified in *Pestalotia*. The reclassification was incomplete, however, and many additional fungi (e.g., *Pestalotia macrochaeta*, listed below) ought to reclassified. Most of these will be placed in *Pestalotiopsis*. More than 50 species of *Pestalotia* and *Pestalotiopsis* occur in the USA and Canada. Ascigerous states of a few of these fungi are known to be in the genera *Broomella* and *Pestalosphaeria* (Xylariales, Amphisphaeriaceae).

Pestalotiopsis species produce conidia in fruit bodies that develop beneath the epidermis of leaves, stems, fruit, and some flower parts. The fruit bodies vary with species from acervular to pycnidial, and most are 75–300 μm in diameter. When plant tissue containing fruit bodies is moistened, massed conidia in a sticky matrix rupture the epidermis and push upward as glistening black droplets or, under drier conditions, as tiny black tongue- or hornlike projections (cirrhi) that may attain lengths of 2 mm (Plate 94F). When wet by dew or rain, the matrix dissolves and a dark film of spores spreads out around the fruit body. The spores are dispersed by splashing or running water and presumably also by insects.

Conidia of most *Pestalotiopsis* species are five celled (some are four celled) and pointed at the ends (Fig. 94). They appear black in mass because the two or three central cells of each spore have dark brown walls. The spores have several colorless appendages (setulae) at the apex and a basal pedicel that provides attachment during spore formation.

Pestalotiopsis species associate with plants at three levels of interaction: some are endophytes, residing in apparently healthy leaves and bark; some are pathogens; and many are colonists of dying of recently dead plant parts. Some species have all three roles in succession. *P. funerea*, an endophyte and pathogen of many gymnosperms, and *P. neglecta* (syn. *P. microspora*), an endophyte of *Taxus* (yew) and *Taxodium* (bald cypress) among other plants, are examples. *P. funerea*, a proven pathogen, occurs worldwide, inhabiting gymnosperms primarily. It has been associated with damping-off, root and collar rot of seedlings, needle blight, shoot or tip blight, twig dieback, and stem cankers. Host genera include *Araucaria, Calocedrus, Cedrus, Chamaecyparis, Cryptomeria,* ×*Cupressocyparis, Cupressus, Gingko, Juniperus, Picea, Pinus, Pseudotsuga, Sequoia, Taxus, Thuja,* and *Tsuga*. *P. funerea* is usually found on plants stressed or damaged by other agents or environmental conditions. *P. neglecta* attracted attention in the 1990s because it produces taxol, a substance utilized for chemotherapy of some human cancers. This fungus also inhabits various plants, including healthy and declining *Torreya taxifolia* in Florida. It has caused leaf spots, chlorosis, and cankers when inoculated to *Torreya* seedlings. The phytotoxins pestalpyrone and hydroxypestalpyrone, known to be produced by the fungus, were detected in symptomatic seedlings.

Opportunistic behavior notwithstanding, *Pestalotia* and *Pestalotiopsis* apparently cause hundreds of plant diseases. The following worldwide sample of host associations of pathogenic species shows the diversity of symptoms and substrates.

Pestalotia macrochaeta: flower blight of *Camellia japonica* (common camellia)

Pestalotiopsis apiculata: shoot dieback of *Cunninghamia lanceolata* (China-fir)

P. disseminata: fruit rot of *Psidium guajava* (guava), seedling blight of *Eucalyptus pellita*, leaf spot of *Rhizophora mangle* (red mangrove) and *Terminalia arjuna*

P. foedans: leaf spots and shoot blight of *Metasequoia glyptostroboides* (dawn-redwood); also associated with needle blight of various pines, seedling blight of *Chamaecyparis obtusa* (Hinoki-cypress), and dieback of *Thuja plicata* (western red cedar)

P. longiseta: gray blight of *Camellia sinensis* (tea)

P. maculans (syn. *P. guepinii*, *P. macrotricha*): leaf spot of *Caesalpinia echinata* (Brazilwood), gray leaf spot and twig dieback of *Camellia japonica* and *C. sinensis*, leaf spot and twig canker of *Rhododendron* species; twig blight of *Corylus* (hazelnut) and *Juglans* (walnut); also occurs on *Cercis* (redbud), *Citrus, Diospyros* (persimmon), *Magnolia, Pyrus* (pear), and *Vaccinium* (blueberry)

P. malicola: fruit rot of *Chaenomeles* (quince), *Malus* (apple), and *Pyrus* (pear)

P. mangiferae: gray leaf spot of *Mangifera indica* (mango)

P. menezesiana: gray leaf blight of *Actinidia deliciosa* (kiwi-fruit)

P. palmarum: leaf spots on many palms, including species of *Bismarckia Borassus, Butia, Caryota, Chamaedorea, Chamaerops, Cocos, Dypsis, Pandanus, Phoenix, Rhapis, Roystonea, Sabal, Syagrus, Veitchia,* and *Washingtonia*; also enlarges lesions initiated by other agents

P. populi-nigrae: cane blight of hybrid *Rosa*, shoot blight and leaf spot of *Populus nigra* (black poplar)

P. sydowiana: leaf spots and twig dieback of ericaceous plants such as *Gaultheria shallon* (salal), *Kalmia latifolia* (mountain-laurel), *Pieris japonica* (Japanese-andromeda), and *Rhododendron*

P. versicolor: cankers on *Eucalyptus* twigs; leaf spots of *Diospyros melanoxylon* (tendu), *Cliftonia* (buckwheat-tree), *Coccoloba uvifera* (sea-grape), and *Nerium oleander* (oleander); gray leaf blight of *Syzygium aromaticum* (clove)

The biology and pathogenicity of *Pestalotiopsis* species on *Ligustrum* (privet) and *Ixora* (symptoms shown in Plate 94) have not been studied. *P. ixorae* was described from lesions in living leaves of *Ixora* sp. in Brazil.

References: 68, 800, 1402, 1871, 2067, 2109, 2271, 2308, 2415, 2779, 3895, 3897, 3910, 3911, 3913, 4290, 4327, 4328

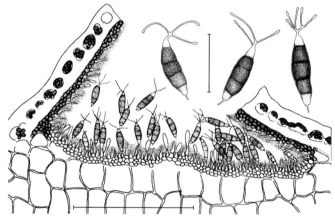

Figure 94. A subepidermal acervulus and conidia of *Pestalotiopsis maculans*. Scale bars = 100 μm for acervulus, 20 μm for conidia. Adapted from reference 2779 by permission of T. R. Nag Raj.

A, B. Tip blight of *Thuja occidentalis* (eastern arborvitae). A. Tan to brown twig tips colonized by *Pestalotiopsis funerea*. B. Black acervuli of *P. funerea* breaking through the epidermis of dead scale-leaves (NY, Aug).

C, D. *Pestalotiopsis* sp. fruiting on a lesion caused by *Phytophthora* sp. on a *Rhododendron* branch. C. Overview of lesion. D. Close view of young acervuli breaking through the epidermis (NY, Oct).

E, F. *Pestalotiopsis* sp. in lesions in year-old leaves of *Ixora coccinea*. Magnified view (F) shows how acervuli break the epidermis and produce black spore horns (arrows) (FL, Mar).

G, H. *Pestalotiopsis* sp. in lesions on leaves of *Ligustrum japonicum* (wax-leaf privet). Viewed under magnification (H), acervuli of this fungus are minute black dots quite inconspicuous in comparison with the fungus on *Ixora* (FL, Mar).

191

Seiridium Cankers of Cypress (Plate 95)

Seiridium cankers, also called cypress cankers, on species of Cupressaceae and Taxodiaceae are caused by at least three *Seiridium* species (coelomycetes) that occur around the world: *S. cardinale*, *S. cupressi*, and *S. unicorne*. The sexual state of *S. cupressi* is *Lepteutypa cupressi* (Xylariales, Amphisphaeriaceae). *S. cardinale* has an undescribed ascigerous state, observed only in California. No sexual state of *S. unicorne* is known. These fungi have caused severe damage in Africa, Japan, the Mediterranean region, New Zealand, and western USA. *S. cardinale*, the most aggressive of the three, began destroying populations of *Cupressus macrocarpa* (Monterey cypress) and *C. sempervirens* (Italian cypress) in California in the 1920s and was transported globally from there. It later began devastating *C. sempervirens* plantings in the Mediterranean region. *S. unicorne*, also globally distributed in temperate regions, causes noteworthy damage to various species of Cupressaceae in Japan and New Zealand and sporadic damage in south-central and southeastern USA. *S. cupressi* occurs in Florida on ×*Cupressocyparis leylandii* (Leyland cypress) but has caused damage mainly in East Africa and New Zealand on *Cupressus*. The three pathogens are distinguishable by analysis of a β-tubulin gene sequence.

Hosts of these fungi, grouped by susceptibility class, are as follows. *Cupressus macrocarpa*, *C. sempervirens*, and ×*Cupressocyparis leylandii* are most susceptible. *Chamaecyparis lawsoniana* (Lawson-cypress), *C. obtusa* (Hinoki-cypress), *C. pisifera* (Sawara-cypress), *Cupressus arizonica* (Arizona cypress), *C. goveniana* (Gowen cypress), *C. lusitanica* (Portuguese cypress), *Juniperus virginiana* (eastern red-cedar), and *Platycladus orientalis* (Oriental arborvitae) are damaged by pathogen strains that prevail in some regions. Plants occasionally infected in nature or relatively resistant in tests include *Calocedrus decurrens* (incense cedar), *Chamaecyparis formosensis*, *C. funebris* (mourning-cypress), *C. nootkatensis* (Alaska cedar), *Cupressus bakeri* (Modoc cypress), *C. forbesii* (tecate cypress), *C. macnabiana* (MacNab cypress), *C. goveniana* subsp. *pygmaea* (Mendocino cypress), *C. torulosa* (Bhutan cypress), *Juniperus chinensis* (Chinese juniper), *J. rigida* (needle juniper), *J. sabiniana* (savin juniper), *J. scopulorum* (Rocky Mountain juniper), *Taxodium distichum* (bald cypress), *Thuja plicata* (western red cedar), and *Thujopsis dolabrata* (hiba-arborvitae). The following are resistant: *Cryptomeria japonica* (Japanese cedar), *Juniperus occidentalis* (Sierra juniper), and *Thuja occidentalis* (eastern arborvitae). Resistant trees halt infection rapidly and then form a cork barrier that isolates a lesion.

Symptoms and signs. Symptoms caused by *S. cardinale* are emphasized here. Girdling cankers on twigs, branches, or mainstems cause fading and death of foliage. Typical cankers are lens shaped and become sunken as diseased bark dries and surrounding healthy tissues grow. Resin usually exudes from active lesions, although old or slowly growing trees tend to exude little resin. Cankers on trunks of young, rapidly growing trees of highly susceptible species may elongate 10–20 cm per year and attain lengths of 60–90 cm before the stems are girdled. Small trees affected by aggressive strains may be killed within a year. Large trees may decline over many years from the cumulative effects of multiple cankers. Death is hastened by water shortage and sometimes by insects, notably *Phloeosinus* species (bark beetles), that attack weakened trees.

Signs of the three pathogens are similar. Fruit bodies (here called pycnidia although they are intermediate in form between pycnidia and acervuli) develop in cankers, and those of *S. cardinale* also form on cone scales. They appear as scattered black pustules 0.3–1.5 mm in diameter that break through the bark surface and open widely during wet weather. Conidia exude while bark is moist. Conidia are six celled, and most measure 17–37 × 6–10 μm; the length/width ratios are 2.5–3 in *S. cardinale*, 4–8 in *S. cupressi*, and 3.2–3.9 in *S. unicorne*. The end cells are pointed and colorless; the four central cells are olive brown to dark brown. Conidia of *S. cupressi* and *S. unicorne* have a colorless appendage at each end. In California the undescribed ascigerous state of *S. cardinale* is common in cypress bark dead for a year or longer.

Disease cycle. The pathogens survive winter and dry summer in colonized bark, where pycnidia with spores are present throughout the year. Most infections are presumed to be started during wet weather by conidia. Wounds on twig or branch bases are common sites of infection, but *S. cardinale* can also infect intact young green leaves and shoots. These fungi colonize bark and outer sapwood

intracellularly, killing parenchyma cells as much as 1 mm in advance of hyphal extension. They produce pycnidia and conidia within several weeks after death of the bark if mild, moist weather prevails. Pycnidial development is interrupted during hot, dry weather. Conidia are dispersed locally in splashing or running water. Those of *S. cardinale* are capable of surviving several weeks on exposed cypress foliage and for more than a year in dry, protected situations. Where perithecia occur, they develop during the winter after the substrate dies. Ascospores, when produced, are presumed to be dispersed by air and to account for some long-distance spread. Transmission of *S. cardinale* on pruning tools and transportation in infested seed and on planting stock have been recorded. Insect transmission is probable also, inasmuch as *Phloeosinus* species visit stressed cypress and related trees and have habits consistent with vector roles. In California the cypress bark moth, *Laspeyresia cupressana*, visits diseased and wounded bark, becomes contaminated with *Seiridium* conidia, and is therefore also a possible vector. A seed bug, *Orsillus maculatus*, carries spores of *S. cardinale* in Italy.

Although wet weather is required for spore dispersal and infection, severe damage is most likely on sites with dry weather and high daytime temperature in summer. *C. macrocarpa* planted on such sites in California was decimated by *S. cardinale* while natural groves of this species along the Pacific Coast in areas of moderate temperature and frequent fog remained unaffected.

Several toxins produced by *Seiridium* species in culture have been characterized, and some caused bark lesions and killed cypress stems into which they were injected. Seiridin and iso-seiridin are the major toxins from *S. cardinale*. Cyclopaldic acid is the major toxin from *S. cupressi*. Variation in aggressiveness of *S. cardinale* isolates has been correlated with their production of seiridins, and host resistance to the fungus has been correlated with ability to tolerate the toxins.

Several clones of *Cupressus arizonica* and *C. sempervirens* with resistance to *S. cardinale* have been deployed in southern Europe. Resistance in *C. sempervirens* is controlled by several genes with additive effects.

References: 216, 643, 729, 1175, 1352, 1353, 1356, 1402, 1859, 2151a, 2771, 2779, 3407, 3409, 3413, 3768, 3928, 3949, 4044, 4161, 4177, 4456

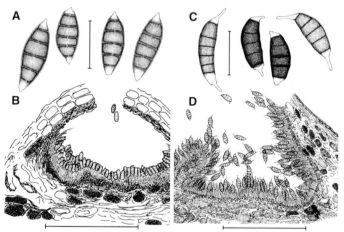

Figure 95. A, B. Conidia and a section of a pycnidium of *Seiridium cardinale*. C, D. Conidia and a partial section of a pycnidium of *S. unicorne*. Scale bars = 20 μm for conidia, 200 μm for pycnidia. Adapted from reference 2779 by permission of T. R. Nag Raj.

A, C, D, F, G. Seiridium canker of *Cupressus macrocarpa* (Monterey cypress), caused by *Seiridium cardinale*. A. Fading foliage in the upper three-fourths of a young tree due to a girdling canker on the trunk (CA, Sep). C. Dripping resin marks the location of a canker on a young trunk (CA, Sep). D, F, G. Successively closer views of bark of a diseased branch with exuded resin and black pycnidia of *S. cardinale* (CA, Jul).

B, E. Seiridium canker of *Platycladus orientalis* (Oriental arborvitae) caused by *S. unicorne*. B. Brown foliage on a branch girdled by a lesion (KS, May). E. A branch canker that originated in a twig crotch (KA, Aug).

Photo credits: B, E—N. A. Tisserat

Cryptosphaeria Canker of Aspen (Plate 96)

Cryptosphaeria canker is caused by *Cryptosphaeria lignyota* (syn. *C. populina*; Diatrypales, Diatrypaceae). The disease is destructive to *Populus tremuloides* (trembling aspen) in parts of western North America, notably Colorado. It occurs from Alaska to Arizona and New Mexico, also in Minnesota and New York. Elsewhere the causal fungus has usually been found growing as a saprobe in branches and trunks of *Populus* species. In this role it occurs in central and eastern North America from the Dakotas and Iowa to Quebec and New Jersey. It also occurs in Europe, where it has been known since early in the 19th century, and China.

Hosts of *C. lignyota* include *P. alba* (white poplar), *P. angustifolia* (narrowleaf cottonwood), *P. balsamifera* (balsam poplar) and its subspecies *trichocarpa* (black cottonwood), *P. ×berolinensis* (Berlin poplar), *P. deltoides* (eastern cottonwood), *P. grandidentata* (bigtooth aspen), *P. grandidentata × tremuloides* hybrids, *P. nigra* (black poplar, including cv. Italica (Lombardy poplar), *P. tremula* (European aspen), *Salix purpurea* (purple willow), and *S. viminalis* (basket willow). Pathogenicity of *C. lignyota* has been demonstrated only in *P. tremuloides*.

In Colorado Cryptosphaeria canker is one of the most common canker diseases of *P. tremuloides*. It was found at more than 80% of the survey locations examined, affecting only 1% of the trees but accounting for 26% of aspen mortality. Pole-size stems were most frequently affected. In New York, variation in canker incidence among aspen families in an experimental plantation indicated genetic variation in susceptibility. Experiments have confirmed that *C. lignyota* can kill *P. tremuloides* and that the principal site of pathogenesis is sapwood. Bark overlying colonized sapwood is invaded secondarily. Thus, Cryptosphaeria canker is similar to the canker-rots caused by basidiomycetes (Plates 155–158).

Symptoms and signs. The cankers usually originate at trunk wounds or branch stubs. They are elongate, often only 5–10 cm wide but up to 3 m long (usually less than 1 m in the East), and may tend to spiral around the stem. Some cankers girdle branches and spread into the mainstem. Cankers expand several millimeters tangentially per year and usually do not girdle trunks. Trees up to 15 cm in diameter may be killed within 3 years after the onset of infection, however, because the pathogen kills a large volume of sapwood.

Young cankers and the margins of older ones are orange-brown on the surface; they become light gray-brown with time. Dead bark adheres tightly to the wood. Internal bark turns black and sooty except for prominent fibers and small light marks 0.5–2 mm in size that give the exposed tissue a mottled appearance (Plate 96D). Woundwood ridges sometimes become noticeable at the edges of cankers after 2 or more years, but these do not halt canker expansion. Reddish brown fluid sometimes bleeds from canker margins.

Dead sapwood, discolored shades of gray, brown, yellow, orange, or pink, extends a meter or more above and below the limits of a canker. This sapwood, especially near the interface with normal wood, fluoresces yellow when viewed under ultraviolet radiation. The fluorescence is not a specific diagnostic character, however, since several other fungi that cause trunk decay in aspen also induce formation of fluorescent substances. Brownish, mottled decay develops in the central part of the column of discolored wood. Decay columns apparently caused by *C. lignyota* averaged 4.6 m in length in one study in Colorado. Most such columns were found in the middle and upper parts of trunks.

Light orange acervuli of the *Libertella* anamorph of *C. lignyota* (*Cytosporina* according to some authors) develop near the edge of a canker and produce colorless, elongate conidia that are single celled, slightly curved to crescent shaped, have a flattened base, and measure 18–32 × 1.2–1.6 μm. The conidia do not germinate on culture media, and their function in nature is unknown. Perhaps they function as spermatia. Perithecia develop singly in a grayish pseudostroma (a tissue composed of compacted hyphae and the remains of dead bark cells) 1–2 cm wide and up to 30 cm long beneath the surface of bark that has been dead at least a year. The pseudostroma and black perithecia impart a gray color to bark in the center of a canker. Pseudostromata vary in size in relation to the size of the canker zone that died in a given year. Perithecia are nearly spherical, 0.4–0.6 mm in diameter, and their ostioles open above the bark surface. The ascospores are yellowish, curved with rounded ends, single celled, and measure 7–12 × 2–2.5 μm.

Disease cycle. The cycle has not been described; its minimum duration is likely to be at least 2 years. Ascospores are presumed to be dispersed by wind, and conidia (spermatia) mainly by water. Ascospores are presumed to cause most infections. The fungus enters trunks via wounds or colonized branches, and it invades sapwood rapidly. In Colorado, columns of discolored wood up to 4 m long developed within 24 months after inoculation, while associated cankers extended 33 cm or less. *C. lignyota* is also able to establish quiescent colonies in healthy sapwood and persist there as an endophyte; it was among the fungi detected in apparently healthy sapwood of *P. tremuloides* in western Canada and New York State. Endophytic fungi can presumably enter sapwood of healthy trees by various routes, including wounds and senescent or dead twigs.

The fungus can readily be isolated from cankers and from the associated discolored and decaying wood, and the *Libertella* state forms in culture. The ability of *C. lignyota* to decay both wood and bark has been confirmed experimentally.

References: 218, 609, 690, 1056, 1213, 1303, 1671, 1672, 1948, 1989, 1994, 3221

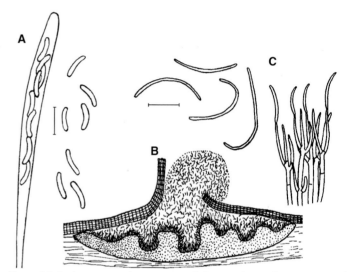

Figure 96. A. An ascus and ascospores of *Cryptosphaeria lignyota*. Asci of this species have long stems; part of the stem of this ascus was omitted from the drawing. B, C. *Libertella* sp., similar to the anamorph of *C. lignyota*. B. An acervulus in bark. C. Conidiophores and conidia, illustrating the typical shape of conidia in this genus. Scale bars = 10 μm. A adapted from reference 1056; B and C adapted from reference 218 by permission of the American Phytopathological Society.

Crytosphaeria canker and trunk rot of *Populus tremuloides* (trembling aspen) (CO, Jun).

A. Coalescing, elongate cankers on a trunk. Note bleeding at canker margins.

B. Close view of coalescing cankers on a pole-size tree. A large canker on the right side of the trunk has joined with a smaller spindle-shaped canker centered on the branch stub.

C, D. A portion of a canker with periderm removed to show blackened cortex, tiny white spots (enlarged in D), and the gray surface of dead sapwood. Gray color of the bark surface at bottom center in C indicates the presence of a pseudostroma where perithecia are developing.

E, F. Close view and magnified view, respectively, of bark with perithecia of *Cryptosphaeria lignyota*. Black dots in E, magnified in F, are the protruding tips of perithecia. Round, black embedded perithecial bodies are revealed where the bark surface was removed in F.

G. Cross section of a diseased trunk showing the relationship between cankers and internal stain and decay caused by *C. populina*.

Eutypella Canker of Maples (Plate 97)

Eutypella parasitica (Diatrypales, Diatrypaceae), the cause of Eutypella canker, attacks *Acer* species (maples) in forests and landscapes. The disease, recognized since the mid-1930s, occurs from Minnesota eastward to Quebec, Maine, and Pennsylvania. It is unknown in Eurasia. Affected species include *Acer campestre*, *A. nigrum*, *A. platanoides*, *A. pseudoplatanus*, *A. rubrum*, *A. saccharinum*, and *A. saccharum* (hedge, black, Norway, sycamore, red, silver, and sugar maples, respectively), and *A. negundo* (box-elder). Eutypella canker is most common on *A. saccharum* in forests and on *A. platanoides* in developed landscapes. *A. pensylvanicum* (striped maple) is susceptible when inoculated. Two reports exist of *E. parasitica* on nonmaple hosts: *Betula alleghaniensis* (yellow birch) and *Prunus pensylvanica* (pin cherry), both in Quebec.

The incidence of Eutypella canker in *A. saccharum* stands is typically 0–4%, but much higher levels are common. Many small trees less than about 10 cm in diameter are girdled and killed, and wood decay associated with the cankers leads to breakage of many trunks. Because it affects the valuable butt log, the canker causes economic loss. By one assessment in Ontario, defective wood in the average tree with Eutypella canker amounted to 12% of total wood volume but 49% of potentially salable wood volume. On a study site in Vermont the loss of salable *A. saccharum* sawtimber volume due to this disease was 3.4%.

Symptoms and signs. Cankers are initiated primarily on young trees, mostly within 3 m of the ground, and are perennial, often remaining active for decades. Their appearance and shape vary with age and host species. During the first several years of development cankers appear as depressed or flattened areas tightly covered with bark, usually with the remains of a branch stub in the center. The margin of an expanding canker has white to buff-colored mycelial fans in recently killed inner bark closely adjacent to healthy tissue. These fans are diagnostic. Old cankers may have a rough exposed face of decaying wood and are often outlined partially or completely by a large ridge of woundwood. The margins of large old cankers that are still active may be outlined by cracks in thick outer bark, as in Plate 97D. Cankers on *A. negundo*, *A. platanoides*, and *A. pseudoplatanus* tend to be oval to nearly circular, while those on *A. rubrum* and *A. saccharum* are elliptic to elongate, sometimes as long as 1.5 m.

Alternation of tree growth and lesion expansion leads to concentric zones of dead tissue in perennial cankers induced by *E. parasitica* and by many other fungal pathogens. A cork barrier and woundwood form around a lesion during each growing season, and the parasite breaches or evades the barrier and expands the lesion during host dormancy. How *E. parasitica* passes the cork barrier is not known. Woundwood development at active margins of Eutypella cankers in any single year is relatively slight, but many affected trunks become swollen around cankers because of the aggregate effect of woundwood production at lesion edges for many years. A prominent woundwood ridge develops only where the parasite has died. Elongate cankers lengthen 1–2 cm annually and widen at a slower rate. *E. parasitica* colonizes sapwood beneath cankers and causes a slow brownish decay of the white-rot type (decay types, page 306). The decay commonly leads to stem breakage, although the internal defect seldom extends more than 30 cm beyond a canker.

Bark at the center of Eutypella cankers at least 6–8 years old is darkened by the black necks of numerous perithecia just beneath the surface, initially in groups a few millimeters in diameter as in Plate 97G, but later almost continuous over the center of a canker face. They seldom form on woundwood ridges dead for less than 4–5 years. Sometimes perithecia form in the wood of the central branch stub. Young jet black perithecia are interspersed with old dull black ones. Perithecia with viable ascospores can be found throughout the year. Their round bodies are 0.6–1.0 mm in diameter, and their necks vary in length from less than 1 mm to as long as 5 mm, depending on depth of the body below the bark surface. They eject ascospores in groups of eight. The ascospores are single celled, 7–11 × 2–2.5 µm, slightly curved, and brown.

E. parasitica also produces an inconspicuous conidial state referable to *Libertella*, a coelomycete, on stromata that break the bark surface in cankers (not shown here). Conidia, produced in irregular fruit bodies intermediate in form between acervuli and pycnidia, are colorless, crescent shaped, and measure 17–32 × 1.2–1.8 µm. These did not germinate when tested on a variety of media, including host tissues. Perhaps they function as spermatia.

Disease cycle. The pathogen perennates in bark and wood of cankers. Its ascospores are released annually and initiate new infections. These spores are ejected from perithecia several millimeters into the air soon after the onset of rain at moderate temperatures (e.g., 12–25°C) and are dispersed by wind. Ascospore octads have been trapped 25 m downwind from cankers. Given adequate moisture ascospores are released from a given portion of a canker for several years, regardless of whether the tree remains alive. They germinate within as few as 10 hours on a moist substrate if the temperature is near the optimum for mycelial growth (28°C).

Small branches, if bruised, broken, or wounded, are susceptible to infection by germinating ascospores. Many infections centered at branch stubs are thought to begin on small lateral branches. Cankers on trunks are initiated at either branch stubs or wounds that penetrate to xylem. *E. parasitica* grows intracellularly in both wood and bark and can apparently grow from wood into bark at the edges of wounds. It advances in inner phloem, where host cells are killed slightly ahead of the mycelial fans. A second layer of mycelial fans sometimes forms in old phloem closer to the bark surface and slightly behind the advancing fans in the inner phloem. Cambial necrosis occurs slightly behind the limit of hyphal advance in phloem. The cycle is completed with perithecial formation and ascospore release beginning several years after the initial infection.

Pure cultures of *E. parasitica* can be isolated readily from diseased bark or wood. The fungus grows well on solid media at 20–35°C and produces irregular pycnidia with crescent-shaped conidia within a month.

Cankers on landscape trees can be arrested at least temporarily by removing a narrow strip of bark all around the canker. This measure denies the fungus its customary route from dead bark into living bark. Long-term observations to learn whether such cankers might be reactivated by growth of the parasite through wood to adjacent bark have not been made.

Numerous additional *Eutypella* species occur on *Acer* and other woody plants in North America, but information about these associations is scant.

References: 898, 1189, 1301, 1392, 1949, 1950, 2098, 2199–2202, 3221, 3539

Eutypella canker of maples, caused by *Eutypella parasitica*.

A. A canker about 9 years old on an *Acer rubrum* (red maple) stem about 8 cm in diameter. Indistinct woundwood ridges and a branch stub in the center of the canker are typical (NY, Apr).

B. Cankers on *A. negundo* (box-elder). The blackened center of the lower canker indicates the presence of perithecia of *E. parasitica* (NY, Apr).

C. Prominent woundwood ridges at the edges of a canker on *A. saccharum* (sugar maple) indicate portions of the canker margin where the parasite is no longer active (NY, Jul).

D. An old canker on *A. platanoides* (Norway maple). The approximate location of the canker margin is indicated by cracks in the bark. Old woundwood ridges within the canker indicate that it was inactive for several years and then resumed expansion (NY, Jul).

E. An old canker on *A. pseudoplatanus* (sycamore maple) (NY, Jan).

F. Close view of white to buff-colored mycelial fans, diagnostic for Eutypella canker, at the margin of a canker on *A. rubrum*. Diagonal parallel lines in dead bark show the limits of successive annual advances of the parasite. Two age classes of mycelial fans, current and 1 year old, are visible (NY, Apr).

G. Magnified view of necks of perithecia of *E. parasitica* protruding from *A. rubrum* bark (NY, Apr).

H. Magnified view of perithecial bodies embedded in *A. rubrum* bark, revealed where surface bark was shaved (NY, Apr).

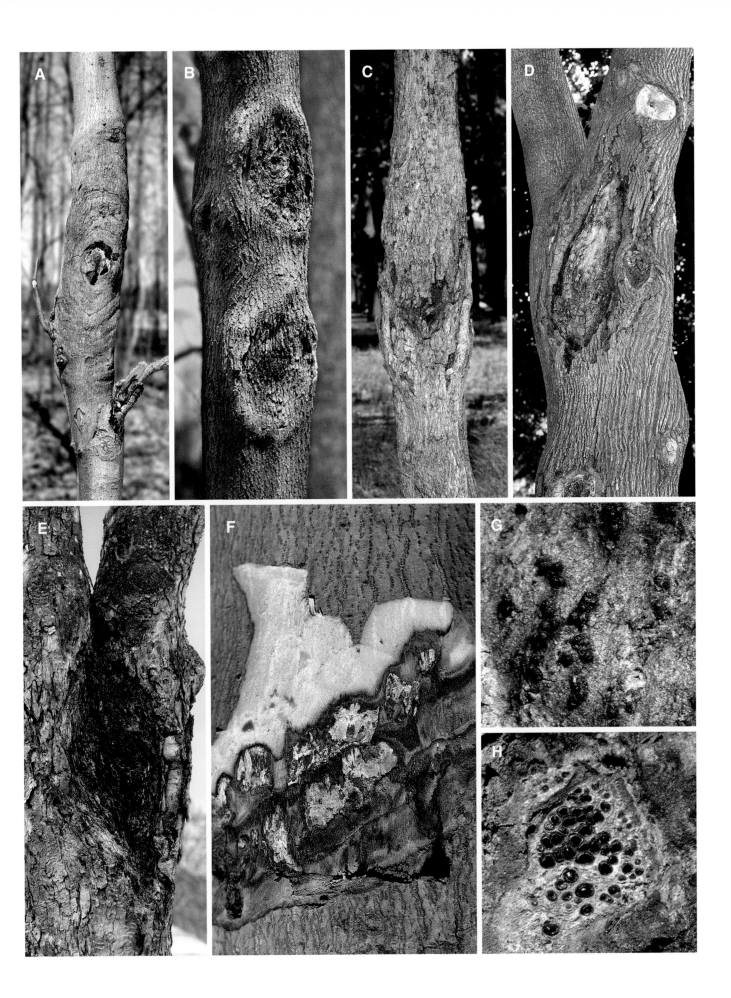

197

Entoleuca (Hypoxylon) Canker of Aspen (Plate 98)

Entoleuca mammata (syn. *Hypoxylon mammatum*; Xylariales, Xylariaceae) causes one of the most destructive diseases of *Populus tremuloides* (trembling aspen) and other aspens (*Populus* section *Populus*). Hypoxylon canker is important mainly in forests, but aspens in open landscapes and residential areas are also affected. The causal fungus occurs from British Columbia and Arizona in the West to Newfoundland and Pennsylvania in the East, but it is common and destructive only east of the Great Plains. It is also widespread in Europe in *P. tremula* (European aspen) and aspen hybrids. Hypoxylon canker is a principal cause of premature death of *P. tremuloides* in the Great Lakes region, where 12% of these trees have been estimated to be infected at a given time. The annual loss is estimated to be 30% of the net growth of aspen in that region. The disease is nearly as severe in eastern North America, where aspen stands have been found with up to 70% of the stems bearing cankers. Cankers develop occasionally on *P. grandidentata* (bigtooth aspen) and rarely on *P. balsamifera* (balsam poplar). Other species sometimes affected are *P. adenopoda* (Chinese aspen), *P. alba* (white poplar), and *Salix daphnoides* (violet willow). *E. mammata* also occurs as a saprobe or innocuous pathogen on trees in several other genera: *Acer, Alnus, Betula, Carpinus, Fagus, Malus, Platanus, Pyrus, Quercus, Sorbus,* and *Ulmus*.

Symptoms and signs. Cankers develop on branches and trunks. Dying branches with yellow undersized leaves or dead branches with brown adhering leaves often call attention to diseased trees. Typical cankers several years old are irregularly elongate and are usually centered at branch bases or stubs or galls induced by *Saperda inornata* (poplar-gall saperda). They have a rough central area with irregular yellowish gray lines and patches on a dull dark gray to nearly black background. From a distance a diseased trunk appears mottled. The outer zone of a canker consists of blistered yellowish orange to orange-brown bark with a smooth, irregularly lobed margin. Some cankers have one or more woundwood ridges and cracked bark at the edges. The wood beneath a canker several years old is always decayed and invaded by boring insects. Some trees with trunk cankers are girdled and die within 3–8 years; others persist for many years. Death is often the result of trunk breakage due to wood decay (by secondary fungi as well as by *E. mammata*) and borer activity.

Cankers developing on young stems at first appear yellowish orange to orange-brown in contrast to normal green to gray-green or light gray bark. They enlarge more rapidly along than around stems and become lobed. White mycelial fans (not shown) sometimes occur in the cambial region approaching but not reaching the edges of a canker. The most distinctive characteristic of young lesions is black-and-cream mottling of phloem, cambium, and outer sapwood. In spring and summer of the second year of lesion enlargement, the dead bark becomes blistered and the periderm (the thin outermost layer) raises slightly and begins to flake or peel off. This process exposes a gray mat of fungal tissue on which tiny pillars (also called hyphal pegs) composed of parallel hyphae have formed. The pillars, $0.2–0.5 \times 0.5–2.5$ mm, raise the periderm and cause the blistered appearance. One-celled, nearly colorless conidia, $5.5–8 \times 1.5–4$ μm in size, form in great numbers on the mat and pillars. The conidia are capable of germination but are not infectious.

The conidial mat and pillars are eventually replaced by perithecial stromata that develop on blackened, cracked bark from which the periderm has disappeared. Stromata are more or less round in outline, 2–5 mm in diameter and 1–2 mm thick, whitish to gray when young, and hard and black when mature. A stroma contains up to 30 perithecia, each 0.7–1.0 mm in diameter with a prominent nipplelike projection where ascospores emerge. The ascospores are unicellular, oblong-elliptic, dark brown, and measure $20–33 \times 9–12$ μm.

Disease cycle. The Hypoxylon canker cycle requires at least four growing seasons. One or 2 years (22–26 months on average) may elapse between inoculation and the onset of canker enlargement, apparently because *E. mammata* remains latent for a time in healthy-appearing stems. No signs are produced during the first year of visible canker formation. Hyphal pillars appear during the second year, and stromata with perithecia begin to appear during the third year.

Ascospores are infectious. They are expelled from perithecia into the air throughout the year after rain or when the stromata are moist, even at air temperatures as low as –4°C. Wet plant surfaces or saturated air for 24–48 hours at temperatures above 16°C are necessary for ascospore germination. Germination is most rapid and complete at 28–32°C. Infection begins on dying year-old twigs at their junctions with young mainstems or branches, and perhaps in wounds that penetrate to the wood. Living bark of many aspen trees inhibits growth of *E. mammata*. Aspens are most susceptible during spring and summer.

Several insects cause wounds that may serve as infection courts. Cankers have been found associated with wounds made by the roundheaded borers *Saperda inornata, S. calcarata* (poplar borer), and *Oberea schaumii* (aspen branch borer), and with oviposition wounds made by *Telamona tremulata* (aspen treehopper), *Magicicada septendecim* (periodical cicada), and *Tibicen linnei* (a dogday cicada). Woodpeckers searching for larvae in galls induced by *S. inornata* also make wounds suitable for infection. Insect population fluctuations may be a reason why many Hypoxylon canker infections begin in occasional wave years. An association between pruning wounds and cankers has been noted, but prior infection of the pruned branch base may account for the observed association.

After lesion development commences, the parasite grows intracellularly in the outer sapwood, cambial region, and bark. Its advance in sapwood precedes that in bark, and it kills host cells in advance of its own progress. Toxins may be involved in this process, but their precise role remains unclear. Cankers expand throughout the year, elongating 7–8 cm per month in summer and a few millimeters per month in winter. Mycelium of the fungus grows most rapidly at 25–30°C. Branch cankers initiated close to the trunk commonly spread into it, but those at a distance are inconsequential because girdled branches soon die and *E. mammata* does not grow far in dead wood.

Aspen clones and families vary in susceptibility to *E. mammata*, and strains of the pathogen vary in aggressiveness. Resistant trees slow the pathogen's advance or may halt it in branches by reacting with rapid branch death, or they limit infection by forming lignified barrier zones and producing callus rich in phenolic substances around lesions. Most strains of the pathogen that cause typical cankers on aspen are apparently host specialized, capable of infecting aspens under a wide range of conditions but unable to infect other plants except when inoculated into water-stressed tissues. An isolate from *Salix daphnoides* caused typical cankers after inoculation into *P. tremuloides*, however.

Environmental and cultural factors affecting incidence and severity of Hypoxylon canker have been much studied, but few reliable statements can be made. Water stress enhances aspen susceptibility under experimental conditions, but neither wet nor dry sites seem unusually favorable for the disease in the field. Understocked stands are likely to be damaged more than fully stocked ones, but cankers enlarge more rapidly on shaded stems than on sunlit ones. Nitrogen fertilization or prior defoliation by insects may enhance susceptibility. Thinning leads to increased disease, but pruning diminishes the incidence of trunk cankers.

References: 74, 78, 87, 162, 291, 292, 322, 362, 363, 400, 549, 557, 609, 1061, 1093, 1184, 1258, 1370, 1672, 1840, 1859, 1977, 2152–2154, 2448, 2454, 2522, 2596, 2920, 3322, 3449, 3852, 4312, 4426

Populus tremuloides (trembling aspen) infected by *Entoleuca mammata*.

A. Dying and dead trees at a forest edge (NY, Jul).

B. A canker that originated at a branch axil (NY, Apr).

C. Close view of part of a canker with dark inner bark visible where the periderm has peeled off. Arrows point to perithecial stromata (NY, Apr).

D. Close view of a canker margin showing the extension of fingerlike necrotic areas in the cambial zone beyond the edge of the lesion in outer bark (NY, Apr).

E. Magnified view of a gray fungal mat and conidial pillars exposed by peeling back periderm in the year-old portion of a canker (NY, Jul).

F, G. Close (F) and magnified (G) views of perithecial stromata of *E. mammata*. Stromata are grayish when young, black when mature. Perithecia (G) have a prominent, black nipplelike tip or papilla with an opening through which ascospores are ejected (NY, Jul).

Biscogniauxia Cankers and Diebacks (Plates 99, 100)

Biscogniauxia (Xylariales, Xylariaceae) is a globally distributed genus with more than 35 known species, a few of which are documented pathogens of angiosperm trees and shrubs. The pathogens are endophytes of healthy trees and colonists of sapwood and bark of stressed trees, causing limb dieback and elongate cankers (strip cankers). *Biscogniauxia* species are often called weak pathogens because they cause severe damage only to drought-stressed or otherwise predisposed trees. This assessment understates their role in tree health, however. They are secondary in the temporal sequence of stressing factors but may, along with so-called secondary insects, be direct causes of dieback or tree death. Some pathogenic *Biscogniauxia* species were formerly classified in *Hypoxylon;* accordingly, the diseases that they cause are commonly called Hypoxylon cankers or diebacks.

The *Biscogniauxia* species considered here are adapted to dry or seasonally dry habitats, cause sapwood decay in limbs and trunks, and can degrade drier wood than can typical basidiomycete decay fungi. The decay is of the white-rot type, in which both cellulose and lignin decompose. Fungi in this group also produce distinctive plates (pseudosclerotial plates) composed of black hyphae and stained host cells within decaying wood. These structures, appearing as black lines on cut or broken wood surfaces, are commonly called zone lines.

Some *Biscogniauxia* species have clear host preferences, although they are not host specific. *B. atropunctata* and *B. mediterranea*, featured in Plate 99, are usually found on *Quercus* (oaks), while *B. marginata* (Plate 100) is usually found on rosaceous trees.

Biscogniauxia species produce dark stromata in or beneath bark. The stromata have an outer layer that opens or disintegrates to expose asexual and later sexual spore-producing structures and permit spore dispersal. The outer layer is usually shed along with overlying bark. The conidial states of *Biscogniauxia* are hyphomycetes with stout branched conidiophores and dry conidia, classified in the genera *Nodulisporium* and *Periconiella*. Mature stromata of species considered here are hard and black internally when mature, and they contain perithecia.
References: 4, 1099, 1350, 1602, 1859, 1972, 1977, 1978, 2522, 2596, 2639, 4312

Biscogniauxia (Hypoxylon) Diebacks of Oaks

Oaks stressed by heat and drought primarily—but also by wounds, root damage, toxic chemicals, or other diseases—often develop dieback that is caused or exacerbated by *Biscogniauxia* species. Two of the common species in North America are *B. atropunctata* (syn. *Hypoxylon atropunctatum*) and *B. mediterranea* (syn. *H. mediterraneum*).

B. atropunctata is found across the eastern half of the USA, most commonly in the South, also in Oregon, southern Ontario, Mexico, and Central America. It contributes to dieback and death of oaks during and after prolonged acute drought. Many oak species are affected, more commonly those in the red oak group (subgenus *Quercus*, section *Lobatae*) than those in the white oak group (section *Quercus*). *B. atropunctata* is also found on *Acer* (maple), *Carpinus* (hornbeam), *Carya* (hickory), *Fagus* (beech), *Platanus* (sycamore), and *Tilia* (basswood).

B. mediterranea is distributed around the world, mainly in warm regions but also as far north as Ontario in North America. It is best known for its association with decline of drought-stressed *Q. suber* (cork oak) and other oaks in the Mediterranean region. (*Phytophthora* species are also associated with the decline and are considered by some researchers to be causal.) The name coal canker has been applied to the association of *B. mediterranea* with oaks in Texas. It occurs on many *Quercus* species and occasionally on *Acer, Alnus, Betula, Castanea, Ceanothus, Celtis, Cornus, Diospyros, Eucalyptus, Fagus, Holodiscus, Juglans, Lithocarpus, Malus, Ostrya, Paulownia, Platanus, Populus, Ribes, Salix, Tilia, Ulmus,* and *Vitis.*

Symptoms and signs. Branches may produce undersized leaves and then die or fail to produce leaves in spring. Foliar yellowing or wilting may also signal the onset of dieback. These symptoms are general indicators of distress and are not necessarily caused by fungi. Stromata of the causal fungi are diagnostic, however. They typically appear beginning in the year after the onset of drought or injury, but those of *B. atropunctata* may form within 2 months after a severe predisposing event. Pressure from a stroma and decay of the inner bark by either of these fungi cause the outer bark to slough off and expose the stroma. Small patches of bark drop away from

stromata of *B. mediterranea*, but *B. atropunctata* commonly displaces strips ranging in length from several centimeters to several meters. Both fungi initially induce a brownish discoloration of sapwood, then a yellow decay with black zone lines.

Stromata of *B. atropunctata* are mostly 2–50 cm long and 2–25 cm wide, but may be larger, and are 0.4–0.7 cm thick. They are initially tan due to a powdery covering of conidiophores and conidia. Conidia may be released as a cloud if a stroma is disturbed. The conidia are nearly colorless individually but are tan in mass, smooth, subspheric, and measure $4–5.5 \times 3–4.5$ µm. As perithecia mature in a stroma, it hardens and the surface color turns to silver gray. Interior parts are black. On close inspection the black tips of perithecia can be seen on the surface. Perithecia liberate dark brown, elliptic to somewhat spindle-shaped, single-celled ascospores that are mostly $23–30 \times 11.5–14.5$ µm in size, each with a straight spore-length groove (germ slit). Old stromata eventually lose the silvery surface and appear black. This fungus has two varieties in addition to the typical form: *B. atropunctata* var. *intermedia* and var. *maritima*. They differ from the typical form in ascospore size and shape.

Stromata of *B. mediterranea* are 2–7 cm long × 0.5–4 cm broad × 0.6–1.5 mm thick. Those with perithecia are coal black and shiny, with raised edges that promote cracking and loss of overlying bark. The stromal surface is studded with pimplelike tips of perithecia. The single-celled ascospores are dark brown, unicellular, oblong to elliptic, usually $16–21 \times 7–10$ µm, with a straight spore-length germ slit. This fungus also has two varieties—*B. mediterranea* var. *macrospora* and var. *microspora*—that differ from the typical form in ascospore size. The latter variety is the common form in Pacific Coast states.

Disease cycle. Hyphae growing from germinated ascospores (and probably conidia also) enter tree stems through minor wounds and natural openings at an early stage of tree growth and form small colonies in apparently healthy bark and sapwood. Both fungi can be isolated from healthy twigs, branches, and some trunks of oaks of all ages. Healthy trees suppress them through a combination of active defense and the inhibitory effect of high water potential on their hyphal growth. Colonies begin rapid growth in water-depleted sapwood and inner bark of trunks and limbs, decay these tissues, form stromata in the cambial zone, and release spores that start new colonies.

Both fungi are favored by a warm, dry or seasonally dry climate. As examples, *B. atropunctata* grows at temperatures up to 40°C with most rapid growth near 30°C. Ascospore germination of *B. mediterranea* is most rapid near 35°C.
References: 114, 252, 379, 424, 733, 771, 1099, 1859, 1977, 1978, 2302, 2390, 2638, 2639, 3960, 4018, 4136, 4138–4140, 4223, 4312

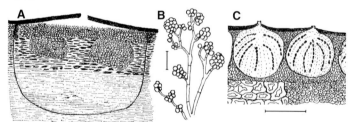

Figure 99. Biscogniauxia atropunctata. A. A young stroma producing conidia beneath broken periderm. The black line beneath the stroma is a pseudosclerotial plate. B. Conidiophores and conidia. C. Section of a perithecial stroma; brown ascospores are arranged linearly within asci. Scale bars = 200 µm for perithecia, 25 µm for conidiophores. Adapted from reference 2638 by permission of the Mycological Society of America.

A. Dieback of a large *Quercus nigra* (water oak) (MS, Apr).
B–E. *Biscogniauxia atropunctata*. B. A tan conidial stroma on *Quercus velutina* (black oak) (AR, winter). C. A silver gray perithecial stroma on *Q. alba* (white oak) (AR, winter). D, E. Young and old mature perithecial stromata, respectively, on *Q. nigra* (MS, Apr).
F, G. *B. mediterranea* on *Q. texana* (Texas red oak). Close (F) and magnified (G) views of a perithecial stroma. Black pimplelike projections are the tips of perithecia (TX, Mar).
H, I. Sapwood of *Q. texana* colonized by *B. atropunctata*. Black zone lines (pseudosclerotial plates) are typical (TX, Mar).
Photo credits: B, C—P. Fenn

201

Blister Canker

Blister canker, also called nailhead canker, commonly affects trees in the apple tribe of the Rosaceae and less commonly affects various other trees. The disease is characterized by strip cankers—elongate zones of dead bark—and limb dieback. The causal fungus, *Biscogniauxia marginata* (Xylariales, Xylariaceae), is an opportunist that causes little damage unless trees are stressed by water shortage. In North America it occurs from Minnesota and Quebec to Oklahoma and Georgia and also in New Mexico. It is widespread in Europe and has also been found in Cuba, French Guiana, and India. It destroyed hundreds of thousands of apple trees in central USA during the first third of the 20th century, but its importance diminished with changes in apple cultivars and orchard practices. Hosts of *B. marginata* include *Amelanchier arborea* (downy serviceberry), *A. canadensis* (eastern serviceberry), *Betula* species (birches), *Carpinus* (hornbeam), *Carya ovata* (shagbark hickory), *C. tomentosa* (mockernut hickory), *Cercis* (redbud), *Gleditsia tria-canthos* (honeylocust), *Magnolia virginiana* (sweet bay), *Malus coronaria* (American crabapple), *M. pumila* (apple), *Pyrus communis* (common pear), *Quercus* sp. (oak), *Sorbus americana* (American mountain-ash), *S. aucuparia* (European mountain-ash), *S. hybrida* (oak-leaf mountain-ash), *Ulmus americana* (American elm), and *Viburnum* sp. When found on plants other than *Amelanchier*, *Malus*, and *Sorbus*, the fungus usually seems to be an inconsequential inhabitant of dead branches.

Symptoms and signs. Symptoms appear in years that follow drought. An elongate canker or a dying or dead stem calls attention to the disease, but the primary symptom is internal—a long column of dead, reddish brown sapwood extending in both directions from a wound. Its expansion toward the cambium leads to dieback or canker formation. The bark and vascular cambium initially possess high resistance to *B. marginata*, but they are colonized after the underlying sapwood dies. Cankers develop on stems that offer some resistance to colonization, and they are usually associated with wounds. Infection that begins at a small wound or crack may become enclosed by woundwood, and a canker may appear above or below it years later. Some cankers become delimited by woundwood and cracked bark at margins.

Blisterlike mounds 3–6 mm in diameter appear in late spring or summer on previously smooth bark in newly evident strip cankers or at the bases of recently dead branches. Fungal stromata, developing beneath the periderm in bark dead for less than a year, cause these swellings and rupture the periderm, usually at lenticels. On some hosts angular flaps of torn periderm roll back from the stromata, as shown in Plate 100F. Stromata become more abundant where the substrate is shaded, and water loss slower, than on surfaces exposed to full sun. If an entire limb dies in one season, it usually becomes too dry to sustain formation of stromata except near the base. *B. marginata* causes a light-colored, relatively dry decay of the wood beneath stromata. It can be isolated readily from spores, stromata, or discolored wood, and its conidial state forms in culture.

The conidial state, a hyphomycete in the genus *Nodulisporium*, develops within the upper part of a young stroma and is exposed as a gray to nearly white powdery mass when the periderm ruptures. Branched conidiophores produce dry conidia, mostly 8×5 μm, that are rounded at one end and pointed at the attached end. Several conidia, honey colored as viewed microscopically, form on each branch tip. One stroma may produce conidia for several years if protected from sunlight.

Mature stromata have a concave, circular to elliptic surface. They vary from 3×3 to 9×5 mm in length and width (sometimes longer on *Sorbus*), rise 2–3 mm above the wood, and are sometimes broader at the top than at the base (resembling nail heads). The raised margins are dark brown to black. Eventually the entire surface turns nearly black and by then is dotted with the ostioles of perithecia.

Perithecia develop in year-old and older stromata, beginning while the conidial state is present. They are shaped like tapered flasks, broadest at the somewhat flattened bases, 0.3–0.5 mm in diameter and 0.5–1 mm high, with narrow necks extending to the surface. The ascospores are brown to dark brown, unicellular, smooth, subglobose, measure $13–16.5 \times 9.5–13$ μm, and have a sigmoid spore-length mark called a germ slit. These ascospore characteristics are diagnostic.

Sometimes *B. marginata* is found along with other opportunistic pathogens on declining trees. We have observed it with another sapwood parasite, *Schizophyllum commune* (Plate 158), on dying limbs of *Amelanchier arborea*.

Disease cycle. A cycle several years in length begins with infection of recently wounded sapwood by ascospores or conidia. Cracked and broken limbs and branch stubs are common sites of infection, which can occur whenever temperature and moisture permit fungal growth. Ascospores are available throughout the year because some stromata produce them in more than one year and the spores do not all reach maturity at once. They are discharged from perithecia after rains throughout the growing season. Peak discharge occurs a day or two after thorough wetting of stromata. Some ascospores are ejected up to 40 mm above stromata. Many ascospores accumulate on the tops of stromata, temporarily trapped there among conidiophores and conidia, and are gradually dispersed during later rains. Both types of spores are dispersed locally by water and potentially also by insects. Ascospores retain germinability for up to 3 years when stored in dry stromata. Conidia lack the longevity of ascospores.

B. marginata may begin to kill bark or cause dieback a year or more after becoming established in sapwood, but external symptoms can be delayed indefinitely unless a tree's defenses are impaired by drought stress. The fungus grows more rapidly in older (e.g., 5 years old) sapwood than in younger wood near the vascular cambium. Sapwood colonization is retarded in well-watered trees. The minimum interval from infection to conidial production on new stromata is probably 2 years, with ascospore production usually requiring an additional year.

Blister canker outbreaks in apple orchards were most severe in regions where hot, dry weather occurs annually. Outbreaks elsewhere occurred only after drought. The disease was more severe on poorly drained than on well-drained sites.

References: 79, 80, 610, 797, 1099, 1298, 1305, 1971, 1978, 4313

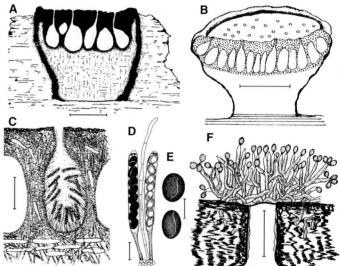

Figure 100. Biscogniauxia marginata. A, B. Vertical sections of perithecial stromata (A) immersed in the substrate and (B) standing on it. C. Section of a perithecium. D, E. Asci and ascospores. F. Conidiophores and conidia on a perithecial stroma. Gap in stroma is a perithecial ostiole. Scale bars: A and B = 1 mm, C = 200 μm, D = 20 μm, E = 10 μm, F = 50 μm. A adapted from reference 79; B–D from reference 1971 by permission of the Mycological Society of America; E from reference 1978 by permission of Mycotaxon, Ltd.; F from reference 797.

A. Dieback of *Amelanchier arborea* (downy serviceberry) caused by *Biscogniauxia marginata* (NY, Jul).

B. Bark of *Amelanchier* sp. with the periderm breaking open where stromata of *B. marginata* are developing (NY, May).

C, E. Year-old broken "blisters" and stromata of *B. marginata* on *Amelanchier* sp. The disclike tops of the stromata have a light color due to presence of conidiophores and conidia (NY, May).

D, F. Mature stromata of *B. marginata* on a branch of *Viburnum* sp. The raised margin around the black, perithecia-bearing center of each stroma is typical. In close view (F), flaps of periderm have rolled back from the margin of each stroma (NY, Jun).

Camillea- and Hypoxylon-Associated Cankers and Diebacks (Plate 101)

Camillea and *Hypoxylon* species (Xylariales, Xylariaceae), like *Biscogniauxia* (Plates 99, 100), are typically found on dead, decaying bark or wood. Some of these fungi are plant parasites, but most are primarily or exclusively saprobic. They colonize trunks, limbs, and sometimes roots weakened or killed by drought, insects, lightning, and other diseases. Some species are endophytes that establish quiescent colonies in healthy stems and are thus in position to grow rapidly in wood and bark after the stems are wounded or severely stressed by water shortage. Species that have been tested, including *C. punctulata* (Plate 101E), cause decay of wood, although the process is much slower than that caused by typical basidiomycete decay fungi.

Camillea dieback of planetree and sycamore. *Camillea tinctor* (syn. *Hypoxylon tinctor*; Plate 101A–D, G) is apparently an opportunistic parasite that colonizes stressed or declining *Platanus* ×*acerifolia* (London planetree) and *P. occidentalis* (American sycamore), causing strip cankers and dieback. Pathogenicity of this fungus has not been proven, but circumstantial evidence of it is provided by the perennial expansion of cankers bearing its stromata. *C. tinctor* has been known for many years as a saprobe on dead *Platanus* and occasionally on other trees, including *Acer* (maple), *Carpinus* (hornbeam), *Fraxinus* (ash), *Liriodendron* (tuliptree), *Magnolia*, *Morus* (mulberry), *Nyssa sylvatica* (black-gum), *Quercus* (oak), *Salix* (willow), and *Vitis* (grapevine), in forests and landscapes of central and southeastern USA and Mexico. It occurs on various other angiosperms in the Caribbean region, South America, Africa, and parts of the western Pacific region (Thailand, New Guinea, Singapore).

Cankers with stromata of *Camillea tinctor* on *Platanus* were first reported from Georgia, USA, where they developed on residual *P. occidentalis* trees in a thinned forest. Heat stress caused by the loss of shade on trunks was presumed to have predisposed them to infection. Urban shade trees, shown in Plate 101, may be affected when predisposed by adverse site conditions and drought. The cankers are perennial and develop on trunks or large limbs as long bands that follow the grain of the wood. If the wood grain turns in a helix, so does a canker. Canker margins are often imperceptible beneath scaly bark. Successive woundwood ridges sometimes mark annual increments of canker expansion, and the centers of cankers may appear sunken. Bark or exposed wood in cankers may have longitudinal cracks. Woundwood at canker margins is barely perceptible on declining trees, but it may become prominent on trees that regain some vitality after a predisposing stress is removed and lesion expansion is halted.

Stromata of *C. tinctor*, 2–15 cm long × 1–2 cm wide and 1–1.5 mm thick, develop beneath the bark surface and cause the overlying bark to fall away, or sometimes are found beneath the outer scales of rough bark near a canker margin. Immature stromata have two layers. The outer layer is shed with the overlying bark, exposing on the surface of the inner layer a conidial anamorph in the genus *Xylocladium*. The inner layer at maturity is studded with dark raised dots, which are the ostioles of embedded perithecia. The perithecia produce ascospores that are pale yellow to light brown, single celled, mostly 16–22 × 6–8 μm, and taper toward each end. Sometimes stromata of several age classes are visible. Beneath the youngest, which are the darkest in color, the inner bark and cambial region are usually discolored orange to orange-brown. As stromata age, the underlying tissue turns darker brown, as may be seen where portions of stromata were removed in Plate 101D. Wood in the vicinity of stromata is usually stained bright yellow or orange and contains prominent black zone lines similar to those shown in Plate 99. Mexican collections are inconsistent in pigment production, however. Cankers associated with *C. tinctor* are superficially similar to those of the canker-stain disease of sycamore (Plate 116), but the latter lack dark stromata.

Two prominent saprobes. *C. punctulata* (syn. *Hypoxylon punctulatum*; Plate 101E, H) is among the more prominent fungi that appear on recently dead parts of *Quercus* (oak) trees in eastern USA

and Mexico. It is an aggressive primary colonist of bark and sapwood of dying trees. This behavior indicates that it may reside as an endophyte in healthy oak tissues, as do several other fungi in the Xylariaceae. Occasional hosts of this fungus include *Betula* (birch), *Carya* (hickory), *Castanea* (chestnut), *Fagus* (beech), *Halesia* (silverbell), and *Ulmus* (elm).

Stromata of *Camillea punctulata* begin to appear as soon as 6–7 months after the death of a limb or tree. Mature stromata are dark brown (black at a distance), less than 1 mm thick, usually somewhat elongate, and 5–20 × 3–15 cm in size. They develop in bark just beneath the corky outer part and have two layers when young. The outer layer, or ectostroma, raises the outer bark, which breaks and falls away, taking the ectostroma with it. This shedding is due to force applied by a ridge of gelatinous tissue around the edge of an ectostroma, augmented by hyphal pillars on its surface. The mature endostromal surface is dotted with ostioles (openings where spores are liberated) of embedded perithecia that produce brown, elliptic, unicellular ascospores measuring 7–9 × 3–4 μm. A gray mat of the conidial state (in genus *Xylocladium*) precedes the perithecial stromata. The conidia form on the opposing faces of the endo- and ectostroma and on hyphal pillars 1–1.5 mm tall composed of parallel cells that arise from the stromal surfaces. This mode of conidial production and the action of hyphal pillars in causing overlying tissue to break away, exposing stromata, are similar to the habits of *Entoleuca mammata* (Plate 98).

In the Appalachian region, oaks dying from oak wilt (Plate 118) are among the substrates of *C. punctulata*. The fungus suppresses saprobic growth and sporulation of the oak wilt pathogen through competition for nitrogenous compounds and sugars, but not to a degree sufficient for practical biological control of the wilt disease.

Hypoxylon cohaerens (Plate 101F, I) occurs in North America and Europe as a common colonist of *Fagus grandifolia* and *F. sylvatica* (American and European beech, respectively). It also occurs rarely on *Acer* (maple) in eastern North America and occasionally on *Corylus* (filbert) and *Quercus* in the West. A small-spored variety, *H. cohaerens* var. *microsporum*, occurs on *Quercus* in Europe. *H. cohaerens* is one of the common invaders of beech trunks dead or dying from beech bark disease (Plate 91). It is sometimes associated with discrete cankers (Plate 101F), but no direct evidence for its pathogenicity has been reported. Perithecial stromata form on the bark surface as reddish brown to pale brown cushions, 2–4 mm in diameter and 1–1.5 mm high, often in clusters over large areas. They darken with age, becoming purplish black. Each stroma contains 6–15 perithecia that produce dark brown, spindle-shaped, single-celled ascospores measuring 9–12 × 4–5 μm.

References: 217, 911, 1099, 1311, 1972, 1977, 2211, 2286, 2527, 2596, 2639, 3328, 4312

A–D, G. Association of *Camillea tinctor* with dieback and strip canker of *Platanus occidentalis* (American sycamore). A. Dead branches and others with sparse tufted foliage on a stressed tree. B, C. Strip cankers, one with a dark brown stroma, on a tree with severe dieback. D. Parallel arrangement of stromata indicates tangential enlargement of a canker from right to left over several years. Orange-brown discoloration of the inner bark and cambial region beneath active stromata, exposed by removing parts of stromata, is typical of tissues colonized by *C. tinctor*. G. Magnified view of the surface of a perithecial stroma. Dark dots are ostioles of perithecia (IL, Jul).

E, H. Perithecial stromata of *C. punctulata* on a recently dead *Quercus* (oak) branch. Dots on the surface in magnified view (H) are ostioles of perithecia (NY, Oct).

F, I. Perithecial stromata of *Hypoxylon cohaerens* on a canker on *Fagus grandifolia* (American beech). Stromata of this species are cushionlike and reddish brown when young and become darker with age. Each stroma contains several perithecia. The cause of this canker was unknown (NY, May).

Basal Canker and Butt Rot Caused by *Kretzschmaria deusta* (Plate 102)

Kretzschmaria deusta (syn. *Hypoxylon deustum*, *Ustulina deusta*; Xylariales, Xylariaceae) kills sapwood and causes a white rot of the major roots, butt, and trunk of angiosperm trees and shrubs in numerous genera around the world, mainly in the north temperate zone. Reported hosts of *K. deusta* in North America include species of *Acer* (maple), *Aesculus* (horse-chestnut), *Alnus* (alder), *Betula* (birch), *Carpinus* (hornbeam), *Carya* (hickory), *Celtis* (sugarberry), *Citrus*, *Fagus* (beech), *Fraxinus* (ash), *Ilex* (holly), *Liriodendron* (tuliptree), *Nyssa* (black-gum), *Platanus* (planetree), *Quercus* (oak), and *Ulmus* (elm). There is one record of *K. deusta* on a gymnosperm—*Chamaecyparis thyoides*, or Atlantic white cedar. Records from elsewhere include additional genera that are common in temperate North America: *Castanea* (chestnut), *Casuarina* (Australian-pine), *Populus* (poplar), *Prunus* (peach and others), *Pyrus* (pear), *Sambucus* (elderberry), *Salix* (willow), and *Tilia* (linden). In the tropics, other *Kretzschmaria* species, notably *K. zonata*, that were formerly confused with *K. deusta* cause tree decline and consequent economic damage in plantations of *Camellia sinensis* (tea), *Citrus*, *Hevea brasiliensis* (rubber), and *Tectona grandis* (teak). *K. clavus* causes root rot and decline of *Macadamia integrifolia* (macadamia) in Hawaii and Taiwan.

K. deusta is usually found as an apparent saprobe growing in and on wood, but its ability to colonize wood in living trees is well documented. It causes basal cankers associated with butt rot on *Acer rubrum* (red maple), *A. saccharum* (sugar maple), *Celtis laevigata* (sugarberry), *Fagus sylvatica* (European beech), and probably other species. Wood decay by *K. deusta* is sometimes associated with tree failure during storms. For example, root and butt rot and basal cankers associated with *K. deusta* occur on mature *Celtis laevigata* in southern Louisiana on sites where infection is presumed to be fostered by the high water table. Prolonged soil saturation and accompanying hypoxia in this situation may weaken or kill roots and dispose them to attack. Diseased trees, if not cut first, eventually topple because of decay in buttress roots. Similar damage due to *K. deusta* has been noted in *Aesculus hippocastanum* and *Tilia* species in Great Britain.

Symptoms and signs. Symptoms caused by *K. deusta* are primarily internal and are not diagnostic. The fungus causes reddish brown wood discoloration and then a light-colored decay, usually associated with an obvious wound on the butt or a buttress root. The reddish brown color of recently colonized sapwood is due to deposition of chemicals associated with wound isolation and defense against fungi and bacteria. These materials are produced by xylem parenchyma cells before they die. Decaying wood, viewed on a cut or broken face, often has alternating zones of discolored wood (former reaction zones in sapwood) and white areas with advanced decay, a pattern typical of several canker-rot diseases. The decay in its early stages is of the soft-rot type, inasmuch as hyphae grow into the thick middle, or S_2, layer of the secondary walls of fiber tracheids and there form elongate cavities by enzymatic digestion of wall substances. Both cellulose and lignin are degraded. The cavities merge and greatly weaken the walls. The advanced decay is light in color, brittle to crumbly, and contains numerous irregular black sheets (pseudosclerotial plates) that appear as lines (zone lines) on broken or cut faces. The plates, refractory to microbial degradation, are composed of black hyphae with swollen bladderlike cells and host cells with lumens and walls impregnated with dark material. Colonized stumps and logs eventually decompose except for a persistent mass of pseudosclerotial plates that appear as brittle black sheets. Pseudosclerotial plates are produced by many wood-decaying fungi, especially members of the Xylariaceae, but those produced by *K. deusta* are unusual for their size, abundance, and persistence. Pseudosclerotial plates within oak wood are shown in Plate 99.

K. deusta also produces large lumpy groups of stromata on the butts and dead roots of living and dead trees and stumps. Stromata develop on both bark-covered and decorticated wood in association with wounds, cankers, and decaying roots, and may persist for several years. They are grayish white and leathery at first, then turn copper brown to brown and eventually black and brittle. Old dead stromata can readily be detached from the substrate. Young stromata produce conidia. In North America the conidial stroma forms in winter to spring, depending on latitude, and the perithecial stroma ripens in early summer to autumn. The conidial stroma is an effuse grayish white to grayish brown layer, 1.5–4 mm thick, consisting of vertically oriented tufts of branched conidiophores on a thick mycelial base. Conidia are nearly colorless, single celled, and wider at one end than the other, measuring $5–9 \times 2–3.5$ μm. If the fungus is isolated from wood, the conidial state also forms on agar media.

Perithecial stromata of *K. deusta* develop on the same area previously occupied by the conidial state, often in massive groups. They may appear as discrete cushions or a sheet of undulating to lumpy tissue, its topography determined in part by that of the host surface. Stromatal groups often become perennial and then develop concentric zones. For example, young gray-brown stromata extend beyond older black stromata on the *Fagus* butt shown in Plate 102D. Scattered large black perithecia (0.7–1.5 mm in diameter and 1–2 mm high) develop in the stroma, immersed in its whitish interior tissue, their nipplelike tips protruding slightly above the surface. The ascospores at maturity are single celled, dark brown to black, and irregularly elliptic with pointed ends, most measuring $27–35 \times 7–9$ μm.

Disease cycle. *K. deusta* is dispersed as airborne ascospores and conidia. Infection and decay initiated by spores commonly begin at wounds caused by fire or mechanical damage. Both types of spores are infectious to freshly wounded sapwood, but information on their relative importance as inoculum is lacking. The ascospores germinate readily in water at 15–30°C. Germination and mycelial growth are most rapid at 20–30°C, at which temperatures most spores germinate within 24 hours.

During the early stages of wood decay the fungus grows from cell to cell mainly through bordered pits. The progress of attack by *K. deusta* is relatively slow, but colonization may be extensive nonetheless. When ascospores were introduced to fresh wounds in sapwood of *Fagus sylvatica* and *Tilia* ×*europaea* (European linden), the fungus could be isolated a year later from stained wood near the wound. Observations of dissected *Acer* and *Fagus* with basal cankers indicated that *K. deusta* may have caused the cankers by spreading outward from a central column of decayed wood. In northeastern USA *K. deusta* infects sprout maples via the parent stump or the stumps of companion sprouts, and the decay column may eventually extend more than 2 m above ground in trunks only 20–25 cm in diameter. Probably many years elapse from the time of infection until stromata with a new generation of spores are produced. Tree-to-tree movement of *Kretzschmaria* species in *Macadamia* and *Camellia sinensis* plantations has been suggested to occur via root contacts and by hyphal growth through soil, but these possibilities have not been studied for *K. deusta* in trees of the temperate zones.

References: 631, 1365, 1859, 1972, 2103, 2639, 2865, 2992, 2997, 3151, 3323, 3508, 3510, 4312, 4366

A–C. Butt rot and basal canker of *Celtis laevigata* (sugarberry). A. The butt of a mature landscape tree is dead around nearly half its circumference and has a large aggregation of rough black stromata of *Kretzschmaria deusta* at the ground line. B. Closer view of the stromata. C. Close view of part of a young stroma; black dots are ostioles of perithecia (LA, Dec).

D–F. Root and butt rot and basal canker of *Fagus sylvatica* (European beech). D. Close view of the edge of a large basal canker. Stromata of *K. deusta* (gray-brown to dark brown, their left edges marked by arrows) lie at the junction of bark and exposed wood. Reddish brown stains on bark result from bleeding at points of attack by ambrosia beetles that tunnel into weakened trees. E, F. A large group of dark lumpy stromata on the soil between brace roots of a tree, growing from roots; arrow in E marks an exposed patch of root bark that appears at left center in F (NY, Sep).

Sooty-bark Canker of Aspen and Poplar (Plate 103)

Sooty-bark canker, caused by *Encoelia pruinosa* (Helotiales, Sclerotiniaceae), is the most damaging canker disease of *Populus tremuloides* (trembling aspen) in the Rocky Mountain region. It affects mature trees primarily, on sites of both high and low quality for aspen growth, and is lethal within several years. Surveys in Colorado in the 1970s revealed occurrence of the disease on more than 90% of the sites examined; 1–2% of living aspens were affected. The causal fungus was also found fruiting on more than half the dead trees; thus it was apparently the most common cause of aspen mortality. It had been known as a saprobe on aspen bark for nearly 70 years before a description of the associated disease appeared in 1956. Sooty-bark canker occurs from Arizona and New Mexico to Alaska in the west, usually on *P. tremuloides*. *P. grandidentata* (bigtooth aspen), *P. balsamifera* (balsam poplar), and *P. balsamifera* subsp. *trichocarpa* (black cottonwood) are occasionally affected.

E. pruinosa occurs also in the Great Lakes region and northeastern USA on *P. grandidentata* and *P. tremuloides*, and in northern Europe on *P. tremula* (European aspen); however, the fungus seems inconsequential in those areas. Experiments in Minnesota revealed the ability of local isolates to cause only small cankers, most of which ceased elongation within 3 years. Inoculation of cut segments of aspen trunks from Colorado and Minnesota with isolates from both states resulted in larger and more numerous lesions induced by Colorado isolates than by those from Minnesota. Also, more lesions formed in inoculated stem segments from Colorado than in segments from Minnesota. Thus, regional differences in importance of sooty-bark canker may be linked to regional population differences in aggressiveness of the pathogen and susceptibility of *P. tremuloides*.

Symptoms and signs. Young cankers appear as slightly sunken elliptic areas with nearly normal surface color and blackened internal tissues. Marginal woundwood development is slight or absent in most cases, but variations in bark color reveal zones of lesion expansion. After bark has been dead 2–3 years, the periderm begins to slough off, revealing blackened inner bark. A band of light gray periderm usually remains attached at the margin of each annual zone of expansion. These bands provide a record of the age of the canker and the rate of its enlargement. Sooty-bark cankers in Colorado elongate an average 45 cm and increase an average 16 cm in width per year. Canker expansion up to a meter in length and 35 cm in width in the first year has been recorded. Blackened inner bark, crumbly (sooty) on the surface, usually remains tightly attached to the wood for several years, held in place by mycelial mats of *E. pruinosa*. Eventually, however, it begins to break away, revealing distinctive patterns of gray to black spots on the wood where fungal mats formerly adhered. The spots remain visible for years.

Apothecia develop in great numbers on blackened bark and are present throughout the year, either exposed or beneath loose periderm. They are leathery, silver gray on the surface with dark brown to black interior tissue, cup shaped, and about 3–4 mm in diameter when moist. They curl into angular or elongate shapes when dry. The outer surface is encrusted with crystals. Ascospores are forcibly ejected and are dispersed by air. These spores are colorless, ellipsoid and slightly curved, single celled, and measure 8–12 × 2.5–3.5 μm.

E. pruinosa can be isolated readily in pure culture from cankers or ascospores. It grows most rapidly at temperatures near 22°C. Ascospores germinate at 5–30°C and do so most rapidly at 20–25°C. The fungus produces tiny spherical microconidia in culture. These do not germinate, and their function is unknown. Perhaps they are spermatia.

Disease cycle. Chronological details of the cycle have not been published. Accounts of canker enlargement and apothecial formation beneath loose periderm indicate that the minimum duration may be 2–3 years. Infection is initiated by germinating ascospores in wounds of various kinds at any height on the trunk. Superficial wounds as well as those extending into xylem are suitable. The fungus advances in the inner bark and cambium and often girdles a tree within 4–5 years. Some cankers resulting from experimental inoculations have ceased expansion after a year or longer, as evidenced by a woundwood ridge at the margin. No comparable behavior has been reported for naturally occurring cankers.

Disturbances such as selective logging and campsite construction and use in aspen stands in the Rocky Mountain region have been reported to cause increased incidence of sooty-bark canker and also Ceratocystis canker of aspen (Plate 115). Incidence of trunk damage by elk has also been related to incidence of sooty-bark canker.
References: 897, 1099, 1668, 1672, 1674, 1948, 1990, 1992, 1994, 3523, 4064, 4209

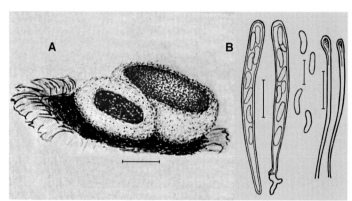

Figure 103. Encoelia pruinosa. A. Apothecia on bark. B. Asci, ascospores, and paraphyses (sterile hyphae that grow among asci). Scale bars = 1 mm for apothecia, 20 μm for asci and paraphyses, 10 μm for ascospores. A adapted from reference 3523; B adapted from reference 4064 by permission of Universitetsforlaget.

Sooty-bark canker of *Populus tremuloides* (trembling aspen), caused by *Encoelia pruinosa.*

A. A mature tree, girdled and killed by the disease 8 years after infection. Light tan arcs mark the limits of annual increments of canker expansion.
B. Part of the canker shown in A. Bark here has been dead 4–6 years. The light gray periderm is absent except in arcs between annual zones of expansion. Sooty-appearing inner bark is beginning to crack and fall away.
C. Close view of sooty-appearing dead bark killed about 3 years earlier. The large white patch is periderm. The tiny silver gray structures on the black surface are apothecia of *E. pruinosa.*
D. Characteristic dark spots on the surface of wood where mycelial mats of *E. pruinosa* held dead inner bark in place for several years (A–D: CO, Jun).
E. Magnified view of dry apothecia (CO, Oct).

Cenangium Dieback of Pines (Plate 104)

Pine branches with brown needles may signal Cenangium dieback. The disease becomes prominent in scattered localities in occasional years in the northern temperate and boreal zones. It is common after drought or an unusually cold winter, especially if a severe winter is preceded by unusually mild autumn weather. It is caused by the ubiquitous, opportunistic fungus *Cenangium ferruginosum* (syn. *C. abietis*; Helotiales, Helotiaceae), which kills bark and cambium of twigs and branches weakened by environmental factors, other pests and pathogens, or natural senescence. *C. ferruginosum* is an endophyte of pine needles, a proven pathogen of conifer twigs and branches, and a competent saprobe that fruits on twigs and branches. It is found occasionally on *Abies* (fir) and *Picea* (spruce) in addition to *Pinus*.

Disorders attributed to *C. ferruginosum* include twig blight of *Pinus contorta* (lodgepole pine), *P. ponderosa* (ponderosa pine), and *P. strobus* (eastern white pine); girdling cankers and dieback on branches of *P. densiflora* (Japanese red pine), *P. koraiensis* (Korean pine), *P. nigra* (Austrian pine), and *P. thunbergii* (Japanese black pine); and death of *P. strobus* saplings in the understory of Appalachian forests after severe winters. Disease observations have also indicated probable pathogenicity of the fungus to *P. cembra* (Swiss stone pine), *P. halepensis* (Aleppo pine), *P. jeffreyi* (Jeffrey pine), *P. lambertiana* (sugar pine), *P. monticola* (western white pine), *P. pinaster* (cluster pine), *P. pinea* (Italian stone pine), and *P. sabiniana* (digger pine). In pathogenicity tests *C. ferruginosum* caused cankers or twig dieback on *Pinus banksiana* (jack pine), *P. contorta*, *P. densiflora*, *P. koraiensis*, *P. ponderosa*, *P. resinosa* (red pine), *P. sylvestris* (Scots pine), and *P. strobus*, as well as on *Picea mariana* (black spruce) and *P. glauca* (white spruce). There are records of the fungus on dead twigs of numerous additional conifers.

Infestation of pines by certain insects may predispose them to attack by *C. ferruginosum*. The fungus has killed *P. ponderosa* branches following infestation by *Matsucoccus vexillorum* (Prescott pine scale) in western USA, and has caused dieback of *P. densiflora* following infestation by *M. matsumurae* in China.

Trees chronically affected by Cenangium dieback are exceptional but may be found on stressful sites. The *P. densiflora* specimen shown in Plate 104A lost branches every year after construction of the adjacent building until its ornamental value was destroyed. Cenangium dieback of *P. sylvestris* sometimes appears in Christmas tree plantations after severe winters and makes trees unsalable. Random individuals are affected, and usually only a minority are severely damaged. Outbreaks of the disease in North American forests, although sometimes conspicuous, have been inconsequential. In some areas of southern and eastern Europe, on the other hand, *C. ferruginosum* and secondary insects have caused considerable damage in drought-stressed forests.

Symptoms and signs. Foliage on scattered twigs and branches turns drab green, then yellowish, and finally brown in late winter and spring. If the outer bark is removed at the base of the dead part, a sharply delimited boundary between brown dead bark and normal tissue is revealed. Needle browning proceeds from bases toward tips, and the needles loosen and drop during the summer. Little or no resin exudes on surfaces of infected parts, and resin infusion of tissue occurs only at margins of killed bark. *C. ferruginosum* causes neither the dark staining characteristic of Atropellis cankers (Plate 106) nor the yellow-green discoloration that sometimes distinguishes Scleroderris cankers (Plate 105). Also unlike *Atropellis* or *Gremmeniella*, *C. ferruginosum* produces apothecia on bark during the first summer after the bark dies.

By early summer, apothecia are visible on killed twigs or branches up to about 5 cm in diameter. They arise from black stromata that form during spring in newly killed bark just beneath the corky outer layer. Foliage distal to the canker may still be green when these form. Dark brown, globose tubercles of fungal tissue then break through the bark surface singly or, usually, in clusters and become larger and lighter in color until they mature during summer. When on twigs, the apothecia commonly emerge at needle bases or scars. The surface of bark that is normally smooth, as on five-needle pines, becomes shriveled during the period of apothecial development. Mature apothecia of *C. ferruginosum* are yellowish brown, deeply cup shaped when moist, 2–5 (mostly 2–3) mm in diameter, with a

yellow to buff interior surface (concealed by an inrolled margin when dry) from which ascospores are discharged during rainy weather. The ascospores are broadly elliptic, single celled, colorless, and somewhat narrowed toward the ends, measure 12–14 × 5–6 μm, and have one or more guttules (presumed oil droplets). After discharging spores, apothecia slowly become black and brittle.

Disease cycle. C. ferruginosum produces germinable conidia in culture but is not known to do so in nature. Therefore, infections are believed to be initiated by ascospores, presumably in summer or early autumn. The saprobic existence of *C. ferruginosum* ensures a supply of spores. The early stages of infection have not been studied, but it seems likely that infections begin in needles and possibly in minor bark wounds, and are held in check by host defenses unless these are weakened or defeated by environmental damage or by other pests. This scheme reconciles the endophytic habit of the fungus with the sporadic appearance of disease symptoms. Extensive infection of green needles is indicated by isolation data. *C. ferruginosum* was the most frequently isolated needle endophyte (64% of all isolates) in *P. sylvestris* in Finland. Similarly, it was among those most frequently isolated from *P. mugo* in Germany and Switzerland. In Japan this fungus was detected in an average of 13% of basal segments of *P. thunbergii* needles and 2.4% of corresponding segments of *P. densiflora* needles. It grows and its spores germinate at temperatures between at least 10° and 25°C on common agar media.

Another fungus in the Helotiaceae, *Crumenulopsis atropurpurea* (syn. *Cenangium atropurpureum*), superficially similar to *C. ferruginosum* and possibly also pathogenic, is widespread on pines in North America. Its apothecia have a purplish outer surface when fresh. It is mentioned here because records of its occurrence were at one time confused with those for *C. ferruginosum*. It fruits on dead twigs of at least *Pinus caribaea*, *P. densiflora*, *P. elliottii*, *P. mugo*, *P. nigra*, *P. ponderosa*, *P. pungens*, *P. resinosa*, *P. rigida*, *P. sylvestris*, and *P. taeda*.

References: 668, 1099, 1213, 1298, 1478, 1546, 1591, 1834, 1981, 2111, 2213, 2274, 2362, 2565, 3196, 3623, 3649, 3701, 4064, 4279

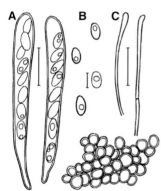

Figure 104. Cenangium ferruginosum. A. Asci. B. Ascospores. C. Paraphyses (sterile hyphae that grow up among asci). D. Rounded cells in the outer tissue of the apothecium, a diagnostic character. Scale bars = 10 μm for ascospores, 25 μm for other structures. Adapted from reference 4064 by permission of Universitetsforlaget.

A–C. Cenangium dieback of *Pinus densiflora*, *P. nigra*, and *P. thunbergii* (Japanese red, Austrian, and Japanese black pines, respectively) (NY, May–Jun).

D–F. A branch of *P. densiflora* partly killed by *Cenangium ferruginosum*. The bark surface appears normal at the branch whorl beyond which needles are brown, but necrosis is revealed by removing the outer bark. Young fruiting bodies and black stromata of the fungus are located by arrows in E and F, respectively (NY, Jul).

G. Apothecia of *C. ferruginosum* in an early stage of development on *P. densiflora* (NY, Jul).

H, I. Apothecia on *P. thunbergii* just before and just after opening (NY, Jun).

Scleroderris Canker of Conifers (Plate 105)

Scleroderris canker, also known as Brunchorstia dieback, is a major disease of *Pinus* species (pines) in snowy regions of Europe and eastern North America. It kills seedlings and saplings, stunts and deforms survivors as the result of trunk cankers or killed leaders, and reduces log quality because of deformities and delayed self-pruning of surviving trees. The pathogen is *Gremmeniella abietina* (syn. *Ascocalyx abietina*, *Scleroderris lagerbergii*; Helotiales, Helotiaceae). It comprises several varieties and recognized subgroups, some of which will likely be found to be distinct at the species level. In 2004 they were recognized as follows. *G. abietina* var. *abietina* is the principal pathogen. *G. abietina* var. *balsamea* is an inconsequential pathogen of *Abies balsamea* (balsam fir) and *Picea* species (spruce) in eastern Canada. A third group, allied with the others on the basis of DNA sequence data, attacks *Larix* (larch) in Europe and North America. The larch colonist has been named *G. laricina*.

G. abietina var. *abietina* includes several races and ecotypes that occupy different habitats, differ in aggressiveness on particular tree species, and are distinguishable by DNA analysis. The Asian race causes damage to *Abies sachalinensis* (Todo fir) in Japan. The North American race occurs from Ontario and Minnesota to Nova Scotia and also in western Canada. It attacks primarily *Pinus contorta* (lodgepole pine) in the West and *P. banksiana* (jack pine), *P. resinosa* (red pine), and less often *P. nigra* (Austrian pine) and *P. strobus* (eastern white pine) in the East. The European race occurs on various pines and inconsequentially on *Picea abies* (Norway spruce) and other conifers throughout most of Europe. It has also been found from eastern Ontario and northern New York to New Brunswick and in Newfoundland. In that region it causes damage mainly to *Pinus resinosa*, and to *P. sylvestris* where the pathogen population builds up on *P. resinosa*. It is occasionally found on *P. nigra* and *P. strobus*. The European and North American races are both capable of causing twig dieback on numerous other coniferous species if plants are massively inoculated and grown under disease-conducive conditions. The European race is more aggressive than the North American race in most of their common host species. Two additional differentiated subgroups of *G. abietina* var. *abietina* occur in Europe: an ecotype at high latitudes in Fennoscandia on *P. contorta* and *P. sylvestris* (Scots pine), and a similar ecotype at high elevations in the Alps, most often on *P. cembra* (Swiss stone pine), *P. mugo* (mugo pine), *P. sylvestris*, and *Larix lyallii* (subalpine larch).

The remainder of this discussion deals with diseases caused by *G. abietina* var. *abietina* in North America. Two- and three-needle pines are most susceptible, five-needle pines and spruces are much less so, and all other conifers are resistant or immune. The North American race infects terminal and lateral shoots and kills seedlings and young trees up to about 2 m tall in nurseries, plantations, and natural stands. It kills low branches of larger saplings and occasionally spreads from these branches into mainstems, causing perennial cankers. The European race kills shoots and causes cankers and dieback on trees of all sizes. The cumulative effects of this damage can kill tall trees. It has destroyed many plantations of *P. resinosa* in Newfoundland, southern Quebec, and adjacent USA, but its spread southward is limited by climatic factors. Its spread is also limited by resistance in *P. banksiana*, which is the dominant pine species in boreal forests of eastern Canada.

Symptoms and signs. The North American and European races cause similar symptoms on small pines of susceptible species. Resinous lesions form primarily during host dormancy in cortical tissue of first- or second-year twigs, extend into needle bases and buds, and girdle twigs. Needle bases turn orange-brown by early spring, and the discoloration spreads toward needle tips. Symptomatic needle clusters can be pulled from twigs readily, and they drop during late spring and summer, leaving bare twigs. Buds on twigs with discolored needles do not open. Lesions in twigs can be found by cutting into the bark near the junction of green and brown foliage. The cambial region in a lesion caused by the North American race is often discolored yellow-green by a pigment secreted by the pathogen. Trunk cankers caused by the North American race are elongate, resinous, and usually have greenish discoloration. Infection by the European race on *P. banksiana* and *P. contorta* causes only tip blight (death of apical buds and adjacent needles).

The pycnidial anamorph of *G. abietina*, *Brunchorstia pinea*, develops after foliar symptoms appear. Pycnidia form singly or in clusters in the bark of killed twigs or sometimes in needle bases. They are most common at the former locations of needle clusters. They are irregular in shape with one or more fertile cavities, nearly black on the surface and white within, and up to 1 mm wide. Pycnidia that have not yet broken the bark surface may be found by shaving superficial bark. They produce conidia in a mucilaginous matrix. Conidia are individually colorless, usually four celled (some have five–seven cells), measure about 30×3 µm, and have pointed ends. Microconidia 5×1.5 µm also form in some pycnidia; their function is undetermined.

Brown cuplike apothecia appear in spring on twigs dead 1–2 years. They are about 1 mm in diameter and slightly taller than wide, with short stalks. The fertile inner surface is cream colored. The ascospores are colorless, ellipsoid, four celled with rounded ends, often slightly curved, and measure $15–22 \times 3–5$ µm. Apothecia of the North American race are common; those of the European race in North America are rare.

Disease cycle. Conidia and ascospores are both infectious, although the latter are relatively unimportant in disease buildup. Both are dispersed during wet weather in late spring and summer, conidia mainly in water and ascospores in air. Conidia may be carried many meters by mist droplets and can survive up to 20 days on dry plant surfaces. Infection occurs primarily in spring and summer, sometimes as late as October. Bracts (scales) at the bases of short shoots (needle clusters) are a common site of infection, with hyphae usually penetrating via stomata. Direct penetration of the epidermis of long and short shoots beneath bracts and bud scales has also been reported. Colonization proceeds both inter- and intracellularly, facilitated by hyphal secretion of pectolytic, cellulolytic, and proteolytic enzymes and perhaps a toxin (scleroderolide). Symptoms sometimes appear in autumn after spring inoculation but usually appear during the next spring. Sometimes the pathogen remains quiescent and causes no symptoms until the second year after inoculation. Pycnidia develop during the first and second years of symptom display and release conidia, completing the disease cycle. Sexual reproduction and completion of the pathogen's life cycle require 2 years.

Snowy winters promote damage from Scleroderris canker. *G. abietina* can grow at temperatures as low as –6°C, which facilitates host colonization while defense reactions are minimal. Symptom induction by the North American race of *G. abietina* var. *abietina* requires at least 44 days during the first winter or first plus second winters after inoculation when temperatures are in the range –6° to +5°C or snow covers tree parts that have incipient infections. Disease builds up on sites such as frost pockets and cold air drainages where plant surfaces stay wet for long periods and airborne spores become concentrated, and where snow lingers in spring. Large-scale outbreaks may occur in years with long periods of cool, moist weather in spring. Damage to new plantations on sites conducive to infection can be minimized by planting resistant species.

References: 202, 203, 338, 643, 987, 989, 990, 1014, 1213, 1461, 1462, 1595, 1596, 1683a, 1801, 1859, 2000, 2213–2216, 2232, 2451, 2473, 2980, 3025, 3026, 3055, 3634, 3685, 3989, 4121, 4122, 4465, 4466, 4485

Scleroderris canker caused by *Gremmeniella abietina* var. *abietina*.

A, B. Symptoms on *Pinus banksiana* (jack pine) caused by the North American race.

A. Death of low branches that were beneath snow in winter (QC, Aug). B. A perennial canker on a mainstem (ON, Jul).

C. A perennial canker with resin-infiltrated, dark-green-stained wood, induced by the North American race on the mainstem of a small *P. resinosa* (red pine) (ON, summer).

D, E. Symptoms on *P. resinosa* caused by the European race. C. A devastated stand of pole-size trees. E. Dieback on a branch on which, within 2 years, all growing points were killed (NY, Jun).

F. Foliar symptoms typical of those caused by either the North American or the European race: orange-brown discoloration beginning at needle bases and spreading outward, shown on *P. sylvestris* (Scots pine) (NY, Jun).

G. A resin-infiltrated canker on a small branch of *P. resinosa*, exposed by removing outer bark. Yellow-green pigment occurs in the cambial zone of the lesion (arrows) (NY, Jun).

H, I. Signs of *G. abietina* var. *abietina*. H. Magnified view of pycnidia of the *Brunchorstia* state on scars where short shoots were shed from a diseased twig of *P. sylvestris* (NY, Jun). I. Magnified view of apothecia on a *P. resinosa* twig (NY, Jun).

Photo credit: C—Canadian Forest Service Great Lakes Forestry Centre

213

Atropellis Cankers of Pines (Plate 106)

Four species of *Atropellis* (Helotiales, Dermateaceae) cause perennial cankers on branches or trunks of native and introduced *Pinus* species (pines) in North America.

A. tingens kills small branches and mainstems of at least 17 pine species. It occurs throughout the eastern half of the USA, most commonly in Appalachian and Piedmont areas, and has been found in Colorado, British Columbia, and Nova Scotia as well. It causes economic damage by disfiguring pines in Christmas tree plantations. Its hosts include *P. banksiana* (jack pine), *P. clausa* (sand pine), *P. contorta* (lodgepole pine), *P. densiflora* (Japanese red pine), *P. echinata* (shortleaf pine), *P. elliottii* (slash pine), *P. nigra* (Austrian pine), *P. pinaster* (cluster pine), *P. ponderosa* (ponderosa pine), *P. pungens* (table-mountain pine), *P. resinosa* (red pine), *P. rigida* (pitch pine), *P. serotina* (pond pine), *P. strobus* (eastern white pine), *P. sylvestris* (Scots pine), *P. taeda* (loblolly pine), and *P. virginiana* (Virginia pine).

A. piniphila, the most important pathogen of the group in the West, occurs from British Columbia and Saskatchewan to California and New Mexico, excepting the central Rocky Mountain region. It also occurs in South Dakota and occasionally in southeastern USA. It attacks branches and mainstems, and is best known for causing elongate resinous cankers on trunks of *P. contorta* and *P. ponderosa*. Trunk cankers caused by this fungus on *P. contorta* reduce the value of trees for lumber and paper pulp because the bark in cankers adheres tightly to the wood and diseased wood is darkly stained and its elasticity reduced. It requires extra cooking and bleaching during papermaking. Additional hosts of *A. piniphila* include *P. albicaulis* (whitebark pine), *P. banksiana*, *P. densiflora*, *P. echinata*, *P. monticola* (western white pine), *P. nigra*, *P. taeda*, and *P. virginiana*.

A. pinicola occurs in northwestern USA and British Columbia. It kills branches and occasionally small trees of *P. albicaulis*, *P. contorta*, *P. lambertiana* (sugar pine), *P. monticola*, *P. nigra*, *P. strobus*, and *P. sylvestris*. *A. apiculata*, known only in North Carolina and Virginia, kills twigs and branches of *P. virginiana*.

Symptoms and signs. Girdling cankers form on branches and small mainstems, and resin drips down the bark of stem cankers. All *Atropellis* species cause dark stains in wood beneath lesions. *A. apiculata* causes a chocolate brown stain, and the other species cause gray to bluish black discoloration that is detectable by cutting into the edge of a canker. These fungi also produce numerous small but conspicuous black apothecia in cankers on bark that has been dead 2–3 years.

Cankers caused by *A. tingens* originate at the bases of needle fascicles and remain small and elliptic. They enlarge rapidly in the first year but slowly thereafter and become inactive after about 10 years. Typical mature cankers on surviving branches may be only about 2 cm long. Multiple lesions girdle twigs and small branches, causing foliage to fade and turn brown in spring and early summer. Dead needle clusters glued in place by resin remain on cankers until they weather away. Cankers several years old have sunken centers and raised margins due to woundwood formation. Woundwood sometimes overgrows old inactive cankers on mainstems.

Damage caused by *A. piniphila* is most common in young dense stands of *P. contorta*. The fungus kills some trees and deforms many others through the effects of multiple cankers. Trunk cankers usually start at branch axils on stems 5–30 years old and thus have branch stubs in their centers, but some cankers start on internodes. The first symptoms are tiny brown lesions in bark, each with a resin drop on its surface. Older lesions appear as elongate depressions covered with bark and, in *P. contorta*, with dripping resin. They elongate about 5 cm per year on *P. contorta* in Alberta but much faster on *P. ponderosa* in the Southwest, where they sometimes attain lengths of 3 m. New resin appears at the margin of colonized bark as long as the canker is active. Sapwood beneath killed bark is stained bluish black. The stained sector is initially wedge shaped in cross section but becomes irregular with age. The stain does not enter previously formed heartwood, although stained wood may be surrounded by heartwood in later years. Multiple cankers on a given stem are sometimes joined internally by stained wood. Cankers caused by *A. piniphila* on *P. contorta* resemble those caused by rusts (Plate 149), but the stained xylem distinguishes Atropellis cankers.

Cankers caused by *A. pinicola* originate at leaf scars or, in the case of trunk cankers, at branch bases. They appear as elongate, flattened depressions when in smooth bark such as that of *P. monticola*. They are most common on trees less than about 13 m tall.

Atropellis species other than *A. apiculata* produce pycnidial stromata on killed bark before their apothecia appear. These stromata are black, about 1 mm across, and contain multiple chambers where spores form. The spores are noninfectious and perhaps function as gametes. They are rod shaped and measure $4–6 \times 1–1.5\ \mu m$ in *A. piniphila* and $4–8 \times 0.5–0.7\ \mu m$ in *A. pinicola* and *A. tingens*. Apothecia form soon afterward, arising initially on cankers 2 or more years old and developing annually thereafter on bark killed 2–3 years previously. On *P. contorta*, the larger the diseased stem, the longer the fruiting of *A. piniphila* is delayed, from 4–5 years after infection on small stems to 25 years or more.

Apothecia of all *Atropellis* species are cuplike and black on the outside, have a brown interior, and are 2–4 mm in diameter when moist and expanded but smaller and globose when dry. They arise singly or in groups and are present throughout the year. They can be distinguished from dark-colored apothecia of other fungi on pine bark by a simple diagnostic test. If tissue from an *Atropellis* apothecium is placed in a drop of 5% potassium hydroxide solution, it releases blue-green—or, from *A. apiculata*, chocolate brown—pigment. *Atropellis* species can be distinguished from one another by characteristics of their colorless ascospores. Those of *A. tingens* are one to four celled, usually $24–40 \times 2.5–3.5\ \mu m$, and narrowed toward one or both ends. Those of *A. piniphila* are fusiform, one or two celled, and mostly $16–28 \times 4–7\ \mu m$. Those of *A. pinicola* are nearly linear, one to six celled, and $32–63 \times 1.5–3.5\ \mu m$. Those of *A. apiculata* are one to three celled, fusiform with sharply pointed ends, and measure $20–24 \times 5–6.5\ \mu m$.

Disease cycles. The cycle of disease caused by *A. piniphila*, partly known from experiments, requires at least 2 years but is usually longer, based on the age of lesions on which apothecia develop. Ascospores are dispersed by wind, mainly in summer to early autumn. They are capable of infecting wounds or bruised bark and perhaps undamaged bark on very young stems. As a lesion develops, dark brown mycelium permeates the bark and wood and causes the typical staining. The fungus grows longitudinally through sapwood far beyond canker margins. It produces pycnidial stromata in the second year of infection (rarely in the first year) and apothecia the next year. Cycles of the other Atropellis diseases are probably similar to that caused by *A. piniphila*.

Epidemics caused by *A. piniphila* and *A. tingens* are documented. Those caused by *A. piniphila* in *P. contorta* occur in slowly growing stands on dry sites. Disease incidence tends to be positively correlated with stand density and is also related to fire history. Sapling stands growing up after a fire are in some cases uniformly infected as the result of spore dispersal from cankers on older trees that survived the fire. Outbreaks of disease caused by *A. tingens* have not been related to particular site factors. Weather patterns that favor outbreaks of these diseases are likewise unknown.

Resistance to *A. piniphila* probably occurs in *P. contorta*, based on the observation of individual trees with remarkably fewer and smaller cankers than occur on their associates in severely damaged stands. No information is available about possible resistance of other pines to *Atropellis* species.

References: 197, 461, 963, 1213, 1614, 1683, 1751, 2005, 2356, 2357, 2829, 3254, 3828, 4278

A–E. Branch cankers on *Pinus nigra* (Austrian pine) caused by *Atropellis tingens*. A. Disease in low branches has destroyed the value of a tree in a Christmas tree plantation. B, C. Cankers that arose at bases of needle fascicles have encircled this branch. The bluish gray to black stain in diseased bark and wood is diagnostic. D, E. Black apothecia 2–3 mm across, in clusters on bark of cankers 2 or more years old (VA, May).

F–H. Atropellis canker of *P. contorta* (lodgepole pine) caused by *A. piniphila*. F. Resin dripping profusely from an old trunk canker (AB, summer). G. Bark in a canker on a small mainstem is darkened by apothecia and underlain by darkly stained wood (BC, Jul). H. Cross section of a small stem with stained wood (AB, summer).

Photo credits: F, H—J. Hopkins; G—E. A. Allen

215

Lachnellula Cankers of Conifers and Strumella Canker of Hardwoods (Plate 107)

Lachnellula Cankers

These cankers are, with one exception, inconsequential disorders of conifer twigs, caused by opportunistic fungi that produce small, usually white apothecia on dead parts. European larch canker, caused by *Lachnellula willkommii* (Helotiales, Hyaloscyphaceae), is the damaging exception. The first North American record of it came in 1927 from Massachusetts, where it was eradicated. In the 1980s, and 1990s, however, it was found on *Larix laricina* (eastern larch) in the Canadian Maritime Provinces and Maine. Incidence of infection averaged 59% of trees in a large sample of natural *L. laricina* stands in coastal New Brunswick. The disease in eastern North America, as in Europe, is most common and severe in coastal areas with oceanic climate, abundant moisture, and moderate winter temperatures. It also occurs in China.

All *Larix* species as well as *Pseudolarix amabilis* (golden-larch) are at least somewhat susceptible. *L. decidua* (European larch), *L. laricina*, *L. occidentalis* (western larch), and *L. sibirica* (Siberian larch) are rated highly susceptible; *L. gmelinii* (Dahurian larch) and *L. kaempferi* (Japanese larch) are less so. The hybrids *L. ×marschlinsii* (Dunkeld larch) and *L. ×leptoeuropaea* are resistant.

Symptoms and signs. Single or multiple cankers occur on branches and mainstems, killing stems up to 10 cm in diameter. The cankers are perennial and expand primarily during host dormancy. They begin as circular to elliptic depressions in the bark of small stems, nearly always surrounding a dead dwarf shoot or twig. Resin exudes from recently infected bark and established cankers. The stem appears swollen at the site of a canker because growth is stimulated adjacent to it. Distal parts of girdled stems die in spring, or the needles turn yellow prematurely in late summer. Old cankers are rough, dark, and resinous, with concentric woundwood ridges and cracked bark at the margins. Dead bark eventually falls from the center of a canker, but the wood remains more or less sound. Many cankers cease enlargement after a few years, apparently because competitive microorganisms displace *Lachnellula willkommii*.

Apothecia of *L. willkommii* form on cankers and are visible throughout the year. They are 1–6 mm broad, cup shaped, hairy, and white with an orange-yellow to pale buff fertile layer within. They discharge ascospores to the air throughout the year when moistened by rain. The ascospores are colorless, single celled, oblong with somewhat pointed ends, and measure 15–28 × 6–9 μm. Apothecia are preceded in spring by small yellowish white stromata that produce colorless spermatia 2–8 × 1–2 μm.

Disease cycle. Ascospores cause infections, probably in autumn. Sites of infection include dwarf shoots and insect feeding injuries, notably those made by the weevil *Hylobius abietis* in Europe. Moribund dwarf shoots in which the periderm is absent or only partially formed are suspected to be important sites of infection. Frost injury may aggravate damage by *L. willkommii* but is not required for it. Infection progresses slowly, the first symptoms appearing 1–4 years after inoculation. Apothecia appear on lesions 1 or more years after infection, completing the cycle.

Many more *Lachnellula* species occur on conifers. Those known to be pathogenic, and hosts on which they apparently cause cankers or twig dieback, are *L. agassizii* on *Abies balsamea* (balsam fir), *A. fraseri* (Fraser fir), and *Pinus strobus* (eastern white pine); *L. arida* and *L. gallica* on *A. balsamea;* and *L. laricis* and *L. occidentalis* on *Larix laricina*. *Lachnellula pini* is common on *P. sylvestris* (Scots pine) in northern Europe. Each of these fungi occurs as a saprobe on several additional conifers.
References: 185, 392, 561, 562, 949, 1213, 1437, 1438, 1859, 1875, 2418, 2420, 2458, 2915, 2955, 3065, 3701, 3942–3945, 4435, 4477–4480

Strumella Canker

This perennial canker is common on *Quercus* species (oaks) and sometimes occurs on other hardwoods. The causal fungus is *Urnula craterium* (Pezizales, Sarcosomataceae). Its conidial state is *Conoplea globosa* (syn. *Strumella coryneoidea*), a hyphomycete. The pathogen, native to North America, occurs from Manitoba and Missouri to northern New England and North Carolina, and in the Pacific Northwest. It has been found also in northern Europe and Japan. Strumella canker occurs commonly on *Q. coccinea* (scarlet oak), *Q. ilicifolia* (bear oak), *Q. rubra* (northern red oak), and *Q. velutina* (black oak), and less commonly on a dozen or so other oaks of both the red oak and white oak groups. The disease has also been found on species of *Acer* (maple), *Carya* (hickory), *Castanea* (chestnut), *Fagus* (beech), *Nyssa* (sour-gum), *Ostrya* (hop-hornbeam), and *Tilia* (basswood), and the causal fungus has been noted on additional hardwoods. Strumella canker seldom affects more than 2–3% of the oaks in a forest stand but has been reported to cause death or severe damage to up to 63% of trees in some young oak plantations.

Symptoms and signs. The cankers develop within 4 m above the ground, mainly on trees less than 25 years old. Trees less than about 10 cm in diameter may be girdled, and pole-size stems may break because of decay at cankers. Infection begins in branch axils or possibly on branches. The first symptom is yellowish brown discoloration of bark around a dead branch or branch stub. Typical perennial cankers enlarge slowly (1–3 cm in width per year), becoming irregular depressed areas with concentric woundwood ridges. The parasite kills bark during host dormancy, and the tree halts this advance and produces callus and woundwood around the lesion each growing season. Diseased trunks are often flattened, widened, and concave at the locations of cankers because growth is stimulated near the lateral margins of the lesions. Strumella cankers up to 1.5 m long and 60 cm wide have been recorded. Killed bark remains attached to the wood for several years but eventually decays and sloughs off, revealing a yellowish decay caused by the same fungus. Adventitious sprouts often develop just below large cankers.

Dark brown to black nodules of sterile mycelium a few millimeters in length and width form on the bark during the first year after its death and persist for several years. White mycelial strands or mats can be seen by cutting into dead bark near a canker margin. Sometimes dark brown to black sporodochia with great numbers of tiny brown spores form in cankers, but sporulation is more often restricted to killed sprouts or branches or to a girdled trunk in and above a canker. The sporodochia are rounded and 1–3 mm in diameter. The spores are spiny, irregularly globose to pear shaped, and measure 6.7–8.1 × 4.7–5.8 μm. They are windborne but have not been observed to germinate. Their role is unknown.

Cuplike, stalked apothecia of *U. craterium* form in early spring on decaying wood of stumps and fallen trunks or limbs. The apothecia are 3–4 cm wide and 4–6 cm deep, grayish on the surface, and black inside. The colorless ascospores, which are windborne and germinate readily, are one celled, smooth, and measure 12–14 × 24–35 μm.

The disease cycle of Strumella canker has not been elucidated.
References: 185, 358, 461, 1099, 1298, 1572, 1765, 1807, 1808, 2085, 4413, 4435

A–C. European larch canker, caused by *Lachnellula willkommii*, on *Larix laricina* (eastern larch). A. Dead and dying branches on an affected tree (NB, Sep). B. Swelling, resin exudation, and tiny white apothecia associated with a branch canker (ME, Jul). C. Close view of a canker with apothecia (NB, Sep).

D, E. A canker with apothecia of *L. agassizii* on an *Abies fraseri* (Fraser fir) seedling. In magnified view (E) the apothecia have an orange-yellow interior and hairy white surface (NY, Aug).

F–I. Strumella canker of *Quercus rubra* (red oak). F. A canker girdling a pole-size tree. Concentric woundwood ridges, adhering bark, and a branch stub in the center are typical. G. A canker that has become inactive around most of its margin. H. Transverse section through a canker showing decayed wood and bark, successive dead woundwood ridges, and the canker margin in bark at the upper right (arrow). I. Close view of sterile nodules of fungal tissue on bark in a canker (NH, Mar).

J. Apothecia of *Urnula craterium* on a fallen *Quercus* branch (NH, spring).

Photo credits: A—C. L. Burlock; B—D. A. Stark; C—K. J. Harrison; J—A. L. Shigo

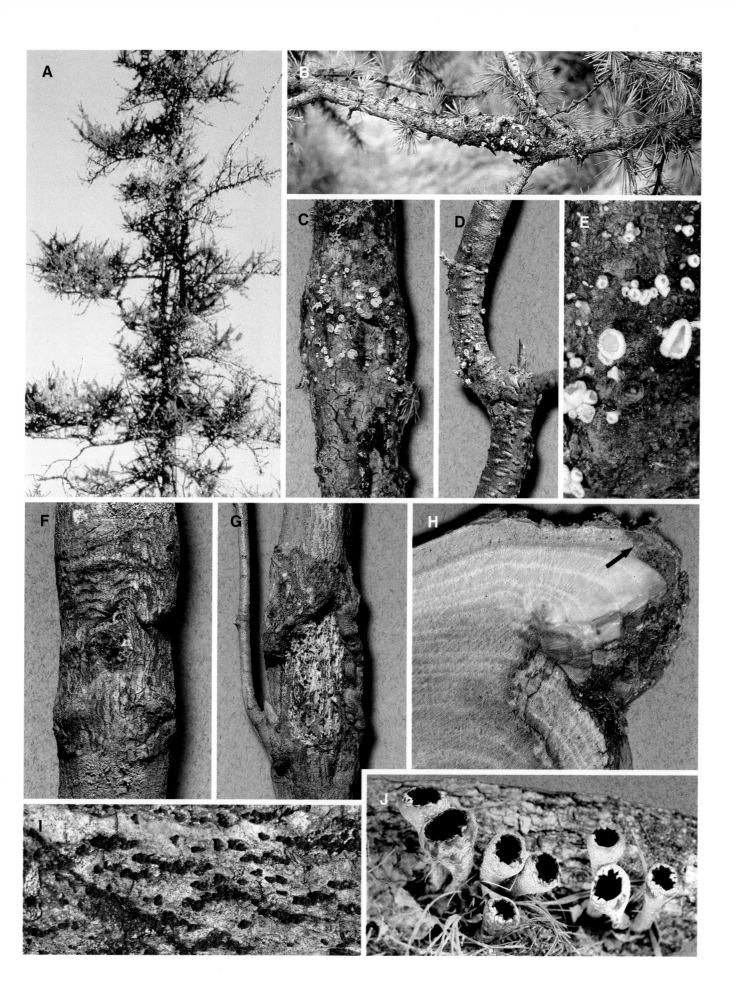

217

Charcoal Root Rot and Fusarium Root Rots (Plate 108)

Charcoal Root Rot

Macrophomina phaseolina (syn. *Sclerotium bataticola*), a coelomycete, causes charcoal root rot. The disease affects more than 300 species, herbaceous and woody, including monocots, dicots, and gymnosperms. It occurs globally in warm-temperate and tropical regions, especially in semiarid and arid places. Damage occurs in nurseries, agricultural fields, and uncultivated soils. Trees are damaged by *M. phaseolina* while in the seedling stage in nurseries, where this pathogen is often accompanied by *Fusarium* species and other root killers. Susceptible woody plants and plant groups include *Abies* (fir), some species of Arecaceae (palms), *Actinidia deliciosa* (kiwi-fruit), *Cedrus* (true cedar), *Eucalyptus*, *Gingko biloba*, *Picea engelmannii* (Engelmann spruce), *Pinus* (pine), *Prunus armeniaca* (apricot), *Pseudotsuga menziesii* (Douglas-fir), and *Sequoiadendron giganteum* (giant sequoia). Reports from California indicate that *Pinus jeffreyi* and *P. ponderosa* (Jeffrey and ponderosa pines) are somewhat resistant, and *Calocedrus decurrens* (incense cedar) is highly so.

Charcoal root rot is named for its primary symptoms: necrosis and blackening of roots that lead in severe cases to stunting, chlorosis, and death. Infection progresses up the root system, resulting in distinctive roughening, loosening, and darkening of bark on affected parts of the taproot and major laterals. The woody cylinder, although dead, remains light colored. Bark roughening results from production of abnormally abundant cork and phelloderm cells. The fungus secretes several cell-wall-degrading enzymes that contribute to necrosis and decay. It then produces numerous tiny black sclerotia (microsclerotia), which are resting structures, beneath the loose periderm on roots and the lower part of the stem. Microsclerotia are often so numerous that they blacken the surface of the cortex. They remain in soil after dead roots decay, germinate when stimulated by root exudates as new roots grow close to them, and cause new infections. The microsclerotia are so heat tolerant that solar heating of soil under polyethylene does not fully eliminate them. Some strains of *M. phaseolina* also produce pycnidia and conidia, but these are not reliably present for diagnostic purposes.

Although infection occurs during periods of moderate temperature, charcoal root rot symptoms intensify during hot summer weather and are exacerbated by drought. The pathogen can cause severe damage at soil temperatures above 15°C. It grows most rapidly at 28–32°C and causes most damage where the temperature of infested soil rises above 30°C. It can grow and produce large quantities of microsclerotia under relatively dry conditions. Infections that begin in conifer nurseries sometimes lead to failure of young trees after outplanting.

References: 773, 800, 950–952, 1132, 1696, 1697, 2414, 2532, 2629, 2749, 2900, 3359, 3546, 3573, 3733, 3744

Fusarium Root Rots

Fusarium species, hyphomycetes, are significant pathogens in forest nurseries, causing seed rot, damping-off, hypocotyl rot, cotyledon blight, stem rot, and root rot. Coniferous seedlings are commonly damaged by these fungi in North America and Europe; angiosperms are significantly affected elsewhere. Damage is most common in soil beds, although container-grown seedlings are also affected. Most of the causal fungi are anamorphs of ascomycetes in the genera *Gibberella* and *Haematonectria* (Hypocreales, Nectriaceae), but some fusaria, notably *F. oxysporum*, lack known teleomorphs. Pathogens of tree seedlings include *Fusarium acuminatum* (anamorph of *Gibberella acuminata*), *F. avenaceum* (*G. avenacea*), *F. circinatum* (*G. circinata*), *F. oxysporum*, *F. proliferatum* (*G. fujikoroi* complex), *F. sambucinum* (*G. pulicaris*), and strains of the *F. solani* complex (some of which are anamorphic *Haematonectria haematococca*). *F. oxysporum* and *F. solani* are illustrated in Figs. 123 and 92, respectively. The diseases mentioned above each tend to be caused by particular fungal strains that display specialized behavior. This specialization occurs even within pathogen species. The pathogens are dispersed by air and with soil, diseased plants, and seed.

Fusarium root rots of coniferous seedlings in North America are caused most often by strains of *F. oxysporum* and the *F. solani* complex, often acting in concert with other pathogenic fungi or with nematodes. Therefore the symptoms (stunting, chlorosis, root necrosis, death) are responses to combinations of organisms and are conditioned by local environment and cultural practices. Many coniferous species are affected, primarily in the first growing season, but root rot and death of seedlings up to 3 years old have occurred in some nurseries in eastern USA. Fortunately, most forest planting sites are unfavorable for root-infecting fusaria that cause damage in nurseries, and those carried from nurseries to forest sites do not persist.

Root rot of *Pseudotsuga menziesii* (Douglas-fir) seedlings caused by *F. oxysporum*, featured here, is important in British Columbia and northwestern USA. Infection occurs near root tips during the first 6 weeks after seed is sown, and hyphae advance both inter- and intracellularly, killing the taproot and lateral roots soon after they emerge. Adventitious roots may form close to the root collar after death of the taproot. Seedlings with severe root rot either die during hot weather in summer or survive stunted until winter and then perish as the result of frost heaving or infection by other opportunistic pathogens during mild weather in late winter. Some strains of the pathogen infect or advance into the hypocotyl and kill the seedling during its first summer. Mortality of *Fusarium*-infected seedlings seems to be triggered by transient water stress on hot summer days. Diseased seedlings that survive to the second growing season usually outgrow the effects of first-year disease, although their roots still harbor the pathogen. Many such seedlings are culled, however, because they do not attain sufficient size for outplanting with others of the same age.

F. oxysporum produces chlamydospores in dead root and hypocotyl tissue, and elongate, curved, multicelled conidia with tapered ends on dead stems near the soil line. The conidia are dispersed locally but do not cause much disease increase in the same crop. Some cells in conidia often convert to chlamydospores by developing thick walls and rounded contours. Chlamydospores persist in decaying root fragments and in soil, and germinate in the presence of host roots. Persistence in soil, as well as introduction with seed, contaminated implements, and apparently as airborne conidia, all ensure presence of the pathogen in the next crop. Other fusaria that attack conifer seedlings have similar habits, except that some species do not produce chlamydospores.

Fusarium root rots are promoted by urea fertilizer or unrotted organic amendments. Therefore the diseases are managed in part by regulating nitrogen sources and applying only the minimum necessary for seedling growth during the first year. Crop rotations that include leaving soil fallow for a year have also been helpful. Nonpathogenic fusaria that associate with seedling roots in nursery soils have shown promise for biological control of Fusarium root rots, as have strains of *Trichoderma harzianum* and certain ectomycorrhizal fungi. Mechanisms of biological control that have been demonstrated in conifer seedlings include competition for substrate, antagonism through antibiosis or parasitism, and induction of resistance in the plants. Biological control is more feasible for seedlings grown in containers than for those in soil beds.

References: 136, 137, 148, 415, 416, 419, 443, 547, 685, 686, 800, 1007, 1031, 1096, 1347, 1487, 1991, 1996, 2128, 2602, 2819, 2889, 2890, 3345, 3385, 3641, 3891, 3899, 3909

A–C. Symptoms and signs of charcoal root rot, caused by *Macrophomina phaseolina*. A. Dead and dying 2-year-old *Abies magnifica* (red fir) in a nursery seedbed (CA, Aug). B. Microsclerotia beneath the periderm on a stem base of *Euphorbia lathyris* (caper spurge) killed by the fungus. Areas blackened by aggregated microsclerotia account for the disease name (AZ, Jul). C. Microsclerotia on the dead hypocotyl of a *Pinus elliottii* (slash pine) seedling (GA, Jan).

D–G. Fusarium root rot of first-year seedlings of *Pseudotsuga menziesii* (Douglas-fir), caused by *Fusarium oxysporum*. D. Dead and distressed seedlings. Curling of top needles indicates distress but is not diagnostic. E. Part of the hypocotyl of a killed seedling on which pinkish sporodochia and white mycelium of *F. oxysporum* have formed at the soil line. F. Seedlings stunted by severe root rot with a normal seedling of the same age for comparison. G. Enlarged view of the severely damaged root system of a seedling. Major lateral roots are absent, and short roots as well as the taproot are decayed (all WA, Aug).

Photo credits: A—J. T. Kliejunas; C—C. E. Cordell

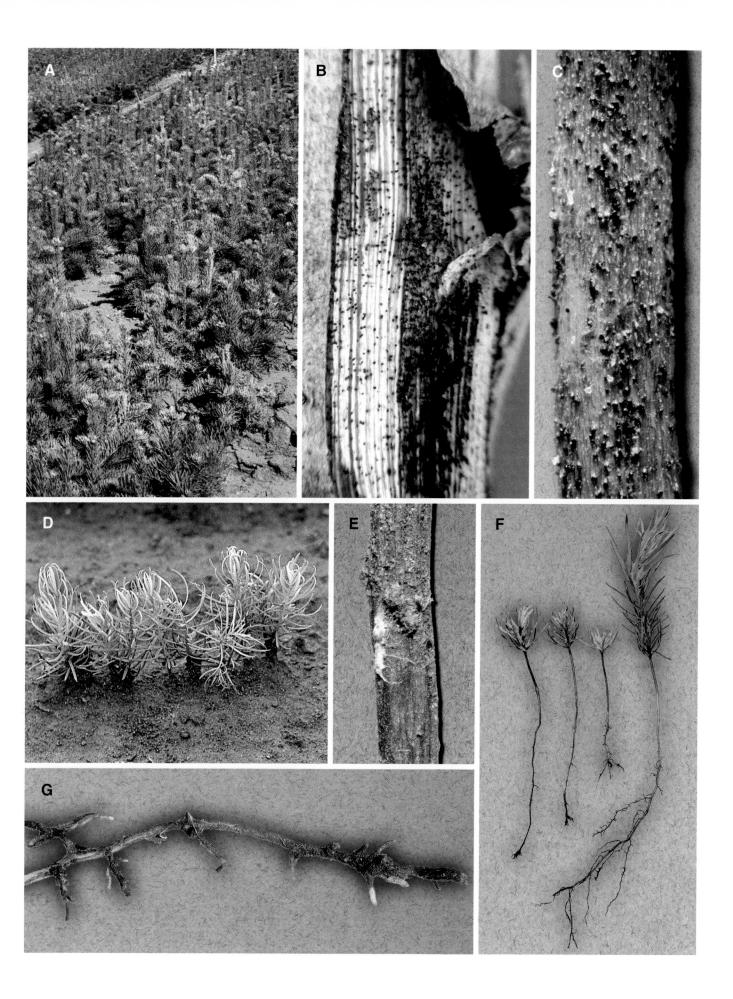

Root Rots and Blights Caused by *Cylindrocladium* and *Cylindrocladiella* (Plate 109)

Cylindrocladium and *Cylindrocladiella* species, hyphomycetes, attack herbaceous and woody plants in more than 130 genera around the world, especially in tropical and subtropical regions. These fungi are anamorphs of *Calonectria* and *Nectricladiella* species, respectively (Hypocreales, Nectriaceae). Food and fiber crops, ornamentals, and seedlings and small saplings of trees and shrubs are affected. Attacks occur in cultivated fields, greenhouses, nurseries, orchards, plantations, and forest stands. Damage to woody plants in North America is usually confined to nurseries. About 35 *Cylindrocladium* species are recognized, of which about 25 are anamorphs of known *Calonectria* species. The *Nectricladiella-Cylindrocladiella* group comprises fewer than 10 species. Some fungi in these groups, such as *Calonectria avesiculata* on *Ilex* (holly) and *Leucothoe*, cause damage mainly to leaves and young stems, but many species have the ability to kill roots, stems, leaves, and sometimes fruit.

Host lists of individual *Cylindrocladium* and *Cylindrocladiella* species overlap broadly. An abridged host genus list for these fungi collectively includes the following:

Abies, Acca, Acer, Amelanchier, Araucaria, Arctostaphylos, Asimina, Bauhinia, Bixa, Bucida, Buddleja, Buxus, Callistemon, Calluna, Camellia, Carica, Carya, Cassia, Castanea, Cedrus, Ceratonia, Cercis, Chamaedorea, Chionanthus, Cissus, Citrus, Cocos, Conocarpus, Cornus, Cryptomeria, Cunninghamia, Cycas, Diospyros, Dypsis, Elaeis, Erica, Eucalyptus, Euonymus, Euphorbia, Ficus, Fuchsia, Gardenia, Howea, Ilex, Jasminum, Juglans, Juniperus, Lagerstroemia, Larix, Leucothoe, Ligustrum, Liquidambar, Liriodendron, Magnolia, Mahonia, Malus, Melaleuca, Morella, Morus, Myrtus, Nandina, Nerium, Osmanthus, Persea, Phellodendron, Photinia, Picea, Pieris, Pinus, Pittosporum, Platanus, Platycladus, Populus, Prosopis, Prunus, Pseudotsuga, Quercus, Rhododendron, Rhus, Ribes, Robinia, Rosa, Rubus, Salix, Sambucus, Swietenia, Syagrus, Syringa, Taxus, Thuja, Tsuga, Vaccinium, Ulmus, Vitis, and *Washingtonia.*

The most widespread and important *Cylindrocladium* species on woody plants in North America are *Cy. scoparium* (teleomorph *Ca. morganii*) and *Cy. floridanum* (a species complex, teleomorph *Ca. kyotensis*). These fungi have hosts in at least 70 and 30 genera, respectively; attack various plant parts; and occur in both warm and cool regions. They are widespread in the eastern half of North America and in scattered other places around the world. *Cy. scoparium* has been found not only in nurseries but also associated with root rot and decline of *Prunus persica* (peach) trees in orchards, pole-size *Liriodendron tulipifera* (tuliptree) in a plantation, and *Liquidambar styraciflua* (sweetgum) in a natural forest stand. Additional species encountered in the USA are *Cy. colhounii* (teleomorph *Ca. colhounii*), *Cy. parasiticum* (syn. *Cy. crotalariae*, teleomorph *Ca. ilicicola*), and *Cy. theae* (teleomorph *Ca. indusiata*).

Symptoms and signs. Symptoms vary with pathogen, host, and environmental conditions but include damping-off, root rot, hypocotyl rot, cutting rot, leaf spots, foliar and shoot blights, lesions on green stems, twig dieback, wilting, and death. Root infections are characterized by multiple dark brown to nearly black lesions, often with longitudinal cracks, on young roots. Coalescing lesions may girdle the taproot or root collar. Severe root disease leads to stunting, yellowing, wilting, and death. If damage is sublethal, new roots may proliferate above the killed part of a root system. Stem infections of both conifers and broadleaved plants arise at leaf bases and may girdle small stems. Diseased conifer needles turn yellow and then, in some pines, red-brown. Dark lesions form in leaves of broadleaved plants. Leaf blight leads to defoliation.

Wefts of brown mycelium and/or tiny white tufts of conidia may be produced on lesions on leaves or stems under warm, humid conditions. The conidia are colorless, long-cylindric, and two to several celled, the septation varying among species. (Conidia are two celled in *Cy. floridanum* and *Cy. scoparium.*) Conidiophores of *Cylindrocladium* species are unusual in that one or more branches elongate markedly and produce sterile swellings called vesicles at their tips. Vesicle shape is one of the diagnostic characters for specific identification. *Cylindrocladium* species also produce brown irregularly shaped microsclerotia, 40–200 µm in the longest dimension, in killed plant parts. As a rule, microsclerotia can be detected only by microscopic examination. The *Calonectria* states of these fungi produce orange to yellow-orange warty perithecia, 0.25–0.45 mm in diameter and height, on killed leaves and bark

under moist conditions. Blobs of colorless ascospores extrude from perithecia when they are wet. (Note: the teleomorph of *Cy. scoparium* is known only from cultures.)

Disease cycles. Cylindrocladium species that affect roots persist in plant debris and soil as microsclerotia. These structures tolerate environmental extremes and are resistant to microorganisms. Microsclerotia of *Cy. scoparium* have been known to survive for 7 years in fallow soil. Microsclerotia germinate when stimulated by root exudates and produce mycelium that infects roots. New microsclerotia form in diseased tissues and are later dispersed with soil, plants, and water. Conidia and sometimes ascospores are dispersed by water and infect roots or aerial parts. Ascospores of most *Calonectria* species are probably much less important than microsclerotia and conidia as inoculum. Conidia are capable of causing secondary cycles of disease. Germ tubes from conidia penetrate leaves via stomata. Penetration and colonization of stems and root collars have not been described in detail. Plants that have been studied seem most susceptible while growing rapidly. Some *Cylindrocladium* species produce toxic metabolites in culture, which indicates a possible role of toxins in symptom induction. Infection of aboveground parts is fostered by warm, moist conditions and close spacing of plants. *Cylindrocladium* species grow most rapidly at 24–28°C.

Cy. scoparium (and probably related species as well) is capable of competitive saprobic colonization of plant materials in soil, and it can persist in association with roots of plants that show no symptoms. A high population of *Cy. floridanum* was found in the soil and associated with clover roots in an abandoned nursery in Minnesota 14 years after production of tree seedlings ceased.

References: 204, 433, 563, 800–802, 824, 848, 855, 1153, 1185, 1289, 1457, 1859, 1967, 1993, 2034, 2167, 2325–2327, 2590, 3003, 3113, 3337, 3341, 3417, 3473, 3474, 3496, 3497, 3755, 3996, 4033, 4156, 4471

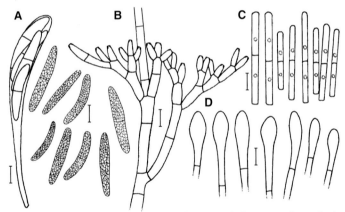

Figure 109. Calonectria morganii and its conidial anamorph, *Cylindrocladium scoparium.* A. Ascus and ascospores. B. A conidiophore with an elongate branch extending upward toward a vesicle that is beyond view. C. Conidia. D. Vesicles that form on long branches of conidiophores. Scale bars = 10 µm. Adapted from reference 848 by permission of the American Phytopathological Society.

A, B. Wilting and dieback of *Rhododendron* sp. (azalea) caused by *Cylindrocladiella parva.* Shoots on diseased branches wilt, shoot bases turn brown, and buds remain closed (NY, greenhouse).

C. Root rot of a *Liriodendron tulipifera* (tuliptree) seedling caused by *Cylindrocladium scoparium.* Dark brown lesions on otherwise light-colored roots are typical (MS, greenhouse).

D–F. Foliar symptoms associated with root rot caused by *C. scoparium* in *Pinus resinosa* (red pine) and *P. strobus* (eastern white pine) seedlings. D, E. Diseased *P. resinosa* in nursery beds. The healthy-appearing seedlings at left in D were growing in fungicide-treated soil. F. Chlorosis associated with root rot in *P. strobus* seedlings (WI, Sep–Oct).

G–J. Blight of *Pinus palustris* (longleaf pine) seedlings caused by *Calonectria kyotensis.* G, H, Needle blight. I, J. Perithecia on needle bases of a killed seedling (FL, Sep–Nov).

Photo credits: D–F—W. G. Thies; G–J—E. L. Barnard

221

Phymatotrichum and Thielaviopsis Root Rots (Plate 110)

Phymatotrichum Root Rot

Phymatotrichopsis omnivora (syn. *Phymatotrichum omnivorum*), a hyphomycete, causes Phymatotrichum root rot, also known as Texas root rot, in the region extending from western Louisiana and Arkansas to southern California and Nevada, also in Utah and Mexico. This pathogen, which attacks more than 2000 plant species, causes significant economic damage to crop plants, especially cotton, and is lethal to many species of woody plants in nurseries, orchards, landscapes, and shelterbelts. Genera with highly susceptible woody plants include *Acacia, Ailanthus, Broussonetia, Carya, Catalpa, Cydonia, Euonymus, Ficus, Gingko, Gleditsia, Grevillea, Ligustrum, Malus, Melia, Morus, Nandina, Pinus, Pittosporum, Populus, Prunus, Pyrus, Quercus, Robinia, Rosa, Salix, Schinus, Syringa, Ulmus,* and *Vitis.*

Symptoms and signs. The disease tends to occur in expanding patches that involve many plants. Multiple coalescing necrotic lesions develop on roots of all sizes. The cortex and phloem become brown and mushy and are eventually decomposed by soil saprobes. Infection at the root collar or on the taproot within the upper 30 cm of soil is usually lethal. Living roots at the edge of a patch of diseased plants typically have small lesions characteristic of early stages of the disease. Only after root damage is extensive do symptoms appear on aboveground parts. Leaves may then turn yellowish or bronze before they wither. Small highly susceptible plants often die within several weeks after first infection. Diseased trees, according to their level of susceptibility, may live for several years and show stunted growth, yellowing, and dieback for a year or longer before death. Resistant plants, although infected, may persist indefinitely.

Brown mycelial strands grow on diseased roots and tend to form networks on the surfaces. Each strand consists of a large central hypha surrounded by many small interwoven hyphae. Viewed microscopically, the strands possess diagnostic cross-shaped brown hyphal structures (cruciform hyphae) attached perpendicular to the strand axis. The tip and arms of each cross are pointed (acicular). *P. omnivora* also produces sclerotia (resting structures) that are resistant to environmental extremes and microbial attack. These form along mycelial strands in soil at depths to 2 m or more. The sclerotia are 1–3 mm in diameter, initially light colored but eventually black. Under warm, moist conditions the pathogen may produce on the soil surface around the stem of a diseased plant a buff-colored mat of mycelium bearing conidiophores and conidia. Such mats are more common beneath the canopies of field crops than around the bases of woody plants. The function of the conidia in nature has not been learned.

Disease cycle. Sclerotia of *P. omnivora* can survive several years in the absence of hosts. Sclerotia produce new hyphae and mycelial strands when stimulated by root exudates in moist soil. The tips of mycelial strands differentiate into individual hyphae that penetrate roots through lenticels, wounds, and other breaks in the periderm. Host cells are killed in advance of hyphae, apparently by a heat-labile toxin. The fungus also secretes cell-wall-degrading enzymes. It spreads as mycelial strands along roots and through soil at rates up to 9 m per year. In soils conducive to its survival *P. omnivora* inhabits the roots of many kinds of plants that do not exhibit aboveground symptoms. Mycelium growing from killed tissue produces more sclerotia.

P. omnivora is favored by warm, alkaline (pH 7.2–8.5), calcareous soils with high content of montmorillonite clay, conditions that in North America exist only in southwestern USA and in Mexico. Acid soils or those with high sodium content inhibit the fungus. It grows and produces sclerotia at 15–35°C, most rapidly at 28°C. The sclerotia can be killed by solar heating of moist soil beneath clear plastic for several weeks.

Monocots, some dicots native to the Southwest, and assorted plants from elsewhere are resistant to *P. omnivora*. These include the following:

Acacia farnesiana, Agave americana, Arecaceae (palms), *Bauhinia variegata, Caragana arborescens, Bambusa* species, *Carya* species except *C. illinoinensis, Celtis* species, *Chilopsis linearis, Clematis, Elaeagnus angustifolia, Eucalyptus camaldulensis, Gymnocladus dioica, Ilex* species, *Juniperus* species, *Maclura pomifera, Malvaviscus arboreus, Nerium oleander, Paulownia tomentosa, Philadelphus* species, *Phyllostachys* species, *Platanus wrightii, Prosopis glandulosa, Punica granatum, Pyracantha atalantioides, Sambucus cerulea, Sapindus saponaria, Styphnolobium japonicum,* and *Yucca* species.
References: 32, 511, 1941, 2399, 3017, 3888, 3969, 4253, 4444

Thielaviopsis Root Rot

Thielaviopsis basicola (syn. *Chalara elegans*) causes Thielaviopsis root rot, also known as black root rot. The fungus is a hyphomycete related by DNA sequence similarity to the ascomycete genus *Ceratocystis*. It occurs in soils and roots all around the temperate zones, associated with plants in more than 130 genera, mainly herbaceous species. The majority of its plant associations are apparently inconsequential, but it kills rootlets of many plants and sometimes infects graft unions close to the soil. It grows on recently cut surfaces of grafted parts, producing a gray mat of mycelium and conidiophores, apparently killing enough tissue to prevent union. *T. basicola* has been blamed for graft failure in *Camellia japonica* (common camellia), *Lonicera* (honeysuckle), and *Paeonia suffruticosa* (tree peony).

Relatively few woody plants are damaged by Thielaviopsis root rot. *Ilex* species (hollies) in nurseries and landscapes may sustain growth loss and exhibit sparse foliage, poor foliar color, and dieback after the death of fibrous roots. Hollies affected include *I. crenata* (Japanese holly), *I. glabra* (inkberry), *I.* ×*meserveae* (Meserve hybrid or blue holly), and *I. opaca* (American holly). Other woody hosts that sustain damage include *Catalpa speciosa* (northern catalpa), *Catharanthus roseus* (periwinkle), *Citrus, Euphorbia pulcherrima* (poinsettia), *Daphne cneorum* (garland-flower), *Olea europaea* (olive), *Robinia pseudoacacia* (black locust), and *Ulmus americana* (American elm). The fungus grows intracellularly and causes small dark lesions that may coalesce and blacken young roots. Colorless conidia and brown cylindrical aleurioconidia (formerly called chlamydospores) form on the lesions, and aleurioconidia also form within lesions. They serve as survival structures in soil. When stimulated by root exudates, aleurioconidia germinate and may cause new infections. *T. basicola* spreads with infested plants, occurs in peat moss, and can be transmitted aerially by shore flies (Diptera: Ephydridae) under moist conditions such as in greenhouses.

References: 799, 984, 1194, 1348, 1739, 1740, 1859, 2220, 2221, 2612, 2654, 2780, 2981, 2982, 3292, 3829, 3901, 4350, 4372, 4475

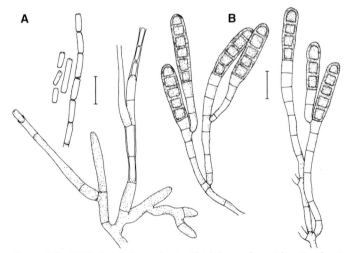

Figure 110. Thielaviopsis basicola. A. Phialides and conidia. B. Aleurioconidia. Scale bars = 20 μm. Adapted from reference 984 by permission of Elsevier.

A. Death of *Citrus aurantium* (sour orange) trees along a residential street typifies the patchy occurrence of damage caused by *Phymatotrichopsis omnivora* (AZ, Jul).

B–D. Symptoms and signs of Phymatotrichum root rot on woody taproots of *Gossypium* (cotton). B. A diseased taproot with the periderm shaved off to reveal numerous small lesions that would have coalesced. C. A taproot in which the pathogen has advanced nearly to the soil line; killed tissue is brown. D. Brown strands of intertwined hyphae on the root surface (AZ, Jul).

E–G. Thielaviopsis root rot of *Ilex crenata* (Japanese holly). E. A diseased plant, its growth suppressed by root infection. F. Close view of the root system. G. Numerous brown lesions revealed in a magnified view of fibrous roots of the plant shown in F (VA, greenhouse).

223

Xylaria and Rosellinia Root Rots (Plate 111)

Xylaria Root Rots

Xylaria (Sphaeriales, Xylariaceae) is a large, globally distributed genus of wood-decaying ascomycetes. They occur as endophytes in leaves and stems and as decomposers of stem and root wood. Two of the more than 25 species in North America are recognized pathogens. *X. mali* and *X. polymorpha* cause black root rot of *Malus* (apple, crabapple) and various other angiosperm trees. Black root rot is named for a black stromal sheath that develops on decaying roots. *X. mali* and *X. polymorpha* each cause an off-white decay of wood beneath the black sheath, and both produce clusters of fingerlike to irregularly club-shaped fruit bodies (perithecial stromata) that originate on the butt or major roots and extend a few centimeters above the soil line. These structures are the basis of the common name of *X. polymorpha*: dead man's fingers.

In common with other members of their genus, *X. mali* and *X. polymorpha* colonize diverse substrates. *X. mali*, the more aggressive parasite, is known only in eastern and south-central USA. It is a proven pathogen of *Acer platanoides* (Norway maple), *Gleditsia triacanthos* (honeylocust), *Malus pumila* (apple), *Prunus mahaleb* (Mahaleb cherry), *Pyrus communis* (common pear), *P. pyrifolia* (Chinese pear), and *Ulmus americana* (American elm). It also occurs on other *Malus*, *Prunus*, and *Pyrus* species.

X. polymorpha occurs in Europe, Africa, India, and Japan as well as North America. It is common from the Gulf Coast states northward to the Great Lakes region but is rare in the West. It is a proven pathogen only of *M. pumila* but is also found on *Acer* (maple), *Betula* (birch), *Carya* (hickory), *Cercis* (redbud), *Citrus*, *Cladrastis* (yellowwood), *Fagus* (beech), *Gleditsia* (honeylocust), *Juglans* (butternut, walnut), *Platanus* (plane), *Poncirus* (trifoliate orange), *Quercus* (oak), *Ribes* (currant), *Robinia* (locust), *Sassafras*, *Ulmus* (elm), and *Vitis* (grapevine), among other plants. This fungus has often been confused with other large xylarias.

Aboveground symptoms of Xylaria root rots are those of stress and decline: slow growth, undersized organs, premature autumn leaf color, dieback, and sometimes basal cankers. These symptoms do not appear until root damage is extensive. Apple trees stressed by root damage may set an abnormally large crop of small fruit. Roots are killed and then decayed. A mycelial encrustation, white at first but turning black, develops on the surface, and a network of tiny black rhizomorphs extends from the edge of the encrustation several centimeters along the root. Decay may extend into the butt.

Stromata of *X. polymorpha* arise in spring and mature in summer. They are nearly white at first, then become tan with a covering of brown conidiophores and unicellular conidia. Mature stromata are mostly 5–8 × 0.5–2 cm in size with a thin black crust and white to pale buff fleshy internal mycelium. They often have a central cavity. Black perithecia up to 1 mm in diameter form just beneath the crust, and at maturity discharge to the air smooth, dark brown, single-celled ascospores through minute pores in the crust. *X. mali* is similar.

Wood undergoing decay by Xylaria, especially *X. polymorpha*, contains an irregular lattice of black sheets (pseudostromatal plates) that appear as black lines (zone lines) on cut or broken faces. Many other wood-decaying fungi also produce these structures (e.g., see Plates 99, 152). Pseudostromatal plates are composed of dark, thick-walled, bladderlike cells of the fungus within wood cells plus a dark substance that impregnates the wood cell walls. The plates are refractory to microbes that decompose wood.

If susceptible trees are planted in soil where others have been killed by *X. mali*, severe disease in the new planting is likely to develop within 2–3 years. Otherwise, cultural practices and fertility have little influence on Xylaria root rot.
References: 611, 623, 996, 1200, 1652, 1859, 3152, 3321, 4108, 4312

Rosellinia Root Rot

Rosellinia necatrix (Sphaeriales, Xylariaceae) in its conidial state, *Dematophora necatrix*, causes a white root rot of more than 170 woody and herbaceous species in 30 families around the world. *Malus pumila* (apple) orchards in many places and *Populus* (poplar) plantations in parts of southern Europe have sustained severe damage. In North America the pathogen occurs in Mexico and California. Woody plants affected in California include species of *Berberis* (barberry), *Castanea* (chestnut), *Ceanothus*, *Cotoneaster*, *Cydonia* (quince), *Ficus* (fig), *Frangula* (buckthorn), *Juglans* (walnut), *Ligustrum* (privet), *Malus*, *Persea* (avocado), *Osmanthus*, *Populus*, *Prunus* (almond, apricot cherry, peach), *Pyrus* (pear), *Ribes* (currant), *Rubus* (brambles), *Viburnum*, and *Vitis* (grapevine). Several other *Rosellinia* species, notably *R. desmazierii*, also cause root rot of woody plants.

Aboveground symptoms of Rosellinia root rot include slow growth and sparse, chlorotic, undersized foliage for a year or more before the plant dies. Symptoms may involve the entire plant or initially just one side. Plants killed quickly often retain dry foliage for some time. During wet weather, white mycelium may appear at the soil surface above killed roots, or the fungus may form at the stem base a dark mat on which synnemata (aggregates of conidiophores) and single-celled conidia develop. The conidia, although germinable, have no known role in the disease cycle. White mycelium and synnemata also form on diseased root segments that are incubated under moist conditions for diagnosis. Viewed microscopically, each hyphal cell is distinctly swollen at its distal end. The perithecial state of *R. necatrix* has not been found in North America.

Infection begins on small roots and progresses to larger ones, eventually killing a plant by girdling at the root collar. Hyphae invade both bark and wood, growing inter- and intracellularly. The phytotoxin cytochalasin E and several other toxic or possibly toxic metabolites of the fungus are produced in infected plants, but their importance in pathogenesis is not established. Fibrous roots rot off, often leaving a residue of white mycelium in soil. White mycelium also grows on or beneath the bark of larger roots, forming plaques and loosely aggregated white strands that grow into adjacent soil. A dark crust sometimes forms on a dead root or the root collar. *R. necatrix* also produces dark microsclerotia about 0.1 mm in size. These and mycelium in root fragments allow the fungus to survive in soil. Mycelial strands growing from decayed roots and hyphae growing from microsclerotia penetrate small roots. The fungus spreads readily in loose soil with high sand content. Transmission also occurs where roots touch one another. Infected planting stock is responsible for long-distance dispersal.

R. necatrix is favored by relatively cool (<20°C), moist soils. Its mycelium grows most rapidly at 22–25°C and is intolerant of heat. Some apple growers delay the loss of diseased trees by removing soil from around the root collar and root buttresses, which allows exposed bark to be heated by sun and air sufficiently to inhibit the pathogen. Solar heating of orchard soil under clear plastic has killed *R. necatrix* in experiments in several countries where summer weather is hot and sunny.

Little attention has been given to deploying rootstocks resistant to *R. necatrix* in North America, although diverse plants, including various *Malus* species and clones, *Carya illinoinensis* (pecan), and *Diospyros* (persimmon), possess resistance.
References: 92, 590, 1859, 2011, 2277, 2366, 2484, 2894, 2896, 3445, 3556, 3764, 3947, 3948, 4108, 4312

A–F. Xylaria root rot. A, B. *Carya ovata* (shagbark hickory) dying of root and butt rot caused by *Xylaria polymorpha* and *Ganoderma applanatum*. The tree had previously been stressed by drought. A. Sparse foliage and dieback. B. Perithecial stromata of *Xylaria polymorpha* at the trunk base (NY, Jul). C. Mature perithecial stromata of *Xylaria* sp. growing from a *Pyrus communis* (common pear) stump (NY, Aug). D. Young stromata of *Xylaria* sp. visible as tan objects at ground level adjacent to bare wood of a basal canker on *Cladrastis lutea* (yellowwood) (NY, Jun). E. Close view of the surface and interior of a stroma cut to expose soft white interior tissue and the perithecial layer beneath the black surface. F. Magnified view of perithecia (NY, Aug).

G–I. Rosellinia root rot of *Malus pumila* (apple). G. Sparse, chlorotic, undersized leaves with uprolled edges on a diseased tree. H. Close view of the root-collar area, where superficial bark has been shaved to reveal the margin of necrosis. Dead bark contains pockets of white mycelium. I. White mycelium of the *Dematophora* state of *Rosellinia necatrix* on a dead root extending from the butt (CA, Jul).

225

Rhizina Root Rot and Southern Blight (Plate 112)

Rhizina Root Rot

The ascomycete fungus *Rhizina undulata* (Pezizales, Rhizinaceae) kills coniferous seedlings and sometimes mature trees on and adjacent to burned areas. It inhabits coniferous forests, both natural and planted, around the Northern Hemisphere and in southern Africa. In North America it has been reported as far south as California and Florida. It has been called a fire fungus because it is activated by heat from fires but then remains active for only a few years. Many tree species in at least seven genera are affected: *Abies* (fir), *Larix* (larch), *Picea* (spruce), *Pinus* (pine), *Pseudotsuga* (Douglas-fir), *Thuja* (arborvitae), and *Tsuga* (hemlock). The disease occurs only on sites that have supported growth of conifers for at least several years. Severe losses occur only on acidic sandy or peat soils where fire was extensive, such as on sites burned over by wildfire or in preparation for planting. Mortality rates above 75% have occurred in planted or natural seedling populations on such sites in western Canada, Europe, and South Africa.

Killing of seedlings or transplants by *R. undulata* usually begins in the first year after a fire and subsides after 2 years. Symptoms progress rapidly. Roots are killed and matted with white to yellowish mycelial strands that extend into the bark. Death of a plant usually results from girdling at the root collar. Nonwoody shoot tips wilt, and older foliage fades or turns yellow and then brown. The fungus grows radially through soil, as mycelial strands from colonized roots, at rates up to 2.5 m per year. On encountering a host root, strands grow on the surface and enter through lenticels. Hyphae then ramify and kill the root cortex.

Larger trees are attacked at margins of fire sites and sometimes many meters beyond, typically in radially expanding groups, to which an alternative disease name, group dying, refers. Symptoms include coalescing cankers on roots and the root collar, depressed growth, resin exudation at the butt, production of a large cone crop, then fading or wilting. Disease foci typically expand for 2–3 years, but longer periods have often been recorded, notably in England. *R. undulata* has been found associated with mature *Picea rubens* (red spruce) in Vermont, *Tsuga heterophylla* (western hemlock) in Washington, and pole-size to mature *Pinus* and *Picea* on other continents.

Apothecia of *R. undulata* form on soil or charred litter in summer and persist until freezing weather. They are connected to roots by cream to yellow mycelial strands, sometimes called rhizoids, and occasionally grow on exposed roots or tree butts. Apothecia initially appear as brownish buttons but soon expand to convex, irregularly lobed, convoluted, tough-fleshy structures 3–8 cm across, sometimes aggregated into irregular crusts up to 15 cm or more across. They have a yellow-brown to reddish brown upper surface with cream to light yellow edges when young. The undersurface is tan-ochre. Mature specimens are dark brown to nearly black. They produce colorless, fusiform ascospores that are usually 28–32 × 8–10 μm, including a protective gelatinous sheath.

Ascospores of *R. undulata* are dispersed in air. After settling on the forest floor they probably are carried by water to the mineral soil or a short distance into sandy or peat soil. Some survive, dormant, until stimulated by heat shock (e.g., 37–40°C for 24 hours). Fire not only provides the heat to break spore dormancy but also creates for the fungus an advantage in competition for substrates. Its mycelium grows most rapidly at 25–30°C.
References: 53, 1859, 1902, 2388, 2724, 2776, 2787, 2892, 2988, 3065, 3415, 3521, 3625, 4019, 4142

Southern Blight

This disease is caused by the basidiomycete fungus *Athelia rolfsii* (syn. *Corticium rolfsii*; Stereales, Atheliaceae) in its asexual state, *Sclerotium rolfsii*. The latter produces only mycelium and sclerotia. It attacks herbaceous and woody plants in more than 250 genera, typically causing root and crown rot, but also damping-off, stem rot, and leaf blight. It occurs globally in tropical and temperate regions. In the USA it is occasionally seen as far north as Washington and New York states. Although capable of killing woody plants up to several years old, *S. rolfsii* is important mainly on those not yet protected by corky bark in the root-collar region. They are affected while in nurseries or, on occasion, shortly after outplanting. Woody hosts include species of the following genera, among others:

Actinidia, Araucaria, Arctostaphylos, Asimina, Aucuba, Berberis, Buddleja, Camellia, Carica, Cassia, Catalpa, Catharanthus, Citrus, Codiaeum, Cornus, Crataegus, Cryptomeria, Cydonia, Daphne, Elaeagnus, Eriobotrya, Eucalyptus, Ficus, Forsythia, Gardenia, Hydrangea, Jasminum, Juglans, Koelreuteria, Ligustrum, Malus, Morus, Murraya, Myrtus, Olea, Persea, Pittosporum, Populus, Prunus, Punica, Rhaphiolepis, Rhododendron, Rosa, Schefflera, Syringa, Thuja, Torreya, Viburnum, Vitis, and *Washingtonia.*

Symptoms and signs. Southern blight typically appears on scattered plants among healthy ones. A plant suddenly wilts and dies as the result of girdling necrosis at the stem base or root collar. Bark a few centimeters above and below the soil line is killed. The lesion is often visible as a darkly discolored area. As attack begins, a coarse web of white mycelium grows on the stem base and on and in the surrounding soil. At about the time of plant death sclerotia form on the surface mycelium, which then disappears. The sclerotia, spheric and 1–2 mm in diameter, are white at first but soon become brown. They have a brown rind and white internal tissue that stores organic compounds to be used later for growth. The surface mycelium and sclerotia are diagnostic.

Disease cycle. Sclerotia persist on crop debris and weed hosts and are dispersed with soil or flowing water. They germinate when moist and stimulated by chemicals that emanate from plant materials. Germination occurs either by growth of individual hyphae through the rind or by eruption of a white mass of mycelium that grows toward the stimulus. Mycelium from an eruptively germinated sclerotium may infect plant tissue as far as 3.5 cm distant. The fungus penetrates succulent or wounded tissues by hyphal growth from appressoria that develop a short distance behind the advancing edge of the mycelium. It produces oxalic acid and an array of macerating enzymes that quickly kill and decompose host tissues. The interval from infection until plant death and formation of a new generation of sclerotia is usually less than a month. Basidia of *Athelia rolfsii* occasionally develop on mycelium on killed leaves of herbaceous plants near the soil line. The basidiospores, although infectious, are considered to play only a minor role in epidemiology of southern blight, but their formation via meiosis sustains variability in the species.

Southern blight is favored by hot (30–35°C), wet weather and acidic (pH 3–6), well-aerated soil. As wet soil begins to dry, *A. rolfsii* may attack plants below the soil line. It is capable of saprobic growth and may spread somewhat in surface soil. In some circumstances dead plant material (presumably a food base) seems to favor infection. Alternate wetting and drying tend to stimulate sclerotial germination in the absence of susceptible plant material. Volatile organic chemicals such as methanol emanating from decomposing plant debris also stimulate sclerotial germination. This can be detrimental to the fungus because its mycelium may be killed by other microorganisms in soil. Some of these organisms also provide a measure of biological control by killing sclerotia. Flooding or solar heating of soil under clear plastic causes rapid decline in viable sclerotial populations.
References: 46, 149, 150, 353, 789, 1099, 1153, 1505, 1859, 1967, 2328, 2432, 2702, 3166–3168, 3377, 3660, 3740, 4054, 4099

A–C. Rhizina root rot, caused by *Rhizina undulata.* A. Young brown and old black apothecia of *R. undulata* associated with diseased and dead *Picea* (spruce) seedlings (ON, summer). B. Apothecia associated with a recently killed *Pinus contorta* (lodgepole pine) seedling (BC, Oct). C. Close view of young apothecia with light-colored edges (ON, summer).

D–G. Southern blight, caused by *Sclerotium rolfsii.* D. Small sclerotia, white at first, brown when mature, at the base of a *Malus pumila* (apple) sapling. E. White mycelium, a diagnostic sign, on the stem base of a diseased *Schefflera actinophylla* and on surrounding soil. F. Close view of mycelium and young sclerotia at the base of a dying *Ajuga reptans* (bugleweed). G. Magnified view of sclerotia (FL, summer).

Photo credits: A, C—Canadian Forest Service; B—R. Garbutt; D–G—C. L. Schoulties

227

Procerum Root Disease of Pines (Plate 113)

Leptographium (syn. *Verticicladiella*) *procerum*, a hyphomycete, infects the bark and sapwood of roots and the stem base of *Pinus* species, causing root lesions, basal cankers, and sometimes decline and death. The fungus is known from DNA sequence data to be a member of the Ophiostomatales and Ophiostomataceae. The decline syndrome, called procerum root disease or white pine root decline, is restricted to young trees, most often *P. strobus* (eastern white pine). *P. nigra* (Austrian pine) and *P. sylvestris* (Scots pine) are sometimes affected. The disease occurs in eastern USA from New England into the Appalachian region and Ohio River basin. It is commonly associated with infestation of major roots and the root collar by weevils (Coleoptera: Curculionidae) and occurs primarily in plantations, especially Christmas tree plantations. Abundant stumps in plantations favor the buildup of weevil populations. Mortality in some plantations has been in the 20–50% range. Landscape trees on adverse sites, particularly on wet sites, are also affected.

The geographic distribution of *L. procerum* in North America is much broader than that of the decline syndrome in *P. strobus*. The fungus also occurs in Europe, South Africa, and New Zealand. It is one of several beetle-associated sapstain fungi that invade roots and stem bases of stressed and dying pines, and it is associated with root lesions on pines growing on unfavorable sites. Accordingly, *L. procerum* is widely viewed as a contributing pathogen rather than as the primary cause of the pine decline syndrome. A related beetle-associated sapstain fungus, *L. terebrantis*, has proven more aggressive than *L. procerum* in comparative pathogenicity tests, although it is not consistently associated with white pine root decline.

Hosts of *L. procerum* in North America include at least the following pines in addition to those named above: *P. banksiana* (jack pine), *P. clausa* (sand pine), *P. contorta* (lodgepole pine), *P. echinata* (shortleaf pine), *P. elliottii* (slash pine), *P. ponderosa*, *P. taeda* (loblolly pine), and *P. virginiana* (Virginia pine). Rare hosts include at least one species of *Abies* (fir), two of *Picea* (spruce), and *Pseudotsuga menziesii* (Douglas-fir). In most situations *L. procerum* is not aggressive. In pathogenicity tests on various pines it usually caused small cankers or localized resin soaking of sapwood in roots. On the other hand, Appalachian and French strains have killed seedlings of *Pinus strobus*, *P. sylvestris*, and *P. taeda*.

Symptoms and signs. Most infections are probably inconsequential. This account deals with those apparently leading to basal cankers, decline, or death of *P. strobus*. Trees up to 20 years old are most commonly affected. Height growth declines 1–3 years before other symptoms appear. In the year of death, buds open late; shoot growth is arrested; and the mature foliage fades, droops, and turns brown, usually in May–July. Some trees discolor before the time of bud break. Dead foliage clings to the twigs for a year or longer. A tree with fading foliage usually has a resinous, girdling canker extending from below soil level to a height of 10–45 cm on the butt. Bark in the canker and wood beneath the canker face are soaked with resin, and the bark is chocolate brown to black. Sapwood in and adjacent to the canker does not conduct water. Resin often exudes from the roots, root collar, and trunk base for a few years before the onset of foliar symptoms. A girdling canker is often subtended by smaller resinous cankers and dieback on main roots. In some cases, especially in the Appalachian region, blackish streaks form in the sapwood of roots and the butt. Viewed in cross section they tend to be wedge shaped because the fungus grows inward along rays. *L. procerum* can easily be isolated from diseased bark and from black-streaked wood, but less readily from resin-soaked wood.

P. nigra (Austrian pine) and *P. sylvestris* (Scots pine) are sometimes affected in a manner similar to *P. strobus*, but the stiff needles of these species do not droop before death. *P. sylvestris* may be diseased for several years without showing external symptoms. In *P. resinosa*, which is not killed by *L. procerum*, the sapwood beneath cankers on roots may be resin soaked to the center of a root but is not darkly stained.

L. procerum produces minute (3–5 × 1–3 μm), colorless conidia in cream-colored drops of mucilaginous fluid on brushlike multiple branches atop stout black-stalked conidiophores (Fig. 113). Conidiophores develop in wounds and tunnels made by insects in cankers and on dead roots and also form quickly, sometimes in concentric rings, when the pathogen is isolated on laboratory media. The sticky spores are adapted for transmission by insects.

Disease cycle. The cycle is incompletely known. *L. procerum* overwinters and perennates in diseased trees and in stumps. Its conidia contaminate insects that breed in diseased bark, and some of these insects later introduce the fungus into healthy trees. *Hylobius pales* (pales weevil) and *Pissodes nemorensis* (deodar weevil) are demonstrated vectors. *H. radicis* (pine root-collar weevil) and *Pachylobius picivorus* (pitch-eating weevil), as well as several species of bark beetles, carry the fungus and are potential vectors. Propagules presumed to be conidia also become numerous in soil adjacent to diseased roots and are most abundant in late summer. These conidia, where transported in water, possibly account for some tree-to-tree spread of the fungus. *L. procerum* can infect small roots, wounded or not, but such infections on trees larger than seedlings probably result in only inconsequential cankers and killed root ends.

L. procerum is often found in pines stressed by adverse sites, root wounds, insect attack, polluted air, or other root pathogens. It is one of the sapstain fungi involved in beetle-associated decline of *P. resinosa* in Wisconsin and *P. palustris* in South Carolina. It causes resin-soaked cankers on *P. resinosa* roots in poorly drained soils where trees undergo slow decline in New York, and was one of three fungi shown to cause cankers near ground level on *P. strobus* saplings damaged by mound-building ants. The fungus has been found associated with wounds made during root pruning and cultivation of several *Pinus* species in seed orchards in southeastern USA, but it has not been credited with significant damage in that circumstance. It was commonly isolated from *P. strobus* showing symptoms of air pollutant injury in Virginia but was not detected in pollution-tolerant trees to which the symptomatic group was compared. *L. procerum* has also been found associated with root-rotting fungi such as *Heterobasidion annosum* in plantations and natural stands of *P. clausa* and *P. taeda*.

References: 591, 1074, 1490, 1507, 1512, 1549, 1650, 1766, 1891, 1892, 2036, 2095–2097, 2206–2208, 2826–2828, 3081, 3218, 3649, 3927, 4390–4392

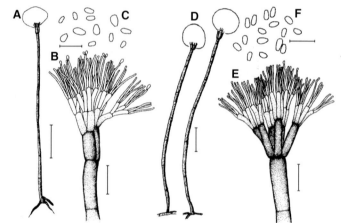

Figure 113. A–C. *Leptographium procerum.* D–F. *L. wageneri* (discussed with Plate 114). A and D, conidiophores, each topped by a mucilaginous drop of liquid containing conidia. B and E, brushlike apical parts of conidiophores. C and F, conidia. Scale bars = 100 μm for conidiophores, 20 μm for their apical parts, 10 μm for conidia. Adapted from reference 1891 by permission of the American Phytopathological Society.

Symptoms caused by *Leptographium procerum* in *Pinus strobus* (eastern white pine).

A, B. Faded and brown foliage in spring. A. In a natural old-field stand (NY, Jun). B. In a Christmas tree plantation (VA, May).

C. An upper branch of a diseased tree showing reduced height growth before other external symptoms. The current-year and 1-year-old branch segments together are not as long as the 2-year-old segment (NY, Sep).

D. Part of a tree that died the previous year, amid others apparently unaffected in a hedge. Dwarfed shoots on the dead tree indicate arrested growth early in the season of death (NY, Apr).

E. Exuded resin and dead, resin-soaked inner bark exposed in a girdling canker on the base of a tree (NY, Jun).

F. Sapwood darkly stained by *L. procerum* in the stump of a recently killed tree (VA, May).

G, H. Resin-infiltrated lesions and dieback on roots of an affected tree (NY, Sep).

Black Stain Root Disease of Conifers (Plate 114)

Black stain root disease is a debilitating, usually lethal disorder characterized by extensive chocolate brown to black staining and death of sapwood in roots and the lower stem of conifers. It is caused by the hyphomycete *Leptographium wageneri* and occurs only in western North America. The fungus is known from DNA sequence data to be a member of the Ophiostomatales and Ophiostomataceae. Three host-specialized varieties of the pathogen with overlapping but spotty distributions are recognized. *L. wageneri* var. *wageneri* attacks *Pinus edulis* and *P. monophylla* (pinyons) across the Southwest and as far north as Idaho. This variety has killed thousands of pinyons in Mesa Verde National Park. *L. wageneri* var. *ponderosum* colonizes *Pinus contorta* (lodgepole pine), *P. jeffreyi* (Jeffrey pine), and *P. ponderosa* (ponderosa pine). It is perhaps the anamorph of the ascomycete *Ophiostoma wageneri* (Ophiostomatales, Ophiostomataceae), which has been seen only once. *L. wageneri* var. *pseudotsugae* attacks *Pseudotsuga menziesii* (Douglas-fir). The latter two varieties cause losses in forests from British Columbia and Montana southward. Other conifers occasionally infected, primarily by *L. wageneri* var. *ponderosum*, include several species of *Picea* (spruce), *Pinus*, *Tsuga* (hemlock), and possibly *Abies* (fir). Two European conifers found to be susceptible when tested were *Pinus sylvestris* (Scots pine, highly susceptible) and *P. nigra* (Austrian pine).

Black stain root disease occurs at elevations ranging from sea level to nearly 3000 m and in habitats ranging from semiarid to well-watered. The disease is most common on relatively cool, moist microsites, especially in areas where summers are hot and dry. It tends to increase in forests disturbed by thinning, logging, and road construction. Bark beetles often compound the problem by breeding in diseased trees and then attacking other trees nearby.

Symptoms and signs. Diseased trees typically occur in groups. Infection begins in one tree and spreads to neighbors at rates up to 7 m (average 1–2 m) per year. *Pseudotsuga* up to 80 years old (but usually less than 30) and *Pinus ponderosa* up to 100 years old may be killed, but not all trees within a disease patch die. Some patches stabilize after several years; others expand, coalesce, and may eventually cover several hectares. *Pseudotsuga* seedlings planted among the stumps of former diseased trees may become infected and begin to die as soon as a year after planting.

External symptoms develop over several years and include reduced height growth, subnormal needle size, premature abscission of old needles, and increased transparency of the crown. A declining tree may produce an abnormally large crop of small cones and may exude resin along the lower trunk. Resin exudation varies with tree species and vigor, bark thickness, and insect attack. Foliage turns yellowish and then, after one or more summers, brown. Before foliar symptoms appear, the hydraulic conductivity of diseased roots diminishes to a small fraction of normal. This change leads to water stress and reduced transpiration and photosynthesis. Bark beetles attacking diseased pines are often the direct cause of their death.

Diseased trees have longitudinal streaks of stained sapwood, sometimes widest at the root collar and tapering into the trunk and roots. Multiple streaks may form, appearing in cross section as arcs in one or more annual rings. In pines the stained wood is sometimes impregnated with resin. Stained sapwood and noninfected sapwood adjacent to it become dysfunctional as hyphae, resin, tyloses, and/or gas emboli block tracheids. The dark stain is due to large, amber to brown, much-branched hyphae, each surrounded by a thick amber sheath that sometimes fills the lumen of a tracheid, and to amber to brown discoloration of tracheid walls. Cells at a distance from hyphae may also be discolored. Stain patterns caused by *L. wageneri* are unusual because the fungus colonizes only tracheids and only those within particular annual rings. Most other fungi that cause stain in conifer sapwood colonize ray parenchyma and grow radially across several rings, staining the wood in sectors that appear wedge shaped in cross section (see Plate 117). Cankers may form near the base of a tree where infection approaches the vascular cambium. *L. wageneri* apparently causes some of these cankers, but others are caused by *Armillaria*, which is often found in them.

Roots and root collars of diseased and recently dead trees and stumps are attacked by beetles that act as vectors of *L. wageneri*. All three varieties sporulate in beetle galleries, producing tiny, colorless, sticky conidia, 4–8 × 1–3 µm, in droplets on brushlike much-branched conidiophores (Fig. 113). They probably also sporulate in other wounds that penetrate diseased tissue.

Disease cycle. *L. wageneri* perennates in diseased trees. Infection most often begins in small roots, the pathogen entering via wounds (made by vectors or other agents) and natural openings that provide access to xylem. It grows along a root in both directions and up the stem to a height of 2 m or more in *Pinus* and as high as 10–15 m in *Pseudotsuga* before trees die. In *Pinus ponderosa* infection spreads from root to stem in the youngest layer of sapwood, but in *Pseudotsuga* the infection can spread from a root into stem sapwood of various ages. While a tree is alive the fungus is restricted to axial and ray tracheids in xylem, growing from one cell to another through bordered pits. It follows the grain of the wood and also grows tangentially but not much radially. It spreads from tree to tree by hyphal growth across root grafts and contacts and through soil for distances of a few centimeters, also as conidia carried by vectors and perhaps flowing water. Vectors account for the establishment of new disease patches. Vectors in *Pseudotsuga* forests include the weevils *Steremnius carinatus* and *Pissodes fasciatus* and the root-feeding bark beetle *Hylastes nigrinus* (all in the Curculionidae). *H. macer* is probably a vector in *Pinus ponderosa* stands. These insects, attracted by volatile chemicals from diseased or otherwise stressed trees or new stumps, attack roots and root collars and breed there. The beetles may introduce *L. wageneri* whether or not the funugus is already present. New adults of the next generation become contaminated with conidia, exit the now-dead roots, and then may feed in the roots of healthy trees, incidentally transferring the pathogen into them.

The three varieties of *L. wageneri* differ somewhat in form, appearance in culture, temperature optima for growth, and tolerance of high temperatures. All grow most rapidly and are most infectious at 15–21°C and are inhibited or killed at temperatures above 21–24°C.

L. wageneri often precedes or occurs along with other root rot pathogens, especially *Armillaria* (Plate 163). Trees killed by *L. wageneri* alone or in collaboration with other pathogens are soon colonized by bluestain and sap-rot fungi that obscure black stain symptoms. Additional *Leptographium* species attack conifer roots in North America. Some cause resinous lesions and/or dark staining in wounded roots under some circumstances, but none causes symptoms closely similar to those caused by *L. wageneri*.

References: 754, 955, 1308, 1484, 1490, 1507, 1508, 1510–1513, 1597, 1625–1628, 1891, 1975, 2036, 3734, 4182, 4261, 4368, 4396, 4404, 4405, 4506

Black stain root disease, caused by *Leptographium wageneri*.

A. *Pinus monophylla* (singleleaf pinyon) killed by *L. wageneri* var. *wageneri*. Trees with tan foliage wilted in early summer; those without foliage died in previous years (CA, Jul).

B, E–I. Symptoms caused by *L. wageneri* var. *pseudotsugae* in *Pseudotsuga menziesii* (Douglas-fir). B. A group of diseased and dead young trees (OR, autumn). E–G. Trunks of young trees showing resin exudation and, where bark was cut away, streaks of black stain ascending in sapwood from roots (OR, summer). H, I. Successively closer cross-sectional views of the trunk of a dying tree. Stained xylem appears black in the youngest sapwood but has faded to gray in older wood. The three youngest growth rings are very narrow. Stain is present in rings up to 7 years old. Infection near or at the cambium has caused a canker at top left in H. Growth abnormalities are evident as wavy lines in outermost wood (CA, Jul).

C, D. Symptoms caused by *L. wageneri* var. *ponderosum* in *Pinus ponderosa* (ponderosa pine). C. Tufted foliage on a declining tree, a symptom noted in a small minority of trees affected by *L. wageneri*. D. A streak of discolored sapwood extending upward from roots (OR, summer).

Photo credits: B—E. Goheen; C–E—W. G. Thies

Ceratocystis Cankers (Plates 115, 116)

Ceratocystis fimbriata (Microascales, Ceratocystidaceae), a cause of cankers, dieback, and wilt diseases, occurs worldwide in tropical and temperate regions and infects diverse plant species, both woody and herbaceous. Woody hosts represent at least 21 genera, including *Acacia, Citrus, Coffea* (coffee), *Eucalyptus, Gmelina, Hevea* (rubber tree), *Mangifera* (mango), *Pimenta* (pimento), *Platanus* (planetree, sycamore), *Populus* (aspen, poplar), *Prunus* (stone-fruit trees: almond, apricot, peach, prune), *Punica* (pomegranate), and *Theobroma* (cacao). Lesions caused by this fungus usually involve both bark and wood, and infected tissues are darkly stained. Sapwood lesions in plants such as *Citrus, Eucalyptus, Platanus,* and *Punica* are accompanied by wilt and dieback. The principal tree diseases caused by *C. fimbriata* in North America are cankers on *Populus* and *Prunus* and canker-stain of *Platanus*. The former two diseases are featured with Plate 115; that of *Platanus* is described with Plate 116.

Ceratocystis Cankers of *Populus* and *Prunus*

Ceratocystis cankers of *Populus* and *Prunus* in North America affect primarily *Populus tremuloides* (trembling aspen) and *Prunus dulcis* (almond). The aspen disease occurs throughout the range of the species in the mountainous West and eastward to Quebec and Pennsylvania. *P. balsamifera* (balsam poplar) is occasionally affected in Quebec. Various *Populus* species are affected in Europe, and *P. deltoides* (eastern cottonwood) is a reported host in India. The almond disease occurs in California.

Symptoms and signs. Cankers on both *Populus* and *Prunus* originate at trunk wounds and branch crotches and are perennial. Those on *Populus tremuloides* are similar to cankers caused on that tree by *Neonectria galligena* (Plate 90). The two diseases cannot be distinguished visually where both occur from Minnesota eastward. Ceratocystis cankers on *Populus* are circular to oval when young but later tend to become diamond shaped to irregular in outline because of flaring dead bark at the sides. Cankers initiated at edges of large wounds such as those caused by logging equipment or falling trees tend to follow the outline of the wound. Old cankers are concave and have concentric ridges of dead woundwood. Killed bark eventually sloughs from the center of a canker. Wood beneath the canker face and for a few centimeters above and below is darkly stained. Amber to reddish brown fluid exudes from a canker margin and may cover the canker surface. This substance plus perithecia of various species of *Ceratocystis* and *Ophiostoma* and dark hyphae and spores of saprobic fungi cause bark and exposed wood in the canker to become dark brown to black. In the West cankers on *P. tremuloides* commonly enlarge for 30–40 years, expanding 1.3 cm in width and 2.8 cm in length per year on average. Individual cankers rarely girdle trees, but multiple coalescing cankers do so, and some diseased trees break because of weakness at cankers.

Ceratocystis cankers on *Prunus dulcis* develop primarily in branch crotches. They were formerly most common at wounds made by mechanical harvesters, but changes in harvest practices solved that problem. The pathogen induces profuse gum production in *P. dulcis*.

Black perithecia of *C. fimbriata*, visible with a hand lens, can often be found in spring near the edges of cankers on bark killed in the previous year. Conidia and perithecia also form within 7–10 days on newly infected trunk wounds. The perithecia are shaped like long-necked flasks with bodies 100–200 μm diameter and necks up to 0.8 mm long. Since several species of *Ceratocystis* and *Ophiostoma* occur in aspen cankers, however, specific identification requires microscopic examination. The ascospores are colorless and flattened on one side, with an outer wall sheath encircling the flat area like the brim on a hat. They measure 6–8 × 3–5 μm and often are found stuck together in clumps at tips of perithecial necks. The stickiness of the spores, a characteristic of *Ceratocystis* in general, facilitates transmission by insects. *C. fimbriata* also produces three types of asexual single-celled spores: thick-walled, brown, barrel-shaped aleurioconidia (also called chlamydospores) 9–18 × 6–13 μm, that form in chains; colorless, cylindric endoconidia, 10–27 × 3–5 μm, characteristic of the genus *Thielaviopsis* (syn. *Chalara*) that are extruded from the tips of tubelike conidiophores; and subglobose endoconidia, mostly 6–10 × 5–9 μm, that also are extruded from tubelike conidiophores. All spore types are infectious. Aleurioconidia and cylindric endoconidia have been

found in and on diseased tissues of both *Populus* and *Prunus*. *C. fimbriata* grows readily in culture, producing dark olive to brown colonies with a fruity aroma, in which all spore states often form. Strains of *C. fimbriata* associated with cankers on *Populus* and *Prunus* have been shown to be host specialized, as is also true for strains from *Platanus* (Plate 116).

Disease cycle. C. fimbriata overwinters in lesions in both bark and sapwood in perennial hosts and also on certain insects that act as vectors. It apparently sporulates on diseased plants throughout the year as temperature and moisture permit. Its fruity odors attract insects, notably sap-feeding beetles (Nitidulidae), which visit cankers and wounds where the fungus is sporulating, become contaminated, and may later transmit spores to wounds in healthy trees. Susceptibility of wounded tissues declines as wounds age. Lesions expand intermittently as callus and woundwood develop at a canker margin during the growing season, and *C. fimbriata* kills part or all of the new tissues during the next dormant season. Colonization proceeds in both bark and wood. Degeneration of tissues in *Populus* begins several cells behind the advancing tips of hyphae. Resistant plants localize infections.

Insects implicated as possible vectors in the aspen canker system in Colorado are *Epuraea* species, *Colopterus truncatus* (both in the Nitidulidae), and *Nudobius corticalis* (Staphylinidae). Some of the contaminated insects overwinter as adults in the ground, and spores of *C. fimbriata* remain viable on them. Nitidulids also transmit *C. fimbriata* among stone-fruit trees.

References: 123, 172, 215, 449, 596, 609, 745–747, 1367, 1669, 1670, 1672, 1674, 1948, 1994, 2452, 2896, 2981, 2982, 3159, 3160, 3358, 3991, 3992, 4123, 4267, 4396, 4425, 4497

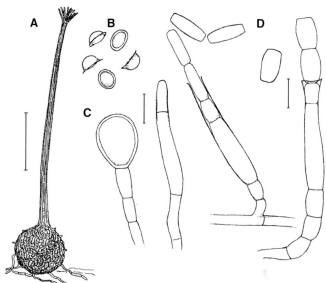

Figure 115. Ceratocystis fimbriata. A. Perithecium. B. Hat-shaped ascospores. C. An aleurioconidium. D. Conidiophores and conidia. Scale bars = 200 μm for perithecium, 10 μm for other structures. Adapted from reference 123 by permission of Gebrüder Borntraeger Verlagsbuchhandlung.

A–E. Ceratocystis canker of *Populus tremuloides* (trembling aspen), caused by *Ceratocystis fimbriata*. A. High incidence of the disease in a young aspen stand (UT, May). B, C. Close views of old cankers with flaring bark associated with callus and woundwood growth at the margins (UT, May; CO, Jun). D. A canker several years old that apparently began at a branch stub. The lesion is expanding upward (canker margin near top of photo); large lateral woundwood ridges indicate inactivity at the sides. E. Rust brown fluid bleeding from a canker approximately 4 years old on a young stem. Woundwood nearly encloses the original site of infection. Zones of lesion expansion during the previous and current growing seasons are discolored orange (D, E: CO, Jun).

F, G. Ceratocystis canker of *Prunus dulcis* (almond). F. Gum exuding from the active parts of a canker in a branch crotch. G. An old canker centered on a wound made by a harvesting machine (CA, summer).

Photo credits: F, G—B. L. Teviotdale

233

Canker-stain of Planetree and Sycamore

Canker-stain of *Platanus* (planetree, sycamore), sometimes called colored canker in Europe, is a lethal disease caused by *Ceratocystis fimbriata* f. *platani* (Microascales, Ceratocystidaceae), herein abbreviated *C.f. platani*. The pathogen is a host-specialized form of *C. fimbriata* that is considered to be indigenous to the USA. It occurs from New Jersey to Georgia, Arkansas, and Missouri, also in California and southern Europe. *P. ×acerifolia* (London planetree) apparently sustains the most damage. *P. orientalis* (Oriental planetree) is also highly susceptible, while *P. occidentalis* (eastern sycamore) is somewhat less so. The disease occurs in both landscape and forest trees. Up to 30% of the trees examined in some plantations of *P. occidentalis* in southern USA have been found infected.

Canker-stain has been known since the 1920s, when outbreaks began in urban plantings of *P. ×acerifolia* in Pennsylvania and New Jersey. These epidemics were eventually halted by sanitary and hygienic procedures: destroying diseased trees, disinfesting equipment after work on any *Platanus* within an area of outbreak, and covering pruning wounds with fungicidal dressings. Outbreaks began in Italy and France in the 1970s, and the disease later appeared in Spain, Switzerland, and Armenia. The pathogen is thought to have been transported from North America to Europe in green wood of *P. occidentalis* used for crates. The Italian population of *C.f. platani* is known to be genetically homogeneous, which is consistent with recent introduction of a small founding population.

Symptoms and signs. Elongate, spreading lesions with reddish brown to blue-black discoloration develop first in sapwood and later involve the cambium and phloem of trunks and limbs. Pole-size trees may die within 2 years after infection, but large trees tend to decline over several years. Diseased inner bark and cambial tissues turn reddish brown to nearly black. Cankers in young smooth bark appear as elongate dark areas and usually have only slight ridges at the edges. Cankers beneath scaly bark are invisible until they become sunken or bark begins to slough, as it does from old cankers. Old cankers on *P. ×acerifolia* appear more or less blackened and cracked. Active lesions in sapwood as viewed in cross section are at first lens shaped and later wedge shaped, pointed toward the center of the stem. Some lesions extend through the pith and expand toward the other side of the stem. Xylem dysfunction leads to wilting on branches above xylem lesions or to decline symptoms: slow twig growth; sparse, stunted, and prematurely yellow foliage; and dieback. Epicormic sprouts often form below cankers, but these die as the infection spreads. The dark stain in the wood fades as secondary organisms, including wood-decay fungi, replace the pathogen. In southern USA dead or dying bark is often colonized by *Botryosphaeria rhodina* (Plate 63), an opportunistic pathogen of stressed trees. *B. rhodina* can also cause cankers and dark discoloration of xylem.

C.f. platani produces ascospores and three types of asexual spores as illustrated for *C. fimbriata* (Fig. 115). It begins to produce endoconidia (anamorph genus *Thielaviopsis*, syn. *Chalara*) in gray colonies in wounds within a few days after infection. These are soon followed by long-necked perithecia that are visible with a hand lens. Ascospores extrude in a tiny blob of mucilaginous fluid from the tips of perithecia. Sticky ascospores and conidia contaminate insects and other animals. Aleurioconidia, also called chlamydospores (Fig. 115), form in diseased xylem and remain viable there for at least 2 years. The fungus produces all four types of spores on cut or broken surfaces of diseased wood or inner bark after a tree is further wounded or cut down.

Disease cycle. C.f. platani perennates in diseased trees and persists in killed wood as noted above. It sporulates on recently cut or wounded wood throughout the year, temperature and moisture permitting. Natural transmission is due primarily to insects, also to rodents, water, and root grafts. Splashing, wind-driven, or flowing water can carry spores. The pathogen has been detected in river water in Europe. Transmission by humans involves spores carried in sawdust; in nonfungicidal tree wound dressings applied by brushing; and on pruning saws, ropes, boots, and other equipment used for tree care. Spores on equipment may remain infectious for a month or more. Infection begins in fresh wounds on any part of a tree, including roots. Fresh wounds are susceptible throughout the year, although least so in winter. Wounds on leaves and soft shoots can become infected, but resistance responses in those organs

prevent systemic spread of infection. The ability of the pathogen to survive for several years in wood of diseased trees entrains a risk of infection of *Platanus* planted on sites where diseased trees formerly grew, because woody residues usually remain after trees are removed.

After wound colonization the pathogen spreads in sapwood. External symptoms may appear as soon as 7 weeks after infection. Stained wood extends far above and below and several millimeters to each side of a canker. Cankers widen several centimeters per year and lengthen up to 3 m per year in southern USA. Mechanisms of pathogenicity have not been elucidated in detail. Circumstantial evidence indicates that toxins, pectolytic enzymes that degrade primary cell walls, and elicitors of host defense are involved. A phytotoxic protein of the hydrophobin group, named cerato-platanin, is produced by the fungus and can elicit defense reactions, but its role in symptom induction has not been established. The usual duration of the disease cycle has not been reported; however, a diseased tree can serve as a source of infectious spores beginning in the first year of infection if new wounds expose tissue colonized by *C.f. platani*. Diseased trees resist infection by various mechanisms. They produce several phytoalexins; infuse tissues around lesions with polymerized phenolics; and produce tyloses, gels, and wall coatings in xylem vessels. These responses halt sap conduction and usually proceed too slowly to halt the pathogen.

Ceratocystis species, including *C.f. platani*, emit fruity odors attractive to insects that act as vectors, notably sap-feeding beetles of the Nitidulidae. Vectors that visit wounds on diseased trees transmit spores to healthy trees. The beetles *Cryptarcha ampla* (Nitidulidae) and *Laemophloeus biguttatus* (Cucujidae) transmitted *C.f. platani* from diseased to healthy trees in experiments in eastern USA. *Morimus asper* (Cerambycidae) has been implicated in Italy. Additional insect vectors are likely to be identified. A vole, *Clethrionomys glareolus*, is also a demonstrated vector in Italy. *C.f. platani* can be transported long distances in green (unseasoned) wood from diseased trees or on green logs or lumber from healthy trees if the wood becomes contaminated or is infested when cut. Spores produced on such wood at new locations can then be spread locally by insects or humans. By most accounts *C.f. platani* is capable of infecting only *Platanus* species. In one test, however, an isolate from diseased *P. occidentalis* caused cankers in *Populus tremuloides* (trembling aspen).

C.f. platani grows and reproduces at temperatures between 10° and 35°C; the optimum is near 25°C. The fungus cannot tolerate 40°C for longer than 48 hours, and it is quickly killed at 80°C. Wood products that potentially harbor the pathogen can be sanitized by heat.

References: 5, 128, 172, 596, 745–747, 845, 880, 1044, 1045, 1385–1387, 1859, 2286, 2950, 2982, 3021, 3263, 3408, 4206, 4207, 4396

Canker-stain in *Platanus occidentalis* (sycamore), caused by *Ceratocystis fimbriata* f. *platani* (all MS, Apr).

A. Sparse foliage of a diseased tree in contrast to healthy trees.

B–D. Cankers on pole-size trees. B. A long canker with wood exposed (upper arrows) but margins imperceptible except for a slight linear depression near its base (lowest arrow). C. A canker delimited by woundwood ridges at margins. Growth of epicormic sprouts below cankers is typical. D. Painted margins of a canker illustrate its vertical extent from an infection site near ground level.

E. Part of the trunk of a mature tree half girdled by canker-stain. Wood long dead is decaying. The pathogen is active in dark brown and black discolored streaks at the canker margin.

F, G. Portions of elongate cankers on pole-size trees. Edges of the inconspicuous canker in F were exposed by removing superficial bark scales. Bark in the canker in G has been colonized by dark secondary fungi.

H. Cross section of the trunk of a young tree with a darkly stained sector of wood extending from a canker face to the pith.

I. Part of a transverse section above the canker shown in E. Discolored sapwood indicates longitudinal spread of the pathogen and secondary organisms.

235

Sapstreak Disease and Bluestain (Plate 117)

Sapstreak Disease

Sapstreak is a disease of angiosperm sapwood caused by *Ceratocystis virescens* (Microascales, Ceratocystidaceae). It occurs in *Acer saccharum* (sugar maple), *Liriodendron tulipifera* (tuliptree), and perhaps other hardwoods. *A. saccharum* is affected in scattered forests and sugar bushes from Wisconsin to Vermont and in North Carolina and Tennessee. *L. tulipifera* is also affected in the latter region. Diseased maples cease sap production and die, and logs and lumber from them are greatly devalued by the stained wood. Stain patterns apparently caused by *C. virescens* also occur in *A. rubrum* (red maple) and *Betula alleghaniensis* (yellow birch), but the disease in these species is little known. *C. virescens* is also a widespread colonist of logs, stumps, and lumber of various angiosperm trees in eastern North America. Strains that cause sapstreak possibly have special pathogenic capability. They differ genetically from strains of the same species associated with stain in *Fagus* (beech) and *Quercus* (oak) logs. *C. virescens* was formerly confused with a conifer-attacking bluestain fungus, *C. coerulescens*.

Symptoms and signs. Symptoms are well known only in *A. saccharum*. The primary symptoms are water-soaked lesions that develop in sapwood and eventually involve nearly the entire cross section of the lower trunk and major roots and may extend far up into branches. Sapwood necrosis leads to externally visible symptoms: slow growth, chlorosis, undersized leaves, and dieback on one or more major branches or throughout the crown. Symptoms intensify from year to year. Xylem lesions begin as dark green to black or sometimes reddish streaks that follow the wood grain and expand along the medullary rays toward the pith and cambium, producing radiate patterns as viewed in cross section. Sapwood surrounded by dark streaks turns yellow-green to yellow-brown and water-soaked, and dark streaks within it turn brown or reddish brown. Elongate cankers form where sapwood lesions approach the cambium. Most affected trees die within 2–4 years, but some survive much longer with intervals of up to several years in which their condition is static before decline resumes. Maples affected by sapstreak are usually attacked by *Armillaria* (Plate 162) or *Xylaria* (Plate 111), which may hasten death. Apparent recovery, associated with compartmentalization of diseased sapwood, has also been noted. Most diseased trees occur adjacent to roads and skid tracks, and nearly all have old basal wounds resulting from logging and driving equipment over roots.

If diseased wood is wounded or a diseased tree cut down, *C. virescens* may sporulate on the moist wood surface, producing within several days a dark gray mat of mycelium, conidiophores, and long-necked black perithecia. The perithecia, ascospores, and two types of endoconidia (cylindric and barrel shaped) are similar to those of *C. fimbriata* (Fig. 115).

Disease cycle. Some elements of the cycle are known; others can be inferred from knowledge of similar diseases. The pathogen perennates in diseased and recently killed trees and also in cut logs of various hardwoods. Sap beetles that visit wounds are suspected to be vectors because they are involved with similar diseases. Occasional transmission occurs through natural root grafts. *A. saccharum* is most susceptible in spring and early summer. Infection occurs primarily through fresh wounds near the base of the trunk and on exposed roots. *C. virescens* is adapted for rapid spread in sapwood and usually avoids being restricted by wound-associated compartmentalization. Foliar and branch symptoms appear 1–6 years after infection. Saplings of this species have died as soon as 2 months after experimental inoculation.

References: 1517, 1612, 1770, 1772, 2982, 3348, 3724, 4396, 4407

Bluestain

Bluestain, as treated here, is fungus-induced dark staining of sapwood in conifer logs, lumber, and tree trunks. Stain colors range from gray or blue-gray to black. Bluestain fungi are primary colonists of stressed, dying, and recently killed trees. Dark melanin pigments in their hyphae cause the discoloration. The fungi are carried into tree trunks and limbs by insects, mainly bark beetles. The principal bluestain fungi are species of *Ceratocystis* (Microascales, Ceratocystidaceae), *Ophiostoma* (Ophiostomatales, Ophiostomataceae), and related anamorphs in the genera *Leptographium*, *Pesotum*, and others. These fungi have evolved

forms and habits that are linked to dispersal by beetles. Both *Ceratocystis* and *Ophiostoma* produce long-necked perithecia and sticky ascospores. The asexual states of *Ceratocystis* species are typified by those of *C. fimbriata* (Fig. 115); those of *Ophiostoma* are similar to those of *O. novo-ulmi* (Plate 119) and *Leptographium procerum* (Fig. 113). Fruiting structures and spores of these fungi typically form in galleries chewed by insects in the cambial region and inner bark. Some beetles feed on bluestain fungi.

The beetles involved most often are species of *Dendroctonus*, *Ips*, and *Scolytus* (Curculionidae, subfamily Scolytinae). *D. frontalis* (southern pine beetle), for example, carries *O. minus*, and *D. ponderosae* (mountain pine beetle) carries *O. clavigerum* and *O. ips* into *Pinus* species in southeastern and western USA, respectively. Beetles carrying bluestain fungi typically attack stressed and dying trees, but when attacking in great numbers may kill initially healthy trees as well. Fire, drought, and root damage by fungi such as *Armillaria* or *Heterobasidion* species are among the stress-inducing factors that initially make trees attractive to bark beetles. Once attack begins, more beetles arrive in response to aggregation pheromones. Although pathogenicity of various bluestain fungi has been demonstrated, the effects of bark beetle attack and sapwood colonization by these fungi cannot be separated in natural circumstances.

External symptoms of bark beetle attack and bluestain are fading and browning foliage and sawdust associated with holes in the bark. Internal symptoms vary according to a tree's vitality. A vigorous tree can halt the attack by infusing attack sites with resin. The resin and the reaction zones around beetle wounds contain terpenes, resin acids, and phenolic compounds that inhibit both insects and fungi. Stressed trees cannot mount effective defenses. Beetles construct galleries in the cambial region and reproduce there. Bluestain fungi spread up and down from the galleries and toward the center of the stem, stopping at the heartwood. They grow primarily in nutrient-rich phloem, xylem ray parenchyma, and the cells that line resin ducts. Colonized zones in the wood are linear to wedge shaped as viewed in cross section. When numerous they coalesce and involve the entire sapwood cylinder. The process may be completed within several weeks in warm weather.

Bark beetles become contaminated with spores of bluestain fungi before they emerge from the stems in which their parents bred. The spores adhere to beetles' bodies and are carried in cavities or pockets in the exoskeleton and sometimes internally, and so accompany the beetles to new breeding sites. Once there, germinating spores establish new colonies.

Other wood-staining fungi. Additional fungi, diverse in classification, pathogenicity, and ecological associations, cause dark stains in wood, primarily in sapwood (sapstain) and primarily in logs and lumber. Both hardwoods and softwoods are affected. The fungi range from saprobes to unspecialized opportunistic pathogens such as *Botryosphaeria rhodina* (Plate 63) and *Diplodia pinea* (Plate 64). Like bluestain fungi they have dark melanin pigments in their hyphae that cause the discoloration. They are dispersed in various ways, mainly by air and water. Some are carried in an incidental manner on insects. Unspecialized wood-staining pathogens enter limbs and trunks of stressed or moribund trees through wounds or by the extension of twig or bark infections. These fungi do not spread far within the wood until it is dying or dead.

References: 400, 461, 732, 844, 1169, 1519, 1891, 2094, 2945, 3608, 4396, 4407

A–D. Sapstreak of *Acer saccharum* (sugar maple), caused by *Ceratocystis virescens*. A. Premature foliar color, defoliation, and dieback in diseased trees. B. An old wound on a surface root that was possibly a site of infection. C, D. Internal symptoms revealed by chopping into exposed roots at the trunk base. Brown-stained sapwood is dead. Darker streaks adjacent to normal white sapwood are sites of current infection. The radiate pattern of discoloration results from growth of the parasite in rays (NY, Jul).

E–G. Bluestain in *Pinus* stems. E, F. Stain patterns indicating centripetal spread of bluestain in medullary rays of a *P. monticola* (western white pine) log (ID, Aug). G. Wedge-shaped stained sectors in a bolt of *P. contorta* (lodgepole pine) (BC, Mar).

Photo credits: A—G. W. Hudler; G—E. A. Allen

237

Oak Wilt (Plate 118)

Oak wilt is a systemic, usually lethal disease caused by *Cerato-cystis fagacearum* (Microascales, Ceratocystidaceae). It occurs only in the USA in a region delimited by Minnesota, Pennsylvania, South Carolina, and Texas. It has been most destructive in the upper Midwest and in Texas, killing millions of trees in the latter state. More than 20 *Quercus* (oak) species, all in the subgenus *Quercus*, have been found naturally infected, and all species that have been tested, including those indigenous to western North America and Europe, are susceptible to some extent. Oaks in section *Lobatae* (red oak group) are generally more susceptible than those in section *Quercus* (white oak group). Affected species in the former group include *Q. coccinea* (scarlet oak), *Q. ellipsoidalis* (northern pin oak), *Q. rubra* (northern red oak), and *Q. velutina* (black oak), among others. Affected species in the white oak group include *Q. alba* (white oak), *Q. fusiformis* (Texas live or plateau oak), *Q. macrocarpa* (bur oak), *Q. lyrata* (overcup oak), *Q. stellata* (post oak), and *Q. virginiana* (southern live oak). *Q. fusiformis*, *Q. virginiana*, and some European oaks in section *Quercus* are highly susceptible exceptions to the rule of partial resistance in white oaks. Infection in trees with partial resistance progresses slowly and is often localized by host defense reactions. The threat posed by oak wilt to Eurasian oaks has led to stringent restrictions on importation of oak materials from the USA.

Symptoms and signs. Oak wilt typically affects groups of trees because the pathogen spreads through root grafts and, in *Q. virginiana* and *Q. fusiformis*, through clonal root systems that support multiple stems. Symptoms appear in spring and summer and vary with oak species and region. In the North trees of the red oak group typically die within a year, often within 6 weeks, after symptoms appear. Wilt is first evident at or near the top of a tree. Leaves turn dull green, bronze, or tan, beginning along the tips and edges, often with an abrupt transition from dying to green tissue. Or leaves may droop, curl lengthwise, and wilt. Yellow to brown color sometimes develops along the veins. Leaf cast begins soon after symptoms appear, and some leaves drop while still green. In species of the red oak group, foliar symptoms often progress through the crown within a few weeks. Brown streaks develop in outer sapwood of some species coincident with foliar symptoms. The streaks are discolored clusters of dysfunctional xylem vessels and associated dead parenchyma cells. Symptoms in trees with partial resistance may appear on single or scattered branches and sometimes resemble normal autumn color changes.

In southern USA the symptoms and progress of the disease differ from the description above. Leaves of *Q. virginiana* and *Q. fusiformis* in Texas become chlorotic or bronze, often with brown tips and yellow to brown color along the veins. Streaks in xylem are not closely associated with foliar symptoms. Diseased trees may defoliate and die quickly, but some survive several years, displaying progressive dieback and producing adventitious sprouts with small leaves on trunks and large limbs. Some trees enter remission. Slow progressive dieback or remission also occurs in *Q. laevis* (turkey oak) in South Carolina.

In the autumn or the next spring after a diseased tree defoliates, *C. fagacearum* may form opposing pads (mats) of mycelium on the surface of the wood and the inner surface of the bark of the trunk. Mats are elliptic, 2.5–20 × 1–10 cm in size, initially gray or infrequently buff colored with white edges, and turn black with age. Compact cushions called pressure pads in the center of a mat cause the bark to crack open, allowing access to the mat by insects. Cylindrical endoconidia (anamorph genus *Thielaviopsis*, syn. *Chalara*) are produced on the mats. A mat initially represents one mating type of the fungus. If conidia of a compatible mating type are brought to the mat by insects, long-necked black perithecia soon form and produce colorless, elongate-ellipsoid ascospores. Mat formation is suppressed in dry years. Mats are rare on trees in the white oak group and have not been observed on *Q. virginiana* or *Q. fusiformis*. In Texas, mats on *Q. texana* (Texas red oak) and *Q. marilandica* (blackjack oak) are the sources of spores that infect live oaks.

Disease cycle. *C. fagacearum* overwinters in diseased and recently killed trees. Sap-feeding beetles in several genera of the Nitidulidae (*Carpophilus*, *Colopterus*, *Cryptarcha*, *Epuraea*, *Glischrochilus*, and *Lobiopa*) visit mats of *C. fagacearum* on killed trees, attracted there by a fruity odor that the fungus emits, and become contaminated with conidia and ascospores. Later, contaminated insects visiting fresh wounds of various kinds on healthy trees transmit the pathogen, thus accounting for the majority of new oak wilt infection foci. Probably not all the carrier species act as vectors. Most insect-mediated infections begin in spring, when oaks are most susceptible. Pruning wounds on shade trees are common sites of infection where *C. fagacearum* mats occur nearby. Wounds more than a few days old or covered with paint are not suitable, however. Infections in trees of the white oak group are mostly dead ends with respect to further transmission by nitidulids. Two species of oak bark beetles, *Pseudopityophthorus minutissimus* and *P. pruinosus* (Curculionidae, subfamily Scolytinae), are also vectors of *C. fagacearum* and could account for some long-distance spread of oak wilt, but their roles are thought to be minor compared with those of nitidulids.

Systemic infection results from passive movement of spores in the sapstream in xylem vessels and cell-to-cell mycelial growth between vessels and through intervening cells. Spores produced in xylem fluid are carried upward in some vessels and downward in others. Mycelial colonies develop after spores lodge in pits and at vessel endwalls. These vessels soon become plugged with tyloses and dark gummy substance, and adjacent parenchyma cells die and darken. The mechanism of killing is not fully understood, although it is clear that water conduction ceases in diseased xylem. Toxins and growth-regulating chemicals are apparently involved in abscission and death of leaves and in formation of abnormal xylem outside the infected part of a xylem ring.

When a diseased oak wilts, *C. fagacearum* is present throughout the symptomatic parts and in other parts as well. After leaves die, the pathogen in twigs and small branches often dies too, apparently because of heating, drying, and competition from secondary fungi. (*C. fagacearum* is intolerant of temperatures above 32°C.) It may remain alive until the next year in the trunk, however, initially growing both toward inner sapwood and outward to the cambial region and inner bark, where mats form and spores become available to vectors. *C. fagacearum* survives up to 4 years in roots.

Isolates of *C. fagacearum* vary in aggressiveness, but no evidence of strains specialized for particular host species has been reported. Geographic variation in frequency of mat formation is due to environment.

After the first tree in a new location becomes diseased, *C. fagacearum* may spread to adjacent stems on the same root system or to oaks of the same taxonomic section via naturally grafted roots. Adjacent trees wilt 1–6 years after infection of the original tree, and disease centers commonly expand 1–15 m per year (up to 40 m per year recorded in Texas). Large foci of disease (tens of hectares) may thus develop in the upper Midwest and in Texas; smaller foci prevail in the East. Cutting roots (with either a vibratory plow or a trench digger) or chemically killing roots between diseased and healthy trees can prevent transmission through roots.

The hazard of oak wilt infection at pruning wounds can be minimized by pruning during the dormant season or during summer. Oak wilt hazard can also be reduced by girdling or felling and cutting newly diseased trees into firewood. This hastens drying and suppresses or prevents mat formation. Systemic fungicide injections are beneficial for oak wilt suppression in live oaks.

References: 93–97, 128, 246, 681, 753, 1186, 1188, 1264, 1265, 1610, 1774, 1889, 2187, 2196, 2303, 2316, 2552, 2579, 2678, 2981, 3109, 3225, 3264, 3892, 3952, 3953, 3955–3957, 4090, 4364, 4375

Oak wilt, caused by *Ceratocystis fagacearum*.

A. Dead and dying *Quercus ellipsoidalis* (northern pin oak) (WI, Jul).

B–E. Foliar symptoms in: B, *Q. ellipsoidalis* (WI, Jul); C, *Q. rubra* (northern red oak) (PA, Aug); D, *Q. velutina* (black oak) (MO, Jul); and E, *Q. virginiana* (TX, Apr).

F, G. Brown streaks in outer sapwood of *Q. velutina*. F. Longitudinal view in a small branch. G. Viewed in cross section of a small trunk (MN, Aug).

H–J. Mats and pressure pads produced by *C. fagacearum* beneath the bark of recently killed *Q. velutina*. H. Cracks in bark (arrows) indicate presence of pressure pads. I. Opposing pressure pads on bark and wood. J. Gray-black pressure pads on a gray mycelial mat where conidia are abundant (MN, Jul).

Photo credits: I, J—D. W. French

239

Dutch Elm Disease (Plate 119)

This famous disease is caused primarily by *Ophiostoma novo-ulmi* (Ophiostomatales, Ophiostomataceae), the most aggressive of three closely related species specialized for infection of *Ulmus* (elm) sapwood. The other two are the original Dutch elm disease pathogen, *O. ulmi*; and a Himalayan species, *O. himal-ulmi*. Epidemics caused by *O. ulmi* began in western Europe early in the 20th century and in North America in the 1920s. Vectored by elm bark beetles and carried long distances in elm wood by people, *O. ulmi* spread to most places where elms grow on both continents and in western Asia. *O. novo-ulmi* began to replace the original pathogen in those regions beginning in the 1940s and was later introduced to New Zealand. The remainder of this account emphasizes disease caused by *O. novo-ulmi*. European and North American subspecies of this pathogen are recognized; their effects on elms are similar.

Symptoms and signs. Leaves wilt and shrivel early in the season or turn yellow to brown and drop during summer. Symptoms usually begin on one branch and gradually involve the entire crown. A tree may die within a year or survive in declining condition for several years. Symptoms in highly susceptible elms develop rapidly during a period of 4–6 weeks beginning when leaves reach full size and then spread through a tree during the remainder of the season or during the next year. Primary symptoms are internal: death of xylem parenchyma, loss of water-conducting ability of xylem vessels, and brown streaks that follow the grain in the springwood of the youngest xylem sheath. The streaks are discolored clusters of dysfunctional vessels and associated parenchyma. Some diseased branches die during the growing season, others during dormancy. Infection in a resistant tree is often localized in small branches. If a tree survives infection, normal sapwood forms outside the sheath containing the brown streaks. The pathogen remains alive but sequestered in these streaks for years. In areas where summers are dry, symptoms induced by heat stress and water shortage (localized yellowing, premature defoliation, and branch dieback) can mimic or mask those of Dutch elm disease. Diagnosis usually depends on isolation and identification of the fungus in culture from streaked xylem. Segments of small (e.g., 1 cm diameter) symptomatic branches, collected while bark and wood are still moist, are convenient for assay.

Foliar and xylem symptoms of Dutch elm disease are caused in part by toxins secreted by the fungus and in part by xylem dysfunction. At least two proteinaceous toxins and several hydrolytic enzymes are produced by *O. novo-ulmi* in diseased elms. Xylem dysfunction is caused in part by enzymatic degradation of pectic cell wall constituents and in part by trees' defense processes, which when expressed too slowly to halt infection ensure extensive plugging of xylem vessels.

Structures of the pathogen vary with its life stage and environment. In xylem vessels in early stages of disease, *O. novo-ulmi* produces spores mainly by budding. Later, during saprobic growth, it produces colorless, unicellular conidia on simple conidiophores and on black-stalked synnemata in galleries made by elm bark beetles. The synnemata consist of bundled conidiophores that produce spores in slimy drops up to 1 mm across. Sometimes, long-necked black perithecia develop among or following synnemata. They produce colorless ascospores that, like the conidia, are in a mucilaginous drop and are thus well suited to adhere to insects.

Disease cycle. *O. novo-ulmi* overwinters in diseased and recently dead elms and in freshly cut elm logs and firewood. It grows as a saprobe in recently dead elm wood and bark and sporulates in galleries excavated by elm bark beetles and their larvae. The smaller European elm bark beetle, *Scolytus multistriatus*, and the native elm bark beetle, *Hylurgopinus rufipes* (both in Curculionidae, subfamily Scolytinae), are vectors of Dutch elm disease fungi in North America. *H. rufipes* is important in Canada and in the northern parts of border states. *S. multistriatus* predominates in most areas of the USA and parts of southern Canada. It and several other scolytid beetles are vectors in Eurasia. Adult beetles become contaminated with spores of *O. novo-ulmi* as they emerge from pupation. They inadvertently carry viable spores for distances up to several kilometers, both to feeding sites, where new infections begin, and to breeding sites, where saprobic growth resumes. Attracted by volatile chemicals that elms emit, the insects bore into the inner bark, cambial region, and outermost sapwood and there incidentally

leave spores. *S. multistriatus* feeds in twig crotches; thus most infections begin in twigs. *H. rufipes* bores in branches and small trunks, initiating infections that involve major branches from the outset. Both beetle species breed in dying and recently dead stems or cut pieces 3 cm or more in diameter.

From a point of inoculation in a healthy elm, the pathogen moves upward and downward by two alternating modes: passive transport of spores in xylem sap and hyphal growth between vessels after spores lodge and germinate in pits and at vessel endwalls. *O. novo-ulmi* regularly reaches the root of large trees within the first season of infection. It proliferates in the roots and then ascends the trunk in a wave of systemic infection that kills the tree or a major part of it the next year. Infection may spread from tree to tree through root grafts where elms are spaced 10 m or less apart.

The severity and pace of a Dutch elm disease epidemic depend on the strain of pathogen, reproductive success of vectors, level of elm susceptibility, and elm population density. The aggressive species *O. novo-ulmi* kills highly susceptible elms rapidly and thus promotes vector population increase and fast epidemic development. The vector population rises to high levels in early years of an epidemic and remains high until nearly all large elms in a locality are gone. Natural enemies of the vectors are not effective while breeding sites are abundant. Reproduction of *S. multistriatus* is limited by severe winters in parts of northern USA and in Canada, however.

Elm species vary from highly susceptible to highly resistant. Highly susceptible species include *U. alata* (winged elm), *U. americana* (American or white elm), *U. ×hollandica* (Dutch elm), *U. laevis* (European white elm), *U. rubra* (red or slippery elm), *U. serotina* (September elm), and *U. thomasii* (rock elm). Those of intermediate susceptibility under North American conditions include *U. crassifolia* (cedar elm), *U. glabra* (Scotch or wych elm), and *U. minor* (European field or smooth-leaf elm, including English elm). Resistant species include *U. davidiana* (David or Japanese elm), *U. parvifolia* (Chinese or lace-bark elm), and *U. pumila* (Siberian elm), as well as many lesser-known Asiatic elms. Susceptibility varies widely within some species, notably *U. glabra*, *U. minor*, and *U. pumila*. Tree breeders have developed many Eurasian elm cultivars with resistance to *O. novo-ulmi* that are suitable for temperate regions of North America. In addition, four cultivars of *U. americana*—Jefferson, New Harmony, Princeton, and Valley Forge—have moderate resistance and may be useful in areas where elm yellows (Plate 197) does not occur.

Control of Dutch elm disease in susceptible elms depends mainly on denying elm bark beetles places to breed. This means a continuing program of detection, prompt removal of diseased limbs or trees, and burial or destruction of all dying or recently dead elm wood found in and near the elms that are to be preserved. Once this basic requirement is met, additional techniques, such as injection with systemic fungicides, can be applied for higher levels of protection. Elms resistant to *O. novo-ulmi* do not require protection. *References:* 128, 308, 440, 456, 477, 478, 480, 482–484, 663, 639, 1006, 1013, 1515, 1727, 2123, 2124, 2419, 2636, 2830, 3280, 3427, 3581, 3603, 3640, 3654, 3692, 3763, 3866, 3871, 3961, 3962, 3984, 4070, 4071, 4364, 4449

A–E. Foliar symptoms of Dutch elm disease caused by *Ophiostoma novo-ulmi*. A. Wilting and defoliation of *Ulmus americana* (American elm) that became infected a year before the photo was made. B. Wilting leaves on branches of *U. glabra* 'Camperdownii.' C–E. Wilting, browning, and shriveling leaves of *U. americana* in midsummer (all NY, Jun–Jul).

F–H. Brown streaks in outer sapwood of twigs and a small trunk of *U. americana*. F. A slant cut through a twig that became infected in its fourth year of growth. Discolored streaks appear as brown dots in the most recently formed wood. G, H. Longitudinal views of brown streaks in outermost xylem exposed by removing bark of a small trunk and a twig, respectively (NY, Jul–Oct).

I. Magnified view of *Ophiostoma novo-ulmi* on elm wood. Synnemata are up to 1.5 mm tall. Hyphae have a granular appearance because of the presence of conidiophores with clusters of colorless conidia (NY, Oct).

Photo credit: B—G. W. Hudler

Verticillium Wilt (Plates 120, 121)

The soilborne fungi *Verticillium dahliae* and *V. albo-atrum* (hyphomycetes) cause yellowing, wilt, scorch, dieback, and decline or death of woody and herbaceous plants in many families in temperate regions around the world. *V. dahliae* is the species that most commonly attacks woody plants in North America. It has been distributed widely in diseased but asymptomatic plants and plant parts and in soil. Damage is common in nurseries, orchards, and landscapes. The few reports of Verticillium wilt in North American woodlands include isolation of *V. albo-atrum* from wilting seedlings of *Liriodendron* (tuliptree) in Delaware and from wilting *Ceanothus* in California, and of *V. dahliae* from dying *Ailanthus altissima* (tree-of-heaven) in Pennsylvania.

Woody genera with species susceptible to *Verticillium* include the following, among others (some listed genera have both susceptible and resistant species):

Acacia, Acer (maple), *Actinidia* (kiwi-fruit), *Aesculus* (buckeye, horse-chestnut), *Ailanthus* (tree-of-heaven), *Albizia* (mimosa), *Amelanchier* (serviceberry), *Aralia, Artemisia* (sagebrush), *Atriplex* (saltbush), *Aucuba, Bauhinia* (butterfly-tree), *Berberis* (barberry), *Blighia* (akee), *Buxus* (boxwood), *Camellia, Carica* (fig), *Catalpa, Ceanothus, Ceratonia* (St. John's bread), *Cercis* (redbud), *Chrysothamnus* (rabbitbrush), *Cinnamomum* (camphor tree), *Cistus* (rock-rose), *Cladrastis* (yellowwood), *Cotinus* (smoke tree), *Cupaniopsis* (carrotwood), *Cytisus* (broom), *Daphne, Diospyros* (persimmon), *Dodonaea* (Florida hopbush), *Elaeagnus* (Russian-olive), *Erica* (heath), *Fatsia* (paperplant), *Ficus* (fig), *Fraxinus* (ash), *Fremontodendron* (flannelbush), *Fuchsia, Gymnocladus* (coffeetree), *Hebe, Hedera* (English ivy), *Hibiscus, Jasminum* (jasmine), *Koelreuteria* (golden-rain tree), *Krascheninnikovia* (winterfat), *Larrea* (creosote bush), *Ligustrum* (privet), *Liriodendron* (tuliptree), *Lonicera* (honeysuckle), *Maclura* (osage-orange), *Magnolia, Malus* (apple, crabapple), *Mangifera* (mango), *Nandina* (sacred-bamboo), *Nyssa* (tupelo), *Olea* (olive), *Osmanthus, Paeonia* (peony), *Parthenium* (guayule), *Parthenocissus* (Virginia creeper), *Persea* (avocado), *Phellodendron* (cork tree), *Photinia, Pistacia* (pistache), *Pittosporum, Populus* (aspen, poplar), *Prunus* (stone-fruit trees), *Pyrus* (pear), *Quercus* (oak), *Rhaphiolepis* (India-hawthorn), *Rhododendron* (azalea), *Rhus* (sumac), *Ribes* (currant), *Robinia* (locust), *Rosa* (rose), *Rubus* (brambles), *Sambucus* (elderberry), *Sassafras, Schinus* (pepper-tree), *Spiraea, Styphnolobium* (Japanese pagoda tree), *Syringa* (lilac), *Tilia* (linden), *Ulmus* (elm), *Viburnum, Vitis* (grapevine), and *Weigela*.

Woody species or genera resistant or immune to *Verticillium* as indicated by testing or practical experience include all gymnosperms, all monocots, cacti, and the following dicots:

Arctostaphylos (manzanita), *Asimina triloba* (pawpaw), *Betula* (birch), *Buxus* (boxwood), *Carya* (hickory, pecan), *Castanea* (chestnut), *Ceanothus* (buckbrush, redroot, snowbrush), *Celtis* (hackberry, sugarberry), *Cercidiphyllum japonicum* (katsura tree), *Chaenomeles japonica* (flowering quince), *Cistus* (rock-rose, particular species), *Citrus, Cornus* (dogwood), *Crataegus* (hawthorn), *Cydonia oblonga* (quince), *Eucalyptus, Fagus* (beech), *Gleditsia triacanthos* (honeylocust), *Hebe* (particular species), *Ilex* (holly), *Juglans* (butternut, walnut), *Liquidambar styraciflua* (sweetgum), *Malus* (apple, crabapple), *Morus* (mulberry), *Nerium oleander* (oleander), *Platanus* (planetree, sycamore), *Populus* (aspen, poplar), *Pyracantha* (firethorn), *Pyrus* (pear), *Quercus* (oak), *Rhododendron, Salix* (willow), *Sorbus* (mountain-ash), *Tilia* (linden), *Umbellularia californica* (California laurel), and *Zelkova serrata* (Japanese zelkova).

Some genera with both susceptible and resistant species appear in both lists. Resistance may vary from one region to another and according to the virulence of local *Verticillium* strains. Rare reports of Verticillium wilt in some resistant species—*Quercus palustris* (pin oak), for example—do not offset their usefulness for sites where the disease has occurred.

Symptoms and signs. Verticillium wilt has acute and chronic phases. Acute symptoms include wilt, leaf curling or drying, abnormal red or yellow color of entire leaves or interveinal areas, defoliation, dieback, and death. These symptoms, which indicate infection of the current year's sapwood, may be restricted to particular branches or may involve an entire plant. They appear from early summer into autumn, depending on the type of plant and the geographic region. Plants may die suddenly, but they more often show progressive or intermittent symptoms. Yellowing and defoliation often progress upward. *Fraxinus pennsylvanica* (green ash) sometimes drops green leaves as well as symptomatic ones in response to infection. Plants may have acute symptoms in consecutive growing seasons or may skip one to several years. In the interim, depending on the severity of infection, the plant may appear normal, or one or more branches may have chronic symptoms: slow growth, sparse foliage, stunted leaves and twigs,

leaf scorch, abnormally heavy seed crop, and dieback. Chronic symptoms indicate stress caused by the death of sapwood in previous years. Acute and chronic symptoms often occur simultaneously.

Infected sapwood may have dark streaks or bands that follow the grain, but this discoloration may be absent during the earliest stages of infection. It develops infrequently in *F. pennsylvanica* and does not develop in some susceptible plants such as *Olea europaea* (olive). Diseased sapwood in most hosts first appears slightly discolored and water-soaked, then assumes a darker color characteristic of the plant species. Seen in cross or slant cuts, the necrotic tissue appears as arcs that follow growth rings. Stems infected for longer than a year may have these symptoms in more than one growth ring. Color of the streaks varies from light tan in *Fraxinus* to dark reddish brown in *Robinia* (black locust), yellowish brown in *Prunus* (cherry) and *Cotinus* (smoke tree), and greenish to nearly black in *Acer* (maple). Hyphae are microscopically visible in vessels of the discolored wood. Wood beneath the dark streaks toward the center of the stem or root dies and turns brown later. Stem cracks sometimes develop in saplings that have a core of dead wood. Microsclerotia of *V. dahliae* form in dying parts, notably roots, leaves, and fruits. Microsclerotia are black cellular aggregates 15–100 µm in diameter. Several additional diagnostic structures develop in *Verticillium* cultures (Fig. 121).

Disease cycle. Verticillium species that cause plant diseases are soilborne. When roots in infested soil grow close to or into contact with *Verticillium* propagules (microsclerotia or resting hyphae, discussed below), these structures germinate and hyphae from them penetrate feeder roots and wounds. Hyphae grow in the root cortex and, in susceptible species, enter the stele. Hyphal growth may continue there, or more important, the fungus may produce spores that move passively in the sap stream. Mycelium of *Verticillium* grows only a few millimeters per day, but spores in the sap stream are quickly carried the length of a xylem vessel—a few centimeters to more than 5 m, depending on the kind of plant and the location within it. Where the spores lodge and germinate, new hyphae grow and intensify the infection. Repetition of these processes can result in the spread of infection throughout large plants. Necrosis develops at loci of spore germination and hyphal growth. Expansion and coalescence of necrotic tissue give rise to the streaks and bands of discolored sapwood.

Wood killed by *Verticillium* does not conduct water or nutrients to leaves. Therefore, noninfected twigs and leaves above necrotic xylem are subject to water stress, nutrient shortage, abrupt cessation of growth, and perhaps death. Wilting and death of all leaves on a

---

Verticillium wilt of *Acer* species (maples) caused by *Verticillium dahliae* (all NY, Aug–Oct).

A, B. Acute symptoms in *A. saccharum* (sugar maple) in a nursery.

C. Withering of scattered leaves at the top of a young *A. saccharum*. These branches remained alive but stunted, and the tree later became asymptomatic.

D. Acute (left) and chronic symptoms in landscape specimens of *A. platanoides* (Norway maple).

E. Severe chronic symptoms in *A. saccharum*.

F. Foliar symptoms on *A. saccharum*: dwarfing, mottling, scorch at tips and margins, water-soaking and then necrosis of interveinal areas.

G. Symptoms on *A. rubrum* (red maple) sprout leaves: dwarfing, downward cupping, yellowing, and scorch.

H. *A. saccharum* with leafless branches killed by *Verticillium* the previous year. No new symptoms appeared during a 5-year observation period, and growing branches partially filled the canopy space where diseased branches died.

I–K. Internal symptoms. I. Light gray-green streaks in outmost sapwood of *A. saccharum*. J. Segments of diseased stems of *A. rubrum* (center) and *A. saccharum* saplings. Segments at top and center were from trees with acute symptoms. Segments at bottom were from a tree with severe chronic symptoms; all but a thin sheath of sapwood was dead. K. Cracks that developed as secondary defects in the mainstem of an *A. rubrum* sapling infected by *V. dahliae*.

Photo credits: A, B, I, K—M. L. Daughtrey

branch indicate infection of the current season's wood near the branch base or in the trunk below. The infection may extend into twigs, buds, and leaf bases in a small plant or into sprouts low on the trunk and main branches of a large plant.

Additional aspects of tree colonization by *Verticillium* are known mainly from studies of *Acer*. *Verticillium* progresses around a growth ring by a combination of upward spread and tangential growth; thus the cross-sectional extent of necrotic xylem increases with distance above its origin. Downward movement is negligible, but the fungus often grows outward to the vicinity of the vascular cambium and bud traces. If the stem remains alive, the growth of *Verticillium* close to these meristematic regions allows continuous infection from the wood of one growing season to that of the next. In *Acer* and *Liriodendron* elongate strips of the vascular cambium sometimes die, resulting in cankers. *Cytospora* sp. (Plate 86) and *Nectria cinnabarina* (Plate 87) are common in the bark of these cankers in *Acer*. If the pathogen fails to cross from one season's wood to the next, the result is remission of acute symptoms and compartmentalization of the diseased wood. The severity of chronic symptoms then depends on the extent of damage in the old wood. Acute symptoms that recur after a year or more of remission indicate a new upward thrust of infection from roots. Infections usually do not extend to the ends of major branches.

Verticillium species produce assorted enzymes and toxins that degrade primary cell walls, kill xylem parenchyma, and cause xylem vessels to become occluded with interior surface coatings and cell-wall breakdown products. Sap flow to distal parts is thus reduced or halted. Foliar symptoms can be due to local pathogenesis or to stress from water and nutrient shortage and possibly to toxins translocated from lesions below. Xylem dysfunction in susceptible plants is aggravated by host defense reactions that occur too slowly to halt infection. In resistant plants these reactions either prevent the pathogen from entering xylem or isolate xylem lesions.

Verticillium species persist within plants and can be transmitted in scions, buds, and rootstocks. After diseased parts die, *V. dahliae* survives as microsclerotia. Microsclerotia that form in leaves or fruit are dispersed locally when these parts drop. When infested parts decompose, the microsclerotia remain. Microsclerotia form in susceptible weeds and in roots of many kinds of resistant plants that have no aboveground symptoms. Thus the pathogen can persist indefinitely in the absence of susceptible plants, and crop rotation is not effective for control.

Microsclerotia are transported along with soil particles in water or wind, on equipment, with plants, and in wood chips from diseased plants. They are resistant to most environmental stresses, toxic chemicals, and microbial attack, and can survive in soil for years in the absence of plants. Only in warm, waterlogged soils do microsclerotia, like roots, die rapidly. They germinate and produce infectious hyphae when stimulated by nutrients that exude from roots or decaying organic matter. Each microsclerotium is capable of germinating several times. If a microsclerotium germinates but conditions favorable for hyphal growth are not sustained, it retains the ability to germinate again upon the return of favorable conditions.

V. albo-atrum depends on specialized dark resting hyphae for survival. These hyphae seem to have attributes and functions similar to those of microsclerotia, although much less is known about them. Conidia of *Verticillium* lack survival value. They cannot tolerate sunlight or drying, and they are rapidly destroyed by other microorganisms in soil.

Isolates of *Verticillium* vary in aggressiveness and virulence. In general a given isolate is capable of causing symptoms in various plants and is likely to be most aggressive in the original host. Isolates of *V. albo-atrum* tend to cause more severe disease than do isolates of *V. dahliae* when these fungi are tested in plants of the same type. *V. dahliae* comprises several vegetative-compatibility (VC) groups that, by definition, are incapable or sparingly capable of exchanging genetic information between groups by hyphal anastomosis. Particular VC groups predominate in some areas, but host specificity of VC groups in woody plants has not been detected. Isolates within VC groups vary in virulence.

Environmental conditions influence the severity of symptoms caused by *Verticillium*. For example, other factors aside, generously watered plants are invaded less extensively than those under water

stress. Stress induced by salt or transplanting also has an adverse effect. Trees that become infected while in nurseries often develop symptoms during the first 1–2 years after transplanting but later enter long-term remission. Soil nutrient conditions that seem to optimize a tree's intrinsic resistance are moderate fertility with the balance of major nutrients tipped slightly toward high potassium and low nitrogen.

Verticillium species do not tolerate high temperatures. *V. dahliae* in branches and trunks of orchard trees in California dies during the hot dry season. Recurrence of disease in these trees depends on reinfection of roots during winter and spring. In hot, sunny regions, *V. dahliae* can be suppressed by allowing soil to heat beneath a film of transparent plastic, but solar heat is insufficient for disease suppression in most regions where Verticillium wilt is a problem.

Nematodes that attack plant roots may aggravate Verticillium wilt. *Pratylenchus penetrans* (a lesion nematode) and *Meloidogyne* species (root-knot nematodes) are noteworthy in this respect. Accordingly, control of nematode pests helps to suppress Verticillium wilt.

The most practical approach to control of Verticillium wilt in landscapes and orchards is to use resistant plant types. Resistant individuals or cultivars occur even within plant groups such as *Acer* that are known for their susceptibility.

References: 123, 127, 128, 131, 133, 188, 354, 447, 650, 651, 718, 1015, 1095, 1337, 1509, 1554, 1647, 1665, 1725, 1865, 2309, 2375, 2440, 2531, 2719, 2720, 2757, 2840, 2896, 2963, 3004, 3293, 3364, 3402, 3487, 3543, 3651, 3726, 3727, 4046–4048, 4072, 4108, 4365

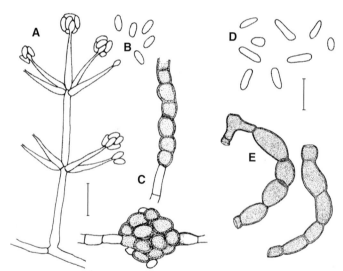

Figure 121. Verticillium speices. A. A conidiophore with conidia in slime drops on branches arranged in whorls, typical of the genus. A–C. *V. dahliae*: conidiophore, conidia, and microsclerotia. D, E. *V. albo-atrum*: conidia and dark resting hyphae. Scale bars = 10 µm. A–C adapted from reference 123 by permission of Gebrüder Borntraeger Verlagsbuchhandlung; D, E from reference 1554 by permission of CAB International.

Verticillium wilt, continued.

A–D. Symptoms caused by *Verticillium dahliae* in *Fraxinus pennsylvanica* (green ash). A, B. One-sided defoliation and dieback in young trees. C, D. Chlorosis, mottling, and marginal and tip necrosis on leaflets (MN, Jul–Sep).

E. Wilted sprouts on a scaffold limb of *Pistacia chinensis* (pistachio) (CA, Aug).

F. Wilt and dieback caused by *V. dahliae* in *Rhus aromatica* (fragrant sumac) (NY, Jul).

G. Brown streaks and central discoloration in a young branch of *Prunus cerasus* (sour cherry) infected by *V. dahliae*. The core of the branch died after being surrounded by a sheath of infected sapwood (NY, Sep).

H–K. Foliar and xylem symptoms caused by *V. dahliae* in *Cotinus coggygria* (smoke tree). Brown streaks lie in outer sapwood between white inner bark and white inner sapwood. The yellowish to brown color of heartwood is normal (NY, Aug).

Photo credits: A–D—C. L. Ash

Persimmon Wilt (Plate 122)

Persimmon wilt is a systemic lethal disease of *Diospyros virginiana* (common persimmon) caused by *Nalanthamala diospyri* (syn. *Acremonium diospyri*), a hyphomycete known from DNA sequence analyses to be allied with *Rubrinectria* (Hypocreales, Nectriaceae). The disease was first reported when an epidemic occurred in the mid-1930s in central Tennessee. The majority of large persimmon trees in that area soon died. Eventually the disease was found from North Carolina to Kansas and south to Florida and Texas. Now it is endemic in that region but is unknown elsewhere.

Symptoms and signs. Symptoms may appear at any time during the growing season but usually develop in late spring. Leaves at branch tips wilt and shrivel or turn yellow and drop. These symptoms soon involve the entire plant. After the onset of hot weather the plant may stand with branches nearly bare except for partly grown fruit. Entire trees sometimes die within 2 months. Trees wilting at the end of a growing season fail to grow the next year or may produce sparse undersized, chlorotic leaves. Such trees either die during the summer or persist in a declining state for 1–2 more years. Dying or recently dead trees are soon colonized by wood-decay fungi, notably *Schizophyllum commune* (Plate 158).

Internal symptoms are narrow dark brown to black streaks that follow the grain of the wood in roots, trunks, and branches. Sapwood of the youngest age class in a wilting plant is always streaked. Numerous streaks may nearly coalesce into bands that appear as arcs within the annual rings in cross section. Streaks are also usually apparent in several older annual rings. Over time the streaks fade to gray.

In autumn of the year when a stem dies the bark surface cracks and loosens, exposing irregularly shaped spore-bearing patches covered with pinkish orange masses of conidia. The patches develop either at the depth of the cork cambium or at that of the vascular cambium and are often many square centimeters in extent. They form beneath dark reddish blisters on smooth-barked stems but cannot be detected beneath rough bark until the cork layer cracks and falls away. Spore-bearing patches may form on branches but are most common near the base of the trunk. Clouds of spores rise when dry spore masses are disturbed.

N. diospyri can be isolated readily from diseased plants. It grows most rapidly at temperatures near 30°C and grows only slowly at 20°C. It sporulates after 7–10 days in culture on common media. The conidia, individually colorless but orange-pink in mass, are unicellular and of two types: ovoid, 3.5–6 × 2–4 μm, with one blunt end; and cylindric, 4.5–20 × 2–4.5 μm, with rounded ends. They are borne in slime droplets atop simple or once-branched conidiophores.

Disease cycle. The first infection in a persimmon grove occurs in wounds made by insects, animals such as cattle, or storm breakage. Thereafter the pathogen spreads perennially among adjacent trees through their common root system and also by airborne spores. Insects that create wounds suitable for infection include a twig girdler, *Oncideres cingulata* (Coleoptera, Cerambycidae); and a powder post beetle, *Xylobiops basilaris* (Bostrichidae). Stubs left by the twig girdler have been found infected before the fungus could be isolated from other parts of trees.

Persimmon trees are susceptible throughout the year, but the interval from infection to wilting varies. Trees inoculated from late winter to early summer may wilt and produce spore masses in the same year. Trees inoculated in summer and autumn wilt and produce spores a year later. Most natural infections by spores presumably occur in late summer and autumn.

Once in the xylem the parasite produces spores that spread in water-conducting vessels throughout the tree. Parenchyma cells that surround colonized vessels die and are colonized and turn brown. Vessels become plugged with tyloses, and both vessels and nearby intercellular spaces also fill with dark, gumlike deposits that are responsible for the streaks seen in diseased wood. Mechanisms of symptom induction are unknown. Affected stems die when the pathogen approaches the vascular cambium around most of the stem's circumference.

Factors that triggered the persimmon wilt epidemics of the 1930s and 1940s remain unknown. An epidemic occurred in Mississippi in 1970 among trees damaged the previous summer by a hurricane. The disease had not been prominent where the outbreak occurred.

Among persimmons and other *Diospyros* species tested for reaction to *N. diospyri, D. virginiana* was most susceptible and *D. texana* (Texas or black persimmon) was somewhat less so. *D. kaki* (Japanese persimmon), the species usually cultivated for fruit, is highly resistant unless grafted on *D. virginiana* rootstock. *Diospyros* species known as ebony and naturalized in Mexico and Central America are highly susceptible. The Asian species *D. lotus* is resistant. Inoculations to *D. blancoi, D. montana,* and *D. rosei,* which are native to the Philippines, India, and Mexico, respectively, resulted in no symptoms.

D. virginiana has both friends and foes, and persimmon wilt has been important to both. The tree is useful for ornament. Its fruit is eaten or made into preserves, and its wood is used for golf club heads and specialty furniture items. On the other hand, it is an aggressive colonist of open land. Persimmon groves encroaching on pastureland in Oklahoma and Arkansas prompted farmers to attempt to spread the wilt by transporting wood from diseased groves to healthy ones. Plant pathologists showed farmers how to inoculate unwanted persimmons with spore suspensions of the pathogen. The procedure was simple: squirt a spore suspension into ax frills. If half the trees in a grove were thus treated in spring at about the time of bud burst, all or nearly all the trees would eventually die, because the pathogen would spread from tree to tree via roots. Entire groves could be killed within 3 years in this way.

N. diospyri is also allied with pathogens of Arecaceae (palms) and *Psidium guajava* (guava). *N. vermoesenii* causes pink rot (necrosis of buds, fronds, trunks) of various palms in California, Florida, and other warm-temperate and tropical regions. *N. guajavae* causes guava wilt in South Africa and Malaysia. *N. diospyri* under the name *Cephalosporium diospyri* was erroneously reported to cause wilt diseases of *Quercus virginiana* (live oak) and *Ulmus* (elms). In fact, the decline and dieback of *Quercus* species once attributed to *C. diospyri* in Texas is caused primarily by the oak wilt fungus, *Ceratocystis fagacearum* (Plate 118). The elm fungus, now known as *Dothiorella ulmi,* does cause a wilt and dieback disease (symptoms similar to Dutch elm disease and Verticillium wilt) but is not related to *N. diospyri. D. ulmi* produces pycnidia on killed branches and in cultures on laboratory media.

References: 128, 829, 830, 837, 1226, 1324, 1553, 2306, 3493, 4154, 4377, 4378

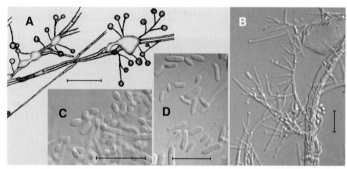

Figure 122. A, B. *Nalanthamala diospyri.* A. Hyphae and conidiophores with slime droplets. B. Conidiophores mounted in water. C, D. Conidia of two sizes. Scale bars: A = 50 μm, others = 20 μm. A adapted from reference 829 by permission of the Mycological Society of America; B–D courtesy of H.-J. Schroers.

Persimmon wilt, caused by *Nalanthamala diospyri,* in *Diospyros virginiana* (common persimmon).

A. Dead and diseased trees in a grove (OK, Jul).

B–D. Foliar symptoms. B. Suppressed twig growth and stunted leaves with uprolled edges cause foliage of a recently infected tree to appear sparse. C, D. Wilted, yellow, and dead leaves on small diseased trees. Shriveled brown leaves that wilted in late spring cling to dead twigs. Leaves becoming symptomatic at midsummer turn yellow and drop, leaving bare twigs (OK & MS, Jul).

E–G. Gray to dark brown or black streaks in outer sapwood. Discoloration caused by *N. diospyri* is visible in several annual rings of a small trunk viewed in cross section. The orange line between bark and wood in E and F is the cambial zone, not a symptom of disease (OK, Jul).

H. A pink sporulating patch on the trunk of a small tree killed by the disease (KS, autumn).

Photo credit: H—N. A. Tisserat

Fusarium Wilts (Plate 123)

Fusarium wilts are caused by host-specialized populations of *Fusarium oxysporum*. This species name is applied to a globally distributed complex of morphologically indistinguishable fungi that includes both pathogens and saprobes. Some of the pathogens cause root rots; others cause systemic infections of xylem known as wilt diseases or vascular wilts. These fungi reproduce only asexually, but DNA analyses have shown them to be related to ascomycetes in the Nectriaceae. Fusarium wilt diseases affect diverse woody and herbaceous angiosperms and monocots, mainly in warm-temperate and tropical regions. The pathogens are soilborne and have been distributed widely in contaminated soil and plant materials. They produce chlamydospores (thick-walled spores adapted for survival) and two types of conidia (macro and micro). The conidia sometimes form on above ground plant parts. Thus these fungi can also be dispersed by water, air, and insects.

Wilt-inducing strains of *F. oxysporum* have differentiated into numerous host-specialized populations called formae speciales, each of which attacks plants in one genus or a few related genera. The best-studied formae speciales each comprise several vegetative-compatibility groups that reproduce as clonal lines. Strains of a given vegetative-compatibility group are generally incapable of hyphal anastomosis and genetic interaction with strains of other groups in the same or other formae speciales. Races able to attack resistant cultivars of susceptible species occur within many of the formae speciales. Strains assignable to a given race may represent more than one vegetative-compatibility group.

Formae speciales of *F. oxysporum* that cause wilt diseases of woody plants in North America include f.sp. *canariensis* in *Phoenix canariensis* (Canary Island date palm), f.sp. *rhois* in *Rhus* (sumac), f.sp. *citri* in *Citrus aurantifolia* (Mexican lime), f.sp. *hebae* in *Hebe*, f.sp. *perniciosum* in *Albizia* (mimosa or silktree) and *Ailanthus altissima* (tree-of-heaven), and f.sp. *pyracanthae* in *Pyracantha* (firethorn). An unnamed forma specialis causes wilt of *Morella cerifera* (wax-myrtle). In Hawaii *F. oxysporum* f.sp. *koae* causes a wilt disease of *Acacia koa* (koa) and is associated with a decline syndrome of that tree in forest stands.
References: 86, 128, 155, 984, 1031, 1109, 1320, 2019, 2020, 2087, 2578, 2819, 2909, 3193, 3899, 4029, 4046, 4055

Mimosa wilt. Wilt caused by *F. oxysporum* f.sp. *perniciosum* is the most serious disease of *Albizia julibrissin* (mimosa or silktree). It has been known since the 1930s and occurs from New York to Florida, Louisiana, and Arkansas, also in California. *Albizia lebbek* (lebbek) and *A. procera* (tall albizia) are affected by the same pathogen in Puerto Rico and Argentina. The disease also occurs in Greece, Iran, and Russia.
Symptoms and signs. Leaves on one or more branches turn chlorotic and droop in early to mid summer. Thereafter, leaves either turn yellow or shrivel and then drop. Trees sometimes die within a month, but symptoms typically progress branch by branch over several months. Gum may ooze from small cracks that open in the bark of trees with advanced symptoms. Adventitious sprouts often develop along the trunk but are killed later (by frost in northern areas). Trees sometimes survive until the next year, produce dwarfed chlorotic leaves, and then die. Parts of the root system may survive longer, however, and sprouts may grow from the root collar.

Brown streaks develop in outermost sapwood beginning before leaves wilt. The streaks are most prominent in the roots and lower trunk on the side where branches wilt first. Viewed in cross section they appear as spots or, after coalescing, as arcs or a ring. Streaks eventually extend into small branches. White froth (Plate 191) with a sweet fermentative odor occasionally exudes from cracks in bark on the trunk and major limbs. Whether *F.o.* f.sp. *perniciosum* or secondary organisms induce this "alcoholic flux" is unknown.
Disease cycle. The pathogen resides in soil as chlamydospores that germinate when stimulated by exudates from host rootlets. Hyphae penetrate intact or wounded rootlets and grow into the xylem. There the pathogen produces microconidia (6–12 × 2–3.5 μm) that are carried upward in the sap stream. They lodge in pits and at vessel endwalls, where new colonies grow and hyphae penetrate adjacent vessels and parenchyma cells. Systemic colonization results from the alternation of hyphal growth where spores lodge and passive distribution of spores in the sap stream in a series of vessels. Parenchyma cells around colonized xylem

vessels and in nearby medullary rays are soon killed and turn brown. Gummy brown material appears in pit apertures and on vessel walls. This material is derived in part from degraded pectic substances from cell walls and in part from defense compounds produced by parenchyma cells before their death. It completely coats or occludes vessel lumens at the time of wilting. Mycelium develops in vessels that have ceased conducting sap, and hyphae then grow from cell to cell via pits, colonizing xylem parenchyma. After a diseased stem has defoliated, the fungus grows from wood into bark and produces orange-pink sporodochia at lenticels on the surface. Colorless macroconidia form on the sporodochia and are dispersed primarily by water. This sporulation continues for up to 2 years after a diseased tree dies. Macroconidia of *F.o.* f.sp. *perniciosum* are colorless, multicellular, canoe shaped in outline, and measure 23–60 × 3–4.5 μm. Other fusaria, capable of causing diagnostic confusion but incapable of causing wilt, also sporulate on the bark of recently killed *Albizia*. Three fusaria are illustrated in Figs. 92 and 93.

Macroconidia of *F.o.* f.sp. *perniciosum* are washed to the soil, where one or more cells per spore develop into chlamydospores with thick brownish walls. Infested soil may be transported by water, earth-moving equipment, and other agents. Transportation of the pathogen in seeds is also possible because it sometimes enters fruits. Seedlings grown from infested seed soon become diseased.

When Fusarium wilt of *Albizia procera* was discovered in 1945 in Puerto Rico, the pathogen there was considered to be *F.o.* f.sp. *perniciosum*, but isolates from *A. procera* could not infect *A. julibrissin*, and vice versa. Strains pathogenic in *A. julibrissin* were designated race 1, and those pathogenic in *A. procera* race 2. The strain that infects *Ailanthus altissima* is pathogenic to *Albizia julibrissin*.

Three clonal cultivars with resistance to *F.o.* f.sp. *perniciosum* have been released: Charlotte, Tryon, and Union. Unfortunately, pathogen strains capable of overcoming the resistance of the first two cultivars have appeared. *Albizia polyphylla* has proven resistant in seedling assays, and several other *Albizia* species seem less susceptible than *A. julibrissin*.

Mimosa wilt can be exacerbated by root-knot nematodes, *Meloidogyne incognita* and *M. javanica* (Plate 214). Trees that normally resist systemic colonization by *F.o.* f.sp. *perniciosum* become susceptible when their roots are damaged by the nematodes.
References: 107, 128, 205, 1285–1288, 1374, 1611, 2528, 2823, 3068, 3679, 3870, 4056

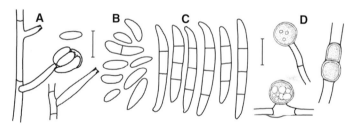

Figure 123. Fusarium oxysporum: A, conidiophore. B, microconidia. C, macroconidia. D, chlamydospores. Scale bars = 10 μm. Adapted from reference 984 by permission of Elsevier.

A–F. Fusarium wilt of *Albizia julibrissin* (mimosa), caused by *Fusarium oxysporum* f.sp. *perniciosum*. A, B. One-sided wilting, defoliation, and dieback (NY, Jun). C. Brown streaks in the outermost wood of a small branch (MD, Sep). D–F. Sporodochia and masses of conidia at lenticels and on adjacent bark. These are visible macroscopically (D), but magnification (E, F) is needed for interpretation. Sporodochia erupting from lenticels are orange-pink, and massed colorless conidia washed from the sporodochia appear as white deposits. Brown eruptions in D and F are normal lenticels (MD, Sep).

G, H. Wilt, decline, and death of *Ailanthus altissima* (tree-of-heaven) caused by *F. oxysporum* f.sp. *perniciosum*. G. Dead and dying trees. H. Light brown discoloration in diseased xylem (VA, Oct).

I. Stained xylem in a stem of *Morella cerifera* (wax-myrtle) severely affected by Fusarium wilt (FL, Mar).

Photo credits: G, H—R. J. Stipes; I—E. L. Barnard

249

Diseases Caused by Basidiomycota (Plates 124–174)
Exobasidium Galls and Blisters (Plate 124)

Exobasidium species (Exobasidiales, Exobasidiaceae) cause colorful leaf spots and swollen deformities on leaves and stems of many plants in the Ericaceae and some in the Empetraceae, Lauraceae, Symplocaceae, and Theaceae. About 60 species of *Exobasidium* are known worldwide, including perhaps 30 in the USA and Canada, all on woody plants.

Signs and symptoms. *Exobasidium* species produce basidia without a fruit body or covering (Fig. 124). Numerous basidia and spores impart a light-colored to white velvety bloom to a sporulating surface (hymenium). The number of basidiospores produced per basidium is variable and usually exceeds four. Most *Exobasidium* species sporulate on the lower surfaces of diseased leaves, but some of these fungi sporulate all over diseased organs. Many of them also produce conidia by budding of basidiospores. Conidia can be found in the same hymenia that produce basidia. *Exobasidium* species grow and sporulate in yeastlike fashion on laboratory media.

A given *Exobasidium* species induces primarily one type of symptom on one or a few closely related plant species. A given plant species, on the other hand, may be host to two or more *Exobasidium* species. Symptom types are: yellow or red leaf spots, thickened or not; annual galls involving buds, leaves, floral parts, or shoot tips; scattered elongated shoots with enlarged, usually reddish leaves, resulting from annual infections that become systemic; clusters of deformed shoots arising annually from perennial mycelium in stems or rhizomes; and perennial stem galls, either spindle shaped or cylindric, which on some plants give rise to brooms. Chlorophyll synthesis and photosynthesis in diseased parts are depressed, respiration and transpiration increase, and phenolic substances accumulate. Some *Exobasidium* species produce indoleacetic acid, which is presumed responsible for swelling and gall formation. Severe infection of *Rhododendron calendulaceum* (flame azalea) by a gall-inducing *Exobasidium* has been shown to reduce seed production and lead to twig and branch dieback.

Knowledge of the identities, host ranges, and distributions of *Exobasidium* species in North America is incomplete because contemporary criteria for identification have not been applied to the majority of old collections and records. The species currently recognized are distinguished on the basis of morphology, host range, symptoms induced, annual versus perennial infection, mode of basidiospore germination (long germ tubes giving rise to mycelium versus short germ tubes giving rise to conidia), upper temperature limit for growth, and physiological characteristics in culture. The trend in *Exobasidium* taxonomy is toward recognition of numerous host-specialized species in place of single species with broad host ranges. This scheme is supported by results of DNA analyses and is summarized for North America as follows.

Species known primarily on cultivated plants include the following. *E. camelliae* and its variety *gracile* cause leaf and shoot galls on *Camellia japonica*, *C. sasanqua*, and hybrids; and *E. monosporum* causes leaf curl of *C. japonica*, all in southeastern USA. Leaf galls on *Rhododendron* species of the azalea group are caused by *E. japonicum* and occur wherever azaleas are grown. Yellow or red spots on other *Rhododendron* species are caused by *E. burtii*, *E. canadense*, and *E. decolorans*. Cranberries (*Vaccinium macrocarpon* and *V. oxycoccos*) are subject to three diseases. Rose bloom, caused by *E. oxycocci*, is characterized by compact, deformed, pink shoots that resemble rose blossoms. Red leaf spot and red shoot are caused by *E. rostrupii* and *E. perenne*, respectively. Red leaf of *V. angustifolium* and *V. myrtilloides* (lowbush blueberries), caused by *E. vaccinii*, appears as oversized red shoots that grow from systemically infected rhizomes. This disease also occurs on *V. corymbosum* (highbush blueberry), usually on terminal leaves.

Species known primarily on wild plants include the following. *E. angustisporum* and *E. uvae-ursi* induce clusters of deformed shoots, and *E. sydowianum* induces red leaf spots, on *Arctostaphylos* (bearberry, manzanita). *E. cassandrae* induces red-leaf spots and *E. savilei* a red-shoot syndrome on *Chamaedaphne calyculata* (cassandra, leatherleaf). *E. cassiopes* and *E. hypogeum* cause shoot deformity and a red-shoot syndrome, respectively, on *Cassiope* species (mountain-heather). *E. empetri* causes red-shoot syndrome on *Empetrum nigrum* (black crowberry). *E. phyllodoces* causes shoot deformity of *Phyllodoce empetriformis* (mountainheath). *E. symploci* induces a bud gall on *Symplocos tinctoria* (sweetleaf). Wild *Vaccinium* species (bilberry, huckleberry, blueberries,

mountain cranberry) sustain shoot and branch infections by *E. dimorphosporum*, *E. expansum*, *E. juelianum*, *E. parvifolii*, and *E. vaccinii-uliginosi*; or they may develop yellow or pink leaf spots caused by *E. cordilleranum*, *E. pachysporum*, and *E. vaccinii* (strict sense).

Exobasidium species formerly identified as *E. vaccinii* (broad sense) but not yet reidentified occur on *Andromeda* species (bog-rosemary), *Arbutus menziesii* (madrone), *Gaultheria ovatifolia* (western teaberry), *Gaylusaccia* species (huckleberries), *Kalmia polifolia* (bog-laurel), *Ledum* species (Labrador-tea), *Leiophyllum buxifolium* (box sandmyrtle), *Leucothoe* species, *Lyonia* species, *Menziesia* species, and *Pieris japonica* (Japanese-andromeda).

Disease cycles. *Exobasidium* species have two general schemes. The more common is a 1-year cycle in which the fungi overwinter in buds from which diseased plant parts develop. Whether the buds are infected before winter or merely provide shelter for spores is unknown. Infection is initiated by basidiospores or conidia from which hyphae either enter stomata or penetrate directly and ramify in young, growing organs. Intercellular mycelium predominates and is nourished in most species by lobed haustoria (absorbing structures) produced within host cells. In late spring to midsummer, the time varying with latitude and elevation, basidia grow out between epidermal cells, rupture the cuticle (if the infected organ is a leaf or stem), and release spores. Basidiospores are dispersed by air, conidia by water. In a variation of this scheme *E. camelliae* var. *gracile* forms its basidia beneath the lower leaf surface and exposes them by causing the overlying leaf tissue to slough off. In some Exobasidium diseases spores of either type may initiate a secondary cycle on young, late-developing stems and leaves. Diseased parts degenerate after sporulation.

In the second type of disease cycle infection begins as described above, and mycelium becomes perennial in stems. It may be localized, inducing galls and brooms, or systemic. Infected buds give rise to deformed organs. The spore-producing layer arises on the lower surfaces of diseased leaves or on hornlike projections that develop annually from stem galls.

Removal of Exobasidium galls from ornamental plants by pruning or hand picking before they sporulate has often been suggested for disease control, but the effectiveness of this practice is unproven. *References:* 287, 403, 404, 665, 779, 825, 1099, 1102, 1298, 1342, 1658, 2657, 2658, 2784, 3222, 3420, 4417, 4418

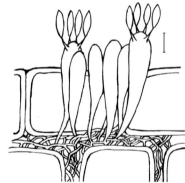

Figure 124. Basidia of *Exobasidium vaccinii* emerging from between epidermal cells of a leaf and producing basidiospores. Scale bar = 10 μm.

A–D. Leaf galls caused by *Exobasidium japonicum* on *Rhododendron* species. A, B. Galls on 'Anchorite' azalea. Sporulating surfaces are white (WA, Jun). C. Galls on *R. periclymenoides* (pinxter-bloom) (NY, Jun). D. Galls on *R. obtusum* (Hiryu azalea) (NY, May).

E. Red leaf spots caused by *Exobasidium* sp. on *Vaccinium ovatum* (evergreen huckleberry) (OR, Jul).

F. A red-shoot syndrome caused by *Exobasidium* sp. on *Arctostaphylos* sp. (manzanita). Diseased leaves are reddish brown and shriveled at the end of winter. Infected buds would soon give rise to new leaves with red upper surfaces (CA, Mar).

G. Leaf and bud galls on *Camellia sasanqua*, caused by *Exobasidium camelliae* var. *gracile*. Sporulation occurs on white areas. Dark spots on the diseased leaf at right are mycelium and spores of a secondary fungus (MS, May).

251

Articularia and Microstroma Leaf Spots and Witches'-brooms (Plate 125)

Articularia Leaf Spot and Witches'-broom

Articularia quercina, a hyphomycete, causes diffuse chlorotic spots on *Quercus* (oak) leaves, and *A. quercina* var. *minor* causes witches'-brooms and white leaf mold on *Carya* (hickory) and *Quercus* species. These little-known fungi, which appear as tiny white stellate tufts on the undersurfaces of leaves, apparently have the ability to parasitize leaves for months and branches for years while causing little damage to host tissues.

When *A. quercina* was first described, it was indicated to be the asexual state of an ascomycete now known as *Cookella microscopica* that apparently followed *Articularia* on the same spots. The connection between these forms was not made through cultural studies, however, and the relationships of *Articularia* to other fungi have not been studied using modern techniques. The similarities of *Articularia* to *Microstroma* with respect to hosts, mode of sporulation, symptoms induced, and growth in culture (yeastlike budding as well as hyphal growth) indicate a need for study of their possible relationship.

A. quercina occurs from Illinois, where it was first recorded, southward to Mississippi, Oklahoma, and Texas. Its hosts include *Carya illinoinensis* (pecan), *Quercus lyrata* (overcup or swamp post oak), *Q. macrocarpa* (bur oak), *Q. prinus* (chestnut oak), and *Q. velutina* (black oak). The variety *minor* occurs on *Quercus gambelii* (Gambel oak), *Q. dunnii* (Palmer oak), and *Q. ×pauciloba* (Rocky Mountain scrub oak) in the southern Rocky Mountain region and on *C. illinoinensis* in Oklahoma and Texas.

Foliar lesions (not shown) are indistinct chlorotic areas as viewed from above. The white tufts of spore-bearing cells on the lower surface develop as multicellular columns growing from globose stromata in the mesophyll beneath stomata. A column consists of tiers of fertile cells with enlarged apices, so that each tier is broader at the top than at the base. Spores are produced in barrel-shaped clumps of eight at the apices of the cells in each tier (Fig. 125). When the spore-bearing tufts are numerous, as on leaves in witches'-brooms, the entire undersurface of a leaf has a white cast. The spores of *A. quercina* are colorless, unicellular, oblong to slightly curved, and measure $12.5-15 \times 4-5$ µm; those of the variety *minor* measure $6-8 \times 2.5-3$ µm.

Articularia disease cycles are unknown but may be like those suggested below for white mold and witches'-brooms caused by *Microstroma juglandis*. *A. quercina* var. *minor* has been isolated from the inner bark and outer sapwood of branches in brooms, which gives credence to the suggestion that it induces brooms and perennates in them.

References: 597, 697, 783, 1099, 1560, 2317, 3001

Microstroma Leaf Spot and Witches'-broom

Microstroma is a small genus (four species) of plant-parasitic basidiomycetes (Microstromatales, Microstromataceae). All produce spore-bearing cells in columns or clusters that issue from stomata of colonized leaves, each cell then giving rise to several spores. No fruit bodies are formed. The classification of these fungi as basidiomycetes is supported by electron microscopic studies of cell wall structure but awaits corroboration from DNA analyses.

M. juglandis causes downy spot, also called white leaf spot or white mold, on *Carya* (hickory) and *Juglans* (walnut). The fungus occurs wherever these plants grow in North America and Europe. It occurs also in northern India and New Zealand. The disease is insignificant in most circumstances. Hosts include *Carya aquatica* (water hickory), *C. cordiformis* (bitternut hickory), *C. glabra* (pignut hickory), *C. illinoinensis* (pecan), *C. ovata* (shagbark hickory), *C. pallida* (sand hickory), *C. texana* (black or Texas hickory), *C. tomentosa* (mockernut hickory), *Juglans ailanthifolia* (Japanese walnut), *J. californica* (California black walnut), *J. cinerea* (butternut), *J. hindsii* (Hinds walnut), *J. major* (Arizona walnut), *J. microcarpa* (little walnut), *J. nigra* (black walnut), and *J. regia* (Persian walnut).

The most common symptoms are leaf spots of various sizes that are pale yellow on the upper surface and downy or powdery white to pale yellow below. Lesions often coalesce to form large angular areas and eventually become brown and dry. Downy lesions sometimes form on growing fruit of *J. regia*. *M. juglandis* var. *robustum* causes blight of *C. illinoinensis* catkins and leaflets.

On some hosts in both *Carya* and *Juglans*, *M. juglandis* has been found associated with witches'-brooms that arise near branch tips

and attain sizes up to about 2 m in diameter. *Microstroma*-associated brooms have been reported on *C. illinoinensis*, *C. ovata*, *C. texana*, *J. major*, and *J. microcarpa*. Viewed from above, leaves in brooms are somewhat dwarfed and yellowish. They tend to shrivel and drop beginning in midsummer. The association of *Microstroma* with brooms has not been critically evaluated. It is possible that the fungus preferentially colonizes succulent leaves in brooms induced by some other agent, such as a phytoplasma (Plate 195).

Microstroma species sporulate on the lower surfaces of lesions while the leaf tissues are alive. *M. juglandis* first produces hyphal aggregations (stromata) in the mesophyll beneath stomata. A cluster of fertile cells interpreted as basidia grows from a stroma through the stomatal pore, and the apex of each cell then bends away from the pore, the cluster thus forming a rosette. Several colorless, single-celled spores develop atop each fertile cell. When sporulation is abundant the leaf surface appears white, or pale yellowish with time. The fungus has been reported to appear in late spring on nearly all leaves in brooms. Sporulating patches darken with age, degeneration, and invasion by secondary organisms, but some may remain until leaf drop. The spores of *M. juglandis* are unicellular, colorless, and $6-8 \times 3-5$ µm in size. Its variety *robustum* has spores up to 14 µm long.

The annual cycle of downy spot has not been worked out, but authors in both North America and Europe have suggested that *M. juglandis* overwinters as stromata in fallen leaves, from which infectious spores are liberated to the air in spring. The fungus is also believed to perennate in witches'-brooms, which would explain the early and uniform sporulation on leaves in brooms.

M. album, a parasite of oak leaves, causes a downy leaf spot that is also known as frosty mildew. This fungus occurs in southeastern and central USA and also in the Pacific Northwest. It is widely distributed in Europe as well. It is similar to *M. juglandis* except that its spore-bearing structures are columnar. Hosts in North America include *Quercus alba* (white oak), *Q. gambelii* (Gambel oak), *Q. garryana* (Oregon oak), *Q. imbricaria* (laurel or shingle oak), *Q. lobata* (California white, or valley oak), *Q. lyrata* (overcup or swamp post oak), and *Q. macrocarpa* (bur oak).

References: 122, 126, 597, 697, 783, 1099, 1649, 2317, 3859, 4376

Figure 125. A. *Articularia quercina:* a conidiophore bearing bundles of conidia. B. *Microstroma album.* C. *M. juglandis.* Scale bars = 25 µm for conidiophore, 10 µm for other structures. A and C adapted from references 697 and 126, respectively, by permission of the Mycological Society of America; B from reference 122 by permission of Gebrüder Borntraeger Verlagsbuchhandlung.

A–C. Articularia witches'-brooms on *Quercus gambelii* (Gambel oak). A, B. Large brooms on old trees (UT, May). C. Brooms on stunted small trees in a pasture thicket (CO, May).

D, E. White mold, caused by *Microstroma juglandis*, on *Juglans nigra* (black walnut) leaves. Lesions viewed from above are yellow spots. Sporulation, often associated with veins, occurs on lower surfaces (IL, Jul).

F–I. White mold, caused by *M. juglandis*, on *Carya ovata* (shagbark hickory). Yellow lesions with indefinite edges, as viewed from above, correspond to areas of white sporulation on lower surfaces. Diseased tissue soon dies (NY, Jun).

Photo credits: D—USDA Forest Service; E—J. A. Matteoni

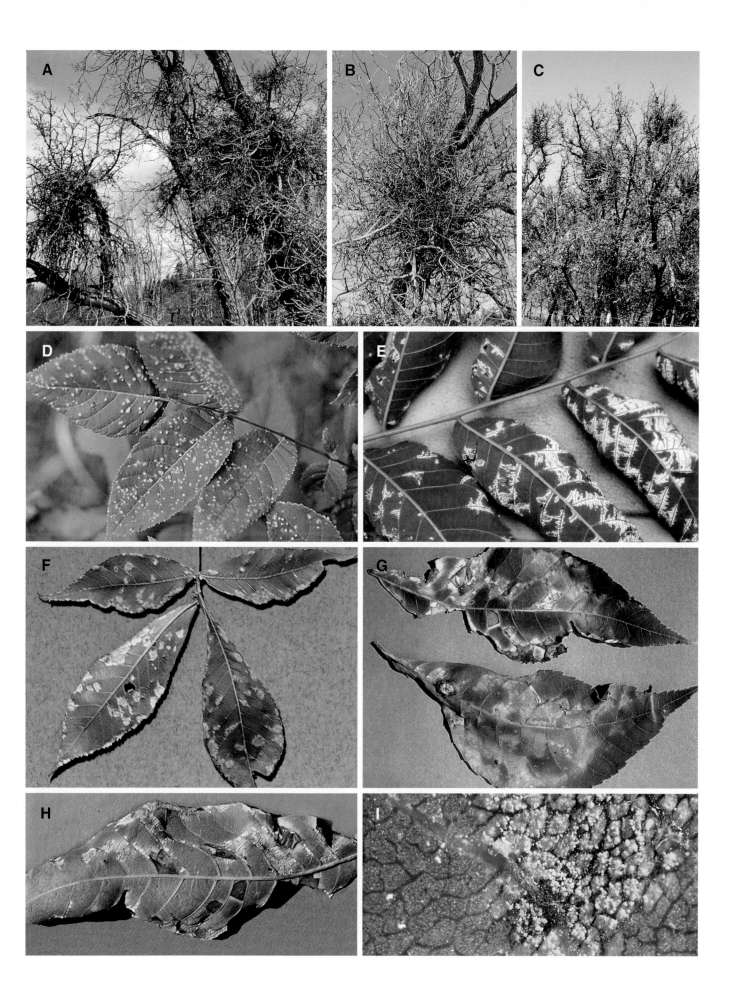

Insolibasidium Blight of Honeysuckle and False Smut of Palms (Plate 126)

Insolibasidium Blight of Honeysuckle

Lonicera species (honeysuckle) are subject to leaf blight caused by *Insolibasidium deformans* (Ustilaginomycetes, Platygloeaceae). The pathogen, found only on *Lonicera*, is common in the Great Plains and eastward in the USA and Canada and occurs also in the Pacific Northwest and in New Zealand. Epidemics in nurseries may cause defoliation, dieback, and lost growth. Susceptible species include *L.* ×*bella* (Bell honeysuckle), *L. caerulea* (sweetberry honeysuckle), *L. canadensis* (fly honeysuckle), *L. dioica* (limber honeysuckle), *L. discolor*, *L. gracilipes*, *L. involucrata* (twinberry), *L. japonica* (Japanese honeysuckle), *L. korolkowii* (blueleaf honeysuckle), *L. maackii* (Amur honeysuckle), *L.* ×*minutiflora*, *L. morrowii* (Morrow honeysuckle), *L. muendeniensis*, *L. nervosa*, *L.* ×*notha*, *L. oblongifolia* (swamp fly honeysuckle), *L. orientalis*, *L. prolifera* (grape honeysuckle), *L. prostrata*, *L. quinquelocularis*, *L. ruprechtiana* (Manchurian honeysuckle), *L. sempervirens* (trumpet honeysuckle), *L. tatarica* (Tatarian honeysuckle), *L. tatsiensis*, and *L.* ×*vilmorinii*. *Symphoricarpos albus* (snowberry), also in the Caprifoliaceae (honeysuckle family), is susceptible if inoculated but has not been found infected in nature.

Symptoms and signs. Blight appears in spring, often on some of the first leaves on new shoots. Infected areas are slightly crinkled or rolled, then turn yellowish green and eventually tan to brown. The veins tend to remain green for a time after interveinal tissues turn brown. Diffuse yellow zones develop around young lesions and become more conspicuous with time. As lesions expand and coalesce they become shriveled and brown, and the affected leaves curl and twist. Although stems are not attacked, an appearance of shoot blight may result from infection of all leaves on a shoot. Severely diseased leaves fall prematurely.

I. deformans produces both basidiospores and conidia. The basidial state develops first, during humid weather at moderate temperatures, as a continuous thin white layer on the lower surface of a lesion. This layer develops from hyphae that emerge through stomata. The basidia are distinctive in that each curves in a semicircle and produces four basidiospores along the arc. The conidial state, *Glomopsis lonicerae*, appears in summer as a white powdery mass on or at the margins of the same lesions where basidia previously formed, most often on shaded severely diseased leaves. Conidiophores emerge singly or in clusters from stomata and are more numerous on the lower leaf surface than on the upper one. Conidia are also common on fallen diseased leaves from summer until early winter. Apparently the conidia, which form in groups of six, are not dispersed but serve as resting spores.

Disease cycle. Basidiospores liberated from overwintered leaves on the ground initiate infection in spring and are presumed responsible for secondary infections throughout the growing season as weather permits. Basidiospores but not conidia have caused blight when inoculated to leaves. Basidiospores are discharged in humid air at 5–28°C. Discharge is most abundant at 14–21°C, and germination is maximal at 14–26°C. This sporulation occurs from spring through early autumn during and after rain. If conducive weather prevails, blight develops on young leaves throughout the season, with new symptoms appearing 8 or more days after each infection period. Conidial production and germination are favored at temperatures similar to those for basidiospores. Perhaps the conidial state functions only for survival and gives rise to basidia in spring.

Honeysuckle species that appeared resistant in Iowa were *L. dioica*, *L. gracilipes*, and *L. sempervirens*. *L. japonica* cv. Halliana was immune in Iowa, but the parent species was found naturally infected in New Zealand.

References: 800, 1339, 1600, 2887, 3287, 3290

False Smut of Palms

This disease affects palms of many species around the globe and causes economic damage to date palms in Africa and Asia. It occurs across southern USA, particularly in humid areas from Florida to Texas. It also occurs on palms grown under glass and was originally described from greenhouse specimens in the USA. The causal fungus, *Graphiola phoenicis* (Exobasidiales, Graphiolaceae), belongs to a distinctive group of between 5 and 10 species.

G. phoenicis is restricted to palms. Those affected in the USA include the following:

Arenga pinnata (sugar palm), *Butia capitata* (pindo palm), *Coccothrinax crinita* (thatch palm), *C. argentata* (silver palm), *Cocos nucifera* (coconut palm), *Dypsis lutescens* (Madagascar or areca palm), *Phoenix canariensis* (Canary Island date palm), *P. dactylifera* (date palm), *P. reclinata* (Senegal date palm), *P. roebelenii* (pygmy date palm), *Roystonea regia* (Florida royal palm), *Sabal minor* (dwarf or scrub palmetto), *S. palmetto* (cabbage palmetto), *Syagrus romanzoffiana* (queen palm), *Washingtonia filifera* (California fan palm), and *W. robusta* (desert or thread palm).

Symptoms and signs. The visible symptoms of false smut are premature senescence of leaflets and fronds and tiny yellow to brown spots associated with sori of the causal fungus (see below). The disease suppresses photosynthesis, fruit production, and tree growth. Diseased *P. dactylifera* leaves contain less chlorophyll, soluble sugars, and major nutrient elements (nitrogen, phosphorus, and potassium) and more phenolic compounds and micronutrients (especially iron and manganese) than healthy ones. Activities of the enzymes indoleacetic acid oxidase and amylase are greater than normal. Fronds of this species, which normally live 6–8 years, may die after as few as 3 years.

The causal fungus develops subepidermally and ruptures both leaf surfaces, producing numerous wartlike black projections (sori) 1–3 mm in diameter, each with a central crater from which light yellowish filaments extrude. The sori appear first near leaf tips and subsequently appear on subapical leaflets and sometimes on rachises. They are generally absent from the youngest leaves but become increasingly numerous on leaves 2–3 years old, eventually numbering 20 or more per square centimeter where infection is severe.

The false smut sorus consists of a hard, dark, persistent layer that encloses a yellowish fertile region where powdery spores are produced among the filaments. The filaments extend as much as 2.5 mm beyond the rim of the crater and are often found coated with spores. Filaments are thought to play a role in spore dispersal, perhaps dislodging spores when disturbed by wind or water drops. The spores are small (e.g., 5×7 μm), two celled, and yellowish in mass. They can germinate by germ tubes or by budding in the manner of yeasts. When initial growth is yeastlike, colonies convert to mycelial growth after 18–24 hours.

Disease cycle. The yellowish spores, called sporidia by some authors, are apparently not infectious. They germinate on palm leaflets by budding and then give rise to short hyphae with sticky sheaths and to spore balls containing numerous minuscule (0.5–1.2 μm) cells. These cells, released when spore balls rupture, are believed to be infectious, giving rise to appressoria from which stomata are penetrated by thin infection pegs. The fungus grows among and within leaf cells and, 5–8 weeks after the onset of infection, produces a compact mass of mycelium (young sorus) beneath the epidermis. Sori rupture the epidermis and begin to release spores after 8–10 weeks, completing the cycle. Rough dark craters of old sori remain after spores have dispersed.

Date palms in Africa are sometimes pruned for control of false smut, but whether this practice aids productivity is unreported. Date palm cultivars apparently vary in resistance to false smut, but they have not been well tested for resistance in the USA. In Florida the cultivar Barhee has shown promise for edible fruit, ornamental value, and false smut resistance.

Two additional *Graphiola* species occur in southeastern USA: *G. congesta* on *Sabal minor* and *S. palmetto*, and *G. thaxteri* on *S. megacarpa* and *S. palmetto*. These fungi are poorly known in comparison with *G. phoenicis*.

References: 46, 170, 764, 1052, 2013, 2908, 3664, 3971

A–C. Insolibasidium blight of *Lonicera tatarica* (Tatarian honeysuckle). A. Mature lesions and blighted leaves. B. Young lesions viewed from the upper surface. C. Conidial (resting spore) state on lower surfaces of lesions (NY, Aug–Sep).

D–F. False smut of *Phoenix canariensis* (Canary Island date palm). D, E. Sori projecting from leaflets. F. Sori and associated chlorotic to necrotic spots. Twisted yellowish filaments extrude from a central cavity in each wartlike sorus (FL, Apr).

Diseases Caused by Rhizoctonia-Forming Fungi (Plate 127)

Rhizoctonia-forming fungi are a cosmopolitan heterogeneous array of mostly soilborne basidiomycetes whose anamorphs (asexual states) are in or similar to the genus *Rhizoctonia*. They include plant pathogens, mycorrhizal symbionts of orchids, and saprobes. Pathogenic strains cause damping-off and root rot of plants in hundreds of genera. Some strains also cause leaf and twig blight and crown rot of older but low-growing plants in nurseries and dense landscape plantings in warm, humid regions. Fungi in this group produce distinctive hyphae, as noted below (thus the term *rhizoctonia-forming*), and many strains produce basidia and basidiospores on mats or webs of hyphae on plants or soil. They do not produce conidia or other mitospores.

Symptoms and signs. Common symptoms in seedbeds or cutting beds are damping-off, hypocotyl rot, root rot, stunting and yellowing, and death. Symptoms on aerial parts of larger plants include necrotic spots and blotches on leaves, shoot blight, and dieback. Mycelium of a rhizoctonia-forming fungus may, under warm, moist conditions, grow out from killed tissues onto the soil and/or plant parts near soil, appearing as a light gray to light brown web. The term *web blight* refers to superficial mycelium on aerial parts of plants that binds detached leaves to stems. Superficial hyphae of some rhizoctonia-forming fungi aggregate into strands to which the term *thread blight* refers.

Detection and identification of rhizoctonia-forming fungi usually require isolation from plants. The hyphae are distinctive because they tend to branch at right angles, have a constriction at the base of each branch, and have a septum in the branch close to the constriction (Fig. 127). Many strains produce hyphae of two forms: primary or "runner" hyphae with long, nearly cylindric cells; and monilioid hyphae that consist of chains of short, barrel-shaped to nearly round cells. Many strains also produce sclerotia in dead plant tissues and in culture. Mature sclerotia of most strains are brown to black and vary in size and shape from spheres less than 1 mm across to crusts several millimeters in extent.

Identification of rhizoctonia-forming species is problematic because their taxonomy and classification are still evolving. One useful character for distinguishing genera is number of nuclei per cell. Of the 10 plant-pathogenic basidium-forming species recognized in 2004, there were five species of *Ceratobasidium* (anamorph *Ceratorhiza*; binucleate), two of *Thanatephorus* (anamorph *Rhizoctonia*; multinucleate), and one of *Waitea* (anamorph *Rhizoctonia*; multinucleate), all in the Ceratobasidiales, Ceratobasidiaceae; also two species of *Helicobasidium* (anamorph *Thanatophytum*; binucleate) of the Platygloeales, Platygloeaceae. Although many rhizoctonia-forming strains can be induced to form their sexual states in culture, they are usually encountered as mycelial anamorphs, and numerous strains are known only as anamorphs.

The anamorph genera *Ceratorhiza*, *Rhizoctonia*, and *Thanatophytum* comprise many genetically isolated anastomosis groups (biological species) and subgroups. The main groups are numbered AG-1 through AG-13 in *Rhizoctonia* and lettered AG-A through AG-S in *Ceratorhiza*. Full diagnosis of a disease caused by one of these fungi includes identification of the anastomosis group. Unfortunately, many disease records predate knowledge of anastomosis groups or modern species distinctions. These records mention only unnamed *Rhizoctonia* or *R. solani* in a broad sense. *R. solani* is the anamorph of the cosmopolitan species *Thanatephorus cucumeris*. The other named species recorded from woody plants in North America are *Ceratorhiza ramicola* (recorded as *R. ramicola*, anamorph of *Ceratobasidium ramicola*) and *Thanatophytum crocorum* (recorded as *R. crocorum*, anamorph of *Helicobasidium brebissonii*). The latter fungus causes violet root rot of various woody and herbaceous plants.

Disease cycles. Rhizoctonia-forming fungi persist in soil and plant debris as mycelium and sclerotia that can be stimulated by plant exudates. Hyphae from these sources infect succulent hypocotyls and roots. Many strains produce infection structures consisting of hyphae with short branches that function as appressoria, often clustered as "infection cushions." Hyphae from these structures penetrate plants and then grow inter- and intracellularly, killing tissues. In web blight, aerial plant parts are penetrated through intact surfaces and sometimes through lenticels or stomata. Mycelium on leaves may spread from plant to plant. Rhizoctonia-forming pathogens secrete diverse hydrolytic enzymes that degrade cell walls, membranes, and contents, but the precise mode of killing remains undescribed. When a Rhizoctonia-forming fungus produces basid-

iospores, they are liberated into the air and are capable of initiating new infections on succulent aerial parts (thus accounting for leaf spots or foliar blight on some hosts) or on roots very close to the soil surface. When killed host tissues decay, a rhizoctonia-forming fungus is returned to soil as hyphae and/or sclerotia.

Representative diseases. Damping-off or rot of succulent roots or cuttings by rhizoctonia-forming fungi can affect any plant. Leaf and stem diseases, including web blight, caused by these fungi have been noted on woody plants in the following genera in North America: *Acca, Carissa, Cotoneaster, Elaeagnus, Erythrina, Gardenia, Hedera, Ilex, Juniperus, Lagerstroemia, Ligustrum, Nandina, Nerium, Pinus, Pittosporum, Rhaphiolepis,* and *Rhododendron* (mainly azalea), among others.

Pine seedling blights in nurseries (Plate 127A–E) may be caused either by strains in the *R. solani* complex or by *Ceratorhiza* species. Losses attributed to *Ceratorhiza* have occurred sporadically in *Pinus palustris* (longleaf pine) and *P. taeda* (loblolly pine) during warm, humid weather in southeastern USA. Attack begins near the soil line. Lesions at needle bases cause foliar fading and browning. Diseased foliage may turn gray with a web of mycelium. The pathogen also spreads laterally, killing seedlings in circular to irregular patches. In *P. palustris* the rot also involves stems and buds near the soil line as well as the upper roots.

Leaf blight (thread blight) caused by *Ceratobasidium stevensii* occurs in southern USA as far north as Kentucky. It affects many plants, notably species of *Aucuba, Citrus, Cornus, Ilex, Liquidambar, Lonicera, Malus, Prunus, Pyrus, Rosa,* and *Viburnum*. The fungus produces brown hyphal strands and brown sclerotia on twigs and branches (Plate 127F).

Foliar blight of *Carissa macrocarpa* (Natal-plum; Plate 127 G, H) can be caused by either *Ceratorhiza ramicola* or *Rhizoctonia solani*. Symptoms appear in summer on low branches. Lesions begin as water-soaked areas and turn brown, sometimes with concentric patterns. The blight spreads upward if promoted by high temperature and humidity. Killed leaves either drop or are held by mycelial strands. Severely affected plants often also have decayed roots.

References: 46, 48, 71, 72, 123, 578, 800, 863, 1066, 1099, 1153, 1193, 1313, 1536, 1789, 1967, 1995, 1996, 2238, 2342, 2523, 2711, 3313, 3371, 3745, 3746, 4159, 4264, 4269

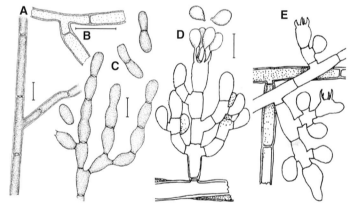

Figure 127. Rhizoctonia-forming fungi. A–C. *Rhizoctonia solani.* A. A primary hypha. B, D, E. Right-angle hyphal branching habit. C. Monilioid hyphae with beadlike swollen cells. D. Upright basidial apparatus with one mature basidium and basidiospores of *Thanatephorus*. E. Laterally branching basidial apparatus of *Ceratobasidium*. Scale bars = 10 μm. A and C adapted from reference 123 by permission of Gebrüder Borntraeger Verlagsbuchhandlung; B, D, and E adapted from reference 3313 by permission of Royal Botanic Gardens, Kew.

A–E. Seedling blight of *Pinus palustris* (longleaf pine) caused by *Ceratobasidium* sp. A, B. Patches of dead, dying, and missing seedlings. C, D. Close views of foliar symptoms on seedlings in the "grass stage." Infection began when sand covered the needle bases and apical bud. E. Mycelium of the pathogen on needles (FL, Aug).

F. Brown mycelial strands and sclerotia of the thread blight fungus *Ceratobasidium stevensii* on a *Malus pumila* (apple) twig (KY, Aug).

G, H. Rhizoctonia blight of *Carissa macrocarpa* (Natal-plum): foliar lesions and death of whole leaves and shoots (FL, Mar).

I. Rhizoctonia rot of a *Peperomia* cutting. Grayish mycelium is visible on the soil and the base of the plant (FL, Apr).

Photo credits: A–E—E. L. Barnard; F—J. R. Hartman

Overview of Rusts, and Phragmidium Rusts (Plate 128)

Overview of Rusts

Rust diseases are caused by obligately parasitic, host-specialized fungi of the order Uredinales in the Basidiomycota. Rust fungi infect all plant parts except roots. Their hyphae grow inter- and intracellularly, and they produce intracellular haustoria. They have complex life cycles that may involve alternate host plants and may include up to four different spore-producing forms and five functionally different kinds of spores. A rust fungus that completes its life cycle on one host is termed *autoecious*. Fungi requiring two host species are termed *heteroecious*. The listed references offer more information on rust morphology, biology, and terminology. The names and functions of life stages of a generalized rust fungus are described below in the order of their appearance in the life cycle.

Spermagonia (syn. pycnia) are small structures of various forms that produce spermatia—noninfectious spores that function as gametes. Spermagonia develop after infection by basidiospores.

Aecia are cuplike, blisterlike, or cylindric asexual fruiting structures that follow spermagonia on the same host and produce nonrepeating, dry, light yellow to orange, wind-dispersed aeciospores. In a typical heteroecious rust, aeciospores do not infect the plant species on which they were produced but are infectious to the species on which teliospores will be produced.

Uredinia (syn. uredia) are asexual, blisterlike pustules that produce urediniospores (syn. uredospores). These spores are dry, orange to rust-colored (purplish in some species), wind-dispersed, and sometimes long-lived. They cause repeating (secondary) cycles on the host on which they form.

Telia are sexual structures of various forms that follow uredinia on the same host and produce teliospores that vary greatly in color and form. Teliospores of most rusts are not dispersed or are dispersed only short distances and in most species are not infectious. Many rusts overwinter as teliospores. Those of most rusts germinate to produce basidia, on which basidiospores are produced after meiosis. Basidiospores are nonrepeating, colorless, short-lived, and wind dispersed. Basidiospores of a heteroecious rust infect the aecial host but do not infect the host on which teliospores were formed.

Rusts that have all the stages above are termed *macrocyclic*. Many rusts omit one or more stages, however, and some rusts produce fruiting structures and spores with the appearance of one state and the function of another. Rust structures with massed spores are often called pustules or sori.

References: 40, 119, 866, 870, 1246, 2077, 1679, 1681, 2344, 2345, 2429, 3028, 3145, 3519, 4387, 4526

Phragmidium Rusts

Phragmidium (Phragmidiaceae) is a genus of autoecious, mostly macrocyclic rust fungi that occur on rosaceous hosts, principally *Potentilla* (cinquefoil), *Rosa* (rose), and *Rubus* (raspberry and other brambles). These rusts produce aecia with yellow to bright orange powdery masses of thick-walled aeciospores, and later uredinia with orange urediniospores, that break through the epidermis of various plant parts. Spores of these types are subglobose to somewhat ellipsoid, spiny or verrucose, with dimensions usually in the range 15–35 μm. Teliospores follow urediniospores, often in the same pustules. Teliospores of most *Phragmidium* species are shades of brown to nearly black, ovoid to cylindric with thick walls and rough surfaces, multicellular with transverse walls, and in most species have a pale to colorless pointed apex. Each teliospore forms atop a colorless stalk cell (pedicel) that in most species is 1–1.5 times the length of the spore.

Rusts on *Rosa*. Ten *Phragmidium* species attack roses in North America: *P. americanum, P. fusiforme, P. montivagum, P. mucronatum, P. rosae-arkansanae, P. rosae-californicae, P. rosae-pimpinellifoliae, P. rosicola, P. speciosum,* and *P. tuberculatum.* Each species is specialized for parasitism of certain rose species, hybrids, or cultivars. *P. mucronatum* and *P. tuberculatum* are the best known and most common rusts on cultivated roses. The former occurs globally wherever roses flourish; the latter occurs in eastern and western North America, Eurasia, and Africa. These two species have similar host ranges, are often confused with each other, and have been suggested to be variants of a single species.

The following description applies particularly to *P. mucronatum.* Rust occurs throughout the year in areas with mild winters.

Elsewhere it appears in spring and intensifies on mature leaves. Spermagonia form on stems and upper surfaces of leaves. Aecia and uredinia develop on lower surfaces of leaves and on other green parts. Orange urediniospores may discolor the plant. Yellow to orange-brown or dark reddish brown lesions develop on upper surfaces of leaves above the rust pustules. Young stems and sepals may be distorted by rust. Severely diseased leaves of highly susceptible plants may turn yellow or brown and then drop. Growth of such plants may be severely retarded.

Black telial pustules supplant uredinia beginning in late summer in cool-temperate areas or in winter in warm regions. Teliospores, 64–95 × 22–33 μm with five–nine cells, are readily dislodged from leaves, alight on various surfaces, and then adhere by means of a gelatinous substance released from the stalk cell after it imbibes water. Teliospores overwinter and germinate in spring to produce basidiospores. In warm regions the rust fungus may also overwinter in leaves and stems. Thus new infections in spring may be initiated by basidiospores, by aeciospores derived from overwintered mycelium in stems, or by urediniospores from pustules that persist through winter on green leaves. Conditions most favorable for development of *P. mucronatum* are frequent wet periods (optimum 12 hours) and temperatures of 18–21°C. Rose rust is not severe in areas with hot or dry summers.

Rusts on *Potentilla*. Five *Phragmidium* species occur on *Potentilla* (cinquefoil) in North America. *Phragmidium andersonii, P. ivesiae,* and *P. potentillae* have transcontinental distributions in the USA and Canada, and *P. andersonii* also occurs in Europe and Asia. *P. biloculare* and *P. boreale* occur in northwestern USA and western Canada. None of these species has been studied beyond description and naming. They have life cycles and spore states as described for *Phragmidium* in general and for *P. mucronatum.* Each *Phragmidium* on *Potentilla* has distinctive teliospores. Those of *P. andersonii* are nearly black; three to five celled; verrucose with a small, colorless, pointed tip; and measure 44–74 × 25–35 μm. The basal part of the stalk cell swells markedly when wet. Severe rust on *Potentilla* suppresses flowering and causes foliar browning and defoliation, so that by late summer the affected plant may appear dead except for a few green leaves near shoot tips. A severely affected plant is likely to be disfigured by rust in subsequent years.

References: 119, 283, 757, 870, 1099, 1298, 1758, 1782, 1852, 1859, 2249, 3422, 3558, 3559, 3570, 4186, 4187

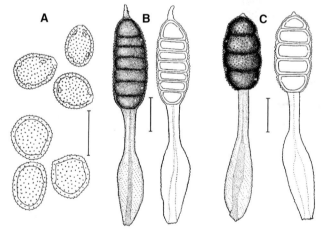

Figure 128. A, B. Aeciospores and teliospores, respectively, of *Phragmidium mucronatum.* C. Teliospores of *P. andersonii.* In each pair of teliospores, left image is surface view, right is medial section. Scale bars = 25 μm. Aeciospores from reference 2249 by permission of CAB International; teliospores from reference 119.

A–E. Rust caused by *Phragmidium mucronatum* on *Rosa.* A, B. Rust-induced necrotic lesions and chlorosis on the upper surface and uredinia on the lower surface of leaves of *Rosa* cv. Sparta (NY, Sep). C, D. Aecia on rose canes (NY, May). E. Magnified view of uredinia on the undersurface of a leaf (NY, Sep).

F–I. Rust caused by *P. andersonii* on *Potentilla* sp. (NY, Aug). F. Foliage discolored by rust. G. Yellow rust-induced lesions, enlarged about 10 times. Dark centers are locations of telia and formerly of uredinia. H, I. Magnified views of black teliospores on rust lesions.

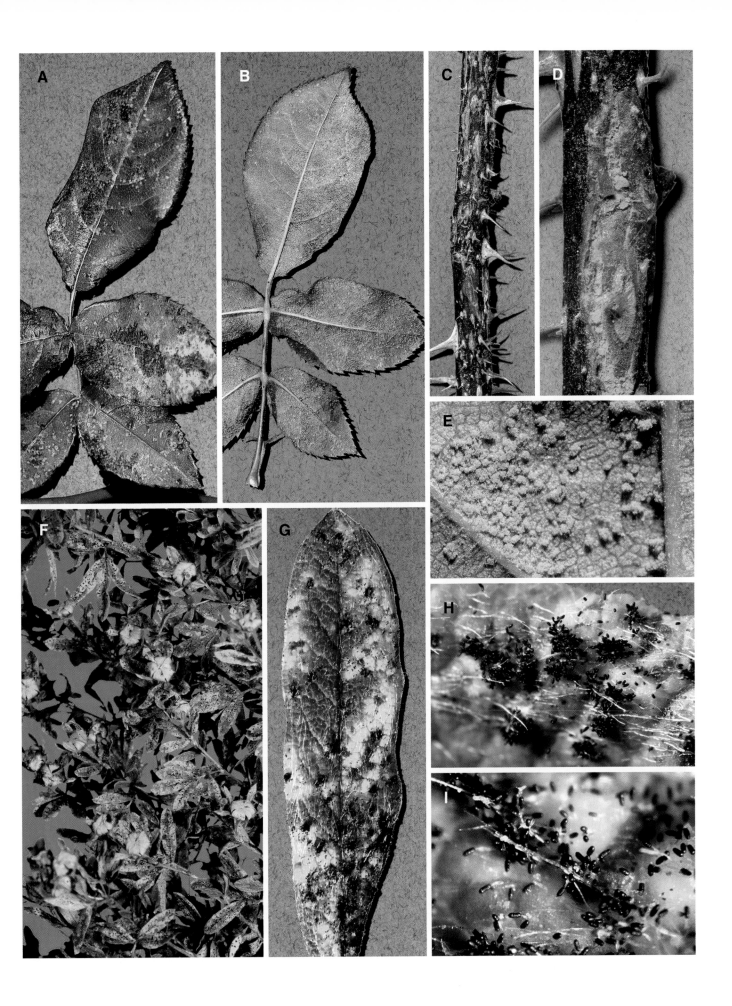

Gymnosporangium Rusts (Plates 129–133)

Overview

Rust fungi of the genus *Gymnosporangium* infect trees and shrubs in the Cupressaceae, and most of them have alternate hosts in the Rosaceae. They cause galls, stem swellings, witches'-brooms, and dieback of twigs and branches on their evergreen hosts. Rosaceous hosts develop colorful spots and localized swellings on leaves, fruit, and green twigs, followed often by casting or distortion and death of these parts.

The genus *Gymnosporangium* (Uredinales, Pucciniaceae) includes about 60 species worldwide, of which about 40 occur in North America. Their telia form on *Juniperus* (juniper, red-cedar) and other members of the Cupressaceae. Nearly all of these fungi lack a uredinial (repeating) state and require an alternate host to complete their life cycles and perpetuate themselves. Two exceptions that form uredinia are *G. nootkatense* on *Chamaecyparis nootkatensis* (Alaska cedar) in the Pacific Northwest and *G. gaeumannii* on *J. communis* (common juniper) in the Canadian Rockies. Spermagonia and aecia of *Gymnosporangium* usually form on rosaceous plants. The only known exceptions are *G. speciosum*, which produces aecia on *Fendlera* and *Philadelphus* (mock-orange) in southwestern USA; *G. ellisii*, which produces aecia on *Comptonia peregrina* (sweetfern) and *Myrica* species (sweetgale, wax-myrtle) in Atlantic Coast states; and the autoecious species *G. bermudianum*, which produces aecia and telia on *J. virginiana* (eastern red-cedar) in southeastern USA. A list of noteworthy *Gymnosporangium* rusts other than those illustrated appears on page 268.

Generalized life cycle and structures. Some *Gymnosporangium* species complete their life cycles in 1 year, but many require 2 years. Spermagonial and aecial states form in spring through late summer in yellow to reddish orange lesions on leaves or in chlorotic to brownish spots on green stems and fruits of angiosperm hosts. Spermagonia in leaves open to the upper surface, and aecia usually form on the undersurface. Spermagonia are flask shaped, immersed in host tissue, and visible only as tiny yellow to black dots on the lesions. Aecia are white or light-colored hornlike or cylindric (rarely cup-shaped) structures that produce powdery yellow to brown aeciospores. Aeciospores are windborne and cause infection of the telial (evergreen) host during summer and autumn. The rust fungus in leaves or fruit of its aecial host is shed with these parts in autumn. If in a twig, the fungus perishes during winter in most cases.

Gymnosporangium species overwinter in their gymnosperm hosts, where they produce telia in spring. In rusts with a 2-year cycle, 21–22 months elapse between infection and telial formation. Telia appear as orange to reddish brown hornlike or cushionlike projections on galls, bark, or leaves. They consist primarily of long-stalked teliospores. Telia become swollen and gelatinous during wet weather in spring, and teliospores then germinate in place, producing basidia and basidiospores. The basidiospores are dispersed by air and infect succulent parts of aecial (angiosperm) hosts. Many *Gymnosporangium* species become perennial in juniper or cedar branches and produce telia annually on galls, swellings, or witches'-brooms.
References: 119, 870, 1099, 1298, 1859, 2044, 2275, 2276, 2429, 2787, 2965, 2967, 4526

Cedar-apple Rust

Cedar-apple rust, caused by *Gymnosporangium juniperi-virginianae*, is the best-known and most economically important of the Gymnosporangium rusts in North America. In apple-growing areas where *Juniperus virginiana* (eastern red-cedar) or *J. scopulorum* (Rocky Mountain juniper) is abundant, this disease has the potential to cause severe fruit infection, premature defoliation, and crop reduction in highly susceptible cultivars. The pathogen is one of the few to have triggered legislation for the purpose of disease suppression. Several states in eastern USA enacted laws requiring the destruction of the principal telial host, *J. virginiana*, near commercial apple orchards. The red-cedars in some areas were valued for ornament and for their fragrant lumber, however, and the eradication efforts often led to hard feelings among neighbors.

Cedar-apple rust, indigenous to North America, is widespread throughout the USA east of the Rocky Mountains and occurs in southern Ontario and Quebec. It has been reported from California

and Washington but is uncommon in the West. Plants most commonly affected are *Malus* species (apple and crabapple), *J. scopulorum*, and *J. virginiana*. Other susceptible plants include *J. chinensis* (Chinese juniper), *J. communis* var. *depressa* (low juniper), *J. horizontalis* (creeping juniper), *J. osteosperma* (Utah juniper), and *J. pinchotii* (Pinchot or redberry juniper). Infection of *Crataegus* (hawthorn) has been reported but is uncommon. There is also a report of this disease on *Cedrus* sp. (cedar) in Florida.

Symptoms and signs. Globose galls and associated twig dieback occur on *Juniperus*. Dieback is most prominent on *J. scopulorum*. Galls begin growth in spring as minute green swellings on leaves, usually on the upper or inner surface. When full grown in autumn the galls are greenish brown, globose to kidney shaped, and usually 10–30 mm in diameter (extremes 2 and 50 mm). Galls consist mainly of parenchyma and abundant intercellular hyphae. Circular depressions appear on a gall in autumn, and telia emerge from these the next spring. The telia are golden brown hornlike projections 10–20 mm long and 1–2 mm wide, consisting primarily of teliospores. During warm spring rains they swell to two to three times their former size, becoming gelatinous and bright orange-yellow. Telia gelatinize and dry several times in response to intermittent rains. They finally disintegrate into a slimy mass that may obscure a gall and adjacent plant parts. The slime eventually dries and falls away. Galls soon die, but they remain attached to twigs for a year or more.

Swollen lesions appear on *Malus* leaves and fruit in late spring and early summer. They are at first greenish yellow, then orange-yellow, often bordered by a red band. Lesion size varies with host susceptibility and inversely with frequency on the infected part. Spermagonia appear in the center of each lesion (on the upper surface of a foliar lesion) as yellow dots that exude tiny droplets of yellow fluid containing spermatia. After several days, spermagonia darken and exudation ceases. Aecia appear in summer, 2–4 weeks after spermagonia, and may be scattered or arranged in a circle (on the lower surface of a foliar lesion). If the lesion is on a fruit or twig, aecia appear among the spermagonia. The aecia, similar to those shown in Plate 131, are off-white cylindric structures 1–2 mm long, with walls that split into strips. The strips flare out and their tips curve toward the base of the aecium, exposing a reddish brown mass of aeciospores. Aecial wall cells are curved in side view. The aeciospores are globoid, thick walled, and measure 21–31 × 16–24 μm.

Disease cycle. Basidiospores, released in spring from telia on galls on *Juniperus*, infect young leaves and sometimes young fruit and green stems of *Malus*. Aecia form on these organs in summer and release aeciospores that infect leaves of *Juniperus* from midsummer to early autumn. Galls form during the next year, and telia are produced on galls 21–22 months after infection, completing the 2-year cycle.

Cedar-apple rust, caused by *Gymnosporangium juniperi-virginianae*.
A. Gelatinized telia on *Juniperus scopulorum* (Rocky Mountain juniper) (NE, Jun).
B. Gelatinized telia on two small galls on *J. virginiana* (eastern red-cedar) (NY, May).
C. Collapsed telia on *J. virginiana* after final gelatinization (NY, Jun).
D, E. Rust galls on *J. virginiana* in winter (D) and early spring (E) at the onset of telial growth (NY, Feb & Apr).
F. Dead rust galls and associated twig dieback on *J. scopulorum* (IL, Jun).
G–L. Rust lesions on *Malus pumila* (apple). G. Young lesions with spermagonia on a 'Stark's Golden Delicious' leaf (NY, Jun). H. Rust spots viewed on upper and lower surfaces of 'Ginger Gold' leaves; dark dots on the lower surface are aecia (NY, Jul). I. Rust-induced necrotic lesions on 'McIntosh' leaves; the necrosis is a resistance response (NY, Jun). J. A rust lesion on 'Ginger Gold' fruit; black dots are the tips of spent spermagonia (NY, Aug). K, L. Magnified views of the upper (left) and lower surfaces of a rust lesion on a 'Runkel' leaf. Black dots in K are spent spermagonia; tattered aecia are arranged in a ring in L (NY, Aug).
Photo credits: E—G. W. Hudler; H–L—D. A. Rosenberger

261

An overnight period of high humidity after rain is sufficient for telial gelatinization, germination of teliospores, formation and dispersal of basidiospores, and infection of a rosaceous host. A given telium releases basidiospores several times because only a portion of the teliospores germinate during each gelatinization. Basidiospores, liberated primarily at night, may be carried several kilometers in air, but most infections occur within a few hundred meters of the spore source. Basidiospores deposited on dry foliage remain germinable for a day or more if humidity remains high and the spores are not exposed to sunlight. Given wet plant surfaces, infection can occur at 2–24°C. A wet period of 4–6 hours at 10–24°C is sufficient for severe infection.

Aeciospores are dry and are dispersed by wind. They drop out of aecia chiefly during morning hours in response to drying. Wetness or high humidity inhibits their dispersal because the wall of an aecium absorbs water and swells, restricting the opening. Wetness on juniper leaves is required for infection, however.

Races of G. juniperi-virginianae that vary in virulence to Malus cultivars are known, and both apple and crabapple cultivars vary in rust resistance. Contemporary apple cultivars resistant to Gymnosporangium rusts include Liberty, Macfree, Moira, Murray, Nova Easygro, Novamac, Priscilla, Redfree, and Trent. Resistance is expressed as restriction of lesions to small flecks or mottled areas or to small spots where only spermagonia develop. Crabapples observed extensively and found rust free or only slightly affected by cedar-apple rust are listed with Plate 42. Junipers whose field performance has indicated resistance to any of four indicated diseases are listed below.

Disease-resistant junipers

Key: † Phomopsis blight, ‡ Kabatina blight, * cedar-apple rust, ** both cedar-apple rust and hawthorn rust.

*Juniperus ashei***	*J. horizontalis* cvs. Admirabilis**,
J. chinensis cvs. Ames‡*, Aurea‡,	Adpressa**, Argenteus**,
Blue Point*, Foemina†**,	Douglasii**, Eximius**,
Fortunei**, Gold Coast‡, Hetzii**,	Filicina**, Glomerata**,
Hetzii Columnaris‡*, Hetzii	Livida**, Petraea**, Plumosa**,
Glauca‡, Iowa†, Japonica**,	Procumbens†, Prostrata Glauca‡
Keteleeri†‡*, Leeana**, Maney‡*,	*J.* x*pfitzeriana*†‡** cvs. Aurea†‡,
Mas**, Mint Julep‡,	Compacta**, Glauca**
Mountbatten‡*, Perfecta‡*,	*J. rigida***
Plumosa‡, Plumosa Aurea**,	*J. sabina*†** cvs. Arcadia†,
Pyramidalis**, Robusta Green†‡*,	Broadmoor†**, Buffalo†,
Shoosmith†, Spartan*‡	Calgary†, Fastigiata**, Knap
J. chinensis var. *procumbens***	Hill†**, Skandia†
cvs. Nana‡, Variegata‡	*J. sabina* var. *tamariscifolia*‡**
J. chinensis var. *sargentii*†**	*J. scopulorum* cvs. Medora*,
cvs. Glauca†‡, Variegata*,	Moonglow*, Silver King†
Viridis‡, Wateri**, Wintergreen‡*	*J. squamata* cvs. Albo-variegata**,
J. communis cvs. Ashfordii†,	Campbellii†, Meyeri**,
Aurea**, Aureospica†**,	Prostrata†, Pumila†, Wilsonii**
Cracovia**, Depressa†**,	*J. squamata* var. *fargesii*†**
Depressa Aurea†, Hibernica‡**,	*J. virginiana* cvs. Aurea**,
Hornbrooki‡, Hulkjaerhus†,	Berg's Rust Resistant**, Blue
Oblonga Pendula**, Repanda†,	Mountain‡*, Burkii**,
Saxatilis†**, Suecica†**,	Globosa**, Grey Owl‡*, Hillii†,
Suecica Nana**	Hillspire*, Kosteri†,
J. communis f. *oblonga***	Pseudocupressus**,
J. conferta†**	Pyramidalis**, Skyrocket**,
J. davurica cv. Expansa‡	Tripartita†**, Venusta**

References: 34, 35, 37, 38, 859, 1099, 1667, 1762, 2655, 2787, 2849, 2942, 2943, 2998, 3000, 4009, 4043, 4228, 4229, 4274, 4275

Hawthorn Rust

Hawthorn rust, caused by *Gymnosporangium globosum*, is similar to cedar-apple rust in most respects. We emphasize the differences. The pathogen, indigenous to North America, is distributed from Saskatchewan to Quebec and southward to Texas and Florida. The most commonly affected evergreen hosts are *Juniperus virginiana* (eastern red-cedar) and *J. scopulorum* (Rocky Mountain

juniper). Others affected are *J. chinensis* (Chinese juniper), *J. communis* var. *depressa* (low juniper), *J. excelsa* (Greek juniper), *J. horizontalis* (creeping juniper), and *J. sabina* (savin juniper). *G. globosum* infects *Malus* (apple and crabapple); many *Crataegus* species (hawthorns); and occasionally *Amelanchier* (serviceberry), *Cydonia* (quince), *Pyrus* (pear), and *Sorbus* (mountain-ash). *Mespilus germanica* (medlar) has become infected when inoculated.

Hawthorn rust most often affects leaves, causing yellow spots. When this rust is severe on *Crataegus*, the foliage may turn bright yellow and drop prematurely. Hawthorn rust occasionally occurs also on fruit or green stems, causing deformity. Black flattened lesions, usually lacking aecia, form on *Cydonia* fruit. On *Pyrus*, foliar lesions are dark brown to nearly black with a reddish border on the upper surface; aecia form on the lower surfaces of the largest lesions and on diseased petioles.

Spermagonia and aecia of *G. globosum* are similar to those of *G. juniperi-virginianae*, but the aecia of *G. globosum* may be distinguished by their greater length (often 3–4 mm), irregular mode of splitting, and relatively straight strips of wall cells after the walls split (in contrast to strips with their tips curved back toward the aecial base). Aecia of this fungus tend to retain cylindric form, at least near the base. Viewed microscopically, *G. globosum* has smaller aeciospores (globoid and 18–25 × 15–19 μm). Also, the aecial wall cells adhere to one another rather than separating in a water mount, and the tip cells of the walls are nearly straight rather than strongly curved. The interval from infection to aecial maturation of hawthorn rust in northern areas is 80–95 days, about 10 days longer than for cedar-apple rust. Precipitation causes immediate release of aeciospores, but in the absence of rain these spores are also released during morning hours as humidity diminishes.

G. globosum induces spheroid galls on *Juniperus* but causes little or no dieback. The galls originate on leaves, grow slowly, become firmly attached to the twigs, and often become flattened on the side next to the twig. They differ from galls of cedar-apple rust in being perennial, reddish brown (rather than greenish) when young, and smaller (3–15 mm diameter) when mature. Hawthorn rust galls in winter display domelike swellings where telia will emerge in spring. Old galls turn grayish brown and have scars where telia were produced previously. Telia of *G. globosum* are chestnut brown, conic with blunt tips, and 3–12 mm long and 1.3 mm wide before gelatinization. Sometimes the telia emerge from the upper surfaces of discolored, slightly swollen leaves. Telia double in size when they gelatinize, and in moist air their teliospores produce and liberate basidiospores within a few hours. When humid weather enhances survival of basidiospores, they can infect alternate hosts at

---

Hawthorn rust, caused by Gymnosporangium globosum.

A. A living gall, several years old, on *Juniperus virginiana* (eastern red-cedar) (NY, Mar).

B. Young, growing galls on *J. communis* (common juniper). A dead awn-shaped leaf remains attached to each gall (NY, Sep).

C–E. Gelatinizing telia—short, blunt, and reddish brown—on large and small galls (NY, May).

F. Close view of telial scars (light areas) on a dormant gall on *J. virginiana* (NY, Feb).

G, H. Yellow foliar lesions on *Crataegus* (hawthorn). Black dots in lesions in H are spermagonia (IL, Jun; NY, Jul).

I–K. Foliar and petiole lesions on *Pyrus calleryana* (callery pear) cv. Bradford. I. Magnified view of the upper surface of a foliar lesion showing spermagonia (yellow to dark dots) on swollen tissue. Resistance of this plant is indicated by small lesions and the necrosis of some infected tissue (dark brown area). J. Magnified view of aecia on the undersurface of a foliar lesion. Aecial side walls are torn from the top toward the base, characteristic of *G. globosum*. Orange-brown aeciospores color the leaf surface among the aecia (NY, Jun). K. Magnified view of a petiole with spermagonia (dark dots) and aecia on the swollen infected portion. Pimplelike bulges near edges of the swelling are developing aecia.

considerable distances from the source. In Ontario severe infection of *Crataegus* was recorded as far as 24 km from telia-bearing *Juniperus*.

Junipers resistant to *G. globosum* are listed on page 262. Hawthorns reported to be resistant include *C. crus-galli* (cockspur thorn), *C. flava* (yellow hawthorn), *C. intricata* (Biltmore hawthorn), *C. laevigata* 'Autumn Glory,' *C. monogyna* (English hawthorn), *C. phaenopyrum* (Washington hawthorn), *C. pruinosa* (frosted hawthorn), and *C. viridis* 'Winter King.'

G. globosum can be confused with *G. bethelii*, which induces similar symptoms and produces similar structures on junipers and hawthorns. Telia of the latter fungus, however, occur primarily on *Juniperus horizontalis* (creeping juniper) and *J. scopulorum* (Rocky Mountain juniper), and the aecia mature 2–4 weeks before those of *G. globosum*. The two fungi also differ microscopically.

References: 406, 1099, 1667, 2412, 2942, 2943, 2966, 4011, 4228, 4229, 4274, 4275

Quince Rust

This disease is caused by *Gymnosporangium clavipes*, which alternates between *Juniperus* and various rosaceous plants. The pathogen occurs only in North America but is distributed throughout the continent; it is common only in eastern regions. It is the most damaging of the Gymnosporangium rusts to rosaceous plants because it affects primarily fruit, green stems, and petioles, stunting and killing fruits and causing swelling, distortion, and death of twigs and petioles. It sometimes causes cankers on small branches. *G. clavipes* has a much broader host range than most members of its group. Rosaceous hosts include more than 480 species in 11 genera: *Amelanchier* (serviceberry), *Aronia* (chokeberry), *Chaenomeles* (flowering quince), *Cotoneaster*, *Crataegus* (hawthorn), *Cydonia* (quince), *Malus* (apple, crabapple), *Mespilus* (medlar), *Photinia*, *Pyrus* (pear), and *Sorbus* (mountain-ash). Apples are somewhat resistant, but their fruits become distorted when infected.

Telia of *G. clavipes* form on *Juniperus* species in section *Juniperus* and section *Sabina*: *J. communis* (common juniper), *J. horizontalis* (creeping juniper), *J. sabina* (savin juniper), *J. scopulorum* (Rocky Mountain juniper), and *J. virginiana* (eastern red-cedar). A host-specialized strain, *G. clavipes* f.sp. *cupressi*, infects *Cupressus lusitanica* (Portuguese cypress) and *Crataegus* in Mexico. Infection by the quince rust fungus apparently impairs the winter hardiness of *J. virginiana*. Many severely infected trees of this species in northeastern USA died during the intensely cold winter of 1933–1934 while adjacent lightly infected or noninfected trees survived.

Symptoms and signs. *G. clavipes* infects leaves and soft stems of *Juniperus*, causing mild swelling of diseased parts before telia form on them. Diseased leaves and many diseased twigs die during the season after first telial formation. Some twigs survive, however, and the fungus becomes perennial in their bark, colonizing a band of tissue that encircles a twig and elongates a few centimeters each year. Branches of increasing size and age thus become involved. The swollen zone becomes spindle shaped and dark brown with flaky bark. Twig infections, when numerous, are associated with gradual loss of vigor and thin foliage on *J. virginiana*, especially on trees in partial shade. Most diseased twigs die within 4–6 years, but some infections that reach larger stems persist many years, causing circular to elliptic swollen patches up to 15–20 cm in diameter. Telia develop annually on the swollen areas.

Chlorotic spots or swellings develop on succulent parts of rosaceous hosts. Leaf veins commonly become infected, and spermagonia form in them, but aecial development on veins is infrequent and then only on large veins. Diseased leaf veins of *Cydonia* swell to nearly twice their original size, and the leaves curl and eventually drop. Diseased *Crataegus* leaves curl and die. Swollen spots with spermagonia and aecia develop on *Cydonia* fruit. Rust development on *Malus* fruit varies with host susceptibility. Some cultivars show no reaction or only flecks; others develop swollen or sunken spots and spermagonia but no aecia; and some support normal aecial development.

Petioles and green twigs or thorns begin to swell soon after infection. The rust fungus commonly spreads up to 3 cm in each

direction from the point of infection, causing a spindle-shaped swelling or, uncommonly, a gall on which spermagonia and aecia develop. Newly formed buds become infected via internal mycelium and begin growth when they would normally remain closed. The resulting shoots swell and elongate only a few centimeters, producing stunted, distorted leaves. Spermagonia and aecia may develop on the shoot axis. Stems or shoots encircled by rust die back to a healthy bud or twig after aecial production. In highly susceptible hosts *G. clavipes* sometimes grows from a thorn or stunted shoot into a year-old branch, causing a canker where aecia form until mid-autumn. If the fungus survives winter at the edge of the colonized area, a new crop of aecia forms there in spring.

Spermagonia of *G. clavipes* are flask shaped and immersed in host tissue except for an opening at the apex. Viewed from above, they appear as reddish dots that blacken with time. Aecia form in cortical tissues, embedded to a depth of about 1 mm. They are white, tubular, up to 3 mm long and 0.3–0.5 mm wide, and become torn at the apex or split along the sides as aeciospores mature. Aecia on twigs always precede aecia on fruit. Fruits of *Amelanchier* and *Crataegus* often become uniformly covered with aecia and then die. The aeciospores are bright orange or orange-yellow in mass. This character and the preference of the fungus for stems and fruit distinguish quince rust from cedar-apple and hawthorn rusts on rosaceous plants.

Telia of *G. clavipes* are cushionlike and arise in spring on *Juniperus* leaves and young stems. Before the first gelatinization, telia are 1–5 mm broad and high, brick red when young, and orange-brown when mature. They turn bright orange and swell to 5–10 times their initial size when wet and gelatinized. Gelatinized telia sometimes coalesce and form an irregular ring around a diseased twig. The teliospores are two celled with long stalks that swell greatly when wet. This swelling is the principal basis of telial gelatinization. Telia may form annually for 20 years or longer at a given site on a branch or trunk. Unless telia are present, trunk and branch infections by *G. clavipes* cannot be distinguished from those caused by several other *Gymnosporangium* species, especially *G. nidus-avis* and *G. effusum*.

Disease cycle. Accounts differ as to whether the quince rust cycle requires one year or two. The pathogen overwinters and perennates in *Juniperus*, where it produces telia beginning in the first or second spring after infection. Teliospores germinate in place (on telia) within several hours after thorough wetting and produce basidiospores that are released into the air. Basidiospores infect rosaceous hosts, causing chlorotic spots or swellings that become visible 7–10 days after infection. Spermagonia appear after 10–15

Quince rust, caused by *Gymnosporangium clavipes*.

A–F. Spindle-shaped stem swellings and stages of telial development and degeneration on *Juniperus virginiana* (eastern red-cedar). A. Swelling with roughened bark on a branch in summer; no signs of the pathogen are visible from summer through winter (NY, Aug). B. Orange-red cushionlike telia protruding from roughened bark in early spring (NY, Apr). C. Mature telia just before gelatinization (NY, May). D, E. Gelatinized telia (NY, Jun). F. Old telia, depleted of germinable teliospores, visible as a dry orange scum on a branch. Dead twigs apparently killed by rust extend from the diseased region (NY, Jun).

G–K. Quince rust on *Crataegus* (hawthorn). G. Young, white, cylindric aecia on fruit (NY, Jul). H–K. Aecial state on stems of *C. laevigata* (English hawthorn) (NY, Aug). H. Aecia on a canker caused by *G. clavipes*. The canker is active for a second year, as indicated by orange aecia at the margins. I. Shoot deformity and dieback. Shoots infected the previous year (gray) became swollen and distorted and died back after aecial production. The pathogen overwintered in the lower shoot. J. Magnified view of a cross section of a swollen first-year stem with aecia. K. Magnified view of a second-year stem segment at the margin of a rust canker. Gray tissue is dead. Orange color is that of aeciospores among tattered remains of aecia.

days, and aecia after 30–60 days (45–60 days in the North). Apple fruits, and presumably fruits of other hosts, are highly susceptible for only about 2 weeks when young. Aeciospores from rosaceous plants cause infection of *Juniperus* leaves and green stems from midsummer to early autumn. These infection sites remain asymptomatic until spring. Mycelium is confined mainly to epidermal cells in leaves and to the outermost living bark of twigs.

After telial production each spring, a cork layer forms beneath the telial stroma, isolating most of the infected tissue and causing both fungal and host cells to die by attrition. Before this barrier forms, however, hyphae of the parasite grow inward to the region of the cork cambium, where they persist among living host cells. These hyphae produce a new telial stroma the next spring. Tissues isolated by the cork barrier become loose superficial bark scales.

Environmental conditions influencing quince rust are relatively well studied. Aeciospores are released not only in response to rain but also in response to aecial drying as humidity diminishes during morning hours. They are dispersed by wind. Given free water, these spores germinate at 3–30°C; the optimum temperature is near 16°C. The optimum temperature for teliospore germination and basidiospore formation is 22–24°C. Basidiospore dispersal is approximately synchronized with the onset of growth in rosaceous hosts, and the spores are capable of germinating within 2 hours on a wet surface. An overnight period of wet weather when telia are mature in spring is sufficient for infection. Individual telia gelatinize six–eight times before collapsing as an orange slime that soon weathers away.

References: 34, 860, 976, 977, 2138, 2653, 2787, 2943, 2965, 4009, 4011, 4229, 4274, 4275, 4526

Juniper Broom Rust

The causal fungus is *Gymnosporangium nidus-avis*, a North American species that occurs from coast to coast and from near the limits of its hosts in Canada southward to Arizona and Florida. Its telial state is associated with witches'-brooms and stem swellings on *Juniperus horizontalis* (prostrate juniper), *J. scopulorum* (Rocky Mountain juniper), *J. virginiana* (eastern red-cedar), and *J. virginiana* var. *silicicola* (southern red-cedar). The spermagonial and aecial states occur on leaves, fruits, and green stems of *Amelanchier* (serviceberry) and on *Chaenomeles* (flowering quince), *Cydonia* (quince), and *Malus* (apple). *Crataegus* (hawthorn) and *Sorbus* (mountain-ash) have also proven susceptible in inoculation tests. Severe infection of *Amelanchier* may kill fruit or leaves and shoots as lesions encircle petioles and green twigs. Foliar infections cause red-bordered brown lesions that enlarge from points of infection to leaf edges.

Infection of *Juniperus* occurs on leaves or green twigs and becomes perennial if the host part remains alive. Growth of rust mycelium near host growing points leads often to formation of witches'-brooms, while infection of a stem at a distance from its apex may lead to swelling but no broom formation. Brooms are most conspicuous on *J. scopulorum* and *J. virginiana*. Most brooms die while small, but some attain diameters of 50–60 cm or more and live 15 years or longer. Branches within brooms tend to be somewhat enlarged, have roughened bark, and often bear only awn-shaped (juvenile) leaves.

The rust fungus often grows many centimeters in living bark along a branch from a broom toward or into the mainstem. Its growth rate in *J. virginiana* has been measured at 5 cm per year. Swellings on nonbroomed branches may be spindle shaped, encircling small branches; or elliptic, appearing as bulges on large branches or trunks. Some infections persist and enlarge for 3 decades or more. Diseased bark usually becomes roughened and in time develops deep longitudinal fissures and corky parallel ridges where the fungus produces telia annually. The ridges are smooth in contrast with the fibrous texture of normal outer bark. Occasionally, long strips of diseased bark die and the fungus produces telia beyond the margins of the dead area. Whether *G. nidus-avis* or a secondary pathogen causes such cankers has not been ascertained.

Orange-brown telia somewhat similar to those of *G. clavipes* (Plate 130) protrude in spring from abnormally roughened bark in brooms and on swollen parts of nonbroomed twigs, branches, and even trunks. Telia may also form in the axils of awn-shaped leaves or between scale-leaves. Telia vary in shape and size from tiny cushions to elongate mounds 3–5 mm long × 3–6 mm wide × 2–9 mm high. They may rise as high as 16 mm when gelatinized. Teliospores germinate in place and liberate basidiospores. Old telia become cinnamon brown and eventually disappear, leaving oval scars that are at first yellowish orange, then brownish.

Basidiospores of *G. nidus-avis* germinate within a few hours on wet surfaces of succulent parts of rosaceous hosts. The fungus then penetrates directly through the cuticle, forming intercellular hyphae that in turn penetrate host cells to produce haustoria. Spermagonia develop within 9–14 days after infection in swollen yellow spots on soft twigs, petioles, or leaf veins, and on yellow to yellow-orange leaf spots with red borders. Rust spots on leaves are 1–5 mm in diameter, depending on host susceptibility. Vein infections cause leaf distortion. Spermagonia appear as yellow dots from which droplets containing yellowish spermatia exude for 5–10 days, after which the spermagonia and often the lesions turn black. Aecia 0.3–0.6 mm in diameter and up to 5 mm long form after an additional 3–4 weeks and open by lateral slits. They tend to retain their cylindric shape, and their wall cells remain straight when wet. Aecia on leaf veins sometimes develop on upper as well as on lower surfaces. The aeciospores are golden brown, thick walled, verrucose, globoid, and measure 20–33 × 18–25 μm. In the North aecia maturing on *Amelanchier* and associated with leaf-and-shoot blight in early to mid-July are likely to be those of *G. nidus-avis*.

Juniper broom rust on *Amelanchier* is somewhat similar to European hawthorn rust caused by *G. clavariiforme*, aecia of which mature at about the same time. The latter rust, however, does not usually cause prominent foliar necrosis. Its aecia are short (to 1.5 mm high), and they split open to the base. The telia of *G. clavariiforme* occur on *Juniperus communis* (common juniper) and are cylindric.

References: 975, 976, 1099, 1298, 2787, 2965, 2967, 3149, 3150, 4213, 4526

Broom Rust of Incense-cedar and Rust Galls on Southwestern Junipers

Gymnosporangium libocedri causes witches'-brooms on *Calocedrus decurrens* (incense cedar) in Nevada and the Pacific Coast states. The spermagonial and aecial states of the fungus develop on leaves and fruit of various rosaceous plants—mostly *Amelanchier* (serviceberry) and *Crataegus* (hawthorn), but also *Chaenomeles* (flowering quince), *Cydonia* (quince), *Malus* (apple and crabapple), *Pyrus* (pear), and *Sorbus* (mountain-ash). *G. libocedri* has occasionally caused severe infection of pear and quince fruit in orchards in the Northwest where *Calocedrus* was growing nearby. Foliar symptoms and signs on rosaceous hosts are similar to those described for other Gymnosporangium rusts. Spermagonia occur on the upper sides of colorful swollen leaf spots, and white aecia develop mostly on the undersides and on fruit. The aecia are cuplike with lacerated rims. The aeciospores are globoid, verrucose, yellow, and 12–20 × 14–23 μm in size.

Most infections on *Calocedrus* are confined to leaves and are scarcely noticeable. Telia develop only on the leaves. Before gelatinization they are cushionlike, reddish brown, and 0.8–1.5 mm in diameter. The long-stalked brown teliospores of *G. libocedri* are distinctive for having up to five cells. Germinated teliospores produce basidiospores that can cause infection of rosaceous hosts at distances as great as 12–16 km from the spore source.

Mycelium of *G. libocedri* sometimes becomes perennial in the wood of twigs and then causes spindle-shaped swellings or witches'-brooms. Telia develop on leaves of the brooms. As

A. Juniper broom rust, caused by *Gymnosporangium nidus-avis*, on *Juniperus scopulorum* (Rocky Mountain juniper) (CO, Jun).
B–E. Leaf- and shoot-blight of *Amelanchier* sp. (serviceberry) caused by *G. nidus-avis*. B. Killed leaves and twigs. C–E. Necrotic foliar lesions and aecia on swollen stems and petioles (ON, Jul).
F. A spindle-shaped gall caused by *G. speciosum* on a branch of *Juniperus deppeana* (alligator juniper) (AZ, Jul).

267

branches grow, and the rust fungus with them, swellings or witches'-brooms come to be located on large branches and sometimes on the mainstem. Diseased wood in swellings or brooms has numerous dark brown flecks arranged in radiate patterns as viewed in transverse section. By one estimate from the study of such wood, living mycelium in incense cedar may persist more than 200 years. Trees with many brooms eventually decline and die.

G. speciosum causes witches'-brooms, spindle-shaped swellings (Plate 132F), and sometimes large woody galls on branches of *Juniperus deppeana* (alligator juniper), *J. monosperma* (cherrystone juniper), *J. occidentalis* (Sierra juniper), and *J. osteosperma* (Utah juniper) in southwestern USA. Brooms caused by this fungus are most common on *J. osteosperma*. They are associated with stem swellings on *J. deppeana*. Galls on *J. deppeana* are usually spindle shaped or oblong but sometimes nearly globose, three to four times the diameter of the branch, and may develop to 35 cm or more in diameter. They have rough corrugated surfaces. Telia form in spring in more or less distinct rows on the swollen parts. Occasional specimens of *J. osteosperma* have been observed that seem to be completely infected, with telia appearing in small patches all over the trunk and branches. The telia are orange-yellow, elongate in basal outline, and 3–4 mm high before gelatinization. The teliospores have either two or three cells.

G. speciosum is unusual in that it alternates not to rosaceous plants but rather to plants in two genera of the Hydrangeaceae: *Fendlera*, notably *F. rupicola* (cliff fendlerbush), and *Philadelphus*, notably *P. microphyllus* (littleleaf mock-orange). Its spermagonia open to the upper surfaces and aecia to the lower surfaces of yellow leaf spots. The aecia are cylindric, white, and 2–3 mm high. Their side walls split open at the top, facilitating the release of yellow aeciospores. The aecial hosts do not sustain significant damage. In New Mexico the anamorphic rust fungus *Uredo apacheca* also infects *J. deppeana*. It produces uredinia on rough galls similar to those caused by *G. speciosum*.
References: 119, 461, 1859, 1880, 2044, 3048

Medlar Rust

Gymnosporangium confusum, a fungus native to Eurasia, causes the disease known as medlar rust. This name is used to avoid confusion with diseases caused by *G. clavipes* and *G. globosum*, which have several of the same hosts. *G. confusum* displays broad host adaptation. Its telial state occurs on eight species of *Juniperus* in both sections of the genus (*Juniperus* and *Sabina*), and its spermagonia and aecia occur on plants in seven rosaceous genera: *Chaenomeles* (flowering quince), *Cotoneaster*, *Crataegus* (hawthorn), *Cydonia* (quince), *Mespilus* (medlar), *Pyrus* (pear), and *Sorbus* (mountain-ash). In North America *G. confusum* has been reported from California, where it was first found on *J. sabina* (savin juniper) and *Crataegus oxycantha* (English hawthorn), and from North Carolina. Medlar rust causes little damage to junipers, but it can destroy the foliage of highly susceptible hawthorns. In California severe infection of hawthorn in spring may lead to browning and defoliation followed by production of a second complement of shoots and leaves. The new foliage remains rust free because inoculum from junipers is available only in spring.

G. confusum infects current-season shoots of *J. sabina* and persists for many years as perennial mycelium. No symptoms or signs are apparent until telia emerge from year-old twigs in spring (early April in California). In subsequent years the infected region develops into a spindle-shaped swelling with abnormally rough bark, from which telia protrude annually in spring. The telia are horn shaped, velvety, dark reddish brown, and up to 8 mm long before gelatinization. When wet they expand and gelatinize, their teliospores germinate, and basidiospores are released. After being depleted of germinable teliospores, telia deliquesce to a slime that coats the bark and surrounding foliage.

Orange spermagonia of *G. confusum* develop on *Crataegus* about 2 weeks after infection (late April in California) in brown thickened spots with yellow margins on both leaf surfaces and in swellings on succulent stems, fruit, pedicels, and calyxes. Aecia develop among spermagonia except on leaves, where they are restricted to the lower surface. They mature about a month after the spermagonia do.

Aecia are cylindric, 2–4 mm long, and dull white, becoming tan and tattered. Sometimes they cover the surfaces of diseased fruit. The aeciospores, infectious to *Juniperus*, are reddish brown, globoid and somewhat angular, thick walled, finely verrucose, and 18–27 μm in diameter. Aeciospores on a wet surface may germinate within 4 hours at 24°C.
References: 198, 1099, 1859, 2044, 4387

Additional Gymnosporangium Rusts

Those significant or conspicuous in North America include the following, among others (host data include records from abroad):

G. asiaticum, Japanese pear rust: widespread in eastern Asia, introduced to eastern and western USA; telia on nonswollen stems and leaves of *Juniperus chinensis* (Chinese juniper), *J. procumbens* (creeping juniper), and *J. virginiana* (eastern red-cedar); aecia on *Chaenomeles* (flowering quince), *Cydonia* (quince), *Photinia*, and *Pyrus* (pear).

G. bethelii: across cool-temperate North America and in Mexico; causes dieback associated with irregular galls along spindle-shaped swellings on stems of *J. horizontalis* (creeping juniper), *J. occidentalis* (Sierra juniper), *J. scopulorum* (Rocky Mountain juniper), and additional *Juniperus* species in Mexico; aecia on fruit and leaves of *Crataegus* (hawthorn).

G. biseptatum: in Atlantic Coast states and California; causes spindle-shaped galls on *Chamaecyparis thyoides* (Atlantic white cedar) and occurs on *Calocedrus decurrens* (incense cedar); aecia on leaves of *Amelanchier* (serviceberry).

G. clavariiforme: across North America and in Eurasia, northern Africa, New Zealand; causes fusiform swellings on stems of *J. communis* (common juniper); aecia on leaves, fruits, and green stems of *Amelanchier*, *Aronia* (chokeberry), *Crataegus*, *Cydonia*, *Pyrus*, and *Sorbus* (mountain-ash).

G. cornutum: around the Northern Hemisphere; causes fusiform swellings on branches of *J. communis* and *J. rigida* (needle juniper); aecia on *Amelanchier*, *Sorbus*, and rarely *Malus* (apple, crabapple).

G. cunninghamianum: in southwestern USA and Asia; causes fusiform swellings and subglobose galls on branches and trunks of *Cupressus arizonica* (Arizona cypress) and *C. bakeri* (Modoc cypress); aecia on *Amelanchier* and *Pyrus*.

G. effusum: in eastern USA and Oklahoma; causes long slender swellings on small branches and occasionally trunks of *J. virginiana*; aecial state unknown.

G. ellisii: along the eastern seaboard of the USA; causes witches'-brooms on *Chamaecyparis thyoides*; aecia on leaves of *Comptonia* (sweetfern) and *Myrica* (bayberry, sweetgale).

G. fuscum (syn. *G. sabinae*), pear trellis rust: in California and British Columbia, Eurasia, and northern Africa; causes fusiform swellings on branches of *J. communis*, *J. oxycedrus* (prickly juniper), *J. phoenicea*, and *J. sabina* (savin juniper); aecia on *Pyrus*.

G. japonicum: in northeastern and northwestern coastal USA and eastern Asia; telia on fusiform swellings on stems of *J. chinensis*; aecia on *Heteromeles* (Christmas-berry, toyon) and *Photinia*.

G. kernianum: in western USA and Mexico; causes witches'-brooms on *J. californica* (California juniper), *J. deppeana* (alligator juniper), *J. monosperma* (cherrystone juniper), *J. occidentalis*, and *J. osteosperma* (Utah juniper); aecia on leaves and fruit of *Amelanchier*, *Cydonia*, *Crataegus*, and *Pyrus*.

G. nelsonii: widespread in temperate North America; broad host range; causes globoid galls on *Juniperus*; telia on *J. californica*, *J. deppeana*, *J. flaccida* (Mexican drooping juniper), *J. horizontalis*, *J. monosperma*, *J. occidentalis*, *J. osteosperma*, *J. scopulorum*, and *J. virginiana*; aecia on *Amelanchier*, *Crataegus*, *Cydonia*, *Malus*, *Peraphyllum* (squaw-apple), *Pyrus*, and *Sorbus*.

G. tremelloides: in western North America and Europe; causes subglobose galls on small branches and hemispheric galls on larger branches of *J. communis*; aecia on leaves of *Cydonia*, *Crataegus*, *Malus*, and *Sorbus*.

References: 119, 461, 566, 1859, 2044, 2275, 2276, 2787, 2965, 2967, 3048, 4108, 4213

A–C. Witches'-brooms on *Calocedrus decurrens* (incense cedar) caused by *Gymnosporangium libocedri*. A, B. Brooms in dying trees are prominent because normal foliage is scarce. C. A rust-induced broom in which many branches radiate from a gall (CA, Jul).

D–F. Medlar rust, caused by *Gymnosporangium confusum*. D. Stem swelling with roughened bark on *Juniperus sabina* (savin juniper). E. Foliage on *Crataegus laevigata* (English Hawthorn) cv. Paul's Scarlet shriveled and brown as the result of severe infection. F. Aecia on fruit (CA, Aug).

269

Puccinia and Cumminsiella Rusts (Plates 134, 135)

Rust fungi of the genus *Puccinia* (Uredinales, Pucciniaceae) are known primarily for damage caused to cereal grain crops. The aecial states of some agriculturally important *Puccinia* species occur on woody plants, usually causing little damage. For example, aecia of *P. graminis*, which causes black stem rust of wheat and other cereals, develop on *Berberis* (barberry) and *Mahonia*; and aecia of *P. coronata*, which causes crown rust of cereals, form on *Frangula* and *Rhamnus* (buckthorns) (Plate 135). A few species of *Puccinia* cause noteworthy diseases of trees or shrubs. Ash rust, considered here, is one example. Cluster-cup rust of *Ribes* (currant and gooseberry), caused by *P. caricina* (Plate 135), is another.

Ash Rust

Puccinia sparganioides causes ash rust, one of the most spectacular and destructive rust diseases of deciduous trees or shrubs. The fungus occurs from Saskatchewan to Nova Scotia and southward to Mexico and the Gulf Coast states. It has also been reported from Brazil. Susceptible tree species include *Fraxinus americana* (white ash), *F. berlandierana* (Mexican ash), *F. caroliniana* (Carolina ash), *F. nigra* (black ash), *F. pennsylvanica* (green or red ash), *F. profunda* (pumpkin ash), *F. quadrangulata* (blue ash), and *F. velutina* (velvet ash). *Forestiera acuminata* (eastern swamp-privet) and *F. segregata* (Florida swamp-privet) are also affected in southern USA.

The spermagonial and aecial states of this rust develop on *Fraxinus* and *Forestiera* in spring and early summer, and the uredinia and telia occur on various species of *Spartina* (cordgrass) during summer and autumn. In marshes along the New England coast this rust infects both *Spartina* and *Distichlis spicata* (saltgrass, spike grass). Epidemics on the grass hosts are scarcely noticed, but those on *Fraxinus* result in striking foliar distortion, browning, and death of green twigs. Severe rust in successive years weakens trees and makes them susceptible to winter damage, causes dieback, and has been reported to kill small *F. americana* in Nova Scotia.

Symptoms and signs. Leaves, petioles, and green twigs of *Fraxinus* are attacked. Spermagonia appear from mid-April in the South to mid-June in the North on yellow to yellow-orange spots on upper surfaces of leaves and on chlorotic spots on petioles and stems. Clusters of aecia become abundant 10–14 days later as bright orange spots on petioles, stems, and undersurfaces of leaves. Diseased tissues swell markedly, causing leaf distortion, sharp bends in petioles, and elliptic to hemispheric wartlike galls on twigs. Spots on leaves may enlarge to several millimeters in diameter, and swellings with aecia on stems and petioles may grow to 2–3 cm long. Individual aecia are cuplike with white walls around a bright orange mass of aeciospores. Tissues bearing aecia, and often the surrounding zone as well, die soon after aecial maturation; thus cankers form on twigs, and petiole infections lead to withering and browning of leaves in early summer. Severely affected trees look scorched. Brown leaves drop during mid to late summer. Rust infections of stems often remain confined to outer tissues and, after aecial production, become separated from underlying normal tissues by a cork layer. This accounts for the wartlike appearance of the swellings on stems. Symptoms on *Spartina* are elliptic yellowish spots or, when severe, general foliar yellowing. Uredinia form beneath the epidermis, oriented with the veins of the leaf as light-colored pustules 2–3 mm long that split open to reveal a mass of yellowish urediniospores. Brownish black telia replace the uredinia in autumn. Severely diseased leaves die prematurely.

Disease cycle. P. sparganioides normally overwinters as teliospores on *Spartina*. During warm, wet weather in spring, teliospores germinate and produce basidiospores. In the North this process requires only 3–4 hours at 12–21°C. Infection of *Fraxinus* occurs within another 3–4 hours if moist air moves basidiospores from marsh grasses to wet surfaces on *Fraxinus*. Beginning in late spring to early summer aeciospores are windborne from *Fraxinus* to *Spartina* and *Distichlis*. Urediniospores are produced on the grasses in May to July, depending on latitude, and cause secondary cycles of rust during mid to late summer. Annual rust activity ends with the production of teliospores.

Ash rust is most severe in coastal regions. In spring severe rust may develop on *Fraxinus* after a period of foggy weather with onshore airflow. Clouds of basidiospores rise from the marshes and move inland under these conditions. Severe ash rust has been noticed as far as 48 km inland in some years. In Texas the optimum temperature for germination of all spore types except spermatia was found to be near 25°C: aeciospores germinated at 10–30°C, urediniospores at 15–30°C, and teliospores and basidiospores at 15–25°C. Rust increase on *Spartina* in coastal marshes is also influenced by water salinity. Salt deposition on the grass apparently inhibits spore germination and rust reproduction.
References: 61, 117, 119, 839, 1026, 2971, 2973, 4526

Cluster-cup Rust of Currant and Gooseberry

Puccinia caricina, the cause of this disease, is globally distributed in the temperate zones and occurs from coast to coast in Canada and the USA. Its spermagonia and aecia develop on *Ribes* (currant and gooseberry) and *Urtica* (nettle), and its uredinia and telia

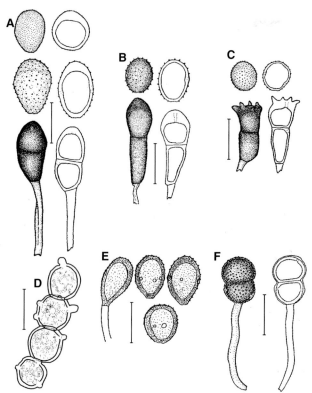

Figure 134. A. *Puccinia sparganioides*: top to bottom, aeciospores, urediniospores, teliospores. B. *P. caricina* urediniospores and teliospores. C. *P. coronata* urediniospores and teliospores. D–F. *Cumminsiella mirabilissima* (discussion and photographs with Plate 135). D. A chain of aeciospores with protrusions at pores in their walls where germ tubes may later emerge. E. Urediniospores. F. Teliospores. In each pair of spores the left image represents surface view, the right sectional view. Scale bars = 25 µm. A–C and F adapted from reference 119; D drawn from reference 2968 by permission of Public Works & Government Services Canada; E adapted from reference 3667 by permission of CAB International.

A–H. Ash rust, caused by *Puccinia sparganioides*, on *Fraxinus americana* (white ash). A. Foliage killed by rust on a large tree; green foliage of a smaller tree in the foreground provides contrast. B. Close view of foliage that is wilting or dead because of rust lesions on petioles. C. Twig cankers and petiole lesions caused by the rust fungus. The canker at left is a year old, and the rust is no longer active in it. D. Wartlike rust-induced swellings on a green twig. E, F. Rust sori viewed on the upper (E) and lower (F) surfaces of leaves. G, H. Magnified views of aecia with orange aeciospores on swollen, bent midveins of leaflets (CT, Jun).

I, J. *Puccinia* sp., similar to the ash rust fungus, on *Spartina alterniflora* (smooth cordgrass). I. Yellow spots and general yellowing associated with uredinial sori. J. Magnified view of uredinia with rust-colored urediniospores (CT, Jun).

271

(Fig. 134B) on many *Carex* species (sedges). *Ribes* affected in North America include the following:

R. acerifolium (mapleleaf currant), *R. alpinum* (alpine currant), *R. americanum* (eastern black currant), *R. aureum* (buffalo or golden currant), *R. bracteosum* (California black currant, stink currant), *R. cereum* (wax or white-flowered currant), *R. cynosbati* (dogberry, prickly gooseberry), *R. divaricatum* (white-stem gooseberry), *R. glandulosum* (skunk currant), *R. ×gordonianum*, *R. hirtellum* (hairy-stem gooseberry), *R. hudsonianum* (northern black currant), *R. lacustre* (bristly black currant, swamp gooseberry), *R. laxiflorum*, *R. missouriense* (Missouri gooseberry), *R. nigrum* (black or European black currant), *R. oxycanthoides* (hawthorn-leaved, inland, or northern gooseberry), *R. rubrum* (northern red currant), *R. sanguineum* (blood, red-flower, or winter currant), *R. triste* (swamp red currant), and *R. uva-crispa* (English or European gooseberry).

P. caricina causes yellow to orange spots on *Ribes* leaf blades, petioles, green stems, fruits, and fruit stalks. Spermagonia form on these spots and are followed several days later by clusters of cup-like aecia (thus the name cluster-cup rust). When on leaves, spermagonia open to the upper surface and aecia to the lower surface. Severe infection kills affected parts. Currant crop losses as great as 75% have been reported from scattered locations in Europe and parts of eastern Canada where winter weather permits overwintering of *P. caricina*. The fungus overwinters on *Carex* as teliospores that can tolerate temperatures down to about –30°C. In spring the teliospores germinate and liberate basidiospores at about the time *Ribes* leaves unfold. Infection of *Ribes* by basidiospores leads to formation of spermagonia and aecia. The aeciospores, yellow in mass, are globoid, 15–23 × 13–19 μm, and have colorless finely verrucose walls. Like other rusts, *P. caricina* has strains that vary in host specificity. A *Ribes* species or genotype resistant to one strain may be highly susceptible to another. *P. caricina* is the only rust fungus that forms spermagonia and aecia on *Ribes*. The structures of other rusts found on *Ribes*, including *Cronartium ribicola*, the cause of white pine blister rust (Plate 146), are uredinia and telia.

References: 119, 1099, 1246, 1298, 2088, 4108

Crown Rust of Buckthorn

Puccinia coronata, a cosmopolitan species, produces spermagonia and aecia on *Frangula* and *Rhamnus* (buckthorns) and also on *Elaeagnus commutata* (silverberry) and *Shepherdia canadensis* (russet buffaloberry). *Frangula* and *Rhamnus* species affected in North America or Europe include. *F. alnus* (European alder-buckthorn, glossy buckthorn), *F. californica* (California buckthorn), *F. caroliniana* (Carolina buckthorn), *F. purshiana* (cascara buckthorn), *R. alnifolia* (alder-leaf buckthorn), *R. alaternus* (Italian buckthorn), *R. cathartica* (common or European buckthorn), *R. crocea* (redberry), *R. davurica* (Dahurian buckthorn), *R. lanceolata* (lance-leaf buckthorn), *R. saxatilis* (rock buckthorn), *R. smithii* (Smith buckthorn), and *R. utilis* (Chinese buckthorn). Additional species are affected in Eurasia.

P. coronata induces yellowish green to yellow swollen spots on buckthorn leaves, petioles, and green stems. Severe infection causes distortion. These infections are initiated by basidiospores liberated in spring from germinated teliospores that overwintered on cereal crop stubble or on any of hundreds of other species of grasses. The common name, crown rust, refers to a feature of the teliospores, which are two celled and club shaped with several projections at the apex like points on a crown (Fig. 134C). A basidiospore germ tube penetrates the epidermis directly, leading to establishment of a mostly intercellular mycelium that gives rise to spermagonia and, several days later, aecia. In leaves spermagonia open to the upper surface and aecia to the lower surface. The aecia develop in clusters as short open cylinders with white sidewalls enclosing powdery orange aeciospores. The aeciospores are globoid to ellipsoid, 16–24 × 15–19 μm, and have colorless finely verrucose walls. The uredinial state of the rust on cereals and other grasses causes yellowing and premature necrosis of leaves. *P. coronata* overwinters as uredinia on grasses in some areas of mild climate.

P. coronata includes several varieties (formae speciales of many authors) that differ in teliospore form and in having different grass

host genera, and hundreds of races distinguished by differing virulence patterns on arrays of host plants, both grasses and buckthorns. All strains appear nearly identical on *Frangula* and *Rhamnus*, where their hybridization results in reassortment of virulence factors.

References: 51, 56, 119, 866, 867, 962, 1079, 1099, 1298, 1942, 3057, 3638, 4108, 4526

Rusts on *Berberis* and *Mahonia*

Several rust fungi in the genera *Puccinia* and *Cumminsiella* of the Pucciniaceae attack *Berberis* (barberry) and *Mahonia* (Oregon-grape and related plants). Heteroecious rusts such as *Puccinia graminis* and *P. brachypodii* produce spermagonia and aecia on *Berberis* and *Mahonia* and their uredinia and telia on cereals and grasses. The seven known *Cumminsiella* species, all native to the Americas, are autoecious, producing their four developmental states (spermagonia, aecia, uredinia, and telia) on *Berberis* or *Mahonia*. *C. mirabilissima* (Plates 135H–K; Fig. 134D–F) attacks leaves and fruits. This fungus is indigenous and widespread in western North America on *B. atrocarpa*, *M. aquifolium* (Oregon-grape), *M. bealei* (Beale mahonia), *M. dictyota* (shining netvein barberry), *M. fortunei*, *M. nervosa* (Cascades mahonia), *M. pinnata* (cluster mahonia), and *M. repens* (creeping mahonia). It occurs in scattered other North American locations on *M. aquifolium* and has become widespread on *Mahonia* in Europe. It also occurs in Australia and New Zealand.

The spermagonial and aecial states of *C. mirabilissima* are apparently uncommon in North America, having been documented only in California and British Columbia. They occur on yellow swollen spots on leaves and fruits of *M. aquifolium*. Aecia, bright yellow when young, occur on reddish purple lesions on fruits and the undersurfaces of leaves. Uredinial and telial pustules also develop on the undersurfaces. Uredinia form on small angular spots that become reddish brown and visible on both leaf surfaces. Uredinia are reddish brown and telia are dark brown. Both uredinia and telia can be found on overwintered leaves. The teliospores are ellipsoid, 30–35 × 20–25 μm, two celled and constricted at the septum, with brown finely verrucose walls and a persistent pedicel (stalk). Infection (usually by urediniospores) occurs on young leaves and may occur late in the season if a host plant is growing at that time. This rust fungus is capable of developing at temperatures only a few degrees above 0°C; the optimum range for its growth is only 8–12°C, and it does not develop at temperatures above 20°C. Urediniospores are capable of withstanding temperatures of –10° to –20°C for at least a month.

References: 119, 267, 870, 1099, 1246, 1298, 1468, 1859, 2089, 2377, 2533, 2968, 3087, 3169, 3667, 4108

A–C. Cluster-cup rust, caused by *Puccinia caricina*, on *Ribes glandulosum* (skunk currant). A. Yellow to orange rust spots and associated yellowing and necrosis on leaves. B. Close view of the upper surface of a rust spot with tips of spermagonia visible as brown dots. C. The lower surface of a leaf with an aecial sorus (NY, Jun).

D–G. Crown rust, caused by *P. coronata*, on *Rhamnus cathartica* (common buckthorn). D. Aecial sori on leaves and on a green stem. E, F. Spermagonia visible as orange dots in the raised centers of the upper surfaces of greenish yellow spots. G. An aecial sorus with clustered cuplike aecia containing orange aeciospores. Aeciospores are also scattered adjacent to the sorus (NY, May).

H–K. Rust caused by *Cumminsiella mirabilissima* on *Mahonia aquifolium* (Oregon-grape). H. Dark rust spots and rust-induced discoloration on year-old leaves. I. Rust-colored uredinial sori on the lower surface of a diseased leaf. J. Magnified view of uredinia with powdery rust-colored urediniospores. K. Magnified view of the undersurface of a leaf with dark brown telia (WA, Jun).

Fig Rust and Birch Rust (Plate 136)

Fig Rust

Rust caused by *Cerotelium fici* (Uredinales, Phakopsoraceae) is found around the world in tropical to warm-temperate regions on plants of the Moraceae (mulberry family). In the USA it occurs from Texas and Arkansas eastward and in California and Hawaii. Reported hosts in the USA include *Ficus aurea* (Florida strangler fig), *F. carica* (common fig), *F. citrifolia* (short-leaf fig), *F. elastica* (Indian rubber-tree, rubberplant), *F. lyrata* (fiddle-leaf fig), *F. pumila* (climbing or creeping ficus), *Maclura pomifera* (osage-orange), and *Morus rubra* (red mulberry). Many more species are affected elsewhere.

C. fici develops on leaves and sometimes fruit of *F. carica*. Symptoms on leaves begin as small yellowish green spots on the undersurface. These progress to angular rusty brown lesions that extend to the upper surface, where they appear purplish to brown. When severe the rust causes yellowing and defoliation in late summer or early autumn. Fruit drop sometimes follows defoliation. Defoliated trees may produce new leaves in the same year, and these are at risk of being killed by frost.

Only the uredinial (repeating) state of *C. fici* is known in North America. The telial state, apparently rare, has been reported from India. The uredinia, initially salmon colored and later golden yellow due to spore color, are found scattered or in irregular groups on the lower surfaces of the lesions. The urediniospores are ellipsoid, measure 18–32 × 14–23 μm, and have short spines. The disease cycle is imperfectly known. In southern USA the fungus is believed to overwinter as urediniospores that are dispersed to developing leaves in spring. About 14 days elapse between infection and the production of new uredinia.

References: 119, 870, 1034, 1099, 1859, 2151, 2564, 3142, 3998

Birch Rust

Melampsoridium betulinum (syn. *M. alni*; Uredinales, Puccinias-traceae) causes birch rust. The disease affects *Betula* species in forests and nurseries around the Northern Hemisphere and in New Zealand and Australia. It occurs across northern North America and as far south as California, Indiana, and New Jersey. It can be destructive to highly susceptible birches, causing defoliation that predisposes them to winter damage and to twig killing by opportunistic fungi. Recorded hosts of the uredinial state (and telial state in many collections) in North America include *Betula alleghaniensis* (yellow birch), *B. nana* (alpine, bog, or dwarf birch), *B. neoalaskana* (Alaska or resin birch), *B. occidentalis* (mountain or water birch), *B. papyrifera* (paper birch) and its varieties, *B. pendula* (European white birch), *B. populifolia* (gray birch), *B. pubescens* (downy or mountain birch), and *B. pumila* (bog, low, or swamp birch). Additional birch species are affected in Eurasia. A rust fungus identified as *M. betulinum* has also been found on *Ostrya virginiana* (hop-hornbeam) in the USA and Japan. It should be compared with *M. carpini*, which occurs on *O. virginiana* and *Carpinus* (hornbeam) in North America and Europe. The aecial state of *M. betulinum*, apparently uncommon in North America, has been recorded on *Larix laricina* (eastern larch, tamarack) in Ontario, Quebec, Connecticut, and Wisconsin. *L. decidua* (European larch) and *L. kaempferi* (Japanese larch) are affected in Europe.

M. betulinum is considered to comprise strains that infect both *Alnus* (alder) and *Betula*. This concept is based on experiments in Norway, where urediniospores from *Betula* infected *Alnus*, urediniospores from *Alnus* infected *Betula*, and basidiospores derived from telia on either *Betula* or *Alnus* infected *Larix* needles. In the 1990s, however, epidemics of rust on *Alnus* in northern Europe were found to be caused by *M. hiratsukanum*, which is morphologically and genetically distinct from *M. betulinum*. Thus more than one alder-infecting rust fungus may occur there; the issue requires clarification. *M. hiratsukanum* occurs on *A. viridis* subsp. *crispa* (American green, or mountain alder) in Alberta but has not been reported from the USA. A *Melampsoridium* recorded as *M. alni* occurs on *Alnus rhombifolia* (Sierra or white alder) and *A. rubra* (red alder) in California and Oregon and on *A. incana* subsp. *tenuifolia* (thinleaf alder) in Texas. The identity of this rust should be reviewed. North American records support the idea that distinct fungi, or at least host-specialized strains, infect birch and alder, because rust on alder is unreported from eastern and western forests where rust on birch is common and alders and birches grow close together.

Symptoms and signs. Chlorotic spots develop on *Larix* needles in spring, and aecia form on the undersides of the spots, appearing as white blisters about 1.5 × 1 mm, 0.5 mm high. The aeciospores have nearly colorless walls but appear orange-yellow because of their contents. Severe disease causes premature yellowing of needles.

Small, yellow, angular spots form on birch leaves. Brown lesions with yellow borders, several millimeters in diameter, may develop by coalescence of small spots. Severe rust causes leaves to turn brown and fall prematurely. Yellow to orange uredinia become noticeable in mid to late summer (although a few form earlier) as tiny pustules 0.5 mm in diameter on the undersides (occasionally on the upper surface) of the spots, on green stems, and sometimes on catkins. Uredinia open by a central pore from which urediniospores issue in chains. These spores are oblong-elliptic, 22–38 × 8–14 μm in size, and spiny except for the apical end. Their walls are colorless, and their contents orange-yellow. Telia develop in autumn as waxy, yellow, flat, crustlike areas among the uredinia. Telia gradually turn dark brown and sometimes cover nearly the entire undersurface of a leaf. The teliospores are arranged in a single palisade-like layer beneath the epidermis, raising it slightly. The epidermis ruptures when teliospores in fallen leaves germinate in spring.

Disease cycle. M. betulinum can persist indefinitely in its uredinial state on *Betula*. This persistence accounts for its occurrence in areas far removed from *Larix*. Urediniospores dispersed from diseased leaves in autumn overwinter among bud scales, especially on low branches, and contaminate young leaves that emerge from the infested buds in early spring. Given cool weather, foliage wet for several hours by rain or dew is suitable for spore germination and infection. These processes are retarded or halted during hot weather. Therefore, although the first infections occur in spring, significant rust buildup is usually delayed until late summer and early autumn. In areas where the rust also alternates between birch and larch, infection of larch needles in spring is caused by windborne basidiospores from telia on fallen birch leaves. After aecial formation, windborne aeciospores cause infections leading to uredinial development on birch leaves.

Data from Oregon and Pennsylvania show that *M. betulinum* prefers cool weather. Urediniospores from Oregon collections germinated at 1–20°C, most rapidly (within 3 hours) near 10°C. Uredinia developed at 10–20°C. When birch leaves were inoculated and held at 12°C, uredinia developed within 13–14 days. Urediniospores did not germinate at 25°C, and they were killed by exposure to 30°C for 6 hours. In Pennsylvania, although uredinia were found as early as June 20, the major period of spore release was from mid-August to mid-October.

Birches vary in susceptibility to rust. In northern Europe moderate to high resistance has been noted in the American species *B. alleghaniensis*, *B. lenta* (black or sweet birch), and *B. papyrifera*; and in the Asian species *B. ermanii*, *B. platyphylla* (Asian white birch), and *B. utilis*. Heritable variation in susceptibility has been noted in *B. pendula* and *B. pubescens*. Rust susceptibility is also conditioned by plant nutrition, with high fertility favoring rust buildup. Rust severity on a given birch is further conditioned by host specificity of the rust strain. *M. betulinum* strains with differing abilities to colonize *B. pubescens* but similar ability to colonize *B. pendula* were found in Finland.

References: 119, 643, 870, 988, 1592, 1903, 2027, 2189, 2988, 3127, 3128, 3326, 4468, 4526

A–C. Rust caused by *Cerotelium fici* on *Ficus carica* (common fig). A, B. The upper and lower surfaces of a diseased leaf with rust-induced necrotic lesions and uredinia in them on the lower surface. C. Magnified view of lesions with uredinia (FL, Mar).

D–H. Birch rust caused by *Melampsoridium betulinum*. D–G. On *Betula occidentalis* (water birch) (CA, Sep). D–F. Angular yellow to brown spots on upper and lower surfaces of leaves. G. Magnified view of uredinia (yellow) and telia (brown) on an undersurface. H. Magnified view of uredinia (yellow) and telia (brown) on the undersurface of a *B. pendula* (European white birch) leaf (NY, Oct).

Pucciniastrum and Allied Rusts (Plates 137, 138)

Pucciniastrum (Uredinales, Pucciniastraceae) and closely related genera accommodate heteroecious rust fungi whose spermagonia and aecia, if known, develop on needles of *Abies* (fir), *Picea* (spruce), or *Tsuga* (hemlock), and whose uredinia and telia develop on leaves or stems of diverse woody and nonwoody dicots. Eight species that until the 1990s were classified in *Pucciniastrum* are known on conifers in North America. All induce premature needle yellowing and shedding. They overwinter in living needles of aecial hosts or as telia in fallen leaves of telial hosts or as perennial mycelium in stems or rhizomes of telial hosts. Some species can continue indefinitely in the uredinial state without host alternation. The aeciospores and urediniospores of fungi in this group are yellow to orange-yellow. Contemporary authors recognize four genera in place of the former broadly conceived *Pucciniastrum*: *Calyptospora*, *Naohidemyces*, *Thekopsora*, and a narrowly construed *Pucciniastrum*.

Plate 137 presents the aecial states of two species: *T. hydrangeae* on *Tsuga* and *P. epilobii* on *Abies*. Plate 138 shows symptoms and signs of the uredinial state of *P. epilobii* on *Fuchsia*, the telial state of *C. goeppertiana* on *Vaccinium ovatum* (California or evergreen huckleberry), and symptoms caused by *N. vaccinii* on *Vaccinium* sp. (blueberry). Several other rusts, not illustrated, are discussed briefly.

Hemlock-Hydrangea Rust

Thekopsora hydrangeae (syn. *Pucciniastrum hydrangeae*) causes rust on *Tsuga canadensis* (eastern hemlock), *T. caroliniana* (Carolina hemlock), *Hydrangea arborescens* (American or wild hydrangea), and *H. paniculata* (panicle hydrangea). Aside from descriptions of the fungus and records of occurrence, the pathogen and disease are little known. This rust occurs from New York to Illinois, Arkansas, and Georgia, and has been more often reported on *Hydrangea* than on *Tsuga*.

Inconspicuous spermagonia develop in spring on green current-season hemlock needles, mostly on the undersides, and are followed there in late spring to early summer by prominent cylindric aecia, light cream in color, arranged along the entire length of the needle in the white rows of stomata. The aecia are 160–220 μm in diameter and 1.0–1.5 mm long. The white covering (peridium) of an aecium splits open from the tip and sometimes along the sides. The aeciospores are orange-yellow in mass, broadly ellipsoid to ovoid, finely verrucose, measure 19–26 × 14–20 μm, and have an elongate smooth spot on one side that distinguishes them from aeciospores of *Thekopsora minima* (the only similar rust on *Tsuga* in eastern North America). Diseased needles turn chlorotic to distinctly yellow by the time aecia form, and die, shrivel, and drop soon after aeciospores are liberated. Thus except for missing needles, hemlock is asymptomatic most of the year.

The aeciospores infect *Hydrangea* leaves, and uredinia develop within 2 weeks. These appear on the undersides of indefinite yellowish spots as tiny blisters barely protruding above the epidermis. They open by an apical pore and release broadly ellipsoid or obovate urediniospores, 16–24 × 12–18 μm, that are sparsely adorned with spines and orange-yellow in mass. Urediniospores cause repeating rust cycles. Premature yellowing and browning occur where uredinia are numerous on a leaf.

Flat, reddish brown telia develop within the epidermal cells of both leaf surfaces of *Hydrangea* in late summer and autumn. The teliospores are brown and several celled, globoid to disc shaped, with the cells of each spore conjoined somewhat like the sections of a citrus fruit. In spring they germinate in place, producing a septate basidium on which four basidiospores develop. The basidiospores infect hemlock needles.

Hemlock-Blueberry Rusts

Three or more rust fungi that produce aecia on *Tsuga* (hemlock) needles and uredinia and telia on leaves of various ericaceous plants around the Northern Hemisphere were formerly classified as forms of *Pucciniastrum vaccinii*. Two of these fungi, *Naohidemyces vaccinii* and *Thekopsora minima*, occur in North America, apparently in separate regions. Both cause yellowing and premature shedding of hemlock needles. *N. vaccinii* occurs in western North America and across Eurasia including Japan. Its aecia develop in

two rows on the undersides of needles of *Tsuga diversifolia* (northern Japanese hemlock), *T. heterophylla* (western hemlock), and *T. mertensiana* (mountain hemlock). These aecia resemble uredinia in that they are tiny domes that release spiny aeciospores through an ostiole (pore) at the top. The aeciospores are subglobose, echinulate, and 20–29 × 14–20 μm in size. In North America the uredinia and telia of *N. vaccinii* are associated with yellow leaf spot and leaf blight of *Lyonia ligustrina* (he-huckleberry, maleberry), *Menziesia ferruginea* (mock-azalea, rusty menziesia), and *Vaccinium* species (bilberry, small cranberry, blueberry, huckleberry). Additional ericaceous plants are infected in Eurasia and Japan. *N. vaccinii* on the Eurasian continent is known only on its telial hosts.

Thekopsora minima occurs in eastern North America, with aecia on *Tsuga canadensis* (eastern hemlock) and *T. caroliniana* (Carolina hemlock), and uredinia and telia on various species of *Gaylussacia*, *Lyonia*, *Rhododendron*, and *Vaccinium*. Its range extends from Canada into the Gulf Coast states, and it also occurs in Japan. Its aecia are similar to those of *T. hydrangeae* in cylindric shape, color, and arrangement in rows on both sides of the midvein on undersides of hemlock needles. The aeciospores are subglobose, 18–26 × 15–18 μm in size, and densely studded with short, blunt, rodlike protuberances. These spore ornaments and the lack of a smooth spot distinguish aeciospores of *T. minima* from those of *T. hydrangeae*. The uredinia are similar to those described above for *T. hydrangeae*. The urediniospores are ellipsoid to oblong, spiny, and 20–24 × 12–18 μm in size. Telia develop intraepidermally with teliospores arranged and functioning as described for *T. hydrangeae*.

References: 8, 53, 119, 334, 635, 779, 870, 1099, 1298, 1677, 1678, 2429, 2533, 3058, 3416, 4108, 4268, 4387, 4526

Fir-Fireweed and Fuchsia Rusts

Pucciniastrum epilobii (syn. *P. pustulatum*), which occurs around the Northern Hemisphere and in Central and South America, Australia, and New Zealand, is the cause of these diseases. In North America it occurs across Canada and as far south as Tennessee and New Mexico. It produces spermagonia and aecia on the undersides of first-year *Abies* (fir) needles and causes bending or curling, yellowing, and death of these needles. Fir hosts in North America include *A. amabilis* (Pacific silver fir), *A. balsamea* (balsam fir), *A. concolor* (white fir), *A. grandis* (grand fir), *A. lasiocarpa* (subalpine fir), and *A. procera* (noble fir). Severe foliar browning of *A. balsamea* and *A. lasiocarpa* has been caused by this rust in localities where *Chamerion* (fireweed) or *Epilobium* (willow-herb) became abundant after fire or logging. Fortunately, the intensity of infection varies from year to year, and older needles are not susceptible. Thus fir trees escape extensive defoliation.

Aecia arise in late spring to midsummer along the lines of stomata on the undersides of the needles in indefinite yellowish spots or along the length of the needle. Aecia occasionally form on cone scales of *A. lasiocarpa*. Aecia are cylindric or nearly so with a white covering (peridium), up to 0.5 mm in diameter, and approximately 1 mm high. They produce yellow verrucose aeciospores in chains. The aeciospores are obovate to globoid, measure 13–21 × 11–15 μm, and have a smooth area along one side. The aecial state of *P. epilobii* is similar to that of *Calyptospora goeppertiana*; however, aecia of the latter fungus do not mature until autumn.

Aeciospores of *P. epilobii* infect *Chamerion* and *Epilobium*—notably *C. angustifolium* (fireweed) and *E. ciliatum* (fringed willow-

---

A–D. Hemlock-hydrangea rust, caused by *Thekopsora hydrangeae*, on *Tsuga canadensis* (eastern hemlock). A, B. Views of upper and lower surfaces of a small branch with scattered diseased current-year needles. White cylindric aecia are visible on yellowish needles. C. A diseased needle with aecia, dying back from the tip. D. Mature aecia extending from the lower surface of a needle. Aeciospores are responsible for the yellowish color (NY, Jun).

E, F. Rust caused by *Pucciniastrum* sp., probably *P. epilobii*, on current-season needles of *Abies grandis* (grand fir). E. Diseased needles are chlorotic or have indefinite chlorotic bands and are slightly bent. F. Close view of the undersurfaces of diseased needles with chlorotic bands and young aecia (ID, Jul).

277

herb)—and also *Clarkia* (farewell-to-spring) and *Fuchsia*, causing yellow-brown leaf spot or leaf blight. Uredinia appear about 2 weeks after infection in small irregular groups on the undersurfaces of pale yellow to reddish spots on leaves, and sometimes appear later on the upper surfaces of *Fuchsia* leaves. These uredinia are minute, yellow, conic or hemispheric swellings, 0.1–0.2 mm in diameter, which at first open by an apical pore, then break open widely. They become distinctly yellow and powdery in appearance as urediniospores are released. On *Fuchsia* the discolored spots may progress rapidly to tan or brown necrotic spots or blotches. In such cases uredinia mature only at the edges of necrotic areas and may easily be overlooked. Severely diseased leaves dry and drop. If the combination of rust strain and *Fuchsia* cultivar is somewhat more compatible, uredinia may appear all over the leaves, but such leaves also die and fall prematurely. Old leaves are more severely affected than young ones. Low light intensity favors this rust.

Brown subepidermal telia, confluent or scattered, develop late in the growing season on angiosperm hosts other than *Fuchsia*. Telia once developed on rusted *Fuchsia* leaves in a refrigerator but have not been reported to occur on these plants in the field in North America. In spring overwintered teliospores germinate and produce basidiospores that infect *Abies* needles. *P. epilobii* is also capable of overwintering in the uredinial state. Uredinia have been observed on *C. angustifolium* shoots as early as mid-April, long before aeciospores mature on *Abies*.

The rust on *Fuchsia* has been classified by some experts as a separate species, *P. pustulatum*. It occurs around the world and affects many *Fuchsia* species and cultivars. *P. pustulatum*, compared with *P. epilobii* in the narrow sense, is reported to have larger aecia (although the size ranges overlap broadly), to produce scattered telia (versus confluent), to possess a column of empty cells beneath each mature teliospore in the host leaf (not present in the latter fungus), and to infect *E. ciliatum* but not *C. angustifolium*. Fuchsia rust is common in warm regions far from fir trees. It is locally severe in California every summer, and it breaks out sporadically in greenhouses around the world as the result of transportation of the pathogen in fuchsia cuttings.

References: 119, 870, 1105, 1677, 1678, 1684, 2368, 2533, 2576, 2940, 3421, 3889, 4281, 4387, 4526

Fir-Blueberry Rust

Calyptospora goeppertiana (syn. *Pucciniastrum goeppertianum*) causes fir-blueberry rust, also known as blueberry witches'-broom rust, around the Northern Hemisphere. It is found all across Canada and northern USA, and its range extends south to Mexico in the West. The fungus causes yellowing and premature shedding of *Abies* needles and induces witches'-brooms that produce no fruit on various species of *Vaccinium* (bilberry, blueberry, cranberry, and huckleberry). The disease has occasionally caused economic damage in lowbush blueberry fields in eastern Canada.

Hosts of the aecial state of *C. goeppertiana* in North America include *Abies amabilis* (Pacific silver fir), *A. balsamea* (balsam fir), *A. concolor* (white fir), *A. grandis* (grand fir), *A. lasiocarpa* (subalpine fir), *A. magnifica* (California red fir), and *A. procera* (noble fir). Damage to these trees is usually inconsequential, but severe rust in Christmas tree plantations has occasionally caused such needle loss that crowns appeared thin and tree value was reduced. Fir-blueberry rust has suppressed growth of young *A. balsamea* under experimental conditions, but the rust ordinarily does not affect enough foliage to retard tree growth.

Spermagonia of *C. goeppertiana* develop in summer on both surfaces of the current year's needles as tiny dome-shaped yellow pustules that turn brown with age. Aecia, whitish and more or less cylindric, mature on the undersurfaces of needles in late summer or autumn. In the West, aecia also develop on second-year needles in spring and early summer. Thus this rust often overwinters in diseased needles. The aeciospores are yellow, ellipsoid in outline, verrucose, 18–30 × 10–18 μm in size, and are produced in chains. They infect *Vaccinium* beginning in late summer and continuing the next spring and early summer.

C. goeppertiana has no uredinial state. Hosts of the telial state include *V. angustifolium* (lowbush blueberry), *V. caespitosum* (dwarf bilberry), *V. corymbosum* (highbush blueberry), *V. deliciosum* (Cascade bilberry), *V. membranaceum* (thinleaf huckleberry), *V.*

myrtilloides (velvetleaf huckleberry), *V. myrtillus* (bilberry), *V. ovalifolium* (blue whortleberry), *V. ovatum* (evergreen huckleberry), *V. parvifolium* (red huckleberry), *V. scoparium* (grouseberry), *V. uliginosum* (bog bilberry), *V. vitis-idaea* (mountain cranberry), and several less familiar species. Infection leads to extensive perennial growth of the fungus within stem phloem, swelling of current-year stems, and formation of witches'-brooms. Leaves on brooms are stunted, and the branches are swollen, spongy, and distinctly yellowish to reddish brown in contrast to the greenish color of normal twigs and young branches.

Telia mature in spring as a conspicuous reddish brown layer in the epidermis, encircling the swollen stems of shoots colonized the previous year. The teliospores have one–four cells with dark brown walls. They germinate in place under wet conditions, producing a septate basidium and basidiospores that infect current-year *Abies* needles during the time of shoot growth. Peaks of basidiospore discharge occur just after rain. Discharge occurs at 10–25°C and is most rapid near 18°C. At this temperature under moist conditions, 18 hours is sufficient for basidiospore production, germination, and infection of fir needles. *A. balsamea* needles are most susceptible during the first 2 weeks of shoot elongation, and susceptibility declines rapidly after 3–5 weeks.

C. goeppertiana becomes perennial in rhizomes of lowbush blueberries. Although many blueberry fields are burned at intervals of 3 years as a cultural practice, the heat is insufficient to kill the rust fungus in rhizomes.

Host alternation of *C. goeppertiana* would seem necessary in the absence of a repeating (uredinial) state. However, observers in both Europe and North America have noted abundant broom rust on *Vaccinium* species far removed from fir trees and have suggested that basidiospores may be capable of infecting the host on which they originate.

References: 119, 334, 870, 901, 1677, 1678, 1684, 2149, 3614–3616, 3618, 3619

Other *Pucciniastrum* and *Thekopsora* Rusts in North America

P. americanum (American spruce-raspberry rust) alternates between *Picea glauca* (white spruce) or uncommonly *P. engelmannii* (Engelmann spruce) and *Rubus* species (raspberries). It affects both foliage and cones of spruce, occasionally reducing seed crops. It occurs across Canada on spruce and raspberries and across northern USA on raspberries. Its uredinial state is also known as late leaf rust or late yellow rust of raspberry.

Pucciniastrum arcticum (Arctic spruce-raspberry rust) also alternates between *Picea* and *Rubus* in cool-temperate regions around the Northern Hemisphere. *P. glauca* is its conifer host across Canada and northern USA. This rust causes yellow-brown leaf spots on *Rubus* species.

Thekopsora sparsa (syn. *Pucciniastrum sparsum*; spruce-bearberry rust) alternates between *Picea* and *Arbutus menziesii* (madrone) or *Arctostaphylos* species (bearberry, kinnikinick, manzanita). Its aecia occur on *P. glauca*, *P. mariana* (black spruce), and *P. pungens* (blue spruce) in northwestern Canada and on additional spruces in Eurasia. Its uredinial and telial states occur in Arctic and alpine areas over a much wider region: Alaska to Mexico and Guatemala and east to Ontario, as well as in Eurasia. It causes brown lesions in purple areas on leaves of its ericaceous hosts.

References: 119, 1099, 1298, 1677, 1678, 4108, 4526

A–D. Fuchsia rust, caused by *Pucciniastrum epilobii*, on *Fuchsia* sp. A. Spreading lesions caused by the rust fungus. B. Close view of the lower surface of a diseased leaf with tiny, inconspicuous uredinia (arrow) at one edge of a lesion. C, D. Magnified views of uredinia. Immature uredinia (D, left) are blisterlike (CA, Jul).

E–G. Fir-blueberry rust, caused by *Calyptospora goeppertiana*, on *Vaccinium ovatum* (evergreen huckleberry). E. Swollen reddish brown twigs and stunted leaves on a rust-induced witches'-broom. F. Close view of a branch on which the disease has deformed all but one shoot. G. Normal branch for comparison (OR, Jul).

H. Yellow leaf spot on *Vaccinium* sp. (blueberry), caused by *Naohidemyces vaccinii* (WA, Jul).

Fir-Fern Rusts (Plate 139)

Rust fungi that infect *Abies* (fir) needles in North America are classified in six or seven genera of the Pucciniastraceae, the number depending on which authorities are followed. These fungi all produce aecia on fir and telia on angiosperms or ferns. Some of the fir-fern rusts occasionally cause economic damage in Christmas tree plantations where trees are planted on sites conducive to fern growth. Needle browning and (or) defoliation by these rusts have been problematic in plantations of *Abies concolor* (white fir), *A. grandis* (grand fir), and *A. magnifica* (California red fir) in the West, and *A. balsamea* (balsam fir) and *A. concolor* in the East.

The aecial states of the genera of rust fungi that occur on *Abies* (fir-fern rusts and others) can be separated on the following characteristics, which for the purpose of generic identification should be considered in order. *Milesina* and *Uredinopsis* produce aeciospores that are white in mass; all other rusts on *Abies* have yellow aeciospores. All of the yellow-spored rusts except *Melampsora* have aecia that are initially covered by a layer of white cells, the peridium. The mycelium of most rusts on *Abies* is localized in needles; mycelia of *Melampsorella* (Plate 140) and one species of *Milesina* become systemic in brooms. Aecia of *Pucciniastrum* mature on current-year or year-old needles; aecia of *Hyalopsora* mature on 2-year-old needles.

The genera *Hyalopsora*, *Milesina*, and *Uredinopsis* have species that alternate between fir and ferns. Only one species of *Hyalopsora*, *H. aspidotus*, is known on fir in North America. It occurs across Canada and in northwestern USA, also in Eurasia. Its uredinia and telia form on *Gymnocarpium dryopteris* (oak fern). *Hyalopsora* species have pigmented spore contents, whereas spores of *Milesina* and *Uredinopsis* lack pigmentation. The white-spored rusts on fir, comprising at least 10 species of *Uredinopsis* and 4 of *Milesina*, cannot be clearly differentiated in their aecial states on *Abies*. All are characterized by white cylindric or tongue-shaped aecia that form on green to yellow needles. All produce their uredinia and telia on ferns. Those with acutely pointed spindle-shaped urediniospores and subepidermal telia are classified in *Uredinopsis*. *Milesina* species produce telia within the epidermis, and their urediniospores are broadly rounded at the apex or, if pointed, are broadest above the middle.

The white-spored rusts known on *Abies* in North America and their alternate hosts and regions of occurrence are as follows:

Milesina fructuosa on *Dryopteris* (wood fern), east
M. laeviuscula on *Polypodium* (polypody), west
M. marginalis on *Dryopteris*, east
M. pycnograndis on *Polypodium*, east; causes witches'-brooms on *Abies*
Uredinopsis americana (syn. *U. mirabilis*) on *Onoclea* (sensitive fern), central and east
U. arthurii on *Woodwardia* (chain fern), east
U. atkinsonii on *Dryopteris*, central and east

U. ceratophora on *Cystopteris* (bladder fern), central and east
U. hashiokai on *Pteridium* (bracken fern), west
U. longimucronata on *Athyrium* (lady fern), transcontinental
U. osmundae on *Osmunda* (flowering fern), central and east
U. filicina (syn. *U. phegopteridis*) on *Dryopteris* and *Gymnocarpium* (oak fern), transcontinental
U. pteridis on *Pteridium*, east and west
U. struthiopteridis on *Athyrium* and *Matteuccia* (ostrich fern), east and west

Each of the fir-fern rust fungi infects, or is probably capable of infecting, several species of *Abies*. The fungi considered here all infect current-year needles, and most species produce aecia on these needles in summer to early autumn. Diseased needles then die and drop. Three species—*U. hashiokai*, *U. pteridis*, and *M. pycnograndis*—become perennial in fir needles, producing aecia on the same needles each spring for several years until the needles die. The aeciospores infect ferns, which develop pale green to yellow spots that later become necrotic and are typically bounded by veins. Severe rust may kill fern leaves.

Fir-fern rust fungi overwinter in various combinations of four survival modes: as dormant telia in dead fir fronds, as aecial mycelium in living fir needles, as uredinial mycelium in living fern fronds, and on fern as resting spores called amphispores that reinfect fern in spring. Amphispores are essentially urediniospores specialized for winter survival. The first two modes, singly or together, lead to production of spores that infect the alternate host—basidiospores from telia on fern to fir and aeciospores from fir to fern. The second two modes of overwintering allow these fungi to persist indefinitely on ferns. Many additional rust fungi related to those named here are known only on ferns.

References: 333, 870, 1104, 1106, 1107, 1177, 2524, 3676, 4268, 4526

A–C. Needle rust caused by *Uredinopsis* sp. on *Abies balsamea* (balsam fir). A, B. Yellow diseased needles contrast with normal green foliage (ON, Aug; NH, Jun) C. Magnified view of white cylindric aecia along the lines of stomata on the undersurface of a needle (ON, Aug).

D–G. Fir-fern rust caused by *U. americana*. D. Severe rust on an *A. concolor* (white fir) seedling that was growing among rust-infected *Onoclea sensibilis* (sensitive fern) (NY, Jul). E, F. Close and magnified views, respectively, of diffuse yellowing and aecia on *A. concolor* needles. Most aecia form on the undersides of the needles, but exceptions occur. White particles on the needle in F are aeciospores (NY, Jul). G. Yellow leaf spots typical of those caused by the uredinial state of *U. americana* on *O. sensibilis* (NY, Jul).

Photo credits: B—W. Merrill and N. Wenner; G—G. W. Hudler

Broom Rust of Fir (Plate 140)

Melampsorella caryophyllacearum (Uredinales, Pucciniastraceae) causes broom rust, or yellow witches'-broom, of *Abies* species (fir). The disease occurs all around the Northern Hemisphere where firs grow in cool, moist places. The pathogen can also be found outside the range of *Abies* on its alternate hosts, species of *Cerastium* (mouse-ear chickweed) and *Stellaria* (chickweed). Recorded fir hosts in North America are *Abies amabilis* (Pacific silver fir), *A. balsamea* (balsam fir), *A. cephalonica* (Greek fir), *A. concolor* (white fir), *A. fraseri* (Fraser fir), *A. grandis* (grand fir), *A. lasiocarpa* (subalpine fir), *A. magnifica* (California red fir), *A. procera* (noble fir), and *A. religiosa* (sacred fir). Additional *Abies* species are affected in Eurasia. In North America the disease reaches greatest severity on *A. lasiocarpa* in the northern Rocky Mountains, where in some localities 70–95% of the trees may be diseased and individual trees may bear several dozen brooms.

Symptoms and signs. Brooms form on branches or mainstems and grow slowly. Old brooms on dominant trees occasionally attain lengths approaching 2 m, but most remain less than half that size. The growth of trees with multiple brooms is retarded. Such trees often decline and die prematurely. Brooms bear only the current year's needles, which are stunted, thickened, and pale green initially but turn yellow during summer. Needles on each twig are arranged in a helix. The needles die and drop in autumn, and a new crop of pale green needles forms in spring. The branch or trunk at the base of a broom swells and becomes a spindle-shaped to globose gall on which cankers may develop. In western North America wind or snow often causes trunk breakage at broom-rust cankers. Orange spermagonia and light yellow blisterlike aecia develop annually on all needles of a broom. The spermagonia arise between the cuticle and epidermis in spring. Aecia develop in summer on the undersurfaces along the lines of stomata and produce orange-yellow aeciospores in chains. The spores are verrucose with rodlike warts.

Uredinia and telia of *M. caryophyllacearum* form on *Cerastium* and *Stellaria*. The pathogen infects perennial species in these genera systemically and perennates in them. Leaves of systemically infected plants are discolored orange. Although uredinia precede telia in the pathogen's life cycle, these structures appear in reverse order each year on systemically infected plants (see below). Teliospores develop in spring in a layer one cell thick within the lower epidermis. Uredinia develop on both sides of the leaves as orange-red pustules that open by a central pore, releasing orange-yellow urediniospores that have spiny walls. Severe rust kills leaves or shoots.

Disease cycle. *M. caryophyllacearum* overwinters as perennial mycelium in *Abies* stems and in systemically infected *Cerastium* and *Stellaria*. In spring it grows up with chickweed shoots and forms teliospores. These germinate in place, producing windborne basidiospores that infect young fir shoots as buds open. The stem of an affected fir shoot swells slightly during the first year. The next spring buds on that stem produce upright shoots that are thicker and shorter than normal, with needles as described above. This aberrant growth over several years results in a broom. Intercellular mycelium of the pathogen is abundant in needles but sparse in stems of brooms. Aecia develop in summer beginning a year after infection, and wind carries aeciospores to leaves of *Cerastium* and *Stellaria*. New and perennial rust colonies on chickweeds produce urediniospores that cause repeating cycles of rust on these plants. Production of basidiospores the next spring completes the 2-year life cycle of the fungus. It persists in chickweeds in the absence of fir trees in some areas.

Aeciospores and urediniospores of *M. caryophyllacearum* are capable of germinating at temperatures of 5–30°C and germinate most rapidly at 20–25°C. Free moisture is required.

References: 335, 643, 870, 1246, 2606, 2941, 3045, 4268, 4319, 4526

Broom rust of fir, caused by *Melampsorella caryophyllacearum*.

A. *Abies magnifica* (California red fir) with a large old broom on which most branches have died (CA, Jul).

B. A typical compact broom on *A. balsamea* (balsam fir) (NY, Jun).

C. Yellow and dying foliage on a broom on *A. balsamea* in late summer (QC, Aug).

D. Yellow witches'-broom on *A. lasiocarpa* (subalpine fir) (UT, Aug).

E. Part of a broom from *A. balsamea*. The needles are dwarfed, thickened, and borne only on current-year twigs (QC, Aug).

F, G. Young aecia with yellow aeciospores on the undersurfaces of stunted, abnormally thickened leaves of a broom on *A. balsamea* (NY, Jun).

H. Scars where aecia were produced earlier in the season on the lower surfaces of diseased *A. balsamea* needles (QC, Aug).

283

Melampsora Rusts (Plates 141, 142)

Rust fungi of the genus *Melampsora* (Uredinales, Melampsoraceae) attack diverse plants, both herbaceous and woody. The most important tree diseases caused by fungi in this group affect *Populus* (aspen, poplar), *Salix* (willow), and conifers in the Pinaceae. Aecial states of two of these fungi on conifers are shown in Plate 141. Uredinial and telial states of two species on poplar and willow are shown in Plate 142.

Leaf rusts of *Populus* and *Salix* have become important, especially in Europe, because of the intensive planting of poplar clones for lumber and fiber and of willow clones for biomass to be used in power generation. Severe rust causes premature leaf senescence and defoliation, which leads to retarded growth the next year. Severe rust in several consecutive years reduces fiber yield and leads to dieback or death associated with attack by opportunistic pathogens. Yield losses of 30% to more than 50% in volume and dry weight of poplar wood and willow biomass have been documented. *Melampsora* rusts seldom cause economically significant damage to conifers in North America, although they occasionally induce severe needle loss or kill shoot tips or cones.

Symptoms and signs. On aecial hosts (conifers emphasized here) diffuse yellowing appears on needles and/or on green stems shortly before aecia appear. Leaves with numerous rust sori turn yellow, shrivel soon after aecial maturation, and drop. Infection on green stems leads to distortion and tip dieback. Rust on cones kills the affected parts. Spermagonia and aecia form on leaves and, in some rusts, on succulent stems or cones. The spermagonia appear as minute blisters. Yellow-orange moundlike aecia arise on indefinite yellowish areas along the rows of stomata. The lack of a prominent white covering layer (peridium) distinguishes these aecia from those of other rusts on conifers. The aeciospores are globose to ellipsoid or obovate, verrucose, and yellow or orange-yellow.

On telial hosts (*Populus* and *Salix* emphasized here) angular to diffuse yellow spots or zones form in leaves and later die. Severe infection leads to shriveled leaves that drop prematurely. Infection of green stems sometimes causes small cankers. Golden yellow uredinia appear mainly on the undersides of the yellow leaf spots or sometimes on green stems beginning in spring and continuing throughout summer. They break open and release yellow to golden urediniospores. Brown crustlike telia form beneath the cuticle or in or beneath the epidermis in late summer and early autumn, and the surrounding tissue dies. Telia are composed of a single layer of oblong brown teliospores oriented perpendicular to the leaf surface and stuck together at their sides. Telia appear nearly black when found in dead fallen leaves.

Generalized disease cycle. Most of the tree pathogens in *Melampsora* are heteroecious and complete their life cycles in 1 year. Exceptions are noted below. In the most common scheme for host-alternating poplar or willow rusts, teliospores in dead leaves on the ground germinate in spring, producing basidiospores that become airborne and infect conifer needles. Spermagonia and aecia then develop within about 2 weeks. Windborne aeciospores infect poplar or willow, resulting in formation of uredinia after about 2 weeks. Urediniospores cause repeating cycles and buildup of rust on these hosts. Leaves are penetrated via stomata. Rust hyphae grow intercellularly, sending penetration pegs into leaf cells for production of haustoria, which absorb nutrients. Telia replace uredinia in late summer and early autumn, and then overwinter, completing the cycle. The presence of coniferous hosts near *Salix* or *Populus* plantations advances the timing of the annual buildup of rust.

Poplar Rusts

About 11 species that cause poplar rusts are recognized; 7 are well known in North America. *M. abietis-canadensis*, considered by some authors to be a form of *M. medusae*, is widespread in the eastern half of the continent. Its yellow aecia appear in late spring on needles, green cones, and slightly swollen, curled green stems of *Tsuga canadensis* (eastern hemlock) and *T. caroliniana* (Carolina hemlock). The aeciospores are ellipsoid to obovate, measure 17–26 × 13–18 µm, and have a finely verrucose surface. Diseased cones attract boring insects. Diseased shoot tips die during summer. This rust and *M. farlowii* occasionally damage hemlock hedges and nursery plants. Uredinia and telia of *M. abietis-canadensis* form on *Populus alba* (white poplar), *P. balsamifera* (balsam poplar), *P. deltoides* and its subspecies *monilifera* (eastern and plains cottonwoods), *P. grandidentata* (bigtooth aspen, a preferred host), and *P. tremuloides* (trembling aspen). The urediniospores are

broadly ellipsoid to globose, sparsely verrucose, sometimes flattened laterally with walls thickened on the flat side, and measure 16–24 × 13–18 µm. The cycle of this rust apparently conforms to the scheme described above. Also, since it occurs far to the west of hemlocks, it perhaps survives winter as urediniospores or uredinial mycelium on or in *Populus*.

(The autoecious rust fungus, *M. farlowii*, also infects leaves, green stems, and cones of *T. canadensis* and *T. caroliniana* and discolors leaves yellow and stems orange, but *M. farlowii* produces only telia and basidiospores. The linear telia are reddish and waxy. They occur on the stomatal surface of needles and also on stems and cones. Telia overwinter in rust-killed twigs, and teliospores germinate there in spring, producing basidiospores that reinfect *Tsuga*.)

M. medusae, the most important of the North American poplar rusts, has spread around the world to most regions where poplars grow. In North America it has been considered to comprise two formae speciales: *M. medusae* f.sp. *deltoidis* and f.sp. *tremuloidis*. The former occurs throughout eastern and central regions on *P. deltoides* (eastern cottonwood) and its subspecies. Records of *M. medusae* on other plants in *Populus* section *Aigeiros* (black poplars) and various hybrids require corroboration. Reports of *M. medusae* in western North America on species in *Populus* section *Tacamahaca* (balsam poplars) and related hybrids would now be referable to *M. occidentalis* or *M. ×columbiana* (page 286). *M. medusae* f.sp. *deltoidis* does not infect *Populus tremuloides* or other aspens. Its principal aecial host is *Larix laricina* (eastern larch, tamarack), on which it occasionally causes severe defoliation.

M. medusae f.sp. *tremuloidis* (syn. *M. albertensis*, conifer-aspen rust) is widespread in western North America, alternating between *P. tremuloides* and various conifers. It causes slight to moderate defoliation of *P. tremuloides* and sometimes severe defoliation of young conifers, notably *L. occidentalis* (western larch) and *Pseudotsuga menziesii* (Douglas-fir). It does not infect *Populus deltoides*. This form was recognized as a distinct species throughout the first half of the 20th century, a distinction that seems likely to be revalidated.

Additional hosts and distinctions between the two forms of *M. medusae* require further study. Their conifer hosts (some known only from inoculation tests) are listed together here and include *Abies concolor* (white fir), *A. grandis* (grand fir), *Larix decidua* (European larch), *L. kaempferi* (Japanese larch), *L. lyallii* (subalpine larch), *L. occidentalis*, *Picea sitchensis* (Sitka spruce), *Pinus banksiana* (jack pine), *P. contorta* (lodgepole pine), *P. lambertiana* (sugar pine), *P. ponderosa* (ponderosa pine), *P. radiata* (Monterey pine), *P. resinosa* (red pine), *P. sylvestris* (Scots pine), and *Tsuga mertensiana* (mountain hemlock).

Aeciospores of *M. medusae* f.sp. *deltoidis* are globoid, verrucose, usually 17–24 × 17–22 µm, and have bilaterally thickened walls. The urediniospores are obovate-ellipsoid, verrucose, 26–35 × 16–23 µm, and have bilaterally thickened walls and an equatorial smooth spot. Aeciospores of this rust are broadly ellipsoid, verrucose, and usually 27–32 × 20–24 µm. The urediniospores are broadly ellipsoid, verrucose, and usually 23–32 × 15–22 µm.

Both formae speciales have the life cycle described for the genus. In addition, *M. medusae* f.sp. *deltoidis* overwinters as uredinia or uredinial mycelium in poplar where winters are mild. This habit permits it to reinfect poplar in spring. In southern USA, where coniferous hosts are absent, rust appears on poplar in April. In the North, where only telia survive, rust appears on larch in May and on poplar soon thereafter. Hot, dry summer weather limits rust buildup, so in most areas it does not become prominent until late summer or early autumn.

A–D. Twig and cone rust of *Tsuga canadensis* (eastern hemlock) caused by *Melampsora abietis-canadensis*. A, B. Views of a deformed shoot that would soon have died. Aecia and aeciospores give the stem a powdery yellow appearance. C. Magnified view of part of a diseased shoot with aecia on the undersurface of a dwarfed needle at left and aeciospores on the stem. Aeciospores in mass are responsible for the yellowish appearance. D. Cones killed by rust. Frass on the cone at left indicates tunneling by a borer (NY, Jun).

E, F. Close and magnified views of young aecia of *M. occidentalis* on current-year needles of *Pseudotsuga menziesii* (Douglas-fir). Aecia like these would soon open and appear powdery with aeciospores. Distortion of needles was caused by *Contarinia* sp., a Douglas-fir needle midge (ID, Jul).

285

Melampsora occidentalis occurs from Alaska and California as far inland as Montana and Colorado and has been carried with its poplar hosts to central USA. It causes yellow leaf spot and premature leaf browning of *Populus balsamifera* subsp. *trichocarpa* (black cottonwood) and premature yellowing and death of needles of *Larix occidentalis* (western larch) and *Pseudotsuga menziesii* (Douglas-fir). Records of its occurrence on other poplars are in error or are referrable to *M. ×columbiana* (see below). Additional aecial hosts (some by inoculation only) include *Abies concolor* (white fir), *L. decidua* (European larch), *L. kaempferi* (Japanese larch), *L. lyallii* (subalpine larch), *Pinus contorta* (lodgepole pine), *P. ponderosa* (ponderosa pine), and *P. radiata* (Monterey pine). Slight susceptibility has been recorded for *Picea sitchensis* (Sitka spruce), *Pinus lambertiana* (sugar pine), and *P. monticola* (western white pine). *M. occidentalis* is usually found only in areas where its coniferous hosts grow. The aecia and uredinia of *M. occidentalis* are macroscopically indistinguishable from those of *M. medusae*, but the former fungus has larger spores. Its aeciospores are hyaline, subglobose, 26–35 × 22–27 µm, finely warty, and have bilaterally thickened walls. The urediniospores are ellipsoid to pyriform, 32–50 × 16–30 µm, with evenly distributed spines and bilaterally thickened walls.

M. ×columbiana, a natural hybrid species derived from *M. medusae* and *M. occidentalis*, is the predominant poplar rust in the Pacific Northwest, where it is more common than *M. occidentalis*. Although it was not recognized until the late 1990s, examination of preserved collections indicates that it had been present for nearly a century. It has morphologic characters intermediate between the parent species. It infects *P. balsamifera* subsp. *trichocarpa* and *P. deltoides*, also *P. ×acuminata* (lanceleaf cottonwood), *P. angustifolia* (narrowleaf cottonwood), *P. fremontii* (Fremont cottonwood), and various hybrid poplars.

M. larici-populina, the most important of the Eurasian poplar rusts, has been introduced to northwestern North America, South America, South Africa, Australia, and New Zealand. Its spermagonia and aecia form on *Larix*. *Pinus contorta* and *P. ponderosa* are also aecial hosts in North America. The uredinia and telia form on many *Populus* species and hybrids in sections *Aigeiros* and *Tacamahaca* (cottonwoods, black and balsam poplars). Aeciospores of *M. larici-populina* are globose or subglobose, finely verrucose, and measure 16–24 × 15–20 µm. The urediniospores are elongate, echinulate, 30–50 × 13–22 µm, and have a smooth spine-free apex and bilaterally thickened walls.

M. populnea comprises several morphologically indistinguishable formae speciales that occur on many *Populus* species but differ in their aecial hosts. These formae are presumed to have originated in Eurasia and to have been introduced elsewhere. Aeciospores of *M. populnea* are more or less globose, finely verrucose, and measure 17–25 × 14–19 µm. The urediniospores are ellipsoid to ovate, 20–40 × 15–22 µm, echinulate, and lack lateral thickenings. This rust, although not often reported in North America, is widely distributed on *P. alba* (white poplar) and its hybrids in the West and occurs in Mexico. Eastern records are rare. The formae speciales and their aecial hosts and distributions are as follows:

f.sp. *allii-populina* with spermagonia and aecia on *Allium* (onion and relatives), *Arum*, and *Muscari*; USA, Eurasia, South Africa

f.sp. *laricis* with spermagonia and aecia on *Larix*; USA, Eurasia, Japan

f.sp. *magnusiana* with spermagonia and aecia on *Chelidonium*, *Corydalis*, *Fumaria*, and *Papaver*; USA, Canada, Eurasia

f.sp. *pinitorqua* with spermagonia and aecia on *Pinus*; Eurasia, Japan; infects pine shoots, causing a serious disease, pine twisting rust, of *Pinus pinaster*, *P. sylvestris*, and other two- and three-needle pines in Europe

f.sp. *rostrupii* with spermagonia and aecia on *Mercurialis*; USA and Eurasia.

References: 119, 274, 609, 870, 1246, 1463, 1618, 1786, 1903, 2250, 2280, 2281, 2379, 2548, 2668, 2832, 2833, 2837, 2841, 2842, 2921, 2925, 3069, 3140, 3141, 3551, 3553, 3661, 3662, 3788, 3789, 3791, 3794, 3797, 3994, 4108, 4211, 4356, 4357, 4387, 4524, 4526

Willow Rusts

Many *Melampsora* species that infect *Salix* (willow) have been recognized, the number depending on whether morphologically similar forms with differing hosts are counted as single species with formae speciales or as multiple species. The formae speciales within a species cannot be differentiated in their uredinial or telial states. Most willow rusts infect only leaves, but strains that infect green stems have been noted in both North America and Europe. North American willow rusts in the multiple-species concept include *M. epitea* f.sp. *tsugae* with aecia on *Tsuga* (hemlock), *M. abieti-caprearum* with aecia on *Abies* (fir), *M. arctica* with aecia on *Saxifraga* (saxifrage), *M. paradoxa* with aecia on *Larix* (larch), and *M. ribesii-purpureae* with aecia on *Ribes* (currant, gooseberry). Some authors have grouped all of these within *M. epitea*.

Numerous additional willow rusts are recognized on other continents, based on morphology of the uredinial and telial states as well as on the results of inoculations of aecial hosts. Willow rusts in Europe include *M. amygdalinae*, *M. capraearum*, *M. epitea* in a narrow sense, *M. larici-pentandrae*, *M. ribesii-viminalis*, and *M. salicis-albae*. These rusts differ in willow host preferences. *M. amygdalinae* is autoecious, its basidiospores infectious to *Salix triandra*; the other species are heteroecious. *M. capraearum*, *M. epitea*, and *M. larici-pentandrae* produce spermagonia and aecia on *Larix*. Larch-alternating forms of *M. epitea* are also designated *M. larici-epitea* or *M. epitea* var. *epitea* and comprise several formae speciales and pathotypes. The infraspecific groups are based on differing preferences for particular willows and, in one pathotype, preference for young stems instead of leaves of *S. viminalis* (basket willow). *M. ribesii-viminalis* and some strains of *M. epitea* infect *Ribes* species. At least six additional species of willow rust fungi occur in Japan and New Zealand.

Melampsora Hybrids and Races, Host Resistance, and Cultural Control

Several of the *Melampsora* species and formae speciales that attack poplar or willow are evolving rapidly through hybridization and buildup of populations specialized for attack of particular tree genotypes. Deployment of poplar and willow clones selected or bred for disease resistance has exerted selection for ability of rust fungi to attack these genotypes and thus has led to emergence of new races and pathotypes.

Known *Melampsora* hybrids include *M. ×columbiana*, discussed above, *M. ×medusae-populina* (from *M. medusae* and *M. larici-populina*), and hybrids between formae speciales of the larch-alternating willow rust, *M. epitea*. Hybridization is important because hybrid strains may have host ranges or virulence characteristics not expressed in parental strains. Hybridization can occur as a result of mixed infections on either aecial or uredinial hosts.

Rust-resistant *Populus* and *Salix* limit colonization by rust fungi and thus limit or prevent spore production and disease buildup. They accomplish this primarily by sacrificing parasitized cells and other cells closely adjacent to them. The speed and scale of this reaction vary. If it occurs within the first day and involves only a few cells at an infection site, the plant is considered to be resistant by virtue of hypersensitive reaction to the rust. Rust resistance may be either race-specific or nonspecific. Race-specific resistance tends to be governed by single genes or gene clusters; nonspecific resistance is governed by multiple factors.

Planting mixtures of poplar or willow clones that differ in resistance to various rust strains has been shown to delay the buildup of rust and result in less severe rust and higher yields than occur in monoclonal plantations. Mixtures of four–six willow clones were as effective as more complex mixtures in experiments.

References: 119, 666, 1246, 1903, 2541, 2542, 2783, 3005–3010, 3217, 3389, 3788, 3789, 3793, 4108, 4281, 4387, 4524, 4526

A–E. Rust caused by *Melampsora occidentalis* on *Populus balsamifera* subsp. *trichocarpa* (black cottonwood). A. Young trees with foliage turned yellow by rust. The nearby conifers include an alternate host, *Pseudotsuga menziesii* (Douglas-fir). B, C. Upper- and lower-surface views of angular yellow lesions associated with rust sori. Yellow spots on the lower surface are uredinia; brown spots are telia (CA, Sep). D. Necrotic blotches developing where telia have formed (OR, Sep). E. Enlarged view of uredinia (yellow) and telia (brown) on the undersurface of a leaf (CA, Sep).

F. Uredinia and telia of *M. medusae* on the undersurface of a *P. deltoides* (eastern cottonwood) leaf. Leaf tissue where uredinia most numerous is beginning to die (brown spots) (NY, Aug).

G, H. Rust caused by *Melampsora* sp. on *Salix* sp. (willow). G. Yellow spots on apical leaves and browning and withering of severely diseased older leaves, which will soon drop. H. Magnified view of uredinia on the undersurface of a leaf (OR, Sep).

287

Chrysomyxa Rusts of Spruce (Plate 143)

Seventeen species of *Chrysomyxa* (Uredinales, Coleosporiaceae) occur on trees and shrubs in North America. Nearly all are heteroecious, producing spermagonia and aecia on *Picea* (spruces) and uredinia and telia on evergreen dicots, mainly in the Ericaceae (heath family). Most *Chrysomyxa* species found on *Picea* seem capable of utilizing more than one host species.

Symptoms, signs, and cycles. Most *Chrysomyxa* rusts on *Picea* cause yellowing and premature needle fall or damage to cones. The yellowing is due not only to chlorophyll breakdown but also to carotene in the fungi. *C. arctostaphyli* causes witches'-brooms, and *C. woroninii* causes stunting and death of shoots. All of the spruce pathogens except the autoecious *C. weirii* produce spermagonia and aecia on needles or cones. The aecia are whitish and blister- or bubblelike or tongue shaped. They release ellipsoid to globoid, verrucose, orange-yellow aeciospores that infect dicot hosts in late spring to summer. Common symptoms on dicots are purple-brown to brown spots or blotches on leaves. Uredinial pustules form on the upper or lower surfaces of overwintered persistent leaves (sometimes on first-year leaves in late summer) and rupture irregularly to release orange-yellow urediniospores that resemble the aeciospores. Urediniospores infect first-year leaves of dicot hosts. Host alternation is optional in *Chrysomyxa* species that produce urediniospores. Telia of species other than *C. weirii* mature on overwintered persistent leaves of dicots, break through the epidermis as cushionlike or conic structures with teliospores arranged in columns, and produce basidiospores that infect *Picea*.

C. arctostaphyli causes broom rust, or yellow witches'-broom, of spruces. It occurs across Canada and from Alaska to Mexico in the West and affects the following spruces: *P. abies* (Norway), *P. engelmannii* (Engelmann), *P. glauca* (white), *P. mariana* (black), *P. pungens* (blue), *P. rubens* (red), and *P. sitchensis* (Sitka). Brooms range in height to about 2 m. Trees with numerous large brooms often have dead or broken tops, grow slowly, and die prematurely. Brooms shed their needles in autumn; new pale green needles emerge in spring. The fungus perennates in stems of brooms and produces spermagonia and aecia in summer on the needles, which soon turn yellow. Spermagonia appear as tiny reddish eruptions and emit a putrid odor. This rust does not produce uredinia. Telia form on undersurfaces of year-old leaves of *Arctostaphylos nevadensis* (pine-mat manzanita), *A. patula* (greenleaf manzanita), and *A. uva-ursi* (bearberry).

C. cassandrae (syn. *C. ledi* var. *cassandrae*) occurs on *Chamaedaphne calyculata* (leatherleaf) around the Northern Hemisphere. Spruce hosts are *P. abies*, *P. mariana*, *P. pungens*, and *P. glauca*.

C. chiogenis, an uncommon rust, occurs in British Columbia and the eastern half of Canada and adjacent parts of the USA. Uredinia and telia form on undersurfaces of year-old leaves of *Gaultheria hispidula* (creeping snowberry). It is unknown on *Picea* in the wild but has produced aecia on inoculated *P. glauca* and *P. mariana*.

C. empetri occurs around the Northern Hemisphere and in the Falkland Islands. It affects *P. engelmannii, P. glauca, P. mariana, P. pungens, P. rubens*, and *P. sitchensis* in Canada and northeastern USA. Uredinia and (uncommonly) telia develop on upper surfaces of year-old leaves of *Empetrum eamesii* (purple crowberry) and *E. nigrum* (black crowberry), associated with leaf blight.

C. ilicina produces uredinia and telia on the undersurfaces of leaves of *Ilex opaca* (American holly) in North Carolina, Tennessee, and West Virginia. This rust has not been found on spruce.

C. ledi, although well represented in North American records, is now considered to be a Eurasian species unknown in North America. It was recently differentiated from six other taxa within a *C. ledi* complex.

C. ledicola occurs across North America and in Japan and Russia. In North America it is common on *P. engelmannii, P. glauca, P. mariana, P. pungens, P. rubens*, and *P. sitchensis*. Needle yellowing during epidemics of this rust is conspicuous. Spermagonia and aecia develop on pale green to yellow current-year needles and rarely on cones. Diseased needles are shed in autumn. Uredinia and telia develop on upper surfaces of leaves of *Ledum glandulosum* (trapper's-tea), *L. groenlandicum* (Labrador-tea), and *L. palustre* (crystal-tea). This rust is distinguished from *C. nagodhii* and *C. neoglandulosi* on *Ledum* by its larger aeciospores and urediniospores.

C. monesis and *C. pirolata* cause coastal and inland spruce cone rusts, respectively. The former occurs from Alaska to Washington on *P. sitchensis*. The latter occurs on *P. abies, P. engelmannii, P. glauca,*

P. mariana, P. pungens, P. rubens, and *P. sitchensis* across North America as far south as Colorado and on additional spruces in Eurasia. These rusts kill seeds or suppress seed size and cause cones to turn brown and open prematurely. Cone rust has severely reduced the spruce seed crop in occasional years and localities. Uredinia and telia of these rusts develop from systemic perennial mycelium, *C. monesis* in *Moneses uniflora* (single-delight), and *C. pirolata* in *M. uniflora, Orthilia secunda* (sidebells-wintergreen), *Pyrola* species (shinleaf), often far from places where spruces grow.

C. nagodhii (syn. *C. ledi* var. *groenlandicum*) infects *Ledum decumbens* (marsh Labrador-tea) and *L. groenlandicum* (bog Labrador-tea) across North America. Its aecia occur on *P. engelmannii, P. glauca, P. mariana, P. pungens*, and *P. rubens*.

C. neoglandulosi (syn. *C. ledi* var. *glandulosi*) is found on *Ledum glandulosum* (trapper's-tea) and *P. engelmannii* in the Northwest. It is capable of infecting rhododendrons.

C. piperiana, the long-spored rhododendron rust, infects needles of *P. sitchensis* and leaves of *Rhododendron macrophyllum* (Pacific rhododendron) and its hybrids in Oregon and California. It occurs as far north as British Columbia on *Rhododendron*.

C. reticulata (syn. *C. ledi* var. *rhododendri*) occurs on *Ledum decumbens* and *L. groenlandicum* in eastern and western North America. It has infected cultivated rhododendrons in nurseries. Its aecia are known only from inoculations to *P. glauca*.

C. rhododendri (syn. *C. ledi* var. *rhododendri*) occurs in Arctic and alpine regions around the Northern Hemisphere, causing brown leaf spot on *Rhododendron* species. In North America this rust is known only in its uredinial state. *Picea* species bear its aecia in Eurasia. In the past, *C. rhododendri* was probably confused with the recently differentiated *C. nagodhii* and *C. reticulata* on cultivated rhododendrons, on which it has been also been reported from Australia and New Zealand.

C. roanensis occurs in North Carolina and Tennessee on *P. rubens, Rhododendron catawbiense*, and *R. minus* (catawba and Piedmont rhododendrons).

C. vaccinii (syn. *C. ledi* var. *vaccinii*) occurs on *Vaccinium parvifolium* (red huckleberry) in British Columbia. Its aecial state, if any, is unknown.

C. weirii (Weir's spruce cushion rust) is autoecious on *P. engelmannii, P. glauca, P. mariana, P. pungens, P. rubens*, and *P. sitchensis*. It occurs across North America and in Asia. Its waxy orange cushionlike telia form in spring on yellow bands on year-old needles, producing teliospores and then basidiospores that infect new needles. Spermagonial, aecial, and uredinial states are unknown. This rust sometimes causes economic damage to *P. pungens* in nurseries and Christmas tree plantations.

C. woroninii (spruce bud rust) affects *P. glauca* and *P. mariana* in western and eastern Canada and additional spruces in Eurasia. It infects needles, grows into buds, and produces aeciospores the next year on stunted needles of yellowish to reddish shoots grown from diseased buds. The shoots then die. Its telia form on current-year leaves of systemically infected *Ledum groenlandicum* and *L. palustre*, on which it induces perennial witches'-brooms.

The anamorphic species, *Peridermium zilleri*, thought to be allied with *Chrysomyxa*, occurs on *Picea sitchensis* in British Columbia.

References: 119, 332, 779, 832–835, 870, 1099, 1298, 1683, 1683a, 1859, 1903, 2520, 2530, 2886, 3044, 3045, 3098, 3418, 3419, 3657, 3659, 3907, 4108, 4526

A, B. Broom rust caused by *Chrysomyxa arctostaphyli* on *Picea pungens* (blue spruce) (CO, Jun) and *P. engelmannii* (Engelmann spruce) (UT, Aug), respectively. The broom in A is about 2 m tall.

C–F. Needle rust caused by *Chrysomyxa ledicola*. C. *Picea mariana* (black spruce) with nearly all the current year's foliage diseased (QC, Aug). D. Close view of *P. rubens* (red spruce) needles with aecia (NY, Aug). E. Enlarged view of aecia on *P. mariana* (QC, Aug). F. Dying foliage of *P. mariana* invaded by secondary fungi after aeciospores dispersed (QC, Aug).

G. Cone rust caused by *C. pirolata* on *P. glauca* (white spruce) (ON, summer).

H, I. Weir's spruce cushion rust, caused by *C. weirii*, on *P. pungens*. H. Severely diseased needles on a year-old twig segment. I. Magnified view of orange cushionlike telia (NY, Jun).

Photo credit: G—Canadian Forest Service

289

Needle Rusts of Pines (Plate 144)

About 15 species of *Coleosporium* (Uredinales, Coleosporiaceae) cause needle rusts of *Pinus* (pines) in North America. Many additional *Coleosporium* species are autoecious on angiosperms and/or occur on other continents. Needle rusts of pines occasionally destroy enough foliage to disfigure or retard the growth of young trees, but usually they cause little damage.

Symptoms and signs. Chlorotic to yellow spots appear in winter to spring, and spermagonia develop beneath the epidermis in the spots in spring. White, bubblelike, columnar or tongue-shaped aecia follow the spermagonia and split open, exposing orange aeciospores. Aecia disappear by the end of summer, leaving scars on yellow to brown spots or bands on green or partly yellowed needles. Severe rust on either *Pinus* or other hosts causes foliar browning and premature leaf drop, notably on low branches of young pines. Orange uredinial pustules develop on the undersides of leaves of angiosperm hosts throughout summer and early autumn. Telia develop subepidermally among the uredinia in late summer and autumn. Telia appear as dark crusts or cushions somewhat larger than uredinia and are composed of a single layer of obovoid teliospores.

Disease cycles. The generalized cycle requires 1 year. Basidiospores from telial hosts infect pine needles of various ages in late summer to early autumn, causing no symptoms until the next season. The pathogen overwinters in living needles. In spring or summer, aeciospores are dispersed from pines by wind and infect leaves of the telial hosts. Uredinia appear on these plants several days later. Urediniospores cause repeating cycles and buildup of rust on the same plants throughout summer. Teliospores germinate in place without a rest period, producing basidiospores and thus completing the cycle. Some *Coleosporium* species persist 2–3 years as mycelium in living pine needles, and some also survive winter in the uredinial state in perennial herbaceous plants. Host alternation is optional for rusts that overwinter in the uredinial state.

Exemplary rusts. The best known rusts of this group in North America are members of the *C. asterum* (syn. *C. solidaginis*) complex, which occurs around the Northern Hemisphere. Many two- and three-needle pines in North America are susceptible, including *P. banksiana* (jack pine), *P. contorta* (lodgepole pine), *P. echinata* (shortleaf pine), *P. mugo* (mugo pine), *P. nigra* (Austrian pine), *P. palustris* (longleaf pine), *P. ponderosa* (ponderosa pine), *P. pungens* (table-mountain pine), *P. resinosa* (red pine), *P. rigida* (pitch pine), *P. sylvestris* (Scots pine), *P. taeda* (loblolly pine), *P. thunbergii* (Japanese black pine), and *P. virginiana* (Virginia pine). Needles of all age classes become infected. Spermagonia appear as reddish orange dots on yellow spots or bands. The aecia develop about a month later as flat-sided white columns, 0.5–1 mm high, which break open at the sides, releasing ellipsoid verrucose aeciospores that measure 28–40 × 20–25 μm. Many diseased 1- and 2-year-old needles persist, and perennial rust mycelium in them produces aecia during one or two more summers.

Uredinial pustules, initially bright orange but fading to yellow, develop 10–15 days after infection of many plants in the Asteraceae: notably species of *Aster* (aster) and *Solidago* (goldenrod), but also *Callistephus* (China-aster), *Chrysoma*, *Chrysopsis* (goldenaster), *Erigeron* (daisy fleabane), *Grindelia* (gumweed), *Heterotheca* (false-goldenaster), *Machaeranthera* (tansyaster), and *Pyrrocoma* (goldenweed). Plants in additional genera are affected in Eurasia. Reddish brown telia, about 0.5 mm in diameter when dry, appear in early autumn. This rust overwinters in both pine needles and the rosettes of perennial herbaceous hosts. An eastern and a western North American form of *C. asterum* are differentiated on the basis of microscopic characters of aecia and aeciospores. In the Great Lakes region three races have been differentiated on the basis of differing *Aster* and *Solidago* hosts. Specialists suspect that East Asian and North American strains of this rust are different species.

Pine-tarweed rust, caused by *C. pacificum*, is common in California and Oregon, where it attacks *P. coulteri* (Coulter pine), *P. jeffreyi* (Jeffrey pine), and *P. radiata* (Monterey pine). Its spermagonia and tongue-shaped aecia occur on all needle surfaces in late January to May, depending on locality and year. The aeciospores are large (40–45 × 25–29 μm). Uredinia and telia develop on species of *Arnica*, *Gaillardia* (blanketflower), *Hemizonia* (hayfield-tarweed, spike-weed), *Madia* (tarweed), *Pyrrocoma*, and *Tagetes* (marigold), all in the Asteraceae. This rust occurs on *Madia* as far north as British Columbia.

Other pine needle rusts in North America. *C. apocynaceum* occurs in southeastern USA and the Caribbean region on *P. caribaea* (Caribbean pine), *P. palustris*, and *P. taeda*. Its telial hosts are *Amsonia* (bluestar) and *Catharanthus roseus* (periwinkle).

C. crowellii occurs on *P. cembroides* (Mexican stone pine), *P. discolor* (border pinyon), *P. edulis* (pinyon pine), *P. flexilis* (limber pine), *P. monophylla* (singleleaf pinyon), and *P. strobiformis* (southwestern white pine) in southwestern USA and Mexico. This autoecious fungus produces only telia, from which basidiospores are dispersed that reinfect pine needles.

C. delicatulum occurs on various two- and three-needle pines in eastern and central USA. Telial hosts are *Euthamia graminifolia* (goldentop) and *Solidago* species.

C. helianthi affects several two- and three-needle pines in eastern and central USA and California. Telial hosts are *Helianthus* (Jerusalem-artichoke, sunflower), *Iva axillaris* (povertyweed), *Madia*, and *Silphium* (rosinweed).

C. inconspicuum occurs on *P. echinata*, *P. palustris*, and *P. virginiana* from Maryland and Ohio to Georgia and Alabama. Its telial hosts are *Coreopsis* species (coreopsis, tickseed).

C. ipomoeae affects several two- and three-needle pines in eastern USA, Arizona, and Central America. Its telia form on *Ipomoea* (morning-glory) and other members of the Convolvulaceae in the Americas and in South Africa.

C. jonesii occurs in north-central and interior western USA and Mexico with aecia on *P. cembroides* and *P. edulis*, and uredinia and telia on *Ribes* species (currant and gooseberry).

C. laciniariae attacks *P. palustris*, *P. rigida*, and *P. taeda* from New Jersey to Florida and Texas. Its uredinia and telia develop on *Liatris* species (gayfeather).

C. minutum occurs on *P. glabra* (spruce pine) and *P. taeda* in Florida and Texas. The telial host is *Forestiera acuminata* (swamp-privet).

C. pinicola occurs on *P. banksiana*, *P. rigida*, *P. taeda*, and *P. virginiana* in eastern USA, Ontario, and Quebec, and on additional pines in Eurasia. This autoecious rust produces telia and then basidiospores on year-old yellow needles in spring.

C. tussilaginis (syn. *C. campanulae*) is widespread in temperate regions, including the eastern half of the USA. It attacks *P. banksiana*, *P. resinosa*, *P. rigida*, *P. sylvestris*, and *P. virginiana*, and additional pines elsewhere. This rust occurs on telial hosts in many places where it is unknown on *Pinus*. Telial hosts in North America include species of *Campanula* (bluebell), *Lysimachia* (loosestrife), *Senecio* (groundsel), and *Triodanis* (Venus' looking-glass).

C. vernoniae affects many two- and three-needle pines in eastern and central USA, Mexico, the Caribbean region, and South America. Alternate hosts are species of *Elephantopus* (elephant's-foot) and *Vernonia* (ironweed).

C. viburni occurs on *P. banksiana* from Quebec to Manitoba and on *Viburnum* species in north-central USA, Mexico, Central and South America, and Japan.

References: 119, 331, 460, 868, 870, 1502, 1576, 1679, 2008, 2010, 2431, 2533, 2628, 2845, 2929, 4526

A, B. *Coleosporium asterum* on *Pinus resinosa* (red pine). A. Aecia on yellow spots on year-old needles. B. Enlarged view of aecia from which the white coverings (peridia) have been mostly removed by wind and abrasion (NY, Jun).

C, D. Needle rust of *P. radiata* (Monterey pine) caused by *C. pacificum*. C. Foliage with yellow to brown bands. D. Rust lesions with the remains of aecia being colonized by secondary molds (CA, Apr).

E. Tongue-shaped aecia of *Coleosporium* sp. on *P. taeda* (loblolly pine) (MS, May).

F, G. *Coleosporium* sp. on *P. palustris* (longleaf pine). F. Elongate swellings in epidermis splitting open before emergence of peridia. G. Aeciospores on a mature aecium (FL, Feb).

H, I. Yellow uredinia and rust-brown telia of *C. asterum* on undersurfaces of *Solidago* (goldenrod) leaves, magnified in I (NY, Sep).

Photo credits: F, G—E. L. Barnard

291

Stem and Cone Rusts of Pines (Plates 145–151)

Overview of North American *Cronartium* Rusts

The stem and cone rusts of pines, caused by species of *Cronartium* (Uredinales, Cronartiaceae) and its anamorphs in the genus *Peridermium*, are the most important rusts of forest trees. Worldwide, at least 16 *Cronartium* species and numerous autoecious *Peridermium* taxa that apparently lack teleomorphs are known. At least 11 *Cronartium* and 6 *Peridermium* species occur in North America. Spermagonial and aecial (*Peridermium*) states of these rust fungi occur on pines. Symptoms on pines differ among diseases but include galls, cankers, dieback of branches and mainstems, tree death, or deformity and death of cones. The stem rusts that cause cankers are also known as blister rusts. Stem rusts that induce gall formation, but usually no cankers, are called gall rusts. Rust infections that become systemic and cause dieback of successive branches, but without trunk cankers, are called limb rusts. Both galls and cankers are induced by some rusts (e.g., fusiform rust; Plate 147). The uredinial and telial states of *Cronartium* develop on leaves of various dicotyledonous plants in the Fagaceae, Grossulariaceae, Myricaceae, Santalaceae, and Scrophulariaceae. *Cronartium* rusts on their telial hosts cause yellow leaf spots or yellow to necrotic blotches and premature defoliation.

The defining structure of *Cronartium* is a columnar or hairlike telium, which emerges usually from the lower surface of a leaf. The infected area may be chlorotic or asymptomatic. These telia are composed of chains of one-celled, thin-walled teliospores, the chains conjoined laterally. Telia swell somewhat when wet, and the teliospores germinate in place, each producing a basidium and four colorless single-celled basidiospores. These short-lived spores, produced and released to the air mostly at night, are intolerant of drying or solar radiation. They infect pine needles or young cones. The interval between infection and formation of spermagonia on pine varies from several weeks in the cone-rust fungus *Cronartium strobilinum* to 2 or more years in some of the canker rusts and gall rusts. Spermagonia develop beneath the periderm as flat, irregularly shaped fertile zones of interwoven rust hyphae up to a few millimeters in diameter. They produce spermatia in a fluid matrix that exudes in drops from minute cracks in the bark or cone surface. Insects and/or hyphal anastomoses cross-fertilize spermagonia. In spring, nearly a year after spermagonial activity (but only several weeks afterward in *C. strobilinum*), prominent light yellow or cream to white, blisterlike aecia of the *Peridermium* state break through the surface where spermagonia appeared previously. The wall (peridium) of an aecium is fragile and soon breaks, releasing powdery yellow to orange (rarely white) aeciospores with thick verrucose walls. Aeciospores tolerate dry air, and those of some species can germinate after being windborne for long distances. Aecia gradually disappear during spring or summer. They may be produced in the same bark area on stem galls in each of several years, but cones die after aecial fruiting. In a canker-rust disease, bark broken by aecial fruiting soon dies, but by then the pathogen has colonized bark beyond the zone of sporulation.

Most species of *Cronartium* and *Peridermium* perennate in living pine tissues. They grow as intercellular mycelium in the cortex and secondary phloem of stems and produce simple haustoria in parenchyma cells. Most species also grow into the outer rings of sapwood, mainly in the medullary rays. Limb rust fungi also invade deep-lying sapwood of the trunk.

Aeciospores of host-alternating taxa infect leaves of their telial hosts (angiosperms) through stomata; aeciospores of autoecious blister or gall rusts infect succulent stems. Uredinia appear on undersurfaces of angiosperm leaves, and on stems of some herbaceous hosts, within approximately 2 weeks after infection. Uredinia are tiny yellow dome-shaped structures that open by a central pore to release orange-yellow urediniospores with thick, spiny walls. Urediniospores cause repeating cycles of infection and rust buildup on leaves of telial hosts during spring or summer. Telia follow the uredinia.

Representative uredinial and telial states of *Cronartium* are shown in Plate 145. *Peridermium* states are presented in Plates 146–151. The following named taxa occur in North America.

C. appalachianum causes Appalachian blister rust on *Pinus virginiana* (Virginia pine) in the Appalachian Mountains. Elongate cankers form on branches and small trunks. The telial host is the parasitic plant *Buckleya distichophylla* (piratebush; Santalaceae).

C. arizonicum causes Coronado limb rust on *P. ponderosa* (ponderosa pine) throughout southwestern USA and in South

Dakota and on additional hard pines (two- and three-needle pines) in Mexico and Guatemala. Its aecial state is *Peridermium filamentosum*. Its telial hosts are species of *Castilleja* (paintbrush), *Orthocarpus* (owl's-clover), and *Pedicularis* (lousewort), all in the Scrophulariaceae.

C. coleosporioides causes stalactiform blister rust (Plate 149), characterized by elongate trunk cankers, on *P. banksiana* (jack pine), *P. contorta* (lodgepole pine), and *P. sylvestris* (Scots pine) across Canada and in central and western USA. Telial hosts are species of *Castilleja*, *Melampyrum* (cow-wheat), and probably *Orthocarpus*, *Pedicularis,* and *Rhinanthus* (yellowrattle), all in the Scrophulariaceae.

C. comandrae causes comandra blister rust, characterized by branch and trunk cankers (Plate 149), on numerous hard pine species across Canada and the USA, most commonly in western states, and in Mexico. Telial hosts are *Comandra* species (comandra, bastard-toadflax) and *Geocaulon lividum* (northern comandra, false-toadflax) (Santalaceae).

C. comptoniae causes sweetfern blister rust (Plate 148), also a canker rust, on several species of hard pines across Canada and northern USA. Telial hosts are in *Comptonia* (sweetfern) and *Myrica* (sweetgale) (Myricaceae).

C. conigenum causes southwestern cone rust (Plate 150) on *P. leiophylla* (Chihuahuan pine) and several additional hard pines in Arizona and southward into Central America. It also causes apical galls on some of its pine hosts. Its telia form on *Quercus* species (oaks; Fagaceae).

C. occidentale causes pinyon blister rust, characterized by stem cankers, on *P. cembroides* (Mexican stone pine), *P. edulis* (pinyon pine), and *P. monophylla* (singleleaf pinyon) in western USA. Telial hosts are *Ribes* species (currant, gooseberry; Grossulariaceae).

C. quercuum is a complex comprising several formae speciales that cause pine-oak gall rusts and fusiform rust (Plate 147) on many hard pines in North America. Telial hosts are *Quercus* species.

C. ribicola causes white pine blister rust (Plate 146) throughout most of the North American and Eurasian ranges of white (five-needle) pines. Telial hosts in North America are *Ribes* species.

C. strobilinum causes southern cone rust (Plate 150) on *P. caribaea* (Caribbean pine), *P. elliottii* (slash pine), and *P. palustris* (longleaf pine) in southeastern USA and Cuba. *Quercus* species are telial hosts.

Peridermium bethelii causes mistletoe blister rust on *P. contorta* and rarely *P. ponderosa* from California and Colorado to Alberta. This rust, closely related to *C. comandrae* but thought to be autoecious, is consistently associated with *Arceuthobium* (dwarf mistletoe) infection on the pines. Its mycelium grows within both the pine and the dwarf mistletoe.

Peridermium harknessii (syn. *Endocronartium harknessii*) causes western gall rust (syn. pine-pine gall rust; Plate 151) on many hard pines across Canada and the USA except in the Southeast. The pathogen is autoecious.

Several additional *Peridermium* species associated with limb rusts and gall rusts have been distinguished based on differences in form and ribosomal DNA sequences but have not been described or named. An unnamed *Cronartium* causes California oak rust on *Quercus,* notably *Q. agrifolia* (California live oak), and *Lithocarpus densiflorus* (tanoak). This rust has not been connected to an aecial state.

A–C. *Cronartium ribicola* on *Ribes* sp. (currant). A. Yellow leaf spots and necrotic blotches caused by rust. B. Uredinia (powdery yellow pustules) and immature telia on the undersurface of a leaf. C. Magnified view of mature telia on a rust-induced lesion (NY, Sep).

D–F. *C. quercuum* on *Quercus shumardii* (Shumard oak). B. Small yellow spots with brown centers are the only symptoms on the upper surface of a leaf. E. Hairlike brown telia on the undersurface. F. Magnified view of telia (FL, Apr).

G. Magnified view of young uredinia of *C. comptoniae* on the undersurface of a *Comptonia peregrina* (sweetfern) leaf (NY, Jun).

H, I. Brown hairlike telia of *C. conigenum*, the southwestern cone-rust fungus, on the undersurfaces of leaves of *Q. hypoleucoides* (silverleaf oak). The telia in I are magnified (AZ, Jul).

J. *Castilleja* sp. (paintbrush), a host of *Cronartium coleosporioides* and *C. arizonicum* (AZ, Jul).

293

References. 119, 460, 461, 870, 1490, 1685, 1903, 2009, 2010, 2148, 2656, 3049–3052, 4167, 4420, 4526

White Pine Blister Rust

This devastating disease of soft pines is caused by an Asian fungus, *Cronartium ribicola,* that now occurs around the Northern Hemisphere. It was introduced to Europe and from there to both eastern and western North America in diseased planting stock near the beginning of the 20th century. It has spread through nearly the entire range of susceptible *Pinus* species (five-needle and stone pines) in Canada and the USA. North American pines affected in nature include *P. albicaulis* (whitebark), *P. aristata* (Rocky Mountain bristlecone), *P. balfouriana* (foxtail), *P. flexilis* (limber), *P. lambertiana* (sugar), *P. monticola* (western white), *P. strobiformis* (southwestern white), and *P. strobus* (eastern white). Additional species are susceptible by inoculation.

Many *Ribes* species (currant and gooseberry; Grossulariaceae; Plate 144) are telial hosts of *C. ribicola* in North America. In Asia, where *C. ribicola* has greater genetic diversity, some strains produce telia on *Pedicularis* (lousewort) and *Castilleja* (paintbrush) of the Scrophulariaceae, some produce telia on *Ribes,* and some strains have both capabilities. Several variants of *C. ribicola* exist that differ in virulence to pines that have some form of resistance. Two autoecious blister rusts (*Peridermium* species; see discussion with Plate 151) closely related to *C. ribicola* occur on five-needle pines in eastern Asia.

Symptoms and signs. Dead or dying tree tops or branches are noticeable at a distance. The dead or dying part is subtended by a girdling lesion that is resinous internally and may exude resin. Lesions on limbs or trunks are usually centered at dead twigs or branches that have swollen, roughened bark at their bases. Nongirdling cankers are diamond shaped to elliptic, 2.5 to 3 times longer than broad; with a rough, dead, often resinous central area. Lesions in young bark, which is normally greenish brown, are surrounded by yellowish green infected bark several centimeters wide. Canker margins in rough bark are indistinct. Rust-infected bark is often gnawed by rodents and slugs.

In spring (or summer at high elevation or latitude), active blister rust cankers have prominent blisterlike aecia, light yellow-orange in color, that arise deep in living bark, rupture the periderm, break open irregularly, and release powdery yellow-orange aeciospores. White remnants of aecial walls (peridia) persist for several weeks. Bark in the zone of aecial production dies soon after spore release and remains swollen, cracked, and rough. Spermagonia develop in summer to early autumn as shallow, irregular to circular structures a few millimeters in diameter, just beneath the periderm surrounding the zone of recent aecial production. Infected bark extends beyond the zone of spermagonial production. Spermagonia produce drops of sticky fluid with yellow-orange spermatia, attractive to insects, then collapse, leaving numerous small brown superficial scars on living bark.

Active canker margins are indistinct, but callus and woundwood may form beyond lesion edges where the rust fungus has been inhibited by mycoparasitic fungi or host resistance or has been removed by rodents. The purple mold fungus, *Tuberculina maxima,* frequently parasitizes *C. ribicola* and other blister rust fungi in pines, suppressing aecial production and inhibiting canker enlargement.

Disease cycle. The cycle of white pine blister rust requires 3–6 years. Infection begins on pine needles of all age classes from midsummer to early autumn (early summer to late autumn in northwestern areas of mild climate) when cool, moisture-laden nocturnal air currents carry basidiospores from *Ribes.* Most pines that become infected are located near the source plants, but air currents around lakes and in mountain valleys can carry infectious spores several kilometers. Basidiospores germinate soon after landing on pine needles in water-saturated air, and the parasite enters through stomata. The initial symptom is a yellow to reddish spot at the infection site, which may appear in autumn of the same year in *P. strobus* but often not until spring in *P. monticola.* Diseased needles may turn yellow and drop prematurely, but often not before hyphae of the rust fungus grow into twigs. Twigs become infected late in the year of needle infection or by summer of the next year. The fungus grows intercellularly in living bark and in rays of the outermost sapwood. Hyphal branches penetrate parenchyma cells and there develop into simple haustoria. Infection progresses along stems at rates up to about 8 cm per year. Rust mycelium can be found in bark 1–5 cm beyond surface discoloration. Infected bark swells because of the volume of rust mycelium and also because of mild hypertrophy and hyperplasia. Swelling is most apparent on twigs and small branches.

Spermagonia develop in summer to autumn of the first or second year after infection of bark, and aecia develop the following spring (or summer at high elevations). Wind carries aeciospores locally and over long distances, and they can be deposited on both upper and lower surfaces of leaves. Aeciospore germ tubes enter stomata on the undersides of *Ribes* leaves. Yellow-orange uredinia form on the undersides of the leaves within about 2 weeks, producing urediniospores that can reinfect *Ribes* and cause a buildup of rust during the summer. Telia usually replace uredinia as day length and temperature diminish from midsummer into early autumn, although telia may appear in late spring in the Pacific Northwest. Teliospores germinate in place during wet weather at night and produce basidiospores that infect pines, completing the cycle.

Not all infections are lethal. Rust progress is often halted when branches are killed by secondary organisms or die naturally. Low branches in fully stocked stands senesce and die after tree crowns close, and the hazard of lethal infection declines markedly. Pruning diseased limbs or all low branches of young pines protects against trunk infections. After trees attain crown closure, infection on branches above the canopy may kill tree tops.

Regions of high blister-rust hazard are mainly in northern and high-elevation areas. After telial formation, at least 60 hours of wet weather with temperatures not exceeding 20°C are required for basidiospore formation, dispersal, and infection of pine.

The eradication of *Ribes* for control of white pine blister rust, although shown to reduce rust incidence markedly over a 70-year term in the Great Lakes region and Maine, was gradually abandoned across North America. It was ineffective in some zones of high rust hazard because *Ribes* plants soon recolonized treated sites, because *C. ribicola* basidiospores were carried to pines from sources beyond treated sites, or both. Unnecessary *Ribes* eradication was performed in many zones of low rust hazard. Regulations against growing currants are being abandoned or made less stringent for similar reasons and because of pressure from currant growers. Several currant cultivars, both red and black fruited, possess resistance to *C. ribicola* strains that prevail in North America.

Several mechanisms of resistance to *C. ribicola* are known in pines, and many trees within susceptible species are resistant. Resistant trees halt or slow the growth of the rust fungus in needles, bark, or both. Resistance screening and breeding programs have been underway for several decades and are producing planting stock of *P. albicaulis, P. lambertiana, P. monticola,* and *P. strobus* with higher frequency of resistance than occurs in unscreened stock.

References: 112, 113, 118, 284, 376, 377, 411, 698, 772, 1085, 1300, 1432, 1459, 1460, 1689, 1705, 1814, 1817, 1820–1824,

---

A, B. Dead and dying limbs and tops: typical white pine blister rust damage on mature *Pinus strobus* (eastern white pine) (ON, Jul) and young *P. albicaulis* (whitebark pine) (OR, Sep), respectively.

C, D. Blister rust on small branches of *P. strobiformis* (southwestern white pine) and *P. strobus,* respectively. Infection began 2–3 years earlier in needles and spread into the stems. Stem swelling is typical (NY, May).

E. Swollen, roughened, cracked bark at the location of a blister rust canker that girdled a branch of *P. albicaulis* (OR, Sep). This symptom is useful for postmortem diagnosis and in the absence of the pathogen's fruiting structures.

F. Blister rust girdling a young *P. strobus* near its base (NY, May).

G, H. Close views of typical aecia of *Cronartium ribicola* on *P. strobiformis.* Spermagonial scars (arrows in H) are visible among the aecia (NY, May).

I. A typical blister-rust canker on *P. strobus* in summer. Colorless dripping resin and white crystallized resin are prominent. Small brown scars (arrows) where spermagonia were fed on by slugs are visible near the canker margin (NY, Aug).

J. Close view of the margin of a blister-rust canker on *P. strobus* in summer. Drops of spermagonial fluid and small blisters (arrows) where periderm is raised above spermagonia are visible on recently infected bark. Remnants of aecia remain on dying bark at right (NY, Jul).

1903, 2010, 2070–2075, 2148, 2316, 2444, 2553, 2622, 2919, 2976, 2979, 3052, 3770, 3850, 3893, 3952, 4167, 4284, 4352, 4420, 4422, 4457, 4505, 4526

Fusiform Rust and Pine-Oak Gall Rusts

Cronartium quercuum, a species complex, causes North American pine-oak gall rusts and fusiform rust. The aecial states of these rusts occur on many *Pinus* species in the hard pine group, and the telial states on many *Quercus* (oak) species, especially those in section *Lobatae* the red-black oak group. *C. quercuum* comprises at least four formae speciales in North America. *C. quercuum* f.sp. *fusiforme* causes fusiform rust on numerous pines. *C.q.* f.sp. *banksianae* and f.sp. *echinatae* induce globose galls on *Pinus banksiana* (jack pine) and *P. echinata* (shortleaf pine), respectively. *C.q.* f.sp. *virginianae* induces galls on *P. clausa* (sand pine) and *P. virginiana* (Virginia pine). The formae have distinct ribosomal DNA sequences and esterase isozyme patterns. *C. quercuum* strains causing globose galls on other pines in North America have not been named as specialized forms. *C. quercuum* has been reported from numerous locations in eastern Asia, but the Asian collections would now be referable to *C. orientale*.

Fusiform rust. Among most economically important tree diseases, fusiform rust occurs from Maryland to Florida, Texas, and Mexico. It is most severe on *Pinus elliottii* (slash pine), followed by *P. taeda* (loblolly pine). Other pines naturally affected include *P. arizonica* var. *cooperi* (Cooper), *P. caribaea* (Caribbean), *P. echinata* (shortleaf), *P. nigra* (Austrian), *P. palustris* (longleaf), *P. pseudostrobus* (smoothbark Mexican), *P. rigida* (pitch), and *P. serotina* (pond). Many other hard pines are susceptible by inoculation. *P. echinata* and *P. palustris* are resistant. Oaks that are most important as sources of spores that infect pines are *Q. falcata* (southern red), *Q. incana* (bluejack), *Q. laevis* (turkey), *Q. laurifolia* (laurel), *Q. marilandica* (blackjack), *Q. nigra* (water), and *Q. phellos* (willow).

Symptoms and signs. Fusiform rust kills seedlings and young trees, and causes galls, cankers, and bushiness that destroy the form and value of older trees. Galls are typically spindle shaped, occasionally oblong, and enlarge slowly. Adventitious branches often arise at galls, sometimes in clusters. Infections that begin on branches within 40 cm of mainstems often progress into them. Insects and wood-decay fungi invade rust cankers, leading to dieback and trunk breakage. Light yellow blisterlike aecia with powdery yellow aeciospores break through the bark surface annually in spring on living parts of rust galls and on swollen tissue at margins of cankers. Pine bark often dies after aecial production. Diseased oak leaves may have chlorotic spots or blotches associated with numerous uredinia or telia (see Plate 145) or may have no symptoms.

Disease cycle. *C. quercuum* f.sp. *fusiforme* requires 2 years or longer to complete its life cycle. Pine needles and succulent stems are infected in spring (March–May) by basidiospores released from telia on oaks. Purplish spots usually appear at sites of infection within a few weeks. The fungus grows from needles (or cotyledons of first-year seedlings) into stems. Stem swelling, mainly the result of stimulated wood production, begins 4–6 months after infection. The pathogen overwinters in pines. Spermagonia form in October–December, occasionally in the first year after infection but usually 1 year later. The next spring (March in most areas) aecia liberate aeciospores that infect expanding oak leaves. Orange uredinia develop on the undersurfaces of oak leaves within 7–10 days and produce urediniospores that are capable of reinfecting oaks. Telia develop 2–3 weeks after the first uredinia, while pine shoots are succulent, or telia may form in the absence of uredinia. Teliospores germinate in place, producing basidiospores that infect pine, completing the cycle.

The severity of damage by fusiform rust varies with pine species and genotype (amount of resistance), abundance of susceptible oaks, site characteristics, and local climate. Fertilization to promote pine growth tends to increase susceptibility to rust. Conditions that promote dispersal of basidiospores and infection of pines are relative humidity above 97%, temperatures of 15–29°C, and wet pine leaves and shoots. These conditions occur frequently while pines are highly susceptible in spring. Eighteen hours under conducive conditions is sufficient time for severe infection. *C.*

quercuum in oak leaves becomes inactive or dies during hot weather (e.g., several days at temperatures above 29°C).

Rust-resistant planting stock of *P. elliottii* and other pines, produced through selection and breeding, is being deployed. However, rust control through the use of resistant trees is complicated by variation in virulence of the rust fungus. A pine selection resistant to one rust strain may be quite susceptible to another. Severe losses are expected to continue for many years because of the time required to convert southern pine forests to stably resistant trees.

References: 89, 499, 506, 569, 875, 946, 964, 1016, 1197, 1243, 1458, 1490, 1695, 1722, 2165, 2166, 2170–2173, 2562, 2651, 2652, 2656, 3049, 3136–3138, 3457, 3458, 3766, 3810, 3846, 3951, 4167, 4362

Pine-oak gall rusts. These rusts occur in Asia, eastern and central North America, western USA, and Mexico. *C. quercuum* f.sp. *banksianae* (syn. eastern gall rust) occurs from Nova Scotia to Ontario and in north-central USA on *P. banksiana* (jack pine). *C.q.* f.sp. *echinatae* and *C.q.* f.sp. *virginianae* occur in southeastern USA, the former on *P. echinata* and the latter on *P. clausa* (sand pine) and *P. virginiana*. The pine-oak gall rust fungus in eastern Asia is *C. orientale*. Galls induced by these fungi are woody and vary in shape from globose to fusiform. Globose or oblong galls often have bark collars at the proximal and distal ends. Yellowish aecia form an anastomosing or cerebroid network on the gall surface in spring. The host and geographic ranges of the named formae speciales of *C. quercuum* and possibly others not yet described that cause globose galls remain to be learned in detail, so their hosts are mentioned collectively here.

Pines affected by host-alternating gall-rust fungi in North America include *P. banksiana*, *P. caribaea*, *P. clausa*, *P. echinata*, *P. elliottii*, *P. lawsonii*, *P. leiophylla* (Chihuahuan), *P. nigra*, *P. ponderosa*, *P. pungens* (table-mountain), *P. resinosa* (red), *P. glabra* (spruce), *P. rigida*, *P. serotina*, *P. sylvestris* (Scots), *P. taeda*, *P. teocote* (twisted-leaf), and *P. virginiana*. The array of oak hosts of pine-oak gall rusts is similar to that of fusiform rust. Common oak hosts north of the range of fusiform rust include *Q. ellipsoidalis* (northern pin oak), *Q. macrocarpa* (bur oak), *Q. palustris* (pin oak), *Q. prinoides* (dwarf chinkapin), *Q. prinus* (chestnut oak), and *Q. rubra*. The cycles of pine-oak gall rusts are similar to that of fusiform rust. In the North, however, aecia and telia form in late spring and summer, respectively.

Damage by pine-oak gall rusts varies with pine species and age. Galls that form on the mainstems of seedlings are usually lethal within 4 years. These rusts have caused severe losses in nursery stock and young plantations of *P. banksiana* and *P. sylvestris*. Galls on *P. banksiana*, *P. rigida*, and *P. virginiana* commonly become cankerous, and those on mainstems stunt tree growth. In Ontario, however, trunk galls induced by *C. quercuum* f.sp. *banksianae* on *P. banksiana* were found to have no effect on productivity in fully stocked stands because healthy trees became dominant in competition for growing space.

References: 75, 569, 1391, 1903, 1939, 2009, 2010, 2170, 2863, 3049, 3136, 3139, 4167

A–F. Fusiform rust, caused by *Cronartium quercuum* f.sp. *fusiforme*, on *Pinus elliottii* (slash pine). A. Damage in a young plantation (GA, Aug). B, C. Young trees on which rust has induced fusiform swellings, multiple leaders, and dieback (GA, Aug & Mar). D. Multiple shoots growing from a rust gall (MS, May). E, F. Fusiform swellings with aecia on small stems (GA, Mar).

G. A fusiform rust canker on a pole-size *P. taeda* (loblolly pine). The woundwood ridge at the right margin indicates inactivation of the pathogen there (GA, Aug).

H. An ovoid gall induced by the fusiform rust fungus on a *P. elliottii* branch (GA, Mar).

I. Pine-oak gall rust, caused by *C. quercuum* f.sp. *virginianae*, on a branch of *P. virginiana* (Virginia pine). Blisterlike aecia have matured and their spores have been liberated (TN, Apr).

Photo credits: F, H—L. D. Dwinell

297

Sweetfern Blister Rust

Sweetfern blister rust, known only in North America, came to prominence early in the 20th century when it began to cause damage in young plantations of hard pines in the Great Lakes region. Foresters had established experimental plantations of pines native to western North America or Europe—*P. contorta* (lodgepole pine), *P. nigra* (Austrian pine), and *P. ponderosa* (ponderosa pine)—on sandy soils that were carpeted with *Comptonia peregrina* (sweetfern). Pines soon began to die, and some plantations failed. *P. banksiana* (jack pine), which is native across the northern part of the Great Lakes region, was also affected. The trees were killed by stem-girdling cankers induced by *Cronartium comptoniae*. Similar epidemic occurrences on other pine species were reported subsequently from the Canadian Maritime Provinces, northeastern USA, and British Columbia. The rust occurs across Canada and the USA, as far south as Georgia in the East, wherever *Comptonia peregrina* or *Myrica gale* (sweetgale), both in the Myricaceae, grows in association with hard pines, especially *P. banksiana* and *P. contorta*. Plantations tend to be more severely affected than natural stands. The disease has caused severe losses in some nurseries.

The pathogen produces spermagonia and aecia on hard pines, and uredinia (Plate 145) and telia on the undersides of leaves of *C. peregrina* and *M. gale*. *Morella cerifera* (wax-myrtle) is susceptible by artificial inoculation. Host alternation is obligatory. Aecial hosts include *P. banksiana*, *P. contorta*, *P. coulteri* (Coulter pine), *P. densiflora* (Japanese red pine), *P. echinata* (shortleaf pine), *P. jeffreyi* (Jeffrey pine), *P. mugo* (mugo pine), *P. muricata* (Bishop pine), *P. nigra*, *P. pinaster* (cluster pine), *P. ponderosa*, *P. radiata* (Monterey pine), *P. rigida* (pitch pine), *P. sylvestris* (Scots pine), *P. taeda* (loblolly pine), *P. pungens* (table-mountain pine), and *P. virginiana* (Virginia pine). Telial hosts are absent from most of the natural range of *P. ponderosa*, so this highly susceptible species escapes severe damage.

Symptoms and signs. Irregular stem swellings, perennial cankers that girdle branches and mainstems of young pines, and yellow or brown foliage on girdled stems are the most prominent symptoms. Nearly all infections of pines begin before trees reach age 10, usually before age 5. Seedlings and the trunks and branches of saplings develop spindle-shaped swellings and often, small galls. Spermagonia similar to those of *Cronartium ribicola* (Plate 146) develop on swollen bark in late summer or autumn beginning the year after infection. They are noticeable on small smooth-barked stems, appearing as yellowish patches up to about 3 mm in diameter, from which orange droplets exude for a few days. Later they darken. Blisterlike aecia, orange to cream colored or nearly white and often confluent, break through the bark in the same area the next spring and rupture irregularly. Aecia on stems with scaly bark are often hidden by loose bark scales. Bark in the area of aecial production dies soon after aeciospores are released. Seedlings that become infected in their first year usually die within 3–4 years. Canker margins on surviving trees often have small irregular galls with dead areas between them. Sometimes the trunk of a small tree is girdled above a whorl of living branches, leading to survival of a shrubby multistemmed individual. Trees infected at a somewhat older age may persist for decades with elongate trunk cankers. Some cankers, hidden by scaly bark, escape detection until trees are more than 25 years old, but most become apparent sooner as sunken areas up to 2 m long near or extending to ground level. The sunken appearance is accentuated by swollen ridges along the sides of the cankers. Insects, particularly larvae of *Dioryctria zimmermani* (Zimmerman pine moth), tunnel in infected bark, and resin oozes from the insect wounds. Trunk deformity, resin-soaked wood beneath cankers, and wood decay associated with cankers devalue the butt log for pulpwood or lumber. Height and diameter growth of jack pine are reduced by sweetfern rust, and trees with trunk cankers eventually die prematurely.

Yellow uredinia develop on undersides of leaves of telial hosts and are followed by hairlike orange to cinnamon telia about 2 mm long, similar to those of other *Cronartium* species. These spore-producing structures are often associated with inconspicuous leaf spots.

Disease cycle. Infection of pine foliage by basidiospores from telial hosts occurs in late summer and autumn, and swellings develop beginning 12–15 months later at branch whorls or on segments between whorls. Spermagonia develop after another year, and aecia form the next spring. Orange aeciospores are dispersed by wind at about the time when leaves of telial hosts are expanding. Aeciospores require only a few hours to infect the lower surfaces of *Comptonia* or *Myrica* leaves under wet conditions at the most favorable temperatures, 12–16°C, but infection can occur (although more slowly) within the temperature range 0–28°C. Uredinia develop about 2 weeks later and are present beginning in June in most areas. Given free water urediniospores and teliospores germinate most rapidly at 16–20°C. Urediniospores cause repeating cycles on *Comptonia* and *Myrica* during summer. Telia develop 6–7 weeks after the first uredinia. Teliospores germinate in place, producing basidiospores that are dispersed to pines.

The rust fungus perennates in pines. Intercellular hyphae become abundant in bark and extend in rays across the cambium into the outer sapwood. Hyphal branches penetrate parenchyma cells in cortex, phloem, and xylem, and differentiate into simple or twisted haustoria. Mycelium in bark and wood extends about 1 cm and 2 cm, respectively, beyond externally visible symptoms. In *P. contorta* the infected area extends 5–15 cm per year. Swelling is due to hyperplasia and hypertrophy in the cambial region, resulting in aberrant wood production. In *P. banksiana* and *P. contorta* the infected tissue contains abnormally large and numerous rays.

In eastern and central regions sweetfern rust is most common on sandy or droughty, relatively infertile soils, reflecting the habitat of *Comptonia*. In the West the disease is most common in moist habitats, reflecting the site preference of *Myrica gale*. Abundance of telial host plants is the strongest determinant of severity of the rust on pines.

P. banksiana and *P. contorta*, the species in which sweetfern rust is most important, vary considerably in susceptibility. This variation offers the possibility of disease management through selection and breeding.

Sweetfern rust and stalactiform rust (Plate 149) are similar in diagnostic characters and pine hosts, and their geographic ranges overlap broadly. When found in branches or in young cankers, the causal fungi cannot be differentiated with certainty except by DNA analysis or inoculation of alternate hosts with aeciospores. For tentative diagnosis, however, stalactiform rust cankers originate at any height on the trunk, whereas sweetfern rust cankers are always near the ground. Swollen ridges and irregular bumps are common at the edges of sweetfern rust cankers but are not features of stalactiform rust. A blister rust on hard pines associated with *Comptonia* or *Myrica* is likely to be the sweetfern rust because *Cronartium comptoniae* usually spreads only a short distance (a few tens of meters) between alternate hosts.

References: 81, 266, 395, 461, 901, 1187, 1394–1396, 1685, 1687, 1829, 1859, 2148, 2205, 2692, 3047, 3049, 3613, 3695, 3772, 3970, 4167, 4524, 4526

A–E, G. Sweetfern blister rust on *Pinus rigida* (pitch pine). A. Dead and dying saplings. B, C. Elongate, sunken, bark-covered cankers outlined by ridges of swollen tissue, aligned with the grain of the underlying wood. Arrow in C locates a small gall induced by the rust fungus. D. Close view of the gall located in C, breaking through the outer bark scales and bearing aecia. E. Close view of a portion of a rust canker on a small stem. Aecia at left were exposed by removing outer bark scales. Rodents gnawed bark from the portion at right. G. Cross section of a young stem showing marked swelling and yellowish discoloration of wood caused by a rust infection that reached this part of the stem when it was in its second year of growth (NY, Jun).

F. Aecia of *Cronartium comptoniae* on a small branch of *P. sylvestris* (Scots pine) (VT, May).

Stalactiform and Comandra Blister Rusts

Cronartium coleosporioides, the cause of stalactiform blister rust, and *C. comandrae,* the cause of comandra blister rust, are widespread on many two- and three-needle pines in Canada and the USA. These native fungi kill seedlings and cause branch dieback and resinous perennial trunk cankers that girdle and kill trees and reduce usable wood volume in living diseased trees. This damage occurs mainly in western forests. Stalactiform blister rust is important primarily on *Pinus contorta* (lodgepole pine), and comandra blister rust is a major disease of *P. contorta* and *P. ponderosa* (ponderosa pine). Stalactiform rust cankers are long (up to several meters) and narrow; comandra rust cankers are usually only two to three times as long as wide and seldom extend more than 1.2 m. Pitch moths (*Dioryctria* and *Synanthedon*) often attack canker margins. Squirrels commonly remove infected bark, and trees respond by producing copious amounts of yellowish resin. This bark removal sometimes halts canker enlargement. Host alternation is obligatory for both rust fungi; thus they occur only where pines grow in association with alternate hosts.

Stalactiform rust. *C. coleosporioides* occurs across Canada and in north-central and western USA as far south as Colorado and southern California. The term *stalactiform* refers to the form of specialized hyphal strands found in aecia along with the aeciospores. The aecial state is *Peridermium stalactiforme.* Outbreaks of this rust in nurseries and young plantations of *P. contorta* and *P. banksiana* (jack pine) have caused losses of 50–80% in extreme cases. Most young trees with trunk cankers are eventually killed by them. On older trees girdling is usually confined to limbs. Stalactiform rust can be confused with sweetfern rust (Plate 148) when cankers are near the ground, and with some limb rusts (Plate 150).

Stalactiform rust affects *P. banksiana* and *P. contorta* most often, but is also known on various native and introduced hard pines: *P. attenuata* (knobcone), *P. coulteri* (Coulter), *P. densiflora* (Japanese red), *P. jeffreyi* (Jeffrey), *P. mugo* (mugo), *P. muricata* (Bishop), *P. nigra* (Austrian), *P. ponderosa, P. radiata* (Monterey), *P. sabiniana* (digger), *P. sylvestris* (Scots), and others by inoculation. Telial hosts are *Castilleja* species (paintbrush), *Melampyrum lineare* (cowwheat), *Orthocarpus* (owl's-clover), *Pedicularis* (lousewort), and *Rhinanthus* (yellowrattle), all in the Scrophulariaceae. *M. lineare* is the principal telial host in the East, *Castilleja* in the West.

The disease cycle requires 2 years or longer. Basidiospores from telia on angiosperm hosts infect pine needles in summer, and the rust fungus grows from needles into branches or young mainstems of seedlings and young trees. Infected bark swells the next year, and spermagonia develop in late summer and autumn, producing orange droplets that contain spermatia. Orange to cream-colored aecia break through the bark surface the following spring and disperse orange aeciospores. (A white-spored form of this rust occurs in Alberta.) Bark on which aecia have matured soon dies and remains swollen and rough. Aecial sporulation is often suppressed or absent at the edges of old trunk cankers. Aeciospores infect telial hosts, resulting in uredinial formation after 2–3 weeks and telia after 5 weeks on both surfaces (mainly the undersurface) of leaves and on stems. All spore types require wet plant surfaces and moderate temperatures (15–21°C) for germination and infection.

C. coleosporioides invades sapwood as well as phloem. In *P. contorta* its hyphae may be found in tracheids as far as 20 cm beyond the point of a canker and 2–5 cm in advance of hyphae in bark. Apparently it grows in the rays from sapwood into bark, which hastens the extension of cankers. Cankers elongate 16 cm or more per year and may reach lengths up to 9 m before girdling occurs. *References:* 53, 77, 461, 1683, 1683a, 1685–1687, 1859, 2004, 2010, 2624, 2829, 3049, 3133, 3318, 4167, 4180, 4447, 4524, 4526

Comandra rust. This rust occurs from New Brunswick to British Columbia in the North and extends southward, mainly in eastern and western mountains, to Alabama, Arkansas, and southern California. It kills young pines of several species and, in western forests, causes loss of growth and volume by killing the tops of *P. contorta* and *P. ponderosa.* Other reported pine hosts include *P. attenuata, P. banksiana, P. brutia* (Calabrian or Turkish), *P. echinata* (shortleaf), *P. elliottii* (slash), *P. glabra* (spruce), *P. jeffreyi, P. mugo,*

P. rigida (pitch), *P. serotina* (pond), *P. sylvestris, P. taeda* (loblolly), *P. virginiana* (Virginia), and additional species by inoculation. Telial hosts are the herbaceous plants *Comandra umbellata* (comandra or bastard-toadflax) and *Geocaulon lividum* (northern comandra) of the Santalaceae, on which the rust causes pale yellow leaf and stem spots and premature leaf abscission.

Symptoms on pines begin with spindle-shaped swellings or occasionally galls on branches and mainstems. Trunk cankers result from infection through needles while stems are young or from growth of the fungus along branches.

The disease cycle requires 1 or 2 years, the duration depending on latitude. Pines become infected in summer by basidiospores from telia on angiosperm hosts. The spores germinate on wet needles, and the parasite enters through stomata. It then grows into the stem, the bark of which swells to two–four times normal thickness. Intercellular mycelium becomes abundant in bark and the outermost sapwood, extending up to 2 cm beyond the swollen area. In *P. contorta* in the Rocky Mountain region, rust mycelium advances about 2 cm per year toward the base of the branch or trunk. Spermagonia form in swollen bark in late summer or autumn of the year of infection or after a lapse of 1–2 years, and aecia develop during the next spring (and summer in the North). Aecial blisters rupture irregularly, revealing orange to red-orange aeciospores that slowly fade to yellow. Bark becomes roughened as the result of aecial production. The pear- or teardrop-shaped aeciospores distinguish *C. comandrae* from all other pine stem rusts except *Peridermium bethelii.* The latter, however, is consistently associated with dwarf mistletoe infections. Aeciospores of *C. comandrae,* dispersed in air during daylight, can survive several days on dry surfaces of *Comandra* or *Geocaulon* and then infect within as few as 6 hours if the surface becomes wet and temperature is in the range of 5–22°C (optimum 15–20°C). Uredinia develop within 1–2 weeks, and telia after 2–6 weeks, depending on temperature. Rust increases on telial hosts during summer as the result of reinfection by urediniospores. Teliospores germinate in place, liberating basidiospores that infect pines. Wet weather at 13–23°C for 24 hours or longer is sufficient for pine infection.

Basidiospores may infect pines at distances greater than 1.6 km from telial hosts in some areas, but aeciospores usually infect *Comandra* or *Geocaulon* within a few hundred meters of source pines. In the South the humid climate is suitable for pine infection every year, but in the West suitable conditions over large areas occur infrequently. High incidence of new infection on pines occurs in only a few years each century. Residual damage continues for decades, however, because of perennial canker activity.

The purple mold *Tuberculina maxima* parasitizes both *C. coleosporioides* and *C. comandrae,* suppressing or inactivating aeciospore production in some cankers, but it does not prevent significant losses.

References: 323, 324, 461, 980, 1250, 1251, 1580, 1683, 1683a, 1685–1687, 1859, 1888, 1903, 1947, 2010, 2022, 2147, 2148, 2580, 3049, 3132–3135, 3318, 3950, 4167, 4526

A–E. Stalactiform rust, caused by *Cronartium coleosporioides,* on *Pinus contorta* (lodgepole pine). A. A mature tree killed by the rust. The yellow zone on the trunk is part of a long canker (OR, Sep). B. Resin exudation identifies the location of a canker (CO, Jun). C. An old canker from which bark was removed by rodents (CA, Jul). D. Aecia of the rust fungus on a small stem; dead pine needles provide a size scale (BC, Mar). E. Part of a stalactiform rust canker that has engulfed a gall. Crystallized resin colors the wood yellow where bark was removed by rodents. Linear ridges indicate previous annual expansion of the canker (OR, Sep).

F, G. Comandra rust cankers on *Pinus ponderosa* (ponderosa pine) (WA, Jul). F. A canker is visible as the yellowish area where bark was removed by rodents and branches are dying. G. Bark of a young tree is roughened where broken by aecia of *C. comandrae.* Massed aeciospores impart yellow color. Branch bases at the infected whorl are swollen.

H. Drops of fluid containing spermatia, oozing from spermagonia of *C. comandrae* on a seedling of *Pinus banksiana* 6 months after it was inoculated with basidiospores (MN, greenhouse).

Photo credits: D—Canadian Forest Service; H—D. R. Bergdahl

301

Limb Rusts and Cone Rusts

Limb rusts. These diseases occur from the Black Hills of South Dakota to northern California and south into Mexico. The term *limb rust* refers to the progressive invasion and killing of branches by rust mycelium that is perennial and systemic in a tree trunk. Limb rusts are among the most destructive rust diseases of *Pinus jeffreyi* (Jeffrey pine) and *P. ponderosa* (ponderosa pine). Additional pine hosts in southwestern USA and Mexico include *P. arizonica* (Arizona pine) and its variety *cooperi*, *P. devoniana*, *P. durangensis* (Durango pine), *P. engelmannii* (Apache pine), and *P. montezumae* (rough-bark Mexican pine).

Trees of all ages and sizes are affected. Infection begins on needle-bearing stems. The causal fungi grow intercellularly in rays into the xylem and then grow longitudinally within tracheids. Hyphal branches enter ray parenchyma and differentiate into haustoria. Hyphae spread from initially infected branches into the trunk, and then advance as much as 20–25 cm per year up and down in the sapwood, growing out into each branch encountered. In transverse view, infected sapwood is discontinuous around the trunk. Limb rust mycelium has been detected in sapwood more than 60 years old. It does not approach the cambium or grow into bark except in small branches and twigs, where aecia are produced. Aecia appear in spring to midsummer, depending on rust strain and locality, and disperse spores for several weeks. Twigs and branches that have borne aecia die. Progressive twig and branch mortality over many years suppresses tree growth and leads eventually to death. Secondary agents such as bark beetles are often the ultimate cause of death.

Limb rusts are caused by five or more distinct taxa, only one of which is fully described and named. *Cronartium arizonicum,* with aecia on *P. arizonica* and *P. ponderosa,* occurs from South Dakota through the Rocky Mountain region to Mexico and Guatemala. Its aecial state is *Peridermium filamentosum.* Its aeciospores infect leaves of *Castilleja* (paintbrush), *Orthocarpus* (cow-wheat), and *Pedicularis* (lousewort). When its aeciospores were inoculated to *C. miniata,* uredinia formed after 18–20 days and telia formed after 39–50 days. The host range and distribution of *C. arizonicum* overlap those of other lesser-known and unnamed limb rust fungi on hard pines in southwestern USA, Mexico, and Central America. One of them is heteroecious, alternating between *P. jeffreyi* and *Castilleja* in the Sierra Nevada of California and Nevada, and is called stalactiform limb rust because its aecia and aeciospores are indistinguishable from those of *Peridermium stalactiforme,* the aecial state of *Cronartium coleosporioides* (Plate 149). It is genetically distinct from *C. coleosporioides,* however. It kills seedlings and saplings and causes a limb rust syndrome in older trees. Aecia of this rust, like those of *C. coleosporioides,* typically measure 1 × 2 mm but often 5 × 6 mm or more where they are confluent, and they have a low profile. Stalactiform limb rust causes more conspicuous roughening of bark than do other limb rust fungi where their aecia are produced.

Two other limb rusts known only in their *Peridermium* states are apparently autoecious, spreading from pine to pine. They differ from each other in host specialization, seasonal timing of aecial maturation, and temperature optima for aeciospore germination. One of these taxa attacks *P. ponderosa* in Arizona, Colorado, and Utah; the other infects *P. jeffreyi* in California and Nevada. Aecia of these fungi are similar to those of *C. arizonicum*—tongue shaped or narrowly conic, not confluent, often 8 mm or more tall, and usually less than 2 × 4 mm at the base. The germ tubes of their aeciospores are shorter and stouter than those of *C. arizonicum* or other host-alternating *Peridermium* species. Peridia (aecial walls) of these strains are tough, and remnants of peridia persist over winter or occasionally for 2–3 years. Additional limb rust taxa occur in Mexico and Guatemala.
References: 869, 1490, 2623, 3046, 3049, 3050, 3053, 4167, 4180, 4524

Cone rusts. Southwestern cone rust, caused by *Cronartium conigenum,* occurs in southern Arizona and southward through Mexico into Central America. Its pine hosts include *Pinus arizonica* var. *cooperi*, *P. caribaea* (Caribbean), *P. devoniana*, *P. douglasiana,* *P. durangensis* (Durango), *P. engelmannii* (Apache), *P. lawsonii, P. leiophylla* (Chihuahuan), *P. lumholtzii* (Mexican weeping), *P. montezumae* (rough-bark Mexican), *P. oocarpa, P. ponderosa* (ponderosa), *P. pseudostrobus* (smooth-bark Mexican), and *P. teocote* (twisted-leaf). The cone rust fungus causes stalked, lobed stem galls on several of these species. Cones become infected during their first year of development, swell into misshapen galls of various sizes, produce no seeds, and do not open. Aecia develop 2–3 years after infection. Galls die after aecial production but remain on the trees. Large galls such as that in Plate 150 F usually terminate the branches that bear them. During periods of outbreak cone rust may kill more than 50% of the cones on groups of trees. In Arizona *C. conigenum* produces spermagonia and aecia in July and August on swollen deformed cones of *P. leiophylla*, and uredinia and telia throughout the summer on the undersides of oak leaves. Oak hosts include *Quercus arizonica* (Arizona white), *Q. chrysolepis* (canyon live), *Q. dunnii* (Palmer), *Q. emoryi* (Emory), *Q. grisea* (gray), *Q. hypoleucoides* (silverleaf), *Q. oblongifolia* (Mexican blue), and *Q. rugosa* (netleaf). Telia are up to 6 mm long and may be so numerous that the lower leaf surface appears covered with loose brown wool (Plate 145).

Southern cone rust, caused by *C. strobilinum,* occurs from North Carolina to Florida and Louisiana and in Cuba. It kills first-year cones of *P. caribaea, P. elliottii* (slash pine) and its variety *densa* (Florida slash pine), and *P. palustris* (longleaf pine). In wet years, which are conducive to infection, it may destroy as much as 90% of the cone crop in some localities. Young cones become infected during the period of pollination in winter, swell during the spring to several times the volume of healthy cones of the same age, turn reddish during the period of spermagonial activity in March, then turn yellow as aecia mature in April and May. As aecia develop, the cone surface ruptures irregularly and falls away, exposing a layer of aeciospores that are dispersed by wind and infect various oaks. After spore dispersal the cones die, dry, and fall. Cone moths (*Dioryctria* species) are attracted to diseased cones at the time aecia develop. Their larvae feed on both fungus and host tissues, and the expanding insect population later attacks healthy cones.

Oak hosts in nature include *Quercus alba* (white), *Q. bicolor* (swamp white), *Q. chapmanii* (Chapman), *Q. incana* (bluejack), *Q. laevis* (turkey), *Q. laurifolia* (laurel), *Q. macrocarpa* (bur), *Q. minima* (dwarf live), *Q. myrtifolia* (myrtle), *Q. nigra* (water), *Q. prinus* (chestnut), *Q. pumila* (running), *Q. stellata* (post), and *Q. virginiana* (southern live). Uredinia develop on oaks in autumn, and telia develop from December to February in the Gulf Coast states. Teliospores germinate in place during wet weather and liberate basidiospores that infect young pine cones. Although both evergreen and deciduous oaks become infected, the latter are not important in the disease cycle because their leaves drop before pine cones become susceptible.

Uredinia and telia diagnosed as *C. strobilinum* have been found on oaks in areas remote from pines with infected cones (e.g., Missouri and Iowa). On this basis it has been suggested that *C. strobilinum* has the ability to overwinter in the uredinial state on green sprout stems of oaks. This idea has not been corroborated through research.
References: 211, 1579, 1581, 3049, 3223, 3224, 3609, 4167

A–C. Limb rust on *Pinus ponderosa* (ponderosa pine). A. Dead and dying branches on a diseased tree. B, C. Successively closer views of aecia on a small branch (CO, Jun).

D, E. Southern cone rust, caused by *C. strobilinum*, on *P. elliottii* var. *densa* (Florida slash pine). D. A diseased cone (left) is greatly enlarged in comparison with its healthy companion (FL, Apr). E. A diseased cone from which aeciospores are being liberated (FL, May).

F. Southwestern cone rust, caused by *C. conigenum*, on *P. leiophylla* (Chihuahuan pine). A healthy cone is at left and a diseased cone, swollen and deformed, at right (AZ, Jun).

Photo credit: E—E. L. Barnard

303

Western Gall Rust

Western gall rust, also known as pine-pine gall rust, occurs across Canada and the USA on two- and three-needle pines. It has been reported as far south as Virginia, Missouri, and northern Mexico. The pathogen, *Peridermium harknessii* (syn. *Endocronartium harknessii*), is autoecious and, as yet, restricted to North America. Its closest known relative is *Cronartium quercuum* f.sp. *banksianae*, which causes eastern gall rust (syn. pine-oak gall rust) on various hard pines.

Pinus contorta (lodgepole pine) and *P. ponderosa* (ponderosa pine) sustain the greatest damage by western gall rust in forests. Some *P. sylvestris* (Scots pine) Christmas tree plantations in central and eastern USA have sustained tree quality losses. Severe disease (numerous galls and dieback) develops only on scattered trees or small groups of them in plantations of this species. Other pines affected include *P. attenuata* (knobcone), *P. banksiana* (jack), *P. canariensis* (Canary Island), *P. coulteri* (Coulter), *P. halepensis* (Aleppo), *P. jeffreyi* (Jeffrey), *P. mugo* (mugo), *P. muricata* (Bishop), *P. nigra* (Austrian), *P. pinaster* (cluster), *P. radiata* (Monterey), and *P. sabiniana* (digger). Several other pines are susceptible by inoculation.

Symptoms and signs. Peridermium harknessii induces the formation of globose or sometimes pear-shaped or somewhat elongate woody galls. They usually form on branches but are common on mainstems of some species, especially *Pinus banksiana* and *P. contorta* (lodgepole and shore pines). Galls are often fissured and typically enlarge to diameters of 1–10 cm before dying. Those on mainstems are hemispheric to globose and sometimes attain diameters of 20–30 cm. Beginning in the second or third year, galls may develop bark collars at one or both ends where thick periderm has been raised by gall expansion along the stem (Plate 151G). The xylem in galls consists of hyperplastic and hypoplastic aggregates of unusually short tracheids and abundant rays. In *P. sylvestris* the relatively thin sapwood region of a gall encloses a resinous reddish brown core. Aecia break through the surfaces of galls annually during the period of pine shoot elongation in spring. Aecia are low in profile, confluent, and often cerebroid in surface view but are more or less continuous beneath the periderm (Plate 151H). They rupture irregularly, and remnants of the peridia (aecial walls) soon disappear. The aeciospores, pale yellow-orange when fresh (except for an albino strain that occurs in north-central USA), are verrucose and variable in shape. Spermagonia (of the sort typical of *Cronartium*, Plate 146) are absent or rare.

Trees with numerous galls grow slowly, often develop witches'-brooms (this varies with host) and dieback of twigs and branches, and in some cases are killed. Cankers sometimes develop at loci of infection in certain hosts, notably *P. contorta*. The cankers may originate at galls or in their absence at the bases of small branches that become infected close to the mainstem. In some trees wood formation at lateral margins of a canker is stimulated, so that the stem expands and flattens. Cankers with bulging flanks are called hip cankers. Western gall rust is often lethal to seedlings. Unfortunately, many diseased seedlings survive. Thus *Peridermium harknessii* has been dispersed widely in seedlings shipped from nurseries, leading to buildup of gall rust in young plantations.

Disease cycle. P. harknessii perennates in living galls. New infections are initiated annually in spring on wet, succulent, current-year stems by aeciospores from galls 2 years old or older. Red pigment appears at infection sites, and stem swelling appears late in the growing season or the next spring. In *Pinus contorta* and *P. muricata*, the peduncles of young female cones also become infected, and the pathogen grows down the peduncle into the stem, where a gall forms. Aecia first form on year-old or 2-year-old galls (2–3 years after infection), and then typically appear in spring for several years (usually < 10, rarely to > 100 years).

Aecia release dry aeciospores that become airborne and cause new infections. The spores are liberated primarily during periods of decreasing atmospheric humidity each morning for 2–3 weeks in spring. Germination and infection occur at 10–30°C. The optimum temperature varies with strain of the pathogen in the range of 15–28°C. Prolonged cool, wet weather during the period of pine shoot growth is conducive to severe infection. In the East these conditions occur in most years. In central and western regions highly conducive weather occurs at intervals of several years, resulting in "waves" of infection.

Gall mortality, cankers, and death of limbs or trees are caused mainly by secondary fungi and insects that invade rust-infected tissues. Several hyperparasitic fungi, especially *Scytalidium uredinicola*, exert some natural control of sporulation of *Peridermium harknessii* and are potentially useful for practical biological control, but secondary fungi probably suppress the disease more than hyperparasites do.

Susceptibility to western gall rust varies among and within host species. Healthy trees within severely damaged populations are common. Intraspecific susceptibility also varies with geographic origin of seed. Species resistance has been verified by inoculations of *Pinus clausa* (sand pine), *P. densiflora* (Japanese red pine), *P. echinata* (shortleaf pine), *P. elliottii* (slash pine), *P. palustris* (longleaf pine), *P. taeda* (loblolly pine), *P. thunbergii* (Japanese black pine), and *P. virginiana* (Virginia pine).

Galls caused by *Peridermium harknessii* are indistinguishable from those of pine-oak gall rust (Plate 147) and other pine gall rusts that are not yet fully described or named. A white-spored gall-rust fungus formerly thought to be a strain of *P. harknessii* occurs on *Pinus ponderosa* in southwestern USA, and a yellow-spored gall rust occurs on *P. jeffreyi* in California. Both of these fungi have been shown to differ genetically from *Peridermium harknessii* and from the *Cronartium quercuum* group. Some strains of pine gall-rust fungi in central and western USA apparently can spread from pine to *Castilleja* (paintbrush), resulting in production of uredinia and (or) telia resembling those of *Cronartium coleosporioides*. In two studies aeciospores from single gall-rust sources infected both pine and angiosperm hosts.

Heteroecious versus autoecious habit of a pine gall-rust fungus can be ascertained by observing germination behavior of aeciospores. Those of heteroecious fungi such as *C. quercuum* are able to infect leaves of angiosperm hosts, and when allowed to germinate on water agar produce long (>500 μm) slender germ tubes with a tendency toward apical branching. Aeciospores of *P. harknessii* and other autoecious gall-rust fungi can infect succulent pine shoots, and their germ tubes are short (average 230 μm in one study), stout, and septate, with a tendency to produce branches near the proximal end.

The life cycle of *P. harknessii* and other autoecious pine stem rust fungi is a subject of scientific disagreement based on cytological studies. One interpretation is that these fungi have a sexual cycle with karyogamy (nuclear fusion) and meiosis occurring in germinating aeciospores and germ tubes, respectively. Therefore these fungi are said to be endocyclic, and the genus name *Endocronartium* is used for them. The alternative interpretation, based on extensive study of *P. harknessii*, is that autoecious pine stem rusts reproduce only mitotically (clonally) and so should be classified in the genus *Peridermium* with the aecial states of *Cronartium* species. DNA sequence comparisons have established that *P. harknessii* and several other autoecious short-cycled pine stem-rust fungi are anamorphic *Cronartium* species and are not a monophyletic group; that is, they are not unified by having a common ancestor within *Cronartium*. Rather, autoecious habit and abbreviated life cycles seem to have evolved several times within *Cronartium*.

References: 52, 75, 76, 593, 594, 870, 871, 1528, 1680, 1681, 1685, 1687, 1903, 2010, 2586, 2597, 2694, 3030, 3035, 3042, 3049, 3990, 4089, 4097, 4167–4169, 4201, 4202, 4320, 4448, 4502, 4526

Western gall rust, caused by *Peridermium harknessii*.

A, B. On *Pinus contorta* (lodgepole pine). A. Galls on the trunk of a young tree (OR, Sep). B. A "hip canker," several decades old, on a large tree (CA, Jul).

C. Branch dieback caused by western gall rust and secondary organisms on *P. ponderosa* (ponderosa pine) (ID, Jul).

D. Swelling and upturned multiple shoots induced by the rust at a branch tip of *Pinus halepensis* (Aleppo pine) (CA, Feb).

E–I. On *Pinus sylvestris* (Scots pine). E. Infections close together on twigs have resulted in nearly continuous swelling, induction of multiple terminal shoots, and dieback (NY, Apr). F. A loose witches'-broom resulting from multiple infections on a limb (NY, Sep). G. Aecia breaking through the bark surface (NY, May). H. Galls in which a layer of yellow aeciospores has developed beneath the periderm, causing the bark scales to break away (NY, May). I. A mature tree killed by gall rust and secondary insects (NY, Sep).

305

Trunk and Limb Rots of Hardwoods (Plates 152–154)

Overview of Wood Types and Decay Processes

We consider first the nature of sapwood, heartwood, and wood decay, and some terms used to describe and classify decaying wood. Normal sapwood is alive and involved in sap conduction, storage of compounds that serve as energy reserves, compartmentalization of wounds, and defense against invading microorganisms (Plates 254, 255). It is light in color and highly resistant to most wood-decay fungi while alive. The vitality and functional importance of sapwood diminish from youngest to oldest layers. Dead sapwood has little resistance to wood-decay fungi.

Heartwood is dead, brightly to darkly colored xylem that develops normally in the centers of trunks and large branches of many species of trees as parenchyma cells in old sapwood die. Heartwood provides mechanical strength but is not involved in conduction or storage. It contains toxic chemicals produced by parenchyma cells before their death that make it resistant to many wood-decay fungi. Heartwood differs in origin and chemical properties from wound-associated discolored wood. Light-colored heartwood that cannot be visibly distinguished from old sapwood is sometimes called ripewood. Some important groups of trees with diffuse-porous xylem, such as *Acer* (maple), *Betula* (birch), *Fagus* (beech), and *Populus* (poplar), do not produce heartwood. Their sapwood extends to the pith of healthy stems.

Most wood decay is caused by fungi and is classified in three categories: white rot, soft rot, and brown rot. During decay of the first two types, wood constituents are depolymerized enzymatically. White rot is typified by more rapid breakdown of lignin than of cellulose and hemicelluloses. When a white rot reaches an advanced stage, the predominantly cellulosic residue is light in color and may be laminated, spongy, stringy, or brittle and crumbly. White mottled rots, white pocket rots, and white stringy rots are characterized by relatively slow reduction in the average length of cellulose polymers, so wood in early stages of decay does not lose strength rapidly. In advanced stages, however, the wood is soft and weak. Some white-rot fungi cause simultaneous degradation of all cell-wall components and rapid reduction in the average length of cellulose molecules, resulting in much strength loss during early stages of decay. Soft rots are biochemically similar to white rots but differ in microscopic aspects. Hyphae grow within the thick S_2 layer of the secondary walls of wood cells and there cause lysigenous cavities that weaken the walls (soft rot type 1); or hyphae grow on secondary walls within wood cells and cause erosion of the walls without degrading the middle lamellae (type 2). Some white rots have soft-rot characteristics while in early stages. In brown rots, cellulose molecules are cleaved randomly by nonenzymatic oxidative and hydrolytic systems (thought to involve organic acids, peroxide, and ferrous ion), causing depolymerization with great loss of wood strength and onset of brittleness at an early stage. Cellulose-degrading enzymes are also produced, and lignin is altered but not digested, so dark-colored breakdown products remain. Wood undergoing brown rot shrinks and cracks into more or less cubical chunks.
References: 17, 27, 398–401, 425, 427, 461, 548, 740, 1359, 1601, 1651, 2241, 2486, 2595, 2596, 2994, 3228, 3499, 3443, 3510, 3600, 3952, 4181, 4493

Representative Decay Fungi

Stereum gausapatum (Russulales, Stereaceae) is the most common heartwood pathogen of *Quercus* (oak) in North America. It occurs from coast to coast in Canada and the USA and in Europe, and causes a white mottled rot of sapwood and/or heartwood. It occasionally colonizes other plants, mainly *Castanea* (chestnut). Reports of *S. gausapatum* on other hardwoods are questionable because it has been confused with the widespread *S. hirsutum*. *S. gausapatum* enters branch stubs and wounds such as fire scars. It also colonizes stumps and spreads into sprouts after these make heartwood connections with stumps. Decay columns elongate approximately 10 cm per year in the South. Basidiocarps arise annually in dense clusters as thin (0.5–1 mm), gray-brown, shelflike structures up to 6 cm or more wide on scars, stumps, and logging slash. The basidiocarps have a smooth undersurface and bleed red fluid where injured.
References: 344, 426, 429, 687, 893, 894, 896, 1075, 1614, 1615, 2293, 2996, 3349, 4059

Hericium erinaceus (Russulales, Hericiaceae), the hedgehog fungus, is a common cause of butt rot in hardwoods in the eastern half of the USA. It occurs also in Pacific Coast states and widely in Eurasia. Reports of *H. erinaceus* in Canada are assignable to *H. americanum*. *H. erinaceus* is most often found on *Quercus* (oak) but attacks many other angiosperms and, rarely, gymnosperms: *Acer* (maple), *Carya* (hickory), *Castanea* (chestnut), *Diospyros* (persimmon), *Fagus* (beech), *Liquidambar* (sweetgum), *Liriodendron* (tuliptree), *Nyssa* (sour-gum), *Platanus* (sycamore), *Pseudotsuga* (Douglas-fir), *Salix* (willow), and additional plants in Eurasia. The decay is a white pocket rot that becomes soft and spongy before its disintegration results in a cavity. Fire scars, stem cracks, and branch stubs are common sites of infection. The decay advances approximately 5 cm per year in *Carya aquatica* (water hickory). *H. erinaceus* is named for the long (2–10 cm) slender teeth that extend from the lower surface of its soft annual basidiocarps when they are mature. Each tooth is covered with fertile tissue where basidiospores are produced. The basidiocarps develop in autumn and winter in the South as white globoid masses, commonly 10–12 cm but sometimes up to 30 cm across. Stumps, logging slash, and branch scars on living trunks are common sites of fruiting. Basidiocarps turn pale yellow or pale brown with age.
References: 344, 1297, 1615, 1614, 2543, 4059

Phellinus igniarius **and relatives**, a group of fungi formerly united in a species complex under the name *Phellinus igniarius* (Hymenochaetales, Hymenochaetaceae), cause white heart rots of hardwoods throughout the cooler parts of the north temperate zone. At least eight named fungi have been sorted out, of which *P. igniarius*, *P. arctostaphyli*, *P. laevigatus*, and *P. tremulae* are recognized in North America. Additional North American segregates are likely to be named. *P. igniarius* in the narrow sense is among the most important trunk decay pathogens in northern hardwood forests, causing extensive damage in *Acer* (maple), *Betula* (birch), and *Fagus* (beech), and occurring on other angiosperms in more than 25 genera. It succeeds pioneer organisms in wounds to sapwood but is not itself an aggressive colonist of fresh wounds. Decay by *P. igniarius* typically contains black pseudosclerotial plates that appear as lines (zone lines, described with Plates 99 and 102) on cut or broken surfaces. *P. arctostaphyli* is restricted to *Arctostaphylos* (manzanita) and *Cercocarpus* (mountain-mahogany). *P. laevigatus* occurs mainly on *Acer*, *Alnus* (alder), *Betula*, and *Fagus*; it causes a canker-rot in *B. alleghaniensis* (yellow birch). *P. tremulae* is restricted to *Populus* (see below). Basidiocarps (conks) of *P. igniarius* are perennial, more or less hoof shaped, up to $11 \times 20 \times 8$ cm in size, and have a gray to blackish upper surface that becomes deeply cracked with age. The margin during growth is sometimes gray-white but soon becomes yellowish brown, and the poroid (lower) surface is cinnamon brown to dark purplish brown with five–six circular pores per millimeter.
References: 632, 1143, 1277, 1280, 1608, 1827, 1859, 2858, 3373, 3510, 3588, 3591, 4092, 4108, 4183, 4517

Phellinus tremulae attacks *Populus* species of the aspen group wherever they grow in the Northern Hemisphere. It is their most important decay pathogen. The yellowish white spongy decay, a white rot, is usually confined to a central core, although *P. tremulae*

A, B. *Stereum gausapatum* on a basal scar on *Quercus nigra* (water oak) (MS, May).

C, D. *Hericium erinaceus*, the hedgehog fungus, fruiting at a branch scar on *Quercus texana* (Nuttall oak) (MS, May).

E–H. Trunk rot caused by *Phellinus igniarius*. E. A young basidiocarp developing at a wound scar on *Juglans cinerea* (butternut) (NY, Apr). F. A typical perennial basidiocarp growing from a canker on a pole-size *Fagus grandifolia* (American beech) (NY, May). G. Zone lines associated with decay by *P. igniarius* and other fungi in a freshly cut *F. grandifolia* butt. Sapwood at left was alive when cut (NY, May). H. Close view of zone lines on a transverse surface of an *Acer saccharum* (sugar maple) log (NY, Oct).

I–K. Heart rot caused by *Phellinus tremulae* in *Populus tremuloides* (trembling aspen). I, J. Perennial basidiocarps on living trunks (NY, Aug). K. Transverse view of a decay column with black zone lines (CO, Jun).

306

307

has been reported capable of colonizing aspen sapwood. It enters trunks via dead branches and stubs and perhaps via fresh wounds. The decay is sweet smelling and contains black pseudosclerotial plates that appear as lines (zone lines) on cut or broken surfaces. Decay columns may extend 2–3 m above and below the perennial basidiocarps, which develop at branch stubs, branch scars, and wound scars. They grow up to 20 cm wide and 15 cm thick, with the upper and lower surfaces at angles nearly 45° from the horizontal, and are usually attached by a granular core of tissue that extends into the trunk. The upper surface is pale brown near the margin; older parts become blackened, crustlike, and cracked. The lower, poroid surface is purplish brown. Basidiospores are released throughout the growing season. Sometimes *P. tremulae* produces hard blackish sterile masses of mycelium (sterile conks, punk knots) at branch scars.
References: 250, 1277, 1280, 1315, 1608, 1672, 1723, 1827, 2332, 2441, 2453, 2858, 2937, 3295, 3374, 4183, 4193, 4360, 4361

Phellinus robiniae is restricted to *Robinia pseudoacacia* (black locust), *R. neomexicana* (New Mexico locust), *Gleditsia triacanthos* (honeylocust), and *Prosopis* (mesquite) in North America, where it occurs from the Atlantic Coast to Idaho and the desert Southwest. It occurs on diverse other angiosperms in Eurasia, South Africa, and Australia. *P. robiniae* is one of the few fungi able to decay the heartwood of *R. pseudoacacia*. It infects trunks through branch stubs and wounds, especially those made by *Megacyllene robiniae*, the locust borer, and causes a spongy yellow heart rot of the white-rot type. Its perennial shelflike basidiocarps grow up to 30 cm wide with concentric ridges. The upper surfaces are at first yellowish brown, but blacken and crack with age; the poroid surfaces are yellowish to reddish brown.

Phellinus everhartii is a common heart-rot pathogen of *Quercus* species (oaks), and is reportedly able to kill their sapwood. It occurs occasionally on other hardwoods: *Acer* (maple), *Betula* (birch), *Carya* (hickory), *Castanea* (chestnut), *Fagus* (beech), *Juglans* (walnut), *Liriodendron* (tuliptree), *Malus* (apple), *Ostrya* (hop-hornbeam), *Populus* (cottonwood), and *Ulmus* (elm). This fungus has been confused with others such as *P. badius* on *Prosopis* (mesquite), *P. igniarius* on various hardwoods, and *P. weirianus* on *Juglans* (walnut). *P. everhartii* seems to be restricted to North America. It occurs in the West from Idaho and Montana to California and Mexico and in the East from the western Great Lakes region and Maritime Provinces to Arkansas and Georgia. The decay is of the white-rot type, appearing golden brown or lighter and typically associated with trunk cracks. The basidiocarps are perennial and hoof shaped, and most are less than 6 × 13 × 8 cm. The upper surface is yellowish brown when young but becomes nearly black and extensively cracked with age. The lower, poroid surface is yellowish brown to reddish brown. Basidiocarps develop at cracks, old wounds, and occasionally at cankers or cankerous swellings. Sometimes sterile blackish cracked masses of fungal tissue (sterile conks) develop instead of fertile basidiocarps.

Phellinus weirianus, a white-rot fungus restricted to mountainous areas in the Southwest (except for rare occurrence in Texas and Oklahoma), mimics the appearance of *P. everhartii*, but the two fungi differ in microscopic characteristics and hosts. *P. weirianus* is restricted to *Juglans major* (Arizona walnut); *P. everhartii* in the Southwest is restricted to *Quercus*.
References: 1277, 1280, 1615, 2545, 2937, 3489, 3510, 4108, 4183

Climacodon septentrionalis (Polyporales, Meruliaceae) causes a spongy white trunk decay in various angiosperms in North America and Eurasia. It is common east of the Great Plains from Canada to the Gulf Coast states and occurs in Alberta. It is most destructive in *Acer* species (maples), which it invades via frost cracks and wound scars, less commonly via branches and branch stubs. Decay columns in *A. saccharum* (sugar maple) extend to 4 m above and 1.5 m below basidiocarps. The advanced decay usually contains thin black pseudosclerotial plates and is surrounded by a zone of brownish wood. Other host genera in North America include *Betula* (birch), *Carya* (hickory), *Fagus* (beech), *Liquidambar* (sweetgum), *Liriodendron* (tuliptree), *Malus* (apple), *Nyssa* (tupelo), *Populus* (aspen), *Quercus* (oak), *Tilia* (basswood), and *Ulmus* (elm). Elaborate cream-colored basidiocarps of *C. septentrionalis* develop annually in summer. They consist of fleshy to fibrous shelflike projections, one above another, joined along the tree trunk, each with the lower surface covered with slender teeth about 1 cm long. The compound structure is commonly 20–30 cm wide and 50–80 cm long.
References: 263, 1299, 2644, 2868, 4108

Bjerkandera adusta (Polyporales, Hapalopilaceae) is a common sapwood pathogen and saprobe that causes a white rot in many angiosperms and occasionally gymnosperms in the temperate zones. It colonizes plants in at least 40 genera in North America and is one of the major pathogens of shade trees in Europe. This fungus aggressively attacks living and dying stems that have been wounded, injured by freezing, or debilitated by other pathogens. It annually produces thin, shelflike basidiocarps 1–6 × 3–10 × 0.1–0.8 cm, usually overlapping one another in dense clusters. The upper surfaces are usually slightly velvety and vary in color from nearly white through shades of tan or gray, sometimes with reddish areas. The lower surface is gray.
References: 982, 1280, 1383, 1388, 2868, 2937, 3374, 4108, 4517

Fomes fomentarius (Polyporales, Polyporaceae) causes white mottled trunk rot of many angiosperm trees from coast to coast in Canada and the northern half of the USA; it is widespread also in Eurasia. This fungus is a secondary colonist of wounded sapwood, in which it follows bacteria and ascomycetes. Its woody, perennial, hoof-shaped basidiocarps are up to 20 cm broad and 15 cm thick. In North America they are most abundant on *Betula* (birch) and *Fagus* (beech), and are also found on living or dead *Acer* (maple), *Alnus* (alder), *Carya* (hickory), *Malus* (apple), *Populus* (aspen, poplar), *Prunus* (cherry), *Quercus* (oak), *Salix* (willow), *Umbellularia* (California-laurel), and rarely on conifers such as *Pseudotsuga* (Douglas-fir) and *Tsuga* (hemlock). It discharges basidiospores primarily in spring, although some specimens may continue throughout the growing season. A typical basidiocarp may release on the order of 10^{11} basidiospores per season—more than its own weight in spores. *F. fomentarius* varies in form and host preferences from one region to another, and isozyme differences have been associated with two forms in Asia.
References: 719, 1280, 1383, 1388, 1653, 1864, 2595, 2937, 3095, 3319, 3374, 3483, 3510, 4098, 4108, 4517

Globifomes graveolens (Polyporales, Polyporaceae) causes a straw-colored to yellowish brown heart rot in *Quercus* species, notably *Q. coccinea* (scarlet oak), *Q. falcata* (southern red oak), *Q. nigra* (water oak), *Q. phellos* (willow oak), *Q. rubra* (northern red oak), and *Q. velutina* (black oak). It rarely colonizes *Acer* (red maple), *Carya* (hickory), *Fagus* (beech), *Liquidambar* (sweetgum), and *Liriodendron* (tuliptree). Found only in the USA from Wisconsin and New York southward, this fungus is unimportant in forests but causes trunk failure in street-side and landscape oaks. Decay begins at wounds, usually low on the trunk. *G. graveolens* is apparently able to grow outward from heartwood, perhaps along small cracks, and kill areas of sapwood and vascular cambium, resulting in the linear depressed cankers where it fruits. Basidiocarps also grow on dead trees. They are globular to ovate, woody in the center, and covered with many downturned scalelike outgrowths, each of which has a layer of tubes on the lower surface. Fresh basidiocarps have varying reddish to brown and gray colors and emit a fruity odor. Old basidiocarps are gray-brown and lack the odor.
References: 896, 1280, 1357, 2769, 2937, 4108

A. *Phellinus robiniae* on *Robinia pseudoacacia* (black locust) (MS, Apr).

B. *P. everhartii* on *Quercus arizonica* (Arizona white oak) (AZ, Jul).

C. *P. weirianus* on *Juglans major* (Arizona walnut) (AZ, Jul).

D. *Climacodon septentrionalis* on *Acer saccharum* (sugar maple) (NY, Aug).

E. *Bjerkandera adusta* on a canker on *Salix caprea* (goat willow) (NY, May).

F. *Fomes fomentarius* on a dead trunk of *Betula alleghaniensis* (yellow birch) (NY, Aug).

G–I. Trunk decay of *Quercus* (oak) caused by *Globifomes graveolens*. G. Old basidiocarps on an elongate canker on *Q. nigra* (water oak). H, I. Young and old basidiocarps, respectively, on *Q. falcata* (southern red oak) (GA, May).

309

Perenniporia fraxinophila (Polyporales, Polyporaceae) causes a white mottled heart rot of *Fraxinus* (ash), notably *F. americana, F. latifolia, F. nigra, F. pennsylvanica, F. quadrangulata,* and *F. velutina.* It uncommonly affects *Acer, Crataegus, Fagus, Garrya, Juniperus, Platanus, Prunus, Quercus, Salix, Shepherdia,* and *Ulmus.* This fungus occurs across North America, most commonly in eastern and central regions, and also in China. It contributes to deterioration of *F. pennsylvanica* (green ash) in shelterbelts in central USA. Decay develops in trunks and major limbs. The advanced stage is straw yellow to yellowish white, soft, and crumbly. Branch stubs are the usual sites of infection and fruiting. The perennial poroid basidiocarps are bracket shaped when fully developed and are dirty white, darkening and becoming cracked on the upper surface with age. They range in size up to 30 × 20 × 10 cm or larger and sometimes arise in groups on dead areas of bark that remains attached to the wood.
References: 1154, 1280, 2937, 3291, 3490, 4108, 4198

Oxyporus populinus (Polyporales, Polyporaceae) is a major heart-rot fungus of *Acer* (maple), especially *A. rubrum, A. negundo,* and *A. saccharum.* It also colonizes *Aesculus, Carya, Cornus, Ginkgo, Gleditsia, Halesia, Liquidambar, Liriodendron, Nyssa, Ostrya, Platanus, Populus, Prunus, Quercus, Salix, Sambucus,* and *Ulmus.* It occurs around the temperate Northern Hemisphere, although uncommonly in western North America, and also in South America and South Africa. *O. populinus* is a wound invader that often follows canker-causing fungi. It causes a spongy straw-colored white rot in heartwood and sapwood. Decay columns usually extend less than 1 m above and below the perennial white basidiocarps. The basidiocarps develop singly or in clusters along cracks, in basal wounds, in cankers caused by other fungi, and in knot holes. Moss usually grows on the upper surface. Well-developed specimens are shelflike and up to 15 cm wide and 10 cm thick, but an irregular mass such as that shown in Plate 154D is also typical. They are soft-corky when dry, spongy when wet.
References: 185, 265, 3591, 3601, 4108

Trametes versicolor (Polyporales, Polyporaceae) occurs through-out the temperate zones of the world as a saprobe and pathogen on woody angiosperms and occasionally on conifers. Its hosts represent more than 70 genera in the USA alone. Living plants on which it has been identified include species of *Acer, Castanea, Catalpa, Malus, Platanus, Prunus, Salix, Sorbus, Syringa, Tilia,* and *Umbellularia,* among others. *T. versicolor* invades wounds, broken limbs, frost cracks, and freezing or sunscald injuries. It then kills sapwood in plants stressed by water shortage, freeze damage, or severe wounding, and causes a spongy white decay. In *Malus* and probably other trees it also kills areas of cambium, causing cankers and dieback. The bark in cankers on *Malus* has a loose papery surface. The thin, tough, bracket-shaped basidiocarps of *T. versicolor,* usually 2–5 × 2–7 × 0.2–0.5 cm, develop annually in summer and early autumn, always one above another in clusters. The upper surface is slightly velvety, with concentric zones of various colors. The lower, poroid surface is white to cream.
References: 17, 325, 630, 910, 1280, 1383, 2059, 2060, 2937, 3319, 3374, 3491, 4108, 4369

Polyporus squamosus (Polyporales, Polyporaceae) causes a spongy white rot leading to cavity formation in trunks of living angiosperm trees; it also decays logs and stumps. It occurs across Canada and the northern half of the USA, in the Southwest, and also in Eurasia and the Southern Hemisphere. In North America trunk decay by this fungus is common in *Acer, Populus, Salix,* and *Ulmus.* Other hosts include species of *Aesculus, Alnus, Betula, Celtis, Fagus, Liriodendron, Pyrus, Tilia,* and *Umbellularia.* Trees in additional genera are affected in Eurasia. *P. squamosus* enters trees via wounds and produces large, fan-shaped, fleshy-fibrous annual basidiocarps at wounds, branch stubs, and cankers. The basidiocarps grow rapidly in spring and early summer from a short, stout, dark-colored lateral stalk and attain widths of 20–30 cm, occasionally 45 cm. The upper surface is yellow-tan with prominent dark brown scales. The lower surface is cream colored and perforated by large tubes with angular mouths measuring 1–3 × 1.4–6 mm.
References: 565, 571, 624, 808, 1280, 1343, 2937, 3374, 4108

Laetiporus sulphureus (Polyporales, Polyporaceae) is the name formerly applied to a complex of fungi that occur worldwide in temperate regions, causing brown cubical rot in the trunks, butts, and sometimes roots of living and dead trees. At least five species occur in North America: *L. sulphureus* in the strict sense, *L. cincinnatus, L. conifericola, L. gilbertsonii,* and *L. huroniensis.* The first three colonize angiosperms; the latter two are found on gymnosperms. *L. cincinnatus* and *L. sulphureus* are widespread in eastern and midwestern North America. *L. gilbertsonii* occurs from California to Washington, and its variety *pallidus* from Florida to Texas. *L. conifericola* occurs in western North America; *L. huroniensis* is found in the East. Most published host records apply to the group as a whole, for which host genera in North America include *Abies, Acer, Betula, Castanea, Celtis, Eucalyptus, Fagus, Fraxinus, Gleditsia, Juglans, Larix, Liriodendron, Malus, Persea, Picea, Pinus, Populus, Prunus, Pseudotsuga, Quercus, Robinia, Schinus, Sequoia, Tamarix, Thuja, Tsuga,* and *Ulmus.*

Laetiporus species enter tree trunks through branch stubs and wounds or via roots. Wood undergoing decay by these fungi develops shrinkage cracks in which sheets of white to pale yellow mycelium form. The basidiocarps, which are edible, arise annually in summer and autumn on stumps, logs, living trees, or the ground at the base of a trunk. They grow in clusters 20–60 cm broad, varying from sulfur yellow to salmon or bright orange on the upper surface and sulfur yellow (or white in *L. cincinnatus* and in the southern variety of *L. gilbertsonii*) on the poroid lower surface. They bleach with age. Individual basidiocarps measure 5–25 × 4–30 × 0.5–2.5 cm. Those of *L. sulphureus* often issue from elongate bark-covered cankers that appear as depressed strips, an indication that the fungus has spread outward killing a zone of sapwood and vascular cambium. Basidiocarps die with the onset of freezing weather in the North but persist longer and disperse spores during winter in the South. Some strains in the *L. sulphureus* complex produce conidia in culture and in decaying wood and tuberlike conidial fruiting structures in nature; the latter are short-lived and collapse into a mass of chlamydospores.
References: 17, 190, 191, 344, 568, 808, 896, 1161, 1266, 1278, 1280, 1615, 1859, 2212, 2545, 2868, 2937, 3324, 3374, 3510, 4108

Piptoporus betulinus (Polyporales, Fomitopsidaceae) occurs only on *Betula* (birch) but is found throughout the range of this genus in North America and Eurasia. Hosts include *B. alleghaniensis, B. lenta, B. occidentalis, B. papyrifera, B. pendula, B. populifolia,* and other species in Eurasia. This fungus causes a red-brown cubical rot. It is usually found on dead trees, rarely on living ones. Its distinctive pale brown to nearly white annual basidiocarps with smooth rounded upper surfaces up to 25 cm wide are unlike those of any other fungus on *Betula.*
References: 9, 808, 1278, 1280, 2405, 2937, 3319, 3374, 3507, 4108

A, B. *Perenniporia fraxinophila* fruiting at old wounds and branch scars on a mature *Fraxinus pennsylvanica* (green ash) (MI, Aug).

C, D. *Oxyporus populinus* in a Nectria canker on *Acer saccharum* (sugar maple). Moss on the upper surface of the basidiocarp is typical (NH, Sep).

E–G. Sapwood rot caused by *Trametes versicolor.* E. Basidiocarps on a *Salix purpurea* (purple osier) stump. Repeated harvesting of shoots disposed the plant to infection (NY, Dec). F. Upper surfaces of basidiocarps on a *Quercus rubra* (red oak) branch (NY, Sep). G. White undersurfaces of basidiocarps and typical white rot in *Q. rubra* sapwood (NY, Sep).

H, I. *Polyporus squamosus* fruiting at a large branch stub on *Acer saccharinum* (silver maple) (NY, Jun), and on an *Ulmus americana* (American elm) stump (NY, Jul).

J, K. *Laetiporus sulphureus.* J. Associated with heart rot in *Prunus serotina* (black cherry). This trunk was alive around half its circumference (NY, Jun). K. Collected from *Quercus* sp. (NY, Jun).

L. *Piptoporus betulinus* on a dead trunk of *Betula populifolia* (gray birch) (NY, Aug).

Photo credit: H—G. W. Hudler

311

Canker-rots of Hardwoods (Plates 155–158)

Canker-rots are caused by wood-decaying fungi that kill sapwood, phloem, and the vascular cambium, resulting in perennial cankers. Some canker-rot fungi, such as *Trametes versicolor* (Plate 154) and *Schizophyllum commune* (Plate 158), enter wounds, parasitize sapwood around the wound, and subsequently kill the overlying vascular cambium and bark. Others, exemplified by *Phellinus spiculosus* here and by other pathogens shown in Plates 156–158, enter trunk wounds or branch stubs, colonize and decay wood within wound-associated compartments (Plates 254, 255), and eventually produce at the original wound or branch stub a mass of mycelium that kills adjacent cambium and bark, resulting in a canker. This lesion serves as a new site of entry into sapwood, which in turn leads to more decay and further compartmentalizing responses by the tree. Trunk decay and cankers expand by repetition of these processes. Internal cracks and new wounds that break barrier zones also contribute to spread of decay fungi outward from initially compartmentalized cores of colonized wood.

Spiculosa Canker

Phellinus spiculosus (Hymenochaetales, Hymenochaetaceae) causes white trunk rot and perennial cankers on *Carya* (hickory), *Gleditsia* (honeylocust), and *Quercus* (oak). Specific hosts include *C. glabra* (pignut hickory), *C. ovata* (shagbark hickory), *C. tomentosa* (mockernut hickory), *G. triacanthos*, *Q. falcata* (southern red oak), *Q. marilandica* (blackjack oak), *Q. nigra* (water oak), *Q. phellos* (willow oak), *Q. rubra* (northern red oak), and *Q. texana* (Texas red oak). The pathogen is apparently restricted to eastern North America from Pennsylvania and Delaware south to Florida and Texas.

Symptoms and signs. Cankers on *Quercus* are usually circular, 5–15 cm in diameter, and within 4 m of the ground. Rough, bulging cankers with irregular woundwood ridges on the surface form occasionally and are similar to those found on *Carya*, described below. Open cankers with depressed centers and moderate woundwood development around the margins also form occasionally, especially on *Q. phellos*. A canker in an early stage of development may look like the site of a branch stub being enclosed by growth of the trunk. Sterile brown fungal tissue, sometimes referred to as a punk knot, occupies the position of the former stub. Often the only indicator of infection is a black or bark-colored punk knot. These can be diagnosed by cutting into them, exposing brown fungal tissue. Punk knots, it has been suggested, maintain a connection for gas exchange between the central decay column and the exterior, thus allowing decay to progress. (Growth of trunk decay fungi slows drastically when trunk wounds close, because the oxygen supply is insufficient for normal activity.) Multiple cankers or punk knots on a tree do not necessarily indicate separate infections; all may be associated with one column of decay.

The cankers on *Carya* are often more conspicuous than those on *Quercus*. They may appear as rough circular swellings where woundwood seems to have enclosed or nearly enclosed wounds. A diseased tree usually has only one canker, but some trees bear several. Sometimes a canker bulges out as a burl with one or more seams where woundwood ridges meet but apparently do not fuse. Open cankers, usually with depressed centers, develop occasionally. Sterile brown fungal tissue forms at branch stubs and often occupies cracks between woundwood ridges at the surface of a swelling. Mycelium extends from the punk knot or crack into a large column of decayed wood in the center of the trunk. Cutting into the swollen area exposes the brown punk knot and brown mycelium between woundwood ridges.

The advanced decay caused by *P. spiculosus* in all hosts is a soft, crumbly, white or slightly yellowish rot, sharply delimited from sound wood, typically confined to heartwood and extending from ground level to a height of 3–4 m or more. The decay columns closely resemble those caused by *Inonotus hispidus* (Plate 157).

P. spiculosus occasionally fruits on dead parts of living trees, but usually the perennial basidiocarps develop on dead standing trees and logs on the ground. They arise in patches or as a continuous sheet up to 1.5 m long between the bark and wood of decayed stems, pushing off the bark as they mature. They are similar in form and behavior to the basidiocarps of *Inonotus obliquus* (Plate 156). The pore surface and interior are brown initially; the surface becomes whitish or grayish, cracked, and dry with age.

Disease cycle. The cycle has not been elucidated, but some elements of it can be deduced from results of tree dissection and isolation studies. Infection, presumably initiated by airborne basidiospores, begins at branch stubs, and cankers eventually form there. Decay columns caused by *P. spiculosus* in *Quercus* species in the South have been estimated to elongate at an average rate of 20 cm per year. Living diseased trees typically have long decay columns before basidiocarps form; therefore, completion of the cycle probably requires decades. Since the decay occurs mainly in heartwood, and since those portions of the decay column that involve sapwood are compartmentalized (Plate 155E–G), this fungus probably kills few trees by itself. More likely it contributes to tree decline and death by diminishing the thickness of functional sapwood and predisposing the trunk to mechanical failure. Possibly it kills trees in which compartmentalization responses are impaired by other stressing factors such as drought or defoliation.

References: 633, 1280, 2544, 2858, 4057

A, E. Spiculosa canker and associated trunk rot in *Quercus nigra* (water oak). A. A perennial targetlike canker, prominently raised because of woundwood formation around the margin, has developed at the site of a former branch stub. E. A section through the trunk several centimeters below the canker reveals advanced white crumbly rot throughout the heart of the tree, plus spokelike extensions of decay into outer sapwood. The extensions are the tips of secondary columns of decay that began as the canker enlarged. They would eventually have merged. The column of advanced decay in this tree was 3 m long (GA, May).

B, D. Spiculosa cankers on *Q. texana* (Texas red oak) and *Q. phellos* (willow oak), respectively. Sterile brown fungal structures (punk knots) occupy the locations of former branch stubs; brown interior tissue is revealed by cutting into the raised knots (MS, Apr).

C, F, G. Dissection of the *Q. texana* shown in B. C. The brown punk knot has grown out from a central column of advanced white rot bounded by a thin purple-brown zone of discolored wood. The decay column has expanded periodically and involves sapwood as the result of canker enlargement adjacent to the punk knot. F. The decay column also expands as a result of boring by insects that breach the barrier zone between decayed and sound sapwood. Two such extensions of decay into the sapwood zone are shown. G. The butt log, approximately 4 m long, with the canker at the midpoint and advanced decay throughout the heart of the log (MS, Apr).

Canker-rots of Birch

Canker-rot caused by *Inonotus obliquus*. Trunk rot associated with prominent sterile conks on *Betula* species (birches) is usually caused by *Inonotus obliquus* (Hymenochaetales, Hymeno-chaetaceae), a fungus that occurs around the Northern Hemisphere. In North America it most commonly infects *B. alleghaniensis* (yellow birch), *B. occidentalis* (water birch), and *B. papyrifera* (paper birch). It also occurs in *B. lenta* (black or sweet birch) and *B. populifolia* (gray birch). Occasional to rare records exist for *Alnus rubra* (red alder), *Fagus grandifolia* (American beech), *F. sylvatica* (European beech), *Ostrya virginiana* (hop-hornbeam), and *Populus balsamifera* subsp. *trichocarpa* (black cottonwood). One report of *I. obliquus* on *Fagus* is of special interest because it involved a basidiocarp on a living tree. When on *Betula*, the fungus produces basidiocarps only after the trees are dead. In Eurasia, where the fungus is widely utilized for medicinal purposes, *I. obliquus* has a similar set of host genera with the addition of *Quercus*.

Symptoms and signs. The disease is easy to diagnose on the basis of the sterile conks. These are hard black perennial masses of fungal tissue, rough and cracked on the surface, that extrude from internal columns of decayed wood at old Nectria cankers, branch stubs, cracks, and wound scars. The internal tissue of a sterile conk is yellow-brown to rust-brown and punky and contains small dark brown fragments of partly decayed bark that was killed and displaced by the growing mass. Diseased trees usually have one to six sterile conks; their presence indicates a decay column so large that at least half the trunk—often the entire trunk—is useless for lumber. The trunk is often swollen adjacent to a sterile conk because of increased xylem and phloem production apparently induced by the pathogen. The bark sometimes thickens to as much as three times normal. Trunks often break at the locations of sterile conks. The decay is of the white-rot type, appearing light reddish brown and mottled and with veins of white mycelium near the sterile conks. It becomes spongy in advanced stages.

Basidiocarps of *I. obliquus* form as sheets of brown tissue in the outer sapwood (1–25 mm deep) 3–4 years after a tree with canker-rot dies, usually while the trunk is still erect. The interval between tree death and sporulation may be longer for trees that break or are felled while alive; it was 7–12 years in one study. The onset of fruiting is signaled by formation of an undulating brown sheet of mycelium at the boundary of the decay column. Elongate ridges of sterile fungal tissue develop parallel to the trunk axis at the edges of the mycelial sheet, exerting radial and tangential force that causes the covering wood and bark to crack along the axis of the trunk and then lift away from the fungal tissue. This tissue soon develops a gray poroid surface from which basidiospores are dispersed, then turns dark brown, dries, and begins to disintegrate. The entire basidiocarp may be 30–150 cm long and 15–30 cm wide. Basidiocarps may be found in July–November but are most abundant during summer. They are short-lived because insects destroy them. A given tree usually produces basidiocarps in only one year.

Disease cycle. Infection is presumed to be initiated by airborne basidiospores. Insect transmission has also been suggested based on the observed destruction of basidiocarps by insects. Dissection studies have indicated that infection occurs at old Nectria cankers, branch stubs, cracks, and wound scars (listed in descending order of frequency). *I. obliquus* is apparently a primary colonist of sapwood, requiring no prior succession of microorganisms. It kills and then degrades sapwood, which extends to the pith in *Betula* and *Fagus*. Eventually it produces sterile conks at the sites of initial infection. A canker forms as massed hyphae at the edge of a sterile conk intermittently grow tangentially into the inner bark, killing it and the vascular cambium. The canker is thus fully occupied by the sterile conk. After each episode of canker expansion the tree reacts with compartmentalizing responses, including formation of a barrier zone at the sapwood surface. Thus the defective portion of the trunk eventually consists of nested columns of decayed wood delimited by discolored wood containing old barrier zones. Decay columns have been reported to elongate as much as 37 cm per year up or down the trunk—fast development compared with the decay caused by most other heart-rot fungi. Over many years the canker and compound column of decaying wood enlarge so much that insufficient normal sapwood remains to sustain the tree. Death ensues, but whether it is caused directly by *I. obliquus* or by opportunistic organisms is unreported.

Possible control of *I. obliquus* by treatments designed to interfere with its reproduction has been tested with no promising result. Felling diseased trees, deep girdling or poisoning them, and/or cutting the trunks into bolts did not prevent the eventual formation of basidiocarps.

Canker-rot caused by *Phellinus laevigatus*. In the Appalachian region *Betula alleghaniensis* (yellow birch) is subject to canker-rot caused by *Phellinus laevigatus* as well as to that caused by *I. obliquus*. The diseases and pathogens differ in several respects. *P. laevigatus* induces rough, irregular, sunken, bark-covered perennial cankers, often with a mat of dark brown to black sterile fungal tissue on the surface. A branch stub or the remains of one is usually visible in the canker. Projecting sterile conks are not present. The internal appearance of the decay column differs from that of decay caused by *I. obliquus*, and the two fungi differ in cultural characteristics. References: 396, 627, 1280, 1298, 3374, 3590, 4092, 4108, 4492

Canker rot of *Betula alleghaniensis* (yellow birch) caused by *Inonotus obliquus*.

A–C. Sterile conks (A) at a canker caused initially by *Neonectria galligena* and (B) emerging from old wound scars. Slight swelling of the trunk at the level of the sterile conks is typical. The edge of each canker, except for part of the old Nectria canker, is within a few millimeters of the edge of the sterile conk. A close view (C) reveals concentric patterns and adhering bark fragments that were carried outward as the sterile conk grew (NY, Sep).

D–G. Internal and exterior views of canker rot. D. Cross-section of a diseased trunk; dark lines around and within the decay column are current and former reaction zones, sites of defense processes in sapwood. E. Sterile conks at each end of a crack where woundwood enclosed an old trunk wound. F. The same trunk, split through the crack, revealing the rust-brown interior of one sterile conk and a large column of mottled rot extending up and down the trunk. Dark brown wood beneath the old wound site is occupied by organisms other than *I. obliquus*. G. Close view of the edge of the canker showing normal bark and sapwood at left, the margin of necrosis in bark (arrow), a zone of discolored sapwood being invaded from right to left by *I. obliquus*, decayed wood laced with white mycelium, and rust-brown tissue of the sterile conk. Bark is destroyed except at the canker margin. Decayed bark is recognizable only as small dark brown fragments within the fungal mass. The cut surface of healthy bark at left, normally light in color, quickly became rust colored where it was exposed to air after being cut (NY, Sep).

H, I. A basidiocarp of *I. obliquus* on a dead trunk. H. A large gray fertile surface following the contour of the trunk is exposed where the bark and outermost sapwood have separated from underlying wood adjacent to an old canker. Dark brown internal tissue of the basidiocarp is revealed where the surface was bruised in four spots near the lower end of the canker. Remnants of sterile conks remain in the canker. I. Close view of the edge of the basidiocarp, showing vertically oriented tubes (NY, Sep).

315

Hispidus Canker

Inonotus hispidus (Hymenochaetales, Hymenochaetaceae) attacks angiosperm trees around the Northern Hemisphere. In North America it occurs throughout eastern, southern, and Pacific coastal USA and in Mexico. It is one of the major trunk decay pathogens of shade trees in Europe. *Quercus* species (oaks) in the red/black group are most commonly affected in North America. They include *Q. coccinea* (scarlet oak), *Q. falcata* (southern red oak), *Q. laevis* (turkey oak), *Q. laurifolia* (laurel oak), *Q. nigra* (water oak), *Q. pagoda* (cherrybark oak), *Q. phellos* (willow oak), *Q. rubra* (northern red oak), *Q. texana* (Nuttall or Texas red oak), and *Q. velutina* (black oak). Species in the white oak group, *Q. alba* (white oak) and *Q. prinus* (chestnut oak) in the East and *Q. garryana* (Oregon oak) in the West, are occasionally affected. Other trees affected in North America represent *Acer* (maple), *Betula* (birch), *Carya* (hickory), *Citrus*, *Fagus* (beech), *Fraxinus* (ash), *Gleditsia* (honeylocust), *Juglans* (walnut), *Liquidambar* (sweetgum), *Morus* (mulberry), *Rhamnus* (buckthorn), *Schinus* (pepper-tree), and *Salix* (willow). *I. hispidus* has been found rarely on *Pinus* (unspecified) and *Abies grandis* (grand fir).

I. hispidus causes large, elongate, bark-covered lesions on which it fruits. Wounds such as pruning cuts and small branch stubs (less than 2.5 cm in diameter) within 5 m of the ground are common sites of infection. The fungus first rots heartwood and then begins to kill sapwood and the vascular cambium, enlarging lesions at the margins. The perennial cankers usually increase to 1–1.5 m in length, and the bark remains firmly attached. Cankers up to 4 m long and 30 years old have been reported. Large woundwood ridges develop along the sides of a canker but are eventually invaded and killed. More woundwood forms at the new canker edge, resulting eventually in a spindle-shaped swelling of the trunk. Cankers lengthen approximately 15 cm per year in the South, and heart rot extends, on average, 35 cm above and below a canker. Trunks of diseased trees often break.

The decay is a white rot that also has soft-rot characteristics because the fungus causes cavities within the secondary cell walls of fiber tracheids. The wood becomes soft, spongy, and straw yellow to pale yellow-brown. Often a narrow, conspicuous black or dark purple-brown zone marks the abrupt transition between decayed and sound wood. Decay columns resemble those caused by *Phellinus spiculosus* (Plate 155).

Basidiocarps of *I. hispidus* form annually in summer to early autumn, usually singly but occasionally in groups of two or three. They grow to full size within 1–2 weeks and are more or less shelflike and irregularly semicircular in outline as viewed from below. They are spongy, somewhat hairy, and yellowish brown to rust-red on the upper surface, and fawn (at first) to rust colored on the lower, poroid surface. They measure 5–30 × 8–25 × 2–10 cm. Growing basidiocarps exude drops of liquid on the lower surface. They disperse spores for about 3 weeks, then slowly shrink and dry to a dark brown or black mass and eventually fall from the tree. Shriveled basidiocarps on cankers or on the ground beneath cankers are diagnostic aids. The fungus may continue to fruit for up to 5 years after diseased trees are felled.
References: 896, 1153, 1276, 1280, 1859, 2544, 2546, 2547, 2549, 2937, 3374, 3508, 3509–3512, 3688, 4058, 4108, 4517

Heart Rot and Canker-rot Caused by *Inonotus andersonii*

Inonotus andersonii, one of the major decay pathogens of *Quercus*, is found across the USA from Pennsylvania and Oregon southward, in scattered European locations, and in Asia. It is an important cause of mortality of oaks in southwestern USA. Hosts include *Q. agrifolia* (California live oak), *Q. alba* (white oak), *Q. arizonica* (Arizona white oak), *Q. coccinea* (scarlet oak), *Q. douglasii* (blue oak), *Q. emoryi* (Emory oak), *Q. gambelii* (Gambel oak), *Q. garryana* (Oregon oak), *Q. hypoleucoides* (silverleaf oak), *Q. lobata* (valley oak), *Q. marilandica* (blackjack oak), *Q. nigra* (water oak), *Q. oblongifolia* (Mexican blue oak), *Q. rubra* (northern red oak), *Q. texana* (Texas red oak), *Q. velutina* (black oak), *Lithocarpus densiflorus* (tanoak), and rarely *Carya tomentosa* (mockernut hickory), *Populus* sp. (poplar), and *Salix nigra* (black willow).

Branch stubs are the most common sites of infection, but the fungus also enters through fire scars and other injuries. When acting as a canker-rot pathogen *I. andersonii* first causes a white rot of the heartwood and then moves outward, killing the sapwood and cambium. It evades effective compartmentalization by virtue of its ability to kill these tissues. Trunks and limbs often break at cankers. In the Southwest, broken limbs with advanced decay or cavities provide habitat for cavity-nesting birds. The fungus produces annual basidiocarps in the form of thick sheets of rusty orange to dull dark brown poroid tissue that follow the contour of colonized stems and have no upper surface (resupinate form). They may exceed 50 cm in length and often have very rough surfaces with peglike outgrowths. They form beneath the bark or outermost wood on killed parts and cause the overlying host tissues to fall away.
References: 344, 628, 1092, 1276, 1280, 2546, 2937, 4108, 4517

Canker-rot Caused by *Inonotus glomeratus*

I. glomeratus is possibly restricted to North America, although there are two reports from offshore locations. It occurs across Canada and northern USA and extends southward to Florida in the East. The fungus causes white to light brown spongy heart rot of *Acer negundo* (box-elder), *A. rubrum* (red maple), *A. saccharum* (sugar maple), *Fagus grandifolia* (American beech), and *F. sylvatica* (European beech), and occasionally to rarely colonizes *Betula papyrifera* (paper birch), *Castanea* (chestnut), *Populus balsamifera* (balsam poplar), *Quercus laurifolia* (laurel oak), *Tsuga* sp. (hemlock), and *Viburnum* sp. In the West it occurs on *B. occidentalis* (water birch), *P. balsamifera* subsp. *trichocarpa* (black cottonwood), and *Q. garryana* (Oregon oak). *I. glomeratus* is thought to be the most important trunk and butt decay pathogen of *A. saccharum* in Ontario, where it accounted for 36% of the infections and 40% of decay volume in one study area. It causes cankers in the manner described and illustrated for canker-rot of birch (Plate 156). Most infections begin at dead branches and stubs, although other wounds are also suitable. After decay is advanced this fungus forms sterile punky masses (sterile conks) that replace branch traces. They extend from a central column of white rot to the trunk surface or, in *Fagus*, as much as 7–8 cm beyond the trunk. The sterile conks are diagnostic. Their surfaces turn black, crusty, and cracked with age. Cankers caused by *I. glomeratus* on *Fagus* tend to be inconspicuous, but woundwood ridges sometimes form at the margins. Cankers on *Acer* are usually irregular in shape with raised margins and inconspicuous sterile conks. Most of these cankers are localized, but some become elongate. *I. glomeratus* produces yellow-brown sheetlike basidiocarps on dead trees or tree parts. These appear 3–4 years or more after tree death, usually on fallen trunks, and sometimes have shelflike protrusions.
References: 628, 629, 1280, 1314, 2868, 2937, 3590, 4108

A–E. Hispidus canker on *Quercus* species (oaks). A. A large canker on *Q. nigra* (water oak) bears the dark remains of basidiocarps of *Inonotus hispidus* that formed the previous year (GA, May). B. Large woundwood ridges flank a canker on *Q. phellos* (willow oak). A rust-brown basidiocarp remains from the previous autumn (MS, Apr). C. Large basidiocarps on a young canker not yet delimited by woundwood ridges on *Q. phellos* (MS, Aug). D. A canker on *Q. phellos* nearly enclosed by woundwood, appearing as a fissure (MS, Apr). E. A blackened dead basidiocarp at the base of a diseased *Q. laurifolia* (laurel oak) (FL, Apr).

F, G. Canker-rot of *Quercus douglasii* (blue oak) caused by *Inonotus andersonii*. F. A partly killed tree with the dead top removed, showing white heart rot and a large brown basidiocarp that has caused the bark to fall away. G. Close view of part of the basidiocarp (AZ, Apr).

H, I. *Fagus grandifolia* (American beech) with trunk rot caused by *Inonotus glomeratus*. Black sterile conks protruding at the locations of former branch stubs are diagnostic. H. An elongate wound probably made by a falling tree, now evident only as a seam beside a large woundwood ridge on the trunk, was possibly the site of infection. I. Close view of a sterile conk (NY, Aug).
Photo credit: C—J. A. Matteoni

317

Sapwood Rot and Cankers Caused by *Cerrena unicolor*

C. unicolor (Polyporales, Polyporaceae), an opportunistic colonist of sapwood of many trees and shrubs, causes canker-rot in trees weakened by wounds or environmental stress. It occurs around the north temperate zone and in South Africa.

Hosts of this fungus are mainly angiosperms; it occurs rarely on gymnosperms. It fruits on cankers and dead branches and trunks. Host genera recorded in North America include the following:

Abies (fir), *Acer* (maple), *Aesculus* (horse-chestnut), *Ailanthus* (tree-of-heaven), *Alnus* (alder), *Amelanchier* (serviceberry), *Betula* (birch), *Carpinus* (hornbeam), *Carya* (hickory), *Castanea* (chestnut), *Celtis* (hackberry), *Cornus* (dogwood), *Crataegus* (hawthorn), *Fagus* (beech), *Fraxinus* (ash), *Ilex* (holly), *Juniperus* (juniper), *Liquidambar* (sweetgum), *Liriodendron* (tuliptree), *Magnolia*, *Malus* (apple), *Nyssa* (tupelo), *Ostrya* (hop-hornbeam), *Populus* (poplar), *Prunus* (cherry), *Pyrus* (pear), *Quercus* (oak), *Rhamnus* (buckthorn), *Robinia* (black locust), *Salix* (willow), *Tilia* (linden), *Tsuga* (hemlock), and *Ulmus* (elm).

Symptoms and signs. Discrete or diffuse cankers and dieback, usually associated with wounds, are the external symptoms and are accompanied by a crumbly white rot of sapwood. In some hosts (e.g., *Acer*) the decaying wood has distinctive color patterns—yellowish white advanced decay alternating with dark reddish brown, less decayed bands where living tissues once resisted invasion. *C. unicolor* is a common secondary factor in the decline of *Acer saccharum* (sugar maple) (Plate 228), in which it sometimes kills a major limb or the tree top. The predisposing injury or stress is often unknown. Occasionally the fungus causes discrete cankers on healthy-appearing *A. saccharum*. Advanced decay may be present 1–1.5 m above a canker. *A. rubrum* (red maple) overtopped by neighbor trees in the forest is also commonly attacked; sprout clumps seem particularly vulnerable. The fungus spreads from one sprout stem to another at their common base, eventually killing all. On fruit trees *C. unicolor* often produces basidiocarps on perennial cankers centered on wounds.

The basidiocarps are small, bracketlike, 0.5–6 × 2–8 × 0.15–0.5 cm, and usually in clusters on bark dead for less than a year. They live for only one season. The upper surfaces are typically hairy, nearly white to greenish or greenish gray, often with concentric ridges. The pores on the lower surface are irregular in outline, and in old specimens the walls of the tubes break up into teeth. When weathered to nearly white, basidiocarps of *C. unicolor* can be confused with those of *Irpex lacteus*, which has similar hosts and habits.

Disease cycle. C. unicolor is a wound parasite and a proven pathogen of *Acer* and *Betula*. Most infections are presumed to be started by airborne basidiospores in broken branches, cracks, or pruning cuts. Insect transmission is also possible because *C. unicolor* is carried by certain insects, notably *Tremex columba* (pigeon tremex), a wood wasp that makes oviposition wounds into sapwood. Once established in sapwood *C. unicolor* kills the overlying cambium and bark. Although capable of infecting healthy trees, it causes the most damage in trees stressed by fire damage, sunscald, dense shade, or top breakage by ice. It readily overcomes defense responses in such trees and evades wound-associated compartmentalization by killing bark and the vascular cambium at the edge of a lesion. From this position it invades sapwood lying outside the wound-associated or lesion-associated barrier zone. Repetition of this process results in an enlarging perennial canker on which it soon fruits. The fungus sometimes kills a large area of cambium in a single year, girdling a stem or causing a large diffuse canker.

References: 17, 401, 625, 1060, 1280, 2937, 3374, 3771, 3868

Sapwood Rot Caused by *Schizophyllum commune*

S. commune (Agaricales, Schizophyllaceae) is a globally distributed saprobe and pathogen on a great many woody and herbaceous plants—monocots, angiosperms, and gymnosperms. The fungus is apparently a generalized opportunistic pathogen inasmuch as some strains parasitize other fungi and some infect dogs and humans. In woody plants *S. commune* causes a white rot of sapwood and kills cambium and bark that overlie sapwood lesions. It aggressively colonizes trees stressed by heat, drought, sunscald

lesions, fire scars, or major wounds. An abridged list of genera with woody hosts follows.

Abies, Acacia, Acer, Aesculus, Ailanthus, Albizia, Alnus, Amelanchier, Arecaceae, Baccharis, Betula, Carpinus, Carya, Catalpa, Ceanothus, Cedrela, Cercis, Chamaecyparis, Cinnamomum, Citrus, Corylus, Diospyros, Eucalyptus, Fagus, Ficus, Fouquieria, Fraxinus, Gleditsia, Ilex, Jacaranda, Juglans, Juniperus, Larix, Liquidambar, Liriodendron, Maclura, Magnolia, Malus, Melia, Morus, Nerium, Nyssa, Olea, Paulownia, Persea, Phyllostachys, Picea, Pinus, Pistacia, Platanus, Poncirus, Populus, Prosopis, Prunus, Pseudotsuga, Pyrus, Quercus, Rhamnus, Rhus, Robinia, Sabal, Salix, Sambucus, Sassafras, Schinus, Sequoia, Sorbus, Tamarix, Taxodium, Thuja, Tilia, Tsuga, Ulmus, Umbellularia, Vaccinium, and *Vitis.*

S. commune is a primary colonist of stems with wounds or other injuries, apparently entering mainstems through branches and stubs. It is among the fungi that compound the damage caused by ice storms. Common sites of infection include bark cracks, broken branches, lesions caused by fire, sunscald lesions, and frost cankers ("winter sunscald" lesions). *S. commune* is a proven pathogen of *Malus pumila* (apple), *Populus alba* (white poplar), and *Prunus* species (nectarine, peach), as well as various tropical species. It has caused noteworthy damage to *Acer saccharum* (sugar maple), *Carya* (hickory), *Fagus sylvatica* (European beech), *M. pumila*, *Populus alba*, *Prunus*, and *Tilia cordata* (littleleaf linden). *S. commune* also infects tree fruits—apples, acorns, and *Pseudotsuga* (Douglas-fir) cones, for example—if they are not stored properly.

Basidiocarps of *S. commune* begin to form only several weeks after it colonizes a substrate. Their presence on a lesion is not a reliable indication of pathogenic activity, however, because the fungus may colonize tissues killed by other organisms. The hairy white annual basidiocarps, 0.5–5 cm wide, usually develop in clusters, are distinctly convex, broadly attached or sometimes with a short lateral stalk, and have bifurcate gills on the lower surface. White to pale brown when young, they turn grayish or darker with age.

S. commune grows at temperatures from near freezing to above 40°C and grows most rapidly at 30–35°C, which helps explain its aggressiveness in heat-stressed trees. The optimum temperature for spore discharge is lower than that for growth. This fungus can also grow in drier substrates than can most other wood-decaying basidiomycetes. It has been much studied as a representative of higher fungi because it fruits readily in culture and is amenable to genetic analysis. Considerable understanding of the genetics and development of higher basidiomycetes is based on knowledge of *S. commune*.

References: 17, 325, 787, 910, 1082, 1909, 1910, 2131, 2135, 2787, 2868, 3187, 3771, 4298, 4371

A–G. Canker-rots caused by *Cerrena unicolor*. A. Basidiocarps on a canker that has nearly girdled a scaffold limb of *Prunus cerasus* (sour cherry) (OR, Jul). B. Basidiocarps on an elongate canker that originated at a branch stub on *Acer saccharum* (sugar maple) (MI, Aug). C, D. Basidiocarps on a large basal canker on a living *A. rubrum* (red maple) (ON, Jul). E. Close views of basidiocarps. Hairy, greenish, concentrically ridged upper surfaces are typical of *C. unicolor* in moist, shady situations. The lower surface has irregularly shaped pores. F, G. Dissection of *A. saccharum* with canker-rot. In longitudinal view (F), decaying wood is bounded by a dark reaction zone. At the canker margin (arrow) the fungus has killed bark overlying sound wood, which will soon be invaded and decayed. In cross section (G) a recently killed stem has concentric reaction zones that formed as the pathogen advanced into younger sapwood. Part of a canker is visible at left (NY, Sep).

H–L. Canker-rots caused by *Schizophyllum commune*. H. Dieback of *Tilia cordata* (littleleaf linden) associated with sapwood colonization by *S. commune* following freeze damage to the trunk (NY, May). I. Basidiocarps in a canker on *Malus pumila* (apple) (ND, Jul). J. A compartmentalized canker inhabited by *S. commune* and *Diplodia* sp. on *Gleditsia triacanthos* (honeylocust) (NY, Mar). K, L. Clusters of hairy white basidiocarps in cankers on *T. cordata* and *M. pumila*, respectively. Viewed from below (L), the basidiocarps have bifurcate gills (NY, May; ND, Jul).

Sapwood Decay and Silverleaf Caused by *Chondrostereum purpureum* (Plate 159)

Chondrostereum purpureum (Polyporales, Meruliaceae) is a wound-invading white-rot fungus that kills and decays sapwood of a wide array of plants. It is globally distributed in the temperate zones and is usually found on angiosperms, occasionally on gymnosperms. During colonization of sapwood in living plants it releases toxins that are carried with the transpiration stream to leaves and there cause silverleaf symptoms in some hosts, notably *Malus* (apple) and *Prunus* (stone-fruit trees). The term *silverleaf* refers to a change in light reflection from leaves due to separation of the epidermis from the palisade. Affected leaves appear somewhat bleached and silvery to leaden as viewed from above.

Recorded hosts of *C. purpureum* include about 190 species, mostly woody, in more than 55 genera. North American records represent the genera listed below.

Abies (fir), *Acer* (maple), *Aesculus* (horse-chestnut), *Alnus* (alder), *Amelanchier* (serviceberry), *Arbutus* (madrone), *Betula* (birch), *Carpinus* (hornbeam), *Carya* (hickory), *Castanea* (chestnut), *Celtis* (sugarberry), *Cornus* (dogwood), *Crataegus* (hawthorn), *Cydonia* (quince), *Fagus* (beech), *Liriodendron* (tuliptree), *Malus* (apple), *Nyssa* (tupelo), *Ostrya* (hop-hornbeam), *Platanus* (planetree, sycamore), *Populus* (aspen, poplar), *Prunus* (apricot, cherry, peach, plum), *Pseudotsuga* (Douglas-fir), *Pyrus* (pear), *Quercus* (oak), *Salix* (willow), *Sorbus* (mountain-ash), *Syringa* (lilac), *Thuja* (western red cedar), *Tsuga* (hemlock), *Ulmus* (elm), and *Vitis* (grapevine). Records from elsewhere indicate susceptibility of many additional plants that grow in North America.

C. purpureum is so aggressive in the sapwood of highly susceptible hosts that it has been formulated as a bioherbicide for suppressing sprout growth from stumps, notably of *Alnus*, *Populus*, and *Prunus serotina* (black cherry), the last-named being a weed species in Europe. *C. purpureum* is associated with decline of *Betula papyrifera* (paper birch) and *B. pendula* (European white birch) in some areas. Its ability to cause decline in initially healthy *B. papyrifera* is untested, however. When inoculated into trunks of *B. alleghaniensis* (yellow birch) it caused cankers and wood decay but failed to kill trees or induce crown decline within 6 years. The fungus is not considered a significant threat to healthy trees in forests, but its possible role as an opportunistic colonist of wounded and stressed *Betula* deserves study.

Symptoms and signs. Retarded growth, undersized leaves, dieback, diffuse cankers, and tree death are the main symptoms in highly susceptible trees. Silvering of leaves occurs in only some of the woody plants (e.g., *Malus*, *Populus*, *Prunus*, *Sorbus*) subject to lethal attack. Other highly susceptible plants (e.g., *Betula*, *Crataegus*, *Fagus*, *Laburnum*) may wilt and die back without silvering. Leaves affected by the silverleaf toxin turn yellow and then brown at midribs and margins, whether or not they first show the silvery sheen. Wilting accompanies the color changes. Anatomical and physiological aspects of silverleaf include abnormally loose arrangement of palisade cells, air space between palisade and epidermis, starch accumulation, degeneration of chloroplasts, and diminished transpiration due to permanent stomatal closure. Symptoms appear on one or a few branches, then spread to other branches. Small highly susceptible trees are killed by the disease. Some infections apparently become compartmentalized, which halts the progression of dieback. Infected sapwood develops a reddish brown stain that, in *Prunus*, is associated with gum occlusion of xylem vessels. The color bleaches as decay develops. Late-stage decay by *C. purpureum* is dry, pale yellow to nearly white, and often mottled.

Basidiocarps of *C. purpureum* develop mainly in autumn (and winter if temperature permits) on recently dead materials—trees, logs, and branches. They usually form in groups and are annual, but may overwinter in areas of mild climate. They are 2–2.5 mm thick and either remain flat against the substrate (resupinate) or extend outward as small brackets mostly 2–6 cm long and 2–4 cm wide. The upper surface is at first hairy, later smooth and wrinkled, purplish to buff or dark yellowish buff with a light-colored margin. The outer or lower surface is smooth to slightly wrinkled, bright purple and covered with basidia when fresh, and brown-violet when old. The basidiospores are smooth, colorless, and elliptic with a point at one end, and measure 6.5–8 × 2.5–3.5 μm. Basidiocarps readily withstand alternate drying and wetting; they release basidiospores intermittently for several months in response to rain.

Disease cycle. Most infections are started by airborne basidiospores in fresh wounds to sapwood. Wounds a month or more old are not suitable. Some spores germinate on the wound surface, but others move passively several millimeters into xylem vessels and then germinate. Infection can occur in any season if spores are available. Considerable seasonal variation in susceptibility occurs, but the patterns seem to vary among species and regions. *C. purpureum* kills xylem parenchyma and utilizes the carbohydrates and nitrogenous compounds therein. It grows rapidly in highly susceptible plants; 4–11 cm per month, for example, in sapwood of *Prunus domestica* (plum). Eventually a sapwood lesion extends to the roots and out to the cambium, resulting in dieback. Infected wood becomes darkly stained, and in gum-producing plants such as *Prunus*, infused with gum. Silvering of foliage begins within a year after infection of fruit trees. The silverleaf toxins are a set of three highly similar endopolygalacturonases that degrade pectin in and between cell walls. The same enzymes are presumed to participate in killing sapwood. Endopolygalacturonases from other sources do not cause silverleaf. Proteinase activity has also been demonstrated in lesions induced by *C. purpureum*.

The fungus fruits in the first and second years after death of its substrate, so the minimum length of a disease cycle in highly susceptible plants is probably 2 years. Infection of fruit trees can begin in nurseries or orchards, and *C. purpureum* can thus be spread with nursery stock. Tree-to-tree spread via grafted roots can occur in trees such as *Prunus mahaleb* (cherry) that naturally make root grafts.

C. purpureum displays considerable genetic variability within and between geographic regions, but strains from various regions and hosts are cross-fertile and display no particular host specialization. Differences in virulence are not associated with particular hosts or regions.

Epidemiology of silverleaf has been studied for keys to disease control. The optimum temperature range for spore germination and growth of the fungus is 24–26°C, and it is severely inhibited or killed at 35°C. Spore discharge occurs at 4–21°C beginning a few hours after rain and continues while the relative humidity remains above 90%. The same basidiocarps produce repeated crops of spores during mild, wet weather for several months to more than a year if not killed by freezing. *C. purpureum* is not an aggressive competitor in dead wood; secondary decay fungi eventually replace it.

Some cultivars of *Malus* and *Prunus* have considerable resistance to *C. purpureum*. Resistance is governed by multiple genes. Fruit trees can be protected against silverleaf to some extent by treating fresh pruning wounds with fungicidal preparations or with biological formulations containing antagonistic and competitive molds, notably *Trichoderma* species.

References: 53, 382, 527–531, 1011, 1023, 1024, 1298, 1325, 1545, 1586, 1969, 1970, 2293, 2569, 2687, 2896, 3088, 3131, 3216, 3541, 3782, 3785, 3786, 3790, 3792, 4108, 4192, 4194

A, B. Silverleaf symptoms in apple, *Malus pumila* 'McIntosh.' A. Silvered foliage related to sapwood parasitism and decay by *Chondrostereum purpureum* on one limb of an otherwise green tree. B. Silvered leaves, smaller than a normal leaf (NY, Aug).

C–E. Symptoms and signs of infection by *C. purpureum* in *Betula papyrifera* (paper birch). C. Dead and dying branches on a shade tree (ON, Aug). D. Mycelium and small basidiocarps revealed by pulling outer bark away from the edge of a trunk wound on a declining tree (ON, Oct). E. Basidiocarps that formed on a log cut earlier in the same season from a tree with severe dieback (ON, Sep).

F–H. *C. purpureum* in *Populus*. F. Young basidiocarps associated with sapwood discoloration in a stump of a shade tree that was cut 2 weeks earlier because of limb dieback (England, autumn). G, H. Basidiocarps on a turned-over log of *P. tremuloides* (trembling aspen) (ON, summer).

Photo credits: C, D, G, H—E. C. Setliff; E—J. McLaughlin and E. C. Setliff; F—E. C. Setliff and A. D. Rayner

321

Trunk Rots of Conifers (Plates 160, 161)

Heart Rots Caused by *Fomitopsis pinicola* and *Echinodontium tinctorium*

Fomitopsis pinicola (Polyporales, Fomitopsidaceae), also known as the red belt fungus, is among the minority of significant heart-rot pathogens that commonly colonize both angiosperms and gymnosperms. It occurs throughout the coniferous forest regions of the Northern Hemisphere, although it is rare or absent in the southern pine region of the USA. It is one of the most important brown-rot pathogens of old-growth western conifers, but it acts slowly and perhaps for this reason is not among the major decay pathogens in second-growth forests. It usually fruits on dead trees and logging slash, and it causes decay in many living mature trees without producing basidiocarps. Wood colonized by this fungus turns dark red-brown. The decay is a brown cubical rot with shrinkage cracks in which white sheets of mycelium develop. In North America gymnosperm host groups include *Abies* (fir), *Calocedrus* (incense-cedar), *Larix* (larch), *Picea* (spruce), *Pinus* (pine), *Pseudotsuga* (Douglas-fir), *Sequoia*, *Thuja* (arborvitae and western red cedar), *Tsuga* (hemlock), and uncommonly *Chamaecyparis* (false-cypress). The most common angiosperm hosts are *Acer* (maple) and *Betula* (birch). Others include *Alnus* (alder), *Carya* (hickory), *Castanea* (chestnut), *Fagus* (beech), *Magnolia*, *Malus* (apple), *Populus* (aspen, poplar), *Prunus* (cherry, plum), *Quercus* (oak), and *Salix* (willow). Isolates from angiosperms are able to cause decay in living gymnosperms, and vice versa.

The perennial basidiocarps of *F. pinicola* are hard-corky to woody, shelflike to rounded on the upper surface, or sometimes somewhat hoof shaped and encrusted, with a red-brown band near the white to cream-colored edge. They vary in size up to 30 × 40 × 20 cm. The oldest parts of the upper surface turn gray to dark brown or black. The white edge may be thick and prominent during growth. The poroid undersurface is white to cream-yellow.

F. pinicola can be found in wounded tree trunks, and it fruits at wounds, but whether these are required for entry is unclear. Decay by this fungus in *Abies* species, *Picea sitchensis* (Sitka spruce), and *Tsuga heterophylla* (western hemlock) is most commonly associated with scars such as those made by falling trees. Branch stubs are also common sites of decay. In *A. concolor* (white fir), *A. magnifica* (red fir), and *T. heterophylla*, decay by *F. pinicola* is often associated with broken tops and with cankers that develop at sites of dwarf mistletoe infection. *F. pinicola* has been isolated from pith of suppressed *A. grandis* (grand fir). Knowledge of the infection biology of *Echinodontium tinctorium*, outlined below, raises the possibility that *F. pinicola* enters young trees through dead twigs or other small natural avenues and remains quiescent until a later wound triggers growth and wood decay. This fungus is probably also introduced to dying or dead trees and to wounded healthy trees by insects. Spores of *F. pinicola* have been detected on various beetles that visit the basidiocarps.

The affinity of *F. pinicola* for dying or dead trees was confirmed in a study of decay in old-growth *Pseudotsuga menziesii* in coastal British Columbia. The fungus accounted for only 5% of the decay occurrences and 3% of the decay volume found in living trees, but was associated with 47% of the decay occurrences and 69% of the decay volume found in dead trees.

F. pinicola and other brown-rot fungi are important litter decomposers in coniferous forest ecosystems. Their decay of dead trees and logging slash leaves a stable lignin-rich residue that is a major component of the organic matter on the forest floor and in the upper layers of soil. This residue enhances the water-holding and cation exchange capacities of soil and is a favorable habitat for the development of ectomycorrhizae and for nitrogen-fixing bacteria. Brown rot caused by *F. pinicola* is difficult to distinguish from that caused by *Laetiporus* species (Plate 154), which may occur in trees of the same species.

References: 559, 1083, 1161, 1278, 1280, 1514, 1690, 1710, 1987, 2068, 2069, 2241, 2742, 2937, 2972, 3056, 3374, 3508–3510, 4108

Echinodontium tinctorium (Russulales, Echinodontiaceae), the Indian paint fungus, is an important heartwood decay pathogen of conifers in northwestern USA and western Canada. It occurs from Alaska to California and New Mexico, primarily in inland forests. The common name of the fungus refers to historic use of the bright red-orange internal tissue of basidiocarps by Native Americans as a source of pigment. *E. tinctorium* causes a yellow to golden brown or rust-red rot that even in early stages weakens the wood in a manner that causes separation along the annual rings (laminated rot). Wood in the early stage of decay is water-soaked and discolored brown. Later it becomes stringy and often contains small white pockets with black centers. Narrow brown to red zone lines are often present. Typically, the entire center of the trunk decays to a height of several meters and may eventually become hollow. Decay extends 3–5 m beyond basidiocarps and may enter main roots and large limbs. The decay is of the delignifying type and is classified as a white rot. It is most common in *Abies concolor* (white fir), *A. grandis* (grand fir), and *Tsuga heterophylla* (western hemlock), and somewhat less common in *A. amabilis* (Pacific silver fir), *A. lasiocarpa* (subalpine fir), *A. magnifica* (red fir), *A. procera* (noble fir), and *Tsuga mertensiana* (mountain hemlock). Reports of this fungus on *Larix occidentalis* (western larch), *Picea engelmannii* (Engelmann spruce), *P. glauca* (white spruce), *Pseudotsuga menziesii* (Douglas-fir), and *Thuja plicata* (western red cedar) are uncommon to rare.

E. tinctorium produces woody, hoof-shaped perennial basidiocarps 4–20 cm in diameter on diseased trunks, usually at junctions with the undersides of branch stubs. Basidiocarps sometimes form at wounds or on limbs and in the latter circumstance are bell shaped. The upper surface is rough and cracked, nearly black. The lower surface is gray-brown to nearly black and is composed of thick blunt spines. The interior tissue is brick-red to rust-red. This pigment also occurs in adjacent decayed wood. Basidiospores of *E. tinctorium* are dispersed during cool, wet weather in autumn but germinate well only after a period of freezing temperature. They remain viable during winter. Infection is believed to occur in spring.

E. tinctorium apparently has two routes of entry into conifer trunks. It enters via small wounds and—in *Tsuga*, at least—via twigs along the trunk. It has been found in wood adjacent to wounds of various ages in *Abies* and *Tsuga*. If an infected wound closes as the result of tree growth, the fungus may become quiescent for an indefinite period. It has often been isolated from apparently normal wood. Eventually it resumes colonization and causes heart rot. Entry into *T. heterophylla* trunks occurs via the stubs of tiny shade-killed twigs only about 1 mm in diameter after trees attain age 40 or more. After 2–3 years these stubs are enclosed by the growing trunk. The fungus then becomes quiescent and may remain so for up to 50 years or sometimes longer, but eventually begins to spread. Events that trigger renewed colonization are unknown, but trunk wounds are suspected to be involved.

References: 25, 26, 1083, 1084, 1389, 1542, 1827, 2443, 2972, 4008, 4108, 4373, 4374

A, B. *Fomitopsis pinicola* and associated wood decay. A. Basidiocarps on a wind-broken *Picea rubens* (red spruce) trunk. The tree sustained extensive trunk decay that led to breakage while it was alive. The transverse fracture without splintering is typical of wood undergoing brown rot (NY, Oct). B. Basidiocarps at the base of a decaying *Tsuga mertensiana* (mountain hemlock) trunk (OR, Sep).

C–H. *Echinodontium tinctorium* and associated wood decay. C. An intact hoof-shaped basidiocarp several years old on *Tsuga heterophylla* (western hemlock). D. The same basidiocarp viewed in section, showing red-orange interior tissue and toothlike projections from the lower surface. E. Close view of the lower surface. F. Stringy, golden brown decayed wood occupies nearly the entire cross section of the trunk of a pole-size tree. Wood at the periphery in the early stage of decay is water-soaked and discolored brown. G. Part of the diseased trunk in longitudinal section, showing at top the delamination (separation of annual rings) of decayed wood. H. Close view of wood in an advanced stage of decay near a basidiocarp, showing typical rust red color and white pockets with black centers (ID, Jul).

323

Heart Rot Caused by *Phellinus pini* and Canker Rot Caused by *P. cancriformans*

Phellinus pini (syn. *Porodaedalea pini, Phellinus gilbertsonii, P. vorax, P. yamanoi, Cryptoderma pini*; Hymenochaetales, Hymenochaetaceae) is the name applied to a complex of important trunk decay pathogens that attack conifers around the Northern Hemisphere. Most conifers that grow in temperate regions are affected by them. In North America *Larix* (larch), *Picea* (spruce), *Pinus* (pine), and *Pseudotsuga* (Douglas-fir) sustain the greatest damage. *Abies* (fir), *Thuja plicata* (western red cedar), and *Tsuga* (hemlock) are commonly affected as well. Occasional infections develop in *Calocedrus* (incense-cedar), *Chamaecyparis* (false-cypress), *Juniperus* (juniper), *Taxus* (yew), and *Thuja occidentalis* (arborvitae). Several intersterility groups (reproductively isolated populations) have been detected in the *Phellinus pini* complex, but most of them have not been differentiated sufficiently to warrant adoption of separate names here. Readers are forewarned, however, that when species of the *P. pini* complex are sorted out, *P. pini* in a strict sense may be a European fungus not represented in North America.

Fungi of the *P. pini* complex cause red heart, red rot, ring scale, red ring rot, and white pocket rot. These names indicate decay features that vary with host and site of infection. Damage is much greater in old trees in virgin and unmanaged forests than it is in young managed forests. Decay typically begins well above ground level, but butt rot is common, and the rot sometimes extends into major roots. Decay columns commonly extend 10 m or more, sometimes 20 m, within mature trees. The decay is usually confined to heartwood, either in one central column or in several columns that extend from branch stubs. It often begins near the junction of heartwood and sapwood and may extend into sapwood adjacent to wounds and along the trunks of trees nearing death. Young trees become infected in special circumstances. For example, *P. pini* infects young *Pinus banksiana* (jack pine) trunks at cankers caused by sweetfern rust (Plate 148).

Wood in early stages of decay appears pinkish to reddish in *Pinus* and purplish in *Picea* and *Pseudotsuga*. Small elongate pockets lined with white fibers appear later and slowly expand and merge. Wood in advanced decay is soft, usually light colored, and fibrous, often with irregular narrow black lines as seen on cut or broken surfaces. Sometimes the decay spreads tangentially, causing a crescent or ring of defect as seen in cross section. It advances slowly, only 5–10 cm per year up or down in *P. strobus* and other white pines. The specimen of decayed wood shown in Plate 161G, while exhibiting pockets and the ultimate fibrous texture of decayed wood, lacks the typical white lining of pockets that might have been seen at an earlier stage of decay. The latter feature, as caused by *Onnia circinata*, is shown in Plate 167.

External indicators of decay include swollen knots where branches were shed many years earlier; irregular bulges with exuding resin; resin flow from knots; and brown basidiocarps at branch bases, branch stubs, knots, wounds, cracks, and sometimes on seemingly normal bark. Punk knots are common in *Larix occidentalis* (western larch), some *Picea* species, *Pinus*, and *Pseudotsuga*. A punk knot is a mass of tightly packed sterile brown hyphae that extends from a decayed branch stub within the trunk to a local swelling on the surface and there appears like a blackened knot. Swelling results from overgrowth of wood around a punk knot. Dark yellowish brown to reddish brown fungal tissue is exposed where a *P. pini* punk knot is cut. Small cankers, up to 9 × 24 cm, apparently caused by *P. pini* are occasionally noted at branch stubs on *Abies balsamea* (balsam fir) in the Great Lakes region. This fungus does not usually act as a canker-rot pathogen in other conifers, however, although it does apparently kill sapwood and cambium and produce basidiocarps on large areas of the bark of moribund trees with extensive heart rot.

Basidiocarps of *P. pini* are annual or perennial and vary from yellowish brown encrustations 2–5 cm in diameter to large bracket- or hoof-shaped forms, or sometimes conchlike specimens with wavy edges, up to 25 cm or more in diameter and 8 cm thick. Bracket forms often occur in clusters. The upper surface is at first reddish brown but becomes gray-brown to brownish black, roughened by concentric ridges and sometimes also by radial cracks. The margin and nearby upper surface are slightly to distinctly velvety. The margin, poroid lower surface, and interior tissue are yellowish brown to reddish brown. The pores are round to radially elongate and average 2–3 per millimeter. Many diseased trees lack basidiocarps until they are near death, at which time numerous specimens may appear. Few or no new basidiocarps form on dead trees. Peak dispersal of basidiospores occurs in spring and autumn in eastern Canada and northeastern USA.

P. pini apparently colonizes wounds such as broken limbs or tops or felling scars in some hosts—for example, *Picea mariana* (black spruce) and *Pinus banksiana*—as well as rust cankers in the latter. The association of basidiocarps with branch stubs, however, is thought to represent outward spread of decay rather than evidence of infection at those locations. Observers have speculated that the infection biology of *Phellinus pini* may be similar to that of *Echinodontium tinctorium*. Decay in *Pinus strobus* often develops at points in heartwood where leaders were killed many years before by *Pissodes strobi* (white pine weevil) or where small branches near the trunk base died. Whether the decay fungus colonizes killed leaders soon after their death or only after several years (as is presumed to occur in lateral branch stubs) is unknown. No evidence for early colonization of weevil-killed leaders was found in *Pinus* or *Picea* in western Canada.

Decay by *Phellinus pini* in old-growth conifers provides habitat for cavity-nesting animals, but forest harvests have greatly diminished the supply of suitably decayed trees. Inoculation of conifers with *P. pini* is practiced in some places as a means of providing potential nest trees for woodpeckers and other cavity-using birds.

References: 251, 393, 395, 461, 1142, 1144, 1280, 1382, 1423, 1424, 1542, 1659, 1819, 1826, 1827, 2240, 2972, 3016, 3374, 3425

Phellinus cancriformans (syn. *P. chrysoloma, P. pini* var. *cancriformans*) causes canker rot in *Abies* (fir) in western North America. This fungus may have a wider distribution and more host genera than that, because it has been confused with other taxa in the *P. pini* complex. *P. cancriformans* causes large, flat, bark-covered cankers on *A. amabilis* (Pacific silver fir), *A. concolor* (white fir), *A. grandis* (grand fir), *A. procera* (noble fir), and *A. shastensis* (Shasta fir). The cankers form as the result of successive advances of the fungus in inner bark, leading to cambial death. This process repeatedly opens new wood near canker margins to invasion and decay. The cankers often attain dimensions of 40 × 60 cm and typically bear numerous basidiocarps. An affected trunk tends to swell at the level of the canker as the result of woundwood formation around the margins. The associated decay is a white rot that at least sometimes lacks distinct white pockets.

Basidiocarps of *P. cancriformans* are small (up to 5.5 × 4.5 × 1.7 cm), crustlike to more or less shelflike, annual or perennial, yellowish brown to reddish brown with a velvety upper surface, and usually form in clusters. A thin black layer separates the velvety surface from the yellow-brown interior tissue. Pores on the lower surface are angular to elongate, 1–6 per millimeter.

P. cancriformans has been confounded with *P. chrysoloma* (syn. *P. pini* var. *abietis, Porodaedalea chrysoloma*). The latter has been reported to colonize *Abies, Larix* (larch), *Picea* (spruce), *Pinus* (pine), *Pseudotsuga* (Douglas-fir), and *Tsuga* (hemlock) around the Northern Hemisphere, but some authorities hold that it is not present in North America. In *A. concolor*, bacteria, including nitrogen-fixing types, are associated with all stages of wood colonization and decay by *P. cancriformans*, but the role of these microbes in diseases caused by the fungus is unknown.

References: 27, 394, 1142, 1144, 1280, 2242, 2938, 3374

A–G. Trunk rot of *Pinus* species caused by *Phellinus pini*. A, B. Basidiocarps on trunks of *Pinus monticola* (western white pine), indicating extensive decay. A. Young basidiocarps growing at locations of old branch stubs. B. Perennial basidiocarps on a flattened dead area (ID, Jul). C. A perennial basidiocarp on *Pinus clausa* (sand pine) (FL, Apr). D, E. Upper and lower surfaces of a basidiocarp from *P. monticola*, the poroid lower surface viewed with magnification (ID, Jul). F. Heart rot in an old-growth *P. monticola* log. Bluestain fungi invaded the sapwood after the tree was felled (ID, Jul). G. Pocket rot caused by *Phellinus pini* in *Pinus clausa* (FL, Apr).

H, I. *Phellinus cancriformans* fruiting on a large canker on *Abies grandis* (grand fir). H. Woundwood formation at canker margins has resulted in trunk swelling. I. Closer view of basidiocarps on the canker (OR, Jul).

Armillaria Root Rots (Plates 162–164)

Armillaria root rots, also called shoestring root rot, mushroom root rot, and "oak root fungus," affect forest, shade, fruit, and ornamental trees and shrubs. *Armillaria* (Agaricales, Tricholomataceae) is a genus of about 40 mushroom-forming basidiomycetes that occur in tropical to boreal regions around the world. They parasitize hundreds of woody and some nonwoody plant species and are ubiquitous decomposers of roots and butts of dead trees and stumps. Armillarias are also called honey mushrooms, honey agarics, and shoestring fungi. These names refer to the color of the mushrooms of some species and the appearance of the rhizomorphs (rootlike organs, described below) by which most species spread in the vegetative state. Most *Armillaria* species are generalists, each having many hosts. In this account we often refer to these fungi collectively using the genus name.

Several *Armillaria* species are aggressive primary pathogens; others are opportunists that cause damage mainly to plants weakened by other malefactors. Armillarias often attack in concert with other root-infecting fungi and secondary insects. Damage caused by *Armillaria* in forests is mainly of four sorts: tree killing, windthrow due to root and butt rot, growth suppression due to loss of functional roots, and predisposition of diseased trees to lethal attack by secondary insects. In some coniferous forests in western North America up to 35% of annual tree mortality is caused by *Armillaria*. In northern coniferous forests many saplings are killed by *Armillaria* during the early years of stand development after harvests and other disturbances that kill larger trees. This mortality commonly averages more than 1% per year for several years, but it does not usually presage stand destruction. Tree killing is often severe, however, in young plantations of conifers growing on sites formerly occupied by broadleaved trees, especially *Quercus* species (oaks). Young trees are at highest risk of infection for the first 10–15 years because stumps and roots of the previous stand serve as food bases from which an *Armillaria* initiates parasitic attack. The pathogen eventually depletes these food reserves, and the trees become less prone to lethal attack as they increase in age and size. Endemic, mostly nonlethal root and butt rot is ensured by root infection and tree-to-tree spread, however. In exceptional situations, tree killing continues as the result of tree-to-tree spread of a virulent pathogen. Severe mortality in patches up to 3–4 hectares in area has occurred in some inland forests affected by *A. ostoyae* in Oregon and Washington. Tree killing by *Armillaria* also occurs around stumps and roots of previously infected trees in orchards and landscape plantings. Damage by *Armillaria* in hardwood forests and urban landscapes of eastern USA often follows defoliation by insects such as *Malacosoma disstria* (forest tent caterpillar) or *Lymantria dispar* (gypsy moth). Armillaria root rot is favored by overwatering in urban landscapes of California.

Ten known species of *Armillaria* occur in North America. Their host ranges overlap broadly, and differences in their virulence are imperfectly known. The two most virulent pathogens, *A. mellea* and *A. ostoyae*, occur throughout the north temperate zone on both angiosperms and gymnosperms. *A. ostoyae* is the more common and damaging species on gymnosperms and is the most common *Armillaria* in boreal and montane coniferous forests in North America. *A. mellea* is the more damaging to angiosperms but also attacks gymnosperms. This species has been credited as a major killer of ornamental trees and shrubs and orchard trees in California. Other species often encountered in North America include (asterisk indicates occurrence also in Europe and/or Asia) *A. calvescens* (east and north-central, hardwoods), *A. gallica* (transcontinental, hardwoods), *A. sinapina* (transcontinental), and *A. tabescens* (southeast). Those least often identified are *A. nabsnona* (west, hardwoods), *A. gemina* (northeast, hardwoods), *A. cepistipes* (northeast and northwest, hardwoods), and an as yet unnamed species that attacks conifers in the West. A different set of Armillarias occurs in the tropics and Southern Hemisphere.

Many reports about *Armillaria* dated before the 1980s cannot be reliably connected to particular species because the name *A. mellea* was used for all North American species except *A. tabescens* (Plate 164). Most of the known species were distinguished—and some were named for the first time—in the late 1970s to 1990s, but field identification remains problematic. Only experts identify the species

reliably. Most *Armillaria* species were differentiated and subsequently have been identified primarily by observing interactions of unknown and standard known strains in culture. Identification by DNA analysis is now possible also. The mushrooms of each species have distinguishing characteristics, but mushrooms are produced for only a few weeks each year, and not on every diseased plant or in every year.

Symptoms and signs. Diseased plants occur singly or in groups. *Armillaria* causes lesions that involve the phloem, cambium, and sapwood of roots; the root collar; and sometimes the trunk base. The lesions may girdle infected parts or remain discrete as cankers. Fanlike veined white sheets of mycelium, called fans or plaques, develop in the cambial zone of lesions. Presence of robust fans indicates current or recent pathogenic activity. Some infections are arrested and die out, however, even when extensive cambial killing has occurred in the root-collar region. Vigorous trees may confine *Armillaria* within cankers and the wood underlying them for many years. Trees of this sort produce woundwood at canker margins. Woundwood may eventually enclose small cankers and associated pockets of decaying wood beneath a layer of sound wood. When a large lesion on the trunk base is arrested, it is likely to remain exposed and become difficult to distinguish from a scar that began with injury by fire or frost. Some old Armillaria cankers can be identified by the images of former mycelial fans etched darkly on the wood.

Symptoms above the trunk base include growth reduction, which may be abrupt or gradual; undersized and/or yellowish leaves; premature autumnal color and leaf drop; branch dieback in the upper crown; or rapid browning and death of the plant. Growth reduction becomes noticeable after more than half the root system is dead. Small trees or shrubs often die quickly, but large ones may be chronically diseased, sustaining decay of part of the root system and butt. Symptoms in conifers usually develop over several years. Dying conifers sometimes produce a distress crop of numerous small cones. Infection stimulates resin production. Resin may saturate the bark and wood near points of attack, accumulate in pockets between bark and wood, and seep out onto the surface and into surrounding soil and litter. Similarly, gum-producing angiosperms such as *Prunus* species (cherries and their relatives) respond by producing gum.

Wood decay by *Armillaria* is usually confined to sapwood of killed roots and the butt until after a tree dies. The decaying wood is at first slightly water-soaked and light brown. With time it becomes light yellow to nearly white, soft and spongy, often stringy in conifers, and marked on the surfaces by black lines (zone lines; see Plate 152). Decay in the butt and major roots of *Abies* (fir), *Betula* (birch), and perhaps other trees results in vertical cracks at the root collar that arise when trees weakened by internal decay are stressed by wind or the weight of snow or ice. Decay caused by an *Armillaria* seldom extends more than 2 m above ground.

Most Armillarias produce branched brown to black rhizomorphs that grow from decaying wood along the surfaces of living or dead

A. *Acer platanoides* (Norway maple) dying of root and butt infection by *Armillaria* sp. and *Phytophthora* sp. (NY, Sep).
B. *Calluna vulgaris* (heather) attacked by *Armillaria* sp. (WA, Sep).
C, D. Mycelial fans of *Armillaria* sp. in the cambial zone of the butt of a recently killed *Quercus prinus* (chestnut oak), and a large lesion in the inner bark and cambial region of a living *Salix babylonica* (weeping willow) butt, respectively. Arrow indicates the upper edge of the lesion in D (NY, Aug-Sep).
E. Rhizomorphs between bark and wood on the butt of a dead *Acer saccharum* (sugar maple) (NY, May).
F, G. Young mushrooms, probably of *Armillaria calvescens*, on the butt of a dead *Ulmus americana* (American elm), and on a *Betula alleghaniensis* (yellow birch) log, respectively (NY, Oct).
H. Mushrooms believed to be of *A. calvescens* associated with *B. alleghaniensis* (NH, Sep).
I. Mushrooms believed to be of *A. mellea* associated with *Quercus* sp. (PA, Sep).
Photo credits: H—H. G. Eno; I—P. M. Wargo

327

roots, out into soil or humus, and up between the bark and wood of dead tree trunks. Rhizomorphs under the bark are flattened and anastomose frequently, but they are cylindric where free in soil, litter, or decomposed wood. They are 1–3 mm in diameter with a white interior and a rind that is reddish brown when young and black when mature. Rhizomorphs grow and branch by the extension of massed hyphae at their tips. Their functions are exploration of the soil and litter within several meters of their food bases, translocation of water and nutrients, and infection of plants that will serve as new substrates. Rhizomorphs on roots or between dead bark and wood of a tree indicate the presence of an *Armillaria*, but not necessarily pathogenic activity. Although some species and strains of *Armillaria* seldom or never form rhizomorphs in the field, most strains can produce them in laboratory cultures.

Armillaria mushrooms usually develop in autumn. In regions with mild climate the fruiting season may extend into winter. Fruiting in spring has been noted occasionally. The mushrooms form on or near decaying wood and most often grow from rhizomorphs connected to decaying wood. The following description is generic. The mushrooms grow in groups of several to 100 or more. The cap, initially globose, expands to a dome or disk 4–15 cm across, often with a depressed center, varying from honey colored to tawny or brown, sometimes dotted with brown scales, and gelatinous when moist. The flesh is white and, in some species, luminescent when fresh (as is wood being decayed by *Armillaria*). The gills are white to flesh colored, and the spores in mass are light cream color. The stem, 1–2 × 5–15 cm long, often has a persistent yellowish to whitish ring encircling its upper part. The basidiospores are small (8–10 × 5–6 μm), colorless, single celled, and oblong to ellipsoid. Mushrooms do not develop consistently in areas of low rainfall, but rhizomorphs (when present) and mycelial plaques between the bark and the wood at the root collar or in roots permit generic diagnosis.

Disease cycles. Armillaria species perpetuate themselves by growth of mycelium and rhizomorphs and by basidiospores. Alternation of parasitic and saprobic activity permits perennial exploitation of a given habitat. Rhizomorphs, supplied with organic nutrients provided primarily by mycelium in wood, are responsible for local spread in most parts of the temperate zones. On encountering a healthy or declining root, rhizomorphs grow on its surface, become attached at intervals, and produce short penetrating branches at points of attachment. The branch tips penetrate intact bark by a combination of mechanical force and enzymatic action. Hyphae growing from rhizomorph tips proliferate beneath the periderm and form mycelial fans that kill cambial tissue in advance of their own progress. Strains that seldom produce rhizomorphs grow as massed hyphae across contacts between infested and healthy roots. *A. tabescens* in southeastern USA and some strains of *A. ostoyae* in inland areas of the Northwest have this habit, although the latter fungus is typically a rhizomorph producer.

Airborne basidiospores initiate occasional *Armillaria* infections. These haploid mononucleate spores can germinate and establish colonies in recently wounded sapwood of stumps and roots. Where hyphae of two compatible haploid monokaryotic colonies mingle and fuse, a heterokaryotic mycelium (having nuclei of different genotypes) or often a diploid mycelium develops. A diploid colony may persist for decades or centuries, colonizing successive woody substrates by means of parasitic and then saprobic activity. A clone may remain continuous (one individual) or become discontinuous, and may occupy a territory ranging from a few square meters to hundreds of hectares. Exceptional clones grow indefinitely. A clone of *A. ostoyae* found in Oregon was occupying 1020 hectares and was estimated to have been growing for 2400 years! A typical infested area, on the other hand, is likely to contain multiple clones in a sort of subterranean mosaic. In the deciduous forests of eastern North America these mosaics include multiple species as well as multiple genotypes (called genets) of a given species.

Epidemiology and disease management. Armillaria species grow much more rapidly toward the tips of disease-girdled roots than toward the butts of living infected trees. When these differential growth rates are averaged, or when growth of rhizomorphs is considered, these fungi spread at rates between 0.2 and 2 m per year in many temperate areas. They commonly colonize the declining root systems of felled trees or those killed by other agents. Some strains are primary pathogens able to kill vigorous plants close

to relatively large food bases, and all pathogenic strains are opportunists that colonize small plants or those predisposed by mechanical damage, defoliation, overirrigation, freeze damage, drought, inadequate light, or severely polluted air. Rhizomorphs commonly colonize root surfaces of healthy trees and penetrate roots after stress-induced chemical changes cause a shift from resistance to susceptibility. Defoliation of broadleaved trees encourages such attack by causing depletion of starch, increased levels of glucose and fructose, changes in amino acids, and production of ethanol in roots. Ethanol and elevated glucose levels stimulate growth of the pathogen, and the energy derived from free sugar enables it to tolerate or metabolize host-produced phenolic compounds at concentrations that would otherwise be inhibitory.

Rhizomorph production typically reaches a peak 10–15 years after a new substrate such as a stump is colonized (sooner if the substrate is small) and then declines markedly. Although rhizomorphs require a woody food base for sustained growth, they derive some nourishment from the decomposing organic matter through which they grow. Rhizomorphs are invariably more abundant and more highly branched in the humus layer than in mineral soil. They are sensitive to variation in oxygen and carbon dioxide concentrations and perhaps for this reason are usually not found more than about 30 cm deep in soil. Where soil conditions permit deep rooting, however, *Armillaria* grows readily. Diseased roots have been found at soil depths greater than 2 m in California. North American species other than *A. tabescens* are generally inhibited by soil temperatures above 26°C, so plants that grow where soil temperature commonly exceeds this level may escape damage.

Armillaria species vary considerably in aggressiveness, as reflected by speed of attack and killing. Strains from particular hosts are generally capable of infecting several kinds of plants. Reports of differing host preferences of species and strains are confounded by the fact that wood of broadleaved trees is a better medium than gymnospenm wood for hyphal growth and rhizomorph production by most strains. Thus, if other conditions are equal, an infested root of *Acer* (maple) or *Quercus* (oak) may be more effective inoculum than a conifer root of the same size.

As mentioned, *Armillaria* often acts in concert with other pathogens and secondary pests. In western North America *Armillaria* is frequently associated with other root-infecting fungi, particularly *Heterobasidion* species, *Phellinus weirii*, and *Leptographium wageneri*. *Pinus ponderosa* (ponderosa pine) with root rot caused by *Armillaria* or other fungi is likely to be attacked by *Dendroctonus brevicomis* (western pine beetle) and *D. ponderosae* (mountain pine beetle). *Abies concolor* (white fir) with root rot becomes attractive to *Scolytus ventralis* (fir engraver). In the East *Quercus* species with root rot are attacked by *Agrilus bilineatus* (two-lined chestnut borer).

Prevention of Armillaria root rot is difficult because the causal fungi are ubiquitous in woody debris in soil and have broad host ranges. Important strategies for protection of shade and ornamental trees and shrubs are to provide care that minimizes stress and to deny the fungi large food bases by removing infested stumps and roots or promoting growth of antagonistic organisms in them. In orchards where summers are hot and dry, diseased trees have sometimes been saved by removing soil from around the root

A–D. Root rot of *Pinus ponderosa* (ponderosa pine) caused by *Armillaria ostoyae*. A, B. Young trees dying after infection by root contact with older, previously killed trees (WA, Jul). C, D. Mycelial fans of *A. ostoyae*. C. Extending about 40 cm above soil line in the cambial zone of a recently killed tree (WA, Jul). D. In the root collar region of a wilted sapling (NM, Aug).

E–G. Armillaria root rot in a *Pinus nigra* (Austrian pine) plantation. E. A recently killed tree. F, G. Trees with girdling lesions at the root collar; necrosis was preceded by resin infiltration of bark and nearby soil. Mycelial fans are visible in the cambial zone where resin-encrusted soil and bark have been removed. White crystallized resin occurs between bark and wood (arrow) above the fan in G (NY, May).

H. *Armillaria ostoyae* mushrooms associated with *Larix occidentalis* (western larch) (BC, Oct).

Photo credit: H—D. J. Morrison

328

buttresses and root collar. The pathogen ceases growth in tissues thus exposed. In California various fruit and ornamental trees and shrubs were observed to escape severe damage by A. mellea—and thus were thought to be resistant—when grown on infested sites that were not irrigated during the normally dry summer season. Many of the same species, however, especially those adapted to the Mediterranean climate of California, were later observed to be infected when growing on irrigated sites. Reliable information about truly resistant plants is not available. Species selection to lessen damage by Armillaria is a management option in some forested areas. For example, Pinus (pine) and Larix (larch) sustain less damage than Abies (fir) or Pseudotsuga (Douglas-fir) in some interior western forests.

Both resistant and susceptible plants have an array of resistance factors, some passive and others induced during infection. Passive resistance is related primarily to the storage of carbohydrate reserves on which active defense processes depend, and to normal production of resins, tannins, and phenolic compounds in root bark, which inhibit attacking fungi. Induced processes include secretion of resins, gum, enzyme inhibitors, phenolic compounds, and phytoalexins at infection sites; infusion of cell walls with fungitoxic compounds; production of enzymes that degrade fungal cell walls; and production of cork barriers at the edges of lesions. These are most effective in well-nourished plants and at soil temperatures most favorable for root growth. Among conifers, species that are somewhat resistant to A. ostoyae produce relatively higher concentrations of phenolic compounds and less sugar in root bark than are produced by more susceptible species. The resistant species thus seem to limit the energy available to the fungus for degrading phenolic compounds and overcoming host defenses.
References: 189, 261, 262, 351, 352, 407, 726, 759, 931, 1067, 1068, 1112, 1166, 1201, 1408, 1516, 1518, 1529–1531, 1683a, 1895, 2090, 2212, 2568, 2690, 2729–2734, 2896, 2975, 3192, 3300–3302, 3304, 3306, 3315, 3346, 3505, 3510, 3565, 3567, 3604, 3627, 3729, 3760, 3836, 3986, 4010, 4170, 4185, 4221, 4224, 4246, 4321, 4437

Mushroom Root Rot

This topic is presented apart from the main treatment of Armillaria diseases because mycologists and plant pathologists have long distinguished Armillaria tabescens from other species in its genus. This fungus occurs in the eastern half of the USA, the Caribbean region, Europe, India, Africa, Japan, and Malaysia. In the USA its mushrooms have been collected as far north as Michigan and New York, but the northernmost records of its pathogenic behavior are from southern Illinois eastward to Maryland. The European and American populations may be different species, but this matter is as yet unresolved. If the American form is found to be genetically distinct and sexually incompatible with European A. tabescens, it will require a new name.

In southern USA from Oklahoma and Texas eastward, A. tabescens causes root rot of more than 200 plant species in about 60 families. The disease is called either mushroom root rot, because clusters of mushrooms often grow at the bases of diseased plants, or Clitocybe root rot, reflecting an earlier classification of the pathogen. The pathogen and the disease fit the general descriptions on the preceding pages, except that A. tabescens rarely produces rhizomorphs in the field.

A. tabescens is an opportunist, best able to attack plants weakened or recently killed by other agents. In Pinus clausa (sand pine) in Florida, for example, it is commonly found in roots also infected by Onnia circinata (Plate 167). Thus, although it has been credited with killing up to 25% of the trees in some P. clausa plantations, the amount of damage it does alone is open to question. In highly susceptible plants, however, it seems to be an aggressive killer. A. tabescens, like other Armillaria species, is a common colonist of Quercus (oak) roots and often causes damage to other plants on sites where oaks grow or once grew.

Some commonly affected plants in the USA are Camellia japonica (camellia), Casuarina species (Australian-pine), Citrus limon (lemon), Cornus florida (flowering dogwood), Eriobotrya japonica (loquat), Ficus elastica (Indian rubber-tree), Hibiscus rosa-sinensis (Chinese hibiscus), Malus pumila (apple), Photinia glabra (Japanese photinia), Pittosporum tobira (Japanese pittosporum), Pinus clausa,

Psidium guajava (guava), Prunus species (cherry, peach), Quercus species (oaks), Rhododendron species (azaleas), Ulmus species (elms), and Vitis species (grapevines).

Symptoms and signs. Trees and shrubs of all ages are attacked, singly or in groups, and trees with severely decayed roots are subject to windthrow. Symptoms before death include yellowing, sparse or undersized leaves, premature defoliation, branch dieback, decayed roots, and lesions that partly or completely girdle the plant at the root collar. Basal lesions often extend 10–30 cm, rarely 60 cm, above soil level and are contiguous with dead roots. In some species of Casuarina basal girdling by A. tabescens leads to pronounced stem swelling above the girdle before the tree dies. Diseased Pinus clausa often has sunken cankers at the root collar and produces resin profusely at sites of infection. The resin saturates bark and wood, accumulates in pockets in the cambial region, and exudes around diseased roots and the root collar, causing encrustations as it hardens. Symptoms on aboveground parts usually do not appear until the fungus has invaded the root collar; then an entire plant may wither and die quickly.

Fanlike mats of white, cream, or chamois-colored mycelium grow beneath the bark of infected roots or the root collar and kill cambial tissue in advance of their own progress. Elliptic perforations 0.5–1.0 mm long occur in the mats, but may be lacking in very young and very old mats. The surfaces of colonized roots often have narrow black ridges of fungal tissue that develops in bark fissures above mycelial mats.

In autumn clusters of mushrooms grow at the bases of diseased plants or occasionally emerge from cracks in the bark on the butt of a tree. Their remains sometimes persist for several weeks. On dry sites and in dry years, however, mushrooms form infrequently. They resemble those of other Armillaria species in growth habit, color, texture, and general appearance, but lack a ring around the stem. North American strains of A. tabescens form mushrooms readily in pure culture, possess binucleate cells and clamp connections in the hyphae of mushrooms, are not luminescent, grow at temperatures up to 32–34°C, and grow optimally at 25–30°C. A. tabescens produces rhizomorphs in culture but does not commonly do so in nature. Its rhizomorphs have been noted under the bark of dying Q. nigra (water oak), and they were infective in pathogenicity tests.

Disease cycle. A. tabescens persists in dead, decaying roots and stumps for many years, and it attacks new hosts by growing as hyphae across root contacts. It grows rapidly in a new host if attack begins in root tissue already dying or recently dead. Symptoms in woody ornamentals frequently arise 3–4 years after transplanting, an indication that wounded roots may be important avenues of entry. Vigorous plants with undamaged roots resist attack. Experimental inoculations of Prunus persica (peach) roots, for example, resulted in only localized lesions after 4 years. A role of basidiospores in establishment of A. tabescens in wounded sapwood is presumed but has not been demonstrated.

Plants found to be infected in only a few roots or a small part of the root collar can be saved for a time by exposing the root collar to aeration and drying, which halts the pathogen.
References: 279, 551, 726, 1166, 1271, 1381, 1518, 2627, 3276, 3277, 3338, 3565, 3627, 4170, 4246, 4258

Mushroom root rot, caused by Armillaria tabescens (FL, Mar–Apr).
A–E. Pinus clausa (sand pine). A. Dying and recently killed 9-year-old trees. Shoots are not elongating and foliage is thin on the dying tree at center. B. Roots of a dying tree; the deep roots are dead and decayed. Resin exudation at the root collar and the pocket of yellow resin beneath the bark on the butt are responses to the pathogen. C–E. Mycelial fans in the cambial region of a dead root. Perforations 0.5–1.0 mm long are diagnostic.
F. Necrotic bark and cambium at the base of a dying Hibiscus rosa-sinensis (Chinese hibiscus). Necrotic tissue extends about 5 cm above the limit of the mycelial fan (arrow). Recently killed sapwood exposed by cutting through the mycelium is darker than normal (arrow in insert).
G. Mushrooms of A. tabescens at the base of a living Quercus sp. (oak) (MI, autumn).
Photo credit: G—T. J. Volk

331

Root and Butt Rots Caused by *Heterobasidion* Species (Plates 165, 166)

Root and butt rots caused by *Heterobasidion* species (Russulales, Bondarzewiaceae) are among the most important and most studied forest diseases. They occur in coniferous and mixed gymnosperm-angiosperm forests around the Northern Hemisphere. Economic damage occurs mainly on sites where gymnosperms have been cut during thinning or harvests. Root rot, windthrow due to rotted roots, butt rot, suppressed growth, and tree mortality are all significant loss factors. In addition, diseased trees in recreational areas are hazards for people and property. Large *Abies* (fir), *Pinus ponderosa* (ponderosa pine), and *Sequoiadendron giganteum* (giant sequoia) have posed notable hazards in California. Root damage by *Heterobasidion* also predisposes trees to pests such as bark beetles and to opportunistic pathogens such as *Armillaria*. In North America diseases caused by *Heterobasidion* have been called annosum root rot.

Heterobasidion species have been found on more than 200 species of plants in the Northern Hemisphere; about two-thirds of these are gymnosperms in at least 17 genera: *Abies, Araucaria, Calocedrus* (incense-cedar), *Cedrus* (cedar), *Chamaecyparis* (false-cypress), *Cryptomeria* (Japanese-cedar), *Cupressus* (cypress), *Juniperus* (juniper), *Larix* (larch), *Picea* (spruce), *Pinus, Pseudotsuga* (Douglas-fir), *Sequoia* (redwood), *Sequoiadendron, Taxus* (yew), *Thuja* (arborvitae), and *Tsuga* (hemlock). Woody angiosperms growing among gymnosperms are occasionally colonized, accounting for at least 30 additional host genera. *Betula* (birch) seems particularly susceptible in northern Europe.

Heterobasidion in the Northern Hemisphere comprises five or six species that are reproductively separated except for rare hybrids. For more than a century, however, they were all united under one name, most recently *H. annosum*. Considered in the strict sense, *H. annosum* is a set of pine-preferring strains found on many gymnosperms and angiosperms in Europe. These strains are sparingly interfertile with North American pine-preferring strains and differ from them at several genetic loci. Some researchers have suggested that the North American pine-preferring strains, which occur across the continent on various gymnosperms and angiosperms, may represent a distinct species, but in 2004 the name *H. annosum* was still used for both European and American pine-preferring strains. A second species, *H. parviporum*, occurs in northern and montane regions of Europe and China. It is named for the small-diameter tubes in its basidiocarps. Its preferred hosts are *Picea abies* (Norway spruce) and *Abies sibirica* (Siberian fir). A third species, *H. abietinum*, is found only in southern and central Europe and displays preference for *Abies*. *H. parviporum* is partly interfertile with *H. abietinum*, and both species are interfertile with an as yet unnamed taxon that occurs on *Abies, Picea, Pseudotsuga*, and *Tsuga* in central and western North America. Five additional named species of *Heterobasidion* are known, but only *H. insulare*, which occurs in Asia on stems of conifers, is found in the Northern Hemisphere. Those five are not considered here.

The range of *Heterobasidion* in North America extends from Alaska to southern Mexico in the West and from southern Quebec and Wisconsin to the Gulf Coast states in the East. It has not been reported from boreal forests of central and eastern Canada. *H. annosum* has caused major damage in plantations of *Pinus elliottii* (slash pine), *P. resinosa* (red pine), and *P. taeda* (loblolly pine) in eastern USA. In the West *H. annosum* or the unnamed species often acts in concert with other root pathogens such as *Armillaria ostoyae, Phellinus weirii*, and *Leptographium wageneri* that collectively cause tree mortality or growth loss. Landscape and ornamental plantings are seldom affected.

All *Heterobasidion* hosts are subject to root killing and decay, but the nature of further damage varies with host group. *Pinus* and *Juniperus* may die from cambial and sapwood necrosis of roots and the butt, but their heartwood usually resists decay. Young *Abies, Calocedrus*, and *Pseudotsuga* may also be killed. *Abies, Larix, Picea*, and *Sequoiadendron* usually escape early death, but root decay leads to windthrow and to butt and heart rot that may extend several meters up the trunk. Extensive infection of sapwood in the butt causes decline and death. *Calocedrus* is affected similarly if it escapes early death. *Tsuga heterophylla* (western hemlock) is subject to butt rot and wound-associated trunk rot. *Heterobasidion* sometimes kills angiosperms whose roots contact infested gymnosperm roots. The only report of independent pathogenicity on an angiosperm concerns *Arbutus menziesii* (madrone) in California.

Symptoms and signs. Lesions most often develop in roots and may spread into the butt. Resinous species often exude resin from infected roots and the butt. Resin-soaked or stained wood extends beneath and beyond the point of advance of the fungus in sapwood. Normal-appearing sapwood adjacent to discolored or resin-soaked wood is abnormally dry and does not conduct sap. Trees with diseased roots often have no symptoms above ground. Pines can tolerate loss of half their roots before diminished growth becomes noticeable. In species commonly killed by *Heterobasidion*, lesions that involve sapwood, cambium, and phloem girdle trees at the root collar. Foliage of young trees may fade and turn brown rapidly with no prior indication of distress. Dead, declining, and windthrown trees with decayed roots occur alone or in groups. Older trees usually decline for several years, producing abnormally short twigs and needles, and may become chlorotic. Bark beetles often kill trees weakened by root disease.

Wood undergoing decay by *Heterobasidion* is initially discolored shades of reddish to purplish brown, or light yellowish brown in some pines. Delicate wefts of white mycelium form between bark and wood, except where tissues are infiltrated with resin. Elliptic to elongate white pockets, often containing single black specks of manganese dioxide, form in the cambial zone. The wood gradually turns weak, soft, and stringy; tan to light brown; and may also have tiny elongate pockets that sometimes contain black specks. The decay is called a white pocket rot, a white stringy rot, or sometimes (in *Abies*) a laminated rot.

Basidiocarps develop on decaying logs, stumps, or butts near the soil line; on superficial roots and those exposed in cavities and burrows; in hollow butts or stumps (often below ground level); and under the bark of stumps. They commonly form on stumps beginning 1.5–3 years after trees are felled, but seldom form on exposed surfaces where summer is hot and dry. They vary in size and shape from tiny white plaques consisting of little more than a tube layer to shelflike conks more than 30 cm across. Large specimens develop in regions of moist climate. Some are perennial. Fresh specimens are leathery and are white when young or growing in darkness. When exposed to light, basidiocarps of *H. annosum* turn buff to reddish brown or dark brown with a white margin (cream colored later) as viewed from above. The lower surface, poroid where a layer of tiny tubes opens to the surface, is initially light cream colored and becomes light buff with age. Basidiocarps are preceded by white mycelial pustules or plaques that form between bark scales and on root surfaces. Growing basidiocarps tend to envelop rather than push aside needles and other litter. Basidiospores, liberated from a fertile layer that lines the tubes, drop out and are carried away by air currents. The basidiospores are colorless, single celled, ovoid, and 4–6 × 3–4.5 μm in size.

Root and butt rot of pine caused by *Heterobasidion annosum*.

A–E. In *Pinus resinosa* (red pine). A. Part of a ruined plantation. Living infected trees in and at the edge of a disease gap display slow growth and tufted foliage (RI, Aug). B. Windthrow following root decay (NY, Sep). C. Young basidiocarps up to 9 cm across, revealed by removing needle litter at the base of a recently killed tree (NY, Sep). D. Irregular sterile white masses of packed mycelium that will differentiate into basidiocarps at the base of a stump, about one-fourth natural size (NY, Jul). E. Wood surface exposed by bark removal at bottom center of the stump shown in D has diagnostic white pockets in the cambial zone (about half natural size).

F. A small *P. strobus* root killed by *H. annosum*. Red-brown resin-infiltrated internal tissue and soil-encrusted resin deposits on the root surface were responses to infection (NY, Sep).

G–J. Basidiocarps of *H. annosum*. G. A large overwintered specimen growing appressed to the undersurface of a *P. resinosa* root taken from a cavity beneath a partly windthrown tree (NY, May). H, I. Two views of large bracket-type basidiocarps on a *P. strobus* stump with typical stringy white decay (WV, Oct.). J. New (white) and dead year-old basidiocarps at the duff line on a *P. resinosa* stump (NY, Sep).

Photo credits: H, I—J. M. Staley

Heterobasidion species have distinctive conidial anamorphs in the genus *Spiniger* that develop under moist conditions on colonized wood: on broken roots, under debris on stumps, in insect galleries, and in cavities formed by shrinkage of soft tissue at the bark-wood interface. Colorless, unicellular, ovoid conidia, most measuring 4.5–8 × 3–6 μm, are borne on the swollen heads of unbranched conidiophores. Each head bears many spores and appears as a tiny white sphere when viewed at low magnification (Plate 166F, G). Sporulating colonies, which have a granular appearance, develop within 2–3 days on newly cut surfaces of wood previously colonized by *Heterobasidion* if samples are incubated at room temperature in a moist atmosphere. Recognition of the anamorph is useful for diagnosis.

Disease cycle. Infection may be initiated by basidiospores, conidia, or mycelium. *Heterobasidion* most commonly becomes established in a stand by means of airborne basidiospores that germinate on freshly cut stumps. Other wounds that expose living sapwood in stems or roots sometimes become infected, notably by *H. parviporum*. Fresh sapwood of a host species is a selective medium for these fungi. The selectivity lasts for a few days to a few weeks, depending on temperature and dryness, after which various saprobes compete for the substrate. *Heterobasidion* cannot replace other microorganisms in dead wood. Given the initial advantage associated with parasitic ability, however, its mycelium kills sapwood and phloem and may persist until the wood is reduced to mush.

Basidiospore dispersal and infection of fresh stumps and wounds can occur year-round in regions with mild climate such as the Pacific Northwest. These activities occur mainly in spring and autumn in the Northeast and during autumn and winter in the South. Spore production is inhibited by prolonged freezing and by hot weather. In parts of southern USA the temperature of stump surfaces in summer rises high enough to prevent colonization by *H. annosum*.

Spores of *Heterobasidion* can percolate with rainwater into coarse soil and infect wounded roots of vigorous trees or unwounded roots of stumps and suppressed or otherwise weak trees. Conidia can become airborne in wind, especially in wind-driven mist, and their dispersal by water and insects has been suggested. Conidia can survive up to 10 months in some soils, and relatively few are required to infect nonwounded stump roots. However, their epidemiological role is believed to be small compared with that of basidiospores in most areas.

If infection begins on stump tops or wounds on tree butts, the pathogen grows down into roots. It spreads from tree to tree via root contacts or grafts, but it does not grow free in soil for more than a few millimeters. Its linear growth rate in roots and stems is 0.1–2.0 m per year, depending on strain of pathogen, host material, temperature, and soil conditions. Parasitic growth is slow because of host resistance, but growth along dying or recently dead roots is relatively rapid. Centrifugal growth from a stump or tree results in groups of diseased and dead trees. In some pine plantations affected by *H. annosum* these "disease gaps" may enlarge for 10–30 years before stabilizing. Multiple infection centers and delayed stabilization result in severe damage.

Sites in mixed coniferous forests sometimes are infested with more than one species of *Heterobasidion*. Multiple genotypes, called genets, are always found at a given site and often in the same stump or root system. Individual genets may persist for decades and grow within root systems to occupy areas up to 10 m in diameter, but smaller sizes are the rule. In general, a given strain of *Heterobasidion* can infect a wide array of plants, but strains do vary in host preference and aggressiveness. *H. annosum* tends to be more aggressive than *H. parviporum* in their common hosts, although the latter species is aggressive in *Picea abies*.

Host-pathogen interactions of *H. annosum* and *H. parviporum* have received considerable study. These fungi secrete various toxins and enzymes that facilitate pathogenicity and wood-decaying ability. Noteworthy among the toxins is fomannoxin, which is produced in pure cultures, has been isolated from naturally infected *Picea sitchensis* (Sitka spruce) wood, and was lethal to *P. sitchensis* seedlings in assays. Pectolytic and proteolytic enzymes are secreted, as are oxidizing and hydrolytic enzymes that break down host defense compounds. Lignin- and cellulose-degrading enzymes digest the wood. Pathogen-induced defenses include accelerated production of resins and fungitoxic compounds (e.g., pinosylvin and

related stilbenes in pines, isorhapontin and astringin in spruces) and enzymes that inactivate fungal attack factors or digest fungal cell walls, and deposition of lignin in parenchyma cell walls. Spruces and presumably other host plants can detoxify fomannoxin and other pathogen-produced toxins. Infiltration of resin into infected tissues greatly slows and sometimes halts growth of the parasite. Effective defense leads to compartmentalization in bark and wood (Plates 254–257).

Epidemiology and management. In North America damage by *H. annosum* is most severe on deep, well-drained sand or sandy loam containing little organic matter. Neutral to alkaline soil conditions enhance the damage. Plantations on former agricultural sites sustain more damage than those on continuously forested sites. Soils with relatively high content of clay or organic matter, acidic reaction, poor internal drainage, or a high water table are not conducive to the disease. Infection may occur frequently on such sites, but tree-to-tree spread is impeded and damage is slight.

Trees weakened by adverse environment or suppressed by dominant neighbors are most likely to be killed by *Heterobasidion*. Weak trees, which usually have low starch reserves, are unable to mobilize adequate defenses. For example, *Pinus ponderosa* (ponderosa pine) is abnormally susceptible to *H. annosum* when weakened by oxidant-polluted air. In *Picea abies* infection frequency and mycelial growth rate were found to be greater in trees under moderate water stress (−500 to −800 kPa) than in nonstressed plants. Severe water stress inhibits both plant and pathogen.

Root rot caused by *Heterobasidion* intensifies in a stand after the second and subsequent thinnings because of local production of spores and infection of stumps. The incidence of stump infection is correlated with subsequent incidence and severity of root disease on sites that are conducive to longevity and spread of the pathogen. In many regions *H. annosum* causes more damage in the second crop than in the first on a given site because it spreads from roots of harvested trees into young trees of the next stand. Longevity of the fungus in stumps is greatest in regions of cool climate and in large stumps. It has been found in *Larix* stumps up to 63 years old in Great Britain. Incidence of stump infection can be greatly reduced by inoculating fresh stump tops with spores of a competitive saprobe such as *Phlebiopsis gigantea* or by treating the stump tops with any of several chemicals, including borax or urea, which is relatively harmless.

Second-crop buildup of *Heterobasidion* does not occur in regions where climate and silvicultural practices favor quick decomposition of infested stumps and roots or where soil conditions inhibit root-to-root spread. In southeastern USA the wood of *Pinus* stumps may be completely decayed in less than 10 years, leaving bark-lined holes in the ground—a hazard for humans but not for the next crop. On sites that were examined in eastern Oregon, 89% of *Abies* stumps became infected with the unnamed *Heterobasidion* but mortality of young trees due to the pathogen during the next decade was negligible. Root and butt rot of residual trees after stand thinning is the main hazard for *Abies* on sites occupied by *Heterobasidion*.

References: 165, 305, 400, 707, 931, 1064, 1196, 1230, 1232, 1233, 1280, 1474, 1492, 1529–1531, 1604, 1605, 1623, 1659, 1787, 1905, 1906, 1945, 1975, 1997, 2169, 2212, 2322, 2323, 2430, 2906, 2927, 3144, 3299, 3307, 3311, 3339, 3461, 3504, 3505, 3510, 3549, 3550, 3592, 3760, 3824, 3847, 3933, 3952, 4431, 4433

A–C. *Heterobasidion* sp. (as yet unnamed) decaying *Tsuga heterophylla* (western hemlock). A. A large perennial basidiocarp on a stump. B. Butt rot at stump height in a young tree. Darkly stained wood surrounding advanced decay is typical. C. Advanced decay. Elongate pockets lined with white cellulose and containing black specks are typical (BC, Aug–Oct).

D. Stain, also called incipient decay, in a *Pinus resinosa* butt, caused by *H. annosum* spreading upward from decaying roots (NY, Aug).

E. Magnified view of the characteristic thin weft of white mycelium between bark and wood of a *P. resinosa* (red pine) root colonized by *H. annosum* (NY, Sep).

F, G. Successively magnified views of the conidial state of *H. annosum* growing on transverse surfaces of wood sections from diseased trees. Each conidiophore is surmounted by a white head on which conidia form.

Photo credits: A–C—D. J. Morrison; G—E. A. Allen

Red Root and Butt Rot of Conifers (Plates 167, 168)

Onnia circinata and *O. tomentosa* (syn. *Inonotus circinatus* and *I. tomentosus*, respectively; Hymenochaetales, Hymenochaetaceae) are highly similar fungi that cause stand-opening disease of *Picea* (spruce) and red root and butt rot of various conifers across North America and in Eurasia. Disease caused by *O. tomentosa* is also called tomentosus root disease. These fungi cause four types of direct damage: growth suppression, devaluation of the butt log at harvest, windthrow, and tree mortality. In addition, trees planted on sites of previous disease are subject to attack when their roots grow into contact with infested roots or stumps of the earlier stand. Identities of the two species have been confounded in the literature of forest pathology. Moreover, *O. circinata* is thought by some experts to be a species complex. *O. circinata* is found most often on *Pinus* (pine), and *O. tomentosa* most often on *Picea*.

O. circinata in the broad sense occurs from British Columbia to Arizona in the West and from Newfoundland to Florida in the East. It also occurs in Europe and China. *O. tomentosa* occurs around the Northern Hemisphere. In North America it is distributed from Alaska to Arizona in the West and from Labrador and Saskatchewan to the southern Appalachian region in the East, its southern range thought to be similar to that of *Picea*.

O. tomentosa is the more studied of the two pathogens and is the more important in northern forests. In central Canada and the Great Lakes region *Larix laricina* (eastern larch), *P. glauca* (white spruce), and *P. mariana* (black spruce) are often damaged. Reports of severe damage by *O. tomentosa* to *Pinus*, notably *P. resinosa* (red pine) and *P. contorta* (lodgepole pine), should be viewed with caution unless microscopic diagnosis is mentioned (see *Symptoms and signs*). In spruce-fir forests of central and eastern Canada *O. tomentosa* is the second most common cause of root and butt rot. It was found in 14% of *P. glauca* and 18% of *P. mariana* there, and in the most heavily damaged stands more than 80% of the trees were affected by severe butt rot or mortality. Only Armillaria root rot (Plate 163), which often occurs in the same stands, is more common in that region. In British Columbia the species most affected are *P. glauca*, *P. glauca × engelmannii* (white × Engelmann spruce), and *Pinus contorta*. Stump surveys on harvest sites in that region revealed root rot incidence ranging from 2% to 27%. Root diseases caused by *O. tomentosa*, *Armillaria ostoyae*, and *Phellinus weirii* all cause significant mortality and growth loss in mixed conifer forests of northwestern USA, often in the same stands.

In southeastern USA *O. circinata* is a common pathogen of *Pinus clausa* (sand pine) and *P. elliottii* (slash pine). It sometimes occurs in association with *Armillaria tabescens* (Plate 164) or *Phytophthora cinnamomi* (Plates 181, 182), or less frequently with *Leptographium procerum* (Plate 113) on *P. clausa* in Florida.

Host records for *O. circinata* and *O. tomentosa* are similar, partly because these fungi have numerous hosts in common and partly because the former species was for many years considered to be a variety of the latter. One or both fungi occur on the following plants in North America, in addition to those named above.

Abies amabilis (Pacific silver fir), *A. balsamea* (balsam fir), *A. grandis* (grand fir), *A. procera* (noble fir), *A. lasiocarpa* (alpine fir), *Larix occidentalis* (western larch), *Picea abies* (Norway spruce), *P. engelmannii* (Engelmann spruce), *P. pungens* (blue spruce), *P. rubens* (red spruce), *P. sitchensis* (Sitka spruce), *Pinus banksiana* (jack pine), *P. echinata* (shortleaf pine), *P. monticola* (western white pine), *P. ponderosa* (ponderosa pine), *P. radiata* (Monterey pine), *P. rigida* (pitch pine), *P. strobus* (eastern white pine), *P. taeda* (loblolly pine), *Pseudotsuga menziesii* (Douglas-fir), *Thuja plicata* (western red cedar), *Tsuga canadensis* (eastern hemlock), and *T. heterophylla* (western hemlock).

The following species are also hosts of *O. tomentosa* in Eurasia: *Abies sachalinensis* (Sakhalin or Todo fir), *Larix kaempferi* (Japanese larch), *Picea jezoensis* (Yeddo spruce), *Pinus nigra* (Austrian pine), *P. sylvestris* (Scots pine), and *P. wallichiana* (Himalayan white pine). Records of *O. tomentosa* on angiosperms are rare; it has been found on *Alnus* (alder) and *Betula* (birch) in British Columbia.

Symptoms and signs. *O. tomentosa* and *O. circinata* cause similar symptoms, the most prominent being openings in forest stands where trees have died or fallen in windstorms. Diseased trees occur singly but more often in groups. Spruces affected by stand-opening disease in central Canada occur in patches that may slowly expand and coalesce to involve 0.5 hectare or more. Aboveground symptoms develop over many years in the North and over 2 years or

more in the South. The growth rate of diseased trees diminishes, and lower branches begin to die prematurely. Foliage becomes thin, and that on upper branches turns chlorotic. In time the trees die because of root loss, but they often break at the butt or are blown over while alive because of decayed wood in the major roots or butt. Diseased roots have decayed distal portions and red-brown resin-soaked wood extending into the living proximal portion and the butt. Reddish brown wood during the incipient stage of decay is typical in most or all hosts. The stain evolves to advanced decay and slowly expands to involve both deep-lying wood and sapwood near the cambium. The decays caused by both pathogens are classified as white pocket rots. The aforementioned reddish brown color persists, but it appears lighter in wood with advanced decay because numerous small elongate pockets filled with white fibers are present (Plate 167I). Decaying wood slowly softens and eventually becomes mushy but never brittle. Mycelium and dark zone lines become prominent in the cambial region of killed spruce roots. Resin exudes from the butt or root buttresses where the cambium and bark are attacked. Resin from roots may mix with soil and accumulate as an encrustation. Although infection may occur in young trees, mortality in natural conifer stands in northern areas seldom occurs before trees are more than 30 years old. Diminished growth of diseased trees is related more closely to the extent of root mortality than to the extent of decay in roots and the butt. The disease progresses much more rapidly in the South; the windthrow and mortality in the *Pinus clausa* stand illustrated in Plate 167 began before age 20.

Field diagnosis depends on recognition of the decay and identification of basidiocarps. The latter are annual, developing in summer in the North and autumn in the South. They arise on stumps, butts, or roots of living or dead trees or from soil containing diseased roots. They are nearly white when young, then become yellowish, and finally turn tan to rusty brown. The margin remains light colored during growth. Basidiocarps are somewhat spongy to leathery when fresh and become rigid when dry. Their caps have a slightly velvety surface and are up to 11 cm broad in *O. tomentosa*, and 18 cm broad (usually <10 cm in southeastern USA) in *O. circinata*, often with concentric zones of varying color. The undersurface is buff to yellowish brown or gray-brown, and the mouths of the tubes average two to four per millimeter. Basidiocarps arising from soil or roots have short central or lateral stalks, but those on the butts of trees lack a stalk. Stalked basidiocarps, usually several in close proximity, are typical of *O. tomentosa*, and solitary basidiocarps laterally attached to the substrate at the soil line are typical of *O. circinata*; exceptions are known, however. Old, darkened inactive basidiocarps of *O. tomentosa* may persist through winter beneath snow in the North and decompose during the next summer. In the South basidiocarps disappear in spring. They are scarce in dry years.

Microscopic characters of basidiocarps of *O. tomentosa* and *O. circinata* include brown setae (pointed structures) that protrude from the hymenium (basidial layer) lining the tubes. Setal form is the definitive basis for distinguishing between these fungi. *O. tomentosa* has nearly straight setae, 50–70 (up to 140) × 7–11 μm; those of *O. circinata* are strongly curved or hooked, and 50–80 × 12–20 μm (Fig. 168). Basidiospores of both species are colorless, smooth, ellipsoid, and measure 5–6 × 3–4 μm.

O. circinata and *O. tomentosa* are sometimes called false velvet-

---

Root rot of *Pinus clausa* (sand pine) caused by *Onnia circinata*.

A, B. Dead and dying trees in a 20-year-old stand. One dead tree in A bears a distress crop of cones. Declining trees at the edge of the disease gap have sparse chlorotic foliage (FL, Apr).

C, D. A basidiocarp of *O. circinata* at the base of a diseased tree (FL, Jan).

E, F. The butt of a diseased tree before and after removal of bark from a resinous basal canker (FL, Apr).

G. Excavation and removal of bark show association of diseased roots (wood stained reddish brown) with a resinous basal canker (FL, Apr).

H. Close view of typical reddish brown stain (incipient decay) within a root (FL, Apr).

I. Close view of white pocket rot caused by *O. circinata* (FL, Apr).
Photo credits: C, D—G. M. Blakeslee

337

top fungi because their basidiocarps are similar to those of *Phaeolus schweinitzii* (Plate 169), the true velvet-top fungus. Basidiocarps of *O. circinata* on *Pinus elliottii* are sometimes so similar to those of *Phaeolus schweinitzii* that microscopic examination is necessary to distinguish between them. The latter fungus can also be identified by immunologic tests and by its growth on a selective medium containing the fungicide thiabendazole, which inhibits the two *Onnia* species.

Disease cycles. Information in this category is based primarily on studies of *O. tomentosa* on *Picea* species in Canada. Infection may be initiated by either mycelium or basidiospores. Results of inoculation trials indicate that basidiospores can infect wounds on roots or the root collar that penetrate sapwood or heartwood. Wounds to heartwood are most suitable. Stump tops are not colonized, however. Certain root-tunneling weevils (*Hylobius* species) cause wounds that serve as points of infection in both *Pinus* and *Picea*. Roots as small as 1 cm in diameter may be colonized by *O. tomentosa* following damage by these insects. Declining roots of stumps or weakened trees also seem to serve as sites of entry. The pathogens spread both distally and proximally in roots and may progress into the butt. Both fungi inhabit decaying root systems for many years. *O. tomentosa* may persist in spruce roots for 15–20 years after death of a tree, and in spruce or pine stumps for 30 years or more. The longevity of *O. circinata* in southern pine roots is presumed to be less because of rapid decay in the warm climate.

Mycelium is effective inoculum where healthy roots grow in contact with infested stumps and roots. *O. tomentosa* can penetrate the bark of spruce roots less than about 4 cm in diameter. Bark disruptions, such as those occurring at root branches, facilitate penetration. Feeder roots often become infected and serve as routes by which the pathogen enters larger roots. It then advances along the center of a root. Mycelium of *O. tomentosa* in roots of *Picea glauca* × *engelmannii* less than about 5 cm diameter in British Columbia was observed to colonize bark approximately 20 cm beyond stain and decay in the xylem. *O. tomentosa* grows linearly only 1–10 cm per year in spruce roots in central Canada; thus 15–20 years usually elapse between infection and death of trees there. In British Columbia, however, spread rates approximate 20 cm per year. Growth of the fungus in the distal direction in spruce roots ceases not long after a diseased tree is cut, perhaps reflecting the pathogen's inability to compete with saprobes that invade dying roots. The disease cycle may be considered complete when a basidiocarp forms on a diseased tree and releases spores or when mycelium grows from a root of one tree into a root of its neighbor.

Both *O. tomentosa* and *O. circinata* are proven pathogens to various conifers. When Canadian isolates of these fungi were inoculated to seedlings of 11 coniferous species, isolates of *O. tomentosa* were more virulent than isolates of *O. circinata* on every species. Trees of all vigor classes are at risk, but not equally so. For example, *P. glauca* plantations in Wisconsin, affected by *O. tomentosa* and observed for 4 years, exhibited mortality rates of 2%, 15%, and 48%, respectively, in dominant, intermediate, and suppressed trees, versus 0%, 0%, and 29% in plots not affected by the fungus. The mortality of suppressed trees in the absence of *O. tomentosa* presumably represented normal attrition of small, weak individuals.

O. tomentosa, in common with many other fungi, has a vegetative-compatibility system that ensures the maintenance of many discrete individuals in a population. Hyphae of a given individual (colony) can anastomose (fuse) with other hyphae of the same individual and with those of the small proportion of other individuals that are identical at genetic loci governing somatic compatibility. In stand-opening disease; small discrete infection centers tend to contain single somatic-compatibility genotypes, while large disease gaps represent the work of multiple genotypes. Somatic compatibility in *O. circinata* has not been studied.

Infection by *O. circinata* in southern pines has not been studied experimentally, but warm climate and year-round growth of the pathogen permit more rapid disease development in the South than in the North. On circumstantial evidence, basal cankers caused by the fusiform rust fungus (Plate 147) may be sites of entry for *O. circinata* in *Pinus elliottii*. Root rot caused by *O. circinata* in *P.*

elliottii was consistently associated with basal cankers in one study, and similar cankers without associated root rot were also found. *P. clausa*, on the other hand, is not affected by fusiform rust, but it does often have basal cankers associated with red root and butt rot. Early reports of *O. circinata* on *P. monticola* in the West mentioned an association with fire scars.

Red root and butt rot occurs on a variety of sites. The stand-opening syndrome in *Picea* in central Canada is most severe on acidic (pH 4–5), relatively infertile soils with low water-holding capacity, and also on shallow soils overlying rock or hardpan. In cool, moist forests of British Columbia disease incidence is highest on the driest sites, which usually have sandy soil and lower nutrient content than do moist sites. In mixed coniferous forests in Idaho the disease is most common in stands at elevations above 1500 m. In Florida disease caused by *O. circinata* in *Pinus clausa* is most severe where soil is poorly drained or the water table or an impermeable soil layer lies close to the surface.

Its widespread occurrence in northern forests notwithstanding, *O. tomentosa* seems adapted to moderate rather than low temperatures. Its mycelium grows most rapidly near 20°C. Its basidiospores are discharged most rapidly in moist air (85–100% relative humidity) at 16–22°C. They germinate well at 15–27°C, and germination is most rapid at 20–24°C.

O. tomentosa and *O. circinata* grow readily but slowly on agar media, the mycelium extending only about 2 mm per day. The former species forms thick-walled chlamydospores in culture, but whether similar spores occur in nature is unknown. It does not produce clamp connections (typical of many basidiomycetes) in its mycelium in culture.

References: 207, 208, 434, 463, 1260, 1280, 1298, 1425, 1683a, 1694, 1819, 1825, 1830, 2212, 2298–2300, 2777, 3374, 4049, 4108, 4183, 4341–4346

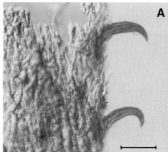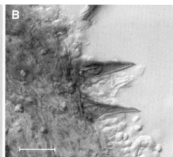

Figure 168. A, B. Microscopic views of setae of *Onnia circinata* and *O. tomentosa*, respectively. Scale bars-approximately 25 μm. Courtesy of B. E. Callan, Canadian Forest Service.

A–C. Root and butt rot of *Picea glauca* (white spruce) said to be caused by *Onnia tomentosa*. A. Basidiocarps in unusual profusion on the butt and buttress roots of a dying tree (ON, Aug). Note the resemblance of these basidiocarps to *O. circinata* in D. B, C. Part of the same scene 2 years later. B. Encrusted resin on a buttress root indicates prior host reaction to the fungus. C. A dead overwintered basidiocarp that grew from a decaying root the previous year (ON, Jul).

D. A mature specimen of *O. circinata* growing from a root of a stump (BC, summer).

E–G. Basidiocarps of *O. tomentosa*. E. Upper and lower surfaces of mature specimens (BC, summer). F. The poroid surface of a stalked specimen (ID, Jul). G. A large young specimen (ON, summer).

H. White pocket rot, typical of that caused by *Onnia* species, in *Pinus monticola* (western white pine) (ID, Jul).

Photo credits: A—Canadian Forestry Service Great Lakes Forestry Centre, D—R. Reich, E—E. A. Allen, G—E. S. Kondo

Brown Root and Butt Rot of Conifers (Plate 169)

Brown root and butt rot, also known as brown cubical butt rot, is caused by the velvet-top fungus, *Phaeolus schweinitzii* (Polyporales, Polyporaceae). This fungus occurs around the world and is one of the most common root and butt pathogens in both natural and planted coniferous forests. It is considered to be the most serious butt rot pathogen of conifers in old-growth forests in the USA. It was associated with butt rot in 5–30% of old-growth *Pseudotsuga menziesii* (Douglas-fir) harvested in various parts of the Northwest, for example, and with 66% of all storm-broken butts of that species in the mountains of Arizona. Damage occurs not only in longtime coniferous forests but also in conifers on sites where hardwoods formerly grew. *P. schweinitzii* also affects conifers in landscapes.

The velvet-top fungus is a generalist on conifers. It commonly causes butt rot in *Abies* (fir, at least 15 species), *Larix* (larch, at least 6 species), *Picea* (spruce, at least 7 species), *Pinus* (pine, more than 50 species), and *Pseudotsuga* (Douglas-fir). Other gymnosperm host genera include *Agathis* (Dammar-pine, kauri), *Calocedrus* (incense-cedar), *Chamaecyparis* (false-cypress), *Sequoiadendron* (giant sequoia), *Taxus* (yew), *Thuja* (arborvitae, western red cedar), and *Tsuga* (hemlock). Angiosperms are rarely infected. There are, however, records of *P. schweinitzii* on *Acacia*, *Betula* (birch), *Eucalyptus* (gum), *Halesia* (silverbell), *Liquidambar* (sweetgum), *Prunus* (cherry), *Quercus* (oak), and *Ulmus* (elm).

Symptoms and signs. Trees of all ages are susceptible to *P. schweinitzii*, and those larger than saplings are commonly attacked. Affected trees are usually scattered rather than in groups. Disease is seldom detected until the pathogen produces basidiocarps or a tree falls or breaks during a storm. Aboveground symptoms are not diagnostic. The growth rate of trees with severely diseased roots diminishes, lower branches begin to die prematurely, foliage becomes thin, and foliage on upper branches turns chlorotic. Trees with these symptoms either die upright or break at the butt.

Trunk rot without a connection to decaying roots is rare. Decay eventually involves most of the cross section of the butt and usually extends less than 3 m up the trunk before the tree breaks or is blown over. Decay columns extending more than 10 m into old *Pinus ponderosa* trunks and as high as 30 m in *Pseudotsuga* have been reported, however. Wood in the incipient stages of decay is stained pale yellow or yellow-brown to reddish brown. It loses strength rapidly and gradually becomes dry and crumbly with a pronounced tendency toward transverse fracture. In late stages it turns darker brown and cracks into more or less cubical pieces with white to yellowish mycelium in the cracks. Decaying wood often has a pungent odor that has been likened to turpentine and to anise.

Basidiocarps grow annually from decaying wood and live for 2–3 months in late summer and autumn, or only in autumn in parts of western North America where summers are dry. They develop on stumps of trees that were diseased before felling, on the bases of living diseased trees near ground level, on roots, or on soil. A basidiocarp on soil is connected by a stout strand of mycelium. Such basidiocarps may arise 1–2 m or more from a trunk, grow to 30 cm across, and stand 15 cm tall on a thick velvety stalk. Specimens growing from trunk bases take the form of brackets lacking a stalk. The upper surface of a mature basidiocarp is dark rust-brown and has a rough-velvety texture when fresh. Young expanding specimens have a thick yellow rim. The underside has one to three angular pores per millimeter, is gray-green to gray-brown when fresh, and turns brown where bruised. Active basidiocarps are water-soaked to soft-corky and often exude droplets on the undersurface. In the North their annual activity is terminated by frost, after which they turn brownish black and corky. After drying they are very light in weight. Dead basidiocarps may persist a year or more. Small or young basidiocarps of *P. schweinitzii* are sometimes confused with those of *Onnia circinata* or *O. tomentosa*.

Disease cycle. P. schweinitzii can infect roots of any age, but many details of its infection biology are unknown. Most infections are believed to be initiated by spores, and the disease cycle is therefore completed when basidiocarps first form on a diseased or killed tree and release a new generation of spores. Basidiospores of *P. schweinitzii* that land on the forest floor can establish persistent infestation of soil; whether the basidiospores persist or germinate to produce resting spores is unknown. *P. schweinitzii* does produce chlamydospores, and they have been found in soil. They also form readily in agar media and furnish one of the diagnostic criteria for identifying the fungus in culture. Although the fungus is not known to grow extensively as a soil saprobe, its mycelium has been detected in soil and it has been isolated from forest soils, including some where no known root rot has occurred. Adjacent trees affected by *P. schweinitzii* usually contain different clones of the fungus, an indication that it does not usually spread from tree to tree through roots. There is, however, one report of butt rot in *Pinus elliottii* (slash pine) 10–22 years old associated with infested stumps of a previous pine crop.

P. schweinitzii is considered to be a weak or opportunistic pathogen. It often enters damaged or perhaps dead superficial roots or wounds such as fire scars at the trunk base. Severe butt rot may occur in trees that lack such external markers, however. The fungus has been reported to infect and kill root tips of *Pseudotsuga menziesii* and induce swelling at the living ends of the resulting root stubs in western USA. Its growth rate from inoculated wounds in woody roots of *P. menziesii* studied in Germany varied from 2 to 20 cm per year. It has caused severe damage to *Pinus resinosa* (red pine) and *P. strobus* (eastern white pine) planted on shallow soils subject to periodic waterlogging and summer drying, which weaken or kill many roots. Roots colonized by *Armillaria* (Plates 162–164) have also been suggested as avenues of infection because the two pathogens are often associated and *Phaeolus schweinitzii* is capable of growing through wood colonized by *Armillaria*. In *Abies grandis* (grand fir) and *Pseudotsuga menziesii* in Idaho, however, *Armillaria* has been considered to be secondary to the velvet-top fungus because it was found at the root collars of trees with severe root rot caused by the latter. In experiments, *P. schweinitzii* penetrated and grew within both suberized (i.e., corky) and nonsuberized rootlets of *Pinus strobus* (eastern white pine) seedlings. On the other hand, it was unable to colonize bark wounds in *Picea sitchensis* (Sitka spruce) roots and was found to be strongly inhibited by constitutively produced antifungal compounds in the root bark.

Site factors that influence infection by *Phaeolus schweinitzii* are poorly understood. Shallow, poorly drained soils apparently predispose *Pinus resinosa* and *P. strobus* to damage in eastern USA. In northwestern forests the fungus attacks various conifers on diverse sites representing a wide range in elevation. In that region it also fruits commonly on old *Pseudotsuga* stumps in young stands that do not seem to be affected by brown root and butt rot.

References: 53, 237–240, 266, 391, 461, 901, 943, 1280, 1694, 2212, 3065, 4183, 4257, 4430, 4431, 4487

A–G. Brown butt rot of *Larix kaempferi* (Japanese larch), caused by *Phaeolus schweinitzii* (NY, Jun–Oct). A. Severe butt rot in a living tree broken by wind. B. Brown cubical rot is visible near ground level in a broken butt. Dark brown basidiocarps at the base of the trunk are diagnostic (Oct). C. Dead basidiocarps of the previous year on a wound scar (Jun). D. Cross section near the base of a stem with butt rot. The decay is associated with a basal wound that occurred when the tree was less than 15 years old (Jun).

E–I. Basidiocarps of *P. schweinitzii*. E, F. Clusters at the bases of a diseased *L. kaempferi*, appearing dark reddish brown above and yellowish tan below after the onset of freezing weather in autumn (NY, Oct). G. A cluster of mature basidiocarps collected from soil at the base of a tree; a stout stalk is apparent at one side of the cluster (NY, Oct). H. Fresh basidiocarps associated with decaying roots of a landscape specimen of *Picea pungens* (NY, Aug). I. The lower, poroid surface of a basidiocarp (NY, Aug).

Laminated Root Rots of Conifers (Plate 170)

Laminated root rots are caused by two fungi that have traditionally been united under the name *Phellinus weirii* (Hymenochaetales, Hymenochaetaceae). The so-called Douglas-fir form of *P. weirii* is a major forest pathogen that occurs from southern British Columbia to northern California and eastward to the continental divide in Montana. It attacks members of the Pinaceae and occasionally Cupressaceae on diverse sites from sea level to the upper limits of managed forests. It is the most economically important pathogen in *Pseudotsuga* (Douglas-fir) forests, especially young managed forests. Annual losses due to laminated root rot caused by the Douglas-fir form in North America have been estimated at more than 4 million cubic meters of wood, and losses continue where susceptible species grow on sites formerly occupied by diseased trees. This pathogen is likely to be recognized as a discrete species and to be given a new name as taxonomic research in the Hymenochaetaceae progresses. In Asia a fungus closely related to the Douglas-fir form of *P. weirii* occurs in several regions from the Urals to Japan.

The cedar form of *P. weirii* preferentially attacks Cupressaceae (mainly *Thuja plicata*, western red cedar), causing laminated root and butt rot, and is occasionally found on species of Pinaceae. Its distribution is similar to that of the Douglas-fir form, but it is found most often in the eastern part of their common range. The cedar form is not considered further here except to clarify information about the Douglas-fir form.

Damage by *P. weirii* is greatest in *Pseudotsuga menziesii*. *Abies amabilis* (Pacific silver fir), *A. concolor* (white fir), *A. grandis* (grand fir), and *Tsuga mertensiana* (mountain hemlock) are also highly susceptible. *A. lasiocarpa* (alpine fir), *A. magnifica* (red fir), *A. procera* (noble fir), *Larix occidentalis* (western larch), *Picea engelmannii* (Engelmann spruce), *P. sitchensis* (Sitka spruce), and *Tsuga heterophylla* (western hemlock) are less susceptible but do sustain butt rot and are sometimes killed. *Calocedrus decurrens* (incense-cedar), *Chamaecyparis nootkatensis* (Alaska cedar) *Pinus contorta* (lodgepole pine), *P. monticola* (western white pine), *P. ponderosa* (ponderosa pine), and *Thuja plicata* are resistant.

Symptoms and signs. Laminated root rot usually affects groups of trees in patches that enlarge for many years, but it is also found in small scattered groups. Trees of any age may be affected, but the disease is most destructive in stands 25–125 years old. Diseased, dead, and fallen trees of highly susceptible species may occupy areas several square meters to a hectare or more in size. Large old disease patches often have green centers where young trees, usually of resistant types, have grown from seed. Trees with diseased roots may show no symptoms above ground for 5–15 years. Eventually their growth slows and the crowns become thin or asymmetric with chlorotic foliage and distress crops of abnormally small cones. Young diseased trees often die before decay extends far into the trunk. Large old trees may have decay columns extending as high as 4 m. Some trees are killed by girdling necrosis in the cambial zone at the butt; others fall while alive because of decay in the roots and butt. Decayed roots typically break at or near the butt and do not lift from the soil as a tree falls. Thin-barked diseased roots of highly susceptible trees are usually ensheathed by prominent superficial light-colored mycelium in which diagnostic reddish brown hairs about 0.5 mm long (setal hyphae) can be seen with a hand lens. This mycelium can be exposed by careful removal of soil from around diseased roots. The mycelium lacks clamp connections. Thin brown fungal crusts often form on root surfaces and on the butt in root crotches. Declining trees are frequently attacked by bark beetles. Additional root pathogens such as *Armillaria* (Plates 162–164) or *Leptographium wageneri* (Plate 114) are often found in the disease patches, especially east of the Cascade Mountains.

Wood in early stages of decay by *P. weirii* is reddish brown to chocolate brown and, when viewed at fresh stump surfaces, usually appears as crescents or irregular patches in the inner sapwood. Wood in late stages of decay is yellowish and tends to separate into sheets corresponding to annual rings. It contains numerous tiny cavities about 0.5 × 1.0 mm in size. Setal hyphae can be seen in cracks and cavities and also between the bark and wood. In addition, thin plaques or tufts of brown mycelium are usually present in cracks in decayed wood. Decaying wood eventually breaks down into a loose stringy mass.

Basidiocarps develop during late summer and early autumn as gray-brown crusts with a poroid surface and white to cream-colored margins while growing. They form on decaying roots and butts, in root crotches, and on the lower sides of fallen trunks, but are neither numerous nor conspicuous in most sites. Those on fallen trunks are typically 30–60 cm long, but larger specimens are common. The tubes are tiny (five to seven pores per millimeter) and oriented vertically. By the next spring most basidiocarps of the Douglas-fir form are dead and moldy. Those of the cedar form on *Thuja plicata* live for several years, producing a new layer of tubes each year.

Disease cycle. Nearly all disease patches caused by *P. weirii* in contemporary forests are considered to be initiated by mycelium from the residue of previous forests. The pathogen persists for decades in decaying roots, butts, and logs, and produces mycelium on root surfaces. Infection begins when healthy roots grow into contact with this mycelium. It grows along a root and penetrates at numerous points through intact or injured bark. Then it spreads within and along the root and infects others at points of contact. Superficial mycelium advances beyond the internal decay in highly susceptible trees. Multiple infections where roots of a single tree cross one another facilitate destruction of the root system. Disease patches expand radially at rates of 20–40 cm per year, which correspond to measured rates of growth of *P. weirii* along roots. Many trees resist the parasite and may halt its advance temporarily. Adventitious roots sometimes grow from callus and woundwood that form where the pathogen has been halted.

Mycelium of *P. weirii* grows as long-lived clones. A given clone tends to be incompatible with most others, so mycelium of this fungus in a disease patch is initially uniform. If patches inhabited by different clones merge, each clone maintains its own sector and contributes to expansion of the composite patch. Mycelial growth occurs at temperatures ranging from below 5° to 30°C, and is most rapid near 25°C.

No role for basidiospores in the disease cycle has been confirmed, although mycelial clones are presumed to be derived from them. Basidiospores were found capable of establishing colonies on surface-sterilized wood of several species, but stump infection by spores has not been demonstrated. Spore infection of roots and butt wounds are unstudied possibilities.

Damage from laminated root rot becomes more severe, and disease patches expand more rapidly, in pure stands of susceptible species than in stands of mixed species. Stump removal and planting of resistant tree species or species mixtures offer possibilities of continued forest productivity on sites where laminated root rot has been severe.

References: 160, 417, 420, 421, 560, 878, 879, 931, 1136, 1309, 1481–1483, 1485, 1694, 1838, 2243, 2535, 2554, 2734, 2818, 2911, 3374, 3995, 3997, 4203–4205

A. An opening caused by root-infecting fungi in a mixed conifer forest. Laminated root rot, caused by *Phellinus weirii*, is the principal disease. *Pseudotsuga menziesii* (Douglas-fir) in the gap have died; *Larix occidentalis* (western larch) and *Pinus monticola* (western white pine) survive in low vigor (ID, Jul).

B, C. Dead and dying young *Pseudotsuga menziesii*. B. Decayed roots remain in the soil where trees have fallen. C. Slow growth, thin foliage, chlorosis, and abnormal cone production are typical aerial symptoms (OR, Jul).

D, E. Excavation at the base of a young *P. menziesii* dying from attack by *Phellinus weirii*. D. Resin exudation from the butt and whitish mycelium plus brown encrustation on surfaces of decaying roots. E. A brown crust of mycelium, diagnostic for *P. weirii* (OR, Jul).

F–I. Discoloration and decay caused by *P. weirii* in *Pseudotsuga*. F. Reddish brown discoloration in inner sapwood in a stump where the fungus has spread upward from roots. G–I. Advanced decay, soft and yellowish, with numerous tiny cavities and a tendency for the annual layers to separate (ID, OR, & WA, summer).

J. Magnified view of reddish brown setal hyphae (sterile hairs) on a mat of mycelium between bark and wood of a recently killed tree (WA, Jun).

K, L. Basidiocarps of *Phellinus weirii* on the lower surfaces of the butts of small fallen *Pseudotsuga menziesii*. K. A fresh growing specimen (OR, autumn). L. A dry specimen that formed the previous autumn (OR, Jun).

Photo credits: B, F, I—W. G. Thies; K—P. A. Angwin

Root and Butt Rots Caused by *Inonotus dryadeus* and *Oxyporus latemarginatus* (Plate 171)

Root Rot Caused by *Inonotus dryadeus*

Inonotus dryadeus (syn. *Fomitiporia dryadea*, *Pseudoinonotus dryadeus*; Hymenochaetales, Hymenochaetaceae), also known as the weeping conk, is among the minority of wood-decaying fungi capable of parasitic colonization of both angiosperm and gymnosperm trees. It occurs around the Northern Hemisphere and in Australasia. In North America it occurs across the USA and in Mexico, British Columbia, and southern Ontario. It is often found as a saprobe on stumps or snags but also causes root and butt rot of various trees, predominantly *Quercus* species (oaks). It is common on conifers in the West. Its North American hosts include the following:

Abies concolor (white fir), *A. grandis* (grand fir), *A. religiosa* (sacred fir), *Acer glabrum* (Rocky Mountain maple), *A. macrophyllum* (bigleaf maple), *A. platanoides* (Norway maple), *A. saccharinum* (silver maple), *Aesculus flava* (yellow buckeye), *Carya* sp. (hickory), *Castanea dentata* (American chestnut), *Ceanothus* sp. (buckbrush), *Eucalyptus* sp., *Fraxinus* sp. (ash), *Liquidambar styraciflua* (sweetgum), *Picea engelmannii* (Engelmann spruce), *P. sitchensis* (Sitka spruce), *Pinus ponderosa* (ponderosa pine), *Platanus wrightii* (Arizona sycamore), *Populus* sp. (poplar), many species of *Quercus*, *Tsuga heterophylla* (western hemlock), *T. mertensiana* (mountain hemlock), and *Ulmus americana* (American elm).

The foregoing list includes records for both living and dead trees. Additional hosts recorded elsewhere include species of *Fagus* (beech), *Populus* (poplar), *Pyrus* (pear), and *Tilia* in Europe; and *Acacia* (wattle) in Australasia. Genetic research to ascertain the degree of relatedness of angiosperm and gymnosperm-inhabiting forms of *I. dryadeus* has not been reported.

The following description of symptoms applies mainly to angiosperms, especially *Quercus*. Root rot caused by *I. dryadeus* is most common in old or suppressed trees and in those on sites unfavorable for vigorous growth. Affected trees may have no aboveground symptoms for many years or until they fall because of decayed roots. Trees that remain standing eventually show symptoms of decline: sparse foliage, poor color, and dieback. Those with many dead roots sometimes die suddenly during periods of summer heat and dryness. Diseased trees ordinarily occur singly rather than in clusters. Roots of trees that fall because of root rot usually break within 30–60 cm of the trunk, and the remainder of the decayed roots stay in the soil. Necrosis of bark and cambium precedes decay of sapwood at a given location on the root system. Irregular white patches of mycelium may form on the surfaces of decaying roots.

Wood decay caused by *I. dryadeus* is of the white mottled type. It spreads toward the butt from initial locations in roots, developing first in sapwood and later in deep-lying wood. In *Quercus* the rot does not spread far above ground level. Superficial roots may be affected only on the lower side or may escape decay altogether. In conifers such as *Abies grandis* the decay spreads into the base of the trunk.

Irregular, often massive basidiocarps grow annually from exposed roots or the trunk at or near ground level, or from the soil, attached to roots. The basidiocarps are light yellow to yellow-brown initially, often with numerous drops of liquid on the surface. They turn brown to chestnut brown at maturity and after several months become blackish with rough, cracked surfaces that eventually weather to gray-brown. The interior is reddish brown to chestnut brown and fibrous-corky in texture. Basidiocarps vary in size up to 40 cm long, 35 cm wide, and 10 cm thick, with rounded margins. The lower surface is poroid, but the tubes in old inactive specimens are usually obscured by mycelium. Mycelium of *I. dryadeus* may permeate sandy soil for several centimeters beneath a fruit body and may hold the soil in a hard mass. Multiple basidiocarps on a tree indicate extensive decay that could cause it to topple during a storm.

Dispersal of airborne basidiospores is assumed to account for establishment of *I. dryadeus* in new substrates, but the site of infection (wounds versus intact roots) is unknown. The rate of invasion is also unknown. No evidence for tree-to-tree spread has been reported. In Mississippi basidiospores are released from November to February when average daily temperatures are in the range 10–16°C. Peak release occurs during daylight and coincides with daily maxima of 22–24°C. This temperature is close to the reported optimum (25°C) for mycelial growth of *I. dryadeus* on agar media.

References: 634, 878, 896, 1117, 1276, 1280, 2361, 2363, 2545, 3374, 3510, 4108, 4183

Rot Caused by *Oxyporus latemarginatus*

Oxyporus latemarginatus (Polyporales, Polyporaceae) causes a white rot in the sapwood of roots and stems of diverse angiosperm trees around the world. It is a common invader of wood behind fire scars on living trees. It is also a common saprobe on logs, snags, and logging slash. In North America it occurs from Nova Scotia and Ontario to Florida and Texas, and from Alberta and British Columbia to California and Arizona, but is uncommon or absent in central regions.

O. latemarginatus is a proven root pathogen. It occasionally causes significant damage in orchards, but its importance in forests and landscapes is unknown. Root and root collar rots caused by it have been noted on at least *Citrus, Juglans regia* (Persian walnut), *Liriodendron* (tuliptree), *Malus pumila* (apple), *Platanus occidentalis* (sycamore), *Populus deltoides* (eastern cottonwood), *Prunus dulcis* (almond), *P. mahaleb* (mahaleb cherry), *P. persica* (peach), and *Pyrus* (pear). Other hosts, mainly those on which the fungus occurs as a saprobe or is associated with trunk decay, include species of the following:

Acacia (acacia, wattle), *Acer* (maple), *Albizia* (mimosa), *Alnus* (alder), *Amelanchier* (serviceberry), *Aralia*, *Betula* (birch), *Carya* (hickory), *Celtis* (hackberry), *Cladrastis* (yellowwood), *Fagus* (beech), *Fraxinus* (ash), *Gleditsia* (honeylocust), *Juglans* (walnut), *Ligustrum* (privet), *Liquidambar* (sweetgum), *Magnolia* (bay), *Melia* (chinaberry), *Nyssa* (sour-gum), *Ostrya* (hop-hornbeam), *Phoenix* (date palm), *Populus* (aspen, cottonwood, poplar), *Quercus, Rhododendron* (azalea), *Robinia* (locust), *Salix* (willow), *Sambucus* (elder), *Syringa* (lilac), *Ulmus* (elm), and *Umbellularia* (California-laurel).

Slow growth and mortality of young trees are the aboveground symptoms of root rot but are not diagnostic. Signs of the pathogen are usually present, however. *O. latemarginatus* produces either a sterile white mat of mycelium or a white to dirty white or pale yellow basidiocarp at the soil line of a dying or recently dead plant. The basidiocarp, active for only one season, may envelop litter and blades of grass as it grows. The fungus also infests soil around diseased roots and the root collar and may form a white crust or network of white mycelial strands on dead parts of roots or on the soil at the stem base.

O. latemarginatus appears to kill small roots first and then grow toward the root collar. It colonizes outer bark and then invades the inner bark, cambial region, and sapwood, causing enlarging cankers and a soft white rot of the wood. Seedlings may be killed within 9 months. Mature (16-year-old) *Prunus* (cherry) have succumbed within 2 years in California. Parasitic attack proceeds most rapidly during warm summer weather.

White mycelium of this fungus grows rapidly on agar media. Growth is maximal at temperatures near 30°C and in one study was 88–90% of maximum at 25–35°C.

Although root rot by *O. latemarginatus* has frequently been linked to waterlogged soil or other conditions that might damage roots, it has also been severe in *Prunus* in orchards with well-drained soils in California. It is sometimes associated with damage by Phytophthora root and crown rot on *Prunus* in that state.

References: 17, 264, 941, 1131, 1280, 2359, 3542, 4108

A, B. Basidiocarps of *Inonotus dryadeus* on *Acer saccharinum* (silver maple) and *A. platanoides* (Norway maple), respectively (ON, Sep & early summer, respectively).

C–E. Indicators of root and butt rot by *I. dryadeus* on *Quercus phellos* (willow oak). C. Irregular dark brown to black old basidiocarps in root crotches of a mature tree. D. Flat, pattylike basidiocarps of two age classes extending from a tree at soil level; the older structures are weathered gray-brown. E. A small clinkerlike basidiocarp that grew from a decaying root beneath the soil surface (MS, Apr).

F, G. Year-old inactive bracket-shaped basidiocarps of *I. dryadeus* on the butt of a dead *Abies grandis* (grand fir) (OR, Jul).

H, I. Root rot of *Prunus persica* (peach) and *Populus deltoides* (cottonwood) seedlings, respectively, caused by *Oxyporus latemarginatus*. The fungus produces white mycelial sheets and poroid basidiocarps at the bases of recently killed small plants (MS, Apr).

Photo credits: A—J. C. Krug; B—P. Dmytrasz

345

Ganoderma Root and Butt Rots and Trunk Decay (Plates 172–174)

Overview

Ganoderma species (Polyporales, Ganodermataceae) are wood-decaying fungi that occur around the world on woody monocots, dicots, and gymnosperms. The genus is estimated to comprise 70–110 species, of which 13–14 occur in North America. *Ganoderma* species cause decay of the white rot type in which lignin and cellulose are degraded together (simultaneous rot) or lignin is removed selectively. Some species are apparently only saprobic, but several are important pathogens of roots, butts, and trunks of living trees and shrubs. Some are consumed by humans as health food or folk remedies and/or are used as sources of pharmacological products. Ganodermas often kill their hosts, especially palms, but often a diseased tree is windthrown or breaks while still alive as a result of decay in the butt and base of the trunk. Tropical Ganodermas cause economic damage in plantations of *Camellia sinensis* (tea), *Cocos nucifera* (coconut palm), *Elaeis guineensis* (oil palm), and *Hevea brasiliensis* (rubber). Temperate species shorten the useful lives and/or timber value of shade, forest, and some orchard trees.

Differentiation of *Ganoderma* species is problematic, even for specialists. Putative species were traditionally distinguished on the basis of color, form, size, and anatomy of basidiocarps; presence/absence and color of a laccate (varnishlike) crust on the upper surface; and size, form, and wall architecture of basidiospores. Additional characters include the presence or absence of chlamydospores in culture, temperature preferences, and host preferences. Some species are thermophilic, growing well at temperatures above 30°C. About three-fourths of the species have annual basidiocarps that are either bracketlike or attached to the substrate by a stalk and have a laccate yellowish to reddish brown or darker crust. The remainder have perennial basidiocarps that lack a stalk and have a hard, dull, gray or gray-brown or darker upper surface. All Ganodermas produce thick-walled reddish brown to yellowish brown or brown basidiospores that often accumulate in conspicuous deposits on and around basidiocarps.

Interfertility studies and analyses of DNA sequences in recent years have revealed several species complexes in *Ganoderma*, each comprising multiple genetically distinct taxa with similar morphological and cultural characteristics. Several of the most familiar species names, such as *G. lucidum* and *G. applanatum*, are associated with collections so genetically heterogeneous as to represent multiple natural species. Therefore, these names will come to be applied more narrowly than they have been heretofore. As this was written, appropriate names were available for only a fraction of the species.

References: 12, 14, 15, 559, 809, 1150, 1280, 1328, 1329, 1859, 2697, 2698, 3374, 3862, 4108

Root and Butt Rot of Palms

At least six *Ganoderma* species attack palms. *G. zonatum* is blamed for damage to numerous palm species in Florida, Georgia, the Caribbean region, and tropical Africa. It has also been reported on various dicots, but such records are probably erroneous. *G. colossum*, a thermophilic species, has been found on *Phoenix canariensis* (Canary Island date palm) in Florida. *G. boninense* is a major pathogen of *Elaeis guineensis* (oil palm) and *Cocos nucifera* (coconut palm) in the region extending from Sri Lanka to Australasia and Japan. *G. tornatum* (of the *G. applanatum* complex, see page 350) has been reported on palms and various dicots throughout the tropics (e.g., on *Casuarina* in Florida), although its true distribution perhaps includes only Southeast Asia and some Pacific islands. The other palm-colonizing Ganodermas are less well known. Some species on palms in the tropics have been confused with temperate species, and this confusion persists. We focus on diseases caused by *G. zonatum*.

Probably most palms in North and Central America are susceptible to *G. zonatum*. It affects palms in more than 30 genera in the USA, including the following:

Acoelorraphe (Everglades palm), *Acrocomia, Adonidia* (Christmas palm), *Aiphanes, Arenga, Attalea, Bactris* (spiny-club palm), *Brahea, Butia* (pindo or jelly palm), *Carpentaria, Caryota* (fishtail palm), *Chamaerops* (fan palm), *Coccothrinax, Cocos, Copernicia, Dictyosperma* (princess palm), *Dypsis* (areca, butterfly, or yellow palm), *Elaeis, Euterpe* (Assai palm), *Gastrococos,* *Hyophorbe* (pignut palm), *Latania, Livistona* (fountain palm), *Nannorrhops, Phoenix* (date palm), *Ptychosperma, Roystonea* (royal palm), *Sabal* (cabbage palm), *Satakentia, Serenoa* (saw or scrub palmetto), *Syagrus* (queen palm, Arikury palm), and *Washingtonia* (fan palm).

Symptoms and signs. Palms of all ages are affected, and the likelihood of visible damage increases with age. Symptoms include loss of vigor, undersized fruit and fronds, premature yellowing and death of the oldest fronds, and wilting. The crown eventually dies and breaks off. Basidiocarps grow out from the base of the trunk or from adventitious roots. Their presence indicates advanced decay, which renders the surface roots dry and brittle and the underlying tissues weak and crumbly. Palm butts long diseased commonly have masses of white mycelium in the decayed tissues and sometimes have hollow centers. Tissues in early stages of decay are dark brown and sometimes are water-soaked.

Basidiocarps of *G. zonatum* are up to 20 × 8 × 5 cm thick at the base; generally lack a stalk; are often fan or kidney shaped or semicircular as viewed from above; and have a hard, thin, varnishlike yellowish brown to reddish brown or gray-brown crust on mature parts of the upper surface. This surface often has irregular lumps and usually has concentric ridges and furrows. The margin is white to cream colored and somewhat swollen while growing. Interior tissues are brown to purplish brown. Most of the lower surface is covered with nearly circular pores, five or six per millimeter, that are the mouths of tubes that extend vertically up to 1.5 cm into the basidiocarp. Each tube is lined with a spore-producing layer, the hymenium, from which thick-walled brown basidiospores 11–13 × 5–6 μm are released. These spores often collect beneath basidiocarps, or on their upper surfaces if dropped from others above, and impart a dusty brown appearance. When growing in culture *G. zonatum* produces hyphal rosettes composed of several hyphal branches projecting from a common origin. It does not produce chlamydospores.

Disease cycle. Diseases caused by *G. boninense* have received the most study. This fungus as well as *G. zonatum* and various unnamed or misidentified isolates from palms are proven pathogens. Infection by *G. boninense* occurs when roots of planted palms grow into contact with infested roots, butts, and stumps. Infection is most likely when the pathogen has a large food base. Decay progresses through a root system and up into the lower part of the trunk. Radial spread from the first infected plant to others nearby is limited, however, because the pathogen does not persist long in small roots. Basidiospores also initiate infections, as indicated by the common occurrence of different genotytpes of the pathogen in adjacent palms. Researchers have suggested that the spores may infect freshly cut stumps or wounds made by insects or other agents, but direct evidence for infection by spores is lacking. Basidiospores probably account for establishment of the pathogen on sites where palms did not previously grow. Stump infection is indicated by the presence of identical isolates of the pathogen in the stump of a formerly healthy palm and in young plants infected by contact with it.

Ganoderma in palm plantations is managed in part by removing stumps. Small residual roots, even if infested with *Ganoderma*, are not a significant threat to new palms if planting is delayed for a few months. During that period the pathogen tends to perish in competition with soil saprobes. This information can be applied to ornamental plantings as well.

References: 10, 46, 106, 888, 1051, 1052, 1150, 1153, 1859, 2793, 3018, 3411, 3863, 3864, 4125

Butt rot of palms caused by *Ganoderma zonatum* (all FL, Mar–Apr).

A, B, D. Yellowing and premature death of fronds, and young basidiocarps on the butt of a dying *Elaeis guineensis* (oil palm).

C. Large inactive basidiocarps on a stump of *Syagrus romanzoffiana* (queen palm).

E. Small inactive basidiocarps on the adventitious roots of a living *S. romanzoffiana*.

F. Upper surfaces of young basidiocarps, about one-third natural size. The **X**-shaped scratch on the white, growing margin of one specimen exposes dark brown interior tissue.

G. Magnified view of the lower surface of a basidiocarp showing the pores, or mouths of tubes, from which spores are released.

347

Diseases of Woody Dicots Caused by Laccate Ganodermas

Many *Ganoderma* species with a varnishlike (laccate) crust attack woody dicots. On molecular evidence (DNA sequences, isozyme analyses) these species are arrayed in two main groups, each with several subdivisions. North American pathogens occur in both main groups. In order to focus on the North American fungi it is necessary to introduce two species, *G. lucidum* and *G. resinaceum*, that were first described in Europe. *G. lucidum*, considered in the strict sense, is possibly restricted to that continent; *G. resinaceum* probably has a wider distribution. Both fungi attack such woody dicots as *Alnus* (alder), *Fagus* (beech), *Populus* (poplar), *Robinia* (black locust), and *Quercus* (oak). *G. lucidum* also infects conifers, particularly *Picea* (spruce). Fungi known historically as *G. lucidum* that attack dicots in North America actually represent at least two species: *G. curtisii* and a taxon closely related to or identical with *G. resinaceum*. The latter taxon, which we refer to as *G. "lucidum,"* occurs in the eastern half of the USA and adjacent Canada and from Oregon to Arizona in the West. Some strains of *G. "lucidum"* are interfertile with *G. resinaceum*. *G. curtisii* is widespread in the eastern half of the USA, especially in the Southeast. It is morphologically and anatomically similar to *G. "lucidum"* (see *Symptoms and signs*), but DNA sequence analyses have revealed marked genetic separation of the two taxa. We use the term laccate *Ganoderma* for information that pertains to both species or to inadequately identified representatives of them.

The geographic and host ranges of *G. curtisii* and *G. "lucidum"* apparently overlap broadly. Their North American host lists are combined here because separate records cannot be sorted out. Some southeastern records for palms and gymnosperms are included provisionally. Hosts include the genera listed below.

Acer (maple), *Albizia* (mimosa), *Betula* (birch), *Buxus* (boxwood), *Calliandra* (powderpuff), *Carpinus* (hornbeam), *Carya* (hickory), *Cassia*, *Castanopsis* (chinkapin), *Casuarina* (Australian-pine), *Celtis* (hackberry, sugarberry), *Cercis* (redbud), *Citrus* (citrus and calamondin), *Dioon* (a cycad), *Diospyros* (persimmon), *Dypsis* (areca palm), *Fagus* (beech), *Fraxinus* (ash), *Gleditsia* (honeylocust), *Gymnocladus* (Kentucky coffee-tree), *Lantana*, *Ligustrum* (privet), *Limonia* (wood-apple), *Liquidambar* (sweetgum), *Lysiloma*, *Magnolia*, *Malus* (apple, crabapple), *Nyssa* (sour-gum), *Parkinsonia* (paloverde), *Persea* (avocado), *Phoenix* (date palm), *Pinus* (pine), *Platanus* (sycamore), *Poncirus* (trifoliate orange), *Populus* (cottonwood), *Prosopis* (mesquite), *Prunus* (cherry, peach, plum), *Quercus* (oak), *Rhus* (sumac), *Robinia* (black locust), *Salix* (willow), *Sassafras*, *Schinus* (pepper-tree), *Swietenia* (mahogany), *Syagrus* (queen palm), *Taxodium* (bald cypress), *Ulmus* (elm), and *Vitis* (grapevine).

Pathogenicity of some North American laccate *Ganoderma* isolates to bark and sapwood of various woody dicots has been proven, and descriptions of naturally occurring symptoms support pathogenicity of both *G. curtisii* and *G. "lucidum."* A laccate *Ganoderma* was found, for example, in about 20% of a sample of more than 300 declining *Acer platanoides* (Norway maple) and *A. rubrum* (red maple) in the New York metropolitan area. *G. curtisii* is often associated with decline and death of landscape trees in the Southeast. In hardwood forests these fungi are usually found growing as saprobes or associated with decay in the butt and trunk of various trees.

Symptoms and signs. Trees affected progressively by *Ganoderma* develop widespread decay of sapwood in the butt and major roots. They eventually lose vigor and have undersized and sometimes yellow or wilting leaves, thin crowns, and dead branches. Some die of the disease; others fall in storms. Wood in advanced stages of decay by *Ganoderma* is light colored and stringy or spongy.

Rapid decline and death of *Gleditsia triacanthos* (honeylocust) and *Sassafras albidum* (sassafras) infected by laccate *Ganoderma* have been reported, but these fungi usually work slowly. Diseased *Quercus rubra* (red oak) sometimes exhibits no change in general health of the crown during 30 years or more after basidiocarps of *Ganoderma "lucidum"* first appear at the ground line. This observation suggests that some infections are arrested or caused by strains of low aggressiveness. In pathogenicity tests 3–5 years elapsed between inoculation and death of *Liquidambar styraciflua* (sweetgum) and *Albizia julibrissin* (mimosa) seedlings, although lesions formed on inoculated roots and were followed by general root necrosis and decay.

Basidiocarps provide the first specific indication of Ganoderma root rot. They grow from roots or butts. Free-standing specimens are connected to roots. Basidiocarps of *G. curtisii* are kidney to fan shaped or sometimes disc shaped, usually $3–12 \times 3–20 \times 0.7–5$ cm in size, with concentric zones of varying topography and a stalk that is attached laterally or rarely near the center. They are leathery to corky when fresh. The crust is yellowish brown to dull red or has reddish yellow areas. The interior tissue is soft and light colored, but brownish near the tube layer. The poroid (lower) surface is white to brownish or pale yellow. The spores are ovoid with a truncate apex, light brown, $9–11$(to 13) $\times 5–7$ μm. The basidiocarps of *G. "lucidum"* are annual, $2–20 \times 3–35 \times 0.5–8$ cm in size, of irregular form, and usually lack a stalk. They develop singly or overlap one another in clusters. The crust varies in color from mahogany to blood red or blackish red. The margin and poroid surface are white during growth and tawny with age. The interior tissue and basidiospores are similar to those of *G. curtisii*. Where basidiocarps are clustered, their upper surfaces and often the surrounding grass or bark may be discolored brown by a deposit of basidiospores.

Disease cycle. Basidiospores are presumed to initiate most infections. Wounds on roots and trunk bases are likely infection courts, and basidiocarps commonly grow from the vicinity of old wounds. Basidiospores, dispersed throughout the summer, are released in greatest numbers during evening hours when the air is humid. Experiments have shown that infection by root contact with previously colonized wood is also possible, but tree-to-tree spread is not indicated by field observations.

Environmental stress and severe wounding seem to predispose trees to damage by laccate Ganodermas. In fern nurseries in Florida, for example, *Quercus* (oak) trees used as a canopy have shown unusual incidence of Ganoderma root rot after the use of herbicides and soil fumigants. *G. curtisii* is among the fungi commonly associated with decline and mortality of drought-stressed *Quercus* in southeastern USA.

References: 11–14, 46, 559, 1150, 1153, 1280, 1328, 1422, 1614, 1859, 2302, 2545, 2937, 3082, 3374, 3510, 3865, 4061, 4108

Ganodermas on Conifers

Three species are known in North America. *G. oregonense* and *G. tsugae* occur on *Abies* (fir), *Picea* (spruce), *Pinus* (pine), *Pseudotsuga* (Douglas-fir), and especially *Tsuga* (hemlock), causing white stringy rots of roots and butts of dead trees. *G. tsugae* is widespread in the East and Southwest, and *G. oregonense* occurs in the Northwest. *G. tsugae* is apparently not pathogenic, but *G. oregonense* occasionally causes heart rot in living *T. heterophylla* (western hemlock). They differ in that the heavily varnished basidiocarps of *G. oregonense* attain larger size, commonly 40 cm or more across, and produce larger spores. The third species, *G. meredithae*, occurs on living and dead *Pinus glabra* (spruce pine) and *P. taeda* (loblolly pine) from Georgia to east Texas. It causes a white root and butt rot. *G. meredithae* is highly similar to *G. curtisii*, and differences between the two have not been adequately delineated.

References: 11–14, 397, 559, 1064, 1150, 1280, 2937, 4108

A, B. Root rot caused by *Ganoderma* sp. in *Gleditsia triacanthos* (honeylocust). A. Dieback associated with the disease (MS, Apr). B. Young basidiocarps of *Ganoderma* sp. developing as irregular white masses at the base of a diseased tree. Several initially discrete masses would later coalesce to form large basidiocarps (MS, Apr).

C, D. Basidiocarps of *G. "lucidum"* associated with root and butt rot of *Quercus rubra* (northern red oak). The cluster of basidiocarps in D was about 30 cm in diameter (NY, Aug–Sep).

E, F. Basidiocarps of *G. curtisii*. E. A growing specimen associated with *Quercus* sp. F. Older specimens with a central stalk (left) or lateral stalk (NC, Sep.)

G. Mature basidiocarps of *G. tsugae* on a decaying *Tsuga canadensis* (eastern hemlock) stump (NY, Sep).

H. A young basidiocarp of *G. oregonense*, about 12 cm across, growing from a decaying root of *T. heterophylla* (western hemlock) (ID, Jul).

I. Stringy white rot caused by *G. oregonense* in *T. heterophylla* (ID, Jul).

Photo credits: E, F—J.-M. Moncalvo

Diseases Caused by *Ganoderma applanatum* and Related Fungi

G. applanatum (syn. *Elfvingia applanata*) is the most common, widespread, and important of the Ganodermas in North America that have perennial unvarnished (nonlaccate) basidiocarps. It causes decay of sapwood and heartwood in butts, trunks, and sometimes roots of a wide array of trees. Both hardwoods and conifers are affected, although usually the former. The fungus is also an important decomposer of logs and stumps. Its basidiocarps are popular with artists for etching because the white lower surface of a fresh specimen turns dark brown where marked by a stylus. When dry, this surface is suitable for painting. *G. applanatum* is apparently widespread in temperate regions of the Northern Hemisphere. Reports of it in the tropics and Southern Hemisphere probably relate to other species, especially those in the *G. australe* complex.

DNA sequence analyses of *G. applanatum* strains have indicated that they represent two morphologically similar but genetically distinct species. If this finding is corroborated by intersterility studies, other distinctions between the two taxa will need to be defined.

When a basidiocarp of *G. applanatum* develops on a living tree, extensive decay is present within and the likelihood of windthrow or trunk breakage is high. For example, 86% of windthrown *Populus tremuloides* (trembling aspen) in a storm-damaged stand in Colorado had basidiocarps of *G. applanatum*. This fungus is the most important cause of root and butt rot of *P. tremuloides* in the central Rocky Mountain region, is a significant cause of trunk rot in hardwoods across North America, and was one of the three most important causes of heart rot in old-growth *Tsuga heterophylla* (western hemlock) in the West. A fungus identified as *G. applanatum* has been associated with perennial cankers on *Citrus* trunks in Florida.

G. applanatum has been recorded on plants in the following genera in North America (several questionable records have been omitted):

Abies (fir), *Acer* (maple), *Aesculus* (horse-chestnut), *Ailanthus* (tree-of-heaven), *Alnus* (alder), *Betula* (birch), *Carpinus* (hornbeam), *Carya* (hickory), *Castanea* (chestnut), *Cercis* (redbud), *Citrus, Crataegus* (hawthorn), *Eucalyptus, Fagus* (beech), *Fraxinus* (ash), *Gleditsia* (honeylocust), *Juglans* (butternut, walnut), *Liquidambar* (sweetgum), *Liriodendron* (tuliptree), *Lonicera* (honeysuckle), *Magnolia* (magnolia), *Malus* (apple), *Morus* (mulberry), *Picea* (spruce), *Pinus* (pine), *Platanus* (sycamore), *Populus* (aspen, cottonwood), *Prunus* (peach, plum), *Pseudotsuga* (Douglas-fir), *Pyrus* (pear), *Quercus* (oak), *Robinia* (black locust), *Salix* (willow), *Thuja* (cedar), *Tilia* (basswood), *Tsuga* (hemlock), and *Ulmus* (elm).

Symptoms and signs. Trees with extensive decay involving sapwood may show slow growth and dieback, but many break or fall without ever showing external symptoms. The decay is a white rot in which lignin and cellulose are degraded together, resulting in great loss of strength at an early stage and leading to transverse fractures. Wood undergoing decay by *G. applanatum* remains light colored and exhibits a mottled appearance. Thin black lines (zone lines or pseudosclerotial plates, Plate 152) develop in it; eventually it becomes soft and spongy. A dark band several millimeters wide separates healthy sapwood from that undergoing decay. Columns of decaying wood usually extend 1–2 m above and below basidiocarps in trunks of *Acer*, *Betula*, and *Fagus*, and were found, on average, 3 m above basidiocarps in old-growth *Tsuga heterophylla*. In roots the fungus is usually restricted to those of relatively large diameter within 1–2 m of the butt.

Basidiocarps of *G. applanatum* usually develop on root buttresses or on the trunk within 3 m of the ground. They are often associated with old wounds and are common on stumps, snags, and fallen trunks. They become hard and woody at an early age and can attain impressive strength and dimensions (Fig. 174). Basidiocarps are roughly semicircular or fan shaped as viewed from above, with a white margin and poroid surface, and a flat to convex upper surface that is initially brown and weathers to gray, often becoming cracked and concentrically ridged. Although they sometimes live 8–10 years, 5–6 years was found to be the norm in southern Ontario. The interior tissues, including one or more layers of tubes 5–10 mm long, are dark brown. Brown, ovoid, thick-walled basidiospores, mostly 9–12 × 6.5–8 μm with a truncate distal end, are released from basidia in the tubes. The tubes must be vertical for effective spore release. If the orientation of a basidiocarp shifts when a tree falls or a log rolls, spore production ceases until the fungus produces a new layer of tubes. Spore release occurs throughout the year as

temperature permits, and the spores collect as a brown deposit on nearby bark, litter, and other basidiocarps. By one estimate a basidiocarp with a poroid surface of only 100 cm^2 may liberate 4.65 billion basidiospores in a 24-hour period!

Disease cycle. G. applanatum colonizes new substrates by means of basidiospores, which are both airborne and carried by insects that visit the vicinity of basidiocarps. The insects become contaminated while foraging, and spores drop off during subsequent activity. *G. applanatum* is apparently a colonist of wounds. It probably follows pioneering fungi into wounded wood, because its spores germinate readily only in the presence of microorganisms. The latter are presumed to provide necessary nutrients or stimulants. Whether the fungus kills sapwood or only exploits wood prepared by earlier organisms in the succession is unknown. The interval between infection and fruit body formation is also unknown but is measured in years and presumably varies with latitude and host species, which differ in their abilities to resist colonization and decay. A relatively large volume of decaying wood is required to support fruiting, because wood has a low nitrogen content while basidiocarps and spores are rich in nitrogen. Nitrogen is recycled from old fungal tissues to young ones in basidiocarps to help sustain production of new tube layers and spores.

Four other nonlaccate *Ganoderma* species in North America deserve brief mention. Their taxonomic status is unclear. *G. annularis* and *G. brownii* have been reported only in California, the former mainly on dead *Prunus* (almond and peach), *Quercus*, and *Umbellularia* (California-laurel); the latter on living and dead *Citrus*, *Hakea* (pincushion-tree), *Prunus* (also almond and peach), *Quercus*, *Schinus* (pepper-tree), and *Umbellularia*. *G. australe*, a fungus resembling *G. applanatum* that occurs widely in Asia and Australasia, has been found in Florida on *Coccoloba diversifolia* (tietongue or pigeon-plum), but apparently is rare there. *G. lobatum* occurs widely in the eastern half of the USA and in the Southwest, mainly on dead hardwoods although it has been found on living *Prunus persica* (peach).

References: 401, 543, 558, 723, 1064, 1150, 1151, 1280, 1859, 2595, 2937, 3340, 3374, 3510, 3540, 4102, 4108, 4322, 4482

Figure 174. Ganoderma applanatum supporting a man on a living *Fagus grandifolia* (American beech).

Butt and trunk decay by *Ganoderma applanatum*.

A, B. A large *Populus deltoides* (eastern cottonwood) with decay caused by *G. applanatum* in the butt and trunk. A. Basidiocarps near ground level (NY, Aug). B. The scene 2 years later after a windstorm; more than half of the transverse area of the trunk was decayed to a height of 5 m (NY, Sep).

C, D. Decay and trunk breakage in *Acer saccharum* (sugar maple). C. Typical basidiocarps on a broken trunk. Brown color of bark near basidiocarps is a deposit of basidiospores. D. Decayed wood shows a transverse fracture (NY, Sep).

E. Basidiocarps about 2 years old on an *A. saccharum* stump. Growing basidiocarps have a prominent white margin (MI, Aug).

F. The trunk of a dead *Fagus grandifolia* (American beech), broken after basal decay. Mottled white rot is typical of that caused by *G. applanatum* (NY, May).

Diseases Caused by Oomycota (Plates 175–182)

Overview of Oomycetes

Oomycetes are organisms formerly thought to be fungi but now classified in Kingdom Straminipila, Phylum Oomycota, Class Oomycetes. They share with fungi the characteristics of being eukaryotic, spore-producing organisms that reproduce both sexually and asexually and whose filamentous, branched bodies lack chlorophyll and obtain nutrients by absorption. They differ from fungi in mitochondrial structure, cell-wall composition, various biochemical and molecular characters, and ancestry as inferred from DNA sequences. Oomycetes have diploid nuclei in their somatic cells (polyploid in some strains), and many produce motile spores (zoospores). Their sexually derived spores are oospores. Important groups of oomycetes that affect woody plants include *Pythium* species, which cause damping-off and rootlet necrosis; *Phytophthora* species, which attack all plant parts; and several genera that attack succulent aboveground parts and cause downy-mildew diseases.

References: 40, 957, 1928, 2713, 3775, 4117

Downy Mildews (Plate 175)

Downy mildews are caused by oomycetes of the Peronosporales (family Peronosporaceae) that produce delicate "downy" tufts or clusters of filamentous spore-bearing structures (sporangiophores) on diseased plant parts. Downy mildews occur globally in humid environments and are destructive to many crops, but the only woody plants that sustain economically significant damage in North America are *Vitis* species (grapevines) and *Rosa* species (roses). Downy mildew of viburnum is common in eastern USA but is usually inconsequential. Other members of the group on woody plants are infrequent to rare. Downy-mildew pathogens display considerable host specificity, usually being restricted to plants in one genus or family. Examples include: *Peronospora grisea* on *Hebe*; *P. sparsa* (syn. *P. rubi*) on *Prunus laurocerasus* (cherry-laurel), *Rosa* (rose), and *Rubus* (brambles); *Plasmopara cercidis* on *Cercis canadensis* (redbud); *P. ribicola* on *Ribes* (currant, gooseberry); *P. viburni* on *Viburnum* (arrowwood, cranberry bush); *P. viticola* on *Ampelopsis, Cissus* (grape ivy), *Parthenocissus* (Virginia creeper, Boston ivy), and *Vitis* (grapevine); and *Pseudoperonospora celtidis* on *Celtis* (hackberry, sugarberry).

Symptoms and signs. Leaves, tender green stems, and fruits are affected. Lesions initiated by spores begin as discolored (usually light green) spots or blotches, mainly on upper surfaces of leaves near ground level. Lesions may darken to reddish brown before turning brown. Large or coalescing lesions on leaves cause defoliation. Leaves, flowers, or fruit infected while young become distorted and may shrivel. Shoots that grow from infected buds are distorted and stunted and develop yellowish or reddish blotches that turn brown. Lesions on which a pathogen is sporulating have a granular to downy appearance due to branched, colorless sporangiophores on which sporangia form (Fig. 175). Sporangiophores on leaves grow out from stomata, typically on the lower surface, and form a loose to dense, white to gray-brown mat, the texture and color varying with pathogen species and lesion age. Oospores form in diseased tissues and can be observed microscopically.

Disease cycles and conducive conditions. Downy-mildew pathogens are strict biotrophs and thus in nature derive sustenance only from living plants. Their life cycles have alternating phases: parasitic mycelial growth accompanied by asexual reproduction, and sexual reproduction followed by quiescent survival. Infection giving rise to the mycelial phase is initiated by any of three types of spores, described below. Hyphae growing from germinated spores penetrate succulent organs, often via stomata. The hyphae ramify intercellularly and obtain nutrients by means of short branches that penetrate host cells and produce haustoria (absorptive structures) in them. The parasitic phase may be perennial if colonized tissues (buds, for example) remain alive through winter.

Asexual reproduction is accomplished by formation of sporangia. Those of *Peronospora* function as conidia, germinating by germ tubes; sporangia of *Plasmopara* and *Pseudoperonospora* liberate zoospores. Water-splashed zoospores and air- or water-dispersed sporangia germinate on wet plants and initiate repeating cycles of disease. Oospores are products of sexual reproduction and function

as survival structures that can withstand environmental extremes and microbial attack. They are released to the soil as dead plant parts decay. Overwintered oospores produce sporangia on germination. New infections begin after these sporangia or zoospores from them are splashed onto leaves or other succulent parts.

Downy mildews of woody plants develop in cool or warm but not hot weather. Sporangium production requires high humidity, and plant surfaces must be wet for several hours for infection to occur. Measures that promote dry foliage suppress these diseases.

Downy mildews of rose, viburnum, and grapevine. *Peronospora sparsa* (syn. *P. rubi*) causes downy mildew of *Rosa* (rose), *Rubus* (brambles), and other rosaceous plants. Epidemics on roses occur sporadically in greenhouses and in nurseries and outdoor plantings. Leaves, stems, buds, and flower parts are attacked. Severe disease causes rapid defoliation and sometimes death of shoots. Bud infection leads to stunting and malformation of shoots and flowers. The pathogen can be transported as dormant mycelium in cuttings and plants. Rose cultivars vary in susceptibility, but most become infected to some extent in sites where the disease builds up on highly susceptible types. *P. sparsa* perennates as mycelium in roots, crowns, and canes of *Rubus* species and causes economic loss in some fruit-growing regions.

Plasmopara viburni, which causes downy mildew of *Viburnum* species, is common east of the Great Plains. Symptoms begin as chlorotic spots on the upper surfaces of leaves, accompanied by sporulation on the lower surfaces. Affected leaves often curl upward. Old lesions turn brown. Severe foliar browning and defoliation may occur.

Plasmopara viticola attacks *Vitis* (grapevine) and related plants. The pathogen is native to North America but has been distributed globally with its hosts. Foliar lesions begin as round light green areas with indefinite margins. These darken and enlarge to irregular patches. Severe infection defoliates vines. Shoots, tendrils, and berry clusters are attacked while young. Grape berries become resistant when partly grown, but if infection is severe, berries that escape infection are likely to be small and of poor quality. Grape yield can be reduced by half or more if vines are not protected. Cultivated grapes derived from North American species are less susceptible than European types.

References: 40, 180, 500, 957, 1445, 1447, 1758, 1859, 3561, 3562, 3775

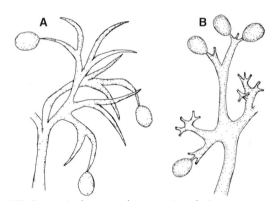

Figure 175. Sporangiophores and sporangia of *Peronospora* (A) and *Plasmopara* (B). From reference 40 by permission of John Wiley & Sons, Inc.

A–C. Downy mildew of rose, caused by *Peronospora sparsa*. A, B. Foliar symptoms on pot-grown plants (NY, Jun). C. Grayish aerial mycelium on the lower surface of a lesion on the cultivar Samantha (CA, Nov).

D–G. Downy mildew of *Viburnum opulus* (European cranberry-bush), caused by *Plasmopara viburni*. D. Browning and partial defoliation in a hedge of *V. opulus* 'Nanum' (NY, Sep). E. Symptoms on mature leaves of *V. opulus* (NY, Aug). F, G. Lower surfaces of *V. opulus* 'Nanum' leaves with areas of white sporangial production (magnified in G). The leaf in F is distorted because infection began while it was expanding (NY, Aug–Sep).

H. Signs and symptoms of downy mildew on *Vitis vinifera* (grape) leaves. The lower surface has white clumps of sporangiophores; the upper surface has yellowish brown spots (NY, Aug).

353

Diseases Caused by *Phytophthora* Species (Plates 176–182)

Overview

Phytophthora is an infamous genus of plant-parasitic oomycetes (Peronosporales, Pythiaceae) whose members attack all major groups of higher plants, causing diverse symptoms: leaf and shoot blight, bud rot, fruit rot, twig and limb dieback, stem cankers, collar rot (bark necrosis at the root collar), root and crown rot (death and decay of roots and root crown), and decline due to root damage. Several of the approximately 60 known species of *Phytophthora* are of global importance, having been widely distributed along with plant materials. Several of them also have broad host ranges, including both herbaceous and woody species, and cause multiple symptoms. A few Phytophthoras have limited distributions and narrow host ranges (see table). Several new species and interspecific hybrids have been discovered in recent years, and more such discoveries are probable. Interspecific hybridization has potential to produce strains with new host and environmental adaptations. Host-specialized strains have been found within some species that have broad host ranges, notably *P. cactorum* and *P. cinnamomi*.

Some Characteristics of Selected Species of *Phytophthora*

Species	Host range	Distribution*	Principal symptoms induced**
alni	narrow	Eur	D, RCR, SC
cactorum	broad	global	D, DB, FR, RCR, RN, SC, SD
cambivora	broad	global except SA	D, RCR, RN, SB, W
cinnamomi	broad	global	D, RCR, RN, SC, W
citricola	broad	global	D, DB, FR, RCR, RN, SB, SC, W
citrophthora	broad	global	D, FR, RCR, RN, SB, SC
cryptogea	broad	global	D, RCR, RN
drechsleri	broad	global	D, RCR, RN, SC
gonapodyides	intermediate	Eur, NA	FR, RCR, RN
heveae	intermediate	global except Eur	D, DB, FR, LB, RCR, SB, SC, W
ilicis	narrow	Eur, NA	DB, FR, LB
inflata	narrow	Eur, NA	RN, SC
lateralis	narrow	Eur, NA	D, RN, RCR, SB, SC
megasperma	broad	global except Afr	D, RCR, RN, SC, W
nemorosa	intermediate?	NA	LB, SC
nicotianae	broad	global	D, DB, LB, RN, RCR, SB, SC
palmivora	broad	global	BR, FR, LB, RCR, SC, SD
pseudosyringae	intermediate?	Eur	RN, SC
pseudotsugae	narrow	NA	RN
quercina	narrow	Eur	RN
ramorum	broad?	Eur, NA	D, DB, LB, SB, SC
syringae	intermediate	Afr, Aus, Eur, NA	DB, FR, RCR, RN, SB, SC

* Afr, Africa; Aus, Australia; Eur, Europe; NA, North America; SA, South America.
** BR, bud rot; D, decline; DB, dieback; FR, fruit rot; LB, leaf blight; RCR, root crown and/or collar rot; RN, rootlet necrosis; SB, shoot blight; SC, stem canker; SD, seedling decay; W, wilt.

Detection and diagnosis. Diagnosis of *Phytophthora* diseases requires symptom interpretation, diagnostic tests, and microscopy. The pathogens produce only microscopic reproductive structures. Field diagnosis is likely to be nonspecific because several *Phytophthora* species as well as unrelated organisms may be capable of causing similar symptoms on a given plant. Detection of *Phytophthora* in diseased or recently killed tissues is often possible by microscopic examination, which may reveal oomycete hyphae and spores, or through the use of immunologic tests or DNA amplification. Commercial test kits adapted for field use are available. Species identification requires examination of a pathogen in pure culture or analysis of DNA from diseased tissues. Bait techniques and selective culture media facilitate the isolation of Phytophthoras from soil and infected plants, but fast-growing fungi often obscure Phytophthoras on nonselective media. Bait techniques involve placing leaves, immature fruit, or seedlings of susceptible plants in contact with moist or flooded soil samples or moist plant material suspected to contain a pathogen.

Rhododendron, *Camellia*, *Eucalyptus*, and *Juniperus* leaves; *Malus* (apple) and *Pyrus* (pear) fruits; and *Lupinus* (lupine) seedlings have been used, among other baits.

Pathogen structures and their functions. *Phytophthora* species produce stout, aseptate, freely branching, colorless hyphae that grow inter- and intracellularly. Some species produce haustoria in host cells. Most species produce three kinds of infective structures: sporangia, zoospores, and oospores; but the oospores of heterothallic species are seldom found in nature because only one mating type is usually present in a site. Some species also produce chlamydospores. The nuclear condition of somatic structures and all mature spores is diploid (or polyploid in some strains). Spores of all types require free water for germination.

Sporangia (Fig. 176) are short-lived structures that under some conditions behave like spores, germinating by a germ tube that gives rise to a new mycelium. More often, however, they "germinate" by converting their contents to wall-less, unicellular, kidney-shaped, motile zoospores that escape through a pore at the apex of the sporangium. Numerous zoospores are produced in each sporangium, thus amplifying the organism's potential to infect host

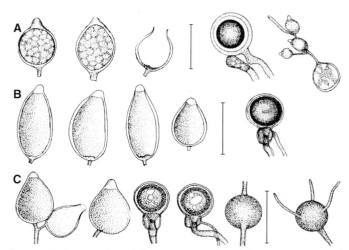

Figure 176. Row A, *Phytophthora cactorum:* from left, two sporangia with zoospores forming, empty sporangium after zoospore release, oospore in an oogonium with an antheridium (male gamete cell) attached laterally, sporangia on a hypha from a germinated oospore. Row B, *P. palmivora:* sporangia and an oogonium with antheridium. Row C, *P. nicotianae:* sporangia, oospores in oogonia, an intercalary chlamydospore, and a germinating chlamydospore. Scale bars = 40 μm. Adapted from reference 1077 by permission of the American Phytopathological Society.

A, B. Shoot blight and dieback of *Rhododendron* sp. caused by *Phytophthora* sp. Wilting in A was caused by the stem lesion at lower right. In B, infection has spread from the stem apex into leaves (NY, Jul).

C. Fruit rot of *Cotoneaster horizontalis* (rock cotoneaster) caused by *P. cactorum.* Dark lesions expand to involve entire fruits, which eventually shrivel (CA, Oct).

D–F. Collar rot (crown canker) of *Cornus nuttallii* (Pacific dogwood) and *C. florida* (flowering dogwood), presumed caused by *P. cactorum* (discussion on p. 358). D. Foliar reddening and dieback of top branches of diseased *C. nuttallii* (OR, Sep). E. Large basal cankers on *C. nuttallii* partially revealed where bark has cracked and fallen away (WA, Jun). F. Crown canker on *C. florida.* Old lesions, previously inactive for a time, as indicated by large woundwood ridges, are expanding beneath the scaly outer bark (NY, May).

G–J. Bleeding canker caused by *P. palmivora* on *Acer rubrum* (red maple). G, H. An affected tree viewed in March (G) and, from the reverse side, in October (H), showing dramatic expansion of stained bark. I. Fluid oozing from bark near the edge of a lesion (FL, Mar). J. The advancing edge of a lesion, exposed by removal of outer bark (FL, Mar).

Photo credits: G–J—E. L. Barnard

tissue. Zoospores move in water by means of flagella, swimming for a few minutes to a few hours at most and changing directions frequently. They are attracted by amino acids and other chemicals that exude from roots or other plant parts, but their net movement in the absence of chemical attraction is seldom more than a few centimeters unless they are in flowing water. Wounds and intact succulent plant parts such as shoot or root tips are suitable for infection. On reaching a suitable surface, a zoospore rounds up, loses its flagella, secretes a cell wall (encysts), and produces a germ tube that penetrates the plant and gives rise to mycelium.

Chlamydospores are somatic cells, within hyphae or at their tips, that develop thick walls and serve as resting structures. Oospores, products of the union of gametangia called oogonia and antheridia, also serve as resting structures. Germinating oospores and chlamydospores give rise to either mycelium or sporangiophores and sporangia, depending on environmental conditions.

Modes of pathogenicity. Knowledge of mechanisms of pathogenicity of Phytophthoras that attack woody plants is fragmentary. That they kill tissues and digest cellular contents is evident. These activities are facilitated by toxins and assorted extracellular enzymes: pectinases, cellulases, proteases, lipases, nucleic-acid-degrading enzymes, and others. *P. cactorum, P. cryptogea, P. megasperma, P. nicotianae,* and *P. quercina* secrete proteinaceous toxins belonging to a group called elicitins that damage plant cell membranes and cause chlorosis, wilting, and necrosis in laboratory assays. They also trigger plant defense processes. Similar proteins are probably secreted by most or all Phytophthoras.

Ecology and epidemiology. Many Phytophthoras are generalists in both hosts and modes of attack, can interchangeably cause similar diseases in a given plant species, and can acquire new virulence characteristics through natural selection and hybridization. The generalists that cause much damage to trees and shrubs are, except for *P. ramorum* (discussed in a later section), soilborne organisms whose activities tend to be favored by wet sites—soils along watercourses or that have impeded internal drainage or are irrigated by flooding. Such conditions not only promote the pathogens' reproduction and dispersal as zoospores but also predispose roots to infection. Roots stressed by oxygen deficiency in waterlogged soil exude more amino acids and other substances attractive to zoospores, and their resistance to pathogens is impaired. Saline soil or previous water shortage also lowers plant resistance.

Phytophthoras are dispersed in water and with soil, diseased plants, infested plant debris, and contaminated tools and equipment. A few species that attack woody plants, notably *P. palmivora* and *P. ramorum,* have deciduous sporangia that can be dispersed aerially, usually in splashing or wind-driven water. *Phytophthora*-contaminated irrigation water causes problems in nurseries and orchards. Once established in a new location, even though the site may be mesic or seasonally dry, a *Phytophthora* may persist and cause damage for years unless the microbial population of the site suppresses it through competition and antagonism. All of the soilborne Phytophthoras infect roots or rootlets whether or not they also attack other organs. Sometimes they persist insidiously, "nibbling" at fine roots without causing noticeable foliar symptoms or growth decline. Multiple Phytophthoras may inhabit a given site, and environmental conditions favorable to one species are likely to favor others. Damage by some Phytophthoras, notably *P. cinnamomi,* tends to be exacerbated when host plants are stressed by factors such as water or nutrient (especially nitrogen) deficiency. Inversely, root damage by Phytophthoras can lead to nutrient deficiency in aerial parts of plants and induce stress that favors attack by opportunistic fungi and insects. Most Phytophthoras, although readily cultivable on laboratory media, do not grow much as saprobes in nature. After death of a colonized plant, if more susceptible plant tissue is not available, these pathogens can persist for several months as chlamydospores or oospores in soil or dead plant material.

Soils vary in suitability as habitat for Phytophthoras. In the Pacific Northwest, for example, *P. cinnamomi* and other Phytophthoras that cause root rot in nurseries are regularly carried with infected seedlings to forest planting sites but do not become established. On the other hand, a soil that is suppressive to one *Phytophthora*

species may fail to suppress others. *P. lateralis,* which causes root rot of *Chamaecyparis lawsoniana* (Lawson-cypress, Port Orford cedar; Plate 180), spreads readily in forests where its host grows in the Pacific Northwest. Strict precautions are necessary to minimize its movement from contaminated sites.

Some noteworthy pathogens and their management. P. cactorum causes root, crown, and collar rot and bleeding cankers on various fruit and shade trees (Plate 176), mainly in temperate zones. *P. inflata* causes pit canker of *Ulmus* (elm; Plate 177). *P. cambivora* causes collar rots of *Castanea* (chestnut) and *Fagus* (beech) in the USA and Europe and of *Prunus dulcis* (almond) in Australia and the USA. *P. ramorum* causes bleeding cankers, decline, and death of *Quercus* and *Lithocarpus* (tanoak), and various symptoms on other woody plants in California, Oregon, and Europe (Plate 178). *P. citricola, P. citrophthora, P. megasperma,* and *P. nicotianae* (syn. *P. parasitica*) cause root, crown, and collar rots and stem cankers of diverse forest, orchard, and shade trees globally (Plate 179). *P. lateralis* causes a devastating root and collar rot of *C. lawsoniana* in northwestern USA and adjacent Canada (Plate 180). *P. cinnamomi* attacks hundreds of species on six continents, causing feeder root necroses; root, crown, and collar rots; stem cankers; and decline (Plates 181, 182).

Management of Phytophthora diseases is difficult. Awareness of the need usually begins with a disease outbreak. Disease spread can be retarded by restricting site disturbance and movement of plants, soil, water, and people. Outbreaks in plantations, orchards, and nurseries can be avoided or minimized by site selection to avoid wet or poorly drained sites and/or by improving drainage, ensuring clean irrigation water, and utilizing drip or trickle irrigation rather than overhead or flood irrigation. In nurseries further benefits come from utilizing soil-less, pathogen-free container mixes and soil cover such as gravel or mulch that minimizes movement of oomycete propagules with soil and splashing water. Sites of severe disease in forests eventually become less conducive to pathogen activity because of changes in plant and microbial populations, increase of resistant plant species, and gradual improvement of soil conditions on previously cleared sites. Moderate resistance to Phytophthoras has been found in various trees, especially fruit and nut trees, but deployment of resistant stock is slow because existing tree plantings tend to be long-lived. In addition, successful selection for resistance to one Phytophthora does not ensure resistance to others. Disease control with pathogen-suppressive rooting substrates and fungicides is possible in nurseries. *Phytophthora* activity in soil can be suppressed somewhat by calcium fertilizers or soil amendment with gypsum. Plants in nurseries, orchards, and landscapes can be treated prophylactically with certain fungicides and resistance-inducing chemicals such as potassium phosphonate. Expansion of basal cankers on a diseased tree can sometimes be halted by heat or surgery. Solar heating of soil beneath clear plastic can suppress Phytophthoras in regions of warm climate.

References: 435, 481, 524, 552, 786, 957, 1003, 1076, 1077, 1118, 1256, 1405, 1453, 1491, 1494, 1590, 1712, 1716, 1859, 1984, 1986, 2181, 2248, 2365, 2378, 2407, 2483, 2609, 2613, 2753, 3278, 3495, 3707, 3825, 4233

A–G. Bleeding canker of *Fagus sylvatica* (European beech), caused by a member of the *Phytophthora citricola* complex (discussion on p. 360). A. Severe dieback due to cankers near ground level. B, C. Discolored bark due to bleeding cankers. D–F. Close views of lesions before and after removing bark. The cambial region is red where recently killed. G. Woundwood ridges at the edges of old cankers indicate that a tree has resisted the pathogen (CT, Nov).

H, I. Pit canker of *Ulmus americana* (American elm), caused by *P. inflata* (discussion, p. 358). H. Coalescing perennial cankers have extended more than halfway around a trunk. I. Deep pit cankers on another tree were exposed by removing outer bark scales (NY, Jun).

Photo credits: A–G—G. W. Hudler

Shoot Blights, Diebacks, and Fruit Rots

Dieback of Rhododendron and other ericaceous plants. Aboveground parts of *Rhododendron* species and *Pieris japonica* (Japanese pieris) may be attacked by *Phytophthora cactorum, P. citricola, P. citrophthora, P. heveae, P. nicotianae, P ramorum,* or *P. syringae* (Plates 176, 181). When these oomycetes infect leaves and twigs, they usually do not advance far down woody stems, although *P. ramorum* sometimes does so. They cause necrotic blotches that are at first olive colored and later brown and may have a dark red margin. Infection by the soilborne Phytophthoras occurs when zoospores or sporangia are delivered in irrigation water or splashed from soil to plant surfaces. Lesions develop within 2–3 days during warm, humid weather. Sporangia and later oospores and chlamydospores form in the dead tissues. Diseased leaves drop, returning the parasites to the soil, where they survive in plant debris. *P. ramorum* spreads from plant to plant aerially as air- or waterborne sporangia or waterborne zoospores. Its propagules can also be splashed onto plants from leaves on the ground. Phytophthoras in *Rhododendron* are sometimes followed by opportunistic fungi such as *Botryosphaeria dothidea* (Plate 59) that mask or aggravate the initial damage.

P. citricola and *P. citrophthora* also cause root and crown rot of *Rhododendron* species, as do *P. cinnamomi* (Plates 181, 182), *P. cryptogea, P. gonapodyides,* and *P. megasperma.* The symptoms include stunting, epinasty, chlorosis, dieback, and eventual death.

Fruit rots. Phytophthoras decay fruits of diverse plants. The usual symptom is a soft brown decay that spreads from one spot to involve an entire fruit. As examples, *P. cactorum* causes fruit rot of *Carya illinoinensis* (pecan), *Cotoneaster* (Plate 176), *Prunus persica* (peach), *Pyracantha* (firethorn), and *Pyrus* (pear), among other plants; *P. citrophthora* and *P. nicotianae* cause brown rot of citrus; *P. heveae* causes bud rot and nut drop of *Cocos nucifera* (coconut palm); *P. ilicis* rots shoots and fruits of *Ilex aquifolium* (English holly); and *P. palmivora* causes pod rot of *Theobroma cacao* (cocoa). Fruit infections are often caused by water-splashed spores or sporangia from soil and other plant surfaces.

References: 318, 1259, 1967, 2112, 2193, 2194, 3257

Cankers, Collar and Crown Rots, and Foot Rots

Cankers on trunks and limbs. Phytophthoras that cause stem cankers include *P. cactorum, P. cinnamomi, P. citricola, P. megasperma, P. palmivora,* and *P. syringae,* as well as *P. inflata, P. lateralis, P. nemorosa,* and *P. ramorum.* The first six species also cause lesions in roots and/or root crowns and trunk collars. Among the many tree genera affected are *Acer* (maple), *Aesculus* (horse-chestnut), *Alnus* (alder), *Arbutus* (madrone), *Betula* (birch), *Castanea* (chestnut), *Elaeagnus* (Russian-olive), *Fagus* (beech), *Juglans* (walnut), *Liquidambar* (sweetgum), *Liriodendron* (tuliptree), *Lithocarpus* (tanoak), *Persea* (avocado), *Prunus* (almond, cherry, peach), *Quercus* (oak), *Salix* (willow), *Tilia* (linden), and *Ulmus* (elm). The lesions are often hidden by outer bark, but their presence may be indicated by bleeding and stained bark where fluid oozes from cracks and wounds; hence the term *bleeding canker.* Multiple lesions, connected within the inner bark, may extend up a trunk from the root crown. Active lesions typically involve the inner bark, cambium, and outermost sapwood. They are water-soaked and discolored yellowish to red or reddish brown or brown with distinct margins. Fluid that exudes is often reddish brown. Gum may also exude from lesions on *Citrus* or *Prunus.* If a lesion enlarges for several years, the marginal area shows the typical color and texture of newly killed tissue while the long-dead center loses its bark and becomes weathered. Some Phytophthoras enter stems via wounds. *P. syringae,* for example, infects pruning wounds on *Prunus dulcis* (almond).

Crown canker of dogwood, caused by *P. cactorum* (Plate 176), is destructive to landscape specimens of *Cornus florida* (flowering dogwood) and *C. nuttallii* (Pacific dogwood). Symptoms first appear as occasional bleeding of reddish brown fluid from bark near the base of the trunk. Eventually, killed bark dries, cracks, and falls away. Portions of a canker margin may become temporarily or permanently inactive, and woundwood may form at the edge. Leaves on branches above a canker become somewhat chlorotic

and smaller than normal, with a tendency to bend downward (epinasty), fold along the midrib, and turn red in late summer; dieback follows. *P. cactorum* also causes root rot of *Cornus* species growing in poorly drained soils.

Pit canker of *Ulmus* (elm; Plate 177; Fig. 178) is caused by *Phytophthora inflata.* The disease occurs in northeastern USA and Quebec in *U. americana* (American elm) and, less commonly, *U. rubra* (red elm). It was often noticed, before elm populations were decimated by Dutch elm disease, in trees growing under adverse conditions, such as in compacted soil. Affected trees develop multiple perennial cankers, often sunken and partly hidden by thick outer bark, on the trunk and scaffold limbs. Trunks thus affected become distorted because of growth in asymmetric bands between cankers. The phloem, cambium, and outermost sapwood at canker margins are discolored red-brown, and dark fluid may bleed from small fissures near canker margins. Narrow bands of diseased tissue often connect closely adjacent cankers. The bark eventually falls away from old cankers, exposing concentric marks on the wood that indicate intermittent enlargement of lesions. The extent of woundwood development at canker margins is variable. Severely affected trees lose vigor, have undersized chlorotic leaves that droop and drop prematurely, and begin to die back. *P. inflata* can be isolated readily from the margins of active cankers. In pure culture it produces distinctively inflated antheridia, for which it was named. *P. inflata* is now obscure. In the 1990s it was blamed for a root rot in *Sambucus racemosa* (European red elderberry) and *Syringa*

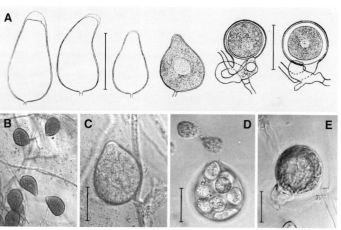

Figure 178. A. *Phytophthora inflata:* from left, four sporangia, an oogonium with antheridium attached, and an oospore within an oogonium, to which an empty antheridium remains attached. B–E. The *Phytophthora* associated with bleeding canker of *Fagus grandifolia.* B. Sporangia viewed at low magnification. C. A mature sporangium. D. Zoospores escaping from a sporangium. E. An oogonium with antheridium. Scale bars = 30 μm in A, 20 μm in B–E. A adapted from reference 1077 by permission of the American Phytopathological Society; B–E courtesy of G. W. Hudler.

Diseases caused by *Phytophthora ramorum* (discussion on p. 360).
A–C. *Quercus agrifolia* (California live oak, with spreading crowns) and *Lithocarpus densiflorus* (tanoak, with upright or conical crowns) affected by "sudden oak death" (CA, summer).
D–H. Cankers on *Q. agrifolia.* D, F. Dark reddish fluid bleeding from bark. E. Lesions apparent after removing outer bark of the specimen shown in D. G. Close view of the typical dark line found between healthy (left) and necrotic tissues at the edge of a lesion in secondary phloem (D–G: CA, summer). H. Cross section of a young stem half girdled by a canker (CA, Nov).
I, J. Twig blight on *L. densiflorus* and leaf blight on *Umbellularia californica* (California bay), respectively (CA, summer).
K. Branch dieback on *Vaccinium ovatum* (evergreen huckleberry) (CA, Feb).
L, M. Shoot blight on *Sequoia sempervirens* (redwood) and *Pseudotsuga menziesii* (Douglas-fir) (CA, Aug & May, respectively).
Photo credits: A—S. J. Frankel; B—P. Svihra; C, I—K. E. Loeffler; D, E—K. Julin; F, J—M. Garbelotto; G, H, L—D. M. Rizzo; K, M—J. M. Davidson

vulgaris (common lilac) in a nursery in England, but its identification could not be corroborated.

In the late 1990s a *Phytophthora* strain allied with *P. citricola* (Fig. 178) was found associated with bleeding cankers in declining *Fagus sylvatica* in northeastern USA (Plate 177). The cankers, occurring from the ground line up to 3 m on some trees, are easily detected because a dark stain spreads on the bark from sites of bleeding. Recently infected areas on purple-leaved cultivars have distinctive red discoloration in the cambial zone, which can be viewed by cutting away dying bark near the edges of lesions. *Phytophthora* isolates from the lesions in *Fagus* have proven pathogenic to plants in several unrelated genera.

References: 212, 387, 435, 485, 546, 652, 993, 1003, 1077, 1446, 1490a, 1494, 1763, 1778, 1859, 2101, 2318, 2669, 2896, 3105, 3894, 4423

Diseases caused by *Phytophthora ramorum*. A syndrome of bleeding cankers and decline of *Q. agrifolia* (coast live oak) and *Lithocarpus densiflorus* (tanoak) (Plate 178) was noticed in the San Francisco Bay area of California beginning in the mid-1990s and was soon dubbed "sudden oak death." Trees of all sizes were dying in large numbers. *Q. kelloggii* (California black oak) was affected also but less often. Drought, insects, and opportunistic fungi were considered as possible causes or contributors. In 2000, however, *Phytophthora ramorum* was detected and soon was shown to cause the syndrome. This species, first found in Europe in the 1990s, was previously unknown in North America. It was later found causing foliar lesions, shoot blight, and/or cankers and dieback on *Q. chrysolepis* (canyon live oak), *Q. parvula* var. *shrevei* (Shreve oak), and, by 2004, more than 40 additional species representing 12 families. Only *Lithocarpus* and *Quercus* regularly incur lethal damage. *Arbutus menziesii* (madrone) is sometimes killed. Several additional *Phytophthora* species occasionally cause cankers on *Lithocarpus* or *Quercus*, but none is a significant contributor to large-scale tree mortality. Tree killing by *P. ramorum* occurs on mesic to dry sites, whereas cankers and crown and collar rots caused by soilborne Phytophthoras tend to be found on wet or poorly drained sites.

Lithocarpus is affected most severely. Lesions develop on leaves, twigs, branches, and trunks, and the trees soon die. New shoots often droop or turn yellow to brown before cankers become noticeable. Cankers caused by *P. ramorum* on *Quercus* usually develop on the lower trunk but may occur much higher, and they often coalesce. They do not extend below the soil surface. Recently killed phloem darkens slightly or becomes reddish brown to dark purplish brown. A dark brown to black line marks the edge of each lesion, and the sapwood surface is similarly blackened. Dark red to reddish brown fluid oozes through cracks in bark overlying lesions on *Quercus* and less commonly on *Lithocarpus* during the wet season. Individual oaks vary in susceptibility. Oaks debilitated by *P. ramorum* are colonized by secondary fungi, notably *Hypoxylon thouarsianum*, and insects, and they typically die after one to several (>5) years. Dead leaves remain attached for 6 months to a year.

Most of the known hosts of *P. ramorum* sustain only foliar and/or twig damage, for which the name ramorum blight has been coined. Plants found naturally infected in the USA represent, in addition to genera already mentioned, *Acer* (maple), *Aesculus* (California buckeye), *Arbutus* (madrone), *Arctostaphylos* (manzanita), *Camellia*, *Corylus* (hazelnut), *Frangula* (buckthorn, coffeeberry), *Heteromeles* (toyon), *Lonicera* (honeysuckle), *Pieris* (pieris), *Pittosporum*, *Pseudotsuga* (Douglas-fir), *Rhododendron*, *Toxicodendron* (Pacific poison-oak), *Rubus* (salmonberry), *Sequoia* (redwood), *Trientalis* (starflower), *Umbellularia* (California bay), *Vaccinium* (huckleberry), and *Viburnum*. Numerous additional species, especially of Ericaceae and Fagaceae, that are common in the nursery trade and in forests of eastern USA and Europe have proven susceptible when inoculated. *P. ramorum* has been found widely in nurseries in Europe and Great Britain, and discoveries in landscapes were increasing as this was written. The pathogen occurs there on *Aesculus* (horse-chestnut), *Camellia*, *Castanea* (chestnut), *Fagus* (beech), *Hamamelis* (witch-hazel), *Syringa* (lilac), *Quercus*, *Taxus* (yew), *Viburnum*, and various ericaceous plants, most commonly *Rhododendron*. Stem cankers caused by *P. ramorum* have been found on *Aesculus*, *Fagus*, and both European and American *Quercus* species.

P. ramorum is favored by relatively cool temperatures (near 20°C). Its sporangia are deciduous; that is, they become detached from sporangiophores. Therefore they can be dispersed with splashing and windblown rain. Chlamydospores of *P. ramorum* can be transported with soil and plant debris. It sporulates on leaves of several hosts, and the spores are dispersed in splashing and dripping water. It occurs in nurseries, in soil associated with diseased plants, in streams and nursery irrigation water, and on the shoes of people leaving infested sites. By mid-2004 it had been transported with nursery stock from California and Oregon to nurseries across the USA and had also been discovered in a few eastern landscapes. It is considered to be a significant threat to oaks in Appalachian forests.

The origin of *P. ramorum* is unknown. DNA analyses have revealed differences between European and Californian strains, and they are of different mating types, so they are presumed to be descended from different founding populations. In western North America *P. ramorum* has been found in forests from the central coast of California to Oregon and in nurseries as far north as British Columbia. It has high potential for distribution in horticultural commerce because various ornamental plants, mainly in the Ericaceae, can support abundant sporulation. *P. ramorum* also has the potential to cause great damage to *Quercus* species in regions where temperature and rainfall regimes during part of the year are similar to those of coastal California. Regulations designed to reduce the hazard of spread have been enacted, and eradication has been

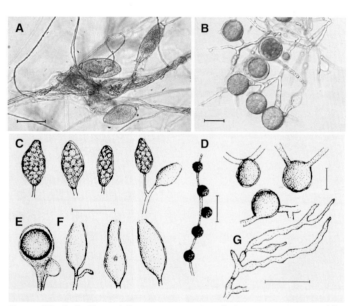

Figure 179. A, B. *Phytophthora ramorum:* A, colorless sporangia; B, golden brown chlamydospores. C–G. *P. lateralis:* C, sporangia; D, chlamydospores; E, an oospore in an oogonium with antheridium still attached; F, empty sporangia; G, typical gnarled hyphae. Scale bars = 50 μm in A, B; 40 μm in C–G. A and B courtesy of D. M. Rizzo and L. Englander, respectively. C–G adapted from reference 1077 by permission of the American Phytopathological Society.

A–F. Foot rot of *Citrus* caused by *Phytophthora nicotianae* (discussion, p. 362). A. Sparse, chlorotic foliage on severely affected *C.* ×*paradisi* (grapefruit). B. Dieback in *C. reticulata* (mandarin orange) associated with foot rot. C. A bleeding lesion near soil level on *C.* ×*paradisi*, partly exposed by removal of bark. D, E. Foot rot revealed by excavation at the root crown of *C. reticulata*. The sunken canker on the subterranean part of the butt extends up to the soil line (D). The canker margin was exposed by shaving surface bark (E) (TX, May). F. An old foot rot lesion on *C.* ×*paradisi*, becoming sunken due to growth of adjacent healthy stem (FL, Jan).

G. Decline in *Malus pumila* (apple), a result of crown rot caused by *P. cactorum* (NY, Aug) (discussion, p. 362).

H. Crown rot of *M. pumila* caused by *P. megasperma*. Excavation at the base of a young declining tree revealed that all the major roots on one side of the tree were dead. The swelling just above ground level is the graft union (NY, Aug).

Photo credits: F—J. H. Graham; G, H—S. N. Jeffers

attempted in each place the pathogen has been detected outside the main outbreak area.

References: 892, 1231, 1877, 2483, 3305, 3826, 4296

Collar, crown, and foot rots. The terms *collar rot, crown rot, basal canker,* and *foot rot* refer to lesions that develop at the trunk base and on the adjacent parts of major roots. The infections usually are perennial. Oomycetes that cause such lesions on multiple plant species include *P. cactorum, P. cambivora, P. cinnamomi, P. citricola, P. citrophthora, P. cryptogea, P. drechsleri, P. megasperma, P. nicotianae, P. palmivora,* and *P. syringae.* Some of these pathogens also cause bleeding trunk cankers and other symptoms such as tip blight, fruit rot, and necrosis of feeder roots. Pathogens or combinations of them that cause particular diseases vary with host and geographic area. Multiple species of *Phytophthora* have been detected in particular disease situations. For example, seven species were found associated with root rot and basal cankers on *Abies procera* (noble fir) in Christmas tree plantations in northwestern USA. Collar rots are among the most important lethal disorders of fruit and nut trees in orchards. *Castanea* (chestnut), *Citrus, Juglans* (walnut), *Malus* (apple, crabapple), *Persea americana* (avocado), and *Prunus* (apricot, peach, cherry, plum) are among the tree taxa affected. Many shade and ornamental trees also are attacked. A partial list includes *Abies* (fir), *Acer* (maple), *Aesculus* (horsechestnut), *Alnus* (alder), *Arbutus* (madrone), *Betula* (birch), *Cornus* (dogwood), *Fagus* (beech), *Liquidambar* (sweetgum), *Liriodendron* (tuliptree), *Platanus* (planetree), *Populus* (poplar), *Quercus* (oak), *Salix* (willow), *Sorbus* (mountain-ash), and *Tilia* (linden). Lesions often go unnoticed until foliar symptoms develop, by which time necrotic bark may extend halfway around the butt. Leaves become stunted, sparse, and chlorotic. Leaves of deciduous trees develop premature autumn color, and twigs and branches begin to die.

Foot rot of citrus (Plate 179A–F), also called collar rot or gummosis because gum exudes from fissures in dying or recently killed bark, is caused primarily by *P. nicotianae* and occasionally by *P. citrophthora.* Lesions usually begin near the soil line and spread upward more than downward. Oozing gum marks their location until the bark cracks and begins to fall away. The aboveground parts of a lesion may cease expansion and become delimited by wound wood, but continued expansion below the soil line often leads to girdling and death. Partial girdling causes one-sided stunting and dieback. *P. nicotianae* is favored by high temperatures (30–32°C) and waterlogged soil. It becomes inactive as soil dries, even within a moisture range that is still favorable for root growth. Thus trees may grow satisfactorily in infested soil much of the time. Foot rot in citrus has been suppressed by using resistant rootstocks, but highly susceptible scions become infected through wounds near the graft union.

Crown rot of *Malus pumila* (apple; Plate 179G, H) can be caused by any of several Phytophthoras. *P. cactorum* is the most common pathogen. *P. megasperma* caused the canker shown in Plate 179H. Nearly all commonly used apple rootstocks are susceptible to some extent. Crown-rot lesions expand rapidly during spring and slowly during the remainder of the warm season. The period of spring susceptibility coincides with renewed activity of the pathogen after winter quiescence. Wet soil and moderate temperatures at that time favor germination of oospores and chlamydospores and dispersal of zoospores. *Malus* species vary in susceptibility to Phytophthoras that cause crown rot. The crabapples *Malus halliana* and *M. sargentii* apparently possess resistance to multiple Phytophthoras.

Surgery and heat treatment to halt the enlargement of basal cankers on orchard and shade trees have been demonstrated but are not often used because they are laborious, destroy healthy tissues around cankers, and may fail to halt tree decline. Some orchardists utilize resistance-inducing or *Phytophthora*-suppressive fungicides to limit damage.

References: 169, 295, 359, 435, 544, 545, 836, 1001, 1003, 1077, 1494, 1754, 1859, 1918, 1919, 1979, 2101, 2181, 2248, 2318, 2492–2494, 2672, 2896, 3105, 3544, 3545, 3827, 4333, 4363

Root Rots and Feeder Root Necroses

Root and crown rot of Port Orford cedar. *Phytophthora lateralis,* a host-specialized pathogen, has depleted the population of *Chamaecyparis lawsoniana* (Lawson-cypress or Port Orford cedar) on sites favorable for the pathogen in a large part of the host's native range. The tree is native to a small region in southwestern Oregon and northern California but grows satisfactorily in a much larger region and is an internationally popular ornamental plant. Various Phytophthoras attack it with destructive effect. *P. lateralis,* origin unknown, was found killing nursery stock in Washington State in the 1920s, and it soon spread north and south. The epidemic intensified in both forests and landscapes, spreading along watercourses and roads, reflecting movement of infected plants or contaminated soil and water. The pathogen now occurs from northern California to southern British Columbia. *C. lawsoniana* has escaped damage in some northwestern forests where it grows on sites unfavorable for the pathogen and relatively free of human activity.

P. lateralis attacks trees of any size or age. It kills small seedlings within a few weeks and large trees within 2–4 years. Large diseased trees are often attacked by *Phloeosinus* species (cedar bark beetles). Diseased inner bark turns from its normal cream color to cinnamon brown. The usual progression of infection and bark necrosis is from tiny roots to increasingly large ones until the tree is girdled at its butt. The foliage withers and turns dull yellow, then bronze, red-brown, and finally dull brown. Infection can also begin on leaves or stems, as on low branches where foliage brushes against wet infested soil. Foliar lesions proliferate as zoospores from initial lesions are splashed about and initiate new infections. More branches become involved gradually, and the tree is slowly killed. Mycelial growth through root grafts between trees is thought to contribute to limited upslope movement of the pathogen and local intensification of the disease.

In the mild climate of the Pacific Northwest, *P. lateralis* is active throughout the year except during hot, dry weather in summer. Chlamydospores in killed roots and foliage act as survival structures during summer. They require a low-temperature stimulus for germination. During cool, wet weather they germinate and give rise to sporangia and zoospores. Both aerial and root infections occur beginning in late autumn but are most frequent in early spring. Hyphae from germinated spores enter nonsuberized roots, foliage, or wounds to inner bark, and spread in the inner bark and cambial region. Foliage that is wet for 2 hours or less may become infected if zoospores are present. Lesions expand at rates averaging 5 cm per month during late autumn through early spring, but growth ceases and chlamydospores form as the temperature rises above 25°C. Lesions resume expansion in autumn.

Infection can occur at temperatures of 3–25°C; the optimum range is 15–20°C. At 10–20°C sporangia form on diseased foliage 36–48 hours after infection. Sporangia of *P. lateralis* (Fig. 179), unlike those of most Phytophthoras, become detached from the sporangiophores and can serve as propagules. Detached sporangia and zoospores are dispersed. Isolates of *P. lateralis* from the Pacific Northwest vary in growth rate, temperature preference, and aggressiveness. Assays of ribosomal DNA sequences have revealed relatively little variation, however, a finding consistent with the concept of a population derived recently from a small, presumably introduced, founding stock.

P. lateralis affects several additional tree and shrub species. *Taxus brevifolia* (Pacific yew) is occasionally infected and killed where it grows in close proximity to diseased *C. lawsoniana. C. nootkatensis* (Alaska cedar), native to coastal forests north of the range of the pathogen, has proven less susceptible than *C. lawsoniana* when tested. Elsewhere, oomycetes identified as *P. lateralis* have been isolated from roots of *Actinidia deliciosa* (kiwi-fruit), *C. lawsoniana, C. obtusa* (Hinoki-cypress), *Juniperus horizontalis* (creeping

---

Root rot of *Chamaecyparis lawsoniana* (Lawson-cypress, Port Orford cedar) caused by *Phytophthora lateralis.*

A–C. Dead and dying trees in three forest situations.

D. Necrotic bark exposed by shaving off the outer layer at the base of a stem that grew from a layered branch in the forest litter layer.

E–G. Basal cankers girdling small trees (E, F) and a mature tree (G). Bark has been shaved to reveal orange-brown phloem and cambium in the cankers. Brown tissue above the healthy white inner bark in G is normal dead, corky outer bark (OR, summer).

Photo credits: C–F—D. J. Goheen

363

juniper), *Kalmia latifolia* (mountain-laurel), *Photinia ×fraseri*, and *Rhododendron* sp. The identities of isolates from *Juniperus* and dicot hosts have not been verified.

Despite the damage caused by *P. lateralis*, there are prospects for growing *C. lawsoniana* in forests of the Pacific Northwest if resistant planting stock and forest management practices that minimize spread of the pathogen are used. Resistant trees have been detected, although they formed only a tiny fraction of those tested. First-generation plants screened for resistance are already available. Measures for pathogen suppression and containment include regulating nontimber uses of forests where *C. lawsoniana* grows, prohibiting access to these forests during wet weather, washing equipment that moves from infested to other areas, eliminating *C. lawsoniana* in buffer zones along roads, growing this tree species on sites unfavorable for infection (well-drained sites away from roads and streams), identifying waters containing *P. lateralis*, and then avoiding them for uses that could spread the pathogen. Forest crop rotation may also be applicable because in one study *P. lateralis* became undetectable in decomposing material from diseased trees within about 7 years.

References: 1486, 1488, 1494, 2767, 2918, 3355, 4083, 4100, 4400

Diseases caused by *Phytophthora cinnamomi*. The most widely respected *Phytophthora* species among those attacking woody plants is probably *P. cinnamomi* (Plates 181, 182). This organism occurs globally in tropical and temperate regions and infects more than 900 plant species in many families. It is common along the Pacific Coast of North America as far north as British Columbia, and it occurs in eastern areas as far north as Ohio and Massachusetts. In the more northerly areas of its North American distribution it sometimes occurs in nurseries but does not persist in forests. Its hosts are primarily trees and shrubs. Major host groups include *Abies* (fir), *Castanea* (chestnut), *Eucalyptus*, *Persea* (avocado), *Pinus* (pine), and ericaceous plants, especially *Rhododendron*. *P. cinnamomi* has devastated forests in parts of southeastern and southwestern Australia, killing both trees (notably *Eucalyptus*) and understory plants. It is associated with the following diseases, among others: the decline syndrome called littleleaf disease of *Pinus echinata* and *P. taeda* (shortleaf and loblolly pines) in southeastern USA; feeder root necrosis, bleeding stem cankers, and decline of *Quercus* (oaks) in Mexico and southern Europe; root, crown, and collar rots of diverse plants, such as *Abies* (fir) in USA and *Persea americana* (avocado) on six continents; and root rot and wilt of various woody ornamentals in American, Australian, and European nurseries. Pathogenic specialization has been detected in strains from various hosts, but most strains can infect diverse plants.

P. cinnamomi attacks the small absorbing roots and in many hosts also invades larger roots or the butt and mainstem, growing mainly in the inner bark and cambial zone. It derives nourishment from recently killed as well as living tissues, and it reproduces in the dead tissues. It produces oospores, chlamydospores, sporangia, and zoospores. Oospores of *P. cinnamomi* are rarely found in nature, however, because only one mating type of this species is present in most areas that it inhabits. Hyphae of different mating types must contact each other to initiate the sexual process that results in oospore formation. Oospores or chlamydospores allow the pathogen to survive environmental adversity or persist several years in soil in the absence of hosts. Under favorable conditions in the presence of host roots, oospores or chlamydospores give rise to mycelium or to sporangia and zoospores. When the water content of soil is insufficient for zoospore activity, hyphae may infect roots.

P. cinnamomi requires warm, wet soil for significant pathogenic activity. It does not persist where soils regularly freeze deeply in winter. Temperature optima for sporulation, infection, and mycelial growth usually are similar within an isolate but may vary between isolates in the range 20–32°C. Temperature minima for these functions are 5–15°C, and maxima are 30–35°C. Plants that become infected at low temperatures may lack foliar symptoms, or symptoms may be delayed until after a period of higher temperature or an environmental change stressful to the plant. The interval between infection and sporulation of *P. cinnamomi* is only a few days, so disease can increase explosively if warm infested soil is flooded or waterlogged. Microbial populations influence the

conduciveness or suppressiveness of soils to disease caused by *P. cinnamomi*. Most nursery soils, soil-less container mixes, and soils formerly used for agricultural crops are likely to be conducive.

P. cinnamomi causes root rot, dieback, and death of various plants in nurseries. Ericaceous plants such as *Pieris japonica* (Japanese pieris) and *Rhododendron* species (azalea and rhododendron) are often affected. Additional species of *Phytophthora* often are involved too. Many plants become infected in propagation beds and remain symptomless until after transplanting, which leads to spot outbreaks in ground beds and container-grown crops. Symptoms first appear on one plant and subsequently on those surrounding it. In typical cases fine roots become infected and turn brown. If infection is confined to the roots, plants become stunted and chlorotic with dieback of scattered branches. The pathogen usually spreads toward the root crown and there either girdles the plant or advances up the stem in the cambial zone and phloem, eventually colonizing xylem and pith and reaching green shoots. Affected plants or branches cease growth and the leaves droop, turn chlorotic to dull yellowish green, and wilt permanently. Colonized tissues turn brown, and chlamydospores form in dead cortex and phloem.

Infection of nursery plants by *P. cinnamomi* can lead to problems in landscape plantings. In one study with *P. cinnamomi* in *Rhododendron obtusum* (Hinodegiri azalea), plants that were deliberately inoculated before outplanting in a landscape situation failed to thrive, and many died. In contrast, similarly inoculated plants that were treated with fungicides before outplanting performed normally.

Damage by *P. cinnamomi* in orchard and nursery situations can be lessened somewhat by amending surface soil with gypsum (hydrated calcium sulfate). This substance and other calcium compounds interfere with sporangium formation and zoospore function and thus suppress infection. Additional management measures are noted in the section on other Phytophthora root rots, p. 366.

Phytophthora root rot of *Abies fraseri* (Fraser fir; Plate 182) is typical of root, crown, and collar rots caused by *P. cinnamomi*. *A. fraseri*, native at high elevations in the southern Appalachian Mountains, is grown as a Christmas tree in nurseries and plantations

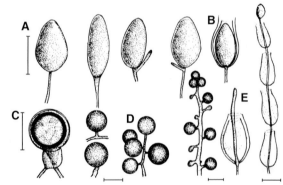

Figure 181. Phytophthora cinnamomi. A. Sporangia. B, E. Internal proliferation of sporangia. C. An oospore in an oogonium to which an antheridium remains attached. D. Chlamydospores. Scale bars = 40 μm. Adapted from reference 1077 by permission of the American Phytopathological Society.

A–C. Phytophthora root rot and wilt of *Pieris japonica* (Japanese pieris). A. A chlorotic plant among apparently healthy plants in a nursery. B. One-sided wilt and chlorosis reflect initially one-sided root necrosis. C. The stem base of a diseased plant, with bark shaved to show brown necrotic inner bark and wood (NY, Jul).

D, E. Root rot of *Rhododendron* sp. (azalea) caused by *Phytophthora cinnamomi*. D. Three diseased plants in contrast to a healthy 'Hershey' azalea. E. The stem base of a diseased plant with bark shaved to reveal necrotic inner bark and wood (NY, Aug).

F, G. Root rot of *Juglans nigra* (black walnut) seedlings caused by *P. citricola*. F. Dead blackened taproots (WI, Apr). G. Killed seedlings, split to show lesions involving the upper taproot and stem base (IN, Apr).

Photo credits: F—N. A. Tisserat; G—R. J. Green

365

at lower elevations. Seedlings there are attacked by *P. cinnamomi* and other *Phytophthora* species. Symptoms begin with reddish brown decay of rootlets and of the cortex and secondary phloem of woody roots. Root loss leads to cessation of growth and then chlorosis, drooping, and browning of foliage. Symptomatic plants occur singly and in enlarging groups. Shipment of infected seedlings from nurseries leads to disease outbreaks in plantations. Once established at a site conducive to infection, this pathogen may persist indefinitely by means of its resting spores and its ability to infect roots of diverse plants.

Littleleaf disease of pines in southeastern USA exemplifies decline diseases caused by *P. cinnamomi* in combination with stressful environment. Littleleaf disease became important during the 1930s and 1940s, when pole-size and mature *Pinus echinata* (shortleaf pine) and less often *P. taeda* (loblolly pine) began to decline on sites previously used for field crops. These sites had been severely eroded and were characterized by low fertility, poor internal drainage, periodic waterlogging, and periodic water shortage. *Phytophthora cinnamomi* was eventually found to cause severe feeder root necrosis on *Pinus echinata* in those conditions. *P. taeda* was less affected by the pathogen, and its decline on some sites was ascribed primarily to the poor soil conditions. Aboveground symptoms of littleleaf disease are slow growth; chlorotic, stunted, tufted foliage; premature needle loss; and progressive dieback. The chlorosis is due mainly to impaired uptake of nutrients, especially nitrogen. Affected trees usually signal distress by producing an abnormally large number of cones. Diseased trees live an average of 6 years after the onset of symptoms. Ectomycorrhizal fungi protect some pine rootlets from *P. cinnamomi*, but this is not sufficient to prevent littleleaf disease on soils with conducive physical characteristics. The disease still occurs, although the magnitude of the problem has diminished. This change reflects tree planters' recognition and avoidance of high-risk soils, replacement of susceptible trees with tolerant ones on many sites, and improvement of soil structure and microbial activity during several decades of forest regrowth.

References: 141, 312–314, 317, 388, 389, 479, 570, 601, 602, 626, 758, 831, 913, 1008, 1032, 1077, 1466, 1489, 1491, 1494, 1624, 1714, 1715, 1717, 1859, 2038, 2039, 2181, 2365, 2406, 2464, 2465, 2488, 2515, 2609, 2610, 2677, 2753, 2882, 3106, 3107, 3272, 3314, 3390, 3486, 3495, 3586, 3707, 3958, 4300–4302, 4329, 4406, 4510

Other Phytophthora root rots. Several additional soilborne *Phytophthora* species kill feeder roots and/or cause girdling lesions on larger roots. Some of these pathogens also cause foot rots. The root problems in North America are best known in nurseries, Christmas tree plantations, and orchards—situations where nutritional, microbial, and physical properties of soils (and soil-less container media in nurseries) are presumed to be more favorable than undisturbed forest soils for survival and activity of oomycetes. Reports of root damage in forests by Phytophthoras other than *P. cinnamomi* and *P. lateralis* began to increase in the late 20th century, however, as researchers in Europe employed more selective and sensitive detection procedures in the diagnosis of declining *Quercus* (oaks). Several well-known *Phytophthora* species and others not previously known were obtained from soil and water using host leaves as baits. More recently, the combination of baiting and identification of *Phytophthora* and *Pythium* species by PCR (polymerase chain reaction) techniques has facilitated the study of oomycete associations with diseases of woody plants.

Nursery problems. Pythiaceous oomycetes (*Pythium* and *Phytophthora*) that kill feeder roots and/or cause girdling lesions on larger roots are perennial problems in nurseries because the pathogens are transported with plants or soil, their dispersal is often favored by irrigation practices, and their spores can contaminate nursery water supplies or recycled irrigation water. *Pythium* species cause damping-off and feeder root necrosis, and *Phytophthora* species may kill feeder roots and cause girdling lesions on larger roots or the root crown. *P. cactorum, P. cinnamomi, P. citricola, P. citrophthora, P. cryptogea, P. gonapodyides, P. lateralis, P. megasperma, P. nicotianae, P. palmivora,* and *P. pseudotsugae* are documented root pathogens in nurseries, causing damage to diverse plants. *Abies* (fir),

Juniperus (juniper), and *Rhododendron* are examples. *P. cactorum* and *P. citricola* also commonly cause leaf blight and shoot dieback of *Rhododendron* and other woody ornamentals. Slow growth, chlorosis, and wilting are common aboveground symptoms in plants not killed by root disease. Leaf baits of *Camellia, Eucalyptus, Juniperus conferta* (shore juniper), and other hosts have proven effective for detecting Phytophthoras in soil-less media used for growing plants in containers. Damage in nurseries can be minimized by using pathogen-suppressive container mixes; irrigating with pathogen-free water; covering soil in container nurseries with woven ground fabric, gravel, or coarse mulch that prevents transfer of pathogens in splashing or running water from soil into pots; and adding calcium chloride or calcium nitrate to water or liquid fertilizer. Calcium ion (Ca^{++}) at nutritional concentrations was shown to cause *P. nicotianae* zoospores to encyst without contacting susceptible tissue and to interfere with infection.

Phytophthoras in Christmas tree plantations and orchards. Although *P. cinnamomi* is the most important of these pathogens in regions of warm climate, several other species cause damage in cooler areas. In Oregon and Washington, for example, *P. cambivora, P. cactorum, P. citricola, P. cryptogea, P. gonapodyides,* and *P. megasperma* were found associated with root rot and stem cankers in planted *Abies procera* (noble fir). *P. citricola* also occurs in *A. fraseri* (Fraser fir) Christmas trees in eastern USA, and *P. megasperma* has killed *Pinus lambertiana* (sugar pine) trees in a seed orchard in California. All of these oomycetes and others are documented causes of tree mortality and crop loss in fruit and nut orchards. Wet soils and irrigation by flooding are common themes in orchard damage. Solar heating of soil (solarization) under clear plastic has shown promise for suppressing *Phytophthora* species in soil-less container mixes and in some nursery and orchard situations, including soils where trees already are growing. This procedure is feasible only in warm, sunny regions, however.

Phytophthora in forests. Several *Phytophthora* species are associated with root necrosis and decline of deciduous oaks, mainly *Q. robur* (English or pedunculate oak) and *Q. petraea* (sessile oak), in forests across Europe. Thirteen species, including four that were previously unknown, were found during surveys. Those with asterisks in the following list caused the most root rot in seedling tests with *Q. robur*: **P. cactorum, *P. cambivora, *P. cinnamomi, *P. citricola, P. drechsleri, P. gonapodyides, P. megasperma,* and *P. syringae*; also *P. europaea, P. pseudosyringae, P. psychrophila, *P. quercina,* and *P. uliginosa.* The latter five species are recently described. Surveys in Bavaria indicated *Phytophthora* species to be strongly involved in oak decline on sites with sandy loam to clay soils and soil pH of 3.5 or greater.

Significant forest and landscape damage in North America by Phytophthoras other than *P. cinnamomi* and *P. lateralis* has not been documented. As noted in an earlier section, root-infecting Phytophthoras that cause damage in nurseries and orchards may fail to thrive in forests, presumably because of less favorable temperature and moisture regimes and inhibitory microbial populations on forest sites. Phytophthora root rots do occur in horticultural landscapes, but they are poorly documented because diagnosis requires excavation that further damages on kills affected plants.

References: 184a, 184b, 524, 552, 710, 720, 1077, 1118, 1149, 1466, 1467, 1491, 1494, 1713, 1714, 1982, 1986, 2365, 2406, 2407, 2493, 2494, 2800, 3495, 3585, 3827, 4155

A, B. Littleleaf disease of pines. A. *Pinus echinata* (shortleaf pine): declining trees next to one relatively healthy individual. B. *P. taeda* (loblolly pine) with chlorotic, stunted, tufted foliage and a distress crop of cones (GA, Aug).

C–F. Root rot of *Abies fraseri* (Fraser fir) seedlings caused by *Phytophthora cinnamomi.* C. Mortality in a nursery bed of year-old seedlings. D. Dead and dying 5-year-old transplants. E. Dead foliage associated with dead roots on one side of a 5-year-old plant. F. A lesion extending up the mainstem from roots. The tree at left was already dead when collected; the pathogen apparently spread from it into the living tree (NC, May).

367

Bacterial Diseases (Plates 183–200)

Overview of Bacterial Diseases

Plant-pathogenic bacteria are classified in more than 20 genera in the phyla Proteobacteria and Posibacteria. Most known bacterial diseases of woody plants are caused by species in six genera of proteobacteria and two genera of posibacteria. The six are *Pseudomonas*, *Xanthomonas*, *Erwinia*, and *Brenneria* (Plates 183–189); *Agrobacterium* (Plate 190); and *Xylella* (Plates 192, 193). Posibacteria in the class Mollicutes are considered with Plates 194–200.

Pseudomonas, *Xanthomonas*, and *Erwinia* include the majority of plant-pathogenic bacteria (phytobacteria) and share many characteristics. Their cells are Gram-negative, mostly short rods with rounded ends, motile by flagella, and culturable. They require free water for multiplication and pathogenic activity. Individual bacterial cells are innocuous; many cells acting together are required to initiate disease. Wet plant surfaces and water-saturated air, which promote accumulation of liquid water in intercellular spaces, favor bacterial movement into plants and thus favor infection. Plant-pathogenic bacteria secrete proteins that degrade plant cell walls and membranes, elicit other damaging processes in susceptible plants, and may induce effective defense in resistant plants. Some bacteria produce toxins that kill plant cells. Their nutrition during pathogenesis is provided by substances leaked from damaged and newly killed cells. The bacteria also secrete slimy extracellular polysaccharides that protect them from drying and from some types of host defenses. Most lesions caused by bacteria in woody plants are localized, although internal spread is a feature of some diseases. The temperature optima for multiplication of most plant-pathogenic bacteria lie between 20° and 30°C. These bacteria are sensitive to heat and drying. They are dispersed passively. Local dispersal is mainly by water, and long-distance transport occurs in or on plant materials. Some phytobacteria are also dispersed by insects and other animals and on tools and equipment. Most are poor saprobic competitors and therefore survive poorly in the absence of hosts, but some soilborne plant pathogens and ice-nucleating strains of *P. syringae*, discussed below, are exceptions to this rule.

The life cycle of a generalized plant pathogen in the groups considered with Plates 183–189 has three phases. Quiescent cells in protected places on or in host plants survive while conditions are unsuitable for multiplication. Then follows a resident, or epiphytic, phase characterized by population increase on surfaces of healthy organs, population density being governed by temperature, availability of water and nutrients, and competition from other microbes. Nutrients sufficient for this phase leak from healthy plants. If wet conditions prevail during the epiphytic phase, cells sufficiently numerous to initiate disease enter a plant through natural openings and wounds, including fresh leaf scars. The pathogenic phase is marked by dramatic population increase in intercellular spaces, then collapse of the colonized tissue and decline of the pathogen population. This phase produces cells that ooze to the plant surface in a slimy matrix when the water content of diseased parts and the surrounding air is high. The bacteria may then be dispersed either to new sites of infection, leading to secondary cycles of disease, or to survival sites. Bacteria also survive for a time in lesions.

Many species of plant-pathogenic bacteria include subpopulations called *pathovars* (abbreviated pv.) that are specialized for attack of particular genera or families of plants. Pathovars within a species can be distinguished by bioassays and in some cases by physiological or biochemical tests.

References: 41, 159, 275, 523, 679, 774, 935, 1022, 1153, 1327, 1547, 1675, 1733, 1734, 1859, 2049, 2272, 2329, 2330, 2712, 2946, 3429, 3628, 3683, 3817, 4490

Leaf Spots, Blights, and Cankers Caused by *Pseudomonas* and *Xanthomonas* (Plates 183–186)

Diseases Caused by Pseudomonas Species

The genus *Pseudomonas* (Pseudomonadaceae) was for many years a broadly defined group that included plant and animal pathogens as well as free-living epiphytic and soil-dwelling forms. The genus is now circumscribed more narrowly. Authentic pseudomonads that are pathogens of woody plants are listed below with representative diseases. Records for the former pseudomonad *Burkholderia andropogonis* are appended.

P. amygdali causes cankers on *Prunus dulcis* (almond).
P. avellanae causes cankers and dieback of *Corylus avellana* (European filbert).
P. cichorii causes leaf spot or leaf blight of *Coffea* (coffee), *Ficus lyrata* (fiddle-leaf fig), *Hedera helix* (English ivy), *Hibiscus* species, *Magnolia grandiflora* (southern magnolia), *Rhododendron*, and *Schefflera arboricola* (dwarf schefflera), also fruit lesions on *Prunus persica* var. *nucipersica* (nectarine).
P. ficuserectae causes leaf spot and shoot blight of *Ficus erecta*.
P. meliae causes bacterial gall of *Melia azedarach* (chinaberry).
P. savastanoi causes galls and cankers on *Fraxinus* (ash), *Olea* (olive), *Nerium oleander* (oleander), and a few other plants (Plate 189).
P. syringae causes many plant diseases (see below).
P. viridiflava causes bud blight of *Hydrangea*; bud rot, blossom blight, and leaf spot of *Actinidia deliciosa* (kiwi-fruit); leaf spot, bud and bract blight, and stem canker of *Euphorbia pulcherrima* (poinsettia); and is associated with bacterial canker of *Prunus armeniaca* (apricot).
Burkholderia andropogonis. (syn. *P. andropogonis*) causes leaf spot of *Bougainvillea*, *Ceratonia siliqua* (carob), and *Vaccinium corymbosum* (highbush blueberry).

Diseases Caused by *Pseudomonas syringae*. The name *P. syringae* is applied to a globally distributed heterogeneous group of opportunistic pathogens. Although capable of inflicting great damage, they commonly inhabit plants without causing noticeable disease. Strains of *P. syringae* cause leaf spots, leaf blight, bud and blossom blight (often called blast), cankers, shoot blight, and dieback of many plants. A few strains cause galls. The economic importance of these bacteria is greatest on orchard trees: *Citrus*, *Prunus* (stone-fruit trees), and *Pyrus* (pear). Symptoms vary with host, strain of pathogen, host predisposition by factors such as frost, and weather conditions.

Two features of *P. syringae*—phytotoxin production and ice nucleation ability—enhance its ability to cause plant damage. *P. syringae* pv. *syringae* produces several broad-spectrum toxins—syringomycin and others—that destroy host cell membranes. Strains vary in the specific toxins produced. Toxin production is stimulated by signal molecules released from plants. The toxins are not the sole determinants of pathogenicity, however, since plants susceptible and resistant to *P. syringae* are sensitive to the toxins.

Ice nucleation refers to the formation of ice crystals around microscopic particles. Ice-nucleating strains of *P. syringae* and certain other bacteria can trigger ice formation in plant tissues cooled to between −2 and −5°C but not acclimated to low temperature. Ice then disrupts cells, causing symptoms of frost damage. In the absence of an ice-nucleating factor, frost-sensitive plants may tolerate brief cooling to these temperatures because water in their tissues remains in liquid form, supercooled. Most

A, B, E, F. Diseases caused by *Pseudomonas syringae* pv. *syringae*. A, B. Bacterial blight of *Syringa vulgaris* (common lilac). A. Blighted shoots and foliar lesions. B. An elongate canker on a current-season twig (NY, Jun). E. Bacterial canker on *Prunus avium* (sweet cherry). Arrows show approximate limits of the necrosis (OR, Jul). F. Pear blast on *Pyrus communis* 'Bartlett.' One-year-old twigs were killed in spring; green shoots grew subsequently (OR, Jul).

C, D, G, H. Diseases caused by undetermined pathovars of *P. syringae*. C, G. Bacterial spot and shoot blight on *Ginkgo biloba*. Lesions on shoots either kill shoot tips, as shown, or cause elongate cankers. Foliar lesions (G) have chlorotic halos (CA, Nov). D, H. Bacterial canker and leaf spot on *Liquidambar styraciflua* (sweetgum). The canker in D has ceased enlarging, as indicated by callus at its edges (CA, Sep).

Photo credits: C, D, G, H—A. L. Bishop

369

plants that are native in temperate zones can tolerate a few degrees of frost for this reason. *P. syringae* in combination with frost causes severe damage to plants that would not be harmed by either agent alone. In addition, the bacterial population rises rapidly in host tissues that have been chilled but not visibly damaged by temperatures below 0°C. Freezing and untimely pruning (in autumn or winter) often predispose plants to bud blight or cankers caused by this pathogen.

Pseudomonas syringae at one time had more than 40 named pathovars. New methods for differentiating pseudomonads have led to recognition of some pathovars as species, a process likely to occur for additional pathovars. *P. syringae* pv. *syringae*, the most widespread and important pathovar, apparently has an unusual set of pathogenicity attributes that permit it to attack diverse plants. Known host genera for pathovar *syringae* and for other or unspecified pathovars of *P. syringae* are discussed below.

Hosts of *P. syringae* pv. *syringae* include species of *Acer* (maple), *Alnus* (alder), *Atriplex* (saltbush), *Carissa*, *Citrus*, *Cornus* (dogwood), *Forsythia*, *Fraxinus* (ash), *Hibiscus*, *Jasminum* (jasmine), *Juglans* (walnut), *Laburnum* (bean-tree), *Magnolia*, *Malus* (apple, crabapple), *Nerium* (oleander), *Persea* (avocado), *Philadelphus* (mock-orange), *Pinus* (pine), *Populus* (aspen, poplar), *Prunus* (ornamental cherries and plums, stone-fruit trees), *Pyrus* (pear), *Quercus* (oak), *Rosa* (rose), *Salix* (willow), *Syringa* (lilac), *Tilia* (linden), and *Wisteria*. Many herbaceous plants are attacked also. Strains of pathovar *syringae* vary in host specificity; no strain is known to infect all recorded hosts.

Bacterial blight of lilac (Plate 183), caused by *P. syringae* pv. *syringae*, is characterized by death and shriveling of leaves, shoots, and sometimes flower clusters. Species affected include *Syringa ×chinensis*, *S. reticulata*, *S. ×persica*, and *S. vulgaris* (Chinese, Japanese, Persian, and common lilacs, respectively). Irregular to circular dark brown spots with yellowish halos form on leaves, then coalesce to form larger lesions; or entire leaves die. Lesions on petioles and shoots cause them to bend or droop, and parts distal to the lesions may then wither and turn brown. Parts invaded by the bacteria turn black. Flower buds are killed by early-season infections. If a stem becomes woody before infection, it remains erect and may escape girdling. Lesions on woody green twigs may appear as black streaks. Lesion formation in stems a year or more old is uncommon.

P. syringae pv. *syringae* causes severe damage to various *Prunus* species, killing buds in winter and blossoms in spring, causing necrotic spots on leaves and fruit, and causing cankers that enlarge on branches and trunks during host dormancy (Plate 183). *P. armeniaca* (apricot), *P. avium* (sweet cherry), *P. domestica* (plum), *P. dulcis* (almond), and *P. persica* (peach) are often affected. *Pseudomonas syringae* pv. *morsprunorum* also causes cankers on these plants. Blossom blight is most severe in years when frost occurs during bloom. Branch cankers form when infection spreads from diseased spurs. Narrow brown streaks extend into normal tissue at the upper and lower margins of the lesions. Cankers tend to be roughly elliptic and to exude gum or sour-smelling liquid. The extent of gummosis varies with host species. It is most profuse on apricot.

Pathovar *syringae* also causes bacterial blast of apple and pear. Brown to black lesions develop on flower parts, green stems, and the bases of fruit clusters. Blossoms are infected most often on trees predisposed by frost or a cold period in spring. Infection is usually restricted to current growth but sometimes extends into fruit spurs and year-old twigs, causing dieback or small cankers. Brown streaks on the surface of the sapwood may extend several centimeters beyond the margin of necrosis in the bark of apple twigs. Blast cankers on pear twigs are usually tan to light brown. The periderm of diseased apple or pear shoots and spurs tends to separate from underlying tissues, so that a superficial layer appears papery, a diagnostic symptom. Diseased pear fruits have black depressed lesions.

Diseases caused by other or undetermined pathovars of *P. syringae* occur on many plants, including *Acer*, *Actinidia* (kiwi-fruit), *Amelanchier* (serviceberry), *Berberis* (barberry), *Broussonetia* (paper-mulberry), *Camellia* (tea), *Castanea* (chestnut), *Coffea* (coffee), *Cunninghamia* (China-fir), *Eriobotrya* (loquat), *Ginkgo*, *Hibiscus*, *Liquidambar* (sweetgum), *Morus* (mulberry), *Philadelphus*, *Photinia*, *Prunus*, *Rhaphiolepis* (India-hawthorn), *Ribes* (currant), *Ulmus* (elm), and *Viburnum*.

Nursery plants of *Ginkgo biloba* and *Liquidambar styraciflua* in California (Plate 183), *Prunus laurocerasus* (cherry-laurel) in British Columbia (Plate 184C), and *Ulmus* species (elm) in northeastern USA (Plate 184A, B) occasionally develop a syndrome that includes shoot blight, twig cankers, and leaf spots. Tender shoot tips may be killed, and nongirdling lesions on young twigs may cause them to bend severely. Lesions that form in growing leaves cause distortion. Lesions on expanded *Ginkgo* leaves have chlorotic halos, and the centers of some old lesions weather away. Lesions in *Liquidambar* leaves are at first water-soaked, then fade to olive green with nearly black margins. The disease of *Ulmus* develops in early spring and sometimes follows bud necrosis on the same branches. Lesions in leaves are tiny (up to 2 mm), dark brown to black, and have chlorotic halos. If numerous they cause yellowing and necrosis of larger areas, followed by leaf cast. Shot holes form in leaves of *Prunus laurocerasus* (cherry laurel; Plate 184C), beginning as water-soaked lesions that enlarge to several millimeters in diameter and turn tan with chlorotic halos. After leaf defenses halt the enlargement of a lesion, an abscission layer develops around it, and the dead part eventually falls out.

References: 46, 303, 381, 432, 556, 618, 619, 636, 701, 846, 904, 1153, 1675, 1859, 2002, 2040, 2049, 2093, 2234, 2688, 2705, 2714, 2896, 2897, 2946, 3330, 3365, 3803

Diseases caused by other pseudomonads. Pseudomonad pathogens and associated diseases are listed with Plate 183. *Pseudomonas cichorii*, which has a wide host range, causes necrotic spots or blotches on leaves of *Ficus lyrata* (fiddle-leaf fig), *Rhododendron catawbiense* (Plate 184D), *Magnolia grandiflora* (southern magnolia; Plate 184E, F), and *Schefflera arboricola* (dwarf schefflera). These diseases were discovered in nursery plants in southeastern USA. Newly unfolded leaves of *M. grandiflora* had numerous dark brown lesions that occupied up to two-thirds of the leaf area. Symptoms were less severe on older foliage. *M. grandiflora*, *M. macrophylla*, *M. ×soulangeana*, and *M. tripetala* all developed foliar lesions after inoculation with the strain from *M. grandiflora*. Lesions on *R. catawbiense* develop on young leaves as irregular water-soaked spots, 1–5 mm across, that enlarge and become red-brown to dark brown and angular. They may coalesce into large necrotic areas. Test plants inoculated with a strain isolated from such lesions developed symptoms after 3–7 days.

A bacterium once named *Pseudomonas lauraceum* causes leaf spot (Plate 184G) and occasional severe leaf blight of *Umbellularia californica* (California laurel). Outbreaks have occurred during rainy winters in California. Small, black, angular spots delimited by veins,

---

A, B. Bacterial spot of an *Ulmus* (elm) hybrid, caused by *Pseudomonas syringae*. A. Leaf distortion as a consequence of lesion formation in growing tissues. B. Close view of angular lesions with chlorotic halos (NY, Jun).

C. Bacterial spot and shot hole caused by *P. syringae* on *Prunus laurocerasus* (cherry laurel) (BC, Apr).

D. Bacterial spot caused by *Pseudomonas cichorii* on *Rhododendron catawbiense* 'Album' (GA, Sep).

E, F. Bacterial spot caused by *P. cichorii* on *Magnolia grandiflora* (southern magnolia). Lesions resulting from infection of a newly unfolded leaf and a slightly older leaf, respectively (AL, May).

G. Bacterial spot of *Umbellularia californica* (California laurel), caused by *Pseudomonas* sp. (CA, Mar).

H. Bacterial spot caused by *Xanthomonas* sp. on *Hydrangea quercifolia* (oak-leaf hydrangea) (GA, Jun) (discussion, p. 372).

I. Bacterial spot caused by *Xanthomonas axonopodis* pv. *poinsettiicola* on *Euphorbia pulcherrima* (poinsettia) (FL, Sep).

J. Bacterial spot caused by *X. codiaei* on *Codiaeum variegatum* (croton) (FL, Aug).

Photo credits: C—S. H. De Boer; D, H—W. Uddin; E, F—J. M. Mullen; I, J—A. R. Chase

often with yellow halos, arise in mature leaves and may enlarge through coalescence. Young lesions often have a swollen central portion that, when broken, exudes bacteria-laden fluid. Minor wounds seem to be common sites of infection, because lesions have been noticed particularly on leaves injured in storms. Pathogenicity of the bacterium to *U. californica* and to *Persea americana* (avocado) was proven by inoculations. Unfortunately, reference cultures were not maintained, and the relationship of "*P. lauraceum*" to any contemporary species or pathovar is unknown.
References: 46, 701, 702, 705, 846, 1063, 1544, 1859, 2049, 2705, 2755, 2896, 2897, 2946, 2969, 3514, 3516, 4112

Diseases Caused by *Xanthomonas* Species

All known *Xanthomonas* species (xanthomonads; *Pseudomonadaceae*) inhabit plants. These bacteria are motile by single polar flagella. Most strains form yellow, smooth, mucoid colonies on solid media. The yellow pigments, called xanthomonadins, and a secreted polysaccharide called xanthan that accounts for the mucoid texture are typical of the genus. Practical taxonomy of xanthomonads has been problematic because traditional criteria for separating bacterial species do not differentiate strains that cause different diseases. From 1980 until the mid-1990s the majority of xanthomonads were classified as pathovars of one heterogeneous species, *X. campestris*. By 2004, however, more than 20 species had been differentiated on the basis of DNA homology, DNA fingerprinting, fatty acid methyl ester profiles, and correlated physiological and biochemical characters. At that time about 25 pathovars of *X. campestris* that attack woody plants had not yet been reclassified, or specialists had not agreed on their classification. Most of them were unrecorded in North America, however. The majority of xanthomonads known to attack trees and shrubs are considered to be pathovars (or subgroups within pathovars) of three species: *X. arboricola*, *X. axonopodis*, and *X. hortorum*. Other pathogens of woody plants include *X. codiaei*, which causes leaf spots of *Codiaeum variegatum* (croton; Plate 184J) and *Euphorbia pulcherrima* (poinsettia); *X. populi*, which causes bacterial canker of *Populus* (poplar) in Europe; and *X. theicola*, which causes a pustular leaf spot of *Camellia sinensis* (tea) in Asia. *X. axonopodis* is heterogeneous, so additional named species will likely be differentiated from it as well as from *X. campestris*. In the discussions that follow we provide synonyms as appropriate. *X. populi* is apparently restricted to *Populus* and is widespread in Europe but is not known to occur elsewhere. North American countries have regulations designed to prevent its importation.
References: 46, 567, 1022, 1153, 1859, 2824, 3429, 3683, 3940, 4144, 4145, 4488, 4490

Some diseases of woody ornamental plants. *Hydrangea quercifolia* is affected by a bacterial leaf spot in Georgia. Water-soaked spots, initially 1–4 mm across, become angular and turn dark brown as they enlarge (Plate 184H). Numerous or coalescing lesions sometimes cause death of mature leaves. In diagnostic tests the bacterium caused symptoms 3–5 days after inoculation and was found to be similar to *X. campestris* pv. *carotae*, a pathovar that, at this writing, had not been assigned to a more appropriate species.

At least four types of bacteria cause leaf spots of *Euphorbia pulcherrima* (poinsettia): *Pseudomonas viridiflava*, *X. arboricola* pv. *poinsettiicola*, *X. axonopodis* pv. *poinsettiicola*, and *X. codiaei*. The latter three were all formerly known as different types of *X. campestris* pv. *poinsettiicola*. They cause angular spots and blotches that sometimes spread along veins (Plate 184I).

Codiaeum variegatum (garden croton) is subject to leaf spots caused by *X. codiaei* (syn. *X. campestris* pv. *poinsettiicola* type B). The lesions (Plate 184J) are 1–10 mm in diameter, tan when fully developed, have irregular margins, and often expand along major veins but do not cross the midvein. Those on undersurfaces have a corky raised border. Of seven cultivars tested for response to the pathogen, Norma and Petra were the most resistant; Apple Leaf and Yellow Icing were intermediate; and Gold Finger, Gold Dust, and Gold Star were the most susceptible. This pathogen also causes leaf spots on *E. pulcherrima*, which is in the same family.

Xanthomonas hortorum pv. *hederae* (syn. *X. campestris* pv. *hederae*) causes foliar lesions on plants in the Araliaceae: *Fatsia japonica* (Japanese fatsia), *Hedera helix* (English ivy), *Schefflera actinophylla* (schefflera), and *S. arboricola* (dwarf schefflera). *S. elegantissima* (false-aralia) and *Polyscias fruticosa* were susceptible when tested. Symptoms on schefflera (Plate 185A, B) range from dark necrotic blotches in young expanding leaves to pinpoint yellow or tan lesions in mature leaves. Bacterial spot of *H. helix* (Plate 185C, F) occurs wherever the plant is grown outdoors and receives frequent rain or sprinkler irrigation. Severe infection results in leaf distortion, blight, and premature defoliation. Lesions, visible first as greenish brown water-soaked spots, expand to 2–10 mm in diameter with rust-brown or darker centers and frequently with chlorotic halos. Lesions bounded by veins appear angular, and some extend into petioles, killing the leaves. Under moist conditions some lesions exude orange-red fluid containing bacteria. Old lesions dry and crack. Stems may be invaded from petioles. A soft dark brown decay develops in succulent stems, but in mature stems the bacteria are restricted to small, dark brown, nongirdling cankers. Symptoms develop 5–21 days after inoculation, depending on temperature, humidity, and the age of plant tissue. The bacteria multiply most rapidly at 22–26°C. Epidemics have developed after propagation of ivy by cuttings from infected stock plants. Accordingly, careful hygiene is important for disease prevention. Ivy cultivars Gold Dust, Perfection, Sweet Heart, Eva, and California were relatively resistant in tests conducted in Florida. Symptoms caused on ivy by *X. hortorum* pv. *hederae* can be confused with those of anthracnose caused by *Colletotrichum trichellum*.

Alnus rhombifolia (white alder) and *A. cordata* (Italian alder) grown under sprinkler irrigation in a California nursery were found affected by a *Xanthomonas* strain that causes leaf blight and necrotic spots (Plate 185I, J). Lesions are irregular in shape, initially water-soaked, then dry and turn brown to tan, partly delimited by veins. Diseased leaf edges become tattered as dead tissue disintegrates. Large lesions have alternating dark brown and tan zones. Bacteria isolated from lesions and adjacent asymptomatic tissue of leaves and stems caused similar symptoms on inoculated test plants after 3–4 weeks. The pathogen was also detected in surface-sterilized peeled twigs, indicating systemic infection. This bacterium, a member of the *X. campestris* complex, has not yet been reclassified.
References: 574, 699, 700, 703, 1021, 1153, 2643, 2870, 4113, 4330

Citrus canker and related diseases. *X. axonopodis* pv. *citri* (syn. *X. citri*, *X. campestris* pv. *citri*) causes Asiatic citrus canker, also called citrus bacterial canker, a disease characterized by pustular lesions on leaves, young twigs, and fruit (Plate 185D, E, G, H). It occurs in Asia, on Pacific and Indian Ocean islands, in South America, and in Florida, where it has repeatedly been the object of eradication

---

A, B. Bacterial spot of *Schefflera actinophylla* (schefflera) and *S. arboricola* (dwarf schefflera), respectively, caused by *Xanthomonas hortorum* pv. *hederae* (FL, Jan & Aug).

C, F. Bacterial spot of *Hedera helix* (English ivy), caused by *X. hortorum* pv. *hederae*. C. Lesions involving leaf blades and petioles. F. Closer view of typical lesions with rust-brown centers and water-soaked margins (MS, May).

D, E, G, H. Asiatic citrus canker, caused by *Xanthomonas axonopodis* pv. *citri*. D. Foliar lesions on *Citrus sinensis* (sweet orange). Typical lesions have prominent yellow halos (Argentina, Mar). E. Halos around lesions on a *C.* ×*paradisi* (grapefruit) leaf accentuated by viewing it in transmitted light (FL, Sep). G. Lower surfaces of leaf lesions, here on *C. limon* (lemon), are raised and corky (FL, Oct). H. Raised corky lesions on grapefruit fruits (FL, Oct).

I, J. Bacterial leaf spot and leaf blight of *Alnus rhombifolia* (white alder), caused by *Xanthomonas* sp. (CA, Jul).

Photo credits: A, B—A. R. Chase; D, E, G, H—T. R. Gottwald; I, J—A. L. Bishop

373

campaigns. Many types of citrus are susceptible to some degree; *Citrus ×paradisi* (grapefruit), *C. aurantifolia* (lime), and *Poncirus trifoliata* (trifoliate orange) are highly susceptible. Although the disease can cause premature leaf and fruit drop when severe, its importance derives mainly from the diminished value of blemished fruit. It was present in southeastern USA for many years in the early 1900s but was eradicated. Reintroduced to Florida again in the 1980s and 1990s despite exclusionary measures, the pathogen is again the subject of an eradication campaign and quarantines.

Pathovar *citri* can induce lesions in tender growing parts within 7 days after inoculation. Young lesions are light green spots, raised in the center and often surrounded by a yellow halo. The center ruptures, leaving a brown depression with raised corky edges. Lesions attain diameters of 2–10 mm, depending on host susceptibility and tissue age, and often are aggregated at leaf tips or margins. Those on stems become 1–3 mm deep. Tree-to-tree spread is by wind-driven rain. Conditions that promote vigorous growth and delay tissue maturation enhance disease severity. Mature tissues are resistant. The disease is favored by frequent rainfall during shoot growth and the early phase of fruit growth and by insects such as leaf miners that wound leaves. Long-distance spread of pathovar *citri* occurs via diseased propagating material and infested citrus debris in shipping containers.

X. axonopodis pathovars *aurantifolii* and *citrumelo* (syn. *X. citri* pvs., *X. campestris* pvs.) cause diseases similar to, but less severe than, Asiatic citrus canker. Pathovar *aurantifolii* occurs in South America and Mexico. Pathovar *citrumelo*, which occurs in Florida and causes bacterial spot, elicited a flurry of action aimed at its eradication when it was mistaken for Asiatic citrus canker in nurseries in 1984. The lesions are not raised, and the necrotic centers often fall out. Although many types of citrus are susceptible, this disease occurs primarily on *C. trifoliata* (trifoliate orange).
References: 1331–1336, 1344–1346, 1540, 1541, 2181, 3238, 3484, 4030, 4146, 4337, 4338

Bacterial spot of *Prunus*. *Xanthomonas arboricola* pv. *pruni* (syn. *X. campestris* pv. *pruni*) infects leaves, twigs, and fruit of all stone-fruit species grown in orchards and several ornamental and wild *Prunus* species. The disease occurs worldwide where stone-fruit crops are grown in warm humid climates. It is common on sandy sites in southeastern USA but is unknown near the Pacific Coast. Ornamental and wild hosts include *P. americana* (American plum), *P. angustifolia* (chickasaw plum), *P. ×blireiana* (flowering plum), *P. caroliniana* (Carolina cherry-laurel), *P. davidiana* (Chinese wild peach), *P. hortulana* (wild goose plum), *P. laurocerasus* (cherry-laurel), *P. maritima* (beach plum), *P. mume* (Japanese flowering apricot), *P. pumila* (sand cherry), *P. pumila* var. *besseyi* (western sand cherry), *P. serotina* (black cherry), *P. serrulata* (Japanese flowering cherry), *P. tenella* (dwarf Russian almond), *P. texana* (peachbush), *P. triloba* (flowering almond), and some lesser-known species. Wild species tend to be less affected than domesticated ones.

The symptoms described here occur on highly susceptible species such as *P. persica* (peach) (Plate 186C–F) and *P. domestica* (common plum). A seasonal succession of symptoms begins with dark, elongate, blisterlike lesions on year-old twigs in early spring as the result of infection through leaf scars the previous autumn. Angular lesions 1–3 mm in diameter then appear on leaves, often aggregated near tips or margins. Numerous lesions cause yellowing and leaf cast. Some lesions, after being isolated by a cork layer, may drop out, leaving shot holes. Small cankers develop on green twigs in late spring. Small corky lesions form on fruit and may coalesce into large necrotic areas. Terminal buds and the adjacent parts of twigs may be killed during winter, resulting in "black tip." The pathogen overwinters in buds and twig cankers and is an epiphyte on both diseased and asymptomatic plants throughout the year. It is dispersed primarily by water and transmitted on pruning tools and by propagation from diseased or infested plants. Frequent wet

periods early in the growing season, especially those lasting more than 18 hours, favor epidemics. Symptoms develop most rapidly at temperatures near 30°C. Although infection in most hosts is believed to be localized, internal spread occurs in *P. domestica*, in which the pathogen moves in fruit stalks from twigs to fruit and in petioles from leaf blades to twigs. Fruit tree cultivars vary widely in susceptibility; many are resistant. Strains of *X. arboricola* pv. *pruni* vary in aggressiveness but have shown little evidence of host specificity within *Prunus*.
References: 2896, 2897, 3103, 3104, 3515, 3574, 4295, 4508

Bacterial blight of walnut. This disease (Plate 186G–L), also known as walnut blight or walnut bacteriosis, is caused by *X. arboricola* pv. *juglandis* (syn. *X. campestris* pv. *juglandis*). The causal agent occurs around the world where *Juglans regia* (Persian walnut) grows. Walnut blight is widespread in eastern and southern USA and is economically important in orchards in western USA. *J. regia* is most susceptible; *J. hindsii* (Hinds or northern California walnut), *J. ailanthifolia* (Japanese walnut), and 'Paradox' hybrid walnuts are less so. *J. cinerea* (butternut) and *J. nigra* (black walnut) are resistant.

Symptoms on *J. regia* are described here. The first infections each year are usually on catkins and young leaves. Diseased florets turn black, and bacteria from them contaminate pollen produced on normal parts of catkins. Lesions on leaflets begin as water-soaked spots that enlarge up to 4 mm and at maturity are angular and brown with yellowish green margins. They are usually most numerous near leaflet edges and often cause deformity but seldom cause abscission. Lesions also occur on leaflet veins, petioles, and rachises. Diseased leaves are a source of inoculum for late-season infection of fruits and buds. Flower and vegetative buds and succulent twigs become infected, leading either to tiny lesions or to death of the buds. Fruit infection is the main cause of economic loss. The first symptom on a fruit is usually a black spot at the blossom end, which may enlarge to involve an entire young fruit. Lesions in partly grown fruit are black and depressed. Drops of black liquid containing bacteria may exude where the epidermis breaks. Fruits infected while very young usually drop; those attacked later remain attached, but their nuts frequently fail to fill. The pathogen overwinters on plant surfaces and in and on diseased and healthy buds. The bacteria are dispersed by water and contaminated pollen. The incubation period of walnut blight is 5–34 days (normally 10–15 days), depending on environmental conditions and age of host parts. Disease develops most rapidly at 20–30°C. Epidemics are favored by rainfall and/or fog from the time of bud opening until nuts are about half grown.
References: 299, 2649, 2759, 2896, 3992

A, B. Bacterial spot, caused by *Xanthomonas arboricola* pv. *pruni*, on *Prunus laurocerasus* (cherry-laurel). A, B. Lesions on leaves and a close view of various stages of shot-hole formation (AL, Jul).

C–F. Bacterial spot and twig canker, caused by *X. arboricola* pv. *pruni*, on *Prunus persica* (peach). C. Damage to twigs. Left: lesions induced in spring on young twigs; right: dead buds resulting from previous year's infections on a year-old twig (NC, May). D, E. Foliar lesions (SC, May; NC, Jun). F. Lesions on a ripening fruit (SC, Jun).

G–L. Walnut blight, caused by *X. arboricola* pv. *juglandis*, on *Juglans regia* (Persian walnut). G, H. Lesions on leaflets (CA, May & Apr, respectively). I. Blighted florets on a catkin (CA, Apr). J. A diseased pedicel with shriveling fruit (CA, Apr). K. A young lesion at the blossom end of a fruit (CA, Jul). L. Shriveled fruits (CA, Jul).

Photo credits: A, B—A. K. Hagan; C, E—D. F. Ritchie; D, F—E. I. Zehr; G—S. E. Lindow; H–J—B. L. Teviotdale.

Fire Blight (Plate 187)

Fire blight, the most important bacterial disease of rosaceous plants, is destructive in apple and pear orchards and ornamental plantings in regions where warm, wet weather occurs during spring. The pathogen, *Erwinia amylovora* of the Enterobacteriaceae, is indigenous to North America but has spread to many parts of the world. Its host list includes more than 130 species in nearly 40 genera, all in the Rosaceae. Species of *Cotoneaster*, *Crataegus* (hawthorn), *Cydonia* (quince), *Malus* (apple, crabapple), *Pyracantha* (firethorn), *Pyrus* (pear), *Rubus* (brambles), and *Sorbus* (mountain-ash) are most commonly damaged. Diseases similar to fire blight but caused by the closely related *Erwinia pyrifoliae* and similar bacteria occur in the Far East.

Symptoms and signs. The principal symptoms are blight of blossoms, fruitlets, fruit spurs, and leafy shoots, as well as cankers and dieback. Blossom blight is usually the first new symptom each year. Flowers and flower clusters appear water-soaked and darkened, then quickly droop, shrivel, and turn brown or black. Lesions progress into peduncles and fruit spurs, and thence often into branches. Leaves and green twigs also become infected through wounds and natural openings. The tips of diseased shoots droop and darken to brown or black. If many shoots are blighted simultaneously the plant may appear scorched; thus the name fire blight. The cambium and secondary phloem in diseased shoots and at canker margins become discolored reddish brown.

Cankers form in twigs and branches. During the growing season the margins of active (expanding) cankers may appear slightly raised or blistered. Inactive cankers, their expansion checked by host defenses, become isolated from healthy tissues as periderms (cork layers) develop in the surrounding bark. Bark at the edges of lesions may crack during the dormant season, and where the edges remain inactive, cankers appear sunken during the next year as adjacent healthy tissues grow. Cankers on trunks or root collars result from infection of basal shoots or wounds by bacteria carried in water from diseased parts above. These cankers may girdle and kill entire trees.

During humid weather or after rain, drops of cream- to honey-colored or orange slimy fluid called ooze emerge from lenticels and tiny wounds on recently infected parts. Ooze is a suspension of bacteria in a polysaccharide matrix. Under somewhat drier conditions, strands of bacteria in semisolid matrix may extrude from diseased tissues.

Disease cycle. General aspects of bacterial disease cycles and pathogenicity are summarized on page 368. Fire blight bacteria overwinter in diseased tissue at the margins of cankers where periderm produced by the host during the previous season did not effectively separate healthy from diseased tissue. The bacteria also overwinter in asymptomatic buds and twigs on trees that had blight the preceding season. Bacterial activity resumes during warm weather in spring, and ooze drops form at canker margins and on newly infected tissues. Bacteria in ooze or extruded strands are transmitted by insects, splashing or running water, birds, and humans. Wind-driven rain splash and aerosols also disperse them. The most common vectors are pollinators (bees and flies), but insects in more than 75 genera have been implicated. Bacterial populations build up on plants during warm, wet weather. Infections occur most often in nectaries and on stigmas of flowers, but fresh wounds on any growing plant part are susceptible. *E. amylovora* also enters stomata and hydathodes in leaves and lenticels in succulent twigs. Bacteria in any of these places multiply rapidly, doubling in number in as little as 1 hour under favorable conditions, and move intercellularly in a watery matrix of bacterial polysaccharide and degraded host material. Within several days the bacteria may spread 15–30 cm in shoots. Blight appears 1–3 weeks after the onset of infection, the interval depending on temperature and moisture. Internal spread continues throughout the growing season, sometimes down stems and into rootstocks in highly susceptible apple cultivars.

The principal extracellular polysaccharides of *E. amylovora* are amylovoran and levan. The former compound is essential for pathogenicity. The polysaccharide mixture is hygroscopic; it absorbs water and nutrients from host cells, permitting further bacterial multiplication and polysaccharide production and resulting in pressure that forces host cells apart and ruptures them. This force accounts for most bacterial movement within the plant, exudation of ooze, and extrusion of bacterial strands. Extracellular polysaccharide also plugs xylem of shoots and thus may have a role in inducing the wilt portion of the fire-blight syndrome.

E. amylovora occurs as an epiphyte on the stigmas of symptomless flowers, also on leaves, twigs, and immature fruit. The population builds up during suitable weather (see below), so new infections can occur any time during host growth. Perhaps fire-blight bacteria in dormant, symptomless apple and pear buds and shoots contribute to population buildup in spring, but this has not been established.

E. amylovora is a Gram-negative, rod-shaped bacterium that is motile by peritrichous flagella. Strains of *E. amylovora* vary in virulence, and some strains exhibit specialization on *Rubus*. However, isolates from any species of the subfamily Maloideae (*Cotoneaster*, *Crataegus*, *Malus*, *Pyrus*, among other genera) seem able to cause typical disease in diverse species of that subfamily.

Fire blight epidemics are favored by warm, humid (relative humidity >60%) weather during bloom. The causal bacteria can multiply at temperatures of 15–32°C; the optimum range is 27–29°C. Predictive systems that utilize weather data are used by fruit growers to identify periods of *E. amylovora* population buildup so that antibacterial sprays can be applied before the onset of infection. Outbreaks of shoot blight often occur after storms with heavy rain, hail, and strong winds. Devastating epidemics sometimes follow hail storms because bacteria enter the wounds. Damage by insects such as cicadas that wound stems also promotes infection. Cultural practices that prolong succulent growth—sprinkler irrigation, heavy fertilization, severe pruning—favor severe disease. For a given level of inherent susceptibility, plants growing in well-drained, moderately fertile soil with a balanced supply of nutrients are least damaged.

Plants that resist the fire blight bacterium are known within even the most susceptible genera and species. *Pyrus calleryana* 'Bradford,' for example, is seldom affected and then sustains only minor damage. Crabapples observed to be relatively free from fire blight in addition to other diseases include the cultivars Baskatong, Centurion, David, Dolgo, Henry Kohankie, Liset, Professor Sprenger, Sugartyme, and *Malus sargentii* 'Tina.' Additional crabapples that were blight-resistant when tested by inoculation include *Malus scheideckeri* 'Hilleri' and the cultivars Golden Hornet, Manchurian, Profusion, Rosedale, and Thunderchild. Numerous apple and pear cultivars grown for fruit have at least moderate resistance. Resistant firethorns include *Pyracantha coccinea* 'Sensation,' *P. koidzumii* and its cultivar Santa Cruz Prostrata, and the hybrid cultivars Cadange Saphyr Orange, Cadron Saphyr Rouge, San Jose, and Shawnee. Cotoneasters with resistance include *C. adpressus*, *C. apiculatus*, *C. dammeri* (some cultivars), *C. dielsianus*, *C. foveolatus*, *C. franchetii*, *C. integerrimus*, *C. nitens*, and *C. zabelii*.

Destruction of ornamental plants by fire blight can be delayed or halted by carefully pruning out all blighted twigs or branches, making cuts at least 20 cm below cankers or margins of dieback while plant surfaces are dry, preferably during winter. Tools must be disinfested between cuts if plants are not dormant.

References: 36, 225, 369, 371, 438, 439, 598, 909, 925, 1245, 1944, 2061, 2064–2066, 2574, 2695, 2696, 2849, 2869, 3023, 3485, 3532, 3898, 4135, 4388, 4533, 4534

A. Branch dieback in *Pyrus communis* (common pear), a result of fire-blight cankers (NY, Sep).

B. Shoot blight on *Eriobotrya japonica* (loquat) (FL, Apr).

C. Fire blight on *Pyracantha coccinea* (firethorn). Spurs and spur leaves died when bacteria spread from blighted blossoms (NY, Aug).

D, E. Shoot blight in a crabapple, *Malus baccata* 'Gracilis.' Spread of infection below the dead part is revealed by internal discoloration in longitudinal section (NY, Sep).

F. Blossom blight on *Pyrus communis*. Although not yet symptomatic, the twig bearing the diseased spur is already infected (NY, May).

G. Fire blight cankers on the trunk and limbs of *Sorbus aucuparia* (European mountain-ash). Bark at edges of active cankers appears blistered (arrows) (NY, Sep).

H. Bacterial ooze issuing from lenticels in bark of a diseased limb of 'Ida Red' apple (NY, Jun).

I. A fire-blight canker on a young branch of *Malus pumila* 'Ida Red' (apple). Bacteria entered the branch from the diseased twig at upper left (NY, Aug).

J. Winter appearance of a fire-blight canker on 'Ida Red' apple (NY, Feb).

377

Bleeding Bacterial Cankers (Plate 188)

Deep bark canker of walnut. *Juglans regia* (Persian walnut) in California and Spain is subject to deep bark canker, caused by *Brenneria rubrifaciens* (syn. *Erwinia rubrifaciens*), a member of the Enterobacteriaceae. Elongate lesions form in the bark of trunks and scaffold limbs but are not visible on the surface. Lesions often merge into strips a meter or more long and occasionally involve inner bark, cambium, and superficial sapwood. Usually the innermost (conductive) phloem and the cambium remain alive and free from bacteria except in spots 1–2 mm in diameter where infection has spread inward along medullary rays. Localized killing of the cambium there results in small dark brown pits in the sapwood. Dark reddish brown, bacteria-laden fluid escapes from lesions to the bark surface via longitudinal cracks and makes dark streaks. Callus formation results in slight swelling around the cracks. Bacteria in dried streaks on bark remain viable for up to 4 months.

Infection begins in wounds that penetrate to the inner bark. Common infection sites are injuries caused by the pads of machines that shake the trees to dislodge ripe fruit. The bacteria colonize secondary phloem 1 or more years old, migrating within sieve tubes in both directions from the initial site of infection. Sieve tubes of this age are no longer functional in translocation. Adjacent parenchyma cells are then killed, resulting in narrow dark streaks that coalesce into necrotic strips. Bacteria may migrate as far as 3 m beyond visibly infected areas, which accounts for development of cankers on limbs distant from sites of primary infection. When severe the disease destroys a tree's productivity. Only Persian walnut is known to be affected. The cultivar Hartley is most susceptible; Ashley, Chandler, Eureka, Franquette, and Payne are sometimes attacked. Cultivars Gustine and Howe were found to be susceptible when inoculated.

B. rubrifaciens is a Gram-negative, rod-shaped bacterium, motile by peritrichous flagella. It produces a water-soluble red pigment, rubrifacine, in culture media. This characteristic, among others, distinguishes it from the related *B. nigrifluens*, discussed below, which causes extensive superficial necrosis of Persian walnut bark (shallow bark canker). Both diseases may occur in a given tree, and both pathogens in a single lesion.

B. rubrifaciens is dispersed locally by splashing and windblown water, and between orchards by contaminated harvesting equipment. Transmission of the pathogen in or on contaminated budwood during grafting is suspected. Sap-sucking birds are possible vectors, because cankers develop at their feeding wounds. Most infections occur and cankers elongate during summer. Water stress may be associated with susceptibility. Reduced incidence of the disease was noted in trees that received irrigation sufficient to wet the soil deeply during winter as a supplement to standard summer irrigation.

Shallow bark canker of walnut. This disease, caused by *Brenneria nigrifluens* (syn. *Erwinia nigrifluens*), was first discovered in California and was later found in southern Europe and the Middle East. It is less damaging than deep bark canker, although not less conspicuous. It affects trunks and scaffold branches of mature trees without relationship to known wounds. The external symptoms are discolored spots where dark brown fluid spreads out on the bark in summer. The spots overlie irregularly shaped shallow lesions, most of which are confined to the outer half of the bark. Lesions that form along longitudinal cracks in the bark sometimes extend to the cambium. Shallow lesions typically become isolated from living bark as periderm forms beneath them. These lesions dry out, separate from normal bark in the manner of bark scales, and drop off. Tree death by this disease has not been noted in California but has been reported in Iran. Young trees usually are not affected, although the disease has been noted in walnut nursery stock in Italy. Lesions on young stems cause longitudinal grooves and deformity.

In California shallow bark canker most commonly affects the cultivar Hartley. It has occurred also in the cultivars Eureka, Mammoth, Mayette, Meylan, Myrtleford, and Payne, and the Paradox hybrid rootstock (*J. hindsii* × *J. regia*). The host range of *B. nigrifluens* is evidently not restricted to *Juglans* or to trees; the bacterium has been found in *Helianthus* (sunflower) in Iran and in *Cynara cardunculus* (globe artichoke) in Argentina. California strains of *B. nigrifluens* from walnut caused cankers when inoculated to wounds in bark of *Prunus dulcis* (almond) branches in spring. However, these lesions later dried out and remained inactive.

Other bacterial cankers. Bleeding cankers on many shade, ornamental, and forest trees, apparently caused by bacteria, are common but seldom seem to persist or cause much damage. Liquid escaping from a lesion onto the bark surface may be the only external symptom. The lesions are usually localized in inner bark and often involve the cambium. The liquid often has a sour odor due to organic acids produced by bacteria. These organisms are presumed to cause the lesions, but their pathogenicity has seldom been tested. Species of *Brenneria*, *Erwinia*, and *Pseudomonas* are associated with some of these diseases. Bacterial infections that cause bleeding often follow environmental stress or another disease or injury, and some authors have suggested that the bacteria are opportunists able to cause damage only in stressed trees. Galleries made by bark beetles or ambrosia beetles in previously stressed trees are common sites of infection.

Bleeding of sour liquid from small cankers should not be confused with that from wetwood (Plate 191). Some fungi such as *Cryptosporiopsis* sp. on *Acer rubrum* (red maple; Plate 86) and *Phytophthora* species on various trees (Plates 176–178) also cause bleeding cankers, so this symptom should be interpreted cautiously.

A second type of bleeding lesion, also seemingly associated with previous stress, results in production of white froth with an alcoholic odor at cracks in the bark surface (Plate 191). The microorganisms in these lesions apparently produce gas and alcohol during fermentation of sap. This "alcoholic flux" is reported to be common on *Liquidambar styraciflua* (sweetgum), *Quercus* (oak), and *Ulmus* (elm) in the Midwest and on *Morus* (mulberry) and *Prunus* (almond, apricot, peach) in California.

References: 662, 1113, 1238, 1547, 1859, 2236, 2728, 2737, 2896, 2898, 3375, 3428, 3431, 3432, 3993, 4385, 4386

A–C. Deep bark canker of *Juglans regia* (Persian walnut), caused by *Brenneria rubrifaciens*. Brown streaks on bark originated as bacteria-laden fluid that exuded from cracks overlying elongate cankers in inner bark (CA, Jul).

D, E, J. Cankers apparently caused by unidentified bacteria on *Quercus velutina* (black oak). Bleeding spots (D) were associated with tiny wounds made by insects. Lesions involved inner bark (E) and the cambium and outer sapwood (J) (NY, May).

F, I. A bleeding canker apparently caused by bacteria on *Aesculus hippocastanum* (horse-chestnut). Extensive staining of bark (F) was caused by exudation from a canker about 10 cm long. Water-soaked sapwood beneath the canker, exposed by removing bark, is dead and stained dark brown (NY, Jul).

G, H. Shallow bark canker, caused by *Brenneria nigrifluens*, on *J. regia*. G. Multiple discolored spots where brown liquid seeped out, indicating presence of cankers. H. Internal appearance of bark, exposed by removing the periderm, where the surface was covered with numerous bleeding spots (CA, Nov).

Photo credits: G, H—B. L. Teviotdale

Bacterial Galls of Olive, Oleander, and Ash (Plate 189)

Olive knot, oleander gall, and bacterial knot of ash and other plants in the olive family are caused by strains of the bacterium *Pseudomonas savastanoi* pv. *savastanoi* (syn. *P. syringae* subsp. *savastanoi*). Olive knot and oleander gall occur throughout the Mediterranean region and in California, where they are significant diseases of *Olea europaea* (olive) and *Nerium oleander* (oleander), respectively. The oleander disease also occurs in South Africa. The pathogen is believed to have been carried to southern Africa and North America with host plants from the Mediterranean region. Bacterial knot of *Fraxinus excelsior* (European ash) is widespread in western Europe but unknown in North America. *Jasminum mesnyi* (primrose jasmine), *J. nudiflorum* (winter jasmine), *Ligustrum japonicum* (wax-leaf privet), *Phillyrea* sp., and *Retama sphaerocarpa* (retama) are also affected in Europe. Most known hosts of *P.s.* pv. *savastanoi* in nature are in the Oleaceae. Exceptions include *N. oleander* of the Apocynaceae, *R. sphaerocarpa* of the Fabaceae, *Myrtus communis* (myrtle) of the Myrtaceae, and *Rhamnus alaternus* (Italian buckthorn) of the Rhamnaceae.

P.s. pv. *savastanoi* comprises at least three groups of strains that differ from one another in genomic profiles as indicated by DNA fingerprints, in fatty acid composition, in production of hormones and bacteriocins, and in host specialization. Strains of *P.s.* pv. *savastanoi* from *N. oleander* and *O. europaea* produce indole-3-acetic acid (IAA) and cytokinins that induce the hypertrophy and hyperplasia leading to galls and knots. The majority of strains from olive studied in Italy produce bacteriocins that are thought to prevent some oleander strains from infecting olive. Bacteriocins are antibiotics produced by bacteria, to which only closely related but nonidentical bacteria are sensitive. Olive strains have infected several species of *Olea*, *Chionanthus virginicus* (fringe tree), *Forestiera* species, *Forsythia* species, *Fraxinus* species, *J. mesnyi*, *Osmanthus* species, and uncommonly *N. oleander*. Oleander strains studied in Italy do not produce bacteriocins. These strains have infected *C. virginicus*, *Forestiera acuminata*, *Fraxinus excelsior*, and commonly *Olea europaea*. Strains from *F. excelsior* produce bacteriocins, and most produce auxins at low levels but do not produce cytokinins and do not induce knots on olive. These strains have infected *F. americana* (white ash), *F. angustifolia*, *F. pennsylvanica* (green ash), and uncommonly *O. europaea*. Strains from the three source species noted above were found to multiply more rapidly and to reach higher population levels in their original host species than in other susceptible plants. Strains from *R. sphaerocarpa* are specialized for infection of that plant.

Strains isolated originally from *Ligustrum*, *Jasminum*, and *Phillyrea* also vary in the characteristics mentioned above but have not been studied in sufficient numbers to determine whether or not they are aligned with the above groups. A strain from *L. japonicum* infected two other *Ligustrum* species, *N. oleander*, and *O. europaea*. Based on host specificity, the groups of strains discussed here could be designated pathovars; however, appropriate nomenclature had not been proposed when this was written.

Symptoms and signs. Olive trees sustain dieback of twigs and branches as well as formation of galls or knots. Fruit from diseased trees may have a bitter, salty, sour, or rancid taste. The main effects of the disease on most other host plants are disfigurement and dieback. Cankerous knots develop on *F. excelsior*.

Knots on *Olea* are globose to irregular, roughened, eventually fissured tumors that may enlarge for several years and may attain sizes of several centimeters in the largest dimension. They arise anywhere on twigs and branches at wounds, freezing injuries, bark cracks where adventitious shoots emerge, and especially leaf scars. A beadlike arrangement of knots along a twig (Plate 189E) indicates that infections began at successive leaf scars. Initially light tan, knots become dark gray to nearly black as the surface tissues die. Leaves, peduncles, and fruit are also susceptible, but infection on these is usually inconspicuous.

On *Nerium*, galls and lesions form on stems, flower parts, seed pods, peduncles, and young leaves. Mature galls on stems are 1–2 cm in diameter. Galls initiated on succulent twigs may become cankerous, because lesions form in the knots and stems sometimes split open at lesions. Leaves, flowers, and seed pods develop wartlike galls and become distorted. The seed pods often become twisted and stunted. Diseased parts turn brown within the first season, and severely infected parts die back. Lesions in succulent tissue sometimes spread internally, upward as much as 25 cm and, less often, downward as much as 20 cm, causing streaks of swollen tissue or numerous smooth secondary knots or tubercles. This spread occurs via laticifers (latex-containing ducts). Olive and ash, which lack laticifers, show little evidence of internal spread, although *P.s.* pv. *savastanoi* has been detected as an internal resident of apparently healthy olive branches.

Bacterial knot of ash, which to date occurs only in Europe, differs from the diseases of olive and oleander in that lesions with swollen, roughened bark at their margins form on twigs, branches, and trunks. As lesions enlarge, the cankerous area becomes raised, greatly roughened, and fissured. These distinctive symptoms are attributable to the strain of the bacterium rather than to the host. When strains of *P.s.* pv. *savastanoi* from ash, oleander, and olive were inoculated into ash, the ash strains caused cankerous swellings in the bark, while the strains from olive and oleander caused knots similar to those on their original hosts. These differences were later correlated with greater hormone production by olive and oleander strains than by ash strains.

Disease cycle. The pathogen perennates in knots and cankers and as part of the epiphytic microbiota on hosts. Bacteria from knots and lesions renew the epiphytic population, so wounds are likely to become contaminated and infected. Only fresh wounds are suitable as infection courts. The susceptibility of leaf scars, for example, diminishes rapidly during the first day after leaf fall, and by the ninth day the scars are not susceptible. Plants remain susceptible throughout the year, but in areas with distinct wet and dry seasons nearly all infection occurs during the wet season—approximately October–May in California.

If infection occurs during host growth, symptoms appear within 10–14 days. If infection occurs during winter, galls appear in spring. The first symptom is often a small water-soaked lesion. A cavity 2–5 mm long develops and becomes filled with a mixture of bacteria and cellular debris. Abnormal plant cell enlargement and division begin adjacent to the lesion and spread outward. Gall tissue, initially spongy, becomes woody, with a disorganized arrangement of xylem, phloem, and parenchyma. While galls are alive, bacteria can be found in creases formed by the infolding of proliferated tissues. Bacteria also multiply in tiny irregular cavities much like the original infection site that are located near the surface of a gall or close to apparently healthy tissues. Fissures extend from cavities to the surface of the knot. Bacteria thus reach the surface, where they are available for dispersal by water or contaminated hands or implements. Rain followed by high relative humidity and temperature near 23–24°C favors dispersal and infection. In Italy the olive fly, *Daucus olea*, carries olive knot bacteria, but transmission by insects in California is unreported.

Bacterial knot on specimen plants, if not too severe, can be suppressed by pruning infected parts during dry weather, disinfesting the tool between cuts. Copper fungicides, to which most bacteria are sensitive, can reduce the incidence of new infections.

References: 59, 153, 641, 1630, 1842, 1843, 1913, 1914, 2253, 2748, 2896, 2947, 3012, 3708, 3902–3904, 4381, 4383, 4384

A–D. Bacterial knot of *Nerium oleander* (oleander). A. Distortion and necrosis of flower parts (CA, Oct). B. Galls on seed pods and stems (CA, Oct). C. Cankerous galls on stems. The small galls originated at leaf scars. The elongate gall may have arisen from multiple infections along a wound or as the result of internal spread of bacteria from one point of infection (CA, Apr). D. Young galls on a leaf (CA, Apr).

E–H. Olive knot. E. A series of knots that originated at leaf scars, resulting in dieback of the horizontal twig. F. Old dark, apparently dead knots along a branch about 6 cm in diameter. These knots apparently developed at wounds. G. Continuous knots extending several centimeters along a twig. H. Close view of a knot on a year-old twig. Light color indicates recent development (CA, Jul).

I. Bacterial knot, characterized by cankerous galls, on *Fraxinus excelsior* (European ash) in the Netherlands.

Photo credit: I—J. D. Janse

Crown Gall (Plate 190)

Crown gall is the most famous plant tumor and one of the most studied plant diseases. It occurs around the world, causing economically significant damage to fruit and nut trees and some woody ornamental plants. The disease in most plants is caused by *Agrobacterium tumefaciens* (syn. *Rhizobium radiobacter*), a soil-inhabiting bacterium in the Rhizobiaceae. *A. tumefaciens* has the broadest host range of any bacterial plant pathogen. More than 600 plant species in more than 90 families are susceptible, although relatively few sustain significant damage.

A. tumefaciens is part of a cluster of species in *Agrobacterium* and *Rhizobium* that induce galls, root proliferation, or root nodules. They are aerobic, Gram-negative, rod-shaped bacteria, motile by one–five flagella, and are readily cultivable on various media. Other pathogens in the cluster are *A. rubi*, which causes cane gall on brambles; *A. vitis*, which causes galls and root lesions on grapevines; *A. fici*, which causes root and stem galls on *Ficus* species (fig); and *A. rhizogenes*, which causes hairy-root disease. All of the agrobacteria have been proposed for reclassification in *Rhizobium*, but the change had not been widely adopted when this was written. Hairy root is characterized by dense clusters of small roots up to about 25 cm long protruding from roots, stems, or galls. This disease is occasionally important on *Juglans*, *Malus*, *Pyrus*, *Rosa*, and *Rubus* (walnut, apple, pear, rose, and brambles, respectively). In nature *A. tumefaciens* and other members of its genus exist as mixtures of saprobic and pathogenic strains. The latter contain plasmids that carry tumor- or root-inducing genes (see below). Host ranges of these bacteria are also determined by plasmid genes.

Symptoms and signs. Galls form on roots and stems, often at the root collar (root crown). Young plants with large or numerous galls tend to be stunted and predisposed to drought damage or winter injury. Floral display or fruit production may be suppressed. Damage is greatest when galls encircle the root crown, although few plants are killed by crown gall alone. Economic losses occur in fruit and nut tree orchards and in nurseries that produce pome- and stone-fruit trees (*Malus* and *Prunus*, respectively). Cherry trees tolerate the disease with little growth loss. Mature trees of various species seem able to support numerous large, often grotesque, galls without much debilitation. Galled plants in nurseries must be culled, however, because regulations prohibit their shipment and they are not acceptable to customers. Nursery plants with latent infections often develop galls after outplanting.

Galls caused by *A. tumefaciens* enlarge during host growth and vary in diameter from a few millimeters to 30 cm or more. Their ultimate size depends on strain of the pathogen, plant species and size, and growth rate of the infected plant part. Aerial galls are common on highly susceptible plants such as *Populus* (poplar), *Rosa*, *Salix* (willow), and *Euonymus fortunei* (climbing euonymus). Galls are initially the color of the infected plant part but darken with age. Those on woody plants become hard as disorganized clusters of xylem elements differentiate in them. Gall surfaces are typically rough and become fissured with age. Dead surface tissue, especially of galls in soil, decays and sloughs off.

Plants with severe crown gall are attacked by other pathogens such as *Armillaria* species that enter through decaying galls. This problem was severe at one time in walnut orchards in California, but it lessened when growers began using *Juglans californica* (California black walnut) and *J. hindsii* (Hinds walnut) as rootstocks for the highly susceptible *J. regia* (Persian walnut).

Disease cycle. Crown-gall bacteria infect fresh wounds of any sort. Suitable wounds occur during propagation, planting, cultivation, frost heaving of soil, and feeding by soil insects or nematodes, especially *Meloidogyne* and *Pratylenchus* (root-knot nematodes and lesion nematodes, respectively). Surviving plant cells in a wound are, for a time, attractive and receptive to crown-gall bacteria, which attach to the cell walls. Receptivity lasts for a few days during plant growth but may continue for months while plants are dormant. The bacteria, through a process outlined below, transform plant cells genetically so that they produce abnormally high concentrations of auxins and cytokinins (plant hormones), which induce gall formation, and opines, which nourish the bacteria. Opines are α-N-substituted amino acid derivatives not found in normal plants. Gall growth proceeds autonomously after plant cells are transformed. Plant sensitivity to cytokinins is a major factor controlling tumorigenesis.

The process that results in genetic transformation by crown-gall bacteria begins when agrobacteria in a wound attach to plant cells and sense sugars and phenolic compounds leaking from them. Some of these chemicals, notably acetosyringone, act as signals to the DNA of the tumor-inducing (Ti) plasmid to turn on genes (*vir* genes) that control transformation. The protein products of these genes accomplish the removal and transport from the plasmid of a single-stranded piece of DNA (T-DNA) bearing genes that encode the capability to produce auxin, cytokinins, and opines, as well as genes whose functions have not been clearly defined. The T-DNA is transported through the bacterial and plant cell walls and membranes and into the plant cell nucleus, where it is integrated into a chromosome. The process of T-DNA excision from a plasmid, cell-to-cell transport, and integration into a plant chromosome is equivalent to the process by which conjugating bacteria exchange genetic information. Once in place the T-DNA replicates during mitosis and is expressed along with the plant genes. Transformed cells multiply uncontrolled, and they export opines. Agrobacteria that possess Ti or Ri (root-inducing) plasmids can metabolize opines as sources of carbon and nitrogen. Root proliferation in hairy-root disease is induced by plasmid-encoded genes in a manner similar to gall induction in crown gall.

Tumors become visible 2–4 weeks after inoculation if plants are growing, but the onset of gall formation may be delayed for months in dormant plants. Crown-gall bacteria multiply and migrate intercellularly in young galls, creating secondary growth centers that contribute to the irregular shape and rough, fissured surfaces of old galls. *A. tumefaciens* can be isolated from young galls easily but often cannot be isolated from old galls. Secondary galls may encircle stems at the root crown and sometimes form on rose or bramble canes as the result of bacterial movement upward within the plant from points of initial infection. Such galls are arrayed in a row or appear as a linear roughened swelling. Tumor development is most favored by temperatures near 22°C. Temperatures above 30°C prevent transformation of normal cells to tumor cells but do not prevent gall growth after transformation.

As the outer parts of galls weather or decay, crown-gall bacteria are returned to the soil, completing the disease cycle. Soils where diseased plants grew may support high populations of agrobacteria on root fragments left after harvest. The bacteria are dispersed in soil or irrigation water, on horticultural implements, and on or within plants. Populations of *A. tumefaciens* in soil diminish markedly during the first several months after removal of plants but then persist at low levels for extended periods. Long-term survival depends on association with roots. *A. tumefaciens* colonizes the root surfaces of many plants irrespective of their susceptibility to infection and gall formation.

References: 73, 306, 452, 453, 743, 948, 1078, 1241, 2706–2708, 2763, 2825, 2910, 3013, 3121, 3124, 3208, 3233, 3294, 3363, 3423, 3424, 3429, 3494, 3818, 4172, 4489, 4521, 4529

Crown gall, caused by *Agrobacterium tumefaciens*.

A. Old galls thought to be caused by *A. tumefaciens* on *Populus nigra* 'Italica' (Lombardy poplar (WA, Jun).

B, C. Large old galls on the trunk and at the root crown of a mature *Juglans cinerea* (butternut) tree (NY, Jul).

D. Streaks of crown-gall tissue on a stem of *Rosa* sp. The streaks could have arisen either from multiple infections along a scraped area or from internal spread of bacteria (NY, Sep).

E, F. Crown gall on 2-year-old *Prunus avium* (mazzard cherry) seedlings. Numerous galls on lateral roots may only retard growth, while one gall encircling the root collar may be lethal (NY, Jul).

G. Crown gall on *Euonymus fortunei* (climbing euonymus). Galls formed during the current season are light brown and somewhat spongy; those a year old or older are dark brown and hard (NY, Oct).

H. Crown gall on *Wisteria* sp. (CA, Sep).

I, J. Successively closer cutaway views of a stem of *Ligustrum vulgare* (common privet) with crown gall developing at the site of a scrape wound. Gall tissue differs from normal bark and wood in texture and anatomy (NY, Sep).

383

Wetwood, Slime Flux, and Alcoholic Flux (Plate 191)

Wetwood. Wetwood is a water-soaked condition of wood that occurs in tree trunks, roots, and limbs. The affected wood is dead, usually anaerobic or nearly so, and usually discolored and malodorous. It supports various anaerobic and facultatively anaerobic bacteria. The discolorations range from pinkish or pale yellow or yellowish green to olive-green, brown, red-brown, or black, often arranged as streaks or columns. In some species wetwood may lack distinctive color but can be recognized by its sour odor. Compared with normal sapwood, wetwood has elevated contents of mineral and organic solutes, depleted storage carbohydrates, usually has elevated pH, and often contains gas under pressure. Zones of wetwood may occur in heartwood, sapwood, or both; are usually irregular in shape; and tend to be associated with wounds. Wetwood develops first in the lower central parts of tree trunks and major roots, and it may extend far up trunks and into limbs. Wetwood in landscape trees is usually unimportant except for the disfiguring appearance of vertical streaks where liquid seeps out of cracks or old wounds and runs down the bark.

Wetwood is universal in mature trees of some *Populus* species (cottonwood and poplar) and *Ulmus* (elm), and its incidence is high in *Abies* (fir), *Acer* (maple), *Morus* (mulberry), *Platanus* (planetree, sycamore), *Quercus* (oak), *Salix* (willow), and *Tsuga* (hemlock). It affects many other species occasionally to commonly. No controls for wetwood are known.

The presence of wetwood devalues logs and sawn lumber. Lumber with wetwood requires more time and energy than normal wood for drying, because wetwood usually contains more water and is less permeable than sapwood to gas and water, and because wetwood fluid has reduced vapor pressure due to its solute content. Fir, hemlock, and oak boards with wetwood tend to check during kiln drying. Poplar wetwood is inferior to normal wood in crushing strength and toughness. Wetwood bacteria secrete pectolytic and cellulolytic enzymes that degrade intercellular substances and primary walls of wood cells, weakening the wood. On the other hand, the anaerobic or hypoxic nature of most wetwood and the toxicity of some of the bacterial metabolites in it prevent or retard decay of wetwood by fungi in living trees.

Wetwood forms when solutes, especially cations, accumulate in a localized zone, creating a gradient of osmotic potential that causes water to move into the zone from adjacent sapwood and/or from decaying wood. Accordingly, a relatively dry transition zone of sapwood occurs adjacent to wetwood. The water content of wetwood in angiosperms is greater (up to 80% greater in *Populus* and *Ulmus*) than that of normal sapwood, which in turn is wetter than heartwood. In *Abies*, however, sapwood and wetwood sometimes have similar water contents.

Cation concentrations in wetwood rise to several times the concentrations found in normal sapwood, concurrently reducing the osmotic potential (e.g., -1.47 MPa in *Ulmus* wetwood versus -0.05 MPa in sapwood). These ions include Ca^{++}, Mg^{++}, Na^+, and, most notably, K^+. Organic acids and alcohols also accumulate in wetwood. The typical pH of wetwood varies with tree species but is often 1–2 units higher (i.e., more basic) than that of sapwood.

Gas pressure and composition in wetwood have been studied in *Populus* and *Ulmus*. Pressure during summer is often in the range 0.025–0.125 MPa above atmospheric pressure (0.1 MPa). Values up to 0.4 MPa have been recorded. Gas composition has been in the following ranges: CH_4, 46–61%; CO_2, 7–16%; H, 1.0–1.5%; N, 23–34%; O_2, 0–7% (often none). The gas in normal sapwood of *U. americana*, in contrast, consists of about 19% O_2, 75% N, and up to 5% CO_2, with no methane or hydrogen. The malodorous compounds in wetwood are products of fermentative bacterial metabolism.

Bacterial populations in angiosperm wetwood typically number 10^6–10^9 cells per milliliter of wetwood fluid. These bacteria are diverse in form (rods, cocci, spirals, and spore formers), Gram reaction, oxygen tolerance or requirement, gas production, and other metabolic characteristics. Some are strictly anaerobic. Genera represented include, among others, *Bacillus, Bacteroides, Citrobacter, Clostridium, Corynebacterium, Edwardsiella, Enterobacter, Erwinia, Klebsiella, Lactobacillus, Methanobrevibacter, Pantoea, Pseudomonas,* and *Staphylococcus.* Species in several of these genera have been detected in diverse trees. Some wetwood bacteria formerly classified in *Corynebacterium* or *Erwinia* would now be assigned to other genera. No information is available about possible host specialization.

Bacteria associated with wetwood enter xylem through wounds and lesions, especially in roots. Scanning electron microscopy and bacterial isolations from angiosperms have indicated that bacterial populations increase in sapwood vessels as wetwood forms, typically in wood several years old. Wetwood bacteria can be detected in normal sapwood adjacent to wetwood but not usually in young sapwood. Whether the high cation concentrations in wetwood are due to accumulation by bacteria, bacterial interaction with living sapwood parenchyma cells, or host-programmed ion accumulation in senescing sapwood is unclear. Wetwood sometimes contains few bacteria. Wet zones in *Pinus* (pine) wood may be nearly devoid of organisms. Wetwood in *Abies* has been suggested to have a nonmicrobial cause, with bacterial colonization a secondary process. Research to test this suggestion in other genera would be useful.

Few wetwood-associated bacteria have been tested for ability to induce the condition, and most results of those tests have been negative or ambiguous. *Enterobacter nimipressuralis* apparently caused brown streaks when inoculated into sapwood of elm saplings. A coryneform bacterium isolated from *Populus nigra* 'Italica' (Lombardy poplar) apparently induced wetwood in the host of origin. Phytotoxicity of elm wetwood fluid was indicated in bioassays, so it seems likely that bacterial metabolites can contribute to the spread of wetwood within trees.

Dieback and slime flux related to wetwood. Wetwood fluid under pressure sometimes spreads into outer sapwood, where in *Ulmus* and *Populus* it may cause gray-brown streaks and bands extending into branches and twigs. These streaks in *Ulmus*, which resemble symptoms of Dutch elm disease, are sometimes associated with leaf scorch or wilt, yellowing, defoliation, and dieback. Wetwood-associated dieback of *Ulmus* is common in the Great Plains of North America and in Asia. Wetwood is also associated with dieback and premature death of *Populus* species, and in Europe with decline in *Abies alba* (silver fir). In addition, wetwood may predispose tree trunks to frost cracks.

Wetwood fluid may ooze out of wounds or cracks, commonly at crotches. This fluid darkens on exposure to air and supports growth of many kinds of bacteria, yeasts, and filamentous fungi that give the plant surface a slimy texture and often a fetid color and odor. This mixture, *wetwood slime*, flows down the bark and dries as a light grayish crust. Chronic slime formation is called *slime flux*.

"Alcoholic" or white flux. This condition, also called frothy flux, is not related to wetwood. It occurs where microorganisms ferment sap in cracks and other wounds in the bark and cambial region. Alcoholic flux is acidic and nearly colorless, sometimes appears as a white froth (Plate 191F), often emits a fermentative odor, and persists only a short time in summer. The associated microorganisms apparently produce gas and ethanol. Alcoholic flux is reportedly common on stressed trees, notably on *Liquidambar styraciflua* (sweetgum), *Quercus* (oak), and *Ulmus* (elm) in the Midwest; *Salix* (willow) in the Southwest; and *Morus* (mulberry) and *Prunus* (almond, apricot, peach) in California. It has also been seen on *Albizia julibrissin* (mimosa or silktree) affected by Fusarium wilt.

References: 509, 660, 662, 768, 820, 1533, 2006, 2203, 2236, 2764–2766, 2792, 2898, 3376, 3383, 3447, 3448, 3518, 3536, 4208, 4216, 4217, 4439, 4440, 4462, 4463, 4509

A, B. Slime flux on *Ulmus pumila* (Siberian elm). Cambial dieback at the sides of the pruning wound (B) may have been caused by toxic chemicals in the wetwood slime (IL, Jun).

C, D. Slime flux issuing from a knot hole in *Casuarina equisetifolia* (horsetail casuarina) and a frost crack in *Acer saccharum* (sugar maple), respectively (FL, Mar; NY, May).

E. Wetwood in *U. americana* (American elm). Spokelike marks show the pattern of holes bored for chemotherapeutic injections to control Dutch elm disease 2–3 years before the tree was cut. Where the drill bit entered wetwood, the condition began to spread into young sapwood (NY, Aug).

F. Cross section of a 7-year-old trunk of *Populus deltoides* (eastern cottonwood). A core of wetwood occupies portions of five annual layers of wood (MS, Apr).

G. Alcoholic flux at small cracks in bark of *Albizia julibrissin* (mimosa) (NY, Jun).

Bacterial Scorch Diseases (Plates 192, 193)

Overview

Bacterial scorch refers to browning of leaf tips and margins caused by xylem-inhabiting fastidious bacteria that infect plants systemically and interfere with water supply to leaves. These bacteria are found only within plants or insect vectors and can be cultivated only on special media. Fastidious bacteria have been recognized as plant pathogens since the early 1970s. All strains known to cause bacterial scorch of woody plants in the Americas represent one species, *Xylella fastidiosa* (related to *Xanthomonas* in the Proteobacteria). The genus *Xylella* has no other named members. These bacteria are Gram-negative rods and are transmitted by xylem-feeding homopteran insects, mainly species of the Cicadellidae, subfamily Cidadellinae (sharpshooter leafhoppers), and Cercopidae (spittlebugs). Diseases caused by *X. fastidiosa* in grapevines and fruit trees have been given crop-specific names—almond leaf scorch, citrus variegated chlorosis, coffee leaf scorch, Pierce's disease of grapevine, phony disease of peach, plum leaf scald. The generic name bacterial scorch is used for similar diseases of other trees and shrubs. These diseases are known mainly in North and South America in tropical to cool-temperate climates. They occur across the southern half of the USA and northward to northern California and southern Ontario. This distribution probably reflects vector ranges and low tolerance of the pathogen for winter cold. Strains of *X. fastidiosa* occur also in southern Europe, India, and Taiwan.

X. fastidiosa causes leaf scorch in plants belonging to at least the following genera in North America: *Acer* (maple), *Aesculus* (buckeye), *Ampelopsis* (peppervine), *Callicarpa* (beautyberry), *Carya* (pecan), *Celtis* (hackberry), *Cornus* (dogwood), *Liquidambar* (sweetgum), *Morus* (mulberry), *Nerium* (oleander), *Parthenocissus* (Virginia creeper), *Platanus* (sycamore), *Prunus* (almond, plum), *Quercus* (oak), *Sambucus* (elderberry), *Ulmus* (elm), and *Vitis* (grapevine). *Coffea arabica* (coffee) and *Pyrus pyrifolia* (Asian pear) are affected by scorch in South America and Taiwan, respectively. *X. fastidiosa* causes variegated chlorosis of *Citrus* in South America and is associated with citrus blight in Florida and several other regions. It is not the primary cause of the latter disease, however. *X. fastidiosa* also has more than 100 plant hosts, both woody and nonwoody, in at least 30 families, in which it does not induce noticeable symptoms. These plants may serve as reservoirs of the pathogen, from which vectors carry it to other tolerant plants and to scorch-susceptible plants.

Considerable genetic and correlated behavioral variation occurs in *X. fastidiosa*. Strains vary in host range, pathogenicity, antigenic characteristics, antibiotic resistance, and degree of fastidiousness. Several groups of strains have been distinguished based on DNA sequence differences, protein profiles, and host preferences: an almond-grapevine-maple group, an almond-grapevine-elm-peach-plum-sycamore-group, an elderberry group, a mulberry group, an oak group, an oleander group, and a citrus-coffee group, the last in South America. Three *X. fastidiosa* subspecies and several pathovars had been proposed informally by mid-2004, but their names were not in general use. Genomic analysis of a strain that causes citrus variegated chlorosis revealed genes that likely encode toxins, antibiotics, iron-binding molecules, thiol groups that are involved in bacterial aggregation, enzymes that degrade plant cell walls, proteins that interact with host molecules, and an extracellular polysaccharide called fastidian gum that may be involved in bacterial coating of vessel walls and plugging of vessels. All of these factors could contribute to the pathogenicity and virulence of *X. fastidiosa*.

Symptoms and signs. Infection in woody plants is typically perennial. Severely affected plants undergo disruption of water transport and display various combinations of related symptoms: marginal reddening or yellowing followed by browning of leaves, decreased fruit production, decline in vigor, delayed budbreak, stunting, dieback, and sometimes death. Leaves begin development more or less normally each year, and foliar symptoms develop in late spring to summer, most often in mid to late summer. Leaf scorch increases during hot, dry weather. It first appears on one branch or on branches in one part of the crown. Additional branches may show symptoms in subsequent years. Scorch often develops first on the oldest (basal) leaves on a shoot and progresses toward the tip, where some leaves may remain green. Xylem vessels in symptomatic parts of diseased plants are lined or plugged with tyloses and with gummy material containing bacteria. These obstructions impede water flow and thus cause foliar symptoms.

The bacteria in xylem can be visualized in detail by electron microscopy, and they can be extracted from xylem sap. They occur in leaves, stems, and, in many hosts, in roots, mostly within xylem vessels but also sometimes in intercellular spaces around vessels. They measure $1–4 \times 0.25–0.4\,\mu m$ (average near $2.0 \times 0.3\,\mu m$), and their walls are rippled in outline. They can be detected and identified in plants and insects by amplification of diagnostic DNA sequences and by immunological tests.

Disease cycle. Infection by *X. fastidiosa* requires placement of bacterial cells within xylem, either by vectors or by grafts. The bacteria multiply in xylem sap and are carried with it both upward and downward, eventually occurring throughout a plant. The interval between infection and appearance of foliar symptoms varies from several weeks to a year according to bacterial strain, plant species, plant size, and time of year. The pathogen perennates in roots and/or stems, moves into the current year's xylem annually, and then is carried into leaves. It is presumed to secrete enzymes that degrade pit membranes, thus permitting it to pass from vessel to vessel. Details of internal spread downward have not been reported, but the bacteria are presumed to move with receding menisci when water columns break (cavitation) as a result of stem drying or wounds.

X. fastidiosa has many vectors; the most important ones differ from one region to another. The best known and most studied vectors in North America are leafhoppers in the genera *Carneocephala*, *Draeculacephala*, *Graphocephala*, *Homalodisca*, and *Oncometopia*, and the spittlebug *Philaenus spumarius*. Cells of the pathogen are generally found adhering to the cuticular lining of the foregut of an infectious vector. Adults that have acquired *X. fastidiosa* can transmit it for an indefinite time. Nymphs can acquire and transmit the bacteria but lose the latter ability when they molt. This information is evidence that *X. fastidiosa* does not circulate in vectors' bodies. Whether it multiplies within vectors is unknown.

References for overview and Pierce's disease of grapevine: 33, 329, 716, 717, 1122, 1182, 1316, 1568, 1603, 1660, 1661, 1748–1750, 1859, 2287, 2310, 2311, 2474, 2583, 2691, 3174–3176, 3429, 3430, 3639, 3632, 4287, 4398

Representative Diseases

Pierce's disease of grapevine. Pierce's disease is one of the most important diseases of *Vitis* (grapevine) in North America and is the most studied of the bacterial scorch diseases. It is believed to have been endemic in southeastern USA since precolonial times, and it has been known in California since the late 1800s. The symptoms are as described for bacterial scorch diseases in general. In addition, leaves often turn distinctly red or yellow before or in association

---

A–C. Leaf scorch of *Aesculus californica* (California buckeye). A. Scorched, partially defoliated *Aesculus* among normal *Quercus* (oak) in summer. Trees with bacterial leaf scorch cannot be distinguished from those with scorch caused by water shortage. B, C. Leaves with scorch caused by *Xylella fastidiosa*. Brown areas began as reddish discoloration (CA, Jul).

D–F. Bacterial scorch of *Prunus dulcis* (almond). D, E. Symptoms in the cultivar Jordanolo. Foliage of a tree with advanced disease is yellow or scorched, and dieback has begun. A tree in an early stage of disease (E) has yellow foliage mainly on one major branch. F. Close view of affected leaves on cultivar Long IXL. Brown areas have a diffuse yellow border (CA, Jul).

G. Pierce's disease of *Vitis vinifera* (grapevine). Leaves on a diseased shoot vary from nearly normal to severely scorched. Brown areas often have a diffuse yellow border (CA, Jul).

H–J. Bacterial scorch of *Nerium oleander* (oleander). H. Scorched leaves and dieback on a declining plant. I, J. Foliar symptoms on other affected plants (CA, Aug–Oct).

Photo credits: H–J—M. Blua

with marginal scorch. Lower leaves on a shoot may have chlorotic mottling, and part or all of the vine may wilt and the leaves wither.

In southeastern USA the majority of grape cultivars within the range of Pierce's disease are derived from *V. rotundifolia* (muscadine grape), a species that is relatively tolerant of infection. Even so, the disease is important there. Strains of *X. fastidiosa* that infect grapevine have many additional hosts that collectively comprise most of the known host range of the bacterium. Grapevine strains also are associated with most of the known vector genera in North America.

Almond leaf scorch. Almond leaf scorch, which first appeared in California in the 1970s, is caused by strains of *X. fastidiosa* that also cause Pierce's disease of grapevine. The two diseases have common vectors. Marginal yellowing and necrosis of leaves appear annually after mid-June. Dead tissue is tan, often with zonate markings. A yellow band often separates green from brown areas, and tissue along the secondary veins may be yellowish. Symptoms appear initially on single branches, spreading and intensifying for 3–8 years until all are affected and dieback begins. Scorched leaves remain on a tree until the normal time of defoliation in autumn. *X. fastidiosa* has been found in almond leaves and stems but not in roots. Scorch caused by buildup of salt in irrigated soil can mimic the symptoms of bacterial leaf scorch, but all leaves on a salt-injured shoot usually are scorched whereas some leaves on a shoot with bacterial leaf scorch usually remain green.

Bacterial scorch of shade and landscape trees and shrubs. These diseases are noticed primarily in eastern USA. The pathogen strains involved belong to at least three subgroups of *X. fastidiosa* that differ in host preferences, but specific information about relationships of shade tree strains to each other and to grapevine and fruit tree strains is scant.

Bacterial scorch of elm. The disease was first described in Washington, DC, and later was detected from Louisiana to the Niagara Peninsula in Ontario. It affects *U. americana* (American elm) primarily, but also occurs in *U. glabra* (Scotch or wych elm), *U. ×hollandica* (Dutch elm), and *U. pumila* (Siberian elm). Affected leaves have marginal zones of discolored tissue that first fade and dry, then turn brown. A yellow band usually separates brown from green tissue. Scorched leaves often curl up (sometimes down) at the edges and are likely to drop prematurely. Stem water potentials (measured as the pressure needed to equal the tension in water columns of the sap stream) are abnormally negative throughout the growing season, an indication that leaves are not receiving water as rapidly as it is lost through transpiration. Starch storage in stem wood is suppressed. Symptom severity in a given tree fluctuates from year to year. *X. fastidiosa* has been extracted from elm twigs, branches, and trunks. Slow increase in disease incidence (10% in 6 years) was noted in an elm population in Washington. *X. fastidiosa* was detected in two species of *Graphocephala* (sharpshooter leafhoppers) associated with elms there, an indication that these insects may be vectors. In pathogenicity tests an *X. fastidiosa* strain from *U. americana* was able to cause scorch in that species but not in *Platanus occidentalis* (American sycamore), and the reverse was true for a strain from *P. occidentalis*. Bacterial scorch predisposes elms to attack by elm bark beetles, which carry the Dutch elm disease fungus (Plate 119). Dutch elm disease in an elm population observed for several years in the District of Columbia was about 12 times as common in scorch-affected elms as in others, with more than 40% of all cases of Dutch elm disease occurring in trees already affected by bacterial scorch.

Bacterial scorch of oak. This malady occurs from southern New York to Missouri and southward, with up to 50% incidence in some landscape plantings. It affects species in both the red oak and white oak groups, including *Quercus alba* (white oak), *Q. coccinea* (scarlet oak), *Q. falcata* (southern red oak), *Q. imbricaria* (shingle oak), *Q. incana* (bluejack oak), *Q. laevis* (turkey oak), *Q. laurifolia* (laurel oak), *Q. macrocarpa* (bur oak), *Q. nigra* (water oak), *Q. palustris* (pin oak), *Q. rubra* (northern red oak), and *Q. virginiana* (live oak). Chlorosis and then faded green color develop at leaf tips and margins, and the discolored area then dries, fades to drab green, and turns brown. A yellow or reddish brown band may separate scorched from green tissue. Symptoms appear first in the outer and usually upper part of a crown and often involve all leaves on a shoot. Leaves remain attached until autumn. Retarded growth and dieback are often associated with scorch. The foliar symptoms are similar to those of oak wilt (Plate 118), but scorch can be distinguished on the basis of symptoms that develop over several years, retention of symptomatic leaves until autumn, and absence of dark streaks in sapwood.

Bacterial scorch of sycamore. *Platanus occidentalis* (American sycamore) is affected in the vicinity of Washington, DC, west to Kentucky and south to Florida and Texas. Dieback is a common associated symptom. Leaf margins and interveinal areas dry out and turn drab olive-green, then tan, usually with a brown band between tan and green tissue. Severely scorched leaves curl upward from the edges but remain on the twigs until autumn. Diseased trees produce leaves later than normal in spring, and the leaves are sometimes smaller than normal. Leaves near the tips of affected twigs sometimes remain green. *P. ×acerifolia* (London planetree) is also affected in some areas. Declining sycamores in the South are often colonized by opportunistic pathogens such as *Botryosphaeria rhodina* (Plate 63). When inoculated, *P. orientalis* (Oriental planetree) was also found susceptible to scorch.

Oleander leaf scorch. Bacterial scorch appeared on *Nerium oleander* (oleander) in California in the 1990s and was soon detected in Florida and Texas as well. Symptoms include chlorotic spots, marginal scorch, tip burn, decline, and death. Many cultivars are affected. Oleander strains of *X. fastidiosa* are vectored by two species of *Homalodisca* (glassy-winged sharpshooters) and at least one of *Graphocephala* and differ in host preferences from strains that infect grapevine.

Bacterial scorch of California buckeye. Leaf scorch of *Aesculus californica* occurs in central and northern California in both wild and planted trees. Symptoms appear soon after leaves unfold. Leaves on affected branches or entire trees are somewhat dwarfed and may appear crinkled. Tips and margins of leaflets develop purplish pigment that progresses through shades of reddish brown to brown scorch. Symptoms recur yearly, but trees remain alive. This disease was long overlooked because the symptoms caused by *X. fastidiosa* often blend with those caused by injuries (e.g., foliar injury by leafhoppers) and resemble the normal response of *A. californica* to hot, dry summer weather. These trees naturally drop their leaves and enter dormancy before autumn.

Other woody plants affected. Bacterial scorch of *Morus rubra* (red mulberry) occurs from near New York City to Louisiana and also in Missouri and Nebraska. It affects trees of all ages from seedlings to mature specimens. The disease has also been found in *Acer rubrum* (red maple) and *Cornus florida* in the East; in *A. saccharum* (sugar maple), *Celtis occidentalis* (hackberry), and *Liquidambar styraciflua* (sweetgum) in Kentucky; and in *Ampelopsis arborea* (peppervine), *Callicarpa americana* (beautyberry), *Parthenocissus quinquefolia* (Virginia creeper), and *Sambucus canadensis* (American elder) in Florida.

References: 128, 206, 423, 716, 816, 905, 1317, 1534, 1535, 1575, 1749, 1750, 2133, 2134, 2556, 2557, 2671, 3172, 3178, 3179, 3391, 3430, 3576–3580, 3583, 4304, 4305, 4347, 4348

A–C. Bacterial scorch of *Quercus rubra* (red oak). A. Scorched leaves throughout the canopy of a diseased tree remain attached to twigs. B, C. Leaf tips and margins dry and fade to drab light green, then turn brown (DC, Aug).

D–F. Bacterial scorch of *Platanus occidentalis* (American sycamore). D. Leaf scorch and dieback. E, F. Diseased leaves fade to light green and then tan at the edges and between veins, often with a narrow brown band between tan and green tissue. Scorched areas have zonate markings, indicating intermittent enlargement (DC, Aug).

G–I. Bacterial scorch of *Ulmus americana* (American elm). G. Dieback in a tree affected for several years. H, I. Tips and margins of leaves dry, fade to drab light green, then turn brown, often with darkest brown near the inner edge of the scorched area, and have a bright yellow band between brown and green tissue (DC, Aug).

389

Diseases Caused by Mollicutes (Plates 194–200)

Overview

Plate 194 leads a series about diseases caused by insect-vectored fastidious bacteria of the class Mollicutes. Plant-pathogenic mollicutes, unknown before 1967, infect phloem systemically and cause chronic or lethal diseases. They reproduce in their vectors and in phloem sieve tubes of angiosperms and monocots, in which they are translocated systemically with photosynthetic products. Reports of mollicutes in gymnosperms require corroboration. Mollicute infections of woody plants are usually perennial and incurable. Mild infections either induce no visible symptoms or are masked by other disorders.

Two kinds of mollicutes that infect plants are known: those for which the genus name 'Candidatus Phytoplasma' has been proposed, and those in the genus *Spiroplasma* (Entomoplasmatales, Spiroplasmataceae). *Candidatus* taxa are distinguished on the basis of molecular and biological properties but are as yet unculturable and therefore cannot yet be described and named in the conventional manner. Members of both groups are single celled and reproduce by fission. Their cells are delimited by flexible membranes and lack walls. Details of their size, form, and structure have been revealed by electron microscopy. All phytoplasmas and those spiroplasmas that infect plants (three species known) are intracellular parasites. Their vectors are homopteran and a few hemipteran insects that feed by sucking phloem sap. Phytoplasmas are pleomorphic and nonmotile. A mature phytoplasma cell consists of a central body with irregular to tubular appendages. Thin sections through such a cell appear as circular to elongate profiles of various sizes (Fig. 194). Spiroplasmas have helical, motile cells and can be grown in pure culture. Their helical form and motility are due to a contractile membrane protein called spiralin.

Phytoplasmas are known from DNA sequence data to be a unique group related to Gram-positive bacteria, and to possess the smallest genomes known. They are classified on the basis of similarities and differences in base sequences of the 16S ribosomal RNA genes and supporting data from other conserved genes. On this basis phytoplasmas comprise eight main lineages with nearly 30 groups (16S rRNA groups) and more than 50 subgroups. Each group and subgroup is known by the name of a disease caused by one of its members. The main groups are thought to be species, or in some cases clusters of closely related species. Several groups or subgroups have been given *Candidatus* names, for example, '*Ca*. Phytoplasma fraxini' for strains associated with ash yellows and lilac witches'-broom, indicating that this group is considered to be a discrete species. Phytoplasma groups associated with some diseases of woody plants are listed below. The list reveals that some diseases can be induced by more than one kind of phytoplasma. Phytoplasmas associated with some diseases have not been classified.

Some phytoplasma groups and diseases caused by their members (asterisk indicates the disease occurs in North America)

Apple proliferation group: apple (*Malus*) proliferation, caused by '*Candidatus* Phytoplasma mali'; buckthorn (*Rhamnus cathartica*) witches'-broom, caused by '*Ca*. Phytoplasma rhamni'; European stone-fruit yellows, caused by '*Ca*. Phytoplasma prunorum'; oak (*Quercus*) decline; *pear (*Pyrus*) decline and *Prunus* yellow leafroll caused by '*Ca*. Phytoplasma pyri'; witches'-broom of *Spartium junceum* (Spanish broom), caused by '*Ca*. Phytoplasma spartii'

Ash yellows group, '*Ca*. Phytoplasma fraxini': *ash (*Fraxinus*) yellows, *lilac (*Syringa*) witches'-broom

Aster yellows group: *blueberry (*Vaccinium*) stunt, *grapevine (*Vitis*) yellows, *gray dogwood (*Cornus racemosa*) stunt, *Hydrangea* phyllody, *Hydrangea* virescence; Japanese *Hydrangea* phyllody, caused by '*Ca*. Phytoplasma japonicum'; mulberry (*Morus*) dwarf, papaya (*Carica papaya*) decline, sandal (*Santalum*) spike; witches'-brooms of *Paulownia*, *Populus* (poplar), *Rhus* (sumac), *Salix* (willow)

Clover proliferation group: *elm (*Ulmus*) yellows, *witches'-broom of *Potentilla fruticosa* (shrubby cinquefoil)

Elm yellows group: alder (*Alnus*) yellows; *elm yellows, caused by '*Ca*. Phytoplasma ulmi'; *Eucalyptus* little-leaf; grapevine yellows; *Rubus* stunt; *Hovenia* witches'-broom; *Ziziphus* (jujube) witches'-broom, caused by '*Ca*. Phytoplasma ziziphi'

Hibiscus witches'-broom group, '*Ca*. Phytoplasma brasiliense': witches'-broom of *Hibiscus*

Lime witches'-broom group, '*Ca*. Phytoplasma aurantifolia': witches'-broom of *Citrus* species

Mexican periwinkle virescence group: *Virescence of *Catharanthus roseus*

Palm lethal yellowing group: *lethal yellowing and *lethal decline of coconut (*Cocos*) and other palms; witches'-broom of *Castanea crenata* (Japanese chestnut), caused by '*Ca*. Phytoplasma castaneae'

Pigeon-pea witches'-broom group: almond (*Prunus dulcis*) witches'-broom and decline, caused by '*Ca*. Phytoplasma phoenicium'; *Gliricidia sepium* little-leaf

Prunus X-disease group: *Prunus* X-disease, blueberry (*Vaccinium*) witches'-broom, *bunch diseases of pecan (*Carya illinoinensis*) and walnut (*Juglans*), *grapevine yellows, *hazel (*Corylus*) stunt, *pin cherry (*P. pensylvanica*) witches'-broom, *poinsettia (*Euphorbia pulcherrima*) free branching, *Rubus* stunt, *Spiraea* stunt

Phytoplasmas vary in degree of host specialization, which is in turn related to feeding preferences of vectors. Strains in the ash yellows group, for example, have been found only in *Fraxinus* and *Syringa*, while strains of the aster yellows group comprise many subgroups that infect a broad array of plants. An individual plant may become infected by two or more kinds of phytoplasmas, but one organism always seems to be more readily detectable than the other(s). Phytoplasma strains that cause a given disease vary in aggressiveness, which is the relative ability to cause damage. Such variation has been detected, for example, in phytoplasmas that cause apple proliferation, ash yellows, and hydrangea virescence.

Symptoms and diagnosis. Phytoplasmal infections cause various symptoms: slow growth or stunting, dwarfed shoots and leaves, chlorosis, abnormal yellow or reddish leaf color, suppressed photosynthesis, impaired ability to export carbohydrate from leaves, aberrant branching (including loss of apical dominance, abnormally upright and spindly twigs, bushy form, witches'-brooms), precocious and/or dwarfed flowers, diminished flowering and fruiting, low seed viability or no seeds produced, phloem hyperplasia, phloem necrosis, rootlet necrosis, decline, and in some diseases death. No single symptom is diagnostic across a spectrum of plants. Symptom development is closely related to buildup of pathogen

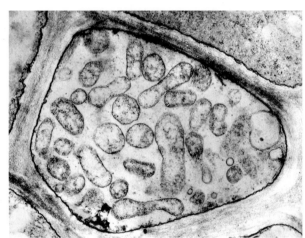

Figure 194. Cell profiles of '*Candidatus* Phytoplasma ulmi,' the elm yellows pathogen, viewed with an electron microscope in a transverse section of a sieve cell of *Ulmus americana* (American elm). The widest phytoplasma cells here are about 0.5 μm in diameter. Courtesy of E. J. Braun.

Phytoplasmal and spiroplasmal diseases of fruit trees.

A–D. Pear decline, caused by '*Candidatus* Phytoplasma pyri' (discussion, p. 392). A, B. Red-leaf symptoms in *Pyrus communis* 'Bartlett' (CA, Sep). C. Declining 'Bartlett' pear trees (CA, Nov). D. 'Bartlett' trees infected with the pear decline phytoplasma but growing on tolerant rootstocks. Rootstocks of the two end trees are more tolerant, and hence support more rapid growth, than rootstocks of the four center trees (CA, Jul).

E–G. Prunus yellow leafroll, also caused by '*Ca*. Phytoplasma pyri' (discussion, p. 392). E. In *Prunus domestica* (plum; CA, Sep). F, G. In *P. persica* (peach; CA, Jul).

H. Stubborn disease of citrus, caused by *Spiroplasma citri* (discussion, p. 392). Stunting and chlorosis of *Citrus limon* (lemon) in foreground contrast with normal trees in the background (CA, Oct).

Photo credits: A, B, E—C. D. Smart; C, H—B. C. Raju

populations in developing organs, which are metabolic sinks to which the parasites are carried along with photosynthetic products. Different phytoplasmas may cause similar or indistinguishable symptoms in some plants. For example, decline of *Prunus* species is caused primarily by phytoplasmas of the X-disease group in North America but by strains of the European stone-fruit yellows group in Europe. Elm yellows is caused by phytoplasmas of the elm yellows group in eastern USA and Europe but by a member of the clover proliferation group in northern Illinois. Accordingly, reliable specific diagnosis requires analysis of DNA from affected plants.

Witches'-brooms and/or decline are features of many phytoplasmal diseases of angiosperms. The severity of brooming and of decline symptoms are generally correlated. The brooms either arise from clustered buds or begin with precocious growth from buds in leaf axils, resulting in a bunch of shoots all produced in the same year as their parent twig. Growth and apical dominance are suppressed, so the shoots remain bunched. Additional generations of shoots form similarly and contribute to a broom's density. Twigs in brooms tend to be dwarfed and abnormally upright, although brooms that weigh heavily on branch ends become prostrate or pendant with only the twig tips turned up. Leaves are undersized or dwarfed and often misshapen. Brooms tend to continue growth or retain leaves later than normal in autumn, and they lack normal cold hardiness. Brooms or branches that support them often die back in winter. Surviving parts resume growth before the normal time in spring. In general, no flowers develop on brooms. Phytoplasmas are more abundant in brooms than in other parts of affected plants.

Environment influences the occurrence and severity of phytoplasmal diseases. They are more common and generally more severe in tropical and warm-temperate areas than in cool regions. This relationship reflects vector distributions and the limiting influence low temperatures have on phytoplasmal survival in phloem. Cold winters and cold storage of small diseased plants in pots have both caused remission of symptoms and permanent cure of phytoplasmal infection in some woody plants, as documented for *Pyrus* (pear) and *Fraxinus* (ash). Drought stress or shading by larger trees can trigger dieback and decline in phytoplasma-infected trees that would otherwise be able to tolerate infection.

Disease cycles. Phytoplasmas perennate in plants and in adult overwintering vectors. Intergenerational transfer in vectors via eggs occurs but is thought to be unusual. The vectors include hundreds of insect species, mostly of Cicadellidae (leafhoppers), but also Cixiidae, Derbidae, and Fulgoridae (planthoppers); Psyllidae (psyllids); and Pentatomidae (stink bugs). Some vector species transmit phytoplasmas associated with several diseases, and some phytoplasmas have several vectors. The vectors of most phytoplasmas are still unknown. Phytoplasmal infection of vector insects does not affect their life spans or reproduction. Other means of phytoplasma transmission include grafting and dodder (*Cuscuta* species). Phytoplasmas are not known to be transmitted through seeds, although they have been detected in some plant embryos.

Mechanisms of phytoplasma pathogenicity are incompletely understood. Photosynthesis and phloem loading of sugar in leaves are suppressed, and hormonal imbalances lead to aberrant growth. Gene functions associated with photosynthesis, sugar transport, response to stress, and hormone synthesis are all affected in plants infected by phytoplasmas. Phloem plugging and necrosis and rootlet necrosis seem to represent host resistance responses that develop slowly and cause great harm without halting plant colonization by phytoplasmas.

Host tolerance and resistance are keys to control of phytoplasmal diseases. *Tolerance* refers to the relative ability of a plant to grow and reproduce while colonized by phytoplasmas. *Resistance* refers to a plant's ability to localize colonization or suppress phytoplasma reproduction. Tolerance has been demonstrated in apple proliferation, ash yellows, elm yellows, and pear decline disease systems, among others. Phytoplasma-tolerant plants are potentially useful as rootstocks or breeding parents if not as cultivars. Resistance occurs in some disease systems. Multiplication of European stone-fruit yellows is inhibited in sweet cherry. Apple proliferation phytoplasmas apparently die out in certain *Malus* cultivars. One form of willow witches'-broom is characterized by localized infections in otherwise healthy, vigorous *Salix* plants, and pathogen-free plants can be propagated by cuttings from the healthy-appearing parts of willows with this disease.

References: 58, 272, 328, 454, 803, 906, 979, 1141, 1360, 1410, 1411, 1692, 1693, 1860, 1901, 1915, 1916, 1980, 2018, 2023, 2078, 2084, 2264, 2265, 2267, 2268, 2282, 2296, 2462, 2466, 2700, 2768, 2791, 2857, 3177, 3225, 3435, 3529, 3694, 3964, 4309, 4317, 4318

Representative Diseases

Pear decline. Pear decline (Plate 194), which killed nearly a million trees during the first few years of its occurrence in western North America in the 1960s, also occurs in northeastern USA and is widespread in Europe. It affects several *Pyrus* species, notably *P. communis* (common pear) when it is grown on rootstocks intolerant of infection. The pear-decline phytoplasma, '*Candidatus* Phytoplasma pyri.' is transmitted by *Cacopsylla* species (psyllids) and is classified in the apple proliferation group. Infected adult psyllids that overwinter in areas of mild climate can begin transmitting phytoplasmas into plants as soon as plant growth begins in spring. Symptoms of pear decline appear in three general patterns determined by the scion-rootstock combination: quick decline, slow decline, or leaf curl. Quick decline occurs in *P. communis* on rootstocks of *Pyrus pyrifolia* (Chinese or Nashi pear) or *P. ussuriensis* in which phloem of the rootstock just below the graft union dies. Affected trees wilt and die during summer or decline and die during the next 1–2 years. Slow decline is characterized by less phloem damage, slow growth, stunted leaves with upturned margins, red leaf color in summer, premature leaf drop, and dieback. Trees with declining tops have many dead rootlets. Trees on tolerant rootstocks such as *P. betulaefolia*, *P. calleryana* (Callery pear), or *P. communis* seedlings may show only slow growth or downward curling, purplish color, and early shedding of leaves. Natural remission of pear decline symptoms sometimes occurs in areas with cold winters. Pear decline is controlled primarily through the use of tolerant cultivars and rootstocks and by suppression of overwintering adult psyllid populations.

Prunus yellow leafroll. This disease (Plate 194) is also caused by '*Ca.* Phytoplasma pyri.' Outbreaks of yellow leafroll in *Prunus persica* (peach) and pear decline in *Pyrus communis* (pear) have occurred coincidentally in adjacent orchards. DNA analyses and vector transmission of the pathogen from pear to peach confirmed that one pathogen causes both diseases. Prunus yellow leafroll is characterized by chlorosis and upturned leaf edges. It also affects *P. domestica* (plum) and *P. dulcis* (almond).

References: 412, 413, 657, 1141, 2083, 2847, 2896, 3435, 3527, 3528, 3530, 3531, 4309

Stubborn disease of citrus. Stubborn disease (Plate 194), caused by *Spiroplasma citri*, is important in several *Citrus* species, notably *C. ×paradisi* (grapefruit), *C. sinensis* (sweet orange), and *C. ×tangelo* (tangelo). It occurs in California and Arizona, and also in

A, B. Bunch disease of *Carya illinoinensis* (pecan) (discussion, p. 394). A. Symptoms in spring. The diseased portions at lower left and top center began growth earlier than normal, and their foliage, although bunched at branch ends, thus appears fuller and darker green than the young foliage on unaffected branches. Brownish catkins are visible on normal branches but are absent from those with bunch symptoms. B. A systemically diseased, stunted tree about 4 m tall. Bunched foliage and multiple spindly stems in place of single branches occur throughout the tree (MS, Apr).

C, D. Bunch disease in *Carya ovata* (shagbark hickory) (AR, May).

E–H. Walnut witches'-broom in *Juglans* species (discussion, p. 394). E. Brooming and decline in *J. microcarpa* (little walnut). Dense brooms, lacking normal cold tolerance, died the previous winter. One branch not yet severely diseased has normal leaves (IL, Jun). F, G. Brooms on *Juglans cinerea* (butternut) (NY, May). H. A young broom on *J. nigra* (eastern black walnut). Spindly shoots have grown from buds in leaf axils (IN, summer).

Photo credits: C, D—P. Fenn; H—R. J. Green

393

Mediterranean and Middle Eastern countries. Stubborn disease has been known since early in the 20th century, but the causal spiroplasma was not discovered until 1970, when it was isolated and its pathogenicity proved. Symptoms include suppressed growth, shortened internodes, abnormally numerous shoots from axillary buds, and a compact bushlike appearance. Shoots tend to grow abnormally upright, and the affected tree grows more during the autumn flush than in spring as normal trees do. Leaves are undersized and show various chlorotic patterns. Fruits and seeds are abnormally small and variously deformed, fruit is bitter or sour, and yield is depressed. Symptoms begin on one branch and spread to all within a few years. Trees infected when young remain stunted. Diseased trees are abnormally sensitive to heat and frost, and dieback occurs in some of them. Several insects harbor *S. citri*, and at least three leafhoppers are natural vectors in the USA. The beet leafhopper (*Circulifer tenellus*) and two species of *Scaphytopius* transmit the pathogen between citrus trees and also to and from herbaceous plants.
References: 884, 1310, 1409, 1859, 2195, 2346, 2463, 2588, 2904

Bunch disease of walnut and butternut. Brooms on *Juglans* (walnut, butternut; Plate 195) are caused by phytoplasma strains in a subgroup of the *Prunus* X-disease group. The disease occurs throughout the range of these species in eastern USA. Its effects vary from slight to lethal, depending on the host species. *J. microcarpa* (little walnut) and *J. ailanthifolia* (Japanese walnut) are highly intolerant of infection; *J. nigra* (eastern black walnut) is tolerant. Other hosts include *J. mandshurica* (Manchurian walnut) and *J. regia* (Persian walnut). Virtually all twigs on small highly susceptible trees may be converted to brooms; large trees tend to retain normal overall form but produce brooms that begin as sprouts along the trunk and limbs. Leaflets in brooms tend to be abnormally narrow, curled or cupped, and often chlorotic. Fruits fall from diseased trees prematurely, or nuts fail to fill normally, the kernels shriveling and turning black. Diseased *J. nigra* often shows no symptoms other than slow growth until a tree is pruned or felled. Then, fast-growing brooms unusual for their long internodes may grow from stubs or the stump. Some diseased trees may live 40 years or longer. Recovery of *J. nigra* from bunch disease has been reported, but this may represent only a temporary increase in growth in response to favorable environment or fertilization rather than a spontaneous cure. Vectors of walnut bunch are unknown.

Bunch disease of pecan and other hickories. Bunch disease of *Carya illinoinensis* (pecan; Plate 195) is caused by phytoplasmas in another subgroup of the *Prunus* X-disease group. It occurs from Kansas and Texas to North Carolina and Georgia, affecting trees in orchards, landscapes, and the wild. Diseased trees typically have one to several brooms on scattered branches or on sprouts on the trunk. Infection may be temporarily localized in particular branches, but eventually the causal phytoplasmas spread throughout the tree and may induce brooms on all branches. Catkins sometimes become broomlike also. Nut production ceases on broomed parts, and nut yield, size, and quality on nonbroomed branches diminish. Foliage on affected branches varies from dark green to chlorotic, and leaflets are sometimes distorted or shorter and broader than normal. Small trees may produce brooms with a profusion of slender sprouts with stunted light green leaves. Leaflets produced on brooms tend to drop prematurely. Systemically diseased trees of highly susceptible cultivars eventually decline and die. Vectors of bunch disease have not been reported. Pecan cultivars Brooks, Caddo, Choctaw, Curtis, Farley, Owens, Seminole, Stuart, and Success are resistant.

Observers have long suspected that bunch diseases of *C. illinoinensis* and other *Carya* species are caused by the same agent. *C. cordiformis* (bitternut hickory), *C. tomentosa* (mockernut hickory), *C. ovata* (shagbark hickory), and *C. aquatica* (water hickory) are affected in the region where bunch disease is common on *C. illinoinensis*, and bunch disease in pecan orchards has developed near diseased wild hickories in several areas. The disease in wild *Carya* species also occurs in areas north of the range of *C. illinoinensis* from Iowa to New York.
References: 715, 767, 1141, 2268, 2404, 2462, 3537, 3538

Lilac witches'-broom. Strains of 'Candidatus Phytoplasma fraxini,' the ash yellows group of phytoplasmas, cause this disease, which is widespread in midwestern and eastern USA. Cultivars intolerant of infection produce dense brooms of short, thin twigs, most often on mainstems near the ground. Leaves on the brooms are variously distorted, usually stunted, and often chlorotic. Leaf scorch and dieback sometimes develop on brooms during the growing season, and brooms often die back during winter. Entire plants sometimes die within a few years after the onset of brooming. Phytoplasmas have been detected in at least 20 *Syringa* species and hybrids. Late-flowering types seem particularly susceptible. Severe disease has been noted in *Syringa josikaea* (Hungarian lilac), *S. reticulata* (Japanese tree lilac), *S. sweginzowii*, and hybrids with these species and *S. komarowii* or *S. villosa* (late or villous lilac) as parents. Many cultivars, notably those of *S. vulgaris* (common lilac), are tolerant of infection and retain nearly normal twig form except for retarded growth and short internodes. Some display only premature swelling of buds and elongation of shoots. Vectors of lilac witches'-broom are unknown.

Willow witches'-broom. This disease has been studied in Alberta, eastern USA, and China, and is reported also from Europe and India. In North America brooms have been noted on *Salix eriocephala* (heartleaf or Missouri River willow) and *S. nigra* (black willow) in scattered eastern localities, and on *S. bebbiana* (Bebb willow), *S. discolor* (pussy willow), *S. exigua* (narrowleaf or sandbar willow), and *S. petiolaris* (meadow willow) in Alberta. The phytoplasmas identified in willow brooms in North America and China are members of the aster yellows group. Two types of brooms occur in *S. nigra*. One type definitely linked to phytoplasmas develops near the ends of scattered branches and remains small (mostly less than 40 cm long) and dense with spindly twigs and stunted leaves (Plate 196F, G). Brooms of this sort typically die in winter. The other, more common, broom type on *S. nigra* is a clump of twigs, often 40–60 cm high and nearly as broad, with twig bases close together and their shafts arching outward and then upward, growing from multiple points on trunks and large limbs. Whether phytoplasmas cause the latter type of broom has not been established.

Dogwood witches'-broom and stunt. Several *Cornus* species (dogwoods) develop phytoplasma-associated witches'-brooms. Phytoplasmas have been detected in brooms on *Cornus amomum* (silky dogwood), *C. florida* (flowering dogwood), *C. racemosa* (gray dogwood), and *C. sericea* (red-osier dogwood) in North America, and on *C. sanguinea* (blood-twig dogwood) in Europe. The pathogen in *C. racemosa* in New York State is a member of the aster yellows group. Identities of phytoplasmas in other *Cornus* species have not been reported.

Other witches'-brooms. Phytoplasmas also cause witches'-brooms on North American *Fraxinus* species (ash), *Gleditsia triacanthos* (honeylocust), *Prunus armeniaca* (apricot), *P. pensylvanica* (pin cherry), *P. persica* (peach), *Robinia pseudoacacia* (black locust), *Sassafras albidum* (sassafras), *Ulmus* species (elms), and *Vaccinium* (blueberry). Many other witches'-brooms possibly caused by phytoplasmas remain undiagnosed.
References: 1379, 1380, 1635, 1637, 1692, 1724, 2462, 3188, 3212, 3225, 3646

A–E. Lilac witches'-broom, caused by a phytoplasma of the ash yellows group. A–C. Symptoms in *Syringa* ×*prestoniae* 'Hiawatha.' A. Chlorosis and stunted growth. B. Close view of small witches'-brooms. C. Chlorosis and mottling of leaves, and fruit abortion. Only seven or eight fruits are maturing among the remnants of dozens of blossoms (NY, Aug). D. Dieback and brooms on branches of *Syringa* ×*josiflexa* 'Royalty' (IL, Jul). E. Dead and dying twigs in a broom on *S. josikaea* (Hungarian lilac). Green-tipped buds survived winter but failed to open (NY, Jul).

F, G. Willow witches'-broom caused by phytoplasmas of the aster yellows group on *Salix* species. F. Brooms and normal twigs on the same branch. Brooms of this sort die during winter. G. Spindly upright twigs with abnormal pigment and dwarfed leaves (NY, Sep).

H. Witches'-broom caused by an unidentified phytoplasma on *Cornus florida* (flowering dogwood) (PA, Jun).

I. Stunted, bushy growth of *Cornus racemosa* (gray or paniceled dogwood) affected by dogwood stunt in comparison with a healthy plant. Both plants, collected from the field and growing in pots, were previously pruned (NY, greenhouse).

Photo credit: D—C. R. Hibben

395

Elm Yellows

Elm yellows is a debilitating or lethal systemic disease of *Ulmus* species (elms) that is usually caused by phytoplasmas of a group provisionally named '*Candidatus* Phytoplasma ulmi,' the elm yellows group, but can also be caused by a strain in the clover proliferation group (see phytoplasma overview with Plate 194). The disease occurs in the eastern half of the USA and in parts of central and southern Europe. An outbreak occurred in the Niagara Peninsula of Canada in the 1980s, but by 2000 the disease was no longer seen there. In the USA elm yellows is known for epidemics that kill nearly all native elms in affected localities. In Europe, by contrast, the disease is of little importance and is usually not lethal. It appears in scattered trees and occasionally increases to epidemic status in northern Italy. Various Eurasian elm species have been found infected. Major differences in symptoms expressed by American and Eurasian elms are indicated by two former names of the disease: phloem necrosis in America and elm witches'-broom in Europe. The remainder of this account emphasizes American experience.

Elm yellows has been found in five native species, one Asian elm species, and an American-Asian hybrid in North America: *Ulmus alata* (winged elm), *U. americana* (American or white elm), *U. crassifolia* (cedar elm), *U. parvifolia* (Chinese elm), *U. rubra* (red or slippery elm), *U. serotina* (September elm), and the hybrid *U. rubra* × *pumila* (red × Siberian elm). All *U. americana* cultivars that have been tested are intolerant of infection. Eurasian elms have remained apparently unaffected when exposed naturally to elm yellows in localities where the disease killed most native elms. On the other hand, several Eurasian species, hybrids, and cultivars were susceptible to North American elm phytoplasmas when inoculated by grafting. The symptoms ranged from depressed growth and broom formation to yellowing and death.

Symptoms and signs. Elms intolerant of infection die within a year or two after foliar symptoms appear. Tolerant elms (Eurasian species) grow slowly, become stunted if infected when small, and may develop chlorosis and witches'-brooms. Recovery is unknown. External symptoms in North American species usually appear in mid to late summer and include yellowing, epinasty (drooping or downward bending of the petioles of turgid leaves), and premature casting of leaves, then death of branches. The sequence often takes only a few weeks. All branches usually show symptoms at once, but sometimes yellowing develops first on one branch and then spreads to others during two or more seasons. Yellow leaves may be interspersed with green ones on a single branch, but more often all leaves become yellowish green, then yellow. When these symptoms arise in late summer or early autumn, only the timing distinguishes them from normal leaf senescence. Diseased trees that survive dormancy may open buds at the normal time but then grow only enough to produce chlorotic, dwarfed shoots and leaves that soon wilt or turn yellow and drop. Trees sometimes wilt and die quickly without prior external symptoms. Shriveled brown leaves adhere to wilted trees for several weeks.

By the time foliar symptoms appear, root mortality and phloem degeneration in living roots and the base of the tree are extensive. Fine roots die first, then successively larger ones succumb. The innermost bark and the cambial zone change color from nearly white to yellow, then butterscotch or tan, sometimes with darker flecks, and finally dark brown. Then other tissues die. The surface of the wood may also be discolored where pigment diffuses from degenerating phloem. On exposure to air, discolored inner phloem and cambium turn brown much more rapidly in diseased elms than in healthy ones. Water movement in xylem and translocation in phloem diminish before any visible symptoms of yellows develop. Stomata close partially or completely and remain closed beginning some weeks before discolored phloem can be found in stems. Living discolored phloem of *U. alata*, *U. americana*, *U. crassifolia*, and *U. serotina* affected by elm yellows produces methyl salicylate (oil of wintergreen), which can be detected by sniffing at the surface of freshly exposed inner bark or at the mouth of a vial enclosing a sample of inner bark. Diseased *U. rubra* produces an aroma somewhat like that of maple syrup. The odor emanates from newly killed bark and leaves.

Witches'-brooms commonly form on small branches of *U. rubra* (red elm) and Eurasian elms such as *U. parvifolia* affected by elm yellows. On *U. rubra* they develop in the second year that a tree shows foliar symptoms, which is usually its final year of life. These brooms usually remain small—a few centimeters long—but some grow much larger. Phytoplasmas can be detected in both symptomatic and apparently normal parts of diseased elms. They become most numerous in the petioles and stems of witches'-brooms.

Disease cycle. Vector insects inoculate elms in summer and early autumn, but few elms develop symptoms in the year of inoculation. The incubation period is at least 3 months in very small trees and 9–10 months or more in large ones. The yellows agent thus overwinters in its plant hosts. Most new infections occur after vector insects develop wings. One vector of elm yellows in North America is the white-banded elm leafhopper, *Scaphoideus luteolus*. Adults of this species are widespread in some areas from early summer until frost in autumn. Additional vectors are probable, moreover, because elm yellows epidemics sometimes develop and persist where *S. luteolus* is scarce. Two possible occasional vectors are *Philaenus spumarius* (meadow spittlebug) and a leafhopper, *Allygus atomarius*. Both species, captured in the vicinity of naturally diseased trees, transmitted yellows to test seedlings. In Italy the leafhopper *Macropsis mendax* is a vector. Elm yellows also spreads among closely spaced trees of the same species via root grafts.

Epidemics of elm yellows, although locally spectacular, do not spread rapidly. The edges of an area of outbreak in New York State, observed for 15 years, advanced only about 6 km per year on average. The disease can be endemic for many years between flare-ups in a given locale. Spot outbreaks and single-tree occurrences beyond the main range of the disease (e.g., in Massachusetts and North Dakota) presumably develop after long-distance transport of vectors by wind.

The foregoing account represents the disease caused by '*Ca.* Phytoplasma ulmi.' As of 2004, the similar disease caused by a strain in the clover proliferation group had been reported only in *U. americana* in one locality in Illinois. It was characterized by foliar symptoms and phloem necrosis like those described above. No information has appeared about vectors or host range of the pathogen.

References: 128, 184, 486, 487, 656, 663, 1013, 1141, 1378, 1894, 2266, 2269, 2508, 2510, 2516, 3654, 3655, 3871, 3938

A–E. Foliar symptoms of elm yellows in *Ulmus americana* (American elm). A–C. Dying trees; symptoms in the tree in B began on one limb. D. Epinastic yellow and green leaves on the same branch. E. Chlorosis and epinasty on the branch at right, contrasted with normal foliage (NY, Aug).

F. Foliar symptoms in *U. rubra* (red elm). Affected leaves first turn yellowish green, then gold or yellow-orange (NY, Aug).

G, H. Phloem symptoms in *U. americana*. G. Normal (left) and discolored inner phloem exposed by peeling bark from small stems. H. Yellow-brown discoloration of the innermost phloem and cambial region (arrow) exposed by a slanting cut into a small stem (NY, Aug).

I. Dead rootlets on a pencil-size *U. americana* root. Rootlet mortality precedes foliar symptoms (NY, Aug).

J. Yellowing, proliferation of tiny upright twigs, and twig dieback in a hybrid elm, *U. rubra* × *U. pumila* (Siberian elm) (NY, Sep).

K, L. Elm yellows in *Ulmus parvifolia* (lacebark or Chinese elm). K. Witches'-brooms with dwarfed shoots at branch tips and red pigment in leaves of normal size that formed before the onset of aberrant growth (NY, Aug). L. Spindly witches'-brooms on a low, shaded branch of a declining tree (NY, Sep).

397

Ash Yellows

Ash yellows causes slow growth or decline and premature death of *Fraxinus* (ash) in parts of the USA, Canada, and Colombia. The causal phytoplasmas are a group provisionally named '*Candidatus* Phytoplasma fraxini,' also known as the ash yellows group. The disease has been known since the 1950s or earlier, but its nature was not well understood until the 1980s (see also ash decline, Plate 229). It occurs in the region extending from southern Quebec, Massachusetts, and North Carolina to Alberta and southwestern Utah. It is common in areas where wooded and open lands are intermixed and uncommon in areas that are primarily forested. Witches'-brooms similar to those caused by the ash yellows agent have been noted on *F. pennsylvanica* (green ash) and *F. berlandieriana* (Mexican ash) in southern USA, but their cause has not been ascertained. *F. uhdei* (shamel ash) is severely damaged by ash yellows in Colombia. Phytoplasmas of the ash yellows group also cause lilac witches'-broom (Plate 196).

F. americana (white ash) sustains the greatest damage from ash yellows in North America. *F. pennsylvanica* is affected throughout the North American range of the disease and is the principal host in the western half of that region. *F. velutina* (velvet ash) is the main host in Utah. The latter two species usually tolerate infection without progressive deterioration. Other ash species that have been found infected in woodlots or arboreta include *F. angustifolia* (narrowleaf ash), *F. bungeana*, *F. excelsior* (European ash), *F. latifolia* (Oregon ash), *F. nigra* (black ash), *F. ornus* (flowering ash), *F. profunda* (pumpkin ash), *F. quadrangulata* (blue ash), and *F. sogdiana*. Their relative abilities to tolerate infection are unknown.

Symptoms and signs. Symptoms vary according to stage of the disease, abilities of trees to tolerate infection, and aggressiveness of strains of the pathogen. This description emphasizes symptoms in *F. americana*. Height and diameter growth diminish, usually to less than half their former rates. Highly susceptible saplings die to ground level within 1–3 years. Slightly more tolerant trees remain alive but grow very slowly and usually develop dieback. Because of slow twig growth, foliage may be arranged in tufts or, in extreme cases, rosettes at branch tips. The foliage is often lighter green than normal and sometimes turns chlorotic. Leaves often fail to attain normal size, and leaflets in tree crowns may be partially folded lengthwise. Slow growth and small folded leaves give an overall appearance of sparse foliage. Occasional trees nearing death produce a set of feeble leaves that all drop during the growing season. Branch dieback occurs in many diseased trees, initially on scattered branches and usually during dormancy. Smooth bark of diseased trees exposed to sun and wind, especially those with dieback, often turns pinkish gray in contrast to the normal drab gray or gray-green. Declining trees die back progressively. On the other hand, many diseased trees, apparently tolerant of infection, continue slow to moderate seasonal growth indefinitely while showing only sparse foliage, dieback of scattered branches, and deliquescent branching. (A deliquescent branch is one on which the terminal shoot has lost dominance, so that the branch axis becomes lost among several twigs of more or less equal length.)

Declining ash trees affected by yellows usually produce witches'-brooms or epicormic sprouts with deliquescent branching, but only a minority have these symptoms at a given time. Brooms grow most often at the root collar, sometimes on the trunk, and rarely on branches. Brooms on branches and rosettes at branch ends occur only on small trees. Brooms at the root collar often do not develop until 1–3 years after the crown has become leafless or a diseased tree has been cut down. Leaves on brooms are usually dwarfed, and many are simple rather than compound. Leaflets are often distorted and narrower than normal. Such leaflets are usually chlorotic, with the greatest yellowing in interveinal areas, and are retained longer than is normal in autumn. Epicormic sprouts, when present, usually occur within about 3 m of ground level. Some evolve into witches'-brooms. Stomata in leaves on brooms or deliquescent branches remain partially to completely closed. Roots of diseased *F. americana* tend to be abnormally short and bushy, often with many dead ends. Diseased saplings commonly sustain rootlet necrosis that leads to sudden wilting and death. More tolerant species do not have these root symptoms.

Shoot growth of many *F. americana* affected by ash yellows begins several days earlier in spring than that in healthy trees. This behavior is most noticeable on witches'-brooms. Diseased trees also tend to display autumn color prematurely and are not as cold hardy as healthy trees. Shoots in brooms usually die back in winter. Freezing damage to the vascular cambium at the base of the tree sometimes occurs and may lead to split bark and rarely to basal cankers extending from the soil line upward 50 cm or more (Plate 248). Damage by freezing is most common in saplings and pole-size trees.

Diagnosis of ash yellows is usually difficult, especially in isolated landscape specimens, because many trees affected by this disease merely grow slowly or decline without producing either of the symptoms most useful for diagnosis—witches'-brooms and deliquescent branching. Phytoplasmas can readily be detected in phloem of diseased trees by electron microscopy or fluorescence microscopy using the DNA-binding fluorochrome known as DAPI (4',6-diamidino-2-phenylindole · 2HCl). Specific diagnosis is possible by a serological test or DNA analysis.

Disease cycle. The ash yellows cycle is presumed to conform to the general model established for phytoplasmal diseases of woody plants in temperate regions: perennation of phytoplasmas in stem and root phloem, colonization of leaf phloem in spring and summer, transmission of phytoplasmas from diseased to healthy plants by vector insects during summer and early autumn, and expression of symptoms (if any) after the next dormant period or after an indefinite longer time. Small trees graft-inoculated in summer show symptoms the next spring. A longer incubation period is likely in large trees. The principal vectors of ash yellows in eastern North America are probably leafhoppers of the genus *Scaphoideus* (Homoptera: Cicadellidae). These insects were the most numerous of many types of leafhoppers collected on sites where the disease occurred in New York State, and they were found to be the most common carriers of ash yellows phytoplasmas.

Ash yellows tends to increase slowly but inexorably within affected tree populations. In many old-field stands and hedgerows from Illinois eastward the majority of *F. americana* become diseased. In New York State the average rate of increase of disease incidence on several sites was found to be near 5% of the initial *Fraxinus* populations per year. Decline associated with ash yellows is aggravated by drought, presumably because trees with rootlet necrosis and stunted root systems undergo more severe or prolonged stress than do healthy trees during drought. Phytoplasma-infected *F. americana* released from drought stress often fail to grow as rapidly as they did before the water shortage, while previously healthy trees regain their former growth rate. Ash yellows is seldom important in nursery-produced shade trees, aside from retarding their growth, because *Fraxinus* cultivars are usually grown on phytoplasma-tolerant *F. pennsylvanica* rootstocks.

References: 128, 505, 1111, 1141, 1379, 1380, 1643, 1662, 2268, 2508, 2509, 3642–3648, 3656, 3693, 3964, 4197

A–I. Ash yellows in *Fraxinus americana* (white ash). A. Young diseased trees in an old-field stand exhibit thin crowns, chlorosis, and dieback. One tree at left center appears healthy (NY, Jul). B. Deliquescent branches and witches'-brooms on a small tree; symptoms as distinct as these are unusual (NY, Apr). C. Deliquescent branches (NY, Apr). D, E. Brooms and deliquescent sprouts on mainstems of young trees (NY, Aug & Apr). F. Undersized leaves clustered on deliquescent twigs at the tip of a slowly growing branch. The stem above the tape marker is 15 years old (NY, Aug). G. Magnified view of a cross section of the trunk of a small tree that had been in decline for about 12 years when cut. Annual layers of wood near the bottom of the image are of normal thickness (NY, Oct). H. The base of a diseased tree with several brooms and split bark due to freezing injury (NY, May). I. Brooms on a dying sapling (NY, Aug).

J. Ash yellows in *F. pennsylvanica* (green ash): sprouts with chlorotic simple leaves on the base of a sapling stem (NY, Aug).

K. Ash yellows in *F. velutina* (velvet ash): brooms at the root collar of a declining tree (UT, May).

X-disease of *Prunus*

X-disease is a lethal decline-type disease of *Prunus* species, mainly cherries and peach. The loss of productivity of diseased fruit trees makes X-disease economically important. Phytoplasma strains that cause X-disease were the first of their group to be studied using tools of molecular biology, so the *Prunus* X-disease group is named for them. Like other phytoplasmas, they have not been cultivated apart from plants or insect vectors. They occur from coast to coast in the northern half of the USA and southern Canada. X-disease is not known elsewhere.

Most of the information about X-disease has come from research on the disease and its vectors in orchards. It was first recognized in the early 1930s in California and Connecticut. Soon thereafter it was noticed in large regions in eastern and western USA and Canada, and it became a limiting factor in peach and cherry production. Major outbreaks in orchards have occurred at irregular intervals, presumably reflecting trends in vector activity.

Known hosts of X-disease phytoplasmas in nature include *Prunus armeniaca* (apricot), *P. avium* (sweet cherry), *P. cerasus* (sour cherry), *P. domestica* (prune), *P. dulcis* (almond), *P. mahaleb* (mahaleb cherry), *P. persica* (peach) and its variety *nucipersica* (nectarine), *P. salicina* (Japanese plum), and *P. virginiana* (chokecherry). X-disease phytoplasmas have been transmitted experimentally by grafts, dodder, or leafhoppers to several other species of *Prunus* and to diverse herbaceous plants, which have then developed stunting, chlorosis, and deformities. *Prunus serrulata* (Japanese flowering cherry) and *P. serotina* (black cherry) are apparently not susceptible.

The geographic range of X-disease is related to that of *P. virginiana*. The disease in this species is important primarily because *P. virginiana* serves as a reservoir from which leafhoppers carry the pathogen to orchard trees. In the northern Great Plains, moreover, *P. virginiana* is planted in windbreaks and for wildlife habitat. X-disease is fatal in wild-type individuals of this species. X-disease occurs in purple-leaved cultivars, but its severity in them has not been determined.

Symptoms. Symptoms vary somewhat with host species, region, and strain of the pathogen. Eastern and western X-disease syndromes are recognized. In this discussion we feature eastern X-disease in *P. virginiana*. Beginning in early summer, leaves turn light green and then chlorotic, often with green tissue remaining along the veins. Many chlorotic leaves drop. Leaves nearest the shoot tip may be thicker than normal, somewhat leathery in texture, and crinkled. Chlorotic spots often develop, turn brown, and drop out of expanded leaves, thus causing voids called shot holes. These holes are not diagnostic, however, because various stresses can induce their formation. Many *Prunus* species exhibit shot-hole formation in response to foliar infections by fungi. Leaves of X-diseased *P. virginiana* exposed to sun often turn reddish purple to bright red or yellow and red beginning in midsummer; stomata begin to close, and transpiration is retarded. Transpiration through red leaves is negligible. Blossom clusters on diseased plants are undersized, and many blossoms abort. Fruits are undersized, less colorful, and more pointed than normal, and seed development is arrested. In one study less than 10% of seeds on diseased plants, versus 99% on healthy plants, had normal contents.

Many diseased *P. virginiana*, instead of setting winter buds in the normal fashion, produce a tuft or rosette of stunted, abnormally upright leaves very close together at the tip of the shoot. These often remain green longer than leaves of normal form below them. Buds in leaf axils of the rosette sometimes open and produce very short secondary shoots. Stems with rosetted shoot tips and red foliage often die during winter. Entire diseased plants sometimes die after the second year, but most die after 3 or more years with symptoms.

Chokecherry propagates itself by root sprouts as well as seeds. Therefore all stems in a clump may be on a common root system, leading to disease in all of them after the first becomes infected. The roots of diseased plants often survive for several years, however, repeatedly sending up new sprouts.

Symptoms of X-disease in peach include formation of shot holes in leaves (often so severe that leaves become tattered), partial folding of leaves, chlorosis, premature defoliation, reduced fruit yield, rosettes at twig tips, dieback, and eventual death. Symptoms usually appear first on one scaffold limb and spread throughout a tree. Affected cherry trees on *P. avium* rootstock (called mazzard rootstock) grow poorly; bloom late; produce light green, undersized, sometimes upward-folded leaves and small, deformed fruit with poor flavor; display dull or sometimes bronzed foliage; and may begin to die back after several years. Infected cherry trees on *P. mahaleb* rootstock either decline slowly or wilt and die within one season. Some *P. avium* cultivars produce enlarged stipules when infected. Other fruit trees are usually less severely affected than cherry and peach.

Disease cycle. Inoculation occurs during summer, when phytoplasma populations in aboveground parts of previously infected plants are high and leafhopper vectors are abundant. Vectors sucking phloem sap from leaves of infected plants acquire phytoplasmas and may later transmit them into healthy plants. A delay of 4–5 weeks occurs between acquisition and transmission. This delay reflects the time required for multiplication and circulation of phytoplasmas into a vector's salivary glands, whence they are dispensed into phloem during feeding. Symptoms of X-disease appear in the year after inoculation. General characteristics of phytoplasmal infections are described in the overview beginning on page 390.

At least 16 species of leafhoppers (Homoptera: Cicadellidae) are vectors of the X-disease phytoplasma. The most common vectors are *Colladonus montanus* and *Fieberiella florii* in the West and *Colladonus clitellarius*, *Paraphlepsius irroratus*, and *Scaphytopius acutus* in the East. Some vectors are capable of transmitting X-disease phytoplasmas among diverse herbaceous and woody plants in several families.

X-disease conforms to the jump-spread model of disease increase, in which symptoms appear in one plant and then in several others nearby. The disease spreads rapidly in *P. virginiana* populations. In an experimental planting in Nebraska disease incidence increased from 5% to over 80% within 5 years, and more than half the plants were dead within 8 years. In eastern stone-fruit orchards most cases of X-disease seem to result from transmission of phytoplasmas from *P. virginiana*. In the West, however, vectors spread the pathogens from tree to tree in orchards of both peach and cherry.

Other phytoplasmal diseases of Prunus *in North America.* Pin cherry witches'-broom is caused by a member of the X-disease group. Peach yellow leafroll and peach rosette are caused by members of the apple proliferation group. Phytoplasmas associated with peach yellows, albino disease of cherry, and cherry buckskin have not been classified. In Europe a set of phytoplasmal diseases collectively called European stone-fruit yellows is caused by strains in the apple proliferation group.

References: 727, 1141, 1412–1414, 1657, 1929, 2079, 2403, 2461, 2462, 2508, 2896, 3029, 3037, 3070, 3173, 3333, 3334, 3665, 4126

A–G. Eastern X-disease in *Prunus virginiana* (chokecherry). A. Red foliage of a diseased plant in contrast with normal vegetation (NY, Aug). B. Plants showing loss of apical dominance and yellow to red leaves with edges rolled upward (PA, Aug). C–E. Yellowish to red leaves of normal size on twigs with rosettes at the tips. Dense foliage in rosettes is a result of short internodes and growth of secondary shoots from buds in leaf axils. Typical rosettes have dwarfed yellow-green leaves (NY & PA, Aug). F. A diseased shoot tip in winter, with dwarfed twigs that grew in leaf axils the previous summer (NY, Mar). G. Shot-hole formation in light green leaves of a plant in early stages of the disease (NY, Jul).

H. *Prunus avium* (sweet cherry) with early symptoms of western X-disease: slow growth and rolled leaves with reddish discoloration on one branch (CA, Sep).

I–K. Eastern X-disease in *P. persica* (peach). I. Yellow, rolled leaves (MI, Jul). J. Purple blotches on rolled leaves (NY, Aug). K. Reddish blotches progressing to necrosis and shot holes (NY, Sep).

L. Enlarged stipules in leaf axils of *P. avium* in an early stage of X-disease (NY, Jun).

M. Undersized yellowish fruit in contrast with a few normal fruit on diseased *P. cerasus* (sour cherry) (MI, Jun).

Photo credits: H—C. D. Smart; I, M—A. L. Jones, J, L—D. A. Rosenberger

401

Lethal Yellowing of Palms

Lethal yellowing became famous in the 1950s–1970s for decimating *Cocos nucifera* (coconut palm) populations in Jamaica and southern Florida. It occurs in the northern Caribbean region and in coastal areas of southern Florida, Mexico, Belize, Guatemala, and Honduras. The pathogen is the type phytoplasma of the palm lethal yellows group. It has killed millions of *C. nucifera* in plantations and landscapes in the above-mentioned areas. Similar diseases caused by other phytoplasmas in the group affect palms in southern Texas and Africa. A lethal decline caused by phytoplasmas of the sugar cane white-leaf group occurs in Indonesia. Lethal yellowing and related diseases have become worldwide threats to susceptible palms in the tropics and subtropical belt. Losses in early phases of the epidemic in the Caribbean region were mainly of one coconut cultivar, Pacific Tall, which had been widely planted because of its desirable horticultural characteristics. Large numbers of *Adonidia merrillii* (Christmas or manila palm), *Phoenix* species (date palms), and *Pritchardia* species were also killed. The initial epidemics of lethal yellowing subsided when the populations of highly susceptible palms had been severely depleted.

More than 30 palm species that grow in the USA are susceptible to lethal yellowing, based on observations in Florida. The disease has been noted in the following, in addition to those mentioned above:

Aiphanes lindeniana, Allagoptera arenaria (seashore palm), *Arenga engleri* (miniature sugar palm), *Borassus flabellifer* (toddy palm), *Caryota mitis* (Burmese fishtail palm), *C. rumphiana* (Albert or giant fishtail palm), *Chelyocarpus chuco, Corypha utan* (buri or gebang palm), *Dypsis cabadae* (cabada palm), *D. decaryi* (triangle palm), *Dictyosperma album* (princess palm), *Gaussia attenuata* (llume palm or Puerto Rican gaussia), *Howea belmoreana* (Belmore sentry palm), *Hyophorbe verschaffeltii* (spindle palm), *Latania* species (Latan palms), *L. rotundifolia* (footstool palm), *Nannorrhops ritchiana* (Mazari palm), *Phoenix canariensis* (Canary Island date palm), *P. dactylifera* (date palm), *P. reclinata* (Senegal date palm), *P. rupicola* (cliff date palm), *P. sylvestris* (wild date palm), *Pritchardia affinis* (Kona palm or luolu), *P. pacifica* (Fiji fan palm), *P. remota* (luolu), *P. thurstonii* (Thurston palm), *Ravenea hildebrandtii* (Hildebrand palm), *Syagrus schizophylla* (Arikury palm), *Trachycarpus fortunei* (windmill palm), *Veitchia arecina, V. mcdanielsi,* and *V. montgomeryana* (Montgomery palm).

All of the palms affected in the USA are introduced species; palms native to Florida and Mexico have remained unaffected. A lethal decline similar to that in palms does occur, however, in *Pandanus utilis* (common screw-pine) in southern Florida and in the palmlike plant *Carludovica palmata* (palma jipijapa) in Mexico.

Progress toward understanding lethal yellowing was slow at first because palms cannot be grafted for experimental transmission of an infectious agent or for resistance tests. Beginning in 1972, however, multiple reports established the association of phytoplasmas with typical symptoms. The vector role of a planthopper, *Myndus crudus* (Homoptera: Cixiidae), was confirmed in the late 1970s.

Symptoms. This description emphasizes symptoms in *Cocos nucifera.* Plants of all ages and sizes are affected. They show four general symptoms in various combinations before death. The usual sequence is premature nut drop, blackening of flower parts, yellowing and browning of mature fronds, and necrosis of the spear leaf before the crown collapses. Fruits in all stages of development fall within 1–4 weeks. Darkened flower parts, which soon shrivel and turn nearly black, are noticeable in recently opened inflorescences. In advanced stages of disease flower parts begin to degenerate while still enclosed in the spathe, and the spathe itself may discolor and split abnormally. Foliar discoloration usually begins on mature fronds, progressing from the tip toward the base and often from the oldest toward the youngest fronds. The discoloration begins with loss of greenness and then, according to palm species and locality, either progresses through shades of yellow, orange-yellow, or bronze to brown, or is followed by withering and browning. Yellowing sometimes begins on only one frond in a green canopy; such a frond is called a flag leaf. While yellowing is the dominant foliar symptom in *C. nucifera*, fronds of some other species, notably *Adonidia merrillii, Borassus flabellifer, Caryota mitis,* and *Phoenix* species, die and turn brown without a yellow phase. Necrosis and browning begin along leaflet margins in these palms. Brown fronds either fall or droop and hang from a dying crown. Finally the youngest extended leaf, or spear leaf, dies,

and necrosis spreads to nearby organs and eventually reaches the growing point. Moist dead tissues there are often invaded by microorganisms that cause a slimy decay. Mature fronds are usually brown by that time. When their sheathing bases no longer strengthen the trunk, it collapses near the former growing point, and the remainder of the crown then topples. Roots begin to die as discoloration develops in the crown, but some roots remain alive until late stages of canopy symptoms. Usually 3–6 months elapse from the onset of overt symptoms until death of highly susceptible palms, but sequences both faster and slower are common. A few cases of natural remission of symptoms have been recorded.

Disturbed water relations are an early, consistent feature of lethal yellowing. Stomata close, and photosynthesis and transpiration diminish while leaves are still green. Leaf yellowing is accompanied by decrease in protein, chlorophyll, and carotenoid contents. Hormonal imbalance is indicated by increase in abscisic acid and formation of ethylene as diseased plants decline.

Disease cycle. The causal phytoplasma is transmitted from diseased to healthy palms only by vectors. The only vector known in Florida, Jamaica, and south Texas is *Myndus crudus.* The interval from infection until onset of overt symptoms is 3–15 months and depends on plant age and size. During this period phytoplasma cells become distributed systemically in the phloem, aggregating at sites of growth such as young leaves and inflorescences, root tips, and the stem apex just below the growing point. They have been detected in palm embryos, but transmission through seed to seedlings has not been demonstrated.

Outbreaks of lethal yellowing typify the jump-spread pattern of disease occurrence and intensification associated with airborne vectors. One tree may become diseased at a distance of several kilometers from known sources of infection. Then new cases develop near the first one if a vector population is present. *Myndus crudus* feeds on palms but breeds among the roots of grasses. In South Florida these grasses are *Cynodon dactylon* (Bermuda grass), *Paspalum notatum* (Bahia grass), *Stenotaphrum secundatum* (St. Augustine grass), and *Zoysia* sp. (zoysia grass), all of which are widely used for turf. Adult vectors often migrate to palms at night and return to the grass by day. Grasses do not become infected by the palm pathogen, so far as is known.

Many types of palms tolerate or resist the lethal yellowing agent or escape infection and can be grown where susceptible trees have died. The *C. nucifera* cultivars Fiji Dwarf, Panama Tall, Malayan Red Dwarf, and various hybrids have been useful. 'Malayan Dwarf' and the hybrid 'Maypan' seemed tolerant initially in comparative trials but eventually became diseased. Palms recorded as unaffected in outbreak areas include *Areca* species (areca palms), *Acoelorraphe wrightii* (Everglades palm), *Butia capitata* (pindo palm), *Dypsis lutescens* (areca palm), *Phoenix roebelenii* (pygmy date palm), *Ptychosperma elegans* (solitaire palm), *P. macarthurii* (MacArthur palm), *Roystonea* species (royal palms), *Sabal palmetto* (cabbage palmetto), *Syagrus romanzoffiana* (queen palm), *Washingtonia* species (Washington and fan palms), and *Wodyetia bifurcata* (foxtail palm). In some areas antibiotic injections have been used to prolong the lives of susceptible palms while resistant or tolerant ones grow large enough to take their places.

References: 532, 644, 684, 803, 1052, 1141, 1524–1527, 1779–1781, 2294, 2462, 2539, 2540, 2585, 2912, 3259, 4003, 4106, 4230

A–F. Symptoms of lethal yellowing in *Cocos nucifera* (coconut palm). A–D. Foliar symptoms progressing from old to young fronds. C, D. Yellowing of the spear leaf accompanying the decline of lower fronds. E. Necrosis of flower parts, a diagnostic symptom. F. A normal inflorescence (left) produced by a diseased palm after injection of an antibiotic, in contrast with a dead inflorescence (right) that was affected before treatment (A, B, E, F: FL, Apr; C, D: Mexico, Feb).

G. A dying spear leaf on an *Adonidia merrillii* (Christmas palm) affected by lethal yellowing (FL, Apr).

H, I. *Phoenix canariensis* (Canary Island date palms) affected by lethal decline, exhibiting withering and browning of foliage but little yellowing (TX, Apr).

403

Viral Diseases (Plates 201–213)

Overview

Viruses are infectious nucleoprotein particles. Each particle, or *virion*, has a nucleic acid core with a protein coat and, in a few virus groups, a membranous outer envelope. Viruses replicate only within living cells, controlling synthetic processes so that the host produces more virus. Hundreds of plant viruses have been characterized, and these infect thousands of species. Many viruses that occur in agricultural crops, herbaceous ornamentals, and weeds also occur in trees and shrubs. Those in woody plants often cause no visible symptoms; or they may cause symptoms ranging in severity from indistinct foliar markings and slight growth suppression to colorful patterns on leaves; distorted leaves, stems, or fruits; stem pitting; cankers; twig and branch dieback; graft union necrosis; or decline ending in death. A plant may simultaneously harbor two or more viruses or a virus plus another type of pathogen, which complicates both symptoms and diagnosis. Furthermore, some herbicides and some environmental factors, notably nutrient deficiencies or imbalances, induce symptoms similar to those of viral diseases. In the discussions that follow, descriptions of particular diseases or symptoms do not usually face the plates that display them.

Various physiological and biochemical disturbances and morphologic symptoms accompany viral infections. Common physiologic changes include decreased photosynthesis, increased respiration, changes in amounts and balances of plant hormones, and altered membrane permeability leading to sucrose accumulation in leaves and starch accumulation in chloroplasts. Photosynthetic changes reflect either inhibition of chlorophyll synthesis or, in mature cells, chlorophyll breakdown in damaged chloroplasts. Hormonal changes include decreased concentrations of auxin and gibberellins associated with decreased growth, increased abscisic acid concentrations, and production of ethylene. Many virus-plant interactions lead to local or general necrosis. Rapidly developing local necrosis is a plant defense reaction that can halt the spread of viruses and other types of pathogens. If a necrosis-inducing virus moves out of infected cells before they die, larger lesions or general necrosis may develop.

Viral and viruslike symptoms are well known and understood in woody species cultivated for orchard and berry crops, but relatively few viral disorders in forest, shade, and ornamental trees and shrubs have been studied, and even fewer well characterized. Viral diseases in nurseries, orchards, and some plantations cause losses in growth, yield, product quality, and plant or product value. These losses are associated with clonal propagation and monoculture of many kinds of trees and shrubs that have been selected or bred for horticultural characteristics. If a stock plant is virus-infected, its vegetative progeny will also be infected unless rigorous procedures are employed to obtain virus-free propagules. Orchard environments favor the spread of viruses because the planting of one or a few genotypes enhances virus transmission by vectors, pollen, and contacts between susceptible plants. The incidence and impact of viral diseases in most forests and landscapes are unknown. In natural forests selection favors plants that resist or tolerate infection with few adverse effects, and the species mixtures and their reproduction by seed do not favor the buildup of virus vector populations or virulent virus strains.

Virus classification and nomenclature. Viruses are classified according to physical and chemical characteristics of the virions: nucleic acid type and content; virion shape, size, and mass; monopartite or multipartite genome (viral nucleic acid entirely within one virion or distributed among two or more virions); presence or absence of a membranous envelope outside the protein coat; and other characters. Plant viruses are classified in about 67 genera, of which about 44 are arranged in 14 families. The remaining genera are as yet unassigned. Viruses representing about 30 genera are known to occur naturally in woody plants, although the commonly reported viruses represent fewer than 10 genera. The nucleic acid of most plant viruses is single-stranded RNA, but double-stranded RNA viruses and both single- and double-stranded DNA viruses also occur in plants. Individual virions are too small to observe with light microscopes but can be visualized by electron microscopy. Virus-induced structures called *inclusions* develop within cells infected by some viruses. Inclusions are generally proteinaceous or consist of aggregated virions. Virions range in shape from polyhedra to rigid or flexuous rods (Fig. 201); those with envelopes appear bacillar or spherical. The genomes of plant viruses

typically comprise only a few genes. These encode structural proteins that form the coat and nonstructural proteins involved in replication, local and long-distance movement within plants, and vector relationships. Determinants of host range and pathogenicity of plant viruses are still poorly understood. Latin nomenclature is not in general use for viruses. Instead, virologists have agreed on a formal set of coined names for virus genera and names of colloquial origin for particular viruses. For example, the genus *Carlavirus* comprises viral species related to *Carnation latent virus*. *Poplar mosaic virus* is a carlavirus that causes the disease known as poplar mosaic. Each formally named virus has an approved abbreviation; for example, PopMV for *Poplar mosaic virus*. Viral order, family, genus, and species names given official standing by taxonomic committees are italicized.

Viroids. Viroids are noncellular, submicroscopic pathogens that infect plants systemically. Each viroid is a small closed loop of infectious RNA without associated protein. Some viroid diseases are important in trees, but none is known to be significant in forest or landscape trees or shrubs in North America. Viroid diseases are not considered further in this book.

Virus host ranges and transmission. The natural host ranges of plant viruses vary widely and are determined in part by modes of transmission. Many viruses that can be transmitted by plant-to-plant contact or by infection of roots growing in contaminated soil have

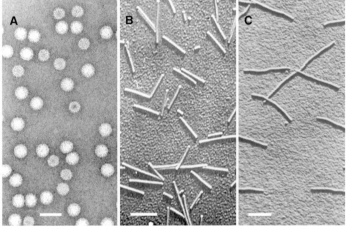

Figure 201. Electron micrographs of three forms of plant virus particles. A. Isometric. B. Rigid rods. C. Flexuous rods. Scale bars = 50 nm in A, 200 nm in B and C.

Some diseases caused by or associated with the nepoviruses *Tomato ringspot virus* and *Tobacco ringspot virus* (discussion, p. 408).

A–D. Stem pitting of *Prunus persica* (peach), caused by *Tomato ringspot virus*. A. Chlorotic foliage on a declining tree (PA, Oct). B–D. Xylem symptoms on rootstocks below the debarked graft unions of three trees. B. Mild stem pitting (MI, summer). C. Union necrosis and severe stem pitting (PA, Aug). D. Severe stem pitting resulting from cambial necrosis (NY, summer).

E–H. Brown-line disease of *Malus pumila* (apple) on MM106 rootstock, caused by *Tomato ringspot virus*. E. Necrosis at a graft union exposed by cutting a diagnostic "window" in the bark of a declining tree (MI, Jul). F. A debarked graft union with a line of small voids where the union was incomplete. Brown color of the wood surface is due to oxidation after the specimen was collected and is unrelated to the disease (NY, summer). G, H. Stem breakage at graft unions that were defective because of cambial and xylem necrosis (NY, Oct; PA, summer).

I, J. Yellow-net disease of *Forsythia* sp. (discussion, p. 410). I. An affected plant with stunted new shoots amid dead and dying canes of the previous year. Foliage of the entire plant has a yellowish cast due to yellow-net symptoms. The dieback, presumed to have been caused by freezing in late winter, did not occur in nearby plants with normal foliage. J. A diseased shoot (NY, Jul). *Tobacco ringspot virus* has been associated with yellow net in eastern USA. *Arabis mosaic virus* has been associated with similar symptoms in Europe.

Photo credits: A, C, H—K. D. Hickey; B, E—D. C. Ramsdell

405

broad natural host ranges. Most plant viruses that are naturally transmitted only by vectors have host ranges that are determined by vectors' feeding preferences. In diagnostic tests, however, most viruses can infect an array of plants more diverse than those found naturally infected. This fact is applied in diagnostic practice by attempting to transmit disease-inducing agents from affected plants (or from nonsymptomatic plants being screened for latent viral infection) to various species of indicator plants.

Plant viruses that cause significant diseases become distributed systemically in their hosts, but this distribution may be erratic and may vary seasonally. Most plant viruses invade phloem and parenchyma, but some are limited to phloem. Viruses are transported in commerce within infected, often symptomless plants or seeds and are transmitted in nature by various insects (mainly those that suck plant sap); certain mites; certain nematodes, protists, and fungi; pollen; grafts; and contact. The most common modes of transmission differ among virus genera. A virus that is vectored by one group of organisms (e.g., aphids) is generally not transmitted by other kinds of organisms. Transmission by a vector may be persistent, semipersistent, or nonpersistent. These terms refer to the time during which a vector, after acquiring a virus, retains and is able to transmit it. Persistent transmission implies that a vector remains viruliferous (able to transmit) until it dies. Some persistently transmitted plant viruses in the family *Rhabdoviridae* (rhabdoviruses) and the genera *Tenuivirus* and *Tospovirus* replicate in their vectors. Some viruses are chemically stable, remaining infectious in water, soil, or plant debris after release from living and decaying host parts; other viruses lose infectivity soon after separation from living tissue.

Plant virus diagnosis. Viral and viruslike symptoms in woody species cultivated for orchard and berry crops are better known and understood than are those in forest, shade, and ornamental trees and shrubs. Symptoms caused by different viruses may be similar, two or more viruses may occur in a given plant, and many nonviral agents or conditions cause symptoms that mimic those caused by viruses. Therefore, diagnostic shortcuts that are useful in agricultural situations tend to be inappropriate for landscape and forest plants. Reliable diagnosis of a newly noticed viruslike syndrome requires multiple procedures, but the importance of such a disorder in a landscape plant is seldom sufficient to justify specialists' time and materials. Full diagnosis is appropriate for disease outbreaks in crops and orchards and occasionally in plantations. The appropriate selection and order of diagnostic procedures vary with circumstance.

• *Tests for known viruses or virus groups.* If plants have symptoms like those of a well-characterized viral disease, then a diagnostician is likely to skip directly to immunological tests or to polymerase chain reaction (PCR)-based diagnosis. These procedures are limited by molecular specificity, however. Although some antibodies and PCR primers are useful for detecting members of virus genera or families, the antiserum or PCR primers for a particular virus may fail to detect unrelated viruses or other types of pathogens. If diseased plants only have symptoms, such as yellowing or stunting, that could be caused by various types of pathogens or abiotic agents, then procedures are needed that can potentially implicate any of diverse causal agents.

• *Tests for unknown potential pathogens.* Virions can often be detected by electron-microscopic examination of diseased tissue or sap squeezed from it. Plants must also be examined for signs of fungi or bacteria that could cause the symptoms. Tests for culturable and unculturable organisms associated with symptoms may be needed.

• *Assessment of possible abiotic factors that could cause or aggravate the symptoms.* Plants' circumstances and history must be considered. The remainder of this discussion deals only with viruses and viruslike agents.

• *Tests for transmissibility of symptoms from diseased to healthy plants.* Tests for graft transmissibility are useful in angiosperms. Most plant viruses and various other pathogens can be transmitted by grafting, but the incubation period (i.e., time from inoculation to appearance of symptoms) is months to years in some tree-virus interactions. Diagnostic transmission of many viruses is possible by rubbing the leaves of herbaceous indicator plants with juice from pulped diseased leaves. If symptoms develop in the assay plants while control plants remain normal, the result furnishes preliminary

information on host range of the pathogen and diagnostic symptoms in indicators. If the original symptoms are then reproduced by inoculating healthy plants of the species in which the disease was found with juice from indicator plants, this result verifies that indicator plants had expressed symptoms caused by the agent(s) associated with the original problem. Some viruses are not transmissible in sap, however. For these, grafts or insects may be used to attempt transmission.

• *Detection, extraction, and purification of virus.* Nonspecific detection of many plant viruses is possible not only by electron microscopy but also by double-stranded RNA assay because dsRNA is an intermediate product in the replication of many viruses. Extraction and purification involve grinding virus-infected tissue in buffer, diluting the suspension, and using differential centrifugation to separate virus particles from plant debris and concentrate them. Concentrated virus particles can readily be visualized with an electron microscope.

• *Identification.* Physical, chemical, and biological properties are determined. Concentrated preparations are tested to determine nucleic acid type and content of particles, whether the genome is entire or divided, base sequence(s) of the nucleic acid(s), and particle mass and density. Virion size and form are ascertained or verified by electron microscopy. Biological properties include retention or loss of infectivity in heated or aged crude plant juice, dilution end point of infectivity in plant sap, host range and symptoms induced in diagnostic plant species, and reactions with antisera to known viruses. Both immunological and PCR-based tests are highly useful for detecting and identifying particular viruses.

• *Inoculation of healthy plants with a purified virus.* Reproduction of the original symptoms in an experiment with appropriate controls is strong evidence for viral cause of a syndrome. A negative result indicates either that a virus associated with the original symptoms did not cause them or conditions were not conducive for symptom expression, or that the virus is not mechanically transmissible.

The following discussions are organized by virus group because most of the well-known viruses that infect woody plants have broad host ranges or at least infect diverse plants. The discussions contain lists of plants affected by particular viruses. The lists include symptoms demonstrated to be caused by the viruses, as well as symptoms observed in plants in which particular viruses were detected whose causal roles were uncertain. Viral diseases of woody plants arranged by plant type are presented in references 792, 1967, and 2861.

References: 554, 555, 777, 792, 1170, 1171, 1191, 1521, 1812, 1863, 1967, 2191, 2514, 2790, 2861, 2905, 3070, 3247

Diseases Caused by Nepoviruses

The name nepovirus indicates nematode transmission and polyhedral particles. Nepoviruses are globally distributed in woody and herbaceous plants, mainly in temperate zones. The genus *Nepovirus* (*Comoviridae*) comprises about 30 species of single-stranded RNA viruses whose genomes are divided between two similar icosahedral particles near 28 nm in diameter. The RNAs of

---

Symptoms in *Fraxinus* (ash) associated with *Tobacco ringspot virus* (TRSV) and unknown viruses (discussion, p. 410).

A–D. Ringspots and chlorotic spots and mottling on leaves of *F. pennsylvanica* (green ash) (A, B) and *F. americana* (white ash) (C, D). These symptoms have been linked to TRSV in inoculated saplings and naturally infected trees. Chlorotic spots on *F. americana* sometimes turn red in late summer (D) or remain yellow after leaves turn red (NY, summer).

E–H. Symptoms caused by unidentified viruses in *F. americana*. E, F. Leaf distortion and yellow color due to suppressed chlorophyll formation near leaflet bases and along midveins. G. Diffuse line patterns and veinal chlorosis. H. Vein-associated yellow line patterns. Viruses associated with these symptoms in northeastern USA include TRSV, *Tomato ringspot virus*, and *Tobacco mosaic virus* (NY, summer).

I. Chlorotic spots and yellow line patterns in deformed leaves of a *F. velutina* (velvet ash) sapling grafted with bark from *F. americana*. The virus(es) were not identified (NY, summer).

Photo credit: A—J. D. Castello

407

both particles are required for infection and replication, which occurs in host cytoplasm. The virions are heat-stable, with temperatures above 55°C required for thermal inactivation. Their coat proteins are strong antigens, and immunological tests for several nepoviruses are available. Nepoviruses are sap-transmissible to indicator plants.

Foliar symptoms induced by nepoviruses include ringspots, mottling, chlorotic spots, and vein yellowing. Some nepoviruses cause localized cambial necrosis or dramatically reduced cambial activity in mainstems and main roots of highly sensitive woody plants, resulting in cankers or pits. Several nepoviruses induce within host cells inclusion bodies consisting of membrane-bound tubules or vesicles containing rows of viruslike particles.

Important nepoviruses with broad host ranges in woody plants in North America are *Tomato ringspot virus*, *Tobacco ringspot virus*, and *Cherry leafroll virus*. In Europe, important nepoviruses in woody plants are *Arabis mosaic virus*, *Cherry leafroll virus*, *Raspberry ringspot virus*, and *Tomato black ring virus*.

About a dozen nepoviruses have known nematode vectors, although nematodes do not transmit all nepoviruses. Some are also carried on pollen and/or in seed. *Cherry leafroll virus* is transmitted primarily by pollen and also with seed, but not by nematodes. *Blackcurrant reversion associated virus* is transmitted by an eriophyid mite. Nepoviruses vectored by *Xiphinema* species (dagger nematodes; Fig. 216) are common in North America. The other nematode vectors of nepoviruses are species of *Longidorus* and *Paralongidorus* (needle nematodes), which along with *Xiphinema* are in the family Longidoridae. These nematodes occur globally in the temperate zones and in some tropical regions. The mode of virus transmission by nematodes is semipersistent. Virus particles are ingested as nematodes feed on succulent root tips. The particles adhere to the surface of the food canal within the stylet of *Longidorus* or the food canal in the esophagus of *Xiphinema*, and are released slowly during subsequent feeding. Both adults and juvenile stages of *Longidorus* and *Xiphinema* transmit viruses. The adults live a year or longer, depending on species, and may remain infective throughout their lives. Juvenile nematodes lose virus particles along with the stylet when they molt. Longidorid nematodes have broad host ranges, which accounts for the presence of nepoviruses in many weeds and wild plants as well as in ornamental and crop plants.
References: 537, 554, 777, 792, 1191, 1471, 1863, 2191, 2896, 3070, 4212

Tomato ringspot virus (ToRSV) is widespread in many herbaceous and woody plants in North America, from where it has apparently been carried in plants to all other vegetated continents. It comprises many strains that differ in serological properties and symptoms induced. Plants in at least 15 families are natural hosts. ToRSV is transmitted by at least four species of *Xiphinema*: *X. americanum*, *X. bricolensis*, *X. californicum*, and *X. rivesi*, all of which were at one time grouped under the name *X. americanum*. Seed transmission of ToRSV occurs in herbaceous plants but is unreported for woody plants except in *Rubus idaeus* (European red raspberry). This virus also spreads with pollen in some herbaceous plants.

ToRSV is associated with the following symptoms in woody plants: leaf mottling of *Daphne mezureum* (February daphne); chlorotic rings, concentric ringspots, and oak-leaf patterns on *Euonymus fortunei* (climbing euonymus, wintercreeper) and *E. kiautschovicus* (creeping strawberry-bush); chlorotic spots, blotches, and line patterns in *Fraxinus americana* (white ash); chlorosis with green blotches and stunting of *Hydrangea macrophylla* (French hydrangea); decline because of necrosis at graft unions (brown-line disease) in *Malus* (apple) and *Prunus* (cherry, peach, prune) (Plate 201); stem pitting and decline of cherry and peach (Plate 201); yellow bud mosaic of peach; mosaic in *Ribes rubrum* (red currant) and *Rosa* (rose); ringspot and decline of *Rubus* (brambles); chlorosis, ringspots, line patterns, and dark green mottling on leaves of *Sambucus canadensis* (American elderberry); and stunting, chlorotic mottle, veinal yellowing, and decline of *Vitis vinifera* (grapevine). ToRSV was detected in nonsymptomatic *Cornus florida* (flowering dogwood) in South Carolina. Foliar symptoms in natural hosts often disappear soon after infection. Naturally infected herbaceous plants

provide reservoirs from which the virus can be transmitted to trees and shrubs planted in *Xiphinema*-infested soils.

ToRSV is a major pathogen in some stone-fruit orchards. For example, the incidence of ToRSV infection in 21 prune orchards in Michigan where trees were declining with brown-line disease averaged 28%, and the incidence of infection in 10 peach orchards in West Virginia where stem pitting was prevalent averaged 48%. ToRSV-induced necrosis at graft unions of fruit trees is due to high sensitivity of certain rootstocks. Two nepoviruses, ToRSV and *Tobacco ringspot virus* (TRSV), were detected in declining *Fraxinus americana* in New York State, as noted with the discussion of TRSV, but their role in slow growth and decline of ash remains unknown.
References: 507, 537, 554, 777, 792, 1158, 1368, 1638, 1655, 1784, 2121, 2670, 3162, 3214, 3236, 3332, 3983, 4104, 4105, 4212

Tobacco ringspot virus (TRSV) is widespread in woody and herbaceous plants in North America and also occurs in Europe, Asia, and Australasia. This virus comprises many strains that differ in serological properties and symptoms induced. Plants in at least 16 families are naturally infected, and representatives of many more families are susceptible by inoculation. The virus is transmitted in nature by *Xiphinema* species (dagger nematodes; Fig. 216), and also in seeds and pollen. Seed transmission is documented only for herbaceous plants, however. TRSV can be transmitted mechanically in plant juice to assay plants and can be detected in all parts of its herbaceous hosts, but it is not readily transmitted by plant-to-plant contact. Natural spread in the USA and Canada has been documented where *Xiphinema americanum*, *X. californicum*, *X. rivesi*, and related species are common, and the vector role of each of these species has been demonstrated.

TRSV has been associated with or may cause the following disorders in woody plants: twig dieback and mild veinal chlorosis of *Cornus florida* (flowering dogwood); yellow net of *Forsythia* (Plate 201); chlorosis, crinkled leaves, and stunting of *Hydrangea macrophylla* (French hydrangea); chlorotic to yellow leaf spots and irregular leaf margins in *Ilex crenata* (Japanese holly); yellow blotches, leaf drop, stunting, and dieback in *Jasminum officinale* (poet's jasmine); necrosis at graft unions (brown line disease) in *Malus pumila* (apple); mosaic in *Rosa* (rose); mottling, mosaic, line patterns, distortion, and stunting of leaves of *Rubus* species (brambles); ringspots, line patterns, and dark green mottling on leaves of *Sambucus canadensis* (American elderberry); ring patterns in *Skimmia*; necrotic ringspot of *Vaccinium corymbosum* (highbush blueberry); and chlorotic rings, mottling, leaf distortion, and general

---

Diseases caused by nepoviruses.

A–C, F, G. Two diseases caused by *Cherry leafroll virus* (CLRV). A–C. Blackline disease of *Juglans regia* (Persian walnut) (discussion, p. 410). A. Dieback and light green foliage bunched at the ends of slowly growing twigs. B. The butt of a diseased tree on *J. hindsii* (northern California black walnut) understock. Profuse sprouting of the understock is a symptom of the disorder. Bark at the graft union has been removed to reveal a transverse black line, the principal diagnostic symptom. C. Close view of the black line, a narrow band of necrotic cambium and phloem at the graft union (CA, Jul). F, G. Mottling, distortion, and chlorotic spots on leaves of *Ulmus americana* (American elm) affected by elm mosaic (discussion, p. 412) (NY, Jun).

D, E. Cherry raspleaf, caused by *Cherry raspleaf virus*, in *Prunus avium* 'Lambert' (sweet cherry): distorted leaves with enations (epidermal outgrowths), typical of the disease (OR, Jul).

H–M. Slow growth and foliar markings associated with *Dogwood mosaic virus* (discussion, p. 412) in *Cornus florida* (flowering dogwood). H. Undersized, pale green leaves give a severely diseased landscape tree an unusual pale color and sparse-appearing canopy. I. Abnormally narrow, twisted bracts. J–M. Symptomatic leaves in comparison with normal ones. J. Narrow leaves with edges upturned and wavy. K. Chlorotic spots and vein-associated line patterns. L. Chlorosis and bleaching of leaf tips. M. Normal leaves (SC, spring).

Photo credits: K, L—O. W. Barnett; K by permission of the American Phytopathological Society

409

stunting of *Vitis* (grapevine). Foliar symptoms in natural hosts tend to be transient.

TRSV has also been detected in diverse woody plants that either lacked symptoms or for which the symptoms were not described: *Cornus racemosa* (gray dogwood), *C. sericea* (red-osier dogwood), *Crataegus* sp. (hawthorn), *Daphne* species, *Diervilla* sp. (bush-honeysuckle), *Hydrangea paniculata* (panicled hydrangea), *Jasminum mesnyi* (Japanese jasmine), *J. nudiflorum* (winter jasmine), *Populus tremuloides* (trembling aspen), *Potentilla* sp. (cinquefoil), *Salix nigra* (black willow), *Ulmus americana* (American elm), and *Vitis aestivalis* (bunch grape).

Woody plants artificially inoculated with TRSV exhibited the following symptoms: mild mosaic in *Acer ginnala* (Amur maple), mosaic and abnormally narrow leaves in *Acer negundo* (box-elder), mild mosaic in *Hibiscus syriacus* (rose-of-sharon), mosaic and stunting in *Prunus avium* (sweet cherry), ringspots and line patterns on *Sambucus canadensis* (American elder), and mosaic and stunting in *Ulmus americana* (American elm).

Plate 202 features symptoms associated with TRSV in *Fraxinus*. Foliar markings in *F. americana* (white ash) or *F. pennsylvanica* (green ash) include faint chlorotic spots and rings, irregular chlorotic sectors and line patterns along veins, mosaic, mottling, red spots and rings in late summer, premature autumn color, and green spots and rings in chlorotic or red leaves in autumn. Symptoms are variable and may be absent from most leaves on a diseased tree. The symptoms are presumed to vary with environment, TRSV strains, host genotype, and presence of additional viruses, notably *Tobacco mosaic virus* (TMV). Both TMV and TRSV have been detected in *F. pennsylvanica* (green ash) and *F. holotricha* 'Moraine,' associated in the latter with chlorosis, mosaic, and puckering of leaflets. Partially purified TRSV inoculated to small *F. americana* and *F. pennsylvanica* suppressed the growth of *F. americana* and caused various combinations of chlorotic spots and blotches, mosaic, ringspots, mottling, veinal yellowing, necrotic or light green line patterns along secondary veins, crinkling, irregular margins on leaflets, and abnormally narrow leaflets in both hosts.

TRSV, ToRSV, and TMV were often detected in *F. americana* during research on the causes of slow growth and dieback of that species in northeastern USA, but the work did not progress far enough to either implicate or rule out virus(es) as significant causal factors in the decline. TRSV and TMV were also detected in seeds of *F. americana*, but the possible seed transmission of these viruses in *Fraxinus* has not been studied.

Yellow-net disorders of *Forsythia* species (Plate 201) are apparently caused by either TRSV or *Arabis mosaic virus* (ArMV); the former nepovirus has been documented in the USA and the latter in Germany. The term *yellow net* connotes yellowing along leaf veins. The disease in the USA is characterized by mild to dramatic yellowing along veins and veinlets. Symptoms diminish in intensity from older to younger leaves on a shoot, and leaves near the tip may appear normal. Yellow net is associated with low vigor and cane dieback that usually occurs during winter. Canes that die during the growing season first produce stunted, chlorotic shoots that wilt and turn brown. Generally all stems of a plant have symptoms, although their nature and intensity vary. A strain of TRSV from *Forsythia* with yellow net in New Jersey induced diagnostic symptoms in herbaceous indicator plants and induced yellow net in previously healthy *Forsythia* inoculated with sap from a diseased indicator plant. TRSV has also been detected in symptomless *Forsythia ovata*. Other viruses reported to occur in *Forsythia*, all in Europe, include *Alfalfa mosaic virus* (genus *Alfamovirus*); *Cucumber mosaic virus* (genus *Cucumovirus*); the nepoviruses *Cherry leafroll virus*, *Raspberry ringspot virus*, and *Tomato black ring virus*; and *Tobacco rattle virus* (genus *Tobravirus*). Symptoms caused by these agents are unreported.

References: 62, 537, 554, 672, 777, 792, 1368, 1633, 1636, 1641, 2224, 2735, 3236, 3451, 3587, 4024, 4212, 4244, 4245

Cherry leafroll virus (CLRV) is among the more common viruses of *Prunus* (cherries and other stone-fruit trees) in orchards and home plantings. It is widespread in Eurasia and North America and occurs also in Australia and New Zealand. Its natural woody hosts represent at least 10 families, and the experimental host range comprises more than 36 families. CLRV induces within host cells inclusion bodies as noted for nepoviruses in general. In some hosts it causes cellular changes such as projections from cell walls and electron-dense globular deposits and poorly defined lamellae in chloroplasts. Some strains of CLRV cause severe symptoms in a given plant, others cause mild ones. Strains from different hosts also differ in serological properties.

Most natural spread of CLRV occurs through seeds and pollen. Transmission rates through seed range from less than 1% to 100%, depending on host and virus strain. Three species of *Xiphinema* (dagger nematodes) are capable of transmitting European strains of the virus that infect *Prunus*, but attempts to demonstrate nematode transmission of other strains have been largely unsuccessful. CLRV is transmissible in sap to many herbaceous plants, among which *Chenopodium*, *Cucumis* (cucumber), *Nicotiana* (tobacco), and *Phaseolus* (bean) are used as diagnostic indicators.

This virus in most of its natural hosts causes either no obvious symptoms or no great damage. Curled leaves or leaf markings such as ringspot, chlorotic spots, or yellow veins (yellow net) are often the only symptoms. In *Prunus avium* (sweet cherry), however, leaves roll upward from margins and trees become stunted and eventually die. Symptoms caused by or associated with this virus in other woody plants include chlorotic spots and mottle evolving to yellow line patterns, rings, and patches in various *Betula* (birch) species; chlorotic mosaic in *Celtis australis* (European hackberry); slow growth, ringspot, mosaic, and foliar deformities in *Cornus florida* (flowering dogwood), discussed below; yellow line patterns in *Cornus sanguinea* (blood-twig dogwood); leaf mottling, stunting, and dieback in *Euonymus europaeus* (European euonymus); chlorotic mottling, mosaic-like flecking, and leaf roll in *Fagus sylvatica* (European beech); mosaic in *Fraxinus americana* (white ash); mosaic, leaf deformity, and ultrastructural aberrations in leaves of *F. excelsior* (European ash); blackline and associated symptoms in *Juglans regia* (Persian walnut), discussed below; yellow spots on leaves of *Ptelea trifoliata* (hop-tree); line patterns on leaves, stunting, and death of "blackberry" (perhaps *Rubus fruticosus*); yellow rings, arcs, oak-leaf patterns, ringspots, and yellow net on *Sambucus nigra* (European elder); chlorotic blotches and necrotic flecks or rings in leaves of *S. canadensis* (American elder) and *S. racemosa* (European red elder); chlorotic ringspots in *Syringa* (lilac); and chlorotic mosaic, ring pattern, and dieback in *Ulmus americana* (American elm), discussed below. CLRV also occurs in *Forsythia*, *Ligustrum* sp. (privet), and *Olea europaea* (olive), but the symptoms it may cause in these plants have not been separated from symptoms caused by other viruses. CLRV-associated diseases of *Cornus*, *Euonymus*, *Juglans*, *Prunus*, *Sambucus*, and *Ulmus* are documented in North America.

The most important disease known to be caused by CLRV in North America is blackline of *Juglans regia* (Plate 203). Blackline, which occurs also in Europe, is characterized by decline and death of trees grafted on rootstocks of certain other *Juglans* species, notably *J. hindsii* (northern California black or Hinds walnut), *J. regia* × *hindsii* ('Paradox' hybrid walnut), *J. hindsii* × *nigra* ('Royal' hybrid walnut), *Pterocarya stenoptera* (Chinese wingnut), and other hybrids with *J. regia* as a parent. *J. regia* on its own roots or on rootstocks of the same species tolerates CLRV infection as noted below. The first indication of blackline disease is gradual loss of vigor and early shedding of leaves in autumn. Symptoms typically appear on one limb and gradually involve others. Slow twig growth leads to bunching of leaves at branch tips and a general openness of the

Diseases caused by carlaviruses: poplar mosaic and lilac mottle (discussion, p. 414).

A–F. Poplar mosaic, caused by *Poplar mosaic virus*. A–E. Symptoms in *Populus deltoides* (eastern cottonwood) saplings. A. Retarded growth and premature loss of apical dominance. B. Dwarfed and distorted terminal leaves and absence of leaves at subterminal nodes. C, D. Chlorotic spots and abnormal vein reddening. E. Necrosis on major veins and the petiole, also abnormal bending of the midrib, causing leaf distortion (ON, Jul). F. Vein-associated chlorosis in 'Dorskamp' hybrid poplar (ON, Jul).

G. Lilac mottle, caused by *Lilac mottle virus*, in *Syringa vulgaris* (common lilac).

Photo credit: G—H. E. Waterworth

411

canopy. Twigs and branches die back, and the entire top of a tree may be dead within 2–5 years. Profuse sprouts grow from the rootstocks of declining trees. Decline coincides with development of a dark brown or black, corky band of dead cambial and phloem tissue at the graft union. This begins to form on one side of the trunk and gradually girdles it. The dead tissue appears as a transverse line 3–6 mm wide in trees on *J. hindsii* rootstock but may extend several centimeters below the graft union on 'Paradox' rootstock. Tissue of the black line extends slightly into sapwood because cambial activity and differentiation of wood continue for a time on both sides of the girdle.

Blackline symptoms are usually delayed until 10–20 years after grafting because most *J. regia* scions grown from healthy seeds remain free from CLRV until they are old enough to produce flowers. After transmission in pollen to healthy trees, CLRV slowly becomes systemic. On reaching a graft union, often several years after the onset of infection, the virus induces such rapid necrosis of rootstock cells that they die before the virus can move beyond them. This hypersensitive response halts downward spread but girdles the tree. The roots remain uninfected. Hypersensitivity to CLRV is a dominant monogenic trait in several *Juglans* species. Prospects for control of CLRV in *Juglans* are linked to further selection of tolerant rootstocks or transfer to *J. regia* of genes encoding hypersensitivity to CLRV. Blackline disease in walnut serves as a model for understanding delayed expression of apparent graft incompatibility in other orchard trees and in shade trees and woody ornamentals.

CLRV is readily transmitted in *J. regia* seeds, with reported incidence of seed infection as high as 32%. The virus suppresses growth and causes chlorotic spots, chlorotic to necrotic rings, yellow blotches, and vein necrosis on undersides of leaflets of occasional trees. The majority of infected trees at a given time are symptomless, however.

Infection of *Cornus florida* by CLRV has been documented in Maryland and South Carolina. Trees harboring the virus may grow slowly and develop twisted bracts; transient chlorotic spots, ringspots, and arcs along the major veins of leaves; abnormally narrow leaves with uprolled wavy margins and light green color; and general thinness of the canopy. Chlorotic spots apparent in spring may disappear during summer. Not all of these symptoms develop on a given tree, and their usefulness for diagnostis of CLRV is not established. *C. florida* seedlings inoculated with CLRV developed chlorotic flecks and ringspots and veinal necrosis. Other viruses that infect flowering dogwood (arabis mosaic, broad bean wilt, cucumber mosaic, tobacco ringspot, and tomato ringspot viruses known in North America) may contribute to these symptoms. *C. florida* infected with CLRV or other viruses often has no symptoms.

Elm mosaic (Plate 203), also caused by a strain of CLRV, is widespread in North America, as indicated by records from Nova Scotia, Wisconsin, Oklahoma, and Virginia. It has been reported only for *Ulmus americana* (American elm). Leaves of affected trees are usually abnormally small and stiff, and some are distorted. Yellow mottling or mosaic, ringspots, and abnormal ridges and wrinkles (rugosity) also appear. Enations (ragged elongate projections as shown on *Prunus* in Plate 203) have been observed along the midvein and between lateral veins on leaves of the cultivar Moline. Twigs may appear spindly and may be produced in abnormal abundance, giving the tree a bushy appearance. Diseased trees lose vigor, and some sustain twig dieback. A tree may persist for decades in this condition. In one test, 1–3.5% of seedlings from diseased *U. americana* seed lots developed mosaic. CLRV can also be transmitted among elms by grafting. The elm strain of CLRV has physical properties similar to those of strains from other hosts but differs from them serologically. Elm mosaic is at times associated with zonate canker on *U. americana*. The agent that causes zonate canker is unknown but is assumed to be viral because it may occur in and be transmitted from symptomless trees.

References: 554, 777, 791, 792, 795, 796, 1157, 1208, 1454, 1455, 1960, 2191, 2244, 2490, 2673, 2674, 2896, 3236, 3360, 3520, 3939, 4242, 4399, 4528

Cherry rasp leaf virus (CRLV) causes ragged or rasplike projections of green tissue (enations) on the undersides of leaves of *Prunus avium* (sweet cherry; Plate 203), *P. mahaleb* (mahaleb cherry), and *P. persica* (peach). This nepovirus is the most common cause of

enations on cherry in North America. It is widespread in western USA and occurs also in Australasia, China, and South Africa. Dark green, rounded to pointed enations grow from the undersides of twisted, abnormally narrow leaves, especially along the midvein and in rows between major lateral veins. Upper surfaces of leaves are rough and have pits that correspond to the enations below. Symptoms tend to be restricted to particular branches low on a tree, and these are abnormally sensitive to frost. Many affected spurs and branches die, giving the canopy an open appearance and reducing fruit yield. Some affected trees undergo general decline. CRLV also affects *Malus pumila* (apple), causing enations and flattened fruit. Other plants in several families are susceptible by inoculation. The virus is transmitted by the dagger nematode, *Xiphinema americanum*, and also is transmitted in seeds of some herbaceous hosts. It has been detected in cherry pollen. Weeds including *Balsamorhiza* (arrowleaf balsamroot), *Plantago* (plantain), and *Taraxacum* (dandelion) have been found asymptomatically infected and may serve as natural reservoirs of CRLV.

References: 430, 554, 777, 1480, 1961, 2881, 3070, 4184

Other nepoviruses. *Dogwood mosaic virus* affects *Cornus florida* (Plate 203). It was discovered in South Carolina, associated with mild yellow mosaic and chlorosis in some leaves. Its geographic distribution is unknown. The virus is transmissible by grafting and mechanically in sap to indicator plants; herbaceous plants in 10 families proved susceptible. Other means of transmission have not been demonstrated.

Arabis mosaic virus (ArMV) and *Tomato black ring virus* (TBRV) occur around the world, except perhaps in South America and Australasia, respectively. ArMV has been found in at least 20 genera of woody plants used as ornamentals: *Acer, Buxus, Caryopteris, Chamaecyparis, Chionanthus, Daphne, Forsythia, Fraxinus, Hedera, Hydrangea, Jasminum, Laburnum, Ligustrum, Populus, Ptelea, Rosa, Sambucus, Spiraea, Staphylea,* and *Syringa*. It is transmitted by at least three *Xiphinema* species and also in seeds. It has been identified in *Vitis* (grapevine) in Missouri but is otherwise unreported in North America. In England ArMV was isolated from *Fraxinus americana* (white ash) with puckered and twisted leaves and chlorotic blotches and ring patterns. It was demonstrated to cause such symptoms in *F. americana, F. excelsior* (European ash), and *F. sieboldiana*, and to suppress growth of *F. americana* on *F. excelsior* rootstock. ArMV is also associated with mottling, chlorotic vein banding, and line patterns in *F. excelsior*. It causes a yellow-net syndrome of *Forsythia intermedia* similar to that associated with *Tobacco ringspot virus* in North America. Tissue along leaf veins turns yellow while areas between veins remain green. ArMV has also been found in *F. europaea* with bright chlorotic leaves. Yellow net caused by ArMV has been transmitted to healthy *Forsythia* by grafting, and a forsythia strain of ArMV from an indicator plant caused yellow net in artificially inoculated *Forsythia*. Various

---

Diseases caused by ilarviruses (discussion, p. 414).

A–C. Sour cherry yellows, caused by *Prune dwarf virus*, in *Prunus cerasus* 'Montmorency' (sour cherry). A. Sparse foliage due to premature abscission, and sparse fruit set due to subnormal number of fruiting spurs. B. The tip of a diseased branch with only one fruit spur on the previous year's wood, and several leaves missing from the current season's shoot before harvest time (OR, Jul). C. Typical yellowing of leaves (NY, summer).

D. Chlorotic to necrotic ringspots in a *P. avium* (mazzard cherry) leaf infected with *Prunus necrotic ringspot virus* (PNRSV) (OR, Jul).

E, F. Symptoms caused by *Apple mosaic virus* in *Malus pumila* (apple). E. Vein clearing. F. Bleached spots and line patterns (NY, summer).

G. Chlorotic to white line patterns in *Corylus avellana* (European filbert) caused by *Apple mosaic virus* (OR, Jul).

H–K. Rose mosaic complex in hybrid roses (discussion, p. 416). H, I. Vein yellowing (yellow net), line patterns and puckering of leaflets possibly caused by *Prunus necrotic ringspot virus* (NY, Jun). J. Mosaic possibly caused by *Apple mosaic virus* (CA, spring). K. Chlorotic ringspots and mottle, similar to symptoms of rose ring pattern (NY, Jul).

symptoms have appeared in herbaceous indicator plants inoculated mechanically or with dodder. ArMV also causes a yellow-net syndrome in *Sambucus* (elderberry). TBRV has been found in at least 15 genera of woody plants, mainly in Europe: *Abeliophyllum, Celastrus, Daphne, Fagus, Forestiera, Forsythia, Fraxinus, Hydrangea, Ligustrum, Picea, Populus, Robinia, Sambucus, Staphylea,* and *Syringa.* It is sometimes associated with foliar symptoms but more often is not. Reported symptoms include chlorosis, chlorotic patches, mottling, and veinal yellowing.

The nepovirus *Hibiscus latent ringspot virus* occurs in *Hibiscus rosa-sinensis* (Chinese hibiscus) in Europe and Africa, sometimes associated with chlorosis.
References: 221, 792, 3451

Diseases Caused by Carlaviruses

About 50 viruses are assigned to the genus *Carlavirus,* which is named for *Carnation latent virus.* A dozen or more carlaviruses have been identified in woody plants. Their particles are filamentous, straight or flexuous, and measure 510–720 × 11–18 nm. Each particle nominally contains the entire viral genome of single-stranded RNA. They replicate in cytoplasm. Carlaviruses are transmitted nonpersistently or semipersistently by insects, mainly aphids, although whiteflies transmit at least two species. Most of these viruses are transmissible by inoculation with sap from diseased plants but are not readily transmitted by plant-to-plant contact. Most carlaviruses seem not to be transmitted in seeds or pollen. Carlarviruses of woody plants include *Blueberry scorch virus, Daphne S virus, Elderberry symptomless virus, Euonymus mosaic virus, Fuchsia latent virus, Honeysuckle latent virus, Lilac mottle virus, Lilac ringspot virus, Mulberry latent virus,* and *Poplar mosaic virus,* among others.

Poplar mosaic virus (PopMV) (Plate 204) was first discovered in Europe and later found in scattered locations around the world except South America. It affects nearly all commercially important cultivars of European-American hybrid poplars as well as *P. balsamifera* (balsam poplar) and its subspecies *trichocarpa* (black cottonwood), *P. deltoides* (eastern cottonwood), and *P. nigra* (black poplar). *P. tremula* (European aspen) and *P. tremuloides* (trembling aspen) have been found infected in Europe. PopMV infects only *Populus* in nature as far as is known. In Europe it has been detected more frequently in *Populus* section *Aigeiros* (cottonwoods and black poplar) than in section *Tacamahaca* (balsam poplars).

PopMV virions are slightly flexuous and measure 675–685 × 15 nm. The virus reaches highest concentrations in symptomatic leaves but can be detected by serological tests in symptomless leaves from the same plants. In winter it is easiest to detect in buds. Indicator plants such as *Vigna unguiculata* (cowpea) and *Nicotiana* species (tobacco) can be successfully inoculated by rubbing leaves with sap from *Populus.*

Symptoms. Height growth may be suppressed 0–50%, depending on tolerance of the cultivar and strain of PopMV. Shoots of highly susceptible plants may die back. These severe symptoms, noted only in young trees, subside within a few years after infection. Other clones tolerate infection with no measurable growth loss. Foliar symptoms usually appear in late spring on the first fully expanded leaves and may disappear during summer. They include diffuse chlorotic spots, mosaics of chlorotic and green tissue, starlike chlorotic spots (asteroid spots), vein-associated chlorosis, veinal reddening and necrosis, and necrotic lesions on petioles and main veins. Leaves with such lesions tend to be deformed and brittle and to abscise prematurely. High temperature tends to mask foliar symptoms. Photosynthesis is retarded and respiration increased in diseased trees. Specific gravity and strength of wood are only slightly affected.

Disease cycle. PopMV perennates in its hosts and is transmitted by grafting and by root contact. It is perpetuated by vegetative propagation of aspen and poplar, in which it invades both parenchyma and phloem. Vectors or other modes of transmission in nature are unknown, although natural spread in plantations has been observed. PopMV does not seem to be borne in seed or pollen.

Hybrid poplar clones differ in tolerance to PopMV. Those derived from crosses of *P. balsamifera* subsp. *trichocarpa* or *P. nigra* with *P. maximowiczii* (Japanese poplar) have been less affected than *P. ×canadensis* (Carolina poplar) clones; *P. balsamifera* subsp.

trichocarpa and hybrids between it and *P. deltoides* develop relatively severe symptoms.

Other carlaviruses. Lilac mottle virus (LiMoV) causes yellow mottling and undersized leaves of *Syringa* (lilac; Plate 204). It was first discovered in *Syringa oblata* in Maryland and is transmitted by aphids (by *Myzus persicae* under experimental conditions). *Honeysuckle latent virus* (HnLV) is associated with a mottled leaf syndrome in *Lonicera periclymenum* in Europe. An unnamed carlavirus was found associated with a chlorotic to yellow variegation in *Jasminum* sp. (jasmine) in Italy. Other carlaviruses have been detected in *Ligustrum* (privet), *Morus* (mulberry), and *Sambucus* (elderberry).
References: 356, 357, 554, 792–794, 2125, 2191, 2794, 2796, 2797, 4241

Diseases Caused by Ilarviruses

Ilarviruses occur in many woody angiosperms and induce economically important diseases of fruit trees and roses. The symptoms are indicated by some of the virus names: *American plum line pattern virus, Apple mosaic virus, Citrus variegation virus, Elm mottle virus, Hydrangea mosaic virus, Lilac ring mottle virus, Prunus necrotic ringspot virus, Prune dwarf virus.* The nucleic acid of an ilarvirus is single-stranded RNA divided into three or more parts that are packaged in separate particles. The particles are asymmetric and vary in shape from roughly spherical to bacillar. Natural spread of ilarviruses is due primarily to their transmission in seeds and with pollen. These viruses are also transmissible by grafting and mechanically in sap to indicator plants, but are not readily transmitted by plant-to-plant contact. Foliar symptoms caused by ilarviruses in trees tend to reappear year after year.

Sour cherry yellows and related diseases. *Prune dwarf virus* (PDV) causes sour cherry yellows (Plate 205), prune dwarf, and peach stunt. The virus occurs worldwide in *Prunus cerasus* (sour cherry), *P. domestica* (plum, prune), *P. persica* (peach), and other stone-fruit trees. It limits sour cherry fruit production in some regions. PDV has been transmitted experimentally to more than 100 *Prunus* species and to many herbaceous plants, but only a small fraction of the woody plants and none of the others have been found naturally diseased. The only reported natural woody hosts of PDV other than *Prunus* species are *Betula nigra* (river birch), *B. papyrifera* (paper birch), and *B. pendula* (European white birch). Effects of PDV on birch have not been described.

P. cerasus infected by PDV shows depressed growth and produces abnormally abundant flower buds instead of vegetative buds on terminal twigs. This conversion leads to fewer fruit spurs and diminished yield. Leaves at basal and intermediate positions along new shoots are mottled bright yellow and green or show various yellow patterns in late spring and early summer. These leaves soon drop, leaving affected twigs bare except for green leaves near the tip. Diseased trees often lose 30–50% of their leaves by midsummer. Leaves on newly infected trees may show acute symptoms consisting of chlorotic mottle or chlorotic rings and flecks. Holes form where tissue within chlorotic rings dies and drops out. Yellowing and leaf cast are promoted by cool weather after bloom.

Diseases caused by *Cucumber mosaic virus* (CMV) in *Nandina,* putative foveaviruses in *Prunus,* and *Camellia yellow mottle virus* in *Camellia.*

A. Stunting, wine-red pigmentation of leaves and flowers, and abnormally narrow leaves in *Nandina domestica* (sacred-bamboo) infected by CMV (SC, May; discussion, p. 418).

B, C. Necrotic spots and leaf distortion caused by *Cherry necrotic rusty mottle virus* in *Prunus avium* 'Sam' (sweet cherry) (OR, Jul; discussion, p. 420).

D. Green ring mottle caused by *Cherry green ring mottle virus* in *P. cerasus* 'Montmorency' (sour cherry) (NY, Sep; discussion, p. 420).

E, F. *Camellia japonica* (common camellia) with mosaic and chlorosis caused by *Camellia yellow mottle virus.* The disease is called yellow mottle or leaf and flower variegation (MS, May; discussion, p. 420).

Photo credit: A—O. W. Barnett by permission of the American Phytopathological Society

415

Yellows is most severe in trees simultaneously infected with PDV and *Prunus necrotic ringspot virus* (PNRSV). PDV causes ringspot or leaf tatter in some *P. avium* (sweet cherry) cultivars, but other cultivars of that species remain symptomless when infected. Diseased plum trees may be dwarfed with narrow, rugose, leathery leaves on some of the branches. *P. persica* infected by the virus also becomes stunted.

PDV is seedborne and is also carried in pollen from tree to tree in orchards. No vector is known, but pollinating insects are suspected to mediate virus transmission in pollen. PDV virions can be found in all parts of infected plants. They vary from isometric and about 19–20 nm in diameter to short rods up to 73 nm long.
References: 554, 777, 1594, 2032, 2181, 2191, 2896, 3070, 3934

***Prunus necrotic ringspot virus* (PNRSV)** (Plate 205) is distributed worldwide in rosaceous fruit and ornamental trees and in other woody plants. Its particles are asymmetric polyhedra of three sizes: 23, 25, and 27 nm in diameter. Besides causing necrotic ringspot or line patterns in many *Prunus* species, PNRSV causes calico disease in *P. dulcis* (almond), rugose mosaic in *P. avium* (sweet cherry), line patterns in *P. domestica* (plum), and part of the rose mosaic syndrome in *Rosa* species. It depresses growth and yield of fruit trees. The virus occurs in *P. pensylvanica* (pin cherry) in eastern USA and in *P. fasciculata* (desert almond) in the Southwest. PNRSV has also been detected in *Betula lenta* (sweet birch) and *Ligustrum japonicum* (wax-leaf privet), but symptoms in these plants are unreported. First-year symptoms (shock symptoms) in *Prunus* occur on single or scattered branches and often consist of chlorotic to necrotic rings from which the centers drop, leaving shot holes. In some cases buds or twigs and branches may die back; this symptom varies with host and strain of the virus. In subsequent years foliar symptoms caused by most strains are mild or obscure, but growth and fruit yield are depressed. PNRSV is transmitted by grafting and spreads on pollen from tree to tree and into seeds via pollinated flowers. The pollen is superficially contaminated, not infected. Accordingly, pollinating insects are suspected to contribute to the virus's spread. PNRSV is also seedborne, notably in *P. pensylvanica*, in which up to 80% incidence of seed infection has been reported. It does not spread by plant-to-plant contact.
References: 554, 620, 763, 777, 792, 2181, 2191, 2896, 3070, 3934

***Apple mosaic virus* (ApMV)** (Plate 205) occurs around the world in the temperate zones in many rosaceous and some nonrosaceous woody plants. ApMV, typical of ilarviruses, is a single-stranded RNA virus with its genome divided among three asymmetrically polyhedral particles 25–29 nm in diameter. It is closely related to PNRSV and occurs in several of the same plant species. In addition to its role in apple mosaic disease, this virus causes line pattern in *Prunus domestica* (plum) and causes or contributes to symptoms of rose mosaic. It also occurs in *Aesculus* ×*carnea* (red horse-chestnut), *A. hippocastanum* (horse-chestnut), *Betula alleghaniensis* (yellow birch), *B. papyrifera* (paper birch), *B. pendula* (European white birch), *Carpinus betulus* (European hornbeam), *Chaenomeles japonica* (flowering quince), *Corylus avellana* (European filbert), *P. dulcis* (almond), *P. triloba* (flowering almond), and *Rubus* (brambles). A virus that caused symptoms like those of apple mosaic in indicator plants was isolated from symptomless *Pyracantha rogersiana* in England, and an agent associated with mosaic in *Prunus virginiana* (chokecherry) in Maine caused vein yellowing and mosaic when graft-inoculated to apple.

Symptoms caused by ApMV vary with strain and host. The virus may infect plants asymptomatically or cause growth or yield reductions in the absence of foliar symptoms. Major yield reductions in apple and filbert have been documented. The most common foliar symptoms in *Malus* are angular cream to yellow spots on a green background, often coalescing to produce large areas devoid of green. Chlorotic bands may develop along major veins. In other instances leaves show a light and dark green mosaic or diffuse yellowish white patches. Severe strains of ApMV induce chlorotic to white areas, whereas mild strains cause only chlorotic spots. Symptoms usually occur on scattered leaves, although all leaves on a branch are sometimes affected. Mild strains of the virus in apple protect against infection by severe strains. Apple mosaic has been described in detail only for commercial apple cultivars.

Many other species of *Malus* are susceptible, however; *M. ioensis* (prairie crabapple) and *M. floribunda* (Japanese flowering crabapple) are highly so. *Aesculus* ×*carnea* and *A. hippocastanum* found naturally diseased in England had a bright yellow and green mosaic. The virus induces primarily line patterns rather than mosaic in several species whose reactions are known from controlled inoculations: *Betula* species, *Chaenomeles japonica* (flowering quince), *Corylus avellana*, *Prunus persica* (peach), and *Rosa* (rose). The symptoms in rose cannot be reliably distinguished from those caused by PNRSV. *Corylus avellana* is subject to infection not only by the common strain of ApMV, which causes line patterns and fleck, but also by *Tulare apple mosaic virus*, another ilarvirus. The latter occurs in California and Europe.

Betula species infected with ApMV may have foliar markings similar to those on *Corylus* and *Malus*: chlorotic to white line patterns, often of the oak-leaf type, and chlorotic flecks. Chlorotic vein banding, concentric ring patterns, and mild mosaic sometimes appear. Symptoms usually occur on only a few leaves and not in every year. The virus has been found in declining *B. pendula* in Europe. The discovery of ApMV in birch was initially of interest because viruslike symptoms had been associated with the destructive birch dieback disease that swept the Maritime Provinces and northeastern USA in the 1930s–1950s. Symptoms of the sort caused by ApMV occurred from Nova Scotia to Wisconsin, however, well beyond the region affected by birch dieback. Moreover, ApMV caused no dieback in experimentally inoculated birch seedlings. Young *B. alleghaniensis* and *B. papyrifera* that developed line patterns within 2 years after inoculation by grafting were symptomless by the fourth year. The possibility remains that ApMV may predispose *Betula* to damage by other factors, but evidence is lacking.

ApMV can be mechanically transmitted in sap from rose or apple leaves, but inadvertent mechanical transmission in the field is unknown. Experimental and diagnostic hosts represent many families. Plants used as diagnostic indicators include *Malus pumila* (apple) cultivars Golden Delicious, Jonathan, and Lord Lambourne; also *Catharanthus roseus* (Madagascar periwinkle), *Cucumis sativus* (cucumber), *Torenia fournieri* (bluewings), and *Vigna unguiculata* (cowpea). *Pyrus calleryana* 'Bradford' (Bradford pear) became infected when inoculated but did not show conspicuous symptoms. ApMV is readily transmitted by grafting and thus spreads in nursery stock. Transmission through natural root grafts also occurs in *Malus*. Symptoms develop after a few weeks to a year, depending on plant size and season of inoculation. Colonization of apple or plum trees is often slow and erratic; symptoms or virus may not be detectable in all branches. Natural spread in orchards is slow and in many instances seems not to occur at all. ApMV in young apple plants can be inactivated by holding them near 36°C for 3–10 weeks.
References: 103, 554, 792, 1191, 1326, 2081, 2181, 2593, 2896, 3112, 3125, 3934, 3935

Rose mosaic. Once thought to be caused by one virus, rose mosaic (Plate 205) is a complex of viral diseases that occur worldwide on wild and cultivated *Rosa* species. The principal

---

Diseases associated with tobamoviruses and a rhabdovirus (discussions, p. 422).

A–E. Symptoms associated with, although not shown to be caused by, *Tobacco mosaic virus* in *Fraxinus* species (ash). A. Wavy margins and abnormally narrow leaflets in *F. americana* (white ash) (NY, May). B–E. Mosaic and chlorosis in *F. pennsylvanica* (green ash; B, E) and *F. americana* (white ash; C, D) that also contained a nepovirus (greenhouse-grown plants).

F, G. Mottling and interveinal chlorosis associated with a rigid, rod-shaped virus similar to tobamoviruses in *Liriodendron tulipifera* (tuliptree) (greenhouse-grown plants).

H. Suppressed growth in *Picea rubens* (red spruce) seedlings resulting from inoculation with a strain of *Tomato mosaic virus* originally obtained from spruce; noninoculated plant at left (NY, greenhouse).

I, J. Symptoms caused by a rhabdovirus in *Populus balsamea* (balsam poplar). I. Chlorosis along veins. J. Close view of chlorosis and necrotic flecks along veins (ON, Jul).
Photo credits: F, G—O. W. Barnett; H—J. D. Castello

417

pathogens in North America are *Apple mosaic virus* (ApMV) and *Prunus necrotic ringspot virus* (PNRSV), both discussed above. Another ilarvirus, *Tobacco streak virus* (TSV), has been isolated from *R. setigera* (climbing or prairie rose) in the Northwest and from other roses in Europe. Rose ring pattern, caused by an uncharacterized viruslike agent, is also part of the mosaic complex in California and Oregon. Several additional viruses, among which the nepovirus *Arabis mosaic virus* (ArMV) is most common, have been isolated from rose with mosaic symptoms in Europe. ArMV is not associated with rose mosaic in North America.

Mosaic disease depresses the vigor of rose plants and renders them abnormally sensitive to winter damage. The quality and number of blossoms are diminished in some cultivars infected with ApMV. PNRSV interferes with establishment of bud grafts of some cultivars. In 'Fragrant Cloud' PNRSV delays the onset of flowering, reduces size and number of blossoms, and increases the proportion of deformed blossoms.

Foliar symptoms vary with strains or combinations of viruses, rose species or cultivar, and environmental conditions. The following have been recorded: puckered leaflets, chlorotic mottle, ringspots, light green or chlorotic line patterns, vein clearing, vein banding (at temperatures above 27°C), and mosaics of green and yellow or white. Some infected plants, even those of highly susceptible cultivars, lack obvious symptoms. In California mosaic symptoms develop during cool (e.g., 15°C) weather in spring, but no new symptoms appear during hot summer weather.

In general, infection caused by a particular virus in *Rosa* cannot be reliably diagnosed from symptoms alone. When isolates of PNRSV and ApMV were transmitted to certain rose cultivars by grafting, however, each virus caused distinct symptoms. ApMV caused chlorotic mottle and a mosaic of white or yellow and green patches or fused chlorotic rings, while PNRSV induced light green to chlorotic line patterns that varied in width from 1–2 mm up to several millimeters. Symptoms caused by ApMV were most distinct during summer when temperatures were above 21°C. Those caused by PNRSV were most distinct during relatively cool weather in early and late summer. PNRSV has also been isolated from rose plants showing vein-associated yellowing (yellow net).

Rose ring pattern is caused by an unknown graft-transmissible agent thought to be viral and distinct from the other viruses of the rose mosaic complex. Foliar symptoms in Hybrid Tea roses include irregular rings, fine line patterns, and chlorotic flecks, but these marks are often indistinct. Color-break rings develop in the petals of some cultivars. 'Queen Anne' plants infected with the ring pattern agent show yellow blotches on leaflets. The agent is transmitted by grafting, but no mode of natural transmission has been reported. *Rosa multiflora* 'Burr' is a reliable indicator; leaflets on diseased plants become stunted, deformed, rugose, and mottled.

Diseased rootstocks are a common source of rose mosaic viruses. PNRSV has been detected in *R. multiflora* rootstocks. It is transmitted by grafting and is also carried on pollen of roses as well as susceptible stone-fruit trees. Vectors other than pollinating insects are unknown. Rose mosaic viruses common in North America can be inactivated in young rooted cuttings by holding them at temperatures near 38°C for 4 weeks. Roses are subject to several additional viral diseases: rose streak, rose rosette, rose wilt, rose spring dwarf, and rose leaf curl. *Tobacco streak virus* (TSV) causes irregular chlorotic areas, vein chlorosis, and twisted leaves in *R. setigera*.

References: 777, 792, 1758, 1850, 2191, 3524, 3934, 3999–4001

Other ilarviruses. Elms in Europe are affected by elm mottle, caused by *Elm mottle virus*. Yellow-and-green mottling and chlorotic spots, lines, and rings that sometimes contain necrotic lesions develop on leaves of *U. glabra* (Scotch elm) and *U. minor* (European field elm). This virus is seedborne in *U. glabra* and also naturally infects *Philadelphus* species (mock-orange) and *Syringa vulgaris* (common lilac). It causes pale yellow to white mosaic, sometimes with chlorotic rings or broad white chevron patterns, and occasional curling of white leaves in the latter plant. It caused white mosaic also when inoculated to *Forsythia intermedia*. *Citrus* species, which tend to have their own set of viruses, are affected by *Citrus leaf rugose virus* and *Citrus variegation virus*.

References: 555, 1962, 3453

Diseases Associated with Cucumoviruses

Cucumoviruses are single-stranded RNA viruses with genomes divided into three parts that are packaged in icosahedral particles 29–30 nm in diameter. These viruses are transmitted in the nonpersistent manner by many species of aphids and are seedborne in some plants. They are known mainly for diseases caused by the globally distributed type species, *Cucumber mosaic virus* (CMV), in vegetable crops. The host range of CMV includes representatives of more than 40 families. It has been detected in at least in the following woody plants in North America (and Europe as indicated). It was found in *Berberis thunbergii* (Japanese barberry) with reddish blotches and undersized distorted leaves and was shown to cause the syndrome. It has been associated with yellow ring and line patterns in *Daphne* species in which *Alfalfa mosaic virus* was also present, and it has been detected in nonsymptomatic *Cornus florida* (flowering dogwood). Leaves of *Euonymus japonicus* (Japanese spindle-tree) with yellow blotches and rings have yielded CMV in the USA and Europe. This virus and others have been associated with chlorotic chevrons, rings, or blotches in *Ligustrum vulgare* (common privet) and *L. ovalifolium* (California privet) in the USA and Europe. CMV in *Nandina domestica* (sacred-bamboo) has been associated with reddening of petals and leaves in spring (Plate 206), distortion and narrowing of leaflets, diminished flower number, and poor fruit set. It causes yellowish green mosaic or chlorotic bands on undersized and deformed leaves in *Rhamnus cathartica* (common buckthorn) in the USA and Europe. CMV along with other viruses occurs in *Sambucus canadensis* (American elder) and *S. nigra* (European elder) on both continents.

Reports from Eurasia add the following to disorders associated with CMV: mild mosaic in *Acer rubrum* (red maple); narrow leaves with chlorotic to yellow arc and ring patterns in *Buddleja davidii* (butterfly bush); chlorotic flecks and vein banding in *Clematis*; mosaic in *Forsythia*; mosaic and distorted leaves in *Hydrangea*; mild mosaic in *Liriodendron tulipifera* (tuliptree); vein yellowing of *Lonicera* (honeysuckle); chlorotic mottling with some necrosis in *Lycium barbarum* (matrimony-vine); mild mosaic in *Magnolia acuminata* (cucumber-tree); chlorotic lines, arcs, and rings in other *Magnolia* species; chlorosis, mosaic, vein banding, chlorotic ringspot, or oak-leaf symptoms and leaf deformity in *Paulownia tomentosa* (empress tree); chlorotic mottle or chevron patterns in *Philadelphus* sp. (mock-orange). In addition, CMV is a demonstrated cause of distorted leaves and dieback in *Piper methysticum* (kava pepper) on several islands in the southwestern Pacific.

The virus has also been detected in woody plants for which no symptoms were reported or in which two or more viruses were associated with observed symptoms: *Chionanthus* (fringe tree), *Jasminum* (jasmine), *Piper* (pepper), *Syringa* (lilac), *Prunus* (plum), and *Ribes* (currant).

The only other known cucumovirus noted in woody plants is *Peanut stunt virus* (PSV) (syn. Robinia mosaic virus), which causes mosaic, mottling, and leaf deformity of *Robinia pseudoacacia* (black locust) in Europe and China and of *R. viscosa* (clammy locust) in Europe. An unidentified cucumovirus was found associated with chlorotic ringspots and yellow-white spots and curved line patterns in *Maclura pomifera* (osage-orange) in Yugoslavia.

References: 220, 554, 777, 792, 907, 1521, 2278, 3102, 3110, 3111, 3191, 3452, 3454, 4367

Rose rosette disease, caused by an unnamed viruslike agent (discussion, p. 424).

A–F. *Rosa multiflora* (multiflora rose). A–D. Overview and close views of a diseased plant with proliferating stunted, thick, red shoots with dwarfed yellow to red leaves (WV, Jun). E. Rosettes of stunted leaves with red pigment on a chlorotic cane (IA, summer). F. Dark red vein-associated discoloration on undersides of leaves, an early symptom (IA, summer).

G. A potted Hybrid Tea rose with abnormally abundant thorns, red-purple pigment, and deformed leaves (IA, Aug).

H. Magnified view of *Phyllocoptes fructiphilus*, an eriophyid mite that is a vector of the causal agent. The specimen is about 200 μm long.

Photo credits: E–H—A. H. Epstein

Diseases Caused by Potyviruses

Potyvirus is the largest plant virus group and is among the most destructive in herbaceous agricultural plants. The group is named for *Potato virus Y*. Potyvirus virions are flexuous filaments that contain single-stranded RNA within a protein coat. They replicate in cytoplasm and induce the formation of microscopically visible intracellular inclusions, notably proteinaceous pinwheel-shaped structures. Particle dimensions vary with particular viruses in the range 620–900 × 11–20 nm. A typical virion contains the entire genome. Potyviruses are transmitted primarily by aphids in the nonpersistent manner. Most are readily transmissible in sap to indicator plants but are not readily transmitted by foliar contact between plants. Several can be transmitted in seed, and a few in pollen.

A virus informally named poplar potyvirus or Populus virus (PV) was first detected in *Populus grandidentata* (bigtooth aspen), *P. tremuloides,* and hybrid poplars affected by bronze leaf disease (Plate 46) in Wisconsin. PV is apparently widespread in the Great Lakes region but causes little or no damage. It was once thought to be a cause of bronze leaf disease, but the fungus *Apioplagiostoma populi* was later shown to induce the leaf bronzing and dieback characteristic of that disease. *P. tremuloides* infected with PV develops chlorotic to necrotic leaf spots early in the growing season. European-American poplar hybrids affected by PV develop chlorotic to necrotic leaf spots similar to those shown in Plate 210D. PV particles measure 800–810 × 13 nm. The virus is transmissible in sap from *Populus* to *Vigna unguiculata* (cowpea), *Phaseolus* (bean), and other indicator plants, in which it induces intracellular inclusions of the pinwheel type. Pathogenicity of PV to *Populus tremuloides* was demonstrated by inoculation with sap from *V. unguiculata.* Assay plants developed chlorotic to necrotic leaf spots similar to those originally observed. The mode of transmission in nature is unknown. Natural hosts other than *Populus* are also unknown.

Other potyviruses that infect trees and shrubs include the following. *Bean yellow mosaic virus* was found to cause veinal chlorosis, mottle, and deformity of *Cladrastis lutea* (yellowwood) leaves in New York State. *Daphne virus Y* causes chlorotic mosaic, leaf distortion, and poor flower quality in *Daphne* species in Australasia and Europe. *Palm mosaic virus,* discovered in California, causes mosaic, ringspots, line patterns, and stunting in *Washingtonia robusta* (Mexican fan palm). *Wisteria vein mosaic virus* causes chlorotic mottled foliage in *Wisteria* species in the USA, Europe, and Australia. The much studied *Plum pox virus* (PPV), dreaded by stone-fruit growers, causes leaf mottling, fruit malformation, fruit drop, and dieback in *Prunus domestica* (plum), and milder symptoms in several other *Prunus* species. It is widespread in Eurasia and also occurs in Nova Scotia, Ontario, and Pennsylvania, where quarantines and eradication efforts were ongoing as this was written. The economically important *Papaya ringspot virus* (PRSV) causes ringspot and decline in *Carica papaya* but is being controlled commercially in some regions through the use of papaya lines genetically engineered to produce the coat protein of the virus constitutively, which prevents replication of PRSV. An unnamed potyvirus was isolated from *Picea abies* (Norway spruce) with yellowed foliage in Bavaria and was shown to be transmitted to herbaceous plants by aphids.
References: 554, 752, 792, 1030, 1159, 1281, 1312, 1427, 2191, 2485, 2519, 2822, 3102, 3156, 4153

Diseases Caused by Foveaviruses

Foveavirus is a genus of linear, single-stranded, monopartite RNA viruses associated with diseases of fruit trees and grapevines. The type member is *Apple stem pitting virus.* The group is named for the pits and cankers that form in infected stems where localized areas of vascular cambium cease activity or die. The virions are flexuous filaments about 800 × 12 nm (perhaps longer in *Cherry green ring mottle virus,* as noted below).

Rusty mottle diseases. Two diseases of *Prunus avium* (sweet cherry) that are widespread in western North America are caused by viruses tentatively assigned to the foveavirus group. Necrotic rusty mottle (Plate 206) is attributed to *Cherry necrotic rusty mottle virus.* It also occurs in Europe. Numerous *Prunus* species are susceptible, but only *P. avium* sustains commercial damage. Angular necrotic

spots form in leaves of susceptible cultivars 3–6 weeks after full bloom and cause distortion of leaves not yet fully expanded. Lesions may remain discrete or may coalesce to involve large areas of the leaf, often inducing leaf drop. Severe symptoms are promoted by prolonged cold weather in spring. Yellow-and-green mottling with green islands or bands develops in summer. Over several years numerous buds and twigs die, leaving branches with terminal tufts of foliage. Cankers and gum blisters form on branches. The causal agent is readily transmitted by grafting, and most cases of the disease in orchards are considered to result from propagating diseased plants in nurseries. The natural mode of transmission is unknown. Other viruses closely related to that associated with necrotic rusty mottle include cherry rusty mottle virus, European isolates of *Cherry mottle leaf virus,* and *Cherry green ring mottle virus.*

Cherry green ring mottle (Plate 206) occurs in *Prunus armeniaca* (apricot), *P. avium* (sweet cherry), *P. serrulata* (Japanese flowering cherry), *P. cerasus* (sour cherry), and *P. persica* (peach and nectarine) in eastern and western North America, the Mediterranean region, and New Zealand. Infection of *P. cerasus* causes economic damage. In 'Montmorency' sour cherry a mottle consisting of persistent green islands or rings on a chlorotic to yellow background appears on scattered leaves 4–6 weeks after flower petals fall. The affected leaves soon drop, and few foliar symptoms are visible during the remainder of the season. Fruit on diseased trees may have mottled and pitted surfaces and bitter flesh. Diseased *P. serrulata* 'Kwanzan' and 'Shirofugen' display epinasty (downward bending of petioles), and parts of the midrib and lateral veins of leaves may die, resulting in twisting and curling. Twig growth is suppressed, and strips of maturing bark on twigs die and crack; dieback follows. The causal agent, *Cherry green ring mottle virus* (CGRMV), is transmitted by grafting and has been distributed widely in propagating materials. Natural spread occurs in orchards, but no vector has been identified.

CGRMV has been tentatively classified as a foveavirus on the basis of RNA sequence similarity although it differs from foveaviruses in certain RNA sequences and genome organization. A strain of CGRMV in Michigan was found to have thin, flexuous, filamentous particles 1000–2000 × 5–6 nm, their length resembling that of closteroviruses. Accordingly, it is possible that CGRMV represents a discrete group not yet characterized.
References: 1255, 3070, 3356, 4494, 4515, 4516

Diseases Caused by Varicosaviruses

Infectious variegation of camellia, caused by *Camellia yellow mottle virus* (Plate 206), affects *Camellia japonica* (common camellia) and *C. sasanqua* (sasanqua camellia). The virus occurs globally and is tentatively classified in genus *Varicosavirus.* The nucleic acid of varicosaviruses is double-stranded RNA packaged in rod-shaped particles. The camellia disease, once confused with genetically controlled variegation, has also been called yellow mottle, yellow spot, or leaf and flower variegation. Symptoms include yellow mottling or circular to irregular or elongate yellow or white zones, sometimes in a mosaic but often concentrated along veins or margins; chlorosis or albinism of entire leaves; roughening or corkiness of the epidermis; and sometimes necrosis of white or chlorotic tissues. Ringspots have also been reported but may represent a different disease. Symptoms are most severe on leaves formed early in the season and may be absent from those formed

---

Viral diseases of *Rhododendron* and *Kalmia* (discussion, p. 426).
A. Rhododendron necrotic ringspot in *Rhododendron* cv. Unique. The causal agent, *Rhododendron necrotic ringspot virus,* is tentatively classified as a potexvirus (OR, summer).
B. Symptoms caused by *Rhododendron necrotic ringspot virus* in *Kalmia latifolia* (mountain-laurel) (OR, summer).
C–G. Viruslike symptoms of unknown cause in cultivated rhododendrons. C–E. A disease of the mosaic type in *R. arborescens* (smooth azalea) causes foliar distortion, narrow leaves, and slow growth. A normal plant of another species is at right in E (NY, Jul). F, G. An unidentified cultivar with chlorotic flecks and yellow-white lines and zones where arrested growth of leaf tissue resulted in distortion (WA, Jun).
Photo credits: A, B—D. L. Coyier

421

late. Flowers show aberrant color patterns that range from scattered white flecks to large, round or irregular white spots or elongate, radiating white zones following veins. Growth of diseased plants is not greatly affected, but leaves with prominent white or yellow zones are abnormally susceptible to sunburn and frost injury. *Camellia hiemalis, C. reticulata,* and *C. saluenensis* are also susceptible. Stems and leaves of *C. reticulata* become distorted when infected. The virions measure 120–170 × 30 nm and have been observed in cytoplasm in leaves, petals, and stamens. The virus is readily transmissible through grafts, including natural root grafts, but is not mechanically transmissible in sap; nor is it transmitted in pollen or by plant-to-plant contact. Modes of natural transmission are unknown. Several strains of the virus have been designated on the basis of different symptoms caused in indicator varieties of *Camellia* to which diseased scions were grafted. Infected rootstocks and attractively variegated plants, although diseased, have been widely dispersed in horticultural trade. Camellia clones free from the virus can be obtained by propagating from the apices of vegetative buds. Up to 67% of plants thus obtained from virus-infected stock plants were virus free in one set of experiments.
References: 554, 792, 840, 1222, 1691, 2630, 3092, 3093, 3094

Diseases Associated with Tobamoviruses

Tobamoviruses comprise a group of about 17 viruses with rigid cylindric particles measuring 297–320 × 15–18 nm, each with a single-stranded monopartite RNA genome. Some of these viruses—notably the type species, *Tobacco mosaic virus* (TMV), and *Tomato mosaic virus* (ToMV)—are ubiquitous, occurring in many plant species and, after release from roots or decaying tissue, in soil and water. They are often detected in asymptomatic plants. Tobamoviruses can spread from plant to plant by contact and are carried on tools, hands, and mouthparts of vagrant insects. These viruses are chemically stable and tolerant of heat and can persist a long time outside plants. Tobamoviruses replicate in cytoplasm and reach high concentrations. Amorphous inclusions and crystalline arrays of tobamovirus virions within cells can be visualized microscopically.

***Tobacco mosaic virus* (TMV) and *Tomato mosaic virus* (ToMV).** These viruses were distinguished from each other beginning in the 1970s, but records of TMV in natural hosts up until the 1980s may have represented either TMV or ToMV, or possibly other tobamoviruses. Therefore we consider tobamoviruses collectively unless virus-specific information is available. TMV and ToMV virions measure about 300 × 18 nm. They differ in RNA sequences and coat proteins, and are readily distinguished by serological assays. These viruses infect many herbaceous and woody plants and have been carried around the world in plants. ToMV has even been detected in fog and cloud water and in ancient glacial ice. Transfer of ToMV from infected foliage to water droplets in fog, and vice versa, are presumed to account for its presence in cloud water and in montane forests. TMV and ToMV are transmitted mainly by contact between plants, on hands or implements during plant care, and through seeds; for example, in *Malus* (apple), *Pyrus* (pear), and *Vitis* (grape). Several types of beetles and grasshoppers, which have chewing-type mouthparts, have been shown to transmit TMV incidentally as they graze from plant to plant. It has been suggested that a tobamovirus spreads with conidia of the powdery mildew fungus *Cystotheca lanestris* among *Quercus* (oaks) in California.

Symptoms associated with tobamoviruses in woody plants around the world include chlorotic spots and mottling of *Acer saccharum* (sugar maple); mottling and leaf deformities in *A. pseudoplatanus* (sycamore maple); chlorotic spots in narrow deformed leaves of *Castanea sativa* (European chestnut) and *Quercus cerris* (European turkey oak); chlorotic spots, mosaic, rings, line patterns, and leaflet distortion in *Fraxinus* species (ash; Plate 207) and mosaic in *F. americana* (white ash) inoculated with ToMV; yellowing, rosetting, and necrotic spotting of *Lithocarpus densiflorus* (tanoak); vein banding and decline in *Olea europaea* (olive) infected with TMV; rolled leaves and interveinal yellowing in *Paulownia tomentosa* (empress tree) infected with TMV; growth reduction, delayed budbreak, and enhanced cold hardiness in *Picea rubens* (red spruce) seedlings inoculated with ToMV (Plate 207); light green diffuse spots in *Pyrus* (pear) leaves; bushy top in *Populus tomentosa* (Chinese

white poplar); chlorotic flecks in leaves of *Quercus petraea* (sessile oak) and chlorotic flecks and star-shaped lesions in deformed leaves of *Q. robur* (English oak); mosaic in *Robinia pseudoacacia* (black locust); chlorotic spots and lines and deformed leaves in *Sambucus canadensis* (American elder); mosaic and chlorotic mottle in *Syringa ×nanceiana* infected with ToMV; and chlorotic spots and mosaic of *Wisteria*.

TMV was often detected—as were the nepoviruses *Tobacco ringspot virus* and *Tomato ringspot virus*—in *Fraxinus americana* (white ash) during studies of ash decline in northeastern USA, but no virus was found consistently associated with it. When the three viruses from *F. americana* were inoculated separately and in combinations to *F. americana* and *F. pennsylvanica* (green ash) saplings, symptoms specific to the tobamovirus could not be distinguished. In other studies, however, a tobamovirus from *F. americana* with chlorotic spots, rings, and line patterns caused chlorotic spots in inoculated *F. americana* and *F. pennsylvanica* seedlings and concentric chlorotic rings in *F. excelsior* (European ash). A tobamovirus was detected in seeds of *F. americana,* but possible transmission of it to seedlings was not investigated.

Tobamoviruses are also widespread in many woody plants that usually show no symptoms or for which the symptoms are generally unknown; for example, *Cornus florida* (flowering dogwood), *Daphne, Eucalyptus, Ficus* (fig), *Malus pumila* (apple), *Persea americana* (avocado), *Picea rubens* (red spruce), *Populus nigra* (black poplar), *Prunus* (cherry, plum), *Pyrus* (pear), *Quercus* (oak), *Ribes* (currant), *Rosa* (rose), *Salix* (willow), *Vitis* (grapevine), and *Wisteria.*

Other tobamoviruses. Florida hibiscus virus and Hibiscus S virus (both names awaiting official sanction) are associated with chlorotic mottle and other symptoms in *Hibiscus rosa-sinensis* (Chinese hibiscus) in Florida and Singapore, respectively. *Frangipani mosaic virus* (FrMV) is associated with mosaic, vein banding, ringspots, bronzing, leaf distortion, and necrosis in *Plumeria* species (frangipani) in Australia and India.
References: 21, 62, 156, 672, 673, 777, 792, 1125, 1137, 1204, 1883, 1884, 2129, 2191, 2225–2227, 2744, 2862, 3235, 3587, 3812, 3900, 4082

Diseases Associated with Rhabdoviruses

The family *Rhabdoviridae* includes both plant and animal pathogens. A typical rhabdovirus virion contains a helically coiled linear molecule of single-stranded RNA associated with several different proteins and covered by a glycoprotein envelope. Rhabdoviruses have the form of short rods or bullets. A few plant rhabdoviruses lack an envelope. Dimensions of plant rhabdovirus particles, including the envelope, are in the range 100–500 × 42–130 nm (average near 260 × 70 nm). These virions are so distinctive that they can be identified to family level when observed in plant cells by electron microscopy. Some are assigned to the genus *Nucleorhabdovirus* because they replicate in plant nuclei; those that replicate in cytoplasm are assigned to *Cytorhabdovirus*. Most of the plant rhabdoviruses have not yet been assigned to a genus. They tend to have restricted host ranges. They are transmitted by grafting and in nature by suctorial insects (aphids, leafhoppers, and piesmids) or mites, and are also transmissible in sap to indicator plants. Many rhabdoviruses occur in herbaceous dicots and in

---

Symptoms caused by unidentified virus(es) in *Populus* species (discussion, p. 426).

A–C. Symptoms in the hybrid *Populus* cultivar Robusta. A. Growth suppression in a diseased plant (right) compared with a healthy plant (left). B. Stunting and distortion of subterminal leaves. C. Vein-associated chlorotic blotches in subterminal leaves (ON, Jul).

D. Small chlorotic spots, similar to those caused by poplar potyvirus, on a leaf of *P. deltoides* (eastern cottonwood) (ON, Jul).

E, F. Vein-associated yellowing (vein banding) in the hybrid poplar cultivar Northwest (ON, Jul).

G, H. Vein-associated yellowing and mosaic, collectively called aspen vein mottling, on *P. tremuloides* (trembling aspen) (ON, Jul).

423

monocots. Most that occur in woody ornamental plants are known from European reports.

A putative rhabdovirus called *Euonymus fasciation virus* is associated with fasciation in *Euonymus japonicus* (Japanese spindletree), and a possibly different rhabdovirus in *E. japonicus* is associated with chlorotic leaf spots and vein yellowing. A strain of *Eggplant mottled dwarf virus*, also known as pittosporum vein-yellowing virus, causes veinal chlorosis in *Pittosporum*. *Sambucus vein clearing virus* causes vein clearing (translucency associated with subnormal greenness along veins) in *Sambucus* (elderberry). Unnamed rhabdoviruses cause apical necrosis in *Carica papaya* (papaya) and malformed leaves with yellow to pink veins in *Codiaeum variegatum* (croton) and *Laburnum anagyroides* (golden-chain tree). A rhabdovirus informally called Hibiscus vein-yellowing virus is associated with vein yellowing in *H. rosa-sinensis* in the Mediterranean region. A rhabdovirus that has not been fully characterized infects *Populus balsamifera* (balsam poplar; Plate 207) in Ontario. It is associated with yellowing and necrosis along leaf veins. Its geographic distribution, host range, and impact on host growth are unknown. A lethal disease called citrus leprosis, caused by *Citrus leprosis virus,* occurs in South America and Central America and is transmitted by mites.
References: 345, 838, 2191, 2354, 2794, 2795, 3099, 3100, 3101

Diseases Associated with Necroviruses

Tobacco necrosis virus (TNV), the type member of the necrovirus group, is a single-stranded monopartite RNA virus. The icosahedral virions are about 28 nm in diameter. TNV occurs around the world. The virus is soilborne, is transmitted by zoospores of the root-infecting fungus *Olpidium brassicae,* and has been detected in roots of many woody and herbaceous plants. It apparently fails to spread into aboveground parts of some plants, and it often causes no noticeable symptoms. TNV has been detected in symptomless flowers, leaves, and fruits of *Malus pumila* (apple) and *Pyrus communis* (pear) and in *Vitis* (grapevine) leaves. In Great Britain it was found in roots and leaves of *Betula pendula* (European white birch), associated with chlorotic foliar markings but not shown to cause them. It was found in *Picea abies* (Norway spruce) roots and caused growth reduction in inoculated seedlings. It has also been detected in roots of *Forsythia, Fraxinus* (ash), *Larix* (larch), *Pinus,* (pine), *Salix* (willow), and *Sambucus* (elderberry), among other woody plants. TNV can be transmitted in sap to many herbaceous plants, in which causes local lesions or systemic necrosis. *Chenopodium, Nicotiana* (tobacco), and *Phaseolus* (bean), are used as indicators. Many strains exist that vary in host range and in the symptoms they induce in assay plants.

TNV has been found in *Populus tremuloides* (trembling aspen) clones affected by a disorder called aspen deterioration in the Rocky Mountain region, but whether it plays a role in the syndrome is unknown. *P. tremuloides* reproduces naturally by root suckers (sprouts from roots) as well as by seed. Therefore it occurs in clonal clusters with the oldest stems in the center. Aspen deterioration primarily affects clones in which the original trees are 80–100 years old. It is characterized by low vigor; crooked, branchy stems; abnormally high mortality rate of stems; and suppressed production of root suckers. Stands damaged by grazing animals are often affected. Both deteriorating and normal trees have been found with viruslike foliar symptoms. These begin in spring as faint chlorotic spots, mottling, and chlorosis along the fine veins, then progress to light green or chlorotic patterns that are often reddish brown by August. When leaves turn yellow in autumn, the patterns may remain orange with brown flecks. Faint chlorotic spots but no vein mottling developed after TNV from *P. tremuloides* leaves was transmitted in sap to leaves of an indicator plant and was then inoculated in sap of that plant into healthy *P. tremuloides* leaves. TNV has also been found in hybrid *Populus* clones in the Great Lakes region.
References: 792, 1634, 2129, 2191, 2860, 2861, 3446

Diseases Caused by Viruslike Agents Not Fully Characterized

Rose rosette. A lethal disease called rose rosette (Plate 208) that affects *Rosa* species has spread widely in North America since its discovery in 1941. It is common in midwestern USA, occurs in scattered localities from California to Maryland and Manitoba to Texas, and is still spreading. It affects predominantly *Rosa multiflora* (multiflora rose) and also occurs in at least the following: *R. bracteata* (Macartney rose), *R. chinensis* (China rose), *R. glauca* (redleaf rose), *R. nutkana* (Nutka rose), *R. odorata* (Tea rose), *R. pisocarpa* (cluster rose), *R. rubiginosa* (sweetbrier), *R. woodsii* (Fendler, interior, or Woods rose), and various hybrids. The following are resistant: *R. acicularis* (prickly rose), *R. arkansana* (prairie rose), *R. blanda* (smooth rose), *R. californica* (California rose), *R. carolina* (Carolina rose), *R. palustris* (swamp rose), *R. pimpinellifolia* (burnett or Scotch rose), and *R. setigera* (prairie rose).

Dying plants or deformed, bushy, stunted, red-discolored apical parts of canes call attention to the disease. The onset of symptoms is more subtle, however, with mosaiclike reddish blotches associated with veins on new leaves of vigorously growing shoots, usually in midsummer. Affected shoots at first elongate rapidly, are pink to magenta, and often become abnormally thorny. Leaves are deformed and show yellow-and-green mosaic as well as red pigment. Growth slows markedly within a few weeks or the next year, and many axillary buds open and produce short reddish shoots with stunted, twisted, misshapen leaves and abnormally narrow leaflets. Repetition of this process results in clusters (rosettes) of stunted organs on canes that become yellowish. Symptoms spread throughout the plant, and it typically dies within 2–4 years. In cultivated roses the first stems affected may elongate rapidly and become thicker and more thorny than normal, then produce short deformed shoots that grow in various directions from the axis. Flowers often abort or are deformed or develop as greenish leaflike organs. Symptomatic shoots on most roses are yellowish red to burgundy-red, but they may be yellowish to greenish on Hybrid Teas and some Tea roses. Diseased parts are abnormally sensitive to frost, which contributes to their inevitable dieback.

The causal agent is presumed to be a virus of a type not yet characterized. Spherical double-membrane-bound particles, 120–150 nm in diameter, that are similar to but larger than known enveloped viruses have been observed by electron microscopy in diseased but not healthy plants. The particles occur singly or in groups in the cytoplasm of parenchyma, phloem, and epidermal cells. Double-stranded RNA, which is the genomic material of some viruses and a replicative intermediate form of many others, can be detected in diseased plants. Symptoms are transmissible by grafting and by the eriophyid mite *Phyllocoptes fructiphilus,* but not mechanically in plant sap except by special procedures using extracts from diseased root tips. These features are consistent with the hypothesis that rose rosette is caused by a viral agent whose particles are unstable and soon degrade outside intact plant cells. *P. fructiphilus* feeds on succulent tissues and overwinters in protected sites such as partly open buds and acute angles between rose stems and petioles. Adults of this four-legged mite are only about 200 μm long. They are dispersed in air as well as on infested plants. Long-distance spread of the vector is presumed to account for the wide distribution of rose rosette in North America. *R. multiflora* inoculated by the mite develops foliar symptoms after 17–30 days or longer, and deformed shoot growth begins after 30–90 days or longer. The pathogen spreads from canes to roots. Incidence of rose

---

Viral diseases of *Cercis canadensis* (redbud) and *Hibiscus rosa-sinensis* (Chinese hibiscus) (discussion, p. 426).

A–C. Undiagnosed dieback and ringspots on *C. canadensis*. A. Undersized deformed leaves on a declining tree. B, C. Necrotic ringspots on leaves from the plant shown in A (GA, May).

D. Redbud yellow ringspot. Diseased plants contain double-membrane-bound particles (not fully characterized) that are transmitted by grafting and eriophyid mites (AR, summer).

E. Half leaves of *C. canadensis* showing progression in severity of an undiagnosed ringspot and line-pattern syndrome from summer to autumn. The affected tree was in severe decline (NY, Jul–Oct).

F–H. Viruslike symptoms on *Hibiscus rosa-sinensis* (Chinese hibiscus). Green blotches and ringspots are prominent on the oldest leaves on the shoot, but leaves formed later show only slight curling and rugosity (FL, Mar).

Photo credit: D—K. S. Kim

425

rosette tends to increase rapidly within local populations of *R. multiflora*, rising to near 90% within a few years under conducive conditions.

R. multiflora, introduced to North America from eastern Asia, is widely used as a rootstock for cultivated roses, but it has become an invasive weed, so rose rosette is welcomed by many landowners. In some midwestern areas where *R. multiflora* in pastures was graft-inoculated for biological control, gardeners expressed concern about the possibility of disease spread from inoculated plants to cultivated roses. Research showed that where rose rosette occurred naturally, deliberate inoculation of wild *R. multiflora* resulted in disease increase only within 100 m of the treated plants.
References: 57, 66, 67, 953, 1069–1072, 1257, 1521, 4012

Rhododendron necrotic ringspot and other viral diseases of *Rhododendron*. One viral disease of *Rhododendron* has been relatively well characterized. It is caused by *Rhododendron necrotic ringspot virus*, tentatively classified in genus *Potexvirus*. The disease occurs in many *Rhododendron* cultivars and in *Kalmia latifolia* (mountain laurel) in northwestern USA and British Columbia and has been noted in the latter plant in Great Britain. Prominent concentric necrotic ring patterns develop in leaves of infected plants (Plate 209). The causal virus is transmissible by grafting and can be detected by electron microscopy in leaf cells as flexuous filaments that measure 500–510 × 13 nm.

Although other viral diseases of *Rhododendron* occur widely, little is known about them. Mosaic-type diseases were noted in Germany and in New Jersey in the early 1930s but were not fully described or shown to be transmissible. *R. ponticum* and catawba hybrids were affected in New Jersey. Two syndromes of unknown cause are shown in Plate 209. In the first of these (Plate 209C–E), symptoms on expanded leaves include mosaic, chlorotic mottling, and rugosity. The leaves are undersized and often somewhat twisted, with edges rolled downward. Typically the entire plant is affected, loses vigor, and may be overgrown by healthy neighbors. In the second syndrome (Plate 209F, G) chlorotic flecks appear on all parts of a leaf. In addition, leaves have chlorotic lines or zones where tissue differentiation ceased prematurely, resulting in wrinkling and deformity. Affected leaves tend to be undersized, and many are abnormally narrow. Severity of symptoms varies from branch to branch.
Reference: 826

Viral diseases of aspen and poplar. Several viruses other than *Poplar mosaic virus* infect *Populus*, but their identities, geographic distributions, and effects on the hosts are not well understood. Known viruses include the nepoviruses *Arabis mosaic virus* and *Raspberry ringspot virus*, which occur in *Populus* only in Europe. In North America a disorder called vein mottling (Plate 210) occurs in *P. tremuloides* (trembling aspen) in Ontario and in the northern Rocky Mountain region. It is characterized by fine chlorotic patterns with diffuse margins along major veins early in the season, followed by light green mottling that may extend over a large part of the leaf blade. Light green areas sometimes turn yellow. The cause is unknown.

P. tremuloides and hybrids between this species and *P. tremula* (European aspen) are also subject to a necrotic leaf spot of undetermined cause. Symptoms consist of chlorotic spots 2–3 mm in diameter that sometimes develop necrotic centers. Marked hypertrophy occurs in certain cells within the chlorotic zone, and unusual viruslike particles have been observed in diseased tissue. The symptoms can be transmitted through grafts and seed, but no pathogen has been characterized. The disease has been noted in Ontario and Quebec.
References: 470, 2794

Redbud diseases. Two undiagnosed viruslike disorders of *Cercis canadensis* (redbud) are shown in Plate 211. The foliar symptoms are associated with progressive twig and branch dieback. The disorder shown in A–C is characterized by chlorotic ringspots and short bands of chlorotic tissue along veins in scattered leaves. These symptoms are visible soon after leaf expansion. Brown flecks of necrotic tissue develop and eventually coalesce in a line near the margin of each chlorotic band or spot. Tissue enclosed by necrotic lines dies and drops out, leaving ragged holes. Some spotted leaves and others on the same branches are deformed by wrinkles and bulges. Branches with foliar lesions produce undersized leaves and die back during winter. A similar disorder, called redbud yellow ringspot (Plate 211D), was noted in Arkansas in the 1970s and persists there. Double-membrane-bound particles 120–150 nm in diameter, resembling those of an enveloped virus but larger than those of any known plant virus, were detected in cytoplasm of cells in diseased leaves. The particles are not sap-transmissible but are readily transmitted by grafting. They are naturally transmitted by unidentified eriophyid mites and are similar to the particles associated with rose rosette.

The disorder shown in Plate 211E also leads to decline and dieback. Foliar symptoms appear in early summer as slightly chlorotic rings or irregular figures on scattered leaves. Reddish brown necrotic flecks develop within the chlorotic tissue, often in two or more concentric lines. Colors intensify, and a greenish black border may develop, resulting in two- or three-tone line patterns by late summer. Symptoms are confined to certain branches, and these branches die back during winters. In Italy *Alfalfa mosaic virus* (genus *Alfalfamovirus*) was isolated from *Cercis siliquastrum* (Judas tree) showing yellowish ringspots, line patterns, and leaf deformities.
References: 1521, 2063, 3195

Hibiscus diseases. The symptoms shown on *Hibiscus rosa-sinensis* (Chinese hibiscus) in Plate 211 were not diagnosed. They appear in spring on the oldest leaves, which develop a mosaic of green rings and blotches on a yellow background. The yellow leaves soon drop, giving the shrub an open appearance. Leaves formed later may have irregular chlorotic spots and be puckered or cupped. Several viral diseases of *H. rosa-sinensis* have been described, but none is reported to cause the dramatic symptoms shown.

The best known viral disease of *Hibiscus* in North America is chlorotic ringspot caused by a carmovirus, *Hibiscus chlorotic ringspot virus*. The virus occurs worldwide in *H. rosa-sinensis*, causing a generalized mottle or numerous tiny chlorotic spots, rings, or vein banding on most leaves of diseased plants. Plants continue to flower, however, and do not become stunted unless infected when very young. The virus is spread in commercial cultivars by grafting and pruning with contaminated shears. The virions are icosahedral, about 28 nm in diameter, and contain single-stranded RNA. *Carnation mosaic virus* (genus *Carmovirus*) also occurs in *H. rosa-sinensis* in the USA and is associated with chlorotic variegation, mottling, and downward rolling of leaf edges. A nepovirus, two tobamoviruses, and a rhabdovirus in *Hibiscus* are mentioned in the discussions of those groups.
References: 2354, 3099, 4243

Apple flatlimb. Flatlimb (Plate 212A) is a graft-transmissible, presumably viral disease that affects many cultivars of *Malus pumila* (apple); Ida Red and Gravenstein are particularly susceptible. The

A. Flatlimb disease in *Malus pumila* 'Gravenstein' (apple). Branches have flattened areas, groovelike depressions, and cankers (WA, Jun).

B, C. Undiagnosed line patterns in *Chaenomeles japonica* (flowering quince). Necrotic flecks have formed within some lines. *Apple mosaic virus* (p. 416) is known to cause line patterns in this species (NY, Jul).

D, E. Undiagnosed mosaic in *C. japonica*. Leaves on spur shoots and the oldest leaves on long shoots develop yellow-and-green mottle or mosaic and drop prematurely, while leaves on terminals remain green (NY, Aug).

F. Undiagnosed line patterns on *Lonicera maackii* (Amur honeysuckle) (GA, Aug).

G. Mosaic and chlorotic rings indicating probable viral infection of a leaf of *Magnolia macrophylla* (bigleaf magnolia) (NY, Jun).

H. Vein banding (vein-associated line patterns) on *Acer platanoides* (Norway maple) (NY, Aug).

I. Necrotic ringspots on *Liquidambar styraciflua* (KS, Aug).
Photo credit: I—N. A. Tisserat

disease occurs around the world where 'Gravenstein' is grown. Flatlimb symptoms have also been observed in *Cydonia* (quince), *Pyrus* (pear), and *Sorbus* (mountain-ash) in Europe. Slight linear depressions or slight flattening develops in strips on shoots and branches where cambial activity slows or ceases. The depressions later become deep furrows or eventually cankers as surrounding tissues grow more or less normally. Some branches become flattened, twisted, and brittle. The causal agent can occur in symptomless rootstocks and symptomless trees, leading to flatlimb in scions after such trees have been "top worked" by grafting susceptible cultivars into them. Symptoms develop beginning a year or more after grafting. The flatlimb agent has been transmitted experimentally to many woody rosaceous species and to *Juglans* (walnut) but has not yet been characterized.
Reference: 1191

Mosaic and line patterns in *Chaenomeles*. The symptoms in *C. japonica* (flowering quince) shown in Plate 212B–D were not diagnosed. Ilarviruses, especially *Apple mosaic virus* (ApMV), would logically be considered during diagnostic work because *C. japonica* is susceptible to natural infection by ApMV and has developed line patterns after inoculation with the virus. The display of mosaic on the oldest quince leaves while young terminal leaves remain unaffected is similar to the apple mosaic syndrome in *Malus*. *Chaenomeles* is also susceptible to one or more viruses that cause line pattern in plum, as well as to another ilarvirus, *Tulare apple mosaic virus*, which has been found in naturally infected plants.
References: 1296, 2081, 3934, 3936

Viruses of *Albizia* and *Ficus*. The symptoms shown in Plate 213A–C on *Albizia julibrissin* (mimosa) and *Ficus carica* (common fig) are caused by incompletely characterized viruses. The diseases, called mimosa striped chlorosis and fig mosaic, respectively, were discovered in Arkansas. Their geographic distributions are unknown. The mimosa virus particles are bacillar and measure 95 × 35 nm. The fig mosaic virus is believed to be transmitted by eriophyid mites.
Reference: 2482

Viruses of *Magnolia* and *Liriodendron*. Viral diseases of *Magnolia* and related plants are poorly known. The cause of the striking green-and-white mosaic and chlorotic rings shown on *M. macrophylla* (bigleaf magnolia) in Plate 212G is unknown. In Europe *M. ×soulangeana* (saucer magnolia) and *M. denudata* (yulan magnolia) are subject to a viral disease characterized by chlorotic lines, arcs, rings, and oak-leaf patterns on leaves. A strain of *Cucumber mosaic virus* (CMV) was isolated from the diseased plants, but its ability to cause the original symptoms was not tested. CMV was also found associated with mild mosaic of *M. acuminata* (cucumber-tree) and *Liriodendron tulipifera* (tuliptree) in Europe. Another graft-transmissible viral syndrome of *L. tulipifera* occurs in southeastern USA. Affected trees have small leaves with chlorotic vein banding, interveinal chlorosis, stippling, and ringspots (Plate 207F, G). Twig growth of affected trees is retarded. Rigid rod particles (average length 272 nm) were associated with the symptoms.
References: 222, 792, 3111, 3455

Oak viruses. Several viral or viruslike diseases of *Quercus* (oak) are known, but none in North America is thought to cause significant deformity or growth loss. In Arkansas, chlorotic ringspots, chlorotic zones along veins (vein banding), and line patterns of the oak-leaf type were found on *Q. marilandica* (blackjack oak; Plate 213F) and *Q. velutina* (black oak) and were shown to be associated with flexuous filamentous particles about 13 nm in diameter. Cells containing the particles also contained abnormal vesicles and cell-wall protrusions. One instance of graft transmission from oak to oak was reported, but attempts to transmit symptoms to herbaceous plants failed. In Europe rod-shaped particles similar to tobamo-viruses and associated with chlorotic flecks or leaf deformities have been observed in *Q. petraea* (sessile oak) and *Q. robur* (English oak). In northern Germany an unidentified dsRNA (replicative form of a virus) was detected in *Q. robur* with graft-transmissible chlorotic

spots, ringspots, and chlorotic mottle. Tests for several viruses known to occur in forest trees were negative. In Denmark a graft-transmissible yellow or chlorotic mottle of *Q. rubra* (northern red oak) was found, but the nature of the causal agent was not reported. Mosaic, chlorotic lesions, and deformed leaves have been noted but not diagnosed in several European oaks.
References: 219, 592, 792, 1760, 2062, 2859, 2862

Viral diseases of conifers. Assorted viruses and viruslike agents have been detected in conifers, sometimes associated with yellowing and slow growth. No viral diseases of conifers are well characterized, however, and no images of them of in North American plants are available other than that in Plate 207. *Tomato mosaic virus* was found in *Picea rubens* (red spruce) trees on a mountain in New York State where the trees were stressed by ozone and acidic cloud water. The virus was detected in cloud water and groundwater at the same site. No foliar symptoms suggestive of viral infection were noted in the trees, but a purified preparation of the virus from *P. rubens* suppressed growth of inoculated seedlings of that species and enhanced the cold hardiness of their aboveground parts. Effects of ToMV on larger plants are unknown. Elsewhere in New York State, two filamentous viruses 12–16 nm in diameter and up to 1260 nm long, with RNA sequences characteristic of tenuiviruses, were detected in *Picea mariana* (black spruce), but the trees had no distinctive symptoms.

In Great Britain and Europe viral disorders of *Pinus sylvestris* (Scots pine) called mosaic (a yellowing syndrome), mottling, and bushy stunt were described in the 1960s. Rod-shaped particles were associated with all three diseases; those of the first measured near 420 × 60 nm, those of the second 300–600 × 15 nm, and those of the third were 25 nm in diameter with variable length. Rod-shaped particles were also detected in *Pinus monticola* (western white pine) in Great Britain, but the particles could not be associated with particular symptoms. Studies of yellowing and decline in *Picea abies* (Norway spruce) in Great Britain and in what is now the Czech Republic also revealed viruslike rod-shaped particles. In the latter country the syndrome and the particles were reported to be transmissible by *Adelges abietis* (eastern spruce gall adelgid) and *Cinara pilicornis* (spruce shoot aphid). A potyvirus closely related to *Potato virus Y* was isolated from *Picea abies* with a yellowing syndrome in Bavaria, and this virus was shown to be aphid-transmissible between herbaceous indicator plants. *Tobacco necrosis virus*, found in *P. abies* roots in Germany, reduced growth in inoculated seedlings of that species. Yellowing and decline in *P. pungens* (blue spruce) and *P. sitchensis* (Sitka spruce) containing rod-shaped viruslike particles have been noted in Great Britain, Germany, and Russia. Several *Abies* (fir) species in Germany and southern France were found to contain isometric viruslike particles with diameters near 20 and 35 nm. These particles were found in both trees and seedlings, indicating probable seed transmission. *Arabis mosaic virus* was found in roots of *Chamaecyparis lawsoniana* (Lawson-cypress) in Great Britain, but no symptoms were noted.
References: 156, 355, 674, 676, 682, 1030, 1146, 2861

Miscellaneous viral and viruslike diseases.

A, B. Striped chlorosis of *Albizia julibrissin* (mimosa), caused by an incompletely characterized DNA virus (AR, summer).

C. *Ficus carica* (common fig) with fig mosaic caused by a double-membrane-bound, mite-transmitted, viruslike agent (AR, summer).

D, E. Yellow ringspots indicating possible viral infection of *Quercus myrtifolia* (myrtle oak) (FL, Mar).

F. Line pattern on *Q. marilandica* (blackjack oak), caused by an unidentified virus (AR, Jul).

G. Line patterns on *Bougainvillea* sp. indicating probable viral infection (FL, Mar).

H, I. Chlorotic spots indicating possible viral infection of *Euonymus* spp. (GA, May; MD, greenhouse).

J. Mosaic indicating probable viral infection of *Celtis occidentalis* (hackberry) (NY, Jun).

Photo credits: A–C—K. S. Kim; F—O. W. Barnett; I—H. E. Waterworth

429

Diseases Caused by Nematodes (Plates 214–216)

Overview of Plant-Parasitic Nematodes

Nematodes (Phylum Nematoda) are soft-bodied tubular animals (Figs. 215, 216) that inhabit many ecological niches: water, soil, plants, and other animals. Many nematodes are free living and consume microorganisms and organic detritus as food, but thousands of species parasitize plants, fungi, or other animals. More than 70 genera of plant-parasitic nematodes in two classes and four orders are recognized. More species occur, and cause more damage, in tropical and warm-temperate regions than in cool regions. Important groups of nematodes affecting woody plants include *Aphelenchoides* (foliar nematodes), *Bursaphelenchus* (pine wood nematodes), *Helicotylenchus* (spiral nematodes), *Hoplolaimus* (lance nematodes), *Longidorus* (needle nematodes), *Meloidogyne* (root-knot nematodes), *Mesocriconema* and other Criconematidae (ring nematodes), *Paratylenchus* (pin nematodes), *Pratylenchus* (lesion nematodes), *Radopholus* (burrowing nematodes), *Tylenchorhynchus* (stunt nematodes), and *Xiphinema* (dagger nematodes), among others. We deal here with representatives of 6 genera.

Plant-parasitic nematodes are ubiquitous. They are dispersed with soil, water, plant materials, cultivation implements, and (for a few types) insect vectors. Most are soilborne and feed on or in roots—succulent roots not protected by periderm. Some nematodes, notably *Aphelenchoides* species, climb and enter succulent aboveground plant parts, and a few species such as the pine wood nematode (Plate 216) invade plants systemically. Some plant-parasitic nematodes (notably species of *Longidorus*, *Xiphinema*, and related genera) are vectors of plant viruses. Damage to woody plants by root-infesting nematodes is often diagnosed in nurseries and orchards but is seldom demonstrated or assessed in other situations.

Nematodes reproduce by eggs and develop through four juvenile stages before reaching adulthood. Each juvenile stage terminates when the nematode molts (sheds its cuticle). Males of some species are unnecessary for reproduction, and some other species apparently lack males. Plant-parasitic nematodes are small (most are less than 2 mm long), usually colorless, worm shaped in one or more stages of development, and have at least one motile stage. The motile stages move sinuously in soil water, on wet plants, or within plants. Nematodes that feed on or in plants puncture cell walls by means of a sclerotized spear, or stylet, and suck out cell contents. During or preparatory to feeding they produce secretions that aid digestion or alter the plant in ways favorable for the nematode. These secretions dissolve intercellular pectic substances, degrade cell walls, suppress cell division, induce cell proliferation, or induce formation of giant cells that serve as long-term feeding sites. Nematodes infest both mycorrhizal and nonmycorrhizal roots. Some plant parasites feed exclusively from outside the root (ectoparasites); others enter and remain within roots for much of their development (endoparasites). Some move from one feeding site to another and are termed migratory; others are sedentary. Females of some sedentary types become swollen and pear shaped to spheroid. The duration of the life cycle varies with species from a few days to a year, and the life span varies from several weeks to 5 years. Most root-feeding nematodes can utilize diverse plants, including weeds; survive in dry or fallow soil; and reproduce rapidly when suitable plants are available.

Some nematode-induced symptoms (e.g., root knot galls) are diagnostic, but most are not. Nematodes may induce various swellings, cause lesions, suppress root elongation or differentiation, or interact with other microorganisms to cause root rot or wilt where neither type of organism alone would do so. If root damage by nematodes is severe, infested plants may grow slowly, become stunted, show chlorosis or other symptoms of nutritional stress, and may be abnormally subject to winter damage or attack by opportunistic fungi or bacteria.

Diagnosis of most nematode diseases requires microscopic examination and enumeration of the pests. This process entails extracting nematodes from soil or plant material, identifying and counting the types present, and interpreting the data. Interpretation of nematode types and numbers is usually essential because plants vary in suitability as hosts and in ability to tolerate particular nematodes, and nematodes vary in ability to damage particular plants. The counting step may be unnecessary for diseases with diagnostic symptoms.

Contemporary approaches to control of root-feeding nematodes involve combinations of crop rotation, heating of soil, amending soil with botanical materials that suppress nematodes, interplanting susceptible plants and those inhibitory to nematodes, cultural procedures that minimize contamination of noninfested soils, growing plants in composted media, and selective use of pesticides. Soil treatments to introduce or encourage particular biological control agents are not yet feasible at field scale. Selection or breeding of nematode-resistant plants has not yet been emphasized for woody ornamental or forest trees.

References: 182, 199, 200, 316, 776, 1002, 1087, 1115, 1152, 1441, 1967, 2421, 2427, 2851, 2875, 3367–3369, 3402, 3610, 3621, 3765

Representative Nematodes That Attack Roots (Plates 214, 215)

Root Knot Nematodes

Meloidogyne species (Tylenchida, Meloidogynidae; Fig. 215A–D) cause root knot and are globally important plant pathogens. More than 55 species are recognized. Several occur in North America, and 4 are widespread and important on woody plants. These are *M. hapla* (northern root knot nematode) and 3 species that inhabit warm regions: *M. arenaria*, *M. incognita*, and *M. javanica*. Woody plants subject to severe root knot include species of *Abelia*, *Aucuba*, *Berberis* (barberry), *Buxus* (boxwood), *Cornus* (dogwood), *Deutzia*, *Ficus* (fig), *Forsythia*, *Gardenia*, *Ilex* (holly), *Prunus* (especially peach), *Rhododendron* (azalea), *Rosa* (rose), *Salix* (willow), *Schefflera*, *Syringa* (lilac), and *Weigela*.

Root knot nematodes are sedentary endoparasites. Only males and second-stage juveniles are motile, and only the second-stage juveniles and adult females feed. Feeding stimulates formation of galls (called knots), in which the nematodes become embedded. Eggs are released at the surfaces of knots by embedded females. The first juvenile stage develops within the egg. Second-stage juveniles hatch out, migrate a short distance in soil, enter succulent roots near the root tip, and migrate to the region of differentiating xylem. They position themselves with bodies in the cortex and heads within the vascular cylinder, where they feed. Their feeding stimulates gall formation. Plant cells in the outermost stele and the cortex undergo abnormal division and growth, and several large, multinucleate cells (giant cells) rich in cytoplasm differentiate around the nematode's head. The giant cells remain alive throughout the life of the nematode. By moving its head only slightly, the animal can ingest food from each cell. It swells to sausage shape, then undergoes three molts within only a few days and becomes an adult male or female. If male the animal is worm shaped. It leaves the root, mates, and remains free in soil. If female the nematode remains in place; her body swells, becoming nearly spherical except for the head region. The swollen white female, more than 0.1 mm in diameter (Fig. 215B), can be exposed by careful dissection. Her posterior is near the root surface, where she deposits several hundred eggs into a gelatinous egg sac.

Under favorable conditions in warm soil (25–30°C), a new generation may be produced in 3–4 weeks. The population increases until it is limited by food supply. Severe infestation may result in conversion of feeder roots to a mass of short, thick, bumpy roots subject to decay by secondary pathogens.

Root knot nematodes interact with other pathogens to induce more damage than either would cause alone. Crown gall (Plate 190) and several fungal root rot and wilt diseases may be aggravated by these nematodes.

A–D. Root knot of *Gardenia* sp. A. Diseased container-grown plants in a nursery; the plant at left is chlorotic and severely stunted. B. Roots of a severely diseased plant are distorted by large numbers of galls (knots). Normal rootlets are scarce. C. Close view of a diseased root. D. Magnified view of knots on a small root (FL, Apr).

E, F. Root knot caused by *Meloidogyne hapla*, the northern root knot nematode, on *Forsythia* sp. E. Distorted, knotted roots. F. Close view of knots. Note brown necrotic tissue and scarcity of normal rootlets (NY, Feb).

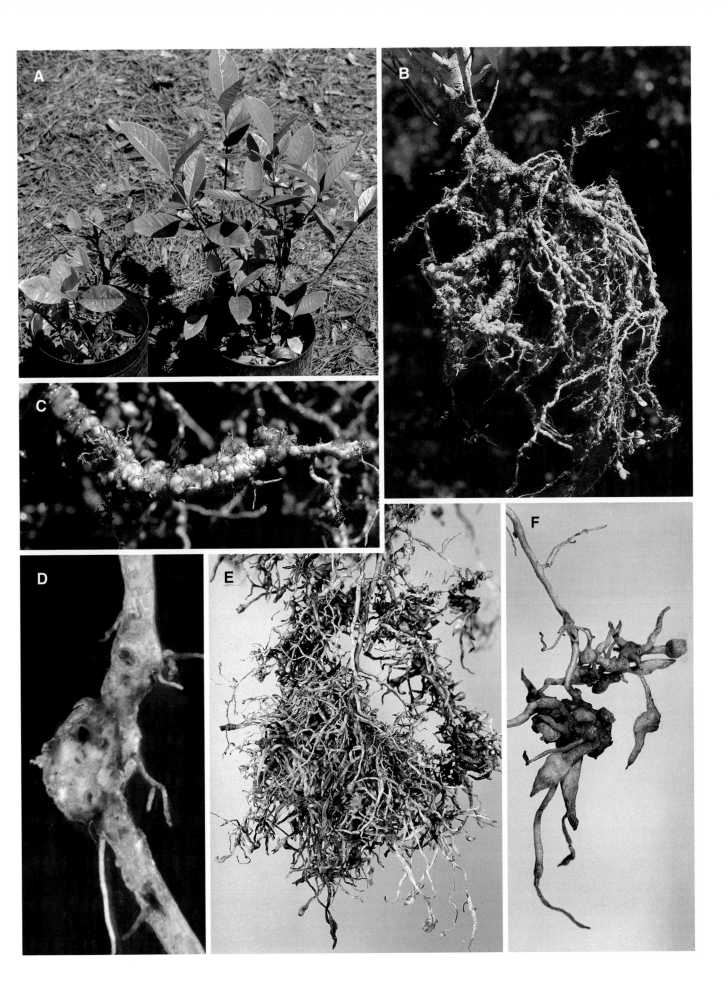

References: 315, 316, 776, 1120, 2017, 2102, 2222, 2262, 3285, 3288, 3363, 3403, 3405, 3414, 3743, 3890

Burrowing, Stunt, Lesion, and Dagger Nematodes

Burrowing nematode. *Radopholus similis* (Tylenchida, Pratylenchidae) is a migratory endoparasite of succulent root tips and other fleshy subterranean organs. It occurs around the world in tropical and subtropical areas, parasitizing plants of more than 250 species. It is best known for damage to *Musa* (banana), *Citrus*, *Cocos nucifera* (coconut palm), and *Piper nigrum* (black pepper). Two races (also called pathotypes) occur, differentiated by the ability of one to infect *Citrus*. The citrus race, known only in the Western Hemisphere, also infects *Carya illinoinensis* (pecan), *Diospyros virginiana* (persimmon), *Mangifera indica* (mango), *Persea americana* (avocado), *Pinus elliottii* (slash pine), *P. taeda* (loblolly pine), *Poncirus trifoliata* (trifoliate orange), *Prunus persica* (peach), and many other herbaceous and woody plants. It kills citrus root tips, inducing the disease called spreading decline. Affected trees lose vigor; foliage becomes somewhat chlorotic, sparse, and undersized; shoots may wilt temporarily during the day; twigs die back; and productivity is lost. Symptoms spread slowly across an orchard. *Citrus ×paradisi* (grapefruit) is more severely affected than *C. sinensis* (sweet orange).

Other host plants of the burrowing nematode in Florida do not show a decline syndrome similar to that in citrus. This difference may be associated with plant rooting habits. *Citrus* is deep rooted, and the nematode prefers deep-lying roots. It causes most damage at soil depths greater than 50 cm and no damage in the topmost 25 cm. Plants with root systems principally in the top 50 cm of soil escape severe damage.

The life cycle of *R. similis* is completed in about 3 weeks. All life stages can be found within roots, and all stages except eggs are motile. The feeding and movement of juveniles and females within a root cause lesions that coalesce and destroy the root tip. The nematodes then leave the root and migrate to another. Males do not feed, however, and after leaving a root do not enter another. Migration of burrowing nematodes, up to 15 m per year, is the basis of spreading decline in orchards. Long-distance dispersal occurs in infested plant materials. Resistant rootstocks appear promising for control in citrus orchards.

References: 182, 316, 776, 1152, 2012, 2130, 2875, 2885, 3130, 4331

Stunt nematodes. *Tylenchorhynchus* species (Tylenchida, Belonolaimidae; Fig. 215E–G) are migratory ectoparasites that feed on epidermal cells and root hairs of various plants, suppressing root growth and often causing sufficient damage to retard top growth. Roots become discolored light to dark brown. *Picea* (spruce), *Robinia pseudoacacia* (black locust), and *Rosa* (rose) are highly favorable hosts for reproduction of these nematodes. *Pieris japonica* (Japanese-andromeda) and *Rhododendron* species (azaleas) become stunted and develop chlorosis, tip burn, and (in azalea) leaf drop when severely affected by stunt nematodes. *Magnolia* and *Pinus* may also be damaged.

References: 201, 316, 776, 2875, 4331

Lesion nematodes. *Pratylenchus* species (Tylenchida, Pratylenchidae; Fig. 216A–C) are migratory endoparasites with a broad plant host range. They occur in cool- and warm-temperate regions. Larvae and adults feed in cortical parenchyma of roots, and females lay eggs singly or in small clusters within the cortex. The nematodes can feed only on living cells and so must move frequently within a root or exit and reenter at another point. In so doing, they leave tracks of ruptured cells in the cortex that are soon colonized by saprobic and pathogenic bacteria and fungi. Feeding and migration result in formation of numerous small lesions. The life cycle is completed in 1–3 months, depending on nematode species, host, and temperature. Root damage by lesion nematodes, usually in concert with fungi and bacteria, causes growth loss of tree seedlings, woody ornamentals in nurseries, fruit trees in orchards, and many agricultural crops. Woody hosts include *Acer* (maple), *Buxus* (boxwood), *Cotinus* (smoke bush), *Cotoneaster*, *Cryptomeria* (Japanese-

cedar), *Ficus* (fig), *Forsythia*, *Ginkgo*, *Juglans* (walnut), *Juniperus* (juniper, eastern red-cedar), *Ligustrum* (privet), *Liquidambar* (sweetgum), *Liriodendron* (tuliptree), *Magnolia*, *Malus* (apple, crabapple), *Olea europaea* (olive), *Persea americana* (avocado), *Picea* (spruce), *Pinus* (pine), *Prunus* (almond, cherry, peach), *Pseudotsuga* (Douglas-fir), *Rosa* (rose), *Salix* (willow), and *Vitis* (grapevine), among others. Diseases such as Fusarium root rot of coniferous seedlings (Plate 108) and Verticillium wilt (Plates 120, 121) are aggravated if roots are infested by *Pratylenchus*.

References: 31, 182, 350, 776, 1015, 1152, 1900, 1967, 2426, 2757, 2875, 3076, 3077, 3840, 4172, 4331

Dagger nematodes. *Xiphinema* species (Dorylaimida, Longidoridae; Fig. 216D, E) are among the largest (to 5 mm) and include some of the slowest-developing and longest-lived plant-parasitic nematodes. Some species undergo only one generation per year, and females may live for a year. Shorter generation times are usual, however. Dagger nematodes feed from outside the root through a hollow spear that may be nearly 200 μm long, enabling the nematode to reach deep-lying root cells. Nematodes feeding at a root tip inhibit root elongation but stimulate cell division. The result is stunted, swollen root tips that are attractive feeding sites for other nematodes of the same species. Formation of short lateral roots may be suppressed also, resulting in an abnormally coarse root system. Secondary pathogens kill many rootlets damaged by the nematodes. *Xiphinema* species are common around the world in horticultural and untended landscapes, orchards, and forests. In North America they have been found associated with decline or lack of productivity of *Acer* (maple), *Celtis* (hackberry), *Elaeagnus angustifolia* (Russian-olive), *Fraxinus* (ash), *Juniperus virginiana* (eastern red-cedar), *Malus* (apple), *Picea* (spruce), *Populus* (cottonwood), *Prunus* (stone fruit trees), *Ulmus* (elm), and *Vitis* (grapevine) among many other woody and herbaceous plants. Root damage by nematodes of the *X. americanum* group to *Picea pungens* (blue spruce) in midwestern nurseries led at one time to winter injury and an outbreak of Leucostoma canker (Plate 83). *X. bakeri* causes corky root disease of *Pseudotsuga menziesii* (Douglas-fir) and other coniferous seedlings in western Canada. *X. index* damages *Ficus carica* (fig) and *Vitis* in California and elsewhere around the world.

Xiphinema is one of three nematode genera (the other two being *Longidorus* and *Paralongidorus*) with species that are vectors of nepoviruses, which damage many woody plants. For example, certain nematodes of the *X. americanum* group (which comprises more than 30 species) transmit tomato ringspot, tobacco ringspot, and/or cherry raspleaf viruses. Tomato ringspot virus causes graft union necrosis and decline in apple and stem pitting and decline in peach and cherry. *X. index* is a vector of *Grapevine fanleaf virus* in California.

References: 316, 776, 1152, 1368, 1371, 1967, 2376, 2435, 2875, 3312, 3410, 3906, 4212, 4331, 4455

A. Sparse foliage and dieback on *Citrus sinensis* 'Valencia' orange on *C. jambhiri* (rough lemon) rootstock damaged by *Radopholus similis*, the burrowing nematode. 'Valencia' orange on nematode-resistant 'Milam' rootstock, at right, is unaffected. Burrowing nematodes enter but do not reproduce in the resistant rootstock (FL, Apr).

B. Chlorosis, tip browning of leaves, and reduction of leaf size associated with damage to roots of *Pieris japonica* (Japanese-andromeda) by a stunt nematode, *Tylenchorhynchus* sp. (NY, Nov).

C, D. Damage to *Malus pumila* (apple) seedlings by a lesion nematode, *Pratylenchus penetrans*. C. Plants growing in nematode-infested soil (right) and noninfested soil. D. Root systems of the two plants shown in C. The damaged roots are stunted and darkened by many small brown lesions (NY, greenhouse).

E, F. *M. pumila* roots damaged by a dagger nematode, *Xiphinema americanum*. E. Proliferation of short lateral roots above the point where two long roots were killed. Rootlets or root tips on which the nematode fed are stunted and swollen. F. Swollen rootlets damaged by the nematode turn dark brown when killed by secondary microbial pathogens (NY, Jun).

433

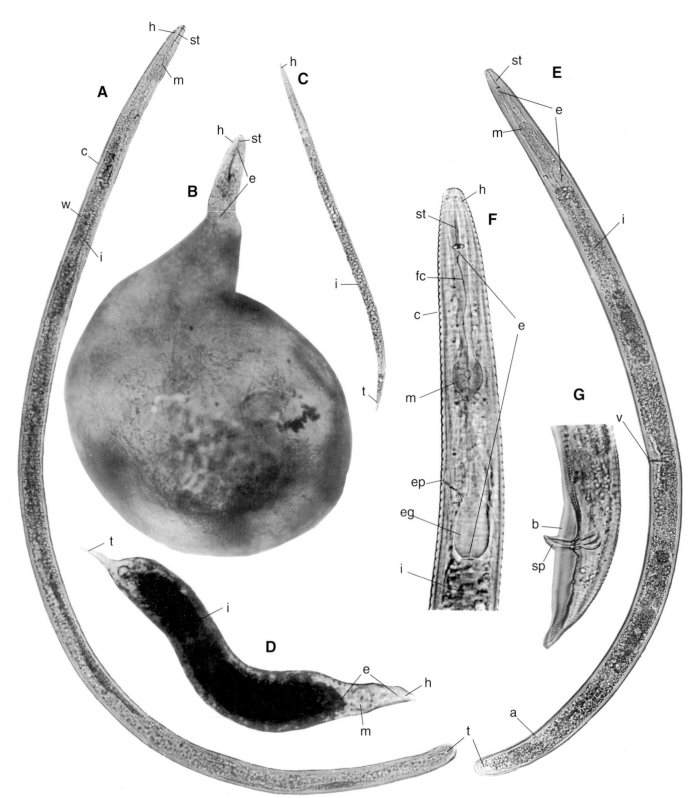

Figure 215. A–D. *Meloidogyne* sp., a root knot nematode. A. Male. B. Female. C. Motile second-stage juvenile. D. Sausage-shaped older juvenile extracted from a root. E–G. *Tylenchorhynchus* sp., a stunt nematode. E. Female. F. Detail of anterior portion. G. Detail of male tail and genital organs. Magnifications differ; *Meloidogyne* female is about 245 μm in diameter; *Tylenchorhynchus* female is about 890 μm long. a = anus, b = bursa (clasping organ), c = cuticle, e = esophageal region, eg = esophageal glands, ep = excretory pore, fc = food canal, h = head region, i = intestine, m = metacorpus (a pumping organ that moves saliva, food), sp = spicules, st = stylet, t = tail, v = vulva, w = body wall.

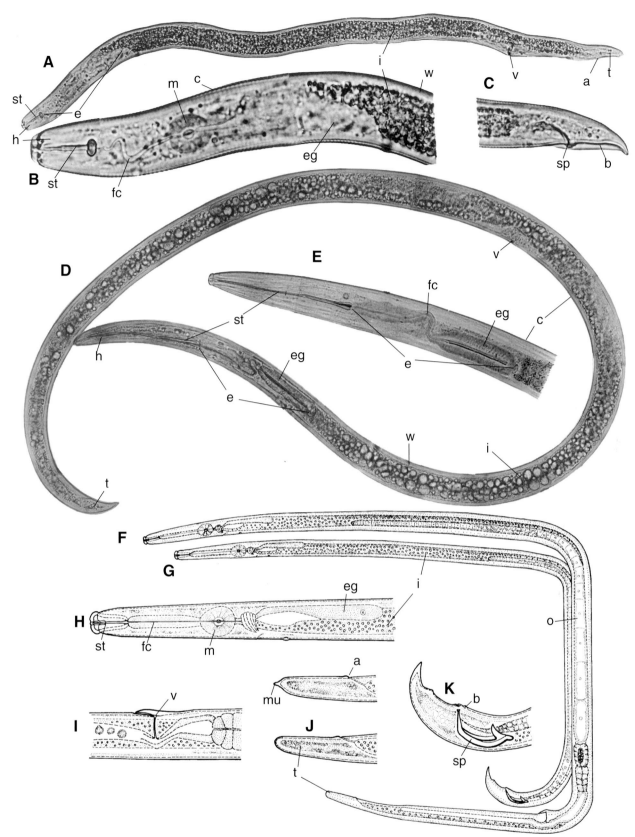

Figure 216. A–C. *Pratylenchus penetrans*, a lesion nematode. A. Female. B. Detail of anterior portion. C. Detail of male tail and genital organs. D, E. *Xiphinema americanum*, a dagger nematode. D. Female. E. Anterior portion. F–K. *Bursaphelenchus xylophilus*, the pine wood nematode. F, G. Adult female and male, respectively. H. Anterior portion. I. Vulvar region of female. J. Female tails, one with mucro (point). K. Male tail region. Magnifications differ; the *Pratylenchus* female is about 455 μm long, the *Xiphinema* female about 1.9 mm long, the *Bursaphelenchus* female about 1.0 mm long. a = anus, b = bursa (clasping organ), c = cuticle, e = esophageal region, eg = esophageal glands, ep = excretory pore, fc = food canal, h = head region, i = intestine, m = metacorpus (a pumping organ that moves saliva, food), mu = mucro, o = oocyte in ovary, sp = spicules, st = stylet, t = tail, v = vulva, w = body wall. F–K adapted from reference 2447 by permission of Brill Academic Publishers.

Wilt of Pines Caused by Pine Wood Nematodes (Plate 216)

Nematodes of the genus *Bursaphelenchus* (Aphelenchida, Parasitaphelenchidae) infest wood and woody plants and some types of beetles that breed in these materials. Plant-pathogenic *Bursaphelenchus* species colonize sapwood of living coniferous trees as well as dying or recently dead trees and logs. They reproduce using either living plant cells or mycelium of wood-invading fungi, especially bluestain fungi, as a food source. They are vectored by beetles of the Cerambycidae (long-horned beetles), primarily species of *Monochamus*. *B. xylophilus* (the pine wood nematode) and *B. mucronatus* are the best known members of their genus. *B. mucronatus* is distributed across Eurasia and is inconsequential in most circumstances. It is commonly found associated with decline of environmentally stressed pines. *B. xylophilus*, however, is a major forest pathogen in some regions. It is indigenous and mostly inconsequential in native conifers across North America, but kills trees of highly susceptible species where they come under stress from heat and drought. *Pinus densiflora* (Japanese red pine), *P. elliottii* (slash pine), *P. nigra* (Austrian pine), *P. sylvestris* (Scots pine), and *P. thunbergii* (Japanese black pine) are among the most susceptible species. Locally destructive outbreaks have occurred in forest and landscape plantings of *P. sylvestris* in central and eastern USA, but the worst damage by far occurs in Japan, where *B. xylophilus* was introduced early in the 20th century. It has decimated populations of *P. densiflora* and *P. thunbergii* in the southern part of that country, especially in coastal areas. More recently damage has begun in Taiwan and eastern China, and the pest has been discovered in Portugal and Nigeria. The threat of damage to native pines in Europe and other as yet uninfested areas has led to severe restrictions on international commerce in coniferous forest products.

B. xylophilus (Fig. 216) has been found in trees or logs of more than 30 species of *Abies* (fir), *Cedrus* (cedar), *Larix* (larch), *Picea* (spruce), *Pinus*, and *Pseudotsuga* (Douglas-fir). Most North American conifers are resistant unless water stressed, however, and most host records reflect discovery of the nematode in logs and declining trees. Wilt and death caused primarily by the nematode is common only in pines.

Symptoms and signs in pines. Single trees or small groups develop symptoms without relationship to previous vigor. One or a few branches or an entire tree may show arrested growth, fading green color, or slight yellowing. Prominent yellowing or wilt and browning soon follow. In lethal attacks trees usually die from early summer to early autumn and retain brown needles for several months. When the disease develops slowly, the oldest needles turn yellow and drop before the remaining foliage turns brown. In large trees, branch dying sometimes begins in autumn and resumes the next spring. Before visible symptoms arise, epithelial cells that line resin ducts die, oleoresin pressure diminishes so that resin no longer flows at wounds, and transpiration slows or ceases. Sectors of wood in the trunk may become resin soaked.

Disease cycle. B. xylophilus, together with assorted fungi, insects, and other nematodes, resides in recently dead coniferous (mostly *Pinus*) trunks, limbs, and logs, in which it feeds on fungi. It is carried to these habitats by *Monochamus* species that breed there. The principal vectors in North America are *M. carolinensis* (the Carolina sawyer) and *M. scutellatus* (whitespotted sawyer). *M. alternatus* (Japanese pine sawyer) is the principal vector in the east Asian region. Several additional cerambycid species are likely to be at least occasional vectors. When young adult beetles of vector species emerge from nematode-infested trees or logs, they carry hundreds to tens of thousands of nematodes in their bodies. The beetles first feed briefly on the bark of healthy pine shoots but are soon attracted to dying or felled trees, where they mate and the females oviposit (lay eggs) beneath the outer bark. Nematodes leave the insects and enter new pine substrates during both feeding and oviposition. Nematode transmission to healthy pines during beetle feeding can lead to the pine wilt syndrome. Transmission during oviposition allows the nematodes to inhabit places where pine wilt is rare or does not occur as well as places where the disease is common. The nematodes, by killing trees, provide additional habitat for the beetles. In central USA all life stages of *M. carolinensis* except pupae are present throughout the year, and the beetles produce two generations per year in southern areas.

In the pine wilt scenario, nematodes entering feeding wounds in shoots move actively and often rapidly (40–50 cm or more per day)

from the site of inoculation. They first colonize cortical resin ducts and kill epithelial cells (parenchyma cells) that line the ducts. Then they move into xylem, reproducing rapidly and causing spreading dysfunction of the tree's resin secretion system. Population density is commonly 10^3–10^4 nematodes per gram of wood (dry weight basis) when foliar symptoms appear, and nematodes occur throughout the trunk, branches, and large roots. Toxins cause or contribute to death of cortical and xylem parenchyma cells and initiate cavitation (displacement of xylem fluid by gas) in xylem tracheids. Spreading catastrophic cavitation in sapwood leads to wilting and tree death.

The toxins are produced during interaction of the nematodes, associated bacteria, and epithelial cells of resin ducts. Bacteria alone do not cause the symptoms. Three of the toxins are 8-hydroxyverbenone and carvone hydrate, derived from pine oleoresin, and phenylacetic acid produced by bacteria. Benzeneacetic acid and 2-methoxycinnamic acid, both also toxic, accumulate in infected tissues. *B. xylophilus* also secretes cellulase and other enzymes that may play a role in the initial damage to parenchyma cells.

Nematode population increase sufficient to initiate systemic failure of the resin-secreting and water-conducting systems requires warm to hot weather (e.g., temperatures above 25°C) and low water potential in the tree. These conditions occur annually in regions where mean summer temperature is above 20°C, such as central USA and much of Japan. In cooler regions trees' defenses appear to keep nematode populations in check, and infected trees seldom wilt. Small populations of *B. xylophilus* may persist in living trees for many years. Long-term infestation of apparently healthy pines by *B. xylophilus* can occur, as was demonstrated by the nematode's recovery from healthy-appearing *P. sylvestris* in Vermont 6 years after the trees were inoculated.

B. xylophilus has two modes of development: propagative and dispersive. During rapid reproduction (propagative mode), whether the nematodes are feeding on fungi or plant cells, four juvenile stages functionally similar to one another precede the adult stage. Given a suitable substrate *B. xylophilus* can complete its life cycle in 12 days at 15°C and in only 4–5 days at 25°C. Each female, after insemination, lays several dozen eggs during a period of 2–4 weeks before she dies. As reproduction diminishes, nematode development switches to the dispersive mode, in which third- and fourth-stage juveniles are adapted to withstand starvation and, in the fourth, or dauerlarval, stage, drying. Third-stage juveniles in the dispersive mode are attracted to the pupal chambers of insect vectors, where they molt to dauerlarvae and enter the bodies of adult beetles just emerging from pupation.

Strains of *B. xylophilus* from different hosts and regions vary in pathogenicity, virulence in various hosts, and host specificity. For example, when strains from *Abies* and *Pinus* were tested for ability to colonize and cause symptoms in seedlings of various conifers, they were most virulent in the species from which they had been isolated and in related species. Resistance to *B. xylophilus* has been detected in *P. densiflora* and *P. thunbergii*.

References: 128, 471, 472, 640, 1018, 1019, 1035, 1127, 1218, 1443, 1844, 1848, 1849, 1867, 1879, 1940, 2026, 2082, 2122, 2313, 2316, 2333–2335, 2436, 2445–2447, 2738, 2899, 3372, 3630, 3678, 3923, 4052, 4231, 4393, 4394

A. *Pinus sylvestris* (Scots pine) dying of wilt caused by *Bursaphelenchus xylophilus* (MO, Jul).

B. *Pinus thunbergii* (Japanese black pine) recently killed by *B. xylophilus*, with foliage still fading on some branches (NY, Aug).

C. Arrested growth of a *P. sylvestris* branch infested by *B. xylophilus*. Some branches on the same tree had died when the photo was made (MO, Jul).

D. Cross-section of a recently killed *P. thunbergii* trunk, showing uniform annual growth until the year of death, stain (bluestain) caused by secondary fungi introduced by bark beetles, and sectors of resin-soaked wood (NY, Aug).

E, F. *Monochamus carolinensis*, a vector of *B. xylophilus*. E. On *P. sylvestris;* young needles beginning growth in foreground provide a size comparison (MO, Apr). F. Pinned male (left) and female specimens.

Photo credits: E, F—M. J. Linit

Plant-Pathogenic Algae and Vascular Plants (Plates 217–227)
Algal Leaf Spot, or Green Scurf (Plate 217)

Algae that parasitize higher plants are widespread in warm, humid regions but represent relatively few taxa. About 15 species belonging to three families are known. A few species in the genus *Cephaleuros* (Phylum Chlorophyta, Order Trentepohliales, Family Trentepohliaceae) are significant pathogens. *C. virescens* causes economic damage to various tree crops, notably *Citrus*, *Coffea arabica* (coffee), *Elaeis guineensis* (oil palm), *Hevea brasiliensis* (rubber), *Mangifera indica* (mango), *Persea americana* (avocado), *Piper nigrum* (pepper), *Camellia sinensis* (tea), and *Theobroma cacao* (cocoa) in warm countries around the globe. It is the only parasitic algal species common in the USA, where it occurs on plants of more than 200 species in more than 60 families in a region extending from North Carolina to Florida and Texas. *C. parasiticus*, an economically significant pathogen of tea abroad, causes a leaf spot of *Magnolia grandiflora* (southern magnolia) in Louisiana but is otherwise unknown in the USA. Plants that have leaves with somewhat leathery surfaces are most conspicuously attacked by parasitic algae.

C. virescens in the USA has been associated with damage to leaves or bark of *Acacia auriculiformis* (northern black wattle), *Bixa orellana* (anatto), *Camellia japonica* (common camellia), *Citrus* species, *Coccoloba* species (pigeon plum and sea grape), *Cotoneaster*, *Dimocarpus longan* (longan), *Eriobotrya japonica* (loquat), *Franklinia*, *Ilex* species (holly), *Magnolia grandiflora*, *Michelia figo* (banana-shrub), *Morella faya* (candleberry-myrtle), *Psidium guajava* (guava), *Rhododendron* (azalea), *Rubus* (blackberry), *Syzygium cumini* (jambolan), and many other plants.

Signs and symptoms. Colonies of *C. virescens* on leaves, twigs, branches, and sometimes fruits become conspicuous as raised spots or blotches before symptoms develop. Colonies are usually less than 1 cm across but may coalesce and form a continuous sheet. Leaf tissue beneath colonies slowly dies, and brown spots appear. Numerous colonies close together cause premature yellowing and defoliation. During algal reproduction in summer, colonies on leaves appear as reddish brown pads with velvety or roughened surfaces. At other times they are grayish green to greenish brown with green margins. Numerous sterile hairs (trichomes) are present on all but the youngest colonies. The reddish brown color is due to a pigment, haematochrome, that forms in trichomes and in the stalks of hairlike asexual spore-producing structures (sporangiophores). This color has led some authors to call the disease red rust.

A layer of cork cells often develops beneath the algal disc, separating healthy leaf or bark tissue from diseased tissue. The cork may impede algal colony growth and cause the disc and killed tissue to project, cushionlike, above the normal plant surface. This isolation response has been observed in *Citrus* leaves, although not in *Camellia*, *Magnolia*, or *Eriobotrya*.

Plant tissue beneath old colonies sometimes shrivels, leaving an air space between the leaf and the center of the algal disc. Columns of algal cells extend across the space, presumably providing anchorage and allowing conduction of water and nutrients to the disc. Whether the alga derives organic nutrients in addition to water and minerals from its hosts is unknown.

Colonies on twigs and branches arise in bark crevices, rarely on green twigs, and induce thickening of bark around the colony. The raised bark eventually cracks into small irregularly shaped plates or becomes somewhat shredded. Colonies that encircle branches may cause stunting or premature yellowing of leaves and eventual death of distal parts. Citrus branches as large as 5 cm in diameter may be killed. The depth of infections in bark and the mechanism by which the alga invades bark are unreported. Mechanical rupture by the pressure of expanding algal cells in bark fissures has been suggested.

Life cycle. Although *C. virescens* can be cultivated readily on artificial media, descriptions of its life history and pathogenicity are all from observations of collected rather than experimental materials. Colonies begin growth in summer and are evident by early autumn. Those that persist on leaves usually develop between the cuticle and epidermis, rarely beneath the epidermis or among mesophyll cells. How the alga penetrates cuticle and causes separation of cells or separation of cuticle from epidermis is unknown, nor has its possible parasitism been studied other than through microscopic observations. Some authors have reported and illustrated haustoria in epidermal cells, but others have not mentioned them. Growing colonies become disc shaped—several cell layers thick at the center and one layer thick at the margin.

Branching chains of algal cells, visible with a hand lens, grow radially at colony margins.

Asexual reproduction begins in summer with development of sporangiophores, each producing several microscopic sporangia at its apex. Sporangia, which look and function like spores, are dispersed by splashing or running water or by air. Some sporangia develop without sporangiophores. They open in a film of water on the plant surface and release motile spores (zoospores) that swarm for a few minutes before settling. These give rise to new algal colonies that begin producing sporangia in late spring or early summer of the next year. Successive annual crops of sporangia develop in the same colony.

Sexual reproduction begins with the formation of sessile gametangia, many in the surface layer of a colony, that liberate motile gametes. These fuse in pairs and give rise to new colonies, either directly or after formation of a dwarf sporophyte generation. Authors differ on this point, and the difference is fundamental. One interpretation is that the algal colony (also called a thallus) is diploid; that meiosis occurs in the sessile gametangia; resulting in haploid gametes; and that after these fuse they give rise to new diploid colonies. The other proposed scenario has alternating haploid and diploid generations that differ in form. The algal colony, supposedly haploid, releases motile gametes that unite to form a diploid zygote that develops into a microscopic "dwarf sporophyte" thallus. This short-lived structure produces sporangia in which meiosis occurs, resulting in the production of haploid zoospores that give rise to new colonies.

The life span of an individual colony is determined mainly by the fate of the infected plant part. Colonies on leaves survive until after the leaves drop. Old colonies often support fungal fruiting structures, some of which are produced by the fungal symbionts of lichenized colonies. Two lichens, *Strigula complanata* and *S. elegans*, formed by fungi in association with *C. virescens* on bark of various plants, are considered to be parasitic. See the discussion with Plate 261 for more information about lichens. Other fungi in old algal colonies may be parasites of weakened leaf tissue or of degenerating parts of the algal disc, or they may simply be saprobic colonists of dead cells.

Abundant rainfall, high temperature, and direct sunlight favor *C. virescens*. The alga is most aggressive and damaging on slowly growing, weak plants. Such plants are also most subject to secondary attack by opportunistic fungi, but the possible role of fungi in damage attributed to *C. virescens* has not been studied.

References: 46, 692–694, 1036, 1037, 1718, 1976, 2101, 2470, 2951, 3122, 3406, 4411

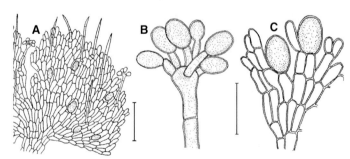

Figure 217. Cephaleuros virescens. A. Part of a colony with sterile hairs and sporangia. B, C. Stalked and sessile sporangia, respectively. Scale bars = 200 μm in A, 50 μm in B and C. Adapted from reference 4411 by permission of North Carolina Academy of Science.

Algal spots caused by *Cephaleuros virescens*.

A. The presence of numerous colonies apparently induces premature senescence of *Magnolia grandiflora* leaves (FL, Mar).

B. Numerous small colonies on a leaf of *M. grandiflora* arose after spore production by the large colony at upper left. The arrangement of the colonies indicates that spores were carried in a film of water (FL, Apr).

C. Young colonies on *Persea borbonia* (red bay) (FL, Mar).

D. Lichenized colonies on *M. grandiflora* display green margins, grayish brown centers, and black perithecia of a fungal partner. The lichen is a species of *Strigula* (FL, Mar).

E. The rusty color of algal colonies on *Gardenia jasminoides* is due to haematochrome pigment (FL, Mar).

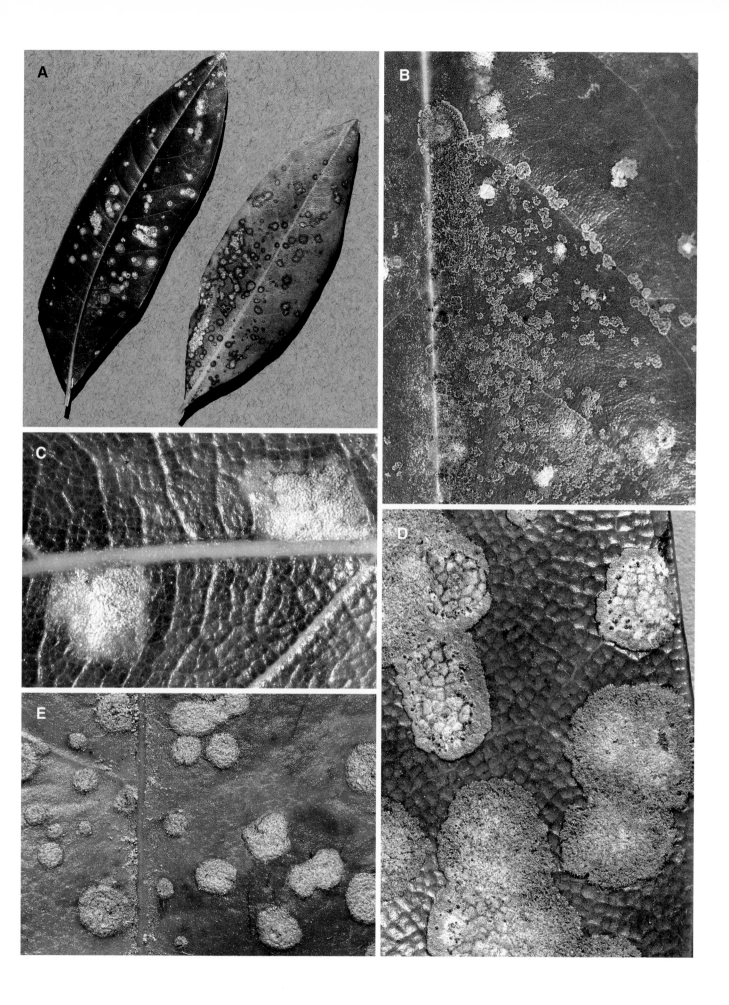

439

North American Leafy Mistletoes (Plates 218–220)

Mistletoes and dwarf mistletoes are shrubby or dwarfed, photosynthetic, plant-parasitic seed plants arrayed in several families of the order Santales. The families Loranthaceae and Viscaceae are represented in North America. Three genera of Viscaceae, but no mistletoes of other families, occur in Canada and the USA: *Arceuthobium* (dwarf mistletoes), *Phoradendron* (American mistletoes), and *Viscum*, the last-named represented by the introduced European species *V. album*. Several additional mistletoe genera occur along with *Arceuthobium* and *Phoradendron* in Mexico. Mistletoes parasitize the stems of trees, shrubs, and sometimes cacti, obtaining water, mineral nutrients, and organic compounds from the hosts' xylem sap. Mistletoes other than *Arceuthobium* are often called xylem-tapping mistletoes and true or leafy mistletoes. They derive a large proportion of their organic nutrients from their own photosynthesis and the remainder (more than half in some mistletoes) from their hosts. *Arceuthobium* species (Plates 221–225), although photosynthetic, extract most of their carbon from their hosts. Both groups cause growth loss, dieback, and sometimes death of trees, the extent of damage being related to severity of infestation. This plate and the following two introduce *Phoradendron* and *V. album*.

Characteristics and Habits of *Phoradendron*

Phoradendron comprises more than 230 species of primarily tropical and subtropical evergreen plants restricted to the Americas. Six species occur in the continental USA, their northern limit describing a skewed arc from Oregon (45°N) through the Southwest, southern Great Plains, and Ohio Valley to southern New York (40°30′N). They parasitize only angiosperms in the East but occur on both angiosperms and gymnosperms in the West. Their northern limits seem to be determined mainly by winter temperature minima, since host plants of a mistletoe species always occur farther north than the mistletoe does. The classification of *Phoradendron* is at present conservative and based entirely on morphology. Species in the USA are *P. bolleanum* (syn. *P. densum*, *P. pauciflorum*), *P. californicum*, *P. capitellatum*, *P. juniperinum* (syn. *P. libocedri*), *P. rubrum*, and *P. serotinum* (syn. *P. leucarpum*, *P. macrophyllum*, *P. tomentosum*, *P. villosum*). *P. bolleanum*, *P. juniperinum*, and *P. serotinum* are considered to be complexes within which some variants have previously been recognized as species. *P. serotinum* comprises four subspecies.

Economic damage by *Phoradendron* is considered to be slight, although these parasites cause decline of many trees, especially in the Southwest. Fruit-bearing branches of *P. serotinum* subsp. *serotinum* (eastern or oak mistletoe) and *P. serotinum* subsp. *tomentosum* (syn. *P. tomentosum*, hairy mistletoe) are collected and sold as Christmas ornaments. The fruit is eaten by birds but is reported to be toxic to humans and livestock. The symptoms caused most often in plants colonized by mistletoe are atrophy and dieback of branch ends beyond the parasite's point of attachment. Galls, elliptic swellings, or clusters of host twigs sometimes develop at loci of infection. Trees that support multiple mistletoe plants tend to lose vigor and decline.

Phoradendron species have simple leaves that are oppositely arranged, fleshy in some species, and either expanded or scalelike. The stem is woody, at least near the base, and brittle. The nodes are closely spaced and somewhat swollen, so that the plants sometimes appear jointed. The stem contains chlorophyll and is the principal site of photosynthesis in species with scalelike leaves. The stems are much more branched and clustered than those of host plants; thus a mistletoe often appears as a nearly spherical bunch of dense vegetation, prominent on a deciduous host during winter. Some mistletoes are monoecious, but all species that occur in the USA are dioecious (male and female flowers on separate plants). The greenish, generally inconspicuous flowers are pollinated by insects and wind. Fruits are nearly spherical berries, white or straw-colored to reddish, 3–6 mm in diameter. Each contains a single seed surrounded by viscid pulp. The first mistletoes on a tree usually grow on small branches. Their numbers increase after the initial female produces fruit. Mistletoe plants on trunks and limbs represent infections that began when the stems were small or that began on sprouts and spread into larger stems.

Mistletoes grow well in full sunlight and develop most extensively high in the crowns of large trees where light and temperature conditions permit them to transpire freely and draw water from the hosts. Trees growing in the open or in disturbed forests with open canopies are infected more frequently than those in undisturbed forests with closed canopies. This difference reflects the roosting preference of birds that eat mistletoe berries and disperse the seeds. These birds include species of *Bombycilla* (waxwings), *Catharus* (thrushes), *Mimus* (mockingbirds), *Sialia* (bluebirds), *Phainopepla* (phainopepla), and *Turdus* (robins), among others. Mistletoe may increase dramatically on trees where birds feed on berries, roost, and deposit seeds on twigs and branches.

Mistletoe life cycle. Birds that eat mistletoe fruit digest the pulp, but the seeds pass through unharmed, retaining a sticky covering with hairlike threads that glue seeds to surfaces on which they fall. The seeds can germinate anywhere if temperature and moisture are suitable, but only those that adhere to thin bark of twigs and small branches of suitable trees can initiate infection and produce new plants. The radicle of a germinating seed flattens itself against the bark, forming an attachment disc, or holdfast. A multicellular projection called the primary haustorium grows from the undersurface of the holdfast and penetrates the bark, often through a lenticel or in the axil of a bud. Once beneath the periderm in living cortex or secondary phloem, the primary haustorium produces a radiating system of branches termed cortical strands or cortical haustoria. Wedge-shaped projections called sinkers grow from cortical strands and pass through the cambium to the outer surface of lignified xylem. Certain cells within the sinker differentiate into water-conducting tracheids and vessels. Some of these come into intimate contact with vessels or tracheids of the host, such that open pits and perforations connect the water-conducting systems of the two plants. This contact permits bulk transport of water, minerals, and nitrogenous organic compounds to the parasite and leads to higher concentrations of mineral elements in mistletoes than in their hosts. Parenchyma cells in mistletoe sinkers, if located in contact with host cells, are also involved in withdrawal of water and minerals, which occurs by diffusion through their cell walls. Concentrations of 10 nutrient elements in *Phoradendron serotinum* examined on seven tree species ranged from 0.97 to 2.88 times greater than those in the hosts.

Sinkers elongate perennially by growth from an intercalary meristem that differentiates where the sinker passes through the host's vascular cambium. Activity of this meristem is synchronized with that of the host so that the sinker elongates as the radius of the host stem increases. Sinkers thus become embedded in wood. The oldest and most deeply embedded sinkers occur beneath the original site of infection. Invasion of a compatible host stem does not trigger compartmentalizing responses; therefore the parasite may persist in relative harmony with its host for many years.

Aerial shoots begin to grow after the system of cortical strands and sinkers is initiated. The first shoots arise from buds on the holdfast, and they grow only a few millimeters during the first year. Additional shoots may grow later from buds that develop on the outer surfaces of cortical strands. The rate of shoot growth increases

---

A. *Phoradendron serotinum* subsp. *serotinum* (eastern or oak mistletoe) on *Nyssa sylvatica* (sour-gum). Adjacent *Fraxinus* (ash) branches are unaffected (WV, Dec).

B. *P. serotinum* subsp. *tomentosum* (hairy mistletoe) on *Quercus* sp. (oak) (TX, Apr).

C–G. *P. serotinum* subsp. *serotinum* on *Quercus* sp. C. An unusually large mistletoe on the trunk of a young tree. D. A vigorous mistletoe growing from a small branch. E, F. Close views of the attachment of the plant shown in D. Bark tissues of host and parasite meet in a convoluted line at the swollen union. Dissection shows continuity of xylem of host and parasite. Greenish mistletoe xylem extends nearly to the center of the oak branch, where infection apparently occurred at a twig axil when the branch was a year old. G. A transverse section through a mistletoe attachment 5 years after infection. The mistletoe, at top, has light-colored wood. Infection occurred on the base of a year-old sprout on a stem that was 7 years old. The parent stem (lower portion) continued normal growth. Abnormally rapid wood production at the site of infection caused the sprout base to swell to nearly the same size as the parent stem. Mistletoe sinkers extend as light-colored wedges to the oldest annual ring of wood in the sprout (FL, Mar).

441

to several centimeters per year after the system of cortical strands and sinkers (endophytic system) is well established. Aerial shoots and stems usually survive only a few years before being broken off or killed by freezing. When a mistletoe stem dies or is removed, multiple new shoots usually develop from adventitious buds on the cortical strands. An endophytic system can remain alive without producing shoots, however. This ability explains the appearance of mistletoe shoots on some tree trunks that have not supported them for years. The longevity of an endophytic system seems limited only by that of the host.

In most circumstances mistletoes draw water and solutes from their hosts by maintaining higher osmotic potentials, higher transpiration rates, and lower water potentials than the hosts do. If water is freely available, a tree is stressed only slightly by mistletoe infestation. When a tree is stressed by drought, however, the parasite's transpiration often intensifies the stress, leading to growth suppression and dieback. Some mistletoes whose hosts grow under conditions of chronic water shortage in arid areas have evolved conservative water use patterns and transpire less rapidly then their hosts. This adaptation is documented for an Australian mistletoe, *Amyema miquelii*, and its host, *Eucalyptus behriana*.

Photosynthetic products from mistletoe leaves and stems are translocated in phloem to the cortical strands and sinkers but do not pass to the host. After mistletoe-induced branch dieback, an infested branch may terminate at a mistletoe plant and possess no leaves of its own. Such a branch receives water, minerals, and xylem-borne organic compounds from other parts of the tree as the result of the mistletoe's transpiration.

Host specialization varies among mistletoes; some have broad host ranges while others colonize only a few tree species. Evidence of adaptation to particular hosts has also been detected within subpopulations of mistletoe species. For example, *P. serotinum* subsp. *tomentosum* seeds from *Celtis* (hackberry), *Prosopis* (mesquite), or *Ulmus* (elm) germinated equally well on all three hosts, but the seedlings produced larger haustorial disks in their original host species. Regional populations of subspecies *tomentosum* in California also display apparent host preferences. *Quercus agrifolia* (California live oak) is commonly parasitized in southern California but not in the San Francisco Bay region, while *Q. douglasii* (blue oak) and *Q. lobata* (valley oak) are colonized in both areas. Some mistletoes parasitize their own kind (autoparasitism) or other mistletoe species (hyperparasitism). For example, the Central American species *Phoradendron fasciculatum* is hyperparasitic on *P. undulatum*, which in turn is parasitic on *Psidium guajava* (guava).

Mistletoe-resistant trees occur within species that are generally susceptible to colonization. Resistance is related to constitutive and parasite-induced production of polyphenolic compounds in secondary phloem and induced periderm formation around the primary haustorium. Genetic determinants of host-parasite incompatibility have not been identified. Constitutive resistance is also related to thickness of collenchyma and fiber layers in young bark, as well as to overall thickness of that bark. A mistletoe haustorium fails to reach the xylem in a resistant interaction, and the mistletoe seedling dies.

References: 134, 135, 147, 192, 605, 612, 613, 615, 742, 1041, 1046, 1140, 1247, 1248, 1253, 1504, 1721, 1813, 1952, 2174–2178, 2186, 2245, 2284, 2295, 2478, 2479, 2770, 2961, 3146, 3220, 3255, 3256, 3283, 3386, 3387, 3503, 3985, 4110, 4266, 4469, 4470

Mistletoes on Angiosperms

Three American mistletoe species and the European *Viscum album* parasitize angiosperms in the USA and Canada. The latter occurs in one California locality and on Vancouver Island. It is the only leafy mistletoe found in the field in Canada.

Phoradendron serotinum subsp. *serotinum* (eastern or oak mistletoe; Plate 218) occurs throughout the eastern and southern states south of an irregular line from southern New York to southeastern Kansas. Over the greater part of its length the line corresponds to the isotherm for mean minimum January temperature of −4.5°C. This mistletoe is most abundant in southern lowlands. It is superficially similar to *V. album* and is the American species used most often for Christmas ornament. It parasitizes more than 100 tree species in at least 50 genera, including the following:

Acer (box-elder, maple), *Aesculus* (buckeye), *Betula* (birch), *Broussonetia* (paper-mulberry), *Carya* (hickory, pecan), *Castanea* (chestnut), *Celtis* (hackberry, sugarberry), *Cinnamomum* (camphor tree), *Cornus* (dogwood), *Crataegus* (hawthorn), *Diospyros* (persimmon), *Fagus* (beech), *Fraxinus* (ash), *Gleditsia* (honeylocust), *Juglans* (walnut), *Liquidambar* (sweetgum), *Liriodendron* (tuliptree), *Maclura* (osage-orange), *Melia* (chinaberry), *Nyssa* (sour gum), *Platanus* (plane tree, sycamore), *Prosopis* (mesquite), *Prunus* (cherry), *Pyrus* (pear), *Quercus* (oak), *Robinia* (locust), *Salix* (willow), *Sapindus* (soapberry), *Sassafras*, *Tilia* (basswood), and *Ulmus* (elm).

In some localities *P. serotinum* subsp. *serotinum* displays host preferences that cannot be explained simply on the basis of host abundance. *Quercus nigra* (water oak) is most commonly parasitized in southern Mississippi; *Nyssa sylvatica* (sour-gum) on the Allegany Plateau in southern Ohio; *Acer saccharinum* (silver maple) and *Ulmus americana* (American elm) in several Ohio River valley drainages; and *Carya ovata* (shagbark hickory) or *Ulmus* species in parts of North Carolina, Tennessee, and Georgia. *P. serotinum* subsp. *serotinum* does not usually kill trees, but severely infested *Celtis laevigata* (sugarberry) and *Q. nigra* on stressful sites sometimes decline and die as branches die back to sites of mistletoe infection successively closer to the trunk. This mistletoe produces flowers from October to late November in the northern part of its range and into January in Florida. The male inflorescence is a jointed spike with a few tiny greenish flowers per segment. White berries 4 mm in diameter mature in autumn or winter a year after pollination. Seed germination occurs in bright light but is inhibited by darkness.

P. serotinum subsp. *macrophyllum* (syn. *P. macrophyllum*; Colorado desert mistletoe) occurs from northern California to Mexico and eastern Texas. Its range overlaps those of subspecies *serotinum* and *tomentosum* in Texas. It parasitizes at least 60 species in about 30 genera, including *Acacia*, *Acer*, *Alnus* (alder), *Celtis*, *Fraxinus*, *Gleditsia*, *Juglans*, *Platanus*, *Populus* (aspen and cottonwood), *Prosopis*, *Quercus*, *Robinia*, *Salix*, *Styphnolobium* (pagoda tree), *Ulmus*, and *Zelkova*. This mistletoe causes branch galls and trunk burls on *Populus fremontii* (Fremont cottonwood) and reportedly kills trees of this species in southern New Mexico. It also causes branch galls on *R. pseudoacacia* (black locust).

P. serotinum subsp. *tomentosum* (hairy mistletoe; Plates 218B, 219A, B) occurs from western Oregon to central Texas and southward to central Mexico. Its range overlaps that of subspecies *macrophyllum* throughout most of that region and overlaps the range of subspecies *serotinum* in Texas. *P. serotinum* subsp. *tomentosum* includes forms previously known as *P. coryae* and *P. villosum*, among others. Its hosts include *Aesculus californica* (California buckeye), *Arctostaphylos* (manzanita), *Celtis laevigata* (sugarberry), *Juglans*, *Maclura pomifera* (osage-orange), *Prosopis*, *Quercus*, and *Ulmus*. It sometimes kills *C. laevigata* in Oklahoma and Texas. It can cause large galls and dieback on *Quercus*, as shown for *Q. emoryi* (Emory oak) in Plate 219B.

P. serotinum subsp. *tomentosum* has moderately to densely pubescent segments in its floral spikes and grows to a diameter of 1 m or more. The variant previously known as *P. coryae*, found in the Southwest, has dense clusters of star-shaped hairs on its leaves.

A, B. *Phoradendron serotinum* subsp. *tomentosum* (hairy mistletoe) on *Quercus oblongifolia* (Mexican blue oak) and *Q. emoryi* (Emory oak), respectively. Galls 6–25 cm or more in diameter, caused by the mistletoe, are visible on the dead limb of *Q. emoryi*. The mistletoe is presumed to have contributed to death of the branch (AZ, Jul).

C. Shoots of *P. serotinum* subsp. *tomentosum* growing from the edges of a large swelling on a trunk of *Q. garryana* (Oregon oak). The infection, many years old, could have begun when the tree was a sapling or when cortical strands of the mistletoe grew from a sprout into the trunk (OR, Jul).

D, E. *Viscum album* (European mistletoe) on *Malus pumila* (apple). The host stem is swollen at the locus of mistletoe attachment (E) (BC, Sep).

F–H. *P. californicum* (desert or mesquite mistletoe) on *Prosopis* sp. (mesquite). F. Dense growth of mistletoe on a deformed host plant. G. A typical mistletoe plant with long, slender, pendant, leafless branches. H. Close view of mistletoe twigs and branches (AZ, Jul).

Photo credits: D, E—C. E. Dorworth

443

P. serotinum subsp. *tomentosum* blooms in winter. The male inflorescence is larger and has more flowers per segment than does that of subspecies *serotinum*. The berries of subspecies *tomentosum* are nearly white and 4–6 mm in diameter. Host-specialized populations of this mistletoe are noted in the discussion of general characteristics of mistletoes.

P. serotinum subsp. *angustifolium* occurs only in Mexico and parasitizes *Quercus*.

P. californicum (mesquite or desert mistletoe; Plate 219F–H) is a nearly leafless species common in southwestern USA and northwestern Mexico on leguminous trees and shrubs such as *Acacia* (acacia), *Parkinsonia* (paloverde), and *Prosopis* (mesquite); and uncommonly on *Baccharis* (baccharis, false-willow), *Condalia* (snakewood), *Larrea* (creosote bush), *Olneya* (desert-ironwood), and *Tamarix* (saltcedar, tamarisk). One collection on *Cupressus macnabiana* (MacNab cypress) is recorded. *P. californicum* produces long, pendulous stems and, in time, dense bushes that attain lengths greater than 1 m. Its leaves are minute and scalelike. The stems vary from green to reddish. Small greenish flowers are produced in December–March. The white to red berries are about 3 mm in diameter. *P. californicum* sometimes parasitizes other individuals of its own species.

P. rubrum (mahogany mistletoe), a West Indian species, parasitizes *Swietenia mahagoni* (mahogany) in southern Florida.

Viscum album (Plate 219D, E), the Christmas mistletoe of Europe, was introduced to Sebastopol, California, in 1900 by horticulturist Luther Burbank. In the ensuing century it spread to more than 20 native and introduced trees and shrubs, all within about 12 km of Burbank's farm. It is most common on *Acer saccharinum* (silver maple), *Alnus rubra* (red alder), *Malus pumila* (apple), *Populus fremontii* (Fremont cottonwood), *Robinia pseudoacacia* (black locust), and *Salix lucida* (Pacific or shining willow). *V. album* was found established on *Malus* sp. in Victoria, British Columbia, late in the 20th century. In Europe *V. album* comprises three subspecies that collectively parasitize angiosperms and gymnosperms in at least 96 genera representing 44 families. Only *V. album* subsp. *album*, which prefers angiosperms, occurs in North America.

References: 135, 223, 579, 612, 613, 615, 742, 991, 1049, 1247, 1561, 1563, 1721, 2174, 2177, 2283, 2284, 2314, 2676, 2953, 3386, 3387, 3438, 3800, 4022

Mistletoes Primarily on Gymnosperms

Three mistletoes—*P. bolleanum*, *P. capitellatum*, and *P. juniperinum*—occur primarily on gymnosperms in the USA, all in western states. All three display a strong preference for the Cupressaceae (cypress family), although *P. bolleanum* (syn. *P. densum*, *P. pauciflorum*) occurs on *Abies concolor* (white fir) as well. They cause only slight economic damage.

P. bolleanum (Bollean mistletoe, fir mistletoe) on *Abies* (Plate 220A–C) grows as prominent dense, dark green, often globose bushes 60–100 cm in diameter high in the crowns of *A. concolor* from central California to Mexico and in one locality in Arizona. Its stems are smooth to slightly hairy, with internodes 6–21 mm long, and they become woody with age. Leaves are 15–30 mm long and up to 8 mm wide. The male inflorescence has one or two segments, each with several tiny flowers. The female inflorescence also has one or two segments, each with two flowers. The fruit is white and smooth, and about 4 mm in diameter. Trees with large or multiple mistletoes often die back from the top and are abnormally susceptible to attack by the beetle *Scolytus ventralis* (fir engraver); cone production is suppressed also. *A. concolor* is the only member of the Pinaceae in the USA that is parasitized by a true mistletoe and the only conifer subject to infection by both leafy and dwarf mistletoes. Cold winters in the northern part of its range prevent *P. bolleanum* from succeeding there.

P. bolleanum on *Cupressus* (cypress) and *Juniperus* (juniper) produces clumps 60–100 cm in diameter with narrow leaves that measure 7–35 × 1–10 mm. It occurs on *Juniperus* from southern Oregon to Mexico and western Texas, and on *Cupressus* in California and Arizona. This species is also found rarely on angiosperms: *Arbutus* (madrone), *Mahonia* (mahonia), *Quercus* (oak), and *Umbellularia* (California-laurel). The variant of *P. bolleanum* previously known as *P. densum* (dense mistletoe) sometimes causes atrophy and death of *Juniperus* branches beyond the point of attachment and thus comes to reside on the clubbed ends of living branches that lack juniper foliage. Its leaves are

smooth and 10–20 × 3–5 mm in size. Observations of *P. bolleanum* on *J. occidentalis* (Sierra juniper) in northern California after the unusually cold winter of 1932 furnished evidence that leafy mistletoes are limited by low temperatures. In an area where this mistletoe was abundant, 90% of the plants, including both aerial shoots and endophytic systems, were killed.

P. capitellatum is a small, densely branched species that occurs from Arizona to western Texas and in northern Mexico on several species of *Juniperus*. Its fleshy leaves measure 10–15 × 1–2 mm, so it is readily distinguished from *P. juniperinum*. A narrow-leaved variant of *P. bolleanum* that occurs in Mexico could be confused with *P. capitellatum*.

P. juniperinum (juniper mistletoe; Plate 220F–H) is a nearly leafless species that parasitizes *Juniperus* and, uncommonly, *Cupressus arizonica* (Arizona cypress) from Oregon and Colorado to Mexico. Plants of this species grow to 40–80 cm in diameter and usually remain erect. The mature stems are woody, with smooth surfaces and internodes usually less than 1 cm long. The scalelike leaves are only about 1 mm long. Flowers are produced in July–September. The male inflorescence consists of one or two segments, each with five–nine flowers. Female inflorescences have one segment with two flowers. Fruits are pinkish white and smooth, about 4 mm in diameter. *P. juniperinum* has been associated with an abnormally high mortality rate of *J. osteosperma* (Utah juniper) and with suppressed mycorrhizal development on *J. monosperma* (cherrystone juniper) on stressful sites. Other hosts include *J. deppeana* (alligator juniper), *J. occidentalis* (Sierra juniper), *J. scopulorum* (Rocky Mountain juniper), and rarely *Chamaebatiaria millefolium* (fernbush) and *Quercus arizonica* (Arizona white oak).

A variant of *P. juniperinum* formerly known as *P. libocedri* (incense-cedar mistletoe; Plate 220D, E) occurs on *Calocedrus decurrens* (incense cedar) discontinuously from Oregon to Mexico. Its internodes are usually more than 1 cm long, and the plants tend to become pendulous with age. This mistletoe is typically found as dark bunches of vegetation throughout the crown of its host and is associated with thin-appearing foliage and occasionally with dieback or tree death. Trunk infections also occur, presumably when the parasite grows as cortical haustoria from a small side branch into the trunk. This leads to spindle- or barrel-shaped swellings up to 2 m long, usually on the upper part of the mainstem, from which mistletoe shoots grow. With time the mistletoe may cease producing shoots, perhaps because it is unable to force its way through the thick corky bark on old stems. Its endophytic system remains alive, however, nourished by the tree. This mistletoe, persisting as an endophytic system in a swelling on the trunk of an old *C. decurrens*, holds the North American record for longevity—over 400 years as estimated by counting annual rings of host wood that developed around sinkers.

Mistletoes on either gymnosperms or angiosperms can be controlled by removing infected branches, cutting at least 30 cm proximal of mistletoe shoots. Mistletoe shoots on limbs or trunks can be broken off or cut off; this reduces water and nutrient loss to the parasite but must be done repeatedly because new sprouts grow from the cortical haustoria. Regrowth from stubs can be reduced by treatment with certain growth regulators or paints. Treatment of mistletoe with certain herbicides while the host is dormant has also proven effective.

References: 135, 616, 1041, 1114, 1249, 1253, 2177, 2315, 2476, 2477, 2479, 4109, 4176, 4179

A–C. *Phoradendron bolleanum* (syn. *P. pauciflorum*; fir mistletoe) on *Abies concolor*. A, B. Foliage of a parasitized tree is thin, and twig dieback is evident. C. The top of a tree has bent under the weight of mistletoe, augmented by snow in winter (CA, Jul).

D, E. *P. juniperinum* (syn. *P. libocedri*) on *Calocedrus decurrens* (incense cedar). D. Pendant mistletoe plants, typical of this species, on low branches. E. A large cluster of shoots on the mainstem in the top of a declining tree. Host foliage is sparse, and dieback has begun (CA, Jul).

F–H. *P. juniperinum* on *Juniperus*. F. Mistletoe (dense olive-green masses) associated with sparse foliage and dieback on *J. scopulorum* (Rocky Mountain juniper). G. Close view of a small mistletoe plant on *J. occidentalis* (Sierra juniper). The branch tip beyond the parasite has died (CA, Jul). H. Close view of mistletoe twigs and fruit. Scalelike leaves appear as pairs of tiny opposing projections at intervals along the twigs (AZ, Apr).

445

Dwarf Mistletoes (Plates 221–225)

Overview

Dwarf mistletoes (*Arceuthobium*, Viscaceae) are small, leafless, chlorophyllous seed plants that parasitize stems of gymnosperms, extracting water, minerals, and organic compounds, especially nitrogenous compounds and sugars. These plants are considered to be an evolutionarily advanced group that has undergone host specialization, size reduction, and form and function changes that distinguish them from other members of Viscaceae. They are distributed around the Northern Hemisphere, with 18 the 26 recognized species in the New World. They occur from Newfoundland to southern Alaska, throughout the coniferous forests of the West, and into Central America. There are 3 species in Canada; these plus 5 more occur in the USA. Mexico and Central America are home to about 15 species.

Hosts of dwarf mistletoes in North America are all members of the Pinaceae (pine family), whereas dwarf mistletoes in Asia infect plants in Pinaceae and Cupressaceae (cypress family). Loss of vigor, dieback, and premature death are the major symptoms. Many species also cause witches'-brooms. Dwarf mistletoes and their seeds provide food for many insects, mammals, and birds, and the witches'-brooms provide shelter.

A dwarf mistletoe plant consists of nonwoody or semiwoody shoots anchored by an endophytic system composed of a rootlike array of cortical strands in the inner bark with extensions (sinkers) into the wood. These plants are distinctive in having stems that lack a central vascular cylinder, fruits that explode when ripe, and radicles with stomata. The seeds contain chlorophyll in the endosperm, a characteristic of Viscaceae. The stems possess vascular bundles with tracheary elements, parenchyma, and phloem fibers. Phloem sieve elements are apparently scarce except in species whose vascular bundles undergo secondary growth.

Dwarf mistletoes cause more damage than any other group of pathogens in forests of western North America. The annual loss of wood due to mortality and growth suppression by these pathogens in western USA was estimated to be 11.3 million cubic meters in the mid-1990s. Height growth is usually suppressed relatively more than radial growth. Direct damage also includes reduction of tree seed quality and quantity and reduction of wood quality. Infected wood has subnormal strength and poor pulping quality due to abnormally short, distorted tracheids and a high proportion of ray tissue. Diseased trees may also have distorted trunks and abnormally large knots. In addition, dwarf mistletoes predispose trees to attack by opportunistic insects and fungi (which are often the direct causes of death) and influence forest succession through effects on tree growth and flammability. Trees damaged or killed by dwarf mistletoes are a major fire hazard. Dwarf mistletoes also weaken, disfigure, and kill landscape trees in residential and resort areas.

Most dwarf mistletoe taxa are host specialized. Each species or subspecies parasitizes one or a few principal and secondary host species but is also capable of infecting other conifers (occasional and rare hosts) growing among severely infected trees of a principal host species. *Arceuthobium* species have common names that reflect their principal hosts or regions of occurrence. The species in Canada and the USA, their principal and secondary hosts (but not occasional or rare hosts), and their distributions are as follows.

Arceuthobium americanum (lodgepole pine dwarf mistletoe; Plate 221) on *Pinus banksiana* (jack pine) and *P. contorta* (lodgepole pine), secondarily on *P. ponderosa* (ponderosa pine); British Columbia to California, Colorado, and Manitoba.

A. blumeri (Blumer's dwarf mistletoe) on *P. pseudostrobus* (smooth-bark Mexican pine) and *P. ayacahuite* (Mexican white pine); Arizona and Mexico.

A. campylopodum on diverse conifers from Alaska to Mexico and throughout the interior West. As presently construed, *A. campylopodum* is a variable species that infects *Abies* (fir), *Larix* (larch), *Picea* (spruce), *Pinus*, and *Tsuga* (hemlock). Its synonyms include 11 names that, before 2004, were considered to represent distinct species. Members of the *A. campylopodum* complex are discussed on pages 450–452.

A. divaricatum (pinyon dwarf mistletoe) on *P. californiarum* (California pine), *P. cembroides* (Mexican pinyon), *P. discolor* (border pinyon), *P. edulis* (Colorado pinyon), *P. monophylla* (singleleaf pinyon), and *P. quadrifolia* (Parry pinyon); California and Baja California to Colorado and Texas.

A. douglasii (Douglas-fir dwarf mistletoe) on *Pseudotsuga menziesii* (Douglas-fir; Plate 223), secondarily on *Abies lasiocarpa* (subalpine fir); British Columbia to Mexico.

A. gillii (Chihuahua pine dwarf mistletoe; Plate 224) on *Pinus herrerae*, *P. leiophylla* (Chihuahua pine), and *P. lumholtzii* (Mexican weeping pine); Arizona, New Mexico, and Mexico.

A. pusillum (eastern dwarf mistletoe; Plate 225) on *Picea glauca* (white spruce), *P. mariana* (black spruce), and *P. rubens* (red spruce); Newfoundland to Pennsylvania and Saskatchewan.

A. vaginatum subsp. *cryptopodum* (southwestern dwarf mistletoe; Plate 224) on *P. arizonica* (Arizona pine), *P. durangensis* (Durango pine), *P. engelmannii* (Apache pine), and *P. ponderosa*; Utah and Colorado to Mexico. *A. vaginatum* subsp. *vaginatum* (Mexican dwarf mistletoe) occurs on *Pinus* species in Mexico.

Most dwarf mistletoe infections remain localized and cause spindle-shaped swelling of host stems. Some species, however, notably *A. americanum*, *A. douglasii*, and *A. pusillum*, grow systemically in host branches and induce little or no swelling. Most species also induce witches'-brooms. Brooms are of two types, called systemic if the endophytic system keeps pace with apical growth of the branch, resulting in systemic infection within the broom; or nonsystemic if the endophytic system remains concentrated near the original site of infection. Aerial shoots develop along the branches of systemically infected brooms, but grow only near the original site of infection on nonsystemic brooms. Cones usually do not form on brooms. If a tree supports multiple infections, the noninfected branches gradually weaken and produce less foliage than normal so the contrast between mistletoe brooms and the remainder of the crown becomes dramatic. Dieback follows.

The aerial shoots of dwarf mistletoes are perennial and branch at least once. They appear jointed because each segment originates in a more or less cuplike depression at a node. Shoots range in height from a few millimeters to more than 70 cm and in habit from scattered to densely clustered. Each shoot grows from a basal cup that remains visible on host bark after the shoot dies and/or breaks off. Flowers on both male and female plants are minute. The fruits, 3–5 mm long in most species, contain viscid pulp and a single seed.

Dwarf mistletoe seeds are dispersed from midsummer to late autumn, the time varying with species. The seeds are expelled as projectiles from ripe fruits that burst with explosive force, those of some species at velocities of 22–26 m per second! Most seeds fall within 5 m of the source tree, but some travel up to 15 m or more. The viscid coating of seeds causes them to stick to twigs, foliage, and other objects. Seeds intercepted by conifer foliage remain there until lubricated by rain, then slide down the needles and either fall to the ground or stick to twigs. Many seeds that initially stick to twigs are later displaced by snow, rain, and presumably birds.

Dwarf mistletoe populations increase, and damage develops most rapidly, in relatively open forests with trees of mixed ages or on steep slopes where seeds are scattered from diseased trees in the overstory onto young trees below. Upward spread within crowns of infected trees is usually gradual and, in relatively young trees, does not keep pace with height growth. Dwarf mistletoe seeds are also dispersed by rodents and birds: species of *Glaucomys*, *Sciurus*, and *Tamiasciurus* (squirrels); *Cyanocitta* and *Perisoreus* (jays), *Junco*

Arceuthobium americanum (lodgepole pine dwarf mistletoe); discussion, p. 448.

A. Witches'-brooms caused by *A. americanum* on a declining *Pinus contorta* (lodgepole pine). The top of this tree, including brooms on upper branches, is dying; brooms on lower branches appear dense and dark green in comparison (OR, Sep).

B. Twig mortality, the result of rodents gnawing infected bark, in brooms on *P. contorta* (CO, Jun).

C. Witches'-brooms caused by *A. americanum* on *P. ponderosa* (ponderosa pine), a secondary host commonly infected where it grows close to infected *P. contorta* (OR, Sep).

D, E. Female and male plants, respectively, of *A. americanum* on *P. contorta*. The branch in E has a spindle-shaped swelling at the site of infection, and part of a female plant is visible at lower right (OR, Sep).

(junco), *Parus* (chickadee), *Picoides* (woodpecker), *Sitta* (nuthatch), *Dendroica* (warbler), and others. The seeds stick temporarily to fur or feathers when animals disturb ripe fruits and cause seed expulsion. Later a seed may be wiped off on a healthy shoot or twig, especially where birds preen. This accounts for dwarf mistletoe infection in the tops of tall trees and for new foci of infection far from seed sources.

The time of seed germination varies with species from autumn to spring. Germinating seeds carry on photosynthesis, which presumably helps the young plant sustain itself until it parasitizes a host. The short radicle produces a holdfast where it contacts a needle base or other obstruction. A penetrating wedge then grows from the holdfast into the stem. Most infections occur on stems less than 5 years old, although some dwarf mistletoes have the ability to penetrate older parts.

A system of cortical strands grows from the penetrating wedge in all directions in the living bark, producing sinkers at irregular intervals. Sinkers are wedge-shaped protrusions that grow along or within medullary rays toward the xylem until their tips pass the vascular cambium and encounter lignified cells. Sinkers consist of parenchyma or parenchyma plus tracheary elements. Some of the latter differentiate after a sinker contacts host tracheids, creating a route for the flow of water and dissolved minerals and organic compounds from host to parasite. Sinkers elongate and increase in width perennially by growth from a meristem located where the sinker passes through the host cambium. Growth of a sinker is synchronized with that of the host, and the tip of a sinker becomes deeply embedded in wood as the radius of a parasitized stem increases.

The endophytic system of a dwarf mistletoe usually develops 2–5 years before aerial shoots begin to form, and it may persist indefinitely without shoots if the host lacks vigor. Aerial shoots begin to produce flowers, male and female on separate plants, after 1–2 years and fruit 5–19 months (usually 12 months) later. Insects and wind pollinate the flowers.

Dwarf mistletoes maintain a higher osmotic potential than that of their hosts. This and the parasite's intrinsically higher rate of transpiration permit it to draw water and continue to transpire even when the host is under severe water stress. The transpiration rate of a dwarf mistletoe may be many times greater (surface area basis) than that of its host. This differential permits it to extract minerals, nitrogenous compounds, and other organic compounds from xylem sap of the host.

Adult dwarf mistletoes assimilate some carbon by photosynthesis, but at low rates compared with their hosts, and they respire at higher rates than their hosts. Therefore they must extract from the hosts much of the carbon necessary for growth and respiration. In the species thus far studied the carbon in cortical strands and sinkers comes mainly from the host. The pathway of carbon from host to dwarf mistletoe is unreported, but carbon translocation within the parasite is presumed to occur via phloem sieve cells and parenchyma.

Some species of dwarf mistletoe exhibit an unexplained mutual exclusion; that is, if the principal parasite of a given host is present in a locality, other dwarf mistletoe species present in that locality tend to be excluded from that host. Conversely, other dwarf mistletoes are likely to parasitize that host species if the principal parasite is absent. In stands containing both *Pinus contorta* (lodgepole pine) and *P. flexilis* (limber pine), for example, if *A. americanum* is present on its principal host, *P. contorta*, then *A. cyanocarpum* (principal host *P. flexilis*) does not parasitize *P. contorta*, although it commonly does so in the absence of *A. americanum*. The reciprocal relationship is also known. Infection of a given tree by two species of dwarf mistletoe is rare.

Resistance to particular dwarf mistletoes occurs not only in their occasional or rare hosts but also in individuals of the principal host species. The bases of resistance are poorly known, however. Rare hosts may be quite susceptible, as evidenced by multiple infections on individual trees.

Several fungal parasites and insects may destroy shoots or fruits of dwarf mistletoes, but no dwarf mistletoe seems significantly limited by biotic agents. Practical control of dwarf mistletoes in forests depends on cutting practices that remove diseased trees and favor even-aged stands. Clear-cutting or controlled burning is often appropriate to sanitize an area where a new stand will be planted or established from seed. In residential areas damage to individual trees of high value can be suppressed or delayed by pruning infected branches.

References: 39, 310, 605, 614, 1248, 1562, 1565, 1566, 1673, 1813, 2178, 2295, 2498, 2852, 2853, 2956, 2964, 3265, 3266, 3378, 3384, 3440, 3813

Representative Dwarf Mistletoes

Arceuthobium americanum (lodgepole pine dwarf mistletoe), a parasite of *Pinus banksiana* (jack pine) and *P. contorta* (lodgepole pine), is the most widespread of the western species, occurring from central British Columbia and northern Alberta to southern California, central Colorado, and eastern Manitoba (and formerly at one location in western Ontario) at elevations 200–3350 m above sea level. In some large forest areas it causes losses equivalent to 25% of annual growth. Tree volume reductions of 28–42% have been recorded in severely infested mature *P. contorta* stands in British Columbia. The annual timber volume yield reduction caused by *A. americanum* in the Intermountain Region was estimated at over 488,000 m³ in the mid-1980s.

Pinus contorta is the principal host of *A. americanum*, except in Saskatchewan and Manitoba, where *P. banksiana* has that role. *A. americanum* frequently parasitizes *P. ponderosa* (ponderosa) pine growing in association with *P. contorta*. Occasional hosts include *P. albicaulis* (whitebark pine), *P. flexilis* (limber pine), and *P. jeffreyi* (Jeffrey pine). Rare hosts include *Abies lasiocarpa* (subalpine fir), *Picea engelmannii* (Engelmann spruce), *P. glauca* (white spruce), *P. mariana* (black spruce), *Pinus aristata* (Rocky Mountain bristlecone pine), *P. attenuata* (knobcone pine), *Pseudotsuga menziesii* (Douglas-fir), and *Pinus sylvestris* (Scots pine) planted near *P. contorta*.

Aerial shoots of *A. americanum* are yellowish to olive-green with branches in whorls. They average 6 cm tall but may grow to 20 cm. The life cycle requires at least 5 years. In the shortest possible cycle, seeds germinate in spring and infection begins during summer of the first year. Aerial shoots appear 2 years later, and flowers open in March–June, depending on locality, of the fourth year. The blue-green fruits mature and seeds are scattered in late summer of the fifth year. The cycle is extended when several years elapse between infection and growth of aerial shoots. Pollen release and seed discharge occur mainly in daylight in response to diurnal fluctuations in humidity and temperature. Wind and nonspecialized insects pollinate the flowers. Birds that visit infected trees account for long-distance dispersal of the seeds.

Infections by *A. americanum* may be localized, inducing stem swellings from which aerial shoots emerge, or may become systemic in branches, resulting in systemic witches'-brooms. Aerial shoots on brooms appear first at branch whorls. Branch segments are usually less than 5 years old when they become infected, but the thin bark of the principal hosts permits infection of older branches—up to 60 years old in *P. contorta*. Severely infected branches often die. Rodents gnawing infected bark are often the direct cause of

---

Dwarf mistletoes of the *Arceuthobium campylopodum* complex (discussion, p. 450).

A–F. *A. campylopodum* in the restricted sense (western dwarf mistletoe) on *Pinus ponderosa* (ponderosa pine). A, B. Witches'-brooms in mature trees. The tree in A has dense foliage on brooms in the lower crown and thin foliage on slow-growing parts above (WA, Jul). B. Proliferation of twigs in a broom is revealed after needles die and drop. C. Aerial shoots growing from an extensive endophytic system on the trunk of a small tree. Infection began many years earlier near the center of the darkened bark. Shoots are sparse near the original site of infection but abundant near the edges of the infected area. D, E. Female shoots bearing fruit, growing from spindle-shaped swellings on branches. Basal cups where previous shoots died and/or broke off are apparent on the branch in D. F. A relatively young male plant growing as a cluster of shoots from a common origin (OR, Sep).

G. *A. occidentale* (digger pine dwarf mistletoe) on *P. sabiniana* (digger pine) (discussion, p. 452). Two plants of different sexes (the male is darker in color) have developed close together, causing an elongate spindle-shaped swelling on the host branch. Brown basal cups are locations of lost shoots (CA, Sep).

branch death. The trunk may also be infected, usually as the result of endophytic systems spreading from branches. Cambial activity slows or ceases at the centers of old trunk infections, resulting in deformity.

Damage increases with time. In young *P. contorta* stands in Alberta where trees were receiving dwarf mistletoe seeds from scattered older trees, infections were found to increase exponentially, with an average doubling time of 1.25 years. The average rate of spread from tree to tree in that region was only about 0.5 m per year, however.

References: 178, 194, 196, 1248, 1273, 1274, 1557, 1559, 1562, 1566, 1707, 1933, 2076, 2080, 2504, 2750, 3170, 3171, 3731, 3808, 3809, 4021, 4214, 4215

Arceuthobium campylopodum **complex.** Taxa of this complex, listed below, differ in host and geographic ranges and morphologic characters but have not been clearly differentiated in DNA sequence analyses. Reproductive isolation among them is largely unstudied, so their appropriate disposition as species, infraspecific taxa, or mere geographic or host-specialized variants is unknown. Their names are retained here as links to published information. Primary and secondary hosts, but not occasional or rare hosts, are indicated.

A. campylopodum in the restricted sense (western dwarf mistletoe; Plate 222) on *Pinus ponderosa* (ponderosa pine) and *P. jeffreyi* (Jeffrey pine) primarily and on *P. attenuata* (knobcone pine) and *P. coulteri* (Coulter pine) as secondary hosts; Idaho and Washington to southern Nevada and Baja California.

A. abietinum (fir dwarf mistletoe) on *Abies*; Washington to northern Mexico and scattered localities in Nevada, Utah, and Arizona. Two host-specialized forms are known, one primarily on *A. concolor* (white fir) and *A. grandis* (grand fir) and secondarily on *Picea breweriana* (Brewer spruce), and the other on *A. magnifica* (California red fir).

A. apachecum (Apache dwarf mistletoe; Plate 223) on *Pinus pseudostrobus* (southwestern white pine); Arizona, New Mexico, and Mexico.

A. californicum (sugar pine dwarf mistletoe) on *P. lambertiana* (sugar pine) and secondarily on *P. monticola* (western white pine); California.

A. cyanocarpum (limber pine dwarf mistletoe) on *P. albicaulis* (whitebark pine), *P. aristata* (Rocky Mountain bristlecone pine), *P. flexilis* (limber pine), and *P. longaeva* (Great Basin bristlecone pine), secondarily on *P. monticola* and *Tsuga mertensiana* (mountain hemlock); California and Oregon to Montana and Colorado.

A. laricis (larch dwarf mistletoe; Plate 223) on *Larix occidentalis* (western larch), secondarily on *P. contorta* and *Tsuga mertensiana*; British Columbia and northwestern states.

A. littorum (coastal dwarf mistletoe) on *P. muricata* (Bishop pine) and *P. radiata* (Monterey pine); California.

A. microcarpum (western spruce dwarf mistletoe) on *Picea engelmannii* (Engelmann spruce), *P. pungens* (blue spruce), and *Pinus aristata*; Arizona and New Mexico.

A. monticola (western white pine dwarf mistletoe) on *P. monticola*, secondarily on *P. lambertiana*; California and Oregon.

A. occidentale (digger pine dwarf mistletoe; Plate 222) on *P. sabiniana* (digger pine), secondarily on *P. attenuata* and *P. coulteri*; California.

A. siskiyouense (knobcone pine dwarf mistletoe) on *P. attenuata*; California and Oregon.

A. tsugense (hemlock dwarf mistletoe; Plate 223) on *T. heterophylla* (western hemlock), *T. mertensiana*, *Abies amabilis* (Pacific silver fir), *A. lasiocarpa* (subalpine fir), *A. procera* (noble fir), and *P. contorta*; coastal Alaska to California. Three subspecies are discussed on page 452.

References: 1566, 2852

Arceuthobium campylopodum in the restricted sense (western dwarf mistletoe; Plate 222) is common on *Pinus ponderosa* (ponderosa pine) and *P. jeffreyi* (Jeffrey pine) at elevations from 30 m above sea level in the Columbia River valley to about 2380 m in the mountains of southern Nevada. *P. attenuata* (knobcone pine) and *P. coulteri* (Coulter pine) are secondary hosts. *P. contorta* (lodgepole pine) and *P. sabiniana* (digger pine) are occasional hosts; *P. lambertiana* (sugar pine) is a rare one. Planted *Picea abies* (Norway spruce), *P. pungens* (blue spruce), *Pinus pinaster* (maritime pine), and *P. sylvestris* (Scots pine) have also been found infected. Several other conifer species have proven susceptible when inoculated with seeds. Damage by *A. campylopodum* is most severe where hosts are stressed by heat and drought. Accordingly, damage is more severe in southern California than in northern areas and more severe on the eastern than the western slopes of the Cascade–Sierra Nevada cordillera.

Aerial shoots of *A. campylopodum* are olive-green to yellow or yellowish brown, up to 13 cm tall (average 8 cm), with branches in a fanlike arrangement (flabellate branching). Within the color range mentioned, male plants tend toward yellow-brown and females toward green, especially in northern areas. Fruits are bluish to olive-green, about 5 mm long.

A. campylopodum has at minimum a 5-year life cycle, as indicated by the fact that swellings with aerial shoots may be found on host branch segments as young as 4 years. Latent infection leading to a much longer life cycle is common, however. For example, on research plots where all aerial shoots and stem swellings were removed by pruning, new swellings or aerial shoots representing more than 350 previously latent infections per hectare appeared during the next 9 years. Flowers of *A. campylopodum* open from August to October, most commonly in mid-September, and seeds mature the next year in late August to late November. The seeds require an after-ripening period and germinate the next spring. After penetration, the endophytic system at first radiates from a compact perennial mass of cells in the outer bark, causing a spindle-shaped swelling. If the branch remains alive, a witches'-broom may develop at the point of initial infection. Brooms often attain large size (>1 m) on old branches. Infections on mainstems often develop an elliptic endophytic system many centimeters long, with aerial shoots arising first at the center but eventually only near the edges, as shown in Plate 222C. This process takes many years, however, because the endophytic system extends less than 2 cm per year on average in each direction.

The susceptibility of *P. ponderosa* to *A. campylopodum* is greatest while trees are young. When scions from trees of various ages were grafted on seedling rootstocks and were inoculated uniformly with dwarf mistletoe seeds, the greatest number of infections developed on scions from the youngest trees, and fewest infections occurred on scions from trees more than 50 years old. Another grafting experiment revealed the extent of growth suppression that can occur in young infected trees. Those with infected scions grew only half as much as those with healthy scions, and one-third of the infected trees—versus none of those that were initially healthy—died within 12 years. Two types of susceptibility associated with damage in old trees have been recognized: susceptibility to infection, allowing the accumulation of many mistletoe plants; and susceptibility to extensive invasion after infection, permitting the development of large endophytic systems in trees that may support relatively few mistletoe plants.

Some strains of *P. ponderosa* are resistant to *A. campylopodum*, as first indicated by absence of infection in certain trees growing in severely infested stands in Oregon. Grafted trees from the non-infected individuals had lower incidence and severity of dwarf mistletoe than did grafted trees from infected parents after a 20-year exposure of both groups to dwarf mistletoe. One trait associated with resistance is drooping needles. Trees with them are less frequently infected by *A. campylopodum* than are trees of typical form, because most seeds of the parasite, although intercepted by foliage, slide off when the needles are wet. Seeds intercepted by twigs or the basal parts of needles above twigs remain on the tree, capable of penetration and infection. Resistance also occurs in *P. jeffreyi*.

References: 725, 760, 1258, 1562, 1566, 2502, 2852, 2956, 3351–3354, 3439, 3441, 3442

Dwarf mistletoes of the *Arceuthobium campylopodum* complex, continued.

A. Witches'-brooms caused by *A. laricis* (larch dwarf mistletoe) on *Larix occidentalis* (western larch) (ID, Jul).

B, C. *A. apachecum* (Apache dwarf mistletoe) on *Pinus strobiformis* (southwestern white pine). B. Male plants growing from a spindle-shaped swelling on a host branch. Small pine shoots, signaling the onset of broom formation, are beginning to grow from the swelling. C. A female plant bearing immature fruit (AZ, Jul).

D–F. *A. cyanocarpum* (limber pine dwarf mistletoe) on *P. flexilis* (limber pine). D, E. Female and male plants, respectively. F. Witches'-brooms on a declining tree (ID, Aug).

G, H. *Arceuthobium tsugense* (hemlock dwarf mistletoe) (discussion p. 452). A. Large dead witches'-brooms on *Tsuga mertensiana* (mountain hemlock) (OR, Sep). B. Typical witches'-brooms in the lower crown of *T. heterophylla* (western hemlock) (WA, Jun).

451

Arceuthobium occidentale (digger pine dwarf mistletoe; Plate 222G) occurs in California at elevations from near sea level to 1200 m, primarily on *Pinus sabiniana* (digger pine) and secondarily on *P. attenuata* (knobcone pine) and *P. coulteri* (Coulter pine). *P. jeffreyi* (Jeffrey pine) and *P. ponderosa* (ponderosa pine) are occasional hosts where they are associated with infected *P. sabiniana*. *P. halepensis* (Aleppo pine), *P. pinea* (Italian stone pine), *P. radiata*, (Monterey pine), and *P. thunbergii* (Japanese black pine) are rare hosts. Several additional pine species have been found susceptible by inoculating them with seeds.

The aerial shoots are straw colored to light brown, range up to 17 cm (average 8 cm) in height, and have branches arranged in a fanlike manner (i.e., flabellate branching). Flowers open in autumn, and seeds mature about 13 months later. The plants produce prominent dense clusters of shoots on the widely spaced branches of *P. sabiniana*. The endophytic system of *A. occidentale* in *P. sabiniana* grows in each direction from a point of infection at an average rate of 2.0–2.3 cm per year. Its extent is nearly the same as the zone of stem swelling. *A. occidentale* does not usually induce witches'-brooms. Broom induction is suspected to be related to the composition of cytokinin-like substances (cell-division factors) produced by dwarf mistletoes. *A. occidentale* and the broom-inducing species *A. vaginatum* differ in these substances.

A. occidentale increases exponentially after a lag period of several years. On *P. sabiniana* deliberately inoculated with seeds, for example, an average of 4 plants with aerial shoots developed per tree within 5–6 years, but the average number rose to 250 plants per tree during the next 10 years.
References: 1556, 1566, 2852, 3433, 3436

Arceuthobium laricis (larch dwarf mistletoe; Plate 223A) occurs on *Larix occidentalis* (western larch) in many localities in northwestern USA and southern British Columbia at elevations of 700–1980 m. *Abies lasiocarpa* (subalpine fir), *Pinus contorta* (lodgepole pine), and *Tsuga mertensiana* (mountain hemlock) are secondary hosts. *Picea engelmannii* (Engelmann spruce), *Pinus albicaulis* (whitebark pine), and *P. ponderosa* (ponderosa pine) are infected occasionally, and *A. amabilis* (Pacific silver fir), *A. grandis* (grand fir), and *P. monticola* (western white pine) rarely. Introduced tree species found infected alongside diseased larch include *Picea abies* (Norway spruce), *Pinus banksiana* (jack pine), *P. resinosa* (red pine), and *P. sylvestris* (Scots pine). This dwarf mistletoe causes witches'-brooms on branches, often infects the trunk, and causes slow growth and premature death of *L. occidentalis*. Its shoots are dark purple and up to 6 cm tall (average 4 cm), with a fanlike arrangement of branches. Flowers are pollinated in late July to early September, and seeds mature 13–14 months later. This species spreads between trees at average rates of 1.2–1.6 m per year, depending on tree spacing. Wide spacing permits longer seed flights and greater spread rates.
References: 276, 1562, 1566, 2499, 2500, 2503, 2852, 3384, 3459, 4353, 4355

Arceuthobium apachecum (Apache dwarf mistletoe; Plate 223B, C) and *A. cyanocarpum* (limber pine dwarf mistletoe; Plate 223D–F) are less economically important species. *A. apachecum* is restricted to *Pinus strobiformis* (southwestern white pine) in the mountains of Arizona, New Mexico, and northern Mexico. Its shoots are yellow-green or sometimes reddish and up to 7 cm tall (average 3–4 cm), with fanlike branching. They develop on spindle-shaped swellings where host buds proliferate and give rise to witches'-brooms. Fruits are blue-green, about 4 mm long, and mature in September, 12 months after pollination.

A. cyanocarpum occurs widely in western USA, primarily on *P. flexilis* (limber pine), *P. albicaulis* (whitebark pine), *P. aristata* (Rocky Mountain bristlecone pine), and *P. longaeva* (Great Basin bristlecone pine); secondarily on *P. monticola* (western white pine) and *T. mertensiana* (mountain hemlock); and occasionally to rarely on *P. balfouriana* (foxtail pine), *P. contorta* (lodgepole pine), and *P. ponderosa* (ponderosa pine). It's small, densely clustered, yellow-green shoots are 3–7 cm tall. Its fruit matures in late summer.
References: 2496, 2852, 3978

Arceuthobium tsugense (hemlock dwarf mistletoe; Plates 223G, H, 224A–D) occurs from southeastern Alaska to central California at elevations from sea level to 2460 m. This species has a broader host range than most dwarf mistletoes and comprises three subspecies. *A. tsugense* subsp. *tsugense* occurs primarily on *Tsuga heterophylla* (western hemlock) and also on *Abies amabilis* (Pacific silver fir), *A. lasiocarpa* (subalpine fir), and *A. procera* (noble fir). *A. grandis* (grand fir) is an occasional host of the subspecies. *Picea engelmannii* (Engelmann spruce), *P. sitchensis* (Sitka spruce), *Pinus monticola* (western white pine), *Pseudotsuga menziesii* (Douglas-fir), and *T. mertensiana* (mountain hemlock) are rarely infected. Several additional conifers have proven susceptible when inoculated with seeds. *Arceuthobium tsugense* subsp. *mertensianae* occurs primarily on *T. mertensiana*, also on *Abies amabilis*, *A. lasiocarpa*, and *A. procera*; and secondarily on *Pinus albicaulis* (whitebark pine), occasionally on *P. monticola*. It occurs rarely on *A. grandis*, *Picea breweriana* (Brewer spruce), *Pinus contorta* var. *latifolia* (lodgepole pine) and *T. heterophylla*. *A. tsugense* subsp. *contortae* occurs on *Pinus contorta* var. *contorta* (shore pine) in the vicinity of Vancouver Island.

Aerial shoots of *A. tsugense* range to 13 cm tall (average 5–7 cm) with fanlike branching. They vary in color from green through straw colored to reddish, darker in winter than in summer. The life cycle requires 4–6 years. Flowers open in late summer to late autumn, depending on locality, and seeds are dispersed 13–14 months later. Seeds germinate in February–May, depending on local climate, and stem swelling appears after 1–2 years. Aerial shoots appear after another 1–2 years, and fruit a year later. Seeds are expelled mainly during daylight and may travel as far as 15 m, although most fall within 5 m.

A. tsugense induces witches'-brooms and suppresses tree growth. Average volume losses of 23% and 39% were recorded for moderate and severe infections, respectively, of *T. heterophylla* on Vancouver Island. Infections occur progressively higher on diseased trees, ascending at 30–65 cm per year on *T. heterophylla*. A dwarf mistletoe population on this host species may double every 2 years or less. More than 4000 infections were recorded on a single tree. Even so, height growth of vigorous trees usually outpaces upward spread of the parasite unless the trees are overtopped by older diseased individuals. Mistletoe-induced brooms that become shaded on low branches, as shown in Plates 223H and 224A, eventually perish without doing much harm. Trunk infections induce swelling, and cankers caused by secondary fungi may develop where endophytic systems grow from branches into trunks. Branch dieback caused by opportunistic fungi also occurs in *Abies* infected by this dwarf mistletoe.
References: 42, 422, 654, 1135, 1562, 1566, 1607, 1828, 2497, 2642, 2852, 2854, 2956, 3281, 3564, 3566, 3568, 3730–3732, 4020, 4232

Arceuthobium douglasii (Douglas-fir dwarf mistletoe; Plate 224) occurs from southern British Columbia into Mexico at elevations from about 275 m in the north to above 3000 m in the south. It does not occur in the coastal forests of the Pacific Northwest, where the principal host, *Pseudotsuga menziesii*, grows most rapidly and attains greatest size. *A. douglasii* occurs occasionally on *Abies amabilis* (Pacific silver fir) and *A. lasiocarpa* (subalpine fir), and rarely on *A. concolor* (white fir), *A. grandis* (grand fir), *Picea engelmannii* (Engelmann spruce), and *P. pungens* (blue spruce). This dwarf mistletoe has a large impact on tree health in some areas. The incidence of dead *P. menziesii* in severely infested stands in five national forests in southwestern USA was found to be three to four times that in healthy stands, and growth of severely infected trees

A–D. *Arceuthobium tsugense* (hemlock dwarf mistletoe) on *Tsuga heterophylla* (western hemlock). A. A typical witches'-broom (WA, Jun). B. A female plant with ripening fruit (WA, Aug). C. A male plant in bloom on a spindle-shaped swelling on a branch (WA, Aug). D. Male shoots as they appear in spring or early summer (WA, Jun).

E–H. *A. douglasii* (Douglas-fir dwarf mistletoe) on *Pseudotsuga menziesii*. E. A large old witches'-broom on a mature tree (OR, Sep). F. A small witches'-broom (AZ, Jul). G. Female shoots on the mainstem of the broom shown in F. H. Shoots of a male plant on a small branch of a broom (AZ, Jul).

I. A female plant of *A. gillii* (Chihuahuan pine dwarf mistletoe) on the trunk of *Pinus leiophylla* (Chihuahuan pine) (AZ, Jul).

J. Witches'-brooms caused by *A. vaginatum* subsp. *cryptopodum* (southwestern dwarf mistletoe) on *Pinus ponderosa* (ponderosa pine) (CO, Jun).

was reduced by more than 50%. Brooms become massive on some trees, and may constitute the entire living part of a crown. Such brooms are associated with dieback and premature death.

A new infection by *A. douglasii* on *P. menziesii* may remain latent for some years, or the branch may begin to swell and a broom develop at the point of infection. Stem swelling, indicating growth of the endophytic system, extends slightly more than 1 cm per year. The parasite becomes systemic within brooms, growing into buds and the primary tissues of young twigs. Aerial shoots arise along host branches at least 3 years old within brooms. Aerial shoots of *A. douglasii* are typically olive-green, sometimes reddish or orange-yellow, and measure up to 8 cm tall but average only 2 cm. Flowers open in spring, and seeds are dispersed 17–18 months later, typically in September. Seeds germinate the next spring. Aerial shoots arise 2 years later and produce flowers after another year; thus the life cycle is at least 6 years long.
References: 1134, 1252, 1428, 1566, 2480, 2501, 2505, 2964, 3384, 3731, 4034, 4035

Arceuthobium gillii (Chihuahuan pine dwarf mistletoe; Plate 224) occurs from southern Arizona and New Mexico into Mexico, primarily on *Pinus leiophylla* (Chihuahuan pine), *P. herrerae* (Herrera pine), and *P. lumholtzii* (Mexican weeping pine). *P. arizonica* (Arizona pine) and *P. arizonica* var. *cooperi* (Cooper pine) are rare hosts. The parasite produces greenish brown shoots up to 25 cm tall (average 11 cm) and as much as 8 mm thick at the base. Flowers are pollinated in spring, and seeds mature on average 19 months later in autumn.
References: 1566, 2852

Arceuthobium vaginatum induces witches'-brooms on various *Pinus* species in southwestern USA and Mexico. Two subspecies, *A. vaginatum* subsp. *vaginatum* (Mexican dwarf mistletoe) and *A.v.* subsp. *cryptopodum* (southwestern dwarf mistletoe), are recognized. The former occurs only in Mexico. The latter occurs from Colorado and Utah southward into Mexico at elevations of 1680–3000 m. Its primary hosts are *Pinus arizonica* (Arizona pine), *P. durangensis* (Durango pine), *P. engelmannii* (Apache pine), and *P. ponderosa* (ponderosa pine). *P. arizonica* var. *cooperi* (Cooper pine) is a secondary host. *P. aristata* (Rocky Mountain bristlecone pine) and *P. contorta* (lodgepole pine) are occasional hosts. *P. flexilis* (limber pine) and *P. strobiformis* (southwestern white pine) are rare hosts.

This dwarf mistletoe is the most important pathogen of *P. ponderosa* in the Southwest, reducing radial growth rates by 35–52% in 55–140-year-old stands in one area studied. It is most common and destructive on the driest sites occupied by *P. ponderosa*. It also predisposes trees to attack by *Dendroctonus ponderosae* (mountain pine beetle). Mortality of *P. ponderosa* observed during 1950–1982 in Grand Canyon National Park was greater than 90% among trees that were severely infected when the study began, but was less than 5% among lightly infected or uninfected trees. Witches'-brooms induced by the parasite often grow for decades and may attain diameters greater than 2 m. Infection is localized in the branch bases of most brooms but becomes systemic in a small minority.

Aerial shoots of *A.v.* subsp. *cryptopodum* are large—up to 10 mm thick at the base (average 4 mm) and up to 27 cm tall (average 10 cm). Usually they are orange to reddish brown with branches in a fanlike arrangement. Stem colors sometimes vary to greenish, red, yellow, or rarely dark purplish. Flowers are pollinated in late spring, and fruits about 5 mm long are dispersed in summer of the next year. The average horizontal distance of unimpeded seed flight is 5 m, but many seeds are intercepted close to the source. For that reason and because 6 years elapse between inoculation and seed production, the average rate of spread of this dwarf mistletoe in forests is only 0.3–0.5 m per year. New centers of infection are started by birds and squirrels that inadvertently carry seeds. Seeds germinate in late summer and may infect stems up to 9 years old.

The induction of witches'-brooms by *A. vaginatum* is believed to be related to the composition of cytokinin-like substances (cell-division factors) produced by the parasite. Cytokinin-like substances have been detected in *A.v.* subsp. *cryptopodum* and in *P. ponderosa* tissues infected by this parasite but not in comparable tissues of healthy *P. ponderosa*. Moreover, these substances differ

from those of *A. occidentale*, which seldom induces brooms in its primary host.
References: 310, 1555, 1558, 1562, 1566, 1803, 2594, 3433, 3751

Arceuthobium pusillum (eastern dwarf mistletoe) parasitizes primarily *Picea glauca* (white spruce), *P. mariana* (black spruce), and *P. rubens* (red spruce) in the region from Newfoundland to northern Pennsylvania, Minnesota, and eastern Saskatchewan. It occurs at elevations from sea level to about 900 m. *Larix laricina* (eastern larch) is frequently infected where it grows with diseased *P. mariana*, but the former species often fails to support production of aerial shoots. *Abies balsamea* (balsam fir), *P. pungens* (blue spruce), *Pinus banksiana* (jack pine), *P. resinosa* (red pine), and *P. strobus* (white pine) are rare hosts. The parasite causes severe growth loss and mortality in *Picea mariana* in many areas and is locally severe on *P. glauca* in Manitoba and along the coasts of Maine and Nova Scotia. It also causes trunk swellings and witches'-brooms on *P. rubens* in the mountains of northeastern USA. *P. rubens* often outgrows the parasite, as indicated by dead brooms on trees with healthy tops. *L. laricina* and *Pinus banksiana* often seem to resist the development of an endophytic system of *A. pusillum* by producing cork barriers that isolate the would-be invader.

A. pusillum, although not the smallest mistletoe (*A. minutissimum* of the Himalayas is smaller), is diminutive in contrast with the other dwarf mistletoes found in western North America. Its aerial shoots occasionally attain a height of 3 cm but average only 1 cm. They are greenish to brown to red and are simple or have short primary branches. The life cycle normally requires at least 4 years. Flowers are pollinated in spring by beetles, flies, wasps, and other insects. Fruits mature and expel seeds in September or October of the same year. Seeds germinate and infection occurs the next spring. Aerial shoots appear beginning in autumn of the second or third year but may be delayed for several years. Shoots develop in the third or fourth year and produce flowers and fruit a year later.

Witches'-brooms induced by *A. pusillum* vary from loose to compact. Host growth is locally stimulated during early stages of infection. The branch swells and the bark often appears reddish. Twigs grow more rapidly than normal and the foliage is dark green. After several years growth slows and the broom loses its verdant appearance. Infection of a mainstem often causes it to be converted to a large erect broom that may persist 40–50 years. Trees with large brooms usually decline and die prematurely, however. The endophytic system of the mistletoe is systemic within a broom, reaching even into the buds. *Picea mariana* twigs infected by *A. pusillum* have higher-than-normal levels of cytokinins and indoleacetic acid and a lower-than-normal concentration of abscisic acid. These imbalances are associated with stem swelling, loss of apical dominance on an affected branch, and diversion of metabolites from host to parasite.

Seeds of *A. pusillum* are dispersed both as projectiles from ripe fruit and by squirrels and birds struck by the sticky seeds when they disturb ripe fruit. Autonomous dispersal occurs primarily during morning hours as temperature rises. Seeds are propelled less than 2 m on average, but some travel more than 10 m. Infection centers in Minnesota were found to expand about 1.2 m per year on average. Animal vectors are responsible for new infections beyond the range of expelled seeds. *Perisoreus canadensis* (gray jay) is perhaps the most common bird vector.
References: 82, 173–177, 738, 1562, 1564, 1566, 1799, 2349, 2923, 3658, 3954

A–C. *Arceuthobium pusillum* (eastern dwarf mistletoe) on *Picea mariana* (black spruce). A, B. Large old witches'-brooms and associated mortality (NY, Jul). C. Two witches'-brooms, illustrating variation in branch proliferation and growth rate (NY, Jun).
D. A witches'-broom induced by *A. pusillum* on *Larix laricina* (eastern larch). Stem swelling and abnormally reddish bark are prominent on the main axis of the branch just below the region of proliferation (NY, Jun).
E–G. Close views of *A. pusillum* on *P. mariana*. E, F. Immature shoots, approximately natural size and enlarged, respectively, on 3-year-old twig segments from a broom (PA, Feb). G. Enlarged view of female shoots (NY, Jun).

455

Cassytha and Dodder (Plate 226)

Cassytha (dodder-laurel) and *Cuscuta* (dodder) are distinct groups of leafless, vinelike, parasitic seed plants. Although they are strikingly similar in general form and habits, they are unrelated. Their similarities are the result of parallel evolution.

Cassytha

The genus *Cassytha* of the Lauraceae is a group of perennial plant-parasitic vines that are widespread in tropical and subtropical regions on diverse hosts. *Cassytha* is the only genus of vinelike plants in its family and the only parasitic one. *C. filiformis*, a cosmopolitan species, is the only representative in the USA. Its colloquial names include devil's gut, dodder-laurel, woe vine, and cassytha. It occurs in Florida, Texas, Puerto Rico, and Hawaii, and is most common in coastal habitats. It develops profusely on low trees and shrubs growing in the open and is considered to be incapable of maintaining itself in closed forest canopies.

C. filiformis is an occasional pest of *Citrus* species in Florida. It also occurs on *Quercus geminata* (sand live oak), *Hibiscus rosa-sinensis* (Chinese hibiscus), *Lagerstroemia indica* (crape-myrtle), *Malpighia coccigera* (Singapore-holly), *Vitex agnus-castus* (lilac chaste-tree), and assorted other plants. It attracts little attention, however, because of its preference for infertile or barren sites where plant damage is seldom noticed. Elsewhere *C. filiformis* parasitizes a great array of plants. The diversity is indicated by records for *Acacia*, *Coccoloba* (sea grape), *Juniperus* (juniper), *Pinus* (pine), and *Rhizophora* (mangrove). On one island in the Bahamas the parasite was found on 81 plant species representing 45 families. Herbaceous plants may serve as initial hosts, but *C. filiformis* flourishes only on woody plants.

C. filiformis has been aptly described as a sprawling vine. It possesses chlorophyll and may appear distinctly green, but this color is often masked by orange pigment that makes the vine appear orange-brown. Each plant consists of a tangle of long, runner-like stems (stolons) 2–3 mm thick with widely separated, tiny, scalelike leaves. It grows initially from a seed that germinates in soil, but the root withers and disappears after several weeks, by which time the parasite has made connections to a host plant. Seedlings grow as much as 10–12 cm per week and reach heights of 25–30 cm while maintained by their own roots, but subsequent growth slows markedly. Mature stolons have been observed to elongate 2.3–6.3 mm per day under dry conditions. Parasitism begins as tips of growing stolons coil, counterclockwise, around host parts and multicellular haustoria grow from the coils into the stems. The stimulus to coil and penetrate is nonspecific; *Cassytha* strands often entwine one another and produce self-parasitic haustoria. Haustorial connections to hosts are made usually to petioles and leaf rachises and to young twigs that lack periderm, but leaf blades may also be parasitized. Haustorial development begins with formation of an attachment disk that sticks to the host by means of polysaccharide adhesive. A wedge-shaped intrusive organ then differentiates and seems to penetrate by a combination of enzymatic action and mechanical force. Connections to twigs seem to be maintained for a time after periderm formation. Stolons spread from one host plant to another, allowing a single *Cassytha* plant to colonize a large area. Host parts intensively colonized often lose vigor and die. A twig bearing a coiled stolon (haustorial coil) may appear swollen between turns of the coil and sunken beneath the coil.

Tiny spikelike flower clusters are produced from haustorial shoots. Insects, perhaps *Thrips* (which have been found in flowers), probably act as pollinators. The fruit is a small drupe, white at maturity, and contains a single seed that may remain viable for up to 2 years. Birds visit the fruit and are suspected of dispersing the seeds.

References: 922, 1585, 2101, 2175, 2176, 2178, 2676, 2770, 2961, 3146, 3209, 4109

Dodder

Cuscuta is a large genus of annual plants in the Convolvulaceae (morning-glory family). About 50 species occur in the USA and southern Canada. They are most abundant and troublesome on herbaceous plants in tropical and warm-temperate areas, but they often occur in cooler regions and may attack succulent parts of low-growing woody plants. Most *Cuscuta* species that have been studied contain chlorophyll and carry on photosynthesis, although inefficiently, but some species have only degenerate chloroplasts and do not fix carbon.

Dodder plants consist of soft, branched, yellowish or reddish strands that grow in a tangle over their hosts. The strands bear scattered minute scales that are vestigial leaves. When a strand encounters a host part, it begins to curl around it and by subsequent growth forms a helical coil from which numerous haustoria penetrate the host. Dodder strands coil as a nonspecific response to contact. Coils progress only counterclockwise and only upward. Dodder is strongly phototropic and does not entwine horizontal objects. The flowers are small, five parted, and are white in most species. They are usually produced in summer and autumn. The fruits are four-seeded capsules a few millimeters in diameter. No special mode of seed dispersal is known. Dodder seeds have been transported globally in seed lots of economic plants.

The seeds germinate in soil, and the young plants are nourished by their own roots for up to several weeks or until a haustorial connection to a host is established. The root then withers and disappears. Seedlings may attain lengths of more than 30 cm on their own roots, and they are attracted to hosts along gradients of water vapor and probably other volatile chemicals. If no suitable host is penetrated, the seedling dies. Incompatible hosts undergo hypersensitive reactions and/or produce defense chemicals (phytoalexins) and anatomical barriers that prevent establishment of haustorial connections.

Dodder can kill herbaceous plants and greatly devitalize woody ones. Sometimes noticeable swelling of a host stem develops beneath dodder strands in response to haustorial penetration. When the strands break as the result of host growth or movement, the separate parts continue as individual plants. Instances of perennial development of dodder by regrowth of shoots from the endophytic system have been reported, but these seem rare.

A *Cuscuta* plant extracts from its host water and nutrients, including organic solutes that travel in host phloem, by means of an intricate haustorial system that establishes connections with the host's parenchyma cells and phloem sieve tubes. Initial penetration is accomplished by numerous intrusive organs that grow from a dodder coil. Once the outer layers of the host are breached, linear structures called search hyphae (not to be confused with hyphae of fungi) differentiate near the tip of the penetrating organ and grow both inter- and intracellularly. Their progress is apparently facilitated by secretion of pectic and cellulolytic enzymes. Some hyphae penetrate parenchyma cells; others extend to the phloem and produce fingerlike branches that seem to clasp sieve tube members. These "contact hyphae" differentiate at maturity into conductive cells that transport solutes from the host to the main body of the parasite. The dodder causes an unloading of sugars, amino acids, and certain ions from host phloem at the site of contact.

Dodder can simultaneously parasitize two or more plants. For this reason and because of its ability to extract substances from phloem, dodder is used for experimental transmission of viruses and phytoplasmas from diseased to healthy plants. In nature, however, the role of dodder as a vector of other pathogens is insignificant.

References: 510, 916, 1217, 1440, 1548, 1644, 1847, 2033, 2126, 2175, 2176, 2178, 2355, 2770, 2843, 2961, 3146, 4109, 4419

A–C. *Cassytha filiformis* on *Quercus geminata* (sand live oak). A. *Cassytha* appearing as a wiry tangle of light green strands. Dead parts of the same vine at upper left are brown. B. Close view of the parasite on a branch tip. Green immature fruits are present. C. *Cassytha* coiling around a twig (FL, Apr).

D–H. Dodder. D. *Cuscuta japonica* (giant Asian dodder) on *Photinia* ×*fraseri* (Fraser or red-tip photinia) (TX, Oct). E. *Cuscuta* sp. on *Nerium oleander* (oleander) (CA, Jul). F. *Cuscuta* sp. on *Senna armata* (desert senna). The host is a nearly leafless leguminous shrub (CA, Jul). G, H. *Cuscuta ceanothi* on *Catharanthus roseus* (Madagascar periwinkle). Flowers and fruits of the parasite are present. Helical lines of scars representing former haustorial connections are visible on the stem in H where dodder strands were displaced (greenhouse plants).

Photo credit: D—K. Camilli and D. N. Appel

Vines That Damage Trees (Plate 227)

Many species of climbing vines are capable of injuring or suppressing the growth of trees and shrubs. *Vitis* species (grapevines), *Celastrus orbiculatus* (Asian bittersweet), *Lonicera japonica* (Japanese honeysuckle), and *Pueraria montana* (syn. *P. lobata*; kudzu) are particularly destructive. The latter three were introduced from Asia. Other vines causing occasional or local damage include *Ampelopsis* species (peppervine), *C. scandens* (American bittersweet), *Euonymus fortunei* (climbing euonymus), *Hedera helix* (English ivy), *Parthenocissus quinquefolia* (Virginia creeper), and *Wisteria sinensis* (Chinese wisteria). Most reports about damage to trees by vines are observational, and some claims of damage are speculative. This page collates examples of damage. Plate 227 shows *P. montana* and the constriction injury caused by *C. scandens*.

Climbing vines in North America are most destructive in the Appalachian and southeastern regions. They cause damage in several ways. Vines compete with trees for water and nutrients. The weight of vines causes twigs and branches to bend, twist, or break. Vines add to the surface area of trees and so increase the likelihood of breakage by wind or the weight of snow or ice. When growing on low vegetation or as ground cover, vines suppress reproduction of other plants, thus maintaining forest canopy gaps and thwarting normal plant succession. By growing over branches atop the canopy, they cause dense shade, which suppresses photosynthesis and may eventually kill trees. Stems or tendrils of some vines grow in tight coils around twigs and young trunks, causing constriction and deformity as the supporting stems increase in girth. In theory, some vines may suppress development of neighboring plants through release of toxic chemicals (allelopathy), but allelopathy probably plays at most a minor role.

Most vines grow well in full sun and are relatively intolerant of shade. Thus they are most common on southern and southwestern slopes and are quick to colonize old fields and forest openings and edges but are less successful or unsuccessful beneath forest canopy. They do not directly suppress tree growth unless they cover tree tops. Growth suppression by vines on trunks and low branches is due to competition for water and nutrients.

The notorious *Pueraria montana* of the Fabaceae was introduced from southeastern Asia to many places in southeastern USA for erosion control and as forage for livestock. The kudzu vine literally became wildly successful and now occurs in many localities from Texas and Florida northward to Kansas, Illinois, and Massachusetts. Although now detested in most localities where it has overgrown other vegetation, *P. montana* still has many advocates because of its usefulness for the purposes noted above. Fanciers of herbs and natural foods are aware of its medicinal and food values for humans. Its leaves are compound with three leaflets. Its stems are hairy, become woody with age, and attain lengths of 20 m or more. They entwine and climb on any object. The fragrant reddish purple flowers are abundant in late spring, but the vine produces pods sparsely. In a given locality *P. montana* spreads mainly by growth of stems that root at nodes when in contact with soil. The roots are swollen, woody, and perennial. Leaves and young stems are sensitive to frost, and the vine defoliates after the first severe frost in autumn. Woody stems, however, including those high in trees, survive winter in the South. *P. montana* dies back to ground level during winter in the coldest parts of its range.

Celastrus scandens (Celastraceae), a native plant, occurs from Montana and Texas eastward. It is common throughout the Appalachian region and northward into Canada. It climbs by coiling in a clockwise helix around stems and other supports. The coils lie at angles 40–70° from the vertical. The ripe fruit is bright orange, and stems with abundant fruit are often used for indoor ornament. Damage caused by *C. scandens* is of two kinds. The vine often climbs to the top of a small tree or shrub and produces a dense foliar canopy that causes deformity and dieback of the shaded host parts. In addition, or as the main effect on plants not overtopped by the vine, the coils of a *Celastrus* stem allow no room for the host stem to increase in girth. As both stems grow, the constriction interferes with the downward translocation of photosynthate in the host, whose stem enlarges markedly above the constriction and very little immediately below it. The host remains alive, however, because conductive elements of xylem and phloem produced by the cambium above the constriction become aligned with coils of the vine's helix. Routes of transport between leaves and roots are thus maintained, although they are lengthened.

C. orbiculatus, from eastern Asia, occurs across the eastern half of the USA north of Florida. It can climb to heights of 17 m or more and can severely damage other plants by strangling and/or overtopping them in the manner described for *C. scandens*. It causes more damage than does *C. scandens*, from which it can be distinguished by the location of female flowers and fruit. They grow in small clusters at nodes of *C. orbiculatus*, whereas they are borne on panicles at the ends of lateral branches on *C. scandens*. The mature capsule (outer fruit covering) is yellow in *C. orbiculatus* and orange in *C. scandens*.

At least 15 *Vitis* species (Vitaceae) grow wild in the USA and Canada. They climb trees and shrubs by means of tendrils that encircle twigs or small branches, providing temporary suspension. The tendrils eventually die and break, and the mainstem of the vine then hangs free from the canopy. Tendrils occasionally cause constriction deformities of young stems, but this injury is insignificant compared with the breakage and growth suppression caused by the vines. Grapevines are a significant obstacle to production of high-quality hardwood timber on many Appalachian sites, where they grow quickly on young trees. In experiments that designated crop trees at a young age and removed surrounding trees to reduce competition, grapevines nullified the benefit of release from other competition. In times of severe drought trees with heavy grapevine growth may perish where trees without vines survive. This tendency was noted in Oklahoma during an intense drought in 1934–1936. At a site in Connecticut breakage by a heavy ice glaze was more severe among trees supporting grapevines than among noninfested trees.

Lonicera japonica (Caprifoliaceae), an evergreen vine from eastern Asia, has become naturalized throughout much of the USA except the northwestern quadrant. It is one of the major weeds that interfere with reestablishment of forests after harvest. In experiments, vine removal from *Liquidambar styraciflua* (sweetgum) resulted in growth increase only if the vines, mainly *Lonicera japonica*, were removed from the trees and from the ground around the trunk. Removal of vines only from the trees, even if the vines extended to three-fourths of tree height, had negligible effect on tree growth. This research demonstrated the importance of competition between trees and vines for water and nutrients.

Hedera helix (Araliaceae), imported from Europe, has escaped from cultivation in scattered locations across North America as far north as British Columbia, Ontario, and Massachusetts, but not in the Great Plains. It may grow as a ground cover occupying many square meters, but it also climbs by means of specialized roots that form along its stem and adhere to rough surfaces. The aerial roots do not penetrate bark other than into natural crevices or cracks, and the vine does little harm unless it grows atop and shades the foliage of the supporting plant. In that event, it may kill small trees or shrubs. *References:* 658, 914, 999, 1000, 1012, 1110, 1128, 2223, 2397, 2575, 2640, 2682, 2880, 2954, 3261, 3611, 3612, 4013, 4025, 4081, 4314

A–D. *Pueraria montana* (kudzu). A. On oak trees at a forest edge (MS, Jul). B. Causing damage to *Liriodendron tulipifera* (tuliptree). Limbs are broken or bent downward. Height growth of the tree has been halted by the weight of the vine and the dense shade it casts above most of the tree's foliage. Many of the bare *P. montana* stems died during winter, but some survived even at the top of the tree, where new leaves are visible (SC, May). C, D. Close views of leaves and young stems, showing the tendency of stems to twine even around themselves (SC, May).

E. An *Ulmus rubra* (red elm) sapling stem entwined but not yet damaged by *Celastrus scandens* (American bittersweet) (NY, Oct).

F. An *Acer negundo* (box-elder) sapling stem deformed by *C. scandens*. The tree stem is swollen above constriction caused by the vine. The *C. scandens* stem is visible in the highest turn of the helix (NY, Sep).

Declines, Environmental Damage, and Unexplained Growth Abnormalities (Plates 228–253)
Decline Diseases with Multiple or Obscure Causal Factors (Plates 228–231)

Concepts and Overview

Decline is progressive loss of vitality and is a natural part of the life cycle of all higher organisms. Premature decline is of concern and, in trees, reflects the influence of stressing factors over periods of years. Causal factors and processes are both biotic and abiotic. The cause of any particular decline syndrome is likely to be consistent with one of four models.

• *Decline caused primarily by continual stress by one factor.* Some phytoplasmas and viruses and many primary insects (i.e., those that attack healthy nonstressed plants) cause their hosts to decline. Some fungal parasites of foliage, roots, or sapwood cause decline. Polluted air or soil can cause decline. Secondary or contributing factors are usually present, but their removal would not reverse the process.

• *Decline caused by drastic injury plus secondary stress.* Severe drought stress, mechanical damage, or defoliation by insects can entrain damage by opportunistic insects and fungi or abiotic factors that would be nearly harmless to a tree of normal vitality. The health of the compromised tree ratchets downward under the influence of transient and persistent factors within its body and its environment.

• *Decline caused by interchangeable predisposing, inciting, and contributing factors.* These factors can be physical (e.g., drought, freezing, mechanical damage, soil compaction), chemical (salinity, nutrient imbalance, soil acidification, misdirected herbicide, polluted air), and biotic (insects, fungi, bacteria, viruses, nematodes). The tree is weakened (predisposed) by one or a combination of factors, such that any additional insult, perhaps of no greater magnitude than the predisposing factor(s), incites decline. If the stressing factors are removed or counterbalanced by favorable factors before the tree becomes moribund, decline is arrested and the tree may recover. This model accommodates slow-decline diseases in which pathogens have major roles.

• *Cohort senescence.* Trees of a cohort attain an aggregate biomass that cannot be sustained by the resources of a particular site. Age and competition are involved. This situation primes a process like that described just above. If the trees are already old for their species, their decline should be considered normal.

The foregoing concepts are based on the following propositions. Tree growth and development vary from site to site and may fluctuate widely on a given site, yet still be judged normal. Trees alter their environment, and they tend to display group behavior. Trees growing in constructed landscapes and disturbed forest sites are stressed by factors not encountered by trees in undisturbed forests. Individual trees vary in exposure to stressing factors and in responses to them. A change in intensity or supply of an environmental factor may cause stress. Introduction of a new biotic or abiotic factor into a tree's environment may cause stress. A factor that causes no damage within a short time may predispose trees to damage by other factors or may cause damage if applied chronically. Repetitive minor injuries may exert cumulative effects. A factor that triggers decline may cease causing stress, but secondary or contributing factors may perpetuate stress and cause decline. Old trees are less resilient than young ones.

Symptoms of decline vary with cause and with tree species. They include slow growth; sparse, undersized or distorted, often chlorotic, nutrient-deficient leaves; browning of leaf margins; premature autumn color; premature leaf drop; abnormally large crops ("distress crops") of fruit; subnormal storage of food reserves (especially starch); and progressive or intermittent dieback of twigs and branches. Adventitious sprouts often develop for a time along the trunks of trees that have sustained branch dieback. Two general sequences of symptoms are recognized. If decline is initiated by an event such as root cutting or severe defoliation, buds and twigs may die as a shock response, and other symptoms follow. If decline results from chronic stress—for example, by salt or water shortage associated with poor root development in compacted soil—foliar symptoms and slow growth are likely to precede dieback. *References:* 142, 144, 1450, 1613, 1767–1769, 1805, 2099, 2450, 2455, 2457, 2567, 2621, 2747, 2974, 3650, 4221

Maple Decline

Decline of *Acer* species (maples) has been studied in forests and urban landscapes. *A. saccharum* (sugar maple) growing in forests and in stands managed for production of syrup and sugar may decline in response to severe defoliation by insects, severe drought,

root freezing during midwinter freeze-thaw cycles or during winters with little snow cover, timber harvesting, root damage by grazing livestock, or overzealous tapping and sap extraction—alone or in combination. Nutrient deficiency and mineral imbalance in trees growing in shallow, poorly buffered soils have also been implicated, as have sapstreak disease, caused by *Ceratocystis virescens* (Plate 117), and damage by *Glycobius speciosus* (sugar maple borer). Opportunistic fungal pathogens attack predisposed trees. *Armillaria* species (Plate 162) attack roots; and fungi such as *Cerrena unicolor* (Plates 158, 228), *Stegonsporium* species (Plate 228), and *Valsa* species (Plate 86) attack stems of weakened trees.

Severe defoliation by insects after shoot growth is complete in summer may cause winter buds to open and produce a new flush of shoots. This process nearly depletes stored food reserves, lowers resistance to fungal pathogens, and delays cold hardiness, thus predisposing the tree to frost damage in autumn. If the tree is further stressed the next season by repeated defoliation or other factors, it may begin to decline. Timber harvesting promotes decline in residual trees where trunks and soil previously shaded are subject to abnormal heating and drying by sunlight. Trunks and roots of residual trees are also often wounded by logging equipment. Grazing cattle cause or contribute to decline because their hoofs break or cut small roots. Sap harvesting may contribute to decline if too many tap wounds are made or if chemicals are used that thwart compartmentalization of tap holes (causing death of a large volume of sapwood) or too much sap is extracted (depleting sugar that would be used as the energy source for growth). Soil mineral imbalances and deficiencies of calcium, potassium, magnesium, and phosphorus have been associated with decline in *A. saccharum*, especially on shallow, poorly buffered soils. Deposition of airborne acidic substances can cause or aggravate such imbalances.

Factors that trigger decline of maples in landscapes include girdling roots (Plate 250), restricted rooting space with associated water shortage or waterlogging, cankers and collar rots caused by fungi or oomycetes (especially *Phytophthora* species), soil compaction leading to water shortage and rootlet mortality, de-icing salt (Plate 232), chronic infection by *Verticillium dahliae* (Plates 120, 121), severe trunk wounds, and root severance during excavation for utility channels, walks, and roadside drainage. The same opportunistic pathogens that occur in forests contribute to decline in urban plantings and are joined there by such fungi as *Ganoderma "lucidum,"* causing root rot (Plate 173), and *Botryosphaeria obtusa* (Plate 62) and *Nectria cinnabarina* (Plate 87), causing cankers and dieback. Decline in urban maples can be avoided or delayed or its incidence suppressed by selecting planting sites to avoid severe environmental stress.

References: 144, 179, 257, 336, 337, 346, 939, 1001, 1025, 1272, 1375, 1449, 1450, 1631, 1632, 1756, 1757, 1767–1769, 1839, 2120, 2197, 2360, 2400, 2416, 2449, 2689, 2787, 2932, 2974, 2986, 3279, 3316, 3317, 3361, 3410, 4221, 4224, 4226, 4227, 4247, 4307

Decline of *Acer saccharum* (sugar maple).

A, B. Premature autumn color and dieback on roadside trees (NY, Sep & Aug).

C. Dieback in a tree that sustained root death due to flooding (NY, Aug).

D. A forest tree, defoliated by insects 6 and 5 years earlier, that never regained vigor (note sparse foliage) and suddenly died during summer drought (ON, Jul).

E, F. Contrast between twig form and growth on a declining tree (E) and a vigorous tree (F). Twigs of the declining tree grow slowly, and many have died, as evidenced by lack of swollen buds. Apical buds on living branches of the declining tree have lost dominance, so all twigs are about the same length (NY, May).

G. A tree declining because roots on one side were buried. Bark has fallen from a dead area that developed on the trunk above dead major roots (NY, Aug).

H. Black spore masses of *Stegonsporium pyriforme* on a recently killed branch of a declining tree. This fungus is believed to be an opportunistic pathogen that kills weak branches (NY, Aug). I, J. Basidiocarps of the canker-rot fungus *Cerrena unicolor,* an opportunistic pathogen, on a dying tree (NY, Aug).

461

Ash Decline

Decline syndromes in *Fraxinus* species are widespread in North America. *F. americana* (white ash) is affected in north-central and northeastern USA and parts of southeastern Canada, *F. pennsylvanica* (green ash, red ash) in the Midwest and Great Plains, and *F. velutina* (velvet ash) in the Southwest. *F. nigra* (black ash) has been affected sporadically in Saskatchewan, northern Minnesota, Quebec, and Maine. The most common environmental circumstance associated with these syndromes is water shortage, but freeze damage (Plate 248) has been a factor in northern locations. Other topics related to ash decline include ash rust (Plate 134), ash yellows (Plate 198), viral diseases of ash (Plates 202, 207), and water stress. Tree decline concepts are presented with Plate 228.

Decline in *F. americana*: "ash dieback." Dieback and decline of *F. americana*, collectively called "ash dieback," have been noted in northeastern USA and adjacent Canada since the 1920s. Several biotic agents as well as drought and freeze damage have been implicated as primary or contributing causes, although the causal factors have differed from place to place. The phytoplasmal disease ash yellows is a major causal factor in areas of mixed land use, but it is absent or uncommon in areas that are primarily forested. "Ash dieback" is broadly characterized by progressive loss of vigor for 2–10 years or longer before trees die. Radial and apical growth diminish, the latter reduction leading to short internodes and tufted foliage at branch ends. Some trees with dieback have normal foliage, but leaves of many trees become undersized and pale green to chlorotic, and the canopy often appears sparse. Premature autumn color and leaf fall are common, as are viruslike symptoms on leaves. The smooth bark of affected trees exposed to sun and wind often turns pinkish in contrast to the normal gray. Annual cankers caused by fungi form on branches and trunks. Twigs and branches die back, usually during dormancy but also during the growing season in drought years. Epicormic sprouts often form below cankers and on trunks of trees with severe dieback. Sprouts that become deliquescently branched or develop into witches'-brooms indicate phytoplasmal infection. Root dieback is common in declining trees. Their death usually occurs during or shortly after drought.

F. americana is an early colonist of many sites ill suited for long-term growth. Its growth rate is highly sensitive to the soil moisture supply in spring and early summer, and it is more sensitive to water shortage than are other trees to which it has been compared. Incidence and severity of dieback in *F. americana* were greater during and after droughts in the 1930s, 1950s, 1960s, and 1990s than in preceding or intervening periods. This information and the association of decline with soils of low water-holding capacity in some areas support an interpretation that ash dieback is often caused primarily by drought and that biotic agents play secondary roles. This interpretation explains decline on stressful sites where ash yellows does not occur and decline in trees that are tolerant of phytoplasmal infection. Ash yellows aggravates drought stress by suppressing root growth and, in highly susceptible trees, causing rootlet necrosis. Infected trees are predisposed to more severe or prolonged water stress than healthy trees experience in similar circumstances. Phytoplasma-infected *F. americana* trees do not recover after drought episodes as well as noninfected trees do. In localities where the causes of decline include ash yellows, declining trees rarely recover, new cases of decline begin annually in young trees, the onset of decline in young trees is random within a stand, and the most common years of onset vary from one stand to another.

Other abiotic factors that may trigger or contribute to ash decline are low temperatures and, in some localities, polluted air. Episodes of dieback in *F. americana* and *F. nigra* in Quebec in the 1920s, for example, were attributed to roots freezing during winters with subnormal snow cover. Frost-induced bark cracks and cankers near ground level are associated with decline in some areas. These symptoms have also been linked to ash yellows, the presumed relationship being that phytoplasmal infection prevents trees from becoming fully cold hardy. Polluted air and atmospheric deposition are probably not significant factors in ash decline. *F. americana* is sensitive to ozone, but foliar symptoms typical of ozone injury are not regularly associated with decline.

Bark-killing fungi are documented contributors to decline in *F. americana*, and viruses are involved but have an unknown role. The most common fungi that cause dieback of twigs and branches are the coelomycetes *Cytospora pruinosa* (syn. *Cytophoma pruinosa*) and *Fusicoccum* sp. (an anamorph of *Botryosphaeria*). Cankers on stems with smooth bark are reddish brown to orange-brown; those in old bark with cork ridges are initially invisible. The bark in cankers begins to split, shred, and drop off 1–2 years after the lesions form. *C. pruinosa* is the more commonly encountered pathogen and was the more virulent in pathogenicity tests. Each fungus, when inoculated into healthy *F. americana* saplings, caused more or larger cankers on trees stressed by water shortage than on trees that received normal rainfall. *C. pruinosa* produces black pycnidia, 450–800 µm in diameter with a simple central fertile cavity, in recently killed bark. Its tiny single-celled conidia ($3–9 \times 1–2.5$ µm) are colorless and curved. Three viruses—*Tobacco mosaic virus*, *Tobacco ringspot virus*, and *Tomato ringspot virus*—were detected in declining *F. americana* in New York State. The last-named virus causes severe damage in fruit and nut trees, but experimental evidence that any virus contributes to ash decline is lacking.

Decline in other *Fraxinus* species. Dieback and decline of *F. pennsylvanica* studied in the upper Midwest and Great Plains seem linked primarily to water stress; harsh environments at planting sites, such as in shelterbelts on the Great Plains; and depredation by opportunistic fungi that cause cankers and trunk decay. Although the ash yellows phytoplasma is widespread in this species and suppresses its growth in some places, the phytoplasma has not been linked to decline. Phloem-inhabiting proteobacteria have been detected in *F. pennsylvanica* in Iowa, but their relationship to tree health is unknown.

F. nigra in normally wet habitats in northern Maine and northern Minnesota has sustained growth decline, chlorosis, and dieback when water tables dropped during drought. In Quebec dieback followed winters with little or no snow cover, when freeze damage to roots was likely. In Saskatchewan young *F. nigra* shade trees have developed dieback after freeze damage to their trunks.

A decline syndrome in *F. velutina* in Zion National Park, Utah, seemed associated primarily with perennially severe foliar damage by insects, particularly *Tropidosteptes pacificus* (ash plant bug), and secondarily with water shortage. Infection by the ash yellows phytoplasma, common in both declining and unaffected trees, was not closely linked to tree condition.

Full recovery of ash trees from decline, as indicated by regrowth of crowns after dieback, is uncommon because the causal factors or circumstances persist or recur. Symptoms often become static, however, if stressing factors diminish. Trees in early stages of decline are usually capable of increased growth when environmental conditions improve.

References: 505, 672, 675, 926, 1111, 1472, 1636, 1638, 1639, 1767, 1768, 2042, 2226, 2350, 2383, 2508, 3116, 3117, 3335, 3336, 3394, 3647, 3648, 4051, 4196–4198, 4428, 4429

A, B. Decline in *Fraxinus americana* (white ash): tufted foliage at twig tips, dwarfed leaves often light green in color, pinkish bark, and dieback of twigs and branches, some (B) with brown leaves still clinging to them (NY, Aug).

C. Decline induced in *Fraxinus velutina* (velvet ash) by the combined effects of water shortage, foliar damage by *Tropidosteptes pacificus* (ash plant bug), and ash yellows (UT, May).

D–J. Cankers in *F. americana* caused by *Cytospora pruinosa*. D, E. Inactive cankers several years old (NY, May). F. A year-old canker about 8 cm long. Infection began at the central wound. Black specks are pycnidia (NY, Oct). G, H. Discrete and diffuse cankers, respectively, on saplings stressed by freeze damage (NY, May). I, J. Magnified views of pycnidia of *C. pruinosa*. I. Maturing in young bark in spring (NY, May). J. Mature pycnidia, some with their tops shaved off to reveal a simple central fertile cavity (NY, Oct).

Birch Decline

A syndrome called birch dieback devastated *Betula alleghaniensis* (yellow birch) and *B. papyrifera* (paper birch) in parts of eastern Canada and northern New England from the 1930s to early 1950s. In the most severely affected areas more than 80% of the mature birch died within 12–13 years, and reproduction was scarce. Death of feeder roots and reduction in growth rate signaled the onset of decline. Most of the fine roots were dead by the time a tree had severe crown damage. Other symptoms included undersized chlorotic leaves near branch tips; shriveled, rolled, or cupped leaves, sometimes with ringspots, vein clearing, or mild mosaic; swollen yellowish buds (versus normal brown); dead buds; distorted shoots that withered early in the season; proliferation of buds and stunted shoots; nonviable seeds; dieback of twigs and branches; and in severe cases, death of a tree after 3–6 years.

Diagnostic research focused initially on *Agrilus anxius* (bronze birch borer), but the borer was not present in trees with early symptoms. No insects or fungi capable of causing severe dieback were found. Viruslike symptoms were common, and some symptoms were transmitted to healthy *B. alleghaniensis* by grafting. *Apple mosaic virus* (Plate 205) was eventually identified, but its role remained unclear. Further research based on knowledge of "postlogging decadence" turned to abiotic factors. *Betula* species and to a lesser extent other northern hardwoods left uncut on logging sites often decline as a result of sudden exposure to full sun, higher air and soil temperatures, lower humidity, and higher wind velocity than occur in uncut stands. Opportunistic organisms such as *Armillaria* species and *Agrilus anxius* attack the disturbed trees.

Climatic warming was considered. Average summer temperatures in the Maritime Provinces rose 1.0–1.4°C during 1920–1950. Dieback was most severe where birch rooting depth was least, as would be expected if soil warming and drying were important. When roots of *B. alleghaniensis* seedlings were subjected to soil temperatures 2°C above normal for one growing season, populations of root-inhabiting fungi were altered and rootlet mortality rose from a normal 6% to 60%. The onset of decline in mature trees, on the other hand, was neither related to measured environmental factors nor correlated with growth perturbation in adjacent healthy trees, as would be expected if environmental change were the primary cause. The outbreak did subside coincident with a temporary downturn in the long-term temperature trend, however. Later research implicated winter freeze-thaw cycles, soil freezing, ice in soil, spring frosts, water stress, and the trunk-decay fungus *Chondrostereum purpureum* (Plate 159) as possible causes or contributing factors. A birch decline in northern Michigan in the late 1980s and early 1990s was linked to increasing temperatures and water shortage, which triggered attack by *A. anxius*.

Leaf browning and dieback of *B. papyrifera* and to a lesser extent *B. alleghaniensis* began in the late 1970s in New Brunswick near the Bay of Fundy. This syndrome was linked to acidic coastal fog, which injures and leaches foliage and hastens acidification of poorly buffered soils. Soil acidification can cause or aggravate soil nutrient imbalances and deficiencies.

Analyses of climatic and historical data related to birch dieback and other hardwood declines of the 20th century have revealed that dieback episodes occurred coincident with forest maturation in several regions. Mature tree populations, having grown to the carrying capacity of their sites, are thought to be more susceptible than younger populations to stresses associated with freezing and drought that trigger decline.

References: 28, 142, 143, 321, 473, 736, 737, 822, 823, 1326, 1363, 1450, 1613, 1964, 2569, 2621, 2747, 3118, 3119, 3246, 3773, 4522

Pine Declines

Well-known decline syndromes in North American pines during the 20th century, other than those caused primarily by insects, included damage to *P. ponderosa* (ponderosa pine) by ozone in southern California (Plate 237); littleleaf of southern pines (Plate 182), attributed to rootlet damage by *Phytophthora cinnamomi* on eroded, growth-limiting sites; procerum root disease of *Pinus strobus* (eastern white pine; Plate 113); and pole blight of *Pinus monticola* (western white pine) in northwestern USA and adjacent British Columbia.

Pole blight of *Pinus monticola*. Pole blight became prominent and then subsided during a 30-year period beginning near 1930. It occurred in the Rocky Mountain and Intermountain regions, causing severe losses in some localities. Examples of it still occur. Pole blight affects primarily dominant and codominant trees of pole size (stem diameter 15–30 cm at breast height). The syndrome begins with rootlet necrosis and continues with declining stem growth, death of larger roots, and crown symptoms. Foliage becomes chlorotic, dwarfed, and tufted on branch ends. Foliar symptoms progress downward, followed by dieback. Distressed trees often produce large cone crops before dying back. Long resinous lesions form on the lower part of the trunk following the grain of the wood. Trees often die within 5–10 years, their demise hastened by secondary insects and bluestain fungi.

Decline of *P. monticola* on sites where other tree species remained healthy was attributed to inferior ability to compete for scarce moisture and adapt to dry conditions. Pole blight was most severe on shallow soils (30–45 cm rooting depth) with limited water-holding capacity. *P. monticola* was found to have fewer root tips, less length of fine roots per unit of soil volume, and a smaller root/shoot ratio than did other conifers under conditions of moderate water shortage. Moreover, *P. monticola* did not curtail top growth under relatively dry conditions that suppressed growth of other conifers. These habits were thought to predispose it to intolerably low water potential during severe drought. Rootlet mortality, cavitation (broken water columns) in sapwood, and external symptoms followed. Pole blight has not been induced experimentally, but the health of many affected trees stabilized when normal rainfall resumed. Some began to produce woundwood at margins of trunk lesions, indicating improving vitality.

No single organism has been consistently associated with pole blight. The fungus *Ophiostoma trinacriforme*, anamorph *Leptographium* (Ophiostomatales, Ophiostomataceae), occurs commonly in trunk lesions. It caused elongate resinous lesions and resin soaking of sapwood when inoculated to healthy trees, but it was considered to be a secondary factor in pole blight because growth reduction always preceded lesion formation. *Armillaria ostoyae* (Plate 163) is also a common invader of roots and root collars of trees with pole blight.

Other pine decline syndromes. Decline followed by premature death of *Pinus strobus* (eastern white pine) in urban landscapes has been studied in Illinois and Virginia and noticed more widely. It occurs on sites with high soil pH, high clay content, shallow and restrictive root space, compacted soil, and/or soil that impedes water movement. Symptoms include chlorosis; diminishing growth rate; undersized needles; premature needle drop, so affected trees have only one age class of needles in winter and crowns appear thin; and shriveling of normally smooth bark as branches or whole trees die. Bark beetles and bluestain fungi enter severely stressed trees and ensure their death. *P. strobus* affected by toxin(s) from roots of *Juglans nigra* undergoes a similar decline syndrome. Pole-size *P. strobus* in dense forest stands in Maine declined and died in the late 1990s in response to drought. The affected trees had colonized sites where the rooting depth was less than 30 cm. Ozone-sensitive individuals of *P. strobus* have declined in some parts of the Appalachian region and northeastern USA. *P. palustris* (longleaf pine) and other southern pines growing on golf courses and residential landscapes in southeastern USA have declined where excess nitrogen and lime or alkaline irrigation water induced nutrient imbalance and chlorosis. Numerous examples of decline in other *Pinus* species have been reported from Europe and Japan.

References: 144, 309, 1218, 1399, 1613, 1624, 2257–2259, 2608, 2747, 2958, 2990, 3460, 3554, 4260

A, B. Birch decline: postlogging decadence of *Betula papyrifera* (paper birch). Gross symptoms are similar to those of "birch dieback" (ON, Jul).

C–G. Pole blight of *Pinus monticola* (western white pine). C. A healthy tree at left, in contrast to a declining tree with chlorotic tufted foliage at right center. A distress crop of cones remains on the top of a dead tree at left center. D, E. Young pole-size trees, normal (D) and diseased (E). F, G. A large resinous canker on the base of a trunk. Bark was removed (G) to show the extent of the lesion and resin-soaked sapwood (ID, July).

465

Oak Decline

Decline of *Quercus* species has caused concern in North America and Europe since the 1950s. Symptoms include slow growth; sparse, undersized, and sometimes distorted or chlorotic leaves; death of scattered twigs; progressive dieback of branches and limbs; strip cankers on trunks; and often growth of adventitious sprouts on the trunk and large limbs following dieback (Plate 253). Chlorosis and/or leaf scorch and premature leaf senescence precede or accompany dieback in some circumstances. We first consider decline in forests and then in urban and landscape plantings. Concepts of tree decline are discussed with Plate 228.

The most common causes of oak decline in forests vary with region and species. Drought stress, defoliation by insects, and attack by secondary insects and fungi are the most frequent causal factors. Severe drought has triggered decline episodes in southern USA and southern Europe. In the USA, species in the red oak group—for example, *Q. coccinea* (scarlet oak), *Q. nigra* (water oak), *Q. rubra* (red oak), and *Q. velutina* (black oak)—are more severely damaged than are members of the white oak group. Drought damage is discussed further with Plates 244 and 245. In Europe frost and freeze damage and root killing by *Phytophthora* species are also involved. Nutrient deficiencies (notably calcium and potassium) and soil nutrient imbalance are involved in some American localities. Trees stressed by primary factors become susceptible to further damage or killing by opportunistic insects and fungi. Borers, notably *Agrilus* species, girdle trunks or limbs. *A. bilineatus* (two-lined chestnut borer) is a common attacker in the USA. Bark-infecting fungi, notably species of *Biscogniauxia* (Plate 99), *Cryphonectria* (Plate 79), and *Endothia* (Plate 81), as well as members of the *Botryosphaeria* group (Plates 59–62), cause dieback and cankers. Root-infecting fungi, notably *Armillaria* species (Plate 162), kill major roots and cause decay of root wood.

Oak decline in forests and urban areas of eastern and central USA is often triggered by defoliating insects, particularly *Lymantria dispar* (gypsy moth) or species of *Archips* or *Croesia* (leaf rollers). Severe defoliation (three-fourths or more of the leaves removed) causes growth loss and can trigger decline if it occurs after leaf expansion and results in a second growth flush in the same season. Such growth depletes carbohydrate reserves and leaves a tree abnormally susceptible to opportunistic insects and fungi. Defoliation by insects or frost in spring before leaves reach full size is of less consequence because regrowth does not exhaust energy reserves and time remains for photosynthetic accumulation of new reserves. Severe defoliation in two or more successive years (or twice in one year, as by frost and then insects) will usually trigger rapid decline and death; trees commonly wilt and die in August. The death rate begins to rise in the second year of defoliation and may exceed 50% in the third year. Similarly, severe defoliation and drought during the same growing season may lead to or accelerate decline.

Common factors that stress *Quercus* species growing as street or landscape trees are drought; defoliation; root wounds caused by excavation equipment; root rots caused by *Ganoderma* species and other fungi; trunk wounds leading to death of sapwood and eventual trunk decay; root death caused by oxygen deficiency in compacted or waterlogged soil; root death caused by soil changes around new structures (change of grade, change in height of water table, heating and drying of surface soil after vegetation is removed); nutrient deficiency, especially iron deficiency induced by alkaline soil (Plate 243); and salinity stress where salt from irrigation water accumulates in the root zone (discussed with Plate 233). De-icing salt is seldom involved, as oaks are relatively tolerant of it. Chronic infestation by *Asterolecanium* species (pit-making scales) often contributes to decline of trees in the white oak group; for example, *Q. alba* (white oak), *Q. garryana* (Oregon white oak), and *Q. prinus* (chestnut oak). *Phytophthora* species that cause cankers on oak trunks and root collars (Plates 176–178) may cause or contribute to decline.

Examples of oak declines in North America include: (1) periodic decline and mortality of oaks in eastern forests, caused by defoliation, drought, borers, *Armillaria* species, and *Biscogniauxia atropunctata*; (2) decline of *Q. virginiana* along the Gulf Coast,

caused by construction damage, soil compaction, trunk wounds, hurricanes, *Endothia gyrosa*, and *Botryosphaeria rhodina*; (3) decline of *Q. palustris* (pin oak) in eastern Virginia, caused by unsuitable soil, drought, and *E. gyrosa*; (4) periodic dieback of *Q. agrifolia* (coast live oak) in California, caused by drought and *Diplodia quercina*; (5) decline of *Quercus* and other trees in irrigated landscapes in Texas, caused by accumulation of soluble salts; (6) decline of residual trees wherever residential developments are constructed in forested landscapes, caused by root disturbances and, in the Midwest, alkalinization of topsoil. Alkalinization occurs when topsoil and alkaline subsoil are mixed and when alkaline substances leach from concrete into topsoil.

References: 98, 252, 340, 380, 479, 551, 933, 969, 986, 1224, 1537, 1578, 1767–1769, 1926, 1982, 1983, 2180, 2301, 2302, 2306, 2380, 2457, 2669, 2715, 2848, 2883, 2884, 2926, 3002, 3298, 3620, 3674, 3675, 3821, 3841, 3959, 3960, 4006, 4007, 4127, 4140, 4219, 4220, 4223, 4225

Citrus Blight

The disease known as citrus blight occurs in several citrus-growing regions of the world, affecting many *Citrus* types on various rootstocks. It causes major losses in Florida (more than one-half million trees per year) but does not occur in western USA. The decline begins any time after trees reach the age of 6 years. In severe situations annual incidence of new disease in a grove is 10–15% of the tree population. Symptoms include slow growth and loss of productivity, chlorotic patterns typical of zinc or manganese deficiency on leaves, wilting despite adequate soil moisture, and branch-by-branch dieback. Trees with dieback usually produce adventitious sprouts on the trunk and scaffold limbs, but the sprouts also eventually die. Aerial symptoms are followed by root decay. Xylem vessels in roots, trunks, and major limbs, but not in small branches, become plugged with an amorphous substance that inhibits water movement. Zinc and water-soluble phenolic substances accumulate to at least three times normal levels in outer sapwood. Affected trees never recover, and no control is known. Blight cannot be diagnosed from visible symptoms because other maladies cause similar decline. Diagnostic criteria include elevated zinc and phenolic concentrations and slow water uptake (less than 10% of the normal rate) during diagnostic injection into the trunk.

No single cause of citrus blight has been found. Some symptoms have been transmitted experimentally by root grafts (not by stem grafts), but epidemiological and experimental evidence indicates that blight is caused primarily by stress-inducing abiotic factors that are mainly nutritional. Biotic agents contribute to symptom severity. Root-infecting, toxin-producing strains of *Fusarium solani* are involved. *Xylella fastidiosa* (Plates 192, 193) often occurs in *Citrus* with blight and may exacerbate the symptoms, but it does not induce them.

References: 327, 519, 520, 762, 938, 1237, 1349, 1750, 2181, 2273, 2321, 2820, 2821, 3260, 3979, 4101, 4451–4453

A, B. *Quercus velutina* (black oak) in decline following repeated severe defoliation by insects. A. A dead tree and one with dead branches have adventitious sprouts that grew on major limbs after peripheral branches died. B. Dwarfed, distorted, chlorotic leaves; stunted twigs; and dieback (IL, Jun).

C–E. Decline in *Q. virginiana* (southern live oak). C. A large tree during spring growth, with sparse-appearing undersized foliage and early-stage dieback attributed to prior heat and drought stress; low branches have somewhat denser foliage. D. Closer view of top branches (MS, Apr). E. Dieback and adventitious sprouts. The cause of decline in this tree was unknown (TX, Nov).

F, G. Late-stage symptoms of citrus blight in *Citrus ×paradisi* (grapefruit): sparse, dwarfed leaves, dieback, and (G) sprouts on the base of the trunk (FL, Apr).

Damage by Salt and Other Inorganic Poisons (Plates 232, 233)

Salt Damage

Mineral salts can retard growth and cause foliar browning, kill vegetative and flower buds, cause twig dieback, and even kill plants if applied in sufficient quantity or frequency. Aerial parts may receive excess salt in aerosols, spray, or dust; roots receive it in soil water. Damage from salt spray, salt dust, and saline meltwater occurs along highways where de-icing salt (mostly sodium chloride) is used. This damage is mostly confined within 50 m of highways. Salt spray damage and saltwater flooding occur during storms along the coasts. Salt spray commonly injures plants within about 300 m of the ocean, and damage from severe storms may extend many kilometers inland. Chronic minor damage occurs as spray and aerosols from bursting bubbles at wave tops are blown to the shore. Detergents in ocean salt spray contribute to chronic damage. Airborne salt may also cause local damage around evaporative cooling towers and potash mines. Saline soils affect plant growth in many western and southwestern localities in North America and in other arid and semiarid regions.

Salt spray and dust. Leaves and twigs collect spray droplets and dust. When a plant is wet, sodium and chloride ions (and assorted other ions from seawater) can enter and accumulate. External symptoms develop quickly during mild weather but are delayed in cold weather. Symptoms on evergreens along highways appear in late winter and intensify in early spring. Conifer needles turn brown, the necrosis beginning at the tips and progressing toward the bases. Damaged needles fall prematurely. Injury in successive years causes branches to become barren and die. Symptoms appear on deciduous species when growth begins in spring. Buds, especially flower buds, fail to open, and twig dieback becomes apparent. New shoots grow from adventitious or previously dormant buds on 2-year-old branch segments. These shoots are often clustered because the source buds are closely spaced at twig bases. The side of a plant away from a highway or shore receives less salt than the side facing the salt source and sustains less damage, so growth and flowering become asymmetric. Repeated twig killing on plants growing near the shore often trains them to shrubby or prostrate form. Foliar browning develops within a few days after a coastal storm during the growing season, and severely damaged leaves drop prematurely. Barren twigs may be apparent the next spring.

Salt absorption and subsequent damage increase with the amount of salt deposited on a plant. The type of salt is relatively unimportant. Damage at the cellular or tissue level is caused mainly by toxicity of chloride ions. Chloride is translocated toward twig tips and toward the tips and margins of leaves, where it accumulates. Sodium and other ions also enter and accumulate, but more slowly, and their concentrations in salt-damaged plants are not as closely correlated with the degree of damage as is the concentration of chloride.

Salt tolerance of many plants seems to be based more on ability to retard the internal accumulation of chloride than on intrinsic ability to tolerate it. Foliar and twig damage appear in many woody species when internal chloride concentrations rise above approximately 0.3% (dry weight basis). Plants with severe damage may have chloride contents of 1.25–2% or more in their twigs, while tolerant species growing under the same conditions but remaining asymptomatic have 0.5–0.9% chloride content. Normal values are under 0.1%. Plants with thick wax layers on leaves and twigs; with leaves that have a low surface-to-volume ratio (e.g., pines with thick needles); with large resinous buds, as in *Populus deltoides* (eastern cottonwood) and *Aesculus hippocastanum* (horse-chestnut); or with tiny buds recessed in the twigs, as in *Robinia pseudoacacia* (black locust) and *Gleditsia triacanthos* (honeylocust), tend to accumulate salt relatively slowly.

Some woody species display moderate tolerance of absorbed salt ions. *Pinus nigra* (Austrian pine), for example, accumulates chloride readily from salt spray but remains visually undamaged by internal concentrations that kill tissues of more sensitive species such as *P. strobus* (eastern white pine). Salt accumulation within leaves and twigs is facilitated by minute injuries where they abrade one another.

Salt accumulation causes hypertrophy (enlargement of cells and tissues) of parenchyma in leaves of coastal plants such as *Ilex vomitoria* (yaupon) and *Quercus virginiana* (southern live oak), leading to abnormal succulence of their leaves. Species that tolerate the highest internal concentrations of chloride show the greatest hypertrophy, but hypertrophied leaves are less tolerant of additional chloride than are normal leaves on the same plant. Salt spray also selects for tolerant species at the shore, limiting the diversity of plant communities and resulting often in nearly pure stands of such trees as *Q. ilicifolia* (bear oak) and *Q. virginiana*.

Salt in soil. Salt in the soil solution may cause damage in several ways. Chloride and other ions absorbed by roots and transported to leaves and twigs can eventually accumulate to toxic levels. As water evaporates or is withdrawn by plants from saline soil, the osmotic concentration of the remaining soil solution rises, which in turn diminishes availability of water to roots and retards metabolic functions, photosynthesis, and growth. Availability of mineral nutrients is also diminished in saline soil because sodium ions replace calcium and other nutrient elements on soil particles, and because excess sodium in soil raises the pH. Alkalinity interferes with solubility of mineral nutrients such as iron. High sodium levels also prevent normal aggregation of soil particles and contribute to compaction, because tiny particles fit together more closely than do aggregates.

Salt pollution of soil along highways is greatest at the end of winter and subsides until salt applications begin the next winter. The degree of pollution also diminishes with distance from the pavement. Chloride dissipates more rapidly than sodium. Over several years, however, both sodium and chloride accumulate in the soil and within plants, causing symptoms in salt-sensitive trees. Salt pollution of coastal soils as the result of storm-caused saline flooding is usually temporary because subsequent rains leach the soil.

In parts of western and southwestern USA naturally saline soil inhibits or prevents growth of various plants that would otherwise be adapted to the region. Native plants in those places have undergone natural selection for salt tolerance. Salt accumulation is also a problem in irrigated landscapes in those regions and along the East Coast where sea water intrudes into groundwater as water tables drop. Mineral salts from ions in irrigation water build up in surface soil unless excess water is applied to leach the soil. (Leaching is usually impractical or impossible because of the amount of water needed or because of impeded soil drainage.) Slow growth, chlorosis, dwarfed or scorched leaves, and dieback then develop. If chloride occurs at concentrations greater than about 40 μg/g in irrigation water, it may accumulate to injurious levels in plants without reaching high levels in soil.

Salt injury has deleterious secondary effects, including impairment of cold hardiness and of resistance to some fungal pathogens. Experiments with *Fraxinus* (ash), *Malus* (crabapple), and *Syringa* (lilac) indicated that much twig dieback attributed to salt may be caused by freeze damage following salt-induced impairment of cold hardiness. Salted soil plus *Verticillium dahliae*

A, B. Foliar damage by salt spray blown 10 km inland by a hurricane. A. Mixed *Quercus* species (oaks) in the background are completely brown, and a small *Acer rubrum* (red maple) in the foreground is defoliated, in contrast to salt-tolerant *Pinus nigra* (Austrian pine) and *P. thunbergii* (Japanese black pine). B. Brown original leaves and green new growth triggered by death of most of the original foliage on a young *Liquidambar styraciflua* (sweetgum) in a nursery 3 weeks after the hurricane (NY, Aug).

C, D. Damage to *Pinus strobus* (eastern white pine) by de-icing salt spray. C. Needle browning and suppressed growth, becoming less severe with distance along a row of trees planted perpendicular to the highway. D. Typical injury to needles, with abrupt transition between dead and green parts (NY, Apr).

E. Damage to *Prunus serotina* (black cherry) by de-icing salt spray from a highway. Repeated killing of buds and twigs has led to branch dieback and to formation of clusters of adventitious shoots resembling witches'-brooms (NY, Apr).

F. Damage to *Thuja occidentalis* (arborvitae) by dog urine (NY, May).

killed experimentally treated young *Acer saccharum* (sugar maple) trees when neither agent alone did so.

Salt-laden soil can be made more satisfactory for plant growth by working gypsum (calcium sulfate) into the surface layer. Appropriate rates vary with local conditions. Fertilizer is also helpful, provided that the nutrient ions potassium and nitrate are available to plants at the same osmotic concentrations as the sodium and chloride.

Selected plants and plant groups are listed below according to their reported abilities to tolerate salt. Data for response to salt spray and to salt in soil have been merged. Plants sensitive in one circumstance but tolerant in another are ranked according to tolerance in the more common type of salt exposure. If this was unknown, we assigned the rank of less tolerance. Genus or family names appear alone if the commonly used species have similar tolerance.

Relative tolerance of plants to salt

High tolerance (symptoms mild or uncommon): *Acer platanoides, Aesculus hippocastanum, Ailanthus altissima, Amelanchier alnifolia, Arecaceae, Atriplex, Avicennia germinans, Baccharis halmifolia, Callicarpa americana, Callistemon viminalis, Caragana arborescens, Citrus, Coccoloba uvifera, Cryptomeria japonica, ×Cupressocyparis leylandii, Elaeagnus angustifolia, Euonymus japonicus, Forestiera segregata, Fraxinus americana, F. pennsylvanica, Ilex cornuta, I. opaca, I. vomitoria, Laguncularia racemosa, Larix decidua, L. kaempferi, Ligustrum amurense, Lonicera korolkowii, L. xylosteum, Morella caroliniensis, M. cerifera, Nerium oleander, Osmanthus americanus, Persea americana, P. borbonia, Philadelphus, Picea pungens, Pinus clausa, P. elliottii, P. mugo, P. nigra, P. palustris, P. pinea, P. thunbergii, Podocarpus macrophyllus, Populus alba, P. balsamifera, P. canescens, P. deltoides, P. nigra, P. tremula, Populus hybrids, Prosopis glandulosa, Prunus virginiana, Quercus alba, Q. geminata, Q. ilicifolia, Q. laurifolia, Q. macrocarpa, Q. myrtifolia, Q. phellos, Q. robur, Q. virginiana, Rhaphiolepis umbellata, Rhizophora mangle, Rhus hirta, Ribes alpinum, Robinia pseudoacacia, Rosa rugosa, Sabal palmetto, Salix alba, S. fragilis, S. nigra, S. purpurea, Sciadopitys verticillata, Serenoa repens, Shepherdia argentea, Spiraea ×vanhouttei, Styphnolobium japonicum, Tamarix, Triadica sebifera,* and *Ulmus pumila.*

Moderate tolerance: *Acer campestre, A. ginnala, A. saccharinum, Alnus glutinosa, A. incana, Araucaria heterophylla, Berberis thunbergii, Betula lenta, B. papyrifera, B. pendula, Carya ovata, Catalpa speciosa, Cercis canadensis, Crataegus monogyna, C. oxycantha, Eucalyptus globulus, E. grandis, Forsythia, Frangula alnus, Fraxinus excelsior, Gleditsia triacanthos, Juglans nigra, J. regia, Juniperus chinensis, J. virginiana, Koelreuteria paniculata, Kolkwitzia amabilis, Lagerstroemia indica, Ligustrum lucidum, L. vulgare, Lonicera japonica, L. tatarica, Maclura pomifera, Magnolia grandiflora, Olea europaea, Oxydendrum arboreum, Picea abies, Pinus banksiana, P. ponderosa, P. rigida, P. sylvestris, P. taeda, Pittosporum tobira, Platanus ×acerifolia, Populus grandidentata, P. tremuloides, Potentilla fruticosa, Prunus avium, P. dulcis, P. salicina, Pyracantha coccinea, Pyrus calleryana, Quercus coccinea, Q. marilandica, Q. rubra, Q. stellata, Rhamnus cathartica, Rhus glabra, Ribes aureum, R. nigrum, Salix capraea, S. pentandra, S. viminalis, Sambucus nigra, Spiraea, Symphoricarpos, Syringa vulgaris, Taxodium distichum, Thuja occidentalis, Tilia americana, T. cordata, T. platyphyllos, Ulmus americana, U. glabra, U. minor, U. parvifolia, Ulmus hybrids,* and *Viburnum opulus.*

Low tolerance (foliar browning, bud death, and twig dieback common and severe): *Abies balsamea, A. concolor, Acer negundo, A. rubrum, A. saccharum, Aesculus ×carnea, Albizia julibrissin, Alnus viridis, Amelanchier laevis, Araucaria araucana, Bauhinia variegata, Buxus sempervirens, Carpinus betulus, C. caroliniana, Carya illinoinensis, Cedrus atlantica, Celtis occidentalis, Cephalanthus occidentalis, Ceratonia siliqua, Chaenomeles japonica, Chamaecyparis pisifera, C. thyoides, Cornus alba, C. amomum, C. florida, C. mas, C. racemosa, C. sanguinea, C. sericea, Corylus americana, C. avellana, C. cornuta, Cotoneaster, Crataegus crus-galli, Euonymus alatus, Fagus grandifolia, F. sylvatica, Ginkgo biloba, Liquidambar styraciflua, Liriodendron tulipifera,* deciduous *Magnolia* species, *Mahonia aquifolium, Malus, Metasequoia, Morus alba, Myrtus communis, Nandina domestica, Nyssa sylvatica, Ostrya virginiana, Physocarpus, Picea glauca, Pinus contorta, P. resinosa, P. strobus, P. virginiana, Platanus occidentalis, Prunus ×blireiana, P. cerasifera, P. persica, P. serotina, P. serrulata* 'Kwanzan,' *Pseudotsuga menziesii, Pyrus communis, Quercus bicolor, Q. falcata, Q. nigra, Q. palustris, Rhamnus alternans, Rhododendron, Rosa canina, R. multiflora, Sambucus canadensis, Sassafras albidum, Sorbus aucuparia, Taxus, Tsuga canadensis, Vaccinium, Viburnum lantana, Vitis vinifera,* and *Zamia.*

References: 30, 54, 101, 115, 116, 171, 179, 242, 340, 466, 653, 778, 966, 971, 1155, 1172, 1267, 1407, 1417, 1448, 1449, 1550, 1570, 1708, 1745, 1855, 1967, 2048, 2140, 2197, 2343, 2384, 2385, 2736, 2872, 3015, 3404, 3555, 3884, 3896, 4069, 4165, 4286, 4308, 4454

Other Inorganic Poisons in Soil

Inorganic chemicals other than common salt that occasionally cause damage to woody plants include various salts and ions of boron, copper, manganese, nickel, zinc, aluminum, arsenic, and lead. Trace amounts of the first five elements are essential for plant growth and are supplied by soil. Soil pollution associated with irrigation, mining, smelter waste, municipal waste, or pesticidal uses of these elements is the usual reason for their occurrence at phytotoxic levels. Symptoms of toxicity (chlorosis and necrosis of leaf tips and margins, depressed growth) are similar for several minerals and are not diagnostic. Diagnosis requires chemical analysis of plant tissues and the rooting medium. Here we illustrate injuries attributed to arsenic and boron.

Salts of arsenic, especially lead arsenate and sodium arsenite, were formerly used as insecticides and herbicides. Arsenic accumulated to phytotoxic levels in some agricultural and nursery soils until the early 1970s. Arsenic residues in soil dissipate slowly. Residues of 40–80 µg/g or more in soil, versus normal values less than 10 µg/g, have been blamed for rootlet mortality and impairment of mycorrhizal function leading to stunting of *Malus pumila* (apple); to rootlet mortality and tip burn of *Pinus resinosa* (red pine) needles in a forest nursery; and to rootlet mortality, needle browning, and defoliation of *Tsuga canadensis* (eastern hemlock) in a landscape nursery. Arsenic does not become highly concentrated in leaves; foliar concentrations less than 20 µg/g may be associated with severe foliar browning.

Excess boron in soil comes from irrigation water and from application or disposal of boron compounds used as herbicides, fire retardants, fertilizers, and laundry products. Toxic concentrations of boron in soil occur naturally in some arid regions where soils are derived from marine sediments. Boron taken up by roots is transported into leaves, where it normally occurs at concentrations of 15–100 µg/g (dry weight basis). Other factors being equal, the more boron available to roots, the greater its accumulation in foliage. Foliar chlorosis and necrosis have been associated with foliar concentrations above 75 µg/g in a wide array of plants. The *P. strobus* (eastern white pine) shown in Plate 233C, D had a foliar boron content of 355 µg/g.

References: 132, 241, 756, 1173, 1292, 1293, 1355, 1574, 1932, 2386, 2572, 3555, 3652, 3653, 3752, 3877, 3878, 3880, 4074, 4434

A, B. De-icing salt spray damage to *Pinus nigra* (Austrian pine). Individual sensitive trees of this normally tolerant species may be severely damaged by the combination of salt, wind, and thaw-freeze cycles in late winter. A. The tree in the foreground has many dead branches near the base, indicating recurrent damage. B. Green needle bases of 1- and 2-year-old needles are nearly hidden by dead needle tips; new green shoots provide contrast (ON, Jul).

C, D. Needle browning of *Pinus strobus* (eastern white pine) caused by borax from laundry wastewater that drained from the pipe at left (arrow) into the root zone (NY, May).

E, F. Needle browning and defoliation of *Tsuga canadensis* 'Sargentii' (Sargent weeping hemlock) associated with high concentrations of arsenic in soil and roots (NY, May).

G. Tip necrosis on leaves of *Prunus dulcis* 'Mission,' a salt-sensitive almond cultivar, caused by accumulation of ions from irrigation water (CA, Jul).

H. Marginal scorch, attributed to ion uptake from salt-polluted soil, on leaves of *Acer platanoides* (Norway maple) (NY, Aug).

Damage by Misapplied Pesticides (Plates 234–236)

Pesticides used in agronomy, forestry, and horticulture are intended to suppress or kill particular plants or plant pests while causing little injury to nontarget organisms. In practice, however, environmental contamination and injury to nontarget organisms occur often, even when normal precautions are taken. Site contamination and/or unwanted plant injury are likely to occur when pesticides are mishandled or applied under improper conditions. Insecticides and fungicides are less likely than herbicides to cause damage to nontarget plants because chemicals in the former two groups are formulated to minimize plant injury, while the intended function of an herbicide is to kill plants. Untrained applicators of herbicides often use more than the recommended amounts, usually because of improperly calibrated equipment, or apply them unevenly or unintentionally to the root zones of woody plants. Herbicide spray drifting to plants outside treated areas also causes injuries. Plates 234–236 show symptoms caused by pesticides, mainly herbicides, on nontarget trees and shrubs. The text deals first with herbicides, then with plant injuries caused by insecticides and fungicides. Photos of damage by a particular pesticide may not face the page with information about it.

Symptoms of injury by herbicides vary with the type of chemical. Hormone-type herbicides and some others cause aberrant growth. Triazines, substituted urea compounds, uracil derivatives, and dichlobenil cause chlorosis, either general or in patterns. Amitrole causes bleaching. Contact herbicides, notably paraquat, cause localized necrosis. Some herbicides cause stunting and death of growing points. Symptoms similar to those of herbicide injury can be caused by some diseases and insects, insufficient or excess water or heat, deficiencies of certain nutrients, poisons not formulated as herbicides, and misapplied pesticides of other types. Therefore, the history and/or chemical analyses of symptomatic plants and/or soil may be required for accurate diagnosis of herbicide injury.

Herbicides that often cause injury to nontarget woody plants in landscapes fall into two general groups as regards their intended purpose. One group is intended to kill established broadleaved weeds selectively. Well-known herbicides in this group include the phenoxyacetic acids (2,4-D, 2,4-DP, MCPA, MCPP, and related compounds) and the benzoic acid derivative dicamba. These act as plant hormones that disrupt growth processes. In sublethal doses they cause growing plant organs to assume aberrant forms. Chemicals in this group may enter nontarget plants by one or more of three modes, listed in descending order of frequency: absorption from spray droplets that drift away from the site of application; absorption from drifting vapor after evaporation of liquid on sprayed surfaces, as occurs with volatile herbicides in warm weather; and absorption by roots.

Herbicides of the hormone type are translocated to growing points and cause aberrations in new leaves and shoots. Nontarget plants usually receive sublethal doses and outgrow the symptoms within 1–2 years. Symptoms develop several days to several weeks after exposure or may appear in spring following an autumn exposure. Symptoms include cupped leaves (convex surface either up or down); abnormal bending or drooping of green twigs, petioles, rachises, or main leaf veins; helical growth of shoot tips ("corkscrewing"), abnormally prominent veins; parallel leaf venation on plants that normally have spreading veins in leaves; underdevelopment of interveinal leaf tissue (resulting in abnormally narrow leaves); wavy, frilled, or curled leaf margins; and, in extreme cases, bud death and dieback. Additional symptoms include abnormally tough or leathery leaves, partial failure of chlorophyll development, swollen buds, delayed bud break in spring, loss of apical dominance, enlarged lenticels, abnormal purple color of normally green stems, lesions in stem bark, and suppressed root growth. Purple stem color induced by phenoxy herbicides has been observed on *Fraxinus* (ash), *Acer saccharum* (sugar maple), *Juglans* (walnut), and *Quercus palustris* (pin oak). Conifers are relatively tolerant of phenoxy herbicides, but their shoots may curl and needles drop in response to damaging doses.

Hormone-type herbicides (phenoxy and benzoic acid derivatives) applied to turf for control of broadleaved weeds often injure trees and shrubs. These chemicals, especially dicamba, are taken up by woody plant roots in and below the root zone of the grass. Dicamba

causes drastic growth suppression, leaf cupping, bending and sometimes coiling ("fiddlenecking") of shoot tips, yellowing of new growth, bud failure, browning or blackening of foliage (depending on plant type), defoliation, and sometimes death. Some woody plants are so sensitive to phenoxy herbicides that they usually show at least mild foliar symptoms of injury if these chemicals are used anywhere in the vicinity. *Acer negundo* (box-elder), *Vitis* species (grapevines), and *Cercis canadensis* (redbud) are among the most sensitive species and are useful indicators if damage to other plants by phenoxy herbicides is suspected.

The other broad group of herbicides comprises chemicals intended either to prevent emergence of weed seedlings (preemergence herbicides) or to kill all vegetation to which they are applied. Some of these chemicals persist in soil for a year or more. Simazine, for example, is a much-used preemergence herbicide that sometimes causes unwanted residual effects after application to nursery soils. Root absorption of simazine is enhanced by mycorrhizae. General vegetation killers that sometimes cause trouble for nontarget trees or shrubs include amitrole, dichlobenil, glyphosate, paraquat, picloram, and substituted urea compounds such as diuron. These chemicals are either sprayed on target plants or applied to soil for absorption by roots. They are used to control vegetation along highways, railroads, fences, power lines, and similar places. Those applied to soil are intended to have residual activity; therefore misapplication leads to persistent effects. As an example, we have seen century-old specimens of *A. saccharum* and *Picea abies* (Norway spruce) killed by picloram that was applied for control of woody weeds around gravestones in a cemetery. The inexperienced applicator did not realize that roots of mature trees extend far from the trunks and may not have appreciated the potency of the herbicide. One general-purpose herbicide sprayed on target plants that does not significantly pollute soil or air is glyphosate. When deposited on soil, glyphosate is adsorbed to soil particles and is soon degraded by microorganisms. This chemical also has low volatility, so it is available to plants only by spray deposition.

Herbicide injuries to coniferous trees.

A. *Abies concolor* (white fir) damaged by dichlobenil applied to the ground in a nursery. *Abies* species are highly sensitive to this herbicide. Variability in damage as shown here could reflect differing sensitivity of individual trees, inconsistent herbicide coverage, and/or soil drainage patterns on the site (NY, Oct).

B. Symptoms caused by simazine to a young *Pinus strobus* (eastern white pine) in a nursery. The tree was planted in contaminated soil. Needles elongated normally but began to droop and turn yellow at the bases as the plant absorbed herbicide. It died subsequently (NY, Aug).

C. Symptoms caused on *Pinus sylvestris* (Scots pine) by a mixture of phenoxy herbicides used for vegetation control beside a highway (NY, Aug).

D. *Pinus thunbergii* (Japanese black pine) injured by a mixture of 2,4-D and MCPP applied to adjacent turf at the beginning of the growing season. These branches are dying. Shoot growth and needle elongation were arrested; growing points are dead (NY, Sep).

E. Bleached needles on *P. sylvestris* attributed to injury by amitrole (NY, Aug).

F. Bleached needle bases on *P. nigra* (Austrian pine) caused by an unidentified herbicide (NY, Aug).

G. *Picea pungens* (blue spruce) injured by dicamba, which was applied to adjacent turf to control broadleaved weeds (NY, Dec).

H. *P. pungens* injured by unidentified herbicide. Shoots of the current season have ceased growth; shoot tips have died. Some apical buds have differentiated abnormally into cones (NY, Jun).

I. *Cedrus atlantica* (Atlas cedar) injured by an unidentified herbicide. Growing points of the twig and of several short shoots have died, and needles are dying. Stem swelling near the twig apex is abnormal (NY, Oct).

Photo credit: F—G. W. Hudler

472

Preemergence herbicides and nonselective vegetation killers tend to halt growth and cause chlorosis of new and old leaves. Most of these chemicals either enter nontarget plants through roots or are carried to plants in dust blown from treated soil. A large enough dose can cause foliar yellowing or browning, leaf cast, and dieback of twigs and branches. Amitrole causes bleaching of leaves or leaf parts to white or pinkish white. Atrazine and simazine cause marginal and interveinal chlorosis of broadleaved plants and yellowing of conifer needles beginning at needle bases. Dichlobenil causes marginal chlorosis or, if injury is severe, general chlorosis and browning. Substituted urea compounds such as diuron cause veinal chlorosis or, at high doses, general chlorosis and death. Picloram in low doses may cause abnormal bending or curling of young shoots, and at higher doses causes foliar browning, leaf cast, and dieback. Trees or shrubs injured by these herbicides are less likely to recover than are plants injured by hormone-type herbicides.

Plants tend to accumulate herbicides from treated or contaminated soils, so the concentration of an herbicide in plant tissue may be several times greater than the concentration in soil. Tissue concentrations associated with injury in plants may be as low as 10–20 parts per billion, depending on the type of herbicide. The minimum damaging dose also varies with plant genotype, stage of growth, time of year, and plant stress caused by other factors. Herbicides generally have greatest effects on plants treated or inadvertently exposed during growth in warm weather. Water shortage or heat stress may enhance toxic effects. Dormant plants or those in a resting phase are less sensitive. Herbicides may also trigger damage by secondary factors. For example, some herbicides reduce cold hardiness; dieback attributed to these chemicals may be caused in part by freeze damage and by opportunistic fungal pathogens that attack tissues weakened by cold injury.

Contact herbicides such as paraquat that kill leaves and succulent shoots but not woody parts are a special diagnostic category. Spray droplets containing paraquat cause brown lesions similar to leaf spots caused by some fungi and superficial lesions on green stems. Stems of green-barked trees in nurseries—for example, *Gleditsia triacanthos* (honeylocust), *Laburnum anagyroides* (golden-chain), *Quercus* species (oaks), *Styphnolobium japonicum* (Japanese pagoda tree), and *Tilia* species (linden)—may develop brown streaks of superficial cork or brown cracked bark after spray treatment with a contact herbicide intended to kill sprouts or to control weeds at the base of the trunk.

A complex association of herbicide injury with decline of broadleaved trees in the Great Plains of Canada and the USA exemplifies the difficulty that diagnosticians may face in fixing blame for tree damage. In that region injury to trees by spray and vapor drift of hormone-type herbicides occurs up to several kilometers from agricultural fields where the chemicals are applied. Foliar symptoms typical of those caused by 2,4-D are common every year on several species, notably *Acer negundo* (box-elder), *Celtis occidentalis* (hackberry), and *Ulmus pumila* (Siberian elm). Branch dieback and premature death of broadleaved trees in hedgerows, shelterbelts, and farm woodlots are also common and widespread. *C. occidentalis* and *U. pumila* are most severely affected. The scene shown in Plate 235A illustrates the problem in *Ulmus*. Such decline has been reported to be more severe in areas where field crops are grown (where more herbicide is used) than in rangeland areas. Some authors have attributed decline of *C. occidentalis* and *U. pumila* to herbicides, but direct evidence for a primary causal role of herbicides is lacking, and alternative explanations are available. Herbicides clearly cause foliar symptoms, but the doses of airborne herbicide that trees receive are far smaller than the doses that cause dieback or death in trees intentionally treated. Herbicides apparently do interact with other stressing factors, especially low temperature, to cause bud failure and dieback, but the climate of the plains is harsh for trees and may itself be sufficient to preclude long life of species planted in shelterbelts. Summer heat and dryness, winter cold and wind, and rapid temperature changes in early and late winter stress or injure trees and also predispose them to attack by opportunistic fungal pathogens. Much dieback in *Ulmus* stressed by drought or freezing

is caused directly by *Sphaeropsis hypodermia* (Plate 65) and *Tubercularia ulmea* (Plate 88). Herbicides may play a role in susceptibility to pathogens, as has been demonstrated in experiments with field crops and coniferous seedlings, but evidence for this role in trees or shrubs has not been presented.

Phenoxy herbicides and dicamba are the weed killers that most commonly injure nontarget trees and shrubs. The following ratings for tolerance to these herbicides were collated from listed references and the authors' and colleagues' observations. Tolerance of glyphosate is also considered because this herbicide is much used for spot treatments in horticultural landscapes.

Relative tolerance of plants to phenoxy herbicides, dicamba, and glyphosate

Plants tolerant of phenoxy herbicides: *Abies* (fir), *Acer macrophyllum* (bigleaf maple), *A. rubrum* (red maple), *A. saccharinum* (silver maple), *Catalpa, Crataegus* (hawthorn), *Fagus grandifolia* (American beech), *Juniperus* (juniper), *J. virginiana* (eastern red-cedar), *Nyssa sylvatica* (sourgum), *Pinus* (pine), *Pseudotsuga menziesii* (Douglas-fir), *Pyrus communis* (common pear), *Quercus virginiana* (southern live oak), *Rhododendron maximum* (great-laurel, rosebay), *R. occidentale* (western azalea), *Taxodium distichum* (bald cypress), *Thuja occidentalis* (eastern arborvitae), and *Tilia americana* (basswood).

Plants of intermediate sensitivity: *Acer saccharum* (sugar maple), *Aesculus* (buckeye), *Alnus rubra* (red alder), *Berberis* (barberry), *Carya* (hickory), *Elaeagnus angustifolia* (Russian-(olive), *Fraxinus americana* (white ash), *F. pennsylvanica* (green ash), *F. quadrangulata* (blue ash), *Gymnocladus dioicus* (Kentucky coffeetree), *Ligustrum* (privet), *Liquidambar styraciflua* (sweetgum), *Morus* (mulberry), *Picea pungens* (blue spruce), *Populus balsamifera* (balsam poplar), *P. deltoides* (eastern cottonwood), *P. tremuloides* (trembling aspen), *Prunus* (cherry, chokecherry, peach), *Quercus alba* (white oak), *Q. kelloggii* (California black oak), *Q. nigra* (water oak), *Q. palustris* (pin oak), *Q. rubra* (red oak), *Rhododendron, Robinia pseudoacacia* (black locust), *Syringa vulgaris* (common lilac), *Taxus* (yew), *Tsuga* (hemlock), *Ulmus alata* (winged elm), and *U. americana* (American elm).

Plants sensitive to phenoxy herbicides: *Acer ginnala* (Amur maple), *A. negundo* (box-elder), *A. platanoides* (Norway maple), *Aesculus hippocastanum* (horse-chestnut), *Ailanthus altissima* (tree-of-heaven), *Albizia julibrissin* (mimosa), *Amelanchier* (serviceberry), *Arbutus menziesii* (ma-

---

Injury to deciduous woody plants by phenoxy herbicides.

A. Dieback and premature death of *Ulmus pumila* (Siberian elm) in a woodlot in a wheat-producing area of the Great Plains. Such damage is associated with chronic exposure to phenoxy herbicides that drift from cultivated fields (CO, Jul).

B. Typical symptoms of injury to *Acer negundo* (box-elder) by 2,4-D aerosol or vapor. Leaves that were fully expanded when exposed appear normal. Leaves just completing growth at the time of exposure have leaflets with midveins bent and edges turned downward. Leaves that were in an early stage of growth became stunted, distorted, and yellowish (NY, Jun).

C. Symptoms of phenoxy herbicide injury in *U. pumila*. The sequence of normal growth, herbicide exposure, and symptom development was as described above for *A. negundo* (ND, Jul).

D. Injury to *Quercus rubra* (northern red oak) by 2,4-D. The tree acquired the herbicide via roots and had begun to outgrow its effects when the photo was made. Deformed leaves have petioles and veins of normal length, but interveinal tissues did not develop, and the veins grew curved (NY, Jul).

E. Leaflet distortion on *Fraxinus latifolia* (Oregon ash) caused by 2,4-D aerosol (WA, Aug).

F. Irregular contours on a trunk of *Gleditsia triacanthos* (honeylocust) injured by a mixture of phenoxy herbicides applied to surrounding turf. The tree had severe foliar symptoms. Trunk abnormalities resulted from localized growth adjacent to areas of inhibited cambial activity (NY, Aug).

G, H. Leaf distortion and lack of chlorophyll formation in the basal parts of leaflets of *Celastrus scandens* (American bittersweet) caused by a phenoxy herbicide mixture applied to nearby turf. The leaflet in H attained nearly normal length before receiving the herbicide. Tissues still developing at that time were in the basal part of the leaf (NY, Jul).

drone), *Betula* (birch), *Celtis occidentalis* (hackberry), *Cercis canadensis* (redbud), *Cladrastis lutea* (yellowwood), *Cornus* (dogwood), *Forsythia, Fraxinus latifolia* (Oregon ash), *Gleditsia triacanthos* (honeylocust), *Juglans* (walnut), *Liriodendron tulipifera* (tuliptree), *Malus pumila* (apple), *Phellodendron amurense* (Amur cork tree), *Platanus* (planetree, sycamore), *Populus deltoides* subsp. *monilifera* (plains cottonwood), *Quercus velutina* (black oak), *Rhododendron canescens* (hoary azalea), *Rhus* (sumac), *Rosa* (rose), *Salix* (willow), *Sambucus* (elderberry), *Tilia cordata* (littleleaf linden), *Ulmus pumila* (Siberian elm), *Vitis* (grapevine), and *Wisteria*.

Plants tolerant of dicamba: *Juniperus virginiana* (eastern red cedar), *Populus deltoides* (eastern cottonwood), *Quercus velutina* (black oak), *Q. virginiana* (southern live oak), and *Rhododendron canescens* (hoary azalea).

Plants of intermediate sensitivity: *Acer platanoides* (Norway maple), *A. rubrum* (red maple), *A. saccharinum* (silver maple), *Aesculus* (buckeye), *Carya* (hickory), *Celtis occidentalis* (hackberry), *Fagus grandifolia* (American beech), *Fraxinus americana* (white ash), *F. pennsylvanica* (green ash), *Juglans nigra* (black walnut), *Quercus rubra* (red oak), *Ulmus alata* (winged elm), and *U. americana* (American elm).

Plants sensitive to dicamba: *Acer negundo* (box-elder), *A. saccharum* (sugar maple), *Ailanthus altissima* (tree-of-heaven), *Alnus rubra* (red alder), *Amelanchier* (serviceberry), *Berberis* (barberry), *Betula* (birch), *Catalpa, Cercis canadensis* (redbud), *Crataegus* (hawthorn), *Gleditsia triacanthos* (honeylocust), *Juglans* (walnut), *Juniperus* (juniper), *Liquidambar styraciflua* (sweetgum), *Liriodendron tulipifera* (tuliptree), *Malus pumila* (apple), *Nyssa sylvatica* (sour-gum), *Picea pungens* (blue spruce), *P. glauca* (white spruce), *Pinus echinata* (shortleaf pine), *Platanus occidentalis* (sycamore), *Populus deltoides* subsp. *monilifera* (plains cottonwood), *P. tremuloides* (trembling aspen), *Pseudotsuga menziesii* (Douglas-fir), *Quercus alba* (white oak), *Q. palustris* (pin oak), *Rhododendron maximum* (great-laurel, rosebay), *Robinia pseudoacacia* (black locust), *Syringa* (lilac), *Taxus cuspidata* (Japanese yew) and related *Taxus* hybrids, *Thuja occidentalis* (eastern arborvitae), *Tilia americana* (basswood), and *Tsuga canadensis* (eastern hemlock).

Variation in plant responses to glyphosate. Conifers are more tolerant than broadleaved evergreens, which are more tolerant than deciduous trees and shrubs. Conifers are most sensitive in the spring, become less so during the warm season, and are insensitive during dormancy, because they do not absorb glyphosate when dormant. The relative tolerance of woody angiosperms also varies seasonally based on ability of plants to absorb the herbicide and to translocate it to stems and roots. Deciduous plants exposed to glyphosate in spring tend to translocate less of it to the stems and roots than they would if exposed in late summer or early autumn. Consequently, spring exposures induce symptoms on shoot tips, but plants recover. Autumn exposures result in systemic damage that becomes apparent the following year. Leaves emerging in spring from buds on plants exposed in autumn are stunted, deformed, chlorotic, and tightly clustered because shoot growth is slight.

References: 50, 158, 289, 455, 655, 936, 937, 973, 1126, 1456, 1645, 1876, 1967, 2353, 2506, 2798, 2799, 2807, 2928, 3067, 3197, 3290, 3472, 3584, 3686, 3713, 3728, 3842

Plant injury by insecticides and fungicides. Such injury is infrequent compared with that by herbicides, because insecticides and fungicides are less toxic to plants than to animals or fungi, and because smaller amounts of them are used in horticultural landscapes. When visible symptoms do occur, a pesticide diluent or solvent is often responsible. The active ingredients in organic insecticides and fungicides seldom cause visible symptoms unless they were deliberately applied at rates higher than specified or under improper environmental conditions or to plants not specified on product labels. Other factors being equal, the likelihood of pesticide injury is greatest for plants stressed by heat (temperatures above 30–32°C) or water shortage or for plants sprayed under poor drying conditions so that liquid remains on their surfaces. Plants sensitive to a particular pesticide are usually identified on the label. An extensive list of plant sensitivities to insecticides and acaracides that were in use in the 1960s (many are still in use) appears in reference 1160.

Symptoms of plant injury by insecticides or fungicides are usually not diagnostic and therefore cannot be interpreted reliably unless the history and treatment of a plant are known. Visible symptoms include yellow to brown leaf spots; chlorosis of leaf tips, margins, or interveinal areas; general chlorosis; browning of leaf margins or interveinal areas; stunted shoots; and abnormal crinkling or curling of leaves. Foliar yellowing or browning is often followed by premature leaf drop. Sprays that contain oil, although among the safest from the standpoint of toxicity to mammals, occasionally cause injury to plants because the oil film interferes with gas and heat exchange at plant surfaces. Symptoms on leaves may range from chlorotic or dark spots to general chlorosis, browning, and leaf cast. Twigs sometimes die back. Many pesticides and antitranspirants temporarily suppress photosynthesis and growth of treated plants without causing visible symptoms, but such effects on woody ornamental plants are usually unimportant.

Before the use of organic pesticides (i.e., those containing carbon) became widespread, inorganic materials such as lead arsenate, lime-sulfur, and copper compounds (including Bordeaux mixture) caused injury to many kinds of plants. Copper compounds are still used, although not often on woody ornamental plants, and may injure sensitive plants such as *Ilex* (holly) or *Acer platanoides* (Norway maple). Red spots with associated hypertrophy may develop on the lower surfaces of *Ilex* leaves. *A. platanoides* leaves sprayed with Bordeaux mixture or fixed copper may develop numerous minute black flecks on the lower surface. If injury is severe, the flecks coalesce into necrotic blotches.

Wood preservatives can also cause plant injury, although in landscapes this is usually restricted to foliage and shoots touching or closely adjacent to freshly treated wood. Plants growing in confined spaces with preservative-treated wood, as in greenhouses or preservative-treated containers, are sometimes injured. Creosote is the wood preservative most often associated with plant injury because it releases toxic volatile chemicals during warm, sunny weather. These can cause uprolled leaf edges, chlorosis, and premature leaf drop. Treated wood in soil usually does no visible harm to nearby plants, however.

References: 151, 899, 1160, 1621, 1646, 1872, 1966, 2893, 3966

Herbicide and insecticide injuries.

A, B. *Platanus ×acerifolia* (London planetree) injured by dicamba applied to surrounding turf. Many buds did not open; leaves are dwarfed, distorted, and chlorotic (NY, May).

C. *Cornus kousa* (Korean dogwood) injured by unidentified herbicide used on adjacent turf (NY, Jul).

D. Leaflet distortion and dwarfing following uptake of picloram by roots of *Juglans nigra* (black walnut).

E. Severe stunting and distortion of new growth on an *Acer macrophyllum* (bigleaf maple) sapling treated the previous year with glyphosate (OR, Jul).

F. *Syringa reticulata* (Japanese tree lilac) injured by amitrole applied to control weeds in its root zone. Bleached leaves or parts of leaf blades, followed by leaf browning, are typical of injury by this chemical (NY, Jun).

G. Veinal chlorosis and general bleaching of *Malus pumila* (apple) leaves caused by the herbicides diuron and terbacil, which were absorbed by roots of nontarget trees in an orchard (NY, Jun).

H, I. *Betula papyrifera* (paper birch) injured by a spray of dimethoate insecticide applied at the beginning of the growing season. Shoot growth and leaf expansion on injured stems were suppressed, and expanded leaves developed marginal and interveinal browning (ON, Jul).

J. Interveinal yellowing and marginal browning on a leaf of *Betula pendula* (European white birch) injured by the systemic insecticide disulfoton. The insecticide was applied to the soil for uptake by roots (NY, Aug).

Photo credit: D—E. Masters and USDA Forest Service

477

Injuries and Diseases Caused by Air Pollutants (Plates 237–241)

Overview

The effects of polluted air on plants and ecosystems are discussed in many books and many thousands of research articles. We first consider the effects of polluted air on plant health, then major phytotoxic air pollutants, and then turn to damage by oxidants, sulfur dioxide, fluorides, and minor air pollutants. Descriptive information in addition to that in the plate captions may not face a plate that displays particular symptoms.

The early examples of plant damage by anthropogenic air pollutants were localized: decline and death of vegetation around smelters, brick works, and other industrial sites; and failure of some trees to thrive in industrial areas. Sulfur dioxide and fluorides were the main phytotoxic pollutants. Later, with burgeoning population and associated needs for power and products, came the continual massive release of oxides of sulfur and nitrogen and other pollutants from the stacks of factories, refineries, power-generating stations, and home chimneys.

Air pollution by ozone and other oxidants became a problem in the second half of the 20th century as the exhaust gases of motor vehicles polluted large areas. In the 1950s–1970s many plant injuries from short-term exposure to air pollutants were described and some diseases caused by chronic exposure were recognized. *Pinus* species in mountain forests of southern California were declining as the result of smog damage, and depressed plant growth in parts of eastern USA was linked to polluted air.

In the 1980s acidic air pollutants and "acid rain" were blamed for "forest decline" in parts of Europe and eastern North America, but research failed to support concepts of wide-scale forest damage by acidic agents. Ozone was found to be the major pollutant affecting trees. By late in the 20th century pollutant emissions from point sources were decreasing in North America and Europe, although such sources still caused significant problems in eastern Europe, China, and developing countries. Ozone levels, however, were not declining. Tree health problems linked to air pollutants exist today in and downwind of metropolitan and industrial regions; in some places where forests growing on poorly buffered soils receive acidic pollutants; and on mountains where trees are often wet by pollutant-laden clouds. Foliar injuries, reduced tree growth, and tree declines in some montane forests in eastern and southwestern USA, Mexico, southern Europe, and parts of Asia have been linked to damage by air pollutants.

The major classes of phytotoxic air pollutants, in descending order of damage caused, are oxidants (ozone, oxides of nitrogen, and peroxyacyl nitrates), sulfur dioxide, and fluorides (hydrogen fluoride and silicon tetrafluoride). In addition, ammonia, chlorine, and hydrogen chloride cause occasional local damage. Metallic elements such as mercury from polluted air accumulate in soils, but their effects on plants at the concentrations encountered in most places are minor or unknown.

Ozone, a constituent of photochemical smog and the most damaging of the oxidants, is a secondary pollutant generated in air by the oxidation of volatile organic compounds in the presence of oxides of nitrogen (NO_x) and sunlight. Most of the NO_x comes from anthropogenic sources, mainly automobiles and coal-fired power plants. Volatile organic compounds come from both natural and anthropogenic sources. Ozone commonly increases to phytotoxic concentrations in urban areas during warm, sunny weather. It may cause damage to plants far from the source of its precursors as masses of polluted air move overland. Ozone from other sources, such as that generated during electrical storms or carried down from the stratosphere by unusual air movements, is generally considered inconsequential for plant health. Peroxyacetyl nitrate and related compounds are secondary pollutants formed within photochemical smog. Although highly toxic to plants, these compounds do not often accumulate to toxic concentrations.

Sulfur dioxide (SO_2) is formed primarily during combustion of coal and oil and roasting of ores. The main sources are power-generating stations, heavy industries, and smelters. Until the 1970s, when air pollution regulations began to take force in North America, damage by sulfur dioxide was primarily localized near such sources, which released great quantities of pollutants near ground level. Today, although the aggregate amount of SO_2 emitted is still huge, most large pollutant emitters have reduced their output and/or have constructed tall smokestacks so that pollutants become greatly diluted before reaching ground level. Diagnostic foliar symptoms of injury by sulfur dioxide have become less common as a result. This gas remains important, however, for its role in acidic deposition.

Fluorides are constituents of many substances mined from the earth. They are released into the air during aluminum reduction; petroleum refining; and the manufacture of steel, phosphate fertilizer, bricks, glass, and ceramics. Plant damage by fluorides occurs near such sources. Damage by chlorine compounds, ammonia, and other minor air pollutants is associated with chemical spills.

Plants sustain injury by pollutants that accumulate in stagnant air during atmospheric inversions in valleys and basins bounded by mountains and in metropolitan regions when they are affected by stationary high-pressure weather systems. As polluted air moves from a source locality or region, plants located downwind (notably in mountain forests) may be injured. Pollutant gases enter plants via stomata and are capable of suppressing photosynthesis, stimulating respiration, and suppressing growth at dose levels insufficient to cause diagnostic foliar symptoms. Sensitive trees may sustain both acute and chronic injuries (explained below), and the same trees tend to be symptomatic year after year. Visible symptoms include foliar lesions, chlorosis, premature foliar senescence or death, diminished growth, and in some cases progressive decline in health. Fortunately, heritable tolerance of air pollutants has been found within most of the species studied.

The severity of air pollutant injury to plants varies with concentration of the pollutant, duration of exposure, intrinsic sensitivity of the plants, and environmental conditions before and during exposure. Injuries are classed as acute or chronic. The former type, associated with brief episodes of greater-than-normal air pollution, results from rapid absorption of enough toxicant to kill leaf cells, causing characteristic markings. Chronic injury results from continuing or repeated absorption of a pollutant in amounts that do not initially kill tissues but cause cumulative physiological disturbance. The minimum concentrations of ozone and sulfur dioxide that cause measurable effects in sensitive plants are on the order of 50 and 500 parts per billion (ppb, volume basis), respectively, for several hours, but higher concentrations are required to cause visible injury or suppress growth in most woody plants. Exposures to pollutants at low concentrations for hours to weeks are required to induce symptoms. Injury by fluorides usually occurs as the result of prolonged foliar accumulation from air containing low concentrations of toxicant—as low as a few parts per billion.

In most places air pollutants' impact on plants is imperceptible to all but trained observers. Most plant species in a given region and most individuals of a given species are either tolerant of pollutants at the concentrations in ambient air or escape conspicuous injury because of their location or growth stage at the time of pollution episodes. Atmospheric turbulence and prevailing winds disperse pollutants and prevent wide-scale conspicuous injury. Growth suppression of various tree and shrub species by chronic exposure to air pollutants has been demonstrated in many experiments, however, and has been detected in studies of tree growth in polluted versus less polluted areas. The most credible quantitative data have

---

A. Photochemical smog in the foothills of the San Bernardino Mountains of southern California (CA, Jul).

B–E. Damage caused by ozone to *Pinus ponderosa* (ponderosa pine) in the San Bernardino Mountains (discussion, p. 480). B. A declining ozone-sensitive tree with stunted shoots and yellow foliage. C. A *P. ponderosa* stand in decline. Normal appearance of the species is shown by the dense crown of one ozone-tolerant tree at right; the others have thin crowns resulting from premature loss of needles and tuftlike arrangement of undersized needles on stunted twigs. D. A branch tip from a diseased tree. The current season's foliage is not yet symptomatic; year-old needles are chlorotic, and needles 2 and 3 years old have fallen prematurely. E. Chlorotic mottling caused by ozone on year-old needles (CA, Jul).

479

come from experiments in which plants were grown for one or more seasons in open-top chambers supplied with carbon-filtered air, ambient air, or air with controlled doses of pollutants.

Several North American tree species have, in particular regions, exhibited reduced growth unexplained by climate but correlated with increasing air pollution, mainly by ozone, since approximately 1950. These species include *Abies concolor* (white fir), *Picea rubens* (red spruce), *Pinus echinata* (shortleaf pine), *P. jeffreyi* (Jeffrey pine), *P. ponderosa* (ponderosa pine), *P. strobus* (eastern white pine), *Quercus alba* (white oak), and *Q. prinus* (chestnut oak), among others. Pollutant-associated growth losses may be partly offset, however, by growth increase associated with the rising carbon dioxide concentration in the earth's atmosphere.

Air pollutants also modify interactions of plants with biotic pathogens. In general, plants injured by air pollutants become more susceptible to opportunistic, nonspecialized pathogens and to secondary insects. Highly specialized pathogens such as rust fungi or tar-spot fungi are themselves adversely affected by air pollutants and cause less disease in polluted air than in clean air.

Acid-forming pollutants, mainly nitrogen dioxide and sulfur dioxide, are associated with tree health problems in some places. Acidic deposition can entrain nutritional imbalances, reducing the ratio of calcium and magnesium to aluminum, for example, in poorly buffered soils and plants growing in them. However, exposures of various woody plants to simulated "acid rain" of moderate acidity (above pH 3.5) have usually caused neither foliar injury nor growth depression. Highly acidic (e.g., pH 3.0) cloud water and coastal fog are implicated in the decline of *Picea rubens* on mountains of eastern USA and in crown damage to *Betula* species (birch) in New Brunswick, respectively. *P. rubens* growing at elevations where trees are often shrouded by clouds display chlorosis, premature needle browning and loss, thin crowns, twig and branch dieback, and premature death. Leaching of calcium and other cations from foliage and soil, depressed photosynthesis, and increased dark respiration are associated with the decline and have been induced experimentally by acidic mist treatments. The syndrome is most severe on mountains in New York and New England, where severe repeated winter injury of *P. rubens* (Plate 246) has also occurred. Observations and experiments revealed that foliage of this species injured by substances in cloud water does not become as cold hardy as normal foliage and thus is predisposed to winter injury. Exclusion of cloud water from *P. rubens* branches during the growing season enhanced subsequent cold hardiness.

References: 161, 247, 340, 580, 648, 695, 821, 1028, 1167, 1192, 1399, 1855, 1856, 1862, 1899, 2007, 2155, 2198, 2251, 2367, 2537, 2567, 2570, 2571, 2646, 2743, 2778, 2907, 3394, 3404, 3677, 3680, 3681, 3741, 3853, 4134

Damage by Ozone

Ozone causes acute and chronic symptoms in sensitive woody plants in areas where daytime average concentrations in summer often exceed 50 ppb or the maximum hourly average exceeds 100 ppb. Typical minimum doses of ozone capable of causing visible injury to intrinsically sensitive plants in a sensitive condition range from 40–100 ppb for 8 hours to 100–250 ppb for 1 hour. This highly reactive gas breaks down during reactions with plant constituents, leaving no residue or diagnostically useful reaction products. Ozone at concentrations near 50 ppb in short-term treatments has initially stimulated growth of young *Fraxinus americana* (white ash), *Liriodendron tulipifera* (tuliptree), *Picea abies* (Norway spruce), and *Populus* (poplar) hybrids. However, growth of woody species tested in longer-term (weeks to years) exposures to ozone at near-ambient levels has been retarded or unaffected, the result varying by species. Most perennial plants are tolerant of ozone and other air pollutants at various times of year or stages of their life cycles. Evergreens as well as deciduous plants are tolerant during dormancy. Also, little injury occurs at night or during severe water stress, when, in most species, stomata are closed and gas exchange inhibited.

Acute foliar injury by ozone results from more or less rapid death of clusters of cells, sometimes preceded or accompanied by pigment accumulation. Membranes (plasmalemma, tonoplast, and chloroplast envelope) in sensitive cells lose integrity, and the cells collapse. Damaged cell clusters are visible externally as dark stipples or, in some species, as white to tan flecks. In conifer needles, mesophyll cells adjacent to stomata are affected first. Ozone-induced cell death is mediated by natural defense systems. Ozone, like many plant pathogens, elicits an oxidative burst in leaves of sensitive plants, resulting in the generation and accumulation of reactive oxygen species—hydrogen peroxide (H_2O_2) and superoxide (O^-)—that cause intracellular damage. Ozone also stimulates phenylpropanoid metabolism, leading to production of phenolic and other toxic compounds, including colored products, and lignin biosynthesis. These reactions, when elicited by biotic pathogens, help protect plants by killing or inhibiting many pathogens and depriving obligate parasites of necessary living substrates, but the same reactions when elicited by ozone merely facilitate damage.

Plants that develop chlorosis and premature senescence in response to chronic ozone exposure do so in part because of oxidative damage to photosynthetic systems in chloroplasts and redirection of metabolic pathways. Photosynthesis is depressed, respiration and the activity of various enzymes involved in biosynthesis are increased, defense compounds are synthesized, and export of sugars diminishes. For the plant as a whole, the result is depressed accumulation of biomass and diminished growth.

Conifers that sustain foliar damage by ozone tend to develop chlorotic mottling; some sustain general chlorosis of mature needles. *Abies religiosa*, which sustains severe damage near Mexico City, is an exception; its needles develop light-colored stipples that progress to reddish blotches. Symptoms on conifers intensify with time, so the oldest needles are the most affected; they senesce and drop prematurely. Ozone injury to conifers is best documented for *Pinus ponderosa* (ponderosa pine) in the San Bernardino and San Gabriel mountains of southern California, which receive photochemical smog from the Los Angeles metropolitan region. Chlorotic spots with diffuse margins develop irregularly along nearly the entire length of needles a year or more old, giving them a mottled appearance, and severely affected trees have only two age classes of needles instead of the normal four or five (Plate 237). *P. jeffreyi* (Jeffrey pine) is affected similarly. Growth diminishes over years, and the trees become progressively less able to withstand drought and to defend against opportunistic fungi and insects. They become more susceptible to attack by the root-rotting fungus *Heterobasidion annosum* (Plates 165, 166) and by bark beetles.

Oxidant injuries and related or mimicking symptoms on *Pinus* species.

A–C. "Semimature tissue needle blight," considered by some to represent acute ozone injury (discussion, p. 482), on a highly susceptible *P. strobus* (eastern white pine). Scattered trees in the field were affected similarly. B. Close view of affected shoots. C. Needles from three fascicles, showing pinkish tan bands that spread toward needle tips (IL, Jun).

D. Ozone injury of *P. strobus:* pinkish yellow lesions with diffuse margins, associated with stomata (NY, Jun).

E. Chronic injury, attributed mainly to oxidants, on shoots of two highly sensitive *P. strobus* (center and right) individuals with a normal shoot for comparison. The shoot at center has undersized yellowish leaves with many lesions similar to those in D. The specimen at right was from a small declining tree that had been affected for many years. Secondary fungi were probably contributing to the needle blight symptoms. Such highly sensitive individuals are now rare (NY, Sep).

F, G. Foliar injuries of *Pinus* species that could be confused with air pollutant injury. F. Enlarged view of chlorotic spots that often arise during winter (winter fleck) on *P. strobus* needles, unrelated to stomata (NY, Jan). G. *Pinus nigra* (Austrian pine) needles with chlorotic lesions of undetermined cause, also called winter fleck (NY, Feb). The specific cause(s) of winter fleck have not been reported.

These biotic agents are often the direct cause of death. Tree and shrub species composition in the affected forests has been shifting, with oxidant-tolerant taxa becoming relatively more prominent as sensitive *P. ponderosa* and *P. jeffreyi* die. The situation is improving, however, as a consequence of air pollution control measures. Although the growth of surviving *P. ponderosa* decreased progressively between 1945 and 1971, growth began to increase thereafter.

Acute injury of conifer foliage by ozone is uncommon. It is best known from the response of highly sensitive individuals of *P. strobus* (eastern white pine). These trees, fortunately, are a minority of the species. Their immature needles have a zone near the needle base (semimature tissue) that is ozone sensitive. Initial symptoms, caused by collapse of mesophyll cells in the semimature zone, are bleached flecks and pinkish yellow spots along the lines of stomata (Plate 238). If injury is severe, many cells collapse and a band of pinkish yellow, eventually brownish, tissue separates the green needle base from the green tip. Necrosis then progresses toward the needle tip. Less sensitive individuals of *P. strobus* undergo yellowing and decline that have been attributed to chronic exposure to ozone. Most trees of this species sustain no visible injury or only chlorotic spotting and premature casting of needles. *P. strobus* tolerant enough to retain needles for the normal maximum of 27 months, however, are now uncommon in areas affected by air pollution.

P. strobus is also subject to "semimature tissue needle blight" (SNB; Plate 238), attributed by various authors to ozone, sulfur dioxide, heat and water stress in succulent needle tissue during the time of needle elongation in late spring, or the needle cast fungus *Canavirgella banfieldii* (Plate 24). The blight affects only particular trees and only in occasional years. Necrosis spreads upward from pinkish yellow bands that develop in semimature tissue near needle bases. Critical tests of the ability of ozone, sulfur dioxide, or *C. banfieldii* to cause symptoms like those of SNB have not been performed. Credible tests would require grafted plants derived from those naturally affected by SNB in comparison with grafted plants derived from unaffected trees.

In typical broadleaved plants the cells most sensitive to ozone are in the palisade parenchyma of recently expanded leaves. Therefore symptoms appear first on subterminal leaves of growing shoots. Groups of cells bounded by the smallest veinlets collapse and appear as dark stipples (light flecks in some species) on the upper surface 24–48 hours after a pollution episode. The tissue along major veins remains green. Severe injury leads to either a bleached or a darkly pigmented appearance of the upper surface. No symptoms are visible on the lower leaf surface unless the injury is unusually severe. Bifacial necrosis may occur in severely injured leaves, which often drop prematurely. Flecks or stipples sometimes fade and may be replaced by general chlorosis.

Ozone can depress the growth of many tree species in the absence of visible symptoms. Such depression has been detected or demonstrated in *Acer saccharinum* (silver maple), *Fagus sylvatica* (European beech), *Fraxinus pennsylvanica* (green ash), *Liriodendron tulipifera* (tuliptree), *Pinus taeda* (loblolly pine), *Platanus occidentalis* (sycamore), *Populus deltoides* (eastern cottonwood), *P. tremuloides* (trembling aspen), *Populus* hybrids, and *Quercus alba* (white oak), among other species.

Although dormant plants are tolerant of polluted air, minor symptoms caused by air pollutants and other agents to conifer needles during the growing season tend to intensify and be augmented by new symptoms during winter in northern and alpine areas. These changes are responses to drying, freeze-thaw stresses, and diminished defense and repair functions during dormancy. Differences in external appearance and microscopic characteristics of injured tissue distinguish needle markings caused by air pollutants from spots, flecks, and tip necrosis caused by stresses of winter weather.

The following lists classify tolerance of woody plant species to ozone according to the most sensitive condition of an average plant. Cultivars and individuals of a species may vary in tolerance, so species may be listed in one category while a particular cultivar is listed in another. The lists were constructed from reports of natural and artificially induced injury. The lists do not necessarily indicate which species are most likely to sustain growth suppression by

chronic exposure to low doses of ozone, as this is not well correlated with sensitivity to acute injury.

Relative tolerance of woody plants to ozone

Tolerant: *Abelia* ×*grandiflora*, *Abies balsamea*, *Acer platanoides*, *A. pseudoplatanus*, *A. rubrum*, *A. saccharum*, *Albizia julibrissin*, *Alnus glutinosa*, *A. viridis*, *Betula nigra*, *Calocedrus decurrens*, *Carpinus betulus*, *Citrus limon*, *Cornus florida*, *C. racemosa*, *Crataegus*, *Fagus sylvatica*, *Fraxinus americana* 'Autumn Purple,' *F. ornus*, *F. pennsylvanica* 'Summit,' *F. profunda*, *F. quadrangulata*, *Fuchsia*, *Ginkgo biloba*, *Gleditsia triacanthos* (several cvs.), *Ilex* ×*attenuata*, *I. cornuta*, *I. crenata*, *I. opaca*, *Juglans nigra*, *Juniperus occidentalis*, *J.* ×*pfitzeriana*, *Kalmia angustifolia*, *Lagerstroemia indica*, *Ligustrum amurense*, *Magnolia* species, *M.* ×*soulangiana*, *M. acuminata*, *Metasequoia glyptostroboides*, *Morella cerifera*, *Nyssa sylvatica*, *Paulownia tomentosa*, *Persea americana*, *Picea abies*, *P. glauca*, *P. pungens*, *Pieris japonica*, *Pinus aristata*, *P. lambertiana*, *P. monophylla*, *P. pseudostrobus*, *P. resinosa*, *P. sabiniana*, *P. thunbergii*, *Prunus persica*, *Pseudotsuga menziesii*, *Pyracantha coccinea* 'Lalandei,' *Pyrus calleryana*, *P. communis* 'Bartlett,' *Quercus acutissima*, *Q. imbricaria*, *Q. macrocarpa*, *Q. palustris* 'Sovereign,' *Q. shumardii*, *Rhododendron carolinianum*, *Robinia pseudoacacia*, *Sequoia sempervirens*, *Sequoiadendron giganteum*, *Sorbus aucuparia*, *Spiraea*, *Styphnolobium japonicum* 'Regent,' *Taxodium distichum*, *Taxus*, *Thuja occidentalis*, *Tilia cordata*, *T. petiolaris*, *Tsuga canadensis*, *Ulmus* hybrids, *Viburnum* species, and *V. carlesii*.

Intermediate: *Abies concolor*, *Acer negundo*, *A. saccharinum*, *Catalpa*, *Cercis canadensis*, *Corylus avellana*, *Elaeagnus commutata*, *Euphorbia pulcherrima*, *Forsythia* ×*intermedia* 'Lynwood,' *Frangula alnus*, *Fraxinus excelsior*, *F. pennsylvanica*, *Gleditsia triacanthos* 'Shademaster,' *Juglans regia*, *Koelreuteria paniculata*, *Larix kaempferi*, *Ligustrum vulgare*, *Liquidambar styraciflua*, *Lonicera korolkowii*, *Malus* species (some cvs. sensitive), *Philadelphus coronarius*, *Pinus attenuata*, *P. contorta*, *P. echinata*, *P. nigra*, *P. rigida*, *P. strobus* (some seedlings sensitive), *P. sylvestris*, *P. torreyana*, *Platanus* ×*acerifolia* (some cvs. sensitive), *Populus nigra*, *Prunus avium* 'Lambert,' *Pseudotsuga macrocarpa*, *Quercus alba* (some cvs. sensitive), *Q. coccinea*, *Q. palustris*, *Q. robur* (tolerance varies), *Q. rubra*, *Q. velutina*, *Rhododendron catawbiense* and its hybrids, *Sambucus nigra*, *Tilia americana* (tolerance varies), *Ulmus glabra*, and *U. parvifolia*.

Sensitive: *Abies religiosa*, *Aesculus glabra*, *Ailanthus altissima*, *Amelanchier alnifolia*, *Betula pendula*, *Buddleja davidii*, *Cercis chinensis*, *Citrus* species, *Cornus sericea*, *Cotoneaster divaricatus*, *C. horizontalis*, *Fraxinus americana*, *Gleditsia triacanthos* 'Imperial,' *Gymnocladus dioicus*, *Larix decidua*, *Liriodendron tulipifera*, *Malus baccata*, *Morus alba*, *M. nigra*, *Parthenocissus quinquefolia*, *Physocarpus opulifolius*, *Pinus banksiana*, *P. coulteri*, *P. halepensis*, *P. hartwegii*, *P. jeffreyi*, *P. parviflora*, *P. ponderosa*, *P. radiata*, *P. taeda*, *P. virginiana*, *Platanus* ×*acerifolia* 'Bloodgood,' *P. occidentalis*, *P. orientalis*, *Populus* hybrids, *P. tremula*, *P. tremuloides*, *Prunus avium* 'Bing,' *P. serotina*, *Quercus gambelii*, *Q. robur* 'Fastigiata,' *Rhamnus cathartica*, *Rhododendron kaempferi* 'Campfire,' *R. mucronatum*, *R. obtusum*, *R. yedoense*, *Rhus hirta*, *Rosa multiflora*, *Salix babylonica*, *S. viminalis*, *Sambucus canadensis*, *S. racemosa*, *Sassafras albidum*, *Spiraea* ×*vanhouttei*, *Symphoricarpos albus*, *S. orbiculatus*, *Syringa reticulata*, *S. vulgaris*, *Tilia* ×*euchlora*, *T. platyphyllos* 'Fastigiata' and 'Orebro,' *Viburnum lantana*, *Vitis* (some types tolerant), and *Zelkova serrata*.

References: 248, 309, 311, 502, 648, 695, 696, 755, 817, 818, 902, 968, 1139, 1180, 1399, 1640, 1654, 1866, 1899, 1905, 1906, 2014–2016, 2150, 2156, 2270, 2337, 2338, 2424, 2428, 2646–2648, 3393, 3394, 3404, 3460, 3677, 3680, 3681, 3716, 4023, 4291, 4409, 4472. See also references for overview.

Oxidant injury on angiosperm leaves.

A–C. Purple pigmentation on upper surfaces of *Fraxinus americana* (white ash) leaves. A. Viewed in the field (PA, Aug). B, C. Close views of purple stippling, slightly enlarged in B. Green zones without injury on the center leaflet were protected by the edges of lateral leaflets. C. Magnified view of purple-brown discoloration in islets delimited by the smallest veins (NY, Sep).

D. Interveinal purple stippling and browning on *Liriodendron tulipifera* (tuliptree) (VA, Aug).

E. Purple stippling on *Prunus serotina* (black cherry) (PA, Aug).

F–H. Symptoms on *Vitis* (grapevine). A. Purple stippling on the cultivar Chambourcin (PA, Sep). G, H. Magnified views of stipples consisting of clusters of bleached cells and, where injury was more severe (H), tiny angular lesions bounded by veinlets (greenhouse plants).

Photo credits: A, D–F—J. M. Skelly

Damage by Sulfur Dioxide

Damage to plants by sulfur dioxide (SO_2) was formerly concentrated around industrial sources and power-generating stations. As one approached a major SO_2 source from the prevailing downwind direction, visible plant injury and mortality increased and the total number of plant species diminished. In the worst cases SO_2 killed all vegetation close to the sources. During the 1970s–1990s, however, many major emitters of SO_2 in developed nations either stopped operating or erected tall stacks and/or built SO_2 scrubbers. Severe vegetation damage in those places ceased, and revegetation of denuded areas has begun. Acute injury by SO_2 has become uncommon because gases from tall stacks become greatly diluted before reaching ground level.

Acute foliar injury is indicated by bleached or pigmented (tan to reddish brown or dark brown, depending on species) necrotic interveinal areas on broadleaved plants and chlorotic spots and bands or brown tips on conifer needles. Young leaves that have just attained full size are generally most sensitive. Typical doses of SO_2 that can cause visible symptoms on sensitive plants are 500 ppb or higher for 1–3 hours. Chronic exposure to low concentrations of SO_2 causes chlorosis, premature leaf senescence, and depressed growth of sensitive plants. Chronic exposure to 100 ppb or less of SO_2 caused no measurable effect on many of the woody species that have been tested. Such concentrations actually stimulated the growth of some plants, at least during the first year.

SO_2 causes injury when it and its sulfite derivative (SO_3^-) are absorbed (via stomata) faster than they are detoxified. Plants oxidize SO_2 successively to sulfite and sulfate, which interfere with photosynthesis and energy metabolism. For equivalent amounts, sulfite is about 30 times more toxic than sulfate. The latter in modest doses is utilized as a nutrient source.

Plants vary in tolerance of SO_2 for several reasons. First, they vary in efficiency of SO_2 absorption; damage is least likely if the gas is absorbed slowly. They also vary in ability to detoxify SO_2 and to dispose of excess sulfur. Sulfur in excess of that utilized as a nutrient is both incorporated into organic compounds and reduced to hydrogen sulfide, which escapes from leaves. Much of the sulfur absorbed by leaves is translocated to other organs, thus diluting it. Tolerance of trees to SO_2 is roughly correlated with shade tolerance. *Acer* species (maples), for example, are generally tolerant, and *Betula* species (birches) are sensitive.

The following classification of plants according to relative tolerance to SO_2 was constructed from reports of naturally and artificially induced injuries. Ratings are for representative plants. Individuals and cultivars within species may differ in tolerance.

Relative tolerance of plants to sulfur dioxide

Tolerant: *Abies amabilis, A. fraseri, Acer campestre, A. ginnala, A. platanoides, A. pseudoplatanus, A. saccharinum, A. saccharum, Aesculus flava, Arctostaphylos uva-ursi, Buxus sempervirens, Carya laciniosa, Ceanothus velutinus, Chamaecyparis nootkatensis, C. pisifera, Citrus, Cornus florida, Crataegus douglasii, Cryptomeria japonica, Elaeagnus commutata, Gingko biloba, Hedera helix, Ilex aquifolium, Juniperus chinensis, J. communis, J. occidentalis, J. osteosperma, J. sabina, J. scopulorum, J. virginiana, Ligustrum ovalifolium, L. vulgare, Lonicera tatarica, Mahonia aquifolium, Metasequoia glyptostroboides, Pinus contorta, P. edulis, P. mugo, P. nigra, Platanus ×acerifolia, Populus balsamifera, P. ×canadensis, Prunus cerasifera, P. serotina, P. virginiana, Quercus alba, Q. coccinea, Q. gambelii, Q. petraea, Q. prinus, Q. rubra, Q. velutina, Q. virginiana, Rhododendron catawbiense, Sambucus nigra, S. racemosa, Shepherdia argentea, Spiraea, Styphnolobium japonicum, Taxodium distichum, Taxus baccata, T. brevifolia, Thuja occidentalis,* and *T. plicata.*

Intermediate: *Abies balsamea, A. concolor, A. grandis, A. lasiocarpa, Acer rubrum, Aesculus hippocastanum, Ailanthus altissima, Alnus glutinosa, Berberis thunbergii, Caragana arborescens, Carpinus caroliniana, Catalpa, Ceanothus sanguineus, Cercidiphyllum japonicum, Cercis canadensis, Chamaecyparis lawsoniana, Cornus mas, C. sanguinea, C. sericea, Corylus avellana, Cotinus coggygria, Cotoneaster divaricatus, Crataegus chryso-carpa, C. laevigata, C. phaenopyrum, Cydonia oblonga, Elaeagnus angusti-folia, Fagus grandifolia, F. sylvatica, Forsythia, Frangula alnus, Fraxinus americana, F. excelsior, F. pennsylvanica, Gleditsia triacanthos, Hamamelis virginiana, Hydrangea paniculata, Juglans nigra, J. regia, Laburnum ana-gyroides, Larix decidua, L. kaempferi, Liquidambar styraciflua, Liriodendron tulipifera, Malus pumila, Morus alba, Nyssa sylvatica, Parthenocissus*

quinquefolia, Philadelphus coronarius, Picea engelmannii, P. glauca, P. pungens, Pinus cembra, P. flexilis, P. monticola, P. peuce, P. strobus, Populus deltoides, P. nigra, Prunus armeniaca, P. avium, P. domestica, P. persica, P. serrulata, Pyracantha coccinea, Pyrus communis, Quercus palustris, Rhamnus cathartica, Rhus glabra, Ribes alpinum, Robinia pseudoacacia, Salix caprea, Sambucus cerulea, Sorbus aucuparia, Spiraea ×vanhouttei, Symphoricarpos albus, Syringa reticulata, S. vulgaris, Tilia americana, T. cordata, T. platyphyllos, Tsuga canadensis, T. heterophylla, Ulmus americana, U. glabra, U. minor, U. parvifolia, Viburnum lantana, V. opulus, and *Weigela.*

Sensitive: *Acer glabrum, A. negundo, A. palmatum, Alnus incana* subsp. *tenuifolia, Amelanchier* species, *Betula alleghaniensis, B. lenta, B. nigra, B. occidentalis, B. papyrifera, B. pendula, B. populifolia, Chaenomeles japonica, Corylus cornuta, Holodiscus discolor, Larix laricina, L. occidentalis, Lonicera morrowii, Morus microphylla, Philadelphus lewisii, Physocarpus capitatus, P. malvaceus, Picea abies, Pinus banksiana, P. caribaea, P. jeffreyi, P. ponderosa, P. resinosa, P. rigida, P. sylvestris, P. virginiana, Platanus occidentalis, Populus angustifolia, P. balsamifera* subsp. *trichocarpa, P. grandidentata, P. tremuloides, Prunus emarginata, Pseudotsuga menziesii, Quercus macrocarpa, Rhus hirta, Ribes rubrum, R. uva-crispa, Salix babylonica, S. fragilis, S. nigra, S. pentandra, S. purpurea, Sassafras albidum, Sorbus scopulina, S. sitchensis, Symphoricarpos oreophilus, Vaccinium angustifolium,* and *Vitis* species.

References: 63, 373, 502, 648, 722, 817, 818, 903, 1180, 1244, 1318, 1451, 1899, 1930, 1931, 2014–2016, 2285, 2337, 2424, 2438, 2439, 2570, 3404, 3444, 3569, 3691, 4118, 4119, 4472. See also references for overview.

Damage by Fluorides and Minor Pollutant Gases

Fluorides are released to the air during the manufacture of various products made from raw materials mined from the earth. Major sources of airborne fluorides include manufacturers of aluminum, phosphate fertilizer, iron and steel, bricks, glass, and ceramics, as well as installations that burn large amounts of coal. Fluoride damage to plants is most common near sites where phosphate fertilizer and aluminum are produced, because these processes employ fluoride-rich materials. Phosphate rock contains fluoride (3–4%) that is released during heating, grinding, or acid treatment. Hydrogen fluoride (HF) is released from molten cryolite (sodium aluminum fluoride) and fluorite (calcium fluoride) during electrolytic reduction of alumina. Hydrofluorosilicic acid is another gaseous fluoride that causes plant injury. Fluorides also occur as natural air contaminants in windborne soil and in ash and gas released from volcanoes and fumaroles, but these natural sources are relatively unimportant. Plants absorb fluorides from air and soil but generally accumulate toxic quantities only from air. Excess fluoride acquired from plants may cause disease in grazing animals whether or not the plants show symptoms.

Gaseous fluorides enter leaves primarily through stomata and accumulate near needle tips of conifers, near leaf tips of plants with parallel leaf veins, and in interveinal areas as well as at leaf tips and

---

Sulfur dioxide damage.

A. Devegetation around a smelter, caused mainly by SO_2. Damage close to the smelter ceased after the tall stack was erected, because toxic gases were then greatly diluted before returning to ground level. Railroad cars in the foreground provide a size standard for the stack (ID, Jul).

B, C. Acute and chronic SO_2 injuries to *Pinus strobus* (eastern white pine). B. Sensitive and tolerant trees, the former growing slowly and bearing sparse, faded foliage. The source of pollutant was a steam plant 1 km upwind. C. Tufted foliage with dead and dying needle tips on a branch from the sensitive tree (NY, Apr).

D, E. Acute injury to leaves of *Betula papyrifera* (paper birch) growing near an iron sintering plant. D. A young plant with more than half its foliar area killed. E. Close view of interveinal necrosis in original leaves and green new growth produced after an episode of severe SO_2 pollution (ON, Jul).

F, G. Acute foliar injury (bifacial necrosis between major veins) on *Diospyros virginiana* (common persimmon) and *Fraxinus americana* (white ash), respectively (MO, Jul).

Photo credits: F, G—J. M. Skelly

485

margins of broadleaved plants. Fluoride at toxic concentrations in plants causes breakdown of chlorophyll, inhibits utilization of calcium and magnesium, and interferes with various metabolic processes. Exposure of sensitive plants to fluoride at 5–100 ppb in air (volume basis) for several hours may cause acute symptoms such as interveinal necrosis on broad leaves or tip burn of conifer needles. Tissues that accumulate an injurious amount turn yellowish, then tan to brown or reddish brown, often with a narrow darker band at the boundary of living tissue. Dead tissue no longer accumulates toxicant, so the site of accumulation shifts to living tissue adjacent to that previously killed; in time this too dies. Thus symptoms of fluoride toxicity may progress incrementally from leaf tips and margins toward main veins and leaf bases. Slow accumulation over days or weeks leads to chronic symptoms— chlorosis at leaf tips and margins. Pollution control devices and tall smokestacks have greatly diminished the atmospheric fluoride concentration at ground level near many pollution sources, so accumulation usually occurs slowly. Given sufficient time—10 days to a year or longer—sensitive plants can accumulate an injurious amount of fluoride from air containing less than 1 ppb. In conifers the youngest needles are most sensitive, and new symptoms arise only on current-season needles. Symptoms may slowly intensify on older needles, however, as these continue to accumulate fluoride for two or more seasons. Such needles become senescent and drop prematurely. The growth of *Pseudotsuga menziesii* (Douglas-fir) and presumably other species may be depressed by fluoride at concentrations in tissue that do not cause visible symptoms.

Plant species and individuals within species vary widely in fluoride tolerance. Normal or expected concentrations of fluoride in plant foliage are on the order of 5–20 ppm for many species. Elevated fluoride content (50 to several hundred ppm) of leaves with chlorotic or necrotic patterns in species known to be sensitive is presumptive evidence of fluoride injury. Since plants vary in tolerance and fluoride is leached from leaves by rain, however, no clear-cut quantitative relationship between fluoride content and severity of symptoms has been found. Plants tolerant of fluoride may accumulate it to concentrations of several hundred ppm in leaves that appear healthy. Also, some plants such as *Camellia japonica* (common camellia) seem to accumulate a large amount of fluoride from soil. Woody plants that have been ranked for sensitivity or tolerance to atmospheric fluoride are listed below.

Relative tolerance of plants to fluorides

Tolerant: *Ailanthus altissima, Alnus glutinosa, Camellia japonica, Cornus florida, C. mas, Elaeagnus angustifolia, Fraxinus velutina* var. *glabra, Juniperus* species, *Kalmia latifolia, Ligustrum* species, *Liquidambar styraciflua, Magnolia* species, *Morus* species, *Parthenocissus quinquefolia, Philadelphus coronarius, Platanus* ×*acerifolia, P. occidentalis, Populus balsamifera, Prunus cerasifera, P. serrulata, Pyracantha coccinea, Pyrus communis, Ribes* species, *Salix* species, *Sambucus nigra, S. racemosa, Spiraea* ×*vanhouttei, Thuja occidentalis, Tilia americana, Ulmus americana,* and *U. parvifolia.*

Intermediate: *Abies grandis, Acer campestre, A. platanoides, A. saccharinum, Betula pendula, Carpinus betulus, Citrus limon, C.* ×*paradisi, C. reticulata, C. sinensis, Corylus avellana, Fraxinus excelsior, F. pennsylvanica,*

Fagus sylvatica, Ilex aquifolium, Juglans nigra, J. regia, Malus pumila, Morus rubra, Picea glauca, Pinus contorta, P. mugo, Populus ×*canadensis, P. nigra, P. tremuloides, Prunus avium, P. persica, P. virginiana, Quercus* species, *Rhododendron* species, *Rhus glabra, R. hirta, Robinia pseudoacacia, Rosa* species, *Sorbus aucuparia, Syringa vulgaris, Taxus cuspidata,* and *Tilia cordata.*

Sensitive: *Acer negundo, Amelanchier* species, *Cercis canadensis, Ilex opaca, Larix occidentalis, Mahonia aquifolium, Paulownia tomentosa, Picea pungens, Pinus ponderosa, P. strobus, P. sylvestris, P. taeda, Prunus armeniaca* (some cvs. intermediate), *Pseudotsuga menziesii, Vaccinium* species, and *Vitis* species (some cvs. intermediate).

Minor air pollutants that cause injury to plants include chlorine, hydrogen chloride, ammonia, ethylene, hydrogen sulfide, and an array of metallic elements (arsenic, copper, lead, mercury, nickel, zinc, and others). Metallic pollutants are discharged in smoke from coal-fired power plants, smelters, and other industrial sources and slowly accumulate in vegetation downwind, but their impact on plant health has not been separated from that of major air pollutants. Gaseous minor pollutants are most likely to cause plant injury near sites of accidental leaks and spills from tanks or pipes. In such cases prominent foliar injury occurs to many kinds of plants. Those nearest the leak or spill may lose all their foliage. The ammonia leak that caused the injury shown in Plate 241 occurred at a distribution facility for anhydrous ammonia, which is used as a fertilizer.
References: 647, 648, 973, 1240, 1755, 1899, 2198, 2438, 4078, 4079, 4158

Damage by fluoride and by ammonia.

A. Foliar browning and death of *Pinus elliottii* (slash pine) caused by fluoride accumulated from severely polluted air near a phosphate fertilizer plant. Tree-to-tree differences in pollutant tolerance are evident (FL, May).

B, C. Needle tip necrosis on *Picea pungens* (blue spruce) caused by a fluoride emission event at a ceramics factory (NY, Aug).

D, E. Needle tip necrosis on *P. mugo* (mugo pine), associated with chronic exposure to low concentration of fluoride in air near a brick works. Only old symptoms on 2- and 3-year-old needles are visible (NE, Jun).

F. Marginal chlorosis on *Acer saccharum* (sugar maple) leaves caused by fluoride accumulation from air near a brick works (VA, summer).

G. Necrosis caused by fluoride at margins of *Cercis canadensis* (redbud) leaves. The tree was growing near an aluminum reduction plant (SC, Apr).

H, I. Damage to *Populus deltoides* (eastern cottonwood), caused by an ammonia spill. The incident occurred 3 weeks before these photographs were made. Most foliage present at the time of the spill was killed; leaves produced subsequently are green (MN, Jul).

Photo credits: B, C—G. W. Hudler; F—J. M. Skelly; G—J. A. Laurence

Mineral Nutrient Deficiencies (Plates 242, 243)

Twenty elements are known or suspected to be essential nutrients for plants, although not all of them have been demonstrated to be essential for woody perennials. All essential elements except carbon, hydrogen, and oxygen are normally supplied by soil. Those required in relatively large amounts include nitrogen, phosphorus, potassium, calcium, magnesium, and sulfur. The others (micronutrients) are needed in small to trace amounts. Growth of healthy plants is usually limited by the supply or availability of one or more nutrient elements. A nutrient deficiency exists if an increase in the effective supply of the nutrient(s) would correct or prevent abnormally slow growth or small size, deformity, depressed yield, abnormal color, or necrosis. Plates 242 and 243 show symptoms caused by deficiencies of some nutrient elements. A pictorial survey of the subject is beyond the intended scope of this book. The references contain more information and illustrations.

Nutritional deficiencies in trees and shrubs cannot usually be diagnosed reliably on the basis of symptoms alone. Deficiencies of different nutrients may cause similar symptoms, and symptoms can be complicated by simultaneous deficiency of two or more elements. Also, many symptoms of deficiency are similar to those caused by other factors such as excess heat, drought, waterlogged soil, chemical injuries, or damage by insects or pathogens. Chemical analyses of plant tissues and/or the rooting medium are needed for accurate diagnosis and prescriptive treatment of deficiencies. Analytical results are most useful if normal concentration ranges and deficiency levels of nutrients in tissues of the affected plant species are known. Diagnosticians who identify nutrient deficiencies accurately by observation alone are relying on personal experience with the species or cultivar in the locality where the disorder occurs. Mineral nutrient elements, some of their functions, and some typical symptoms of deficiency are summarized below—major nutrients first.

Nitrogen (N) is a component of amino acids, structural proteins, enzymes, nucleic acids, chlorophyll, growth regulators, and alkaloids. It is required for diverse functions. Normal concentrations in tree foliage are 12–35 mg/g dry matter. The nitrogen supply limits growth of woody plants on most sites. Deficiency causes slow growth, diminished leaf size or number of leaflets on compound leaves, subnormal chlorophyll synthesis, and collapse of chloroplasts, resulting in pale green color or chlorosis. Symptoms are usually more severe on old leaves than on young ones because nitrogen is withdrawn from old or senescing parts to supply the needs of physiologically active organs.

Phosphorus (P) is a constituent of nucleic acids, nucleoproteins, phospholipids, energy transfer compounds, and buffer systems. It is mobile and occurs in both organic and inorganic forms within plants. Normal and deficient concentrations in tree leaves are 1.2–2.5 and <1.0 mg/g dry matter, respectively. Deficiency causes slow growth, general chlorosis, or abnormal reddish purple pigment on petioles or in interveinal areas of broadleaved plants. Seedlings deficient in phosphorus are stunted, often are chlorotic, and may have abnormal reddish purple color. Purple pigmentation presages browning in coniferous seedlings. Young plants lacking mycorrhizae often have a phosphorus deficiency.

Potassium (K) is required in polypeptide synthesis by ribosomes and thus is essential in protein synthesis. It also regulates proton pumps; mediates activity of some enzymes; and is involved in nitrogen metabolism, carbohydrate translocation, turgor maintenance, and regulation of transpiration. It remains in inorganic form and is highly mobile in plants; thus deficiency symptoms appear first in old tissues. Normal and deficient concentrations in conifer leaves are 40–100 and <20–40 mg/g dry matter, respectively. Corresponding values for broadleaved woody plants are 100–150 and <60–100 mg/g. Deficiency causes slow growth; chlorosis and browning of leaf tips, margins, and interveinal areas; sparse-appearing foliage; and shoot dieback. Red or purple pigment may appear before tissues turn brown. Potassium-deficient conifer needles display general chlorosis, subnormal length, tip browning in severe cases, and premature senescence and abscission. Potassium deficiency also promotes susceptibility to various pathogens, such as *Verticillium* species (Plates 120, 121).

Calcium (Ca) stabilizes pectate polymers in and between primary cell walls. It also binds to membranes and is essential for their integrity and functions. It is important in intracellular ionic balance,

gene expression, mitosis, and secretion. Calcium is a component of the messenger polypeptide calmodulin, an activator of several enzyme systems. The cation Ca^{++} is a buffer component. The calcium concentration in leaves is usually in the range of 5–30 mg/g dry matter, a typical value in trees being near 15 mg/g. Calcium is immobile after incorporation into structural polymers, so deficiency symptoms appear first in young shoots. Deficiency is indicated by chlorosis and subnormal size of youngest leaves and marginal and sometimes veinal necrosis that progresses inward and toward the bases of chlorotic leaves. Severe deficiency results in death of shoot tips. Lime or calcined magnesite is beneficial for amelioration of calcium deficiency.

Magnesium (Mg) is a constituent of chlorophyll and an activator of enzymes involved in phosphate transfer and photosynthesis. It is essential for nucleic acid and polypeptide synthesis and for the functions of ribozymes (catalytic RNAs that cleave RNA). The cation Mg^{++} is a buffer component. Magnesium is relatively mobile in plants, so deficiency symptoms appear first in old tissues. Normal and deficient concentrations in tree leaves are 10–20 and <8 mg/g dry matter, respectively. Deficiency symptoms in angiosperms and palms include chlorosis that develops first at margins and on apices of the oldest leaves, then extends between the veins and to younger leaves. Leaf bases often remain green. Severe deficiency results in general chlorosis, irregular marginal necrosis, and premature leaf drop. Some magnesium-deficient plants produce purplish red pigment in interveinal areas of leaves. Deficiency in conifers causes premature yellowing and browning of second-year or older needles, yellow discoloration of current-year needle tips in pines, and general yellowing of spruce needles. Needle tips turn brown in cases of severe deficiency.

Sulfur (S) is a constituent of certain amino acids and proteins, vitamins, and coenzyme A. Disulphide bonds in polypeptides stabilize their structure and thus contribute to protein conformation. Sulfhydryl (SH) groups participate in some enzyme reactions in sugar and lipid metabolism and in detoxification of oxygen radicals, and they are constituents of chelating molecules that detoxify heavy metals. Typical foliar concentration of sulfur in trees is near 20 mg/g dry matter, and deficiency symptoms are associated with concentrations below about 15 mg/g. Sulfur is relatively immobile within plants, and deficiency symptoms are thus likely to appear first in young shoots. Deficiency causes diminished protein synthesis, accumulation of amino acids, general chlorosis (more severe in interveinal areas than along leaf veins in some species), and subnormal leaf size.

Boron (B) is involved in uracil synthesis and thus indirectly in RNA synthesis and in carbohydrate utilization. It is required for

---

Nutrient deficiencies in monocots and gymnosperms.
A. "Frizzle top" of *Syagrus romanzoffiana* (queen palm), a distortion of leaflets attributed to manganese deficiency (FL, Apr).
B. Chlorosis attributed to iron deficiency in *S. romanzoffiana* (TX, Apr).
C. Chlorosis induced primarily by nitrogen deficiency in young *Pinus taeda* (loblolly pine). These trees grew from seed at a site where topsoil had been removed. Normal *P. taeda* in the background provide contrast (GA, May).
D. Lime-induced chlorosis of *P. elliottii* (slash pine), attributed to iron deficiency. Fertilizer, lime, and irrigation water containing carbonate were applied to turf in the trees' root zone. This raised the soil pH, and the pines developed deficiency symptoms (FL, Apr).
E, F. Chlorosis of needle tips, induced primarily by magnesium deficiency, in young *P. sylvestris* (Scots pine) (NY, Sep).
G. Premature senescence of year-old needles on a young *P. strobus* (eastern white pine) affected by deficiencies of magnesium and potassium (NY, Sep).
H, I. Premature senescence of second-year needles, induced by magnesium deficiency, in *Pseudotsuga menziesii* (Douglas-fir) and *Abies balsamea* (balsam fir), respectively. A normal *P. menziesii* twig is at right in H (NY, Sep).
J. Potassium deficiency symptoms in year-old *Pinus resinosa* (red pine) needles. Current-year needles are green in response to a potassium fertilizer application (NY, summer).
K. Severe magnesium deficiency symptoms in *P. resinosa* (NY, Sep).

489

normal cell wall structure, plasma membrane function, and meristematic activity. Normal and deficient concentrations in tree leaves are 15–100 and <15 µg/g dry matter, respectively. Boron-deficient plants may have chlorotic, undersized leaves; fused leaves or leaflets that would normally be separate; shoot dieback; bud proliferation at bases of dead shoots; and blunt, much-branched roots. Boron deficiency in *Malus* (apple) causes, in addition to some of the symptoms above, brittle deformed leaves and cambial lesions. Deficiency can be alleviated by applying boron compounds to soil.

Chlorine (Cl) and calcium are together required for photolysis (the water-splitting reaction) in photosystem II. Chlorine also enhances photophosphorylation and is important in turgor regulation. Normal and deficient concentrations in tree leaves are 2–20 mg/g and 70–700 µg/g dry matter, respectively. Deficiency symptoms, known only in some in herbaceous plants, include wilting of leaves at the edges, chlorosis, and undersized leaves. Surplus chloride ion can inhibit nitrate uptake.

Cobalt (Co) is essential for symbiotic nitrogen fixation. It is a component of the coenzyme cobalamine (vitamin B_{12}). Evidence of essentiality in non-nitrogen-fixing plants is scant, although chlorosis in young leaves of cobalt-deprived plants has been reported.

Copper (Cu) is an activator of several oxidative enzymes and a constituent of phenolases and enzymes that are involved in photosynthesis, electron transport in mitochondria, and detoxification of free radicals. It is required for lignification, protein utilization, and chlorophyll synthesis. Normal and deficient concentrations in tree leaves are 5–20 and <4 µg/g dry matter, respectively. Deficiency symptoms may appear in plants growing in poorly buffered sandy soils, peat moss–vermiculate root media, or organic soil where copper is bound in insoluble complexes. The symptoms include marginal chlorosis; general light green or gray-green color, sometimes with contrasting chlorotic veins; marginal or interveinal necrosis; cupping of leaves; stunting of leaves and shoots; premature leaf abscission; and twig dieback. Conifers may grow prostrate or have drooping shoots and brown needle tips. Small shoots may proliferate below dead tips, and the new shoots may die in turn. Copper is relatively immobile in plants; thus young tissues show deficiency symptoms first. Applying nitrogen to copper-deficient plants may accentuate the symptoms. Deficiency can be corrected by acidifying neutral or alkaline soil, fertilizing with copper sulfate, or spraying plants with copper-containing products.

Iron (Fe) functions as a prosthetic group or coenzyme in certain enzyme systems. It is involved in the synthesis of chloroplast proteins, chlorophyll, and haem-containing enzymes, notably oxidases and respiratory enzymes. Normal and deficient concentrations in tree leaves are 40–100 and 25–40 µg/g dry matter, respectively. Iron is available from soils at pH 6 or less, but it is bound in insoluble forms in alkaline or calcareous soils. It is relatively immobile within plants; thus deficiency symptoms tend to occur on young shoots. Deficiency causes interveinal chlorosis and slow growth that is at first irregular from branch to branch. Affected leaves tend to be undersized and in severe cases develop irregular marginal browning. Branches with severe symptoms eventually die back. Deficiencies of manganese and zinc may cause similar symptoms, and subnormal levels of two or all of these elements may occur in chlorotic foliage. Application of a chelated iron compound or acidifying substance (e.g., elemental sulfur) to soil may be beneficial. Injection or implantation of iron salts or chelated iron into holes drilled in sapwood of affected trees also often helps. *Quercus palustris* (pin oak) and *Q. ellipsoidalis* (northern pin oak) are notoriously prone to "iron chlorosis" resulting from iron deficiency. This proneness varies with seed source, however.

Manganese (Mn) is an activator of enzyme systems and an electron donor in the water-splitting system in photosynthesis. It assists in respiration and is a component of an enzyme that detoxifies superoxide radicals. Normal and deficient concentrations in tree leaves are 50 to several hundred and 5–40 (usually <20) µg/g, respectively. Manganese is most readily available from soils at pH 6 or less, and deficiency is most likely to occur in plants growing on alkaline or calcareous soils. Symptoms include interveinal chlorosis; undersized leaves; and, in some plants, wavy, crinkled, or curled leaf margins. Deficiency can be corrected by application of manganese sulfate to infertile acidic soil, but

implantation of this substance into sapwood of the trunk may be more effective—although a short-term fix—where deficiency symptoms occur on alkaline or calcareous soil. Manganese can be toxic at relatively low concentrations. Therefore suspected deficiency should be confirmed by tissue analysis before manganese is added.

Molybdenum (Mo), associated with nitrate reductase, is essential for conversion of nitrate, the principal form of nitrogen absorbed by plants, to reduced forms that can be incorporated into organic compounds. It is also involved in purine metabolism in some legumes. Normal and deficient concentrations in tree leaves are 0.3–1.0 and <0.2 µg/g dry matter, respectively. Natural deficiency has been described for only a few trees and shrubs. Molybolenum-deficient *Citrus* develops large yellow spots between veins on leaves. *Hibiscus rosa-sinensis* (Chinese hibiscus) displays foliar stunting and deformity resembling that caused by phenoxy herbicides.

Nickel (Ni) is better known for toxicity than for beneficial effects, probably because it can displace other heavy metal nutrients from physiologically important sites. It is considered to be essential for some cereal crops, in which it is necessary for seed germination. It must be present for urease to convert urea into usable nitrogen and for iron absorption. Deficiency is associated with dwarfed, cupped leaflets (mouse-ear symptom) in *Carya illinoinensis* (pecan). Symptoms in other woody plants are unknown.

Silicon (Si) is bound in pectin polymers in primary cell walls and contributes to their strength. It also forms complexes with orthophenols that are lignin precursors and is therefore important for lignin synthesis. Normal concentrations in plant tissues range from 5 to 75 mg/g dry matter, depending on whether or not a plant is a silicon-accumulating species. The Pinaceae are silicon accumulators. Deficiency symptoms, documented in herbaceous plants, include leaf deformities, premature death of old leaves, and wilting due to reduced control of transpiration. An abundant supply enhances resistance to foliar pathogens and some leaf-feeding insects. Silicon can also mitigate aluminum toxicity.

Vanadium (V) is reported to enhance growth, and is thus probably beneficial for many plants, but its function is unknown and its essentiality has not been proven. Its concentration in healthy plants is usually in the range 0.2–4.0 µg/g dry matter.

Zinc (Zn) is a catalytic component of many enzyme systems. It is involved, for example, in synthesis of tryptophan, which is a precursor of auxin, and in detoxification of superoxide radicals. It is readily available from soils at pH 6 or less. Normal and deficient concentrations in tree leaves are 12–80 and 5–20 µg/g dry matter. Zinc-deficient *Carya illinoinensis* may contain up to 40 µg/g. Deficiency causes interveinal chlorosis or mottling, dwarfed leaves, crinkled leaf margins, and stunted or rosetted shoots. Symptoms tend to be variable within a tree and between adjacent trees. Zinc deficiency can be corrected by injection of zinc sulfate solution or, in some cases, by driving a few zinc-coated nails into a stem.

References: 132, 224, 336, 337, 342, 378, 405, 704, 933, 959, 1052, 1062, 1153, 1421, 1583, 1666, 1841, 1845, 1904, 1912, 2255, 2475, 2591, 2607, 2932, 3014, 3262, 3714, 3875, 3876, 4166, 4190, 4297, 4306, 4315, 4436

Nutrient deficiencies in angiosperms.

A, B. Chlorosis and twig dieback in an iron-deficient *Triadica sebifera* (Chinese tallowtree) (TX, Apr).

C, D. Chlorosis induced by iron deficiency in *Quercus palustris* (pin oak) and *Q. rubra* (northern red oak), respectively. Tissue along the veins typically retains green color longest (NY, Jun).

E, F. Interveinal chlorosis, indicating iron deficiency, in *Rhododendron* species (TN, Apr).

G, H. Marginal yellow and red pigmentation in *Prunus serotina* (black cherry), and marginal chlorosis in *Betula papyrifera* (paper birch), induced by deficiencies of both magnesium and potassium (NY, Sep).

I. Interveinal chlorosis attributed to manganese deficiency in *Acer rubrum* (red maple) (MI, Aug).

J. Chlorosis, stunted leaves and shoots, and twig dieback induced by copper deficiency in variegated *Ligustrum sinense* (Chinese privet), compared with a normal plant (right) (FL, Apr).

Damage by Drought, Heat, and Freezing (Plates 244–248)

Injuries caused by water shortage and temperature extremes are inextricably linked. The order of text presentation is water shortage, heat stress, frost during the growing season, frost rings in wood, freezing in winter, root freezing, and predisposition to attack by fungal pathogens. Captions for some images are expanded to convey diagnostic features or circumstances of injury. Text information relevant to a particular illustration may not face it. For cracks caused in tree trunks by freezing or drying, see Plate 256.

Water Shortage

The amount and seasonal distribution of precipitation determine which woody species grow and how fast they grow at a given latitude and elevation. Native and naturalized plants in a particular locality are adapted to its seasonal and year-to-year variations in water supply; only severe drought is likely to cause noticeable injury to them. Planted trees and shrubs, on the other hand, often show symptoms of water stress. Perhaps more important, severe water deficit predisposes trees to infection by opportunistic pathogens and attack by secondary insects.

Water deficits develop normally in plants when loss by transpiration exceeds absorption from soil, as it does almost daily during the growing season. Water is replenished at night and during periods of rain or dew formation, when transpiration slows or ceases. As soil dries, however, roots fail to extract as much water as has been lost, and physiological stress develops. If the water potential of plant tissues decreases sufficiently, they degenerate and die. Leaves of drought-stressed plants may droop, wilt, turn yellow, turn brown at the tips and margins (leaf scorch), curl, or show combinations of these symptoms. Water deficits also develop in dormant plants, both evergreen and deciduous, during winter as water evaporates from leaves and stems. Injury by freezing exacerbates water loss. Plants vary in ability to tolerate low water potential. The minimum tolerated by most woody species that grow in mesic environments is −1.6 to −2.4 megapascals (MPa), but water potentials of −6 MPa have been measured in desert plants such as *Prosopis glandulosa* (mesquite), and −9 MPa in *Picea engelmannii* (Engelmann spruce) during late winter. Physiological and cytological symptoms of water shortage include growth cessation; loss of turgor; stomatal closure; disorganization of chloroplasts; diminished respiration; and greatly reduced photosynthesis, protein synthesis, and physiologic defenses against pests and pathogens. When leaves lose turgor and die, the oldest leaves usually succumb first. When extreme water stress causes cavitation (fracture of xylem water columns), this occurs first in twigs and then in successively larger stems.

In eastern USA the drought-sensitive European species *Aesculus hippocastanum* (horse-chestnut) shows leaf scorch nearly every year. *Acer platanoides* (Norway maple) often develops leaf scorch where it is planted in restricted rooting spaces. *Cornus florida* (flowering dogwood) and *Oxydendrum arboreum* (sourwood) are among the native species most likely to wilt and to develop foliar browning. Leaves of other deciduous trees such as *Acer rubrum* (red maple), *Carya* species (hickories), *Populus* species (poplars), *Robinia pseudoacacia* (black locust), and *Tilia* species (basswood, linden) become prematurely senescent (usually yellow) and drop. Severely stressed deciduous species may drop all their leaves. This commonly occurs where trees have colonized pockets of soil on ledges or in any soil after severe mechanical damage to roots. Similarly, the oldest age class of needles on coniferous trees may turn yellow and drop prematurely. Dieback of twigs and then limbs may follow foliar symptoms. Drought damage often appears on trees in groups because they share common soil conditions and because root grafts may form among trees of the same species, as in various species of *Quercus* and *Pinus*. These factors promote a similar level of water stress in most members of a group, but small individuals usually sustain the greatest injury.

Severe water deficit in *Pinus* during summer causes needles to lose turgor and bend or droop at a point near the needle base where lignification is incomplete. Needles then either fade and turn brown or remain green and permanently bent. Drooping and bending of water-stressed needles have been noted in *P. monticola* (western white pine), *P. ponderosa* (ponderosa pine), *P. resinosa* (red pine),

and *P. strobus* (eastern white pine). Similar symptoms can be caused by frost (Plate 245D).

Intense short-term water stress may develop in immature leaves during bright, hot, windy weather in late spring or early summer, even when soil moisture is abundant. Young leaves of *A. saccharum* (sugar maple), for example, may suddenly die or develop interveinal patches of dead light brown tissue. Similarly, dramatic marginal browning of leaves of *Quercus kelloggii* (California black oak) and to a lesser extent *Q. chrysolepis* (canyon live oak) and *Q. wislizeni* (interior live oak) has been noted in early summer in northern California. Drooping and death of young *Abies concolor* (white fir) and *P. menziesii* shoots have been observed during hot weather in late May and early June. Such injuries were at one time attributed solely to heat, but the relative roles of heat and drying in these disorders are unknown.

Many instances of drought-related dieback of native trees across large areas have been reported. Appalachian and southern *Quercus* species (oaks) incurred damage intermittently throughout the 20th century. Species commonly affected included *Q. coccinea* (scarlet oak), *Q. falcata* (southern red oak), *Q. laurifolia* (laurel oak), *Q. nigra* (water oak), *Q. phellos* (willow oak), *Q. rubra* (northern red oak), and *Q. velutina* (black oak). The trees sustained damage not only from water stress, but also from opportunistic organisms, notably the fungus *Biscogniauxia atropunctata* (Plate 99) and the buprestid beetle *Agrilus bilineatus* (two-lined chestnut borer), both of which colonize weakened and dying oaks. Drought-associated oak decline also became prominent in parts of Europe during the late 20th century.

Western conifers, notably *Abies grandis* (grand fir), *Calocedrus decurrens* (incense cedar), *Larix occidentalis* (western larch), *Pinus contorta* (lodgepole pine), *P. monticola* (western white pine), and *Pseudotsuga menziesii* (Douglas-fir), undergo foliar browning and dieback in years of severe heat and dryness. Drought-stressed *Picea glauca* (white spruce) may become chlorotic. Pole blight of *Pinus monticola* (Plate 230), which was important in the Inland Empire region in 1930s–1950s, was considered to be primarily a syndrome of water stress.

The risk of damage or death by drying is greatest for seedlings and recently transplanted plants. Seedlings lack a significant internal reservoir such as that represented by sapwood in larger plants. Trees

A–E. Injuries caused by water shortage in summer. A. Dieback and mortality of *Juniperus ashei* (Ashe juniper) and *Quercus* sp. (oak) a year after an intense heat wave and drought. Plants at the base of the slope where drying was least severe are green (OK, Jul). B. Interveinal chlorosis and necrotic spots on leaves of *Liriodendron tulipifera* (tuliptree). This disorder develops during hot, dry weather after midsummer. Affected leaves drop prematurely (NY, Jul). C. Drought-induced leaf scorch on *Ixora coccinea* (FL, Mar). D. Drought-induced leaf scorch on *Aesculus hippocastanum* (horse-chestnut) (NY, Aug). E. Scorch of *Acer platanoides* (Norway maple) leaves (NY, Aug).

F–J. Injuries associated with freezing and drying. F. Dieback of *Prunus lusitanica* (Portugal-laurel) exposed during winter to warm air discharged through a ground-level ventilator from an adjacent building. Defoliation and dieback followed freezing and desiccation (WA, Jun). G. Freeze damage to leaves of *Mahonia aquifolium* (Oregon-grape). Injured leaves died back from tips and margins. Tissues with sublethal injury produced purple pigment (NY, Apr). H. Winter injury to a recently transplanted *Platanus* ×*acerifolia* (London planetree). The root-pruned tree was planted in autumn in sandy soil where deep frost occurred in absence of snow cover during winter. Dieback of twigs and branches was followed by growth of adventitious sprouts (NY, Aug). I. Yellowing and reddish pigment in winter-injured leaves of *Buxus sempervirens* (common boxwood) (NY, May). J. Twig dieback and leaf reddening in *Cotoneaster divaricatus* (spreading cotoneaster) injured by freezing the previous winter. Some buds failed to open; other buds produced normal-appearing shoots that came under water stress and changed color during summer when injured sapwood failed to conduct sufficient water (NY, Aug).

and shrubs lose many absorbing roots during transplanting, so the creation of an abnormal water deficit is unavoidable. If the root ball contains a highly porous rooting medium, it is likely to dry out even though the surrounding soil may contain water sufficient for plant growth. The hazard of desiccation continues until roots grow beyond the ball.

Drying of leaves and twigs in winter and early spring is often linked to prior freezing injury. Symptoms include chlorosis and foliar browning (in evergreens) and twig dieback. Severe desiccation can cause radial cracks in sapwood of the trunk. Foliar browning is common in *Picea rubens* (red spruce) and *Tsuga canadensis* (eastern hemlock) in eastern North America and in *P. glauca, P. pungens* (blue spruce), *Pinus contorta, P. ponderosa,* and *Pseudotsuga menziesii* in the West. The symptoms commonly appear after warm, dry, windy weather in late winter or early spring. The terms *parch blight* and *red belt injury,* referring to reddish brown foliage, have been applied to this syndrome. The latter term refers to injury within a narrow altitudinal stratum on mountainsides.

Established trees and shrubs respond to and cope with water deficits in various ways. Stomata in many species close after development of significant internal stress. Chlorophyll formation, photosynthesis, and growth slow or cease. Chlorosis may develop. In trees that produce latex or resin the exudation pressure of the substance diminishes. Green leaves, stems, roots, and fruits shrink. Stem shrinkage may continue for several weeks during drought, and radial cracks may form in the sapwood (Plate 256). The cambium in the lower part of the trunk of a severely stressed tree may fail to produce any new wood in a year of severe drought, or patches of cambium may die, resulting in cankers. Roots in drying soil become less permeable to water, and root tips may die.

Absorption begins as soon as water is added to soil, and shrunken organs expand quickly, but there is a lag time of up to several days before roots become as permeable to water as they were before the onset of the dry period. Stomata that closed during drought may recover normal function slowly or not at all, and the rate of photosynthesis may remain below the prestress level for a time. These behaviors vary with plant species and the severity of water stress.

Woody plants have evolved many adaptations to avoid or cope with water shortage. Constitutive adaptations include reduced leaf size, heavily cutinized leaves, deep rooting, high water-use efficiency (ratio of biomass accumulated to water transpired), high leaf mass per unit area (which enhances water use efficiency), small leaf-cell size (which facilitates a higher osmotic pressure with a given amount of solutes), and high elasticity of leaf tissue and stem xylem. Drought-tolerant species can maintain turgor and avoid embolisms in xylem at lower water potentials than can less tolerant species. For example, *Pinus banksiana* (jack pine), which typically colonizes sandy sites, loses turgor in needles when their water content diminishes to 75–65% of the amount in saturated needles (relative water content), while *Picea mariana* (black spruce), a colonist of wet sites, loses needle turgor at 86–76% relative water content. *Larrea tridentata* (creosote bush) leaves can tolerate relative water content as low as 50%. The drought-tolerant Chinese species *Acer truncatum* (Shantung maple) can maintain turgor at leaf water potentials as low as −4.1 MPa. *Juniperus* (juniper) is less vulnerable than other evergreens are to xylem embolism.

Inducible adaptations include accumulation of osmotically active molecules in leaves, elastic adjustment (cell and tissue shrinkage), stomatal closure in dry air or when leaves are water stressed, accumulation of wax in stomatal antechambers, and shedding of stressed leaves. The ability to maintain turgor through osmotic and/or elastic adjustments permits plants to avoid chloroplast damage and continue gas exchange even at low leaf water potentials. Water-stress tolerance of *Fraxinus excelsior* (European ash), for example, is related to accumulation of malate and mannitol, which are osmotically active compounds. Proline and monosaccharides (fructose, glucose) accumulate in drought-stressed needles of *Pinus taeda.*

Drought tolerance and heat tolerance are linked. Some trees develop enhanced heat tolerance during the onset of water stress. For example, *Cedrus* species (cedar) preconditioned by water

shortage were scarcely affected by exposure to air temperatures above 45°C for 5 hours in experimental conditions.
References: 286, 468, 1094, 1192, 1406, 1465, 1836, 2139, 2141, 2142, 2144, 2145, 2210, 2256, 2584, 2626, 2760, 2926, 2949, 2977, 3331, 3960, 4004, 4091, 4107, 4140, 4262, 4358

Heat Stress
Few claims of heat injury to trees or shrubs, other than by fire, are well supported by evidence from experiments or temperature measurements. Solar heat sufficient to kill succulent plant tissue (approximately 54°C or higher) can occur at the surface of dry soils and can cause the collapse of soft tissues in young coniferous seedlings in nurseries and forest openings. Temperatures up to 75°C at soil level have been related to low survival of *Pinus ponderosa* seedlings in a sparse forest in Idaho. Heat lesions in nonlignified hypocotyls cause them to bend under the weight of the tops, and the tops soon die. Older seedlings remain erect but girdled. An affected stem with lignified hypocotyl cell walls swells above the girdle; the roots cease growing and the top of the seedling turns chlorotic. The plant dies later in the season, in winter, or rarely the next year. Seedlings with adequate water usually avoid damage because water flowing through their stems absorbs sufficient heat energy to maintain stem temperature as much as 30°C lower than that of the soil surface in bright sunlight.

Heat injury to immature leaf tissues not yet actively transpiring has been suggested as the cause of sudden foliar collapse during exceptionally hot weather in late spring or early summer (as in *Quercus* species in California, mentioned on page 482), but there is no evidence to confirm that heat rather than desiccation causes such symptoms.

Injury to tree trunks by solar heat (sunburn) probably occurs in some circumstances, but the details have not been established by

A, B. Sunscald on the southwest side of the trunk and foliar symptoms of decline due to exposure of a mature *Pinus strobus* (eastern white) pine by removal of adjacent trees. Mild injury caused reddish discoloration of smooth bark that was formerly gray or gray-green. Severe injury caused roughening of bark as outer layers of secondary phloem died, shrank, split, and peeled back in ragged scales. Slow growth and sparse off-color foliage are presumed to result partly from injury along the trunk and partly from the changed root environment. A normal *P. strobus* in left background provides contrast (NY, Jun).
C. Heat injury to first-year seedlings of *Tsuga canadensis* (eastern hemlock). The left seedling of each pair has a girdling band of dead tissue at the former location of the soil line. Swelling above the girdle (arrows) resulted from growth stimulation associated with blocked downward translocation of organic compounds. Chlorosis reflects inadequate supply of minerals from starving roots (NY, Aug).
D. Injury to *Pinus resinosa* (red pine) by summer frosts that occurred in successive years. Immature or recently matured tissue near needle bases was most sensitive. Injured tissues collapsed, and needle droop and death resulted. The current season's shoot is stunted and dying; year-old needles are drooping; older needles are missing (NY, Oct).
E. Freeze injury to mature first-year needles of *Pinus thunbergii* (Japanese black pine). Injured tissues near needle bases became bleached and turned brown, but the injury was not lethal and the needle tips remained green. Needles injured in this fashion usually become senescent and drop a year or more prematurely (NY, Jun).
F. Injury to *Abies grandis* (grand fir) by a late spring frost. New shoots were several centimeters long at the time of freezing. Year-old foliage and the basal parts of current shoots survived (WA, Jul).
G. Foliar browning and twig mortality of *Taxus cuspidata* (Japanese yew), caused by freezing in late winter. Foliage deacclimated during mild, sunny weather and then froze when the temperature dropped to a nighttime low that was normal for the season. The symptoms appeared during the next warm period (NY, May).

critical research. Temperature at the surface of tree bark in bright sunlight occasionally rises to 50°C or higher. Living tissue beneath the rhytidome (dead corky layers) normally remains much cooler, however, because cork is an effective insulator and because heat that penetrates to living bank is conducted inward and upward by the transpiration stream. A speculative scenario for injury to thin-barked tree stems by solar radiation is that living bark beneath a thin rhytidome may experience heat stress sufficient to trigger periderm formation if the cooling effect of transpiration is diminished during drought. Periderms isolate heat-stressed superficial tissues, which die of attrition and become coarse unnatural bark scales.

In cool-temperate regions tree trunks subject to intense heat in summer are also subject to freeze damage (winter sunscald) in winter. Living bark and cambium are injured when deacclimated tissue freezes after warming by the afternoon sun. Injury similar to that shown on *Pinus strobus* (eastern white pine) in Plate 245 is common on trees of many species that have smooth, thin bark. The damage occurs on the south to southwest sides of stems (north to northwest in the Southern Hemisphere) that were formerly shaded by their own branches or those of adjacent trees. Bark becomes pinkish or reddish, and the surface slowly roughens and darkens as patches of outer bark die, dry, and crack. Some of this roughening is presumed to be due to killing of the outer secondary phloem by heat or freezing. Much roughening is caused by formation of patches of periderm beneath damaged outer bark (details with Plate 257), which isolates the superficial tissues. Severe damage includes cambial necrosis, resulting in long cankers. Similar injuries occur on upper surfaces of limbs when sheltering branches above them are lost.

Heat tolerance and drought tolerance determine which plant species colonize hot sites. Heat stress depresses photosynthesis. The temperature threshold for photosynthetic depression is higher, and recovery from heat stress occurs quicker, in heat-tolerant than in less tolerant species. Heat tolerance of *Quercus muehlenbergii* (chinkapin oak), for example, has been demonstrated and is suggested to be a determinant of its ability to grow naturally in the prairie region of eastern Kansas where few other tree species persist.

Some aspects of damage to tree trunks by fire should be mentioned. The inner bark and sapwood of living trees are incombustible because of their water content. A fire may char the surface bark and kill underlying tissues, or heat insufficient to char the surface may kill bark and cambium of a thin-barked tree, but the first such fire usually does not burn into a living trunk. Damage becomes apparent after a year or more as dead bark cracks and begins to expose the wood. Trunk expansion and woundwood formation at the edges of the damaged area eventually cause the dead bark to loosen and fall. A fire occurring after the formation of a zone of dead, dry bark and wood may burn into the trunk, especially in resinous wood of conifers (Plate 254).

References: 143, 872, 1465, 1532, 1593, 1793, 2119, 2204, 2626, 4133, 4262

Frost and Freeze Damage

Frost during the growing season. Most trees and shrubs can withstand air temperatures of −1° to −2°C without injury, even during the time of shoot growth, because solutes in water within the tissues depress the freezing point somewhat below that of pure water. At lower temperatures water either becomes supercooled or freezes. Pure water can supercool to −38°C, but most water in plants freezes at much higher temperatures because plants contain substances that act as nuclei for formation of ice crystals. When crystals begin to form, ice spreads rapidly. Intercellular ice (between cells) causes relatively little harm, but ice within cells causes lethal disruption of membranes and organelles.

Leaves and stems killed by spring frost are usually small and succulent at the time of injury. After thawing, they at first appear water-soaked, then shrivel and turn reddish brown to dark brown or nearly black, depending on the species. Dead leaves or shoots break off or abscise during the ensuing several weeks. New shoots begin to grow from dormant or adventitious buds almost immediately and soon mask the early-season damage. By midsummer only a trained observer is likely to notice the stubs of frost-killed spring shoots.

The likelihood that a particular plant will be damaged by spring frost is related to the date when it begins growth. Conifers that break bud earliest are most likely to sustain damage, which may involve browning of year-old needles or collapse of new shoots. *Larix decidua* (European larch), *L. kaempferi* (Japanese larch), and *Pseudotsuga menziesii* (Douglas-fir) planted in the Appalachian region begin growth earlier than other conifers planted there and are more frequently damaged by spring frost. On the other hand, deciduous trees that begin growth earliest are the least frequently damaged. More than simple escape is involved; species such as *Acer saccharum* (sugar maple) and *Prunus serotina* (black cherry) that begin growth early may have a large amount of relatively succulent tissue exposed, yet escape damage by a frost that kills all new growth of plants that begin growth later, such as *Fraxinus americana* (white ash) and *Robinia pseudoacacia* (black locust). Of 29 deciduous species studied in West Virginia, only *Fagus grandifolia* (American beech) and *Magnolia fraseri* (Fraser magnolia) ranked high in susceptibility to damage by spring frost while still being among the earliest species to begin spring growth.

Damage by summer frost occurs in the boreal region and some high-elevation areas. *Pinus* (pines) in plantations on flat land and in topographic depressions (frost pockets) in the boreal forest are most often affected. Trees that become established naturally in locations prone to summer frost are survivors from seedling populations that were thinned by frost injury as well as other agents. Injury occurs during clear nights when, due to radiational cooling, a layer of cold

---

A, B. Winter freeze injury to needles of *Pinus strobus* (eastern white pine). Needles nearest the apical buds become chlorotic or mottled yellow to yellow-green in late winter or early spring and drop at the beginning of the growing season. Buds are usually not injured, and new shoots develop normally. Injury of this type commonly follows warm weather in late winter or early spring during which the youngest needles lose cold hardiness. A sudden return to freezing temperatures normal for the season causes the injury (VT, May).

C, D. Winter freeze injury to needles and twigs of *Picea pungens* (blue spruce). This type of injury develops in late winter or early spring when freezing follows warm weather. C. Symptoms of mild injury resembling those of a needle cast disease. Scattered brown lesions appear on green needles, or entire needles turn brown while others remain green. Xylem at the base of the apical bud may turn light brown in contrast to the nearly white pith and green cortex, but the bud remains alive. Injured needles later become masked by growth of new shoots and drop prematurely (NY, Mar). D. If injury is more severe, most needles on a twig turn brown, and xylem in the twig, including xylem traces leading into needles, becomes discolored (NY, May).

E. Freeze damage to *Picea rubens* (red spruce). The circumstances of this injury were similar to those described above: normal winter freezing following winter thaw. The base of the tree was protected by snow and sustained little injury (VT, Jun).

F. Winter freeze damage to *Pinus thunbergii* (Japanese black pine) following untimely pruning. The tree from which this twig came was pruned in August, which promoted growth of new shoots that did not acclimate to cold weather before winter. Foliage and stems produced during the previous spring remained unharmed (NY, Jun).

G–I. Damage to *Pseudotsuga menziesii* (Douglas-fir) by freezing in early spring. G. Top killing. H. Deformity of the trunk of a small tree, caused by freeze injury in previous years. The scar of an irregular canker, now covered by woundwood, is visible at center, and a lens-shaped split in the bark is present near the base of the stem portion shown. I. Magnified cross-section of a young stem at the base of a portion killed by freezing. The injury occurred after 2 years of normal growth. In the upper part of the photo, phloem, cambium, and sapwood were all killed. A prominent frost ring extends from the necrotic zone around the stem beneath living bark and cambium. Sapwood produced after the injury appears as a narrow light-colored band between the bark and the frost ring (WA, Jun).

air forms at ground level and accumulates to a meter or more in depth, killing needles and succulent stem tissue. Needle tissue near the basal sheath collapses, and the needles droop and may later turn brown. If frost kills entire shoots, drooping needles may remain in place for a year or more, slowly weathering to gray. The plant resumes growth the next year from adventitious buds and develops bushy form. If the stem of a shoot survives, the injured needles abscise, leaving a bare twig. Shoots of the next season may grow from surviving terminal buds or from adventitious buds near the stem apex, but this growth is likely to be slight, producing tufted foliage. In unusual cases light frost during shoot elongation may damage stems enough to cause bending or drooping but not death, and the stems remain permanently deformed. Injury does not occur every year, even on frost-prone sites, so some trees planted there eventually grow tall enough to escape top killing.

Frost injury to many plants, both woody and herbaceous, is aggravated by certain bacteria that act as nuclei for the formation of ice crystals. In the absence of these bacteria nonacclimated plants can supercool several degrees and may thus escape damage from light frosts. Several species of plant-inhabiting bacteria have ice-nucleating ability and contribute to frost damage at temperatures above −5°C. The pathogen *Pseudomonas syringae* is the most common; others are *P. viridiflava*, *P. fluorescens*, *Pantoea agglomerans*, *P. ananatis,* and *Xanthomonas translucens.* Ice nucleation is the basis of the reported interaction of *P. syringae* with freeze damage in bacterial canker of stone-fruit trees (Plate 183) and frost-induced shoot dieback of deciduous nursery stock. Large populations of this bacterium may also be present on plant surfaces in the absence of disease. Frost hardiness is enhanced if ice-nucleating bacteria are suppressed by bactericides or by competition from other bacteria that lack ice nucleation properties.

Frost rings in wood. Freezing injury to sapwood or cambium during growth leads to aberrations in wood that is forming at the time and in several tiers of wood cells formed subsequently. Injury sustained between growing seasons leads to aberrant wood formation in the early part of the next growing season. The aberrant tissue extends as a sheet partly or completely around the stem, appearing as a ring (frost ring) in cross section (Plate 246I). Frost rings have different structures, depending on plant species and the severity and timing of injury. Freezing during the growing season may kill differentiating xylem elements, leaving a sheet of dead undifferentiated cells overlying crumpled, partly lignified xylem elements. The vascular cambium usually survives and produces an abnormally large amount of xylem parenchyma and abnormally shaped tracheary elements. Gum or resin and dark pigment may be deposited. Injury during cambial inactivity causes frost rings that lack remnants of undifferentiated or partly differentiated xylem. Slight freezing injury to the cambium and/or differentiating xylem commonly causes frost rings in the absence of external symptoms. More severe injury, usually accompanied by foliar symptoms or dieback, leads to formation of small pockets of parenchyma (which may eventually turn brown) along the frost ring or to tangential separation of wood rings. Frost rings may also form in phloem, but they disappear because of rearrangement of cell positions as stems grow.

Injury by freezing during dormancy. Woody plants in temperate zones undergo seasonal changes in ability to tolerate low temperature. In autumn, as temperature and daylength diminish, they become acclimated to withstand frost. Acclimation is typically complete before freezing weather begins. The degree of cold acclimation varies during winter, more or less in relation to ambient temperature. The hardiest trees or shrubs, fully acclimated, can withstand temperatures lower than ever occur where they grow. All freezable water is extracted from their living cells and crystalizes intercellularly. Only chemically bound (unfreezable) water remains within living cells. These plants include *Betula* species (birches), *Cornus sericea* (red-osier dogwood), *Populus tremuloides* (trembling aspen), and *Salix* species (willows). The majority of woody plants of the temperate zones become acclimated to withstand temperatures between −20° and −40°C. They avoid lethal freezing by passive export of intracellular water to intercellular space and supercooling of the remaining intracellular water. Intercellular ice has a lower

water potential than liquid water and therefore draws water out of cells. Tissues in trees as diverse as *Carya* (hickory) and *Malus* (apple) may then supercool as low as −40°C without harm. Apparently the process of cold acclimation involves reduction or elimination of ice-nucleating centers within cells and/or development of barriers to ice crystal formation. As temperatures rise in late winter and early spring, plants deacclimate until, by the time growth begins, they can no longer tolerate more than a few degrees of frost.

Sapwood, cambium, and phloem acclimate to different degrees. During midwinter the sapwood is the least tolerant of cold; that is, intracellular freezing of sapwood parenchyma occurs at a higher temperature than does freezing of cambial or phloem cells. During autumn and early spring sapwood is relatively more tolerant than cambium and phloem. Unusually warm weather in autumn can delay cold acclimation, and warm weather in winter or early spring can induce partial deacclimation. Most damage to woody plants by freezing in late autumn or winter follows delayed acclimation or partial deacclimation. The damage is usually caused by temperatures that are normal for the season and is observed mostly in planted trees and shrubs growing in areas where winters are colder than those in the plants' natural ranges. Deep midwinter cold waves cause damage infrequently unless preceded by unseasonable warmth.

The most common external symptoms caused by winter freezing are dieback, foliar browning, sunscald, and bark splitting near the base of the trunk. Dieback of twigs and branches and foliar browning in evergreens commonly follow injury that occurs when

---

A, F, G. Freeze damage to *Prunus serrulata* 'Kwanzan' (flowering cherry). A. Stunted, wilting, and dead foliage on a tree that sustained trunk damage of the sort shown in panels F and G. Water stress caused by death of sapwood resulted in wilting after leaf expansion on some branches and failure of leaves to attain full size on other branches (NY, Jun). F, G. Trunks of young trees damaged by freezing in late winter. The stems have lens-shaped cracks in the bark near the ground line where callus and woundwood formed at the edges of killed areas and caused the rigid outer bark to split. Long splits developed in bark above the cracks, and the edges flared back. Flaring bark results from mechanical stress imposed by drying of killed tissues, coupled with growth of tissues in surviving parts of the phloem (WA, Jun). The dying stem in G has discolored dead sapwood, and the flaring bark has a reddish brown inner surface where a periderm formed after the bark separated from the wood (NY, Jul).

B, D. Freeze damage to *Prunus persica* (peach). B. An injured twig has dead buds, stunted growth from one bud that survived, and freeze-induced reddish discoloration in sapwood (NY, Jun). D. Enlarged view of a cross-section of a 2-year-old stem that sustained freeze damage after its first year of growth. Wood that froze is dead, discolored brown, and enclosed within a dark brown sheet (frost ring, also a barrier zone, as explained with Plates 254 and 255) that was produced by the cambium after it sustained sublethal injury (NY, May).

C, E. Winter freeze damage to *Ilex aquifolium* (English holly). C. Most leaves and many buds present during winter were killed, and most leaves had dropped when the photograph was made. Scattered new shoots, stunted and containing red pigment, had grown from surviving buds. The tree died the following summer of water stress resulting from damage to sapwood. E. Cross section of an injured branch about 1 cm in diameter, with xylem discolored pinkish while the cambial zone and bark appear normal. The discolored xylem is dying, its water-conducting capacity already impaired (WA, Jun).

H, I. Internal symptoms of previous freeze injury to *Euonymus fortunei* (wintercreeper). H. The boundary between living 2-year-old wood and older wood killed by freezing is marked by a dark brown line. The injured xylem temporarily conducted enough water to allow leaf functions, which in turn allowed formation of new sapwood outside the damaged tissues. I. Tissues have been separated at the brown line to show in tangential view that it is a sheet of dead cells (NH, Aug).

499

winter temperatures arrive suddenly after warm autumn weather. Severe twig dieback (and foliar browning in conifers) caused by freezing may also follow late summer pruning, which retards cold acclimation.

Sunscald develops on the south and southwest sides of trunks and on the upper surfaces of limbs exposed to sun. The cambial temperature of the sun-warmed bark may exceed 20°C in late winter when the ambient temperature barely exceeds 0°C. This heating causes deacclimation and can be followed by lethal freezing at night. Damaged bark and cambium dry out, crack, separate from the wood, and eventually fall away. Small trees with thin bark are most susceptible to this type of injury. Sunscald has been noted, for example, on trunks or limbs of *Acer* species (maples), *Betula alleghaniensis, Catalpa, Fagus* species (beech), *Pinus strobus* (eastern white pine), *Prunus serrulata* (flowering cherry), *Pyrus calleryana* (callery pear), *Quercus rubra* (red oak), *Salix babylonica* (weeping willow), *Tilia cordata* (littleleaf linden), and many kinds of fruit trees. In a variation of this type of injury the bark remains alive after the cambium dies. A pocket then forms between living bark and wood (and can be detected by tapping) as the bark is lifted by growth of woundwood at the edges of the injured area. Usually the covering of this pocket splits as shown in Plates 247 and 248 (further illustration and discussion with Plate 256), but sometimes, as in "blister shake" of *Liriodendron tulipifera* (tuliptree), the bark remains intact over a pocket between two age classes of wood. Less severe cambial injuries result eventually in weakness between annual layers of wood, which causes them to separate under mechanical stress, as when drying. The defect is called ring shake.

Freezing that kills the sapwood or cambium leads to cankers, dieback, and often wilting and death during the next growing season because damaged sapwood conducts insufficient water. A frost ring forms in the wood of surviving parts of a damaged plant. Freeze-killed sapwood in surviving stems becomes compartmentalized (Plates 254, 255) and usually turns dark. Sapwood exposed by bark splitting or branch dieback is soon invaded by decay fungi. In trees prone to wetwood—*Ulmus* (elm) and *Platanus* (sycamore), for example—this condition becomes more widespread within a tree after damage by freezing.

Root freezing. Freeze damage to roots can entrain dieback and foliar symptoms of distress. Roots become less acclimated to cold than do parts above ground, and may freeze if not protected by snow, ground cover, or mulch. Thaw-freeze and root-freeze events in winter and early spring are believed to have been key factors triggering and synchronizing episodes of hardwood dieback in eastern Canada and northeastern USA during the 20th century. Species affected included *Acer saccharum* (sugar maple), *Betula alleghaniensis* (yellow birch), *B. papyrifera* (paper birch), and *Fraxinus americana* (white ash). The importance of snow cover for tree health in northern forests was demonstrated in Quebec by installing screens above the root zones of mature *A. saccharum* trees so that rain or melt water but not snow could reach the root zone. Roots froze, and severe top dieback developed within the first year. *References:* 130, 143, 286, 301, 346, 465, 572, 608, 621, 642, 695, 821, 828, 872, 910, 918, 927, 1088, 1192, 1205, 1304, 1429, 1520, 1616, 1688, 1761, 1793, 1870, 1946, 2058, 2239, 2272, 2291, 2329, 2331, 2387, 2442, 2587, 2684, 2962, 2988, 3011, 3120, 3189, 3211, 3251, 3274, 3303, 3331, 3382, 3575, 3845, 3874, 3941, 4005, 4050, 4093, 4094, 4116, 4134, 4178, 4263, 4282, 4500, 4501, 4504, 4522, 4523, 4532

Predisposition to Attack by Opportunistic Pathogens

Drought or freezing insufficient to cause evident stress in plants can nevertheless impair their defense against fungi that cause cankers, dieback, sapwood decay, or root rot. Diplodia dieback of *Pinus* (Plate 64) increases during drought years for this reason. Cankers caused by *Valsa sordida* (Plate 85), *Fusarium lateritium* (Plate 92), or *Cryptodiaporthe salicella* (Plate 77) in *Populus balsamifera* subsp. *trichocarpa* (black cottonwood) or *Salix* species were found to develop in dormant bark with water content less than about 80% of that at saturation but not in bark with greater water content. Cankers were caused by *Botryosphaeria dothidea* (Plate 59) in young *Sorbus aucuparia* (European mountain ash), *Betula*

pendula (European white birch), and *Cornus sericea* (red-osier dogwood) that were inoculated and then allowed to dry to internal water potentials below −1.2 MPa, but not so low as to cause wilting. Canker expansion ceased a few days after plants were rewatered, and well-watered plants were resistant throughout the experiments. *Picea abies* (Norway spruce) roots inoculated with *Heterobasidion annosum* (Plates 165, 166) resisted infection when their xylem water potential was above −0.5 MPa, but not when it was in the range −0.5 to −1.0 MPa.

When partially cold-acclimated plants of *B. pendula, C. sericea, Liquidambar styraciflua* (sweetgum), and *S. aucuparia* were inoculated with *Botryosphaeria dothidea,* cooled to between −20° and −30°C (which caused no measurable injury), and then slowly thawed, lesions formed within 2 weeks; but plants cooled to temperatures above −20°C before thawing remained resistant. Susceptibility of *Euonymus alatus* (winged euonymus) and *Frangula alnus* (alder buckthorn) to *Nectria cinnabarina* (Plate 87) and *Tubercularia ulmea* (Plate 88) was conditioned similarly by freezing.

Additional diseases for which predisposition is important include other Cytospora and Valsa cankers (Plates 82–86), dieback of *Platanus occidentalis* (sycamore) and *Quercus* species caused by *Botryosphaeria rhodina* (Plate 63), sooty canker of *Juglans* (walnut) and fruit trees (Plate 75), Seiridium canker of *Cupressus macrocarpa* (Monterey cypress; Plate 95), Thyronectria canker of *Gleditsia triacanthos* (honeylocust; Plate 89), and canker-rots of various trees caused by *Cerrena unicolor, Trametes versicolor,* and *Schizophyllum commune* (Plates 154, 158).

Many fungal pathogens of bark and sapwood enter stems and reside quiescently as endophytes until the hosts undergo water stress or senescence. Then the pathogens begin aggressive colonization. This behavior has been demonstrated for *Cryptosphaeria populina* (Plate 96), *Entoleuca mammata* (Plate 98), and *Valsa sordida* (Plate 85) in *Populus tremuloides* (trembling aspen), and for *Biscogniauxia atropunctata* in *Quercus* species (oaks; Plate 99). Secondary insects often join opportunistic fungi in the attack on stressed trees. Drought-stressed conifers, for example, may be attacked by bark beetles and bluestain fungi. The buprestid beetle *Agrilus bilineatus* (two-lined chestnut borer) and *Biscogniauxia* species collaborate in killing drought-stressed oaks in southern USA.

References: 157, 252, 363, 408, 468, 690, 828, 844, 910, 1953, 2323, 3478–3482, 4288

A. Twig dieback and cankers on *Morus rubra* (red mulberry), believed caused by freeze damage followed by opportunistic fungi (NY, Jun).

B. Frost injury to *Liriodendron tulipifera* (tuliptree). A light frost when leaves were nearly full grown killed most leaves on the limb. Petioles were not killed except adjacent to the leaf blades, so the dead leaf blades droop at the tips of green petioles (NC, May).

C–E. Winter freeze damage to *Syringa vulgaris* (common lilac). C. Wilting, foliar browning, and dieback in young landscape plants. These symptoms developed in early summer because damaged sapwood could not conduct enough water to new shoots. D. Curled chlorotic leaves on a plant stressed by freeze damage to sapwood at the stem base. E. Symptoms at a stem base. Lens-shaped splits in bark are evident externally. Removal of bark and outermost sapwood reveals a frost canker and numerous small lesions resulting from localized cambial necrosis (NY, Jul).

F. Tattered appearance of an *Acer platanoides* 'Crimson King' (Norway maple) leaf in which interveinal tissues failed to develop. Injury of this type occurs while leaves are still enclosed in buds and is attributed to freeze damage, although proof of cause is lacking (NY, Jun).

G–I. Frost cankers on pole-size *Fraxinus americana* (white ash). The trees lacked normal cold hardiness because of infection with the ash yellows phytoplasma (Plate 198), which induced the witches'-brooms shown in Plate 248G and H. G. Bark has split as the result of pressure exerted by woundwood ridges at edges of an area where the cambium was killed. H. Removal of loose bark reveals extensive dead sapwood and a woundwood ridge at its edge. I. An old frost canker from which the bark has long since fallen (NY, May).

501

Damage by Flooding or Waterlogged Soil (Plate 249)

Roots in flooded or waterlogged soils often die of anoxia (oxygen deficiency). Such damage occurs not only to plants on obviously wet sites but also to those in planting holes along city streets and in landscapes where soil drainage is impeded by high clay content. Most trees and shrubs cannot grow for long in waterlogged soil, and some die if flooded for only a few days during the growing season. Plant roots and soil organisms quickly use up the gaseous oxygen in waterlogged soil. Carbon dioxide increases; redox potential diminishes; iron and manganese are reduced to soluble, toxic forms; and sulphides, ethylene, fatty acids, phenolic compounds, and other toxic organic products accumulate. Anoxia leads to hydrolysis of cyanogenic glycosides and release of cyanide in flooded roots of *Prunus* and other plants that produce cyanogenic compounds.

Oxygen deficiency in roots causes a switch from aerobic to anaerobic respiration, which is much less efficient in utilization of organic compounds as energy sources and results in accumulation of toxic end products such as ethanol. Roots soon lose some of their permeability, retarding uptake of water and minerals. After a few days internal water shortage, accumulation of amino acids in leaf sap, stomatal closure, and depressed photosynthesis and translocation of organic compounds occur. Stomata either remain closed or reopen only after root absorption resumes. If the plant remains alive, diminished root function soon causes foliar nitrogen deficiency. Synthesis and translocation of growth regulators (gibberellins and cytokinins) in roots slow, and concentrations of auxin and ethylene in stems increase. Mycorrhizal fungi, which associate with plant roots symbiotically (Plates 259, 260), are also adversely affected, further suppressing plant uptake of mineral nutrients, especially phosphorus. Internal water deficit in some plants increases until they die, but many kinds of plants regain normal hydration while their stomata remain closed during flooding. Stomata of some tolerant plants reopen as the plants adapt to flooding.

External symptoms associated with root damage in waterlogged soil include various of the following: epinasty (downward bending of leaf petioles), stem swelling (particularly in small plants), chlorosis, edema (eruptive parenchymatous masses that resemble lenticels) on leaves, red or purple pigment in leaves, browning of leaf margins, reduction or cessation of growth (particularly in roots), twig dieback, absence or diminished yield of fruit, death of roots, wilting, leaf drop, and death of an entire plant. Seedlings develop symptoms more quickly than do older plants. Plants with roots injured by waterlogged soil may die or experience drought stress when, after the soil drains, the root system is unable to meet transpirational demands of the top. This situation has been noted in plantations of *Pinus resinosa* (red pine) and *P. taeda* (loblolly pine) on poorly drained sites, and it occurs in trees planted in holes in concrete landscapes.

Plants stressed or injured by waterlogging also become abnormally susceptible to certain soilborne pathogens. *Phytophthora* species (Plates 176–182), for example, cause root rot most often in soils that are periodically waterlogged. Excess water not only promotes susceptibility of roots but also acts as a dispersal medium for zoospores of these oomycetes. Root damage to conifers in periodically waterlogged soils seems to predispose them to root rot by *Phaeolus schweinitzii* (Plate 169).

Dormant woody plants can tolerate flooding or waterlogging for several weeks without lasting harm. This tolerance is usually associated with low oxygen demands of roots and organisms in cold soil, but dormant roots also display tolerance at moderate temperatures (e.g., 20°C).

Growing plants that tolerate flooding or waterlogged soil do so by means of adaptive changes in form or function. Lenticels on submerged roots and stems become enlarged; edematous eruptions may develop on the lower sides of leaves; adventitious roots may grow from the lower stem or the upper main roots at the level of the water table; and intercellular spaces may form in the root cortex. These changes facilitate gas exchange. Edema develops in such plants as *Camellia*, *Hibiscus*, *Ligustrum* (privet), *Schefflera* (schefflera, octopus tree), and *Taxus* (yew) when soil is waterlogged and

transpiration impaired. Root tissue with intercellular air spaces is normally present in some plants, such as *Nyssa biflora* (swamp tupelo) and *Salix* species (willow), that inhabit wet sites, and is formed by other plants, such as *Pinus contorta* (lodgepole pine) and *Populus* species (poplar), in response to flooding. When induced by flooding, air spaces are created by the dissolution or collapse of certain cells and the swelling of others. Functional adaptations include increased anaerobic metabolism, avoidance of ethanol accumulation, and tolerance of high carbon dioxide concentration.

Most angiosperms tolerate flooding better than most gymnosperms do. The following classification of trees and shrubs according to tolerance of flooded or waterlogged soil has been compiled from several sources. Where available, the ratings are for established plants rather than for seedlings.

Relative tolerance of plants to flooded or waterlogged soil

Tolerant: *Acer rubrum* (some ecotypes), *Alnus cordata*, *A. japonica*, *Carya aquatica*, *Casuarina equisetifolia*, *Cephalanthus occidentalis*, *Cornus sericea*, *Eucalyptus camaldulensis*, *Forestiera acuminata*, *Fraxinus mandshurica*, *F. nigra*, *F. pennsylvanica*, *F. profunda*, *Gleditsia aquatica*, *Ilex decidua*, *Larix laricina*, *Ligustrum sinense*, *Nyssa aquatica*, *N. biflora*, *Quercus lyrata*, *Rhizophora* species, *Salix interior*, *S. nigra*, *S. viminalis*, and *Taxodium distichum*.

Intermediate: *Abies balsamea*, *A. grandis*, *Acer negundo*, *A. saccharinum*, *Alnus incana*, *Berberis thunbergii*, *Betula nigra*, *Celtis occidentalis*, *Chamaecyparis thyoides*, *Cornus mas*, *C. racemosa*, *Diospyros virginiana*, *Fraxinus americana*, *Gleditsia triacanthos*, *Juniperus chinensis* 'Pfitzeriana,' *Larix kaempferi*, *Ligustrum obtusifolium*, *Liquidambar styraciflua*, *Maclura pomifera*, *Magnolia virginiana*, *Malus* 'Dolgo', *Persea borbonia*, *Picea mariana*, *Pyrus betulaefolia*, *P. calleryana*, *P. communis*, *Pinus contorta*, *P. elliottii*, *P. serotina*, *P. taeda*, *Platanus occidentalis*, *Populus balsamifera*, *P. deltoides*, *P. tremuloides*, *Quercus macrocarpa*, *Q. michauxii*, *Q. nigra*, *Q. palustris*, *Q. petraea*, *Q. phellos*, *Q. robur*, *Q. texana*, *Salix alba*, *S. discolor*, *Thuja occidentalis*, *Triadica sebifera*, *Ulmus americana*, *U. crassifolia*, and *Viburnum* species.

Intolerant: *Acer platanoides*, *A. saccharum*, *Aesculus flava*, *Asimina triloba*, *Betula papyrifera*, *B. pendula*, *B. populifolia*, *Carpinus caroliniana*, *Carya glabra*, *C. illinoinensis*, *C. ovata*, *C. tomentosa*, *Celastrus orbiculatus*, *Celtis laevigata*, *Cercis canadensis*, *Cladrastis lutea*, *Cornus florida*, *Crataegus* ×*lavallei*, *C. phaenopyrum*, *Euonymus fortunei*, *Fagus*, *Forsythia*, *Ilex opaca*, *Juglans nigra*, *J. regia*, *Juniperus virginiana*, *Ligustrum amurense*, *L. vulgare*, *Liriodendron tulipifera*, *Lonicera morrowii*, *L. tatarica*, *Magnolia grandiflora*, *M.* ×*soulangiana*, *Malus* species, *Melia azedarach*, *Morus rubra*, *Nyssa sylvatica*, *Ostrya virginiana*, *Oxydendrum arboreum*, *Picea abies*, *P. glauca*, *P. pungens*, *P. sitchensis*, *Pinus banksiana*, *P. caribaea*, *P. echinata*, *P. resinosa*, *P. strobus*, *P. virginiana*, *Philadelphus coronarius*, *Physocarpus opulifolius*, *Prunus persica*, *P. serotina*, *P. subhirtella*, *Quercus alba*, *Q. falcata*, *Q. laurifolia*, *Q. marilandica*, *Q. muehlenbergii*, *Q. pagoda*, *Q. rubra*, *Q. shumardii*, *Q. stellata*, *Q. virginiana*, *Robinia pseudoacacia*, *Sassafras*, *Sorbus aucuparia*, *Taxus cuspidata*, *Taxus* hybrids, *Tilia americana*, *Tsuga canadensis*, *Ulmus alata*, *U. pumila*, and *U. rubra*.

References: 83, 90, 120, 186, 241, 266, 341, 521, 778, 798, 819, 997, 1145, 1153, 1202, 1236, 1284, 1742, 1882, 2024, 2025, 2140, 2143–2145, 2209, 2406, 3015, 3062, 3464, 3486, 3881, 4063, 4084, 4324, 4461, 4467

A. *Acer saccharum* (sugar maple) killed by flooding. Water covered the root zone for several days in summer. Vegetation in the background at slightly higher elevation escaped damage (NY, Jul).

B, C. Damage caused by waterlogged soil to a hedge of *Taxus cuspidata* (Japanese yew). B. Chlorosis, dieback, and stunted shoots. C. Dead decaying roots on a declining plant. Living fibrous rootlets were found only near the soil surface (NY, May).

D, E. Edema on lower surfaces of leaves of *T. cuspidata*. Upper surfaces of leaves with edema are normal in form but often are chlorotic (NY, Sep).

F, G. Edema on the lower surfaces of leaves of *Camellia sasanqua* (sasanqua camellia) (FL, Apr).

H. Root necrosis and butt swelling of *Abies* ×*shastensis* (Shasta red fir), caused by waterlogged soil. When the tree's foliage began to fade, all roots and internal tissue below the center of the basal swelling were dead (WA, Jun).

503

Damage by Girdling Roots, Hail, Ice Glaze, and Sheet Ice (Plate 250)

Girdling Roots

Roots that constrict part or all of the butt of a tree or its primary root axis are called girdling roots. The constriction is usually self-inflicted and is confined mainly to planted trees grown originally in nurseries. Nursery practices such as cultivating close to the stem and growing trees in containers cause some secondary and tertiary roots to grow tangentially with respect to a tree's axis. Planting trees too deep contributes to the problem by encouraging root growth above the level of the initial buttress roots. Constriction begins years later, when the butt has enlarged enough to contact a tangential root. Multiple girdling roots may encircle a butt. Trunk diameter is usually 16–20 cm or more when constriction begins to cause symptoms in branches and foliage. One or more branches then begin to decline. The damage is irreversible. Symptoms usually result from the action of multiple girdling roots rather than just one. Trees that grow from seed in nature are rarely self-girdled.

Damage by girdling roots is most common in *Acer* (maples), especially *A. platanoides* (Norway maple), *A. rubrum* (red maple), and *A. saccharum* (sugar maple). Root-girdled trees in other genera seem to tolerate the constriction better than maples do. *Fraxinus pennsylvanica* (green ash), *Gleditsia triacanthos* (honeylocust), and *Tilia cordata* (littleleaf linden) often have girdling roots. A tree with such roots may become mechanically unstable and either break at the level of constriction below the soil line or tip over during a storm. We have observed trees as diverse as *Pinus nigra* (Austrian pine) and *Ulmus americana* (American elm) blown over by wind, exposing roots coiled around greatly constricted butts. The constricted zone and coiled roots together resembled the shapes of the containers in which the trees grew when young. Similar damage has been observed in *Pinus* species that as seedlings were carelessly planted or had their roots deliberately wound in a coil. Breakage at the level of constriction is common and is partly the result of wood decay, presumably because the butt and roots below a constriction, deprived of the normal supply of organic nutrients from above, lose the ability to defend against decay fungi.

Symptoms caused by girdling roots include slow growth, subnormal leaf size, premature autumn color, branch dieback, and abnormal butt shape. Some branches may die while others appear normal for many years, because constriction usually affects only one sector of the butt or becomes severe on one side before affecting the remainder. Symptoms similar to those caused by girdling roots can be induced by diverse injuries and some diseases, so reliable diagnosis depends on detecting the offending root(s), usually by excavation. Girdling roots below the soil line commonly occur at depths less than 8 cm but are sometimes much deeper. Their presence is often indicated by absence of normal flaring of the butt at the root collar; instead the trunk ascends more or less vertically from the soil. A girdling root at the soil line usually causes some swelling where the stem begins to overgrow the root. This swelling is unlike normal flaring of the butt.

Most details of the process of injury by girdling roots await study; we therefore present an interim hypothesis. Pressure from a girdling root is believed to disrupt and eventually halt cambial activity in a band partway around the butt. The resulting aberrations in phloem and outer sapwood inhibit or block translocation of carbohydrates and growth regulators in phloem and conduction of sap and nutrients in xylem. Roots below the girdle decline in growth and absorptive function, and the branches most directly served by the isolated roots decline in vigor.

Damage to xylem beneath a girdling root has been documented for *A. platanoides*. In one tree examined in detail sapwood beneath a girdling root contained abnormally small xylem vessels whose lumens occupied only 10% as much cross-sectional area as those of vessels in normal wood. The abnormal vessels were aligned more with the constricting root than with the trunk axis. This aberration was thought likely to disrupt sap flow and induce distress in leaves and branches. Phloem formation and function in relation to girdling roots have not been reported. Surgical removal of girdling roots or severance of those found deeply embedded in the root collar or main roots of a tree has been practiced widely with the aim of preventing further damage, but there is little evidence that it does so. In one study, girdling roots began to grow again at wound sites after

surgical removal of the original offenders. Severance or removal of circling roots on container-grown plants just before planting has been advocated as a preventive measure but is impractical in many circumstances and has not been put to long-term test. Several measures for producing planting stock without circling roots have been reported but also are not yet known to prevent later development of girdling roots. These measures include growing young trees in containers with grooved sides or corners that train roots up or down rather than around a plant's axis; growing trees in in-ground fabric containers or in foam-lined baskets or shallow, wide containers on the ground; or growing them in copper-treated containers that inhibit root growth at the container surface.
References: 100, 1338, 1726, 1798, 1951, 3967, 3968, 4036, 4249–4251

Damage by Hail, Ice Glaze, and Sheet Ice

Injuries by hail like those shown in Plate 250 occur often, although in small geographic areas. Hailstones can lacerate or remove leaves, bruise or break the bark of twigs and small branches, remove twigs, and kill small trees. Severe hail wounds close together may kill all the bark on one side of a stem. In extreme cases the bark may be pounded off the windward sides of stems. Bruises and wounds tend to be elliptic and vary in length from a few millimeters to 10 cm or more. All occur on the upper side of branches and on the windward sides of upright stems. Bruised bark may crack after a storm as the result of drying and mechanical stress from woundwood growth at the edges of injured areas. All tree species in the locality of a hail storm display similar symptoms, although the severity varies among species. Bruises or wounds result in dieback of twigs and branches if tissues around the injuries dry out. Dieback is more likely if injury occurs during dormancy than if it occurs during the growing season. Secondary damage by canker fungi is also more likely if the injury occurs during dormancy.

Injury by sheet ice, as shown in panels H and I, where stems worked against a stationary ice sheet, probably occurs only to young trees with thin bark. In such cases some degree of injury can be found at the same level on all trees where water was impounded. Ice crusts on snow do not cause this symptom because snow usually melts away from stems. Moving sheets or blocks of ice remove bark from trees and deform or destroy saplings along riverbanks. Similar injury occurs occasionally on the shores of northern lakes where flooding occurs in late winter or early spring and ice is driven onshore by wind.

A, B. Injury to *Acer rubrum* (red maple) by girdling roots. A. Dieback of twigs and branches. B. Girdling at the soil line. A tree such as this one often also has several additional girdling roots below the soil surface (NY, Apr).

C–E. Ice glaze and associated damage. C, D. Glaze approximately 5 mm thick on *Quercus* sp. E. Broken tops and branches, and regrowth after one year, on *Populus deltoides* (eastern cottonwood) (NY, Mar).

F, G. Hail injury on twigs of *A. saccharum* (sugar maple) and *Malus* sp. (crabapple), respectively. The storm that caused these wounds occurred early in the growing season. The initial injury consisted mainly of bruises. During the growing season bruised tissue died, dried, and then split because of the mechanical stress imposed by woundwood formation at the edges of dead tissue. Small bruises caused only minor splits in the outer bark, and new bark soon filled these injuries. The largest bruises killed bark and cambium, resulting in exposure of sapwood. By autumn, when these photos were made, woundwood had overgrown many small wounds but the largest remained open. Had such severe injury occurred at the end of the growing season the injuries would have resulted in twig death from desiccation during winter (NY, Oct).

H, I. Girdling of *Acer platanoides* (Norway maple) by bruises made by sheet ice. These trees grew in a nursery in a depression where a pool of water occasionally formed and froze during winter. Wind action caused bruises where the stems worked against the ice. The trees began growth normally the next spring but wilted after the leaves expanded (NY, Aug).

Ice glaze on trees, the result of freezing rain, can be enormously destructive. The weight of a glaze more than about 1 cm thick breaks twigs, branches, and trunks or uproots trees. The largest trees suffer the most prominent damage, as limbs of all sizes may break. Saplings may be bent to the ground or flattened, and some that survive remain bent. Breakage increases if wind rises before the ice melts. Some effects of the great ice storm of 1998 were measured in southern Quebec, where in one forest the ice brought down 19.9 metric tons (air dry), or 33.6 m³, of woody debris per hectare, 7–10% of the aboveground biomass in the forest.

Only a slight correlation has been detected between wood strength of a tree species and its susceptibility to breakage by ice glaze, but flexibility is an obvious asset. Conifers sometimes escape with slight damage while hardwoods break. Observations of *Pseudotsuga menziesii* (Douglas-fir) have indicated that glaze damage in coniferous stands with closed canopies may be less severe than in stands opened by thinning or partial harvest.

Severe crown damage in ice storms leads in turn to severe decay. Many hardwood stands in eastern USA and southern Canada contain trees with grotesquely deformed tops and large internal defects resulting from breakage by ice. Long-term effects of ice storms on forest health (as distinguished from health of damaged trees) may be less than observers suppose, however, because the accumulation of biomass is only temporarily retarded, and growing trees close canopy gaps within a few decades.

Species' susceptibilies to breakage by ice glaze vary greatly with region and circumstance but have been compiled here as a general guide. Species judged highly susceptible by one observer and resistant by another are ranked intermediate in the list that follows.

Relative susceptibility of trees to breakage by ice glaze

Most damaged: *Acer negundo, A. rubrum, A. saccharinum, Betula papyrifera, Celtis occidentalis, Fagus grandifolia, Fraxinus nigra, Juglans cinerea, Larix laricina, Liriodendron tulipifera, Pinus rigida. P. strobus, P. virginiana, Populus deltoides, P. grandidentata, P. tremuloides, Prunus serotina, Pyrus calleryana, Quercus prinus, Q. rubra, Q. velutina, Salix* species, *Sassafras albidum, Tilia americana, Ulmus americana, U. pumila,* and *U. rubra*.

Intermediate: *Acer platanoides, A. saccharum, Betula alleghaniensis, B. pendula, B. populifolia, Fraxinus americana, F. pennsylvanica, Gleditsia triacanthos, Juniperus virginiana, Magnolia acuminata, Nyssa sylvatica, Pinus sylvestris, Platanus occidentalis, Quercus coccinea, Q. ellipsoidalis, Robinia pseudoacacia, Thuja occidentalis, Tilia cordata,* and *Zelkova serrata*.

Least damaged: *Abies balsamea, Ailanthus altissima, Carya* species, *Catalpa, Cornus florida, Diospyros virginiana, Juglans nigra, Magnolia tripetala, Malus pumila, Ostrya virginiana, Picea abies, P. rubens, Pinus nigra, P. resinosa, Pseudotsuga menziesii, Pyrus communis, Quercus alba,* and *Tsuga canadensis*.

References: 464, 550, 630, 667, 861, 995, 1009, 1027, 1133, 1744, 1747, 2292, 2456, 2550, 2563, 2739, 3232, 3234, 3273, 3296, 3362, 3535, 3612, 3666, 3709, 3742, 3771, 4340, 4507

Lightning Damage (Plate 251)

Lightning causes significant damage to forest and landscape trees even in the absence of fire. For example, lightning was responsible for one-third of all mortality in *Pinus ponderosa* (ponderosa pine) stands in northern Arizona in the 1940s and was the most common cause of tree death in *P. palustris* (longleaf pine) stands in southern Georgia in the 1990s. Trees struck but not killed are likely to be disfigured by death of limbs or the top. In addition, lightning-caused wounds and killed parts provide entry for borers and wood-decay fungi. Conifers weakened by lightning strikes may be killed by bark beetles and associated bluestain fungi. Lightning scars on living trees in forests devalue logs at harvest time.

Lone trees, tall trees on the windward edges of groups or stands, and dominant trees in woodlands are struck most often. An electrical current at exceedingly high voltage follows the most conductive path between the top and the roots, sometimes along the surface but often in the outer sapwood. Steam or possibly gas from electrolytic dissociation of water is apparently generated instantly in the conducting tissue, because a strip of bark and sapwood is often blown off the trunk, leaving a continuous or intermittent rough

groove that follows the wood grain. Sometimes the trunk shatters and blows apart. Sometimes the bark and covering soil are blown off a root before the charge dissipates in soil. A high proportion of lightning-struck trees eventually die prematurely because of the injury or because of attack by opportunistic organisms, but precise figures are not available.

Trees that survive a lightning strike develop various abnormalities. Injury during the annual period of cambial activity may cause the formation of a sheath of abnormal wood (a ring as viewed in cross section). Since such rings occur in nonwounded branches as well as grooved trunks, they probably reflect a cambial response to the electrical charge or associated heat. Lightning rings in resinous conifers contain an abnormally large number of resin ducts, and injured conifers may exude resin. Galls or short ridges, sometimes several centimeters in diameter, may develop along the trunks of lightning-damaged angiosperms. Long seams are common. If roots on one side of a tree have been killed, adventitious roots sometimes grow from the trunk above, usually beneath the protective cover of dead bark.

Lightning occasionally causes group death. In such cases tree crowns in patches up to 30 m in diameter suddenly turn brown. Sometimes no external marks implicate lightning as the cause, but more often, typical grooves may be found in the trunk of one or more central trees. The roots of the unmarked trees are presumed to have been killed where the electrical charge dissipatesd in soil. Similar damage has been noted in nurseries, where all plants in the center of a patch die and those around the edges remain stunted for a time. The deepest roots of surviving seedlings may be found dead.

The frequency of lightning strikes varies among tree species, but the reasons are unknown. No investigation has been reported in which frequency of lightning strike could be related to tree species apart from tree location, tree size, and frequency of each species in the local tree population. In eastern USA the frequency of lightning strikes has been reported to diminish among tree genera in the following order: *Quercus, Ulmus, Pinus, Liriodendron, Populus, Fraxinus, Acer, Platanus, Tsuga,* and *Picea. Juglans nigra,* although its genus is not listed above, is often struck. *Aesculus, Betula,* and *Fagus* are not often struck.

Lightning rods are sometimes installed on large valuable trees, with copper cables leading down the trunk and through the soil to grounding rods driven into soil beyond the branch spread of the tree. The degree of protection thus given has not been assessed objectively.

References: 635, 841, 842, 981, 1181, 1683a, 1881, 2164, 2179, 2372, 2944, 2987, 3275, 4175

A. *Quercus rubra* (red oak), dying after a lightning strike (MI, Aug).

B. *Ulmus americana* (American elm) with strips of bark and sapwood blown off the trunk where a lightning charge traveled through the cambial region and outer sapwood (NY, Jul).

C. A lightning survivor. This *Q. rubra*, with killed sapwood being enclosed by woundwood, will have a decayed core and trunk cracks that open repeatedly, leading to more decay (NY, May).

D. *Pinus strobus* (eastern white pine) with a lightning-killed top (NY, Apr).

E, F. Branch mortality on *P. resinosa* (red pine). A strip of exposed sapwood where bark was blown off the trunk (F) implicates lightning as the cause of branch death (NY, Aug).

G. Growth abnormalities on *Acer platanoides* (Norway maple), resulting from lightning injury many years earlier. The electrical charge killed broad strips of cambium and roots on one side of the tree. Dead bark remained in place temporarily, and numerous adventitious roots grew from above the killed area downward beneath the killed bark. The dead bark eventually broke away, exposing the roots. An adventitious root about 8 cm in diameter is visible at right (NY, Aug).

H. The base of a lightning-struck *Pinus palustris* (longleaf pine). The electrical charge grounded through one major root, blowing away its bark and the covering soil (MS, May).

Noninfectious and Unexplained Growth Abnormalities (Plates 252, 253)

Fasciation

This term connotes abnormal flattening of plant organs, usually stems, as the result of a change in form of the apical meristem from a minute dome to a jagged row of generative cells perpendicular to the stem axis. The stem generated by this meristematic ridge appears ribbed, as if several stems had fused laterally, because the width of the generative zone varies along the ridge. Foliage on fasciated stems is denser than normal because each rib bears leaves. Leaves are usually normal in shape but may be dwarfed. Unequal elongation of young stem tissues below the row meristem may lead to grotesque bending and coiling. Often the row meristem cleaves into several parts, so that the ribs separate into discrete stems, some fasciated, others normal. Dwarfed, malformed shoots that abort at an early stage may develop along the row, and fasciated branch tips often die during winter. The term *fasciation* has also been used less specifically in horticulture to refer to not only to flattened stems but also to leafy galls and bunches of shoots arising from them, mainly on herbaceous plants.

Fasciation occurs in both woody and herbaceous plants. In some instances it results from mutation in reproductive cells and is transmitted to progeny as a heritable character. In other cases mutations occur in meristematic cells and can be perpetuated by vegetative propagation but not through seed. In still other instances a single plant will alternately produce normal and fasciated twigs, one growing from the other, reflecting fluctuations in hormonal balances. The hormonal herbicide 2,4-D can induce fasciation. Zinc deficiency can lead to fasciation in *Pinus caribaea* and some herbaceous plants. A rhabdovirus is believed to cause fasciation in *Euonymus japonicus* (evergreen euonymus). Fasciation in *Diospyros* (persimmon) in China and in *Olea europaea* (olive) in Italy is associated with phytoplasmal infection. The bacterium *Rhodococcus fascians*, which causes fasciation in many herbaceous plants, possesses a plasmid with genes that enable it to synthesize cytokininlike compounds that mediate cell division. Bacterial cells cured of the plasmid are avirulent. Woody plants affected by *R. fascians* include *Buddleja davidii* (butterfly bush), *Forsythia*, *Hebe*, and *Viburnum opulus* (European cranberry-bush). Fasciation of *Betula pendula* (European white birch) has been induced by treatment with napthaleneacetic acid (an auxin analog) and cytokinins. Fasciation in *Pyrus calleryana* (callery pear) has been induced by indolebutyric acid. Mutations induced in *Abies balsamea* (balsam fir) by gamma radiation have resulted in fasciation.

Chimeras

A chimera in botanical context is a plant or plant organ consisting of tissues of more than one genotype. Natural chimeras apparently arise from mutations in certain cells at growing points or in the vascular cambium such that their vegetative progeny express a form or color different from that of the rest of the plant. Freezing can induce chimeras that originate in the vascular cambium, as exemplified by galls on trunks of *Picea* species (spruce) in cold regions, but the inducing factors for most natural chimeras are undetermined. Chimeras can be induced artificially by radiation, freezing, and certain chemicals such as colchicine. Chimeras can be created by grafting within meristems. Variegated plants represent chimeras resulting from mutations (or in some cases grafting) within apical meristems. Unfortunately, variegated trees or shrubs tend to be poorly suited for horticultural use because the achlorophyllous tissues often die prematurely. Variegated *Acer platanoides* (Norway maple), for example, has a pronounced tendency to develop leaf scorch. Some viral infections cause foliar or flower symptoms resembling chimeras, but a viral disease usually presents additional symptoms.

Graft Union Abnormalities

When freshly cut tissues of compatible plants are grafted together by appropriate methods, a strong union forms. Initially, callus derived from parenchyma cells of both stock and scion forms at the interface. This process is followed by formation of a cork cambium and periderm at the outer surface of the union and by differentiation of vascular cambium, leading to the formation of phloem and xylem across the interface. A strong union is assured by interlocking growth of xylem cells derived from stock and scion. The entire process requires several weeks.

Many variations of abnormal growth and growth failure at graft unions are lumped under the general term *graft incompatibility*. In the simplest case no tissues of stock and scion unite, despite adequate technique and suitable environmental conditions. Delayed or partial incompatibility is of greater concern. Stock and scion apparently unite, but subsequent growth is aberrant, leading to mechanical failure or symptoms of physiological distress. In one type of incompatible union stock and scion become united by callus but not by conducting tissues. The scion grows, but the union remains weak and usually fails within 1–2 years. In another type of delayed incompatibility a xylem union forms between stock and scion, but phloem connections are not established. The result is swelling of the scion above the union and slow starvation of the stock. In another type no interlocking growth of xylem cells occurs; instead xylem elements of stock and scion grow parallel to one another at the union. Enough water moves across the union to satisfy mineral nutrient needs and transpirational demand of the scion, but the union remains physically weak and eventually breaks. There is a well-documented report, for example, of an *Abies concolor* (white fir) tree that broke cleanly off its rootstock at ground level after 40 years of apparently normal growth. Similar incompatibilities have been reported in various fruit and nut trees and in *Acer rubrum* (red maple).

Graft incompatibility may be indicated by one or more of the following symptoms: markedly greater growth rate of the stock than the scion, or vice versa; difference between stock and scion in the onset or cessation of growth; local overgrowth above, below, or at the graft union; breakage of the grafted plant at the graft union; declining vigor, stunting, chlorosis, premature leaf shedding, or death of the scion or the entire plant.

Graft incompatibility may be due to intrinsic factors, but often it is caused by viral infection. One member of the grafted pair becomes infected but tolerates the virus, whereas the other member responds hypersensitively, resulting in necrosis at the graft union. Blackline disease of *Juglans* (walnut; Plate 203) and brownline disease of *Malus domestica* (apple; Plate 201) exemplify delayed incompatibility caused by viruses.

If a scion merely grows more rapidly than its understock, and no symptoms of distress occur in either grafted member, the two are not necessarily incompatible. Continuation of this inequality results in a strangely shaped trunk, however.

References: 49, 273, 1029, 1307, 1538, 1620, 1908, 2092, 2307, 2347, 2538, 2710, 2816, 2864, 3180, 3210, 3703, 3761, 3814, 4027, 4077, 4323, 4483, 4500, 4501

A, B. Fasciation in *Prunus serrulata* 'Kwanzan' (flowering cherry). A. An affected branch from an otherwise normal tree. B. The same branch with leaves and normal twigs removed (NY, Aug).

C. Fasciation of *Myrtus communis* (myrtle); one normal twig is at upper right (CA, Oct).

D, E. Fasciation of *Daphne* sp. D. Transition from normal to fasciated stem form and clusters of dwarfed distorted shoots along the apical row of growing points are apparent. A fasciated branch has grown from the right end and a normal branch from the left end of the row. E. Dieback on a fasciated branch (CA, Aug).

F. Fasciation of *Betula pendula* (European white birch). Two normal twigs developed from the fasciated stem. Bud opening on the normal twigs was delayed, perhaps because of cold injury (NY, May).

G. A chimera characterized by one-sided dwarfing and albinism in a *Juglans nigra* (black walnut) leaf. A mutation apparently occurred early in the formation of the leaf primordium such that cells differentiating on one side had no ability to synthesize chlorophyll (NY, Jun).

H. One-sided albinism in a leaf of *Rosa* sp. Whether the symptom resulted from a somatic mutation or from viral infection (rose mosaic complex) was not known (NY, Jun).

I. A graft union between two species of *Prunus*. Radial growth rate of the stock and scion are dissimilar, so swelling appears to have occurred above the union (NY, Aug).

J. Gross swelling, indicating incompatibility, at a graft union of *Acacia pubescens* (hairy wattle) on an unidentified *Acacia* (PA, Jun).

K. Necrosis at a 3-year-old *Acer rubrum* (red maple) graft union associated with death of the rootstock. Dead xylem is light gray. The cause of the failure was not ascertained (NY, Jun).

509

Adventitious Shoots and Roots

Adventitious shoots or roots are those growing at unexpected sites. They arise in response to injury, disease, or stress that interferes with the normal dominance of apical meristems. Adventitious shoots, also called water sprouts, grow from either epicormic buds or from buds that arise along the trunk and limbs when apical buds die or lose dominance. Severe breakage of branches, dieback resulting from disease or defoliation, and altered hormone balance in tissues subtending wounds and lesions are among the conditions that trigger formation of adventitious shoots. Sprout formation at the base of a tree often indicates chronic distress (for examples, see Plates 198, 203, 231). Sprout formation along a trunk often follows dieback and often presages death. Such was the case with the *Quercus prinus* (chestnut oak) shown in Plate 253A. The trees were defoliated by *Alsophila pometaria* (fall cankerworm) in two successive years, resulting in dieback followed by sprouting. They were killed by opportunistic insects and fungi within a year after the photograph was made.

Acer platanoides (Norway maple) occasionally develops more or less profuse adventitious sprouts from trunk swellings (Plate 253B). A virus was found associated with similar disorders in *A. negundo* (box-elder) and *A. pseudoplatanus* (sycamore maple) in Hungary, but it was not shown to cause the sprouting. Viral association with such symptoms in *Acer* in North America has not been reported.

Adventitious roots form on diverse plants in response to wounding, flooding, or death of the roots below. Such roots often form at the upper edges of stem wounds near ground level, on the upper main roots or lower stems of intermittently flooded plants or those growing in periodically waterlogged soil, on branches continually pressed against moist soil, beneath the loosened bark of trees damaged by lightning (Plate 251) or freezing, and on the lower trunks of trees whose original roots were buried beneath additional soil. The propensity to form adventitious roots varies greatly among woody species and is correlated with ability to be successfully propagated by cuttings. *Salix* species (willows) are well known for their rooting ability, although they usually produce adventitious roots of more or less normal form, unlike the dense mat shown in panels C and D. A dense mat of adventitious roots is likely to form where they grow in a confined space such as beneath bark. Although most adventitious roots form in response to stimuli or damage by noninfectious agents, there is one infectious disorder called hairy root that affects various plants and could be confused with adventitious roots induced by other stimuli. Hairy root is caused by the bacterium *Agrobacterium rhizogenes*. More information about it is presented with Plate 190.

Galls and Burls

The terms *gall*, *tumor*, and *burl* are all used to denote swellings, usually abnormal ones. Galls and tumors are swellings of various sizes and shapes, woody or not, on any plant organ. Galls and various other stem swellings may be induced by bacteria, fungi, insects, mistletoes, environmental insults, and mutations in somatic cells. *Burl* usually means a large woody swelling that is more or less hemispheric. Burls often bear many buds or in some cases sprouts. When such burls are sawn and the cut face smoothed, the wood grain is seen to swirl around each bud trace. This highly figured wood (burlwood) is prized by woodworkers. Burls on *Prunus serotina* (black cherry), *Acer saccharum* (sugar maple), *Sequoia sempervirens* (coast redwood), and *Juglans nigra* (black walnut) are highly valued. The term *burl* has also been used to refer to any stem swelling with swirling wood grain. Tissue called burlwood may form, for example, at the edges of cankers caused by freezing. Some woody species normally produce burls with many buds (bud burls) at the root collar, partly or wholly subterranean, as an adaptation for regrowth after injury or death of the stem, as commonly occurs in wildfires. These burls are also called lignotubers. They are common on *Arbutus menziesii* (madrone), *Arctostaphylos* (manzanita), *Ceanothus*, *Betula* (birch), *Eucalyptus*, *Kalmia* (mountain-laurel), *Quercus gambelii* (Gambel oak), *Rhododendron*, and *Sequoia*. Bud burls induced by unknown stimuli occur occasionally on *Acer macrophyllum* (bigleaf maple), *A. platanoides* (Norway maple), and *Tilia cordata* (littleleaf linden). Burls on *T. cordata* trunks become fissured and are invaded by secondary agents. Trees with such galls grow slowly and decline prematurely. Many stem swellings, including all of those shown here, remain unexplained or incompletely explained.

Stem galls also occur on various conifers; for example, *Picea glauca* (white spruce) in interior Alaska and along the coasts of Maine and the Maritime Provinces, *P. sitchensis* (Sitka spruce) in western Canada, and *Pinus contorta* (lodgepole pine) and other conifers in the northern Rocky Mountains. The galls vary from tiny protuberances to globose masses many times wider than the stems that bear them. They apparently originate in single cells near the pith in first-year stems and enlarge for many years. No evidence for an infectious cause has been presented. Somatic mutations associated with low-temperature injury are suspected. Small stem galls on *Pseudotsuga menziesii* (Douglas-fir) are caused by unnamed bacteria in the genus *Erwinia*. A gall disease of *P. menziesii* described in the 1930s was attributed to a bacterium that was later called *Agrobacterium pseudotsugae*, but cultures with that name were neither pathogenic nor related to *Agrobacterium* when tested in later years.

Witches'-brooms

Brooms may be caused by fungi, mites, mistletoes, phytoplasmas, mutations, and environmental insults that kill growing points, resulting in the proliferation of new shoots. Many witches'-brooms on conifers and some on angiosperms apparently result from mutations in meristematic cells. Some of these have been propagated as dwarf cultivars. The brooms on *Pinus* species shown in Plate 253 illustrate the compact growth habit and dense foliage desired in dwarf cultivars. Some dwarf conifers are aneuploids; that is, they have an abnormal number of chromosomes as the result of gain or loss during mitotic accidents. Brooms and dwarfism have been induced in *P. sylvestris* (Scots pine) by treatment of seeds with the mutagen N-nitrosodimethylurea. Pines with large brooms near their tops often decline and die prematurely even though no infectious agent is known to be present. The *P. sylvestris* shown in panel J died a few years after the photograph was made. Large brooms are presumed to be metabolic sinks that retain or divert photosynthetic products that would normally be used for growth of other parts of the tree. Seeds from some witches'-brooms that originate as mutations in conifers produce dwarf and normal seedlings in equal proportions, indicating inheritance of a dominant characteristic from the maternal parent.

References: 90, 461, 744, 895, 945, 1497, 1835, 1908, 2347, 2693, 2816, 2930, 3043, 3143, 3879, 3946, 4026, 4076, 4096, 4255, 4256, 4325, 4326, 4402, 4467

A. Dieback and adventitious sprouts on *Quercus prinus* (chestnut oak). Defoliation by insects triggered the dieback, which in turn led to production of sprouts along the trunks (NY, Sep).

B. Burls and adventitious sprouts on *Acer platanoides* (Norway maple). The cause of these abnormalities is unknown (NY, Mar).

C, D. Adventitious roots on *Salix babylonica* (weeping willow). The roots began to develop between bark and sapwood after an unknown agent (probably lightning) killed the vascular cambium around most of the butt of the tree. After the bark split open or was removed, the root tips dried and stopped elongating (NY, Aug).

E. Large burls of unknown cause on a butt and trunk of *Schinus molle* (California pepper-tree) (CA, May).

F. Galls of unknown cause on a *Tilia cordata* (littleleaf linden) trunk. Growth of *T. cordata* with trunk galls of this sort is retarded, and the trees decline prematurely (NY, May).

G. A burl developing on the trunk of a young *Acer saccharum* (sugar maple) (NY, May).

H. Multiple burls, cause unknown, on the trunk of a mature *Ostrya virginiana* (hop-hornbeam) (NY, Aug).

I. "Tissue proliferation," a noninfectious tumorous condition on the base of a year-old *Rhododendron* 'Nova Zembla' cutting (NY, Jul). This condition in rhododendrons was sometimes misdiagnosed as crown gall (and was so mislabeled in the first edition of this book) until research revealed that it develops spontaneously and cannot be induced by inoculation with known gall-inducing microbes. The condition develops more frequently in clonal lines maintained by micropropagation than in clones propagated by cuttings.

J. A large witches'-broom, presumed to have resulted from a somatic mutation, on a mature *Pinus sylvestris* (Scots pine). The broom is apparently being nourished at the expense of the rest of the tree, which has begun to decline (NY, Apr).

K. A witches'-broom, viewed from below to reveal its form, on *Pinus strobus* (eastern white pine). Slight swelling of the supporting branch near the base of the broom indicates the original location of a presumed mutation in a meristem. Year-old needles in the interior of the broom had turned brown (natural for the season) and were being shed when the photograph was made (NY, Oct).

Restoration of Sapwood and Bark after Injury or Infection (Plates 254–257)
Wounds, Microbial Colonization, and Compartmentalization (Plates 254, 255)

This unit summarizes concepts that apply generally to trees and shrubs. Details are presented in expanded captions of the photographs. Alternative interpretations, exceptions, fine points, and special cases are treated in the references.

1. Healthy woody stems and roots, in common with other organs, harbor numerous quiescent microorganisms internally. These include both bacteria and fungi, some of which are pathogens and others benign organisms. They are collectively called endophytes. They enter plants by way of tiny wounds and natural openings or penetrate normal surfaces such as cuticle and cracks in periderm and then persist as microcolonies. Pathogens living as endophytes are normally held in check by the physical and chemical conditions of healthy bark and sapwood, which include high water content, constitutive chemical resistance factors, and probably induced resistance responses. When plant organs senesce or are wounded or otherwise injured, these endophytes can resume growth or multiplication. Some of them become aggressive colonists of damaged plant parts and, together with newly arriving organisms (see point 5), kill tissues and participate later in litter breakdown. Examples of plant-pathogenic bacteria and fungi that have endophytic capability in stems include *Erwinia amylovora* (Plate 187), *Pseudomonas syringae* (Plate 183), *Apiognomonia quercina* (Plate 49), *Colletotrichum gloeosporioides* (Plate 56), *Valsa sordida* (Plate 85), *Biscogniauxia atropunctata* (Plate 99), and *Echinodontium tinctorium* (Plate 160).

2. The sapwood of a living tree is capable of isolating damaged portions of itself and of interacting dynamically with invading microorganisms. These capabilities reside in xylem parenchyma cells and are sustained by the transport of energy-bearing organic compounds inward from the phloem via medullary rays. The youngest sapwood is the most competent with respect to defense and repair.

3. Normal wood is divided anatomically into numerous elongate compartments that are delimited by the vascular cambium and aggregations of parenchyma cells. The wood in trees of the temperate zones consists of nested sets of compartments. A set of compartments is produced annually in each new sheath of xylem. The vascular cambium is the outer boundary of the youngest set. Medullary rays, where parenchyma cells are concentrated, are the radial boundaries, or side walls, of compartments. The inner and outer (tangential) boundaries of each compartment other than the outermost one consist of dense latewood plus axial parenchyma cells concentrated near the junctions of annual layers of wood. Axial parenchyma cells are associated with vessels in angiosperm wood and with resin canals in many gymnosperm woods. The inner tangential boundaries are poorly developed in many trees, especially gymnosperms, and are indistinct or absent in tropical trees that grow more or less continuously. Axial parenchyma in stems that lack regular growth rings does participate in compartmentalization of wounds and infections, however.

4. When sapwood is wounded, the tree initiates processes, collectively called compartmentalization, that isolate the wound from normal tissues. Parenchyma cells near the wound produce and secrete substances that make a zone of wood around the wound impervious to air and water, and toxic or inhibitory to microorganisms. These substances include gums, phenolic compounds, and suberin in angiosperms; and resins, terpenoids, and polyphenolics in gymnosperms. Vessels or tracheids that would normally conduct sap become plugged with these substances above and below the wound and near the wound edges. Bordered pits (channels between tracheids) in conifer wood become blocked by aspiration, the tori (discs on pit membranes) acting like valves. Parenchyma cells also produce tyloses that plug vessels and tracheids. (A tylosis is a balloonlike protrusion of the elastic wall of a parenchyma cell into a vessel or tracheid via a pit in the thick wall of the latter cell.) The parenchyma cells that produce antimicrobial substances become infused with the inhibitory compounds, die, and thus become part of a zone of altered wood. The altered wood darkens as phenolic compounds are oxidized. This region of sapwood response is called the reaction zone. It extends much farther up and down the stem than inward or to either side because of the way wood elements are arranged. Elongate conduits (vessels or tracheids) extend up and down, and parenchyma cells usually are less abundant above and below any point in the wound than beside it or toward the center of the stem.

Wounding also stimulates the vascular cambium around the wound to produce abnormal xylem with reduced numbers and sizes

---

A. Trunk decay and a cavity that originated at a fire scar in a mature *Pinus elliottii* (slash pine). The tree survived repeated fires, one of which burned elliptic craters into the decaying wood. Nearly the entire cross section of the trunk is contained within the compartments defined by wound responses after part of the cambium was killed by heat. Decay is proceeding within wood formed before the final episode of cambial killing (within the cylinder defined by the youngest barrier zone) and will continue as long as the wound remains open. Wood formed after the last fire, outside the barrier zone, will remain sound (FL, Mar).

B. A small *Fagus grandifolia* (American beech) with trunk decay that originated at a wound where a falling neighbor tree smashed off a strip of bark many years earlier. Wood that was present at the time of wounding is disappearing and a cavity forming as the result of decay followed by insect, bird, and rodent activity. Wood produced since the time of wounding remains sound. This wound is slowly being enclosed by healthy wood and bark produced by the vascular cambium in rolls (ribs) of woundwood that originated in undamaged tissues at the edges of the wound. These advancing ribs will eventually fuse, restoring the vascular cambium around the entire trunk. Decay within the compartment will cease or be greatly retarded when wound closure is complete. Note that where decayed wood has been removed, woundwood rolls wrap around into the cavity. After the ribs fuse, bark and sapwood facing the interior of the tree will slowly die of attrition. The outlines of the original wound and a line where woundwood ribs met will remain on the bark surface. After restoration of a continuous vascular cambium across the former wound site, trunk growth will result in an increasingly thick layer of normal sapwood between the superficial scar and the compartmentalized cavity (NY, May).

C–E. Compartmentalization of wound-associated discoloration and decay in an *Acer rubrum* (red maple) trunk. The tree, growing slowly in the forest understory, was about 70 years old when cut. C. Longitudinal section at the base of an open cavity resulting from an old wound, about one-half natural size. The wound was caused by a falling tree that damaged the cambium on the right side of the trunk 31 years before the tree was cut. At the level of the wound, discoloration and decay expanded to involve all wood present at the time of wounding, as delimited by the barrier zone (bz). A cavity (cav) developed at the wound site, and a column of defective wood extended above and below it. The cavity is lined with decayed wood. Discolored wood (di), the precursor of decay, lies adjacent to the barrier zone. Woundwood (w) that grew from the edge of the original wound has wrapped around the edge of the cavity. A small wound (x) occurred 9 years before the tree was cut and was completely covered by woundwood after 7 years. During that time little discoloration and no decay originated at the small wound, and further development of discoloration was precluded by wound closure. The point of fusion of woundwood rolls covering the small wound is marked by f. D. An enlarged view of the region where woundwood rolls fused to close the small wound in C. The year-to-year advance of the woundwood rolls that closed the wound is marked by faint lines that delimit annual layers of wood. Discoloration extended only three annual layers into the wood at the time of wound closure. E. The column of defective wood at the top of the cavity, approximately 90 cm above the location shown in C and about one-half natural size. The tree was not wounded at this level, but a barrier zone (bz) corresponding to the location of the vascular cambium at the time of wounding restricts the defect to wood present at the time of wounding. Decaying wood is surrounded by discolored wood (di). The darkly stained wood at the bulge in the stem is a compartmentalized defect near a branch stub that was overgrown by the expanding trunk (NY, Apr).

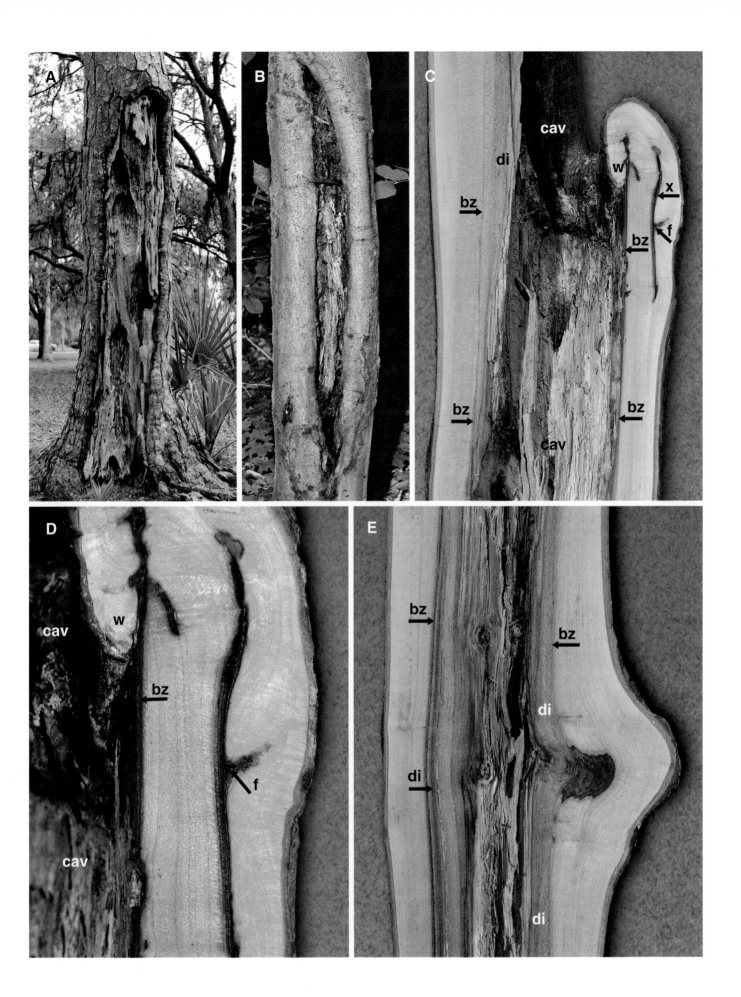

of fibers and conducting elements and a high proportion of parenchyma cells. This parenchyma produces chemicals like those mentioned above, resulting in a thin layer of new xylem more impervious than preexisting wood altered merely by infusion. The newly formed abnormal xylem, called the barrier zone, separates sapwood present before wounding from that produced subsequently. If wounding occurs during annual radial growth, the barrier zone forms within a growth ring. If wounding occurs between growth periods, the barrier zone forms at the beginning of the next annual ring.

A tree utilizes energy from stored compounds and sacrifices some healthy tissue to form the perimeter around a wound. The wound-isolation response within preexisting wood is strongest in young sapwood and diminishes with increasing depth into older wood. Similarly, the degree of physical and chemical aberration of xylem in the barrier zone is greatest near the wound and diminishes with circumferential and vertical distance from the wound. Sapwood beneath a fully formed barrier zone slowly dies because the barrier zone interrupts the normal inward flow of nutrients in rays to xylem parenchyma. Probably the barrier zone also inhibits gas exchange.

5. As soon as a wound occurs, various microorganisms colonize it and begin to spread into the surrounding wood. Microorganisms cannot be kept out of tree wounds. Therefore, a tree's reaction to a wound and its reaction to invading organisms cannot be separated in practice. Some of the invaders arrive from outside the tree, carried by air, water, or arthropods. Some are former endophytes activated by the sudden availability of nutrients, diminishing water potential, and lack of chemical regulation by host cells. Bacteria, yeasts, and molds that grow rapidly by utilizing simple carbon compounds in the wound are the first, or primary, colonists. They include or are followed by organisms, chiefly ascomycetes, that are able to grow into the reaction zone and detoxify inhibitory chemicals, utilizing some of them as nutrients. These organisms are resisted but not halted. Their advance causes continual expansion of the reaction zone up, down, and inward from the wound. They discolor the wood by converting chemicals in the reaction zone to colored products, but they do not cause decay. After staining fungi detoxify the wood, decay fungi move in, months to years after the wounding event. Most decay fungi are relatively intolerant of fungitoxic compounds produced by parenchyma and do not rapidly detoxify these chemicals. But decay fungi are able to degrade complex polymers—lignin, cellulose, and hemicelluloses—and use them as carbon sources. This is the main process in wood decay. Microorganisms in succession, continually confronted by the reaction of sapwood parenchyma, slowly advance and cause discoloration and decay within wood that was present before wounding. Microorganisms do not breach the barrier zone; therefore sapwood formed after wounding remains healthy. If a wound penetrates heartwood (normal, colored dead wood in the center of the tree) or inward-spreading microbes enter heartwood, the invaders are resisted by means of chemicals previously deposited in the wood. These chemicals are produced by parenchyma cells as the final process encoded by their genetic program before they die and become part of the heartwood.

Some pathogenic wood-decay fungi, both ascomycetes and basidiomycetes, are able to invade sapwood through fresh wounds. These fungi, tolerant of chemicals in the reaction zone, play all the roles that primary colonists, detoxifiers, and decay fungi play in the successions mentioned above. *Cryptosphaeria lignyota* (Plate 96), *Entoleuca mammata* (Plate 98), *Trametes versicolor* (Plate 154), *Cerrena unicolor* (Plate 158), and *Heterobasidion annosum* (Plates 165, 166), are examples. Many wood-decay fungi that parasitize sapwood are poor competitive saprobes and are unable to replace staining fungi or other decay fungi in established successions. *E. mammata* and *H. annosum* are in this group.

6. Wound closure proceeds as the result of *woundwood* (bark plus sapwood produced at the edge of a wound) formation. Activity of the vascular cambium is stimulated and remains so near a wound, leading to more rapid wood formation than occurs elsewhere at the same level on a stem. Woundwood expands faster tangentially than radially; thus a wound closes while the stem expands in girth. When a wound closes by fusion of woundwood rolls, the advance of

discoloration and decay from that wound ceases. Cessation is thought to be caused by the increased water potential and altered gas concentrations (increased CO_2 and decreased O_2) in discolored and decayed wood sealed beneath sound wood. These conditions may inhibit staining and decay fungi.

7. A tree may sustain wounds at various times. Therefore it may contain multiple, overlapping regions of discoloration and decay.

8. When a barrier zone is breached by a new wound or by a crack developing from within the column of defective wood, sapwood outside the original barrier zone is subject to discoloration and decay caused by organisms that spread outward from the previously compartmentalized column. These organisms are resisted by the same processes described above.

9. Trees vary in ability to compartmentalize wound-associated discoloration and decay. This ability is under genetic control and is conditioned by environmental factors.

References: 159, 260, 400, 402, 425, 427, 450, 691, 1010, 1223, 1242, 1283, 1377, 1651, 1858, 2260, 2374, 2512, 2513, 2661, 2754, 2805, 2806, 2991–2996, 3227, 3228, 3380, 3401, 3462, 3463, 3499, 3500, 3509, 3510, 3589, 3593–3602, 3607, 3725, 4037, 4038, 4272

A, B. Cross section of the *Acer rubrum* (red maple) trunk shown in Plate 254, about 1.5 m below the wound. The column of discolored and decaying wood at this level occupies nearly the entire cross section of the stem as it existed 31 years earlier at the time of wounding. The annual ring containing the barrier zone associated with the old wound above is indicated by the dashed line (bz). The barrier zone is attenuated and barely perceptible at this distance from the wound. This section shows how discoloration and decay spread outward when a barrier zone is breached by a new wound or crack. Discoloration (di) beyond the barrier zone is extensive at three cracks and is beginning at two others. Tangential spread of discoloration from the three large cracks has been resisted by ray parenchyma. Therefore the defects associated with cracks extend spokelike from the central core as viewed in cross section and are compartmentalized near the plane of the crack. The crack at right never extended to the vascular cambium. The crack at left reached the vascular cambium 8 years before the tree was cut but did not break the bark. This injury to the cambium temporarily stimulated sapwood formation that resulted in a bulge on the side of the tree. When the tree was cut this crack extended to within four annual rings of the cambium. The largest crack, shown approximately 1.5 times actual size in B, extended intermittently, as indicated by dark color patterns in the wood. The crack first broke through the cambium and bark 4 years before the tree was cut, and it opened and closed annually thereafter. The opening induced wound-wood growth and the development of a prominent rib on the trunk (NY, Apr).

C. An *Acer rubrum* with a crack similar in external appearance to that through which the section in A was cut. Woundwood growth at the edges of the crack resulted in formation of a prominent rib that follows the grain of the wood (MI, Aug).

D, E. Part of a cross section of an *A. saccharum* (sugar maple) trunk repeatedly tapped (bored) to extract sap. The "shadows" of eight tap holes slightly above and below the plane of the section are visible. Wound-associated discoloration and decay are extensive. White mycelium of a wood-decay fungus grew from colonized wood onto the surface of the section during several days of incubation. Columns of discoloration and decay, initially compartmentalized behind closed wounds, expanded and coalesced because repeated tapping broke compartment boundaries. Part of a closed tap wound is shown close up in E. A T-shaped mark, with the leg of the T pointing out, remains where woundwood rolls closed the wound and then fused to produce an uninterrupted covering of sapwood (NY, Sep).

F, G. Examples of basal wounds that lead to extensive decay in the butts of trees. F. *Gleditsia triacanthos* (honeylocust) repeatedly assaulted by automobile bumpers. G. *Quercus velutina* (black oak) attacked by a beaver (GA, May).

Frost Cracks, Drought Cracks, and Related Defects (Plate 256)

Plate 256 completes a series of three about relationships between wounds and the development of discoloration and decay in tree trunks and roots. Cracks in wood are often called checks or shakes. The former term is usually applied to small radial separations such as those in the ends of drying logs and pieces of lumber. Radial cracks that originate within standing trunks are also called ray or radial shakes. Tangential cracks that separate annual layers of wood or that develop at sites of former cambial damage by heat or freezing, or at barrier zones associated with wounds, are called ring shakes. Freezing or drying of wood in a stem can cause cracks and contribute to their enlargement; hence the terms *frost crack* and *drought crack*.

Cracks result from mechanical stress induced by freezing, drying, wind force, or weight of snow or ice. Cracks may form in previously sound wood, but they are most likely to start at sites of former injury, thus usually in wood undergoing discoloration and decay. Cracks in wood and bark of shade and ornamental trees often develop as a consequence of supposedly benign or beneficial treatments. For example, cracks may start at pruning wounds and injection wounds.

Discoloration and decay spread along cracks. After a radial crack either breaks the cambium and bark or breaks a barrier zone (Plate 255), the wood to either side of the crack is subject to discoloration and decay as microorganisms spread along the crack and tangentially away from it. Woundwood formation along cracks that break the vascular cambium may result in formation of prominent vertical ribs. Many people have heard sharp reports as cracks in hardwood trees pop open on frosty winter mornings, but gradual reopening of cracks as trunks undergo thermal shrinkage is the usual situation.

Possible mechanisms of frost crack formation include internal freeze-drying of wood through migration of water out of cell walls into cell lumina and intercellular spaces; accumulation of ice lenses between wood cells, forcing them apart; expansion of water during freezing (in waterlogged stems); and, most likely, mechanical stress due to shrinkage of the outer part of the stem during cooling. Frost cracks can be induced in sound wood by experimental freezing treatments; therefore it seems likely that freezing may induce cracks in sound wood in nature.

Radial cracks apparently caused by drought in otherwise normal wood of conifers have been reliably reported from Alaska, Great Britain, and continental Europe. Such cracks originate in recently formed layers of sapwood and often extend only across the width of one annual layer of wood, appearing in transverse section as diamond-shaped clefts amid partially crushed tracheids. Some cracks extend across several annual rings or through the bark, appearing on the surface as long lines that ascend the trunk in a high helix corresponding to the grain of the wood. Most drought cracks are entirely internal and are discovered only when trees are felled. Microscopic study of drought cracks in *Picea abies* (Norway spruce) revealed that they start within low-density wood of single annual rings and are apparently induced by prolonged hot, dry weather beginning in midsummer. Cracks were most severe in trees that had been growing rapidly and had wood of lower density than normal. The tendency toward formation of drought cracks varied among clones, which indicated that it is under genetic control in *P. abies*.

Tangential and radial cracks often develop near the bases of tree trunks that have sustained cambial damage by freezing. The sequence begins when a localized area of cambium or outermost sapwood is killed but the phloem survives. During the next growing season surviving cambial cells or new vascular cambium differentiated from phloem parenchyma overlying the damaged zone begin to produce callus and woundwood, and swelling results. Pressure from callus and woundwood splits the bark and any new xylem to the depth of the original injury, and may cause the split to flare open widely. The cambium at each side of the split then produces callus and woundwood that contribute to swelling of the damaged part of the stem. After wound closure, the wood may separate tangentially along the plane of weakness where the cambium or sapwood was previously killed, or a radial crack may open at the line of fusion of woundwood ribs. The woody cylinder

present at the time of damage is subject to discoloration and decay. Several of the points above are illustrated and explained in more detail with Plates 247, 248, 254, 255, and 257.

Trunk cracks can also be caused by the force of wind on the crown of a tree. The tension in wood on the windward side of the trunk is transmitted across the stem-root boundary and causes a delamination (layer separation) in a root. This crack later elongates up into the trunk and may eventually be a cause of trunk breakage. *References:* 451, 589, 670, 721, 919, 2162, 2163, 2239, 2373, 2398, 2511, 2513, 2812, 3024, 3027, 3060, 3397, 4349

A–C. Relationships among drought stress, internal defect, and stem cracks in *Platanus* ×*acerifolia* (London planetree). A. Trees in this young landscape planting, subject to severe water stress during summer, display dieback, death of occasional individuals, and basal sprouting. Basal sprouts developed most often on trees with weakly growing tops and cracked, defective trunks. B. The trunk of the tree with basal sprouts in A has 2-year-old drought cracks and associated cankers, accentuated by woundwood formation. C. The trunk of a similar tree with discolored bark that indicates cambial dieback along drought cracks. If this tree had not been cut for examination, it would have survived despite an elongate canker associated with the trunk cracks and extensive internal defects that are visible on the cut surface. The tree's history was interpreted from examination of growth rings and internal defects. At the height of the cut the tree had completed 5 years' growth. It was pruned and transplanted before the third growing season. The pruning wounds were poorly compartmentalized, leading to death and discoloration of all wood present at the time of pruning. One pruning wound is still open. Three radial cracks formed in discolored wood associated with pruning wounds. During dry weather after the current year's growth was complete, one radial crack, pointing upward in the photo, opened to the stem surface, exposing sapwood that quickly became discolored. A second crack, at right, broke the barrier zone surrounding the wood of the first 2 years but did not break the vascular cambium or bark. Discoloration spread from the defective core outward along the crack. A third radial crack, pointing down in the photo, remained confined to the compartmentalized wood of the first 2 years. Closure of the two pruning wounds intersected by the diagonal cut became complete during the year when the tree was cut, but the wound on the near side of the stem reopened along the line of woundwood fusion. The pruning wound that remained closed, at top right, lay beneath the short vertical seam that is visible on the bark surface at the top of the left segment (NY, Aug).

D. Vertical cracks and swelling of the butt of a young recently transplanted *Quercus rubra* (red oak) that sustained freeze damage to the cambium during the previous winter. The short cracks shown here closed completely within a year after the photograph was made, but one crack reopened after 6 years and then was 1.3 m long. Cracks that begin and then close in young trees such as this one may reopen when the trees are much older, as happened in the *Q. velutina* (black oak) shown in F. The volume of wood potentially subject to discoloration and decay expands each year that a crack reopens (NY, Aug).

E. *Casuarina* sp. (Australian-pine, she-oak) with a lens-shaped wound resulting from freeze damage to the vascular cambium several years earlier. The wound began as a bark crack soon after freeze damage. Growth, as explained in the text, caused the crack to widen and the bark on the right side to flare out from the wood. The exposed woody cylinder is being colonized by microorganisms that cause discoloration and decay (FL, Mar).

F. A frost rib on a mature *Quercus velutina*. The "rib" is woundwood that developed where a crack opened through the cambium and bark. The tree had sustained a wound at the root collar many years before (scar visible as irregular bark surface below the rib), resulting in internal defect within which a radial crack formed and eventually extended to the surface (NY, Jul).

Bark Formation and Restoration (Plate 257)

Bark comprises all tissues outside the vascular cambium. The inner bark (young phloem) close to the cambium is the location of conducting elements involved in transport of photosynthetic products and growth regulators. The corky outer bark (rhytidome) is an inert protective covering. The zone between conductive phloem and rhytidome consists primarily of phloem parenchyma, crushed conductive cells (sieve cells in gymnosperms, sieve tube elements in angiosperms), phloem fibers in some plants, phelloderm, and the cork cambium (phellogen). The cork cambium and its derivatives, collectively called periderm, build up the thickness of corky bark. Wounds and infections in bark elicit responses that, when complete, isolate damaged tissues and restore the continuity of corky bark. These responses involve production of antimicrobial, water-proofing, and wall-strengthening compounds; regeneration of cork cambium; and production of new periderm. The processes of normal periderm formation and restoration of periderm around and beneath damaged tissues are similar. Responses to wounding and infection in bark are homologous with compartmentalization of wounds and infections in xylem (Plates 254, 255). Plate 257 presents an overview of bark formation and responses of living bark to wounding and infection.

We first consider bark like that pictured in panels A–C. Viewed in cross section (C) the successive major tissues outside the xylem (x) are vascular cambium (vc); conductive secondary phloem (csp) consisting of sieve tubes and companion cells in angiosperms and files of sieve cells in gymnosperms, phloem parenchyma, and phloem fibers (fibers absent in the species shown); nonconductive secondary phloem (nsp) formed in previous years, consisting of parenchyma cells, phloem fibers, and crushed remains of conductive cells (and companion cells in angiosperms); cortex (co), a parenchymatous tissue that formed during the first year of stem growth and served for support and protection before wood and bark formed; the cork cambium (cc); and a layer of cork (ck), or phellem, derived from the cork cambium. (The term *cortex* is often used loosely to refer to living bark tissues other than conductive phloem. In this discussion, however, *cortex* is used only as defined above.) The cork cambium first arises in the cortex where a sheet of mature parenchyma cells returns to meristematic condition and the cells begin to divide and produce cork. The periderm thus formed is thin and more or less impervious to water and microorganisms. It is covered by imperceptible remains of the original epidermis.

Now consider brown patches on green bark and the transition to mature bark (Plate 257B–D). Patches of brown bark on otherwise smooth green stems may be symptoms of abrasion or disease but often represent an early stage of normal development of corky bark. Moving outward from the vascular cambium as before, we encounter a new cork cambium that formed from parenchyma cells in the nonconductive secondary phloem beneath the first periderm. The edges of this new sheet of cork cambium meet the original one at the edges of a brown patch. A new periderm, *exophylactic periderm*, generated by this cork cambium isolates the tissues beyond it (secondary phloem, phloem fibers in many species, and the first periderm and epidermis), which then constitute a plate or scale of dead bark. As successive bark plates form, their edges overlapping, the stem becomes encircled by corky bark (rhytidome). Thereafter, new exophylactic periderms form more or less regularly in the nonconductive secondary phloem, preventing indefinite thickening of that tissue and causing indefinite thickening of the outer corky bark. Bark plates are relatively inelastic and crack or flake off as a stem expands, producing the characteristic texture of mature bark. Root bark forms similarly. Some trees, such as *Fagus* (beech), produce rhytidome only in response to wounding or infection.

Whenever a portion of the cork cambium becomes nonfunctional through wounding (as in panel E) or infection, this dysfunction triggers processes that lead to formation of *necrophylactic periderm*, which restores continuity of the protective layer over living tissues. When the cork cambium is breached or senesces, the underlying parenchyma cells begin to senesce and die, producing a fungitoxic environment as their final function. Parenchyma cells somewhat more distant secrete lignin and suberin in their walls before they too senesce and die. These changes occur within a few days after wounding (longer in cold weather) and create a temporary protective tissue that impedes water loss and is refractory to organisms (impervious tissue, labeled *it*). The new cork cambium differentiates after 2 to many weeks, depending on the season, at the inner edge of the impervious tissue. If, after wounding or infection, periderm is not renewed fast enough to prevent further damage by drying, or the process is disrupted by an insect or pathogen, it begins again in tissues beneath and adjacent to the new damage.

If a wound or infection reaches the vicinity of the vascular cambium, this meristem generates abnormally abundant parenchyma cells that produce toxic and impervious substances and participate in restoration of phloem and cork cambium. If the vascular cambium is impaired or breached, impervious tissue and periderm formation extend to the sapwood, and impervious tissue extends into the wood. There, through the activities of xylem parenchyma cells, wood elements become plugged with gums, toxic metabolites, and tyloses, and the walls of wood parenchyma cells become impregnated with suberin. Impervious tissue in xylem constitutes the boundary of a compartment within which non-functional xylem is isolated from normal tissues.

Leaf scars are natural wounds that are sealed off by the processes discussed above. Xylem tracheary elements in a leaf trace become plugged, and parenchyma cells beneath the separation zone (where the leaf breaks off) produce first an impervious tissue and then a periderm that separates the surface of the scar from underlying normal tissues.

References: 52, 364–366, 371, 400, 450, 1169, 1552, 1796, 2162, 2163, 2341, 2758, 2933, 2934, 3595, 3597, 4014, 4086, 4185, 4432

A. Stages in formation of mature bark on *Platanus* ×*acerifolia* (London planetree). Smooth gray bark has only a thin periderm overlying living secondary phloem. Brown patches are zones where a deeper periderm has formed, separating senescing superficial tissue from inner bark. Tissue outside the cork sheet died and turned brown. This is the beginning of a bark scale. Tan patches are where bark scales formed and sloughed (NY, Dec).

B, C. Early stages of formation of mature bark in *Pinus strobus* (eastern white pine). B. Surface view. C. Magnified view of a cross section through the bark shown in B. Beneath each brown zone an exophylactic periderm separates living from dead secondary phloem. Abbreviations are identified below (NY, Dec).

D. Magnified view of a cross section of mature *P. strobus* bark. A thick plate of dead bark consisting of alternating zones of cork and secondary phloem has built up from successive events of periderm formation that isolated zones of nonconductive secondary phloem. The photograph was made at the edge of a bark crevice (NY, Oct).

E. Microscopic cross section of bark of a young *P. strobus* stem, showing the isolation of a wound by formation of necrophylactic periderm in the cortex and secondary phloem. The stem was wounded in midsummer and was sectioned 3 weeks later. Tissues near the wound are dead, as revealed by red staining. A cork layer, visible as a thin band of cells in regular tiers, was forming at the margin of the dead area, isolating it from normal tissues. Abbreviations are identified below. Magnification 60× (NY, Aug).

F–H. Microscopic cross sections of *Pinus resinosa* (red pine) roots responding to attack by the pathogen *Heterobasidion annosum*. Diseased tissue is at upper right in each image. Healthy tissues are stained cyan except for phloem parenchyma cells containing phenolic materials, which have reddish purple contents. F. The edge of a spreading lesion involving bark and wood. No effective resistance is being expressed. G. The edge of a lesion where defense reactions in root bark have halted the tangential spread of infection. A reaction zone with impervious tissue is apparent at the edge of the lesion, and a cork cambium is differentiating in the adjacent living secondary phloem. Defense in xylem will depend on reactions of parenchyma cells in medullary rays and lining resin ducts. H. The edge of a lesion that is being isolated by necrophylactic periderm formation. Cork has begin to form beneath impervious tissue at left. Parenchyma is being produced in place of normal phloem and xylem where necrosis approaches the vascular cambium at right. Magnification 90× (NY, Sep).

Abbreviations: cc, cork cambium; ck, cork; co, cortex; csp, conducting secondary phloem; fp, first periderm; it, impervious tissue; mr, medullary ray; nsp, nonconductive secondary phloem; p, parenchyma; rc, resin canal; rp, resin pocket; sp, secondary phloem; vc, vascular cambium; x, xylem.

Photo credit: E—G. W. Hudler

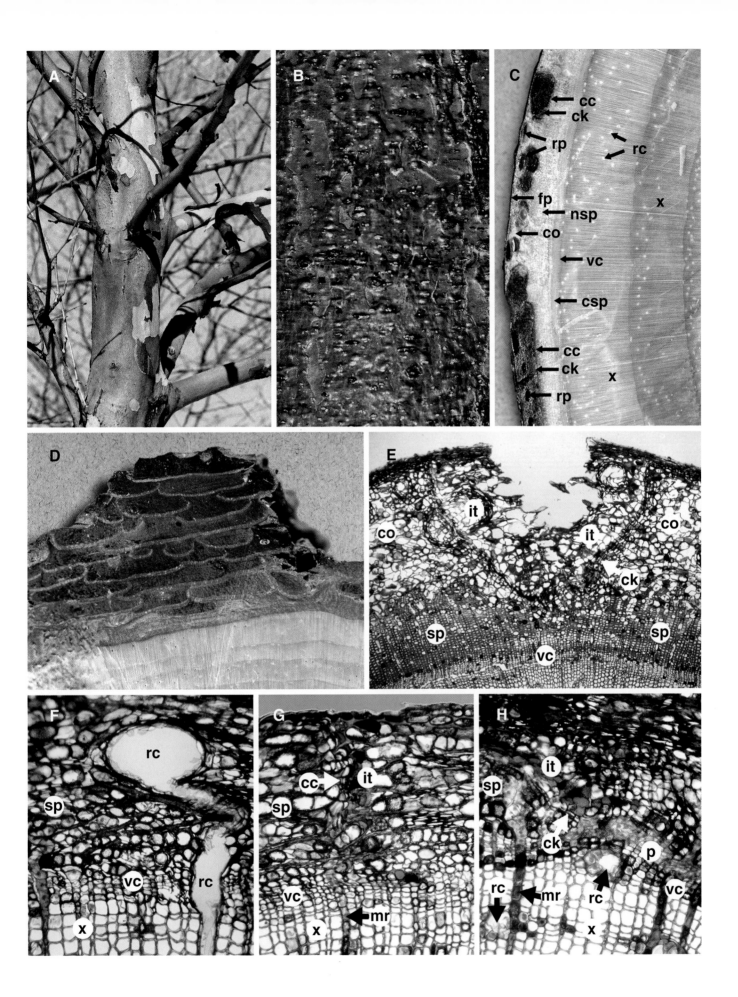

Associations of Normal Woody Plants with Other Organisms (Plates 258–261)
Smooth Patch and Bark Rot, and Normal Foliar Shedding (Plate 258)

Two unrelated topics are presented here: the decay of bark on living trees by saprobic fungi, and normal foliar color changes and shedding.

Smooth Patch and Bark Rot

Many fungi colonize and decompose the dead, corky outer layers of bark on living tree trunks. They cause no known harm to the trees, but their basidiocarps on the bark may be misinterpreted as indicating the presence of a canker or an extensive dead area with decaying wood beneath.

Smooth patch, also known as white patch and bark patch, is caused by species of *Aleurodiscus* (Aleurodiscaceae), *Dendrothele* (Corticiaceae), and *Hyphoderma* (Hyphodermataceae), all of which are basidiomycetes in the order Stereales. Fungi in these groups produce smooth, more or less disclike, light-colored basidiocarps that may aggregate or coalesce into larger structures. Except for two species mentioned below, these fungi are saprobic colonists of bark and wood, common on trunks, branches, and twigs. *A. oakesii* is a noteworthy member of the group in North America; it occurs also in eastern Asia and the Pyrenees in Europe. It causes smooth patch on *Acer* (maple), *Betula* (birch), *Carpinus* (hornbeam), *Carya* (hickory), *Fraxinus* (ash), *Ostrya* (hop-hornbeam), *Quercus* (oak), *Rhododendron*, *Salix* (willow), *Tilia* (basswood, linden), and *Ulmus* (elm). Smooth patch results from the decomposition and sloughing of rough or fissured outer bark in a slowly expanding area. The resulting slightly sunken, relatively smooth, light gray area contrasts with normal bark texture and color. The largest patches apparently form by confluence of small ones. Basidiocarps of *Aleurodiscus oakesii* are visible all year, but in dry weather they may curl or shrivel and become inconspicuous. Individual basidiocarps are 1.5–5.5 mm in diameter but appear much larger where they are confluent or growing in clusters.

Other smooth-patch fungi that occur on bark of living and dead hardwood trees in the USA or Canada include *Dendrothele acerina*, *D. alliacea*, *D. candida*, *D. dryina*, *D. griseocana*, *D. macrodens*, *D. microspora*, *D. nivosa*, *D. strumosa*, and *Hyphoderma baculorubrense*. The last-named fungus occurs on *Juniperus* species and *Q. virginiana* (live oak). Several of the *Dendrothele* species have overlapping host ranges. Some common substrate genera and the number of the above-named *Dendrothele* species reported to occur on them include *Acer* (maple) 6, *Carya* (hickory) 6, *Castanea* (chestnut) 2, *Fagus* (beech) 1, *Fraxinus* 7, *Liquidambar* (sweetgum) 2, *Nyssa* (tupelo) 2, *Ostrya* 2, *Platanus* (sycamore, planetree) 2, *Quercus* 3, *Salix* 3, *Tilia* 4, and *Ulmus* 5. Only two species, out of several dozen in the genera to which North American smooth-patch fungi belong, are thought to be pathogenic. *Aleurodiscus amorphus* causes cankers on trunks and branches of unthrifty *Abies* (fir) and other conifers, and *A. canadensis* possibly causes twig blight of *Picea rubens* (red spruce). Both fungi also occur harmlessly on various trees.

Bark rot is caused by several fungi of the Hymenochaetales and Polyporales that decay the outer bark of living trees but do not cause smooth patches. *Perenniporia phloiophila* (Polyporales, Polyporaceae) is one of these. It apparently occurs only on trunks and large limbs of *Quercus virginiana* and is found from South Carolina to Texas and Mexico. Its perennial basidiocarps develop more or less flat against the bark as units up to 6 cm wide. They may become confluent and up to 1 m long or wide. The pore surface is cream colored to pale buff, and the circular to angular pore mouths number three to five per millimeter. This fungus is easily confused with *P. medulla-panis*, which decays the bark and wood of various deciduous trees as well as fence posts and structural timbers

throughout eastern North America and in California. Substrates of the latter fungus include *Acer* (maple), *Artemisia* (sagebrush), *Castanea* (chestnut), *Fagus* (beech), *Fraxinus* (ash), *Gleditsia* (honeylocust), *Liriodendron* (tuliptree), *Ostrya* (hop-hornbeam), *Populus* (poplar), *Prunus* (cherry), *Quercus* (oak), *Rhus* (sumac), *Robinia* (black locust), *Tilia* (linden), and *Ulmus* (elm). Microscopic examination of spores and anatomical characters is necessary to distinguish *P. medulla-panis* from related fungi.

References: 1099, 1154, 1227, 1279, 1280, 1299, 1479, 2218, 2289, 2290, 2879, 2935, 3699, 3981

Normal Foliar Senescence and Shedding

All trees and shrubs renew part or all of their foliage annually, producing new leaves and shedding old ones. The leaves of deciduous plants live only a few months, but those of evergreens live from one to several years, depending on the species. Most evergreens shed their oldest leaves after producing new ones, although some shed old leaves throughout the year. Usually the old leaves turn yellow and then straw colored to brown before dropping. This display of normal senescence is sometimes mistaken for malfunction.

Conifers normally shed their oldest foliage in autumn. Nearly all *Pinus* species (pines) bear their needles in fascicles of two–five, and the needles remain together when they drop. If the preceding growing season or winter was particularly stressful, some needles may be shed in spring, but most drop later. Trees of the Cupressaceae such as *Juniperus* (juniper), *Chamaecyparis* (false-cypress), and *Thuja* (arborvitae, western red cedar) bear scalelike leaves covering tiny branchlets and shed the oldest branchlets in autumn. Needle shedding in conifers such as *Abies* (fir) and *Picea* (spruce) that bear several age classes of needles is not rigidly restricted to the oldest age class, although it is concentrated there. The foliage on a given branch segment may thin progressively over 2–3 years.

Broadleaved evergreens vary in their habits of foliar shedding. Some drop their old leaves as soon as new ones grow; other species wait until autumn. The timing is often influenced by environmental stress. Early shedding may occur after a severe winter.

Endophytic fungi in normal leaves often begin to grow and cause discoloration as the leaves undergo senescence. These organisms reproduce in fallen leaves and participate in the recycling of carbon and minerals.

A–C. Smooth patch and bark rot. A, B. Smooth patch caused by *Aleurodiscus oakesii* on *Quercus arizonica* (Arizona white oak). A. Smooth patch on a mature tree. B. Close view of basidiocarps of a young colony of *A. oakesii,* approximately natural size (AZ, Jul). C. Basidiocarps of *Perenniporia phloiophila,* shown about one-half natural size, on bark of *Q. virginiana* (live oak) (TX, May).

D–G. Normal leaf senescence and shedding. D. Autumn browning of *Thuja occidentalis* (eastern arborvitae). Instead of shedding the oldest individual scale-leaves, this species sheds the oldest branchlets (NY, Oct). E, F. Autumn browning of old needles on *Pinus resinosa* (red pine). Third-year (2-year-old) needles have died and are beginning to drop (NY, Oct). G. Yellowing and dropping of second-year (1-year-old) leaves of *Magnolia grandiflora* (southern magnolia). This species normally drops its second-year leaves throughout spring and into summer. The yellow leaves in this case had been stressed by an unusually cold winter and were dropping somewhat prematurely (GA, May).

Symbiotic Relationships of Roots (Plates 259, 260)

Mycorrhizae

The term *mycorrhiza* means fungal root. A mycorrhiza is an absorbing organ composed of both root and fungal cells in symbiosis. The fungal partner (mycobiont) obtains carbohydrates, lipids, vitamins, and other organic compounds from the plant. The plant receives from the fungus various mineral nutrients, especially phosphorus, that would be less available to it from soil via its own absorptive capability. The mycorrhizal fungus also may enhance the plant's tolerance of environmental extremes and its resistance to or tolerance of pathogens. Most plants in the wild, including all trees and shrubs, are mycorrhizal, and many species are obligately so; that is, they cannot grow for long or complete their life cycles in a nonmycorrhizal state. Similarly, most mycorrhizal fungi are unable to persist in nature as saprobes. Some are obligately symbiotic and cannot be cultivated apart from living rootlets. Several thousand fungal species enter mycorrhizal associations.

All mycorrhizae consist of a root or rootlike structure plus external (extramatrical) mycelium that extends into the soil or other root medium. Mycorrhizal fungi colonize only primary tissues, usually the cortex of young rootlets. The stele and the apical meristem of a mycorrhiza remain uninfected. On many woody plants only short roots, as opposed to those that grow extensively, become mycorrhizal. The symbiosis extends the life of cortical cells. Mycorrhizal roots of some types are not perceptibly different in form from uninfected rootlets; other types become more or less swollen and highly branched. The changes in form are induced, at least in part, by hormones produced by the fungi.

Mycorrhizae are classified according to their anatomy, which in some types is linked to a particular group of plants or fungi. Ectomycorrhizae are characterized by intercellular colonization and by envelopment of short roots by fungal hyphae. Living cortical cells become separated (except possibly for connections via plasmodesmata) by a network of hyphae called the Hartig net. This network is continuous with a mantle of hyphae over the root surface. Ectomycorrhizae become swollen in comparison with nonmycorrhizal short roots. Some develop elaborate branching, and many of those with thick mantles are distinctively colored. Many or perhaps most ectomycorrhizal fungi can be induced to grow in pure culture, although they tend to grow more slowly than saprobic fungi.

Endomycorrhizae are those in which the fungal partner colonizes rootlets intracellularly, forming either hyphal coils or arbuscules (highly branched absorptive structures) within living cortical cells. The intracellular structures mediate the exchange of materials between the symbionts and eventually senesce along with the plant cells. Arbuscular mycorrhizal fungi often also produce large sporelike vesicles within cortical cells. Usually the endomycorrhizal root differs little from uninfected roots in form or color, and its mycorrhizal condition can only be discerned microscopically. Major subgroups of endomycorrhizae include the arbuscular type, possessed by most plants, and the ericoid type, formed by many plants in the Ericales. Ericoid mycorrhizae have intracellular hyphal coils that presumably serve the same functions as arbuscules. Two additional mycorrhizal types—ectendomycorrhizae and arbutoid (for *Arbutus* and related plants) mycorrhizae—have characteristics of both ecto- and endomycorrhizae: sheathing mycelium and a Hartig net as well as intracellular hyphae. The mycobionts of ericoid and arbutoid mycorrhizae are ascomycetes and basidiomycetes, respectively, and are cultivable apart from their plant partners.

Most plants are endomycorrhizal, and the great majority have arbuscular mycorrhizae. The arbuscular mycobionts are classified in the Zygomycota, Order Glomales. Their hyphae are aseptate, and they reproduce asexually by chlamydospores or azygospores. Many species produce globose chlamydospores that are gigantic in comparison with spores of other fungi, attaining diameters of 100–300 μm or more. Arbuscular mycorrhizal fungi are obligately symbiotic in nature and have not been induced to grow on laboratory substrates. Their spores germinate readily on various media, but no colony develops. These fungi are dispersed with infected plants and as spores or hyphae with soil. In general they are unspecialized with respect to the identity of the plant partner. Any arbuscular mycorrhizal fungus can form mycorrhizae with many different plants, both herbaceous and woody, and a plant species in which arbuscular mycorrhizae are normal is receptive to various arbuscular mycorrhizal fungi. These fungi vary widely in the degree to which they promote plant growth and health.

Ectomycorrhizae form only on woody plants, but these represent at least 30 families, notably Betulaceae, Fagaceae, Pinaceae, Tiliaceae, and Salicaceae (birch, beech, pine, linden, and willow, respectively). Some tree groups such as *Populus* (poplar) and *Salix* (willow) may form either ecto- or endomycorrhizae. The mycobionts in ectomycorrhizae are mostly basidiomycetes—mushroom-forming agarics and boletes (Plate 260) and corticioid fungi whose basidiocarps grow as fertile sheets on wood or bark. They are dispersed in nature as airborne basidiospores. Mycobionts are also transported with plants in commerce. Some ectomycorrhizal fungi are suitable partners for only a few species of trees; others enter symbiosis with many. Similarly, some trees are capable of forming ectomycorrhizae with hundreds of fungi. Mature trees usually host several ectomycorrhizal fungi simultaneously. The diversity of mycobionts in the roots increases with a tree's age.

Production of fruit bodies or large resting spores by a mycorrhizal fungus requires it first to develop an extensive mycelium and then to extract enough carbon and nitrogen from the plant to support reproductive growth. Fungi that are ectomycorrhizal with seedlings begin to fruit as early as the second year of seedling growth and may fruit annually thereafter, the abundance of sporocarps varying with weather.

Mycorrhizal fungi absorb mineral nutrients from soil and translocate them to roots. The accumulation of phosphorus by mycorrhizae is striking. Although the total phosphorus in most soils is adequate for plant growth, much of it is bound in relatively insoluble forms that are scarcely available to nonmycorrhizal roots. Extramatrical mycelium of a mycorrhiza has a much greater absorbing surface, exploits a greater volume of soil, and thus obtains phosphorus more readily than a nonmycorrhizal root can. Mycorrhizae enhance the uptake of other mineral nutrients less dramatically. They also mingle

A. Six-month-old *Pseudotsuga menziesii* (Douglas-fir) seedlings mycorrhizal with the fungus *Laccaria bicolor* (left) in comparison with nonmycorrhizal stunted seedlings (greenhouse plants).

B–G. Magnified views of ectomycorrhizae and nonmycorrhizal short roots. B, C. Pinnately branched ectomycorrhizae formed by *L. bicolor* and *Amanita muscaria,* respectively, with *P. menziesii* (WA, Aug; NY, Aug). D, E. *Pinus resinosa* (red pine) roots with unidentified ectomycorrhizae (D) exhibiting the dichotomous branching typical of mycorrhizae on pines, in comparison with nonmycorrhizal short roots (E) (NY, Aug). F, G. Ectomycorrhizae formed by *Cenococcum geophilum* and *Pisolithus tinctorius,* respectively, with *Pinus mugo* (mugo pine). Stout black hyphae extend from the mantles of *C. geophilum* mycorrhizae. Buff-brown mycelium extends from the *P. tinctorius* mycorrhizae into the root medium; the white object is a coarse grain of quartz sand (greenhouse plants).

H, I. Microscopic views of the mantle and Hartig net of unidentified ectomycorrhizae from *Pseudotsuga menziesii.* H. Cross section: fungal mantle is at upper left; large round cells are cortical parenchyma; red-stained root cells, some collapsed, are tannin cells rich in phenolic compounds. I. Longitudinal section: the mantle at left contains collapsed epidermal and tannin cells; two cortical cells, one with red-stained tanniferous contents, are at center; the mazelike aggregation of hyphae at right is part of the Hartig net; the red-stained object below the plane of focus is a cortical cell nucleus.

J–N. Microscopic aspects of endomycorrhizae. J. An azygospore of *Glomus* sp. on a *Malus pumila* (apple) rootlet. K. Arbuscules of *G. fasciculatus* in cortical cells of an *M. pumila* root. L. Highly magnified view of one arbuscule of *G. mosseae;* colorless cortical cell wall is at right. M. Vesicles and arbuscules in a *Vitis* (grapevine) root. N. Intracellular hyphae of an unidentified mycorrhizal fungus in an *Acer saccharum* (sugar maple) rootlet. Scale bars = 25 μm in H, I, L, N; 50 μm in J, K, M.

Photo credits: C, N—R. F. Lucas; J, K—T. J. Stephens; L—M. Brundrett; M—D. R. Deal

with litter-decomposing organisms in the humus layer and there obtain nutrients from organic sources that would be unavailable to a nonmycorrhizal plant.

Mycorrhizae also have an important role in the carbon cycle because the hyphae of mycorrhizal fungi release carbon compounds into soil. The glycoprotein glomalin, secreted by arbuscular mycorrhizal fungi, is a major component of soil organic matter. This substance, discovered in the 1990s, enhances the aggregation of soil particles and is thus important for maintenance of soil structure and water-holding capacity.

Fertilizer can be substituted for the nutritional benefits of mycorrhizae in many circumstances. High fertility levels such as those commonly used in nurseries often mask any mycorrhizal contribution to plant nutrition and in some cases inhibit mycorrhizal formation or select for mycorrhizal fungi that tolerate high levels.

Mycorrhizae enhance plants' tolerance of adverse conditions such as drought, high temperature, salinity, acidity, or toxic elements in soil. The degree of enhancement varies with the mycobiont; it has been demonstrated most extensively in *Pinus* ectomycorrhizal with the puffball fungus *Pisolithus tinctorius*. Enhanced environmental tolerance is important for successful revegetation of disturbed or polluted sites such as mine spoils and landfills and also for many planting sites in horticultural landscapes. Regarding drought tolerance, mycobionts mediate resistance to water flow from soil into plants. A mycorrhizal plant maintains less resistance than does a nonmycorrhizal plant and is therefore able to utilize soil water more effectively. Mycorrhizal hyphae may also have access to small water-filled pores that are inaccessible to roots.

Many reports exist of mycorrhiza-mediated tolerance or resistance of plants to pathogens, although cases of increased susceptibility or damage have also been noted. Enhanced tolerance presumably follows from the superior nutrition and water economy of the mycorrhizal plant. In addition, some ectomycorrhizal fungi produce antibiotics that are toxic to fungal pathogens, and various mycorrhizal fungi induce resistance to pathogens.

Dramatic benefits of mycorrhizal inoculation have been documented in certain situations: soils that have been fumigated but not adequately recolonized by mycorrhizal fungi, regions where ectomycorrhizal trees have never grown, and sites where topsoil has been removed or buried. One of the famous examples is the story of an early planting of *Pinus elliottii* (slash pine) in Puerto Rico. Seedlings grown from seed that had been imported to the island remained tiny and chlorotic and began to die. Ectomycorrhizal fungi were apparently absent on the site. A group of the declining trees was treated with a small amount of crude mycorrhizal inoculum in the form of soil from a pine stand on the mainland and began to grow normally, while control trees continued to decline. This result corroborated earlier evidence that pines require ectomycorrhizal fungi. Similar accounts exist regarding the necessity for mycorrhizae in establishment of pines and other trees in biocide-treated soils, prairie soils, mine spoils, and bare eroded soils at sites of severe air pollution damage. Benefits have accrued from inoculation with natural materials (topsoil, roots, duff) containing mycorrhizal fungi, as well as with pure cultures of ectomycorrhizal fungi, chopped roots containing endomycorrhizal fungi, and spores of either ectomycorrhizal or endomycorrhizal fungi.

Despite the documented benefits of mycorrhizae, inoculation of plants, soils, or seeds with mycorrhizal fungi has not led to sustained cost-effective enhancement of plant performance or disease control in most field situations. The reason is that indigenous fungi enter mycorrhizal association with woody plants in most places, so a mycorrhizal condition is nearly assured without special arrangements. Moreover, strains of mycorrhiza-forming fungi adapted to a particular locale tend to replace introduced strains, even though the latter may initially promote superior plant growth. Commercial inocula of mycorrhizal fungi have been developed but are not widely utilized. Claimed benefits of inoculation of the root zones of established landscape trees and shrubs with mycorrhizal fungi are plausible but not based on published research performed in landscape situations.
References: 20, 55, 146, 400, 437, 553, 603, 739, 800, 967, 983, 1007, 1263, 1306, 1420, 1523, 1932, 2043, 2229, 2468, 2487–2489, 2751, 2815, 3059, 3108, 3229, 3230, 3297, 3395, 3557, 3715, 3736, 3737, 3886, 3890, 3891, 4066, 4073, 4141, 4446, 4473

Nitrogen-Fixing Associations

Actinorhizae. Nitrogen-fixing nodules called actinorhizae develop on the roots of plants in more than 25 genera of angiosperms as the result of infection by bacteria of the genus *Frankia* (Actinomycetales, Frankiaceae). These nodules are perennial, multilobed, coralloid structures as much as 3–4 cm in diameter. Actinorhizae have been studied mainly on *Alnus* (alder) but also occur on such plants as *Casuarina, Ceanothus, Cercocarpus* (mountain-mahogany), *Elaeagnus* (Russian-olive), *Myrica* (bayberry, sweetfern), *Purshia* (bitterbrush), and *Shepherdia* (buffaloberry). Nodules on *Alnus* may live 3–8 years as the bacterial endophyte perennates within the lobes. Actinorhizae fix much more nitrogen than is used for their own growth and maintenance; the remainder is translocated to other parts of the plant.

Infection leading to formation of actinorhizae on most hosts occurs via root hairs, within which thin actinomycete hyphae develop. Some hosts are penetrated intercellularly. Cortical cells near the site of infection are stimulated to begin division, and the endophyte soon enters these cells. The root swells at the infection site, and the swelling (called a primary nodule) appears reddish if viewed in transmitted light. A lateral root is induced at the swelling, and it is immediately transformed into a true nodule, its cortex infected by the endophyte. The apical meristem of the nodule remains uninfected, and branching from this point results in the complex actinorhizal form. The endophyte forms intracellular hyphae and vesicles, and some strains produce spores. Vesicles are the actual sites of nitrogen fixation. Spores often persist in old parts of a nodule where hyphae and vesicles have been digested by host cells. As lobes of a nodule decay, spores are released into the soil, where they are capable of initiating new infections. Strains of *Frankia* show considerable plant specificity, and they vary in infectivity and efficiency of nitrogen fixation.

Nitrogen-fixing nodules on legumes. Symbiotic bacteria in six genera of the Rhizobiaceae (a family of soilborne Proteobacteria) induce root nodules on plants in the Fabaceae (legumes) and a few other known plants. These bacteria, collectively called rhizobia, were all at one time classified in the genus *Rhizobium*. They enter root hairs or wounds on young roots and produce strands that extend into cortical cells, stimulating the latter to begin meristematic activity that results in nodule formation. Bacteria released within cortical cells of the developing nodule differentiate into swollen bacteroids, which are the sites of nitrogen fixation. Nodules on woody legumes apparently live more than one season. Strains of rhizobia are specialized for particular plant species and vary in infectivity and efficiency of nitrogen fixation. A strain symbiotic with an agricultural plant, for example, would not be suitable for a tree. Trees with bacterial nodules are capable of fixing a significant quantity of nitrogen, from about 20 kg to more than 200 kg per hectare annually. Virtually all of this fixation in temperate zones occurs during the growing season, as fixation is inhibited at low temperatures.
References: 446, 533, 885, 917, 1057, 1138, 1219, 1282, 1291, 1319, 1837, 2115, 2261, 2437, 2871, 2983, 3513, 3767, 3815, 3937, 4065, 4066, 4085, 4294

A–F. Basidiocarps of ectomycorrhizal fungi. A. *Laccaria* sp. in association with *Pinus ponderosa* (ponderosa pine) and *Pseudotsuga menziesii* (Douglas-fir). One mushroom is upended to reveal the lower surface of its cap. B, C. *Amanita muscaria* (B) and *A. pantherina* associated with the same two tree species. The former fungus is known as the fly agaric, the latter as the panther fungus. Both are poisonous to humans if eaten (AZ, Jul). D, E. *Suillus* sp. in association with *P. ponderosa*. The poroid lower surface of a basidiocarp is shown in E (AZ, Jul). F. *Suillus subaureus* fruiting in a lawn near *Pinus strobus* (eastern white pine), *Quercus alba* (white oak), and *Tsuga canadensis* (eastern hemlock) (NY, Sep).

G. Actinorhizae on an *Alnus* (alder) root. Feeder roots of the plant, at lower left, provide a size standard (NY, Jul).

H. Nitrogen-fixing nodules on rootlets of *Robinia pseudoacacia* (black locust). Typical nodules are 3 mm long (NY, Oct).

Epiphytes, Lichens, and Moss (Plate 261)

Epiphytes

Epiphytes are photosynthetic plants that grow on, but do not parasitize, other plants. They usually attach to the bark of trees by means of roots or rhizomes and use rain or water vapor in saturated air to meet their water needs. Essential nutrients reach an epiphyte in water that has leached leaves and branches above, from air contaminants, and through microbial activity in organic debris that accumulates at the site of attachment. Some epiphytes harbor nitrogen-fixing bacteria. Except for the mosses that are abundant on trees in the North, epiphytes occur mainly in warm to tropical regions in light exposures ranging from full sunlight to deep shade. Epiphytic species have evolved in several plant families, notably the Bromeliaceae (pineapple family) and Orchidaceae (orchid family). Epiphytes that can grow in full sunlight sometimes develop luxuriantly on dead or dying trees and for this reason have often been assumed to be parasitic or pathogenic. Two representatives of the New World genus *Tillandsia* (airplant) of the Bromeliaceae are featured here.

Tillandsia comprises more than 350 species; most occur in Mesoamerica and South America. About 18 species are native in southern USA, distributed from Arizona eastward. They are sometimes misconstrued as parasites because they are unlike most seed plants in form, habit, and color. They are adapted to collect and conserve water, and some species can withstand prolonged drying. The adaptations vary by species and include reduced leaf size; water-impounding foliar structures; stomatal opening only at night, when least water is lost by evaporation; and Crassulacean acid metabolism. Plants with the last-named characteristic capture CO_2 at night and convert it to malate, then recover it during daylight by decarboxylation of malate, and fix the carbon in the photosynthetic carbon reduction cycle.

T. usneoides (Spanish moss, or graybeard) is a prominent representative of its genus that occurs from southern Maryland to Mexico and southward into South America. Its stems are slender, branched, and pendant, attaining lengths of 6 m or more. Its leaves are scattered, linear, and up to 5 cm long. The flowers are borne inconspicuously in leaf axils. Some observers have noted diminished growth or subnormal leaf size on tree branches draped with Spanish moss and on this basis have suggested that the plant is injurious, but proof is lacking.

T. recurvata (small ballmoss, bunch moss), which occurs from Arizona eastward in the USA and in Central America and South America, attaches to tree bark, twigs, wires, poles, and even chain-link fences. It appears as gray-green tufts that develop into dense clusters composed of numerous individual plants. Its stems are branched but short, and the ash-colored to reddish scaly leaves grow to lengths of about 5 cm. When it grows on bark *T. recurvata* produces rootlike holdfasts for attachment. Its small blue flowers and later the seeds are produced on stalks 8–10 cm long. The seeds are windborne. Plants found on wires have likely been dislodged from some other surface, for wires would seem to be inhospitable sites for seed germination. *T. recurvata* is considered a weed species in many circumstances.

Pathogenicity (but not parasitism) is claimed for some species of *Tillandsia*, including *T. recurvata*, based on observations that large populations on certain hosts create dense shade and thus suppress photosynthesis and growth and cause premature decline. Evidence for pathogenicity based on experiments or careful comparative observations has not appeared. Nevertheless, herbicides, copper fungicides, or sodium bicarbonate have been used to suppress these plants.

References: 102, 249, 319, 320, 508, 606, 678, 741, 1080, 1153, 2391, 2481, 2716, 2877, 3161, 3231, 3320, 3867, 4124, 4530

Lichens and Moss

A lichen is a perennial compound organism consisting of a fungus (usually an ascomycete) and a green alga or cyanobacterium in symbiotic union. The terms *mycobiont* and *phycobiont* are often used for the fungal and the algal partner, respectively. The entire structure, called a thallus, is long-lived and morphologically unlike either of its constituent organisms. Fungal and algal or bacterial components in the vegetative part of a lichen thallus can be distinguished from one another only by microscopic examination.

The alga or bacterium supplies the fungus with photosynthetic products and vitamins, and the fungus obtains water and minerals essential for both symbionts from the air and from the substrate. Some cyanobacteria in or associated with lichen thalli fix nitrogen that is utilized by both symbionts. Other lichens obtain nitrogen from organic detritus, bird excrement, and/or leachate from plants. Lichens colonize various surfaces, often rocks and bark.

Lichen thalli have various forms. Crustose types are appressed to the substrate. Foliose lichens have leaflike lobes borne above the substrate. Fruticose lichens have linear, branched thalli that may be fingerlike, strap shaped, or hairlike. The mycobiont usually dominates a thallus, at least in terms of relative mass and abundance. The surface layer of a thallus, called the cortex, consists of gelatinous fungal hyphae that in some species are covered with an amorphous sheet of polysaccharide, perforated to permit gas exchange. Beneath the cortex is a layer of phycobiont or bacterial cells and, beneath it, the medulla, where hyphae are relatively loosely arranged and less gelatinized. The medulla is a storage site for carbohydrate, lichen substances (secondary metabolites peculiar to lichens), and water. Many lichens have surface protrusions consisting of both algal and fungal cells.

Lichens reproduce vegetatively. Various substructures or fragments of thalli containing both symbionts are dispersed by wind or water and have the ability to colonize new substrates. Some lichens also reproduce by reassociation of the separate partners after spore dispersal, but this is a little-known phenomenon.

Lichens are classified primarily on the basis of morphological characteristics of the mycobionts and secondarily on the basis of thallus form and chemical components. Lichens bear the Latin names of their mycobionts; the identity of the alga or bacterium is not considered. Indeed, one species of alga can be the phycobiont in taxonomically unrelated lichens, and some lichens contain two different phycobionts.

Inexperienced observers often wonder if lichens growing profusely on trees are parasitic or pathogenic. Slowly growing trees, including those in decline, are likely to support lichen colonies. One early observer, on finding hyphae beneath the cork cambium on a tree populated by *Usnea* sp., suggested the lichen was parasitic, but contemporary authorities do not count parasitism among lichen attributes. Except for lichenized forms (*Strigula* species) of plant-parasitic algae (*Cephaleuros* species; Plate 217), no claim of plant pathogenicity has been well supported. Many lichens grow most rapidly when fully exposed to light, which may explain their profusion on dead and moribund trees.

Mosses (Bryophyta) are seldom blamed for ill health of trees, but they acidify soil, apparently kill feeder roots, and play a role in forest-to-peatland succession. Their role in tree health deserves further study.

References: 24, 247, 522, 536, 970, 1225, 1275, 1442, 2099, 2100, 2254, 2351, 2752, 2786, 3066, 3181

A, B. *Tillandsia usneoides* (Spanish moss) on *Pinus elliottii* (slash pine). A. Abundant growth of the epiphyte on a dying tree in contrast to slight colonization of a healthier tree. B. Close view of the epiphyte on a small branch (FL, Mar).

C. *T. recurvata*, an epiphyte known as small ballmoss, on a twig of *Casuarina equisetifolia* (Australian-pine) (FL, Mar).

D. Lichens on dead branches of an unidentified plant. The large foliose lichen along the left branch is *Parmotrema perforatum*. The lichens with many fine branches are primarily *Ramalina montagnei* and *R. willeyi* (cartilage lichens) (MS, Aug).

E. Lichens on a young *Quercus* sp. (oak). The red crustose lichen, *Chiodecton rubrocincta*, is common on bark of many trees in swamp and floodplain habitats in southern USA and has been collected along the Atlantic Coast as far north as Delaware (FL, Mar).

F. Two unidentified lichens on bark of living *Prunus pensylvanica* (pin cherry) (ON, Jul).

G. *Usnea* sp. (beard lichen) hanging in festoons on a dying *Pseudotsuga menziesii* (Douglas-fir) (ID, Jul).

H. *Hypogymnia physodes* (tube lichen) on a small branch of *Pinus resinosa* (red pine) (NY, May).

abaxial—away (facing away) from the axis of an organ. See *adaxial*.

abiotic—lacking life or never having been alive.

abscission—of plants: the shedding of leaves or other parts as the result of physical weakness in a specialized layer of cells (abscission layer) that develops at the base.

achlorophyllous—lacking chlorophyll.

acervulus (pl. -li)—a fruiting structure of certain microfungi, consisting of a layer of hyphae that bear conidiophores, lacking specialized wall structure, arising subcuticularly, subepidermally, or deeper in plant tissue, colorless to dark, usually visible with a hand lens and sometimes with the unaided eye, and often appearing like a tiny blister that opens wide at the plant surface (illustrated in Figs. 47–56).

adaxial—on or facing the axis of an organ. Examples: the adaxial surface of a pine needle faces the other needle(s) in a fascicle; the adaxial surface of a palm frond once faced the axis of the trunk and is usually uppermost.

adelgid—an insect belonging to the homopteran family Adelgidae.

adventitious shoots or roots—those formed at unexpected sites.

aeciospore—of rust fungi: a nonrepeating, asexual spore. Aeciospores are often yellow to orange in mass, are borne in chains in an aecium, and are usually incapable of infecting the host on which they are produced.

aecium (pl. -ia)—a fruiting structure that produces aeciospores of a rust fungus; it forms after the spermagonium and before the uredinium in the life cycle, is often cup shaped, in some species blisterlike or tubular, often with colorless (white) walls.

agent of disease—an organism or abiotic factor that causes disease; a pathogen.

aggressiveness—of a plant pathogen: relative ability to colonize and cause damage to plants. An aggressive strain causes more damage to each of an array of hosts than does a less aggressive strain. See also *virulence*.

aleurioconidium (pl. -ia)—a thick-walled, usually darkly pigmented spore that develops by wall extension at the end of a conidiogenous (conidium-generating) cell or hyphal branch. See also *chlamydospore*.

anaerobic—living, acting, or occurring in the absence of oxygen.

anamorph—a mitotically reproducing form of a fungus, usually an asexual state of an ascomycete or basidiomycete.

annual canker—see *canker*.

antheridium—a male gamete of an oomycete. Antheridia unite with oogonia to produce oospores.

apothecium (pl. -ia)—the fruit body of certain Ascomycota that are commonly called discomycetes or cup fungi, in which the fertile layer of asci is exposed by a wide opening. Apothecia vary from linear and embedded in the substrate to sessile or stalked structures that may be shaped like a cup, saucer, or wineglass. In size they vary from nearly microscopic to several centimeters in diameter and height (see Figs. 32, 33).

appressorium (pl. -ia)—of fungi and some parasitic seed plants: a swollen or flattened portion of a germ tube (of a fungus) or radicle (of a seed plant), serving for attachment to a host, from which often the parasite penetrates the host.

arbuscule—of arbuscular mycorrhizae: a much-branched, microscopic haustorial structure of a fungal symbiont that forms within a living cortical cell of a root. The interface of arbuscule with plant protoplast is a site of exchange of nutrients and growth-regulating chemicals.

arthrospore—of fungi: a spore resulting from the division of a hypha into separate cells.

ascigerous—of fungi: having asci.

ascocarp—a fruit body containing asci; the sexual fruit body of a fungus belonging to the Ascomycota.

ascomycetes—fungi of the Ascomycota.

Ascomycota—fungi that produce asci and ascospores; one of the major divisions of fungi; sac fungi.

ascospore—a spore produced within an ascus.

ascostroma (pl. -ata)—of certain Ascomycota: a mass of hyphae within or on which asci form; often microscopic, dark in color, and partially to wholly embedded in dead plant tissue.

ascus (pl. -ci)—a saclike cell of a fungus belonging to Ascomycota in which ascospores form.

autoecious—of a rust fungus: completing its life cycle on one host. See also *heteroecious*.

auxin—any of a group of plant hormones related to indoleacetic acid that promote cell enlargement.

axil—of plants: the cavity or angle formed at the upper side of the junction of a petiole or stem with its parent stem.

axillary—pertaining to an axil, in an axil.

azygospore—a parthenogenetic zygospore; formed by some arbuscular mycorrhizal fungi (Zygomycota, Glomales).

bacillar—shaped like a short rod.

bacteriocin—a bacterial antibiotic to which bacteria closely related to the producer are sensitive. Bacteriocins are chemically diverse.

basidiocarp—a fruit body of a fungus belonging to Basidiomycota that bears or contains basidia.

basidiomycetes—fungi of the Basidiomycota.

Basidiomycota—a major division of the fungi, whose members bear their sexual spores on basidia.

basidium (pl. -ia)—the cell or organ of the sexual state of a fungus belonging to Basidiomycota, on which basidiospores form.

biological species—a population of interbreeding or interfertile individuals.

biotroph—an organism that derives its nutrition from other living organisms.

biotype—a subpopulation of a species that has distinctive genetic markers and associated phenotypic properties. Biotypes of plant pathogens often are adapted to special environments or have the ability to attack particular plants that are resistant to other strains of the species.

bracket—of fungi: a shelflike basidiocarp.

broom—in plant pathology: short for witches'-broom.

brown rot—in decay of wood by fungi: degradation characterized by selective removal of cellulose and hemicelluloses that leaves a colored residue of altered lignin.

bud trace—the microscopic vascular connection between stem and bud.

callus—of plants: undifferentiated parenchymatous tissue that proliferates at cut surfaces or edges of wounds or lesions; see also *woundwood*.

cambium, cork—a lateral meristem that produces cork.

cambium, vascular—a lateral meristem by growth from which a stem or root increases in girth.

canker—a necrotic lesion in bark of a stem or root, often extending to the xylem; also, the scar left after shedding of bark killed by localized disease or injury.

canker, annual—a canker that enlarges only once, and does so within an interval briefer than the growth cycle of the plant, usually less than a year.

canker, diffuse—a canker that enlarges without characteristic shape or noticeable woundwood formation at its margins.

canker, perennial—a canker that enlarges during more than 1 year.

canker, target—a canker that includes concentric ridges of woundwood.

carbonaceous—hard, black, and brittle; resembling charcoal.

cellulolytic—of certain enzymes: able to digest cellulose.

cerebroid—having a convoluted surface similar to that of the cerebrum of the brain.

chitin—the principal structural polymer in fungal cell walls and arthropod exoskeletons.

chitinase—a type of enzyme that degrades chitin.

chlamydospore—a thick-walled, usually pigmented, nondeciduous spore that arises within a hypha or conidium and functions as a resting or survival structure.

chlorotic—abnormally yellowish.

cirrus (or cirrhus, pl. -ri or -hi)—of fungi: a mass of spores in the form of a ribbon or tendril, forced from the fruit body of a microfungus.

clamp connection—in many basidiomycetes, a clamplike outgrowth of an apical hyphal cell that forms during cell division and connects the apical and subapical cells by fusing with the latter.

cleistothecium (pl. -ia)—an ascocarp that lacks an opening.

cm—centimeter.

coelomycete—a fungus that produces conidia within a fruiting structure (acervulus, pycnidium, or intermediate form). Such fungi are anamorphs of ascomycetes or have no known sexual state.

collar—see *trunk collar*.

collenchyma—a plant tissue composed of somewhat elongate living cells with thickened, nonlignified primary walls. Its function is support.

compartmentalization—in trees: the processes that results in isolation of wounded or diseased xylem from normal xylem by the formation of chemically and anatomically specialized tissue around the damaged zone.

conidiogenous—producing conidia.

conidioma (pl. -ata)—a multicellular structure (acervulus, pycnidium, sporodochium, or synnema) that bears or contains conidia.

conidiophore—of fungi: a specialized hypha that bears or produces conidia.

conidium (pl. -ia)—an asexual spore that, when mature, separates from the cell that produced it.

conk—a basidiocarp of a wood-decaying fungus, usually of a polypore.

conserved gene—one in which evolutionary changes in the nucleotide base sequence occur rarely; a gene that occurs in diverse organisms.

cortex—in plants: the primary tissue of a stem or root between the vascular cylinder and the epidermis, composed mainly of parenchyma. The term *cortex* is also used to refer to living secondary phloem.

cortical haustoria—of mistletoes: cortical strands.

cortical strands—of mistletoes: radiating strands of mistletoe tissue that grow through cortex and secondary phloem of a parasitized tree stem.

crown—the branches of an intact tree; see also *root crown*.

cultivar—cultivated variety.

cuticle—of plants: the layer of waxy substance on the surface of a leaf, fruit, or young stem.

cutinolytic—of certain enzymes: able to digest cutin.

cv.—cultivar.

cytokinin—any of a group of plant hormones related to zeatin that promote cell division.

damping-off—of germinating seedlings: disease characterized by a lesion at or near the soil line that prevents a seedling from growing above the soil or causes a newly emerged seedling to fall over.

delignification—chemical, usually enzymatic, removal of lignin from xylem, leaving a cellulosic residue.

dieback—dead apical parts, usually twigs or limbs; also the process of dying back.

disc—of *Valsa* and related fungi: a more or less flat apical part of a stroma that protrudes above the bark surface. Also, of fungi with apothecia that open broadly (discomycetes): the exposed fertile portion of an apothecium.

discomycete—an ascomycetous fungus whose mature ascocarps are open and flat to cup shaped; a cup fungus.

disease cycle—of a disease caused by a biotic agent: the cyclical sequence of host and parasite development and interaction that result in disease, in reproduction or replication of the pathogen, and in the readying of a new generation of the parasite for infection.

distal—of a stem or root: away from the base or origin.

DNA—deoxyribonucleic acid.

dwarf shoot—see *shoot, dwarf*.

ecotype—of any species: a subpopulation specialized to inhabit a particular place or type of place; a subpopulation that inhabits a distinctive ecosystem.

effuse—spreading out loosely or flat.

enation—a small, abnormal outgrowth from a plant organ, usually from the epidermis of a leaf. Some enations are small galls.

endemic—of a disease or pathogen: common in a place during an extended time and causing little damage.

endoconidiophore—a conidiophore that produces conidia within itself; a phialide.

endoconidium—a conidium produced within its conidiophore. Endoconidia are extruded from the tips of their conidiophores.

endophyte—a microorganism that resides in a plant, usually without giving evidence of its presence.

endophytic—within a plant.

endophytic system—of mistletoes: the part of the parasite that grows within the host; cortical strands and sinkers.

epicormic sprouts—sprouts arising abnormally along a trunk or limb as the result of release of dormant buds or the differentiation of buds from callus; water sprouts.

epidermis—of plants: the outermost cell layer.

epinasty—downward bending of a turgid plant part, usually the petiole or midvein of a leaf.

epiparasite—an organism that parasitizes a parasite.

epiphyte—a plant that grows nonparasitically on another plant.

epiphytic—on a plant.

ericaceous—of the plant family Ericaceae.

erumpent—breaking through the surface; bursting forth.

eukaryotic (or eucaryotic)—of cells, possessing membrane-delimited nuclei.

exophylactic periderm—a periderm that develops as a normal aspect of bark formation and contributes to corky outer bark by isolating part of the outermost secondary phloem. The isolated tissue and cork in the periderm become a bark plate. See *periderm*.

extractive—any chemical that may be extracted by means of solvents.

fastigiate—of plants: having leaves or branches more or less parallel and pointed upward.

feeder root—a rootlet that absorbs water and minerals.

filamentous fungus—a fungus that produces hyphae.

first-year needles—of conifers: the age class of needles formed most recently; needles of the current year.

flag—a dying or dead shoot or small branch with faded or brown foliage prominent against a contrasting background.

forma specialis—in certain fungi: an infraspecific population of a plant-pathogenic species, distinguished by host preference (usually at the genus level) but scarcely or not at all by morphological criteria.

frass—particles and perhaps excrement remaining where insects, especially those boring in wood or bark, have chewed plant tissues.

frond—the leaf of a palm or fern.

fruit body—of fungi: a multicellular structure that produces spores.

fusiform—spindle shaped, tapered toward the ends.

f.sp.—forma specialis.

gall—a localized overgrowth.

gamete—sex cell; a reproductive cell capable of uniting with another, resulting in a mingling of their cytoplasm and nuclei.

genome—the entire genetic code of an organism or virus.

germ tube—a hypha produced by a germinating spore or microsclerotium.

germ slit—a linear, thin zone in a spore wall, often appearing as a groove, where a germ tube will emerge.

glaucous—having a bluish gray, waxy surface.

green-island effect—retention of chlorophyll around an infection site or lesion in an otherwise senescing leaf.

gummosis—exudation of gum.

haematochrome—a red-orange pigment produced by certain algae.

hard pine—a pine with two or three needles per cluster.

haustorium (pl. -ia)—of plant-parasitic fungi: a specialized hyphal branch that differentiates within a host cell for absorption of nutrients. Of mistletoes and other parasitic plants: a multicellular, usually highly branched structure that differentiates in the cortex and/or secondary phloem of the host for anchorage and nutrient absorption.

heteroecious—of rust fungi: requiring two host species for completion of the life cycle.

heterokaryon (or heterocaryon)—a cellular structure containing nuclei of different genotypes.

heterokaryotic—having nuclei of different genotypes.

heterothallic—of fungi and oomycetes: requiring gametes from genetically different thalli to complete the sexual cycle.

holomorph—all of the forms and developmental phases of a fungus considered collectively.

homokaryotic—having nuclei of one genotype.

homothallic—of fungi: self-fertile in the sexual cycle.

hydathode—a tiny gland at the edge of a leaf from which droplets of water may issue.

hymenium—of fungi: a layer of spore-producing cells, usually in or on a fruit body.

hyperplasia—abnormal increase in number of cells in an organ or part, as in a gall.

hypertrophy—abnormal enlargement, overgrowth.

hypha (pl. -ae)—usually, a filament of a fungus or oomycete. Also, a linear extension of the haustorium of a dodder that makes intimate contact with phloem sieve cells of a host.

hyphomycete—a fungus that produces conidiophores and/or conidia on sporodochia or on mycelium. Such fungi are anamorphs of ascomycetes or have no known sexual state.

hyphopodium (pl. -ia)—a short branch, one or two cells long, of the epiphytic mycelium of a black mildew fungus.

hypovirulence—of a pathogen: abnormally low virulence; diminished ability to damage its host(s).

hysterothecium (pl. -ia)—an ascocarp that opens by a slit.

incubation period—in plant pathology: the interval between inoculation and display of symptoms.

infect—of a parasite: to begin or continue an interactive, usually pathogenic, relationship with a host.

infection—the interaction of parasite with host; according to some authors, the beginning of that interaction.

infection court—in plant pathology: the site of the onset of infection; the site of inoculation.

infest—of a pest or pathogen: to populate or inhabit a thing or place.

infestation—the occurrence or development of an infesting population.

inflorescence—a flower cluster.

inoculation—in plant pathology: the placement of inoculum at a site (the infection court) where infection is possible.

inoculum—of a plant parasite: a unit or units capable of infecting.

isolate—in plant pathology: a culture or subpopulation of a microorganism separated from its parent population and maintained in a controlled circumstance; also, to effect such separation and control; for example, to isolate a pathogen from diseased plant tissue.

isometric—possessing equal dimensions in all aspects.

isozyme—any one enzyme of a family of enzymes that act on a particular substrate.

kg—kilogram.

knot (as symptom)—a localized abnormal swelling; a gall.

latent infection—infection unaccompanied by visible symptoms.

latewood—see *summerwood*.

leader (of terminal shoots)—the topmost or dominant shoot of a tree or branch.

lenticel—a passage, loosely filled with parenchyma cells, between the cortex of a stem or root and the outside environment; visible on the surface as a small, lens-shaped eruption.

lesion—a localized zone of dead or moribund tissue.

lignification—in plant tissue: deposition of lignin within cell walls.

lignin—a complex structural polymer that imparts rigidity to certain plant cell walls, especially fibers, tracheids, and vessels.

lysigenous—created by or pertaining to lysis or dissolution.

megapascal (MPa)—10^6 pascals; approximately 7.5×10^3 mm mercury.

meiospore—a spore whose nucleus resulted from meiosis, as an ascospore or basidiospore.

meristem—a layer or zone of undifferentiated cells capable of division to produce more generative cells as well as cells that enlarge and differentiate, resulting in growth.

meristem, intercalary—a meristem located between differentiated tissues.

microbiota—microorganisms, the microorganisms in an environment.

microsclerotium—a tiny, mass of thick-walled, dark-colored fungal cells, specialized for survival, capable of germination to produce mycelium (see Fig. 121).

microzoospore—of certain algae: a tiny, motile spore.

mitospore—a spore whose nucleus or nuclei resulted from mitosis (see also *meiospore*).

mollicute—a wall-less, prokaryotic, usually pleomorphic microorganism of the class Mollicutes.

monokaryotic—having one nucleus per cell.

monopartite—of a virus: each virion typically containing the entire genome.

morphologic—pertaining to form.

multipartite—of a virus: its genome divided among two or more virions.

mycelium—an aggregation of hyphae, usually of a fungus or oomycete.

mycobiont—the fungal symbiont in a lichen or mycorrhiza.

mycoplasmalike organism—(obsolete): see *phytoplasma*.

nanometer—1×10^{-9} meters.

necrophylactic periderm—a periderm that separates wounded or diseased tissues from normal tissues in bark; also known as wound periderm. See *periderm*.

necrosis—death.

nm—nanometer.

normal length—of virus particles: the most common length, believed to be the intrinsic natural length of a virus that consists of rod-shaped or linear particles.

oleoresin—resin, pitch; a viscous, aromatic mixture of terpenes, resin acids, and fatty acids produced by various conifers.

olivaceous—olive colored.

oogonium—a cell of an oomycete that functions as a female gamete, within which, after fertilization by an antheridium, an oospore develops.

Oomycota—a phylum of funguslike organisms in Kingdom Straminipila (syn. Chromista)

oospore—a resting spore produced as the result of fertilization or a similar parthenogenetic process in a member of the Oomycota.

ostiole—a more or less circular, differentiated pore in a fruit body of a microfungus, from which spores issue.

paraphysis (pl. -es)—a specialized sterile hypha that grows up among asci or basidia from the base of a hymenium, thought to function as a spacer between fertile elements.

pathogen—an agent capable of causing disease. The term usually refers to biotic agents.

pathogenicity—ability to cause disease.

pathotype—of plant-pathogenic fungi: an infraspecific population distinguishable by its virulence pattern on particular host genotypes. See also *race*.

pathovar—in plant-pathogenic bacteria: an infraspecific population distinguishable mainly or only on the basis of pathogenicity to certain plants.

pectolytic—of certain enzymes: capable of digesting pectic compounds.

peduncle—the stalk of a flower or fruit.

perennial canker—see *canker*.

periderm—a tissue external to the cortex and/or phloem of a stem or root, composed of cork (phellem), the cork cambium, and phelloderm; more or less impervious to water, solutes, and organisms; protects underlying tissues. See also *exophylactic periderm, necrophylactic periderm*.

perithecium (pl. -ia)—a more or less globose to flasklike ascocarp with a wall of its own and an opening for the release of ascospores.

peritrichous—of flagella of a microorganism: occurring all over the cell surface.

petiole—leaf stalk.

phellem—cork; suberized tissue differentiated from phellogen on its abaxial (outer) side.

phelloderm—the thin layer of living tissue differentiated from phellogen on its adaxial (inner) side.

phellogen—cork cambium.

phenotype—the set of characteristics encoded by an organism's genes.

phialide—a type of conidiophore or portion thereof that is open at one end and produces conidia (phialoconidia, endoconidia) from within itself, often in chains.

phloem—a plant tissue specialized for conduction of food and growth regulators. In stems and roots with secondary thickening, phloem lies primarily or entirely outside the vascular cambium.

phloem, secondary—phloem formed by the vascular cambium.

phylogeny—the scheme of evolutionary relationships among life forms. Phylogenies are often represented graphically by branching diagrams that show ancestors and relatives of a given type of organism.

phytoalexin—an antimicrobial compound of low molecular weight, synthesized and accumulated by a plant in response to infection or other stress. Phytoalexins are chemically diverse.

phytoplasma—any of a group of a prokaryotic, wall-less, pleomorphic, biotrophic microorganisms that inhabit phloem sieve tubes. Phytoplasmas are not yet isolable in pure culture, and most have not been given Latin names.

phytotoxin—a substance toxic to plants. Selective, specific, or host-specific toxins affect particular plants, usually hosts of a toxin-producing pathogen.

pine, hard—any *Pinus* species with two or three needles per fascicle, except pinyon pines.

pine, soft—any *Pinus* species with five needles per fascicle (excepting *P. torreyana* and five-needle strains of *P. ponderosa*), and also pinyon pines, which have one to four needles per fascicle.

pinna (pl. -ae)—of palm fronds: a leaflet.

pit membrane—in xylem: the apposed flexible primary walls or porous cellulosic skeletons of them in a pit pair (a channel between two adjacent xylem cells with thickened walls).

pl.—plural.

plasmid—a circular piece of cytoplasmic DNA.

polypore—one of a large group of wood-decaying fungi that produce basidiocarps with poroid lower surfaces. The pores are mouths of tubes within which basidiospores are produced.

poroid layer—of polypores: the layer of fertile tubes.

primary cycle—of plant disease: the first cycle that begins in a given year.

primary infection—the first infection in a sequence or in a year.

prokaryote—a unicellular microorganism lacking an organized nucleus and organelles.

prokaryotic—having the characteristics of a prokaryote.

propagule—a structure by which an organism is propagated or multiplies.

proximal—toward the origin, toward the base of a plant organ.

pseudosclerotial plate—of certain fungi that colonize wood: a hard, dark plate formed within decaying wood; more or less impervious to water, solutes, and organisms; composed of large, thick-walled or encrusted fungal cells; affording protection to mycelium behind it; appearing as a black line (zone line) in transverse view.

pseudostroma—of fungi: a stroma that includes host tissue.

pseudothecium (pl. -ia)—of certain microfungi: a pseudoperithecium; an ascostroma that resembles a perithecium.

punk knot—of a wood-decaying fungus in a trunk or limb: a dense sterile mass of mycelium issuing from the interior of the stem to the surface along a channel once occupied by a branch.

pv.—pathovar.

pycnidium (pl. -ia)—of fungi: a subglobose to cup- or flask-shaped fruit body containing conidiophores, producing conidia.

pycnothyrium—a tiny, disc- or shield-shaped fruiting body with radiate upper surface, usually superficial on a leaf, in which conidia form.

race—of a plant pathogen: an infraspecific population distinguishable only or mainly on the basis of its host range or (uncommonly) its aggressiveness or geographic origin.

rachis—the stemlike axis of a compound leaf or inflorescence.

ramify—of filamentous fungi: to produce branching hyphae in a substrate.

ray, medullary ray—in woody stems and roots: a radial sheet of cells, mainly parenchyma but sometimes also tracheary cells, with their long axes (if any) perpendicular to the stem axis; serving for food storage and conduction from phloem to inner sapwood.

resistance—of a plant to a pathogen: the ability to retard, suppress, or prevent infection or colonization; also the functioning of attributes or processes that do so.

resupinate—of a basidiocarp: pressed against the substrate with the hymenium or tube layer on the outer side, often crustlike.

rhizomorph—a thick strand of fungal hyphae organized and capable of growth as a rootlike unit.

rhytidome—dead, corky bark; includes cork and remnants of secondary phloem.

root crown—the region near the soil line where roots emerge from the axis of a plant.

root hair—a hairlike projection from an epidermal cell of a root.

rot, brown—of wood: decay characterized by selective degradation of cellulose and hemicelluloses leaving a crumbly brown residue rich in undigested lignin.

rot, white—of wood: decay characterized by degradation of lignin at a rate equal to or usually greater than the rate of degradation of cellulose and hemicelluloses, leaving a light-colored residue that usually contains relatively more cellulose and hemicelluloses than did the original wood.

rugose—of a surface: rough or wrinkled.

samara—a one-seeded winged fruit, as of *Acer, Fraxinus*, or *Ulmus*.

saprobe—a microorganism that lives on and degrades dead organic matter.

saprobic—pertaining to utilization of dead organic matter for nutrition.

saprophagous—feeding on dead, usually decaying, organic matter; saprobic.

saprophyte—see *saprobe*.

secondary cycle—of plant disease: any cycle initiated by inoculum generated during the same season.

second-year, third-year, etc., needles—of conifers: needles in their second, third, etc., year of life. A second-year needle is between 1 and 2 years old.

septum (pl. -ta)—in fungi: the wall or partition between two cells.

seta (pl. -ae)—a stiff hair or bristle, macro- or microscopic.

shoot, dwarf—of gymnosperms such as *Pinus* or *Larix:* a lateral shoot consisting primarily of a cluster of needles at the apex of a very short stem.

shoot, short—a dwarf shoot.

sign—of a plant pathogen: any visible structure of the pathogen.

simultaneous rot—in decay of wood by fungi: degradation characterized by removal of both lignin and cellulosic constituents leading to rapid loss of strength, brittleness, and eventually crumbly texture.

sinker—of a mistletoe: a wedgelike structure embedded in the xylem of a host.

soft rot—in decay of wood by fungi: degradation characterized by formation of cavities within the thick secondary walls of wood cells, resulting in loss of strength.

somatic—of or pertaining to cells or tissues other than those involved in reproduction.

sorus (pl. -ri)—of rusts, smuts, and false smut: a mass of spores and the fruit body that produced them.

spathe—of palms and other plants: a large bract or pair of bracts that sheath a flower cluster.

spermagonium (pl. -ia)—a fungal fruit body that produces spermatia.

spermatium (pl. -ia)—of fungi: a nonmotile gamete (sex cell) produced and dispersed in the manner of a spore that is capable of emptying its contents into a receptive structure, thereby establishing a dikaryon or diploid.

spermatization—the placing of spermatia on a receptive structure.

sporangiophore—a sporangium-bearing hypha.

sporangium (pl. -ia)—of fungi and oomycetes: a microscopic saclike structure the contents of which become converted into spores. The term *sporangium* is also used for asexual spores that function as conidia in oomycetes.

spore—a minute propagule that functions in the manner of a seed but lacks an embryo. Most spores are microscopic.

sporocarp—a fruit body that produces spores.

sporodochium (pl. -ia)—a tiny, cushionlike fungal structure that protrudes above the surface of a substrate, consisting of a more or less thick pad of hyphae that bear conidiophores; white to pink, reddish, orange, or dark in color.

sporophore—a structure that bears spores.

sporophyte—of plants such as ferns that reproduce by alternation of generations: the stage that produces spores; also, the diploid part of the plant's life cycle.

sporulate—to produce spores.

sp.—unspecified or unidentified species (singular).

spp.—species (plural).

springwood—in a woody perennial, the portion of an annual sheath of xylem formed during the first part of the growth period; also called earlywood.

sterile conk—a hard or tough, dense mass of sterile mycelium of a wood-decaying fungus on a trunk or limb.

sterile fungus—a colony or life stage not producing spores.

stipule—an appendage at the base of a petiole.

stoma (pl. -ata)—a microscopic pore in a plant surface, for gas exchange.

strain—of a pathogen or other organism: an infraspecific population defined on some arbitrary basis, such as geographic or host origin.

stroma (pl. -ata)—of fungi: a mass of hyphae, with or without plant tissue or other substrate, in or on which spores or fruit bodies are usually produced; microscopic to macroscopic and colorless to bright or dark colored, forming within or on a substrate.

subcuticular—beneath the cuticle of a plant, on the epidermis.

suberin—a hydrophobic substance found in walls of some types of plant cells, especially cork cells, that makes them impermeable to water.

suberized—of plant cell walls: impregnated with suberin; corky.

subsp.—subspecies.

subspecies—an infraspecific population defined on the basis of one or more characters (morphologic for most organisms) that distinguish its members from typical representatives of the species.

suctorial—of certain insects: adapted for sucking host fluids by means of specialized mouthparts that pierce cells.

sugar bush—a grove of sugar maple trees from which sap is harvested.

summerwood—in a woody perennial, the portion of an annual sheath of xylem formed during summer; also called latewood.

sunscald—a lesion attributed to excess solar heat, or the process of formation of such a lesion. Sunscald lesions on stems commonly result from freezing of bark in which cold hardiness was lost during solar heating.

suscept—in plant pathology: a plant susceptible to a given pathogen.

susceptibility—of a plant to a pathogen: a condition in which the plant is capable of interacting with the pathogen such that disease results.

symptom—abnormal appearance or function.

syn.—synonym.

synnema (pl. -ata)—a group of conidiophores joined together at their sides to form an elongate spore-bearing structure, often large enough to be seen with the unaided eye.

teleomorph—the sexually (meiotically) reproducing state of a fungus.

teliospore—of rust and smut fungi: a spore that gives rise to a basidium and basidiospores on germination. Teliospores of most rust fungi are thick-walled resting spores.

telium (pl. -ia)—of rust fungi: a fruiting structure consisting of teliospores, produced after the uredinial stage in the life cycle.

thallospore—an asexual spore that has no conidiophore or is not separate from the hypha or conidiophore that produced it.

thallus—of a fungus: the vegetative body (somatic phase).

toruloid (also torulose)—cylindric but with swellings at intervals.

tracheary element—water-conducting cell: a vessel member or tracheid.

transpiration—in plants: movement of water from the root medium through a plant into the atmosphere as the consequence of evaporation.

trichome—a tiny projection from a plant epidermis; may be hairlike, spiny, or glandular.

trunk collar—the stem region just above roots; term used primarily in fruit tree culture.

urediniospore (also urediospore)—of rust fungi: an asexual, thick-walled spore capable of infecting the host on which it is produced.

uredinium (also uredium, pl. -ia,)—of rust fungi: a fruit body that produces urediniospores, formed after the aecium and before the telium in the life cycle.

Ustilaginomycetes—a class of fungi in the phylum Basidiomycota, comprising smut fungi and their allies.

var.—variety.

vector—of a disease or pathogen: an agent, usually another organism, that transmits a pathogen.

vegetative compatibility—of fungi: ability of two strains to undergo hyphal anastomosis resulting in viable cells containing nuclei of both strains in a common cytoplasm.

verrucose—possessing wartlike projections.

vesicle—a microscopic, bladderlike sac.

virion—a virus particle.

virulence—the relative competence of a pathogen to cause disease in a given plant. A particular strain of a pathogen may be virulent on one host cultivar but avirulent on another cultivar that possesses resistance to it. See also *aggressiveness*.

white rot—in decay of wood by fungi: degradation characterized by selective removal of lignin, leaving a light-colored residue rich in cellulose and hemicelluloses.

witches'-broom—an abnormally dense cluster of twigs growing from a common locus or resulting from the proliferation of buds.

woundwood—organized tissues including xylem, phloem, and periderms that grow at the edges of wounds or lesions and gradually cover them. Woundwood is preceded by callus and grows more rapidly than normal parts of stems or roots.

xylem—water-conducting tissue; wood.

yeast—a unicellular fungus that reproduces by budding, usually one of a group of Ascomycota that form asci directly from zygotes or single cells; also, one of a group of Basidiomycota related to smut fungi.

zone line—a black line visible in decaying wood where a cut or fracture crosses a pseudosclerotial plate.

zoospore—a spore that is motile by means of flagella.

zygospore—a thick-walled resting spore formed by the fusion of two gametes borne on somatic hyphae, characteristic of Zygomycota.

1. Aa, H. A. van der. 1973. Studies in *Phyllosticta* I. Stud. Mycol. No. 5. 110 pp.

2. Aa, H. A. van der, and Vanev, S. 2002. A revision of the species described in *Phyllosticta*. Centralbureau voor Schimmelcultures, Baarn. 510 pp.

3. Abe, K., and Kurihara, A. 1993. Species and varietal differences in scab resistance of pear. J. Jap. Soc. Hortic. Sci. 61:789–794.

4. Abe, Y. 1989. Effect of moisture on decay of wood by xylariaceous and diatrypaceous fungi and quantitative changes in the chemical components of decayed woods. Trans. Mycol. Soc. Jap. 30:169–181.

5. Accordi, S. M. 1986. Spread of *Ceratocystis fimbriata* f. *platani* through root anastomoses. Inf. Fitopatol. 36:53–58.

6. Adams, G., Hammer, S., and Proffer, T. 1990. Vegetative compatibility in *Leucostoma persoonii*. Phytopathology 80:287–291.

7. Adams, G. C. J., Gottwald, T. R., and Leach, C. M. 1986. Environmental factors initiating liberation of conidia of powdery mildews. Phytopathology 76:1239–1245.

8. Adams, J. F. 1920. The alternate stage of *Pucciniastrum hydrangeae*. Mycologia 12:33–35.

9. Adams, T. J. H., Todd, N. K., and Rayner, A. D. M. 1981. Antagonism between dikaryons of *Piptoporus betulinus*. Trans. Br. Mycol. Soc. 76:510–513.

10. Adaskaveg, J. E., Blanchette, R. A., and Gilbertson, R. L. 1991. Decay of date palm wood by white-rot and brown-rot fungi. Can. J. Bot. 69:615–629.

11. Adaskaveg, J. E., and Gilbertson, R. L. 1986. Cultural studies and genetics of sexuality of *Ganoderma lucidum* and *G. tsugae* in relation to the taxonomy of the *G. lucidum* complex. Mycologia 78:694–705.

12. Adaskaveg, J. E., and Gilbertson, R. L. 1988. Basidiospores, pilocystidia and other basidiocarp characters in several species of the *Ganoderma lucidum* complex. Mycologia 80:493–507.

13. Adaskaveg, J. E., and Gilbertson, R. L. 1988. *Ganoderma meredithae*, a new species on pines in the southeastern United States. Mycotaxon 31:251–257.

14. Adaskaveg, J. E., and Gilbertson, R. L. 1989. Cultural studies of four North American species in the *Ganoderma lucidum* complex with comparisons to *G. lucidum* and *G. tsugae*. Mycol. Res. 92:182–191.

15. Adaskaveg, J. E., Gilbertson, R. L., and Blanchette, R. A. 1990. Comparative studies on delignification caused by *Ganoderma* species. Appl. Environ. Microbiol. 56:1932–1943.

16. Adaskaveg, J. E., and Hartin, R. J. 1997. Characterization of *Colletotrichum acutatum* isolates causing anthracnose of almond and peach in California. Phytopathology 87:979–987.

17. Adaskaveg, J. E., and Ogawa, J. M. 1990. Wood decay pathology of fruit and nut trees in California. Plant Dis. 74:341–352.

18. Ades, P. K., and Simpson, J. A. 1991. Variation in susceptibility to Dothistroma needle blight among provenances of *Pinus radiata* var. *radiata*. Silvae Genet. 40:6–13.

19. Ades, P. K., Simpson, J. A., and Eldridge, K. G. 1992. Genetic variation in susceptibility of Dothistroma needle blight among provenances and families of *Pinus muricata*. Can. J. For. Res. 22:1111–1117.

20. Adholeya, A., and Singh, S., eds. 1995. Mycorrhizae, biofertilizers for the future. Tata Energy Res. Inst., New Delhi. 548 pp.

21. Adkins, S., Kamenova, I., Achor, D., and Lewandowski, D. J. 2003. Biological and molecular characterization of a novel tobamovirus with a unique host range. Plant Dis. 87:1190–1196.

22. Aducci, P., Ballio, A., Federico, R., Marra, M., Graniti, A., and Frisullo, S. 1983. Translocation of fusicoccin in almond plants infected by *Fusicoccum amygdali* Del. Phytopathol. Medit. 22:100–102.

23. Agostini, J. P., and Timmer, L. W. 1994. Population dynamics and survival of strains of *Colletotrichum gloeosporioides* on citrus in Florida. Phytopathology 84:420–425.

24. Ahmadjian, V. 1993. The lichen symbiosis. Wiley, New York. 250 pp.

25. Aho, P. E., and Filip, G. M. 1982. Incidence of wounding and *Echinodontium tinctorium* infection in advanced white fir regeneration. Can. J. For. Res. 12:705–708.

26. Aho, P. E., Filip, G. M., and Lombard, F. F. 1987. Decay fungi and wounding in advance grand and white fir regeneration. For. Sci. 33:347–355.

27. Aho, P. E., Seidler, R. J., Evans, H. J., and Raju, P. N. 1974. Distribution, enumeration, and identification of nitrogen-fixing bacteria associated with decay in living white fir trees. Phytopathology 64:1413–1420.

28. Ahokas, H. 1997. Acidification of forest top soils in 60 years to the southwest of Helsinki. For. Ecol. Manag. 94:187–193.

29. Ainsworth, G. C., Sparrow, F. K., and Sussman, A. S. 1973. The fungi. An advanced treatise. Vols. 4A and 4B. Academic Press, New York. 621 and 504 pp.

30. Alaoui, S. B., Sehmer, L., Barnola, P., and Dizengremel, P. 1998. Effect of NaCl salinity on growth and mineral partitioning in *Quercus robur* L., a rhythmically growing species. Trees Struct. Funct. 12:424–430.

31. Al-Banna, L., Williamson, V. M., and Gardner, S. L. 1997. Phylogenetic analysis of nematodes of the genus *Pratylenchus* using nuclear 26S rDNA. Mol. Phylogenet. Evol. 7:94–102.

32. Alderman, S. C., and Stowell, L. J. 1986. Strand ontogeny in *Phymatotrichum omnivorum*. Trans. Br. Mycol. Soc. 86:207–211.

33. Aldrich, J. H., Gould, A. B., and Martin, F. G. 1992. Distribution of *Xylella fastidiosa* within roots of peach. Plant Dis. 76:885–888.

34. Aldwinckle, H. S. 1974. Field susceptibility of 41 apple cultivars to cedar apple rust and quince rust. Plant Dis. Rep. 58:696–699.

35. Aldwinckle, H. S. 1975. Pathogenic races of *Gymnosporangium juniperi-virginianae* on apple. Phytopathology 65:958–961.

36. Aldwinckle, H. S., Borejsza-Wysocka, E. E., Malnoy, M., Brown, S. K., Norelli, J. L., Beer, S. V., Meng, X., He, S. Y., Jin, Q. L., and Janick, J. 2003. Development of fire blight resistant apple cultivars by genetic engineering. Acta Hortic. No. 622:105–111.

37. Aldwinckle, H. S., Lamb, C., and Gustafson, H. L. 1977. Nature and inheritance of resistance to *Gymnosporangium juniperi-virginianae* in apple cultivars. Phytopathology 67:259–266.

38. Aldwinckle, H. S., Pearson, R. C., and Seem, R. C. 1980. Infection periods of *Gymnosporangium juniperi-virginianae* on apple. Phytopathology 70:1070–1073.

39. Alexander, M. E., and Hawksworth, F. G. 1975. Wildland fires and dwarf mistletoes: a literature review of ecology and prescribed burning. USDA For. Serv. Gen. Tech. Rep. RM-14. 12 pp.

40. Alexopoulos, C. J., Mims, C. W., and Blackwell, M. 1996. Introductory mycology. 4th ed. Wiley, New York. 868 pp.

41. Alfano, J. R., and Collmer, A. 1996. Bacterial pathogens in plants: life up against the wall. Plant Cell 8:1683–1698.

42. Alfaro, R. I., Bloomberg, W. J., Smith, R. B., and Thomson, A. J. 1985. Epidemiology of dwarf mistletoe in western hemlock stands in south coastal British Columbia. Can. J. For. Res. 15:909–913.

43. Alfenas, A. C., Jeng, R., and Hubbes, M. 1983. Virulence of *Cryphonectria cubensis* on *Eucalyptus* species differing in resistance. Eur. J. For. Pathol. 13:197–205.

44. Alfieri, S. A. Jr. 1979. *Kutilakesa pironii* sp. nov., a stem gall- and canker-inciting fungus new to the United States. Mycotaxon 10:217–218.

45. Alfieri, S. A. Jr., Knauss, J. F., and Wehlburg, C. 1979. A stem gall- and canker-inciting fungus new to the United States. Plant Dis. Rep. 63:1016–1020.

46. Alfieri, S. A. Jr., Langdon, K. R., Kimbrough, J. W., El-Gholl, N. E., and Wehlburg, C. 1994. Diseases and disorders of plants in Florida. Fla. Dep. Agric. Consum. Serv., Div. Plant Indus., Gainesville. Bull. 14. 1114 pp. Also Suppl. No. 1 by N. E. El-Gholl, T. S. Schubert, and N. C. Coile. 1997. 138 pp.

47. Alfieri, S. A. Jr., and Samuels, G. J. 1979. *Nectriella pironii* and its *Kutilakesa*-like anamorph, a parasite of ornamental shrubs. Mycologia 71:1178–1185.

48. Alfieri, S. A. Jr., Seymour, C. P., and Denmark, J. C. 1972. Aerial blight of *Carissa grandiflora* caused by *Rhizoctonia solani*. Plant Dis. Rep. 56:511–514.

49. Alioto, D., Marcone, C., and Ragozzino, E. 2000. Witches'-broom and fasciation diseases of Spanish broom (*Spartium junceum*). Inf. Fitopatol. 50:61–64.

50. Al-Khatib, K., and Parker, P. 1993. Herbicide injury symptoms on cherry, grape, alfalfa and rose. Wash. State Univ. Coop. Exten. Bull. EB1748. 16 pp.

51. Al-Kherb, S. M., Roelfs, A. P., and Groth, J. V. 1987. Diversity for virulence in a sexually reproducing population of *Puccinia coronata*. Can. J. Bot. 65:994–998.

52. Allen, E. A., Blenis, P. V., and Hiratsuka, Y. 1990. Histological evidence of resistance to *Endocronartium harknessii* in *Pinus contorta* var. *latifolia*. Can. J. Bot. 68:1728–1737.

53. Allen, E. A., Morrison, D. J., and Wallis, G. W. 1996. Common tree diseases of British Columbia. Nat. Resour. Can., Can. For. Serv., Victoria, BC. 178 pp.

54. Allen, J. A., Chambers, J. L., and Stine, M. 1994. Prospects for increasing the salt tolerance of forest trees: a review. Tree Physiol. 14:843–853.

55. Allen, M. F. 1991. The ecology of mycorrhizae. Cambridge Univ. Press, Cambridge, New York. 184 pp.

56. Allen, R. F. 1932. A cytological study of heterothallism in *Puccinia coronata*. J. Agric. Res. 45:513–541.

57. Allington, W. B., Staples, R., and Viehmeyer, G. 1968. Transmission of rose rosette virus by the eriophyid mite *Phyllocoptes fructiphilus*. J. Econ. Entomol. 61:1137–1140.

58. Alma, A., Bosco, D., Danielli, A., Bertaccini, A., Vibio, M., and Arzone, A. 1997. Identification of phytoplasmas in eggs, nymphs and adults of *Scaphoideus titanus* Ball reared on healthy plants. Insect Mol. Biol. 6:115–121.

59. Alvarez, F., Rios, J. E. G. de los, Jimenez, P., Rojas, A., Reche, P., and Troya, M. T. 1998. Phenotypic variability in different strains of *Pseudomonas syringae* subsp. *savastanoi* isolated from different hosts. Eur. J. Plant Pathol. 104:603–609.

60. Amano, K. 1986. Host range and geographical distribution of the powdery mildew fungi. Jap. Sci. Soc. Press, Tokyo. 741 pp.

61. Amerson, H. V. 1976. A structural characterization of the uredial stage of *Puccinia sparganioides* and its associated interactions with *Spartina alterniflora*. Ph.D. diss., N.C. State Univ., Raleigh. 116 pp.

62. Amico, L. A., O'Shea, M. T., and Castello, J. D. 1985. Transmission of tobacco mosaic and tobacco ringspot viruses from Moraine ash in New York. Plant Dis. 69:542.

63. Amiro, B. D., and Courtin, G. M. 1981. Patterns of vegetation in the vicinity of an industrially disturbed ecosystem, Sudbury, Ontario. Can. J. Bot. 59:1623–1629.

64. Ammon, V., Graves, B., and Griffin, D. 1990. Influence of environment and inoculum density on penetration and colonization of sycamore leaves by *Apiognomonia veneta*. Plant Dis. 74:989–991.

65. Ammon, V., Vann, S., and McCracken, F. 1991. Incidence of anthracnose and twig canker in two Mississippi sycamore plantations. Can. J. For. Res. 21:1253–1256.

66. Amrine, J. W. Jr., Hindal, D. F., Stasny, T. A., Williams, R. L., and Coffman, C. C. 1988. Transmission of the rose rosette disease agent to *Rosa multiflora* by *Phyllocoptes fructiphilus* (Acari: Eriophyidae). Entomol. News 99:239–252.

67. Amrine, J. W. Jr., and Stasny, T. A. 1994. Catalog of the Eriophyoidea (Acarina: Prostigmata) of the world. Indira Publishing House, West Bloomfield, MI. 804 pp.

68. An, X., and Han, L. 1994. On the occurrence and control of coconut Pestalotia leaf spot. Plant Prot. 20:16–17.

69. Anagnostakis, S. L., and Hillman, B. 1992. Evolution of the chestnut tree and its blight. Arnoldia 52:2–10.

70. Andersen, P. C., Aldrich, J. H., and Gould, A. B. 1990. Impact of pecan leaf blotch on gas exchange of pecan leaves. Plant Dis. 74:203–207.

71. Andersen, T. F. 1996. A comparative taxonomic study of *Rhizoctonia sensu lato* employing morphological, ultrastructural and molecular methods. Mycol. Res. 100:1117–1128.

72. Andersen, T. F., and Stalpers, J. A. 1994. A checklist of *Rhizoctonia* epithets. Mycotaxon 51:437–457.

73. Anderson, A. R., and Moore, L. W. 1979. Host specificity in the genus *Agrobacterium*. Phytopathology 69:320–323.

74. Anderson, D. L., and French, D. W. 1972. Isolation of *Hypoxylon mammatum* from aspen stem sections. Can. J. Bot. 50:1971–1972.

75. Anderson, G. W., and French, D. W. 1965. Differentiation of *Cronartium quercuum* and *Cronartium coleosporioides* on the basis of aeciospore germ tubes. Phytopathology 55:171–173.

76. Anderson, G. W., and French, D. W. 1965. Western gall rust in the Lake States. For. Sci. 11:139–141.

77. Anderson, N. A., French, D. W., and Anderson, R. L. 1967. The stalactiform rust of jack pine. J. For. 65:398–402.

78. Anderson, G. W., and Martin, M. P. 1981. Factors related to incidence of Hypoxylon cankers in aspen and survival of cankered trees. For. Sci. 27:461–476.

79. Anderson, H. W. 1922. Orchard practice for the control of blister canker of apple trees. Ill. Agric. Exp. Stn. Circ. 258. 16 pp.

80. Anderson, H. W. 1930. Experiments with blister canker of apple trees. Ill. Agric. Exp. Stn. Bull. 340:55–90.

81. Anderson, N. A., and French, D. W. 1964. Sweetfern rust on jack pine. J. For. 62:467–471.

82. Anderson, N. A., and Kaufert, F. H. 1959. Brooming response of black spruce to dwarfmistletoe infection. For. Sci. 5:356–364.

83. Anderson, P. H., and Pezeshki, S. R. 2001. Effects of flood preconditioning on responses of three bottomland tree species to soil waterlogging. J. Plant Physiol. 158:227–233.

84. Anderson, P. J. 1914. The morphology and life history of the chestnut blight fungus. Penn. Chestnut Tree Blight Comm. Bull. 7. 44 pp.

85. Anderson, P. J., and Rankin, W. H. 1914. Endothia canker of chestnut. Cornell Univ. Agric. Exp. Stn. Bull. 347:529–618.

86. Anderson, R. C., Gardner, D. E., Daehler, C. C., and Meinzer, F. C. 2002. Dieback of *Acacia koa* in Hawaii: ecological and pathological characteristics of affected stands. For. Ecol. Manag. 162:273–286.

87. Anderson, R. L., Anderson, G. W., and Schipper, A. L. 1979. Hypoxylon canker of aspen. USDA For. Serv. For. Insect Dis. Leafl. 6. 7 pp. Internet pub.

88. Anderson, R. L., and Hoffard, W. H. 1978. Fusarium canker–ambrosia beetle complex on tulip poplar in Ohio. Plant Dis. Rep. 62:751.

89. Anderson, R. L., McClure, J. P., Cost, N., and Uhler, R. J. 1986. Estimating fusiform rust losses in five southeast states. South. J. Appl. For. 10:237–240.

90. Angeles, G., Evert, R. F., and Kozlowski, T. T. 1986. Development of lenticels and adventitious roots in flooded *Ulmus americana* seedlings. Can. J. For. Res. 16:585–590.

91. Anselmi, N. 1986. Resurgence of *Cryptodiaporthe populea* in Italy. Bull. OEPP 16:571–583.

92. Anselmi, N., and Giorcelli, A. 1990. Factors influencing the incidence of *Rosellinia necatrix* Prill. in poplars. Eur. J. For. Pathol. 20:175–183.

93. Appel, D. N. 1994. Identification and control of oak wilt in Texas urban forests. J. Arboric. 20:250–258.

94. Appel, D. N. 1995. The oak wilt enigma: perspectives from the Texas epidemic. Annu. Rev. Phytopathol. 33:103–118.

95. Appel, D. N., Kurdyla, T., and Lewis, R. Jr. 1990. Nitidulids as vectors of the oak wilt fungus and other *Ceratocystis* spp. in Texas. Eur. J. For. Pathol. 20:412–417.

96. Appel, D. N., Maggio, R. C., Nelson, E. L., and Jeger, M. J. 1989. Measurement of expanding oak wilt centers in live oak. Phytopathology 79:1318–1322.

97. Appel, D. N., Peters, R., and Lewis, R. Jr. 1987. Tree susceptibility, inoculum availability, and potential vectors in a Texas oak wilt center. J. Arboric. 13:169–173.

98. Appel, D. N., and Stipes, R. J. 1984. Canker expansion on water-stressed pin oaks colonized by *Endothia gyrosa*. Plant Dis. 68:851–853.

99. Appel, D. N., and Stipes, R. J. 1986. A description of declining and blighted pin oaks in eastern Virginia. J. Arboric. 12:155–158.

100. Appleton, B. L. 1993. Nursery production alternatives for reduction or elimination of circling tree roots. J. Arboric. 19:383–388.

101. Appleton, B. L., Huff, R. R., and French, S. C. 1999. Evaluating trees for saltwater spray tolerance for oceanfront sites. J. Arboric. 25:205–210.

102. Arambarri, A. M., and Mandrile, E. L. 1999. *Tillandsia* L. (Bromeliaceae): anatomy and ethno-pharmacology. Acta Hortic. No. 503:133–139.

103. Aramburu, J., and Rovira, M. 1998. The effects of apple mosaic ilarvirus (ApMV) on hazelnut (*Corylus avellana* L.). J. Hortic. Sci. Biotechnol. 73:97–101.

104. Arauz, L. F., and Sutton, T. B. 1989. Influence of temperature and moisture on germination of ascospores and conidia of *Botryosphaeria obtusa*. Phytopathology 79:667–674.

105. Arauz, L. F., and Sutton, T. B. 1989. Temperature and wetness duration requirements for apple infection by *Botryosphaeria obtusa*. Phytopathology 79:440–444.

106. Ariffin, D., Idris, A. S., and Abdul, H. H. 1989. Significance of the black line within oil palm tissue decayed by *Ganoderma boninense*. Elaeis 1:11–16.

107. Armel, G. R., Richardson, R. J., Stipes, R. J., and Hipkins, P. L. 1997. Biocontrol of the invasive *Ailanthus altissima* with fungal weaponry. (Abstr.) Va. J. Sci. 48:94.

108. Arnett, J. D., and Witcher, W. 1974. Histochemical studies of yellow poplar infected with *Fusarium solani*. Phytopathology 64:414–418.

109. Arnold, R. H. 1967. A canker and foliage disease of yellow birch. I. Description of the causal fungus, *Diaporthe alleghaniensis* sp. nov., and the symptoms on the host. Can. J. Bot. 45:783–801.

110. Arnold, R. H., and Carter, J. C. 1974. *Fusicoccum elaeagni*, the cause of canker and dieback of Russian olive, redescribed and redisposed to the genus *Phomopsis*. Mycologia 66:191–197.

111. Arnold, R. H., and Straby, A. E. 1973. *Phomopsis elaeagni* on Russian olive (*Elaeagnus angustifolia*) in Canada. Can. Plant Dis. Surv. 53:183–186.

112. Arsdel, E. P. Van. 1961. Growing white pine in the Lake States to avoid blister rust. USDA For. Serv. Lake States For. Exp. Stn. Stn. Pap. 92. 11 pp.

113. Arsdel, E. P. Van. 1967. The nocturnal diffusion and transport of spores. Phytopathology 57:1221–1229.

114. Arsdel, E. P. Van. 1972. Some cankers on oaks in Texas. Plant Dis. Rep. 56:300–304.

115. Arsdel, E. P. Van. 1979. Symptoms and conditions of environmental tree disease. Weeds Trees Turf 18(6):16–20, 23, 26, 28, 29.

116. Arsdel, E. P. Van. 1980. Managing trees to reduce damage from low-level saline irrigation. Weeds Trees Turf 19(6):26–28, 61.

117. Arsdel, E. P. Van, and Chitzanidis, A. 1970. Life cycle and spread of ash rust in Texas. (Abstr.) Phytopathology 60:1317.

118. Arsdel, E. P. Van, Riker, A. J., and Patton, R. F. 1956. Effects of temperature and moisture on the spread of white pine blister rust. Phytopathology 46:307–318.

119. Arthur, J. C. 1934. Manual of the rusts in the United States and Canada. Purdue Res. Found. Lafayette, IN. 438 pp.

120. Arthur, J. J., Leone, I. A., and Flower, F. B. 1981. Flooding and landfill gas effects on red and sugar maples. J. Environ. Qual. 10:431–433.

121. Arx, J. A. von 1957. Schurft op Pyracantha. Tijdschr. Plantenz. 63: 198–199.

122. Arx, J. A. von 1970. A revision of the fungi classified as Gloeosporium. 2nd ed. Bibl. Mycol. Vol. 24. 203 pp.

123. Arx, J. A. von. 1987. Plant pathogenic fungi. Cramer, Berlin. 288 pp.

124. Arx, J. A. von. 1981. The genera of fungi sporulating in pure culture. 3rd ed. J. Cramer, Lehre, Germany. 424 pp.

125. Arx, J. A. von, and Müller, E. 1954. Die Gattungen der amerosporen Pyrenomyceten. Beitr. Kryptogamenfl. Schweiz. Vol. 11. 434 pp.

126. Arx, J. A. von, Walt, J. P. van der, and Liebenberg, N. V. D. M. 1982. The classification of Taphrina and other fungi with yeast-like cultural states. Mycologia 74:285–296.

127. Ash, C. L. 1999. The ecology of Verticillium dahliae in Fraxinus and Acer species. Ph.D. diss., Univ. Minnesota, St. Paul. 112 pp.

128. Ash, C. L., ed. 2001. Shade tree wilt diseases. APS Press, St. Paul, MN. 257 pp.

129. Ashcroft, J. M. 1934. European canker of black walnut and other trees. W. Va. Agric. Exp. Stn. Bull. 261. 52 pp.

130. Ashworth, E. N. 1993. Formation and spread of ice in plant tissues. Hortic. Rev. 13:215–255.

131. Ashworth, L. J. Jr., and Gaona, S. A. 1982. Evaluation of clear polyethylene mulch for controlling Verticillium wilt in established pistachio nut groves. Phytopathology 72:243–246.

132. Ashworth, L. J. Jr., Gaona, S. A., and Surber, E. 1985. Nutritional diseases of pistachio trees: potassium and phosphorus deficiencies and chloride and boron toxicities. Phytopathology 75:1084–1091.

133. Ashworth, L. J. Jr., Gaona, S. A., and Surber, E. 1985. Verticillium wilt of pistachio: the influence of potassium nutrition on susceptibility to infection by Verticillium dahliae. Phytopathology 75:1091–1093.

134. Ashworth, V. E. T. M. 2000. Phylogenetic relationships in Phoradendreae (Viscaceae) inferred from three regions of the nuclear ribosomal cistron. I. Major lineages and paraphyly of Phoradendron. II. The North American species of Phoradendron. Syst. Bot. 25:349–370; El Aliso 19:41–53.

135. Ashworth, V. E. T. M., and Santos, G. dos. 1997. Wood anatomy of four Californian mistletoe species (Phoradendron, Viscaceae). IAWA J. 18:229–245.

136. Asiegbu, F. O., Johansson, M., and Daniel, G. 1997. Cytopathological responses of Picea abies seedling roots challenged in vitro by a necrotrophic Fusarium avenaceum (Fr.) Sacc. J. Phytopathol. 145:239–244.

137. Asiegbu, F. O., Kacprzak, M., Daniel, G., Johansson, M., Stenlid, J., and Manka, M. 1999. Biochemical interactions of conifer seedling roots with Fusarium spp. Can. J. Microbiol. 45:923–935.

138. Astrom, B., and Ramstedt, M. 1994. Stem cankers on Swedish biomass willows caused by Cryptodiaporthe salicella and other fungi. Eur. J. For. Pathol. 24:264–276.

139. Atilano, R. A. 1983. A foliar blight of Ming aralia caused by Alternaria panax. Plant Dis. 67:224–226.

140. Atilano, R. A. 1983. Alternaria leaf spot of Schefflera arboricola. Plant Dis. 67:64–66.

141. Atkinson, R. G. 1965. Phytophthora species inciting root rot of Chamaecyparis lawsoniana and other ornamentals in coastal British Columbia. Can. J. Bot. 43:1471–1475.

142. Auclair, A. N. D., Eglinton, P. D., and Minnemeyer, S. L. 1997. Principal forest dieback episodes in northern hardwoods: development of numeric indices of areal extent and severity. Water Air Soil Pollut. 93:175–198.

143. Auclair, A. N. D., Lill, J. T., and Revenga, C. 1996. The role of climate variability and global warming in the dieback of northern hardwoods. Water Air Soil Pollut. 91:163–186.

144. Auclair, A. N. D., Martin, H. C., and Walker, S. L. 1990. A case study of forest decline in western Canada and the adjacent United States. Water Air Soil Pollut. 53:13–31.

145. Audley, D. E., and Skelly, J. M. 1994. A Phomopsis species associated with nonlethal adelgid galls on upper crown branchlets of red spruce in West Virginia. Plant Dis. 78:569–571.

146. Auge, R. M. 2001. Water relations, drought and vesicular-arbuscular mycorrhizal symbiosis. Mycorrhiza 11:3–42.

147. Aukema, J. E. 2003. Vectors, viscin, and Viscaceae: mistletoes as parasites, mutualists, and resources. Front. Ecol. Environ. 1:212–219.

148. Axelrood, P. E., Chapman, W. K., Seifert, K. A., Trotter, D. B., and Shrimpton, G. 1998. Cylindrocarpon and Fusarium root colonization of Douglas-fir seedlings from British Columbia reforestation sites. Can. J. For. Res. 28:1198–1206.

149. Aycock, R. 1966. Stem rot and other diseases caused by Sclerotium rolfsii. N.C. Agric. Exp. Stn. Tech. Bull. 174. 202 pp.

150. Aycock, R., West, E., Watkins, G. M., Cooper, W. E., Wilson, C., Boyle, L. W., Garren, K. H., and Harrison, A. L. 1961. Symposium on Sclerotium rolfsii. Phytopathology 51:107–128.

151. Ayers, J. C. Jr., and Barden, V. A. 1975. Net photosynthesis and dark respiration of apple leaves as affected by pesticides. J. Am. Soc. Hortic. Sci. 100:24–28.

152. Aylor, D. E. 1998. The aerobiology of apple scab. Plant Dis. 82:838–849.

153. Azad, H. R., and Cooksey, D. A. 1995. A semiselective medium for detecting epiphytic and systemic populations of Pseudomonas savastanoi from oleander. Phytopathology 85:740–745.

154. Baayen, R. P., Bonants, P. J. M., Verkley, G., Carroll, G. C., Aa, H. A. van der, Weerdt, M. de, Brouwershaven, I. R. van, Schutte, G. C., Maccheroni, W. Jr., Blanco, C. G. de, and Azevedo, J. L. 2002. Nonpathogenic isolates of the citrus black spot fungus, Guignardia citricarpa, identified as a cosmopolitan endophyte of woody plants, G. mangiferae (Phyllosticta capitalensis). Phytopathology 92:464–477.

155. Baayen, R. P., O'Donnell, K., Bonants, P. J. M., Cigelnik, E., Kroon, L. P. N. M., Roebroeck, E. J. A., and Waalwijk, C. 2000. Gene genealogies and AFLP analyses in the Fusarium oxysporum complex identify monophyletic and nonmonophyletic formae speciales causing wilt and rot disease. Phytopathology 90:891–900.

156. Bachand, G. D., Castello, J. D., Schaedle, M., Stehman, S. V., and Livingston, W. H. 1996. Effects of tomato mosaic tobamovirus infection on red spruce seedlings. Can. J. For. Res. 26:973–981.

157. Bachi, P. R., and Peterson, J. L. 1985. Enhancement of Sphaeropsis sapinea stem invasion of pines by water deficits. Plant Dis. 69:798–799.

158. Bachman, G., Wilson, C., and Whitwell, T. 1995. Tolerance of containerized landscape plants to the postemergence herbicides Stinger, Manage and Basagran. J. Environ. Hortic. 13:129–132.

159. Bacon, C. W., and White, J. F., eds. 2000. Microbial endophytes. Marcel Dekker, New York. 487 pp.

160. Bae, H. H., Hansen, E. M., and Strauss, S. H. 1994. Restriction fragment length polymorphisms demonstrate single origin of infection centers in Phellinus weirii. Can. J. Bot. 72:440–447.

161. Baes, C. F. III, and McLaughlin, S. B. 1984. Trace elements in tree rings: evidence of recent and historical air pollution. Science 224:494–496.

162. Bagga, D. K., and Smalley, E. B. 1974. The development of Hypoxylon canker of Populus tremuloides: role of ascospores, conidia, and toxins; role of interacting environmental factors. Phytopathology 64:654–658, 658–662.

163. Bagga, H. S., and Boone, D. M. 1968. Genes in Venturia inaequalis controlling pathogenicity to crabapples. Phytopathology 58:1176–1182.

164. Bagga, H. S., and Boone, D. M. 1968. Inheritance of resistance to Venturia inaequalis in crabapples. Phytopathology 58:1183–1187.

165. Bahnweg, G., Moller, E. M., Anegg, S., Langebartels, C., Wienhaus, O., and Sandermann, H. Jr. 2002. Detection of Heterobasidion annosum s.l. [(Fr.) Bref.] in Norway spruce by polymerase chain reaction. J. Phytopathol. 150:382–389.

166. Bahnweg, G., Schubert, R., Kehr, R. D., Muller-Starck, G., Heller, W., Langebartels, C., and Sandermann, H. Jr. 2000. Controlled inoculation of Norway spruce (Picea abies) with Sirococcus conigenus: PCR-based quantification of the pathogen in host tissue and infection-related increase of phenolic metabolites. Trees Struct. Funct. 14:435–441.

167. Baidyaroy, D., and Bertrand, H. 2001. Biological control of the chestnut blight pathogen using mitochondrial agents. Rec. Res. Devel. Genet. 1:93–104.

168. Bailey, J. A., and Jeger, M. J., eds. 1992. Colletotrichum: biology, pathology, and control. CAB International, Wallingford, UK. 388 pp.

169. Baines, R. C. 1939. Phytophthora trunk canker or collar rot of apple trees. J. Agric. Res. 59:159–184.

170. Bains, L. M. 1994. Phoenix dactylifera cultivars with resistance to Graphiola leaf spot. Principes 38:134–137.

171. Bajji, M., Kinet, J. M., and Lutts, S. 1998. Salt stress effects on roots and leaves of Atriplex halimus L. and their corresponding callus cultures. Plant Sci. 137:131–142.

172. Baker, C. J., Harrington, T. C., Krauss, U., and Alfenas, A. C. 2003. Genetic variability and host specialization in the Latin American clade of *Ceratocystis fimbriata*. Phytopathology 93:1274–1284.

173. Baker, F. A., and French, D. W. 1980. Spread of *Arceuthobium pusillum* and rates of infection and mortality in black spruce stands. Plant Dis. 64:1074–1076.

174. Baker, F. A., and French, D. W. 1986. Dispersal of *Arceuthobium pusillum* seeds. Can. J. For. Res. 16:1–5.

175. Baker, F. A., and French, D. W. 1991. Radial enlargement of mortality centers caused by *Arceuthobium pusillum* Peck in black spruce stands. For. Sci. 37:364–367.

176. Baker, F. A., French, D. W., and Hudler, G. W. 1981. Development of *Arceuthobium pusillum* on inoculated black spruce. For. Sci. 27:203–205.

177. Baker, F. A., French, D. W., Kulman, H. M., Davis, O., and Bright, R. C. 1985. Pollination of the eastern dwarf mistletoe. Can. J. For. Res. 15:708–714.

178. Baker, F. A., Slivitsky, M., and Knowles, K. 1992. Impact of dwarf mistletoe on jack pine forests in Manitoba. Plant Dis. 76:1256–1259.

179. Baker, J. H. 1965. Relationship between salt concentrations in leaves and sap and the decline of sugar maples along roadsides. Mass. Agric. Exp. Stn. Bull. 553. 16 pp.

180. Baker, K. F. 1953. Recent epidemics of downy mildew of rose. Plant Dis. Rep. 37:331–337.

181. Baker, K. F., and Dimock, A. W. 1969. Black spot. Pages 172–184 in: Roses. J. W. Mastalerz and R. W. Langhans, eds. Pennsylvania Flower Growers, New York State Flower Growers Assoc., and Roses, Inc. 329 pp.

182. Baker, T. J., and Gowen, S. R. 1996. Staining: nematodes and arbuscular mycorrhizae in the same root sample. Fund. Appl. Nematol. 19:607–608.

183. Baker, W. A., Partridge, E. C., and Morgan-Jones, G. 2000. Notes on Hyphomycetes. LXXVIII. *Asperisporium sequoiae*, the causal organism of conifer needle blight, reclassified in *Cercosporidium*, with comments on the status of the genus. Mycotaxon 76:247–256.

184. Baker, W. L. 1949. Notes on the transmission of the virus causing phloem necrosis of American elm, with notes on the biology of its insect vector. J. Econ. Entomol. 42:729–732.

184a. Balcì, Y., and Halmschlager, E. 2003. Incidence of *Phytophthora* species in oak forests in Austria and their possible involvement in oak decline. For. Pathol. 33:157–174.

184b. Balcì, Y., and Halmschlager, E. 2003. *Phytophthora* species in oak ecosystems in Turkey and their association with declining oak trees. Plant Pathol. 52:694–702.

185. Baldwin, H. I., Boyce, J. S., Brown, R. C., Cline, A. C., Filley, W. O., Reynolds, H. A., and Turner, G. W. C., eds. 1940. Important tree pests of the Northeast. Mass. For. Park Assoc., Boston. 194 pp.

186. Ball, M. C. 1988. Ecophysiology of mangroves. Trees Struct. Funct. 2:129–142.

187. Ballio, A. 1978. Fusicoccin, the vivotoxin of *Fusicoccum amygdali* Del. Chemical properties and biological activity. Ann. Phytopathol. 10:145–156.

188. Banfield, W. M. 1941. Distribution by the sap stream of spores of three fungi that induce vascular wilt diseases of elm. J. Agric. Res. 62:637–681.

189. Banik, M. T., and Burdsall, H. H. Jr. 1998. Assessment of compatibility among *Armillaria cepistipes, A. sinapina,* and North American biological species X and XI, using culture morphology and molecular biology. Mycologia 90:798–805.

190. Banik, M. T., and Burdsall, H. H. Jr. 1999. Incompatibility between *Laetiporus cincinnatus* and *L. sulphureus* in culture. Mycotaxon 70:461–469.

191. Banik, M. T., and Burdsall, H. H. Jr. 2000. Incompatibility groups among North American populations of *Laetiporus sulphureus sensu lato.* Mycologia 92:649–655.

192. Bannister, P., King, W. M., and Strong, G. L. 1999. Aspects of the water relations of *Ileostylus micranthus* (Hook. f.) Tieghem, a New Zealand mistletoe. Ann. Bot. 84:79–86.

193. Barakat, R. M., Johnson, D. A., and Grove, G. G. 1995. Factors affecting conidial exudation and survival, and ascospore germination of *Leucostoma cincta*. Plant Dis. 79:1245–1248.

194. Baranyay, J. A. 1970. Lodgepole pine dwarf mistletoe in Alberta. Can. For. Serv. Pub. 1286. 22 pp.

195. Baranyay, J. A., and Hiratsuka, Y. 1967. Identification and distribution of *Ciborinia whetzelii* (Seaver) Seaver in western Canada. Can. J. Bot. 45:189–191.

196. Baranyay, J. A., and Safranyik, L. 1970. Effect of dwarf mistletoe on growth and mortality of lodgepole pine in Alberta. Can. For. Serv. Pub. 1285. 19 pp.

197. Baranyay, J. A., Szabo, T., and Hunt, K. 1973. Effect of Atropellis canker on growth and utilization of lodgepole pine. Can. For. Serv. Pac. For. Res. Cent., Inf. Rep. BC-X-86. 22 pp.

198. Barbe, G. D., McCartney, W. O., and Rosenberg, D. Y. 1966. *Gymnosporangium confusum* found in California. Plant Dis. Rep. 50:768–769.

199. Barker, K. R., and Davis, E. L. 1996. Assessing plant-nematode infestations and infections. Adv. Bot. Res. Incorp. Adv. Plant Pathol. 23:103–136.

200. Barker, K. R., Pederson, G. A., and Windham, G. L., eds. 1998. Plant and nematode interactions. Am. Soc. Agron. and Crop Sci. Soc. Am., Madison, WI. 771 pp.

201. Barker, K. R., Worf, G. L., and Epstein, A. H. 1965. Nematodes associated with the decline of azaleas in Wisconsin. Plant Dis. Rep. 49:47–49.

202. Barklund, P., Livsey, S., Karlman, M., and Stephan, R., eds. 1993. Shoot diseases of conifers. Proceedings of an international symposium, Garpenberg, Sweden. June 1991. Swedish Univ. Agric. Sci., Uppsala. 175 pp.

203. Barklund, P., and Unestam, T. 1988. Infection experiments with *Gremmeniella abietina* on seedlings of Norway spruce and Scots pine. Eur. J. For. Pathol. 18:409–420.

204. Barnard, E. L. 1984. Occurrence, impact, and fungicidal control of girdling stem cankers caused by *Cylindrocladium scoparium* on eucalyptus seedlings in a South Florida nursery. Plant Dis. 68:471–473.

205. Barnard, E. L., and Ash, E. C. III. 1991. Fusarium wilt of waxmyrtle. Fla. Dep. Agric. Consum. Serv. Plant Pathol. Circ. 350. 3 pp.

206. Barnard, E. L., Ash, E. C., Hopkins, D. L., and McGovern, R. J. 1998. Distribution of *Xylella fastidiosa* in oaks in Florida and its association with growth decline in *Quercus laevis*. Plant Dis. 82:569–572.

207. Barnard, E. L., Blakeslee, G. M., English, J. T., Oak, S. W., and Anderson, R. L. 1985. Pathogenic fungi associated with sand pine root disease in Florida. Plant Dis. 69:196–199.

208. Barnard, E. L., and Dixon, W. N. 1983. Insects and diseases: important problems of Florida's forest and shade tree resources. Fla. Dep. Agric. Consum. Serv. Bull. 196-A. 120 pp.

209. Barnard, E. L., El-Gholl, N. E., and Gilly, S. P. 1988. Comparative spore morphology and pathogenicity of four Florida isolates of *Nectria galligena*. Plant Dis. 72:973–976.

210. Barnard, E. L., Geary, T., English, J. T., and Gilly, S. P. 1987. Basal cankers and coppice failure of *Eucalyptus grandis* in Florida. Plant Dis. 71:358–361.

211. Barnard, E. L., and Miller, T. 1999. Southern cone rust: a growing problem in need of attention? USDA For. Serv. Gen. Tech. Rep. SRS-30:569–572.

212. Barnard, E. L., and Mitchell, D. J. 1993. Phytophthora basal canker of red maple. Fla. Dep. Agric. Consum. Serv. Plant Pathol. Circ. 361. 3 pp.

213. Barnard, E. L., and Schroeder, R. A. 1984. Anthracnose of acacia in Florida: occurrence and fungicidal control. Proc. Fla. State Hortic. Soc. 97:244–247.

214. Barnes, E. H. 1979. Atlas and manual of plant pathology. Kluwer Academic/Plenum, New York. 325 pp.

215. Barnes, I., Roux, J., Wingfield, B. D., O'Neill, M., and Wingfield, M. J. 2003. *Ceratocystis fimbriata* infecting *Eucalyptus grandis* in Uruguay. Australas. Plant Pathol. 32:361–366.

216. Barnes, I., Roux, J., Wingfield, M. J., Coetzee, M. P. A., and Wingfield, B. D. 2001. Characterization of *Seiridium* spp. associated with cypress canker based on β-tubulin and histone sequences. Plant Dis. 85:317–321.

217. Barnett, H. L. 1957. *Hypoxylon punctulatum* and its conidial stage on dead oak trees and in culture. Mycologia 49:588–595.

218. Barnett, H. L., and Hunter, B. B. 1998. Illustrated genera of imperfect fungi. 4th ed. APS Press, St. Paul, MN. 218 pp.

219. Barnett, O. W. 1971. Mosaic of blackjack oak. Plant Dis. Rep. 55:411.

220. Barnett, O. W., and Baxter, L. W. 1973. Cucumber mosaic virus on *Nandina domestica* in South Carolina. Plant Dis. Rep. 57:917–920.

221. Barnett, O. W., Reddick, B. B., Burrows, P. M., and Baxter, L. W. Jr. 1989. Characterization of dogwood mosaic nepovirus from *Cornus florida*. Phytopathology 79:951–958.

222. Barnett, O. W., Shelburne, V. B., Yao, J., and Tainter, F. H. 1996. A new virus disease of yellow-poplar. Plant Dis. 80:1317–1319.

223. Barney, C. W., Hawksworth, F. G., and Geils, B. W. 1998. Hosts of *Viscum album*. Eur. J. For. Pathol. 28:187–208.

224. Barney, D. L., Walser, R. H., Davis, T. D., and Williams, C. F. 1985. Trunk injection of iron compounds as a treatment for overcoming iron chlorosis in apple trees. HortScience 20:236–238.

225. Barny, M. A. 1995. *Erwinia amylovora hrpN* mutants, blocked in harpin synthesis, express a reduced virulence on host plants and elicit variable hypersensitive reactions on tobacco. Eur. J. Plant Pathol. 101:333–340.

226. Barr, M. E. 1955. Species of sooty molds from western North America. Can. J. Bot. 33:497–514.

227. Barr, M. E. 1968. The Venturiaceae in North America. Can. J. Bot. 46:799–864.

228. Barr, M. E. 1972. Preliminary studies on the Dothideales in temperate North America. Contrib. Univ. Mich. Herb. 9:523–638.

229. Barr, M. E. 1978. The Diaporthales in North America with emphasis on *Gnomonia* and its segregates. Mycologia Mem. No. 7. 232 pp.

230. Barr, M. E. 1984. *Herpotrichia* and its segregates. Mycotaxon 20:1–38.

231. Barr, M. E. 1987. Prodromus to Class Loculoascomycetes. Pub. by author, Univ. Mass., Amherst. 168 pp. Distrib. by Luprecht & Cramer, Ltd., Forestburgh, NY.

232. Barr, M. E. 1989. The Venturiaceae in North America: revisions and additions. Sydowia 41:25–40.

233. Barr, M. E. 1990. Prodromus to nonlichenized, pyrenomycetous members of Class Hymenoascomycetes. Mycotaxon 39:43–184.

234. Barr, M. E. 1991. Revisions and additions to the Diaporthales. Mycotaxon 41:287–305.

235. Barr, M. E. 1992. Additions to and notes on the Phaeosphaeriaceae (Pleosporales, Loculoascomycetes). Mycotaxon 43:371–400.

236. Barr, M. E. 1996. Planistromellaceae, a new family in the Dothideales. Mycotaxon 60:433–442.

237. Barrett, D. K. 1970. *Armillaria mellea* as a possible factor predisposing roots to infection by *Polyporus schweinitzii*. Trans. Br. Mycol. Soc. 55:459–462.

238. Barrett, D. K. 1985. Basidiospores of *Phaeolus schweinitzii*: a source of soil infestation. Eur. J. For. Pathol. 15:417–425.

239. Barrett, D. K., and Greig, B. J. W. 1985. The occurrence of *Phaeolus schweinitzii* in the soils of Sitka spruce plantations with broadleaved and non-woodland histories. Eur. J. For. Pathol. 15:412–417.

240. Barrett, D. K., and Uscuplic, M. 1971. The field distribution of interacting strains of *Polyporus schweinitzii* and their origin. New Phytol. 70:581–598.

241. Barrick, K. A., and Noble, M. G. 1993. The iron and manganese status of seven upper montane tree species in Colorado, USA, following long-term waterlogging. J. Ecol. 81:523–531.

242. Barrick, W. E., Flore, J. A., and Davidson, H. 1979. Deicing salt spray injury in selected *Pinus* spp. J. Am. Soc. Hortic. Sci. 104:617–622.

243. Barrows-Broaddus, J. 1990. Colonization of cones and seed of loblolly pine following inoculation with *Fusarium subglutinans*. Plant Dis. 74:1002–1005.

244. Barrows-Broaddus, J., and Dwinell, L. D. 1985. Branch dieback and cone and seed infection caused by *Fusarium moniliforme* var. *subglutinans* in a loblolly pine seed orchard in South Carolina. Phytopathology 75:1104–1108.

245. Barrus, M. F., and Horsfall, J. G. 1928. Preliminary note on snowberry anthracnose. Phytopathology 18:797–801.

246. Bartelt, R. J., Kyhl, J. F., Ambourn, A. K., Juzwik, J., and Seybold, S. J. 2004. Male-produced aggregation pheromone of *Carpophilus sayi*, a nitidulid vector of oak wilt disease, and pheromonal comparison with *Carpophilus lugubris*. Agric. For. Entomol. 6:39–46.

247. Bartholmess, H. 1989. Relations between lichens, air pollution load and forest decline in Baden-Wurttemberg. Forstwiss. Centralbl. 108:188–196.

248. Bartholomay, G. A., Eckert, R. T., and Smith, K. T. 1997. Reductions in tree-ring widths of white pine following ozone exposure at Acadia National Park, Maine, USA. Can. J. For. Res. 27:361–368.

249. Bartoli, C. G., Beltrano, J., Fernandez, L. V., and Caldiz, D. O. 1993. Control of the epiphytic weeds *Tillandsia recurvata* and *Tillandsia aeranthos* with different herbicides. For. Ecol. Manag. 59:289–294.

250. Basham, J. T. 1958. Decay of trembling aspen. Can. J. Bot. 36:491–505.

251. Basham, J. T. 1975. Heart rot of jack pine in Ontario. IV. Heartwood-inhabiting fungi: their entry and interactions within living trees. Can. J. For. Res. 5:706–721.

252. Bassett, E. N., and Fenn, P. 1984. Latent colonization and pathogenicity of *Hypoxylon atropunctatum* on oaks. Plant Dis. 68:317–319.

253. Batko, S. 1956. *Meria laricis* on Japanese and hybrid larch in Britain. Trans. Br. Mycol. Soc. 39:13–16.

254. Batra, L. R. 1960. The species of *Ciborinia* pathogenic to *Salix*, *Magnolia*, and *Quercus*. Am. J. Bot. 47:819–827.

255. Batra, L. R. 1983. *Monilinia vaccinii-corymbosi* (Sclerotiniaceae): its biology on blueberry and comparison with related species. Mycologia 75:131–152.

256. Batra, L. R. 1991. World species of *Monilinia* (Fungi): their ecology, biosystematics and control. Mycol. Mem. No. 16. J. Cramer, Berlin. 246 pp.

257. Bauce, E., and Allen, D. C. 1991. Etiology of a sugar maple decline. Can. J. For. Res. 21:686–693.

258. Baudoin, A. B. A. M. 1986. Environmental conditions required for infection of photinia leaves by *Entomosporium mespili*. Plant Dis. 70:519–521.

259. Baudoin, A. B. A. M. 1986. Infection of photinia leaves by *Entomosporium mespili*. Plant Dis. 70:191–194.

260. Baum, S., and Schwarze, F. 2002. Large-leaved lime (*Tilia platyphyllos*) has a low ability to compartmentalize decay fungi via reaction zone formation. New Phytol. 154:481–490.

261. Baumgartner, K., and Rizzo, D. M. 2001. Distribution of *Armillaria* species in California. Mycologia 93:821–830.

262. Baumgartner, K., and Rizzo, D. M. 2001. Ecology of *Armillaria* spp. in mixed-hardwood forests of California. Plant Dis. 85:947–951.

263. Baxter, D. V. 1925. The biology and pathology of some of the hardwood heart-rotting fungi. Am. J. Bot. 12:522–576.

264. Baxter, D. V. 1936. Some resupinate polypores from the region of the Great Lakes. VII. Pap. Mich. Acad. Sci. 21:243–267.

265. Baxter, D. V. 1948. Some resupinate polypores from the region of the Great Lakes. XVIII. Pap. Mich. Acad. Sci. 32(1946):189–211.

266. Baxter, D. V. 1967. Disease in forest plantations: thief of time. Cranbrook Inst. Sci. Bull. 51. 251 pp.

267. Baxter, J. W. 1957. The genus *Cumminsiella*. Mycologia 49:864–873.

268. Baxter, L. W., and Plakidas, A. G. 1954. Dieback and canker of camellias caused by *Glomerella cingulata*. Phytopathology 44:129–133.

269. Baxter, L. W. Jr., and Fagan, S. G. 1974. A comparison of the relative susceptibility of seedlings of *Camellia japonica* and *C. sasanqua* to dieback and canker caused by a strain of *Glomerella cingulata* pathogenic to camellias. Plant Dis. Rep. 58:139–141.

270. Baxter, L. W. Jr., Fagan, S. G., and Owen, M. G. 1982. Response of selected *Camellia reticulata* cultivars to *Glomerella cingulata*, cause of contagious camellia dieback and canker. Plant Dis. 66:1023–1024.

271. Bazzigher, G. 1976. Der schwarze Schneeschimmel der Koniferen [*Herpotrichia juniperi* (Duby) Petrak und *Herpotrichia coulteri* (Peck) Bose]. Eur. J. For. Pathol. 6:109–122.

272. Beanland, L., Hoy, C. W., Miller, S. A., and Nault, L. R. 2000. Influence of aster yellows phytoplasma on the fitness of aster leafhopper (Homoptera: Cicadellidae). Ann. Entomol. Soc. Am. 93:271–276.

273. Beardsell, D. V., and Considine, J. A. 1987. Lineages, lineage stability and pattern formation in leaves of variegated chimeras of *Lophostemon confertus* (R. Br.) Wilson & Waterhouse and *Tristaniopsis laurina* (Smith) Wilson & Waterhouse (Myrtaceae). Austral. J. Bot. 35:701–714.

274. Beare, J. A., Archer, S. A., and Bell, J. N. B. 1999. Effects of *Melampsora* leaf rust disease and chronic ozone exposure on poplar. Environ. Pollut. 105:419–426.

275. Beattie, G. A., and Lindow, S. E. 1995. The secret life of foliar bacterial pathogens on leaves. Annu. Rev. Phytopathol. 33:145–172.

276. Beatty, J. L., Filip, G. M., and Mathiasen, R. L. 1997. Larch dwarf mistletoe. USDA For. Serv. For. Insect Dis. Leafl. 169. Internet pub.

277. Becker, C. M., Burr, T. J., and Smith, C. A. 1992. Overwintering of conidia of *Venturia inaequalis* in apple buds in New York orchards. Plant Dis. 76:121–126.

278. Becker, C. M., and Pearson, R. C. 1996. Black rot lesions on overwintered canes of *Euvitis* supply conidia of *Guignardia bidwellii* for primary inoculum in spring. Plant Dis. 80:24–27.

279. Beckman, T. G. 1998. Developing *Armillaria* resistant rootstocks for peach. Acta Hortic. No. 465:219–224.

280. Bedker, P. J., and Blanchette, R. A. 1983. Development of cankers caused by *Nectria cinnabarina* on honey locusts after root pruning. Plant Dis. 67:1010–1013.

281. Bedker, P. J., Blanchette, R. A., and French, D. W. 1982. *Nectria cinnabarina*: the cause of a canker disease of honey locust in Minnesota. Plant Dis. 66:1067–1070.

282. Bedker, P. J., and Wingfield, M. J. 1983. Taxonomy of three canker-causing fungi of honey locust in the United States. Trans. Br. Mycol. Soc. 81:179–183.

283. Bedlan, G. 1984. The genus *Phragmidium* Link. with particular reference to the *Phragmidium mucronatum* and *Phragmidium*

potentillae groups in Central Europe. Pflanzenschutzberichte 46:33–60.

284. Bega, R. V. 1959. The capacity and period of maximum production of sporidia in *Cronartium ribicola*. Phytopathology 49:54–57.

285. Bega, R. V. 1978. *Phomopsis lokoyae* outbreak in a California forest nursery. Plant Dis. Rep. 62:567–569.

286. Bega, R. V., tech. coord. 1978. Diseases of Pacific Coast conifers. USDA Agric. Handb. 521. 206 pp.

287. Begerow, D., Bauer, R., and Oberwinkler, F. 2002. The Exobasidiales: an evolutionary hypothesis. Mycol. Progr. 1:187–199.

288. Beig, M. A., and Khan, M. A. 1999. Occurrence of *Venturia acerina*, the perfect state of *Cladosporium humile*, on *Populus deltoides* in India. Appl. Biol. Res. 1:163–165.

289. Beigl, H. J. 1977, 1978. Veränderung der Krankheitsdisposition von Forstpflanzen durch Herbizide. I. Freilandversuche mit *Lophodermium pinastri*. III. Versuche mit Erregern der Umfallkrankheit. Eur. J. For. Pathol. 7:200–219; 8:240–258.

290. Bélanger, R. R., Bushnell, W. R., Dik, A. J., and Carver, T. L. W., eds. 2002. The powdery mildews: a comprehensive treatise. APS Press, St. Paul, MN. 292 pp.

291. Bélanger, R. R., Manion, P. D., and Griffin, D. H. 1989. *Hypoxylon mammatum* ascospore infection of *Populus tremuloides* clones: effects of moisture stress in tissue culture. Phytopathology 79:315–317.

292. Bélanger, R. R., Manion, P. D., and Griffin, D. H. 1990. Amino acid content of water-stressed plantlets of *Populus tremuloides* clones in relation to clonal susceptibility to *Hypoxylon mammatum* in vitro. Can. J. Bot. 68:26–29.

293. Belisario, A. 1991. *Dicarpella dryina* sp. nov., teleomorph of *Tubakia dryina*. Mycotaxon 41:147–155.

294. Belisario, A. 1999. Cultural characteristics and pathogenicity of *Melanconium juglandinum*. Eur. J. For. Pathol. 29:317–322.

295. Belisario, A., Cacciola, S. O., and Magriano di san Lio, G. 1997. *Phytophthora cactorum* on walnut seedlings in Italian nurseries. Eur. J. For. Pathol. 27:137–146.

296. Belisario, A., Forti, E., and Corazza, L. 1999. Collar and root rot of walnut trees, associated with *Fusarium solani*. Petria 9:277–282.

297. Belisario, A., Forti, E., Corazza, L., and Kesteren, H. A. von. 1999. First report of *Alternaria alternata* causing leaf spot on English walnut. Plant Dis. 83:696.

298. Belisario, A., and Onofri, S. 1995. Conidiogenesis and morphology of *Melanconium juglandinum*. Mycol. Res. 99:1059–1062.

299. Belisario, A., Zoina, A., Pezza, L. and Luongo, L. 1999. Susceptibility of species of *Juglans* to pathovars of *Xanthomonas campestris*. Eur. J. For. Pathol. 29:75–80.

300. Bell, R. L., and Zwet, T. van der. 1988. Susceptibility of *Pyrus* germplasm to Fabraea leaf spot. Acta Hortic. No. 224:229–236.

301. Bella, I. E., and Navratil, S. 1987. Growth losses from winter drying (red belt damage) in lodgepole pine stands on the east slopes of the Rockies in Alberta. Can. J. For. Res. 17:1289–1292.

302. Bénaouf, G., and Parisi, L. 2000. Genetics of host-pathogen relationships between *Venturia inaequalis* races 6 and 7 and *Malus* species. Phytopathology 90:236–242.

303. Bender, C. L., Alarcon-Chaidez, F., and Gross, D. C. 1999. *Pseudomonas syringae* phytotoxins: mode of action, regulation, and biosynthesis by peptide and polyketide synthetases. Microbiol. Mol. Biol. Rev. 63:266–292.

304. Bender, C. L., and Coyier, D. L. 1984. Isolation and identification of races of *Sphaerotheca pannosa* var. *rosae*. Phytopathology 74:100–103.

305. Bendz-Hellgren, M., and Stenlid, J. 1997. Decreased volume growth of *Picea abies* in response to *Heterobasidion annosum* infection. Can. J. For. Res. 27:1519–1524.

306. Beneddra, T., Picard, C., Petit, A., and Nesme, X. 1996. Correlation between susceptibility to crown gall and sensitivity to cytokinin in aspen cultivars. Phytopathology 86:225–231.

307. Benet, H., Guries, R. P., Boury, S., and Smalley, E. B. 1995. Identification of RAPD markers linked to a black leaf spot resistance gene in Chinese elm. Theor. Appl. Genet. 90:1068–1073.

308. Benhamou, N., Lafontaine, J. G., Joly, J. R., and Ouellette, G. B. 1985. Ultrastructural localization in host tissues of a toxic glycopeptide produced by *Ophiostoma ulmi*, using monoclonal antibodies. Can. J. Bot. 63:1185–1195.

309. Bennett, J. P., Anderson, R. L., Mielke, M. L., and Ebersole, J. J. 1994. Foliar injury air pollution surveys of eastern white pine (*Pinus strobus* L.): a review. Environ. Monit. Assess. 30:247–274.

310. Bennetts, R. E., White, G. C., Hawksworth, F. G., and Severs, S. E. 1996. The influence of dwarf mistletoe on bird communities in Colorado ponderosa pine forests. Ecol. Applic. 6:899–909.

311. Benoit, L. F., Skelly, J. M., Moore, L. D., and Dochinger, L. S. 1982. Radial growth reductions of *Pinus strobus* L. correlated with foliar ozone sensitivity as an indicator of ozone-induced losses in eastern forests. Can. J. For. Res. 12:673–678.

312. Benson, D. M. 1982. Cold inactivation of *Phytophthora cinnamomi*. Phytopathology 72:560–563.

313. Benson, D. M. 1990. Landscape survival of fungicide-treated azaleas inoculated with *Phytophthora cinnamomi*. Plant Dis. 74:635–637.

314. Benson, D. M. 1991. Detection of *Phytophthora cinnamomi* in azalea with commercial serological assay kits. Plant Dis. 75:478–482.

315. Benson, D. M., and Barker, K. R. 1982. Susceptibility of Japanese boxwood, dwarf gardenia, Compacta (Japanese) holly, spiny Greek and Blue Rug junipers, and nandina to four nematode species. Plant Dis. 66:1176–1179.

316. Benson, D. M., and Barker, K. R. 1985. Nematodes—a threat to ornamental plants in the nursery and landscape. Plant Dis. 69:97–100.

317. Benson, D. M., and Cochran, F. D. 1980. Resistance of evergreen hybrid azaleas to root rot caused by *Phytophthora cinnamomi*. Plant Dis. 64:214–215.

318. Benson, D. M., and Jones, R. K. 1980. Etiology of rhododendron dieback caused by four species of *Phytophthora*. Plant Dis. 64:687–691.

319. Benzing, D. H. 1990. Vascular epiphytes: general biology and related biota. Cambridge Univ. Press, Cambridge, UK. 354 pp.

320. Benzing, D. H. 2000. Bromeliaceae: profile of an adaptive radiation. Cambridge Univ. Press, Cambridge, UK, New York. 690 pp.

321. Berbee, J. G. 1957. Virus symptoms associated with birch dieback. Can. Dep. Agric., Sci. Serv., For. Biol. Div. Bi-mon. Prog. Rep. 13(1):1.

322. Berbee, J. G., and Rogers, J. D. 1964. Life cycle and host range of *Hypoxylon pruinatum* and its pathogenesis on poplars. Phytopathology 54:257–261.

323. Bergdahl, D. R., and French, D. W. 1976. Epidemiology of comandra rust on jack pine and comandra in Minnesota. Can. J. For. Res. 6:326–334.

324. Bergdahl, D. R., and French, D. W. 1976. Relative susceptibility of five pine species, 2 to 36 months of age, to infection by *Cronartium comandrae*. Can. J. For. Res. 6:319–325.

325. Bergdahl, D. R., and French, D. W. 1985. Association of wood decay fungi with decline and mortality of apple trees in Minnesota. Plant Dis. 69:887–890.

326. Bergdahl, D. R., Sendak, P. E., Tobi, D. R., Bove, J. R., and Tritton, L. M. 2002. Incidence of annual canker of sugar maple studied with the aid of a geographic information system. North. J. Appl. For. 19:73–79.

327. Berger, R. D. 1998. Cause and control of citrus blight. Laranja 19:91–105.

328. Berges, R., Rott, M., and Seemüller, E. 2000. Range of phytoplasma concentrations in various plant hosts as determined by competitive polymerase chain reaction. Phytopathology 90:1145–1152.

329. Berisha, B., Chen, Y. D., Zhang, G. Y., Xu, B. Y., and Chen, T. A. 1998. Isolation of Pierce's disease bacteria from grapevines in Europe. Eur. J. Plant Pathol. 104:427–433.

330. Bernatzky, R., and Mulcahy, D. L. 1992. Marker-aided selection in a backcross breeding program for resistance to chestnut blight in the American chestnut. Can. J. For. Res. 22:1031–1035.

331. Berndt, R. 1996. Ultrastructure of the D-haustoria of *Coleosporium* spp. (rust fungi, Uredinales). Sydowia 48:263–272.

332. Berndt, R. 1999. *Chrysomyxa* rust: morphology and ultrastructure of D-haustoria, uredinia, and telia. Can. J. Bot. 77:1469–1484.

333. Berndt, R., Bauer, R., and Oberwinkler, F. 1994. Ultrastructure of the host-parasite interface in the fern rusts *Milesia*, *Uredinopsis*, and *Hyalopsora* (Pucciniastraceae, Uredinales). Can. J. Bot. 72:1084–1094.

334. Berndt, R., and Oberwinkler, F. 1995. Ultrastructure of the parasitic interface of *Pucciniastrum*, *Thekopsora*, *Naohidemyces*, and *Calyptospora* (Uredinales, Pucciniastraceae) in the dikaryotic stage. Mycoscience 36:51–59.

335. Berndt, R., and Oberwinkler, F. 1997. Haustorial ultrastructure and morphology of *Melampsorella* and *Thekopsora areolata*. Mycologia 89:698–705.

336. Bernier, B., and Brazeau, M. 1988. Foliar nutrient status in relation to sugar maple dieback and decline in the Quebec Appalachians. Can. J. For. Res. 18:754–761.

337. Bernier, B., Pare, D., and Brazeau, M. 1989. Natural stresses, nutrient imbalances and forest decline in southeastern Quebec. Water Air Soil Pollut. 48:239–250.

338. Bernier, L., Hamelin, R. C., and Ouellette, G. B. 1994. Comparison of ribosomal DNA length and restriction site polymorphisms in *Gremmeniella* and *Ascocalyx* isolates. Appl. Environ. Microbiol. 60:1279–1286.

339. Bernstein, B., Zehr, E. I., Dean, R. A., and Shabi, E. 1995. Characteristics of *Colletotrichum* from peach, apple, pecan, and other hosts. Plant Dis. 79:478–482.

340. Berrang, P., and Karnosky, D. F. 1983. Street trees for metropolitan New York. N.Y. Bot. Gard., Inst. Urban Hortic. Pub. No. 1. 177 pp.

341. Berrang, P., Karnosky, D. F., and Stanton, B. J. 1985. Environmental factors affecting tree health in New York City. J. Arboric. 11:185–189.

342. Berrang, P., and Steiner, K. C. 1980. Resistance of pin oak progenies to iron chlorosis. J. Am. Soc. Hortic. Sci. 105:519–522.

343. Berry, F. H. 1960. Etiology and control of walnut anthracnose. Md. Agric. Exp. Stn. Bull. A-113. 22 pp.

344. Berry, F. H., and Lombard, F. F. 1978. Basidiomycetes associated with decay of living oak trees. USDA For. Serv. Res. Pap. NE-413. 8 pp.

345. Bertaccini, A., and Bellardi, M. G. 1992. A rhabdovirus inducing vein yellowing in croton. Plant Pathol. 41:79–82.

346. Bertrand, A., Robitaille, G., Nadeau, P., and Boutin, R. 1994. Effects of soil freezing and drought stress on abscisic acid content of sugar maple sap and leaves. Tree Physiol. 14:413–425.

347. Bertrand, P. F., and English, H. 1976. Release and dispersal of conidia and ascospores of *Valsa leucostoma*. Phytopathology 66:987–991.

348. Bertrand, P. F., and English, H. 1976. Virulence and seasonal activity of *Cytospora leucostoma* and *C. cincta* in French prune trees in California. Plant Dis. Rep. 60:106–110.

349. Bertrand, P. F., English, H., and Carlson, R. M. 1976. Relation of soil physical and fertility properties to the occurrence of Cytospora canker in French prune orchards. Phytopathology 66:1321–1324.

350. Bertrums, E. J. 1998. Population decline of root lesion nematode (*Pratylenchus penetrans* Cobb) after a *Taxus baccata* crop. Meded. Fac. Landbouw. Toegepaste Biol. Wetensch., Univ. Gent 63:649–653.

351. Bérubé, J. A., and Dessureault, M. 1988. Morphological characterization of *Armillaria ostoyae* and *Armillaria sinapina* sp. nov. Can. J. Bot. 66:2027–2034.

352. Bérubé, J. A., and Dessureault, M. 1989. Morphological studies on the *Armillaria mellea* complex: two new species, *A. gemina* and *A. calvescens*. Mycologia 81:216–225. [Includes key to North American taxa then known.]

353. Beute, M. K., and Rodriguez-Kabana, R. 1979. Effect of volatile organic compounds from remoistened plant tissues on growth and germination of sclerotia of *Sclerotium rolfsii*. Phytopathology 69:802–805.

354. Bhat, R. G., and Subbarao, K. V. 1999. Host range specificity in *Verticillium dahliae*. Phytopathology 89:1218–1225.

355. Biddle, P. G., and Tinsley, T. W. 1968. Virus diseases of conifers in Great Britain. Nature 219:1387–1388.

356. Biddle, P. G., and Tinsley, T. W. 1971. Poplar mosaic in Great Britain. New Phytol. 70:61–66.

357. Biddle, P. G., and Tinsley, T. W. 1971. Some effects of poplar mosaic virus on the growth of poplar trees. New Phytol. 70:67–75.

358. Bidwell, C. B., and Bramble, W. C. 1934. The Strumella disease in southern Connecticut. J. For. 32:15–23.

359. Bielenin, A., and Jones, A. L. 1988. Prevalence and pathogenicity of *Phytophthora* spp. from sour cherry trees in Michigan. Plant Dis. 72:473–476.

360. Biella, S., Smith, M. L., Aist, J. R., Cortesi, P., and Milgroom, M. G. 2002. Programmed cell death correlates with virus transmission in a filamentous fungus. Proc. R. Soc. Lond. Ser. B, Biol. Sci. 269:2269–2276.

361. Bier, J. E. 1939. Septoria canker of introduced and native hybrid poplars. Can. J. Res. C17:195–204.

362. Bier, J. E. 1940. Studies in forest pathology. III. Hypoxylon canker of poplar. Can. Dep. Agric. Pub. 691. 40 pp.

363. Bier, J. E. 1959–1961. The relation of bark moisture content to the development of canker diseases caused by native, facultative parasites. I. Cryptodiaporthe canker on willow. II. Fusarium canker of black cottonwood. IV. Pathogenicity studies of *Cryptodiaporthe salicella* (Fr.) Petrak, and *Fusarium lateritium* Nees., on *Populus trichocarpa* Torrey and Gray, *P.* 'Robusta', *P. tremuloides* Michx. and *Salix* sp. VI. Pathogenicity studies of *Hypoxylon pruinatum* (Klotzsch) Cke. and *Septoria musiva* Pk. on species of *Acer*, *Populus*, and *Salix*. Can. J. Bot. 37:229–238, 781–788; 39:139–144; 1555–1561.

364. Biggs, A. R. 1984. Boundary-zone formation in peach bark in response to wounds and *Cytospora leucostoma* infection. Can. J. Bot. 62:2814–2821.

365. Biggs, A. R. 1984. Intracellular suberin: occurrence and detection in tree bark. IAWA Bull., n.s., 5:243–248.

366. Biggs, A. R. 1985. Suberized boundary zones and the chronology of wound response in bark. Phytopathology 75:1191–1195.

367. Biggs, A. R. 1990. Managing wound-associated diseases by understanding wound healing in the bark of woody plants. J. Arboric. 16:108–112.

368. Biggs, A. R. 1992. A review of pruning techniques for peach trees in relation to the incidence of dieback and infection by *Leucostoma* spp. Acta Hortic. No. 322:127–134.

369. Biggs, A. R. 1994. Characteristics of fire blight cankers following shoot inoculations of three apple cultivars. HortScience 29:795–797.

370. Biggs, A. R. 1995. Detection of latent infections in apple fruit with paraquat. Plant Dis. 79:1062–1067.

371. Biggs, A. R., ed. 1993. Handbook of cytology, histology, and histochemistry of fruit tree diseases. CRC Press, Boca Raton, FL. 330 pp.

372. Biggs, A. R., and Britton, K. O. 1988. Presymptom histopathology of peach trees inoculated with *Botryosphaeria obtusa* and *B. dothidea*. Phytopathology 78:1109–1118.

373. Biggs, A. R., and Davis, D. D. 1981. Effect of SO_2 on growth and sulfur content of hybrid poplar. Can. J. For. Res. 11:830–833.

374. Biggs, A. R., Davis, D. D., and Merrill, W. 1983. Histopathology of cankers on *Populus* caused by *Cytospora chrysosperma*. Can. J. Bot. 61:563–574.

375. Biggs, A. R., and Northover, J. 1985. Inoculum sources for *Monilinia fructicola* in Ontario peach orchards. Can. J. Plant Pathol. 7:302–307.

376. Bingham, R. T. 1983. Blister rust resistant western white pine for the Inland Empire: the story of the first 25 years of the research and development program. USDA For. Serv. Gen. Tech. Rep. INT-146. 45 pp.

377. Bingham, R. T., Hoff, R. J., and McDonald, G. I., sci. dir. and prog. coord. 1972. Biology of rust resistance in forest trees: Proc. NATO-IUFRO Adv. Study Inst., 1969. USDA Misc. Pub. 1221. 681 pp.

378. Binns, W. O., Mayhead, G. J., and MacKenzie, J. M. 1980. Nutrient deficiencies of conifers in British forests. G.B. For. Comm. Leafl. 76. 23 pp.

379. Biocca, M., and Motta, E. 1995. Aspects of latency of *Hypoxylon mediterraneum* in declining Turkey oaks (*Quercus cerris*). Petria 5:171–175.

380. Biocca, M., Tainter, F. H., Starkey, D. A., Oak, S. W., and Williams, J. G. 1993. The persistence of oak decline in the western North Carolina Nantahala Mountains. Castanea 58:178–184.

381. Bishop, A. L., and Basarich, L. 1989. Leaf spot and stem canker of *Ginkgo biloba* caused by *Pseudomonas syringae*. Plant Dis. 73:368.

382. Bishop, G. C. 1979. Infection of cherry trees and production of a toxin that causes foliar silvering by different isolates of *Chondrostereum purpureum*. Austral. J. Agric. Res. 30:659–665.

383. Bissett, J., and Palm, M. E. 1989. Species of *Phyllosticta* on conifers. Can. J. Bot. 67:3378–3385.

384. Björkman, E. 1948. Studier över snöskyttesvampens (*Phacidium infestans* Karst.) biologi samt metoder för snöskyttets bekämpande. Meddel. fran Statens Skogsforskningsinst. 37(2):1–136.

385. Black, W. M., and Neely, D. 1978. Effects of temperature, free moisture, and relative humidity on the occurrence of walnut anthracnose. Phytopathology 68:1054–1056.

386. Black, W. M., and Neely, D. 1978. Relative resistance of *Juglans* species and hybrids to walnut anthracnose. Plant Dis. Rep. 62:497–499.

387. Blackwell, E. 1943. The life history of *Phytophthora cactorum* (Leb. & Cohn) Schroet. Trans. Br. Mycol. Soc. 26:71–89.

388. Blaker, N. S., and MacDonald, J. D. 1981. Predisposing effects of soil moisture extremes on the susceptibility of rhododendron to Phytophthora root and crown rot. Phytopathology 71:831–834.

389. Blaker, N. S., and MacDonald, J. D. 1983. Influence of container medium pH on sporangium formation, zoospore release, and infection of rhododendron by *Phytophthora cinnamomi*. Plant Dis. 67:259–263.

390. Blakeslee, G. M., Jokela, E. J., Hollis, C. H., Wilson, D. S., Lante, W. D., and Allen, J. E. 1999. Pitch canker in young loblolly pines: influence of precommercial thinning and fertilization on disease incidence and severity. South. J. Appl. For. 23:139–143.

391. Blakeslee, G. M., and Oak, S. W. 1980. Residual naval stores stumps as reservoirs of inoculum for infection of slash pines by *Phaeolus schweinitzii*. Plant Dis. 64:167.

392. Blanchette, J.-Y. 2001. Microscopic observations of the early stages of the infection process by *Lachnellula willkommii* (Hartig) Dennis artificially inoculated on larch (*Larix laricina* (Du Roi) K. Koch) seedlings. Ph.D. diss., Univ. New Brunswick, Fredericton. 94 pp.

393. Blanchette, R. A. 1980. Wood decomposition by *Phellinus* (*Fomes*) *pini:* a scanning electron microscopy study. Can. J. Bot. 58:1496–1503.

394. Blanchette, R. A. 1982. Decay and canker formation by *Phellinus pini* in white and balsam fir. Can. J. For. Res. 12:538–544.

395. Blanchette, R. A. 1982. *Phellinus* (*Fomes*) *pini* decay associated with sweetfern rust in sapwood of jack pine. Can. J. For. Res. 12:304–310.

396. Blanchette, R. A. 1982. Progressive stages of discoloration and decay associated with the canker-rot fungus, *Inonotus obliquus*, in birch. Phytopathology 72:1272–1277.

397. Blanchette, R. A. 1984. Selective delignification of eastern hemlock by *Ganoderma tsugae*. Phytopathology 74:153–160.

398. Blanchette, R. A. 1991. Delignification by wood-decay fungi. Annu. Rev. Phytopath. 29:381–398.

399. Blanchette, R. A. 1995. Degradation of the lignocellulose complex in wood. Can. J. Bot. 73:S999–S1010.

400. Blanchette, R. A., and Biggs, A. R., eds. 1992. Defense mechanisms of woody plants against fungi. Springer-Verlag, Berlin, Heidelberg, New York. 458 pp.

401. Blanchette, R. A., Otjen, L., Effland, M. J., and Eslyn, W. E. 1985. Changes in structural and chemical components of wood delignified by fungi. Wood Sci. Technol. 19:35–46.

402. Blanchette, R. A., Sutherland, J. B., and Crawford, D. L. 1981. Actinomycetes in discolored wood of living silver maple. Can. J. Bot. 59:1–7.

403. Blanz, P., and Doring, H. 1995. Taxonomic relationships in the genus *Exobasidium* (Basidiomycetes) based on ribosomal DNA analysis. Stud. Mycol. No. 38:119–127.

404. Blanz, P., and Oberwinkler, F. 1983. A contribution to the species definition in the genus *Exobasidium* (Basidiomycetes). Syst. Appl. Microbiol. 4:199–206.

405. Blaser, H. W., Marr, C., and Takahashi, D. 1967. Anatomy of boron-deficient *Thuja plicata*. Am. J. Bot. 54:1107–1113.

406. Bliss, D. E. 1933. The pathogenicity and seasonal development of *Gymnosporangium* in Iowa. Iowa Agric. Exp. Stn. Res. Bull. 166: 339–392.

407. Bliss, D. E. 1946. The relation of soil temperature to the development of Armillaria root rot. Phytopathology 36:302–318.

408. Blodgett, J. T., Kruger, E. L., and Stanosz, G. R. 1997. Effects of moderate water stress on disease development by *Sphaeropsis sapinea* on red pine. Phytopathology 87:422–428.

409. Blodgett, J. T., Kruger, E. L., and Stanosz, G. R. 1997. *Sphaeropsis sapinea* and water stress in a red pine plantation in central Wisconsin. Phytopathology 87:429–434.

410. Blodgett, J. T., and Stanosz, G. R. 1999. Differences in aggressiveness of *Sphaeropsis sapinea* RAPD marker group isolates on several conifers. Plant Dis. 83:853–856.

411. Blodgett, J. T., and Sullivan, K. F. 2004. First report of white pine blister rust on Rocky Mountain bristlecone pine. Plant Dis. 84:311.

412. Blomquist, C. L., and Kirkpatrick, B. C. 2002. Frequency and seasonal distribution of pear psylla infected with the pear decline phytoplasma in California pear orchards. Phytopathology 92:1218–1226.

413. Blomquist, C. L., and Kirkpatrick, B. C. 2002. Identification of phytoplasma taxa and insect vectors of peach yellow leaf roll disease in California. Plant Dis. 86:759–763.

414. Bloomberg, W. J. 1962. Cytospora canker of poplars: factors influencing the development of the disease. Can. J. Bot. 40:1272–1280.

415. Bloomberg, W. J. 1971. Diseases of Douglas-fir seedlings caused by *Fusarium oxysporum*. Phytopathology 61:467–470.

416. Bloomberg, W. J. 1973. Fusarium root rot of Douglas-fir seedlings. Phytopathology 63:337–341.

417. Bloomberg, W. J. 1990. Effect of stand conditions on advance of *Phellinus weirii* in Douglas-fir plantations. Phytopathology 80:553–559.

418. Bloomberg, W. J., and Farris, S. H. 1963. Cytospora canker of poplars: bark wounding in relation to canker development. Can. J. Bot. 41:303–310.

419. Bloomberg, W. J., and Lock, W. 1972. Strain differences in *Fusarium oxysporum* causing diseases of Douglas-fir seedlings. Phytopathology 62:481–485.

420. Bloomberg, W. J., and Reynolds, G. 1982. Factors affecting transfer and spread of *Phellinus weirii* mycelium in roots of second-growth Douglas-fir. Can. J. For. Res. 12:424–427.

421. Bloomberg, W. J., and Reynolds, G. 1985. Growth loss and mortality in laminated root rot infection centers in second-growth Douglas-fir on Vancouver Island. For. Sci. 31:497–508.

422. Bloomberg, W. J., Smith, R. B., and Wereld, A. van der. 1980. A model of spread and intensification of dwarf mistletoe infection in young western hemlock stands. Can. J. For. Res. 10:42–52.

423. Blua, M. J., Redak, R. A., Morgan, D. J. W., and Costa, H. S. 2001. Seasonal flight activity of two *Homalodisca* species (Homoptera: Cicadellidae) that spread *Xylella fastidiosa* in southern California. J. Econ. Entomol. 94:1506–1510.

424. Boca, M., and Motta, E. 1995. Aspects of latency of *Hypoxylon mediterraneum* in declining Turkey oaks (*Quercus cerris*). Petria 5:171–175.

425. Boddy, L. 1991. Importance of wood decay fungi in forest ecosystems. Pages 501–540 in: Handbook of applied mycology. Vol. 1: Soil and plants. D. K. Akora, B. Rai, K. G. Mukerji, and G. R. Knudsen, eds. Marcel Dekker, New York.

426. Boddy, L., and Rayner, A. D. M. 1982. Population structure, inter-mycelial interactions, and infection biology of *Stereum gausapatum*. Trans. Br. Mycol. Soc. 78:337–351.

427. Boddy, L., and Rayner, A. D. M. 1983. Origins of decay in living deciduous trees: the role of moisture content and a re-appraisal of the expanded concept of tree decay. New Phytol. 94:623–641.

428. Boddy, L., and Rayner, A. D. M. 1984. Fungi inhabiting oak, *Quercus robur*, twigs before and at fall. Trans. Br. Mycol. Soc. 82:501–506.

429. Boddy, L., and Rayner, A. D. M. 1984. Internal spread of fungi inoculated into attached oak branches. New Phytol. 98:155–164.

430. Bodine, E. W., and Newton, J. H. 1942. The rasp leaf of cherry. Phytopathology 32:333–335.

431. Boelema, B. H. 1973. A Cercospora leaf spot and stem necrosis on *Rosa* spp. in the Transvaal. Phytophylactica 5:7–12.

432. Boer, S. H. De. 1980. Leaf spot of cherry laurel caused by *Pseudomonas syringae*. Can. J. Plant Pathol. 2:235–238.

433. Boesewinkel, H. J. 1982. *Cylindrocladiella*, a new genus to accommodate *Cylindrocladium parvum* and other small-spored species of *Cylindrocladium*. Can. J. Bot. 60:2288–2294.

434. Bohaychuk, W. P., and Whitney, R. D. 1973. Environmental factors influencing basidiospore discharge in *Polyporus tomentosus*. Can. J. Bot. 51:801–815.

435. Bonants, P. J. M., Hagenaar-de Weerdt, M., Man in't Veld, W. A., and Baayen, R. P. 2000. Molecular characterization of natural hybrids of *Phytophthora nicotianae* and *P. cactorum*. Phytopathology 90:867–874.

436. Bonello, P., Gordon, T. R., and Storer, A. J. 2001. Systemic induced resistance in Monterey pine. For. Pathol. 31:99–106.

437. Bonfante, P., and Perotto, S. 1995. Strategies of arbuscular mycorrhizal fungi when infecting host plants. New Phytol. 130:3–21.

438. Bonn, W. G., ed. 1996. Seventh international workshop on fireblight, St. Catharines, Ontario, Canada, 1995. Acta Hortic. No. 411. 431 pp.

439. Bonn, W. G., and Elfving, D. C. 1989. Evaluation of flowering crabapples for resistance to fire blight caused by *Erwinia amylovora*. Can. J. Plant Pathol. 11:57–59.

440. Bonsen, K. J. M., Scheffer, R. J., and Elgersma, D. M. 1985. Barrier zone formation as a resistance mechanism of elms to Dutch elm disease. IAWA Bull. 6:71–77.

441. Booth, C. 1959. Studies of Pyrenomycetes IV. *Nectria* (Part 1). Commonw. Mycol. Inst. Mycol. Pap. No. 73. 115 pp.

442. Booth, C. 1967. *Nectria galligena*. C.M.I. Descr. Pathogenic Fungi Bact. No. 147. 2 pp.

443. Booth, C. 1971. The genus *Fusarium*. Commonw. Mycol. Inst., Kew, Surrey, UK. 237 pp.

444. Booth, C., Gibson, I. A. S., and Sutton, B. C. 1973. *Cryptodiaporthe populea*. C.M.I. Descr. Pathogenic Fungi Bact. No. 364. 2 pp.

445. Boothroyd, C. W. 1951. A new leaf spot of *Celastrus scandens* L., the climbing bittersweet. Mycologia 43:373–375.

446. Boring, L. R., and Swank, W. T. 1984. Symbiotic nitrogen fixation in regenerating black locust (*Robinia pseudoacacia* L.) stands. For. Sci. 30:528–537.

447. Born, G. L. 1974. Root infection of woody hosts with *Verticillium albo-atrum*. Ill. Nat. Hist. Surv. Bull. 31:205–249.

448. Born, G. L., and Crane, J. L. 1972. *Kaskaskia gleditsiae* gen. et sp. nov. parasitic on thornless honey locust in Illinois. Phytopathology 62:926–930.

449. Bostock, R. M., and Middleton, G. E. 1987. Relationship of wound periderm formation to resistance to *Ceratocystis fimbriata* in almond bark. Phytopathology 77:1174–1180.

450. Bostock, R. M., and Stermer, B. A. 1989. Perspectives on wound healing in resistance to pathogens. Annu. Rev. Phytopathol. 27:343–371.

451. Boulet, G. B., and Nepveu, G. 1988. Relations between some wood properties and stem cracks in standing trees of *Abies grandis*. Ann. Sci. For. 45:33–51.

452. Bouzar, H., Chilton, W. S., Nesme, X., Dessaux, Y., Vaudequin, V., Petit, A., Jones, J. B., and Hodge, N. C. 1995. A new *Agrobacterium* strain isolated from aerial tumors on *Ficus benjamina* L. Appl. Environ. Microbiol. 61:65–73.

453. Bouzar, H., Jones, J. B., and Hodge, N. C. 1993. Differential characterization of *Agrobacterium* species using carbon-source utilization patterns and fatty acid profiles. Phytopathology 83:733–739.

454. Bové, J. M., and Garnier, M. 1998. Walled and wall-less eubacteria from plants: sieve-tube-restricted plant pathogens. Plant Cell Tissue Organ Cult. 52:7–16.

455. Bovey, R. W. 1977. Response of selected woody plants in the United States to herbicides. USDA Agric. Handb. 493. 101 pp.

456. Bowden, C. G., Smalley, E., Guries, R. P., Hubbes, M., Temple, B., and Horgen, P. A. 1996. Lack of association between cerato-ulmin production and virulence in *Ophiostoma novo-ulmi*. Mol. Plant-Microbe Interact. 9:556–564.

457. Bowen, P. R. 1930. A maple leaf disease caused by *Cristulariella depraedens*. Conn. Agric. Exp. Stn. Bull. 316:625–647.

458. Boyce, J. S. 1933. A canker of Douglas fir associated with *Phomopsis lokoyae*. J. For. 31:664–672.

459. Boyce, J. S. 1940. A needle cast of Douglas-fir associated with *Adelopus gäumannii*. Phytopathology 30:649–659.

460. Boyce, J. S. 1943. Host relationships and distribution of conifer rusts in the United States and Canada. Trans. Conn. Acad. Arts Sci. 35:329–482.

461. Boyce, J. S. 1961. Forest pathology. 3rd ed. McGraw-Hill, New York. 572 pp.

462. Boyce, J. S. Jr. 1954. Hypoderma needle blight of southern pines. J. For. 52:496–498.

463. Boyce, J. S. Jr. 1967. Red root and butt rot in planted slash pines. J. For. 65:493–494.

464. Boyce, R. L. 1988. Wind direction and fir wave motion. Can. J. For. Res. 18:461–466.

465. Boyce, R. L. 1995. Patterns of foliar injury to red spruce on Whiteface Mountain, New York, during a high-injury winter. Can. J. For. Res. 25:166–169.

466. Boyce, S. G. 1954. The salt spray community. Ecol. Monogr. 24:29–67.

467. Boyd, E. S. 1934. A developmental study of a new species of *Ophiodothella*. Mycologia 26:456–468.

468. Boyer, J. S. 1995. Biochemical and biophysical aspects of water deficits and the predisposition to disease. Annu. Rev. Phytopathol. 33:251–274.

469. Boyer, M. G. 1961. A Fusarium canker disease of *Populus deltoides* Marsh. Can. J. Bot. 39:1195–1204.

470. Boyer, M. G., and Navratil, S. 1970. Some aspects of transmission and electron microscopy of necrotic leaf spot of aspen. Can. J. Bot. 48:1141–1145.

471. Braasch, H. 1996. Pathogenicity tests with *Bursaphelenchus mucronatus* on Scots pine and Norway spruce seedlings in Germany. Eur. J. For. Pathol. 26:205–216.

472. Braasch, H. 2000. Influence of temperature and water supply on mortality of 3-year-old pines inoculated with *Bursaphelenchus xylophilus* and *B. mucronatus*. Nachrichtenbl. Deut. Pflanzenschutzd. 52:244–249.

473. Braathe, P. 1996. Birch dieback—the effects of artificial spring frost on budburst and foliage. Norweg. J. Agric. Sci. 10:1–6.

474. Bradshaw, R. E., Ganley, R. J., Jones, W. T., and Dyer, P. S. 2000. High levels of dothistromin toxin produced by the forest pathogen *Dothistroma pini*. Mycol. Res. 104:325–332.

475. Brako, L., Rossman, A. Y., and Farr, D. F. 1995. Scientific and common names of 7,000 vascular plants in the United States. APS Press, St. Paul, MN. 295 pp.

476. Brandt, R. W. 1960. The Rhabdocline needle cast of Douglas-fir. State Univ. Coll. For. (Syracuse, NY) Tech. Pub. 84. 66 pp.

477. Brasier, C. M. 1986. The population biology of Dutch elm disease: its principal features and some implications for other host-pathogen systems. Adv. Plant Pathol. 5:53–118.

478. Brasier, C. M. 1996. Low genetic diversity of the *Ophiostoma novo-ulmi* population in North America. Mycologia 88:951–964.

479. Brasier, C. M. 1996. *Phytophthora cinnamomi* and oak decline in southern Europe. Environmental constraints including climate change. Ann. Sci. For. 53:347–358.

480. Brasier, C. M. 2001. Rapid evolution of introduced plant pathogens via interspecific hybridization. BioScience 51:123–133.

481. Brasier, C. M., Cooke, D., and Duncan, J. M. 1999. Origin of a new *Phytophthora* pathogen through interspecific hybridization. Proc. Natl. Acad. Sci. USA 96:5878–5883.

482. Brasier, C. M., and Kirk, S. A. 2001. Designation of the EAN and NAN races of *Ophiostoma novo-ulmi* as subspecies. Mycol. Res. 105:547–554.

483. Brasier, C. M., Kirk, S. A., and Tegli, S. 1995. Naturally occurring non cerato-ulmin producing mutants of *Ophiostoma novo-ulmi* are pathogenic but lack aerial mycelium. Mycol. Res. 99:436–440.

484. Brasier, C. M., and Mehrotra, M. D. 1995. *Ophiostoma himal-ulmi* sp. nov., a new species of Dutch elm disease fungus endemic to the Himalayas. Mycol. Res. 99:205–215.

485. Brasier, C. M., and Strouts, R. G. 1976. New records of *Phytophthora* on trees in Britain. I. Phytophthora root rot and bleeding canker of horse chestnut (*Aesculus hippocastanum* L.). Eur. J. For. Pathol. 6:129–136.

486. Braun, E. J., and Sinclair, W. A. 1976. Histopathology of phloem necrosis in *Ulmus americana*. Phytopathology 66:598–607.

487. Braun, E. J., and Sinclair, W. A. 1979. Phloem necrosis of elms: symptoms and histopathological observations in tolerant hosts. Phytopathology 69:354–358.

488. Braun, H. J. 1976. Beech (*Fagus sylvatica* L.) bark disease, caused by *Cryptococcus fagi* Baer. I. Anatomy of bark of *Fagus sylvatica* L. as a basic factor. Eur. J. For. Pathol. 6:136–146.

489. Braun, U. 1987. A monograph of the Erysiphales (powdery mildews). Nova Hedwigia Beih. No. 89. 700 pp.

490. Braun, U. 1993. New genera of phytopathogenic deuteromycetes. Cryptog. Bot. 4:107–114.

491. Braun, U. 1993. Taxonomic notes on some species of the *Cercospora* complex (II). Cryptog. Bot. 2:235–244.

492. Braun, U. 1994. Studies on *Ramularia* and allied genera. Nova Hedwigia 58:191–222.

493. Braun, U. 1995. Miscellaneous notes on phytopathogenic hyphomycetes (II). Mycotaxon 55:223–241.

494. Braun, U. 1995. The powdery mildews (Erysiphales) of Europe. Gustav Fischer, Jena. 337 pp.

495. Braun, U. 1995, 1998. A monograph of *Cercosporella, Ramularia*, and allied genera (phytopathogenic hyphomycetes). Vols. 1, 2. IHW-Verlag, Munich. 333 and 493 pp.

496. Braun, U. 1996. Taxonomic notes on some species of the *Cercospora* complex (IV). Sydowia 48:205–217.

497. Braun, U. 1999. Some critical notes on the classification and generic concept of the Erysiphaceae. Schlechtendalia 3:49–55.

498. Braun, U., and Takamatsu, S. 2000. Phylogeny of *Erysiphe, Microsphaera, Uncinula* (Erysipheae) and *Cystotheca, Podosphaera, Sphaerotheca* (Cystotheceae) inferred from rDNA ITS sequences—some taxonomic consequences. Schlechtendalia 4:1–33.

499. Brawner, J. T., Carter, D. R., Huber, D. A., and White, T. L. 1999. Projected gains in rotation-age volume and value from fusiform rust resistant slash and loblolly pines. Can. J. For. Res. 29:737–742.

500. Breese, W. A., Shattock, R. C., Williamson, B., and Hackett, C. 1994. In vitro spore germination and infection of cultivars of *Rubus* and *Rosa* by downy mildews from both hosts. Ann. Appl. Biol. 125:73–85.

501. Brener, W. D., Setliff, E. C., and Norgren, R. L. 1974. *Sclerophoma pythiophila* associated with a tip dieback of juniper in Wisconsin. Plant Dis. Rep. 58:653–657.

502. Brennan, E., Leone, I., Harkov, R., and Rhoads, A. 1981. Austrian pine injury traced to ozone and sulfur dioxide pollution. Plant Dis. 65:363–364.

503. Brenneman, T. B., and Reilly, C. C. 1989. Recent occurrence of pecan anthracnose caused by *Glomerella cingulata*. Plant Dis. 73:775.

504. Bresadola, J. 1894. Fungi aliquot saxonici novi vel critici a cl. W. Krieger lecti. Hedwigia 33:206–210.

505. Bricker, J. S., and Stutz, J. C. 2004. Phytoplasmas associated with ash decline. J. Arboric. 30:193–199.

506. Bridgwater, F. E., and Smith, W. D. 2002. Decline in values of slash pine stands infected with fusiform rust. South. J. Appl. For. 26:134–139.

507. Brierley, P. 1954. Symptoms in the florist's hydrangea caused by tomato ringspot virus and an unidentified sap-transmissible virus. Phytopathology 44:696–699.

508. Brighigna, L., Montaini, P., Favilli, F., and Carabez, T. A. 1992. Role of the nitrogen-fixing bacterial microflora in the epiphytism of *Tillandsia* (Bromeliaceae). Am. J. Bot. 79:723–727.

509. Brill, H., Bock, E., and Bauch, J. 1981. The significance of microorganisms in the wood of *Abies alba* with regard to silver fir dieback. Forstwiss. Centralbl. 100:195–206.

510. Bringmann, G., Schlauer, J., Ruckert, M., Wiesen, B., Ehrenfeld, K., Proksch, P., and Czygan, F. C. 1999. Host-derived acetogenins involved in the incompatible parasitic relationship between *Cuscuta reflexa* (Convolvulaceae) and *Ancistrocladus heyneanus* (Ancistrocladaceae). Plant Biol. 1:581–584.

511. Brinkerhoff, L. A., and Streets, R. B. 1946. Pathogenicity and pathological histology of *Phymatotrichum omnivorum* (the fungus causing cotton or Texas root rot) in a woody perennial—the pecan. Ariz. Agric. Exp. Stn. Bull. 111:103–126.

512. Britton, K. O., Anderson, R. A., and Redlin, S. C. 1995. *Colletotrichum acutatum* causes anthracnose of black gum (*Nyssa sylvatica*). Plant Dis. 79:1187.

513. Britton, K. O., and Hendrix, F. F. 1982. Three species of *Botryosphaeria* cause peach tree gummosis in Georgia. Plant Dis. 66:1120–1121.

514. Britton, K. O., and Hendrix, F. F. 1989. Infection of peach buds by *Botryosphaeria obtusa*. Plant Dis. 73:65–68.

515. Britton, K. O., Pepper, W. D., Loftis, D. L., and Chellemi, D. O. 1994. Effect of timber harvest practices on populations of *Cornus florida* and severity of dogwood anthracnose in western North Carolina. Plant Dis. 78:398–402.

516. Britton-Jones, H. R. 1925. On the diseases known as "bark canker" and "die-back" in fruit trees. J. Pomol. Hortic. Sci. 4:162–183.

517. Britz, H., Coutinho, T. A., Wingfield, M. J., and Marasas, W. F. O. 2002. Validation of the description of *Gibberella circinata* and morphological differentiation of the anamorph *Fusarium circinatum*. Sydowia 54:9–22.

518. Britz, H., Coutinho, T. A., Wingfield, M. J., Marasas, W. F. O., Gordon, T. R., and Leslie, J. F. 1999. *Fusarium subglutinans* f.sp. *pini* represents a distinct mating population in the *Gibberella fujikuroi* species complex. Appl. Environ. Microbiol. 65:1198–1201.

519. Brlansky, R. H., Lee, R. F., and Collins, M. H. 1985. Structural comparison of xylem occlusions in the trunks of citrus trees with blight and other decline diseases. Phytopathology 75:145–150.

520. Brlansky, R. H., Timmer, L. W., Lee, R. F., and Graham, J. H. 1984. Relationship of xylem plugging to reduced water uptake and symptom development in citrus trees with blight and blightlike declines. Phytopathology 74:1325–1328.

521. Broadfoot, W. M., and Williston, H. L. 1973. Flooding effects on southern forests. J. For. 71:584–587.

522. Brodo, I. M., Sharnoff, S. D., and Sharnoff, S. 2001. Lichens of North America. Yale Univ. Press, New Haven, CT. 795 pp.

523. Broek, A. V. Van de, and Leyden, J. Van der. 1995. The role of bacterial motility, chemotaxis, and attachment in bacteria-plant interactions. Mol. Plant-Microbe Interact. 8:800–810.

524. Broembsen, S. L. von, and Deacon, J. W. 1997. Calcium interference with zoospore biology and infectivity of *Phytophthora parasitica* in nutrient irrigation solutions. Phytopathology 87:522–528.

525. Brookhouser, L. W., and Peterson, G. W. 1971. Infection of Austrian, Scots, and ponderosa pines by *Diplodia pinea*. Phytopathology 61:409–414.

526. Brooks, F. E., and Ferrin, D. M. 1994. Branch dieback of southern California chaparral vegetation caused by *Botryosphaeria dothidea*. Phytopathology 84:78–83.

527. Brooks, F. T. 1911, 1913. Silver leaf disease. I, II. J. Agric. Sci. 4:133–144; 5:288–308.

528. Brooks, F. T., and Bailey, M. A. 1919. Silver leaf disease. III. J. Agric. Sci. 9:189.

529. Brooks, F. T., and Brenchley, G. H. 1931. Silver leaf disease. VI. J. Pomol. 9:1–15.

530. Brooks, F. T., and Moore, W. C. 1926. Silver leaf disease. V. J. Pomol. 5:11–97.

531. Brooks, F. T., and Storey, H. H. 1923. Silver leaf disease. IV. J. Pomol. Hortic. Sci. 3:117–141.

532. Broschat, T. K., Harrison, N. A., and Donselman, H. 2002. Losses to lethal yellowing cast doubt on coconut cultivar resistance. Palms 46:185–189.

533. Broughton, W. J., ed. 1981, 1982. Nitrogen fixation. Vol. 1: Ecology. Vol. 2: *Rhizobium*. Clarendon Press, Oxford. 306 and 353 pp.

534. Brown, A. E., Sreenivasaprasad, S., and Timmer, L. W. 1996. Molecular characterization of slow-growing orange and key lime anthracnose strains of *Colletotrichum* from citrus as *C. acutatum*. Phytopathology 86:523–527.

535. Brown, D. A., Windham, M. T., and Trigiano, R. N. 1996. Resistance to dogwood anthracnose among *Cornus* species. J. Arboric. 22:83–86.

536. Brown, D. H., ed. 1985. Lichen physiology and cell biology. Plenum Press, New York. 362 pp.

537. Brown, D. J. F., Halbrendt, J. M., Jones, A. T., Vrain, T. C., and Robbins, R. T. 1994. Transmission of three North American nepoviruses by populations of four distinct species of the *Xiphinema americanum* group. Phytopathology 84:646–649.

538. Brown, E. A., and Hendrix, F. F. 1981. Pathogenicity and histopathology of *Botryosphaeria dothidea* on apple stems. Phytopathology 71:375–379.

539. Brown, E. A. II, and Britton, K. O. 1986. *Botryosphaeria* diseases of apple and peach in the southeastern United States. Plant Dis. 70:480–484.

540. Brown, G. E. 1971. Pycnidial release and survival of *Diplodia natalensis* spores. Phytopathology 61:559–561.

541. Brown, N. A. 1938. The tumor disease of oak and hickory trees. Phytopathology 28:401–411.

542. Brown, N. A. 1941. Tumors on elm and maple trees. Phytopathology 31:541–548.

543. Brown, T. S. Jr., and Merrill, W. 1973. Germination of basidiospores of *Fomes applanatus*. Phytopathology 63:547–550.

544. Browne, G. T., and Mircetich, S. M. 1996. Effects of month of inoculation on severity of disease caused by *Phytophthora* spp. in apple root crowns and excised shoots. Phytopathology 86:290–294.

545. Browne, G. T., Mircetich, S. M., and Cummins, J. N. 1995. Relative resistance of eighteen selections of *Malus* spp. to three species of *Phytophthora*. Phytopathology 85:72–76.

546. Browne, G. T., and Viveros, M. A. 1999. Lethal cankers caused by *Phytophthora* spp. in almond scions: specific etiology and potential inoculum sources. Plant Dis. 83:739–745.

547. Brownell, K. H., and Schneider, R. W. 1983. Fusarium hypocotyl rot of sugar pine in California forest nurseries. Plant Dis. 67:105–107.

548. Bruce, A., and Palfreyman, J. W., eds. 1998. Forest products biotechnology. Taylor & Francis, London. 326 pp.

549. Bruck, R. I., and Manion, P. D. 1980. Interacting environmental factors associated with the incidence of Hypoxylon canker on trembling aspen. Can. J. For. Res. 10:17–24.

550. Bruederle, L. P., and Stearns, F. W. 1985. Ice storm damage to a southern Wisconsin mesic forest. Bull. Torrey Bot. Club 112:167–175.

551. Bruhn, J. N., Wetteroff, J. J. Jr., Mihail, J. D., Kabrick, J. M., and Pickens, J. B. 2000. Distribution of *Armillaria* species in upland Ozark Mountain forests with respect to site, overstory species composition and oak decline. For. Pathol. 30:43–60.

552. Brummer, M., Arend, M., Fromm, J., Schlenzig, A., and Osswald, W. F. 2002. Ultrastructural changes and immunocytochemical localization of the elicitin quercinin in *Quercus robur* L. roots infected with *Phytophthora quercina*. Physiol. Mol. Plant Pathol. 61:109–120.

553. Brundrett, M., Bougher, N., Dell, B., Grove, T., and Malajczuk, N. 1996. Working with mycorrhizas in forestry and agriculture. Austral. Cent. Int. Agric. Res. ACIAR Monogr. No. 32. 374 pp.

554. Brunt, A., Crabtree, K., Dallwitz, M., Gibbs, A., and Watson, L., eds. 1996. Viruses of plants. CAB International, Wallingford, UK. 1484 pp.

555. Brunt, A. A., Crabtree, K., Dallwitz, M. J., Gibbs, A. J., Watson, L. and Zurcher, E. J., eds. 1996–. Plant viruses online. Descriptions and lists from the VIDE database. Austral. Natl. Univ., Canberra. Internet resource.

556. Bryan, M. K. 1928. Lilac blight in the United States. J. Agric. Res. 36:225–235.

557. Bucciarelli, B., Ostry, M. E., Fulcher, R. G., Anderson, N. A., and Vance, C. P. 1999. Histochemical and microspectrophotometric analyses of early wound responses of resistant and susceptible *Populus tremuloides* inoculated with *Entoleuca mammata* (≡*Hypoxylon mammatum*). Can. J. Bot. 77:548–555.

558. Buchanan, P. K., Hseu, R. S., and Moncalvo, J. M., eds. 1995. *Ganoderma*: systematics, phytopathology and pharmacology. Proceedings of symposium, 5th Int. Mycol. Cong., Vancouver, 1994. Coll. Agric., Natl. Taiwan Univ., Taipei. 135 pp.

559. Buckland, D. C., Foster, R. E., and Nordin, V. J. 1949. Studies in forest pathology. VII. Decay in western hemlock and fir in the Franklin River area, British Columbia. Can. J. Res. C27:312–331.

560. Buckland, D. C., Molnar, A. C., and Wallis, G. W. 1954. Yellow laminated root rot of Douglas-fir. Can. J. Bot. 32:69–81.

561. Buczaki, S. T. 1973. A microecological approach to larch canker biology. Trans. Br. Mycol. Soc. 61:315–329.

562. Buczaki, S. T. 1973. Some factors governing mycelial establishment and lesion extension in the larch canker disease. Eur. J. For. Pathol. 3:39–49.

563. Bugbee, W. M., and Anderson, N. A. 1963. Infection of spruce seedlings by *Cylindrocladium scoparium*. Phytopathology 53:1267–1271.

564. Buisman, C. 1931. Three species of *Botryodiplodia* (Sacc.) on elm trees in the United States. J. Arnold Arbor. 12:289–296.

565. Buller, A. H. R. 1906. The biology of *Polyporus squamosus*, Huds., a timber destroying fungus. J. Econ. Biol. 1:101–138.

566. Bunderson, E. D., Weber, D. J., and Nelson, D. L. 1986. Diseases associated with *Juniperus osteosperma* and a model for predicting their occurrence with environmental site factors. Great Basin Nat. 46:427–440.

567. Burdekin, D. A. 1972. Bacterial canker of poplar. Ann. Appl. Biol. 72:295–299.

568. Burdsall, H. H. Jr., and Banik, M. T. 2001. The genus *Laetiporus* in North America. Harvard Pap. Bot. 6:43–55.

569. Burdsall, H. H. Jr., and Snow, G. A. 1977. Taxonomy of *Cronartium quercuum* and *C. fusiforme*. Mycologia 69:503–508.

570. Burgess, T., McComb, J. A., Colquhoun, I., and Hardy, G. E. S. 1999. Increased susceptibility of *Eucalyptus marginata* to stem infection by *Phytophthora cinnamomi* resulting from root hypoxia. Plant Pathol. 48:797–806.

571. Burk, W. R., and Rex, R. E. 1974. *Polyporus squamosus* in Utah. Mycotaxon 1:135–136.

572. Burke, M. J., Gusta, L. V., Quamme, H. A., Weiser, C. J., and Li, P. H. 1976. Freezing and injury in plants. Annu. Rev. Plant Physiol. 27:507–528.

573. Burkholder, W. H. 1917. The anthracnose disease of the raspberry and related plants. Cornell Univ. Agric. Exp. Stn. Bull. 395:153–183.

574. Burkholder, W. H., and Guterman, C. E. F. 1932. Synergism in a bacterial disease of *Hedera helix*. Phytopathology 22:781–784.

575. Burks, S., Jacobi, W. R., and McIntyre, G. A. 1998. Cytospora canker development on aspen in response to nitrogen fertilization. J. Arboric. 24:28–34.

576. Burnham, C. R. 1988. The restoration of the American chestnut. Am. Scientist 76:478–487.

577. Burnham, C. R., Rutter, P. A., and French, D. W. 1986. Breeding blight-resistant chestnuts. Plant Breed. Rev. 4:347–397.

578. Burr, T. J., Hunter, J. E., Ogawa, J. M., and Abawi, G. S. 1978. A root rot of apple caused by *Rhizoctonia solani* in New York nurseries. Plant Dis. Rep. 62:476–478.

579. Büssing, A. 2000. Mistletoe: the genus *Viscum*. Harwood Academic, Amsterdam. 265 pp.

580. Bussotti, F., and Ferretti, M. 1998. Air pollution, forest condition and forest decline in southern Europe: an overview. Environ. Pollut. 101:49–65.

581. Butin, H. 1958. Untersuchungen über ein Toxin in Kulturfiltraten von *Dothichiza populea* Sacc. et. Br. Phytopathol. Z. 33:135–146.

582. Butin, H. 1970. Zwei neue Arten der Gattung *Phaeocryptopus* Naumov. Phytopathol. Z. 68:269–275.

583. Butin, H. 1973. Morphologische und taxonomische Untersuchungen an *Naemacyclus niveus* (Pers. ex Fr.) Fuck. ex Sacc. und verwandten Arten. Eur. J. For. Pathol. 3:146–163.

584. Butin, H. 1985. Teleomorph- und Anamorph-entwicklung von *Scirrhia pini* Funk & Parker auf Nadeln von *Pinus nigra* Arnold. Sydowia 38:20–27.

585. Butin, H. 1992. Effect of endophytic fungi from oak (*Quercus robur* L.) on mortality of leaf inhabiting gall insects. Eur. J. For. Pathol. 22:237–246.

586. Butin, H., and Kehr, R. 1995. Leaf blotch of lime associated with *Asteromella tiliae* comb. nov. and the latter's connection to *Didymosphaeria petrakiana*. Mycol. Res. 99:1191–1194.

587. Butin, H., and Kehr, R. 1998. *Gloeosporidina platani* sp. nov., the spermatial state of the anthracnose fungus *Apiognomonia veneta* (Sacc. & Speg.) Höhn. Eur. J. For. Pathol. 28:297–305.

588. Butin, H., and Kehr, R. 2000. *Rhizosphaera pseudotsugae* sp. nov. and related species. Mycol. Res. 104:1012–1016.

589. Butin, H., and Shigo, A. L. 1981. Radial shakes and "frost cracks" in living oak trees. USDA For. Serv. Res. Pap. NE-478. 21 pp.

590. Butler, E. J., and Jones, S. G. 1955. Plant pathology. Macmillan, London. 979 pp.

591. Butnor, J. R., Seiler, J. R., and Gray, J. A. 2000. Influence of procerum root disease on the water relations of eastern white pine (*Pinus strobus* L.). J. Sustain. For. 10:95–105.

592. Buttner, C., and Fuhrling, M. 1996. Studies on virus infection of diseased *Quercus robur* (L.) from forest stands in northern Germany. Ann. Sci. For. 53:383–388.

593. Byler, J. W., Cobb, F. W. Jr., and Parmeter, J. R. Jr. 1972. Effects of secondary fungi on the epidemiology of western gall rust. Can. J. Bot. 50:1061–1066.

594. Byler, J. W., Cobb, F. W. Jr., and Parmeter, J. R. Jr. 1972. Occurrence and significance of fungi inhabiting galls caused by *Peridermium harknessii*. Can. J. Bot. 50:1275–1282.

595. Byrde, R. J. W., and Willets, H. A. 1977. The brown rot fungi of fruit: their biology and control. Pergamon Press, New York. 171 pp.

596. CAB International. 2003. Crop protection compendium, 2003 ed. CABI Pub., Wallingford, UK. Internet pub.

597. CAB International. CABI Bioscience Databases. Index Fungorum. The CABI Bioscience and CBS database of fungal names. Internet resource.

598. Cadic, A., Paulin, J. P., and Belin, J. 1990. New *Pyracantha* resistant to scab (*Spilocaea pyracanthae* (Otth.) Rostrup) and to fire blight (*Erwinia amylovora* (Burr.) Winsl. et al.). Acta Hortic. No. 273:303–306.

599. Caetano-Anollés, G., Trigiano, R. N., and Windham, M. T. 1997. Sequence signatures from DNA amplification fingerprints reveal fine population structure of the dogwood pathogen *Discula destructiva*. FEMS Microbiol. Lett. 145:377–383.

600. Caetano-Anollés, G., Trigiano, R. N., and Windham, M. T. 2001. Patterns of evolution in *Discula* fungi and the origin of dogwood anthracnose in North America, studied using arbitrarily amplified and ribosomal DNA. Cur. Genet. 39:346–354.

601. Cahill, D. M., and Hardham, A. R. 1994. A dipstick immunoassay for the specific detection of *Phytophthora cinnamomi* in soils. Phytopathology 84:1284–1292.

602. Cahill, D. [M.], Legge, N., Grant, B., and Weste, G. 1989. Cellular and histological changes induced by *Phytophthora cinnamomi* in a group of plant species ranging from fully susceptible to fully resistant. Phytopathology 79:417–424.

603. Cairney, J. W. G., and Chambers, S. M., eds. 1999. Ectomycorrhizal fungi: key genera in profile. Springer, Berlin, New York. 369 pp.

604. Calavan, E. C., and Wallace, J. M. 1954. *Hendersonula toruloidea* Nattrass on citrus in California. Phytopathology 44:635–639.

605. Calder, D. M., and Bernhardt, P., eds. 1983. The biology of mistletoes. Academic Press, New York. 348 pp.

606. Caldiz, D. O., Beltrano, J., Fernandez, L. V., and Andia, I. 1993. Survey of *Tillandsia recurvata* L.: preference, abundance and its significance for natural forests. For. Ecol. Manag. 57:161–168.

607. Calkins, J. B., and Swanson, B. T. 1997. Susceptibility of 'Skyline' honeylocust to cankers caused by *Nectria cinnabarina* influenced by nursery field management system. J. Environ. Hortic. 15:6–11.

608. Calkins, J. B., and Swanson, B. T. 1998. Plant cold acclimation, hardiness, and winter injury in response to bare soil and groundcover-based nursery field management systems. J. Environ. Hortic. 16:82–89.

609. Callan, B. E. 1998. Diseases of *Populus* in British Columbia: a diagnostic manual. Can. For. Serv. Pac. For. Cent., Victoria, BC. 157 pp.

610. Callan, B. E., and Rogers, J. D. 1986. Cultural characters and anamorphs of *Biscogniauxia marginata*, *Biscogniauxia dennisii*, and *Biscogniauxia repanda*. Can. J. Bot. 64:842–847.

611. Callan, B. E., and Rogers, J. D. 1993. A synoptic key to *Xylaria* species from continental United States and Canada based on cultural and anamorphic features. Mycotaxon 49:141–154.

612. Calvin, C. L. 1967. Anatomy of the endophytic system of the mistletoe *Phoradendron flavescens*. Bot. Gaz. 128:117–137.

613. Calvin, C. L. 1997. Host-formed tyloses in vessels of the mistletoe *Phoradendron* (Viscaceae). IAWA J. 18:117–126.

614. Calvin, C. L., Hawksworth, F. G., and Knutson, D. M. 1984. Phloem in *Arceuthobium globosum* (Viscaceae). Bot. Gaz. 145:461–464.

615. Calvin, C. L., and Wilson, C. A. 1995. Relationship of the mistletoe *Phoradendron macrophyllum* (Viscaceae) to the wood of its host. IAWA J. 16:33–45.

616. Calvin, C. L., Wilson, C. A., and Varughese, G. 1991. Growth of longitudinal stands of *Phoradendron juniperinum* (Viscaceae) in shoots of *Juniperus occidentalis*. Ann. Bot. 67:153–161.

617. Cam, B. Le, Parisi, L., and Arene, L. 2002. Evidence of two formae speciales in *Venturia inaequalis*, responsible for apple and Pyracantha scab. Phytopathology 92:314–320.

618. Cameron, H. R. 1962. Diseases of deciduous fruit trees incited by *Pseudomonas syringae* van Hall. Ore. Agric. Exp. Stn. Tech. Bull. 66. 64 pp.

619. Cameron, H. R. 1970. *Pseudomonas* content of cherry trees. Phytopathology 60:1343–1346.

620. Cameron, H. R., Milbrath, J. A., and Tate, L. A. 1973. Pollen transmission of Prunus ringspot virus in prune and sour cherry orchards. Plant Dis. Rep. 57:241–243.

621. Cameron, R. W. F., and Dixon, G. R. 2000. The influence of temperature, daylength and calendar date on cold tolerance of *Rhododendron*. J. Hortic. Sci. Biotechnol. 75:481–487.

622. Camp, R. R., and Whittingham, W. F. 1974. Ultrastructural alterations in oak leaves parasitized by *Taphrina caerulescens*. Am. J. Bot. 61:964–972.

623. Campbell, A. H. 1933. Zone lines in plant tissues. I. The black lines formed by *Xylaria polymorpha* (Pers.) Grev. in hardwoods. Ann. Appl. Biol. 20:123–145.

624. Campbell, A. H., and Munson, R. G. 1936. Zone lines in plant tissues. III. The black lines formed by *Polyporus squamosus* (Huds.) Fr. Ann. Appl. Biol. 23:453–464.

625. Campbell, W. A. 1939. *Daedalea unicolor* decay and associated cankers of maples and other hardwoods. J. For. 37:974–977.

626. Campbell, W. A., and Copeland, O. L. Jr. 1954. Littleleaf disease of shortleaf and loblolly pines. USDA Circ. 940. 41 pp.

627. Campbell, W. A., and Davidson, R. W. 1938. A *Poria* as the fruiting stage of the fungus causing sterile conks on birch. Mycologia 30:553–560.

628. Campbell, W. A., and Davidson, R. W. 1939. *Poria andersonii* and *Polyporus glomeratus*, two distinct heart-rotting fungi. Mycologia 31:161–168.

629. Campbell, W. A., and Davidson, R. W. 1939. Sterile conks of *Polyporus glomeratus* and associated cankers on beech and red maple. Mycologia 31:606–611.

630. Campbell, W. A., and Davidson, R. W. 1940. Top rot in glaze-damaged black cherry and sugar maple on the Allegany Plateau. J. For. 38:963–965.

631. Campbell, W. A., and Davidson, R. W. 1940. *Ustulina vulgaris* decay in sugar maple and other hardwoods. J. For. 38:474–477.

632. Campbell, W. A., and Davidson, R. W. 1941. Cankers and decay of yellow birch associated with *Fomes igniarius* var. *laevigatus*. J. For. 39:559–560.

633. Campbell, W. A., and Davidson, R. W. 1942. A species of *Poria* causing rot and cankers of hickory and oak. Mycologia 34:17–26.

634. Campbell, W. A., and Miller, J. H. 1952. Windthrow of root-rotted oak shade trees. Plant Dis. Rep. 36:490.

635. Canadian Forest Service, Forest Insect and Disease Survey. 1950–1996. Annual Report of the Forest Insect and Disease Survey, 1950–1979; Forest insect and disease conditions in Canada, 1980–1995. Nat. Resour. Can. Can. For. Serv., Ottawa.

636. Canfield, M. L., Baca, S., and Moore, L. W. 1986. Isolation of *Pseudomonas syringae* from 40 cultivars of diseased woody plants with tip dieback in Pacific Northwest nurseries. Plant Dis. 70:647–650.

637. Cannon, P. F., and Minter, D. W. 1984. *Rhytisma acerinum*. C.M.I. Descr. Pathogenic Fungi Bact. No. 791. 2 pp.

638. Cannon, P. F., and Minter, D. W. 1986. The Rhytismataceae of the Indian subcontinent. CAB Int. Mycol. Pap. No. 155. 123 pp.

639. Cannon, W. N. J., and Worley, D. P. 1980. Dutch elm disease control: performance and costs. USDA For. Serv. Res. Pap. NE-457. 8 pp.

640. Cao, Y., Han, Z., and Li, C. 2001. Studies on wilting toxic substances produced in pines infested by pine wood nematodes. Sci. Silvae Sinicae 37:75–79.

641. Caponero, A., Contesini, A. M., and Iacobellis, N. S. 1995. Population diversity of *Pseudomonas syringae* subsp. *savastanoi* on olive and oleander. Plant Pathol. 44:848–855.

642. Caporn, S. J. M., Ashenden, T. W., and Lee, J. A. 2000. The effect of exposure to NO_2 and SO_2 on frost hardiness in *Calluna vulgaris*. Environ. Exp. Bot. 43:111–119.

643. Capretti, P., Heiniger, U., and Stephan, R., eds. 1995. Shoot and foliage diseases in forest trees. IUFRO Proceedings, Vallombrosa, Firenze, Italy, 1994. Inst. Patol. Zool. For. Agr., Univ. Studi di Firenze, Italy. 309 pp.

644. Cardena, R., Villanueva, M. A., Santamaria, J. M., and Oropeza, C. M. 1991. Presence in Yucatan of mycoplasma-like organisms in *Cocos nucifera* palms showing lethal yellowing disease symptoms. Can. J. Plant Pathol. 13:135–138.

645. Cardwell, N. A., and McDaniel, G. L. 1998. Comparison of chitinases from dogwood anthracnose resistant and susceptible *Cornus* species. HortScience 33:298–301.

646. Carisse, O., and Dewdney, M. 2002. A review of non-fungicidal approaches for the control of apple scab. Phytoprotection 83:1–29.

647. Carlson, C. E. 1978. Fluoride induced impact on a coniferous forest near the Anaconda aluminum plant in northwestern Montana. Ph.D. diss., Univ. Montana. 176 pp.

648. Carlson, C. E., and Gilligan, C. J. 1983. Histological differentiation among abiotic causes of conifer needle necrosis. USDA For. Serv. Res. Pap. INT-298. 16 pp.

649. Carmichael, J. W., Kendrick, W. B., Conners, I. L., and Sigler, L. 1980. Genera of Hyphomycetes. Univ. Alberta Press, Edmonton. 386 pp.

650. Caroselli, N. E. 1957. Verticillium wilt of maples. R. I. Agric. Exp. Stn. Bull. 335. 84 pp.

651. Caroselli, N. E. 1959. The relation of sapwood moisture content to the incidence of maple wilt caused by *Verticillium albo-atrum*. Phytopathology 49:496–498.

652. Caroselli, N. E., and Tucker, C. M. 1949. Pit canker of elm. Phytopathology 39:481–488.

653. Carpenter, E. D. 1970. Salt tolerance of ornamental plants. Am. Nurseryman 131(2):12, 54 . . . 71.

654. Carpenter, L. R., Nelson, E. E., and Stewart, J. L. 1979. Development of dwarf mistletoe infections on western hemlock in coastal Oregon. For. Sci. 25:237–243.

655. Carpenter, P. L. 1972. Dicamba injury to *Taxus*. HortScience 7:573.

656. Carraro, L., Ferrini, F., Ermacora, P., Loi, N., Martini, M., and Osler, R. 2004. *Macropsis mendax* as a vector of elm yellows phytoplasma of *Ulmus* species. Plant Pathol. 53:90–95.

657. Carraro, L., Loi, N., and Ermacora, P. 2001. The 'life cycle' of pear decline phytoplasma in the vector *Cacopsylla pyri*. J. Plant Pathol. 83:87–90.

658. Carter, G. A., and Teramura, A. L. 1988. Vine photosynthesis and relationships to climbing mechanics in a forest understory. Am. J. Bot. 75:1011–1018.

659. Carter, J. C. 1941. Preliminary investigation of oak diseases in Illinois. Ill. Nat. Hist. Surv. Bull. 21:195–230.

660. Carter, J. C. 1945. Wetwood of elms. Ill. Nat. Hist. Surv. Bull. 23:401–448.

661. Carter, J. C. 1947. Tubercularia canker and dieback of Siberian elm (*Ulmus pumila* L.). Phytopathology 37:243–246.

662. Carter, J. C. 1975. Diseases of Midwest trees. Univ. Ill. Coll. Agric. Spec. Pub. 35. 168 pp.

663. Carter, J. C., and Carter, L. R. 1974. An urban epiphytotic of phloem necrosis and Dutch elm disease, 1944–1972. Ill. Nat. Hist. Surv. Bull. 31:113–143.

664. Carter, J. C., and Sacamano, C. M. 1967. Fusicoccum canker, a new disease of Russian olive. Mycologia 59:535–537.

665. Caruso, F. L., and Ramsdell, D. C., eds. 1995. Compendium of blueberry and cranberry diseases. APS Press, St. Paul, MN. 87 pp.

666. Carvalho, A. A. Jr., Martins, E. M. F., and Figueiredo, M. B. 1998. Axenic culture of *Melampsora epitea*, the willow (*Salix babylonica*) rust, from urediniospores. Fitopatol. Brasil. 23:379–385.

667. Carvell, K. L., Tryon, E. H., and True, R. P. 1957. Effects of glaze on the development of Appalachian hardwoods. J. For. 55:130–132.

668. Cash, E. K., and Davidson, R. W. 1940. Some new species of ascomycetes on coniferous hosts. Mycologia 32:728–735.

669. Cash, E. K., and Waterman, A. M. 1957. A new species of *Plagiostoma* associated with a leaf disease of hybrid aspens. Mycologia 49:756–760.

670. Caspari, C. O., and Sachsse, H. 1990. Crack damage on Norway spruce—distribution, symptoms, cause, effects. Forst. Holz 45:685–688.

671. Casteldine, P., Grout, B. W. W., and Roberts, A. V. 1981. Cuticular resistance to *Diplocarpon rosae*. Trans. Br. Mycol. Soc. 77:665–666.

672. Castello, J. D., Amico, L. A., and O'Shea, M. T. 1984. Detection of tobacco mosaic and tobacco ringspot viruses in white ash trees by enzyme-linked immunosorbent assay. Plant Dis. 68:787–790.

673. Castello, J. D., Hibben, C. R., and Jacobi, V. 1992. Isolation of tomato mosaic virus from lilac. Plant Dis. 76:696–699.

674. Castello, J. D., Rogers, S. O., Bachand, G. D., Fillhart, R. C., Murray, J. S., Weidemann, K., Bachand, M., and Almond, M. A. 2000. Detection and partial characterization of tenuiviruses from black spruce. Plant Dis. 84:143–147.

675. Castello, J. D., Silverborg, S. B., and Manion, P. D. 1985. Intensification of ash decline in New York State from 1962 through 1980. Plant Dis. 69:243–246.

676. Castello, J. D., Wargo, P. M., Jacobi, V., Bachand, G. D., Tobi, D. R., and Rogers, M. A. M. 1995. Tomato mosaic virus infection of red spruce on Whiteface Mountain, New York: prevalence and potential impact. Can. J. For. Res. 25:1340–1345.

677. Castlebury, L. A., Rossman, A. Y., Jaklitsch, W. J., and Vasilyeva, L. N. 2002. A preliminary overview of the Diaporthales based on large subunit nuclear ribosomal DNA sequences. Mycologia 94:1017–1031.

678. Castro, H. J. C., Wolf, J. H. D., Garcia, F. J. G., and Gonzalez, E. M. 1999. The influence of humidity, nutrients and light on the establishment of the epiphytic bromeliad *Tillandsia guatemalensis* in the highlands of Chiapas, Mexico. Rev. Biol. Trop. 47:763–773.

679. Cavalier-Smith, T. 2002. The neomuran origin of the Archaebacteria, the negibacterial root of the universal tree and bacterial megaclassification. Int. J. Syst. Evol. Microbiol. 52:7–76.

680. Cayley, D. M. 1923. Fungi associated with "die back" in stone fruit trees. I. Ann. Appl. Biol. 10:253–275.

681. Cease, K. R., and Juzwik, J. 2001. Predominant nitidulid species (Coleoptera: Nitidulidae) associated with spring oak wilt mats in Minnesota. Can. J. For. Res. 31:635–643.

682. Čech, M., Králik, O., and Blattný, C. 1961. Rod-shaped particles associated with virosis of spruce. Phytopathology 51:183–185.

683. Celio, G. J., and Hausbeck, M. K. 1998. Conidial germination, infection structure formation, and early colony development of powdery mildew on poinsettia. Phytopathology 88:105–113.

684. Centre for Information on Coconut Lethal Yellowing (CICLY). 2002. Version of 15 Nov 2002. Internet pub.

685. Chakravarty, P., and Hwang, S. F. 1991. Effect of an ectomycorrhizal fungus, Laccaria laccata, on Fusarium damping-off in Pinus banksiana seedlings. Eur. J. For. Pathol. 21:97–106.

686. Chakravarty, P., Khasa, D., Dancik, B., Sigler, L., Wichlacz, M., Trifonov, L. S., and Ayer, W. A. 1999. Integrated control of Fusarium damping-off in conifer seedlings. Z. Pflanzenkr. Pflanzenschutz 106:342–352.

687. Chamuris, G. P. 1985. On distinguishing Stereum gausapatum from the "S. hirsutum complex." Mycotaxon 22:1–12.

688. Chang, L. S., Iezzoni, A., and Adams, G. 1991. Heritability of Leucostoma persoonii canker resistance among diverse peach genotypes. HortScience 26:60–62.

689. Chang, L. S., Iezzoni, A. F., Adams, G. C., and Ewers, F. W. 1991. Hydraulic conductance in susceptible versus tolerant peach seedlings infected with Leucostoma persoonii. J. Am. Soc. Hortic. Sci. 116:831–834.

690. Chapela, I. H. 1989. Fungi in healthy stems and branches of American beech and aspen: a comparative study. New Phytol. 113:65–75.

691. Chapela, I. H., and Boddy, L. 1988. Fungal colonization of attached beech branches. I. Early stages of development of fungal communities. II. Spatial and temporal organization of communities arising from latent invaders in bark and functional sapwood, under different moisture regimes. New Phytol. 110:39–45, 47–57.

692. Chapman, R. L. 1976, 1981. Ultrastructure of Cephaleuros virescens (Chroolepidaceae; Chlorophyta). I. Scanning electron microscopy of zoosporangia. III. Zoospores. Am. J. Bot. 63:1060–1070; 68:554–556.

693. Chapman, R. L. 1984. An assessment of the current state of our knowledge of the Trentepohliaceae. Pages 233–250 in: Systematics of the green algae. D. E. G. Irvine and D. M. John, eds. Academic Press, London.

694. Chapman, R. L., and Henk, M. C. 1985. Observations on the habit, morphology and ultrastructure of Cephaleuros parasiticus (Chlorophyta) and a comparison with C. virescens. J. Phycol. 21:513–522.

695. Chappelka, A. H., and Freer-Smith, P. H. 1995. Predisposition of trees by air pollutants to low temperatures and moisture stress. Environ. Pollut. 87:105–117.

696. Chappelka, A. H., and Samuelson, L. J. 1998. Ambient ozone effects on forest trees of the eastern United States: a review. New Phytol. 139:91–108.

697. Charles, V. K. 1935. A little known pecan fungus. Mycologia 27:74–82.

698. Charlton, J. W. 1963. Relating climate to eastern white pine blister rust infection hazard. USDA For. Serv., East. Reg., Upper Darby, PA. 38 pp.

699. Chase, A. R. 1984. Xanthomonas campestris pv. hederae causes a leaf spot of five species of Araliaceae. Plant Pathol. 33:439–440.

700. Chase, A. R. 1985. Bacterial leaf spot of Codiaeum variegatum cultivars caused by Xanthomonas campestris pv. poinsettiicola. Plant Pathol. 34:446–448.

701. Chase, A. R. 1986. Comparison of three bacterial leaf spots of Hibiscus rosa-sinensis. Plant Dis. 70:334–336.

702. Chase, A. R. 1987. Leaf and petiole rot of Ficus lyrata cv. Compacta caused by Pseudomonas cichorii. Plant Pathol. 36:219–221.

703. Chase, A. R. 1990. Effect of nitrogen, phosphorus and potassium rates on severity of Xanthomonas leaf spot of Schefflera. J. Environ. Hortic. 8:74–78.

704. Chase, A. R., and Broschat, T. K., eds. 1991. Diseases and disorders of ornamental palms. APS Press, St. Paul, MN. 56 pp.

705. Chase, A. R., and Jones, J. B. 1986. Effects of host nutrition, leaf age, and preinoculation light levels on severity of leaf spot of dwarf schefflera caused by Pseudomonas cichorii. Plant Dis. 70:561–563.

706. Chase, A. R., and Poole, R. T. 1986. Effects of fertilizer rate on severity of Alternaria leaf spot of three plants in the Araliaceae. Plant Dis. 70:1144–1145.

707. Chase, T. E., and Ullrich, R. C. 1990. Five genes determining intersterility in Heterobasidion annosum. Mycologia 82:73–81.

708. Chastagner, G. A., and Byther, R. S. 1983. Infection period of Phaeocryptopus gaeumannii on Douglas-fir needles in western Washington. Plant Dis. 67:811–813.

709. Chastagner, G. A., Byther, R. S., MacDonald, J. D., and Michaels, E. 1984. Impact of Swiss needle cast on postharvest hydration and needle retention of Douglas-fir Christmas trees. Plant Dis. 68:192–195.

710. Chastagner, G. A., Hamm, P. B., and Riley, K. L. 1995. Symptoms and Phytophthora spp. associated with root rot and stem canker of Noble fir Christmas trees in the Pacific Northwest. Plant Dis. 79:290–293.

711. Chatfield, J. A., Draper, E. A., Herms, D. A., and Cochran, K. D. 2002. Apple scab on crabapples at Secrest Arboretum: 2001. Ohio Agric. Res. Devel. Cent. Spec. Circ. No. 186:93–96.

712. Chelkowski, J, and Visconti, A., eds. 1992. Alternaria: biology, plant diseases and metabolites. Elsevier, Amsterdam. 573 pp.

713. Chellemi, D. O., and Britton, K. O. 1992. Influence of canopy microclimate on incidence and severity of dogwood anthracnose. Can. J. Bot. 70:1093–1096.

714. Chellemi, D. O., Knox, G., and Palm, M. E. 1993. Limb dieback of flowering dogwood caused by Colletotrichum acutatum. Plant Dis. 77:100.

715. Chen, J., Chang, C. J., and Jarret, R. L. 1992. DNA probes as molecular markers to monitor the seasonal occurrence of walnut witches'-broom mycoplasmalike organism. Plant Dis. 76:1116–1119.

716. Chen, J., Jarret, R. L., Qin, X., Hartung, J. S., Banks, D., Chang, C. J., and Hopkins, D. L. 2000. 16S rDNA sequence analysis of Xylella fastidiosa strains. Syst. Appl. Microbiol. 23:349–354.

717. Chen, J., Lamikanra, O., Chang, C. J., and Hopkins, D. L. 1995. Randomly amplified polymorphic DNA analysis of Xylella fastidiosa Pierce's disease and oak leaf scorch pathotypes. Appl. Environ. Microbiol. 61:1688–1690.

718. Chen, W. 1994. Vegetative compatibility groups of Verticillium dahliae from ornamental woody plants. Phytopathology 84:214–219.

719. Cheng, D., Yanaguchi, T., Wang, Z., and Pan, X. 2000. Genetic differentiation between two morphological types of Fomes fomentarius based on isozyme analysis. Mycosystema 19:81–86.

720. Chern, L. L., Ann, P. J., and Young, H. R. 1998. Root and foot rot of loquat in Taiwan caused by Phytophthora. Plant Dis. 82:651–656.

721. Cherubini, P., Schweingruber, F. H., and Forster, T. 1997. Morphology and ecological significance of intra-annual radial cracks in living conifers. Trees Struct. Funct. 11:216–222.

722. Chiba, O., and Tanaka, K. 1968. The effect of sulfur dioxide on the development of pine needle blight caused by Rhizosphaera kalkhoffii Bubak (I). J. Jap. For. Soc. 50:135–139.

723. Childs, J. F. L. 1953. Concentric canker and wood rot of citrus associated with Fomes applanatus in Florida. Phytopathology 43:99–100.

724. Childs, T. W. 1968. Elytroderma disease of ponderosa pine in the Pacific Northwest. USDA For. Serv. Res. Pap. PNW-69. 45 pp.

725. Childs, T. W., and Edgren, J. W. 1967. Dwarfmistletoe effects on ponderosa pine growth and trunk form. For. Sci. 13:167–174.

726. Chillali, M., Idder, I. H., Guillaumin, J. J., Mohammed, C., Lung-Escarmant, B., and Botton, B. 1998. Variation in the ITS and IGS regions of ribosomal DNA among the biological species of European Armillaria. Mycol. Res. 102:533–540.

727. Chiykowski, L. N., and Sinha, R. C. 1988. Some factors affecting the transmission of eastern peach X-mycoplasmalike organism by the leafhopper Paraphlepsius irroratus. Can. J. Plant Pathol. 10:85–92.

728. Chou, C. K. S. 1978. Penetration of young stems of Pinus radiata by Diplodia pinea. Physiol. Plant Pathol. 13:189–192.

729. Chou, C. K. S. 1990. Pathogenic variation of Seiridium spp. isolated from cankered Cupressaceae hosts in New Zealand. Eur. J. For. Pathol. 20:32–43.

730. Chou, C. K. S. 1991. Perspectives of disease threat in large-scale Pinus radiata monoculture—the New Zealand experience. Eur. J. For. Pathol. 21:71–81.

731. Christensen, C. M. 1940. Studies on the biology of Valsa sordida and Cytospora chrysosperma. Phytopathology 30:459–475.

732. Christiansen, E., Krokene, P., Berryman, A. A., Franceschi, V. R., Krekling, T., Lieutier, F., Lonneborg, A., and Solheim, H. 1999. Mechanical injury and fungal infection induce acquired resistance in Norway spruce. Tree Physiol. 19:399–403.

733. Chun, S., Fenn, P., and Kim, K. 1999. Characterization of chitinase in oak tissues and changes in its activity related to water stress and

inoculation with *Hypoxylon atropunctatum*. Plant Pathol. J. 15:144–151.

734. Chupp, C. 1954. A monograph of the fungus genus *Cercospora*. Published by author, Ithaca, NY. 667 pp.

735. Cilliers, A. J., Swart, W. J., and Wingfield, M. J. 1994. Selective medium for isolating *Lasiodiplodia theobromae*. Plant Dis. 78:1052–1055.

736. Clark, J. 1961. Birch dieback. Pages 1551–1555 in: Recent advances in botany. Univ. Toronto Press, Toronto.

737. Clark, J., and Barter, G. W. 1958. Growth and climate in relation to dieback of yellow birch. For. Sci. 4:343–364.

738. Clark, J., and Bonga, J. M. 1970. Photosynthesis and respiration in black spruce (*Picea mariana*) parasitized by eastern dwarf mistletoe (*Arceuthobium pusillum*). Can. J. Bot. 48:2029–2031.

739. Clark, R. B. 1997. Arbuscular mycorrhizal adaptation, spore germination, root colonization, and host plant growth and mineral acquisition at low pH. Plant Soil 192:15–22.

740. Clausen, C. A. 1996. Bacterial associations with decaying wood: a review. Int. Biodeter. Biodegrad. 37:101–107.

741. Claver, F. K., Alaniz, J. R., and Caldiz, D. O. 1983. *Tillandsia* spp.: epiphytic weeds of trees and bushes. For. Ecol. Manag. 6:367–372.

742. Clay, K., Dement, D., and Rejmanek, M. 1985. Experimental evidence for host races in mistletoe (*Phoradendron tomentosum*). Am. J. Bot. 72:1225–1231.

743. Cleene, M. De, and Ley, J. De. 1976. The host range of crown gall. Bot. Rev. 42:389–466.

744. Cleene, M. De, and Ley, J. De. 1981. The reevaluation of pathogenicity of "*Agrobacterium pseudotsugae.*" Phytopathol. Z. 101:185–188.

745. Clerivet, A., and Alami, I. 1999. Effects of jasmonic acid and of an elicitor from *Ceratocystis fimbriata* f.sp. *platani* on the accumulation of phytoalexins in leaves of susceptible and resistant plane trees. Plant Sci. 148:105–110.

746. Clerivet, A., Deon, V., Alami, I., Lopez, F., Geiger, J. P., and Nicole, M. 2000. Tyloses and gels associated with cellulose accumulation in vessels are responses of plane tree seedlings (*Platanus* × *acerifolia*) to the vascular fungus *Ceratocystis fimbriata* f.sp. *platani*. Trees Struct. Funct. 15:25–31.

747. Clerivet, A., and El Modafar, C. 1994. Vascular modifications in *Platanus acerifolia* seedlings inoculated with *Ceratocystis fimbriata* f.sp. *platani*. Eur. J. For. Pathol. 24:1–10.

748. Cline, M. N., Crane, J. L., and Cline, S. D. 1983. The teleomorph of *Cristulariella moricola*. Mycologia 75:988–994.

749. Cline, M. N., and Neely, D. 1979. *Cristulariella pyramidalis* and its pathogenesis on black walnut. Plant Dis. Rep. 63:1028–1032.

750. Cline, S., and Neely, D. 1983. Penetration and infection of leaves of black walnut by *Marssonina juglandis* and resulting lesion development. Phytopathology 73:494–497.

751. Clinton, G. P., and McCormick, F. A. 1929. The willow scab fungus. Conn. Agric. Exp. Stn. Bull. 302:443–469.

752. Clover, G. R. G., Tang, Z., Smales, T. E., and Pearson, M. N. 2003. Taxonomy of *Wisteria vein mosaic virus* and extensions to its host range and geographical distribution. Plant Pathol. 52:92–96.

753. Cobb, F. W. Jr., and Fergus, C. L. 1964. Pathogenicity, host specificity, and mat production of seven isolates of the oak wilt fungus. Phytopathology 54:865–866.

754. Cobb, F. W. Jr., Slaughter, G. W., Rowney, D. L., and DeMars, C. J. 1982. Rate of spread of *Ceratocystis wageneri* in ponderosa pine stands in the central Sierra Nevada. Phytopathology 72:1359–1362.

755. Cobb, F. W. Jr., Wood, D. L., Stark, R. W., and Parmeter, J. R. Jr. 1968. Photochemical oxidant injury and bark beetle (Coleoptera: Scolytidae) infestation of ponderosa pine. IV. Theory on the relationships between oxidant injury and bark beetle infestation. Hilgardia 39:141–152.

756. Cobbett, C. S. 2000. Phytochelatin biosynthesis and function in heavy-metal detoxification. Curr. Opin. Plant Biol. 3:211–216.

757. Cochrane, V. W. 1945. The common leaf rust of cultivated roses, caused by *Phragmidium mucronatum* (Fr.) Schlecht. Cornell Univ. Agric. Exp. Stn. Mem. 268. 39 pp.

758. Coelho, A. C., Cravador, A., Bollen, A., Ferraz, J. F. P., Moreira, A. C., Fauconnier, A., and Godfroid, E. 1997. Highly specific and sensitive non-radioactive molecular identification of *Phytophthora cinnamomi*. Mycol. Res. 101:1499–1507.

759. Coetzee, M. P. A., Wingfield, B. D., Harrington, T. C., Dalevi, D., Coutinho, T. A., and Wingfield, M. J. 2000. Geographical diversity of *Armillaria mellea* s.s. based on phylogenetic analysis. Mycologia 92:105–113.

760. Cohen, L. I. 1954. The anatomy of the endophytic system of the dwarf mistletoe, *Arceuthobium campylopodum*. Am. J. Bot. 41:840–847.

761. Cohen, L. I. 1967. The pathology of *Hypodermella laricis* on larch, *Larix occidentalis*. Am. J. Bot. 54:118–124.

762. Cohen, M., Pelosi, R. R., and Brlansky, R. H. 1983. Nature and location of xylem blockage structures in trees with citrus blight. Phytopathology 73:1125–1130.

763. Cole, A., Mink, G. I., and Regev, S. 1982. Location of Prunus necrotic ringspot virus on pollen grains from infected almond and cherry trees. Phytopathology 72:1542–1545.

764. Cole, G. T. 1983. *Graphiola phoenicis*: a taxonomic enigma. Mycologia 75:93–116.

765. Cole, J. R. 1933. Liver-spot disease of pecan foliage caused by *Gnomonia caryae pecanae*, nov. var. J. Agric. Res. 47:869–881.

766. Cole, J. R. 1935. *Gnomonia nerviseda*, the perfect stage of the fungus that causes vein spot disease of pecan foliage. J. Agric. Res. 50:91–96.

767. Cole, J. R. 1937. Bunch disease of pecans. Phytopathology 27:604–612.

768. Coleman, J. S., Murdoch, C. W., Campana, R. J., and Smith, W. H. 1985. Decay resistance of elm wetwood. Can. J. Plant Pathol. 7:151–154.

769. Coley-Smith, J. R., Verhoeff, K., and Jarvis, W. R., eds. 1980. The biology of *Botrytis*. Academic Press, New York. 318 pp.

770. Collado, I. G., Hernandez, G. R., Prieto, V., Hanson, J. R., and Redordinos, L. G. 1996. Biologically active sesquiterpenoid metabolites from the fungus *Botrytis cinerea*. Phytochemistry 41:513–517.

771. Collado, J., Platas, G., and Pelaez, F. 2001. Identification of an endophytic *Nodulisporium* sp. from *Quercus ilex* in central Spain as the anamorph of *Biscogniauxia mediterranea* by rDNA sequence analysis and effect of different ecological factors on distribution of the fungus. Mycologia 93:875–886.

772. Colley, R. H. 1918. Parasitism, morphology, and cytology of *Cronartium ribicola*. J. Agric. Res. 15:619–659.

773. Collins, D. J., Wyllie, T. D., and Anderson, S. H. 1991. Biological activity of *Macrophomina phaseolina* in soil. Soil Biol. Biochem. 23:495–496.

774. Collmer, A., and Bauer, D. W. 1994. *Erwinia chrysanthemi* and *Pseudomonas syringae*: plant pathogens trafficking in extracellular virulence proteins. Pages 43–78 in: Bacterial pathogenesis of plants and animals. J. L. Dangl, ed. Springer-Verlag, Berlin.

775. Colmenares, A. J., Aleu, J., Duran-Patron, R., Collado, I. G., and Hernandez-Galan, R. 2002. The putative role of botrydial and related metabolites in the infection mechanism of *Botrytis cinerea*. J. Chem. Ecol. 28:997–1005.

776. Commonwealth Institute of Helminthology. 1972–present. C.I.H. descriptions of plant parasitic nematodes. Sets 1–. Commonw. Inst. Helminthol., St. Albans, UK.

777. Commonwealth Mycological Institute and Association of Applied Biologists. 1970–1989. CMI/AAB descriptions of plant viruses Sets 1–22, Nos. 1–354. Commonw. Agric. Bur., Farnham Royal, Slough, UK.

778. Conner, W. H. 1994. The effect of salinity and waterlogging on growth and survival of baldcypress and Chinese tallow seedlings. J. Coast. Res. 10:1045–1049.

779. Conners, I. L. 1967. An annotated index of plant diseases in Canada. Can. Dep. Agric. Pub. 1251. 381 pp.

780. Conners, I. L., comp. 1956. Canadian Plant Disease Survey, annual reports 1954, 1955. Vols. 34, 35. Agric. Can., Res. Branch, Ottawa.

781. Conradie, E., Swart, W. J., and Wingfield, M. J. 1992. Susceptibility of *Eucalyptus grandis* to *Cryphonectria cubensis*. Eur. J. For. Pathol. 22:312–315.

782. Conti, G. G., Bassi, M., Maffi, D., and Bonecchi, R. 1985. Host-parasite relationship in a susceptible and a resistant rose cultivar inoculated with *Sphaerotheca pannosa*. I. Fungal growth, mechanical barriers and hypersensitive reaction. Phytopathol. Z. 113:71–80.

783. Converse, R. H. 1953. *Articularia* and *Microstroma* on pecan in Oklahoma. Plant Dis. Rep. 37:511–512.

784. Cook, R. T. A. 1981. Overwintering of *Diplocarpon rosae* at Wisley. Trans. Br. Mycol. Soc. 77:549–556.

785. Cook, R. T. A., Inman, A. J., and Billings, C. 1997. Identification and classification of powdery mildew anamorphs using light and scanning electron microscopy and host range data. Mycol. Res. 101:975–1002.

786. Cooke, D. E. L., Drenth, A., Duncan, J. M., Wagels, G., and Brasier, C. M. 2000. A molecular phylogeny of *Phytophthora* and related oomycetes. Fungal Genet. Biol. 30:17–32.

787. Cooke, W. B. 1961. The genus *Schizophyllum*. Mycologia 53:575–599.

788. Cooke, W. B. 1962. A taxonomic study in the "black yeasts." Mycopathol. Mycol. Appl. 17:1–43.

789. Cooley, J. S. 1936. *Sclerotium rolfsii* as a disease of nursery apple trees. Phytopathology 26:1081–1083.

790. Cooley, S. J. 1984. *Meria laricis* on nursery seedlings of western larch in Washington. Plant Dis. 84:826.

791. Cooper, J. I. 1980. The prevalence of cherry leaf roll virus in *Juglans regia* in the United Kingdom. Acta Phytopathol. Acad. Sci. Hung. 15:139–145.

792. Cooper, J. I. 1993. Virus diseases of trees and shrubs, 2nd ed. Chapman & Hall, London. 205 pp.

793. Cooper, J. I., and Edwards, M. L. 1981. The distribution of poplar mosaic virus in hybrid poplars and virus detection by ELISA. Ann. Appl. Biol. 99:53–61.

794. Cooper, J. I., Edwards, M. L., and Siwecki, R. 1986. The detection of poplar mosaic virus and its occurrence in a range of clones in England and Poland. Eur. J. For. Pathol. 16:116–125.

795. Cooper, J. I., and Massalski, P. R. 1984. Viruses and virus-like diseases affecting *Betula* spp. Proc. R. Soc. Edinb., Sect. B, 85:183.

796. Cooper, J. I., Massalski, P. R., and Edwards, M. L. 1984. Cherry leaf roll virus in the female gametophyte and seed of birch and its relevance to vertical virus transmission. Ann. Appl. Biol. 105:55–64.

797. Cooper, J. R. 1917. Studies of the etiology and control of blister canker on apple trees. Nebr. Agric. Exp. Stn. Res. Bull. 12. 117 pp.

798. Copeland, O. L. Jr., and McAlpine, R. G. 1962. Soil characteristics associated with spot die-out in loblolly pine plantations. For. Sci. 8:12–15.

799. Corbaz, R. 1985. Pathotypes et variations du pouvoir pathogène chez *Chalara elegans* Nag Raj et Kendrick (=*Thielaviopsis basicola*). Phytopathol. Z. 113:289–299.

800. Cordell, C. E., Anderson, R. L., Hoffard, W. H., Landis, T. D., Smith, R. S. Jr., and Toko, H. V., tech. coords. 1989. Forest nursery pests. USDA Agric. Handb. 680. 184 pp.

801. Cordell, C. E., and Matuszewski, M. 1974. *Cylindrocladium scoparium*—damaging black walnut seedlings in Kentucky nurseries. Plant Dis. Rep. 58:188–189.

802. Cordell, C. E., and Rowan, S. J. 1975. *Cylindrocladium scoparium* infection in a natural sweetgum stand. Plant Dis. Rep. 59:775–776.

803. Cordova, I., Jones, P., Harrison N. A., and Oropeza, C. 2003. In situ PCR detection of phytoplasma DNA in embryos from coconut palms with lethal yellowing disease. Mol. Plant Pathol. 4:99–108.

804. Corlett, M. 1976. *Apiosporina collinsii*. Fungi Canadenses No. 76. 2 pp.

805. Corlett, M. 1976. *Apiosporina morbosa*. Fungi Canadenses No. 84. 2 pp.

806. Corlett, M. 1991. An annotated list of the published names in *Mycosphaerella* and *Sphaerella*. J. Cramer, Berlin. 328 pp.

807. Corlett, M. 1995. An annotated list of the published names in *Mycosphaerella* and *Sphaerella*: corrections and additions. Mycotaxon 53:37–56.

808. Corner, E. J. H. 1953. The construction of polypores—1. Introduction: *Polyporus sulphureus, P. squamosus, P. betulinus,* and *Polystictus microcyclus*. Phytomorphology 3:157–167.

809. Corner, E. J. H. 1983. *Ad polyporaceas*. I. *Amauroderma* and *Ganoderma*. Nova Hedwigia 75:1–182.

810. Correll, J. C., Gordon, T. R., and McCain, A. H. 1992. Genetic diversity in California and Florida populations of the pitch canker fungus *Fusarium subglutinans* f.sp. *pini*. Phytopathology 82:415–420.

811. Correll, J. C., Gordon, T. R., McCain, A. H., Fox, J. W., Koehler, C. S., Wood, D. L., and Schultz, M. E. 1991. Pitch canker disease in California: pathogenicity, distribution, and canker development on Monterey pine (*Pinus radiata*). Plant Dis. 75:676–682.

812. Corsaro, M. M., Castro, C. de, Evidente, A., Lanzetta, R., Molinaro, A., Parrilli, M., and Sparapano, L. 1998. Phytotoxic extracellular polysaccharide fractions from *Cryphonectria parasitica* (Murr.) Barr strains. Carbohyd. Polymers 37:167–172.

813. Cortesi, P., Gadoury, D. M., Seem, R. C., and Pearson, R. C. 1995. Distribution and retention of cleistothecia of *Uncinula necator* on the bark of grapevines. Plant Dis. 79:15–19.

814. Cortesi, P., McCulloch, C. E., Song, H. Y., Lin, H. Q., and Milgroom, M. G. 2001. Genetic control of horizontal virus transmission in the chestnut blight fungus, *Cryphonectria parasitica*. Genetics 159:107–118.

815. Cortesi, P., and Milgroom, M. G. 1998. Genetics of vegetative incompatibility in *Cryphonectria parasitica*. Appl. Environ. Microbiol. 64:2988–2994.

816. Costa, H. S., Blua, M. S., Bethke, J. A., and Redak, R. A. 2000. Transmission of *Xylella fastidiosa* to oleander by the glassywinged sharpshooter, *Homalodisca coagulata*. HortScience 35:1265–1267.

817. Costonis, A. C. 1970. Acute foliar injury of eastern white pine induced by sulfur dioxide and ozone. Phytopathology 60:994–999.

818. Costonis, A. C. 1971. Effects of ambient sulfur dioxide and ozone on eastern white pine in a rural environment. Phytopathology 61:717–720.

819. Coutts, M. P., and Philipson, J. J. 1978. Tolerance of tree roots to waterlogging. II. Adaptation of Sitka spruce and lodgepole pine to waterlogged soil. New Phytol. 80:71–77.

820. Coutts, M. P., and Rishbeth, J. 1977. The formation of wetwood in grand fir. Eur. J. For. Pathol. 7:13–22.

821. Cox, R. M., Percy, K. L., Jensen, K. F., and Simpson, C. M., eds. 1996. Air pollution and multiple stresses: IUFRO proceedings, 16th international meeting for specialists in air pollution effects on forest ecosystems, Fredericton, New Brunswick, Canada, 1994. Can. For. Serv. Fredericton, NB. 402 pp.

822. Cox, R. M., Lemieux, G., and Lodin, M. 1996. The assessment and condition of Fundy white birches in relation to ambient exposure to acid marine fogs. Can. J. For. Res. 26:682–688.

823. Cox, R. M., and Malcolm, J. W. 1997. Effects of duration of a simulated winter thaw on dieback and xylem conductivity of *Betula papyrifera*. Tree Physiol. 17:389–396.

824. Cox, R. S. 1954. *Cylindrocladium scoparium* on conifer seedlings. Del. Agric. Exp. Stn. Bull. 301 (Tech.). 40 pp.

825. Coyier, D. L., and Roane, M. K. 1986. Compendium of rhododendron and azalea diseases. APS Press, St. Paul, MN. 65 pp.

826. Coyier, D. L., Stace-Smith, R., Allen T. C., and Leung, E. 1977. Viruslike particles associated with a rhododendron necrotic ringspot disease. Phytopathology 67:1090–1095.

827. Coyne, C. J., Mehlenbacher, S. A., and Smith, D. C. 1998. Sources of resistance to eastern filbert blight in hazelnut. J. Am. Soc. Hortic. Sci. 123:253–257.

828. Crandall, B. S. 1943. Bacterial infection and decay of the inner wood of winter-injured young London plane trees. Phytopathology 33:963–964.

829. Crandall, B. S. 1945. A new species of *Cephalosporium* causing persimmon wilt. Mycologia 37:495–498.

830. Crandall, B. S., and Baker, W. L. 1950. The wilt disease of American persimmon, caused by *Cephalosporium diospyri*. Phytopathology 40:307–325.

831. Crandall, B. S., Gravatt, G. F., and Ryan, M. M. 1945. Root disease of *Castanea* species and some coniferous and broadleaf nursery stocks, caused by *Phytophthora cinnamomi*. Phytopathology 35:162–180.

832. Crane, P. E. 2001. Morphology, taxonomy, and nomenclature of the *Chrysomyxa ledi* complex and related rust fungi on spruce and Ericaceae in North America and Europe. Can. J. Bot. 79:957–982.

833. Crane, P. E., and Hiratsuka, Y. 1999. Evidence for environmental determination of uredinia and telia production in *Chrysomyxa pirolata* (inland spruce cone rust). Can. J. Bot. 78:660–667.

834. Crane, P. E., Hiratsuka, Y., and Currah, R. S. 2000. Clarification of the life-cycle of *Chrysomyxa woroninii* on *Ledum* and *Picea*. Mycol. Res. 104:581–586.

835. Crane, P. E., Hiratsuka, Y., and Currah, R. S. 2000. Reproductive biology and evidence for water dispersal of teliospores in *Chrysomyxa weirii*, a microcyclic spruce needle rust. Mycologia 92:754–763.

836. Creager, D. B. 1937. Phytophthora crown rot of dogwood. J. Arnold Arbor. 18:344–348.

837. Creager, D. B. 1937. The Cephalosporium disease of elms. Contrib. Arnold Arbor. No. 10. 91 pp.

838. Credi, R. 1991. Rhabdovirus-like particles associated with a vein yellowing disease of *Laburnum anagyroides* occurring in Italy. Phytopathol. Medit. 30:77–82.

839. Creelman, D. W. 1956. The occurrence of ash rust in western Nova Scotia. Plant Dis. Rep. 40:580.

840. Creze, J. 1983. Where do we stand in regard to the grafting of apex in camellia? Int. Camellia J. No. 15:56–60.

841. Cripe, R. E. 1979. Lightning protection for trees and related property. J. Arboric. 5:145–149.

842. Cripe, R. E. 1985. Lightning protection for trees. Arbor Age 5(4):13, 14, 16, 18, 20.

843. Croghan, C. F., and Robbins, K. 1986. Cankers caused by *Botryodiplodia gallae* associated with oak sprout mortality in Michigan. Plant Dis. 70:76–77.

844. Croise, L., Lieutier, F., Cochard, H., and Dreyer, E. 2001. Effects of drought stress and high density stem inoculations with *Leptographium wingfieldii* on hydraulic properties of young Scots pine trees. Tree Physiol. 21:427–436.

845. Crone, L. J., and Bachelder, S. 1961. Insect transmission of canker stain fungus, *Ceratocystis fimbriata* f. *platani*. (Abstr.) Phytopathology 51:576.

846. Crosse, J. E. 1966. Epidemiological relations of the pseudomonad pathogens of deciduous fruit trees. Annu. Rev. Phytopathol. 4:291–310.

847. Crous, P. W. 1998. *Mycosphaerella* spp. and their anamorphs associated with leaf spot diseases of *Eucalyptus*. APS Press, St. Paul, MN. 170 pp.

848. Crous, P. W. 2002. Taxonomy and pathology of *Cylindrocladium* (*Calonectria*) and allied genera. APS Press, St. Paul, MN. 294 pp.

849. Crous, P. W., Aptroot, A., Kang, J. C., Braun, U., and Wingfield, M. J. 2000. The genus *Mycosphaerella* and its anamorphs. Stud. Mycol. No. 45:107–121.

850. Crous, P. W., and Braun, U. 2003. *Mycosphaerella* and its anamorphs: 1. Names published in *Cercospora* and *Passalora*. Centraalbur. voor Schimmelcultures, Utrecht. 571 pp.

851. Crous, P. W., and Corlett, M. 1998. Reassessment of *Mycosphaerella* spp. and their anamorphs occurring on *Platanus*. Can. J. Bot. 76:1523–1532.

852. Crous, P. W., Ferreira, F. A., and Sutton, B. C. 1997. A comparison of the fungal genera *Phaeophleospora* and *Kirramyces* (Coelomycetes). S.Afr. J. Bot. 63:111–115.

853. Crous, P. W., Kang, J., Braun, U., and Kang, J. C. 2001. A phylogenetic redefinition of anamorph genera in *Mycosphaerella* based on ITS rDNA sequence and morphology. Mycologia 93:1081–1101.

854. Crous, P. W., and Palm, M. E. 1999. Reassessment of the anamorph genera *Botryodiplodia*, *Dothiorella* and *Fusicoccum*. Sydowia 51:167–175.

855. Crous, P. W., and Wingfield, M. J. 1994. A monograph of *Cylindrocladium*, including anamorphs of *Calonectria*. Mycotaxon 51:341–435.

856. Crous, P. W., and Wingfield, M. J. 1997. New species of *Mycosphaerella* occurring on *Eucalyptus* leaves in Indonesia and Africa. Can. J. Bot. 75:781–790.

857. Crous, P. W., Wingfield, M. J., Marasas, W. F. O., and Sutton, B. C. 1989. *Pseudocercospora eucalyptorum* sp. nov. on *Eucalyptus* leaves. Mycol. Res. 93:394–398.

858. Crowe, F., Starkey, D., and Lengleek, V. 1982. Honeylocust canker in Kansas caused by *Thyronectria austro-americana*. Plant Dis. 66:155–158.

859. Crowell, I. H. 1934. The hosts, life history, and control of the cedar-apple rust fungus *Gymnosporangium juniperi-virginianae* Schw. J. Arnold Arbor. 15:163–232.

860. Crowell, I. H. 1935. The hosts, life history, and control of *Gymnosporangium clavipes* C. & P. J. Arnold Arbor. 16:367–410.

861. Croxton, W. C. 1939. A study of the tolerance of trees to breakage by ice accumulation. Ecology 20:71–73.

862. Cruickshank, R. H., and Wade, G. C. 1992. The activation of latent infections of *Monilinia fructicola* on apricots by volatiles from the ripening fruit. J. Phytopathol. 136:107–112.

863. Cubeta, M. A., and Vilgalys, R. 1997. Population biology of the *Rhizoctonia solani* complex. Phytopathology 87:480–484.

864. Cummings Carlson, J. E., and Kuntz, J. E. 1986. Stem canker on black walnut caused by *Fusarium sporotrichioides*. Annu. Rep. North. Nut Growers Assoc. 77:85–92.

865. Cummings Carlson, J. E., Mielke, M. E., Appleby, J. E., Hatcher, R., Hayes, E. M., Luley, C. J., O'Brien, J. G., and Rugg, D. J. 1993. Survey of black walnut canker in plantations in five central states. North. J. Appl. For. 10:10–13.

866. Cummins, G. B. 1962. Supplement to Arthur's manual of the rusts in the United States and Canada. Hafner, New York. 24 pp.

867. Cummins, G. B. 1971. Rust fungi of cereals, grasses, and bamboos. Springer-Verlag, New York. 570 pp.

868. Cummins, G. B. 1978. Rust fungi on legumes and composites in North America. Univ. Ariz. Press, Tucson. 424 pp.

869. Cummins, G. B. 1984. Two new rust fungi (Uredinales). Mycotaxon 20:617–618.

870. Cummins, G. B., and Hiratsuka, Y. 2003. Illustrated genera of rust fungi. 3rd ed. APS Press, St. Paul, MN. 225 pp.

871. Currie, C. R., and Hiratsuka, Y. 1996. Evaluating insect mediated dispersal of *Scytalidium uredinicola* for biological control of western gall rust. Can. J. For. Res. 26:1754–1760.

872. Curry, J. R., and Church, T. W. Jr. 1952. Observations on winter drying of conifers in the Adirondacks. J. For. 50:114–116.

873. Curzi, M. 1927. Di uno speciale parassitismo dell' *Ascochyta syringae*. Riv. Patol. Veg. 17:22–23.

874. Cutler, H. G., Parker, S. R., Ross, S. A., Crumley, F. G., and Schreiner, P. R. 1996. Homobotcinolide: a biologically active natural homolog of botcinolide from *Botrytis cinerea*. Biosci. Biotech. Biochem. 60:656–658.

875. Czabator, F. J. 1971. Fusiform rust of southern pines—a critical review. USDA For. Serv. Res. Pap. SO-65. 39 pp.

876. Czabator, F. J. 1976. A new species of *Ploioderma* associated with a pine needle blight. Mem. N.Y. Bot. Gard. 28:41–44.

877. Czabator, F. J., Staley, J. M., and Snow, G. A. 1971. Extensive southern pine needle blight during 1970–71 and associated fungi. Plant Dis. Rep. 55:764–766.

878. Dai, Y.-C. 1999. *Phellinus sensu lato* (Aphyllophorales, Hymenochaetaceae) in East Asia. Acta Bot. Fenn. 166:1–115.

879. Dai, Y.-C., and Qin, G.-F. 1998. *Phellinidium sulphurascens*—a forest pathogen in China. Fungal Sci. 13:101–107.

880. D'Ambra, V., Ferrata, M., and Baldan, B. P. 1977. Growth and behaviour of *Ceratocystis fimbriata* in plane-tree tissues. A scanning electron microscope study. Eur. J. For. Pathol. 7:351–357.

881. Dance, B. W. 1957. A fungus associated with blight and dieback of hybrid aspen. Can. Dep. Agric., Sci. Serv., For. Biol. Div. Bi-mon. Prog. Rep. 13(6):1–2.

882. Dance, B. W. 1961. Leaf and shoot blight of poplars (Section *Tacamahaca* Spach) caused by *Venturia populina* (Vuill.) Fabric. Can. J. Bot. 39:875–890.

883. Dance, B. W. 1961. Spore dispersal in *Pollaccia radiosa* (Lib.) Bald. and Cif. Can. J. Bot. 39:1429–1435.

884. Daniels, M. J. 1983. Mechanisms of spiroplasma pathogenicity. Annu. Rev. Phytopathol. 21:29–43.

885. Danso, S. K. A., Bowen, G. D., and Sanginga, N. 1992. Biological nitrogen fixation in trees in agro-ecosystems. Plant Soil 141:177–196.

886. Darker, G. C. 1932. The Hypodermataceae of conifers. Contrib. Arnold Arbor. No. 1. 131 pp.

887. Darker, G. D. 1967. A revision of the genera of the Hypodermataceae. Can. J. Bot. 45:1399–1444.

888. Darus, A., Seman, I. A., and Hassan, A. H. 1991. Histopathological studies on colonization of oil palm root by *Ganoderma boninense*. Elaeis 3:289–293.

889. Daughtrey, M. L., and Hibben, C. R. 1994. Dogwood anthracnose: a new disease threatens two native *Cornus* species. Annu. Rev. Phytopathol. 32:61–73.

890. Daughtrey, M. L., Hibben, C. R., Britton, K. O., Windham, M. T., and Redlin, S. C. 1996. Dogwood anthracnose: understanding a disease new to North America. Plant Dis. 80:349–358.

891. Daughtrey, M. L., Wick, R. L., and Peterson, J. L. 1995. Compendium of flowering potted plant diseases. APS Press, St. Paul, MN. 90 pp.

892. Davidson, J. M., Werres, S., Garbelotto, M., Hansen, E. M., and Rizzo, D. M. 2003. Sudden oak death and associated diseases caused by *Phytophthora ramorum*. Plant Health Progr. 7 July 2003. Internet pub.

893. Davidson, R. W. 1934. *Stereum gausapatum*, cause of heart rot of oaks. Phytopathology 24:831–832.

894. Davidson, R. W. 1935. Decay in living sprout oak trees. Plant Dis. Rep. 19:94–95.

895. Davidson, R. W., and Campbell, W. A. 1944. Observations on a gall of sugar maple. Phytopathology 34:132–135.

896. Davidson, R. W., Campbell, W. A., and Vaughan, D. B. 1942. Fungi causing decay of living oaks in the eastern United States and their cultural identification. USDA Tech. Bull. 765. 65 pp.

897. Davidson, R. W., and Cash, E. K. 1956. A *Cenangium* associated with sooty-bark canker of aspen. Phytopathology 46:34–36.

898. Davidson, R. W., and Lorenz, R. C. 1938. Species of *Eutypella* and *Schizoxylon* associated with cankers of maple. Phytopathology 28:733–745.

899. Davies, W. J., and Kozlowski, T. T. 1974. Short- and long-term effects of antitranspirants on water relations and photosynthesis of woody plants. J. Am. Soc. Hortic. Sci. 99:297–304.

900. Davis, B. H. 1938. The Cercospora leafspot of rose caused by *Mycosphaerella rosicola*. Mycologia 30:282–298.

901. Davis, C., and Meyer, T. 1997. Field guide to tree diseases of Ontario. NODA/NFP Tech. Rep. TR-46. Can. For. Serv. Great Lakes For. Ctr. and Ontario Minist. Nat. Resour., Sault Ste. Marie. 135 pp.

902. Davis, D. D., Umbach, D. M., and Coppolino, J. B. 1981. Susceptibility of tree and shrub species and response of black cherry foliage to ozone. Plant Dis. 65:904–907.

903. Davis, D. D., and Wilhour, R. G. 1976. Susceptibility of woody plants to sulfur dioxide and photochemical oxidants. EPA Ecol. Res. Ser. EPA-600/3-76-102. 71 pp.

904. Davis, J. R., and English, H. 1969. Factors related to the development of bacterial canker in peach. Phytopathology 59:588–595.

905. Davis, M. J., Thomson, S. V., and Purcell, A. H. 1980. Etiological role of the xylem-limited bacterium causing Pierce's disease in almond leaf scorch. Phytopathology 70:472–475.

906. Davis, R. E., and Sinclair, W. A. 1998. Phytoplasma identity and disease etiology. Phytopathology 88:1372–1376.

907. Davis, R. I., Brown, J. F., and Pone, S. P. 1996. Causal relationship between cucumber mosaic cucumovirus and kava dieback in the South Pacific. Plant Dis. 80:194–198.

908. Davis, R. M., Farrald, C. J., and Davila, D. 1987. Botryodiplodia trunk lesions in Texas citrus. Plant Dis. 71:848–849.

909. Davis, S. H. [Jr.], and Peterson, J. L. 1976. Susceptibility of cotoneasters to fire blight. J. Arboric. 2:90–91.

910. Davis, S. H. Jr., and Peterson, J. L. 1980. Trunk decay on Greenspire linden. J. Arboric. 6:258–260.

911. Davis, T. C. 1966. Appraisal of *Hypoxylon punctulatum* as a biological control agent of *Ceratocystis fagacearum* in oak-wilt trees. Phytopathology 56:772–775.

912. Davison, A. D. 1972. Factors affecting development of madrone canker. Plant Dis. Rep. 56:50–52.

913. Davison, E. M., Stukely, M. J. C., Crane, C. E., and Tay, F. C. S. 1994. Invasion of phloem and xylem of woody stems and roots of *Eucalyptus marginata* and *Pinus radiata* by *Phytophthora cinnamomi*. Phytopathology 84:335–340.

914. Davison, S. E., and Forman, R. T. T. 1982. Herb and shrub dynamics in a mature oak forest: a thirty-year study. Bull. Torrey Bot. Club 109:64–73.

915. Dawe, A. L., and Nuss, D. L. 2001. Hypoviruses and chestnut blight: exploiting viruses to understand and modulate fungal pathogenesis. Annu. Rev. Genet. 35:1–29.

916. Dawson, J. H., Musselman, L. J., Wolswinkel, P., and Dorr, I. 1994. Biology and control of *Cuscuta*. Rev. Weed Sci. 6:265–317.

917. Dawson, J. O., Berg, R. H., Paschke, M. W, and Wheeler, C. T., eds. 1999. 11th International conference on *Frankia* and actinorhizal plants. Can. J. Bot. 77:1203–1400.

918. Day, W. R. 1928. Damage by late frost on Douglas fir, Sitka spruce, and other conifers. Forestry 2:19–30.

919. Day, W. R. 1954. Drought crack of conifers. G.B. For. Comm. For. Rec. No. 26. 40 pp.

920. Daykin, M. E., and Milholland, R. D. 1990. Histopathology of blueberry twig blight caused by *Phomopsis vaccinii*. Phytopathology 80:736–740.

921. Dearness, J., and Hansbrough, J. R. 1934. *Cytospora* infection following fire injury in western British Columbia. Can. J. Res. 10:125–128.

922. Debabrata, D. 2000. A preliminary taxonomic survey of host-range of *Cassytha filiformis* L. (Cassythaceae) in the lateritic district of Purulia, West Bengal. J. Econ. Taxon. Bot. 24:467–473.

923. Debener, T., Drewes-Alvarez, R., and Rockstroh, K. 1998. Identification of five physiological races of blackspot, *Diplocarpon rosae* Wolf on roses. Plant Breed. 117:267–270.

924. Deckert, R. J., Hsiang, T., and Peterson, R. L. 2002. Genetic relationships of endophytic *Lophodermium nitens* isolates from needles of *Pinus strobus*. Mycol. Res. 106:305–313.

925. Decourtye, L., and Cadic, A. 1993. Breeding high quality *Pyracantha*. C. R. Acad. Agric. France 79:67–75.

926. Défago, G. 1942. Seconde contribution à la connaissance des Valsées v. Höhnel. Phytopathol. Z. 14:103–147.

927. DeHayes, D. H., Thornton, F. C., Waite, C. E., and Ingle, M. A. 1991. Ambient cloud deposition reduces cold tolerance of red spruce seedlings. Can. J. For. Res. 21:1292–1295.

928. Deighton, F. C. 1973. Five North American *Cercospora*-like fungi. Trans. Br. Mycol. Soc. 61:107–120.

929. Deighton, F. C. 1976. Studies on *Cercospora* and allied genera. VI. *Pseudocercospora* Speg., *Pantospora* Cif. and *Cercoseptoria* Petr. Commonw. Mycol. Inst. Mycol. Pap. 140. 168 pp.

930. Deighton, F. C. 1987. New species of *Pseudocercospora* and *Mycovellosiella*, and new combinations into *Pseudocercospora* and *Phaeoramularia*. Trans. Br. Mycol. Soc. 88:365–391.

931. Delatour, C., Guillaumin, J. J., Lung-Escarmant, B., and Marçais, B. 1998. Root and butt rots of forest trees: 9th international conference on root and butt rots, Carcans-Maubuisson, France, 1997. Inst. Natl. Recher. Agron. (INRA), Paris. 459 pp.

932. Demaree, J. B., and Cole, J. R. 1936. A disporous *Gnomonia* on pecan. Phytopathology 26:1025–1029.

933. Demchik, M. C., and Sharpe, W. E. 2000. The effect of soil nutrition, soil acidity and drought on northern red oak (*Quercus rubra* L.) growth and nutrition on Pennsylvania sites with high and low red oak mortality. For. Ecol. Manag. 136:199–207.

934. Denman, S., Crous, P. W., Taylor, J. E., Kang, J. C., Pascoe, I., and Wingfield, M. J. 2000. An overview of the taxonomic history of *Botryosphaeria*, and a re-evaluation of its anamorphs based on morphology and ITS rDNA phylogeny. Stud. Mycol. No. 45:129–140.

935. Denny, T. P. 1995. Involvement of bacterial polysaccharides in plant pathogenesis. Annu. Rev. Phytopathol. 33:173–197.

936. Derr, J. F., and Appleton, B. L. 1988. Herbicide injury to trees and shrubs: a pictorial guide to symptom diagnosis. Blue Crab Press, Virginia Beach, VA. 72 pp.

937. Derr, J. F., and Appleton, B. L. 1989. Behind herbicide injury. Am. Nurseryman 169(10):56–59.

938. Derrick, K. S. 1998. Evidence indicating citrus blight is an infectious disease. Laranja 19:107–116.

939. Dessureault, M., Lachance, D., Roy, G., Robitaille, L., and Gagnon, G. 1985. Symposium: maple decline in Quebec. Phytoprotection 66:69–99.

940. Deusen, P. C. Van, and Snow, G. A. 1991. Paired-tree study suggests 20-year recurrent slash pine blight. Can. J. For. Res. 21:1145–1147.

941. DeVay, J. E., Sinden, S. L., Lukezic, F. L., Werenfels, L. F., and Backman, P. A. 1968. Poria root and crown rot of cherry trees. Phytopathology 58:1239–1241.

942. Devey, M., Matheson, C., and Gordon, T. 1999. Current and potential impacts of pitch canker in radiata pine. Proceedings of the IMPACT Monterey Workshop, Monterey, CA. CSIRO For. & For. Prod. Tech. Rep. No. 112. 120 pp.

943. Dewey, F. M., Barrett, D. K., Vose, I. R., and Lamb, C. J. 1984. Immuno-fluorescence microscopy for the detection and identification of propagules of *Phaeolus schweinitzii* in infested soil. Phytopathology 74:291–296.

944. Dey, A. N., and Debata, D. K. 2000. Studies on leaf spot disease of *Populus deltoides* Marsh. caused by *Alternaria raphani*. Indian For. 126:1013–1014.

945. DeYoung, R. M., Copeman, R. J., and Hunt, R. S. 1998. Two strains in the genus *Erwinia* cause galls on Douglas-fir in southwestern British Columbia. Can. J. Plant Pathol. 20:194–200.

946. Dhakal, L. P., White, T. L., and Hodge, G. R. 1996. Realized genetic gains from slash pine tree improvement. Silvae Genet. 45:190–197.

947. Dhanvantari, B. N. 1978. Cold predisposition of dormant peach twigs to nodal cankers caused by *Leucostoma* spp. Phytopathology 68:1779–1783.

948. Dhanvantari, B. N., Johnson, P. W., and Dirks, V. A. 1975. The role of nematodes in crown gall infection of peach in southwestern Ontario. Plant Dis. Rep. 59:109–112.

949. Dharne, C. G. 1965. Taxonomic investigations on the discomycetous genus *Lachnellula* Karst. Phytopathol. Z. 53:101–144.

950. Dhingra, O. D., and Chagas, D. 1981. Effect of soil temperature, moisture, and nitrogen on competitive saprophytic ability of *Macrophomina phaseolina*. Trans. Br. Mycol. Soc. 77:15–20.

951. Dhingra, O. D., and Sinclair, J. B. 1977. An annotated bibliography of *Macrophomina phaseolina*, 1905–1975. Univ. Federal de Viçosa, Viçosa, Brazil; and Univ. Ill., Urbana. 244 pp.

952. Dhingra, O. D., and Sinclair, J. B. 1978. Biology and pathology of *Macrophomina phaseolina*. Impresna Universitaria Univ. Federal de Viçosa, Viçosa, Brazil. 166 pp.

953. Di, R., Hill, J. H., and Epstein, A. H. 1990. Double-stranded RNA associated with the rose rosette disease of multiflora rose. Plant Dis. 74:56–58.

954. Diamandis, S. 1978. "Top-dying" of Norway spruce, *Picea abies* (L.) Karst., with special reference to *Rhizosphaera kalkhoffii* Bubak. II. Status of *R. kalkhoffii* in "top-dying" of Norway spruce. Eur. J. For. Pathol. 8:345–356.

955. Diamandis, S., Epstein, L., Cobb, F. W. Jr., Popenuck, T., and Hecht, P. E. 1997. Development of *Leptographium wageneri* on root surfaces and other substrata. Eur. J. For. Pathol. 27:381–390.

956. Diamandis, S., and Minter, D. W. 1980. *Rhizosphaera kalkhoffii*. C.M.I. Descr. Pathogenic Fungi Bact. No. 656. 2 pp.

957. Dick, M. W. 2001. Straminipilous fungi: systematics of the peronosporomycetes, including accounts of the marine straminipilous protists, the plasmodiophorids, and similar organisms. Kluwer Academic, Dordrecht. 670 pp.

958. Dickens, J. S. W., and Cook, R. T. A. 1989. *Glomerella cingulata* on camellia. Plant Pathol. 38:75–85.

959. Dickey, R. D. 1977. Nutritional deficiencies of woody ornamental plants used in Florida landscapes. Fla. Agric. Exp. Stn. Bull. 791. 63 pp.

960. DiCosmo, F., Nag Raj, T. R., and Kendrick, W. B. 1984. A revision of the Phacidiaceae and related anamorphs. Mycotaxon 21:1–234.

961. DiCosmo, F., Peredo, H., and Minter, D. W. 1983. *Cyclaneusma* gen. nov., *Naemacyclus*, and *Lasiostictis*, a nomenclatural problem resolved. Eur. J. For. Pathol. 13:206–212.

962. Dietz, S. M. 1926. The alternate hosts of crown rust, *Puccinia coronata* Corda. J. Agric. Res. 33:953–970.

963. Diller, J. D. 1943. A canker of eastern pines associated with *Atropellis tingens*. J. For. 41:41–52.

964. Dinus, R. J., and Schmidt, R. S., eds. 1977. Management of fusiform rust in southern pines. Symp. Proc., Univ. Fla., Gainesville. 163 pp.

965. Dirr, M. A. 1998. Manual of woody landscape plants. Stipes, Champaign, IL. 1187 pp.

966. Dirr, M. A., and Biedermann, J. 1980. Amelioration of salt damage to cotoneaster by gypsum. J. Arboric. 6:108–110.

967. Dixon, R. K., Pallardy, S. G., Garrett, H. E., and Cox, G. S. 1983. Comparative water relations of container-grown and bare-root ectomycorrhizal and nonmycorrhizal *Quercus velutina* seedlings. Can. J. Bot. 61:1559–1565.

968. Dizengremel, P. 2001. Effects of ozone on the carbon metabolism of forest trees. Plant Physiol. Biochem. 39:729–742.

969. Doane, C. C., and McManus, M. L., eds. 1981. The gypsy moth: research toward integrated pest management. USDA Tech. Bull. 1584. 757 pp.

970. Dobson, F. S. 2000. Lichens. 4th ed. Richmond, Slough, UK. 431 pp.

971. Dobson, M. C. 1991. De-icing salt damage to trees and shrubs. HMSO, London. 64 pp.

972. Dochinger, L. S. 1967. Occurrence of poplar cankers caused by *Fusarium solani* in Iowa. Plant Dis. Rep. 51:900–903.

973. Dochinger, L. S. 1973. Trees for polluted air. USDA Misc. Pub. 1230. 12 pp.

974. Dochinger, L. S., and Seliskar, C. E. 1962. Fusarium canker found on yellow-poplar. J. For. 60:331–333.

975. Dodge, B. O. 1931. A destructive red-cedar rust disease. J. N.Y. Bot. Gard. 32:101–108.

976. Dodge, B. O. 1933. The course of mycelia of Gymnosporangia in the trunks of cedars. Proc. Natl. Shade Tree Conf. 9:94–101.

977. Dodge, B. O. 1933. The orange-rust of hawthorn and quince invades the trunk of red cedar. J. N.Y. Bot. Gard. 34:233–237.

978. Dohm, A., Ludwig, C., Schilling, D., Debener, T., Huylenbroeck, J. van, Bockstaele, E. van, and Debergh, P. 2002. Transformation of roses with genes for antifungal proteins to reduce their susceptibility to fungal diseases. Acta Hortic. No. 572:105–111.

979. Doi, Y., Teranaka, M., Yora, K., and Asuyama, H. 1967. Mycoplasma- or PLT group-like microorganisms found in phloem elements of plants infected with mulberry dwarf, potato witches' broom, aster yellows, or paulownia witches' broom. Ann. Phytopathol. Soc. Jap. 33:259–266.

980. Dolezal, W. E., and Tainter, F. H. 1979. Phenology of comandra rust in Arkansas. Phytopathology 69:41–44.

981. Dolwin, J. A. 1985. Lightning and trees. Arboric. J. 9:251–258.

982. Domanski, S. 1982. *Bjerkandera adusta* on young *Quercus rubra* and *Quercus robur* injured by late spring frosts in the Upper Silesia industrial district of Poland. Eur. J. For. Pathol. 12:406–413.

983. Dommergues, Y. R., and Krupa, S. V., eds. 1978. Interactions between non-pathogenic soil microorganisms and plants. Elsevier, Amsterdam. 475 pp.

984. Domsch, K. H., Gams, W., and Anderson, T.-H. 1980. Compendium of soil fungi. Vols. 1, 2. Academic Press, London. 859 and 405 pp.

985. Donaubauer, E. 1995. Über die *Phomopsis*-Krankheit bei Fichten (*Picea abies* [L.] Karst.). Forstliche Bundesversuchs. Wien Berichte No. 88:29–32.

986. Donaubauer, E. 1998. The role of pathogens in the present oak decline in Europe—a literature review. Eur. J. For. Pathol. 28:91–98.

987. Donaubauer, E., and Stephen, B. R., eds. 1988. Recent research on Scleroderris canker of conifers. IUFRO Proc., Salzburg, Austria, and Ljubljana, Yugoslavia, 1986. 167 pp.

988. Dooley, H. L. 1984. Temperature effects on germination of uredospores of *Melampsoridium betulinum* and on rust development. Plant Dis. 68:686–688.

989. Dorworth, C. E. 1972. Epidemiology of *Scleroderris lagerbergii* in central Ontario. Can. J. Bot. 50:751–765.

990. Dorworth, C. E. 1981. Status of pathogenic and physiologic races of *Gremmeniella abietina*. Plant Dis. 65:927–931.

991. Dorworth, C. E. 1989. European mistletoe (*Viscus album* subsp. *album*) in Canada. Plant Dis. 73:444.

992. Doss, R. P., Potter, S. W., Soeldner, A. H., Christian, J. K., and Fukunaga, L. E. 1995. Adhesion of germlings of *Botrytis cinerea*. Appl. Environ. Microbiol. 61:260–265.

993. Doster, M. A., and Bostock, R. M. 1988. Incidence, distribution, and development of pruning wound cankers caused by *Phytophthora syringae* in almond orchards in California. Phytopathology 78:468–472.

994. Dotzler, M. 1991. Infektionsversuche mit *Rhizosphaera kalkhoffii* und *Lophodermium piceae* an unterschiedlich gestressen Jungfichten (*Picea abies* [L.] Karst.). Eur. J. For. Pathol. 21:107–123.

995. Downs, A. A. 1938. Glaze damage in the birch-beech-maple-hemlock type of Pennsylvania and New York. J. For. 36:63–70.

996. Dozier, W. A. Jr., Latham, A. J., Kouskolekas, C. A., and Mayton, E. L. 1974. Susceptibility of apple rootstocks to black root rot and wooly apple aphids. HortScience 9:35–36.

997. Drew, M. C., and Lynch, J. M. 1980. Soil anaerobiosis, microorganisms, and root function. Annu. Rev. Phytopathol. 18:37–66.

998. Drewes-Alvarez, R. 2003. Blackspot. Pages 148–153 in: Encyclopedia of rose science. Vol. 1. A. Roberts, T. Debener, and S. Gudin, eds. Elsevier, Amsterdam.

999. Dreyer, G. D. 1994. Element stewardship abstract for *Celastrus orbiculatus* Thunb. (*C. articulatus*). The Nature Conservancy, Arlington, VA. Internet pub.

1000. Dreyer, G. D., Baird, L., and Fickler, C. 1987. *Celastrus scandens* and *Celastrus orbiculatus*: Comparisons of reproductive potential between a native and an introduced woody vine. Bull. Torrey Bot. Club 114:260–264.

1001. Drilias, M. J., Kuntz, J. E., and Worf, G. L. 1982. Collar rot and basal canker of sugar maple. J. Arboric. 8:29–33.

1002. Dropkin, V. H. 1989. Introduction to plant nematology. 2nd ed. Wiley, New York. 304 pp.

1003. Dubery, I. A., Meyer, D., and Bothma, C. 1994. Purification and characterization of cactorein, a phytotoxin secreted by *Phytophthora cactorum*. Phytochemistry 35:307–312.

1004. Dubin, H. J., and English, H. 1975. Epidemiology of European apple canker in California. Phytopathology 65:542–550.

1005. Dubin, J., and Staley, J. M. 1969. *Lophodermium* (*Davisomycella*) *ponderosae*. Mycologia 61:396–397.

1006. Duchesne, L. C., Jeng, R. S., Hubbes, M., and Sticklen, M. B. 1990. Accumulation of mansonones E and F in seedlings of *Ulmus americana* in response to inoculation with *Ophiostoma ulmi*. Trees Struct. Funct. 4:187–190.

1007. Duchesne, L. C., Peterson, R. L., and Ellis, B. E. 1989. The time-course of disease suppression and antibiosis by the ectomycorrhizal fungus *Paxillus involutus*. New Phytol. 111:693–698.

1008. Dudzinski, M. J., Old, K. M., and Gibbs, R. J. 1993. Pathogenic variability in Australian isolates of *Phytophthora cinnamomi*. Austral. J. Bot. 41:721–732.

1009. Duguay, S. M., Arii, K., Hooper, M., and Lechowicz, M. J. 2001. Ice storm damage and early recovery in an old-growth forest. Environ. Monit. Assess. 67:97–108.

1010. Dujesiefken, D., Rhaesa, A., Eckstein, D., and Stobbe, H. 1999. Tree wound reactions of differently treated boreholes. J. Arboric. 25:113–123.

1011. Dumas, M. T., Wood, J. E., Mitchell, E. G., and Boyonoski, N. W. 1997. Control of stump sprouting of *Populus tremuloides* and *P. grandidentata* by inoculation with *Chondrostereum purpureum*. Biol. Control 10:37–41.

1012. Duncan, W. H. 1975. Woody vines of the southeastern United States. Univ. Ga. Press, Athens. 76 pp.

1013. Dunn, C. P., ed. 2000. The elms: breeding, conservation, and disease management. Kluwer Academic, Boston, Dordrecht. 361 pp.

1014. Dusabenyagasani, M., Lecours, N., and Hamelin, R. C. 1998. Sequence-tagged sites (STS) for studies of molecular epidemiology of Scleroderris canker of conifers. Theor. Appl. Genet. 97:789–796.

1015. Dwinell, L. D. 1967. Selected interactions of *Verticillium dahliae*, *Pratylenchus penetrans*, major nutrient elements, *Acer saccharum*, and *Ulmus americana*. Ph.D. diss., Cornell Univ., Ithaca, NY. 262 pp.

1016. Dwinell, L. D. 1974. Susceptibility of southern oaks to *Cronartium fusiforme* and *Cronartium quercuum*. Phytopathology 64:400–403.

1017. Dwinell, L. D. 1978. Susceptibility of southern pines to infection by *Fusarium moniliforme* var. *subglutinans*. Plant Dis. Rep. 62:108–111.

1018. Dwinell, L. D. 1985. Relative susceptibilities of five pine species to three populations of the pinewood nematode. Plant Dis. 69:440–442.

1019. Dwinell, L. D. 1997. The pinewood nematode: regulation and mitigation. Annu. Rev. Phytopathol. 35:153–166.

1020. Dwinell, L. D., Barrows-Broaddus, J. B., and Kuhlman, E. G. 1985. Pitch canker: a disease complex of southern pines. Plant Dis. 69:270–276.

1021. Dye, D. W. 1967. Bacterial spot of ivy caused by *Xanthomonas hederae* (Arnaud, 1920) Dowson 1939, in New Zealand. N.Z. J. Sci. 10:481–485.

1022. Dye, D. W., Bradbury, J. F., Goto, M., Hayward, A. C., Lelliott, R. A., and Schroth, M. N. 1980. International standards for naming pathovars of phytopathogenic bacteria and a list of pathovar names and pathotype strains. Rev. Plant Pathol. 59:153–168.

1023. Dye, M. H. 1972. Silver-leaf disease of fruit trees. N.Z. Minist. Agric. Fish. Bull 104. 20 pp.

1024. Dye, M. H. 1974. Basidiocarp development and spore release by *Stereum purpureum* in the field. N.Z. J. Agric. Res. 17:93–100.

1025. Dyer, S. M., and Mader, D. L. 1986. Declined urban sugar maples: growth patterns, nutritional status and site factors. J. Arboric. 12:6–13.

1026. Dyke, C. G. Van, and Amerson, H. V. 1976. Interactions of *Puccinia sparganioides* with smooth cordgrass (*Spartina alterniflora*). Plant Dis. Rep. 60:670–674.

1027. Dyke, O. Van. 1999. A literature review of ice storm impacts on forests in eastern North America. Ontario Minist. Nat. Resour. Southcent. Sci. Sect. Tech. Rep. 112. 29 pp.

1028. Eagar, C., and Adams, M. B., eds. 1992. Ecology and decline of red spruce in the eastern United States. Ecol. Stud. 96. Springer-Verlag, New York. 417 pp.

1029. Eames, A. J., and Cox, L. G. 1945. A remarkable tree fall and an unusual type of graft-union failure. Am. J. Bot. 32:331–335.

1030. Ebrahim, N. F., and Heitefuss, R. 1989. Isolation of a potyvirus from diseased Norway spruce in the Bavarian forest. Eur. J. For. Pathol. 19:222–230.

1031. Edel, V., Steinberg, C., Gautheron, N., and Alabouvette, C. 2000. Ribosomal DNA-targeted oligonucleotide probe and PCR assay specific for *Fusarium oxysporum*. Mycol. Res. 104:518–526.

1032. Eden, M. A., Hill, R. A., and Galpoththage, M. 2000. An efficient baiting assay for quantification of *Phytophthora cinnamomi* in soil. Plant Pathol. 49:515–522.

1033. Edgerton, C. W. 1908. The physiology and development of some anthracnoses. Bot. Gaz. 45:367–408.

1034. Edgerton, C. W. 1911. Diseases of the fig tree and fruit. La. Agric. Exp. Stn. Bull. 126. 20 pp.

1035. Edwards, O. R., and Linit, M. J. 1992. Transmission of *Bursaphelenchus xylophilus* through oviposition wounds of *Monochamus carolinensis* (Coleoptera: Cerambycidae). J. Nematol. 24:133–139.

1036. Eesvelde, S. van, Liau, S. S., and Damme, P. van. 1993. Epiphytic and parasitic algae (Trentepohliaceae) on oil palm (*Elaeis guineensis* Jacq.). I. General description of species. Meded. Fac. Landbouw. Univ. Gent 58:1033–1039.

1037. Eesvelde, S. van, Liau, S. S., and Damme, P. van. 1993. Epiphytic and parasitic algae (Trentepohliaceae) on oil palm (*Elaeis guineensis* Jacq.). II. Experimental work. Meded. Fac. Landbouw. Univ. Gent 58:1040–1050.

1038. Egolf, D. R. 1978. *Pyracantha* 'Navaho' and 'Teton.' HortScience 13:483–484.

1039. Egolf, D. R. 1987. *Pyracantha* × 'Apache.' HortScience 22:173–174.

1040. Egolf, D. R. 1987. *Pyracantha* × 'Pueblo.' HortScience 22:510–511.

1041. Ehleringer, J. R., Cook, C. S., and Tieszen, L. L. 1986. Comparative water use and nitrogen relationships in a mistletoe and its host. Oecologia 68:279–284.

1042. Ehrlich, J. 1934. The beech bark disease: a *Nectria* disease of *Fagus* following *Cryptococcus fagi* (Baer). Can. J. Res. 10:593–692.

1043. Eisensmith, S. P., Sjulin, T. M., Jones, A. L., and Cress, C. E. 1982. Effects of leaf age and inoculum concentration on infection of sour cherry by *Coccomyces hiemalis*. Phytopathology 72:574–577.

1044. El Modafar, C., Clerivet, A., and Macheix, J. J. 1996. Flavan accumulation in stems of *Platanus* × *acerifolia* seedlings inoculated with *Ceratocystis fimbriata* f.sp. *platani,* the canker stain disease agent. Can. J. Bot. 74:1982–1987.

1045. El Modafar, C., Clerivet, A., Vigouroux, A., and Macheix, J. J. 1995. Accumulation of phytoalexins in leaves of plane tree (*Platanus* spp.) expressing susceptibility or resistance to *Ceratocystis fimbriata* f.sp. *platani.* Eur. J. Plant Pathol. 101:503–509.

1046. El Sharkawy, M. A., Cock, J. H., and Pilar-Hernandez, A. del. 1986. Differential response of stomata to air humidity in the parasitic mistletoe (*Phthirusa pyrifolia*) and its host, mandarin orange (*Citrus reticulata*). Photosyn. Res. 9:333–343.

1047. Elad, Y. 1996. Mechanisms involved in the biological control of *Botrytis cinerea* incited diseases. Eur. J. Plant Pathol. 102:719–732.

1048. Elad, Y., and Evensen, K. 1995. Physiological aspects of resistance to *Botrytis cinerea*. Phytopathology 85:637–643.

1049. Eleuterius, L. N. 1976. Observations on the mistletoe (*Phoradendron flavescens*) in south Mississippi, with special reference to the mortality of *Quercus nigra*. Castanea 41:265–268.

1050. El-Gholl, N. E., Schubert, T. S., and Coile, N. C. 1997. Diseases and disorders of plants in Florida. Fla. Dep. Agric. Consum. Serv., Div. Plant Indus. Bull. 14. Suppl. No. 1. 138 pp.

1051. Elliott, M. L., and Broschat, T. K. 2001. Observations and pathogenicity experiments on *Ganoderma zonatum* in Florida. Palms 45:62–72.

1052. Elliott, M. L. Broschat, T. K., Uchida, Y., and Simone, G. W. 2004. Compendium of ornamental palm diseases and disorders. APS Press, St. Paul, MN. 69 pp.

1053. Ellis, J. B., and Everhart, B. M. 1892. The North American Pyrenomycetes. Ellis & Everhart, Newfield, NJ. 793 pp.

1054. Ellis, M. A., Ferree, D. C., and Spring, D. E. 1981. Photosynthesis, transpiration, and carbohydrate content of apple leaves infected by *Podosphaera leucotricha*. Phytopathology 71:392–395.

1055. Ellis, M. B. 1971. Dematiaceous hyphomycetes. Commonw. Mycol. Inst., Kew, Surrey, UK. 608 pp.

1056. Ellis, M. B., and Ellis, J. P. 1997. Microfungi on land plants: an identification handbook. Richmond, Slough, UK. 868 pp.

1057. Elmerich, C., Kondorosi, A., and Newton, W. E., eds. 1998. Biological nitrogen fixation for the 21st century. Kluwer Academic, Dordrecht. 708 pp.

1058. Encinas, O., and Daniel, G. 1995. Wood cell biodegradation by the blue stain fungus *Botryodiplodia theobromae* Pat. Mater. Org. (Berl.) 29:255–272.

1059. Encinas, O., Henningsson, B., and Daniel, G. 1998. Changes in toughness and fracture characteristics of wood attacked by the blue stain fungus *Lasiodiplodia theobromae*. Holzforschung 52:82–88.

1060. Enebak, S. A., and Blanchette, R. A. 1989. Canker formation and decay in sugar maple and paper birch infected by *Cerrena unicolor*. Can. J. For. Res. 19:225–231.

1061. Enebak, S. A., Li, B., and Ostry, M. E. 1996. Seedling response of two trembling aspen (*Populus tremuloides*) families to infection by *Hypoxylon mammatum*. Eur. J. For. Pathol. 26:245–252.

1062. Engelhard, A. W., ed. 1989. Soilborne plant pathogens: management of diseases with macro- and microelements. APS Press, St. Paul, MN. 217 pp.

1063. Engelhard, A. W., and Jones, J. B. 1990. Stem canker and leaf spot of poinsettia caused by *Pseudomonas viridiflava* in Florida. Plant Dis. 74:528–529.

1064. Englerth, G. H. 1942. Decay of western hemlock in western Oregon and Washington. Yale Univ. Sch. For. Bull. 50. 53 pp.

1065. English, H., Davis, J. R., and DeVay, J. E. 1975. Relationship of *Botryosphaeria dothidea* and *Hendersonula toruloidea* to a canker disease of almond. Phytopathology 65:114–122.

1066. English, J. T., Ploetz, R. C., and Barnard, E. L. 1986. Seedling blight of longleaf pine caused by a binucleate *Rhizoctonia solani*–like fungus. Plant Dis. 70:148–150.

1067. Entry, J. A., Cromack, K. Jr., Hansen, E., and Waring, R. 1991. Response of western coniferous seedlings to infection by *Armillaria ostoyae* under limited light and nitrogen. Phytopathology 81:89–94.

1068. Entry, J. A., Martin, N. E., Kelsey, R. G., and Cromack, K. Jr. 1992. Chemical constituents in root bark of five species of western conifer saplings and infection by *Armillaria ostoyae*. Phytopathology 82:393–397.

1069. Epstein, A. H., and Hill, J. H. 1995. The biology of rose rosette disease: a mite-associated disease of uncertain aetiology. J. Phytopathol. 143:353–360.

1070. Epstein, A. H., and Hill, J. H. 1999. Status of rose rosette disease as a biological control for multiflora rose. Plant Dis. 83:92–101.

1071. Epstein, A. H., and Hill, J. H., eds. 1995. Proceedings of the international symposium: Rose rosette and other eriophyid mite–transmitted plant disease agents of uncertain etiology, Ames, IA, 1994. Iowa State Univ., Ames. 82 pp.

1072. Epstein, A. H., Hill, J. H., and Nutter, F. W. Jr. 1997. Augmentation of rose rosette disease for biocontrol of multiflora rose (*Rosa multiflora*). Weed Sci. 45:172–178.

1073. Erbaugh, D. K., Windham, M. T., Stodola, A. J. W., and Auge, R. M. 1995. Light intensity and drought stress as predisposition factors for dogwood anthracnose. J. Environ. Hortic. 13:186–189.

1074. Erbilgin, N., and Raffa, K. F. 2002. Association of declining red pine stands with reduced populations of bark beetle predators, seasonal increases in root colonizing insects, and incidence of root pathogens. For. Ecol. Manag. 164:221–236.

1075. Eriksson, J., Ryvarden, L., Hjortstam, K., and Larsson, K. H. 1973–1988. The Corticiaceae of North Europe. Vols. 1–8. Fungiflora, Oslo.

1076. Erwin, D. C., Bartnicki-Garcia, S., and Tsao, P. H., eds. 1983. *Phytophthora*: its biology, taxonomy, ecology, and pathology. Am. Phytopathol. Soc., St. Paul, MN. 392 pp.

1077. Erwin, D. C., and Ribeiro, O. K. 1996. Phytophthora diseases worldwide. APS Press, St. Paul, MN. 562 pp.

1078. Escobar, M. A., Leslie, C. A., McGranahan, G. H., and Dandekar, A. M. 2002. Silencing crown gall disease in walnut (*Juglans regia* L.). Plant Sci. 163:591–597.

1079. Eshed, N., and Dinoor, A. 1980. Genetics of pathogenicity in *Puccinia coronata:* pathogenic specialization at the host genus level. Phytopathology 70:1042–1046.

1080. Espejo, S. A., and Lopez, F. A. R. 1998. Current floristic and phytogeographic knowledge of Mexican Bromeliaceae. Rev. Biol. Trop. 46:493–513.

1081. Esser, K., and Bennett, J. W., eds. 1994–2003. The Mycota. Vols. 1–12. Springer, New York.

1082. Essig, F. M. 1922. The morphology, development, and economic aspects of *Schizophyllum commune* Fries. Univ. Calif. (Berkeley) Pub. Bot. 7:447–498.

1083. Etheridge, D. E. 1973. Wound parasites causing tree decay in British Columbia. Can. For. Serv. Pac. For. Res. Cent., For. Pest Leafl. 62. 15 pp.

1084. Etheridge, D. E., and Craig, H. M. 1976. Factors influencing infection and initiation of decay by the Indian paint fungus (*Echinodontium tinctorium*) in western hemlock. Can. J. For. Res. 6:299–318.

1085. Et-touil, K., Bernier, L., Beaulieu, J., Bérubé, J. A., Hopkin, A., and Hamelin, R. C. 1999. Genetic structure of *Cronartium ribicola* populations in eastern Canada. Phytopathology 89:915–919.

1086. Evans, H. C. 1984. The genus *Mycosphaerella* and its anamorphs *Cercoseptoria, Dothistroma,* and *Lecanosticta* on pines. Commonw. Mycol. Inst. Mycol. Pap. No. 153. 102 pp.

1087. Evans, K., Trudgill, D. L. and Webster, J. M., eds. 1993. Plant parasitic nematodes in temperate agriculture. CAB International, Wallingford, UK. 648 pp.

1088. Evans, L. S., Cocchiara, J., Jaklitsch, P., Feliciano, J., and Biesemeyer, P. T. 1999. Needle and stem injuries of *Picea rubens* (Pinaceae) in the Adirondack Mountains, USA. Environ. Exp. Bot. 41:267–280.

1089. Evidente, A., Sparapano, L., Fierro, O., Bruno, G., Giordano, F., and Motta, A. 1997. Sphaeropsidins B and C, phytotoxic pimarane diterpenes from *Sphaeropsis sapinea* f.sp. *cupressi* and *Diplodia mutila*. Phytochemistry 45:705–713.

1090. Evidente, A., Sparapano, L., Motta, A., Giordano, F., Fierro, O., and Frisullo, S. 1996. A phytotoxic pimarane diterpene of *Sphaeropsis sapinea* f.sp. *cupressi,* the pathogen of a canker disease of cypress. Phytochemistry 42:1541–1546.

1091. Ewers, F. W., McManus, P. S., Goldman, A., Gucci, R., and Fulbright, D. W. 1989. The effect of virulent and hypovirulent strains of *Endothia parasitica* on hydraulic conductance in American chestnut. Can. J. Bot. 67:1402–1407.

1092. Fairweather, M. L., and Gilbertson, R. L. 1992. *Inonotus andersonii:* a wood decay fungus of oak trees in Arizona. Pages 195–198 in: Ecology and management of oak and associated woodlands: perspectives in the southwestern United States and northern Mexico. USDA For. Serv. Gen. Tech. Rep. RM-218. 224 pp.

1093. Falk, S. P., Griffin, D. H., and Manion, P. D. 1989. Hypoxylon canker incidence and mortality in naturally occurring aspen clones. Plant Dis. 73:394–397.

1094. Fan, S., Blake, T. J., and Blumwald, E. 1994. The relative contribution of elastic and osmotic adjustments to turgor maintenance of woody species. Physiol. Plant. 90:408–413.

1095. Farley, J. D., Wilhelm, S., and Snyder, W. C. 1971. Repeated germination and sporulation of microsclerotia of *Verticillium albo-atrum* in soil. Phytopathology 61:260–264.

1096. Farquhar, M. L., and Peterson, R. L. 1989. Pathogenesis in Fusarium root rot of primary roots of *Pinus resinosa* grown in test tubes. Can. J. Plant Pathol. 11:221–228.

1097. Farr, D. F. 1991. Septoria species on *Cornus*. Mycologia 83:611–623.

1098. Farr, D. F. 1993. *Marssonina obclavata* sp. nov. with observations on *Marssonina* and *Septogloeum*. Mycologia 85:814–824.

1099. Farr, D. F., Bills, G. F., Chamuris, G. P., and Rossman, A. Y. 1989. Fungi on plants and plant products in the United States. APS Press, St. Paul, MN. 1252 pp.

1100. Farr, D. F., Castlebury, L. A., and Pardo-Schultheiss, R. A. 1999. *Phomopsis amygdali* causes peach shoot blight of cultivated peach trees in the southeastern United States. Mycologia 91:1008–1015.

1101. Farr, D. F., Castlebury, L. A., and Rossman, A. Y. 2002. Morphological and molecular characterization of *Phomopsis vaccinii* and additional isolates of *Phomopsis* from blueberry and cranberry in the eastern United States. Mycologia 94:494–504.

1102. Farr, D. F., Esteban, H. B., and Palm, M. E. 1996. Fungi on *Rhododendron:* a world reference. Parkway, Boone, NC. 192 pp.

1103. Faull, J. H. 1930. The spread and control of Phacidium blight in spruce plantations. J. Arnold Arbor. 11:136–147.

1104. Faull, J. H. 1934. The biology of Milesian rusts. J. Arnold Arbor. 15:50–85.

1105. Faull, J. H. 1938. *Pucciniastrum* on *Epilobium* and *Abies*. J. Arnold Arbor. 19:163–173.

1106. Faull, J. H. 1938. Taxonomy and geographical distribution of the genus *Uredinopsis*. Contrib. Arnold Arbor. No. 11. 120 pp.

1107. Faull, J. H. 1938. The biology of rusts of the genus *Uredinopsis*. J. Arnold Arbor. 19:402–436.

1108. Fayret, J. 1967. Action de la température et de la lumière sur la multiplication asexuée et de la reproduction sexuelle de *Gnomonia leptostyla* (Fr.) Cesat. et de Not., en culture pure. C. R. Hebd. Séances Acad. Sci. (Paris) D 265:1897–1900.

1109. Feather, T. V., Ohr, H. D., Munnecke, D. E., and Carpenter, J. B. 1989. The occurrence of *Fusarium oxysporum* on *Phoenix canariensis,* a potential danger to date production in California. Plant Dis. 73:78–80.

1110. Featherly, H. L. 1941. The effect of grapevines on trees. Proc. Okla. Acad. Sci. 21:61–62.

1111. Feeley, C. J., Hart, E. R., Thompson, J. R., and Harrington, T. C. 2001. Occurrence, associated symptoms, and potential insect vectors of the ash yellows phytoplasma in Iowa, U.S. J. Arboric. 27:331–339.

1112. Feiler, S., Tesche, M., Zentsch, W., Schmidt, P. A., and Bellmann, C. 1992. Investigations on water relations of spruce infected by *Armillaria* sp. Eur. J. For. Pathol. 22:329–336.

1113. Feistner, G., Korth, H., Budzikiewicz, H., and Pulverer, G. 1982. Rubrifacine from *Erwinia rubrifaciens*. Curr. Microbiol. 10:169–172.

1114. Felix, L. S., Uhrenholdt, B., and Parmeter, J. R. Jr. 1971. Association of *Scolytus ventralis* (Coleoptera: Scolytidae) and *Phoradendron bolleanum* subspecies *pauciflorum* on *Abies concolor*. Can. Entomol. 103:1697–1703.

1115. Fenoll, C. C., Grundler, F. M. W., and Ohl, S. A., eds. 1997. Cellular and molecular aspects of plant-nematode interactions. Kluwer Academic, Dordrecht. 286 pp.

1116. Fergus, C. L. 1954. An epiphytotic of Phyllosticta leaf spot of maple. Plant Dis. Rep. 38:678–679.

1117. Fergus, C. L. 1956. Some observations about *Polyporus dryadeus* on oak. Plant Dis. Rep. 40:827–829.

1118. Ferguson, A. J., and Jeffers, S. N. 1999. Detecting multiple species of *Phytophthora* in container mixes from ornamental crop nurseries. Plant Dis. 83:1129–1136.

1119. Fermaud, M., and Gaunt, R. E. 1995. *Thrips obscuratus* as a potential vector of *Botrytis cinerea* in kiwifruit. Mycol. Res. 99:267–273.

1120. Fernandez, C., Pinochet, J., and Felipe, A. 1993. Influence of temperature on the expression of resistance in six *Prunus* rootstocks infected with *Meloidogyne incognita*. Nematropica 23:195–202.

1121. Fernando, T., Jayasinghe, C. K., and Wijesundera, R. L. C. 2001. Cell wall degrading enzyme secretion by *Colletotrichum acutatum,* the causative fungus of secondary leaf fall of *Hevea brasiliensis*. Mycol. Res. 105:195–201.

1122. Ferreira, H., Gonçalves, E. R., Rodrigues, N. J., and Rosato, Y. B. 2000. Primers specific for *Xylella fastidiosa* based on RAPD differential fragments. Summa Phytopathol. 26:15–20.

1123. Ferrin, D. M., and Ramsdell, D. C. 1977. Ascospore dispersal and infection of grapes by *Guignardia bidwellii,* the causal agent of grape black rot disease. Phytopathology 67:1501–1505.

1124. Ferrin, D. M., and Ramsdell, D. C. 1978. Influence of conidia dispersal and environment on infection of grape by *Guignardia bidwellii*. Phytopathology 68:892–895.

1125. Ferris, M. A., Castello, J. D., and Sinclair, W. A. 1989. Effects of virus and mycoplasmalike organism infection on green and white ash. Phytopathology 79:579–583.

1126. Feucht, J. R. 1988. Herbicide injuries to trees—symptoms and solutions. J. Arboric. 14:215–219.

1127. Fiedling, N. J., and Evans, H. F. 1996. The pine wood nematode *Bursaphelenchus xylophilus* (Steiner and Buhrer) Nickle (=*B. lignicolus* Mamiya and Kiyohara): an assessment of the current position. Forestry 69:35–46.

1128. Fike, J., and Niering, W. A. 1999. Four decades of old field vegetation development and the role of *Celastrus orbiculatus* in the northeastern United States. J. Veg. Sci. 10:483–492.

1129. Filer, T. H. Jr. 1967. Pathogenicity of *Cytospora, Phomopsis,* and *Hypomyces* on *Populus deltoides*. Phytopathology 57:978–980.

1130. Filer, T. H. Jr. 1969. Sycamore canker caused by *Botryodiplodia theobromae*. Phytopathology 59:76–78.

1131. Filer, T. H. Jr., Davis, R. G., and Hegwood, C. P. 1979. Hardwood hosts for *Poria latemarginata*. (Abstr.) Phytopathology 69:527.

1132. Filho, E. S., and Dhingra, O. D. 1980. Survival of *Macrophomina phaseolina* sclerotia in nitrogen amended soils. Phytopathol. Z. 97:136–143.

1133. Filip, G. M., Bryant, L. D., and Parks, C. A. 1989. Mass movement of river ice causes severe tree wounds along the Grande Ronde River in northeastern Oregon. Northwest Sci. 63:211–213.

1134. Filip, G. M., Colbert, J. J., Shaw, C. G. III, Hessburg, P. F., and Hosman, K. P. 1993. Influence of dwarf mistletoe and western spruce budworm on growth and mortality of Douglas-fir in unmanaged stands. For. Sci. 39:465–477.

1135. Filip, G. M., Hadfield, J. S., and Schmitt, C. 1979. Branch mortality of true firs in west-central Oregon associated with dwarf mistletoe and canker fungi. Plant Dis. Rep. 63:189–193.

1136. Filip, G. M., and Schmitt, C. L. 1979. Susceptibility of native conifers to laminated root rot east of the Cascade range in Oregon and Washington. For. Sci. 25:261–265.

1137. Fillhart, R., Bachand, G. D., and Castello, J. D. 1998. Detection of infectious tobamoviruses in forest soils. Appl. Environ. Microbiol. 64:1430–1435.

1138. Finan, T. M., O'Brian, M. R., Layzell, D. B., Vessey, J. K., and Newton, W., eds. 2002. Nitrogen fixation: global perspectives. CABI Pub., Wallingford, UK, New York. 553 pp.

1139. Findley, D. A., Keever, G. J., Chappelka, A. H., Gilliam, C. H., and Eakes, D. J. 1997. Ozone sensitivity of selected southeastern landscape plants. J. Environ. Hortic. 15:51–55.

1140. Fineran, B. A., and Calvin, C. L. 2000. Transfer cells and flange cells in sinkers of the mistletoe *Phoradendron macrophyllum* (Viscaceae), and their novel combination. Protoplasma 211:76–93.

1141. Firrao, G., Andersen, M., Bertaccini, A., and 23 authors. 2004. 'Candidatus Phytoplasma,' a taxon for the wall-less, non-helical prokaryotes that colonize plant phloem and insects. Int. J. Syst. Evol. Microbiol. 54:1243–1255.

1142. Fischer, M. 1994. Pairing tests in the *Phellinus pini* group. Mycologia 86:524–539.

1143. Fischer, M. 1995. *Phellinus igniarius* and its closest relatives in Europe. Mycol. Res. 99:735–744.

1144. Fischer, M. 1996. Molecular and microscopical studies in the *Phellinus pini* group. Mycologia 88:230–238.

1145. Fisher, H. M., and Stone, E. L. 1990. Air-conducting porosity in slash pine roots from saturated soils. For. Sci. 36:18–33.

1146. Flachmann, M., Lesemann, D. E., Frenzel, B., and Koenig, R. 1990. Isometric virus-like particles in *Abies alba* Mill. and other *Abies* species: partial purification and improved detection by means of immunoelectron microscopy. J. Phytopathol. 129:193–202.

1147. Flack, N. J., and Swinburne, T. R. 1977. Host range of *Nectria galligena* Bres. and the pathogenicity of some Northern Ireland isolates. Trans. Br. Mycol. Soc. 68:185–192.

1148. Flaishman, M. A., and Kolattukudy, P. E. 1994. Timing of fungal invasion using host's ripening hormone as a signal. Proc. Natl. Acad. Sci. USA 91:6579–6583.

1149. Fleischmann, F., Schneider, D., Matyssek, R., and Osswald, W. F. 2002. Investigations on net CO_2 assimilation, transpiration and root growth of *Fagus sylvatica* infested with four different *Phytophthora* species. Plant Biol. 4:144–152.

1150. Flood, J., Bridge, P. D., and Holderness, M., eds. 2000. *Ganoderma* diseases of perennial crops. CABI Pub., Wallingford, UK. 275 pp.

1151. Florance, E. R., and Shaw, C. G. III. 1988. Surface morphology of basidiospores from decay fungi that are common in Pacific Northwest forests. Northwest Sci. 62:233–241.

1152. Florida Department of Agriculture & Consumer Services, Division of Plant Industry. 1962–present. Nematology Circulars 1–.

1153. Florida Department of Agriculture & Consumer Services, Division of Plant Industry. 1962–present. Plant Pathology Circulars 1–.

1154. Flott, J. J., and Gilbertson, R. L. 1991. Cultural studies of four North American species of *Perenniporia* (Aphyllophorales: Polyporaceae). Mycol. Res. 95:1113–1122.

1155. Flückiger, W., and Braun, S. 1981. Perspectives of reducing the deleterious effect of deicing salt upon vegetation. Plant Soil 63:527–529.

1156. Flynn, P. H., and Gleason, M. L. 1993. Isolation of *Botryosphaeria stevensii*, cause of Botryosphaeria canker, from Rocky Mountain juniper in Iowa. Plant Dis. 77:210.

1157. Ford, R. E., Moline, H. E., McDaniel, G. L., Mayhew, D. E., and Epstein, A. H. 1972. Discovery and characterization of elm mosaic virus in Iowa. Phytopathology 62:987–992.

1158. Forer, L. B., Powell, C. A., and Stouffer, R. F. 1984. Transmission of tomato ringspot virus to apple rootstock cuttings and to cherry and peach seedlings by *Xiphinema rivesi*. Plant Dis. 68:1052–1054.

1159. Forster, R. L. S., and Milne, K. S. 1976. Daphne virus Y: a potyvirus from daphne. N.Z. J. Agric. Res. 19:359–371.

1160. Forsyth, J., and Maynard, J. 1969. The sensitivity of ornamental plants to insecticides and acaricides. Commonw. Bur. Hortic. Plantation Crops, Hortic. Rev. No. 1. 66 pp.

1161. Foster, R. E., and Foster, A. T. 1951. Studies in forest pathology. VIII. Decay of western hemlock on the Queen Charlotte Islands, British Columbia. Can. J. Bot. 29:479–521.

1162. Fourie, J. F., and Holz, G. 1995. Initial infection processes by *Botrytis cinerea* on nectarine and plum fruit and the development of decay. Phytopathology 85:82–87.

1163. Fowler, M. E. 1947. Glomerella leaf spot of *Magnolia*. Plant Dis. Rep. 31:298.

1164. Fowler, M. E., and Berry, F. H. 1958. Blossom-end rot of Chinese chestnuts. Plant Dis. Rep. 42:91–96.

1165. Fox, J. W., Wood, D. L., Koehler, C. S., and O'Keefe, S. T. 1991. Engraver beetles (Scolytidae: *Ips* species) as vectors of the pitch canker fungus, *Fusarium subglutinans*. Can. Entomol. 123:1355–1367.

1166. Fox, R. T. V., ed. 2000. Armillaria root rot: biology and control of honey fungus. Intercept, Andover, UK. 222 pp.

1167. Fox, S., and Mickler, R. A., eds. 1996. Impact of air pollutants on southern pine forests. Ecol. Stud. 118. Springer-Verlag, New York. 513 pp.

1168. Fraedrich, S. W., Miller, T., and Zarnoch, S. J. 1994. Factors affecting the incidence of black seed rot in slash pine. Can. J. For. Res. 24:1717–1725.

1169. Franceschi, V. R., Krokene, P., Krekling, T., and Christiansen, E. 2000. Phloem parenchyma cells are involved in local and distant defense responses to fungal inoculation or bark-beetle attack in Norway spruce (Pinaceae). Am. J. Bot. 87:314–326.

1170. Francki, R. I. B., Milne, R. G., and Hatta, T. 1985. Atlas of plant viruses. CRC Press, Boca Raton, FL. Vols. 1, 2.

1171. Francki, R. I. B., Regenmortel, M. H. V. van, Fraenkel-Conrat, H., Koenig, R., Milne, R. G., Harrison, B. D., and Murant, A. F., eds. 1985. The plant viruses. Vols. 1–5. Plenum, New York.

1172. Francois, L. E. 1982. Salt tolerance of eight ornamental tree species. J. Am. Soc. Hortic. Sci. 107:66–68.

1173. Francois, L. E., and Clark, R. A. 1979. Boron tolerance of twenty-five ornamental shrub species. J. Am. Soc. Hortic. Sci. 104:319–322.

1174. Franich, R. A., Carson, M. J., and Carson, S. D. 1986. Synthesis and accumulation of benzoic acid in *Pinus radiata* needles in response to tissue injury by dothistromin, and correlation with resistance of *P. radiata* families to *Dothistroma pini*. Physiol. Mol. Plant Pathol. 28:267–286.

1175. Frankie, G. W., and Parmeter, J. R. Jr. 1972. A preliminary study of the relationship between *Coryneum cardinale* (Fungi Imperfecti) and *Laspeyresia cupressana* (Lepidoptera: Tortricidae). Plant Dis. Rep. 56:992–994.

1176. Franz, F., Grotjahn, R., and Acker, G. 1993. Identification of *Naemacyclus minor* hyphae within needle tissues of *Pinus sylvestris* by immunoelectron microscopy. Arch. Microbiol. 160:265–272.

1177. Fraser, W. P. 1914. Notes on *Uredinopsis mirabilis* and other rusts. Mycologia 6:25–28.

1178. Free, S. J., Holtz, B. A., and Michailides, T. J. 1996. Mating behavior in field populations of *Monilinia fructicola*. Mycologia 88:208–211.

1179. Freeman, S., Minz, D., Maymon, M., and Zveibil, A. 2001. Genetic diversity within *Colletotrichum acutatum* sensu Simmonds. Phytopathology 91:586–592.

1180. Freer-Smith, P. H. 1984. The responses of six broadleaved trees during long-term exposure to SO_2 and NO_2. New Phytol. 97:49–61.

1181. Freier, G. D. 1977. Lightning and trees. J. Arboric. 3:131–137.

1182. Freitag, J. H. 1951. Host range of the Pierce's disease virus of grapes as determined by insect transmission. Phytopathology 41:920–934.

1183. French, D. W., and Bergdahl, D. R. 1983. Phomopsis canker of 'Robusta' poplar. J. Arboric. 9:151–152.

1184. French, D. W., Hodges, C. S. Jr., and Froyd, J. D. 1969. Pathogenicity and taxonomy of *Hypoxylon mammatum*. Can. J. Bot. 47:223–226.

1185. French, D. W., and Menge, J. A. 1978. Survival of *Cylindrocladium floridanum* in naturally and artificially infested forest tree nurseries. Plant Dis. Rep. 62:806–810.

1186. French, D. W., and Schroeder, D. B. 1969. Oak wilt fungus, *Ceratocystis fagacearum*, as a selective silvicide. For. Sci. 15:198–203.

1187. French, D. W., Smeltzer, D. L. K., and Anderson, N. A. 1983. Effect of stem rust fungi on 45-yr-old jack pine. Plant Dis. 67:26–28.

1188. French, D. W., and Stienstra, W. C. 1975. Oak wilt disease. Univ. Minn. Exten. Folder 310. 6 pp.

1189. French, W. J. 1969. Eutypella canker on *Acer* in New York. N.Y. State Coll. For. Tech. Pub. 94. 56 pp.

1190. Freyer, K. 1976. Untersuchungen zur Biologie, Morphologie und Verbreitung von *Herpotrichia parasitica* (Hartig) E. Rostrup (formals *Trichosphaeria parasitica* Hartig). I. Verbreitung und Morphologie. II. Infektionsweg, Wirtsspektrum und Kulturverhalten. Eur. J. For. Pathol. 6:152–166, 222–238.

1191. Fridlund, P. R., ed. 1989. Virus and viruslike diseases of pome fruits and simulating noninfectious disorders. Wash. State Univ. Coop. Exten. SP0003. 330 pp.

1192. Friedland, A. J., Gregory, R. A., Kärenlampi, L., and Johnson, A. H. 1984. Winter damage to foliage as a factor in red spruce decline. Can. J. For. Res. 14:963–965.

1193. Frisina, T. A., and Benson, D. M. 1987. Characterization and pathogenicity of binucleate *Rhizoctonia* spp. from azaleas and other woody ornamental plants with web blight. Plant Dis. 71:977–981.

1194. Frisullo, S., Bruno, G., and Sparapano, L. 1999. A foot and root rot of young olive trees caused by *Thielaviopsis basicola*. Petria 9:249–257.

1195. Frisullo, S., Camele, I., Carlucci, A., and Lops, F. 2000. *Botryosphaeria* and *Botryosphaeria*-like micromycetes on declining oaks in Apulia and Basilicata. Petria 10:1–10.

1196. Froelich, R. C., Cowling, E. B., Collicott, L. V., and Dell, R. R. 1977. *Fomes annosus* reduces height and diameter growth of planted slash pine. For. Sci. 23:299–306.

1197. Froelich, R. C., and Snow, G. A. 1986. Predicting site hazard to fusiform rust. For. Sci. 32:21–35.

1198. Fröhlich, J., and Hyde, K. D. 1995. *Maculatipalma fronsicola* gen. et sp. nov. causing leaf spots on palm species in north Queensland with descriptions of related genera: *Apioplagiostoma* and *Plagiostoma*. Mycol. Res. 99:727–734.

1199. Froidevaux, L., and Müller, E. 1972. *Anisogramma virgultorum* (Fr.) Theiss. & Syd., a pathogenic ascomycete on *Betula pubescens* Ehrhart. Eur. J. For. Pathol. 2:185–187.

1200. Fromme, F. D. 1928. The black rootrot disease of apple. Va. Agric. Exp. Stn. Tech. Bull. 34. 52 pp.

1201. Frontz, T. M., Davis, D. D., Bunyard, B. A., and Royse, D. J. 1998. Identification of *Armillaria* species isolated from bigtooth aspen based on rDNA RFLP analysis. Can. J. For. Res. 28:141–149.

1202. Frye, J., and Grosse, W. 1992. Growth responses to flooding and recovery of deciduous trees. Z. Naturforsch. Sec. C, Biosci. 47:683–689.

1203. Fucikovsky, L. 1994. Tumour caused by *Sphaeropsis tumefaciens* on avocado. Rev. Mex. Micol. 10:81–185.

1204. Fuhrling, M., and Buttner, C. 1998. Detection of tobamoviruses in sycamore maple (*Acer pseudoplatanus*) exhibiting mottling and leaf deformation. Forstwissensch. Centralbl. 117:92–97.

1205. Fujikawa, S., Jitsuyama, Y., and Kuroda, K. 1999. Determination of the role of cold acclimation–induced diverse changes in plant cells from the viewpoint of avoidance of freezing injury. J. Plant Res. 112:237–244.

1206. Fukushi, T. 1921. A willow-canker disease caused by *Physalospora miyabeana* and its conidial form *Gloeosporium*. Ann. Phytopathol. Soc. Jap. 1(4):1–12.

1207. Fulbright, D. 1999. Chestnut blight and hypovirulence. Plant-Microbe Interact. 4:57–79.

1208. Fulton, J. P., and Fulton, R. W. 1970. A comparison of some properties of elm mosaic and tomato ringspot viruses. Phytopathology 60:114–115.

1209. Funk, A. 1968. *Diaporthe lokoyae* n.sp., the perfect state of *Phomopsis lokoyae*. Can. J. Bot. 46:601–603.

1210. Funk, A. 1969. *Potebniamyces* (*Phacidiella*) disease of true firs in British Columbia. Can. J. Bot. 47:751–753.

1211. Funk, A. 1970. Taxonomy of *Phomopsis boycei* and its relationship to *Potebniamyces balsamicola*. Can. J. Bot. 48:1023–1025.

1212. Funk, A. 1972. Sirococcus shoot-blight of western hemlock in British Columbia and Alaska. Plant Dis. Rep. 56:645–647.

1213. Funk, A. 1981. Parasitic microfungi of western trees. Can. For. Serv. Pac. For. Res. Cent. BC-X-222. 190 pp.

1214. Funk, A. 1985. Foliar fungi of western trees. Can. For. Serv. Pac. For. Res. Cent. BC-X-265. 159 pp.

1215. Funk, A. 1989. Observations on an aspen leaf spot disease and associated fungus, *Pollaccia borealis*. Can. J. Plant Pathol. 11:353–356.

1216. Furnier, G. R., Stolz, A. M., Mustaph, R. M., and Ostry, M. E. 1999. Genetic evidence that butternut canker was recently introduced into North America. Can. J. Bot. 77:783–785.

1217. Furuhashi, K., Tada, Y., Okamoto, K., Sugai, M., Kubota, M., and Watanabe, M. 1997. Phytochrome participation in induction of haustoria in *Cuscuta japonica*, a holoparasitic flowering plant. Plant Cell Physiol. 38:935–940.

1218. Futai, K., Togashi, K., and Ikeda, T., eds. 1999. Sustainability of pine forests in relation to pine wilt and decline. Proceedings of international symposium, Tokyo, 1998. Shokado, Tokyo. 336 pp.

1219. Gabriel, W. L. 2000. The actinorhizal symbiosis. J. Plant Growth Regul. 19:167–182.

1220. Gadgil, P. D. 1984. Cyclaneusma (Naemacyclus) needle-cast of *Pinus radiata* in New Zealand. 1: Biology of *Cyclaneusma minus*. N.Z. J. For. Sci. 14:179–196.

1221. Gadoury, D. M., and Pearson, R. C. 1991. Heterothallism and pathogenic specialization in *Uncinula necator*. Phytopathology 81:1287–1293.

1222. Gailhofer, M., Thaler, I., and Miličić, D. 1988. Occurrence of camellia leaf yellow mottle virus (CLYMV) on east Adriatic coast. Acta Hortic. No. 234:385–391.

1223. Gallagher, P. W., and Snydor, T. D. 1983. Variation in wound response among cultivars of red maple. J. Am. Soc. Hortic. Sci. 108:744–746.

1224. Gallego, F. J., Perez de Algaba, A., and Fernandez, E. R. 1999. Etiology of oak decline in Spain. Eur. J. For. Pathol. 29:17–27.

1225. Galun, M., ed. 1988. Handbook of lichenology. Vols. 1–3. CRC Press, Boca Raton, FL. 297, 181, and 147 pp.

1226. Gams, W. 1971. *Cephalosporium*-artige Schimmelpilze (Hyphomycetes). Gustav Fischer, Jena. 262 pp.

1227. Gao, M., and Chamuris, G. P. 1993. Microstructural and histochemical changes in *Acer platanoides* rhytidome caused by *Dendrothele acerina* (Aphyllophorales) and *Mycena meliigena* (Agaricales). Mycologia 85:987–995.

1228. Gao, S., and Shain, L. 1995. Activity of polygalacturonase produced by *Cryphonectria parasitica* in chestnut bark and its inhibition by extracts from American and Chinese chestnut. Physiol. Mol. Plant Pathol. 46:199–213.

1229. Gao, S., and Shain, L. 1995. Effects of water stress on chestnut blight. Can. J. For. Res. 25:1030–1035.

1230. Garbelotto, M., Cobb, F. W., Bruns, T. D., Otrosina, W. J., Popenuck, T., and Slaughter, G. 1999. Genetic structure of *Heterobasidion annosum* in white fir mortality centers in California. Phytopathology 89:546–554.

1231. Garbelotto, M., Davidson, J. M., Ivors, K., Maloney, P. E., Hüberle, D., Koike, S. T., and Rizzo, D. M. 2003. Bay laurel and native plants other than oaks are the main hosts for the sudden oak death pathogen, *P. ramorum*, in California. APSnet feature story, April 2003. Am. Phytopathol. Soc., St. Paul, MN. Internet pub.

1232. Garbelotto, M., Lee, H. K., Slaughter, G., Popenuck, T., Cobb, F. W., and Bruns, T. D. 1997. Heterokaryosis is not required for virulence of *Heterobasidion annosum*. Mycologia 89:92–102.

1233. Garbelotto, M., Otrosina, W. J., Cobb, F. W., and Bruns, T. D. 1998. The European S and F interstility groups of *Heterobasidion annosum* may represent sympatric protospecies. Can. J. Bot. 76:397–409.

1234. Garbutt, R. 1996. Foliage diseases in western larch in British Columbia. Can. For. Serv. Pac. For. Cent. For. Pest Leafl. 71. 4 pp.

1235. Garcia, S. M., and Jones, A. L. 1993. Influence of temperature on apothecial development and ascospore discharge by *Blumeriella jaapii*. Plant Dis. 77:776–779.

1236. Gardiner, E. S., and Hodges, J. D. 1996. Physiological, morphological and growth responses to rhizosphere hypoxia by seedlings of North American bottomland oaks. Ann. Sci. For. 53:303–316.

1237. Gardner, J. M., Feldman, A. W., and Stamper, D. H. 1983. Role and fate of bacteria in vascular occlusions of citrus. Physiol. Plant Pathol. 23:295–309.

1238. Gardner, J. M., and Kado, C. I. 1973. Evidence for systemic movement of *Erwinia rubrifaciens* in Persian walnuts by the use of double-antibiotic markers. Phytopathology 63:1085–1086.

1239. Gardner, M. W., Yarwood, C. E., and Kuafala, T. 1972. Oak mildews. Plant Dis. Rep. 56:313–317.

1240. Garrec, J. P., and Plébin, R. 1981. Étude de la relation entre la pluviosité et l'accumulation du fluor dans les forêts résineuses soumises à une pollution fluorée. Eur. J. For. Pathol. 11:129–136.

1241. Garrett, C. M. E. 1987. The effect of crown gall on growth of cherry trees. Plant Pathol. 36:339–345.

1242. Garrett, P. W., Randall, W. K., Shigo, A. L., and Shortle, W. C. 1979. Inheritance of compartmentalization of wounds in sweetgum (*Liquidambar styraciflua* L.) and eastern cottonwood (*Populus deltoides* Bartr.). USDA For. Serv. Res. Pap. NE-433. 4 pp.

1243. Garrett, P. W., and Trew, F. 1986. Resistance of pitch × loblolly pine hybrids to fusiform rust (*Cronartium quercuum* f.sp. *fusiforme*). Plant Dis. 70:564–565.

1244. Garsed, S. G., and Rutter, A. J. 1982. Relative performance of conifer populations in various tests for sensitivity to SO_2, and the implications for selecting trees for planting in polluted areas. New Phytol. 92:349–367.

1245. Gaudriault, S., Paulin, J. P., and Barny, M. A. 2002. The DspB/F protein of *Erwinia amylovora* is a type III secretion chaperone ensuring efficient intrabacterial production of the Hrp-secreted DspA/E pathogenicity factor. Mol. Plant Pathol. 3:313–320.

1246. Gäumann, E. 1959. Die Rostpilze Mitteleuropas. Beitr. Kryptogamenflora Schweiz. Vol. 12. 1407 pp.

1247. Gedalovich, E., Kuijt, J., and Carpita, N. C. 1988. Chemical composition of viscin, an adhesive involved in dispersal of the parasite *Phoradendron californicum* (Viscaceae). Physiol. Mol. Plant Pathol. 32:61–76.

1248. Gedalovich-Shedletzky, E., Delmer, D. P., and Kuijt, J. 1989. Chemical composition of viscin mucilage from three mistletoe species—a comparison. Ann. Bot. 64:249–252.

1249. Gehring, C. A., and Whitham, T. G. 1992. Reduced mycorrhizae on *Juniperus monosperma* with mistletoe: the influence of environmental stress and tree gender on a plant parasite and a plant-fungal mutualism. Oecologia 89:298–303.

1250. Geils, B. W., and Jacobi, W. R. 1990. Development of comandra blister rust on lodgepole pine. Can. J. For. Res. 20:159–165.

1251. Geils, B. W., and Jacobi, W. R. 1993. Effects of comandra blister rust on growth and survival of lodgepole pine. Phytopathology 83:638–644.

1252. Geils, B. W., and Mathiasen, R. L. 1990. Intensification of dwarf mistletoe on southwestern Douglas-fir. For. Sci. 36:955–969.

1253. Geils, B. W., Tovar, J. C., and Moody, B., tech. coords. 2002. Mistletoes of North American conifers. USDA For. Serv. Gen. Tech. Rep. RMRS-GTR-98. 123 pp.

1254. Gennaro, M., Gonthier, P., Nicolotti, G., and Cellerino, G. P. 2001. First report of *Tubakia dryina* in buds and shoots of *Quercus cerris* and *Quercus robur*. Plant Dis. 85:1289.

1255. Gentit, P., Foissac, X., Svanella-Dumas, L., Peypelut, M., Macquaire, G., and Candresse, T. 2002. Molecular characterization of foveaviruses associated with the cherry necrotic mottle leaf disease and complete sequencing of an European isolate of cherry green ring mottle virus. Arch. Virol. 147:1033–1042.

1256. Gerettson, C. L. 1989. A compendium and classification of the species of the genus *Phytophthora* de Bary by the canons of traditional taxonomy. For. Commiss. New South Wales Tech. Pap. 45. 103 pp.

1257. Gergerich, R. C., and Kim, K. S. 1983. A description of the causal agent of rose rosette disease. Ark. Farm Res. 32:7.

1258. Gerhold, H. D., Schreiner, E. J., McDermott, R. E., and Winieski, J. A., eds. 1966. Breeding pest-resistant trees. Pergamon, New York. 505 pp.

1259. Gerlach, W. W. P., Hoitink, H. A. J., and Ellett, C. W. 1974. Shoot blight and stem dieback of *Pieris japonica* caused by *Phytophthora citricola*, *P. citrophthora*, and *Botryosphaeria dothidea*. Phytopathology 64:1368–1370.

1260. Germain, H., Laflamme, G., Bernier, L., Boulet, B., and Hamelin, R. C. 2002. DNA polymorphism and molecular diagnosis in *Inonotus* spp. Can. J. Plant Pathol. 24:194–199.

1261. Gernandt, D. S., Camacho, F. J., and Stone, J. K. 1997. *Meria laricis*, an anamorph of *Rhabdocline*. Mycologia 89:735–744.

1262. Gernandt, D. S., Platt, J. L., Stone, J. K., Spatafora, J. W., Holst-Jensen, A., Hamelin, R. C., and Kohn, L. M. 2001. Phylogenetics of Helotiales and Rhytismatales based on partial small subunit nuclear ribosomal DNA sequences. Mycologia 93:915–933.

1263. Gianinazzi, P. V. 1996. Plant cell responses to arbuscular mycorrhizal fungi: getting to the roots of the symbiosis. Plant Cell 8:1871–1883.

1264. Gibbs, J. N. 1980. Survival of *Ceratocystis fagacearum* in branches of trees killed by oak wilt in Minnesota. Eur. J. For. Pathol. 10:218–224.

1265. Gibbs, J. N., and French, D. W. 1980. The transmission of oak wilt. USDA For. Serv. Res. Pap. NC-185. 17 pp.

1266. Gibbs, J. N., and Greig, B. J. W. 1990. Survey of parkland trees after the great storm of October 16, 1987. Arboric. J. 14:321–347.

1267. Gibbs, J. N., and Palmer, C. A. 1994. A survey of damage to roadside trees in London caused by the application of de-icing salt during the 1990/91 winter. Arboric. J. 18:321–343.

1268. Gibbs, J. N., and Reffold, T. C. 1982. *Gnomonia platani* and bark killing of London plane. Eur. J. For. Pathol. 12:395–398.

1269. Gibson, I. A. S. 1972. Dothistroma blight of *Pinus radiata*. Annu. Rev. Phytopathol. 10:51–72.

1270. Gibson, I. A. S. 1974. Impact and control of Dothistroma blight of pines. Eur. J. For. Pathol. 4:89–100.

1271. Gibson, I. A. S., and Corbett, D. C. M. 1964. Variation in isolates from Armillaria root disease in Nyasaland. Phytopathology 54:122–123.

1272. Giese, R. L., Houston, D. R., Benjamin, D. M., Kuntz, J. E., and Skilling, D. D. 1964. Studies of maple blight. Wisc. Agric. Exp. Stn. Res. Bull. 250. 129 pp.

1273. Gilbert, J., and Punter, D. 1990. Release and dispersal of pollen from dwarf mistletoe on jack pine in Manitoba in relation to microclimate. Can. J. For. Res. 20:267–273.

1274. Gilbert, J., and Punter, D. 1991. Germination of pollen of the dwarf mistletoe *Arceuthobium americanum*. Can. J. Bot. 69:685–688.

1275. Gilbert, O. L. 2000. Lichens. HarperCollins, London. 288 pp.

1276. Gilbertson, R. L. 1976. The genus *Inonotus* (Aphylloporales: Hymenochaetaceae) in Arizona. Mem. N.Y. Bot. Gard. 28:67–85.

1277. Gilbertson, R. L. 1979. The genus *Phellinus* (Aphyllophorales, Hymenochaetaceae) in western North America. Mycotaxon 9:51–89.

1278. Gilbertson, R. L. 1981. North American wood-rotting fungi that cause brown rots. Mycotaxon 12:372–416.

1279. Gilbertson, R. L., and Blackwell, M. 1984. Two new basidiomycetes on living live oak in the Southeast and Gulf Coast region. Mycotaxon 20:85–93.

1280. Gilbertson, R. L., and Ryvarden, L. 1986, 1987. North American polypores. Vol. 1: *Albatrellus–Lindtneria*. Vol. 2: *Megasporaporia–Wrightoporia*. Fungiflora, Oslo. 885 pp.

1281. Gildow, F. E., and Moorman, G. W. 2002. Plum pox virus (PPV) on ornamentals. Penn. State Univ. Coop. Exten. Plant Dis. Facts. 6 pp. Internet pub.

1282. Giles, K. L., and Atherly, A. G., eds. 1981. Biology of the Rhizobiaceae. Int. Rev. Cytol. Suppl. 13. 368 pp.

1283. Gill, A. M. 1974. Toward an understanding of fire-scar formation: field observation and laboratory simulation. For. Sci. 20:198–205.

1284. Gill, C. J. 1970. The flooding tolerance of woody species—a review. For. Abstr. 31:671–688.

1285. Gill, D. L. 1958. Effect of root-knot nematodes on Fusarium wilt of mimosa. Plant Dis. Rep. 42:587–590.

1286. Gill, D. L. 1967. Fusarium wilt infection of apparently healthy mimosa trees. Plant Dis. 51:148–150.

1287. Gill, D. L. 1968. Mimosa-wilt *Fusarium* carried in seed. Plant Dis. Rep. 52:949–951.

1288. Gill, D. L. 1979. 'Union' mimosa. HortScience 14:644.

1289. Gill, D. L., Alfieri, S. A. Jr., and Sobers, E. K. 1971. A new leaf disease of *Ilex* spp. caused by *Cylindrocladium avesiculatum* sp. nov. Phytopathology 61:58–60.

1290. Gill, D. L., and Sobers, E. K. 1974. Control of *Cercospora* sp. leafspot of *Ligustrum japonicum*. Plant Dis. Rep. 58:1015–1017.

1291. Giller, K. E. 2001. Nitrogen fixation in tropical cropping systems. 2nd ed. CABI Pub., New York. 423 pp.

1292. Gilliam, C. H., and Smith, E. M. 1980. Sources and symptoms of boron toxicity in container grown woody ornamentals. J. Arboric. 6:209–212.

1293. Gilliam, C. H., Smith, E. M., Still, S. M., and Sheppard, W. J. 1981. Treating boron toxicity in *Rhododendron catawbiense*. HortScience 16:764–765.

1294. Gilman, J. C., and McNew, G. L. 1940. Fungi associated with tree cankers in Iowa. II. *Diaporthe*, *Apioporthe*, *Cryptodiaporthe*, *Pseudovalsa* and their related conidial forms. Iowa State Coll. J. Sci. 14:129–153.

1295. Gilman, J. C., and Wadley, B. N. 1952. The ascigerous stage of *Septoria querceti* Theum. Mycologia 44:216–220.

1296. Gilmer, R. M. 1958. Two viruses that induce mosaic of apple. Phytopathology 48:432–434.

1297. Ginns, J. H. 1985. *Hericium* in North America: cultural characteristics and mating behavior. Can. J. Bot. 63:1551–1563.

1298. Ginns, J. H. 1986. A compendium of plant disease and decay fungi in Canada. Can. Gov. Pub. Cent., Ottawa. 416 pp.

1299. Ginns, J. H., and Lefebvre, M. N. L. 1993. Lignicolous, corticioid fungi (Basidiomycota) of North America: systematics, distribution, and ecology. APS Press, St. Paul, MN. 247 pp.

1300. Gitzendanner, M. A., White, E. E., Foord, B. M., Dupper, G. E., Hodgskiss, P. D., and Kinloch, B. B. Jr. 1996. Genetics of *Cronartium ribicola*. III. Mating system. Can. J. Bot. 74:1852–1859.

1301. Glawe, D. A. 1983. Observations on the anamorph of *Eutypella parasitica*. Mycologia 75:742–743.

1302. Glawe, D. A., and Crane, J. L. 1987. Illinois fungi. 13. *Tubakia dryina*. Mycotaxon 29:101–112.

1303. Glawe, D. A., and Jacobs, K. A. 1987. Taxonomic notes on *Eutypella vitis*, *Cryptosphaeria populina*, and *Diatrype stigma*. Mycologia 79:135–139.

1304. Glerum, C., and Farrar, J. L. 1966. Frost ring formation in the stems of some coniferous species. Can. J. Bot. 44:879–886.

1305. Gloyer, W. O. 1921. Blister canker of apple and its control. N.Y. Agric. Exp. Stn. Bull. 485. 100 pp.

1306. Godbold, D. L., Jentschke, G., Wintr, S., and Marschner, P. 1998. Ectomycorrhizas and amelioration of metal stress in forest trees. Chemosphere 36:757–762.

1307. Goethals, K., Vereecke, D., Jaziri, M., Montague, M. Van, and Holsters, M. 2001. Leafy gall formation by *Rhodococcus fascians*. Annu. Rev. Phytopathol. 39:27–52.

1308. Goheen, D. J., Cobb, F. W. Jr., and McKibbin, G. N. 1978. Influence of soil moisture on infection of ponderosa pine by *Verticicladiella wageneri*. Phytopathology 68:913–916.

1309. Goheen, D. J., and Filip. G. M. 1980. Root pathogen complexes in Pacific Northwest forests. Plant Dis. 64:793–794.

1310. Golino, D. A., and Oldfield, G. N. 1990. Plant pathogenic spiroplasmas and their leafhopper vectors. Adv. Dis. Vector Res. 6:267–299.

1311. Gonsález, F. S. M., and Rogers, J. D. 1993. *Biscogniauxia* and *Camillea* in Mexico. Mycotaxon 46:229–258.

1312. Gonsalves, D. 1998. Control of papaya ringspot virus in papaya: a case study. Annu. Rev. Phytopathol. 36:415–437.

1313. Gonzalez, D., Carling, D. E., Kuninaga, S., Vilgalys, R., and Cubeta, M. A. 2001. Ribosomal DNA systematics of *Ceratobasidium* and *Thanatephorus* with *Rhizoctonia* anamorphs. Mycologia 93:1138–1150.

1314. Good, H. M., and Nelson, C. D. 1951. A histological study of sugar maple decayed by *Polyporus glomeratus* Peck. Can. J. Bot. 29:215–223.

1315. Good, H. M., and Nelson, J. I. 1962. Fungi associated with *Fomes igniarius* var. *populinus* in living poplar trees and their probable significance in decay. Can. J. Bot. 40:615–624.

1316. Goodwin, P. H., DeVay, J. E., and Meredith, C. P. 1988. Roles of water stress and phytotoxins in the development of Pierce's disease of the grapevine. Physiol. Mol. Plant Pathol. 32:1–15.

1317. Goodwin, P. H., and Zhang, S. 1997. Distribution of *Xylella fastidiosa* in southern Ontario as determined by the polymerase chain reaction. Can. J. Plant Pathol. 19:13–18.

1318. Gordon, A. G., and Gorham, E. 1963. Ecological aspects of air pollution from an iron-sintering plant at Wawa, Ontario. Can. J. Bot. 41:1063–1078.

1319. Gordon, J. C., and Wheeler, C. T., eds. 1983. Biological nitrogen fixation in forest ecosystems: foundations and applications. M. Nijhoff/W. Junk, The Hague. 342 pp.

1320. Gordon, T. R., and Martyn, R. D. 1997. The evolutionary biology of *Fusarium oxysporum*. Annu. Rev. Phytopathol. 35:111–128.

1321. Gordon, T. R., Okamoto, D., Storer, A. J., and Wood, D. L. 1998. Susceptibility of five landscape pines to pitch canker disease, caused by *Fusarium subglutinans* f.sp. *pini*. HortScience 33:868–871.

1322. Gordon, T. R., Storer, A. J., and Wood, D. L. 2001. The pitch canker epidemic in California. Plant Dis. 85:1128–1139.

1323. Gordon, T. R., Wikler, K. R., Clark, S. L., Okamoto, D., Storer, A. J., and Bonello, P. 1998. Resistance to pitch canker disease, caused by *Fusarium subglutinans* f.sp. *pini*, in Monterey pine (*Pinus radiata*). Plant Pathol. 47:706–711.

1324. Goss, R. W., and Frink, P. R. 1934. Cephalosporium wilt and dieback of the white elm. Nebr. Agric. Exp. Stn. Res. Bull. 70. 24 pp.

1325. Gosselin, L., Jobidon, R., and Bernier, L. 1999. Genetic variability and structure of Canadian populations of *Chondrostereum purpureum*, a potential biophytocide. Mol. Ecol. 8:113–122.

1326. Gotlieb, A. R., and Berbee, J. G. 1973. Line pattern of birch caused by apple mosaic virus. Phytopathology 63:1470–1477.

1327. Goto, M. 1992. Fundamentals of bacterial plant pathology. Academic Press, San Diego. 342 pp.

1328. Gottlieb, A. M., Ferrer, E., and Wright, J. E. 2000. rDNA analyses as an aid to the taxonomy of species of *Ganoderma*. Mycol. Res. 104:1033–1045.

1329. Gottlieb, A. M., and Wright, J. E. 1999. Taxonomy of *Ganoderma* from southern South America: subgenus *Ganoderma*. Mycol. Res. 103:661–673.

1330. Gottwald, T. R., and Cameron, H. R. 1979. Studies in the morphology and life history of *Anisogramma anomala*. Mycologia 71:1107–1126.

1331. Gottwald, T. R., Graham, J. H., and Egel, D. S. 1992. Analysis of foci of Asiatic citrus canker in a Florida citrus orchard. Plant Dis. 76:389–396.

1332. Gottwald, T. R., Graham, J. H., and Schubert, T. S. 1997. An epidemiological analysis of the spread of citrus canker in urban Miami, Florida, and synergistic interaction with the Asian citrus leafminer. Fruits 52:383–390.

1333. Gottwald, T. R., Graham, J. H., and Schubert, T. S. 2002. Citrus canker: the pathogen and its impact. Plant Health Progress, 12 August 2002. Internet pub.

1334. Gottwald, T. R., Reynolds, K. M., Campbell, C. L., and Timmer, L. W. 1992. Spatial and spatiotemporal autocorrelation analysis of citrus canker epidemics in citrus nurseries and groves in Argentina. Phytopathology 82:843–851.

1335. Gottwald, T. R., and Timmer, L. W. 1995. The efficacy of windbreaks in reducing the spread of citrus canker caused by *Xanthomonas campestris* pv. *citri*. Trop. Agric. 72:194–201.

1336. Gottwald, T. R., Timmer, L. W., and McGuire, R. G. 1989. Analysis of disease progress of citrus canker in nurseries in Argentina. Phytopathology 79:1276–1283.

1337. Goud, J. C., and Termorshuizen, A. J. 2002. Pathogenicity and virulence of the two Dutch VCGs of *Verticillium dahliae* to woody ornamentals. Eur. J. Plant Pathol. 108:771–782.

1338. Gouin, F. R. 1983. Girdling roots: fact or fiction? Comb. Proc. Int. Plant Prop. Soc. 33:428–432.

1339. Gould, C. T. 1945. The parasitism of *Glomerularia lonicerae* (Pk.) D. and H. in *Lonicera* species. Iowa State Coll. J. Sci. 19:301–331.

1340. Gourbiére, F., and Morelet, M. 1978. Le genre *Rhizosphaera* Mangin et Hariot 1. *R. oudemansii* et *R. macrospora* sp. nov. Rev. Mycol. 43:81–95.

1341. Gourbiére, F., and Morelet, M. 1980. Le genre *Rhizosphaera* Mangin et Hariot 2. *R. pini, R. kobayashii*, et *R. kalkhoffii*. Cryptog. Mycol. 1:69–81.

1342. Graafland, W. 1960. The parasitism of *Exobasidium japonicum* Shir. on azalea. Acta Bot. Neerl. 9:347–379.

1343. Graff, P. W. 1936. North American polypores—I. *Polyporus squamosus* and its varieties. Mycologia 28:154–170.

1344. Graham, J. H., and Gottwald, T. R. 1990. Variation in aggressiveness of *Xanthomonas campestris* pv. *citrumelo* associated with citrus bacterial spot in Florida citrus nurseries. Phytopathology 80:190–196.

1345. Graham, J. H., Gottwald, T. R., and Fardelmann, D. 1990. Cultivar-specific interactions for strains of *Xanthomonas campestris* from Florida that cause citrus canker and citrus bacterial spot. Plant Dis. 74:753–756.

1346. Graham, J. H., Gottwald, T. R., Riley, T. D., and Bruce, M. A. 1992. Susceptibility of citrus fruit to bacterial spot and citrus canker. Phytopathology 82:452–457.

1347. Graham, J. H., and Linderman, R. G. 1983. Pathogenic seedborne *Fusarium oxysporum* from Douglas-fir. Plant Dis. 67:323–325.

1348. Graham, J. H., and Timmer, N. H. 1991. Peat-based media as a source of *Thielaviopsis basicola* causing black root rot on citrus seedlings. Plant Dis. 75:1246–1249.

1349. Graham, J. H., Timmer, L. W., and Young, R. H. 1983. Necrosis of major roots in relation to citrus blight. Plant Dis. 67:1273–1276.

1350. Granata, G., and Whalley, A. J. S. 1994. Decline of beech associated with *Biscogniauxia nummularia* in Italy. Petria 4:111–115.

1351. Grand, L. F., and Menge, J. A. 1974. Sclerotia of *Cristulariella pyramidalis* in nature. Mycologia 66:712–715.

1352. Graniti, A. 1998. Cypress canker: a pandemic in progress. Annu. Rev. Phytopathol. 36:91–114.

1353. Graniti, A., Sparapano, L., and Evidente, A. 1992. Cyclopaldic acid: a major phytotoxic metabolite of *Seiridium cupressi* the pathogen of a canker disease of cypress. Plant Pathol. 41:563–568.

1354. Grasso, S., and La Rosa, R. 1982. Cancri da *Phomopsis incarcerata* su rosa. Riv. Patol. Veg. 18:143–148.

1355. Grattan, S. R., Shannon, M. C., Grieve, C. M., Poss, J. A., Suarez, D., and Leland, F. 1997. Interactive effects of salinity and boron on the performance and water use of eucalyptus. Acta Hortic. No. 449:607–613.

1356. Graves, A. A., and Witcher, W. 1971. Monochaetia canker of Arizona cypress and redcedar in South Carolina. Plant Dis. Rep. 55:810–813.

1357. Graves, A. H. 1919. Some diseases of trees in greater New York. Mycologia 11:111–124.

1358. Graves, A. H. 1923. The Melanconis disease of the butternut (*Juglans cinerea* L.). Phytopathology 13:411–435.

1359. Green, F. I., and Highley, T. L. 1997. Mechanism of brown-rot decay: paradigm or paradox? Int. Biodeter. Biodegrad. 39:113–124.

1360. Green, M. J., Thompson, D. A., and MacKenzie, D. J. 1999. Easy and efficient DNA extraction from woody plants for the detection of phytoplasmas by polymerase chain reaction. Plant Dis. 83:482–485.

1361. Green, R. J. Jr. 1977. Dieback of black walnut seedlings caused by *Phomopsis elaeagni*. Plant Dis. Rep. 61:582–584.

1362. Greene, G. L. 1962. Physiological basis of gall formation in black-knot disease of *Prunus*. Phytopathology 52:880–884.

1363. Greenidge, K. N. H. 1953. Further studies of birch dieback in Nova Scotia. Can. J. Bot. 31:548–559.

1364. Gregory, N. F., Mulrooney, R. P., Rossman, A. Y., and Castlebury, L. A. 2004. Anthracnose caused by *Discula fraxinea* on the new host Chinese fringetree and white ash in Delaware. Plant Dis. 88:427.

1365. Greig, B. J. W. 1989. Decay in an avenue of horse chestnut (*Aesculus hippocastanum* L.) caused by *Ustulina deusta*. Arboric. J. 13:1–6.

1366. Gremmen, J. 1978. Research on Dothichiza-bark necrosis (*Cryptodiaporthe populea*) in poplar. Eur. J. For. Pathol. 8:362–368.

1367. Gremmen, J., and Kam, M. de. 1977. *Ceratocystis fimbriata*, a fungus associated with poplar canker in Poland. Eur. J. For. Pathol. 7:44–47.

1368. Griesbach, J. A., and Maggenti, A. R. 1989. Vector capability of *Xiphinema americanum sensu lato* in California. J. Nematol. 21:517–523.

1369. Griffin, D. H. 1994. Fungal physiology. 2nd ed. Wiley-Liss, New York. 458 pp.

1370. Griffin, D. H., Quinn, K. E., Gilbert, G. S., Wang, C. J. K., and Rosemarin, S. 1992. The role of ascospores and conidia as propagules in the disease cycle of *Hypoxylon mammatum*. Phytopathology 82:114–119.

1371. Griffin, G. D., and Epstein, A. H. 1964. Association of dagger nematode, *Xiphinema americanum*, with stunting and winterkill of ornamental spruce. Phytopathology 54:177–180.

1372. Griffin, G. J. 1986. Chestnut blight and its control. Hortic. Rev. 8:291–336.

1373. Griffin, G. J., Smith, H. C., Dietz, A., and Elkins, J. R. 1991. Importance of hardwood competition to American chestnut survival, growth, and blight development in forest clearcuts. Can. J. Bot. 69:1804–1809.

1374. Griffin, G. J., and Stipes, R. J. 1975. High populations of *Fusarium oxysporum* f.sp. *perniciosum* in mimosa bark lenticellar sporodochia and in soil. Plant Dis. Rep. 59:787–790.

1375. Griffin, H. D. 1965. Maple dieback in Ontario. For. Chron. 41:295–300.

1376. Griffin, M. S., Sutherland, J. R., and Dennis, J. J. 1987. Blight of conifer seedlings caused by *Colletotrichum gloeosporioides*. New For. 1:81–88.

1377. Griffith, G. S., and Boddy, L. 1990. Fungal decomposition of attached angiosperm twigs. I. Decay community development in ash, beech and oak. New Phytol. 116:407–415.

1378. Griffiths, H. M., Sinclair, W. A., Boudon-Padieu, E., Daire, X., Lee, I.-M., Sfalanga, A., and Bertaccini, A. 1999. Phytoplasmas associated with elm yellows: molecular variability and differentiation from related organisms. Plant Dis. 83:1101–1104.

1379. Griffiths, H. M., Sinclair, W. A., Davis, R. E., Lee, I.-M., Dally, E. L., Guo, Y. H., Chen, T. A., and Hibben, C. R. 1994. Characterization of mycoplasmalike organisms from *Fraxinus*, *Syringa*, and associated plants from geographically diverse sites. Phytopathology 84:119–126.

1380. Griffiths, H. M., Sinclair, W. A., Smart, C. D., and Davis, R. E. 1999. The phytoplasma associated with ash yellows and lilac witches-broom: 'Candidatus Phytoplasma fraxini.' Int. J. Syst. Bacteriol. 49:1605–1614.

1381. Grillo, R., Korhonen, K., Hantula, J., and Hietala, A. M. 2000. Genetic evidence for somatic haploidization in developing fruit bodies of *Armillaria tabescens*. Fungal Genet. Biol. 30:135–145.

1382. Groot, R. C. De. 1966. Phenolic extractives in lateral branches and injured leaders of *Pinus strobus* L. Can. J. Bot. 44:57–61.

1383. Grosclaude, C. 1992. Decay caused by wood destroying fungi on plane trees. Cryptog. Mycol. 13:181–191.

1384. Grosclaude, G., Germain, E., Simone, J., and Leglise, P. 1976. La nécrose des inflorescences mâles du noisetier provoquée par *Gloeosporium coryli* (Desm.) Sacc. Ann. Phytopathol. 8:87–89.

1385. Grosclaude, C., Olivier, R., Pizzuto, J. C., and Romiti, C. 1991. Experimental study on the dissemination of *Ceratocystis fimbriata* f. *platani* inoculum by river water. Eur. J. For. Pathol. 21:168–171.

1386. Grosclaude, C., Olivier, R., Pizzuto, J. C., Romiti, C., and Madec, S. 1988. Detection of *Ceratocystis fimbriata* f. *platani* by trapping. Application to the study of the persistence of the parasite in infected wood. Eur. J. For. Pathol. 18:385–390.

1387. Grosclaude, C., Olivier, R., and Romiti, C. 1995. Canker stain of London plane. Survival of the causal agent in soil. Phytoma No. 479:41–42.

1388. Grosclaude, C., Olivier, R., Romiti, C., and Pizzuto, J. C. 1992. About planes. The xylophagous ability of some basidiomycete fungi. Phytoma No. 436:64–66.

1389. Gross, H. L. 1964. The Echinodontiaceae. Mycopathol. Mycol. Appl. 24:1–26.

1390. Gross, H. L. 1967. Cytospora canker of black cherry. Plant Dis. Rep. 51:941–944.

1391. Gross, H. L. 1983. Negligible cull and growth loss of jack pine associated with globose gall rust. For. Chron. 59:308–311.

1392. Gross, H. L. 1984. Impact of Eutypella canker on the maple resource of the Owen Sound and Wingham forest districts. For. Chron. 60:18–21.

1393. Gross, H. L., and Basham, J. T. 1981. Diseases of aspen suckers in northern Ontario. Can. For. Serv., Great Lakes For. Res. Cent. Inf. Rep. O-X-329. 16 pp.

1394. Gross, H. L., Ek, A. R., and Patton, R. F. 1983. Site character and infection hazard for the sweetfern rust disease in northern Ontario. For. Sci. 29:771–778.

1395. Gross, H. L., Patton, R. F., and Ek, A. R. 1978. Reduced growth, cull, and mortality of jack pine associated with sweetfern rust cankers. Can. J. For. Res. 8:47–53.

1396. Gross, H. L., Patton, R. F., and Ek, A. R. 1980. Spatial aspects of sweetfern rust disease in northern Ontario jack pine–sweetfern stands. Can. J. For. Res. 10:199–208.

1397. Grove, W. B. 1935, 1937. British stem- and leaf-fungi (Coelomycetes). Vols. 1, 2. Cambridge Univ. Press, Cambridge, UK. 488 and 407 pp.

1398. Groves, J. W., and Bowerman, C. A. 1955. The species of *Ciborinia* on *Populus*. Can. J. Bot. 33:577–590.

1399. Grulke, N. E., Andersen, C. P., Fenn, M. E., and Miller, P. R. 1998. Ozone exposure and nitrogen deposition lowers root biomass of ponderosa pine in the San Bernardino Mountains, California. Environ. Pollut. 103:63–73.

1400. Gryzenhout, M., Eisenberg, B. E., Coutinho, T. A., Wingfield, B. D., and Wingfield, M. J. 2003. Pathogenicity of *Cryphonectria eucalypti* to *Eucalyptus* clones in South Africa. For. Ecol. Manag. 176:427–437.

1401. Gryzenhout, M., Myburg, H., Merwe, N. A. van der, Wingfield, B. D., and Wingfield, M. J. 2005. *Chrysoporthe*, a new genus to accommodate *Cryphonectria cubensis*. Stud. Mycol. 50:119–142.

1402. Guba, E. F. 1961. Monograph of *Monochaetia* and *Pestalotia*. Harvard Univ. Press, Cambridge, MA. 342 pp.

1403. Guerber, J. C., and Correll, J. C. 2001. Characterization of *Glomerella acutata*, the teleomorph of *Colletotrichum acutatum*. Mycologia 93:216–229.

1404. Guerber, J. C., Liu, B., Correll, J. C., and Johnston, P. R. 2003. Characterization of diversity in *Colletotrichum acutatum sensu lato* by sequence analysis of two gene introns, mtDNA and intron RFLPs, and mating compatibility. Mycologia 95:872–895.

1405. Guest, D. I., Pegg, K. G., and Whiley, A. W. 1995. Control of Phytophthora diseases of tree crops using trunk-injected phosphonates. Hortic. Rev. 17:299–330.

1406. Guicherd, P., Peltier, J. P., Gout, E., Bligny, R., and Marigo, G. 1997. Osmotic adjustment in *Fraxinus excelsior* L.: malate and mannitol accumulation in leaves under drought conditions. Trees Struct. Funct. 11:155–161.

1407. Guidi, L., Lorenzini, G., and Soldatini, G. F. 1988. Phytotoxicity of sea-water aerosols on forest plants with special reference to the role of surfactants. Environ. Exp. Bot. 28:85–94.

1408. Guillaumin, J. J., Mohammed, C., Anselmi, N., Courtecuisse, R., Gregory, S. C., Holdenrieder, O., Intini, M., Lung-Escarmant, B., Marxmüller, H., Morrison, D., Rishbeth, J., Termorshuizen, A. J., Tirro, A., and Dam, B. van. 1993. Geographical distribution and ecology of the *Armillaria* species in western Europe. Eur. J. For. Pathol. 23:321–341.

1409. Gumpf, D. J. 1988. Stubborn diseases of citrus caused by *Spiroplasma citri*. Pages 327–342 in: Mycoplasma diseases of crops: basic and applied aspects. K Maramorosch and S. P. Raychaudhuri, eds., Springer-Verlag, New York.

1410. Gundersen, D. E., and Lee, I.-M. 1996. Ultrasensitive detection of phytoplasmas by nested-PCR assays using two universal primer pairs. Phytopathol. Medit. 35:144–151.

1411. Gundersen, D. E., Lee, I.-M., Rehner, S. A., Davis, R. E., and Kingsbury, D. T. 1994. Phylogeny of mycoplasmalike organisms (phytoplasmas): a basis for their classification. J. Bacteriol. 176:5244–5254.

1412. Gundersen, D. E., Lee, I.-M., Schaff, D. A., Harrison, N. A., Chang, C. J., Davis, R. E., and Kingsbury, D. T. 1996. Genomic diversity and differentiation among phytoplasma strains in 16S rRNA groups I (aster yellows and related phytoplasmas) and III (X-disease and related phytoplasmas). Int. J. Syst. Bacteriol. 46:64–75.

1413. Guo, Y. H., Cheng, Z.-M., and Walla, J. A. 2000. Characterization of X-disease phytoplasmas in chokecherry from North Dakota by PCR-RFLP and sequence analysis of the rRNA gene region. Plant Dis. 84:1235–1240.

1414. Guo, Y. H., Cheng, Z.-M., Walla, J. A., and Zhang, Z. 1998. Diagnosis of X-disease phytoplasma in stone fruits by a monoclonal antibody developed directly from a woody plant. J. Environ. Hortic. 16:33–37.

1415. Guo, Y. L., and Hsieh, W.-H. 1995. The genus *Pseudocercospora* in China. Mycosystema Monogr. Ser. No. 2. International Academic, Beijing. 388 pp.

1416. Guo, Y. L., and Liu, X. J. 1991. Studies on the genus *Pseudocercospora* in China V. Mycosystema 4:99–118.

1417. Gustafsson, M. E. R. 1997. Raised levels of marine aerosol deposition owing to increased storm frequency; a cause of forest decline in southern Sweden? Agric. For. Meteorol. 84:169–177.

1418. Guyon, J. C., Jacobi, W. R., and McIntyre, G. A. 1996. Effects of environmental stress on the development of Cytospora canker of aspen. Plant Dis. 80:1320–1326.

1419. Haasis, F. A. 1953. Flower blight of camellias. Am. Camellia Yearb. 1953:111–119.

1420. Hacskaylo, E., ed. 1971. Mycorrhizae. USDA Misc. Pub. 1189. 255 pp.

1421. Hacskaylo, J., Finn, R. F., and Vimmerstedt, J. P. 1969. Deficiency symptoms of some forest trees. Ohio Agric. Res. Devel. Cent. Res. Bull. 1015. 68 pp.

1422. Haddow, W. R. 1931. Studies in *Ganoderma*. J. Arnold Arbor. 12:25–46.

1423. Haddow, W. R. 1938. On the classification, nomenclature, hosts, and geographical range of *Trametes pini* (Thore) Fries. Trans. Br. Mycol. Soc. 22:182–193.

1424. Haddow, W. R. 1938. The disease caused by *Trametes pini* (Thore) Fries in white pine (*Pinus strobus* L.). Trans. R. Can. Inst. 29:21–80.

1425. Haddow, W. R. 1941. On the history and diagnosis of *Polyporus tomentosus* Fries, *Polyporus circinatus* Fries, and *Polyporus dualis* Peck. Trans. Br. Mycol. Soc. 25:179–190.

1426. Haddow, W. R., and Newman, F. S. 1942. A disease of the Scots pine (*Pinus sylvestris* L.) caused by the fungus *Diplodia pinea* Kickx associated with the pine spittle-bug (*Aphrophora parallela* Say.). I. Symptoms and etiology. Trans. R. Can. Inst. 24:1–17.

1427. Hadidi, A., ed. 1998. Proceedings of the 17th international symposium on virus and virus-like diseases of temperate fruit crops. Fruit tree diseases. Acta Hortic. No. 472. Vol. 1:1–337, vol. 2:341–794.

1428. Hadfield, J. S., Mathiasen, R. L., and Hawksworth, F. G. 2002. Douglas-fir dwarf mistletoe. USDA For. Serv. For. Insect Dis. Leafl. 54. 10 pp. Internet pub.

1429. Hadley, J. L., and Amundson, R. G. 1992. Effects of radiational heating at low air temperature on water balance, cold tolerance, and visible injury of red spruce foliage. Tree Physiol. 11:1–17.

1430. Haemmerli, U. A., Brandle, U. E., Petrini, O., and McDermott, J. M. 1992. Differentiation of isolates of *Discula umbrinella* (teleomorph: *Apiognomonia errabunda*) from beech, chestnut, and oak using randomly amplified polymorphic DNA markers. Mol. Plant-Microbe Interact. 5:479–483.

1431. Hagan, A. K., Akridge, J. R., Bowen, K. L., Olive, J. W., and Tilt, K. M. 2001. Resistance of selected cultivars of Indian hawthorn to Entomosporium leaf spot, fireblight, and anthracnose in Alabama. J. Environ. Hortic. 19:43–46.

1432. Hagle, S. K., McDonald, G. I., and Norby, E. A. 1989. White pine blister rust in northern Idaho and western Montana: alternatives for integrated management. USDA For. Serv. Gen. Tech. Rep. INT-261. 35 pp.

1433. Hahn, G. G. 1920. *Phomopsis juniperovora*, a new species causing blight of nursery cedars. Phytopathology 10:249–253.

1434. Hahn, G. G. 1928. *Phomopsis conorum* (Sacc.) Died.—an old fungus of the Douglas fir and other conifers. Trans. Br. Mycol. Soc. 13:278–286.

1435. Hahn, G. G. 1930. Life-history studies of the species of *Phomopsis* occurring on conifers. Part I. Trans. Br. Mycol. Soc. 15:32–93.

1436. Hahn, G. G. 1943. Taxonomy, distribution, and pathology of *Phomopsis occulta* and *P. juniperovora*. Mycologia 35:112–129.

1437. Hahn, G. G., and Ayers, T. T. 1934. Dasyscyphae on conifers in North America. I. The large-spored, white-excipled species. Mycologia 26:73–101.

1438. Hahn, G. G., and Ayers, T. T. 1943. Role of *Dasyscypha willkommii* and related fungi in the production of canker and die-back of larches. J. For. 41:483–495.

1439. Hahn, G. G., Hartley, C., and Pierce, R. G. 1917. A nursery blight of cedars. J. Agric. Res. 10:533–539.

1440. Haidar, M. A., Orr, G. L., and Westra, P. 1998. The response of dodder (*Cuscuta* spp.) seedlings to phytohormones under various light regimes. Ann. Appl. Biol. 132:331–338.

1441. Halbrendt, J. M. 1996. Allelopathy in the management of plant-parasitic nematodes. J. Nematol. 28:8–14.

1442. Hale, M. E. 1983. The biology of lichens. 3rd ed. Arnold, Baltimore. 190 pp.

1443. Halik, S., and Bergdahl, D. R. 1994. Long-term survival of *Bursaphelenchus xylophilus* in living *Pinus sylvestris* in an established plantation. Eur. J. For. Pathol. 24:357–363.

1444. Halik, S., and Bergdahl, D. R. 2002. Potential beetle vectors of *Sirococcus clavigignenti-juglandacearum* on butternut. Plant Dis. 86:521–527.

1445. Hall, G., Cook, R. T. A., and Bradshaw, N. J. 1992. First record of *Peronospora sparsa* on *Prunus laurocerasus*. Plant Pathol. 41:224–227.

1446. Hall, G., Dobson, S., and Nicholls, C. 1992. First record of *Phytophthora inflata* in the United Kingdom. Plant Pathol. 41:95–97.

1447. Hall, G. S. 1996. Modern approaches to species concepts in downy mildews. Plant Pathol. 45:1009–1026.

1448. Hall, R., Hofstra, G., and Lumis, G. P. 1972. Effects of deicing salt on eastern white pine: foliar injury, growth suppression, and seasonal changes in foliar concentrations of sodium and chloride. Can. J. For. Res. 2:244–249.

1449. Hall, R., Hofstra, G., and Lumis, G. P. 1973. Leaf necrosis of roadside sugar maple in Ontario in relation to elemental composition of soil and leaves. Phytopathology 63:1426–1427.

1450. Hall, R. C. 1933. Post-logging decadence in northern hardwoods. Univ. Mich. Sch. For. Conserv. Bull. 3. 66 pp.

1451. Hällgren, J. E., and Fredriksson, S.-A. 1982. Emission of hydrogen sulfide from sulfur dioxide–fumigated pine trees. Plant Physiol. 70:456–459.

1452. Halmschlager, E., Butin, H., and Donaubauer, E. 1993. Endophytic fungi in leaves and twigs of *Quercus petraea*. Eur. J. For. Pathol. 23:51–63.

1453. Halsall, D. M. 1982. A forest soil suppressive to *Phytophthora cinnamomi* and conducive to *Phytophthora cryptogea*. I. Survival, germination and infectivity of mycelium and chlamydospores. II. Suppression of sporulation. Austral. J. Bot. 30:11–25, 27–37.

1454. Hamacher, J., and Quadt, A. 1991. Light- and electron microscopic studies of cherry leaf roll virus (CLRV) on European ash (*Fraxinus excelsior* L.). J. Phytopathol. 131:215–226.

1455. Hamacher, J., and Quadt, A. 1994. Isolation of cherry leafroll and brome mosaic viruses from European beech and transmission to beech seedlings. Plant Dis. 78:849–853.

1456. Hamel, C., Morin, F., Fortin, A., Granger, R. L., and Smith, D. L. 1994. Mycorrhizal colonization increases herbicide toxicity in apple. J. Am. Soc. Hortic. Sci. 119:1255–1260.

1457. Hamelin, R. C., Bérubé, P., Gignac, M., and Bourassa, M. 1996. Identification of root rot fungi in nursery seedlings by nested multiplex PCR. Appl. Environ. Microbiol. 62:4026–4031.

1458. Hamelin, R. C., Doudrick, R. L., and Nance, W. L. 1994. Genetic diversity in *Cronartium quercuum* f.sp. *fusiforme* on loblolly pines in southern USDA Curr. Genet. 26:359–363.

1459. Hamelin, R. C., Dusabenyagasani, M., and Et-touil, K. 1998. Fine-level genetic structure of white pine blister rust populations. Phytopathology 88:1187–1191.

1460. Hamelin, R. C., Hunt, R. S., Geils, B. W., Jensen, G. D., Jacobi, V., and Lecours, N. 2000. Barrier to gene flow between eastern and western populations of *Cronartium ribicola* in North America. Phytopathology 90:1073–1078.

1461. Hamelin, R. C., Lecours, N., Hansson, P., Hellgren, M., and Laflamme, G. 1996. Genetic differentiation within the European race of *Gremmeniella abietina*. Mycol. Res. 100:49–56.

1462. Hamelin, R. C., and Rail, J. 1997. Phylogeny of *Gremmeniella* spp. based on sequences of the 5.8S rDNA and internal transcribed spacer region. Can. J. Bot. 75:693–698.

1463. Hamelin, R. C., Shain, L., and Thielges, B. A. 1992. Influence of leaf wetness, temperature, and rain on poplar leaf rust epidemics. Can. J. For. Res. 22:1249–1254.

1464. Hamelin, R. C., and Sutherland, J. R. 1991. Variation in the susceptibility of lodgepole pine provenances to Sirococcus shoot blight: results from artificial inoculations. Eur. J. For. Pathol. 21:189–192.

1465. Hamerlynck, E., and Knapp, A. K. 1996. Photosynthetic and stomatal responses to high temperature and light in two oaks at the western limit of their range. Tree Physiol. 16:557–565.

1466. Hamm, P. B., and Hansen, E. M. 1982. Pathogenicity of *Phytophthora* species to Pacific Northwest conifers. Eur. J. For. Pathol. 12:167–174.

1467. Hamm, P. B., and Hansen, E. M. 1983. *Phytophthora pseudotsugae*, a new species causing root rot of Douglas-fir. Can. J. Bot. 61:2626–2631.

1468. Hammarlund, C. 1930. Rostsvampar pa *Mahonia* (*Puccinia mirabilissima* Peck och *P. graminis* Pers.). Bot. Notis. 1930:380–407.

1469. Hampson, M. C. 1981. Phomopsis canker on weeping fig in Newfoundland. Can. Plant Dis. Surv. 61:1–4.

1470. Hampson, M. C., and Sinclair, W. A. 1973. Xylem dysfunction in peach caused by *Cytospora leucostoma*. Phytopathology 63:676–681.

1471. Han, S. S., Karasev, A. V., Ieki, H., and Iwanami, T. 2002. Nucleotide sequence and taxonomy of *Cycas necrotic stunt virus*. Arch. Virol. 147:2207–2214.

1472. Han, Y., Castello, J. D., and Leopold, D. J. 1991. Ash yellows, drought, and decline in radial growth of white ash. Plant Dis. 75:18–23.

1473. Han, Z. M., Yin, T. M., Li, C. D., Huang, M. R., and Wu, R. L. 2000. Host effect on genetic variation of *Marssonina brunnea* pathogenic to poplars. Theor. Appl. Genet. 100:614–620.

1474. Hanlin, R. T. 1982. Conidiogenesis in *Spiniger meineckellus*. Mycologia 74:236–241.

1475. Hanlin, R. T. 1990. Icones Ascomycetum Georgiae: *Ophiodothella vaccinii*. Mycotaxon 39:1–8.

1476. Hanlin, R. T. 1998. Combined keys to Illustrated genera of ascomycetes. Vols. 1, 2. APS Press, St. Paul, MN. 263 pp.

1477. Hanlin, R. T., Goh, T. K., and Skarshaug, A. J. 1992. A key to and descriptions of species assigned to *Ophiodothella* based on the literature. Mycotaxon 44:103–126.

1478. Hanlin, R. T., Jimenez, B., Chang, L. H., and Brown, E. A. 1992. *Crumenulopsis atropurpurea* comb. nov., from Japanese red pine in Georgia. Mycologia 84:650–658.

1479. Hansbrough, J. R. 1934. Occurrence and parasitism of *Aleurodiscus amorphus* in North America. J. For. 32:452–458.

1480. Hansen, A. J., Nyland, G., McElroy, F. D., and Stace-Smith, R. 1974. Origin, host range and spread of cherry rasp leaf disease in North America. Phytopathology 64:721–727.

1481. Hansen, E. M. 1979. Nuclear condition and vegetative characteristics of homocaryotic and heterocaryotic isolates of *Phellinus weirii*. Can. J. Bot. 57:1579–1582.

1482. Hansen, E. M. 1979. Sexual and vegetative incompatibility reactions in *Phellinus weirii*. Can. J. Bot. 57:1573–1578.

1483. Hansen, E. M. 1986. Inoculation of Douglas-fir roots with isolates of *Phellinus weirii* on sites differing in root rot severity. Can. J. For. Res. 16:619–623.

1484. Hansen, E. M., and Goheen, D. J. 1988. Rate of increase of black-stain root disease in Douglas-fir plantations in Oregon and Washington. Can. J. For. Res. 18:942–946.

1485. Hansen, E. M., and Goheen, E. M. 2000. *Phellinus weirii* and other native root pathogens as determinants of forest structure and process in western North America. Annu. Rev. Phytopathol. 38:515–539.

1486. Hansen, E. M., Goheen, D. J., Jules, E. S., and Ullian, B. 2000. Managing Port-Orford-cedar and the introduced pathogen *Phytophthora lateralis*. Plant Dis. 84:4–14.

1487. Hansen, E. M., and Hamm, P. B. 1988. Canker diseases of Douglas-fir seedlings in Oregon and Washington bareroot nurseries. Can. J. For. Res. 18:1053–1058.

1488. Hansen, E. M., and Hamm, P. B. 1996. Survival of *Phytophthora lateralis* in infected roots of Port Orford cedar. Plant Dis. 80:1075–1078.

1489. Hansen, E. M., Hamm, P. B., Julis, A. J., and Roth, L. F. 1979. Isolation, incidence, and management of *Phytophthora* in forest tree nurseries in the Pacific Northwest. Plant Dis. Rep. 63:607–611.

1490. Hansen, E. M., and Lewis, K. J., eds. 1997. Compendium of conifer diseases. APS Press, St. Paul, MN. 101 pp.

1490a. Hansen, E. M., Reeser, P. W., Davidson, J. M., Garbelotto, M., Ivors, K., Douhan, L., and Rizzo, D. M. 2003. *Phytophthora nemorosa*, a new species causing cankers and leaf blight of forest trees in California and Oregon, U.S.A. Mycotaxon 88:129–138.

1491. Hansen, E. M., Roth, L. F., Hamm, P. B., and Julis, A. J. 1980. Survival, spread, and pathogenicity of *Phytophthora* spp. on Douglas-fir seedlings planted on forest sites. Phytopathology 70:422–425.

1492. Hansen, E. M., Stenlid, J., and Johansson, M. 1993. Genetic control of somatic incompatibility in the root-rotting basidiomycete *Heterobasidion annosum*. Mycol. Res. 97:1229–1233.

1493. Hansen, E. M., Stone, J. K., Capitano, B. R., Rosso, P., Sutton, W., Winton, L., Kanaskie, A., and McWilliams, M. G. 2000. Incidence and impact of Swiss needle cast in forest plantations of Douglas-fir in coastal Oregon. Plant Dis. 84:773–778.

1494. Hansen, E. M., and Sutton, W., eds. 2000. Phytophthora diseases of forest trees. Ore. State Univ., Corvallis. 152 pp.

1495. Hansen, H. N., and Barrett, J. T. 1938. Gardenia canker. Mycologia 30:15–19.

1496. Hansen, H. N., and Scott, C. E. 1934. A canker and gall disease of *Gardenia*. Science 79:18.

1497. Hansen, H. N., and Smith, R. E. 1937. A bacterial gall disease of Douglas fir, *Pseudotsuga taxifolia*. Hilgardia 10:569–577.

1498. Hansen, H. N., and Thomas, H. E. 1940. Flower blight of camellias. Phytopathology 30:166–170.

1499. Hansford, C. G. 1961. The Meliolineae. A monograph. Sydowia Beih. 2. 806 pp.

1500. Hansford, C. G. 1965. Iconographia meliolinearum. Sydowia Beih. 5. Unpaged.

1501. Hanso, M., Torva, A., and Mardiste, M. 1975, 1980, 1986. Black snow mould in Estonia. 1. Ecology and morphology of *Herpotrichia juniperi*. 2. Dispersal biology of *Herpotrichia juniperi* (Duby) Petr. 3. Sporulation of *Herpotrichia juniperi* in relation to season and weather conditions. Metsanduslikud Uurimused 12:262–279; 16:99–119; 21:114–126.

1502. Harada, Y. 1994. Materials for the rust flora of Japan VI. Mycoscience 35:295–299.

1503. Harada, Y., and Noro, S. 1988. *Grovesinia pruni*, sp. nov., the teleomorph of a new zonate leaf spot fungus on *Prunus* in Jap. Trans. Mycol. Soc. Jap. 29:85–92.

1504. Hariri, E. B., Jeune, B., Baudino, S., Urech, K., and Sallé, G. 1992. Development of a coefficient of resistance to mistletoe for oak. Can. J. Bot. 70:1239–1246.

1505. Harlton, C. E., Levesque, C. A., and Punja, Z. K. 1995. Genetic diversity in *Sclerotium* (*Athelia*) *rolfsii* and related species. Phytopathology 85:1269–1281.

1506. Harniss, R. O., and Nelson, D. L. 1984. A severe epidemic of Marssonina leaf blight on quaking aspen in northern Utah. USDA For. Serv. Res. Note INT-339. 6 pp.

1507. Harrington, T. C., and Cobb, F. W. Jr. 1983. Pathogenicity of *Leptographium* and *Verticicladiella* spp. isolated from roots of western North American conifers. Phytopathology 73:596–599.

1508. Harrington, T. C., and Cobb, F. W. Jr. 1984. Host specialization of three morphological variants of *Verticicladiella wageneri*. Phytopathology 74:286–290.

1509. Harrington, T. C., and Cobb, F. W. Jr. 1984. *Verticillium albo-atrum* on *Ceanothus* in a California forest. Plant Dis. 68:1012.

1510. Harrington, T. C., and Cobb, F. W. Jr. 1986. Varieties of *Verticicladiella wageneri*. Mycologia 78:562–567.

1511. Harrington, T. C., and Cobb, F. W. Jr. 1987. *Leptographium wageneri* var. *pseudotsugae*, var. nov., cause of black stain root disease on Douglas-fir. Mycotaxon 30:501–507.

1512. Harrington, T. C., and Cobb, F. W. Jr., eds. 1988. Leptographium root diseases on conifers. APS Press, St. Paul, MN. 148 pp.

1513. Harrington, T. C., Cobb, F. W. Jr., and Lownsbery, J. W. 1985. Activity of *Hylastes nigrinus*, a vector of *Verticicladiella wageneri*, in thinned stands of Douglas-fir. Can. J. For. Res. 15:519–523.

1514. Harrington, T. C., Furniss, M. M., and Shaw, C. G. 1981. Dissemination of hymenomycetes by *Dendroctonus pseudotsugae* (Coleoptera: Scolytidae). Phytopathology 71:551–554.

1515. Harrington, T. C., McNew, D., Steimel, J., Hofstra, D., and Farrell, R. 2001. Phylogeny and taxonomy of the *Ophiostoma piceae* complex and the Dutch elm disease fungi. Mycologia 93:111–136.

1516. Harrington, T. C., and Rizzo, D. M. 1993. Identification of *Armillaria* species from New Hampshire. Mycologia 85:365–368.

1517. Harrington, T. C., Steimel, J. P., Wingfield, M. J., and Kile, G. A. 1996. Isozyme variation and species delimitation in the *Ceratocystis coerulescens* complex. Mycologia 88:104–113.

1518. Harrington, T. C., and Wingfield, B. D. 1995. A PCR-based identification method for species of *Armillaria*. Mycologia 87:280–288.

1519. Harrington, T. C., and Wingfield, M. J. 1998. The *Ceratocystis* species on conifers. Can. J. Bot. 76:1446–1457.

1520. Harris, H. A. 1934. Frost ring formation in some winter-injured deciduous trees and shrubs. Am. J. Bot. 21:485–498.

1521. Harris, K. F., Smith, O. P., and Duffus, J. E., eds. 2001. Virus-insect-plant interactions. Academic Press, San Diego. 376 pp.

1522. Harrison, K. A. 1965. Willow blight and the survival of some *Salix* species in Nova Scotia. Can. Plant Dis. Surv. 45:94–95.

1523. Harrison, M. J. 1999. Molecular and cellular aspects of the arbuscular mycorrhizal symbiosis. Annu. Rev. Plant Physiol. Plant Mol. Biol. 50:361–389.

1524. Harrison, M. J., Myrie, W., Jones, P., Carpio, M. L., Castillo, M., Doyle, M. M., and Oropeza, C. 2002. 16S rRNA interoperon sequence heterogeneity distinguishes strain populations of palm lethal yellowing phytoplasma in the Caribbean region. Ann. Appl. Biol. 141:183–193.

1525. Harrison, N. A., Narvaez, M., Almeyda, H., Cordova, I., Carpio, M. L., and Oropeza, C. 2002. First report of group 16SrIV phytoplasmas infecting coconut palms with leaf yellowing symptoms on the Pacific coast of Mexico. Plant Pathol. 51:808.

1526. Harrison, N. A., Richardson, P. A., Kramer, J. B., and Tsai, J. H. 1994. Detection of the mycoplasma-like organism associated with lethal yellowing disease of palms in Florida by polymerase chain reaction. Plant Pathol. 43:998–1008.

1527. Harrison, N. A., Womack, M., and Carpio, M. L. 2002. Detection and characterization of a lethal yellowing (16SrIV) group phytoplasma in Canary Island date palms affected by lethal decline in Texas. Plant Dis. 86:676–681.

1528. Hart, J. H., Mosher, D. G., and Ajewole, R. 1978. Occurrence of Endocronartium harknessii and Cronartium quercuum on Scots and jack pine in Michigan's Lower Peninsula. Plant Dis. Rep. 62:779–782.

1529. Hartig, R. 1874. Wichtige Krankheiten der Waldbäume. Beiträge zur Mycologie und Phytopathologie für Botaniker und Forstmänner. J. Springer, Berlin. 127 pp. English trans. 1975, W. Liese, D. H. Lambert, and W. Merrill. Am. Phytopathol. Soc., St. Paul, MN. 120 pp.

1530. Hartig, R. 1882. Lehrbuch der Baumkrankheiten. J. Springer, Berlin. 198 pp. English trans. 1894, W. Somerville and H. M. Ward. Macmillan, London. 331 pp.

1531. Hartig, R. 1894. Textbook of the diseases of trees. English trans. W. Somerville and H. M. Ward. Macmillan, London. 331 pp.

1532. Hartley, C. 1918. Stem lesions caused by excessive heat. J. Agric. Res. 14:595–604.

1533. Hartley, C., Davidson, R. W., and Crandall, B. S. 1961. Wetwood, bacteria, and increased pH in trees. USDA For. Serv. For. Prod. Lab. Rep. 2215. 34 pp.

1534. Hartman, J. R., Eshenaur, B. C., and Jarlfors, U. E. 1995. Bacterial leaf scorch caused by Xylella fastidiosa: a Kentucky survey; a unique pathogen; and bur oak, a new host. J. Arboric. 21:77–82.

1535. Hartman, J. R., Jarlfors, U. E., Fountain, W. M., and Thomas, R. 1996. First report of bacterial leaf scorch caused by Xylella fastidiosa on sugar maple and sweetgum. Plant Dis. 80:1302.

1536. Hartman, J. R., Smith, C. A., and Stuckey, R. E. 1981. Thread blight disease in Kentucky. (Abstr.) Phytopathology 71:1004.

1537. Hartmann, G., and Blank, R. 1992. Winter frost, insect defoliation and attack by Agrilus biguttatus as causal factors in the complex of oak decline in northern Germany. Forst Holz 47:443–452.

1538. Hartmann, H. T., and Kester, D. E. 2002. Hartmann and Kester's plant propagation: principles and practices. 7th ed. Prentice-Hall, Upper Saddle River, NJ. 880 pp.

1539. Hartung, J. S., Burton, C. L., and Ramsdell, D. C. 1981. Epidemiological studies of blueberry anthracnose disease caused by Colletotrichum gloeosporioides. Phytopathology 71:449–453.

1540. Hartung, J. S., and Civerolo, E. L. 1989. Restriction fragment length polymorphisms distinguish Xanthomonas campestris strains isolated from Florida citrus nurseries from X. c. pv. citri. Phytopathology 79:793–799.

1541. Hartung, J. S., Pruvost, O. P., Villemot, I., and Alvarez, A. 1996. Rapid and sensitive colorimetric detection of Xanthomonas axonopodis pv. citri by immunocapture and a nested-polymerase chain reaction assay. Phytopathology 86:95–101.

1542. Harvey, A. E., Larsen, M. J., Jurgensen, M. F., and Jones, E. A. 1989. Nitrogenase activity associated with decayed wood of living northern Idaho conifers. Mycologia 81:765–771.

1543. Harvey, G. M. 1976. Epiphytology of a needle cast fungus, Lophodermella morbida, in ponderosa pine plantations in western Oregon. For. Sci. 22:223–230.

1544. Harvey, J. M. 1952. Bacterial leaf spot of Umbellularia californica. Madroño 11:195–198.

1545. Hasui, Y., Fukui, Y., Kikuchi, J., Kato, N., Miyairi, K., and Okuno, T. 1998. Isolation, characterization, and sugar chain structure of endoPG Ia, Ib, and Ic from Stereum purpureum. Biosci. Biotechnol. Biochem. 62:852–857.

1546. Hata, K., Futai, K., and Tsuda, M. 1998. Seasonal and needle age-dependent changes of the endophytic mycobiota in Pinus thunbergii and Pinus densiflora needles. Can. J. Bot. 76:245–250.

1547. Hauben, L., Moore, E. R. B., Vauterin, L., Steenackers, M., Mergaert, J., Verdonck, L., and Swings, J. 1998. Phylogenetic position of phytopathogens within the Enterobacteriaceae. Syst. Appl. Microbiol. 21:384–397.

1548. Haupt, S., Oparka, K. J., Sauer, N., and Neumann, S. 2001. Macromolecular trafficking between Nicotiana tabacum and the holoparasite Cuscuta reflexa. J. Exp. Bot. 52:173–177.

1549. Hausner, G., Reid, J., and Klassen, G. R. 2000. On the phylogeny of members of Ceratocystis s.s. and Ophiostoma that possess different anamorphic states, with emphasis on the anamorph genus Leptographium, based on partial ribosomal DNA sequences. Can. J. Bot. 78:903–916.

1550. Hautala, E. L., Wulff, A., and Oksanen, J. 1992. Effects of deicing salt on visible symptoms, element concentrations and membrane damage in first-year needles of roadside Scots pine (Pinus sylvestris). Ann. Bot. Fenn. 29:179–185.

1551. Havir, E. A., and Anagnostakis, S. L. 1983. Oxalate production by virulent but not by hypovirulent strains of Endothia parasitica. Physiol. Plant Pathol. 23:369–376.

1552. Hawkins, S., and Boudet, A. 1996. Wound-induced lignin and suberin deposition in a woody angiosperm (Eucalyptus gunnii Hook.): histochemistry of early changes in young plants. Protoplasma 191:96–104.

1553. Hawksworth, D. L., Gibson, I. A. S., and Gams, W. 1976. Phialophora parasitica associated with disease conditions in various trees. Trans. Br. Mycol. Soc. 66:427–431.

1554. Hawksworth, D. L., and Talboys, P. W. 1970. Verticillium albo-atrum. C.M.I. Descr. Pathogenic Fungi Bact. No. 255. 2 pp.

1555. Hawksworth, F. G. 1961. Dwarfmistletoe of ponderosa pine in the Southwest. USDA Tech. Bull. 1246. 112 pp.

1556. Hawksworth, F. G. 1969. Rapid intensification and upward spread of dwarf mistletoe in inoculated digger pines. Plant Dis. Rep. 53:615–617.

1557. Hawksworth, F. G., and Dooling, O. J. 1984 Lodgepole pine dwarf mistletoe. USDA For. Serv. For. Insect Dis. Leafl. 18. Internet pub.

1558. Hawksworth, F. G., and Geils, B. W. 1990. How long do mistletoe-infected ponderosa pines live? West. J. Appl. For. 5:47–48.

1559. Hawksworth, F. G., and Johnson, D. W. 1989. Biology and management of dwarf mistletoe in lodgepole pine in the Rocky Mountains. USDA For. Serv. Gen. Tech. Rep. RM-169. 38 pp.

1560. Hawksworth, F. G., and Mielke, J. L. 1962. Witches'-broom of Gambel oak associated with Articularia quercina var. minor. Phytopathology 52:451–454.

1561. Hawksworth, F. G., and Scharpf, R. F. 1986. Spread of European mistletoe (Viscum album) in California, USA. Eur. J. For. Pathol. 16:1–5.

1562. Hawksworth, F. G., and Scharpf, R. F., tech. coords. 1984. Biology of dwarf mistletoes: proceedings of the symposium. USDA For. Serv. Gen. Tech. Rep. RM-111. 131 pp.

1563. Hawksworth, F. G., Scharpf, R. F., and Marosy, M. 1991. European mistletoe continues to spread in Sonoma County. Cal. Agric. 45:39–40.

1564. Hawksworth, F. G., and Shigo, A. L. 1980. Dwarf mistletoe on red spruce in the White Mountains of New Hampshire. Plant Dis. 64:880–882.

1565. Hawksworth, F. G., Wicker, E. F., and Scharpf, R. F. 1977. Fungal parasites of dwarf mistletoes. USDA For. Serv. Gen. Tech. Rep. RM-36. 14 pp.

1566. Hawksworth, F. G., and Wiens, D. 1996. Dwarf mistletoes: biology, pathology, and systematics. USDA Agric. Handb. 709. 410 pp.

1567. Hayes, E. M., and Ocamb, C. M. 1997. Fusarium canker and freeze injury of black walnut. USDA For. Serv. Gen. Tech. Rep. NC-191:184–191.

1568. He, C. X., Li, W. B., Ayres, A. J., Hartung, J. S., Miranda, V. S., and Teixeira, D. C. 2000. Distribution of Xylella fastidiosa in citrus rootstocks and transmission of citrus variegated chlorosis between sweet orange plants through natural root grafts. Plant Dis. 84:622–626.

1569. Headland, J. K., Griffin, G. J., Stipes, R. J., and Elkins, J. R. 1976. Severity of natural Endothia parasitica infection of Chinese chestnut. Plant Dis. Rep. 60:426–429.

1570. Headley, D. B., and Bassuk, N. 1991. Effect of time and application of sodium chloride in the dormant season on selected tree seedlings. J. Environ. Hortic. 9:130–136.

1571. Heald, F. D. 1913. The symptoms of chestnut tree blight and a brief description of the blight fungus. Penn. Chestnut Tree Blight Commiss. Bull. 5. 15 pp.

1572. Heald, F. D., and Studhalter, R. A. 1914. The Strumella disease of oak and chestnut trees. Penn. Dep. For. Bull. 10. 15 pp.

1573. Heald, F. D., and Wolf, F. A. 1911. New species of Texas fungi. Mycologia 3:5–22.

1574. Heale, E. L., and Ormrod, D. P. 1982. Effects of nickel and copper on Acer rubrum, Cornus stolonifera, Lonicera tatarica, and Pinus resinosa. Can. J. Bot. 60:2674–2681.

1575. Hearon, S. S., Sherald, J. L., and Kostka, S. J. 1980. Association of xylem-limited bacteria with elm, sycamore, and oak leaf scorch. Can. J. Bot. 58:1986–1993.

1576. Heath, M. C. 1992. Host species specificity of the goldenrod rust fungus and the existence of rust resistance within some goldenrod species. Can. J. Bot. 70:2461–2466.

1577. Hebard, F. V., Griffin, G. J., and Elkins, J. R. 1984. Developmental histopathology of cankers incited by hypovirulent and virulent isolates of *Endothia parasitica* on susceptible and resistant chestnut trees. Phytopathology 74:140–149.

1578. Hecht-Poinar, E. I., Britton, J. C., and Parmeter, J. R. Jr. 1981. Dieback of oaks in California. Plant Dis. 65:281.

1579. Hedgcock, G. G., and Hahn, G. G. 1922. Two important pine cone rusts and their new cronartial stages. Phytopathology 12:109–122.

1580. Hedgcock, G. G., and Long, W. H. 1915. A disease of pines caused by *Cronartium pyriforme*. USDA Bull. 247. 20 pp.

1581. Hedgcock, G. G., and Siggers, P. V. 1949. A comparison of the pine-oak rusts. USDA Tech. Bull. 978. 30 pp.

1582. Hedges, F., and Tenny, L. S. 1912. A knot of citrus trees caused by *Sphaeropsis tumefaciens*. USDA Bull. 247. 74 pp.

1583. Heiberg, S. O., and White, D. P. 1951. Potassium deficiency of reforested pine and spruce stands in northern New York. Soil Sci. Soc. Am. Proc. 15:369–376.

1584. Heichel, G. H., Turner, N. C., and Walton, G. S. 1972. Anthracnose causes dieback of regrowth on defoliated oak and maple. Plant Dis. Rep. 56:1046–1047.

1585. Heide, J. H. S. 1991. Anatomy and ultrastructure of the haustorium of *Cassytha pubescens* R. Br. I. The adhesive disk. Bot. Gaz. 152:321–334.

1586. Heimann, M. 1976. The spread of the fungal parasite silver leaf (*Stereum purpureum*) by natural root grafting. Deutsche Baumschule 28:34–37.

1587. Heimann, M. F., and Worf, G. L. 1987. Basal stem canker of poinsettia caused by *Fusarium solani* in Wisconsin. Plant Dis. 71:469.

1588. Heiniger, U., ed. 2001. *Castanea sativa:* pathology, genetic resources, ecology, and management. For. Snow Landscape Res. 76:327–517.

1589. Heiniger, U., and Rigling, D. 1994. Biological control of chestnut blight in Europe. Annu. Rev. Phytopathol. 32:581–599.

1590. Heiser, I., Fromm, J., Giefing, M., Koehl, J., Jung, T., and Osswald, W. 1999. Investigations on the action of *Phytophthora quercina*, *P. citricola* and *P. gonapodyides* toxins on tobacco plants. Plant Physiol. Biochem. 37:73–81.

1591. Helander, M. L., Sieber, T. N., Petrini, O., and Neuvonen, S. 1994. Endophytic fungi in Scots pine needles: spatial variation and consequences of simulated acid rain. Can. J. Bot. 72:1108–1113.

1592. Helander, M. L., Vuorinen, P., Saikkonen, K., and Lappalainen, J. 1998. Evidence for resistance of mountain birch (*Betula pubescens* ssp. *czerepanovii*) to birch rust (*Melampsoridium betulinum*). Mycol. Res. 102:63–66.

1593. Helgerson, O. T. 1990. Heat damage in tree seedlings and its prevention. New For. 3:333–358.

1594. Helguera, P. R., Docampo, D. M., Nome, S. F., and Ducasse, D. A. 2002. Enhanced detection of *Prune dwarf virus* in peach leaves by immunocapture-reverse transcription-polymerase chain reaction with nested polymerase chain reaction (IC-RT-PCR nested PCR). J. Phytopathol. 150:94–96.

1595. Hellgren, M., and Barklund, P. 1992. Studies of the life cycle of *Gremmeniella abietina* on Scots pine in southern Sweden. Eur. J. For. Pathol. 22:300–311.

1596. Hellgren, M., and Hogberg, N. 1995. Ecotypic variation of *Gremmeniella abietina* in northern Europe: disease patterns reflected by DNA variation. Can. J. Bot. 73:1531–1539.

1597. Helms, J. A., Cobb, F. W. Jr., and Whitney, H. S. 1971. Effect of infection by *Verticicladiella wageneri* on the physiology of *Pinus ponderosa*. Phytopathology 61:920–925.

1598. Helton, A. W., Johnson, J. B., and Dilbeck, R. D. 1988. Arthropod carriers of *Cytospora* propagules in stone-fruit orchards. Plant Dis. 72:734.

1599. Hemmi, T., and Kurata, S. 1931. Studies on septorioses of plants. II. *Septoria azaleae* Voglino causing the brown-spot disease of the cultivated azaleas in Japan. Mem. Coll. Agric. Kyoto Imp. Univ. No. 13, Art. 1. 22 pp.

1600. Henderson, D. M. 1961. *Glomospora* and *Glomopsis*. R. Bot. Gard. Edinb. Notes 23:497–502.

1601. Hendrickson, O. Q. 1991. Abundance and activity of N_2-fixing bacteria in decaying wood. Can. J. For. Res. 21:1299–1304.

1602. Hendry, S. J., Lonsdale, D., and Boddy, L. 1998. Strip-cankering of beech (*Fagus sylvatica*): pathology and distribution of symptomatic trees. New Phytol. 140:549–565.

1603. Hendson, M., Purcell, A. H., Chen, D. Q., Smart, C., Guilhabert, M., and Kirkpatrick, B. 2001. Genetic diversity of Pierce's disease strains and other pathotypes of *Xylella fastidiosa*. Appl. Environ. Microbiol. 67:895–903.

1604. Heneen, W. K., Gustafsson, M., Brismar, K., and Karlsson, G. 1994. Interactions between Norway spruce (*Picea abies*) and *Heterobasidion annosum*. II. Infection of woody roots. Can. J. Bot. 72:884–889.

1605. Heneen, W. K., Gustafsson, M., Karlsson, G., and Brismar, K. 1994. Interactions between Norway spruce (*Picea abies*) and *Heterobasidion annosum*. I. Infection of nonsuberized and young suberized roots. Can. J. Bot. 72:872–883.

1606. Hennebert, G. L. 1973. *Botrytis* and *Botrytis*-like genera. Persoonia 7:183–204.

1607. Hennon, P. E., Beatty, J. S., and Hildebrand, D. 2001. Hemlock dwarf mistletoe. USDA For. Serv. For. Insect Dis. Leafl. 135. 8 pp. Internet pub.

1608. Hennon, P. E., and Hansen, E. M. 1987. Nuclear behavior of *Phellinus arctostaphyli*, *P. igniarius* and *P. tremulae*. Mycologia 79:501–507.

1609. Henry, B. W. 1951. Oak leaf blister severe in south Mississippi. Plant Dis. Rep. 35:384.

1610. Henry, B. W., Moses, C. S., Richards, C. A., and Riker, A. J. 1944. Oak wilt: its significance, symptoms, and cause. Phytopathology 34:636–647.

1611. Hepting, G. H. 1939. A vascular wilt of the mimosa tree (*Albizia julibrissin*). USDA Circ. 535. 10 pp.

1612. Hepting, G. H. 1944. Sapstreak, a new killing disease of sugar maple. Phytopathology 34:1069–1076.

1613. Hepting, G. H. 1963. Climate and forest diseases. Annu. Rev. Phytopathol. 1:31–50.

1614. Hepting, G. H. 1971. Diseases of forest and shade trees of the United States. USDA Agric. Handb. 386. 658 pp.

1615. Hepting, G. H., and Hedgcock, G. G. 1937. Decay in merchantable oak, yellow poplar, and basswood in the Appalachian region. USDA Tech. Bull. 570. 29 pp.

1616. Hepting, G. H., Miller, J. H., and Campbell, W. A. 1951. Winter of 1950–51 damaging to southeastern woody vegetation. Plant Dis. Rep. 35:502–503.

1617. Hepting, G. H., and Roth, E. R. 1946. Pitch canker, a new disease of some southern pines. J. For. 44:742–744.

1618. Hepting, G. H., and Toole, E. R. 1939. The hemlock rust caused by *Melampsora farlowii*. Phytopathology 29:463–473.

1619. Hermanides-Nijhof, E. J. 1977. *Aureobasidium* and allied genera. Stud. Mycol. No. 15:141–177.

1620. Herrero, J. 1951. Studies of compatible and incompatible graft combinations with special reference to hardy fruit trees. J. Hortic. Sci. 26:186–237.

1621. Herridge, E. A., and Lambe, R. C. 1960. A holly leafspot associated with the use of copper fungicides. (Abstr.) Phytopathology 50:84.

1622. Hesler, L. R. 1916. Black rot, leaf spot, and canker of pomaceous fruits. Cornell Univ. Agric. Exp. Stn. Bull. 379:49–148.

1623. Heslin, M. C., Stuart, M. R., Murchú, P. O., and Donnelly, D. M. X. 1983. Fomannoxin, a phytotoxic metabolite of *Fomes annosus*: in vitro production, host toxicity, and isolation from naturally infected Sitka spruce heartwood. Eur. J. For. Pathol. 13:11–23.

1624. Hess, N. J., Otrosina, W. J., Jones, J. P., Goddard, A. J., Walkinshaw, C. H., and Haywood, J. D. 1999. Reassessment of loblolly pine decline on the Oakmulgee Ranger District, Talladega National Forest, Alabama. USDA For. Serv. Gen. Tech. Rep. SRS-30:560–564.

1625. Hessburg, P. F., Goheen, D. J., and Bega, R. V. 1995. Black stain root disease of conifers. USDA For. Serv. For. Insect Dis. Leafl. 145. 11 pp. Internet pub.

1626. Hessburg, P. F., Goheen, D. J., and Koester, H. 2001. Association of black stain root disease with roads, skid trails, and precommercial thinning in southwest Oregon. West. J. Appl. For. 16:127–135.

1627. Hessburg, P. F., and Hansen, E. M. 1986. Soil temperature and rate of colonization of *Ceratocystis wageneri* in Douglas-fir. Phytopathology 76:627–631.

1628. Hessburg, P. F., and Hansen, E. M. 2000. Infection of Douglas-fir by *Leptographium wageneri*. Can. J. Bot. 78:1254–1261.

1629. Hewitt, H. G., and Ayers, E. G. 1976. Effect of infection by *Microsphaera alphitoides* (powdery mildew) on carbohydrate levels and translocation of seedlings of *Quercus robur*. New Phytol. 77:379–390.

1630. Hewitt, W. B. 1939. Leaf-scar infection in relation to the olive-knot disease. Hilgardia 12:41–66.

1631. Hibben, C. R. 1959. Relations of *Stegonosporium ovatum* (Pers. ex Mérat) Hughes with dieback of sugar maple (*Acer saccharum* Marsh.). M.S. thesis, Cornell Univ., Ithaca, NY. 63 pp.

1632. Hibben, C. R. 1964. Identity and significance of certain organisms associated with sugar maple decline in New York woodlands. Phytopathology 54:1389–1392.

1633. Hibben, C. R., and Bozarth, R. F. 1972. Identification of an ash strain of tobacco ringspot virus. Phytopathology 62:1023–1029.

1634. Hibben, C. R., Bozarth, R. F., and Reese, J. 1979. Identification of tobacco necrosis virus in deteriorating clones of aspen. For. Sci. 25:557–567.

1635. Hibben, C. R., and Franzen, L. M. 1989. Susceptibility of lilacs to mycoplasmalike organisms. J. Environ. Hortic. 7:163–167.

1636. Hibben, C. R., and Hagar, S. S. 1975. Pathogenicity of an ash isolate of tobacco ringspot virus. Plant Dis. Rep. 59:57–60.

1637. Hibben, C. R., Lewis, C. A., and Castello, J. D. 1986. Mycoplasmalike organisms, cause of lilac witches'-broom. Plant Dis. 70:342–345.

1638. Hibben, C. R., Reese, J. A., and Castello, J. D. 1988. Identification of tomato ringspot virus in ash in New York. Plant Dis. 72:175.

1639. Hibben, C. R., and Silverborg, S. B. 1978. Severity and causes of ash dieback. J. Arboric. 4:274–279.

1640. Hibben, C. R., and Walker, J. T. 1966. A leaf roll–necrosis complex of lilacs in an urban environment. Proc. Am. Soc. Hortic. Sci. 89:636–642.

1641. Hibben, C. R., and Walker, J. T. 1971. Nematode transmission of the ash strain of tobacco ringspot virus. Plant Dis. Rep. 55:475–478.

1642. Hibben, C. R., Walker, J. T., and Allison, J. R. 1977. Powdery mildew ratings of lilac species and cultivars. Plant Dis. Rep. 61:192–196.

1643. Hibben, C. R., and Wolanski, B. 1971. Dodder transmission of a mycoplasma from ash witches'-broom. Phytopathology 61:151–156.

1644. Hibberd, J. M., Bungard, R. A., Press, M. C., Jeschke, W. D., Scholes, J. D., and Quick, W. P. 1998. Localization of photosynthetic metabolism in the parasitic angiosperm Cuscuta reflexa. Planta 205:506–513.

1645. Hibbs, R. H. 1976. Decline of hackberry attributed to ambient herbicide drift. Proc. Iowa Acad. Sci. 72:187–190.

1646. Hicock, H. W., and Olson, A. R. 1954. The toxicity to plants of wood preservatives and their solvents. Conn. Agric. Exp. Stn. Circ. 189. 4 pp.

1647. Hiemstra, J. A., and Rataj-Guranowska, M. 2003. Vegetative compatibility groups in Verticillium dahliae isolates from the Netherlands as compared to VCG diversity in Europe and in the USA. Eur. J. Plant Pathol. 109:827–839.

1648. Higgins, B. B. 1914. Contribution to the life history and physiology of Cylindrosporium on stone fruits. Am. J. Bot. 1:145–173.

1649. Higgins, B. B. 1917. A disease of pecan catkins. Phytopathology 7:42–45.

1650. Highley, L., and Tattar, T. A. 1985. Leptographium terebrantis and black turpentine beetles associated with blue stain and mortality of black and Scots pines on Cape Cod, Massachusetts. Plant Dis. 69:528–530.

1651. Highley, T. L., Bar-Lev, S. S., Kirk, T. K., and Larsen, M. J. 1983. Influence of O_2 and CO_2 on wood decay by heartrot and saprot fungi. Phytopathology 73:630–633.

1652. Hilborn, M. T. 1937. The anatomy of a black zone caused by Xylaria polymorpha. Phytopathology 27:1177–1179.

1653. Hilborn, M. T. 1942. The biology of Fomes fomentarius. Maine Agric. Exp. Stn. Bull. 409:161–214.

1654. Hildebrand, E., Skelly, J. M., Fredericksen, T. S., Percy, K. E., Cox, R. M., and Jensen, K. F. 1996. Foliar response of ozone-sensitive hardwood tree species from 1991 to 1993 in the Shenandoah National Park, Virginia. Can. J. For. Res. 26:658–669.

1655. Hildebrand, E. M. 1942. Tomato ringspot on currant. Am. J. Bot. 29:362–366.

1656. Hildebrand, E. M. 1947. Perennial peach canker and the canker complex in New York, with methods of control. Cornell Univ. Agric. Exp. Stn. Mem. 276. 61 pp.

1657. Hildebrand, E. M. 1953. Yellow-red or X-disease of peach. Cornell Univ. Agric. Exp. Stn. Mem. 323. 54 pp.

1658. Hildebrand, P. D., Nickerson, N. L., McRae, K. B., and Lu, X. 2000. Incidence and impact of red leaf disease caused by Exobasidium vaccinii in lowbush blueberry fields in Nova Scotia. Can. J. Plant Pathol. 22:364–367.

1659. Hiley, W. E. 1919. The fungal diseases of the common larch. Oxford Univ. Press, Oxford, UK. 204 pp.

1660. Hill, B. L., and Purcell, A. H. 1995. Acquisition and retention of Xylella fastidiosa by an efficient vector, Graphocephala atropunctata. Phytopathology 85:209–212.

1661. Hill, B. L., and Purcell, A. H. 1997. Populations of Xylella fastidiosa in plants required for transmission by an efficient vector. Phytopathology 87:1197–1201.

1662. Hill, G. T., and Sinclair, W. A. 2000. Taxa of leafhoppers carrying phytoplasmas on sites of ash yellows occurrence in New York State. Plant Dis. 84:134–138.

1663. Hill, S. A. 1975. The importance of wood scab caused by Venturia inaequalis (Cke.) Wint. as a source of infection for apple leaves in the spring. Phytopathol. Z. 82:216–223.

1664. Himelick, E. B. 1961. Sycamore anthracnose. Proc. Natl. Shade Tree Conf. 37:136–143.

1665. Himelick, E. B. 1969. Tree and shrub hosts of Verticillium albo-atrum. Ill. Nat. Hist. Surv. Biol. Notes No. 66. 8 pp.

1666. Himelick, E. B., and Himelick, K. J. 1980. Systemic treatment for chlorotic trees. J. Arboric. 6:192–196.

1667. Himelick, E. B., and Neely, D. 1960. Juniper hosts of cedar-apple rust and cedar-hawthorn rust. Plant Dis. Rep. 44:109–112.

1668. Hinds, T. E. 1962. Inoculations with the sooty-bark canker fungus on aspen. Plant Dis. Rep. 46:57–58.

1669. Hinds, T. E. 1972. Ceratocystis canker of aspen. Phytopathology 62:213–220.

1670. Hinds, T. E. 1972. Insect transmission of Ceratocystis species associated with aspen cankers. Phytopathology 62:221–225.

1671. Hinds, T. E. 1981. Cryptosphaeria canker and Libertella decay of aspen. Phytopathology 71:1137–1145.

1672. Hinds, T. E. 1985. Diseases. Pages 87–106 in: Aspen: ecology and management in the western United States. USDA For. Serv. Gen. Tech. Rep. RM-119. 283 pp.

1673. Hinds, T. E., and Hawksworth, F. G. 1965. Seed dispersal velocity in four dwarf mistletoes. Science 148:517–519.

1674. Hinds, T. E., and Ryan, M. G. 1985. Expansion of sooty-bark and Ceratocystis cankers on aspen. Plant Dis. 69:842–844.

1675. Hirano, S. S., and Upper, C. D. 2000. Bacteria in the leaf ecosystem with emphasis on Pseudomonas syringae—a pathogen, ice nucleus, and epiphyte. Microbiol. Mol. Biol. Rev. 64:624–653.

1676. Hirata, T., Cunnington, J. H., Paksiri, U., Limkaisang, S., Shishkoff, N., Grigaliunaite, B., Sato, Y., Takamatsu, S., Urailak, P., and Saranya, L. 2000. Evolutionary analysis of subsection Magnicellulatae of Podosphaera section Sphaerotheca (Erysiphales) based on the rDNA internal transcribed spacer sequences with special reference to host plants. Can. J. Bot. 78:1521–1530.

1677. Hiratsuka, N. 1936. A monograph of the Pucciniastreae. Mem. Tottori Agric. Coll., Tottori, Jap. Vol. 4. 374 pp.

1678. Hiratsuka, N. 1958. Revision of taxonomy of the Pucciniastreae with special reference to species of the Japanese archipelago. Tokyo Univ. Educ. Faculty Agric., Tokyo. Contrib. No. 31. 167 pp.

1679. Hiratsuka, N., Sato, S., Katsuya, K., Kakishima, M., Hiratsuka, Y., Kaneko, S., Ono, Y., Sato, T., Harada, Y., Hiratsuka, T., and Nakayama, K. 1992. The rust flora of Japan. Tsukuba Shuppankai, Takezono, Ibaraki, Jap. 1205 pp.

1680. Hiratsuka, Y. 1969. Endocronartium, a new genus for autoecious pine stem rusts. Can. J. Bot. 47:1493–1495.

1681. Hiratsuka, Y. 1973. The nuclear cycle and the terminology of spore states in the Uredinales. Mycologia 65:432–443.

1682. Hiratsuka, Y. 1984. New leaf spot fungus, Marssonina balsamiferae, on Populus balsamifera in Manitoba and Ontario. Mycotaxon 19:133–136.

1683. Hiratsuka, Y. 1987. Forest tree diseases of the prairie provinces. Can. For. Serv. Inf. Rep. Nor-X-286. 142 pp.

1683a. Hiratsuka, Y., Langor, D. W., and Crane, P. E. 1995. Forest insects and diseases of the prairie provinces. Can. For. Serv. North. For. Cent. Spec. Rep. 3. 297 pp.

1684. Hiratsuka, Y., McArthur, L. E., and Emond, F. J. 1967. A distinction between Pucciniastrum goeppertianum and P. epilobii, with clarification of status of Peridermium holwayi and P. ornamentale. Can. J. Bot. 45:1913–1915.

1685. Hiratsuka, Y., and Powell, J. M. 1976. Pine stem rusts of Canada. Can. For. Serv. North. For. Res. Cent., Edmonton, Tech. Rep. No. 4. 103 pp.

1686. Hiratsuka, Y., Powell, J. M., and Sickle, G. A. Van.1988. Impact of pine stem rusts of hard pines in Alberta and the Northwest Territories. Can. For. Serv. Inf. Rep. NOR-X-299. 9 pp.

1687. Hiratsuka, Y., Samoil, J. K., Blenis, P. V., Crane, P. E., and Laishley, B. L., eds. 1991. Rusts of pine. Proceedings of the IUFRO rusts of pine working party conference. Can. For. Serv. Inf. Rep. NOR-X-317. 408 pp.

1688. Hiratsuka, Y., and Zalasky, H. 1993. Frost and other climate-related damage of forest trees in the prairie provinces. Can. For. Serv. Northwest Region Inf. Rep. NOR-X-331. 25 pp.

1689. Hirt, R. R. 1964. Cronartium ribicola: its growth and reproduction in the tissues of eastern white pine. State Univ. Coll. For. (Syracuse) Tech. Pub. 86. 30 pp.

1690. Hirt, R. R., and Eliason, E. J. 1938. The development of decay in living trees inoculated with *Fomes pinicola*. J. For. 36:705–709.

1691. Hiruki, C. 1985. A preliminary study on infectious variegation of *Camellia*. Acta Hortic. No. 164:55–59.

1692. Hiruki, C. 1999. Phytoplasma diseases of forest and urban trees and shrubs in Canada and East Asia: biology, genomic diversity and management. Pages 39–52 in: Biotechnology and plant protection in forestry science. S. P. Raychaudhuri and K. Maramorosch, eds. Science Pub., Lebanon, NH.

1693. Hiruki, C., ed. 1988. Tree mycoplasma diseases and epidemiology. Univ. Alberta Press, Edmonton. 245 pp.

1694. Hobbs, S. D., and Partridge, A. D. 1979. Wood decays, root rots, and stand composition along an elevation gradient. For. Sci. 25:31–42.

1695. Hodge, G. R., White, T. L., Schmidt, R. A., and Allen, J. E. 1993. Stability of rust infection ratios for resistant and susceptible slash and loblolly pine across rust hazard levels. South. J. Appl. For. 17:188–192.

1696. Hodges, C. S. Jr. 1962. Black root rot of pine seedlings. Phytopathology 52:210–219.

1697. Hodges, C. S. Jr. 1963. Black root rot of pine. Phytopathology 53:1131–1134.

1698. Hodges, C. S. [Jr.] 1983. Pine mortality in Hawaii associated with *Botryosphaeria dothidea*. Plant Dis. 67:555–556.

1699. Hodges, C. S. Jr., Alfenas, A. C., and Ferreira, F. A. 1986. The conspecificity of *Cryphonectria cubensis* and *Endothia eugeniae*. Mycologia 78:343–350.

1700. Hodges, C. S. [Jr.], and May, L. C. 1972. *Cercospora sequoiae* in Brasil. Fitopatologia 7:32–34.

1701. Hodges, C. S. [Jr.], Geary, T. F., and Cordell, C. E. 1979. The occurrence of *Diaporthe cubensis* in eucalypts in Florida, Hawaii, and Puerto Rico. Plant Dis. Rep. 63:216–220.

1702. Hodges, C. S. [Jr.], and Haasis, F. A. 1964. The perfect stage of *Cercospora magnoliae*. Mycologia 56:53–57.

1703. Hoff, R. J. 1985. Susceptibility of lodgepole pine to the needle cast fungus *Lophodermella concolor*. USDA For. Serv. Res. Note INT-349. 6 pp.

1704. Hoff, R. J. 1987. Susceptibility of inland Douglas-fir to Rhabdocline needle cast. USDA For. Serv. Res. Note INT-375. 3 pp.

1705. Hoff, R. J., Bingham, R. T., and McDonald, G. I. 1980. Relative blister rust resistance of white pines. Eur. J. For. Pathol. 10:307–316.

1706. Hoffman, G. M., and Fliege, F. 1967. *Kabatina juniperi* als Ursache eines Zweigsterbens an verschiedenen Juniperusarten. Z. Pflanzenkr. Pflanzenschutz 74:587–593.

1707. Hoffman, J. T., and Hobbs, E. L. 1985. Lodgepole pine dwarf mistletoe in the Intermountain Region. Plant Dis. 69:429–431.

1708. Hofstra, G., Hall, R., and Lumis, G. P. 1979. Studies of salt-induced damage to roadside plants in Ontario. J. Arboric. 5:25–31.

1709. Hogan, E. P., and Griffin, G. J. 2002. Spread of Cryphonectria hypovirus 1 into 45 vegetative compatibility types of *Cryphonectria parasitica* on grafted American chestnut trees. For. Pathol. 32:73–85.

1710. Hogberg, N., Holdenrieder, O., and Stenlid, J. 1999. Population structure of the wood decay fungus *Fomitopsis pinicola*. Heredity 83:354–360.

1711. Höhnel, F. von. 1916. Über *Cheilaria aceris*. Sitzungsb. K. Akad. Wiss., Math-nat. Kl. I, 125:81–84.

1712. Hoitink, H. A. J., and Boehm, M. J. 1999. Biocontrol within the context of soil microbial communities: a substrate-dependent phenomenon. Annu. Rev. Phytopathol. 37:427–446.

1713. Hoitink, H. A. J., and Schmitthenner, A. F. 1969. Rhododendron wilt caused by *Phytophthora citricola*. Phytopathology 59:708–709.

1714. Hoitink, H. A. J., and Schmitthenner, A. F. 1974. Relative prevalence and virulence of *Phytophthora* species involved in rhododendron root rot. Phytopathology 64:1371–1374.

1715. Hoitink, H. A. J., and Schmitthenner, A. F. 1974. Resistance of *Rhododendron* species and hybrids to Phytophthora root rot. Plant Dis. Rep. 58:650–653.

1716. Hoitink, H. A. J., Stone, A. G., and Han, D. Y. 1997. Suppression of plant diseases by composts. HortScience 32:184–187.

1717. Hoitink, H. A. J., VanDoren, D. M. Jr., and Schmitthenner, A. F. 1977. Suppression of *Phytophthora cinnamomi* in a composted hardwood bark potting medium. Phytopathology 67:561–565.

1718. Holcomb, G. E. 1986. Hosts of the parasitic alga *Cephaleuros virescens* in Louisiana and new host records for the continental United States. Plant Dis. 70:1180–1183.

1719. Holdenrieder, O., and Kowalski, T. 1989. Pycnidial formation and pathogenicity in *Tubakia dryina*. Mycol. Res. 92:166–169.

1720. Holliday, P., and Punithalingam, E. 1970. *Sphaeropsis tumefaciens*. C.M.I. Descr. Pathogenic Fungi Bact. No. 278. 2 pp.

1721. Hollinger, D. Y. 1983. Photosynthesis and water relations of the mistletoe *Phoradendron villosum* and its host, the California valley oak, *Quercus lobata*. Oecologia 60:396–400.

1722. Hollis, C. A., and Schmidt, R. A. 1977. Site factors related to fusiform rust incidence in north Florida slash pine plantations. For. Sci. 23:69–77.

1723. Holmer, L., Nitare, L., and Stenlid, J. 1994. Population structure and decay pattern of *Phellinus tremulae* in *Populus tremula* as determined by somatic incompatibility. Can. J. Bot. 72:1391–1396.

1724. Holmes, F. O., Hirumi, H., and Maramorosch, K. 1972. Witches'-broom of willow: *Salix* yellows. Phytopathology 62:826–828.

1725. Holmes, F. W. 1976. Verticillium wilt of salt-injured sugar maple—preliminary study. (Abstr.) Proc. Am. Phytopathol. Soc. 3:305–306.

1726. Holmes, F. W. 1984. Effects on maples of prolonged exposure by artificial girdling roots. J. Arboric. 10:40–44.

1727. Holmes, F. W., and Heybroek, H. M., trans. 1990. Dutch elm disease—the early papers: selected works of seven Dutch women phytopathologists. APS Press, St. Paul, MN. 154 pp.

1728. Holmes, J., and Rich, A. E. 1970. Factors affecting release and dissemination of *Physalospora obtusa* spores in a New Hampshire apple orchard. Phytopathology 60:1052–1054.

1729. Holst-Jensen, A., Kohn, L. M., Jakobsen, K. S., and Schumacher, T. 1997. Molecular phylogeny and evolution of *Monilinia* (Sclerotiniaceae) based on coding and noncoding rDNA sequences. Am. J. Bot. 84:686–701.

1730. Holst-Jensen, A., Kohn, L. M., and Schumacher, T. 1997. Nuclear rDNA phylogeny of the Sclerotiniaceae. Mycologia 89:885–899.

1731. Holst-Jensen, A., Vaage, M., and Schumacher, T. 1998. An approximation to the phylogeny of *Sclerotinia* and related genera. Nordic J. Bot. 18:705–719.

1732. Holt, H. L., Grant, J. F., and Windham, M. T. 1998. Incidence of arthropods infested with conidia of the dogwood anthracnose fungus, *Discula destructiva* Redlin, on flowering dogwoods in the natural environment. J. Entomol. Sci. 33:329–335.

1733. Holt, J. G., ed. in chief. 1984–1989. Bergey's manual of systematic bacteriology. Vols. 1–4. Williams & Wilkins, Baltimore. 2648 pp.

1734. Holt, J. G., Krieg, N. R., Sneath, P. H. A., Staley, J. T., and Williams, S. T., eds. 1994. Bergey's manual of determinative bacteriology. 9th ed. Williams & Wilkins, Baltimore. 787 pp.

1735. Holtz, B. A., Michailides, T. J., and Hong, C. 1998. Development of apothecia from stone fruit infected and stromatized by *Monilinia fructicola* in California. Plant Dis. 82:1375–1380.

1736. Honey, E. E. 1940. *Monilinia* causing a brown rot and blight of the common azalea. Phytopathology 30:537–538.

1737. Hong, C., and Michailides, T. J. 1998. Effect of temperature on the discharge and germination of ascospores by apothecia of *Monilinia fructicola*. Plant Dis. 82:195–202.

1738. Hood, I. A. 1982. *Phaeocryptopus gaeumannii* on *Pseudotsuga menziesii* in southern British Columbia. N.Z. J. For. Sci. 12:415–424.

1739. Hood, M. E., and Shew, H. D. 1997. Initial cellular interactions between *Thielaviopsis basicola* and tobacco root hairs. Phytopathology 87:228–235.

1740. Hood, M. E., and Shew, H. D. 1997. Reassessment of the role of saprophytic activity in the ecology of *Thielaviopsis basicola*. Phytopathology 87:1214–1219.

1741. Hoog, G. S. de, and Hermanides-Nijhof, E. J. 1977. The black yeasts and allied Hyphomycetes. Stud. Mycol. No. 15. 222 pp.

1742. Hook, D. D. 1984. Waterlogging tolerance of lowland tree species of the South. South. J. Appl. For. 8:136–149.

1743. Hook, J. M. Van, and Busteed, R. C. 1935. Anthracnose of *Betula nigra*. Proc. Ind. Acad. Sci. 44:81.

1744. Hooper, M. C., Arii, K., and Lechowicz, M. J. 2001. Impact of a major ice storm on an old-growth hardwood forest. Can. J. Bot. 79:70–75.

1745. Hootman, R. G., Kelsey, P. D., Reid, R., and Heide-Spravka, K. von der. 1994. Factors affecting accumulation of deicing salts in soils around trees. J. Arboric. 20:196–201.

1746. Hoover, K., Wood, D. L., Storer, A. J., Fox, J. W., and Bros, W. E. 1996. Transmission of the pitch canker fungus, *Fusarium subglutinans* f.sp. *pini*, to Monterey pine, *Pinus radiata*, by cone- and twig-infesting beetles. Can. Entomol. 128:981–994.

1747. Hopkin, A., Griefenhagen, S., and Holland, J. 2001. Decay, stains, and beetles in ice-storm-damaged forests: a review. For. Chron. 77:605–611.

1748. Hopkins, D. L. 1989. *Xylella fastidiosa*: xylem-limited bacterial pathogen of plants. Annu. Rev. Phytopathol. 27:271–290.

1749. Hopkins, D. L., and Purcell, A. H. 2002. *Xylella fastidiosa*: cause of Pierce's disease of grapevine and other emergent diseases. Plant Dis. 86:1056–1066.

1750. Hopkins, D. L., Thompson, C. M., Wichman, R. L., Bistline, F. W., and Russo, L. W. 1995. Effect of inoculation of mature citrus trees in the grove with *Xylella fastidiosa* on citrus blight incidence. Proc. Fla. State Hortic. Soc. 108:103–106.

1751. Hopkins, J. C. 1963. Atropellis canker of lodgepole pine: etiology, symptoms, and canker development rates. Can. J. Bot. 41:1535–1545.

1752. Horie, H., and Kobayashi, T. 1979. Brown leaf spot of mountain-laurel, *Kalmia latifolia* L., caused by *Cercospora kalmiae* Ell. & Ev. Bull. Tokyo Agric. Exp. Stn. 12:10–18.

1753. Horie, H., and Kobayashi, T. 1979–1980. Entomosporium leaf spot of Pomoideae (Rosaceae) in Japan. I. Distribution of the disease, morphology and physiology of the fungus. II. Parasitism and over-wintering of the fungus. III. Additional basis for identification of the fungus and distribution of the disease. Eur. J. For. Pathol. 9:366–379; 10:117–124, 225–235.

1754. Horner, I. J., and Wilcox, W. F. 1996. Temporal changes in activity and dormant spore populations of *Phytophthora cactorum* in New York apple orchard soils. Phytopathology 86:1133–1139.

1755. Horntvedt, R. 1995. Fluoride uptake in conifers related to emissions from aluminium smelters in Norway. Sci. Total Environ. 163:35–37.

1756. Horsley, S. B., Long, R. P., Bailey, S. W., Hallett, R. A., and Hall, T. J. 2000. Factors associated with the decline disease of sugar maple on the Allegheny Plateau. Can. J. For. Res. 30:1365–1378.

1757. Horsley, S. B., Long, R. P., Bailey, S. W., Hallett, R. A., and Wargo, P. M. 2002. Health of eastern North American sugar maple forests and factors affecting decline. North. J. Appl. For. 19:34–44.

1758. Horst, R. K. 1983. Compendium of rose diseases. Am. Phytopathol. Soc., St. Paul, MN. 50 pp.

1759. Horst, R. K. 2001. Westcott's plant disease handbook. 6th ed. Van Nostrand Reinhold, New York. 1008 pp.

1760. Horvath, J., Eke, I., Gal, T., and Dezéry, M. 1975. Demonstration of virus-like particles in sweet chestnut and oak with leaf deformations in Hungary. Z. Pflanzenkr. Pflanzenschutz 82:498–502.

1761. Hoshino, T., Odaira, M., Yoshida, M., and Tsuda, S. 1999. Physiological and biochemical significance of antifreeze substances in plants. J. Plant Res. 112:255–261.

1762. Hotson, H. H., and Cutter, V. M. 1951. The isolation and culture of *Gymnosporangium juniperi-virginianae* Schw. upon artificial media. Proc. Natl. Acad. Sci. USA 37:400–403.

1763. Hough, A., Mulder, N. J., and LaGrange, J. M. 1979. Heat treatment for the control of Phytophthora gummosis in citrus. Plant Dis. Rep. 63:40–43.

1764. Houston, D. B., and Houston, D. R. 2000. Allozyme genetic diversity among *Fagus grandifolia* trees resistant or susceptible to beech bark disease in natural populations. Can. J. For. Res. 30:778–789.

1765. Houston, D. R. 1963. Inoculation of oaks with *Urnula craterium* (Schw.) Fr. produces cankers identical to Strumella cankers. Plant Dis. Rep. 47:867–869.

1766. Houston, D. R. 1969. Basal canker of white pine. For. Sci. 15:66–83.

1767. Houston, D. R. 1981. Stress triggered tree diseases. The diebacks and declines. USDA For. Serv. NS-INF-41-81. 36 pp.

1768. Houston, D. R. 1985. Dieback and declines of urban trees. J. Arboric. 11:65–72.

1769. Houston, D. R. 1985. Diebacks and declines of urban trees. Pages 120–137 in: Improving the quality of urban life with plants. D. F. and S. L. Karnosky, eds. N.Y. Bot. Gard., Inst. Urban Hortic. Pub. No. 2. 200 pp.

1770. Houston, D. R. 1993. Recognizing and managing sapstreak disease of sugar maple. USDA For. Serv. Res. Pap. NE-675. 11 pp.

1771. Houston, D. R. 1994. Major new tree disease epidemics: beech bark disease. Annu. Rev. Phytopathol. 32:75–87.

1772. Houston, D. R. 1994. Sapstreak disease of sugar maple: development over time and space. USDA For. Serv. Res. Pap. NE-687. 19 pp.

1773. Houston, D. R. 1994. Temporal and spatial shift within the *Nectria* pathogen complex associated with beech bark disease of *Fagus grandifolia*. Can. J. For. Res. 24:960–968.

1774. Houston, D. R., Drake, C. R., and Kuntz, J. E. 1965. Effects of environment on oak wilt development. Phytopathology 55:1114–1121.

1775. Houston, D. R., and O'Brien, J. T. 1983. Beech bark disease. USDA For. Serv. For. Insect Dis. Leafl. 75. Internet pub.

1776. Houston, D. R., Parker, E. J., and Lonsdale, D. 1979. Beech bark disease: patterns of spread and development of the inciting agent *Cryptococcus fagisuga*. Can. J. For. Res. 9:336–344.

1777. Houston, D. R., Parker, E. J., Perrin, R., and Lang, K. J. 1979. Beech bark disease: a comparison of the disease in North America, Great Britain, France, and Germany. Eur. J. For. Pathol. 9:199–211.

1778. Howard, F. L. 1941. The bleeding canker disease of hardwoods and possibilities of control. Proc. West. Shade Tree Conf. 8:46–55.

1779. Howard, F. W., and Harrison, N. A. (undated) Lethal yellowing of palms. Univ. Fla. Internet pub.

1780. Howard, F. W., and Thomas, D. L. 1980. Transmission of palm lethal decline to *Veitchia merrillii* by a planthopper *Myndus crudus*. J. Econ. Entomol. 73:715–717.

1781. Howard, F. W., Williams, D. S., and Norris, R. C. 1984. Insect transmission of lethal yellowing to young palms. Int. J. Entomol. 26:331–338.

1782. Howden, J. C. W., and Jacobs, L. 1973. Report on the rust work at Bath. Rose Annu. 1973:113–119.

1783. Howells, J. 1993. Clematis wilt. A review of the literature. Plantsman 15:148–160.

1784. Hoy, J. W., Mircetich, S. M., and Lownsbery, B. F. 1984. Differential transmission of Prunus tomato ringspot virus strains by *Xiphinema californicum*. Phytopathology 74:332–335.

1785. Hrazdina, G., Borejsza, W. W., and Lester, C. 1997. Phytoalexin production in an apple cultivar resistant to *Venturia inaequalis*. Phytopathology 87:868–876.

1786. Hsiang, T., and Chastagner, G. A. 1993. Variation in *Melampsora occidentalis* rust on poplars in the Pacific Northwest. Can. J. Plant Pathol. 15:175–181.

1787. Hsiang, T., Edmonds, R. L., and Driver, C. H. 1989. Conidia of *Heterobasidion annosum* from *Tsuga heterophylla* forests in western Washington. Can. J. Bot. 67:1262–1266.

1788. Hsiang, T., Huang, J., Yang, L., Cook, S., Shen, Y., and Richter, K. J. 2000. Occurrence of *Kabatina juniperi* in Ontario and genetic analysis using RAPD markers. Can. J. Plant Pathol. 22:79–88.

1789. Huang, J. W., and Kuhlman, E. G. 1990. Fungi associated with damping-off of slash pine seedlings in Georgia. Plant Dis. 74:27–30.

1790. Huang, Z. Y., Smalley, E. B., and Guries, R. P. 1995. Differentiation of *Mycosphaerella dearnessii* by cultural characters and RAPD analysis. Phytopathology 85:522–527.

1791. Hubbes, M. 1959. Untersuchungen über *Dothichiza populea* Sacc. et Briard, den Erreger des Rindenbrandes der Pappel. Phytopathol. Z. 35:58–96.

1792. Hubbes, M. 1960. Systematische und physiologische Untersuch-ungen an Valsaceen auf Weiden. Phytopathol. Z. 39:65–93.

1793. Huberman, M. A. 1943. Sunscald of eastern white pine, *Pinus strobus* L. Ecology 24:456–471.

1794. Hubert, E. E. 1931. An outline of forest pathology. Wiley, New York. 543 pp.

1795. Hubert, E. E., and Humphrey, C. J. 1920. A canker of the elm in Wisconsin. Wisc. Dep. Agric. Biennial Rep., Bull. 33:158–163.

1796. Hudler, G. W. 1984. Wound healing in bark of woody plants. J. Arboric. 10:241–245.

1797. Hudler, G. W., Banik, M. T., and Miller, S. G. 1987. Unusual epidemic of tar spot on Norway maple in upstate New York. Plant Dis. 71:65–68.

1798. Hudler, G. W., and Beale, M. A. 1981. Anatomical features of girdling root injury. J. Arboric. 7:29–32.

1799. Hudler, G. W., and French, D. W. 1976. Dispersal and survival of seed of eastern dwarf mistletoe. Can. J. For. Res. 6:335–340.

1800. Hudler, G. W., Jensen-Tracy, S., and Banik, M. T. 1998. *Rhytisma americanum* sp. nov.: a previously undescribed species on maples (*Acer* spp.). Mycotaxon 68:405–416.

1801. Hudler, G. W., Knudsen, G. R., and Beale, M. A. 1983. Dose-response relationships of five conifers to infection by conidia of *Gremmeniella abietina*. Plant Dis. 67:192–194.

1802. Hudler, G. W., and Oshima, N. 1976. The occurrence and distribution of *Thyronectria austro-americana* on honeylocust in Colorado. Plant Dis. Rep. 60:920–922.

1803. Hudler, G. W., Oshima, N., and Hawksworth, F. G. 1979. Bird dissemination of dwarf mistletoe on ponderosa pine in Colorado. Am. Midl. Nat. 102:273–280.

1804. Hudson, H. J. 1987. Guignardia leaf blotch of horsechestnut. Trans. Br. Mycol. Soc. 89:400–401.

1805. Huettl, R. F., and Mueller-Dombois, D., eds. 1993. Forest decline in the Atlantic and Pacific region. Springer-Verlag, Berlin, New York. 366 pp.

1806. Hughes, S. J. 1952. Studies on micro-fungi. XIV. *Stigmella, Stigmina, Camptomeris, Polythrincium,* and *Fusicladiella*. Commonw. Mycol. Inst. Mycol. Pap. No. 49. 25 pp.

1807. Hughes, S. J. 1958. Revisiones hyphomycetum aliquot cum appendice de nominibus rejiciendis. Can. J. Bot. 36:727–836.

1808. Hughes, S. J. 1960. Microfungi V. *Conoplea* Pers. and *Exosporium* Link. Can. J. Bot. 38:659–696.

1809. Hughes, S. J. 1976. Sooty moulds. Mycologia 68:693–820.

1810. Huhndorf, S. M. 1992. Studies in *Leptosphaeria*. Transfer of *Leptosphaeria opuntiae* to *Montagnula* (Ascomycetes). Brittonia 44:208–212.

1811. Huhndorf, S. M. 1992. Systematics of *Leptosphaeria* species found on the Rosaceae. Ill. Nat. Hist. Surv. Bull. 34:479–534.

1812. Hull, R. 2002. Matthews' plant virology. 4th ed. Academic Press, San Diego. 1001 pp.

1813. Hull, R. J., and Leonard, O. A. 1964. Physiological aspects of parasitism in mistletoes (*Arceuthobium* and *Phoradendron*) I. The carbohydrate nutrition of mistletoe. Plant Physiol. 39:996–1007.

1814. Hummer, K. E., and Sniezko, R., conveners. 2000. Ribes, pines, and white pine blister rust. Proceedings of the conference, Corvallis, Oregon, 1999. HortTechnology 10:511–570.

1815. Humphreys-Jones, D. R. 1977. Leaf and shoot death (*Coniothyrium fuckelii* Sacc.) of *Juniperus communis* L. var. *compressa* Carr. Plant Pathol. 26:47–48.

1816. Humphreys-Jones, D. R. 1977. Leaf blotch (*Coniothyrium viburni* Died.) of *Viburnum burkwoodii* L. Plant Pathol. 26:101.

1817. Hunt, R. S. 1984. Inoculations of Scrophulariaceae with *Cronartium ribicola*. Can. J. Bot. 62:2523–2524.

1818. Hunt, R. S. 1995. Common pine needle casts and blights in the Pacific Region. Can. For. Serv. Pac. For. Cent. For. Pest Leafl. 43. 7 pp.

1819. Hunt, R. S. 1997. Differential medium for *Phellinus pini* and *Inonotus tomentosus*. Can. J. Plant Pathol. 19:307–309.

1820. Hunt, R. S. 1997. Relative value of slow-canker growth and bark reactions as resistance responses to white pine blister rust. Can. J. Plant Pathol. 19:352–357.

1821. Hunt, R. S. 1998. Pruning western white pine in British Columbia to reduce white pine blister rust losses: 10-year results. West. J. Appl. For. 13:60–63.

1822. Hunt, R. S. 2002. Relationship between early family-selection traits and natural blister rust cankering in western white pine families. Can. J. Plant Pathol. 24:200–204.

1823. Hunt, R. S. 2004. Blister-rust-resistant white pines for British Columbia. Can. For. Serv. BC-X-397. 18 pp.

1824. Hunt, R. S. 2004. White pine blister rust. Rec. Res. Devel. Mycol. 1:73–85. Transworld Res. Network, Kerala, India.

1825. Hunt, R. S., and Ekramoddoullah, A. K. 1996. Protein patterns distinguish among Canadian isolates of *Inonotus tomentosus, I. circinatus* and *Phellinus pini*. Mycologia 88:395–402.

1826. Hunt, R. S., Ekramoddoullah, A. K. M., and Zamani, A. 1999. Production of a polyclonal antibody to *Phellinus pini* and examination of its potential use in diagnostic assays. Eur. J. For. Pathol. 29:259–272.

1827. Hunt, R. S., and Etheridge, D. E. 1995. True heart-rots of the Pacific Region. Can. For. Serv. Pac. For. Cent. For. Pest Leafl. 55. 8 pp.

1828. Hunt, R. S., Owens, J. N., and Smith, R. B. 1996. Penetration of western hemlock, *Tsuga heterophylla,* by the dwarf mistletoe *Arceuthobium tsugense,* and development of the parasite cortical system. Can. J. Plant Pathol. 18:342–346.

1829. Hunt, R. S., and Sickle, G. A. Van. 1984. Variation in susceptibility to sweet fern rust among *Pinus contorta* and *P. banksiana*. Can. J. For. Res. 14:672–675.

1830. Hunt, R. S., and Unger, L. 1994. Tomentosus root disease. Can. For. Serv. Pac. For. Cent. For. Pest Leafl. 77. 8 pp.

1831. Hunt, R. S., Ying, C. C., and Ashbee, D. 1987. Variation in damage among *Pinus contorta* provenances caused by the needle cast fungus *Lophodermella concolor*. Can. J. For. Res. 17:594–597.

1832. Hunt, R. S., and Ziller, W. G. 1978. Host-genus keys to the Hypodermataceae of conifer leaves. Mycotaxon 6:481–496.

1833. Hunter, P. P., and Stipes, R. J. 1978. The effect of month of inoculation with *Endothia gyrosa* on development of pruned branch canker of pin oak (*Quercus palustris*). Plant Dis. Rep. 62:940–944.

1834. Huo, Y. L., and Li, G. W. 1990. Investigations on decline and death of *Pinus densiflora* in the Yantai region, Shandong Province. For. Pest Dis. 3:21–22.

1835. Huppuch, C. D. 1960. Observations on white oak stem swellings. Plant Dis. Rep. 44:238–239.

1836. Hursh, C. R., and Haasis, F. W. 1931. Effects of 1925 summer drought on southern Appalachian hardwoods. Ecology 12:380–386.

1837. Huss, D. K. 1997. Actinorhizal symbioses and their N_2 fixation. New Phytol. 136:375–405.

1838. Hutchins, A. S., Fay, H., and Knutson, D. 1985. A selective medium for *Phellinus weirii*. Can. J. For. Res. 15:746–748.

1839. Hutchinson, T. C., Watmough, S. A., Sager, E. P. S., and Karagatzides, J. D. 1998. Effects of excess nitrogen deposition and soil acidification on sugar maple (*Acer saccharum*) in Ontario, Canada: an experimental study. Can. J. For. Res. 28:299–310.

1840. Hutchison, L. J. 1999. Wood-inhabiting microfungi isolated from *Populus tremuloides* from Alberta and northeastern British Columbia. Can. J. Bot. 77:898–905.

1841. Hüttl, R. F., and Schaaf, W., eds. 1997. Magnesium deficiency in forest ecosystems. Nutrients in ecosystems. Kluwer Academic, Dordrecht. 362 pp.

1842. Iacobellis, N. S., Caponero, A., and Evidente, A. 1998. Characterization of *Pseudomonas syringae* ssp. *savastanoi* strains isolated from ash. Plant Pathol. 47:73–83.

1843. Iacobellis, N. S., Sisto, A., Surico, G., Evidente, A., and DiMaio, E. 1994. Pathogenicity of *Pseudomonas syringae* subsp. *savastanoi* mutants defective in phytohormone production. J. Phytopathol. 140:238–248.

1844. Ichihara, Y., Fukuda, K., and Suzuki, K. 2000. Early symptom development and histological changes associated with migration of *Bursaphelenchus xylophilus* in seedling tissues of *Pinus thunbergii*. Plant Dis. 84:675–680.

1845. Iglesias, I., Dalmau, R., Marce, X., Campillo, M. C. del, Barron, V., and Torrent, J. 2000. Fertilization with iron(II)-phosphate effectively prevents iron chlorosis in pear trees, *Pyrus communis* L. Acta Hortic. No. 511:65–72.

1846. Igoe, M. J., Peterson, N. C., and Roberts, D. L. 1995. A Phomopsis canker associated with branch dieback of Colorado blue spruce in Michigan. Plant Dis. 79:202–205.

1847. Ihl, B., and Wiese, K. 2000. Studies on *Cuscuta reflexa* Roxb.: VIII. Mechanical induction of haustoria formation in non-twining stems of the parasite. Flora 195:1–8.

1848. Ikeda, T. 1996. Xylem dysfunction in *Bursaphelenchus xylophilus*-infected *Pinus thunbergii* in relation to xylem cavitation and water status. Ann. Phytopathol. Soc. Jap. 62:554–558.

1849. Ikeda, T., and Kiyohara, T. 1995. Water relations, xylem embolism and histological features of *Pinus thunbergii* inoculated with virulent or avirulent pine wood nematode, *Bursaphelenchus xylophilus*. J. Exp. Bot. 46:441–449.

1850. Ikin, R., and Frost, R. R. 1974. Virus diseases of roses I. Their occurrence in the United Kingdom. Phytopathol. Z. 79:160–168.

1851. Illingworth, K. 1973. Variation in the susceptibility of lodgepole pine provenances to Sirococcus blight. Can. J. For. Res. 3:585–589.

1852. Ingold, C. T., Davey, R. A., and Wakley, G. 1981. The teliospore pedicel of *Phragmidium mucronatum*. Trans. Br. Mycol. Soc. 77:439–442.

1853. Ingold, C. T., and Hudson, H. J. 1993. The biology of fungi. 6th ed. Chapman & Hall, New York. 224 pp.

1854. Ingram, D. E. 1914. A twig blight of *Quercus prinus* and related species. J. Agric. Res. 1:339–346.

1855. Innes, J. L., and Haron, A. H., eds. 2000. Air pollution and the forests of developing and rapidly industrializing regions. Report No. 4 of the IUFRO Task Force on Environmental Change. CABI Pub., Wallingford, UK. 262 pp.

1856. Innes, J. L., and Oleksyn, J., eds. 2000. Forest dynamics in heavily polluted regions: Report No. 1 of the IUFRO task force on environmental change. CABI Pub., Oxford, UK. 248 pp.

1857. Institut National de la Recherche Agronomique (Paris). 1980. Colloque sur la maladie de l'écorce du hêtre. Ann. Sci. For. 37:269–392.

1858. International Association of Wood Anatomists. 1984. Special issue on discolored wood. IAWA Bull., n.s., 5:91–154.

1859. International Mycological Institute. 1964–present. IMI descriptions of fungi and bacteria. [Formerly C.M.I. descriptions of pathogenic fungi and bacteria, Nos. 1–1010, and C.M.I. descriptions of fungi and bacteria, Nos. 1011–1030.] Int. Mycol. Inst., Kew, Surrey, UK.

1860. International Organization for Mycoplasmology, International Research Programme on Comparative Mycoplasmology (IRCPM), Phytoplasma Working Team of the IRPCM. 1999. Proceedings, first Internet conference on phytopathogenic mollicutes, 1999. Internet pub.

1861. International Union of Forest Research Organizations [IUFRO]. 1983. Proceedings of beech bark disease working party conference. USDA For. Serv. Gen. Tech. Rep. WO-37. 140 pp.

1862. International Union of Forest Research Organizations Research Group 7.04.00. 1999. Contributions from 18th IUFRO workshop on air pollution stress. Forest growth responses to the pollution climate of the 21st century. Water Air Soil Poll. 116, Nos. 1–2. 448 pp.

1863. International Working Group on Fruit Tree Viruses. 1998. Detection of virus and virus-like diseases of fruit trees. Acta Hortic. No. 472:761–783.

1864. Intini, M., Paoletti, E., and Lombardi, M. 1998. Incidence and characteristics of *Fomes fomentarius* on trees in an urban park. Monti e Boschi 49:29–33.

1865. Ioannou, N., Schneider, R. W., and Grogan, R. G. 1977. Effect of flooding on the soil gas composition and the production of microsclerotia by *Verticillium dahliae* in the field. Phytopathology 67:651–656.

1866. Isebrands, J. G., McDonald, E. P., Kruger, E., Hendrey, G., Percy, K., Pregitzer, K., Sober, J., and Karnosky, D. F. 2001. Growth responses of *Populus tremuloides* clones to interacting elevated carbon dioxide and tropospheric ozone. Environ. Pollut. 115:359–371.

1867. Ishida, K., and Hogetsu, T. 1997. Role of resin canals in the early stages of pine wilt disease caused by the pine wood nematode. Can. J. Bot. 75:346–351.

1868. Ishii, H., Udagawa, H., Nishimoto, S., Tsuda, T., and Nakashima, H. 1992. Scab resistance in pear species and cultivars. Acta Phytopathol. Entomol. Hung. 27:293–298.

1869. Ishii, H., and Yanase, H. 2000. *Venturia nashicola*, the scab fungus of Japanese and Chinese pears: a species distinct from *V. pirina*. Mycol. Res. 104:755–759.

1870. Ishikawa, M., Price, W. S., Ide, H., and Arata, Y. 1997. Visualization of freezing behaviors in leaf and flower buds of full-moon maple by nuclear magnetic resonance microscopy. Plant Physiol. 115:1515–1524.

1871. Ito, K., Chiba, O., Ono, K., and Hosaka, Y. 1954. Pestalotia disease of *Camellia japonica* L. Bull. Gov. For. Exp. Stn., Tokyo 70:103–124.

1872. Ito, K., Kobayashi, T., and Hayashi, H. 1972. Stem gall or stem swelling of larch seedlings caused by a phytotoxicity of benzene hexachloride (BHC). Bull. Gov. For. Exp. Stn., Tokyo, No. 245. 19 pp.

1873. Ito, K., and Shibukawa, K. 1956. Studies on some anthracnoses of woody plants—III. A new anthracnose of acacia with special reference to the life history of the causal fungus. Bull. Gov. For. Exp. Stn., Tokyo 92:51–64.

1874. Ito, K., Shibukawa, K., and Kobayashi, T. 1974. Etiological and pathological studies on the needle blight of *Cryptomeria japonica*—IV. Blight and canker of the tree caused by *Cercospora sequoiae* Ellis et Everhart (*C. cryptomeriae* Shirai). Bull. Gov. For. Exp. Stn., Tokyo No. 268:81–134.

1875. Ito, K., Zinno, Y., and Kobayashi, T. 1963. Larch canker in Japan. Bull. Gov. For. Exp. Stn., Tokyo 155:25–47.

1876. Itoh, M., and Manabe, K. 1997. Effect of leaching of a soil-applied herbicide, diuron, on its phytotoxicity in grape and peach. J. Jap. Soc. Hortic. Sci. 66:221–228.

1877. Ivors, K. L., Hayden, K. J., Bonants, P. J. M., Rizzo, D. M., and Garbelotto, M. 2004. AFLP and phylogenetic analyses of North American and European populations of Phytophthora ramorum. Mycol. Res. 108:378–392.

1878. Ivory, M. H. 1994. Records of foliage pathogens of *Pinus* species in tropical countries. Plant Pathol. 43:511–518.

1879. Iwahori, H., Kanzaki, N., and Futai, K. 2000. A simple, polymerase chain reaction-restriction fragment length polymorphism-aided diagnosis method for pine wilt disease. For. Pathol. 30:157–164.

1880. Jackson, H. S. 1914. A new pomaceous rust of economic importance, *Gymnosporangium blasdaleanum*. Phytopathology 4:261–268.

1881. Jackson, L. W. R. 1940. Lightning injury of black locust seedlings. Phytopathology 30:183–184.

1882. Jackson, M. B., and Attwood, P. A. 1996. Roots of willow (*Salix viminalis* L.) show marked tolerance to oxygen shortage in flooded soils and in solution culture. Plant Soil 187:37–45.

1883. Jacobi, V., and Castello, J. D. 1992. Infection of red spruce, black spruce, and balsam fir seedlings with tomato mosaic virus. Can. J. For. Res. 22:919–924.

1884. Jacobi, V., Castello, J. D., and Flachmann, M. 1992. Isolation of tomato mosaic virus from red spruce. Plant Dis. 76:518–522.

1885. Jacobi, W. R. 1984. Optimal conditions for *in vitro* growth, asexual spore release, and germination of *Thyronectria austro-americana*. Phytopathology 74:566–569.

1886. Jacobi, W. R. 1989. Resistance of honeylocust cultivars to *Thyronectria austro-americana*. Plant Dis. 73:805–807.

1887. Jacobi, W. R. 1992. Seasonal effects on wound susceptibility and canker expansion in honeylocusts inoculated with *Thyronectria austro-americana*. J. Arboric. 18:288–293.

1888. Jacobi, W. R., Geils, B. W., Taylor, J. E., and Zentz, W. R. 1993. Predicting the incidence of comandra blister rust on lodgepole pine: site, stand, and alternate-host influences. Phytopathology 83:630–637.

1889. Jacobi, W. R., and MacDonald, W. L. 1980. Colonization of resistant and susceptible oaks by *Ceratocystis fagacearum*. Phytopathology 70:618–623.

1890. Jacobi, W. R., and Riffle, J. W. 1989. Effects of water stress on Thyronectria canker of honeylocusts. Phytopathology 79:1333–1337.

1891. Jacobs, K., and Wingfield, M. J. 2001. *Leptographium* species: tree pathogens, insect associates, and agents of blue stain. APS Press, St. Paul, MN. 207 pp.

1892. Jacobs, K., Wingfield, M. J., Uzunovic, A., and Frisullo, S. 2001. Three new species of *Leptographium* from pine. Mycol. Res. 105:490–499.

1893. Jacobs, K. A., and Danielson, D. 2002. Variation in field susceptibility of native and exotic ash species to anthracnose. J. Arboric. 28:35–40.

1894. Jacobs, K. A., Lee, I.-M., Griffiths, H. M., Miller, F. D. Jr., and Bottner, K. D. 2003. A new member of the clover proliferation phytoplasma group (16SrVI) associated with elm yellows in Illinois. Plant Dis. 87:241–246.

1895. Jacobs, K. A., MacDonald, J. D., Cobb, F. W. Jr., and Wells, K. 1994. Identification of *Armillaria* species in California. Mycologia 86:113–116.

1896. Jacobs, K. A., and Rehner, S. A. 1998. Comparison of cultural and morphological characters and ITS sequences in anamorphs of *Botryosphaeria* and related taxa. Mycologia 90:601–610.

1897. Jacobs, K. A., Santamour, F. S. Jr., Johnson, G. R., and Dirr, M. A. 1996. Differential resistance to Entomosporium leafspot disease and hydrogen cyanide potential in *Photinia*. J. Environ. Hortic. 14:154–157.

1898. Jacobs, K. A., and Spravka, M. 1996. High- and low-performing crabapples during abnormally severe apple scab conditions. HortTechnology 6:140.

1899. Jacobson, J. S., and Hill, A. C., eds. 1970. Recognition of air pollution injury to vegetation: a pictorial atlas. Air Pollut. Control Assoc. Pittsburgh, PA. 109 pp.

1900. Jaffee, B. A., Abawi, G. S., and Mai, W. F. 1982. Role of soil microflora and *Pratylenchus penetrans* in an apple replant disease. Phytopathology 72:247–251.

1901. Jagoueix, E. S., Tarendeau, F., Guolter, K., Danet, J. L., Bové, J. M., and Garnier, M. 2001. *Catharanthus roseus* genes regulated differentially by mollicute infections. Mol. Plant-Microbe Interact. 14:225–233.

1902. Jalaluddin, M. 1967. Studies on *Rhizina undulata*. I. Mycelial growth and ascospore germination. II. Observations and experiments in East Anglian plantations. Trans. Br. Mycol. Soc. 50:449–459, 461–472.

1903. Jalkanen, R., Crane, P. E., Walla, J. A., and Aalto, T., eds. 1998. Proceedings of the first IUFRO rusts of forest trees working party conference, 1998, Saariselkä, Finland. Finn. For. Res. Inst. Res. Pap. 712. 309 pp.

1904. Jalkanen, R. E., Redfern, D. B., and Sheppard, L. J. 1998. Nutrient deficits increase frost hardiness in Sitka spruce (*Picea sitchensis*) needles. For. Ecol. Manag. 107:191–201.

1905. James, R. L., Cobb, F. W. Jr., Miller, P. R., and Parmeter, J. R. Jr. 1980. Effects of oxidant air pollution on susceptibility of pine roots to *Fomes annosus*. Phytopathology 70:560–563.

1906. James, R. L., Cobb, F. W. Jr., Wilcox, W. W., and Rowney, D. L. 1980. Effects of photochemical oxidant injury of ponderosa and Jeffrey pines on susceptibility of sapwood and freshly cut stumps to *Fomes annosus*. Phytopathology 70:704–708.

1907. James, R. L., Dumroese, R. K., and Wenny, D. L. 1995. *Botrytis cinerea* carried by adult fungus gnats (Diptera: Sciaridae) in container nurseries. Tree Planters' Notes 46:48–53.

1908. James, S. 1984. Lignotubers and burls—their structure, function, and significance in Mediterranean ecosystems. Bot. Rev. 50:225–266.

1909. James, T. Y., Moncalvo, J. M., Li, S., and Vilgalys, R. 2001. Polymorphism at the ribosomal DNA spacers and its relation to breeding structure of the widespread mushroom *Schizophyllum commune*. Genetics 157:149–161.

1910. James, T. Y., Porter, D., Hamrick, J. L., and Vilgalys, R. 1999. Evidence for limited intercontinental gene flow in the cosmopolitan mushroom, *Schizophyllum commune*. Evolution 53:1665–1677.

1911. Jandl, R., Anglberger, H., Reh, M., and Halmschlager, E. 2000. Effect of fertilization of a secondary spruce stand in the Kobernausserwald with shoot blight symptoms. Bodenkultur 51:247–258.

1912. Jandl, R., Glatzel, G., Katzensteiner, K., and Eckmüllner, O. 2001. Amelioration of magnesium deficiency in a Norway spruce stand (*Picea abies*) with calcined magnesite. Water Air Soil Pollut. 125:1–17.

1913. Janse, J. D. 1981–1982. The bacterial disease of ash (*Fraxinus excelsior*) caused by *Pseudomonas syringae* pv. *fraxini*. I. History, occurrence and symptoms. II. Etiology and taxonomic considerations. III. Pathogenesis. Eur. J. For. Pathol. 11:306–315, 425–438; 12:218–231.

1914. Janse, J. D. 1991. Pathovar discrimination within *Pseudomonas syringae* subsp. *savastanoi* using whole cell fatty acids and pathogenicity as criteria. Syst. Appl. Microbiol. 14:79–84.

1915. Jarausch, W., Eyquard, J. P., Mazy, K., Lansac, M., and Dosba, F. 1999. High level of resistance of sweet cherry (*Prunus avium* L.) towards European stone fruit yellows phytoplasmas. Adv. Hortic. Sci. 13:108–112.

1916. Jarausch, W., Lansac, M., and Dosba, F. 1999. Seasonal colonization pattern of European stone fruit yellows phytoplasmas in different *Prunus* species detected by specific PCR. J. Phytopathol. 147:47–54.

1917. Jarvis, W. R. 1977. *Botryotinia* and *Botrytis* species: taxonomy, physiology, and pathogenicity. A guide to the literature. Can. Dep. Agric., Res. Branch, Monogr. 15. 195 pp.

1918. Jeffers, S. N., and Aldwinckle, H. S. 1988. Phytophthora crown rot of apple trees: sources of *Phytophthora cactorum* and *P. cambivora* as primary inoculum. Phytopathology 78:328–335.

1919. Jeffers, S. N., Aldwinckle, H. S., Burr, T. J., and Arneson, P. A. 1982. *Phytophthora* and *Pythium* species associated with crown rot in New York apple orchards. Phytopathology 72:533–538.

1920. Jenkins, A. E. 1930. *Sphaceloma symphoricarpi*. Mycologia 22:106–110.

1921. Jenkins, A. E. 1930. The cherry leaf-spot fungus, *Mycosphaerella cerasella* Aderh., its morphology and life history. Phytopathology 20:329–337.

1922. Jenkins, A. E., and Bitancourt, A. A. 1957. Studies in the Myriangiales. VII. Elsinoaceae on evergreen euonymus, rose, and English ivy. Mycologia 49:95–101.

1923. Jenkins, A. E., Bitancourt, A. A., and Pollack, F. G. 1946. Spot anthracnoses in the United States. J. Wash. Acad. Sci. 36:416–421.

1924. Jenkins, A. E., and Jehle, R. A. 1951. An anthracnose of bittersweet in Maryland. Plant Dis. Rep. 35:413–414.

1925. Jenkins, A. E., Miller, J. H., and Hepting, G. H. 1953. Spot anthracnose and other leaf and petal spots of flowering dogwood. Natl. Hortic. Mag. 32:57–69.

1926. Jenkins, M. A., and Pallardy, S. G. 1995. The influence of drought on red oak group species growth and mortality in the Missouri Ozarks. Can. J. For. Res. 25:1119–1127.

1927. Jennings, D. H. 1995. The physiology of fungal nutrition. Cambridge Univ. Press, Cambridge, UK. 622 pp.

1928. Jennings, D. H., and Lysek, G. 1996. Fungal biology: understanding the fungal lifestyle. BIOS Scientific Pub., Oxford, UK. 156 pp.

1929. Jensen, D. D. 1971. Herbaceous host plants of western X-disease agent. Phytopathology 61:1465–1470.

1930. Jensen, K. F. 1983. Growth relationships in silver maple seedlings fumigated with O_3 and SO_2. Can. J. For. Res. 13:298–302.

1931. Jensen, K. F., and Kozlowski, T. T. 1975. Absorption and transportation of sulfur dioxide by seedlings of four forest tree species. J. Environ. Qual. 4:379–382.

1932. Jentschke, G., and Godbold, D. L. 2000. Metal toxicity and ectomycorrhizas. Physiol. Plant. 109:107–116.

1933. Jerome, C. A., and Ford, B. A. 2002. The discovery of three genetic races of the dwarf mistletoe *Arceuthobium americanum* (Viscaceae) provides insight into the evolution of parasitic angiosperms. Mol. Ecol. 11:387–405.

1934. Jewell, F. F. Sr. 1983. Histopathology of the brown spot fungus on longleaf pine needles. Phytopathology 73:854–858.

1935. Jewell, F. F. Sr. 1990. Histopathology of longleaf pine needles infected by *Ploioderma hedgcockii* (Dearn.) Darker. Eur. J. For. Pathol. 20:24–31.

1936. Jewell, F. F. Sr. 1993. Histopathology of loblolly pine needles infected by *Mycosphaerella dearnessii*, the brown-spot needle disease pathogen. Eur. J. For. Pathol. 23:110–119.

1937. Jewell, F. F. Sr. 1994. Histopathology of *Pinus elliottii* var. *elliottii* needles infected by *Ploioderma hedgcockii*. Eur. J. For. Pathol. 24:323–334.

1938. Jewell, F. F. Sr. 2001. Histopathology of *Pinus taeda* needles infected by *Ploioderma lethale*. For. Pathol. 31:33–42.

1939. Jewell, F. F., and Walker, N. M. 1967. Histology of *Cronartium quercuum* galls on shortleaf pine. Phytopathology 57:545–550.

1940. Jikumaru, S., and Togashi, K. 2000. Temperature effects on the transmission of *Bursaphelenchus xylophilus* (Nemata: Aphelenchoididae) by *Monochamus alternatus* (Coleoptera: Cerambycidae). J. Nematol. 32:110–116.

1941. Jimenez, D. F., and Chew, M. Y. 1996. Periods of solarization and their influence on populations of soil fungi and melon production (*Cucumis melo* L.). Rev. Mex. Fitopatol. 14:38–47.

1942. Jin, Y., and Steffenson, B. J. 1999. *Puccinia coronata* var. *hordei* var. nov.: morphology and pathogenicity. Mycologia 91:877–884.

1943. Jing, X. F., Yang, Z. X., Zhang, Y. X., and Li, X. Z. 1983. Studies on hawthorne blossom blight I. Causal organism. II. Symptoms. Acta Phytopathol. Sinica 12:33–36; 13:42–44.

1944. Jock, S., Donat, V., Lopez, M. M., Bazzi, C., and Geider, K. 2002. Following spread of fire blight in western, central and southern Europe by molecular differentiation of *Erwinia amylovora* strains with PFGE analysis. Environ. Microbiol. 4:106–114.

1945. Johansson, M. 1988. Pectic enzyme activity of spruce (S) and pine (P) strains of *Heterobasidion annosum* (Fr.) Bref. Physiol. Mol. Plant Pathol. 33:333–349.

1946. Johnson, A. H. 1992. The role of abiotic stresses in the decline of red spruce in high elevation forests of the eastern United States. Annu. Rev. Phytopathol. 30:349–367.

1947. Johnson, D. W. 1986. Comandra rust. USDA For. Serv. For. Insect Dis. Leafl. 62. Internet pub.

1948. Johnson, D. W., Beatty, J. S., and Hinds, T. E. 1995. Cankers on western quaking aspen. USDA For. Serv. For. Insect Dis. Leafl. 152. Internet pub.

1949. Johnson, D. W., and Kuntz, J. E. 1979. Eutypella canker of maple: ascospore discharge and dissemination. Phytopathology 69:130–135.

1950. Johnson, D. W., and Kuntz, J. E. 1979. *Eutypella parasitica*: ascospore germination and mycelial growth. Can. J. Bot. 57:624–628.

1951. Johnson, G. R., and Hauer, R. J. 2000. A practitioner's guide to stem girdling roots of trees: Impacts on trees, symptomology, and prevention. Univ. Minn. Exten. Serv. BU-07501-GO. 20 pp.

1952. Johnson, J. M., and Choinski, J. S. Jr. 1993. Photosynthesis in the *Tapinanthus-Diplorhynchus* mistletoe-host relationship. Ann. Bot. 72:117–122.

1953. Johnson, J. W., Gleason, M. L., Parker, S. K., Provin, E. B., Iles, J. K., and Flynn, P. H. 1997. Duration of water stress affects development of Sphaeropsis canker on Scots pine. J. Arboric. 23:73–76.

1954. Johnson, K. B., Mehlenbacher, S. A., Stone, J. K., Pscheidt, J. W., and Pinkerton, J. N. 1996. Eastern filbert blight of European hazelnut: it's becoming a manageable disease. Plant Dis. 80:1308–1316.

1955. Johnson, K. B., Pinkerton, J. N., Gauderault, S. M., and Stone, J. K. 1994. Infection of European hazelnut by *Anisogramma anomala*: site of infection and effect of host developmental stage. Phytopathology 84:1465–1470.

1956. Johnston, J. C., and Trione, E. J. 1974. Cytokinin production by the fungi *Taphrina cerasi* and *T. deformans*. Can. J. Bot. 52:1583–1589.

1957. Johnston, P. R., and Jones, D. 1997. Relationships among *Colletotrichum* isolates from fruit-rots assessed using rDNA sequences. Mycologia 89:420–430.

1958. Jokela, J. J., Paxton, J. D., and Zegar, E. J. 1976. Marssonina leaf spot and rust on eastern cottonwood. Plant Dis. Rep. 60:1020–1024.

1959. Jones, A. L., and Aldwinckle, H. S., eds. 1990. Compendium of apple and pear diseases. APS Press, St. Paul, MN. 100 pp.

1960. Jones, A. T. 1973. A comparison of some properties of four strains of cherry leafroll virus. Ann. Appl. Biol. 74:211–217.

1961. Jones, A. T., Brown, D. J. F., Halbrendt, J. M., Vrain, T. C., and Robbins, R. T. 1995. The transmission of three nepoviruses by populations of four *Xiphinema americanum*–group species. Acta Hortic. No. 385:105–109.

1962. Jones, A. T., and Mayo, M. A. 1973. Purification and properties of elm mottle virus. Ann. Appl. Biol. 75:347–357.

1963. Jones, C., Griffin, G. J., and Elkins, J. R. 1980. Association of climatic stress with blight on Chinese chestnut in the eastern United States. Plant Dis. 64:1001–1004.

1964. Jones, E. A., Reed, D. D., Mroz, G. D., Liechty, H. O., and Cattelino, P. J. 1993. Climate stress as a precursor to forest decline: paper birch in northern Michigan, 1985–1990. Can. J. For. Res. 23:229–233.

1965. Jones, J. P., and Holcomb, G. E. 1978. Conidium ontogeny and cytology of *Tubakia dryina* from Louisiana hardwoods. Mycologia 70:1212–1216.

1966. Jones, L. H. 1943. Creosote injurious to plants. Horticulture, n.s., 21:462–463.

1967. Jones, R. K., and Benson, D. M., eds. 2001. Diseases of woody ornamentals and trees in nurseries. APS Press, St. Paul, MN. 482 pp.

1968. Jones, S. G. 1925. Life history and cytology of *Rhytisma acerinum* (Pers.) Fries. Ann. Bot. 39:41–75.

1969. Jong, M. D. de. 2000. The BioChon story: deployment of *Chondrostereum purpureum* to suppress stump sprouting in hardwoods. Mycologist 14:58–62.

1970. Jong, M. D. de, Scheepens, P. C., and Zadoks, J. C. 1990. Risk analysis for biological control: a Dutch case study in biocontrol of *Prunus serotina* by the fungus *Chondrostereum purpureum*. Plant Dis. 74:189–194.

1971. Jong, S. C., and Benjamin, C. R. 1971. North American species of *Nummularia*. Mycologia 63:862–876.

1972. Jong, S. C., and Rogers, J. D. 1972. Illustrations and descriptions of conidial states of some *Hypoxylon* species. Wash. Agric. Exp. Stn. Tech. Bull. 71. 51 pp.

1973. Jong, S. N. de, Levesque, C. A., Verkley, G. J. M., Abeln, E. C. A., Rahe, J. E., and Braun, P. G. 2001. Phylogenetic relationships among *Neofabraea* species causing tree cankers and bull's-eye rot of apple based on DNA sequencing of ITS nuclear rDNA, mitochondrial rDNA, and the β-tubulin gene. Mycol. Res. 105:658–669.

1974. Jorgensen, E., and Cafley, J. D. 1961. Branch and stem cankers of white and Norway spruces in Ontario. For. Chron. 37:394–404.

1975. Joseph, G., Kelsey, R. G., and Thies, W. G. 1998. Hydraulic conductivity in roots of ponderosa pine infected with black-stain (*Leptographium wageneri*) or annosus (*Heterobasidion annosum*) root disease. Tree Physiol. 18:333–339.

1976. Joubert, J. J., and Rijkenberg, F. H. J. 1971. Parasitic green algae. Annu. Rev. Phytopathol. 9:45–64.

1977. Ju, Y.-M., and Rogers, J. D. 1996. A revision of the genus *Hypoxylon*. Mycologia Mem. No. 20. 365 pp.

1978. Ju, Y.-M., Rogers, J. D., Martin, F. S., and Granmo, A. 1998. The genus *Biscogniauxia*. Mycotaxon 66:1–98.

1979. Julis, A. J., Clayton, C. N., and Sutton, T. B. 1978. Detection and distribution of *Phytophthora cactorum* and *P. cambivora* on apple rootstocks. Plant Dis. Rep. 62:516–520.

1980. Jung, H., Sawayanagi, T., Kakizawa, S., Nishigawa, H., Miyata, S. I., Oshima, K., Ugaki, M., Lee, J., Hibi, T., and Namba, S. 2002. 'Candidatus *Phytoplasma castaneae*,' a novel phytoplasma taxon associated with chestnut witches' broom disease. Int. J. Syst. Evol. Microbiol. 52:1543–1549.

1981. Jung, J., Lee, S., and Lee, J. 2001. Comparison of Cenangium dieback fungus isolated from three different species of pine. Plant Pathol. J. 17:216–221.

1982. Jung, T., Blaschke, H., and Neumann, P. 1996. Isolation, identification and pathogenicity of *Phytophthora* species from declining oak stands. Eur. J. For. Pathol. 26:253–272.

1983. Jung, T., Blaschke, H., and Osswald, W. 2000. Involvement of soilborne *Phytophthora* species in Central European oak decline and the effect of site factors on the disease. Plant Pathol. 49:706–718.

1984. Jung, T., Cooke, D. E. L., Blaschke, H., Duncan, J. M., and Osswald, W. 1999. *Phytophthora quercina* sp. nov., causing root rot of European oaks. Mycol. Res. 103:785–798.

1985. Jung, T., Hansen, E. M., Winton, L., Osswald, W., and Delatour, C. 2002. Three new species of *Phytophthora* from European oak forests. Mycol. Res. 106:397–411.

1986. Jung, T., Nechwatal, J., Cooke, D. E. L., Hartmann, G., Blaschke, M., Osswald, W. F., Duncan, J. M., and Delatour, C. 2003. *Phytophthora pseudosyringae* sp. nov., a new species causing root and collar rot of deciduous tree species in Europe. Mycol. Res. 107:772–789.

1987. Jurgensen, M. G., Larsen, J. J., Spano, S. D., Harvey, A. E., and Gale, M. R. 1984. Nitrogen fixation associated with increased wood decay in Douglas-fir residue. For. Sci. 30:1038–1044.

1988. Juzwik, J. 1993. Morphology, cultural characteristics, and pathogenicity of *Rhizosphaera kalkhoffii* on *Picea* spp. in northern Minnesota and Wisconsin. Plant Dis. 77:630–634.

1989. Juzwik, J., and French, D. W. 1990. Cryptosphaeria canker on *Populus tremuloides* in Minnesota. Can. J. Bot. 68:2044–2045.

1990. Juzwik, J., French, D. W., and Hinds, T. E. 1986. *Encoelia pruinosa* on *Populus tremuloides* in Minnesota: occurrence, pathogenicity, and comparison with Colorado isolates. Can. J. Bot. 64:2728–2731.

1991. Juzwik, J., Gust, K. M., and Allmaras, R. R. 1999. Influence of cultural practices on edaphic factors related to root disease in *Pinus* nursery seedlings. Plant Soil 207:195–208.

1992. Juzwik, J., and Hinds, T. E. 1984. Ascospore germination, mycelial growth, and microconidial anamorphs of *Encoelia pruinosa* in culture. Can. J. Bot. 62:1916–1919.

1993. Juzwik, J., Honhart, C., and Chong, N. 1988. Cylindrocladium root rot in Ontario bare-root nurseries: estimate of spruce seedling losses. Can. J. For. Res. 18:1493–1496.

1994. Juzwik, J., Nishijima, W. T., and Hinds, T. E. 1978. Survey of aspen cankers in Colorado. Plant Dis. Rep. 62:906–910.

1995. Kacprzak, M., Asiegbu, F. O., Daniel, G., and Manka, M. 2000. Changes in root structure of Scots pine seedlings infected with *Rhizoctonia solani* observed under fluorescence and electron microscopes. Phytopathol. Polon. 19:127–140.

1996. Kacprzak, M., Asiegbu, F. O., and Manka, M. 2000. Cell wall lignification in roots of Scots pine, Norway spruce and European larch seedlings after infection with *Rhizoctonia solani* and *Fusarium* spp. Sylwan 144:105–113.

1997. Kadlec, Z., Stary, P., and Zumr, V. 1992. Field evidence for the large pine weevil, *Hylobius abietis* as a vector of *Heterobasidion annosum*. Eur. J. For. Pathol. 22:316–318.

1998. Kais, A. G. 1975. Environmental factors affecting brown spot infection on longleaf pine. Phytopathology 65:1389–1392.

1999. Kais, A. G. 1977. Influence of needle age and inoculum spore density on susceptibility of longleaf pine to *Scirrhia acicola*. Phytopathology 67:686–688.

2000. Kaitera, J., Hantula, J., and Jalkanen, R. 1997. Development of fruiting bodies of large tree type of *Gremmeniella abietina* var. *abietina* and timing of infection on Scots pine in northern Finland. Eur. J. For. Pathol. 27:115–124.

2001. Kam, M. de. 1975. Ascospore discharge in *Drepanopeziza punctiformis* in relation to infection of some poplar clones. Eur. J. For. Pathol. 5:304–309.

2002. Kam, M. de 1982. Damage to poplar caused by *Pseudomonas syringae* in combination with frost and fluctuating temperatures. Eur. J. For. Pathol. 12:203–209.

2003. Kamiri, L. K., and Laemmlen, F. F. 1981. Epidemiology of Cytospora canker caused in Colorado blue spruce by *Valsa kunzei*. Phytopathology 71:941–947.

2004. Kamp, B. J. van der. 1993. Production of uredinia and telia by stalactiform blister rust in British Columbia. Can. J. Bot. 71:519–521.

2005. Kamp, B. J. van der 1994. Lodgepole pine stem diseases and management of stand density in the British Columbia interior. For. Chron. 70:773–779.

2006. Kamp, B. J. van der, Gokhale, A. A., and Smith, R. S. 1979. Decay resistance owing to near-anaerobic conditions in black cottonwood wetwood. Can. J. For. Res. 9:39–44.

2007. Kandler, O., and Innes, J. L. 1995. Air pollution and forest decline in Central Europe. Environ. Pollut. 90:171–180.

2008. Kaneko, S. 1981. The species of *Coleosporium*, the causes of pine needle rusts, in the Japanese Archipelago. Rep. Tottori Mycol. Inst. 19. 151 pp.

2009. Kaneko, S. 2000. *Cronartium orientale*, sp. nov., segregation of the pine gall rust in eastern Asia from *Cronartium quercuum*. Mycoscience 41:115–122.

2010. Kaneko, S., Katsuya, K., Kakishima, M., and Ono, Y., eds. 1995. Proceedings of the fourth IUFRO rusts of pines working party conference, Tsukuba, Japan, 1994. For. & For. Prod. Res. Inst. Univ. Tsukuba, Tsukuba, Ibaraki, Jap. 197 pp.

2011. Kanematsu, S., Hayashi, T., and Kudo, A. 1997. Isolation of *Rosellinia necatrix* mutants with impaired cytochalasin E production and its pathogenicity. Ann. Phytopathol. Soc. Jap. 63:425–431.

2012. Kaplan, D. T., Thomas, W. K., Frisse, L. M., Sarah, J. L., Stanton, J. M., Speijer, P. R., Marin, D. H., and Opperman, C. H. 2000. Phylogenetic analysis of geographically diverse *Radopholus similis* via rDNA sequence reveals a monomorphic motif. J. Nematol. 32:134–142.

2013. Kapur, S., Kapur, S. P., and Rehlia, A. S. 1978. Studies on some of the biochemical changes occurring in date palm leaves infected by smut. Indian Phytopathol. 31:394–395.

2014. Karnosky, D. F. 1977. Evidence for genetic control of response to sulfur dioxide and ozone in *Populus tremuloides*. Can. J. For. Res. 7:437–440.

2015. Karnosky, D. F. 1981. Chamber and field evaluations of air pollution tolerances of urban trees. J. Arboric. 7:99–105.

2016. Karnosky, D. F., and Steiner, K. C. 1981. Provenance and family variation in response of *Fraxinus americana* and *F. pennsylvanica* to ozone and sulfur dioxide. Phytopathology 71:804–807.

2017. Karssen, G., and Hoenselaar, T. Van. 1998. Revision of the genus *Meloidogyne* Goldi, 1892 (Nematoda: Heteroderidae) in Europe. Nematologica 44:713–788.

2018. Kartte, S., and Seemüller, E. 1991. Histopathology of apple proliferation in *Malus* taxa and hybrids of different susceptibility. J. Phytopathol. 131:149–160.

2019. Katan, T. 1999. Current status of vegetative compatibility groups in *Fusarium oxysporum*. Phytoparasitica 27:51–64.

2020. Katan, T., and Primo, P. di. 1999. Current status of vegetative compatibility groups in *Fusarium oxysporum*: supplement (1999). Phytoparasitica 27:273–277.

2021. Katovich, S. A., and Ostry, M. E. 1998. Insects associated with butternut and butternut canker in Minnesota and Wisconsin. Great Lakes Entomol. 31:97–108.

2022. Kauffman, B. W., Applegate, H. W., Cordell, C. E., and Thor, E. 1980. Susceptibility of eight pine species to comandra blister rust in Tennessee. Plant Dis. 64:375–377.

2023. Kawakita, H., Saiki, T., Wei, W., Mitsuhashi, W., Watanabe, K., and Sato, M. 2000. Identification of mulberry dwarf phytoplasmas in the genital organs and eggs of leafhopper *Hishimonoides sellatiformis.* Phytopathology 90:909–914.

2024. Kawase, M. 1981. Anatomical and morphological adaptation of plants to waterlogging. HortScience 16:30–34.

2025. Kawase, M. 1981. Effect of ethylene on aerenchyma development. Am. J. Bot. 68:651–658.

2026. Kawazu, K., and Kaneko, N. 1997. Asepsis of the pine wood nematode isolate OKD-3 causes it to lose its pathogenicity. Jap. J. Nematol. 27:76–80.

2027. Kechel, H. G., and Boden, E. 1984. Differential resistance of North American and European birches to *Melampsoridium betulinum.* Holzzucht 38:27–30.

2028. Kehr, R. D. 1992. Pezicula canker of *Quercus rubra* L., caused by *Pezicula cinnamomea* (DC.) Sacc. II. Morphology and biology of the causal agent. Eur. J. For. Pathol. 22:29–40.

2029. Keifer, H. H., Baker, E. W., Kono, T., Delfinado, M., and Styer, W. E. 1982. An illustrated guide to plant abnormalities caused by eriophyid mites in North America. USDA Agric. Handb. 573. 178 pp.

2030. Keitt, G. W., Blodgett, E. C., Wilson, E. E., and Magie, R. O. 1937. The epidemiology and control of cherry leaf spot. Wisc. Agric. Exp. Stn. Res. Bull. 132. 117 pp.

2031. Kellerman, W. A., and Swingle, W. T. 1889. Branch knot of the hackberry. Kans. Agric. Exp. Stn. Rep. 1888:302–315.

2032. Kelley, R. D., and Cameron, H. R. 1986. Location of prune dwarf and *Prunus* necrotic ringspot viruses associated with sweet cherry pollen and seed. Phytopathology 76:317–322.

2033. Kelly, C. K., Harris, D., and Perez, I. R. 2001. Is breaking up hard to do? Breakage, growth, and survival in the parasitic clonal plant *Cuscuta corymbosa* (Convolvulaceae). Am. J. Bot. 88:1458–1468.

2034. Kelman, A., and Gooding, G. V. Jr. 1965. A root and stem rot of yellow-poplar caused by *Cylindrocladium scoparium.* Plant Dis. Rep. 49:797–801.

2035. Kelman, A., Hodges, C. S., and Garris, H. R. 1960. Needle blight of redcedar, *Juniperus virginiana* L. Plant Dis. Rep. 44:527–531.

2036. Kendrick, W. B. 1962. The *Leptographium* complex. *Verticicladiella* Hughes. Can. J. Bot. 40:771–797.

2037. Kendrick, [W.] B. 1992. The fifth kingdom. 2nd ed. Focus Information Group, Newburyport, MA. 406 pp.

2038. Kenerley, C. M., and Bruck, R. I. 1981. Phytophthora root rot of balsam fir and Norway spruce in North Carolina. Plant Dis. 65:614–615.

2039. Kenerley, C. M., and Bruck, R. I. 1983. Overwintering and survival of *Phytophthora cinnamomi* in Fraser fir and cover cropped nursery beds in North Carolina. Phytopathology 73:1643–1647.

2040. Kennedy, B. W., Froyd, J., and Bowden, R. 1984. Bacterial blight of mock orange (*Philadelphus* spp.) caused by *Pseudomonas syringae.* Plant Dis. 68:916–917.

2041. Kennedy, L. L., and Stewart, A. W. 1967. Development and taxonomy of *Apiosporina collinsii.* Can. J. Bot. 45:1597–1604.

2042. Kepley, J. B., and Jacobi, W. R. 2000. Pathogenicity of *Cytospora* fungi on six hardwood species. J. Arboric. 26:326–333.

2043. Kerley, S. J., and Read, D. J. 1998. The biology of mycorrhiza in the Ericaceae XX. Plant and mycorrhizal necromass as nitrogenous substrates for the ericoid mycorrhizal fungus *Hymenoscyphus ericae* and its host. New Phytol. 139:353–360.

2044. Kern, F. D. 1973. A revised taxonomic account of *Gymnosporangium.* Penn. State Univ. Press, University Park. 134 pp.

2045. Kern, H. 1955. Taxonomic studies in the genus *Leucostoma.* Mich. Acad. Sci. Pap. 40:9–22.

2046. Kern, H. 1957. Untersuchungen über die Umgrenzung der Arten in der Ascomycetengattung *Leucostoma.* Phytopathol. Z. 30:149–180.

2047. Kern, H., and Naef-Roth, S. 1975. Zur Bildung von Auxinen und Cytokininen durch *Taphrina*-Arten. Phytopathol. Z. 83:193–222.

2048. Kerr, A. M. 2000. Defoliation of an island (Guam, Mariana Archipelago, Western Pacific Ocean) following a saltspray-laden "dry" typhoon. J. Trop. Ecol. 16:895–901.

2049. Kersters, K., Ludwig, W., Vancanneyt, M., Vos, P. de, Gillis, M., and Schleifer, K. H. 1996. Recent changes in the classification of the pseudomonads: an overview. Syst. Appl. Microbiol. 19:465–477.

2050. Kessler, K. J. Jr. 1974. An apparent symbiosis between *Fusarium* fungi and ambrosia beetles causes canker on black walnut trees. Plant Dis. Rep. 58:1044–1047.

2051. Kessler, K. J. Jr. 1974. Annual canker of sugar maple in northeastern Minnesota. Plant Dis. Rep. 58:1042–1043.

2052. Kessler, K. J. Jr. 1984. *Mycosphaerella juglandis,* causal agent of a leaf spot of *Juglans nigra.* Mycologia 76:362–366.

2053. Kessler, K. J. Jr. 1985. Mycosphaerella leaf spot of black walnut. Plant Dis. 69:1092–1094.

2054. Kessler, K. J. Jr. 1990. A canker disease of autumn olive caused by *Tubercularia vulgaris.* Eur. J. For. Pathol. 20:148–153.

2055. Kessler, K. J. Jr., and Weber, B. C., coords. 1979. Walnut diseases and insects. USDA For. Serv. Gen. Tech. Rep. NC-52. 100 pp.

2056. Khairi, S. M., and Preece, T. F. 1978. Hawthorn powdery mildew: occurrence, survival, and ascospore productivity of *Podosphaera clandestina* cleistothecia in England. Trans. Br. Mycol. Soc. 71:289–293.

2057. Khairi, S. M., and Preece, T. F. 1978. Hawthorn powdery mildew: overwintering mycelium in buds and the effect of clipping hedges on disease epidemiology. Trans. Br. Mycol. Soc. 71:399–404.

2058. Kienholz, R. 1933. Frost damage to red pine. J. For. 31:392–399.

2059. Kile, G. A. 1976. The effect of seasonal pruning and time since pruning upon changes in apple sapwood and its susceptibility to invasion by *Trametes versicolor.* Phytopathol. Z. 87:231–240.

2060. Kile, G. A., and Wade, G. C. 1974. *Trametes versicolor* on apple. I. Host-pathogen relationship. Phytopathol. Z. 81:328–338.

2061. Kim, J. F., and Beer, S. V. 2001. Molecular basis of the Hrp pathogenicity of the fire blight pathogen *Erwinia amylovora:* a type III protein secretion system encoded in a pathogenicity island. Plant Pathol. J. 17:77–82.

2062. Kim, K. S., and Fulton, J. P. 1973. Association of viruslike particles with a ringspot disease of oak. Plant Dis. Rep. 57:1029–1031.

2063. Kim, K. S., and Martin, E. M. 1978. Viruslike particles associated with yellow ringspot of redbud. (Abstr.) Phytopathol. News 12:199.

2064. Kim, W., Gardan, L., Rhim, S., and Geider, K. 1999. *Erwinia pyrifoliae* sp. nov., a novel pathogen that affects Asian pear trees (*Pyrus pyrifolia* Nakai). Int. J. Syst. Bacteriol. 49:899–906.

2065. Kim, W., Hildebrand, M., Jock, S., and Geider, K. 2001. Molecular comparison of pathogenic bacteria from pear trees in Japan and the fire blight pathogen *Erwinia amylovora.* Microbiology 147:2951–2959.

2066. Kim, W., Jock, S., Paulin, J.-P., Rhim, S.-L., and Geider, K. 2001. Molecular detection and differentiation of *Erwinia pyrifoliae* and host range analysis of the Asian pear pathogen. Plant Dis. 85:1183–1188.

2067. Kimishima, E., Kobayashi, Y., and Kobayashi, T. 1996. Cane blight of rose caused by *Pestalotiopsis populi-nigrae* (Sawada et Ito) Morelet. Ann. Phytopathol. Soc. Jap. 62:502–504.

2068. Kimmey, J. W. 1964. Heart rots of western hemlock. USDA For. Serv. For. Pest Leafl. 90. 7 pp.

2069. Kimmey, J. W., and Bynum, H. H. 1961. Heart rots of red and white firs. USDA For. Serv. For. Pest Leafl. 52. 4 pp.

2070. Kinloch, B. B. Jr. 2003. White pine blister rust in North America: past and prognosis. Phytopathology 93:1044–1047.

2071. Kinloch, B. B. Jr., and Dupper, G. E. 1999. Evidence of cytoplasmic inheritance of virulence in *Cronartium ribicola* to major gene resistance in sugar pine. Phytopathology 89:192–196.

2072. Kinloch, B. B. Jr., and Dupper, G. E. 2002. Genetic specificity in the white pine–blister rust pathosystem. Phytopathology 92:278–280.

2073. Kinloch, B. B. Jr., Sniezko, R. A., Barnes, G. D., and Greathouse, T. E. 1999. A major gene for resistance to white pine blister rust in western white pine from the Western Cascade Range. Phytopathology 89:861–867.

2074. Kinloch, B. B. Jr., Sniezko, R. A., and Dupper, G. E. 2003. Origin and distribution of *Cr2,* a gene for resistance to white pine blister rust in natural populations of western white pine. Phytopathology 93:691–694.

2075. Kinloch, B. B. Jr., Westfall, R. D., White, E. E., Gitzendanner, M. A., Dupper, G. E., Foord, B. M., and Hodgskiss, P. D. 1998. Genetics of *Cronartium ribicola.* IV. Population structure in western North America. Can. J. Bot. 76:91–98.

2076. Kipfmueller, K. F., and Baker, W. L. 1998. Fires and dwarf mistletoe in a Rocky Mountain lodgepole pine ecosystem. For. Ecol. Manag. 108:77–84.

2077. Kirk, P. M., Cannon, P. F., David, J. C., and Stalpers, J. A., eds. 2001. Ainsworth & Bisby's dictionary of the fungi. 9th ed. CABI Pub., New York. 655 pp.

2078. Kirkpatrick, B. C. 1997. Phytoplasmas: the evolution of a molecular taxonomic genus. Nachrichtenbl. Deut. Pflanzenschutzd. 49:126–130.

2079. Kirkpatrick, B. C., Fisher, G. A., Fraser, J. D., and Purcell, A. H. 1990. Epidemiological and phylogenetic studies on western X-disease mycoplasma-like organisms. Pages 288–297 in: Recent advances in mycoplasmology. G. Stanek, G. H. Casell, J. G. Tully, and R. F. Whitcomb, eds. Int. J. Med. Microbiol., Ser. A. Suppl. 20.

2080. Kirkpatrick, L. A., and Tinnin, R. O. 1986. Preliminary investigations of the impact of dwarf mistletoe (*Arceuthobium americanum* Nutt.) on water relations in lodgepole pine (*Pinus contorta* Dougl.). Am. J. Bot. 73:670–671.

2081. Kishi, K., Abiko, K., Takanashi, K., and Yano, R. 1973. Studies on the virus diseases of stone fruit. VII. Line pattern of plum and flowering cherry. Ann. Phytopathol. Soc. Jap. 39:288–296.

2082. Kishi, Y. 1995. Pine wood nematode and the Japanese pine sawyer. Thomas, Tokyo. 302 pp.

2083. Kison, H., Kirkpatrick, B. C., and Seemüller, E. 1997. Genetic comparison of the peach yellow leaf roll agent with European fruit tree phytoplasmas of the apple proliferation group. Plant Pathol. 46:538–544.

2084. Kison, H., and Seemüller, E. 2001. Differences in strain virulence of the European stone fruit yellows phytoplasma and susceptibility of stone fruit trees on various rootstocks to this pathogen. J. Phytopathol. 149:533–541.

2085. Kistler, B. R., and Merrill, W. 1968. Effects of Strumella coryneoidea on oak sapwood. Phytopathology 58:1429–1430.

2086. Kistler, B. R., and Merrill, W. 1978. Etiology, symptomatology, epidemiology, and control of Naemacyclus needlecast of Scotch pine. Phytopathology 68:267–271.

2087. Kistler, H. C. 1997. Genetic diversity in the plant-pathogenic fungus Fusarium oxysporum. Phytopathology 87:474–479.

2088. Klebahn, H. 1931. Kulturversuche und Bemerkungen über Rost-pilze. XVIII Bericht (1925–1930). Mit einem Anhang über Ustilago longissima. Z. Pflanzenkr. Pflanzenschutz 41:209–223.

2089. Klebahn, H. 1935. Einige Beobachtungen und Versuche über den Mahonia-Rost. Z. Pflanzenkr. Pflanzenschutz 45:529–537.

2090. Klein, G. H. W., and Blenis, P. V. 1991. Spread of Armillaria ostoyae in juvenile lodgepole pine stands in west central Alberta. Can. J. For. Res. 21:20–24.

2091. Klein, L. A., Windham, M. T., and Trigiano, R. N. 1998. Natural occurrence of Microsphaera pulchra and Phyllactinia guttata on two Cornus species. Plant Dis. 82:383–385.

2092. Klekowski, E. J. Jr., Lowenfeld, R., and Klekowski, E. H. 1996. Mangrove genetics. IV. Postzygotic mutations fixed as periclinal chimeras. Int. J. Plant Sci. 157:398–405.

2093. Klement, Z., Rozsnyay, D. S., Báló, E., Pánczél, M., and Prileszky, G. 1984. The effect of cold on development of bacterial canker in apricot trees infected with Pseudomonas syringae pv. syringae. Physiol. Plant Pathol. 24:237–246.

2094. Klepzig, K. D., Moser, J. C., Lombardero, F. J., Hofstetter, R. W., and Ayres, M. P. 2001. Symbiosis and competition: complex interactions among beetles, fungi and mites. Symbiosis 30:83–96.

2095. Klepzig, K. D., Raffa, K. F., and Smalley, E. B. 1991. Association of an insect-fungal complex with red pine decline in Wisconsin. For. Sci. 37:1119–1139.

2096. Klepzig, K. D., Smalley, E. B., and Raffa, K. F. 1995. Dendroctonus valens and Hylastes porculus (Coleoptera: Scolytidae): vectors of pathogenic fungi (Ophiostomatales) associated with red pine decline disease. Great Lakes Entomol. 28:81–87.

2097. Klepzig, K. D., Smalley, E. B., and Raffa, K. F. 1996. Interactions of ecologically similar saprogenic fungi with healthy and abiotically stressed conifers. For. Ecol. Manag. 86:163–169.

2098. Kliejunas, J. T., and Kuntz, J. E. 1972. Development of stromata and the imperfect state of Eutypella parasitica in maple. Can. J. Bot. 50:1453–1456.

2099. Klinger, L. F. 1990. Global patterns in community succession: 1. Bryophytes and forest decline. Mem. Torrey Bot. Club 24:1–50.

2100. Klinger, L. F. 1996. Coupling of soils and vegetation in peatland succession. Arctic Alpine Res. 28:380–387.

2101. Knorr, L. C. 1973. Citrus diseases and disorders. Univ. Presses Fla., Gainesville. 163 pp.

2102. Ko, M. P., Schmitt, D. P., and Saxby, M. 1997. Effects of container bases on the spread of Meloidogyne incognita in a Hawaiian ornamental nursery. Plant Dis. 81:607–613.

2103. Ko, W. H., Kunimoto, R. K., and Maedo, I. 1977. Root decay caused by Kretzschmaria clavus: its relation to macadamia decline. Phytopathology 67:18–21.

2104. Kobayashi, T. 1958. Phomopsis disease on Japanese cedar, Cryptomeria japonica D. Don, with special reference to the life-history and taxonomy of the causal fungus. Bull. Gov. For. Exp. Stn., Tokyo. No. 107:3–25.

2105. Kobayashi, T. 1968. Notes on Japanese species of the genus Melanconium. J. Mycol. Soc. Jap. 9:1–11.

2106. Kobayashi, T. 1970. Taxonomic studies of Japanese Diaporthaceae with special reference to their life-histories. Bull. Gov. For. Exp. Stn., Tokyo. No. 226. 242 pp.

2107. Kobayashi, T. 1974. Pathogenicity of Melanconis spp. causing dieback diseases of broad-leaved trees. J. Jap. For. Soc. 56:332–336.

2108. Kobayashi, T., and Chiba, O. 1961. Fungi inhabiting poplars in Japan I. Bull. Gov. For. Exp. Stn., Tokyo. No. 130. 43 pp.

2109. Kobayashi, T., Ishihara, M., and Ono, Y. 2001. A new species of Pestalosphaeria, the teleomorph of Pestalotiopsis neglecta. Mycoscience 42:211–216.

2110. Kobayashi, T., and Ito, K. 1957. Phomopsis and its perfect stage Diaporthe causing a die-back of the pawlonia tree. Bull. Gov. For. Exp. Stn., Meguro. No. 103:57–68.

2111. Kobayashi, T., and Mamiya, Y. 1963. A Cenangium causing dieback of Japanese pines. Bull. Gov. For. Exp. Stn., Tokyo 161:123–150.

2112. Koch, J. 1971. Rhododendron blight caused by Phytophthora cambivora. Friesia 9:418–425.

2113. Koch, L. W. 1933–1935. Investigations on black knot of plums and cherries. I. Development and discharge of spores and experiments in control. II. The occurrence and significance of certain fungi found in association with Dibotryon morbosum (Sch.) T. & S. III. Symptomatology, life history, and cultural studies of Dibotryon morbosum (Sch.) T. & S. IV. Studies in pathogenicity and pathological histology. Sci. Agric. 13:576–590; 15:80–95, 411–423, 729–744.

2114. Koch, L. W. 1934. Studies on the overwintering of certain fungi parasitic and saprophytic on fruit trees. Can. J. Res. 11:190–206.

2115. Kohls, S. J., Thimmapuram, J., Buschena, C. A., Paschke, M. W., and Dawson, J. O. 1994. Nodulation patterns of actinorhizal plants in the family Rosaceae. Plant Soil 162:229–239.

2116. Kohn, L. M. 1979. A monographic revision of the genus Sclerotinia. Mycotaxon 9:365–444.

2117. Kohn, L. M., and Grenville, D. J. 1989. Anatomy and histochemistry of stromatal anamorphs in the Sclerotiniaceae. Can. J. Bot. 67:371–393.

2118. Kojwang, H. O., and Kurkela, T. 1984. Linospora ceuthocarpa on aspen (Populus tremula) in Finland. Karstenia 24:33–40.

2119. Kolb, P. F., and Robberecht, R. 1996. High temperature and drought stress effects on survival of Pinus ponderosa seedlings. Tree Physiol. 16:665–672.

2120. Kolb, T. E., and McCormick, L. H. 1993. Etiology of sugar maple decline in four Pennsylvania stands. Can. J. For. Res. 23:2395–2402.

2121. Kommineni, K. V., and Ramsdell, D. C. 1997. An anatomical study of prune brown line disease and immuno-localization of tomato ringspot virus in plum bark. Plant Dis. 81:855–861.

2122. Kondo, E., Foudin, A., Linit, M., Smith, M., Bolla, R., Winter, R., and Dropkin, V. 1982. Pine wilt disease—nematological, entomological, and biochemical investigations. Univ. Mo. Agric. Exp. Stn. SR282. 56 pp.

2123. Kondo, E. S., Hiratsuka, Y., and Denyer, W. B. G., eds. 1982. Proceedings of the Dutch elm disease symposium and workshop. Environ. Can. and Manitoba Dep. Nat. Res. 517 pp.

2124. Konrad, H., Kirisits, T., Riegler, M., Halmschlager, E., and Stauffer, C. 2002. Genetic evidence for natural hybridization between the Dutch elm disease pathogens Ophiostoma novo-ulmi ssp. novo-ulmi and O. novo-ulmi ssp. americana. Plant Pathol. 51:78–84.

2125. Kontzog, H. G., and Ebrahim, N. F. 1992. Determination and evaluation of quantitative virus resistance in poplars inoculated with poplar mosaic virus (PopMV). Eur. J. For. Pathol. 22:193–203.

2126. Kooij, T. A. W. van der, Krause, K., Dorr, I., and Krupinska, K. 2000. Molecular, functional and ultrastructural characterisation of plastids from six species of the parasitic flowering plant genus Cuscuta. Planta 210:701–707.

2127. Kope, H. H. 2000. Didymascella thujina. Fungi Canadenses No. 343. Can. J. Plant Pathol. 22:407–409.

2128. Kope, H. H., Axelrood, P. E., Sutherland, J., and Reddy, M. S. 1996. Prevalence and incidence of the root-inhabiting fungi, Fusarium, Cylindrocarpon and Pythium, on container-grown Douglas-fir and spruce seedlings in British Columbia. New For. 12:55–67.

2129. Kopp, R. F., Castello, J. D., and Abrahamson, L. P. 1999. Viruses in Salix grown for bioenergy. Eur. J. For. Pathol. 29:117–122.

2130. Koshy, P. K., and Sosamma, V. K. 1996. Effect of Radopholus similis on growth, flowering and yield of coconut. J. Plantation Crops 24:157–165.

2131. Koske, R. E., and Tessier, B. 1986. Growth of some wood and litter-decay basidiomycetes at reduced water potential. Trans. Br. Mycol. Soc. 86:156–168.

2132. Koski, R., and Jacobi, W. R. 2004. Tree pathogen survival in chipped wood mulch. J. Arboric. 30:165–171.

2133. Kostka, S. J., Tattar, T. A., and Sherald, J. L. 1986. Elm leaf scorch: abnormal physiology in American elms infected with fastidious, xylem-inhabiting bacteria. Can. J. For. Res. 16:1088–1091.

2134. Kostka, S. J., Tattar, T. A., Sherald, J. L., and Hurtt, S. S. 1986. Mulberry leaf scorch, new disease caused by a fastidious, xylem-inhabiting bacterium. Plant Dis. 70:690–693.

2135. Kothe, E. 1999. Mating types and pheromone recognition in the homobasidiomycete Schizophyllum commune. Fungal Genet. Biol. 27:146–152.

2136. Kou, K.-C., Kao, C.-W., and Leu, L.-S. 1988. The symptoms, causal organism, and fungicide screening for the control of grape Cristulariella leaf spot. Plant Prot. Bull. (Taiwan) 30:111–124.

2137. Kowalski, T., and Krygier, J. 1996. Mycological study on symptomless and diseased needles in pine stand attacked by *Lophodermella sulcigena* (Rostr.) v. Hohn. Phytopathol. Polon. 11:159–168.

2138. Kozar, F., and Netolitzky, H. J. 1975. Ultrastructure and cytology of pycnia, aecia, and aeciospores of *Gymnosporangium clavipes*. Can. J. Bot. 53:972–977.

2139. Kozlowski, T. T. 1982. Water supply and tree growth. For. Abstr. 43:57–95, 145–161.

2140. Kozlowski, T. T. 1997. Responses of woody plants to flooding and salinity. Tree Physiol. Monogr. No. 1. 29 pp. Internet pub.

2141. Kozlowski, T. T., ed. 1978. Water deficits and plant growth. Vol. 5. Water and plant disease. Academic Press, New York. 323 pp.

2142. Kozlowski, T. T., ed. 1981. Water deficits and plant growth. Vol. 6. Woody plant communities. Academic Press, New York. 582 pp.

2143. Kozlowski, T. T., ed. 1984. Flooding and plant growth. Academic Press, New York. 356 pp.

2144. Kozlowski, T. T., Kramer, P. J., and Pallardy, S. G. 1991. The physiological ecology of woody plants. Academic Press, San Diego. 657 pp.

2145. Kramer, P. J., and Boyer, J. S. 1995. Water relations of plants and soils. Academic Press, San Diego. 495 pp.

2146. Kraus, J. F., and Hunt, D. L. 1971. Inherent variation of resistance to *Hypoderma lethale* in slash and loblolly pines. For. Sci. 17:143–144.

2147. Krebill, R. G. 1968. *Cronartium comandrae* in the Rocky Mountain states. USDA For. Serv. Res. Pap. INT-50. 28 pp.

2148. Krebill, R. G. 1968. Histology of canker rusts in pines. Phytopathology 58:155–164.

2149. Krebill, R. G. 1972. *Pucciniastrum goeppertianum* in *Abies lasiocarpa* in the Rocky Mountain states. Am. Midl. Nat. 87:570–576.

2150. Kress, L. W., and Skelly, J. M. 1982. Response of several eastern forest tree species to chronic doses of ozone and nitrogen dioxide. Plant Dis. 66:1149–1152.

2151. Krezdorn, A. H., and Adriance, G. W. 1961. Fig growing in the South. USDA Agric. Handb. 196. 26 pp.

2151a. Krokene, P., Barnes, I., Wingfield, B. D., and Wingfield, M. J. 2004. A PCR-RFLP based diagnostic technique to rapidly identify *Seiridium* species causing cypress canker. Mycologia 96:1352–1354.

2152. Kruger, B. M., and Manion, P. D. 1993. Genetic control of *Populus tremuloides* sensitivity to metabolites of *Hypoxylon mammatum*. Can. J. Bot. 71:1276–1279.

2153. Kruger, B. M., and Manion, P. D. 1993. Sensitivity of *Populus tremuloides* to toxic metabolites of *Hypoxylon mammatum* and susceptibility to *H. mammatum* infection. Can. J. Bot. 71:1298–1303.

2154. Kruger, B. M., and Manion, P. D. 1994. Antifungal compounds in aspen: effect of water stress. Can. J. Bot. 72:454–460.

2155. Krupa, S. V. 1997. Air pollution, people, and plants. APS Press, St. Paul, MN. 197 pp.

2156. Krupa, S. [V.], McGrath, M. T., Andersen, C. P., Booker, F. L., Burkey, K. O., Chappelka, A. H., Chevone, B. I., Pell, E. J., and Zilinskas, B. A. 2001. Ambient ozone and plant health. Plant Dis. 85:4–12.

2157. Krupinsky, J. M. 1981. *Botryodiplodia hypodermia* and *Tubercularia ulmea* in cankers on Siberian elm in northern Great Plains windbreaks. Plant Dis. 65:677–678.

2158. Krupinsky, J. M. 1983. Variation in virulence of *Botryodiplodia hypodermia* to *Ulmus pumila*. Phytopathology 73:108–110.

2159. Krupinsky, J. M. 1989. Variability in *Septoria musiva* in aggressiveness. Phytopathology 79:413–416.

2160. Krupinsky, J. M., and Cunningham, R. A. 1993. Response of Siberian elm to inoculations with *Sphaeropsis ulmicola*. Plant Dis. 77:678–681.

2161. Kubisiak, T. L., Hebard, F. V., Nelson, C. D., Zhang, J., Bernatzky, R., Huang, H., Anagnostakis, S. L., and Doudrick, R. L. 1997. Molecular mapping of resistance to blight in an interspecific cross in the genus *Castanea*. Phytopathology 87:751–759.

2162. Kubler, H. 1983. Mechanism of frost crack formation in trees—a review and synthesis. For. Sci. 29:559–568.

2163. Kubler, H. 1988. Frost cracks in stems of trees. Arboric. J. 12:163–175.

2164. Kucera, L. J., Eichenberger, B., and Stoll, A. 1985. The lightning gap—cause and development. Pages 127–138 in: Xylorama. Trends in wood research. Birkhauser Verlag, Basel.

2165. Kuhlman, E. G. 1992. Interaction of virulent single-gall rust isolates of *Cronartium quercuum* f.sp. *fusiforme* and resistant families of loblolly pine. For. Sci. 38:641–651.

2166. Kuhlman, E. G., Amerson, H. V., Jordan, A. P., and Pepper, W. D. 1997. Inoculum density and expression of major gene resistance to fusiform rust disease in loblolly pine. Plant Dis. 81:597–600.

2167. Kuhlman, E. G., Cordell, C. E., and Filer, T. H. Jr. 1980. Cylindrocladium root rots of sweetgum seedlings in southern forest tree nurseries. Plant Dis. 64:1079–1080.

2168. Kuhlman, E. G., Dianis, S. D., and Smith, T. K. 1982. Epidemiology of pitch canker disease in a loblolly pine seed orchard in North Carolina. Phytopathology 72:1212–1216.

2169. Kuhlman, E. G., Hodges, C. S. Jr., and Froelich, R. C. 1976. Minimizing losses to *Fomes annosus* in the southern United States. USDA For. Serv. Res. Pap. SE-151. 16 pp.

2170. Kuhlman, E. G., and Kaneko, S. 1991. Comparisons of basidiospores and urediniospores of formae speciales of *Cronartium quercuum*. Mycologia 83:440–445.

2171. Kuhlman, E. G., and Matthews, F. R. 1993. Variation in virulence among single-aeciospore isolates from single-gall isolates of *Cronartium quercuum* f.sp. *fusiforme*. Can. J. For. Res. 23:67–71.

2172. Kuhlman, E. G., and Pepper, W. D. 1994. Temperature effects on basidiospore germination and on infection of slash pine seedlings by *Cronartium quercuum* f.sp. *fusiforme*. Phytopathology 84:735–739.

2173. Kuhlman, E. G., and Powers, H. R. Jr. 1988. Resistance response in half-sib loblolly pine progenies after inoculation with *Cronartium quercuum* f.sp. *fusiforme*. Phytopathology 78:484–487.

2174. Kuijt, J. 1964. Critical observations on the parasitism of New World mistletoes. Can. J. Bot. 42:1243–1278.

2175. Kuijt, J. 1969. The biology of parasitic flowering plants. Univ. Calif. Press, Berkeley. 246 pp.

2176. Kuijt, J. 1977. Haustoria of phanerogamic parasites. Annu. Rev. Phytopathol. 15:91–118.

2177. Kuijt, J. 2003. Monograph of *Phoradendron* (Viscaceae). Syst. Bot. Monogr. 66. 643 pp.

2178. Kuijt, J., and Toth, R. 1976. Ultrastructure of angiosperm haustoria—a review. Ann. Bot. 40:1121–1130.

2179. Kula, E., and Zabecki, W. 1997. Lightning-struck areas as the centres of bark-beetle-infested patches within spruce stands. Sylwan 141:89–97.

2180. Kulman, H. M. 1971. Effects of insect defoliation on growth and mortality of trees. Annu. Rev. Entomol. 16:289–324.

2181. Kumar, J., Chaube, H. S., Singh, U. S., and Mukhopadhyay, A. N., eds. 1992. Plant diseases of international importance. Vol. 3: Diseases of fruit crops. Prentice-Hall, Englewood Cliffs, NJ. 456 pp.

2182. Kumi, J., and Lang, K. J. 1979. Susceptibility of various spruce species to *Rhizosphaera kalkhoffii* and some cultural characteristics of the fungus *in vitro*. Eur. J. For. Pathol. 9:35–46.

2183. Kuo, K., and Hoch, H. C. 1995. Visualization of the extracellular matrix surrounding pycnidiospores, germlings, and appressoria of *Phyllosticta ampelicida*. Mycologia 87:759–771.

2184. Kuo, K., and Hoch, H. C. 1996. Germination of *Phyllosticta ampelicida* pycnidiospores: prerequisite of adhesion to the substratum and the relationship of substratum wettability. Fungal Genet. Biol. 20:18–29.

2185. Kuo, K., and Hoch, H. C. 1996. The parasitic relationship between *Phyllosticta ampelicida* and *Vitis vinifera*. Mycologia 88:626–634.

2186. Kuppers, M. 1992. Carbon discrimination, water-use efficiency, nitrogen and phosphorus nutrition of the host/mistletoe pair *Eucalyptus behriana* F. Muell. and *Amyema miquelii* (Lehm. ex Miq.) Tiegh. at permanently low water status in the field. Trees Struct. Funct. 7:8–11.

2187. Kurdyla, T. M., Guthrie, P. A. I., McDonald, B. A., and Appel, D. N. 1995. RFLPs in mitochondrial and nuclear DNA indicate low levels of genetic diversity in the oak wilt pathogen *Ceratocystis fagacearum*. Curr. Genet. 27:373–378.

2188. Kurian, P., and Stelzig, D. A. 1979. The synergistic role of oxalic acid and endopolygalacturonase in bean leaves infected by *Cristulariella pyramidalis*. Phytopathology 69:1301–1304.

2189. Kurkela, T., Hanso, M., and Hantula, J. 1999. Differentiating characteristics between *Melampsoridium* rusts infecting birch and alder leaves. Mycologia 91:987–992.

2190. Kurkela, T. T. 1996. Ascospore production period of *Phacidium infestans*, a snow blight fungus on *Pinus sylvestris*. Scand. J. For. Res. 11:60–67.

2191. Kurstak, E., ed. 1981. Handbook of plant virus infections. Comparative diagnosis. Elsevier/North Holland Biomedical Press, Amsterdam. 943 pp.

2192. Kusaba, M., and Tsuge, T. 1995. Phylogeny of *Alternaria* fungi known to produce host-specific toxins on the basis of variation in internal transcribed spacers of ribosomal DNA. Curr. Genet. 28:491–498.

2193. Kuske, C. R., and Benson, D. M. 1983. Overwintering and survival of *Phytophthora parasitica*, causing dieback of rhododendron. Phytopathology 73:1192–1196.

2194. Kuske, C. R., and Benson, D. M. 1983. Survival and splash dispersal of *Phytophthora parasitica*, causing dieback of rhododendron. Phytopathology 73:1188–1191.

2195. Kwon, M., Wayadande, A. C., and Fletcher, J. 1999. *Spiroplasma citri* movement into the intestines and salivary glands of its leafhopper vector, *Circulifer tenellus*. Phytopathology 89:1144–1151.

2196. Kyhl, J. F., Bartelt, R. J., Cosse, A., Juzwik, J., and Seybold, S. J. 2002. Semiochemical-mediated flight responses of sap beetle vectors of oak wilt, *Ceratocystis fagacearum*. J. Chem. Ecol. 28:1527–1547.

2197. Lacasse, N. L., and Rich, A. E. 1964. Maple decline in New Hampshire. Phytopathology 54:1071–1075.

2198. Lacasse, N. L., and Treshow, M., eds. 1977. Diagnosing vegetation injury caused by pollution. U.S. EPA Air Pollut. Training Inst. Chapters paginated separately.

2199. Lachance, D. 1971. Discharge and germination of *Eutypella parasitica* ascospores. Can. J. Bot. 49:1111–1118.

2200. Lachance, D. 1971. Inoculation and development of Eutypella canker of maple. Can. J. For. Res. 1:228–234.

2201. Lachance, D. 1994. Eutypella canker of maple. Can. For. Serv. Inf. Leafl. LFC 8. 8 pp.

2202. Lachance, D., and Kuntz, J. E. 1970. Ascocarp development of *Eutypella parasitica*. Can. J. Bot. 48:1977–1979.

2203. Lachance, M.-A., Metcalf, B. J., and Starmer, W. T. 1982. Yeasts from exudates of *Quercus, Ulmus, Populus,* and *Pseudotsuga:* new isolations and elucidation of some factors affecting ecological specificity. Microb. Ecol. 8:191–198.

2204. Lachmund, H. G. 1921. Some phases in the formation of fire scars. J. For. 19:638–640.

2205. Lachmund, H. G. 1929. *Cronartium comptoniae* Arth. in western North America. Phytopathology 19:453–466.

2206. Lackner, A. L., and Alexander, S. A. 1982. Occurrence and pathogenicity of *Verticicladiella procera* in Christmas tree plantations in Virginia. Plant Dis. 66:211–212.

2207. Lackner, A. L., and Alexander, S. A. 1983. Root disease and insect infestations on air-pollution-sensitive *Pinus strobus* and studies of pathogenicity of *Verticicladiella procera*. Plant Dis. 67:679–681.

2208. Lackner, A. L., and Alexander, S. A. 1984. Incidence and development of *Verticicladiella procera* in Virginia Christmas tree plantations. Plant Dis. 68:210–212.

2209. Laderman, A. D., ed. 1998. Coastally restricted forests. Oxford Univ. Press, New York. 334 pp.

2210. Ladjal, M., Epron, D., and Ducrey, M. 2000. Effects of drought preconditioning on thermotolerance of photosystem II and susceptibility of photosynthesis to heat stress in cedar seedlings. Tree Physiol. 20:1235–1241.

2211. Laessøe, T., Rogers, J. D., and Whalley, A. J. S. 1989. *Camillea, Jongiella* and light-spored species of *Hypoxylon*. Mycol. Res. 93:121–155.

2212. Laflamme, G., Bérubé, J. A., and Bussières, G., eds. 2003. Root and butt rots of forest trees. Can. For. Serv. Laurentian For. Cent. Inf. Rep. LAU-X-126. 450 pp.

2213. Laflamme, G., Bérubé, J. A., and Hamelin, R. C., eds. 1998. Foliage, shoot, and stem diseases of trees. Can. For. Serv. Laurentian For. Cent. Inf. Rep. LAU-X-122. 272 pp.

2214. Laflamme, G., and Blais, R. 2000. Resistance of *Pinus banksiana* to the European race of *Gremmeniella abietina*. Phytoprotection 81:49–55.

2215. Laflamme, G., Hopkin, A. A., and Harrison, K. J. 1998. Status of the European race of Scleroderris canker in Canada. For. Chron. 74:561–566.

2216. Laflamme, G., Ylimartimo, A., and Blais, R. 1996. Host preference of two *Gremmeniella abietina* varieties on balsam fir, jack pine, and black spruce in eastern Canada. Can. J. Plant Pathol. 18:330–334.

2217. Lagerberg, T. 1949. Some notes on the Phacidiaceae and a new member of this family, *Lophophacidium hyperboreum* nov. gen. et sp. Svensk Bot. Tidskr. 43:420–437.

2218. Lair, E. D. 1946. Smooth patch, a bark disease of oak. J. Elisha Mitchell Sci. Soc. 62:212–220.

2219. Lalancette, N., and Robison, D. M. 2001. Seasonal availability of inoculum for constriction canker of peach in New Jersey. Phytopathology 91:1109–1115.

2220. Lambe, R. C., and Wills, W. H. 1978. Pathogenicity of *Thielaviopsis basicola* to Japanese holly (*Ilex crenata*). Plant Dis. Rep. 62:859–863.

2221. Lambe, R. C., and Wills, W. H. 1980. Distribution of dieback associated with Thielaviopsis black root rot of Japanese holly. Plant Dis. 64:956.

2222. Lamberti, F., and Taylor, C. E., eds. 1979. Root knot nematodes (*Meloidogyne* species). Systematics, biology and control. Academic Press, New York. 477 pp.

2223. Lamson, N. I., and Smith, H. C. 1978. Response to crop tree release: sugar maple, red oak, black cherry, and yellow poplar saplings in a 9-year-old stand. USDA For. Serv. Res. Pap. NE-394. 8 pp.

2224. Lana, A. F., Peterson, J. F., Rouselle, G. L., and Vrain, T. C. 1983. Association of tobacco ringspot virus with a union incompatibility of apple. Phytopathol. Z. 106:141–148.

2225. Lana, A. F., Thomas, O. T., and Peterson, J. F. 1980. A virus isolated from sugar maple. Phytopathol. Z. 97:214–218.

2226. Lana, A. O., and Agrios, G. N. 1974. Properties of a strain of tobacco mosaic virus isolated from white ash trees. Phytopathology 64:1490–1495.

2227. Lana, A. O., and Agrios, G. N. 1974. Transmission of a mosaic disease of white ash to woody and herbaceous hosts. Plant Dis. Rep. 58:536–540.

2228. Landgraf, F. A., and Zehr, E. I. 1982. Inoculum sources for *Monilinia fructicola* in South Carolina peach orchards. Phytopathology 72:185–190.

2229. Landis, T. D., Tinus, R. W., McDonald, S. E., and Barnett, J. P., eds. 1990. The biological component: nursery pests and mycorrhizae. USDA Agric. Handb. 674. 171 pp.

2230. Landis, W. R., and Hart, J. H. 1967. Cankers of ornamental crabapples. Plant Dis. Rep. 51:230–234.

2231. Lang, K. J. 2000. New hosts of *Cristulariella depraedans*. For. Pathol. 30:117–120.

2232. Lang, K. J., and Schütt, P. 1974. Anatomische Untersuchungen zur Infektionsbiologie von *Scleroderris lagerbergii* Gr. [*Brunchorstia pinea* (Karst.) von Höhn]. Eur. J. For. Pathol. 4:166–174.

2233. Langford, M. H., and Keitt, G. W. 1942. Heterothallism and variability in *Venturia pyrina*. Phytopathology 32:357–369.

2234. Langridge, Y. N., and Dye, D. W. 1982. A bacterial disease of *Pinus radiata* seedlings caused by *Pseudomonas syringae*. N.Z. J. Agric. Res. 25:273–276.

2235. Lanier, L., Joly, P., Bondoux, P., and Bellèmere. 1978. Mycologie et pathologie forestières I. Mycologie forestière. Masson, Paris. 487 pp.

2236. Large, J. R. 1944. Alcoholic flux or white slime flux of tung trees. Plant Dis. Rep. 28:35–36.

2237. Large, J. R. 1948. Canker of tung trees caused by *Physalospora rhodina*. Phytopathology 38:359–363.

2238. Large, J. R., Painter, J. H., and Lewis, W. A. 1950. Thread blight in tung orchards and its control. Phytopathology 40:453–459.

2239. Larsen, F. E., and Higgins, S. S. 1992. Longitudinal bark cracking on trunks of young Asian pear trees in response to a rapid drop in winter temperature. Fruit Var. J. 46:225–229.

2240. Larsen, M. J. 2000. *Phellinus gilbertsonii* sp. nov. from western North America causing heart rot of coastal Douglas-fir. Folia Cryptog. Estonica 37:51–54.

2241. Larsen, M. J., Jurgensen, M. F., and Harvey, A. E. 1978. N₂ fixation associated with wood decayed by some common fungi in western Montana. Can. J. For. Res. 8:341–345.

2242. Larsen, M. J., Lombard, F. F., and Aho, P. E. 1979. A new variety of *Phellinus pini* associated with cankers and decay in white firs in southwestern Oregon and northern California. Can. J. For. Res. 9:31–38.

2243. Larsen, M. J., Lombard, F. F., and Clark, J. W. 1994. *Phellinus sulphurascens* and the closely related *P. weirii* in North America. Mycologia 86:121–130.

2244. Larsen, R. C., Gergerich, R. C., and Kim, K. S. 1990. Characterization and ultrastructural studies of a nepovirus from *Euonymus*. Phytopathology 80:134–140.

2245. Larson, D. L. 1996. Seed dispersal by specialist versus generalist foragers: the plant's perspective. Oikos 76:113–120.

2246. Latham, A. J. 1969. Zonate leafspot of pecan caused by *Cristulariella pyramidalis*. Phytopathology 59:103–107.

2247. Latham, A. J. 1974. Effect of temperature and moisture on *Cristulariella pyramidalis*. Phytopathology 64:635–639.

2248. Latorre, B. A., and Wilcox, W. F. 1996. Detection of *Phytophthora* spp. in fruit trees by immunological assays. Fitopatologia 31:202–206.

2249. Laundon, G. F., and Rainbow, A. F. 1969. *Phragmidium mucronatum*. C.M.I. Descr. Pathogenic Fungi Bact. No. 204. 2 pp.

2250. Laurans, F., and Pilate, G. 1999. Histological aspects of a hypersensitive response in poplar to *Melampsora larici-populina*. Phytopathology 89:233–238.

2251. Laurence, J. A. 1981. Effects of air pollutants on plant-pathogen interactions. Z. Pflanzenkr. Pflanzenschutz 88:156–172.

2252. Lavallée, A. 1964. A larch canker caused by *Leucostoma kunzei* (Fr.) Munk ex Kern. Can. J. Bot. 42:1495–1502.

2253. Lavermicocca, P., and Surico, G. 1987. Epiphytic occurrence of *Pseudomonas syringae* pv. *savastanoi* and other bacteria on olive and oleander. Phytopathol. Medit. 26:137–141.

2254. Lawrey, J. D. 1984. Biology of lichenized fungi. Praeger, New York. 408 pp.

2255. Leaf, A. L. 1968. K, Mg, and S deficiencies in forest trees. Pages 88–122 in: Forest fertilization: theory and practice. Tenn. Valley Authority, Muscle Shoals, AL. 306 pp.

2256. Leaphart, C. D. 1959. Drought damage to western white pine and associated tree species. Plant Dis. Rep. 43:809–813.

2257. Leaphart, C. D., and Gill, L. S. 1959. Effects of inoculations with *Leptographium* spp. on western white pine. Phytopathology 49:350–353.

2258. Leaphart, C. D., and Stage, A. R. 1971. Climate: a factor in the origin of the pole blight disease of *Pinus monticola* Dougl. Ecology 52:229–239.

2259. Leaphart, C. D., and Wicker, E. F. 1966. Explanation of pole blight from responses of seedlings grown in modified environments. Can. J. Bot. 44:121–137.

2260. Leben, C. 1985. Wound occlusion and discoloration columns in red maple. New Phytol. 99:485–490.

2261. Lechevalier, M. P. 1994. Taxonomy of the genus *Frankia* (Actinomycetales). Int. J. Syst. Bacteriol. 44:1–8.

2262. Lecouls, A. C., Salesses, G., Minot, J. C., Voisin, R., Bonnet, A., and Esmenjaud, D. 1997. Spectrum of the *Ma* genes for resistance to *Meloidogyne* spp. in myrobalan plum. Theor. Appl. Genet. 95:1325–1334.

2263. Lee, H., and Yu, S. 1995. Distribution of mycotoxin-producing isolates in the genus *Alternaria*. Korean J. Plant Pathol. 11:151–157.

2264. Lee, I.-M., Bertaccini, A., Vibio, M., and Gundersen, D. E. 1995. Detection of multiple phytoplasmas in perennial fruit trees with decline symptoms in Italy. Phytopathology 85:728–735.

2265. Lee, I.-M., Davis, R. E., and Gundersen-Rindal, D. E. 2000. Phytoplasma: phytopathogenic mollicutes. Annu. Rev. Microbiol. 54:221–255.

2266. Lee, I.-M., Davis, R. E., Sinclair, W. A., DeWitt, N. D., and Conti, M. 1993. Genetic relatedness of mycoplasmalike organisms detected in *Ulmus* spp. in the United States and Italy by means of DNA probes and polymerase chain reactions. Phytopathology 83:829–833.

2267. Lee, I.-M., Gundersen-Rindal, D. E., and Bertaccini, A. 1998. Phytoplasma: ecology and genomic diversity. Phytopathology 88:1359–1366.

2268. Lee, I.-M., Gundersen-Rindal, D. E., Davis, R. E., and Bartoszyk, I. M. 1998. Revised classification scheme of phytoplasmas based on RFLP analyses of 16S rRNA and ribosomal protein gene sequences. Int. J. Syst. Bacteriol. 48:1153–1169.

2269. Lee, I.-M., Martini, M., Marcone, C., and Zhu, S. F. 2004. Classification of phytoplasma strains in the elm yellows group (16SrV) and proposal of 'Candidatus Phytoplasma ulmi' for the phytoplasma associated with elm yellows. Int. J. Syst. Evol. Microbiol. 54:337–347.

2270. Lee, J. C., Skelly, J. M., Steiner, K. C., Zhang, J. W., and Savage, J. E. 1999. Foliar response of black cherry (*Prunus serotina*) clones to ambient ozone exposure in central Pennsylvania. Environ. Pollut. 105:325–331.

2271. Lee, J. C., Yang, X., Schwartz, M., Strobel, G., and Clardy, J. 1995. The relationship between an endangered North American tree and an endophytic fungus. Chem. Biol. 2:721–727.

2272. Lee, R. E. J., Warren, G. J., and Gusta, L. V., eds. 1995. Biological ice nucleation and its applications. APS Press, St. Paul, MN. 420 pp.

2273. Lee, R. F., Marais, L. J., Timmer, L. W., and Graham, J. H. 1984. Syringe injection of water into the trunk: a rapid diagnostic test for citrus blight. Plant Dis. 68:511–513.

2274. Lee, S., Jung, J. H., and Lee, J. K. 1998. Cultural characteristics and pathogenicity test of a die-back fungus, *Cenangium ferruginosum*, isolated from *Pinus koraiensis*. J. Korean For. Soc. 87:557–561.

2275. Lee, S., and Kakishima, M. 1999. Aeciospore surface structures of *Gymnosporangium* and *Roestelia* (Uredinales). Mycoscience 40:109–120.

2276. Lee, S., and Kakishima, M. 1999. Surface structures of peridial cells of *Gymnosporangium* and *Roestelia* (Uredinales). Mycoscience 40:121–131.

2277. Lee, S., Ko, K., and Aldwinckle, H. S. 2000. Resistance of selected *Malus* germplasm to *Rosellinia necatrix*. J. Am. Pomol. Soc. 54:219–228.

2278. Lee, S., Park, S., and Choi, J. 1997. Characterization of an isolate of cucumber mosaic virus from forsythia (*Forsythia koreana* Nakai.). Korean J. Plant Pathol. 13:358–363.

2279. Leeuwen, G. C. M. van, Baayen, R. P., Holb, I. J., and Jeger, M. J. 2002. Distinction of the Asiatic brown rot fungus *Monilia polystroma* sp. nov. from *M. fructigena*. Mycol. Res. 106:444–451.

2280. Lefèvre, F., Goué-Mourier, M. C., Faivre-Rampant, R. P., and Villar, M. 1998. A single gene cluster controls incompatibility and partial resistance to various *Melampsora larici-populina* races in hybrid poplars. Phytopathology 88:156–163.

2281. Lefèvre, F., Pichot, C., and Pinon, J. 1994. Intra- and interspecific inheritance of some components of the resistance to leaf rust (*Melampsora larici-populina* Kleb.) in poplars. Theor. Appl. Genet. 88:501–507.

2282. Lefol, C., Lherminier, J., Noudon, P. E., Larrue, J., Louis, C., and Caudwell, A. 1994. Propagation of flavescence dorée MLO (mycoplasma-like organisms) in the leafhopper vector *Euscelidius variegatus* Kbm. J. Invert. Pathol. 63:285–293.

2283. Lei, S. A. 1999. Age, size and water status of *Acacia greggii* influencing the infection and reproductive success of *Phoradendron californicum*. Am. Midl. Nat. 141:358–365.

2284. Lei, S. A. 2001. Survival and development of *Phoradendron californicum* and *Acacia greggii* during a drought. West. North Am. Nat. 61:78–84.

2285. Leininger, T. D., Miller, P. R., Schilling, S. L., and Dunn, P. H. 1991. Seedling responses of five species of western conifers to simulated ambient sulfur dioxide exposures. For. Sci. 37:1538–1549.

2286. Leininger, T. D., Solomon, J. D., Wilson, A. D., and Schiff, N. M. 1999. A guide to major insects, diseases, air pollution injury, and chemical injury of sycamore. USDA For. Serv. Gen. Tech. Rep. SRS 28. 44 pp.

2287. Leite, B., Ishida, M. L., Alves, E., Carrer, H., Pascholati, S. F., and Kitajima, E. W. 2002. Genomics and X-ray microanalysis indicate that Ca^{2+} and thiols mediate the aggregation and adhesion of *Xylella fastidiosa*. Brazil. J. Med. Biol. Res. 35:645–650.

2288. Leith, I. D., and Fowler, D. 1988. Urban distribution of *Rhytisma acerinum* (Pers.) Fries (tar spot) on sycamore. New Phytol. 108:175–181.

2289. Lemke, P. A. 1964. The genus *Aleurodiscus* (*sensu stricto*) in North America. Can. J. Bot. 42:213–282.

2290. Lemke, P. A. 1965. *Dendrothele* (1907) vs. *Aleurocorticium* (1963). Persoonia 3:365–367.

2291. Lemoine, D., Granier, A., and Cochard, H. 1999. Mechanism of freeze-induced embolism in *Fagus sylvatica* L. Trees Struct. Funct. 13:206–210.

2292. Lemon, P. C. 1961. Forest ecology of ice storms. Bull. Torrey Bot. Club 88:21–29.

2293. Lentz, P. L. 1955. *Stereum* and allied genera of fungi in the upper Mississippi Valley. USDA Monogr. 24. 74 pp.

2294. Leon, R., Santamaria, J. M., Alpizar, L., Escamilla, J. A., and Oropeza, C. 1996. Physiological and biochemical changes in shoots of coconut palms affected by lethal yellowing. New Phytol. 134:227–234.

2295. Leonard, O. A., and Hull, R. J. 1965. Translocation relationships in and between mistletoes and their hosts. Hilgardia 37:115–151.

2296. Lepka, P., Stitt, M., Moll, E., and Seemüller, E. 1999. Effect of phytoplasmal infection on concentration and translocation of carbohydrates and amino acids in periwinkle and tobacco. Physiol. Mol. Plant Pathol. 55:59–68.

2297. Leslie, J. F. 1993. Fungal vegetative compatibility. Annu. Rev. Phytopathol. 31:127–150.

2298. Lewis, K. J., and Hansen, E. M. 1991. Survival of *Inonotus tomentosus* in stumps and subsequent infection of young stands in north central British Columbia. Can. J. For. Res. 21:1049–1057.

2299. Lewis, K. J., and Hansen, E. M. 1991. Vegetative compatibility groups and protein electrophoresis indicate a role for basidiospores in spread of *Inonotus tomentosus* in spruce forests of British Columbia. Can. J. Bot. 69:1756–1763.

2300. Lewis, K. J., Morrison, D. J., and Hansen, E. M. 1992. Spread of *Inonotus tomentosus* from infection centres in spruce forests in British Columbia. Can. J. For. Res. 22:68–72.

2301. Lewis, R. Jr. 1981. Decline symptoms in Mississippi gulf coastal live oaks. (Abstr.) J. Miss. Acad. Sci. 27 (Suppl.):6.

2302. Lewis, R. Jr. 1981. *Hypoxylon* spp., *Ganoderma lucidum*, and *Agrilus bilineatus* in association with drought related oak mortality in the South. (Abstr.) Phytopathology 71:890.

2303. Lewis, R. Jr. 1985. Temperature tolerance and survival of *Ceratocystis fagacearum* in Texas. Plant Dis. 69:443–444.

2304. Lewis, R. Jr., and Arsdel, E. P. Van. 1978. Development of Botryo-diplodia cankers in sycamore at controlled temperatures. Plant Dis. Rep. 62:125–126.

2305. Lewis, R. Jr., and Arsdel, E. P. Van. 1978. Vulnerability of water-stressed sycamores to strains of *Botryodiplodia theobromae*. Plant Dis. Rep. 62:62–63.

2306. Lewis, R. Jr., and Oliveria, F. L. 1979. Live oak decline in Texas. J. Arboric. 5:241–244.

2307. Li, H., Qiu, B., Shi, C., Jin, K., Zhou, Q., and Huang, X. 1997. PCR amplification of 16S rDNA of phytoplasma associated with cherry fasciated disease and RFLP analysis. For. Res. 10:478–481.

2308. Li, J., Strobel, G., Sidhu, R., Hess, W. M., and Ford, E. J. 1996. Endophytic taxol-producing fungi from bald cypress, *Taxodium distichum*. Microbiology 142:2223–2226.

2309. Li, K. N., Rouse, D. I., Eyestone, E. J., and German, T. L. 1999. The generation of specific DNA primers using random amplified polymorphic DNA and its application to *Verticillium dahliae*. Mycol. Res. 103:1361–1368.

2310. Li, W. B., Pria, W. D. Jr., Teixeira, D. C., Miranda, V. S., Ayres, A. J., Franco, C. F., Costa, M. G., He, C. X., Costa, P. I., and Hartung, J. S. 2001. Coffee leaf scorch caused by a strain of *Xylella fastidiosa* from citrus. Plant Dis. 85:501–505.

2311. Li, W. B., Zreik, L., Fernandes, N. G., Miranda, V. S., Teixeira, D. C., Ayres, A. J., Garnier, M., and Bové, J. M. 1999. A triply cloned strain of *Xylella fastidiosa* multiplies and induces symptoms of citrus variegated chlorosis in sweet orange. Curr. Microbiol. 39:106–108.

2312. Liang, H., Catranis, C. M., Maynard, C. A., and Powell, W. A. 2002. Enhanced resistance to the poplar fungal pathogen, *Septoria musiva*, in hybrid poplar clones transformed with genes encoding antimicrobial peptides. Biotechnol. Lett. 24:383–389.

2313. Liao, J. L., Zhang, L. H., and Feng, Z. X. 2001. Reliable identification of *Bursaphelenchus xylophilus* by rDNA amplification. Nematol. Medit. 29:131–135.

2314. Lichter, J. M., and Berry, A. M. 1991. Establishment of the mistletoe *Phoradendron macrophyllum*: phenology of early stages and host compatibility studies. Bot. Gaz. 152:468–475.

2315. Lichter, J. M., Reid, M. S., and Berry, A. M. 1991. New methods for control of leafy mistletoe (*Phoradendron* spp.) on landscape trees. J. Arboric. 17:127–130.

2316. Liebhold, A. M., MacDonald, W. L., Bergdahl, D., and Mastro, V. C. 1995. Invasion by exotic forest pests: a threat to forest ecosystems. For. Sci. Monogr. 30. 49 pp.

2317. Lightle, P. C., and Hawksworth, F. G. 1965. New hosts for broom-causing fungi in the Southwest. Plant Dis. Rep. 49:417–418.

2318. Lilja, A., Karjalainen, R., Parikka, P., Kammiovirta, K., and Nuorteva, H. 1998. Pathogenicity and genetic variation of *Phytophthora cactorum* from silver birch and strawberry. Eur. J. Plant Pathol. 104:529–535.

2319. Limber, D. P., and Cash, E. K. 1945. *Actinopelte dryina*. Mycologia 37:129–137.

2320. Lin, Z. S., and Liang, Z. C. 1988. Comparative resistance of several pine species to brown spot needleblight and the origin of the disease. Sci. Silvae Sinicae 24:25–47.

2321. Lindbeck, A. G. C., and Brlansky, R. H. 2000. Cytology of fibrous roots from citrus blight–affected trees. Plant Dis. 84:164–167.

2322. Lindberg, M., and Johansson, M. 1991. Growth of *Heterobasidion annosum* through bark of *Picea abies*. Eur. J. For. Pathol. 21:377–388.

2323. Lindberg, M., and Johansson, M. 1992. Resistance of *Picea abies* seedlings to infection by *Heterobasidion annosum* in relation to drought stress. Eur. J. For. Pathol. 22:115–124.

2324. Linde, C., Kemp, G. H. J., and Wingfield, M. J. 1997. First report of Sphaeropsis canker on cypress in South Africa. Eur. J. For. Pathol. 27:173–177.

2325. Linderman, R. G. 1973. Formation of microsclerotia of *Cylindrocladium* spp. in infected azalea leaves, flowers, and roots. Phytopathology 63:187–191.

2326. Linderman, R. G. 1974. Ascospore discharge from perithecia of *Calonectria theae*, *C. crotalariae*, and *C. kyotensis*. Phytopathology 64:567–569.

2327. Linderman, R. G. 1974. The role of abscised *Cylindrocladium*-infected azalea leaves in the epidemiology of Cylindrocladium wilt of azalea. Phytopathology 64:481–485.

2328. Linderman, R. G., and Gilbert, R. G. 1969. Stimulation of *Sclerotium rolfsii* in soil by volatile components of alfalfa hay. Phytopathology 59:1366–1372.

2329. Lindow, S. E. 1983. The role of bacterial ice nucleation in frost injury to plants. Annu. Rev. Phytopathol. 21:363–384.

2330. Lindow, S. E. 1993. Biological control of plant frost injury: the ice story. Pages 113–128 in: Advanced engineered pesticides. L. Kim, ed. Marcel Dekker, New York.

2331. Lindow, S. E., and Connell, J. H. 1984. Reduction of frost injury to almond by control of ice nucleation active bacteria. J. Am. Soc. Hortic. Sci. 109:48–53.

2332. Lindsey, J. P., and Gilbertson, R. L. 1978. Basidiomycetes that decay aspen in North America. Cramer, Vaduz, Austria. 406 pp.

2333. Linit, M. J. 1988. Nematode-vector relationships in the pine wilt disease system. J. Nematol. 20:227–235.

2334. Linit, M. J. 1990. Transmission of pinewood nematode through feeding wounds of *Monochamus carolinensis* (Coleoptera: Cerambycidae). J. Nematol. 22:231–236.

2335. Linit, M. J., and Tamura, H. 1987. Relative susceptibility of four pine species to infection by pinewood nematode. J. Nematol. 19:44–50.

2336. Linzon, S. N. 1960. The development of foliar symptoms and the possible cause and origin of white pine needle blight. Can. J. Bot. 38:153–161.

2337. Linzon, S. N. 1966. Damage to eastern white pine by sulfur dioxide, semimature-tissue needle blight, and ozone. J. Air Pollut. Control Assn. 16:140–144.

2338. Linzon, S. N. 1967. Histological studies of semimature-tissue needle blight on eastern white pine. Can. J. Bot. 45:133–143.

2339. Linzon, S. N. 1967. Ozone damage and semimature-tissue needle blight of eastern white pine. Can. J. Bot. 45:2047–2061.

2340. Linzon, S. N. 1968. Etiological studies on needle fungi associated with semimature-tissue needle blight of eastern white pine. Can. J. Bot. 46:1565–1574.

2341. Lipetz, J. 1970. Wound-healing in higher plants. Int. Rev. Cytol. 27:1–28.

2342. Litterick, A. M., McQuilken, M. P., and Holmes, S. J. 1995. Sources of *Rhizoctonia* species in ericaceous plant nurseries. Z. Pflanzenkr. Pflanzenschutz 102:441–444.

2343. Little, S., Mohr, J. J., and Spicer, L. L. 1958. Salt-water storm damage to loblolly pine forests. J. For. 56:27–28.

2344. Littlefield, L. J. 1981. Biology of the plant rusts: an introduction. Iowa State Univ. Press, Ames. 103 pp.

2345. Littlefield, L. J., and Heath, M. C. 1979. Ultrastructure of rust fungi. Academic Press, New York. 277 pp.

2346. Liu, H.-Y., Gumpf, D. J., Oldfield, G. N., and Calavan, E. C. 1983. Transmission of *Spiroplasma citri* by *Circulifer tenellus*. Phytopathology 73:582–585.

2347. Liu, X., Tian, S., Qin, G., and Shen, R. 1999. Examination of persimmon fasciation and black locust witches' broom with the polymerase chain reaction. J. Beijing For. Univ. 21:26–28.

2348. Liu, Y. C., Double, M. L., MacDonald, W. L., and Milgroom, M. G. 2002. Persistence of *Cryphonectria* hypoviruses after their release for biological control of chestnut blight in West Virginia forests. For. Pathol. 32:345–356.

2349. Livingston, W. H., Ritz, G. F., and Rombold, J. S. 1991. Eastern dwarf mistletoe on red spruce in eastern Maine. North. J. Appl. For. 8:123–125.

2350. Livingston, W. H., and White, A. S. 1997. May drought confirmed as likely cause of brown ash dieback in Maine (Abstr.). Phytopathology 87:S59.

2351. Llimona, X., Lumbsch, H. T., and Ott, S. 2000. Progress and problems in lichenology at the turn of the millenium. Cramer, Berlin. 326 pp.

2352. Lo, M. H., Abrahamson, L. P., White, E. H., and Manion, P. D. 1995. Early measures of basal area and canker disease predict growth potential of some hybrid poplar clones. Can. J. For. Res. 25:1113–1118.

2353. Lockerman, R. H., Putnam, A. R., Rice, R. P. Jr., and Meggitt, W. F. 1975. Diagnosis and prevention of herbicide injury. Mich. State Univ., Coop. Exten Serv. Bull. E-809. 19 pp.

2354. Lockhart, B. E. L. 1987. Evidence for identity of plant rhabdoviruses causing vein-yellowing diseases of tomato and *Hibiscus rosa-sinensis*. Plant Dis. 71:731–733.

2355. Loffler, C., Czygan, F. C., and Proksch, P. 1999. Role of indole-3-acetic acid in the interaction of the phanerogamic parasite *Cuscuta* and host plants. Plant Biol. 1:613–617.

2356. Lohman, M. L., and Cash, E. K. 1940. *Atropellis* species from pine cankers in the United States. J. Wash. Acad. Sci. 30:255–262.

2357. Lohman, M. L., Cash, E. K., and Davidson, R. W. 1942. An undescribed *Atropellis* on cankered *Pinus virginiana*. J. Wash. Acad. Sci. 32:296–298.

2358. Lohman, M. L., and Watson, A. J. 1943. Identity and host relations of *Nectria* species associated with diseases of hardwoods in the eastern states. Lloydia 6:77–108.

573

2359. Lombard, F. F., Davidson, R. W., and Lowe, J. L. 1960. Cultural characteristics of *Fomes ulmarius* and *Poria ambigua*. Mycologia 52:280–294.

2360. Long, R. P., Horsley, S. B., and Lilja, P. R. 1997. Impact of forest liming on growth and crown vigor of sugar maple and associated hardwoods. Can. J. For. Res. 27:1560–1573.

2361. Long, W. H. 1913. *Polyporus dryadeus*, a root parasite on the oak. J. Agric. Res. 1:239–250.

2362. Long, W. H. 1924. The self pruning of western yellow pine. Phytopathology 14:336–337.

2363. Long, W. H. 1930. *Polyporus dryadeus*, a root parasite on white fir. Phytopathology 20:758–759.

2364. Lonsdale, D. 1980. *Nectria* infection of beech bark in relation to infestation by *Cryptococcus fagisuga* Lindiger. Eur. J. For. Pathol. 10:161–168.

2365. Lopez, H. C. J., Perez, J. R. M., Basallote, U. M. J., Bonilla, T. Z., and Melero, V. J. M. 1997. Effect of soil solarization on the control of Phytophthora root rot in avocado. Plant Pathol. 46:329–340.

2366. Lopez, H. C. J., Perez, J. R. M., Basallote, U. M. J., Zea, B. T., and Melero, V. J. M. 1999. Loss of viability of *Dematophora necatrix* in solarized soils. Eur. J. Plant Pathol. 105:571–576.

2367. Lorenzini, G., and Soldatini, G. F., eds. 1995. Responses of plants to air pollution: biological and economic aspects. Proceedings of a conference held in Pisa, Italy, 1994. Agric. Medit. Spec. Vol. Pacini Editore, Pisa. 415 pp.

2368. Loring, L. B., and Roth, L. F. 1964. *Pucciniastrum epilobii* on *Fuchsia* in Oregon. Plant Dis. Rep. 48:99.

2369. Lortie, M. 1964. Pathogenesis in cankers caused by *Nectria galligena*. Phytopathology 54:261–263.

2370. Lortie, M., and Kuntz, J. E. 1963. Ascospore discharge and conidium release by *Nectria galligena* Bres. under field and laboratory conditions. Can. J. Bot. 41:1203–1210.

2371. Louis, C., Girard, M., Kuhl, G., and Lopez, F. M. 1996. Persistence of *Botrytis cinerea* in its vector *Drosophila melanogaster*. Phytopathology 86:934–939.

2372. Lovelady, C. N., Pulley, P. E., Coulson, R. N., and Flamm, R. O. 1991. Relation of lightning to herbivory by the southern pine bark beetle guild (Coleoptera: Scolytidae). Environ. Entomol. 20:1279–1284.

2373. Lowell, E. C., and Fahey, T. D. 1994. The effect of frost cracks in sugar pine trees on lumber grade. For. Prod. J. 44:57–60.

2374. Lowerts, G. A., and Kellison, R. C. 1981. Genetically controlled resistance to discoloration and decay in wounded trees of yellow-poplar. Silvae Genet. 30:98–101.

2375. Lu, J. D., Gan, L., and Yan, L. F. 1991. Purification of the toxin of *Verticillium dahliae* and its properties. Acta Phytopathol. Sinica 21:129–133.

2376. Luc, M., Coomans, A., Loof, P. A. A., and Baujard, P. 1998. The *Xiphinema americanum*–group (Nematoda: Longidoridae). 2. Observations on *Xiphinema brevicollum* Lordello & da Costa, 1961 and comments on the group. Fund. Appl. Nematol. 21:475–490.

2377. Lucas, J. A. 1988. An outbreak of mahonia rust, *Cumminsiella mirabilissima*. Mycologist 2:163.

2378. Lucas, J. A., Shattock, R. C., Shaw, D. S., and Cooke, L. R., eds. 1991. *Phytophthora*. Cambridge Univ. Press, Cambridge, UK. 447 pp.

2379. Ludwig, C. A. 1915. Notes on some North American rusts with *Caeoma*-like sori. Phytopathology 5:273–281.

2380. Luisi, N., Lerario, P., and Bianco, M. C. 1996. *Botryosphaeria stevensii*: pathogenicity on oaks and phytotoxic effects. Ital. For. Mont. 51:250–263.

2381. Luisi, N., Manicone, R. P., Trombetta, N. M., and Cusano, G. 1995. Predisposition to forest decline of some Mediterranean oaks in relation to drought resistance. Ital. For. Mont. 50:44–59.

2382. Luley, C. J., and McNabb, H. S. Jr. 1991. Estimation of seasonal ascospore production of *Mycosphaerella populorum*. Can. J. For. Res. 21:1349–1353.

2383. Luley, C. J., Mielke, M. E., Castello, J. D., Cummings Carlson, J., Appleby, J., and Hatcher, R. 1992. Ash crown condition and the incidence of ash yellows and other insects and diseases in Illinois, Iowa, Missouri, and Wisconsin. Plant Dis. 76:1209–1212.

2384. Lumis, G. P., Hofstra, G., and Hall, R. 1973. Sensitivity of roadside trees and shrubs to aerial drift of deicing salt. HortScience 8:475–477.

2385. Lumis, G. P., Hofstra, G., and Hall, R. 1976. Roadside woody plant susceptibility to sodium and chloride accumulation during winter and spring. Can. J. Plant Sci. 56:853–859.

2386. Lumis, G. P., and Johnson, A. G. 1982. Boron toxicity and growth suppression of *Forsythia* and *Thuja* grown in mixes amended with municipal waste compost. HortScience 27:821–822.

2387. Lund, A. E., and Livingston, W. H. 1999. Freezing cycles enhance winter injury in *Picea rubens*. Tree Physiol. 19:65–69.

2388. Lundquist, J. E. 1993. Assessing Rhizina root disease hazard: a case study. S.Afr. For. J. 164:51–53.

2389. Luque, J., and Girbal, J. 1988. Dieback of cork oak (*Quercus suber*) in Catalonia (NE Spain) caused by *Botryosphaeria stevensii*. Eur. J. For. Pathol. 19:7–13.

2390. Luque, J., Parlade, J., and Pera, J. 2000. Pathogenicity of fungi isolated from *Quercus suber* in Catalonia (NE Spain). For. Pathol. 30:247–263.

2391. Lüttge, U., ed. 1989. Vascular plants as epiphytes: evolution and ecophysiology. Springer-Verlag, Berlin, New York. 270 pp.

2392. Luttrell, E. S. 1940. *Morenoella quercina*, cause of leaf spot of oaks. Mycologia 32:652–666.

2393. Luttrell, E. S. 1948. Physiologic specialization in *Guignardia bidwellii*, cause of black rot of *Vitis* and *Parthenocissus* species. Phytopathology 38:716–723.

2394. Luttrell, E. S. 1949. Horse chestnut anthracnose in Missouri. Plant Dis. Rep. 33:324–327.

2395. Luttrell, E. S. 1950. *Botryosphaeria* stem canker of elm. Plant Dis. Rep. 34:138–139.

2396. Luttrell, E. S. 1989. Morphology of *Meliola floridensis*. Mycologia 81:192–204.

2397. Lutz, H. J. 1943. Injuries to trees caused by *Celastrus* and *Vitis*. Bull. Torrey Bot. Club 70:436–439.

2398. Lutz, H. J. 1952. Occurrence of clefts in the wood of living white spruce in Alaska. J. For. 50:99–102.

2399. Lyda, S. D. 1978. Ecology of *Phymatotrichum omnivorum*. Annu. Rev. Phytopathol. 16:193–209.

2400. Lyon, J., and Sharpe, W. E. 1999. An assessment of the Ca:Al ratios of selected Pennsylvania forest soils. Water Air Soil Pollut. 109:53–65.

2401. Ma, Z., and Michailides, T. J. 2002. Characterization of *Botryosphaeria dothidea* isolates collected from pistachio and other plant hosts in California. Phytopathology 92:519–526.

2402. Ma, Z., Morgan, D. P., and Michailides, T. J. 2001. Effects of water stress on Botryosphaeria blight of pistachio caused by *Botryosphaeria dothidea*. Plant Dis. 85:745–749.

2403. MacBeath, J. H., Nyland, G., and Spurr, A. R. 1972. Morphology of mycoplasmalike bodies associated with peach X-disease in *Prunus persica*. Phytopathology 62:935–937.

2404. MacDaniels, L. H., Johnson, W. T., and Braun, E. J. 1976. The black walnut bunch disease syndrome. North. Nut Growers Assoc. Annu. Rep. 66:71–87.

2405. MacDonald, J. A. 1937. A study of *Polyporus betulinus* (Bull.) Fries. Ann. Appl. Biol. 24:289–310.

2406. MacDonald, J. D. 1982. Role of environmental stress in the development of Phytophthora root rots. J. Arboric. 8:217–223.

2407. MacDonald, J. D., Ali, S. M. S., Kabashima, J., and Stites, J. 1994. Occurrence of *Phytophthora* species in recirculated nursery irrigation effluents. Plant Dis. 78:607–611.

2408. MacDonald, W. L., Cech, F. C., Luchok, J., and Smith, C. 1979. Proceedings of the American chestnut symposium. West Va. Univ. Books, Morgantown. 122 pp.

2409. MacDonald, W. L., and Fulbright, D. W. 1991. Biological control of chestnut blight: use and limitations of transmissible hypovirulence. Plant Dis. 75:656–661.

2410. MacHardy, W. E. 1996. Apple scab: biology, epidemiology, and management. APS Press, St. Paul, MN. 545 pp.

2411. MacHardy, W. E., Gadoury, D. M., and Gessler, C. 2001. Parasitic and biological fitness of *Venturia inaequalis*: relationship to disease management strategies. Plant Dis. 85:1036–1051.

2412. MacLachlan, J. D. 1935. The hosts of *Gymnosporangium globosum* Farl. and their relative susceptibility. J. Arnold Arbor. 16:98–142.

2413. Madar, Z., Kimchi, M., and Solel, Z. 1996. Fusarium canker in Italian cypress. Eur. J. For. Pathol. 26:107–112.

2414. Madar, Z., and Reuveni, R. 1988. Mortality of *Cedrus* seedlings caused by the fungus *Macrophomina phaseolina*. Hassadeh 68:1808–1809.

2415. Madar, Z., Solel, Z., and Kimchi, M. 1991. Pestalotiopsis canker of cypress in Israel. Phytoparasitica 19:79–81.

2416. Mader, D. L., and Thompson, B. W. 1969. Foliar and soil nutrients in relation to maple decline. Soil Sci. Soc. Am. Proc. 33:794–800.

2417. Magan, N., Kirkwood, I. A., McLeod, A. R., and Smith, M. K. 1995. Effect of open-air fumigation with sulphur dioxide and ozone on phyllosphere and endophytic fungi of conifer needles. Plant Cell Environ. 18:291–302.

2418. Magasi, L. P., ed. 1983. Proceedings of European larch canker workshop. Can. For. Serv. Maritimes For. Res. Cent. MFRC Workshop Proc. No. 3. 46 pp.

2419. Magasi, L. P., Harrison, K. J., Urquhart, D. A., and Murray, D. M. 1993. Three decades of Dutch elm disease in Fredericton, NB. Can. For. Serv. Inf. Rep. M-X-185E. 39 pp.

2420. Magasi, L. P., and Pond, S. E. 1982. European larch canker: a new disease in Canada and a new North American host record. Plant Dis. 66:339.

2421. Maggenti, A. R. 1981. Nematodes: development as plant parasites. Annu. Rev. Microbiol. 35:135–154.

2422. Maguire, D. A., Kanaskie, A., Voelker, W., Johnson, R., and Johnson, G. 2002. Growth of young Douglas-fir plantations across a gradient in Swiss needle cast severity. West. J. Appl. For. 17:86–95.

2423. Mahoney, E. M., Milgroom, M. G., Sinclair, W. A., and Houston, D. R. 1999. Origin, genetic diversity, and population structure of Nectria coccinea var. faginata in North America. Mycologia 91:583–592.

2424. Mahoney, M. J., Skelly, J. M., Chevone, B. I., and Moore, L. D. 1984. Response of yellow poplar (Liriodendron tulipifera L.) seedling shoot growth to low concentrations of O_3, SO_2, and NO_2. Can. J. For. Res. 14:150–153.

2425. Mahoney, M. J., and Tattar, T. A. 1980. Identification, etiology, and control of Euonymus fortunei anthracnose caused by Colletotrichum gloeosporioides. Plant Dis. 64:854–856.

2426. Mai, W. F., Bloom, J. R., and Chen, T. A., eds. 1977. Biology and ecology of the plant-parasitic nematode Pratylenchus penetrans. Penn. Agric. Exp. Stn. Bull. 815. 64 pp.

2427. Mai, W. F., and Mullin, P. G. 1996. Plant-parasitic nematodes: a pictorial key to genera. 5th ed. Cornell Univ. Press, Ithaca, NY. 277 pp.

2428. Maier, M. U. 1999. Predisposition of trees to drought stress by ozone. Tree Physiol. 19:71–78.

2429. Maier, W., Begerow, D., Weiss, M., and Oberwinkler, F. 2003. Phylogeny of the rust fungi: an approach using nuclear large subunit ribosomal DNA sequences. Can. J. Bot. 81:12–23.

2430. Maijala, P., Harrington, T. C., and Raudaskoski, M. 2003. A peroxidase gene family and gene trees in Heterobasidion and related genera. Mycologia 95:209–221.

2431. Mains, E. B. 1938. Host specialization in Coleosporium solidaginis and C. campanulae. Pap. Mich. Acad. Sci. Arts Lett. 23:171–175.

2432. Maiti, S., and Sen, C. 1988. Effect of moisture and temperature on the survival of sclerotia of Sclerotium rolfsii in soil. J. Phytopathol. 121:175–180.

2433. Malek, B. von, and Debener, T. 1998. Genetic analysis of resistance to blackspot (Diplocarpon rosae) in tetraploid roses. Theor. Appl. Genet. 96:228–231.

2434. Malek, B. von, Weber, W. E., and Debener, T. 2000. Identification of molecular markers linked to Rdr1, a gene conferring resistance to blackspot in roses. Theor. Appl. Genet. 101:977–983.

2435. Malek, R. B. 1968. The dagger nematode, Xiphinema americanum, associated with decline of shelterbelt trees in South Dakota. Plant Dis. Rep. 52:795–798.

2436. Malek, R. B., and Appleby, J. E. 1984. Epidemiology of pine wilt in Illinois. Plant Dis. 68:180–186.

2437. Malek, W., and Sajnaga, E. 1999. Current taxonomy of the rhizobia. Acta Microbiol. Polon. 48:109–122.

2438. Malhotra, S. S., and Blauel, R. A. 1980. Diagnosis of air pollutant and natural stress symptoms on forest vegetation in western Canada. Environ. Can. Inf. Rep. NOR-X-228. 84 pp.

2439. Malhotra, S. S., and Hocking, D. 1976. Biochemical and cytological effects of sulfur dioxide on plant metabolism. New Phytol. 76:227–237.

2440. Malia, M. E., and Tattar, T. A. 1978. Electrical resistance, physical characteristics, and cation concentrations in xylem of sugar maple infected with Verticillium dahliae. Can. J. For. Res. 8:322–327.

2441. Mallett, K. I., and Myrholm, C. L. 1995. The mating system of Phellinus tremulae. Mycologia 87:597–603.

2442. Malone, S. R., and Ashworth, E. N. 1991. Freezing stress response in woody tissues observed using low-temperature scanning electron microscopy and freeze substitution techniques. Plant Physiol. 95:871–881.

2443. Maloy, O. 1991. Review of Echinodontium tinctorium 1895–1990. Wash. State Univ. Exten. Bull. EB1592. 29 pp.

2444. Maloy, O. C. 1997. White pine blister rust control in North America: a case history. Annu. Rev. Phytopathol. 35:87–109.

2445. Mamiya, Y. 1983. Pathology of the pine wilt disease caused by Bursaphelenchus xylophilus. Annu. Rev. Phytopathol. 21:201–220.

2446. Mamiya, Y., and Furukawa, M. 1977. Fecundity and reproductive rate of Bursaphelenchus lignicolus. Jap. J. Nematol. 7:6–9.

2447. Mamiya, Y., and Kiyohara, T. 1972. Description of Bursaphelenchus lignicolus n.sp. (Nematoda: Aphelenchoididae) from pine wood and histopathology of nematode-infested trees. Nematologica 18:120–124.

2448. Manion, P. D. 1975. Two infection sites of Hypoxylon mammatum in trembling aspen (Populus tremuloides). Can. J. Bot. 53:2621–2624.

2449. Manion, P. D. 1981. Norway maple decline. J. Arboric. 7:38–42.

2450. Manion, P. D. 1991. Tree disease concepts. 2nd ed. Prentice-Hall, Englewood Cliffs, NJ. 402 pp.

2451. Manion, P. D., ed. 1984. Scleroderris canker of conifers. Martinus Nijhoff/W. Junk, The Hague. 273 pp.

2452. Manion, P. D., and French, D. W. 1967. Nectria galligena and Ceratocystis fimbriata cankers of aspen in Minnesota. For. Sci. 13:23–28.

2453. Manion, P. D., and French, D. W. 1969. The role of glucose in stimulating germination of Fomes igniarius var. populinus basidiospores. Phytopathology 59:293–296.

2454. Manion, P. D., and Griffin, D. H. 1986. Sixty-five years of research on Hypoxylon canker of aspen. Plant Dis. 70:803–808.

2455. Manion, P. D., and Griffin, D. H. 2001. Large landscape scale analysis of tree death in the Adirondack Park, New York. For. Sci. 47:542–579.

2456. Manion, P. D., Griffin, D. H., and Rubin, B. D. 2001. Ice damage impacts on the health of the northern New York State forest. For. Chron. 77:619–625.

2457. Manion, P. D., and Lachance, D., eds. 1992. Forest decline concepts. APS Press, St. Paul, MN. 249 pp.

2458. Manners, J. G. 1957. Studies on larch canker. II. The incidence and anatomy of cankers produced experimentally either by inoculation or by freezing. Trans. Br. Mycol. Soc. 40:500–508.

2459. Manter, D. K., Bond, B. J., Kavanagh, K. L., Rosso, P. H., and Filip, G. M. 2000. Pseudothecia of Swiss needle cast fungus, Phaeocryptopus gaeumannii, physically block stomata of Douglas-fir, reducing CO_2 assimilation. New Phytol. 148:481–491.

2460. Manter, D. K., and Livingston, W. H. 1996. Influence of thawing rate and fungal infection by Rhizosphaera kalkhoffii on freezing injury in red spruce (Picea rubens) needles. Can. J. For. Res. 26:918–927.

2461. Maramorosch, K., and Harris, K. F. 1979. Leafhopper vectors and plant disease agents. Academic Press, New York. 654 pp.

2462. Maramorosch, K., and Raychaudhuri, S. P., eds. 1981. Mycoplasma diseases of trees and shrubs. Academic Press, New York. 362 pp.

2463. Maramorosch, K., and Raychaudhuri, S. P. eds. 1988. Mycoplasma diseases of crops: basic and applied aspects. Springer-Verlag, New York. 456 pp.

2464. Marcais, B., Dupuis, F., and Desprez-Loustau, M. L. 1993. Influence of water stress on susceptibility of red oak (Quercus rubra) to Phytophthora cinnamomi. Eur. J. For. Pathol. 23:295–305.

2465. Marcais, B., Dupuis, F., and Desprez-Loustau, M. L. 1996. Susceptibility of the Quercus rubra root system to Phytophthora cinnamomi; comparison with chestnut and other oak species. Eur. J. For. Pathol. 26:133–143.

2466. Marcone, C., Neimark, H., Ragozzino, A., Lauer, U., and Seemüller, E. 1999. Chromosome sizes of phytoplasmas composing major phylogenetic groups and subgroups. Phytopathology 89:805–810.

2467. Marks, G. C., Berbee, J. G., and Riker, A. J. 1965. Colletotrichum shoot blight of poplars. For. Sci. 11:204–215.

2468. Marks, G. C., and Kozlowski, T. T., eds. 1973. Ectomycorrhizae: their ecology and physiology. Academic Press, New York. 444 pp.

2469. Marks, G. C., and Minko, G. 1969. The pathogenicity of Diplodia pinea to Pinus radiata D. Don. Austral. J. Bot. 17:1–12.

2470. Marlatt, R. B., and Alfieri, S. A. Jr. 1981. Hosts of a parasitic alga, Cephaleuros Kunze, in Florida. Plant Dis. 65:520–522.

2471. Marlatt, R. B., and Ridings, W. H. 1976. Sphaeropsis disease of Carissa grandiflora. Plant Dis. Rep. 60:842–843.

2472. Marlatt, R. B., and Ridings, W. H. 1979. Sphaeropsis gall of Schinus terebinthifolius, a new host. Plant Dis. Rep. 63:786–787.

2473. Marosy, M., Patton, R. F., and Upper, C. D. 1989. A conducive day concept to explain the effect of low temperature on the development of Scleroderris shoot blight. Phytopathology 79:1293–1301.

2474. Marques, L. L. R., Ceri, H., Manfio, G. P., Reid, D. M., and Olson, M. E. 2002. Characterization of biofilm formation by Xylella fastidiosa in vitro. Plant Dis. 86:633–638.

2475. Marschner, H. 1995. Mineral nutrition of higher plants. 2nd ed. Academic Press, San Diego, London. 889 pp.

2476. Marshall, J. D., Dawson, T. E., and Ehleringer, J. R. 1993. Gender-related differences in gas exchange are not related to host quality in the xylem-tapping mistletoe, Phoradendron juniperinum (Viscaceae). Am. J. Bot. 80:641–645.

2477. Marshall, J. D., Dawson, T. E., and Ehleringer, J. R. 1994. Integrated nitrogen, carbon, and water relations of a xylem-tapping mistletoe following nitrogen fertilization of the host. Oecologia 100:430–438.

2478. Marshall, J. D., and Ehleringer, J. R. 1990. Are xylem-tapping mistletoes partially heterotrophic? Oecologia 84:244–248.

2479. Marshall, J. D., Ehleringer, J. R., Schulze, E. D., and Farquhar, G. 1994. Carbon isotope composition, gas exchange and heterotrophy in Australian mistletoes. Funct. Ecol. 8:237–241.

2480. Marshall, K., and Filip, G. M. 1999. The relationship of Douglas-fir dwarf mistletoe (*Arceuthobium douglasii*) to stand conditions and plant associations in the southern Cascade Mountains, Oregon. Northwest Sci. 73:301–311.

2481. Martin, C. E., Eades, C. A., and Pitner, R. A. 1986. Effects of irradiance on Crassulacean acid metabolism in the epiphyte *Tillandsia usneoides* L. (Bromeliaceae). Plant Physiol. 80:23–26.

2482. Martin, E. M., and Kim, K. S. 1987. A new type of plant virus causing striped chlorosis of mimosa. Phytopathology 77:935–940.

2483. Martin, F. N., and Tooley, P. W. 2003. Phylogenetic relationships of *Phytophthora ramorum*, *P. nemorosa* and *P. pseudosyringae*, three species recovered from areas in California with sudden oak death. Mycol. Res. 107:1379–1391.

2484. Martin, P. 1967. Studies in the Xylariaceae. II. *Rosellinia* and the *Primo Cinera* section of *Hypoxylon*. J. S.Afr. Bot. 33:315–328.

2485. Martin, R. R., Berbee, J. G., and Omuemu, J. O. 1982. Isolation of a potyvirus from declining clones of *Populus*. Phytopathology 72:1158–1162.

2486. Martinez, A. T., Hatakka, A., Lundell, T., and Jeffries, T. W. 2002. Molecular biology and structure-function of lignin-degrading heme peroxidases. Rec. Adv. Lignin Biodegrad. Biosyn. 30:425–444.

2487. Marx, D. H., Cordell, C. E., Kenney, D. S., Mexal, J. G., Artman, J. D., Riffle, J. W., and Molina, R. J. 1984. Commercial vegetative inoculum of *Pisolithus tinctorius* and inoculation techniques for development of ectomycorrhizae on bare-root tree seedlings. For. Sci. Monogr. 25. 101 pp.

2488. Marx, D. H., and Davey, C. B. 1969. The influence of ectotrophic mycorrhizal fungi on the resistance of pine roots to pathogenic infections. IV. Resistance of naturally occurring mycorrhizae to infections by *Phytophthora cinnamomi*. Phytopathology 59:559–565.

2489. Marx, D. H., and Schenck, N. C. 1983. Potential of mycorrhizal symbiosis in agricultural and forest productivity. Pages 334–347 in: Challenging problems in plant health. T. Kommedahl and P. H. Williams, eds. Am. Phytopathol. Soc., St. Paul, MN. 538 pp.

2490. Massalski, P. R., and Cooper, J. I. 1984. The location of virus-like particles in the male gametophyte of birch, walnut, and cherry naturally infected with cherry leaf roll virus and its relevance to vertical transmission of the virus. Plant Pathol. 33:255–262.

2491. Massie, L. B., and Peterson, J. L. 1968. Factors affecting the initiation and development of Fusarium canker on *Sophora japonica* in relation to growth and sporulation of *Fusarium lateritium*. Phytopathology 58:1620–1623.

2492. Matheron, M. E., and Matejka, J. C. 1993. Seasonal differences in susceptibility of three citrus rootstocks to root lesions caused by *Phytophthora citrophthora* and *P. parasitica*. Plant Dis. 77:729–732.

2493. Matheron, M. E., and Mircetich, S. M. 1985. Influence of flooding duration on development of Phytophthora root and crown rot of *Juglans hindsii* and Paradox walnut rootstocks. Phytopathology 75:973–976.

2494. Matheron, M. E., and Mircetich, S. M. 1985. Pathogenicity and relative virulence of *Phytophthora* spp. from walnut and other plants to rootstocks of English walnut trees. Phytopathology 75:977–981.

2495. Matheron, M. E., and Sigler, L. 1994. First report of eucalyptus dieback caused by *Nattrassia mangiferae* in North America. Plant Dis. 78:432.

2496. Mathiasen, R. L. 1979. Distribution and effect of dwarf mistletoes parasitizing *Pinus strobiformis* in Arizona and New Mexico, USA, and northern Mexico. Southwest. Nat. 24:455–461.

2497. Mathiasen, R. L. 1994. Natural infection of new hosts by hemlock dwarf mistletoe. USDA For. Serv. Res. Note RM–530. 6 pp.

2498. Mathiasen, R. L. 1996. Dwarf mistletoes in forest canopies. Northwest Sci. 70:61–71.

2499. Mathiasen, R. L. 1998. Comparative susceptibility of conifers to larch dwarf mistletoe in the Pacific Northwest. For. Sci. 44:559–568.

2500. Mathiasen, R. L. 1998. Infection of young western larch by larch dwarf mistletoe in northern Idaho and western Montana. West. J. Appl. For. 13:41–46.

2501. Mathiasen, R. L. 1999. Comparative susceptibility of subalpine firs to Douglas-fir dwarf mistletoe. Can. J. Plant Pathol. 21:45–51.

2502. Mathiasen, R. L., Allison, J. R., and Geils, B. W. 1998. Western dwarf mistletoe parasitizing Colorado blue spruce and Norway spruce in California. Plant Dis. 82:351.

2503. Mathiasen, R. L., Beatty, J. S., and Hildebrand, D. M. 1995. First report of larch dwarf mistletoe on Pacific silver fir and on mountain hemlock in the Cascade Mountains. Plant Dis. 79:1249.

2504. Mathiasen, R. L., Beatty, J. S., and Hildebrand, D. M. 1996. First report of lodgepole pine dwarf mistletoe on subalpine fir. Plant Dis. 80:342.

2505. Mathiasen, R. L., Hawksworth, F. G., and Edminster, C. B. 1990. Effects of dwarf mistletoe on growth and mortality of Douglas-fir in the Southwest. Great Basin Nat. 50:173–179.

2506. Matschke, J., and Amenda, R. 1995. The dying of root tips in woody plants caused by herbicides. Allg. Forst Z. 50:1100–1104.

2507. Matteoni, J. A., and Neely, D. 1979. *Gnomonia leptostyla*: growth, sporulation, and heterothallism. Mycologia 71:1034–1042.

2508. Matteoni, J. A., and Sinclair, W. A. 1983. Stomatal closure in plants infected with mycoplasmalike organisms. Phytopathology 73:398–402.

2509. Matteoni, J. A., and Sinclair, W. A. 1985. Role of the mycoplasmal disease, ash yellows, in decline of white ash in New York State. Phytopathology 75:355–360.

2510. Matteoni, J. A., and Sinclair, W. A. 1989. A note on the presence of elm yellows in the Niagara Peninsula. Phytoprotection 70:137–139.

2511. Mattheck, C., and Bethge, K. 1990. Wind breakage of trees initiated by root delamination. Trees Struct. Funct. 4:225–227.

2512. Mattheck, C., Bethge, K., and West, P. W. 1994. Breakage of hollow tree stems. Trees Struct. Funct. 9:47–50.

2513. Mattheck, C., and Kubler, H. 1995. Wood—the internal optimization of trees. Springer-Verlag, Berlin, Heidelberg, New York. 129 pp.

2514. Matthews, R. E. F. 1993. Diagnosis of plant virus diseases. CRC Press, Boca Raton, FL. 440 pp.

2515. Maurel, M., Robin, C., Capron, G., and Desprez-Loustau, M. L. 2001. Effects of root damage associated with *Phytophthora cinnamomi* on water relations, biomass accumulation, mineral nutrition and vulnerability to water deficit of five oak and chestnut species. For. Pathol. 31:353–369.

2516. Mäurer, R., Seemüller, E., and Sinclair, W. A. 1993. Genetic relatedness of mycoplasmalike organisms affecting elm, alder, and ash in Europe and North America. Phytopathology 83:971–976.

2517. Maxwell, D. L., Kruger, E. L., and Stanosz, G. R. 1997. Effects of water stress on colonization of poplar stems and excised leaf disks by *Septoria musiva*. Phytopathology 87:381–388.

2518. May, C. 1961. Diseases of shade and ornamental maples. USDA Agric. Handb. 211. 22 pp.

2519. Mayhew, D. E., and Tidwell, T. E. 1978. Palm mosaic. Plant Dis. Rep. 62:803–806.

2520. Mayr, S., Siller, C., Kriss, M., Oberhuber, W., and Bauer, H. 2001. Photosynthesis in rust-infected adult Norway spruce in the field. New Phytol. 151:683–689.

2521. Mayr, U., Michalek, S., Treutter, D., and Feucht, W. 1997. Phenolic compounds of apple and their relationship to scab resistance. J. Phytopathol. 145:69–75.

2522. Mazzaglia, A., Anselmi, N., Vicario, S., and Vannini, A. 2001. Sequence analysis of the 5.8S rDNA and ITS regions in evaluating genetic relationships among some species of *Hypoxylon* and related genera. Mycol. Res. 105:670–675.

2523. Mazzola, M. 1997. Identification and pathogenicity of *Rhizoctonia* spp. isolated from apple roots and orchard soils. Phytopathology 87:582–587.

2524. Mazzola, M., and Bergdahl, D. R. 1989. The phenology and spore dissemination patterns of *Uredinopsis mirabilis* on *Abies balsamea* and *Onoclea sensibilis*. Can. J. For. Res. 19:88–95.

2525. Mmbaga, M. T. 2000. Winter survival and source of primary inoculum of powdery mildew of dogwood in Tennessee. Plant Dis. 84:574–579.

2526. Mmbaga, M. T. 2002. Ascocarp formation and survival and primary inoculum production in *Erysiphe* (sect. *Microsphaera*) *pulchra* in dogwood powdery mildew. Ann. Appl. Biol. 141:153–161.

2527. McAlpine, R. G. 1961. *Hypoxylon tinctor* associated with a canker on American sycamore trees in Georgia. Plant Dis. Rep. 45:196–198.

2528. McArdle, A. J., and Santamour, F. S. 1986. Screening mimosa (*Albizia julibrissin*) seedlings for resistance to nematodes and Fusarium wilt. Plant Dis. 70:249–251.

2529. McArthur, G. W. F. M. 1959. Cercospora leaf spot on *Rhododendron* (*C. handelii* Bubak). N. Z. J. Agric. Res. 2:86–89.

2530. McBeath, J. H. 1984. Symptomology on spruce trees and spore characteristics of a bud rust pathogen. Phytopathology 74:456–461.

2531. McCain, A. H., Raabe, R. D., and Wilhelm, S. 1979. Plants resistant or susceptible to Verticillium wilt. Univ. Calif., Div. Agric. Sci., Leafl. 2703. 10 pp.

2532. McCain, A. H., and Scharpf, R. F. 1989. Effect of inoculum density of *Macrophomina phaseolina* on seedling susceptibility of six conifer species. Eur. J. For. Pathol. 19:119–123.

2533. McCain, J. W., Hennen, J. F., and Ono, Y. 1990. New host species and state distribution records for North American rust fungi (Uredinales). Mycotaxon 39:281–300.

2534. McCarroll, D. R., and Thor, E. 1985. Pectolytic, cellulytic and proteolytic activities expressed by cultures of *Endothia parasitica*, and inhibition of these activities by components extracted from Chinese and American chestnut inner bark. Physiol. Plant Pathol. 26:367–378.

2535. McCauley, K. J., and Cook, S. A. 1980. *Phellinus weirii* infestation of two mountain hemlock forests in the Oregon Cascades. For. Sci. 26:23–29.

2536. McClain, R. L. 1925. Scab of Christmas berry, *Photinia arbutifolia* Lindl., due to *Fusicladium photinicola* n.sp. Phytopathology 15:178–182.

2537. McClenahen, J. R., and Dochinger, L. S. 1985. Tree ring response of white oak to climate and air pollution near the Ohio River valley. J. Environ. Qual. 14:274–280.

2538. McClintock, J. A. 1948. A study of uncongeniality between peaches as scions and the Marianna plum as a stock. J. Agric. Res. 77:253–260.

2539. McCoy, R. E., ed. 1983. Lethal yellowing of palms. Fla. Agric. Exp. Stn. Bull. 834. 100 pp.

2540. McCoy, R. E., Miller, M. E., Thomas, D. L., and Amador, J. 1980. Lethal decline of *Phoenix* palms in Texas associated with mycoplasmalike organisms. Plant Dis. 64:1038–1040.

2541. McCracken, A. R., and Dawson, W. M. 1997. Growing clonal mixtures of willow to reduce effect of *Melampsora epitea* var. *epitea*. Eur. J. For. Pathol. 27:319–329.

2542. McCracken, A. R., and Dawson, W. M. 1998. Short rotation coppice willow in Northern Ireland since 1973: development of the use of mixtures in the control of foliar rust (*Melampsora* spp.). Eur. J. For. Pathol. 28:241–250.

2543. McCracken, F. I. 1970. Spore production of *Hericium erinaceus*. Phytopathology 60:1639–1641.

2544. McCracken, F. I. 1978. Canker-rots in southern hardwoods. USDA For. Serv. For. Insect Dis. Leafl. 33. 4 pp.

2545. McCracken, F. I. 1978. Spore release of some decay fungi of southern hardwoods. Can. J. Bot. 56:426–431.

2546. McCracken, F. I. 1988. Microorganisms associated with canker rots and heart rot of oak. Eur. J. For. Pathol. 18:391–396.

2547. McCracken, F. I., and Toole, E. R. 1969. Sporophore development and sporulation of *Polyporus hispidus*. Phytopathology 59:884–885.

2548. McCracken, F. I., Schipper, A. L., and Widin, K. D. 1984. Observations on occurrence of cottonwood leaf rust in central United States. Eur. J. For. Pathol. 14:226–233.

2549. McCracken, F. I., and Toole, E. R. 1974. Felling infected oaks in natural stands reduces dissemination of *Polyporus hispidus* spores. Phytopathology 64:265–266.

2550. McCulloch, W. F. 1943. Ice breakage in partially cut and uncut second growth Douglas-fir stands. J. For. 41:275–278.

2551. McDermott, J. M., and Robinson, R. A. 1989. Provenance variation for disease resistance in *Pseudotsuga menziesii* to the Swiss needle-cast pathogen, *Phaeocryptopus gaeumannii*. Can. J. For. Res. 19:244–246.

2552. McDonald, B. A., Bellamy, B. K., Zhan, J., and Appel, D. N. 1998. The effect of an oak wilt epidemic on the genetic structure of a Texas live oak population. Can. J. Bot. 76:1900–1907.

2553. McDonald, G. I., and Hoff, R. J. 2001. Blister rust: an introduced plague. Pages 193–220 in: Whitebark pine communities: ecology and restoration. D. F. Tomback, S. F. Arno, and R. E. Keane, eds. Island Press, Washington, DC.

2554. McDougall, D. N., and Blanchette, R. A. 1996. Metal ion adsorption by pseudosclerotial plates of *Phellinus weirii*. Mycologia 88:98–103.

2555. McDowell, J., and Merrill, W. 1985. *Rhabdocline* taxa in Pennsylvania. Plant Dis. 69:714–715.

2556. McElrone, A. J., Sherald, J. L., and Forseth, I. N. 2001. Effects of water stress on symptomatology and growth of *Parthenocissus quinquefolia* infected by *Xylella fastidiosa*. Plant Dis. 85:1160–1164.

2557. McElrone, A. J., Sherald, J. L., and Pooler, M. R. 1999. Identification of alternative hosts of *Xylella fastidiosa* in the Washington, D.C., area using nested polymerase chain reaction (PCR). J. Arboric. 25:258–263.

2558. McFadden-Smith, W., Northover, J., and Sears, W. 2000. Dynamics of ascospore release by *Apiosporina morbosa* from sour cherry black knots. Plant Dis. 84:45–48.

2559. McGranahan, G. H., and Smalley, E. B. 1984. Conidial morphology, axenic growth, and sporulation of *Stegophora ulmea*. Phytopathology 74:1300–1303.

2560. McGranahan, G. H., and Smalley, E. B. 1984. Influence of moisture, temperature, leaf maturity, and host genotype on infection of elms by *Stegophora ulmea*. Phytopathology 74:1296–1300.

2561. McIntyre, G. A., Jacobi, W. R., and Ramaley, A. W. 1996. Factors affecting Cytospora canker occurrence on aspen. J. Arboric. 22:229–233.

2562. McKeand, S. E., Li, B., and Amerson, H. V. 1999. Genetic variation in fusiform rust resistance in loblolly pine across a wide geographic range. Silvae Genet. 48:255–260.

2563. McKellar, A. D. 1942. Ice damage to slash pine, longleaf pine, and loblolly pine plantations in the Piedmont section of Georgia. J. For. 40:794–797.

2564. McKenzie, E. H. C. 1986. New plant disease record in New Zealand: fig rust (*Cerotelium fici*) on *Ficus carica*. N.Z. J. Agric. Res. 29:707–710.

2565. McKenzie, H. L., Gill, L. S., and Ellis, D. E. 1948. The Prescott scale (*Matsucoccus vexillorum*) and associated organisms that cause flagging injury to ponderosa pine in the Southwest. J. Agric. Res. 76:33–51.

2566. McKenzie, M. A., Jones, L. H., and Gilgut, C. J. 1940. *Phomopsis gardeniae* in relation to gardenia culture. Plant Dis. Rep. 24:58–62.

2567. McLaughlin, D. 1998. A decade of forest tree monitoring in Canada: evidence of air pollution effects. Environ. Rev. 6:151–171.

2568. McLaughlin, J. A. 2001. Distribution, hosts, and site relationships of *Armillaria* spp. in central and southern Ontario. Can. J. For. Res. 31:1481–1490.

2569. McLaughlin, J. A., and Setliff, E. C. 1990. *Chondrostereum purpureum* associated with decline of *Betula papyrifera* in Thunder Bay, Ontario. Plant Dis. 74:331.

2570. McLaughlin, S. B., McConathy, R. K., Duvick, D., and Mann, L. K. 1982. Effects of chronic air pollution stress on photosynthesis, carbon allocation, and growth of white pine trees. For. Sci. 28:60–70.

2571. McLaughlin, S. B., and Percy, K. 1999. Forest health in North America: some perspectives on actual and potential roles of climate and air pollution. Water Air Soil Pollut. 116:151–197.

2572. McLeod, K. W., and Ciravolo, T. G. 1998. Boron tolerance and potential boron removal by bottomland tree seedlings. Wetlands 18:431–436.

2573. McManus, P. S., Ewers, F. W., and Fulbright, D. W. 1989. Characterization of the chestnut blight canker and the localization and isolation of the pathogen *Cryphonectria parasitica*. Can. J. Bot. 67:3600–3607.

2574. McManus, P. S., and Jones, A. L. 1994. Role of wind-driven rain, aerosols, and contaminated budwood in incidence and spatial pattern of fire blight in an apple nursery. Plant Dis. 78:1059–1066.

2575. McNab, W. H., and Meeker, M. 1987. Oriental bittersweet: a growing threat to hardwood silviculture in the Appalachians. North. J. Appl. For. 4:174–177.

2576. McNabb, R. F. R., and Laurenson, J. B. 1965. A rust of cultivated fuchsias. N.Z. J. Agric. Res. 8:336–339.

2577. McPartland, J. M., and Schoeneweiss, D. F. 1984. Hyphal morphology of *Botryosphaeria dothidea* in vessels of unstressed and drought-stressed stems of *Betula alba*. Phytopathology 74:358–362.

2578. McRitchie, J. J. 1973. Pathogenicity and control of *Fusarium oxysporum* wilt of variegated pyracantha. Plant Dis. Rep. 57:389–391.

2579. McWain, P., and Gregory, G. F. 1972. A neutral mannan from *Ceratocystis fagacearum* culture filtrate. Phytochemistry 11:2609–2612.

2580. Mead, M. A., Dolezal, W. E., and Tainter, F. H. 1978. Eighteen newly discovered pine hosts of comandra blister rust fungus. Plant Dis. Rep. 62:885–887.

2581. Mehlenbacher, S. A., Azarenko, A. N., Smith, D. C., and McCluskey, R. 2000. 'Lewis' hazelnut. HortScience 35:314–315.

2582. Mehlenbacher, S. A., Azarenko, A. N., Smith, D. C., and McCluskey, R. 2001. 'Clark' hazelnut. HortScience 36:995–996.

2583. Mehta, A., and Rosato, Y. B. 2001. Phylogenetic relationships of *Xylella fastidiosa* strains from different hosts, based on 16S rDNA and 16S-23S intergenic spacer sequences. Int. J. Syst. Evol. Microbiol. 51:311–318.

2584. Meier, C. E., Newton, R. J., Puryear, J. D., and Sen, S. 1992. Physiological responses of loblolly pine (*Pinus taeda* L.) seedlings to drought stress: osmotic adjustment and tissue elasticity. J. Plant Physiol. 140:754–760.

2585. Mejía, F., Palmieri, M., Oropeza, C., Doyle, M., Harrison, N., Aguilar, E., Narváez, M., Estrada, R., and Ortiz, G. 2004. First report of coconut lethal yellowing disease in Guatemala. Br. Soc. Plant Pathol. New Dis. Rep. Internet pub.

2586. Meinecke, E. P. 1929. Experiments with repeating pine rusts. Phytopathology 19:327–342.

2587. Melcher, P. J., Cordell, S., Jones, T. J., Scowcroft, P. G., Niemczura, W., Giambelluca, T. W., and Goldstein, G. 2000. Supercooling capacity increases from sea level to tree line in the Hawaiian tree species *Metrosideros polymorpha*. Int. J. Plant Sci. 161:369–379.

2588. Melcher, U., and Fletcher, J. 1999. Genetic variation in *Spiroplasma citri*. Eur. J. Plant Pathol. 105:519–533.

2589. Mel'nik, V. A. 2000. Key to the fungi of the genus *Ascochyta* Lib. (Coelomycetes). Mitt. Biol. Bundes. Land- u. Forstw. Berlin-Dahlem. Heft 379. Parey Buchverlag, Berlin. 192 pp.

2590. Menge, J. A., and French, D. W. 1976. Effect of plant residue amendments and chemical treatments upon the inoculum potential of *Cylindrocladium floridanum* in soil. Phytopathology 66:1085–1089.

2591. Mengel, K., and Kirby, E. A., eds. 2001. Principles of plant nutrition. 5th ed. Kluwer Academic, Dordrecht. 849 pp.

2592. Menon, R. 1956. Studies on Venturiaceae on rosaceous plants. Phytopathol. Z. 27:117–146.

2593. Menzel, W., Jelkmann, W., and Maiss, E. 2002. Detection of four apple viruses by multiplex RT-PCR assays with coamplification of plant mRNA as internal control. J. Virol. Meth. 99:81–92.

2594. Merrill, L. M., Hawksworth, F. G., and Jacobi, W. R. 1987. Frequency and severity of ponderosa pine dwarf mistletoe in relation to habitat type and topography in Colorado. Plant Dis. 71:342–344.

2595. Merrill, W., and Cowling, E. B. 1966. Role of nitrogen in wood deterioration: amount and distribution of nitrogen in fungi. Phytopathology 56:1083–1090.

2596. Merrill, W., French, D. W., and Wood, F. A. 1964. Decay of wood by species of the Xylariaceae. Phytopathology 54:56–58.

2597. Merrill, W., and Kistler, B. R. 1976. Phenology and control of *Endocronartium harknessii* in Pennsylvania. Phytopathology 66:1246–1248.

2598. Merrill, W., and Kistler, B. R. 1976. Seasonal development and control of *Lophodermium pinastri* in Pennsylvania. Plant Dis. Rep. 60:652–655.

2599. Merrill, W., and Kistler, B. R. 1978. Accelerated development of Rhizosphaera needlecast of blue spruce in Pennsylvania. Plant Dis. Rep. 62:34–35.

2600. Merrill, W., and Kistler, B. R. 1978. Needlecast and weeds interact to cause branch mortality in Scots pine. Plant Dis. Rep. 62:200–202.

2601. Merrill, W., Kistler, B. R., Zang, L., and Bowen, K. 1980. Infection periods in Naemacyclus needlecast of Scots pine. Plant Dis. 64:759–761.

2602. Merrill, W., McCall, K., and Zang, L. 1981. Fusarium root rot of Douglas-fir and Fraser fir seedlings in Pennsylvania. Plant Dis. 65:913–914.

2603. Merrill, W., and Ostry, M. E., eds. 1990. Recent research on foliage diseases. Conf. proc., 1989, Carlisle, PA. USDA For. Serv. Gen. Tech. Rep. WO-56. 145 pp.

2604. Merrill, W., and Wenner, N. G. 1996. Cyclaneusma needlecast and needle retention in Scots pine. Plant Dis. 80:294–298.

2605. Merrill, W., Wenner, N. G., and Dreisbach, T. A. 1996. *Canavirgella banfieldii* gen. and sp. nov.: a needlecast fungus on pine. Can. J. Bot. 74:1476–1481.

2606. Merrill, W., Wenner, N. G., and Peplinski, J. D. 1993. New host and distribution records from Pennsylvania conifers. Plant Dis. 77:430–432.

2607. Messenger, A. S. 1984. Seasonal variations of foliar nutrients in green and chlorotic red maples. J. Environ. Hortic. 2:117–119.

2608. Messenger, A. S., and Stelford, M. W. 1997. White pine chlorosis in northern Illinois: iron deficiency or not? J. Arboric. 23:191–195.

2609. Messenger, B. J., Menge, J. A., and Pond, E. 2000. Effects of gypsum soil amendments on avocado growth, soil drainage, and resistance to *Phytophthora cinnamomi*. Plant Dis. 84:612–616.

2610. Messenger, B. J., Menge, J. A., and Pond, E. 2000. Effects of gypsum on zoospores and sporangia of *Phytophthora cinnamomi* in field soil. Plant Dis. 84:617–621.

2611. Meszka, B., Bielenin, A., and Masny, S. 2002. Development and maturation of pseudothecia of *Venturia inaequalis* in leaves of different apple cultivars. Phytopathol. Polon. 25:5–11.

2612. Meyer, J. R., Shew, H. D., and Harrison, U. J. 1994. Inhibition of germination and growth of *Thielaviopsis basicola* by aluminum. Phytopathology 84:598–602.

2613. Mezzetti, B., Capasso, R., Evidente, A., Hammerschlag, F. A., Zimmerman, R. H., Cristinzio, G., and Rosati, P. 1994. Interaction of partially purified phytotoxins from *Phytophthora cactorum* on apple cell plasma membrane. J. Phytopathol. 142:219–226.

2614. Micales, J. A., and Stipes, R. J. 1986. The differentiation of *Endothia* and *Cryphonectria* species by exposure to selected fungitoxicants. Mycotaxon 26:99–117.

2615. Micales, J. A., and Stipes, R. J. 1987. A reexamination of the fungal genera *Cryphonectria* and *Endothia*. Phytopathology 77:650–654.

2616. Michaels, E., and Chastagner, G. A. 1984. Distribution, severity, and impact of Swiss needle cast in Douglas-fir Christmas trees in western Washington and Oregon. Plant Dis. 68:939–942.

2617. Michaels, E., and Chastagner, G. A. 1984. Seasonal availability of *Phaeocryptopus gaeumannii* ascospores and conditions that influence their release. Plant Dis. 68:942–944.

2618. Michailides, T. J. 1991. Pathogenicity, distribution, sources of inoculum, and infection courts of *Botryosphaeria dothidea* on pistachio. Phytopathology 81:566–573.

2619. Michailides, T. J., Morgan, D. P., Felts, D., and Phillimore, J. 2002. First report of *Botryosphaeria rhodina* causing shoot blight of pistachio in California. Plant Dis. 86:1273.

2620. Michalopoulos-Skarmoutsos, H., and Skarmoutsos, G. 1999. Pathogenicity of fungi affecting black locust (*Robinia pseudoacacia*) in Greece. Phytoparasitica 27:239–240.

2621. Mickler, R. A., Birdsey, R. A., and Hom, J., eds. 2000. Responses of northern U.S. forests to environmental change. Springer-Verlag, New York. 578 pp.

2622. Mielke, J. L. 1943. White pine blister rust in western North America. Yale Univ. Sch. For. Bull. 52. 155 pp.

2623. Mielke, J. L. 1952. The rust fungus *Cronartium filamentosum* in Rocky Mountain ponderosa pine. J. For. 50:365–373.

2624. Mielke, J. L. 1956. The rust fungus (*Cronartium stalactiforme*) in lodgepole pine. J. For. 54:518–521.

2625. Mielke, J. L. 1957. Aspen leaf blight in the Intermountain Region. USDA For. Serv. Intermount. For. Range Exp. Stn. Res. Note 42. 5 pp.

2626. Mielke, J. L., and Kimmey, J. W. 1942. Heat injury to the leaves of California black oak and some other broadleaves. Plant Dis. Rep. 26:116–119.

2627. Mihail, J. D., Bruhn, J. N., and Leininger, T. D. 2002. The effects of moisture and oxygen availability on rhizomorph generation by *Armillaria tabescens* in comparison with *A. gallica* and *A. mellea*. Mycol. Res. 106:697–704.

2628. Mihail, J. D., Bruhn, J. N., Meyer, T. R., and Bell, F. W. 2002. Pine needle rust effect on *Pinus banksiana* in response to interspecific plant competition and telial host density. Can. J. For. Res. 32:1372–1380.

2629. Mihail, J. D., Orum, T. V., Alcorn, S. M., and Stroehlein, J. L. 1989. *Macrophomina phaseolina* in the Sonoran Desert. Can. J. Bot. 67:76–82.

2630. Milbrath, J. A., and McWhorter, F. P. 1946. Yellow-mottle leaf, a virus disease of camellia. Am. Camellia Yearb. 1946:51–53.

2631. Miles, L. E. 1921. Leaf spots of the elm. Bot. Gaz. 71:161–196.

2632. Milgroom, M. G., and Cortesi, P. 1999. Analysis of population structure of the chestnut blight fungus based on vegetative incompatibility genotypes. Proc. Natl. Acad. Sci. USA 96:10518–10523.

2633. Milgroom, M. G., Wang, K., Zhou, Y., Lipari, S. E., and Kaneko, S. 1996. Intercontinental population structure of the chestnut blight fungus, *Cryphonectria parasitica*. Mycologia 88:179–190.

2634. Milholland, R. D. 1972. Histopathology and pathogenicity of *Botryosphaeria dothidea* on blueberry stems. Phytopathology 62:654–660.

2635. Millar, C. S. 1981. Infection processes on conifer needles. Pages 185–207 in: Microbial ecology of the phylloplane. J. P. Blakeman, ed. Academic Press, London. 502 pp.

2636. Millar, J. G., Zhao, C. H., Lanier, G. N., O'Callaghan, D. P., Griggs, M., West, J. R., and Silverstein, R. M. 1986. Components of moribund American elm trees as attractants to elm bark beetles, *Hylurgopinus rufipes* and *Scolytus multistriatus*. J. Chem. Ecol. 12:583–608.

2637. Miller, H. N. 1961. Cause and control of diseases of potted plants. Fla. Agric. Exp. Stn. Annu. Rep. 1961:157–158.

2638. Miller, J. H. 1928. Biologic studies in the Sphaeriales—II. Mycologia 20:305–339.

2639. Miller, J. H. 1961. A monograph of the world species of *Hypoxylon*. Univ. Ga. Press, Athens. 158 pp.

2640. Miller, J. H., and Edwards, B. 1983. Kudzu: where did it come from? And how can we stop it? South. J. Appl. For. 7:165–169.

2641. Miller, J. H., and Wolf, F. A. 1936. A leaf-spot disease of honey locust caused by a new species of *Linospora*. Mycologia 28:171–180.

2642. Miller, J. R., and Tocher, R. D. 1975. Photosynthesis and respiration of *Arceuthobium tsugense* (Loranthaceae). Am. J. Bot. 62:765–769.

2643. Miller, J. W., and Seymour, C. P. 1973. A comparative study of *Corynebacterium poinsettiae* and *Xanthomonas poinsettiaecola* on poinsettia and crown-of-thorns. Proc. Fla. State Hortic. Soc. 85:344–347.

2644. Miller, L. W., and Boyle, J. S. 1943. The Hydnaceae of Iowa. Univ. Iowa Stud. Nat. Hist. 18(2):1–92.

2645. Miller, P. M., and Rich, S. 1968. Reducing spring discharge of *Venturia inaequalis* ascospores by composting overwintering leaves. Plant Dis. Rep. 52:728–730.

2646. Miller, P. R., tech. coord. 1980. Proceedings of symposium on effects of air pollutants on Mediterranean and temperate forest ecosystems. USDA For. Serv. Gen. Tech. Rep. PSW-43. 256 pp.

2647. Miller, P. R., and Evans, L. S. 1974. Histopathology of oxidant injury and winter fleck injury on needles of western pines. Phytopathology 64:801–806.

2648. Miller, P. R., Parmeter, J. R. Jr., Taylor, O. C., and Cardiff, E. A. 1963. Ozone injury to the foliage of *Pinus ponderosa*. Phytopathology 53:1072–1076.

2649. Miller, P. W., and Bollen, W. B. 1946. Walnut bacteriosis and its control. Ore. Agric. Exp. Stn. Tech. Bull. 9. 107 pp.

2650. Miller, S. B., and Baxter, L. W. Jr. 1970. Dieback in azaleas caused by *Phomopsis* species. Phytopathology 60:387–388.

2651. Miller, T., Cowling, E. B., Powers, H. R. Jr., and Blalock, T. E. 1976. Types of resistance and compatibility in slash pine seedlings infected by *Cronartium fusiforme*. Phytopathology 66:1229–1235.

2652. Miller, T., Patton, R. F., and Powers, H. R. Jr. 1980. Mode of infection and early colonization of slash pine seedlings by *Cronartium quercuum* f.sp. *fusiforme*. Phytopathology 70:1206–1208.

2653. Mims, C. W. 1981. Ultrastructure of teliospore germination and basidiospore formation in the rust fungus *Gymnosporangium clavipes*. Can. J. Bot. 59:1041–1049.

2654. Mims, C. W., Copes, W. E., and Richardson, E. A. 2000. Ultrastructure of the penetration and infection of pansy roots by *Thielaviopsis basicola*. Phytopathology 90:843–850.

2655. Mims, C. W., and Glidewell, D. C. 1978. Some ultrastructural observations on the host-pathogen relationship within the telial gall of the rust fungus *Gymnosporangium juniperi-virginianae*. Bot. Gaz. 139:11–17.

2656. Mims, C. W., Liljebjelke, K. A., and Covert, S. F. 1996. Ultrastructure of telia and teliospores of the rust fungus *Cronartium quercuum* f.sp. *fusiforme*. Mycologia 88:47–56.

2657. Mims, C. W., and Nickerson, N. L. 1986. Ultrastructure of the host-pathogen relationship in red leaf disease of low-bush blueberry caused by the fungus *Exobasidium vaccinii*. Can. J. Bot. 64:1338–1343.

2658. Mims, C. W., Richardson, E. A., and Roberson, R. W. 1987. Ultrastructure of basidium and basidiospore development in three species of the fungus *Exobasidium*. Can. J. Bot. 65:1236–1244.

2659. Mims, C. W., Sewall, T. C., and Richardson, E. A. 2000. Ultrastructure of conidiogenesis and mature conidia in the plant pathogenic *Entomosporium mespili*. Mycol. Res. 104:453–462.

2660. Mims, C. W., Sewall, T. C., and Richardson, E. A. 2000. Ultrastructure of the host-pathogen relationship in Entomosporium leaf spot disease of *Photinia*. Int. J. Plant Sci. 161:291–295.

2661. Minardi, P. 1996. Defence mechanisms and microbial diffusion during the decay processes in the xylem of living trees. Petria 6:11–35.

2662. Minter, D. W. 1981. *Lophodermium* on pines. Commonw. Mycol. Inst. Mycol. Pap. No. 147. 54 pp.

2663. Minter, D. W., and Cannon, P. F. 1984. Ascospore discharge in some members of the Rhytismataceae. Trans. Br. Mycol. Soc. 83:65–92.

2664. Minter, D. W., and Gibson, I. A. S. 1978. *Ploioderma lethale*. C.M.I. Descr. Pathogenic Fungi Bact. No. 570. 2 pp.

2665. Minter, D. W., and Millar, C. S. 1978. *Lophodermium seditiosum*. C.M.I. Descr. Pathogenic Fungi Bact. No. 568. 2 pp.

2666. Minter, D. W., and Millar, C. S. 1980. Ecology and biology of three *Lophodermium* species on secondary needles of *Pinus sylvestris*. Eur. J. For. Pathol. 10:169–181.

2667. Minter, D. W., Staley, J. M., and Millar, C. S. 1978. Four species of *Lophodermium* on *Pinus sylvestris*. Trans. Br. Mycol. Soc. 71:295–301.

2668. Miot, S., Frey, P., and Pinon, J. 1999. Varietal mixture of poplar clones: effects on infection by *Melampsora larici-populina* and on plant growth. Eur. J. For. Pathol. 29:411–423.

2669. Mircetich, S. M., Campbell, R. N., and Matheron, M. E. 1977. Phytophthora trunk canker of coast live oak and cork oak trees in California. Plant Dis. Rep. 61:66–70.

2670. Mircetich, S. M., and Hoy, J. W. 1981. Brownline of prune trees, a disease associated with tomato ringspot virus infection of myrobalan and peach rootstocks. Phytopathology 71:30–35.

2671. Mircetich, S. M., Lowe, S. K., Moller, W. J., and Nyland, G. 1976. Etiology of almond leaf scorch disease and transmission of the causal agent. Phytopathology 66:17–24.

2672. Mircetich, S. M., and Matheron, M. E. 1976. Phytophthora root and crown rot of cherry trees. Phytopathology 66:549–558.

2673. Mircetich, S. M., and Rowhani, A. 1984. The relationship of cherry leafroll virus and blackline disease of English walnut trees. Phytopathology 74:423–428.

2674. Mircetich, S. M., Sanborn, R. R., and Ramos, D. E. 1980. Natural spread, graft transmission, and possible etiology of walnut blackline disease. Phytopathology 70:962–968.

2675. Mirzaee, M. R., Mohammadi, M., and Rahimian, H. 2002. *Nattrassia mangiferae*, the cause of die-back and trunk cankers of *Ficus religiosa* and branch wilt of *Psidium guajava* in Iran. J. Phytopathol. 150:244–247.

2676. Missouri Botanical Garden, W³TROPICOS VAST (VAScular Tropicos) database. St. Louis, MO. Internet resource.

2677. Mistretta, P. A. 1984. Littleleaf disease. USDA For. Serv. For. Insect Dis. Leafl. 20. Internet pub.

2678. Mistretta, P. A., Anderson, R. L., MacDonald, W. L., and Lewis, R. Jr. 1984. Annotated bibliography of oak wilt, 1943–80. USDA For. Serv. Gen. Tech. Rep. WO-45. 132 pp.

2679. Mitchell, C. P., Millar, C. S., and Minter, D. W. 1978. Studies on decomposition of Scots pine needles. Trans. Br. Mycol. Soc. 71:343–348.

2680. Mitchell, C. P., Millar, C. S., and Williamson, B. 1978. The biology of *Lophodermella conjuncta* Darker on Corsican pine needles. Eur. J. For. Pathol. 8:108–118.

2681. Mitchell, C. P., Williamson, B., and Millar, C. S. 1976. *Hendersonia acicola* on pine needles infected by *Lophodermella sulcigena*. Eur. J. For. Pathol. 6:92–102.

2682. Mitich, L. W. 2000. Kudzu [*Pueraria lobata* (Willd.) Ohwi]. Weed Technol. 14:231–235.

2683. Mittal, R. K., Pritam, S., and Wang, B. S. P. 1987. *Botrytis*: a hazard to reforestation. A literature review. Eur. J. For. Pathol. 17:369–384.

2684. Mix, A. J. 1916. Sun-scald of fruit trees. A type of winter injury. Cornell Univ. Agric. Exp. Stn. Bull. 382:235–284.

2685. Mix, A. J. 1925. Anthracnose of European privet. Phytopathology 15:261–272.

2686. Mix, A. J. 1949. A monograph of the genus *Taphrina*. Univ. Kans. Sci. Bull. 23, Pt. 1, No. 1. 167 pp.

2687. Miyairi, K., Fujita, K., Okuno, T., and Sawai, K. 1977. A toxic protein causative of silver-leaf disease symptoms on apple trees. Agric. Biol. Chem. 41:1897–1902.

2688. Mo, Y. Y., Geibel, M., Bonsall, R. F., and Gross, D. C. 1995. Analysis of sweet cherry (*Prunus avium* L.) leaves for plant signal molecules that activate the *syrB* gene required for synthesis of the phytotoxin, syringomycin, by *Pseudomonas syringae* pv. *syringae*. Plant Physiol. 107:603–612.

2689. Mohamed, H. K., Pathak, S., Roy, D. N., Hutchinson, T. C., McLaughlin, D. L., and Kinch, J. C. 1997. Relationship between sugar maple decline and corresponding chemical changes in the stem tissue. Water Air Soil Pollut. 96:321–327.

2690. Mohammed, C., Guillaumin, J. J., and Berthelay, S. 1994. *Armillaria* species identified in China and Japan. Mycol. Res. 98:607–613.

2691. Mollenhauer, H. H., and Hopkins, D. L. 1976. Xylem morphology of Pierce's disease–infected grapevines with different levels of tolerance. Physiol. Plant Pathol. 9:95–100.

2692. Molnar, A. C. 1961. An outbreak of *Cronartium comptoniae* on Monterey and Bishop pines on Vancouver Island, British Columbia. Plant Dis. Rep. 45:854–855.

2693. Molotkov, P. I., Kirichenko, O. I., and Bengus, Y. V. 1989. The origin of "witches' brooms" on Scots pine. Tsitologiya i Genet. 24:14–19.

2694. Moltzan, B. D., Blenis, P. V., and Hiratsuka, Y. 2001. Effects of spore availability, spore germinability, and shoot susceptibility on gall rust infection of pine. Plant Dis. 85:1193–1199.

2695. Momol, M. T., Momol, E. A., Lamboy, W. F., Norelli, J. L., Beer, S. V., and Aldwinckle, H. S. 1997. Characterization of *Erwinia amylovora* strains using random amplified polymorphic DNA fragments (RAPDs). J. Appl. Microbiol. 82:389–398.

2696. Momol, M. T., Norelli, J. L., Piccioni, D. E., Momol, E. A., Gustafson, H. L., Cummins, J. N., and Aldwinckle, H. S. 1998. Internal movement of *Erwinia amylovora* through symptomless apple scion tissues into the rootstock. Plant Dis. 82:646–650.

2697. Moncalvo, J. M., and Ryvarden, L. 1997. A nomenclatural study of the Ganodermataceae Donk. Synop. Fungorum. No. 11. 114 pp.

2698. Moncalvo, J. M., Wang, H., and Hseu, R. 1995. Gene phylogeny of the *Ganoderma lucidum* complex based on ribosomal DNA sequences. Comparison with traditional taxonomic characters. Mycol. Res. 99:1489–1499.

2699. Monod, M. 1983. Monographie taxonomique des Gnomoniaceae. Sydowia Ann. Mycol., Ser. 2, Beih. 9. 315 pp.

2700. Montano, H. G., Davis, R. E., Dally, E. L., Hogenhout, S., Pimentel, J., and Brioso, P. S. T. 2001. 'Candidatus Phytoplasma brasiliense,' a new phytoplasma taxon associated with hibiscus witches' broom disease. Int. J. Syst. Evol. Microbiol. 51:1109–1118.

2701. Montealegre, A. J. R. 1992. Asian pear twig canker in Chile caused by *Fusarium lateritium*. Fitopatologia 27:80–84.

2702. Montealegre, J. R., Rojas, M. A., Varnero, M. T., and Aballay, E. 1996. Effect of soil solarization on the control of *Sclerotium rolfsii* and nematodes in the Metropolitan Region of Chile. Fitopatologia 31:70–83.

2703. Moore, D. 1998. Fungal morphogenesis. Cambridge Univ. Press, Cambridge, UK. 469 pp.

2704. Moore, D., and Frazer, L. N. 2002. Essential fungal genetics. Springer-Verlag, New York. 357 pp.

2705. Moore, E. R. B., Mau, M., Arnscheidt, A., Bottger, E. C., Hutson, R. A., Collins, M. D., Peer, Y. van de, Wachter, R. de, and Timmis, K. N. 1996. The determination and comparison of the 16S rRNA gene sequences of species of the genus *Pseudomonas* (sensu stricto) and estimation of the natural intrageneric relationships. Syst. Appl. Microbiol. 19:478–492.

2706. Moore, L. W. 1976. Latent infections and seasonal variability of crown gall development in seedlings of three *Prunus* species. Phytopathology 66:1097–1101.

2707. Moore, L. W., and Allen, J. 1986. Controlled heating of root-pruned dormant *Prunus* spp. seedlings before transplanting to prevent crown gall. Plant Dis. 70:532–536.

2708. Moore, L. W., Chilton, W. S., and Canfield, M. L. 1997. Diversity of opines and opine-catabolizing bacteria isolated from naturally occurring crown gall tumors. Appl. Environ. Microbiol. 63:201–207.

2709. Moore, M. H. 1963. A *Gloeosporium* bud-rot and twig-canker disease of cultivated hazel. J. Hortic. Sci. 38:109–118.

2710. Moore, R., and Walker, D. B. 1981. Studies of vegetative compatibility-incompatibility in higher plants. I. A structural study of a compatible autograft in *Sedum telephoides* (Crassulaceae). II. A structural study of an incompatible heterograft between *Sedum telephoides* (Crassulaceae) and *Solanum pennellii* (Solanaceae). Am. J. Bot. 68:820–830, 831–842.

2711. Moore, R. T. 1987. The genera of *Rhizoctonia*-like fungi—*Ascorhizoctonia*, *Ceratorhiza* gen. nov., *Epulorhiza* gen. nov., *Moniliopsis*, and *Rhizoctonia*. Mycotaxon 29:91–100.

2712. Moore, W. E. C., and Moore, L. V. H. 1992. Index of the bacterial and yeast nomenclatural changes. ASM Press, Herndon, VA. 72 pp.

2713. Moore-Landecker, E. 1996. Fundamentals of the fungi. 4th ed. Prentice Hall, Upper Saddle River, NJ. 574 pp.

2714. Moorman, G. W., Lukezic, F. L., and Levine, R. 1988. *Pseudomonas syringae*, causal agent of a leaf spot on *Ulmus* sp. Plant Dis. 72:801.

2715. Moraal, L. G., and Hilszczanski, J. 2000. The oak buprestid beetle, *Agrilus biguttatus* (F.) (Col., Buprestidae), a recent factor in oak decline in Europe. Anzeiger f. Schadlingsk. 73:134–138.

2716. Morales, J. F. 1998. Bromelias de Costa Rica. Heredia, Santo Domingo, Costa Rica. 175 pp.

2717. Moreau, M., Moreau, C., and Péresse, M. 1971. Une maladie chancreuse des genévriers d'ornement. Ann. Phytopathol. 3:233–241.

2718. Morehart, A. L., Carroll, R. B., and Stuart, M. 1980. *Phomopsis* canker and dieback of *Elaeagnus angustifolia*. Plant Dis. 64:66–69.

2719. Morehart, A. L., Donohue, F. M. III, and Melchior, G. L. 1980. Verticillium wilt of yellow poplar. Phytopathology 70:756–760.

2720. Morehart, A. L., and Melchior, G. L. 1982. Influence of water stress on *Verticillium* wilt of yellow-poplar. Can. J. Bot. 60:201–209.

2721. Morelet, M. 1985. Les *Venturia* des peupliers de la section *Leuce* I.—Taxinomie. Cryptog. Mycol. 6:101–117.

2722. Morelet, M. 1986. Risks of adaptation of *Venturia tremulae* to selected aspens. Bull. OEPP 16:589–592.

2723. Moret, B. A., and Nadal, P. M. 1991. Distribución de *Diplodia mutila* Fr. apud Mont. en Catalunya. Acta Bot. Malacitana 16:93–96.

2724. Morgan, P. D., Wallin, E. K., and Driver, C. H. 1974. Occurrence of Rhizina root rot in an old-growth conifer stand in the Pacific Northwest. Plant Dis. Rep. 58:492–494.

2725. Morgan-Jones, G., and White, J. F. 1987. Notes on Coelomycetes. II. Concerning the *Fusicoccum* anamorph of *Botryosphaeria ribis*. Mycotaxon 30:117–125.

2726. Morgan-Jones, G., and White, J. F. J. 1987. Notes on Coelomycetes. III. Concerning *Microsphaeropsis concentrica*: morphology and ultrastructure. Mycotaxon 30:177–187.

2727. Mori, Y., Sato, Y., and Takamatsu, S. 2000. Evolutionary analysis of the powdery mildew fungi using nucleotide sequences of the nuclear ribosomal DNA. Mycologia 92:74–93.

2728. Morone, C., Janse, J. D., and Scortichini, M. 1998. Bark canker of Persian walnut (*Juglans regia*) trees incited by *Erwinia nigrifluens* in Italy. J. Phytopathol. 146:637–639.

2729. Morrison, D. J. 1981. Armillaria root disease. A guide to disease diagnosis, development, and management in British Columbia. Can. For. Serv. Pac. For. Res. Cent. BC-X-203. 15 pp.

2730. Morrison, D. J. 1982. Effects of soil organic matter on rhizomorph growth by *Armillaria mellea*. Trans. Br. Mycol. Soc. 78:201–207.

2731. Morrison, D. J., Chu, D., and Johnson, A. L. S. 1985. Species of *Armillaria* in British Columbia. Can. J. Plant Pathol. 7:242–246.

2732. Morrison, D. J., and Pellow, K. W. 2002. Variation in virulence among isolates of *Armillaria ostoyae*. For. Pathol. 32:99–107.

2733. Morrison, D. J., Pellow, K. W., Nemec, A. F. L., Norris, D. J., and Semenoff, P. 2001. Effects of selective cutting on the epidemiology of Armillaria root disease in the southern interior of British Columbia. Can. J. For. Res. 31:59–70.

2734. Morrison, D. J., Wallis, G. M., and Weir, L. C. 1988. Control of Armillaria and Phellinus root diseases: 20-year results from the Skimikin stump removal experiment. Can. For. Serv. Pac. For. Cent. Inf. Rep. BC-X-302. 16 pp.

2735. Morton, C. S., Barnett, O. W., and Baxter, L. W. 1977. Damage caused by tobacco ringspot virus to poet's jasmine (*Jasminum officinale*). Proc. South. Nurserymen's Assoc. Res. Conf. Annu. Rep. 22:111–114.

2736. Moss, A. E. 1940. Effect on trees of wind-driven salt water. J. For. 38:421–425.

2737. Mosseler, A. J., and Hubbes, M. 1983. *Erwinia* spp. and a new canker disease of hybrid poplars in Ontario. Eur. J. For. Pathol. 13:261–278.

2738. Mota, M. M., Braasch, H., Bravo, M. A., Penas, A. C., Burgermeister, W., Metge, K., and Sousa, E. 1999. First report of *Bursaphelenchus xylophilus* in Portugal and in Europe. Nematology 1:727–734.

2739. Mou, P., and Warrillow, M. P. 2000. Ice storm damage to a mixed hardwood forest and its impacts on forest regeneration in the ridge and valley region of southwestern Virginia. J. Torrey Bot. Soc. 127:66–82.

2740. Mouchacca, J. 1987. Some interesting micromycetes observed on living or dead leaves of *Carpinus betulus* L. Cryptogam. Mycol. 8:141–158.

2741. Moulton, G. A., King, J., and Maleike, R. 1996. Crabapples for western Washington landscapes. Wash. State Univ. Coop. Exten. Bull. 1809. 12 pp.

2742. Mounce, I. 1929. Studies in forest pathology. II. The biology of *Fomes pinicola* (Sw.) Cooke. Can. Dep. Agric. Bull., n.s., 111. 75 pp.

2743. Mudd, J. B., and Kozlowski, T. T., eds. 1975. Responses of plants to air pollution. Academic Press, New York. 383 pp.

2744. Mueller, W. C. 1967. Tobacco mosaic virus obtained from diseased wisteria and elder. Plant Dis. Rep. 51:1053–1054.

2745. Mueller, W. C., Goos, R. D., Quainoo, J., and Morgham, A. T. 1991. The structure of the phialides (mucronate hyphopodia) of the Meliolaceae. Can. J. Bot. 69:803–807.

2746. Mueller, W. C., Morgham, A. T., and Goos, R. D. 1991. Ultrastructure of the haustoria of *Meliola sandwicensis*. Mycol. Res. 95:1208–1210.

2747. Mueller-Dombois, D., Canfield, J. E., Holt, R. A., and Buelow, G. P. 1983. Tree group death in North American and Hawaiian forests: a pathological problem or a new problem for vegetation ecology? Phytocoenologia 11:117–137.

2748. Mugnai, L., Giovannetti, L., Ventura, S., and Surico, G. 1994. The grouping of strains of *Pseudomonas syringae* subsp. *savastanoi* by DNA restriction fingerprinting. J. Phytopathol. 142:209–218.

2749. Muhammad, I. K., Muhammad, A. K., and Sultan, M. K. 1999. Effect of soil solarization on sclerotial viability of *Macrophomina phaseolina* (Tassi) Goid. Pakistan J. Phytopathol. 11:156–158.

2750. Muir, J. A. 1972. Increase of dwarf mistletoe infections on young lodgepole pine. Can. J. For. Res. 2:413–416.

2751. Mukerji, K. G., ed. 1996. Concepts in mycorrhizal research. Kluwer Academic, Dordrecht, Boston. 374 pp.

2752. Mukerji, K. G., Upreti, D. K., Chamola, B. P., and Upadhyay, R. K. 1999. Biology of lichens. Aravali Books International, New Delhi. 419 pp.

2753. Mukhopadhyay, A. N., Kumar, J., Chaube, H. S., and Singh, U. S., eds. 1992. Plant diseases of international importance. Vol. 4:

Diseases of sugar, forest, and plantation crops. Prentice-Hall, Englewood Cliffs, NJ. 376 pp.

2754. Mulhern, J., Shortle, W., and Shigo, A. 1979. Barrier zones in red maple: an optical and scanning microscope examination. For. Sci. 25:311–316.

2755. Mullen, J. M., and Cobb, G. S. 1984. Leaf spot of southern magnolia caused by *Pseudomonas cichorii*. Plant Dis. 68:1013–1015.

2756. Mullen, J. M., Gilliam, C. H., Hagan, A. K., and Morgan, J. G. 1991. Canker of dogwood caused by *Lasiodiplodia theobromae*, a disease influenced by drought stress or cultivar selection. Plant Dis. 75:886–889.

2757. Müller, J. 1977. Wechselwirkungen zwischen fünf *Pratylenchus*-Arten und *Verticillium albo-atrum*. Z. Pflanzenkr. Pflanzenschutz 84:215–220.

2758. Mullick, D. B. 1977. The non-specific nature of defense in bark and wood during wounding, insect and pathogen attack. Rec. Adv. Phytochem. 11:395–441.

2759. Mulrean, E. N., and Schroth, M. N. 1982. Ecology of *Xanthomonas campestris* pv. *juglandis* on Persian (English) walnuts. Phytopathology 72:434–438.

2760. Munger, T. T. 1916. Parch blight on Douglas fir in the Pacific Northwest. Plant World 19:46–47.

2761. Munkvold, G. P., and Neely, D. 1990. Pathogenicity of *Tubakia dryina*. Plant Dis. 74:518–522.

2762. Munkvold, G. P., and Neely, D. 1991. Development of *Tubakia dryina* on host tissue. Can. J. Bot. 69:1865–1871.

2763. Munnecke, D. E., Chandler, P. A., and Starr, M. P. 1963. Hairy root (*Agrobacterium rhizogenes*) of field roses. Phytopathology 53:788–799.

2764. Murdoch, C. W., Biermann, C. J., and Campana, R. J. 1983. Pressure and composition of intrastem gases produced in wetwood of American elm. Plant Dis. 67:74–76.

2765. Murdoch, C. W., and Campana, R. J. 1983. Bacterial species associated with wetwood of elm. Phytopathology 73:1270–1273.

2766. Murdoch, C. W., Campana, R. J., and Biermann, C. J. 1987. Physical and chemical properties of wetwood in American elm (*Ulmus americana*). Can. J. Plant Pathol. 9:20–23.

2767. Murray, M. S., and Hansen, E. M. 1997. Susceptibility of Pacific yew to *Phytophthora lateralis*. Plant Dis. 81:1400–1404.

2768. Murray, R. G. E., and Stackebrandt, E. 1995. Taxonomic note: implementation of the provisional status *Candidatus* for incompletely described procaryotes. Int. J. Syst. Bacteriol. 45:186–187.

2769. Murrill, W. A. 1904. The Polyporaceae of North America—VIII. *Hapalopilus, Pycnoporus*, and new monotypic genera. Bull. Torrey Bot. Club 31:424–425.

2770. Musselman, L. J., Worsham, A. D., and Eplee, R. E. 1979. Proceedings, the second international symposium on parasitic weeds. N.C. State Univ., Raleigh. 296 pp.

2771. Mutto, S., and Panconesi, A. 1987. Ultrastructural modifications in *Cupressus sempervirens* tissues invaded by *Seiridium cardinale*. Eur. J. For. Pathol. 17:193–204.

2772. Myburg, H., Gryzenhout, M., Wingfield, B. D., Stipes, R. J., and Wingfield, M. J. 2004. Phylogenetic relationships of *Cryphonectria* and *Endothia* species, based on DNA sequence data and morphology. Mycologia 96:988–999.

2773. Myburg, H., Gryzenhout, M., Wingfield, B. D., and Wingfield, M. J. 2002. β-tubulin and histone *H3* gene sequences distinguish *Cryphonectria cubensis* from South Africa, Asia, and South America. Can. J. Bot. 80:590–596.

2774. Myburg, H., Gryzenhout, M., Wingfield, B. D., and Wingfield, M. J. 2003. Conspecificity of *Endothia eugeniae* and *Cryphonectria cubensis*: a re-evaluation based on morphology and DNA sequence data. Mycoscience 44:187–196.

2775. Myburg, H., Wingfield, B. D., and Wingfield, M. J. 1999. Phylogeny of *Cryphonectria cubensis* and allied species inferred from DNA analysis. Mycologia 91:243–250.

2776. Myren, D. T., ed. 1994. Tree diseases of eastern Canada. Nat. Resour. Can., Can. For. Serv., Ottawa. 159 pp.

2777. Myren, D. T., and Patton, R. F. 1971. Establishment and spread of *Polyporus tomentosus* in pine and spruce plantations in Wisconsin. Can. J. Bot. 49:1033–1040.

2778. Naegele, J. A., ed. 1973. Air pollution damage to vegetation. Am. Chem. Soc., Washington, DC. 137 pp.

2779. Nag Raj, T. R. 1993. Coelomycetous anamorphs with appendage-bearing conidia. Mycologue Pub., Waterloo, ON. 1101 pp.

2780. Nag Raj, T. R., and Kendrick, W. B. 1975. A monograph of *Chalara* and allied genera. Wilfred Laurier Univ. Press, Waterloo, ON. 200 pp.

2781. Nair, V. M. G. 1999. Butternut canker—an international concern. Pages 239–252 in: Biotechnology and plant protection in forestry

2782. Nair, V. M. G., Kostichka, C. J., and Kuntz, J. E. 1979. *Sirococcus clavigignenti-juglandacearum*: an undescribed species causing canker on butternut. Mycologia 71:641–646.

2783. Nakamura, H., Kaneko, S., Yamaoka, Y., and Kakishima, M. 1998. Differentiation of *Melampsora* rust species on willows in Japan using PCR-RFLP analysis of ITS regions of ribosomal DNA. Mycoscience 39:105–113.

2784. Nannfeldt, J. A. 1981. *Exobasidium*, a taxonomic reassessment applied to the European species. Symb. Bot. Upsal. 23:1–72.

2785. Nash, B. L., Stanosz, G. R., Taylor, G., and Davis, D. D. 1994. *Discula campestris* infection of sugar maple leaves associated with pear thrips injury. Plant Dis. 78:285–289.

2786. Nash, T. H. III. 1996. Lichen biology. Cambridge Univ. Press, Cambridge, New York. 303 pp.

2787. National Mycological Herbarium (Canada). 1973–1993. Fungi Canadenses Nos. 1–300. Agric. Can., Res. Branch, Biosyst. Res. Inst., Ottawa. (Later issues published in Can. J. Plant Pathol.)

2788. Nattrass, R. M. 1928. The Physalospora disease of the basket willow. Trans. Br. Mycol. Soc. 13:286–304.

2789. Nattrass, R. M. 1933. A new species of *Hendersonula* (*H. toruloidea*) on deciduous trees in Egypt. Trans. Br. Mycol. Soc. 18:189–198.

2790. Nault, L. R. 1997. Arthropod transmission of plant viruses: a new synthesis. Ann. Entomol. Soc. Am. 90:521–541.

2791. Nault, L. R., and Rodriguez, J. G., eds. 1985. The leafhoppers and planthoppers. Wiley, New York. 500 pp.

2792. Naumov, G. I., Naumova, E. S., and Sniegowski, P. D. 1998. *Saccharomyces paradoxus* and *Saccharomyces cerevisiae* are associated with exudates of North American oaks. Can. J. Microbiol. 44:1045–1050.

2793. Navaratam, S. J., and Leong, C. K. 1965. Root inoculation of oil palm seedlings with *Ganoderma* sp. Plant Dis. Rep. 49:1011–1012.

2794. Navratil, S. 1979. Virus and virus-like diseases of poplar: are they threatening diseases? Rep. 19 in: Poplar research, management, and utilization in Canada. D. C. F. Fayle, A. Zsuffa, and H. W. Anderson, eds. Ont. Minist. Nat. Resour., For. Res. Inf. Pap. 102. 17 pp.

2795. Navratil, S. 1981. A rhabdovirus associated with vein yellowing and vein necrosis of balsam poplar. (Abstr.) Phytopathology 71:245.

2796. Navratil, S., and Boyer, M. G. 1968. The identification of poplar mosaic virus in Canada. Can. J. Bot. 46:722–723.

2797. Navratil, S., and Boyer, M. G. 1972. Ultrastructural changes in leaf tissue of *Populus deltoides* subsp. *angulata* Ait. infected with poplar mosaic virus. Can. J. For. Res. 2:308–312.

2798. Neal, J. C., and Skroch, W. A. 1985. Effects of timing and rate of glyphosate application on toxicity to selected woody ornamentals. J. Am. Soc. Hortic. Sci. 110:860–864.

2799. Neal, J. C., and Skroch, W. A. 1987. Influence of timing and rate of glyphosate applications on conifer growth. J. Environ. Hortic. 5:97–101.

2800. Nechwatal, J., Schlenzig, A., Jung, T., Cooke, D. E. L., Duncan, J. M., and Osswald, W. F. 2001. A combination of baiting and PCR techniques for the detection of *Phytophthora quercina* and *P. citricola* in soil samples from oak stands. For. Pathol. 31:85–97.

2801. Neely, D. 1968. Bleeding necrosis of sweet gum in Illinois and Indiana. Plant Dis. Rep. 52:223–225.

2802. Neely, D. 1968. The somatic pressure cushion of *Gnomonia platani*. Mycologia 60:84–89.

2803. Neely, D. 1971. Additional *Aesculus* species and subspecies susceptible to leaf blotch. Plant Dis. Rep. 55:37–38.

2804. Neely, D. 1981. Application of nitrogen fertilizer to control anthracnose of black walnut. Plant Dis. 65:580–581.

2805. Neely, D. 1988. Tree wound closure. J. Arboric. 14:148–152.

2806. Neely, D. 1988. Wound closure rates on trees. J. Arboric. 14:250–254.

2807. Neely, D., and Crowley, W. R. Jr. 1974. Toxicity of soil-applied herbicides to shade trees. HortScience 9:147–149.

2808. Neely, D., and Himelick, E. B. 1963. *Aesculus* species susceptible to leaf blotch. Plant Dis. Rep. 47:170.

2809. Neely, D., and Himelick, E. B. 1963. Temperature and sycamore anthracnose severity. Plant Dis. Rep. 47:171–175.

2810. Neely, D., and Himelick, E. B. 1965. Nomenclature of the sycamore anthracnose fungus. Mycologia 57:834–837.

2811. Neely, D., and Himelick, E. B. 1967. Characteristics and nomenclature of the oak anthracnose fungus. Phytopathology 57:1230–1236.

2812. Neely, D., and Himelick, E. B. 1987. Freeze-crack-related measurements on *Platanus* × *acerifolia* trees. For. Sci. 33:239–244.

2813. Neely, D., and Himelick, E. B. 1989. Susceptibility of honeylocust cultivars to Thyronectria canker. J. Arboric. 15:189–191.

2814. Neely, D., and Nolte, D. S. 1989. Septoria leaf spot on dogwoods. J. Arboric. 15:263–267.

2815. Nehls, U., Mikolajewski, S., Magel, E., and Hampp, R. 2001. Carbohydrate metabolism in ectomycorrhizas: gene expression, monosaccharide transport and metabolic control. New Phytol. 150:533–541.

2816. Neilson-Jones, W. 1969. Plant chimeras. 2nd ed. Methuen, London. 123 pp.

2817. Nelson, E. E., Silen, R. R., and Mandel, N. L. 1988. Effects of Douglas-fir parentage on Swiss needlecast expression. Eur. J. For. Pathol. 19:1–6.

2818. Nelson, E. E., and Sturrock, R. N. 1993. Susceptibility of western conifers to laminated root rot (Phellinus weirii) in Oregon and British Columbia field tests. West. J. Appl. For. 8:67–70.

2819. Nelson, P. E., Toussoun, T. A., and Cook, R. J., eds. 1981. Fusarium: diseases, biology, and taxonomy. Penn. State Univ. Press, University Park. 457 pp.

2820. Nemec, S., and Lee, O. 1996. Management of citrus blight with soil amendments. Proc. Soil Crop Sci. Soc. Fla. 55:12–16.

2821. Nemec, S., Myhre, D. L., and Bryan, D. S. 1995. Nitrogen forms affect citrus blight and alter soil factors affecting root systems. Proc. Soil Crop Sci. Soc. Fla. 54:60–65.

2822. Nemeth, M. 1994. History and importance of plum pox in stone-fruit production. Bull. OEPP 24:525–536.

2823. Nesbitt, R. B., Tidwell, T. E., Stipes, R. J., and Griffin, G. J. 1999. First report of mimosa wilt disease of silk tree (Albizia julibrissin) in California caused by Fusarium oxysporum f.sp. perniciosum. Plant Dis. 83:487.

2824. Nesme, X., Steenackers, M., Steenackers, V., Picard, C., Menard, M., Ride, S., and Ride, M. 1994. Differential host-pathogen interactions among clones of poplar and strains of Xanthomonas populi pv. populi. Phytopathology 84:101–107.

2825. Nester, E. W. 2000. DNA and protein transfer from bacteria to eukaryotes--the Agrobacterium story. Mol. Plant Pathol. 1:87–90.

2826. Nevill, R. J., and Alexander, S. A. 1992. Pathogenicity of three fungal associates of Hylobius pales and Pissodes nemorensis (Coleoptera: Curculionidae) to eastern white pine. Can. J. For. Res. 22:1438–1440.

2827. Nevill, R. J., and Alexander, S. A. 1992. Root- and stem-colonizing insects recovered from eastern white pines with procerum disease. Can. J. For. Res. 22:1712–1716.

2828. Nevill, R. J., and Alexander, S. A. 1992. Transmission of Leptographium procerum to eastern white pine by Hylobius pales and Pissodes nemorensis (Coleoptera: Curculionidae). Plant Dis. 76:307–310.

2829. Nevill, R. J., Merler, H., and Borden, J. H. 1989. Reduced volume, grade and value of lodgepole pine lumber caused by Atropellis canker and stalactiform blister rust. For. Chron. 65:36–41.

2830. Newbanks, D., Bosch, A., and Zimmermann, M. H. 1983. Evidence for xylem dysfunction by embolization in Dutch elm disease. Phytopathology 73:1060–1063.

2831. Newcombe, G. 1996. The specificity of fungal pathogens of Populus. Pages 223–246 in: Biology of Populus and its implications for management and conservation. R. F. Stettler, H. D. Bradshaw Jr., P. E. Heilman, and T. M. Hinckley, eds. NRC Res. Press, Ottawa.

2832. Newcombe, G. 1998. A review of exapted resistance to diseases of Populus. Eur. J. For. Pathol. 28:209–216.

2833. Newcombe, G. 1998. Association of Mmd1, a major gene for resistance to Melampsora medusae f.sp. deltoidae, with quantitative traits in poplar rust. Phytopathology 88:114–121.

2834. Newcombe, G. 1998. Southerly extension of poplar leaf blight (Linospora tetraspora) in the Pacific Northwest. Plant Dis. 82:590.

2835. Newcombe, G. 2003. Native Venturia inopina sp. nov., specific to Populus trichocarpa and its hybrids. Mycol. Res. 107:108–116.

2836. Newcombe, G., and Bradshaw, H. D. Jr. 1996. Quantitative trait loci conferring resistance in hybrid poplar to leaf spot caused by Septoria populicola. Can. J. For. Res. 26:1943–1950.

2837. Newcombe, G., Chastagner, G. A., and McDonald, S. K. 1994. Additional coniferous aecial hosts of the poplar leaf rusts, Melampsora larici-populina and M. medusae f.sp. deltoidae. Plant Dis. 78:1218.

2838. Newcombe, G., and Oosten, C. van. 1997. Variation in resistance to Venturia populina, the cause of poplar leaf and shoot blight in the Pacific Northwest. Can. J. For. Res. 27:883–889.

2839. Newcombe, G., and Ostry, M. 2001. Recessive resistance to Septoria stem canker of hybrid poplar. Phytopathology 91:1081–1084.

2840. Newcombe, G., and Robb, J. 1989. The chronological development of a lipid-to-suberin response at Verticillium trapping sites in alfalfa. Physiol. Mol. Plant Pathol. 34:55–73.

2841. Newcombe, G., Stirling, B., and Bradshaw, H. D. Jr. 2001. Abundant pathogenic variation in the new hybrid rust Melampsora ×columbiana on hybrid poplar. Phytopathology 91:981–985.

2842. Newcombe, G., Stirling, B., McDonald, S., and Bradshaw, H. D. Jr. 2000. Melampsora ×columbiana, a natural hybrid of M. medusae and M. occidentalis. Mycol. Res. 104:261–274.

2843. Neyland, R. 2001. A phylogeny inferred from large ribosomal subunit (26S) rDNA sequences suggests that Cuscuta is a derived member of Convolvulaceae. Brittonia 53:108–115.

2844. Ngo, H. C., Baxter, L. W. Jr., and Fagan, S. G. 1978. The status of our knowledge in 1978 of twig blight, canker, and dieback of camellias caused by a strain of Glomerella cingulata. Pages 75–91 in: American camellia yearbook 1978. Am. Camellia Soc., Fort Valley, GA. 256 pp.

2845. Nicholls, T. H., Patton, R. F., and Arsdel, E. P. Van. 1968. Life cycle and seasonal development of Coleosporium pine needle rust in Wisconsin. Phytopathology 58:822–829.

2846. Nicholls, T. H., Prey, A. J., and Skilling, D. D. 1974. Rhizosphaera kalkhoffii damages blue spruce Christmas tree plantations. Plant Dis. Rep. 58:1094–1096.

2847. Nichols, C. W., Schneider, H., O'Reilly, H. J., Shalla, T. A., and Griggs, W. H. 1960. Pear decline in California. Calif. Dep. Agric. Bull. 49:186–192.

2848. Nichols, J. O. 1968. Oak mortality in Pennsylvania: a ten-year study. J. For. 66:681–684.

2849. Nichols, L. P. 1985. Disease resistant crabapples (results of 1984 survey). Penn. State Univ. Plant Pathol. Contrib. 1502. 7 pp.

2850. Nichols, L. P., and Peterson, D. H. 1971. Overwintering of Fusicladium dendriticum, the imperfect stage of the apple scab fungus, on twigs of flowering crab apples. Plant Dis. Rep. 55:509.

2851. Nickle, W. R., ed. 1984. Plant and insect nematodes. Marcel Dekker, New York. 925 pp.

2852. Nickrent, D. L., García, M. A., Martín, M. P., and Mathiasen, R. L. 2004. A phylogeny of all species of Arceuthobium (Viscaceae) using nuclear and chloroplast DNA sequences. Am. J. Bot. 91:125–138.

2853. Nickrent, D. L., Schuette, K. P., and Starr, E. M. 1994. A molecular phylogeny of Arceuthobium (Viscaceae) based on nuclear ribosomal DNA internal transcribed spacer sequences. Am. J. Bot. 81:1149–1160.

2854. Nickrent, D. L., and Stell, A. L. 1990. Electrophoretic evidence for genetic differentiation in two host races of hemlock dwarf mistletoe (Arceuthobium tsugense). Biochem. Syst. Ecol. 18:267–280.

2855. Niedbalski, M., Crane, J. L., and Neely, D. 1979. Illinois fungi. 10. Development, morphology, and taxonomy of Cristulariella pyramidalis. Mycologia 71:722–730.

2856. Niederleitner, S., and Knoppik, D. 1997. Effects of the cherry leaf spot pathogen Blumeriella jaapii on gas exchange before and after expression of symptoms on cherry leaves. Physiol. Mol. Plant Pathol. 51:145–153.

2857. Nielsen, M. W. 1979. Taxonomic relationships of leafhopper vectors of plant pathogens. Pages 3–27 in: Leafhopper vectors and plant disease agents. K. Maramorosch and K. L. Harris, eds. Academic Press, New York.

2858. Niemelä, T. 1972–1975. On Fennoscandian polypores. II. Phellinus laevigatus (Fr.) Bourd. & Galz. and P. lundellii Niemelä n.sp. III. Phellinus tremulae (Bond.) Bond. & Borisov. IV. Phellinus igniarius, P. nigricans, and P. populicola n.sp. Ann. Bot. Fenn. 9:41–59; 11:202–215; 12:93–122.

2859. Nienhaus, F. 1975. Viren und virusverdächtige Erkrankungen in Eichen (Quercus robur und Quercus sessiliflora). Z. Pflanzenkr. Pflanzenschutz 82:739–749.

2860. Nienhaus, F., Buttner, C., and Hamacher, J. 1990. Virus infection of forest trees by mechanical transmission. J. Phytopathol. 129:141–150.

2861. Nienhaus, F., and Castello, J. D. 1989. Viruses in forest trees. Annu. Rev. Phytopathol. 27:165–186.

2862. Nienhaus, F., and Yarwood, C. E. 1972. Transmission of virus from oak leaves fractionated with Sephadex. Phytopathology 62:313–315.

2863. Nighswander, J. E., and Patton, R. F. 1965. The epidemiology of the jack pine–oak gall rust (Cronartium quercuum) in Wisconsin. Can. J. Bot. 43:1561–1581.

2864. Nilsson, O., Moritz, T., Sundberg, B., Sandberg, G., and Olsson, O. 1996. Expression of the Agrobacterium rhizogenes rolC gene in a deciduous forest tree alters growth and development and leads to stem fasciation. Plant Physiol. 112:493–502.

2865. Nilsson, T., Daniel, G., Kirk, T. K., and Obst, J. R. 1989. Chemistry and microscopy of wood decay by some higher ascomycetes. Holzforschung 43:11–18.

2866. Nirenberg, H. I., and O'Donnell, K. 1998. New *Fusarium* species and combinations within the *Gibberella fujikuroi* species complex. Mycologia 90:434–458.

2867. Nishida, H., and Sugiyama, J. 1994. Archiascomycetes: detection of a major new lineage within the Ascomycota. Mycoscience 35:361–366.

2868. Nordin, V. J. 1954. Studies in forest pathology XIII. Decay in sugar maple in the Ottawa-Huron and Algoma Extension forest region of Ontario. Can. J. Bot. 32:221–258.

2869. Norelli, J. L., Aldwinckle, H. S., and Beer, S. V. 1986. Differential susceptibility of *Malus* spp. cultivars Robusta 5, Novole, and Ottawa 523 to *Erwinia amylovora*. Plant Dis. 70:1017–1019.

2870. Norman, D. J., Chase, A. R., Stall, R. E., and Jones, J. B. 1999. Heterogeneity of *Xanthomonas campestris* pv. *hederae* strains from araliaceous hosts. Phytopathology 89:646–652.

2871. Normand, P., Fernandez, M. Simonet, P., and Domenach, A. M., eds. 1992. *Frankia* and actinorhizal plants. Acta Oecol. 13:365–516.

2872. Northover, J. 1987. Sodium chloride injury to dormant roadside peach trees and its effect on the incidence of infections by *Leucostoma* spp. Phytopathology 77:835–840.

2873. Northover, J., and McFadden-Smith, W. 1995. Control and epidemiology of *Apiosporina morbosa* of plum and sour cherry. Can. J. Plant Pathol. 17:57–68.

2874. Northover, P. R., and Desjardins, M. 2003. First report of bronze leaf disease on hybrid poplar (*Populus ×canescens* 'Tower') caused by *Apioplagiostoma populi* in Manitoba, Canada. Plant Dis. 87:1538.

2875. Norton, D. C. 1978. Ecology of plant-parasitic nematodes. Wiley, New York. 268 pp.

2876. Norton, J. D., and Boyhan, G. E. 1991. Inheritance of resistance to black knot in plums. HortScience 26:1540.

2877. Nowak, E. J., and Martin, C. E. 1997. Physiological and anatomical responses to water deficits in the CAM epiphyte *Tillandsia ionantha* (Bromeliaceae). Int. J. Plant Sci. 158:818–826.

2878. Nüesch, J. 1960. Beitrag zur Kenntnis der weidenbewohnenden Venturiaceae. Phytopathol. Z. 39:329–360.

2879. Núñez, M., and Ryvarden, L. 1997. The genus *Aleurodiscus* (Basidiomycotina). Synop. Fungorum No. 12. Fungiflora, Oslo. 164 pp.

2880. Nuzzo, V. 2002. Element stewardship abstract for *Lonicera japonica*, rev. 2002. The Nature Conservancy, Arlington, VA. Internet pub.

2881. Nyland, G., Lownsbery, B. F., Lowe, S. K., and Mitchell, J. F. 1969. The transmission of cherry rasp leaf virus by *Xiphinema americanum*. Phytopathology 59:1111–1112.

2882. Oak, S. W. 1985. Adoption of littleleaf disease hazard rating for use in forest management in South Carolina national forests. Pages 246–251 in: Integrated Pest Management Research Symposium: the proceedings. S. J. Branham and R. C. Thatcher, eds. USDA For. Serv. Gen. Tech. Rep. SO-56.

2883. Oak, S. W., Huber, C. M., and Sheffield, R. M. 1991. Incidence and impact of oak decline in western Virginia, 1986. USDA For. Serv. Resour. Bull. SE-123. 16 pp.

2884. Oak, S. [W.], Tainter, F., Williams, J., and Starkey, D. 1996. Oak decline risk rating for the southeastern United States. Ann. Sci. For. 53:721–730.

2885. O'Bannon, J. H. 1977. Worldwide dissemination of *Radopholus similis* and its importance in crop production. J. Nematol. 9:16–25.

2886. Oberhuber, W., Thomaser, G., Mayr, S., and Bauer, H. 1999. Radial growth of Norway spruce infected by *Chrysomyxa rhododendri*. Phyton 39:147–154.

2887. Oberwinkler, F., and Bandoni, R. 1984. *Herpobasidium* and allied genera. Trans. Br. Mycol. Soc. 83:639–658.

2888. O'Brien, J. T. 1973. Sirococcus shoot blight of red pine. Plant Dis. Rep. 57:246–247.

2889. Ocamb, C. M., and Juzwik, J. 1995. *Fusarium* species associated with rhizosphere soil and diseased roots of eastern white pine seedlings and associated nursery soil. Can. J. Plant Pathol. 17:325–330.

2890. O'Donnell, K. 1996. Progress towards a phylogenetic classification of *Fusarium*. Sydowia 48:57–70.

2891. O'Donnell, K., Cigelnik, E., and Nirenberg, H. I. 1998. Molecular systematics and phylogeography of the *Gibberella fujikuroi* species complex. Mycologia 90:465–493.

2892. O'Donnell, K., Cigelnik, E., Weber, N. S., and Trappe, J. M. 1997. Phylogenetic relationships among ascomycetous truffles and the true and false morels inferred from 18S and 28S ribosomal DNA sequence analysis. Mycologia 89:48–65.

2893. Oetting, R. D., Morishita, F. S., Helmkamp, A. L., and Bowen, W. R. 1980. Phytotoxicity of eight insecticides to some nursery-grown ornamentals. J. Econ. Entomol. 73:29–31.

2894. Ofong, A. U., Pearce, R. B., and Barrett, D. K. 1991. Biology and pathogenicity of *Rosellinia desmazieresii* [*sic*]. Mycol. Res. 95:189–194.

2895. Ogawa, J. M., Bose, E., Manji, B. T., and Peterson, L. J. 1977. Life cycle and chemical control of Modesto tree anthracnose. Plant Dis. Rep. 61:792–796.

2896. Ogawa, J. M., and English, H. 1991. Diseases of temperate zone tree fruit and nut crops. Univ. Calif., Div. Agric. & Nat. Resour. Pub. 3345. 461 pp.

2897. Ogawa, J. M., Zehr, E. I., Bird, G. W., Ritchie, D. F., Uriu, K., and Uyemoto, J. K., eds. 1995. Compendium of stone fruit diseases. APS Press, St. Paul, MN. 98 pp.

2898. Ogilvie, L. 1924. Observations on the "slime-fluxes" of trees. Trans. Br. Mycol. Soc. 9:167–182.

2899. Oku, H. 1988. Role of phytotoxins in pine wilt disease. J. Nematol. 20:245–251.

2900. Olaya, G., and Abawi, G. S. 1996. Effect of water potential on mycelial growth and on production and germination of sclerotia of *Macrophomina phaseolina*. Plant Dis. 80:1347–1350.

2901. Old, K. M., Gibbs, R., Craig, I., Myers, B. J., and Yuan, Z. Q. 1990. Effect of drought and defoliation on the susceptibility of eucalypts to cankers caused by *Endothia gyrosa* and *Botryosphaeria ribis*. Austral. J. Bot. 38:571–581.

2902. Old, K. M., and Kobayashi, T. 1988. Eucalypts are susceptible to the chestnut blight fungus, *Cryphonectria parasitica*. Austral. J. Bot. 36:599–603.

2903. Old, K. M., Wingfield, M. J., and Yuan, Z. K. 2003. A manual of diseases of eucalypts in South-East Asia. Center for International Forestry Research, Jakarta, Indonesia. 98 pp.

2904. Oldfield, G. N. 1988. Ecological associations of *Spiroplasma citri* with insects, plants and other plant mycoplasmas in the western United States. Pages 175–191 in: Mycoplasma diseases of crops: basic and applied aspects. K. Maramorosch and S. P. Raychaudhuri, eds. Springer-Verlag, New York.

2905. Olmos, A., Bertolini, E., and Cambra, M. 2002. Simultaneous and co-operational amplification (Co-PCR): a new concept for detection of plant viruses. J. Virol. Meth. 106:51–59.

2906. Olson, Å., and Stenlid, J. 2000. Functional units in root diseases: lessons from *Heterobasidion annosum*. Pages 139–156 in: *Ganoderma* diseases of perennial crops. J. Flood, P. D. Bridge, and M. Holderness, eds. CABI Pub., Wallingford, UK.

2907. Olson, R. K., Binkley, D., and Böhm, M., eds. 1992. The response of western forests to air pollution. Ecol. Stud. 97. Springer-Verlag, New York. 532 pp.

2908. Omamor, I. B. 1985. Studies on the infective stage of *Graphiola phoenicis*. Date Palm J. 4:55–68.

2909. O'Mara, J., and Tisserat, N. 1997. A vascular wilt of fragrant sumac caused by *Fusarium oxysporum*. Plant Dis. 81:1333.

2910. Ophel, K., and Kerr, A. 1990. *Agrobacterium vitis* sp. nov. for strains of *Agrobacterium* biovar 3 from grapevines. Int. J. Syst. Bacteriol. 40:236–341.

2911. Oren, R., Thies, W. G., and Waring, R. H. 1985. Tree vigor and stand growth of Douglas-fir as influenced by laminated root rot. Can. J. For. Res. 15:985–988.

2912. Oropeza, C., Howard, F. W., and Ashburner, G. R., eds. 1995. Lethal yellowing: research and practical aspects. Kluwer Academic, Dordrecht. 268 pp.

2913. Osorio, M., and Stephan, B. R. 1989. Ascospore germination and appressorium formation in-vitro of some species of the Rhytismataceae. Mycol. Res. 93:439–451.

2914. Osorio, M., and Stephan, B. R. 1991. Life cycle of *Lophodermium piceae* in Norway spruce needles. Eur. J. For. Pathol. 21:152–163.

2915. Ostaff, D. P. 1985. Age distribution of European larch canker in New Brunswick. Plant Dis. 69:796–798.

2916. Osterbauer, N. K., Johnson, K. B., Mehlenbacher, S. A., and Sawyer, T. L. 1997. Analysis of resistance to eastern filbert blight in *Corylus avellana*. Plant Dis. 81:388–394.

2917. Ostrofsky, A., and Peterson, G. W. 1981. Etiologic and cultural studies of *Kabatina juniperi*. Plant Dis. 65:908–910.

2918. Ostrofsky, W. D., Pratt, R. G., and Roth, L. F. 1977. Detection of *Phytophthora lateralis* in soil organic matter and factors that affect its survival. Phytopathology 67:79–84.

2919. Ostrofsky, W. D., Rumpf, T., Struble, D., and Bradbury, R. 1988. Incidence of white pine blister rust in Maine after 70 years of a *Ribes* eradication program. Plant Dis. 72:967–970.

2920. Ostry, M. E., and Anderson, N. A. 1998. Interactions of insects, woodpeckers, and Hypoxylon canker on aspen. USDA For. Serv. Res. Pap. NC-331. 17 pp.

2921. Ostry, M. E., and McNabb, H. S. Jr. 1985. Susceptibility of *Populus* species and hybrids to disease in the north central United States. Plant Dis. 69:755–757.

2922. Ostry, M. E., Mielke, M. E., and Skilling, D. D. 1994. Butternut—strategies for managing a threatened tree. USDA For. Serv. Gen. Tech. Rep. NC-165. 7 pp.

2923. Ostry, M. E., Nicholls, T. H., and French, D. W. 1983. Animal vectors of eastern dwarf mistletoe of black spruce. USDA For. Serv. Res. Pap. NC-232. 16 pp.

2924. Ostry, M. E., Wilson, L. F., and McNabb, H. S. Jr. 1989. Impact and control of *Septoria musiva* on hybrid poplars. USDA For. Serv. Gen. Tech. Rep. NC-133. 5 pp.

2925. Ostry, M. E., Wilson, L. F., McNabb, H. S. Jr., and Moore, L. M. 1988. A guide to insect, disease, and animal pests of poplars. USDA Agric. Handb. 677. 118 pp.

2926. Oszako, T., and Delatour, C., eds. 2000. Recent advances on oak health in Europe. Selected papers from a conference held in Warsaw, Poland, 1999. For. Res. Inst., Warsaw. 281 pp.

2927. Otrosina, W. J., and Scharpf, R. F., eds. 1989. Proceedings of the symposium on research and management of annosus root disease (*Heterobasidion annosum*) in western North America. Monterey, CA, 1989. USDA For. Serv. Gen. Tech. Rep. PSW-116. 177 pp.

2928. Otta, J. D. 1974. Effects of 2,4-D herbicide on Siberian elm. For. Sci. 20:287–290.

2929. Ouellette, G. B. 1966. *Coleosporium viburni* on jack pine and its relationship with *C. asterum*. Can. J. Bot. 44:1117–1120.

2930. Ouellette, G. B. 1999. Anomalies de croissance et dépérissement du tilleull au Québec: à prévenir dès la production des plantes en pépinière. Phytoprotection 79:71–85.

2931. Ouellette, G. B., Rioux, D., Bussières, G., Simard, M., and Bernier, L. 2000. Extracellular sheath formation by *Sphaeropsis hypodermia* and association with its infection in elm trees. Phytoprotection 81:69–86.

2932. Ouimet, R., and Camire, C. 1995. Foliar deficiencies of sugar maple stands associated with soil cation imbalances in the Quebec Appalachians. Can. J. Soil Sci. 75:169–175.

2933. Oven, P., and Torelli, N. 1994. Wound response of the bark in healthy and declining silver firs (*Abies alba*). IAWA J. 15:407–415.

2934. Oven, P., Torelli, N., Shortle, W. C., and Zupancic, M. 1999. The formation of a ligno-suberised layer and necrophylactic periderm in beech bark (*Fagus sylvatica* L.). Flora 194:137–144.

2935. Overholts, L. O. 1923. Diagnoses of American Porias—II. Bull. Torrey Bot. Club 50:245–253.

2936. Overholts, L. O. 1926. Mycological notes for 1924. Mycologia 18:31–38.

2937. Overholts, L. O. 1953. The Polyporaceae of the United States, Alaska, and Canada. Univ. Mich. Press, Ann Arbor. 466 pp.

2938. Owens, C. E. 1936. Studies on the wood-rotting fungus, *Fomes pini*. I. Variations in morphology and growth habit. Am. J. Bot. 23:144–149.

2939. Paavolainen, L., Hantula, J., and Kurkela, T. 2000. *Pyrenopeziza betulicola* and an anamorphic fungus occurring in leaf spots of birch. Mycol. Res. 104:611–617.

2940. Pady, S. M. 1933. Teliospore development in the Pucciniastraceae. Can. J. Res. 9:458–485.

2941. Pady, S. M. 1946. The development and germination of the intraepidermal teliospores of *Melampsorella cerastii*. Mycologia 38:477–499.

2942. Pady, S. M., and Kramer, C. L. 1971. Basidiospore discharge in *Gymnosporangium*. Phytopathology 61:951–953.

2943. Pady, S. M., Kramer, C. L., and Clary, R. 1969. Aeciospore release in *Gymnosporangium*. Can. J. Bot. 47:1027–1032.

2944. Palik, B. J., and Pederson, N. 1996. Overstory mortality and canopy disturbances in longleaf pine ecosystems. Can. J. For. Res. 26:2035–2047.

2945. Paine, T. D., Raffa, K. F., and Harrington, T. C. 1997. Interactions among scolytid bark beetles, their associated fungi, and live conifers. Annu. Rev. Entomol. 42:179–206.

2946. Palleroni, N. J. 1994. *Pseudomonas* classification. A new case history in the taxonomy of Gram-negative bacteria. Antonie van Leeuw. 64:231–251.

2947. Palm, C. J., Gaffney, T., and Kosuge, T. 1989. Cotranscription of genes encoding indoleacetic acid production in *Pseudomonas syringae* subsp. *savastanoi*. J. Bacteriol. 171:1002–1009.

2948. Palm, M. E., and Civerolo, E. L. 1994. Isolation, pathogenicity, and partial host range of *Alternaria limicola*, causal agent of mancha foliar de los cítricos in Mexico. Plant Dis. 78:879–883.

2949. Palomaki, V., Holopainen, J. K., and Holopainen, T. 1995. Effects of drought and waterlogging on ultrastructure of Scots pine and Norway spruce needles. Trees Struct. Funct. 9:98–105.

2950. Panconesi, A. 1999. Canker stain of plane trees: a serious danger to urban plantings in Europe. J. Plant Pathol. 81:3–15.

2951. Panikkar, M. V. N., Ampili, P., and Chauhan, V. D. 1989. Observations on *Cephaleuros virescens* Kunze from Kerala, India. J. Econ. Taxon. Bot. 13:67–70.

2952. Pantidou, M. E., and Darker, G. D. 1963. The genus *Didymascella* on *Juniperus*. Mycologia 55:415–420.

2953. Panvini, A. D., and Eickmeier, W. G. 1993. Nutrient and water relations of the mistletoe *Phoradendron leucarpum* (Viscaceae): how tightly are they integrated? Am. J. Bot. 80:872–878.

2954. Pappert, R. A., Hamrick, J. L., and Donovan, L. A. 2000. Genetic variation in *Pueraria lobata* (Fabaceae), an introduced, clonal, invasive plant of the southeastern United States. Am. J. Bot. 87:1240–1245.

2955. Paques, L. E., Sylvestre-Guinot, G., and Delatour, C. 1999. Genetic variation among clones of *Larix decidua polonica* for resistance to *Lachnellula willkommii*. Ann. For. Sci. 56:155–166.

2956. Paquet, P. J., Knutson, D. M., Tinnin, R. O., and Tocher, R. D. 1986. Characteristics of viscin from the seeds of dwarf mistletoe. Bot. Gaz. 147:156–158.

2957. Parbery, I. H., and Brown, J. F. 1986. Sooty moulds and black mildews in extra-tropical rainforests. Pages 101–120 in: Microbiology of the phyllosphere. N. J. Fokkema and J. van den Heuvel, eds. Cambridge Univ. Press, Cambridge, UK.

2958. Parker, A. K. 1957. The nature of the association of *Europhium trinacriforme* with pole blight lesions. Can. J. Bot. 35:845–856.

2959. Parker, A. K. 1970. Effect of relative humidity and temperature on needle cast disease of Douglas-fir. Phytopathology 60:1270–1273.

2960. Parker, A. K., and Reid, J. 1969. The genus *Rhabdocline* Syd. Can. J. Bot. 47:1533–1545.

2961. Parker, C., and Riches, C. R. 1993. Parasitic weeds of the world: biology and control. CAB International, Wallingford, UK. 332 pp.

2962. Parker, J. 1963. Cold resistance in woody plants. Bot. Rev. 29:124–201.

2963. Parker, K. G. 1959. Verticillium hadromycosis of deciduous tree fruits. Plant Dis. Rep. Suppl. 255:38–61.

2964. Parks, C. G., Bull, E. L., Tinnin, R. O., Shepherd, J. F., and Blumton, A. K. 1999. Wildlife use of dwarf mistletoe brooms in Douglas-fir in northeast Oregon. West. J. Appl. For. 14:100–105.

2965. Parmelee, J. A. 1965. The genus *Gymnosporangium* in eastern Canada. Can. J. Bot. 43:239–267.

2966. Parmelee, J. A. 1968. Effective range of basidiospores of *Gymnosporangium*. Can. Plant Dis. Surv. 48:150–151.

2967. Parmelee, J. A. 1971. The genus *Gymnosporangium* in western Canada. Can. J. Bot. 49:903–926.

2968. Parmelee, J. A., and Carteret, P. M. de. 1984. *Cumminsiella mirabilissima*. Fungi Canadenses No. 288. 2 pp.

2969. Parmeter, J. R. Jr., Bega, R. V., and Hood, J. R. 1960. Epidemic leaf-blighting of California-laurel. Plant Dis. Rep. 44:669–671.

2970. Parris, G. K., and Byrd, J. 1962. Oak anthracnose in Mississippi. Plant Dis. Rep. 46:677–681.

2971. Partridge, A. D. 1957. The ash leaf rust syndrome in New Hampshire: suscepts, incitant, epidemiology and control. Ph.D. diss., Univ. New Hampshire, Durham.

2972. Partridge, A. D., and Miller, D. L. 1974. Major wood decays in the inland Northwest. Ida. Res. Found., Moscow, Nat. Resour. Ser. No. 3. 125 pp.

2973. Partridge, A. D., and Rich, A. E. 1957. A study of the ash leaf rust syndrome in New Hampshire: suscepts, incitant, epidemiology, and control. (Abstr.) Phytopathology 47:246.

2973a. Partridge, E. C., and Morgan-Jones, G. 2003. Notes on Hypho-mycetes. XC. *Fusicladosporium*, a new genus for *Cladosporium*-like anamorphs of *Venturia*, and the pecan scab–inducing fungus. Mycotaxon 85:357–370.

2974. Patterson, J. C. 1977. Soil compaction—effects on urban vegetation. J. Arboric. 3:161–167.

2975. Patton, R. F. 1978. Penetration and colonization of oak roots by *Armillaria mellea* in Wisconsin. Eur. J. For. Pathol. 8:259–267.

2976. Patton, R. F., and Johnson, D. W. 1970. Mode of penetration of needles of eastern white pine by *Cronartium ribicola*. Phytopathology 60:977–982.

2977. Patton, R. F., and Riker, A. J. 1954. Needle droop and needle blight of red pine. J. For. 52:412–418.

2978. Patton, R. F., and Spear, R. N. 1983. Needle cast of European larch caused by *Mycosphaerella laricina* in Wisconsin and Iowa. Plant Dis. 67:1149–1153.

2979. Patton, R. F., and Spear, R. N. 1989. Histopathology of colonization in leaf tissue of *Castilleja, Pedicularis, Phaseolus,* and *Ribes* species by *Cronartium ribicola*. Phytopathology 79:539–547.

2980. Patton, R. F., Spear, R. N., and Blenis, P. V. 1984. The mode of infection and early stages of colonization of pines by *Gremmeniella abietina*. Eur. J. For. Pathol. 14:193–202.

2981. Paulin, A. E., and Harrington, T. C. 2000. Phylogenetic placement of anamorphic species of *Chalara* among *Ceratocystis* species and other ascomycetes. Stud. Mycol. No. 45:209–222.

2982. Paulin-Mahady, A. E., Harrington, T. C., and McNew, D. 2002. Phylogenetic and taxonomic evaluation of *Chalara, Chalaropsis*, and *Thielaviopsis* anamorphs associated with *Ceratocystis*. Mycologia 94:62–72.

2983. Pawlowski, K., and Bisseling, T. 1996. Rhizobial and actinorhizal symbioses: what are the shared features? Plant Cell 8:1899–1913.

2984. Pawsey, R. G. 1960. An investigation into Keithia disease of *Thuja plicata*. Forestry 33:174–186.

2985. Paxton, J. D., and Wilson, E. E. 1965. Anatomical and physiological aspects of branch wilt disease of Persian walnut. Phytopathology 55:21–26.

2986. Payette, S., Fortin, M. J., and Morneau, C. 1996. The recent sugar maple decline in southern Quebec: probable causes deduced from tree rings. Can. J. For. Res. 26:1069–1078.

2987. Peace, T. R. 1940. An interesting case of lightning damage to a group of trees. Q. J. For. 34:61–63.

2988. Peace, T. R. 1962. Pathology of trees and shrubs. Clarendon Press, Oxford. 722 pp.

2989. Peace, T. R., and Holmes, C. H. 1933. *Meria laricis*. The leaf cast disease of larch. Oxford For. Mem. 15. 29 pp.

2990. Peacock, C. H., and Webb, R. S. 1984. Screening potential causes of pine tree decline on Florida golf courses. Soil Crop Sci. Soc. Fla. Proc. 43:34–35.

2991. Pearce, R. B. 1990. Occurrence of decay-associated xylem suberization in a range of woody species. Eur. J. For. Pathol. 20:275–289.

2992. Pearce, R. B. 1991. Reaction zone relics and the dynamics of fungal spread in the xylem of woody angiosperms. Physiol. Mol. Plant Pathol. 39:41–55.

2993. Pearce, R. B. 1996. Antimicrobial defences in the wood of living trees. New Phytol. 132:203–233.

2994. Pearce, R. B. 2000. Decay development and its restriction in trees. J. Arboric. 26:1–11.

2995. Pearce, R. B., and Holloway, P. J. 1984. Suberin in the sapwood of oak (*Quercus robur* L.): its composition from a compartmentalization barrier and its occurrence in tyloses in undecayed wood. Physiol. Plant Pathol. 24:71–81.

2996. Pearce, R. B., and Rutherford, J. 1981. A wound-associated suberized barrier to the spread of decay in the sapwood of oak (*Quercus robur* L.). Physiol. Plant Pathol. 19:359–369.

2997. Pearce, R. B., Sumer, S., Doran, S. J., Carpenter, T. A., and Hall, L. D. 1994. Non-invasive imaging of fungal colonization and host response in the living sapwood of sycamore (*Acer pseudoplatanus* L.) using nuclear magnetic resonance. Physiol. Mol. Plant Pathol. 45:359–384.

2998. Pearson, R. C., Aldwinckle, H. S., and Seem, R. C. 1977. Teliospore germination and basidiospore formation in *Gymnosporangium juniperi-virginianae*: a regression model of temperature and time effects. Can. J. Bot. 55:2832–2837.

2999. Pearson, R. C., and Goheen, A. C. 1988. Compendium of grape diseases. APS Press, St. Paul, MN. 93 pp.

3000. Pearson, R. C., Seem, R. C., and Meyer, F. W. 1980. Environmental factors influencing the discharge of basidiospores of *Gymnosporangium juniperi-virginianae*. Phytopathology 70:262–266.

3001. Peck, C. H. 1881. Two new species of fungi. Bull. Torrey Bot. Club 8:49–51.

3002. Pedersen, B. S. 1998. The role of stress in the mortality of midwestern oaks as indicated by growth prior to death. Ecology 79:79–93.

3003. Peerally, A. 1991. The classification and phytopathology of *Cylindrocladium* species. Mycotaxon 40:323–366.

3004. Pegg, G. F. 1974 Verticillium diseases. Rev. Plant Pathol. 53:157–182.

3005. Pei, M. H., Hunter, T., and Ruiz, C. 1999. Occurrence of *Melampsora* rusts in biomass willow plantations for renewable energy in the United Kingdom. Biomass Bioenergy 17:153–163.

3006. Pei, M. H., Parker, S. R., Hunter, T., and Royle, D. J. 1997. Variation in populations of *Melampsora* willow rust and the implications for design of short rotation coppice plantations. Aspects Appl. Biol. 49:91–96.

3007. Pei, M. H., Royle, D. J., and Hunter, T. 1993. Identity and host alternation of some willow rusts (*Melampsora* spp.) in England. Mycol. Res. 97:845–851.

3008. Pei, M. H., Royle, D. J., and Hunter, T. 1995. A comparative study of stem- and leaf-infecting forms of *Melampsora* rust on *Salix viminalis* in the UK. Mycol. Res. 99:357–363.

3009. Pei, M. H., Royle, D. J., and Hunter, T. 1996. Pathogenic specialization in *Melampsora epitea* var. *epitea* on *Salix*. Plant Pathol. 45:679–690.

3010. Pei, M. H., Royle, D. J., and Hunter, T. 1999. Hybridization in larch-alternating *Melampsora epitea* (*M. larici-epitea*). Mycol. Res. 103:1440–1446.

3011. Pellett, H., Gearhart, M., and Dirr, M. 1981. Cold hardiness capability of woody ornamental plant taxa. J. Am. Soc. Hortic. Sci. 106:239–243.

3012. Penyalver, R., Garcia, A., Ferrer, A., Bertolini, E., and Lopez, M. M. 2000. Detection of *Pseudomonas savastanoi* pv. *savastanoi* in olive plants by enrichment and PCR. Appl. Environ. Microbiol. 66:2673–2677.

3013. Penyalver, R., Vicedo, B., and Lopez, M. M. 2000. Use of the genetically engineered *Agrobacterium* strain K1026 for biological control of crown gall. Eur. J. Plant Pathol. 106:801–810.

3014. Perala, D. A., and Sucoff, E. 1965. Diagnosing potassium deficiency in American elm, silver maple, Russian olive, hackberry, and box elder. For. Sci. 11:347–352.

3015. Percival, G. C., Biggs, M. P., and Dixon, G. R. 1998. The influence of sodium chloride and waterlogging stresses on *Alnus cordata*. J. Arboric. 24:19–27.

3016. Percival, W. C. 1933. A contribution to the biology of *Fomes pini* (Thore) Lloyd (*Trametes pini* [Thore] Fries). N.Y. State Coll. For. Tech. Pub. 40. 72 pp.

3017. Percy, R. G. 1983. Potential range of *Phymatotrichum omnivorum* as determined by edaphic factors. Plant Dis. 67:981–983.

3018. Peries, O. S. 1974. Ganoderma basal stem rot of coconut: a new record of the disease in Sri Lanka. Plant Dis. Rep. 58:293–295.

3019. Pero, R. W., and Howard, F. L. 1970. Activity of juniper diffusates on spores of *Phomopsis juniperovora*. Phytopathology 60:491–495.

3020. Perrin, R. 1984. Variability of the susceptibility of beech (*Fagus sylvatica* L.) to *Nectria coccinea*, one of the pathogens of bark disease. Eur. J. For. Pathol. 14:321–325.

3021. Perry, E., and McCain, A. H. 1988. Incidence and management of canker stain in London plane trees in Modesto, California. J. Arboric. 14:18–19.

3022. Perry, R. G., and Peterson, J. L. 1982. Susceptibility and response of juniper species to *Kabatina juniperi* infection in New Jersey. Plant Dis. 66:1189–1191.

3023. Persiel, F., and Zeller, W. 1990. Breeding upright growing types of *Cotoneaster* for resistance to fire blight, *Erwinia amylovora* (Burr.) Winslow et al. Acta Hortic. No. 273:297–301.

3024. Persson, A. 1994. Stem cracks in Norway spruce in southern Scandinavia: causes and consequences. Ann. Sci. For. 51:315–327.

3025. Petäistö, R. L., and Kurkela, T. 1993. The susceptibility of Scots pine seedlings to *Gremmeniella abietina*: effect of growth phase, cold and drought stress. Eur. J. For. Pathol. 23:385–399.

3026. Petäistö, R. L., and Laine, A. 1999. Effects of winter storage temperature and age of *Pinus sylvestris* seedlings on the occurrence of disease induced by *Gremmeniella abietina*. Scand. J. For. Res. 14:227–233.

3027. Peters, M., Ossenbruggen, P., and Shigo, A. 1985. Cracking and failure behavior models of defective balsam fir trees. Holzforschung 39:125–135.

3028. Petersen, R. H. 1974. The rust fungus life cycle. Bot. Rev. 40:453–513.

3029. Peterson, G. W. 1966. Western X-disease virus of chokecherry: transmission and seed effects. Plant Dis. Rep. 50:659–660.

3030. Peterson, G. W. 1973. Dispersal of aeciospores of *Peridermium harknessii* in central Nebraska. Phytopathology 63:170–172.

3031. Peterson, G. W. 1973. Infection of *Juniperus virginiana* and *J. scopulorum* by *Phomopsis juniperovora*. Phytopathology 63:246–251.

3032. Peterson, G. W. 1976. Disease of Russian-olive caused by *Botryodiplodia theobromae*. Plant Dis. Rep. 60:490–494.

3033. Peterson, G. W. 1977. Epidemiology and control of a blight of *Juniperus virginiana* caused by *Cercospora sequoiae* var. *juniperi*. Phytopathology 67:234–238.

3034. Peterson, G. W. 1977. Infection, epidemiology, and control of Diplodia blight of Austrian, ponderosa, and Scots pines. Phytopathology 67:511–514.

3035. Peterson, G. W. 1981. Pine and juniper diseases in the Great Plains. USDA For. Serv. Gen. Tech. Rep. RM-86. 47 pp.

3036. Peterson, G. W. 1984. Resistance to *Dothistroma pini* within geographic seed sources of *Pinus ponderosa*. Phytopathology 74:956–960.

3037. Peterson, G. W. 1984. Spread and damage of western X-disease of chokecherry in eastern Nebraska plantings. Plant Dis. 68:103–104.

3038. Peterson, G. W., tech. coord. 1986. Recent research on conifer needle diseases: conference proceedings, 1984, Gulfport, MS. USDA For. Serv. Gen. Tech. Rep. WO-50. 106 pp.

3039. Peterson, J. L., and Davis, S. H. Jr. 1965. A Fusarium canker of *Sophora japonica*. Plant Dis. Rep. 49:835–836.

3040. Peterson, J. L., Davis, S. H. Jr., and Judd, R. W. Jr. 1976. Effect of fungicide and application timing on Cercospora leafspot of mountain laurel. Plant Dis. Rep. 60:138–140.

3041. Peterson, M. J., and Sutherland, J. R. 1990. Controlling gray mold on container-grown Douglas-fir by modified styroblock and under-bench, forced air ventilation. West. J. Appl. For. 5:75–79.

3042. Peterson, R. S. 1960. Development of western gall rust in lodgepole pine. Phytopathology 50:876–881.

3043. Peterson, R. S. 1961. Conifer tumors in the central Rocky Mountains. Plant Dis. Rep. 45:472–474.

3044. Peterson, R. S. 1961. Host alternation of spruce broom rust. Science 134:468–469.

3045. Peterson, R. S. 1963. Effects of broom rusts on spruce and fir. USDA For. Serv. Res. Pap. INT-7. 10 pp.

3046. Peterson, R. S. 1966. Limb rust damage to pine. USDA For. Serv. Res. Pap. INT-31. 10 pp.

3047. Peterson, R. S. 1966. On sweetfern blister rust. Plant Dis. Rep. 50: 744–746.

3048. Peterson, R. S. 1967. Studies of juniper rusts in the West. Madroño 19:79–91.

3049. Peterson, R. S. 1967. The *Peridermium* species on pine stems. Bull. Torrey Bot. Club 94:511–542.

3050. Peterson, R. S. 1968. Limb rusts of pine: the causal fungi. Phytopathology 58:309–315.

3051. Peterson, R. S. 1973. Studies of *Cronartium* (Uredinales). Rep. Tottori Mycol. Inst. 10:203–223.

3052. Peterson, R. S., and Jewell, F. F. 1968. Status of American stem rusts of pine. Annu. Rev. Phytopathol. 6:23–40.

3053. Peterson, R. S., and Shurtleff, R. G. Jr. 1965. Mycelium of limb rust fungi. Am. J. Bot. 52:519–525.

3054. Petrak, F., and Sydow, H. 1926–1927. Die Gattungen der Pyrenomyceten, Sphaeropsideen und Melanconieen. Verlag des Reportoriums (Feddes Reportorium), Berlin-Dahlem. Heft 42. 551 pp.

3055. Petrini, O., Petrini, L., Laflamme, G., and Ouellette, G. B. 1989. Taxonomic position of *Gremmeniella abietina* and related species: a reappraisal. Can. J. Bot. 67:2805–2814.

3056. Pettey, T. M., and Shaw, C. G. 1986. Isolation of *Fomitopsis pinicola* from in-flight bark beetles (Coleoptera: Scolytidae). Can. J. Bot. 64:1507–1509.

3057. Peturson, B. 1954. The relative prevalence of specialized forms of *Puccinia coronata* that occur on *Rhamnus cathartica* in Canada. Can. J. Bot. 32:40–47.

3058. Pfister, S. E., Halik, S., and Bergdahl, D. R. 2004. Effect of temperature on *Thekopsora minima* urediniospores and uredinia. Plant Dis. 88:359–362.

3059. Pfleger, F. L., and Linderman, R. G., eds. 1994. Mycorrhizae and plant health. APS Press, St. Paul, MN. 344 pp.

3060. Phelps, J. E., McGinnes, E. A. Jr., and Lieu, P. J.-Y. 1975. Anatomy of xylem tissue formation associated with radial seams and cracks in black oak. Wood Sci. 8:397–405.

3061. Phelps, W. R. 1974. Live oak in North Carolina and Florida infected by *Endothia parasitica*. Plant Dis. Rep. 58:596–598.

3062. Philipson, J. J., and Coutts, M. P. 1980. The tolerance of tree roots to waterlogging. IV. Oxygen transport in woody roots of Sitka spruce and lodgepole pine. New Phytol. 65:489–494.

3063. Phillips, A. J. L. 2000. *Botryosphaeria populi* sp. nov. and its *Fusicoccum* anamorph from poplar trees in Portugal. Mycotaxon 76:135–140.

3064. Phillips, A. J. L., Fonseca, F., Povoa, V., Castilho, R., and Nolasco, G. 2002. A reassessment of the anamorphic fungus *Fusicoccum luteum* and description of its teleomorph *Botryosphaeria lutea* sp. nov. Sydowia 54:59–77.

3065. Phillips, D. H., and Burdekin, D. A. 1982. Diseases of forest and ornamental trees. Macmillan, London. 435 pp.

3066. Phillips, J. F. V. 1929. The influence of *Usnea* sp. (near *barbata* Fr.) upon the supporting trees. Trans. R. Soc. S.Afr. 17:101–107.

3067. Phipps, H. M. 1963. The role of 2,4-D in the appearance of a leaf blight of some plains tree species. For. Sci. 9:283–288.

3068. Phipps, P. M., and Stipes, R. J. 1976. Histopathology of mimosa infected with *Fusarium oxysporum* f.sp. *perniciosum*. Phytopathology 66:839–843.

3069. Pichot, C., and Teissier, C. E. du. 1993. Susceptibility of *P. deltoides* Bartr. to *Melampsora larici-populina* and *M. alli-populina*. I. Qualitative analysis of a 6 × 6 factorial mating design. II. Quantitative analysis of a 6 × 6 factorial mating design. Silvae Genet. 42:179–188, 188–199.

3070. Pine, T. S., Gilmer, R. M., Moore, J. D., Nyland, G., and Welsh, M. F., eds. 1976. Virus diseases and noninfectious disorders of stone fruits in North America. USDA Agric. Handb. 437. 433 pp.

3071. Pinkerton, J. N., Johnson, K. B., Aylor, D. E., and Stone, J. K. 2001. Spatial and temporal increase of eastern filbert blight in European hazelnut orchards in the Pacific Northwest. Phytopathology 91:1214–1223.

3072. Pinkerton, J. N., Johnson, K. B., Mehlenbacher, S. A., and Pscheidt, J. W. 1993. Susceptibility of European hazelnut clones to eastern filbert blight. Plant Dis. 77:261–266.

3073. Pinkerton, J. N., Johnson, K. B., Stone, J. K., and Ivors, K. L. 1998. Factors affecting the release of ascospores of *Anisogramma anomala*. Phytopathology 88:122–128.

3074. Pinkerton, J. N., Johnson, K. B., Stone, J. K., and Ivors, K. L. 1998. Maturation and seasonal discharge pattern of ascospores of *Anisogramma anomala*. Phytopathology 88:1165–1173.

3075. Pinkerton, J. N., Stone, J. K., Nelson, S. J., and Johnson, K. B. 1995. Infection of European hazelnut by *Anisogramma anomala*: ascospore adhesion, mode of penetration of immature shoots, and host response. Phytopathology 85:1260–1268.

3076. Pinochet, J., Calvet, C., Camprubi, A., and Fernandez, C. 1996. Interactions between migratory endoparasitic nematodes and arbuscular mycorrhizal fungi in perennial crops: a review. Plant Soil 185:183–190.

3077. Pinochet, J., Fernandez, C., Alcaniz, E., and Felipe, A. 1996. Damage by a lesion nematode, *Pratylenchus vulnus*, to *Prunus* rootstocks. Plant Dis. 80:754–757.

3078. Pinon, J., and Morelet, M. 1975. Le *Linospora ceuthocarpa* (Fr.) Munk ex Morelet, parasite foliaire des peupliers. Eur. J. For. Pathol. 5:367–376.

3079. Pinon, J., and Poissonnier, M. 1975. Étude épidemiologique du *Marssonina brunnea* (El. et Ev.) P. Magn. Eur. J. For. Pathol. 5:97–111.

3080. Pinon, J., and Valadon, A. 1997. Comportement des cultivars de peupliers commercialisables dans l'Union européenne vis-à-vis de quelques parasites majeurs. Ann. Sci. For. 54:19–38.

3081. Piou, D. 1993. Possible role of *Hylobius abietis* in the transport of *Leptographium procerum* and in the infection of Scots pine. Ann. Sci. For. 50:297–308.

3082. Pirone, P. P. 1957. *Ganoderma lucidum*, a parasite of shade trees. Bull. Torrey Bot. Club 84:424–428.

3083. Pirozynski, K. A. 1974. *Cercospora handelii*. Fungi Canadenses No. 22. 2 pp.

3084. Pirozynski, K. A. 1974. *Marssonina brunnea*. Fungi Canadenses No. 13. 2 pp.

3085. Pirozynski, K. A. 1974. *Marssonina castagnei*. Fungi Canadenses No. 14. 2 pp.

3086. Pirozynski, K. A. 1974. *Marssonina populi*. Fungi Canadenses No. 15. 2 pp.

3087. Pisi, A., and Bellardi, M. G. 1993. Rust and powdery mildew on *Mahonia aquifolium* Nutt. Phytopathol. Medit. 32:51–54.

3088. Pitt, D. G., Dumas, M. T., Wall, R. E., Thompson, D. G., Lanteigne, L., Hintz, W., Sampson, G., and Wagner, R. G. 1999. *Chondrostereum purpureum* as a biological control agent in forest vegetation management. I. Efficacy on speckled alder, red maple, and aspen in eastern Canada. Can. J. For. Res. 29:841–851.

3089. Plakidas, A. G. 1940. Angular leaf spot of *Pittosporum*. Mycologia 32:601–608.

3090. Plakidas, A. G. 1942. *Venturia acerina*, the perfect stage of *Cladosporium humile*. Mycologia 34:27–37.

3091. Plakidas, A. G. 1945. Blight of Oriental arborvitae. Phytopathology 35:181–190.

3092. Plakidas, A. G. 1954. Transmission of leaf and flower variegation in camellias by grafting. Phytopathology 44:14–18.

3093. Plakidas, A. G. 1962. Strains of the color-breaking virus of camellia. Phytopathology 52:77–79.

3094. Plakidas, A. G. 1991. Transmission of the color-breaking virus through root grafts. Am. Camellia Yearb. 46:51–53.

3095. Plank, S., and Wolkinger, F. 1976. Études du cours des hyphes de *Fomes fomentarius* dans le bois d'*Aesculus hippocastanum* au microscope électronique à balayage. Can. J. Bot. 54:2231–2238.

3096. Plante, F., and Bernier, L. 1997. Variability of virulence of *Nectria galligena* towards northern hardwoods. Eur. J. For. Pathol. 27:261–272.

586

3097. Plante, F., Hamelin, R. C., and Bernier, L. 2002. A comparative study of genetic diversity of populations of *Nectria galligena* and *N. coccinea* var. *faginata* in North America. Mycol. Res. 106:183–193.

3098. Plattner, K., Volgger, W., Oberhuber, W., Mayr, S., and Bauer, H. 1999. Dry mass production in seedlings of Norway spruce infected by the needle rust *Chrysomyxa rhododendri*. Eur. J. For. Pathol. 29:365–370.

3099. Plavšić, B., Miličić, D., and Erić, Z. 1984. Occurrence of a rhabdovirus in *Hibiscus rosa-sinensis* L. (Abstr.). Page 35 in: 6th International Symposium on Virus Diseases of Ornamental Plants, Cornell Univ., Ithaca, NY, 1984.

3100. Plavšić-Banjac, B., Miličić, D., and Erić, Z. 1976. Rhabdovirus in *Pittosporum tobira* plants suffering from vein yellowing disease. Phytopathol. Z. 86:225–232.

3101. Pleše, N., and Erić, Z. 1980. Rhabdovirus in *Euonymus japonica*. Acta Phytopathol. Acad. Sci. Hung. 15:291–295.

3102. Pleše, N., and Miličić, D. 1973. Two viruses isolated from *Maclura pomifera*. Phytopathol. Z. 77:178–183.

3103. Plessis, H. J. du. 1987. Canker development on plum shoots following systemic movement of *Xanthomonas campestris* pv. *pruni* from inoculated leaves. Plant Dis. 71:1078–1080.

3104. Plessis, H. J. du. 1990. Systemic invasion of plum seed and fruit by *Xanthomonas campestris* pv. *pruni* through stalks. J. Phytopathol. 130:37–45.

3105. Plich, M., and Rudnicki, R. M. 1979. Studies of the toxins of *Phytophthora cactorum* pathogenic to apple trees. 1. Isolation, some of the properties, and activities of a toxin produced by the fungus cultured *in vitro*. Phytopathol. Z. 94:270–278.

3106. Ploetz, R. C., and Schaffer, B. 1989. Effects of flooding and Phytophthora root rot on net gas exchange and growth of avocado. Phytopathology 79:204–208.

3107. Podger, F. D. 1972. *Phytophthora cinnamomi*, a cause of lethal disease in indigenous plant communities in Western Australia. Phytopathology 62:972–981.

3108. Podila, G. K., and Douds, D. D. Jr., eds. 2000. Current advances in mycorrhizae research. APS Press, St. Paul, MN. 193 pp.

3109. Pokorny, J. 1999. How to collect field samples and identify the oak wilt fungus in the laboratory. USDA For. Serv. Northeast. Area State & Priv. For. NA-FR-01-99. 12 pp.

3110. Polak, Z. 1996. Scarlet maple—a new host of cucumber mosaic virus. Ochrana Rostlin 32:73–76.

3111. Polak, Z. 1999. Mild mosaic of cucumber and tulip trees caused by cucumber mosaic virus. Zahradnictvi—Hortic. Sci. 26:25–26.

3112. Polak, Z., and Zieglerova, J. 1996. Apple mosaic virus associated with decline of silver birch. Ochrana Rostlin 32:15–18.

3113. Polizzi, G. 1996. Myrtaceae, natural hosts of *Cylindrocladium scoparium*. Inf. Fitopatol. 46:59–64.

3114. Pomerleau, R. 1937–1938. Recherches sur le *Gnomonia ulmea* (Schw.) Thüm. Nat. Can. 64:261–289, 297–318; 65:23–41, 253–279.

3115. Pomerleau, R. 1940. Studies on the ink-spot disease of poplar. Can. J. Res. C18:199–214.

3116. Pomerleau, R. 1944. Observations sur quelques maladies non parasitaires des arbres dans le Québec. Can. J. Res. C22:171–189.

3117. Pomerleau, R. 1953. History of hardwood species dying in Quebec. Can. Dep. Agric., Ottawa, report of Symposium on Birch Dieback, Part 1:10–11.

3118. Pomerleau, R., Auclair, A., and Lachance, D. 1991. Experiments on the causal mechanisms of dieback on deciduous forests in Quebec. For. Can. Inf. Rep. LAU-X-96. 48 pp.

3119. Pomerleau, R., and Lortie, M. 1962. Relationships of dieback to the rooting depth of white birch. For. Sci. 8:219–224.

3120. Pomerleau, R., and Ray, R. G. 1957. Occurrence and effects of summer frost in a conifer plantation. Can. For. Serv. Tech. Note 51. 15 pp.

3121. Poncet, C., Antonini, C., Bettachini, A., Hericher, D., Pionnat, S., Simonini, L., Dessaux, Y., and Nesme, X. 1995. Impact of the crown gall disease on vigour and yield of rose trees. Acta Hortic. No. 424:221–225.

3122. Ponte, J. J. da, Andrade, N. C. de, and Silveira-Filho, J. 1999. Lichen parasitism in ornamental palm of the genus *Caryota*. Rev. Agric. Piracicaba 74:245–248.

3123. Pooler, M. R., Jacobs, K. A., and Kramer, M. 2002. Differential resistance to *Botryosphaeria ribis* among *Cercis* taxa. Plant Dis. 86:880–882.

3124. Porter, J. R. 1991. Host range and implications of plant infection by *Agrobacterium rhizogenes*. Crit. Rev. Plant Sci. 10:387–421.

3125. Posnette, A. F., ed. 1963. Virus diseases of apples and pears. Commonw. Bur. Hortic. Plantation Crops Tech. Communic. 30. 141 pp.

3126. Potebnia, A. von. 1910. Beiträge zur Micromycetenflora Mittel-Russlands. Ann. Mycol. 8:42–93.

3127. Poteri, M. 1992. Screening of clones of *Betula pendula* and *B. pubescens* against two forms of *Melampsoridium betulinum* leaf rust fungus. Eur. J. For. Pathol. 22:166–173.

3128. Poteri, M., and Rousi, M. 1996. Variation in *Melampsoridium* resistance among European white-birch clones grown in different fertilization treatments. Eur. J. For. Pathol. 26:171–181.

3129. Potter, D. A., and Hartman, J. R. 1993. Susceptibility of honeylocust cultivars to *Thyronectria austro-americana* and response of *Agrilus* borers and bagworms to infected and non-infected trees. J. Environ. Hortic. 11:176–181.

3130. Poucher, C., Ford, H. W., Suit, R. F., and DuCharme, E. P. 1967. Burrowing nematode in citrus. Fla. Dep. Agric. Div. Plant Indus. Bull. 7. 63 pp.

3131. Pouzar, Z. 1959. New genera of higher fungi. III. Česká Mykol. 13:10–19.

3132. Powell, J. M. 1971. Incidence and effect of *Tuberculina maxima* on cankers of the pine stem rust, *Cronartium comandrae*. Phytoprotection 52:104–111.

3133. Powell, J. M. 1971. Occurrence of *Tuberculina maxima* on pine stem rusts in western Canada. Can. Plant Dis. Surv. 51:83–85.

3134. Powell, J. M. 1972. Seasonal and diurnal periodicity in the release of *Cronartium comandrae* aeciospores from stem cankers of lodgepole pine. Can. J. For. Res. 2:78–88.

3135. Powers, H. R. Jr. 1972. Comandra rust on southern pines. J. For. 70:18–20.

3136. Powers, H. R. Jr., Lin, D., and Hubbes, M. 1989. Interspecific and intraspecific differentiation within the genus *Cronartium* by isozyme and protein pattern analysis. Plant Dis. 73:691–694.

3137. Powers, H. R. [Jr.], Miller, T., and Belanger, R. P. 1993. Management strategies to reduce losses from fusiform rust. South. J. Appl. For. 17:146–149.

3138. Powers, H. R. Jr., Schmidt, R. A., and Snow, G. A. 1981. Current status and management of fusiform rust on southern pines. Annu. Rev. Phytopathol. 19:353–371.

3139. Powers, H. R. Jr., Snow, G., Lin, D., and Hubbes, M. 1991. Isozyme analysis as an indicator of synonymy of the causal agents of gall rust on sand and Virginia pine. Plant Dis. 75:1225–1227.

3140. Prakash, C. S., and Heather, W. A. 1986. Inheritance of resistance to races of *Melampsora medusae* in *Populus deltoides*. Silvae Genet. 35:74–77.

3141. Prakash, C. S., and Thielges, B. A. 1989. Interaction of geographic isolates of *Melampsora medusae* and *Populus*: effect of temperature. Can. J. Bot. 67:486–490.

3142. Prasad, K. V., Yadav, B. R. D., and Sullia, S. B. 1993. Taxonomic status of rust on mulberry in India. Curr. Sci. 65:424–426.

3143. Prasad, N. S., Rao, A. R., and Rao, G. M. 1997. Fasciation in *Casuarina equisetifolia*. Indian For. 123:773–774.

3144. Pratt, J. E., Gibbs, J. N., and Webber, J. F. 1999. Registration of *Phlebiopsis gigantea* as a forest biocontrol agent in the UK: recent experience. Biocontrol Sci. Technol. 9:113–118.

3145. Preece, T. F., and Hick, A. J. 1990. An introductory scanning electron microscope atlas of rust fungi. Farrand, London. 220 pp.

3146. Press, M. C., and Graves, J. D., eds. 1995. Parasitic plants. Chapman & Hall, London. 292 pp.

3147. Pretorius, M. C., Crous, P. W., Groenewald, J. Z., and Braun, U. 2003. Phylogeny of some cercosporoid fungi from *Citrus*. Sydowia 55:286–305.

3148. Price, T. V. 1970. Epidemiology and control of powdery mildew (*Sphaerotheca pannosa*) on roses. Ann. Appl. Biol. 65:231–248.

3149. Prince, A. E. 1943. Basidium formation and spore discharge in *Gymnosporangium nidus-avis*. Farlowia 1:79–93.

3150. Prince, A. E. 1946. The biology of *Gymnosporangium nidus-avis* Thaxter. Farlowia 2:475–525.

3151. Prljincević, M. B. 1982. Economic significance of the infection of beech forests by *Hypoxylon deustum* (Hoffm. et Fr.) Grev. at Šara mountain. Eur. J. For. Pathol. 12:7–10.

3152. Proffer, T. J. 1988. Xylaria root rot of urban trees caused by *Xylaria polymorpha*. Plant Dis. 72:79.

3153. Proffer, T. J. 1996. First report of *Diplodia mutila* on golden-chain tree (*Laburnum ×watereri*). Plant Dis. 80:1303.

3154. Proffer, T. J., and Hart, J. H. 1988. Vegetative compatibility groups in *Leucocytospora kunzei*. Phytopathology 78:256–260.

3155. Proffer, T. J., and Hart, J. H. 1994. Wound inoculations of Colorado blue spruce with isolates of *Leucostoma* (*Cytospora*) *kunzei* and other *Cytospora* species. J. Arboric. 20:215–221.

3156. Provvidenti, R., and Hunter, J. E. 1975. Bean yellow mosaic virus infection in *Cladrastis lutea*, an ornamental leguminous tree. Plant Dis. Rep. 59:86–87.

3157. Prusky, D., Freeman, S., and Dickman, M. B., eds. 2000. *Colletotrichum*: host specificity, pathology, and host-pathogen interaction. APS Press, St. Paul, MN. 393 pp.

3158. Pryor, B. M., and Michailides, T. J. 2002. Morphological, pathogenic, and molecular characterization of *Alternaria* isolates associated with Alternaria late blight of pistachio. Phytopathology 92:406–416.

3159. Przybyl, K. 1984. Pathological changes and defense responses in poplar tissues caused by *Ceratocystis fimbriata*. Eur. J. For. Pathol. 14:183–191.

3160. Przybyl, K. 1988. The response of *Populus* 'NE 42' (*P. maximowiczii* × *P. trichocarpa*) to infection by *Ceratocystis fimbriata* isolated from cacao-tree and plane. Eur. J. For. Pathol. 18:8–12.

3161. Puente, M. E., and Bashan, Y. 1994. The desert epiphyte *Tillandsia recurvata* harbours the nitrogen-fixing bacterium *Pseudomonas stutzeri*. Can. J. Bot. 72:406–408.

3162. Puffinberger, C. W., and Corbett, M. K. 1985. Euonymus chlorotic ringspot disease caused by tomato ringspot virus. Phytopathology 75:423–428.

3163. Punithalingam, E. 1976. *Botryodiplodia theobromae*. C.M.I. Descr. Pathogenic Fungi Bact. No. 519. 3 pp.

3164. Punithalingham, E., and Gibson, I. A. S. 1973. *Phomopsis juniperovora*. C.M.I. Descr. Pathogenic Fungi Bact. No. 370. 2 pp.

3165. Punithalingam, E., and Waterston, J. M. 1970. *Diplodia pinea*. C.M.I. Descr. Pathogenic Fungi Bact. No. 273. 2 pp.

3166. Punja, Z. K., and Grogan, R. G. 1981. Eruptive germination of sclerotia of *Sclerotium rolfsii*. Phytopathology 71:1092–1099.

3167. Punja, Z. K., and Grogan, R. G. 1981. Mycelial growth and infection without a food base by eruptively germinating sclerotia of *Sclerotium rolfsii*. Phytopathology 71:1099–1103.

3168. Punja, Z. K., and Grogan, R. G. 1983. Germination and infection by basidiospores of *Athelia* (*Sclerotium*) *rolfsii*. Plant Dis. 67:875–878.

3169. Punja, Z. K., and Ormrod, D. J. 1979. New or noteworthy plant diseases in coastal British Columbia, 1975 to 1977. Can. Plant Dis. Surv. 59:22–24.

3170. Punter, D., and Gilbert, J. 1989. Animal vectors of *Arceuthobium americanum* seed in Manitoba. Can. J. For. Res. 19:865–869.

3171. Punter, D., and Gilbert, J. 1991. Explosive discharge of jack pine dwarf mistletoe (*Arceuthobium americanum*) seed in Manitoba. Can. J. For. Res. 21:434–438.

3172. Purcell, A. H. 1980. Almond leaf scorch: leafhopper and spittlebug vectors. J. Econ. Entomol. 73:834–838.

3173. Purcell, A. H. 1987. Comparative epidemiology of X-disease in North America. Pages 175–185 in: Proceedings of second international workshop on leafhoppers and planthoppers of economic importance, Provo, UT, 1986. Commonw. Inst. Entomol., London.

3174. Purcell, A. H. 1990. Homopteran transmission of xylem-inhibiting bacteria. Adv. Dis. Vector Res. 6:243–266.

3175. Purcell, A. H. 1997. *Xylella fastidiosa*, a regional problem or global threat? J. Plant Pathol. 79:99–105.

3176. Purcell, A. H., and Hopkins, D. L. 1996. Fastidious xylem-limited bacterial plant pathogens. Annu. Rev. Phytopathol. 34:131–151.

3177. Purcell, A. H., and Nault, L. R. 1991. Interactions among plant pathogenic prokaryotes, plants, and insect vectors. Pages 383–405 in: Microbial mediation of plant-herbivore interactions. P. Barbosa, V. A. Krischik, and C. G. Jones, eds. Wiley, New York.

3178. Purcell, A. H., and Saunders, S. R. 1999. Fate of Pierce's disease strains of *Xylella fastidiosa* in common riparian plants in California. Plant Dis. 83:825–830.

3179. Purcell, A. H., Saunders, S. R., Hendson, M., Grebus, M. E., and Henry, M. J. 1999. Causal role of *Xylella fastidiosa* in oleander leaf scorch disease. Phytopathology 89:53–58.

3180. Purohit, S. S., ed. 1985. Hormonal regulation of plant growth and development. M. Nijhoff/W. Junk, Dordrecht, Boston. 412 pp.

3181. Purvis, W. 2000. Lichens. Smithsonian Inst. Press, Washington, DC. 112 pp.

3182. Pusey, P. L. 1989. Availability and dispersal of ascospores and conidia of *Botryosphaeria* in peach orchards. Phytopathology 79:635–639.

3183. Pusey, P. L. 1989. Influence of water stress on susceptibility of nonwounded peach bark to *Botryosphaeria dothidea*. Plant Dis. 73:1000–1003.

3184. Pusey, P. L. 1993. Role of *Botryosphaeria* species in peach tree gummosis on the basis of differential isolation from outer and inner bark. Plant Dis. 77:170–174.

3185. Pusey, P. L., and Bertrand, P. F. 1993. Seasonal infection of nonwounded peach bark by *Botryosphaeria dothidea*. Phytopathology 83:825–829.

3186. Puterka, G. J., Scorza, R., and Brown, M. W. 1993. Reduced incidence of lesser peachtree borer and Leucostoma canker in peach-almond hybrids. J. Am. Soc. Hortic. Sci. 118:864–867.

3187. Putterill, V. A. 1922. The biology of *Schizophyllum commune* Fries with special reference to its parasitism. S.Afr. Dep. Agric. Sci. Bull. 25. 35 pp.

3188. Qiu, B., Li, H., Shi, C., and Jin, K. 1998. Amplification of phytoplasma 16S rDNA from 20 infected plants in China and their RFLP analysis. Sci. Silvae Sinicae 34:67–74.

3189. Quamme, H. A. 1985. Avoidance of freezing injury in woody plants by deep supercooling. Acta Hortic. No. 168:11–30.

3190. Quaroni, S., Sardi, P., and Locci, R. 1980. Disseccamenti apicali in *Acer pseudoplatanus* associati a *Diaporthe eres* Nits. Riv. Patol. Veg. 16:109–115.

3191. Quiaoit, A. R., and Fulton, R. W. 1966. A disease of *Rhamnus cathartica* caused by cucumber mosaic virus. Plant Dis. Rep. 50:613–614.

3192. Raabe, R. D. 1979. Resistance or susceptibility of certain plants to Armillaria root rot. Univ. Calif. Div. Agric. Sci. Leafl. 2591. 11 pp.

3193. Raabe, R. D. 1985. Fusarium wilt of *Hebe* species. Plant Dis. 69:450–451.

3194. Raabe, R. D., and Gardner, M. W. 1972. Scab of pyracantha, loquat, toyon, and kageneckia. Phytopathology 62:914–916.

3195. Rabiti, A. L., and Marani, F. 1992. Alfalfa mosaic virus (AMV) isolated from *Cercis siliquastrum*. Phytopathol. Medit. 31:71–76.

3196. Rack, K., and Scheidemann, U. 1987. On the succession and pathogenic properties of fungi inhabiting pine needles. Eur. J. For. Pathol. 17:102–109.

3197. Radosevich, S. R., Roncoroni, E. J., Conrad, S. G., and McHenry, W. B. 1980. Seasonal tolerance of six coniferous species to eight foliage-active herbicides. For. Sci. 26:3–9.

3198. Ragazzi, A., Fedi, I. D., and Moricca, S. 1990. Colonization pattern of *Quercus cerris* by *Diplodia mutila* and *Phomopsis quercina*. Phytopathol. Medit. 29:209–212.

3199. Ragazzi, A., Moricca, S., Capretti, P., and Dellavalle, I. 1999. Endophytic presence of *Discula quercina* on declining *Quercus cerris*. J. Phytopathol. 147:437–440.

3200. Ragazzi, A., Moricca, S., Capretti, P., and Dellavalle, I. 1999. Infection ability of ascospores and conidia of *Apiognomonia quercina* on oaks. Z. Pflanzenkr. Pflanzenschutz 106:490–494.

3201. Ragazzi, A., Moricca, S., Capretti, P., and Dellavalle, I. 2000. Analysis of *Discula quercina* isolates from *Quercus* spp. Z. Pflanzenkr. Pflanzenschutz 107:170–175.

3202. Ragazzi, A., Moricca, S., Capretti, P., Dellavalle, I., Mancini, F., and Turco, E. 2001. Endophytic fungi in *Quercus cerris*: isolation frequency in relation to phenological phase, tree health and the organ affected. Phytopathol. Medit. 40:165–171.

3203. Ragazzi, A., Moricca, S., and Dellavalle, I. 1999. Epidemiological aspects of *Discula quercina* on oak: inoculum density and conidia production. Z. Pflanzenkr. Pflanzenschutz 106:501–506.

3204. Ragazzi, A., Moricca, S., and Dellavalle, I. 1999. Interactions between *Quercus* spp. and *Diplodia mutila* under water stress conditions. Z. Pflanzenkr. Pflanzenschutz 106:495–500.

3205. Ragazzi, A., Moricca, S., and Dellavalle, I. 1999. Water stress and the development of cankers by *Diplodia mutila* on *Quercus robur*. J. Phytopathol. 147:425–428.

3206. Raggi, V. 1987. Water relations in peach leaves infected by *Taphrina deformans* (peach leaf curl)—diffusive resistance, total transpiration and water potential. Physiol. Mol. Plant Pathol. 30:109–120.

3207. Raggi, V. 1995. CO_2 assimilation, respiration and chlorophyll fluorescence in peach leaves infected by *Taphrina deformans*. Physiol. Plant. 93:540–544.

3208. Raio, A., Zoina, A., and Moore, L. W. 1997. The effect of solar heating of soil on natural and inoculated agrobacteria. Plant Pathol. 46:320–328.

3209. Rajanna, L., and Shivamurthy, G. R. 2001. Occurrence of graniferous tracheary elements in the haustorium of *Cassytha filiformis* Linn., a stem parasite of Lauraceae. Taiwania 46:40–48.

3210. Rajaseger, G., Tan, H. T. W., Turner, I. M., Saw, L. G., and Kumar, P. P. 1999. Random amplified polymorphic DNA variation among and within selected *Ixora* (Rubiaceae) populations and mutants. Ann. Bot. 84:253–257.

3211. Rajashekar, C. B., and Burke, M. J. 1996. Freezing characteristics of rigid plant tissues. Development of cell tension during extracellular freezing. Plant Physiol. 111:597–603.

3212. Raju, B. C., Chen, T. A., and Varney, E. H. 1976. Mycoplasmalike organisms associated with a witches'-broom disease of *Cornus amomum*. Plant Dis. Rep. 60:462–464.

588

3213. Ram, C. 1993. Cultural characteristics, sporulation and virulence of *Botryodiplodia theobromae* "strain," causal agent of leaf blight of coconut palm (*Cocos nucifera*). Fitopatol. Brasil. 18:143–146.

3214. Ramsdell, D. C., Adler, V. A., and Kesner, C. R. 1993. Prune brown line disease occurrence in declining 'Stanley' plum orchards in Michigan and factors related to its incidence. HortTechnology 3:325–329.

3215. Ramsdell, D. C., Nelson, J. W., and Myers, R. L. 1975. Mummy berry disease of highbush blueberry: epidemiology and control. Phytopathology 65:229–232.

3216. Ramsfield, T. D., Becker, E. M., Rathlef, S. M., Tang, Y., Vrain, T. C., Shamoun, S. F., and Hintz, W. E. 1996. Geographic variation of *Chondrostereum purpureum* detected by polymorphisms in the ribosomal DNA. Can. J. Bot. 74:1919–1929.

3217. Ramstedt, M., Hurtado, S., and Astrom, B. 2002. Pathotypes of Melampsora rust on *Salix* in short-rotation forestry plantations. Plant Pathol. 51:185–190.

3218. Rane, K. K., and Tattar, T. A. 1987. Pathogenicity of blue-stain fungi associated with *Dendroctonus terebrans*. Plant Dis. 71:879–883.

3219. Ranney, T. G., Grand, L. F., and Knighten, J. L. 1995. Susceptibility of cultivars and hybrids of kousa dogwood to dogwood anthracnose and powdery mildew. J. Arboric. 21:11–16.

3220. Ransom, J. K., Musselman, L. J., Worsham, A. D, and Parker, C., eds. 1991. Proceedings of the 5th international symposium of parasitic weeds, Nairobi, Kenya. CIMMYT, Nairobi. 550 pp.

3221. Rappaz, F. 1987. Taxonomy and nomenclature of Diatrypaceae with eight-spored asci. Mycol. Helvet. 2:285–648.

3222. Rautenberg, E. 1973. Untersuchungen an *Exobasidium*-Gallen von *Rhododendron simsii* Planch. I. Der Erreger *Exobasidium japonicum* Shir. und Begleitpilze. II. Bedingungen der Gallentwicklung und die Wechselbeziehung zwischen Wirtspflanze und Cecidie. III. Fluoreszierende Phenolderivate in Gallen und Blättern. Phytopathol. Z. 78:2–13, 121–133, 214–226.

3223. Rayachhetry, M. B., Webb, R. S., Kimbrough, J. W., and Miller, T. 1995. Haustorial morphology of *Cronartium conigenum* in naturally infected cones of three *Pinus* species from Guatemala. Eur. J. For. Pathol. 25:152–158.

3224. Rayachhetry, M. B., Webb, R. S., Miller, T., and Kimbrough, J. W. 1995. Histology of *Pinus maximinoi* cones infected by *Cronartium conigenum*. Eur. J. For. Pathol. 25:100–108.

3225. Raychaudhuri, S. P., and Maramorosch, K., eds. 1996. Forest trees and palms: diseases and control. Science Pub., Lebanon, NH. 336 pp.

3226. Raymond, F. L., and Reid, J. 1961. Dieback of balsam fir in Ontario. Can. J. Bot. 39:233–251.

3227. Rayner, A. D. M., and Boddy, L. 1988. Fungal communities in the decay of wood. Adv. Microb. Ecol. 10:115–166.

3228. Rayner, A. D. M., and Boddy, L. 1988. Fungal decomposition of wood: its biology and ecology. Wiley, Chichester, UK. 587 pp.

3229. Read, D. J., Lewis, D. H., and Alexander, I. J., eds. 1992. Mycorrhizas in ecosystems. CAB International, Wallingford, UK. 419 pp.

3230. Read, D. J., and Perez-Moreno, J. 2003. Mycorrhizas and nutrient cycling in ecosystems—a journey towards relevance? New Phytol. 157:475–492.

3231. Read, M. 1989. Bromeliads threatened by trade. Kew Mag. 6:22–29.

3232. Read, R. A., and Sprackling, J. A. 1981. Hail damage variation by seed source in a ponderosa pine plantation. USDA For. Serv. Res. Note RM-410. 6 pp.

3233. Ream, W., and Gelvin, S. B., eds. 1996. Crown gall: advances in understanding interkingdom gene transfer. APS Press, St. Paul, MN. 145 pp.

3234. Rebertus, A. J., Shifley, S. R., Richards, R. H., and Roovers, L. M. 1997. Ice storm damage to an old-growth oak-hickory forest in Missouri. Am. Midl. Nat. 137:48–61.

3235. Reddick, B. B. 1989. Isolation and partial characterization of a tobamovirus from flowering dogwood in Tennessee. Plant Dis. 73:174–176.

3236. Reddick, B. B., Barnett, O. W., and Baxter, L. W. Jr. 1979. Isolation of cherry leafroll, tobacco ringspot, and tomato ringspot viruses from dogwood trees in South Carolina. Plant Dis. Rep. 63:529–532.

3237. Reddick, D. 1911. The black rot disease of grapes. Cornell Univ. Agric. Exp. Stn. Bull. 293:287–364.

3238. Reddy, M. R. S. 1997. Sources of resistance to bacterial canker in citrus. J. Mycol. Plant Pathol. 27:80–81.

3239. Reddy, S., Spencer, J. A., and Newman, S. E. 1992. Leaflet surfaces of blackspot-resistant and susceptible roses and their reactions to fungal invasion. HortScience 27:133–135.

3240. Redhead, S. A. 1975. The genus *Cristulariella*. Can. J. Bot. 53:700–707.

3241. Redlin, S. C. 1991. *Discula destructiva* sp. nov., cause of dogwood anthracnose. Mycologia 83:633–642.

3242. Redlin, S. C. 1992. Scanning electron microscopy of the conidioma of *Discula destructiva* (Coelomycetes). Mycologia 84:257–260.

3243. Redlin, S. C., and Rossman, A. Y. 1991. *Cryptodiaporthe corni* (Diaporthales), cause of Cryptodiaporthe canker of pagoda dogwood. Mycologia 83:200–209.

3244. Redlin, S. C., and Stack, R. W. 1988. *Gnomoniella fraxini* sp. nov., teleomorph of the ash anthracnose fungus and its connection to *Discula fraxinea* comb. nov. Mycotaxon 32:175–198.

3245. Redlin, S. C., and Stack, R. W. 2001. Cryptodiaporthe canker of pagoda dogwood in North Dakota. Plant Dis. 85:1290.

3246. Redmond, D. R. 1955. Studies in forest pathology. XV. Rootlets, mycorrhiza, and soil temperature in relation to birch dieback. Can. J. Bot. 33:595–627.

3247. Regenmortel, M. H. V. van, Fauquet, C. M., and Bishop, D. H. L., eds. 2000. Virus taxonomy: classification and nomenclature of viruses. Seventh report of the International Committee on Taxonomy of Viruses. Academic Press, San Diego. 1162 pp.

3248. Rehfeldt, G. E. 1992. Breeding strategies for *Larix occidentalis*: adaptations to the biotic and abiotic environment in relation to improving growth. Can. J. For. Res. 22:5–13.

3249. Rehfeldt, G. E. 1995. Genetic variation, climate models and the ecological genetics of *Larix occidentalis*. For. Ecol. Manag. 78:21–37.

3250. Rehner, S. A., and Uecker, F. A. 1994. Nuclear ribosomal internal transcribed spacer phylogeny and host diversity in the coelomycete *Phomopsis*. Can. J. Bot. 72:1666–1674.

3251. Reich, R. W., and Kamp, B. J. van der. 1993. Frost, canker, and dieback of Douglas-fir in the central interior of British Columbia. Can. J. For. Res. 23:373–379.

3252. Reid, J., and Cain, R. F. 1962. Studies on the organisms associated with "snow-blight" of conifers in North America. I. A new genus of the Helotiales. II. Some species of the genera *Phacidium, Lophophacidium, Sarcotrochila*, and *Hemiphacidium*. Mycologia 54:194–200, 481–497.

3253. Reid, J., and Dowsett, J. A. 1990. On *Dicarpella, Sphaerognomonia*, and *Apiosporopsis*. Can. J. Bot. 68:2398–2407.

3254. Reid, J., and Funk, A. 1966. The genus *Atropellis* and a new genus of the Helotiales associated with branch cankers of western hemlock. Mycologia 58:417–439.

3255. Reid, N., Smith, M. S., and Yan, Z. 1995. Ecology and population biology of mistletoes. Pages 285–310 in: Forest canopies. M. D. Lowman and N. M. Nadkarni, eds. Academic Press, San Diego.

3256. Reid, N., Yan, Z. G., and Fittler, J. 1994. Impact of mistletoes (*Amyema miquelii*) on host (*Eucalyptus blakelyi* and *Eucalyptus melliodora*) survival and growth in temperate Australia. For. Ecol. Manag. 70:55–65.

3257. Reilly, C. C., Hotchkiss, M. W., and Hendrix, F. F. Jr. 1998. Phytophthora shuck and kernel rot, a new disease of pecan caused by *Phytophthora cactorum*. Plant Dis. 82:347–349.

3258. Reilly, C. C., Wood, B. W., and Hotchkiss, M. W. 1996. Relative susceptibility of pecan cultivars to zonate leaf spot. HortScience 31:986–987.

3259. Reinert, J. A. 1977. Field biology and control of *Haplaxius crudus* on St. Augustine grass and Christmas palm. J. Econ. Entomol. 70:54–56.

3260. Rensburg, J. C. J. van, Labuschagne, N., and Nemec, S. 2001. Occurrence of *Fusarium*-produced naphthazarins in citrus trees and sensitivity of rootstocks to isomarticin in relation to citrus blight. Plant Pathol. 50:258–265.

3261. Reuschel, D., Mattheck, C., and Althaus, C. 1998. The mechanical effect of climbing plants on host trees. Allg. Forst Jagdz. 169:87–91.

3262. Reuter, D. J., and Robinson, J. B., eds. 1997. Plant analysis: an interpretation manual. 2nd ed. CSIRO Pub., Collingwood, Australia. 572 pp.

3263. Rex, E. G., and Walter, J. M. 1946. The canker stain disease of planetrees, with recommendations for controlling it in New Jersey. N.J. Dep. Agric. Circ. 360. 23 pp.

3264. Rexrode, C. O. 1976. Insect transmission of oak wilt. J. Arboric. 2:61–66.

3265. Rey, L., Sadik, A., Fer, A., and Renaudin, S. 1991. Trophic relations of the dwarf mistletoe *Arceuthobium oxycedri* with its host *Juniperus oxycedrus*. J. Plant Physiol. 138:411–416.

3266. Rey, L., Sadik, A., Fer, A., and Renaudin, S. 1992. Étude de quelques aspects du métabolisme carboné et azoté chez l'*Arceuthobium oxycedri*, gui nain du genévrier. Can. J. Bot. 70:1709–1716.

3267. Reynolds, D. R. 1978. Foliicolous Ascomycetes 1: The capnodiaceous genus *Scorias*, reproduction. Nat. Hist. Mus. Los Angeles County, Contrib. Sci. No. 288. 16 pp.

3268. Reynolds, D. R. 1979. Foliicolous Ascomycetes. 3. The stalked capnodiaceous species. Mycotaxon 8:417–445.

3269. Reynolds, D. R. 1989. Foliicolous fungi 8. *Capnodium* in California, USA. Mycotaxon 34:197–216.

3270. Reynolds, D. R. 1998. Capnodiaceous sooty mold phylogeny. Can. J. Bot. 76:2125–2130.

3271. Reynolds, D. R. 1999. *Capnodium citri*: the sooty mold fungi comprising the taxon concept. Mycopathologia 148:141–147.

3272. Reynolds, K. M., Benson, D. M., and Bruck, R. I. 1985. Epidemiology of Phytophthora root rot of Fraser fir: root colonization and inoculum production. Phytopathology 75:1004–1009.

3273. Rhoades, R. W. 1999. Ice storm damage in a small valley in southwestern Virginia. Castanea 64:243–251.

3274. Rhoads, A. S. 1923. The formation and pathological anatomy of frost rings in conifers injured by late frosts. USDA Tech. Bull. 1131. 15 pp.

3275. Rhoads, A. S. 1943. Lightning injury to pine and oak trees in Florida. Plant Dis. Rep. 27:556–557.

3276. Rhoads, A. S. 1950. Clitocybe root rot of woody plants in the southeastern United States. USDA Circ. 853. 25 pp.

3277. Rhoads, A. S. 1956. The occurrence and destructiveness of Clitocybe root rot of woody plants in Florida. Lloydia 19:193–240.

3278. Ribeiro, O. K. 1978. A sourcebook of the genus *Phytophthora*. Cramer, Lehre. 417 pp.

3279. Rich, S., and Walton, G. S. 1979. Decline of curbside sugar maples in Connecticut. J. Arboric. 5:265–268.

3280. Richards, W. C. 1994. Cerato-ulmin: a wilt toxin of Dutch elm disease. Can. For. Serv. Inf. Rep. O-X-432. 37 pp.

3281. Richardson, K. S., and Kamp, B. J. van der. 1972. The rate of upward advance and intensification of dwarf mistletoe on immature western hemlock. Can. J. For. Res. 2:313–316.

3282. Richmond, B. G. 1932. A Diaporthe canker of American elm. Science 75:110–111.

3283. Richter, A., Popp, M., Mensen, R., Stewart, G. R., and Willert, D. J. von. 1995. Heterotrophic carbon gain of the parasitic angiosperm *Tapinanthus oleifolius*. Austral. J. Plant Physiol. 22:537–544.

3284. Ridings, W. H., and Marlatt, R. B. 1976. Sphaeropsis witches' broom of *Nerium oleander*. Proc. Fla. State Hortic. Soc. 89:302–303.

3285. Riffle, J. W. 1973. Histopathology of *Pinus ponderosa* ectomycorrhizae infected with a *Meloidogyne* species. Phytopathology 63:1034–1040.

3286. Riffle, J. W. 1978. Development of cankers on *Ulmus pumila* related to month of inoculation with *Botryodiplodia hypodermia*. Phytopathology 68:1115–1119.

3287. Riffle, J. W. 1981. Recovery of *Herpobasidium deformans* basidiospores from *Lonicera tatarica* leaves overwintered in nursery beds. (Abstr.) Phytopathology 71:251–252.

3288. Riffle, J. W., and Kuntz, J. W. 1967. Pathogenicity and host range of *Meloidogyne ovalis*. Phytopathology 57:104–107.

3289. Riffle, J. W., and Peterson, G. W. 1986. Thyronectria canker of honeylocust: influence of temperature and wound age on disease development. Phytopathology 76:313–316.

3290. Riffle, J. W., and Peterson, G. W., tech. coords. 1986. Diseases of trees in the Great Plains. USDA For. Serv. Gen. Tech. Rep. RM-129. 149 pp.

3291. Riffle, J. W., Sharon, E. M., and Harrell, M. O. 1984. Incidence of *Fomes fraxinophilus* on green ash in Nebraska woodlands. Plant Dis. 68:322–324.

3292. Riggs, W., and Mims, C. W. 2000. Ultrastructure of chlamydospore development in the plant pathogenic fungus *Thielaviopsis basicola*. Mycologia 92:123–129.

3293. Rijkers, A. J. M., Hiemstra, J. A., and Bollen, G. J. 1992. Formation of microsclerotia of *Verticillium dahliae* in petioles of infected ash trees. Neth. J. Plant Pathol. 98:261–264.

3294. Riker, A. J., Keitt, G. W., Hildebrand, E. M., and Banfield, W. M. 1934. Hairy root, crown gall, and other malformations at the unions of piece-root-grafted apple trees and their control. J. Agric. Res. 48:913–939.

3295. Riley, C. G. 1952. Studies in forest pathology. IX. *Fomes igniarius* decay of poplar. Can. J. Bot. 30:710–734.

3296. Riley, C. G. 1953. Hail damage in forest stands. For. Chron. 29:139–143.

3297. Rillig, M. C., Wright, S. F., Nichols, K. A., Schmidt, W. F., and Torn, M. S. 2001. Large contribution of arbuscular mycorrhizal fungi to soil carbon pools in tropical forest soils. Plant Soil 233:167–177.

3298. Rink, G., and Budelsky, C. A., eds. 1989. Seventh Central Hardwood Forest Conference. Proceedings. USDA For. Serv. Gen. Tech. Rep. NC-132. 313 pp.

3299. Rishbeth, J. 1950, 1951. Observations on the biology of *Fomes annosus*, with particular reference to East Anglian pine plantations.

I. The outbreaks of disease and ecological status of the fungus. II. Spore production, stump infection, and saprophytic activity in stumps. Ann. Bot., n.s., 14:365–383; 15:1–21.

3300. Rishbeth, J. 1978. Effects of soil temperature and atmosphere on growth of *Armillaria* rhizomorphs. Trans. Br. Mycol. Soc. 70:213–220.

3301. Rishbeth, J. 1978. Infection foci of *Armillaria mellea* in first-rotation hardwoods. Ann. Bot. 42:1131–1139.

3302. Rishbeth, J. 1988. Stump infection by *Armillaria* in first-rotation conifers. Eur. J. For. Pathol. 18:401–408.

3303. Ristic, Z., and Ashworth, E. N. 1995. Response of xylem ray parenchyma cells of supercooling wood tissues to freezing stress: microscopic study. Int. J. Plant Sci. 156:784–792.

3304. Rizzo, D. M., Blanchette, R. A., and May, G. 1995. Distribution of *Armillaria ostoyae* genets in a *Pinus resinosa–Pinus banksiana* forest. Can. J. Bot. 73:776–787.

3305. Rizzo, D. M., Garbelotto, M., Davidson, J. M., Slaughter, G. W., and Koike, S. T. 2002 *Phytophthora ramorum* as the cause of extensive mortality of *Quercus* spp. and *Lithocarpus densiflorus* in California. Plant Dis. 86:205–214.

3306. Rizzo, D. M., and Harrington, T. C. 1992. Nuclear migration in diploid-haploid pairings of *Armillaria ostoyae*. Mycologia 84:863–869.

3307. Rizzo, D. M., and Slaughter, G. W. 2001. Root disease and canopy gaps in developed areas of Yosemite Valley, California. For. Ecol. Manag. 146:159–167.

3308. Roane, M. K., Griffin, G. J., and Elkins, J. R. 1986. Chestnut blight, other Endothia diseases, and the genus *Endothia*. APS Press, St. Paul, MN. 53 pp.

3309. Roane, M. K., Stipes, R. J., Phipps, P. M., and Miller, O. K. Jr. 1974. *Endothia gyrosa*, causal pathogen of pin oak blight. Mycologia 66:1042–1047.

3310. Robbins, J. 2001. The powdery mildews as gall-inducers. Cecidology 16:13–22.

3311. Robbins, K. 1984. Annosus root disease in eastern conifers. USDA For. Serv. For. Insect Dis. Leafl. 76. Internet pub.

3312. Robbins, R. T. 1993. Distribution of *Xiphinema americanum* and related species in North America. J. Nematol. 25:344–348.

3313. Roberts, P. 1999. Rhizoctonia-forming fungi. R. Bot. Gard., Kew, UK. 239 pp.

3314. Robin, C., and Desprez-Loustau, M. L. 1998. Testing variability in pathogenicity of *Phytophthora cinnamomi*. Eur. J. Plant Pathol. 104:465–475.

3315. Robinson, R. M., and Morrison, D. J. 2001. Lesion formation and host response to infection by *Armillaria ostoyae* in the roots of western larch and Douglas-fir. For. Pathol. 31:371–385.

3316. Robitaille, G., Boutin, R., and Lachance, D. 1995. Effects of soil freezing stress on sap flow and sugar content of mature sugar maples (*Acer saccharum*). Can. J. For. Res. 25:577–587.

3317. Robitaille, G., Boutin, R., and Lachance, D. 1996. Effect of deep soil frost on sugar maple (*Acer saccharum* Marsh.) leaf dry weight and area. Pages 35–41 in: Air pollutants and natural stresses. IUFRO proceedings, 16th international meeting for specialists in air pollution effects on forest ecosystems, Fredericton, New Brunswick, Canada, 1994. R. Cox, K. Percy, and K. Jensen, eds. Can. For. Serv., Fredericton, NB.

3318. Rocchini, L. A., Lewis, K. J., Lindgren, B. S., and Bennett, R. G. 1999. Association of pitch moths (Lepidoptera: Sesiidae and Pyralidae) with rust diseases in a lodgepole pine provenance trial. Can. J. For. Res. 29:1610–1614.

3319. Rockett, T. R., and Kramer, C. L. 1974. Periodicity and total spore germination by lignicolous basidiomycetes. Mycologia 66:817–829.

3320. Rodriguez, A. 1995. Airplants: a survey of the genus *Tillandsia*. New Plantsman 2:72–88.

3321. Rogers, J. D., and Callan, B. E. 1986. *Xylaria polymorpha* and its allies in continental United States. Mycologia 78:391–400.

3322. Rogers, J. D., and Ju, Y. 1996. *Entoleuca mammata* comb. nov. for *Hypoxylon mammatum* and the genus *Entoleuca*. Mycotaxon 59:441–448.

3323. Rogers, J. D., and Ju, Y. 1998. The genus *Kretzschmaria*. Mycotaxon 68:345–393.

3324. Rogers, S. O., Holdenrieder, O., and Sieber, T. N. 1999. Intraspecific comparisons of *Laetiporus sulphureus* isolates from broadleaf and coniferous trees in Europe. Mycol. Res. 103:1245–1251.

3325. Roll-Hansen, F. 1989. *Phacidium infestans*. A literature review. Eur. J. For. Pathol. 19:237–250.

3326. Roll-Hansen, F., and Roll-Hansen, H. 1981. *Melampsoridium* on *Alnus* in Europe. *M. alni* conspecific with *M. betulinum*. Eur. J. For. Pathol. 11:77–87.

3327. Ronald, P. S., St. Pierre, R. G., and Bains, P. S. 2001. Resistance to *Entomosporium mespili* among cultivars of saskatoon, *Amelanchier alnifolia*. Can. J. Plant Pathol. 23:391–402.

3328. Roncadori, R. S. 1962. The nutritional competition between *Hypoxylon punctulatum* and *Ceratocystis fagacearum*. Phytopathology 52:498–502.

3329. Roquebert, M.-F., and Fayret, J. 1982. *Marssoniella juglandis*: anamorphe de *Gnomonia leptostyla*. Can. J. Bot. 60:1320–1329.

3330. Rosen, H. R. 1935. Rose blast induced by *Phytomonas syringae*. J. Agric. Res. 51:235–241.

3331. Rosen, P. M., Good, G. L., and Steponkus, P. L. 1983. Desiccation injury and direct freezing injury to evergreen azaleas: a comparison of cultivars. J. Am. Soc. Hortic. Sci. 108:28–31.

3332. Rosenberger, D. A., Cummins, J. N., and Gonsalves, D. 1989. Evidence that tomato ringspot virus causes apple union necrosis and decline: symptom development in inoculated apple trees. Plant Dis. 73:262–265.

3333. Rosenberger, D. A., and Jones, A. L. 1977. Seasonal variation in infectivity of inoculum from X-diseased peach and chokecherry plants. Plant Dis. Rep. 61:1022–1024.

3334. Rosenberger, D. A., and Jones, A. L. 1978. Leafhopper vectors of the peach X-disease pathogen and its seasonal transmission from chokecherry. Phytopathology 68:782–790.

3335. Ross, E. W. 1964. Cankers associated with ash die-back. Phytopathology 54:272–275.

3336. Ross, E. W. 1966. Ash dieback: etiological and developmental studies. State University Coll. For. (Syracuse, NY) Tech. Pub. 88. 80 pp.

3337. Ross, E. W. 1967. Association of *Cylindrocladium scoparium* with mortality in a 27-year-old yellow-poplar plantation. Plant Dis. Rep. 51:38–39.

3338. Ross, E. W. 1970. Sand pine root rot—pathogen: *Clitocybe tabescens*. J. For. 68:156–158.

3339. Ross, E. W. 1973. *Fomes annosus* in the southeastern United States. USDA Tech. Bull. 1459. 26 pp.

3340. Ross, W. D. 1976. Fungi associated with root diseases of aspen in Wyoming. Can. J. Bot. 54:734–744.

3341. Rossman, A. Y. 1979. A preliminary account of the taxa described in *Calonectria*. Mycotaxon 8:485–558.

3342. Rossman, A. Y. 1989. A synopsis of the *Nectria cinnabarina*–group. Mem. N.Y. Bot. Gard. 49:253–265.

3343. Rossman, A. Y., Castlebury, L. A., Adams, G. C., and Putnam, M. L. 2002. *Phlyctema vagabunda* isolated from coin canker of ash trees in Michigan. Plant Dis. 86:442.

3344. Rossman, A. Y., Palm, M. E., and Spielman, L. J. 1987. A literature guide for the identification of plant pathogenic fungi. APS Press, St. Paul, MN. 252 pp.

3345. Rossman, A. Y., Samuels, G. J., Rogerson, C. T. and Lowen, R. 1999. Genera of Bionectriaceae, Hypocreaceae and Nectriaceae (Hypocreales, Ascomycetes). Stud. Mycol. No. 42. 248 pp.

3346. Rosso, P., and Hansen, E. 1998. Tree vigour and the susceptibility of Douglas-fir to Armillaria root disease. Eur. J. For. Pathol. 28:43–52.

3347. Rotem, J. 1994. The genus *Alternaria*: biology, epidemiology, pathogenicity. APS Press, St. Paul, MN. 326 pp.

3348. Roth, E. R., Hepting, G. H., and Toole, E. R. 1959. Sapstreak disease of sugar maple and yellow-poplar in North Carolina. USDA For. Serv. Southeast. For. Exp. Stn. Res. Note 134. 2 pp.

3349. Roth, E. R., and Sleeth, B. 1939. Butt rot in unburned sprout oak stands. USDA Tech. Bull. 684. 42 pp.

3350. Roth, L. F. 1959. Perennial infection of ponderosa pine by *Elytroderma deformans*. For. Sci. 5:182–191.

3351. Roth, L. F. 1971. Dwarf mistletoe damage to small ponderosa pines. For. Sci. 17:373–380.

3352. Roth, L. F. 1974. Juvenile susceptibility of ponderosa pine to dwarf mistletoe. Phytopathology 64:689–692.

3353. Roth, L. F. 2001. Dwarf mistletoe–induced mortality in Northwest ponderosa pine growing stock. West. J. Appl. For. 16:136–141.

3354. Roth, L. F., and Barrett, J. W. 1985. Response of dwarf mistletoe–infested ponderosa pine to thinning: 2. Dwarf mistletoe propagation. USDA For. Serv. Res. Pap. PNW-331. 20 pp.

3355. Roth, L. F., Harvey, R. D. Jr., and Kliejunas, J. T. 1987. Port-Orford-cedar root disease. USDA For. Serv. Pac. Northwest Region R6 FPM PR 010 91. Internet pub.

3356. Rott, M. E., and Jelkmann, W. 2001. Characterization and detection of several filamentous viruses of cherry: adaptation of an alternative cloning method (DOP-PCR), and modification of an RNA extraction protocol. Eur. J. Plant Pathol. 107:411–420.

3357. Roux, J., Myburg, H., Wingfield, B. D., and Wingfield, M. J. 2003. Biological and phylogenetic analyses suggest that two *Cryphonec-tria* spp. cause cankers of *Eucalyptus* in Africa. Plant Dis. 87:1329–1332.

3358. Roux, J., Wingfield, M. J., Bouillet, J. P., Wingfield, B. D., and Alfenas, A. C. 2000. A serious new wilt disease of *Eucalyptus* caused by *Ceratocystis fimbriata* in central Africa. For. Pathol. 30:175–184.

3359. Rowan, S. J. 1960. The susceptibility of twenty-three tree species to black root rot. Plant Dis. Rep. 44:646–647.

3360. Rowhani, A., and Mircetich, S. M. 1988. Pathogenicity on walnut and serological comparisons of cherry leafroll virus strains. Phytopathology 78:817–822.

3361. Ruark, G. A., Mader, D. L., Veneman, P. L. M., and Tattar, T. A. 1983. Soil factors related to urban sugar maple decline. J. Arboric. 9:1–6.

3362. Rubin, B. D., and Manion, P. D. 2001. Landscape-scale forest structure in northern New York and potential successional impacts of the 1998 ice storm. For. Chron. 77:613–618.

3363. Rubio-Cabetas, M. J., Minot, J. C., Voisin, R., and Esmenjaud, D. 2001. Interaction of root-knot nematodes (RKN) and the bacterium *Agrobacterium tumefaciens* in roots of *Prunus cerasifera*: evidence of the protective effect of the *Ma RKN* resistance genes against expression of crown gall symptoms. Eur. J. Plant Pathol. 107:433–441.

3364. Rudolph, B. A. 1931. Verticillium hadromycosis. Hilgardia 5:197–361.

3365. Rudolph, K., Burr, T. J., Mansfield, J. W., Stead, D., Vivian, A., and Kietzell, J. von, eds. 1997. *Pseudomonas syringae* pathovars and related pathogens. Kluwer Academic, Dordrecht. 663 pp.

3366. Ruehle, G. D. 1941. Poinsettia scab caused by *Sphaceloma*. Phytopathology 31:947–948.

3367. Ruehle, J. L. 1964. Plant-parasitic nematodes associated with pine species in southern forests. Plant Dis. Rep. 48:60–61.

3368. Ruehle, J. L. 1967. Distribution of plant-parasitic nematodes associated with forest trees of the world. USDA For. Serv. Southeast. For. Exp. Stn. 156 pp.

3369. Ruehle, J. L. 1968. Plant-parasitic nematodes associated with southern hardwood and coniferous forest trees. Plant Dis. Rep. 52:837–839.

3370. Rulamort, M. 1986. Remarques taxonomiques et nomenclaturales sur quelques micromycètes. Bull. Soc. Bot. Centre-Ouest, n.s., 17: 191–192.

3371. Runion, G. B., and Kelley, W. D. 1993. Characterization of a binucleate *Rhizoctonia* species causing foliar blight of loblolly pine. Plant Dis. 77:754–755.

3372. Rutherford, T. A., Mamiya, Y., and Webster, J. M. 1990. Nematode-induced pine wilt disease: factors influencing its occurrence and distribution. For. Sci. 36:145–155.

3373. Ryvarden, L. 1976, 1978. The Polyporaceae of North Europe. Vol. 1: *Albatrellus–Incrustoporia*. Vol. 2: *Inonotus–Tyromyces*. Fungiflora, Oslo. 577 pp.

3374. Ryvarden, L., and Gilbertson, R. L. 1993, 1994. European polypores. Part 1: *Abortiporus–Lindtneria*. Part 2: *Meripilus–Tyromyces*. Fungiflora, Oslo. 743 pp.

3375. Saccardi, A., Bonetti, V., Melegatti, A., and Cristanini, M. 1998. Occurrence of *Erwinia nigrifluens* on English walnut (*Juglans regia*) in the Veneto region (northern Italy). J. Plant Pathol. 80:63–65.

3376. Sachs, I. B., Ward, J. C., and Kinney, R. E. 1975. Scanning electron microscopy of bacterial wetwood and normal heartwood in poplar trees. Pages 453–459 in: Scanning electron microscopy/1974. O. Johari and I. Corvin, eds. IIT Res. Inst., Chicago.

3377. Sachslehner, A., Nidetzky, B., Kulbe, K. D., and Haltrich, D. 1998. Induction of mannanase, xylanase, and endoglucanase activities in *Sclerotium rolfsii*. Appl. Environ. Microbiol. 64:594–600.

3378. Sadik, A., Rey, L., and Renaudin, S. 1986. Le système endophytique d'*Arceuthobium oxycedri*. I. Organisation, étude cytologique et cytochimique. II. Aspects ultrastructuraux des zones de contact entre les tissus de l'hôte et du parasite. Can. J. Bot. 64:1104–1111, 2778–2784.

3379. Saenz, G. S., and Taylor, J. W. 1999. Phylogeny of the Erysiphales (powdery mildews) inferred from internal transcribed spacer ribosomal DNA sequences. Can. J. Bot. 77:150–168.

3380. Saikkonen, K., Faeth, S. H., Helander, M. L., and Sullivan T. J. 1998. Fungal endophytes: a continuum of interactions with host plants. Annu. Rev. Ecol. Syst. 29:319–343.

3381. Saito, I., Tamura, O., and Takakuwa, M. 1972. Ascospore dispersal in *Valsa ceratosperma*, the causal fungus of Japanese apple canker. Ann. Phytopathol. Soc. Jap. 38:367–374.

3382. Sakai, A. 1982. Freezing resistance of ornamental trees and shrubs. J. Am. Soc. Hortic. Sci 107:572–581.

3383. Sakamoto, Y., and Kato, A. 2002. Some properties of the bacterial wetwood (watermark) in *Salix sachalinensis* caused by *Erwinia salicis*. IAWA J. 23:179–190.

3384. Sala, A., Carey, E. V., and Callaway, R. M. 2001. Dwarf mistletoe affects whole-tree water relations of Douglas fir and western larch primarily through changes in leaf to sapwood ratios. Oecologia 126:42–52.

3385. Salerno, M. I., Gianinazzi, S., and Gianinazzi, P. V. 2000. Effects on growth and comparison of root tissue colonization patterns of *Eucalyptus viminalis* by pathogenic and nonpathogenic strains of *Fusarium oxysporum*. New Phytol. 146:317–324.

3386. Sallé, G. 1979. Le système endophytique du *Viscum album*: anatomie et fonctionnement des suçoirs secondaires. Can. J. Bot. 57:435–449.

3387. Sallé, G. C., Hariri, E. B., and Andary, C. 1994. Polyphenols and resistance of poplar (*Populus* spp.) to mistletoe (*Viscum album* L.). Acta Hortic. No. 381:756–762.

3388. Salmon, E. S., and Ware, W. M. 1927. Leaf scorch of azalea. Gard. Chron. 81:286–288.

3389. Samils, B., Stepien, V., Lagercrantz, U., Lascoux, M., and Gullberg, U. 2001. Genetic diversity in relation to sexual and asexual reproduction in populations of *Melampsora larici-epitea*. Eur. J. Plant Pathol. 107:871–881.

3390. Sanchez, M. E., Caetano, P., Ferraz, J., and Trapero, A. 2002. Phytophthora disease of *Quercus ilex* in south-western Spain. For. Pathol. 32:5–18.

3391. Sanderlin, R. S., and Heyderich, A. K. I. 2000. Evidence that *Xylella fastidiosa* can cause leaf scorch disease of pecan. Plant Dis. 84:1282–1286.

3392. Sanderlin, R. S., and Grauke, L. J. 1987. Vein spot disease caused by *Gnomonia nerviseda* on pecan in Illinois. Plant Dis. 71:1146.

3393. Sandermann, H. J. 1996. Ozone and plant health. Annu. Rev. Phytopathol. 34:347–366.

3394. Sandermann, H. J., Wellburn, A. R., and Heath, R. L., eds. 1997. Forest decline and ozone: a comparison of controlled chamber and field experiments. Springer, New York. 400 pp.

3395. Sanders, F. E., Mosse, B., and Tinker, P. B., eds. 1975. Endomycorrhizas. Academic Press, New York. 626 pp.

3396. Sanderson, P. G., and Worf, G. L. 1986. Phomopsis shoot blight of Colorado blue spruce. J. Environ. Hortic. 4:134–138.

3397. Sano, K., and Fukazawa, K. 1996. Timing of the occurrence of frost cracks in winter. Trees Struct. Funct. 11:47–53.

3398. Santamour, F. S. Jr. 1976. Resistance to sycamore anthracnose disease in hybrid *Platanus*. Plant Dis. Rep. 60:161–162.

3399. Santamour, F. S. Jr. 1977. The selection and breeding of pest-resistant landscape trees. J. Arboric. 3:146–152.

3400. Santamour, F. S. Jr. 1984. 'Columbia' and 'Liberty' planetrees. HortScience 19:901–902.

3401. Santamour, F. S. Jr. 1984. Wound compartmentalization in cultivars of *Acer*, *Gleditsia*, and other genera. J. Environ. Hortic. 2:123–125.

3402. Santamour, F. S. Jr. 1992. Influence of root-knot nematodes on Verticillium wilt of maples. J. Arboric. 18:298–301.

3403. Santamour, F. S. Jr., and Batzli, J. M. 1990. Host checklist of root-knot nematodes on broad-leaved landscape trees. J. Arboric. 16:162–168.

3404. Santamour, F. S. Jr., Gerhold, H. D., and Little, S., eds. 1976. Better trees for metropolitan landscapes. USDA For. Serv. Gen. Tech. Rep. NE-22. 256 pp.

3405. Santamour, F. S. Jr., and Riedel, L. G. H. 1993. Susceptibility of various landscape trees to root-knot nematodes. J. Arboric. 19:257–259.

3406. Santesson, R. 1952. Foliicolous lichens. I. A Revision of the taxonomy of the obligately foliicolous lichenized fungi. Symb. Bot. Upsal. 12:1–590.

3407. Santini, A., Camussi, A., and Raddi, P. 1997. Genetic variability of canker resistance trait in *Cupressus sempervirens* L. progenies. J. Appl. Genet. 38:453–461.

3408. Santini, A., and Capretti, P. 2000. Analysis of the Italian population of *Ceratocystis fimbriata* f.sp. *platani* using RAPD and minisatellite markers. Plant Pathol. 49:461–467.

3409. Santini, A., and Lonardo, V. di. 2000. Genetic variability of the 'bark canker resistance' character in several natural provenances of *Cupressus sempervirens*. For. Pathol. 30:87–96.

3410. Sanzo, C. P. Di, and Rohde, R. A. 1969. *Xiphinema americanum* associated with maple decline in Massachusetts. Phytopathology 59:279–284.

3411. Sariah, M., Hussin, M. Z., Miller, R. N. G., and Holderness, M. 1994. Pathogenicity of *Ganoderma boninense* tested by inoculation of oil palm seedlings. Plant Pathol. 43:507–510.

3412. Sasaki, I., and Nagayama, H. 1996. ß-glucosidase of *Botrytis cinerea*: its involvement in the pathogenicity of this fungus. Biosci. Biotechnol. Biochem. 60:54–56.

3413. Sasaki, K., and Kobayashi, T. 1975. Resinous canker disease of Cupressaceae caused by *Monochaetia unicornis* (Cke. et Ell.) Sacc. (I). The causal fungus and its pathogenicity. Bull. Gov. For. Exp. Stn., Meguro 271:27–38.

3414. Sasser, J. N., Haasis, F. A., and Cannon, T. F. 1966. Pathogenicity of *Meloidogyne* species on *Ilex*. Plant Dis. Rep. 50:664–668.

3415. Sato, K., Yokozawa, Y., and Shoji, T. 1974. Studies on Rhizina root rot causing group dying of pine trees. Bull. Gov. For. Exp. Stn., Meguro 268:13–48.

3416. Sato, S., Katsuya, K., and Hiratsuka, Y. 1993. Morphology, taxonomy and nomenclature of *Tsuga*-Ericaceae rusts. Trans. Mycol. Soc. Jap. 34:47–62.

3417. Saunders, J. E., Juzwik, J., and Hutchison, R. 1992. Outplanting survival of Cylindrocladium root rot affected black spruce seedlings. Can. J. For. Res. 22:1204–1207.

3418. Savile, D. B. O. 1950. North American species of *Chrysomyxa*. Can. J. Res. C28:318–330.

3419. Savile, D. B. O. 1955. *Chrysomyxa* in North America—additions and corrections. Can. J. Bot. 33:487–496.

3420. Savile, D. B. O. 1959. Notes on *Exobasidium*. Can. J. Bot. 37:641–656.

3421. Savile, D. B. O. 1962. Some fungal parasites of Onagraceae. Can. J. Bot. 40:1385–1398.

3422. Savile, D. B. O. 1976. *Phragmidium ivesiae* and its allies in North America. Can. J. Bot. 54:1690–1696.

3423. Savka, M. A., Black, R. C., Binns, A. N., and Farrand, S. K. 1996. Translocation and exudation of tumor metabolites in crown galled plants. Mol. Plant-Microbe Interact. 9:310–313.

3424. Sawada, H., Ieki, H., Oyaizu, H., and Matsumoto, S. 1993. Proposal for rejection of *Agrobacterium tumefaciens* and revised descriptions for the genus *Agrobacterium* and for *Agrobacterium radiobacter* and *Agrobacterium rhizogenes*. Int. J. Syst. Bacteriol. 43:694–702.

3425. Sawahata, T., and Hayashi, Y. 2002. Pairing tests and morphological studies of *Phellinus pini*–complex in Japan. Nippon Kingakukai Kaiho 43:51–59.

3426. Sbaghi, M., Jeandet, P., Bessis, R., and Leroux, P. 1996. Degradation of stilbene-type phytoalexins in relation to the pathogenicity of *Botrytis cinerea* to grapevines. Plant Pathol. 45:139–144.

3427. Scala, A., Pattuelli, M., Coppola, L., Guastini, M., Tegli, S., Sorbo, G. del, Mittempergher, L., and Scala, F. 1997. Dutch elm disease progression and quantitative determination of cerato-ulmin in leaves, stems and branches of elms inoculated with *Ophiostoma novo-ulmi* and *O. ulmi*. Physiol. Mol. Plant Pathol. 50:349–360.

3428. Schaad, N. W., Heskett, M. G., Gardner, J. M., and Kado, C. I. 1973. Influence of inoculum dosage, time after wounding, and season on infection of Persian walnut trees by *Erwinia rubrifaciens*. Phytopathology 63:327–329.

3429. Schaad, N. W., Jones, J. B., and Chun, W., eds. 2001. Laboratory guide for identification of plant pathogenic bacteria. 3rd ed. APS Press, St. Paul, MN. 373 pp.

3430. Schaad, N. W., Opgenorth, D., and Gaush, P. 2002. Real-time polymerase chain reaction for one-hour on-site diagnosis of Pierce's disease of grape in early season asymptomatic vines. Phytopathology 92:721–728.

3431. Schaad, N. W., and Wilson, E. E. 1970. Pathological anatomy of the bacterial phloem canker disease of *Juglans regia*. Can. J. Bot. 48:1055–1060.

3432. Schaad, N. W., and Wilson, E. E. 1971. The ecology of *Erwinia rubrifaciens* and the development of phloem canker of Persian walnut. Ann. Appl. Biol. 69:125–136.

3433. Schaffer, B., Hawksworth, F. G., Wullschleger, S. D., and Reid, C. P. P. 1983. Cytokinin-like activity related to host reactions to dwarf mistletoes (*Arceuthobium* spp.). For. Sci. 29:66–70.

3434. Schafleitner, R., and Wilhelm, E. 1997. Effect of virulent and hypovirulent *Cryphonectria parasitica* (Murr.) Barr on the inter-cellular pathogen related proteins and on total protein pattern of chestnut (*Castanea sativa* Mill.). Physiol. Mol. Plant Pathol. 51:323–332.

3435. Schaper, U., and Seemüller, E. 1982. Condition of the phloem and the persistence of mycoplasmalike organisms associated with apple proliferation and pear decline. Phytopathology 72:736–742.

3436. Scharpf, R. F. 1962. Growth rate of the endophytic system of the dwarf mistletoe on digger pine. USDA For. Serv. Pac. Southwest For. Range Exp. Stn. Res. Note 193. 5 pp.

3437. Scharpf, R. F., and Bynum, H. H. 1975. Cytospora canker of true firs. USDA For. Serv. For. Pest Leafl. 146. 5 pp.

3438. Scharpf, R. F., and Hawksworth, F. G. 1976. Luther Burbank introduced European mistletoe into California. Plant Dis. Rep. 60:740–742.

3439. Scharpf, R. F., Kinloch, B. B., and Jenkinson, J. L. 1991. One seed source of Jeffrey pine shows resistance to dwarf mistletoe. USDA For. Serv. Res. Pap. PSW-RP 207. 8 pp.

3440. Scharpf, R. F., and Parmeter, J. R. Jr., tech. coords. 1978. Proceedings of the symposium on dwarf mistletoe control through forest management. USDA For. Serv. Gen. Tech. Rep. PSW-31. 190 pp.

3441. Scharpf, R. F., and Roth, L. F. 1992. Resistance of ponderosa pine to western dwarf mistletoe in central Oregon. USDA For. Serv. Res. Pap. PSW-RP 208. 9 pp.

3442. Scharpf, R. F., Smith, R. S., and Vogler, D. 1987. Pruning dwarf mistletoe brooms reduces stress on Jeffrey pines, Cleveland National Forest, California. USDA For. Serv. Res. Pap. PSW-186. 7 pp.

3443. Scheffer, T. C., and Cowling, E. B. 1966. Natural resistance of wood to microbial deterioration. Annu. Rev. Phytopathol. 4:147–170.

3444. Scheffer, T. C., and Hedgcock, C. G. 1955. Injury to northwestern forest trees by sulfur dioxide from smelters. USDA Tech. Bull. 1117. 49 pp.

3445. Schena, L., Nigro, F., and Ippolito, A. 2002. Identification and detection of *Rosellinia necatrix* by conventional and real-time Scorpion-PCR. Eur. J. Plant Pathol. 108:355–366.

3446. Schier, G. A. 1975. Deterioration of aspen clones in the middle Rocky Mountains. USDA For. Serv. Res. Paper INT-170. 14 pp.

3447. Schink, B., Ward, J. C., and Zeikus, J. G. 1981. Microbiology of wetwood: importance of pectin degradation and *Clostridium* species in living trees. Appl. Environ. Microbiol. 42:526–532.

3448. Schink, B., Ward, J. C., and Zeikus, J. G. 1981. Microbiology of wetwood: role of anaerobic bacterial populations in living trees. J. Gen. Microbiol. 123:313–322.

3449. Schipper, A. L. Jr. 1978. A *Hypoxylon mammatum* pathotoxin responsible for canker formation in quaking aspen. Phytopathology 68:866–872.

3450. Schlagbauer, H. E., and Holz, G. 1989. Penetration of plums by *Monilinia laxa* and histology of a defence reaction. Phytophylactica 21:39–43.

3451. Schmelzer, K. 1962. Untersuchungen an Viren der Zier- und Wildgehölze. 2. Mitteilung: Virosen an *Forsythia*, *Lonicera*, *Ligustrum*, und *Laburnum*. Phytopathol. Z. 46:105–138.

3452. Schmelzer, K. 1968. Neue Befunde über die geographische Verbreitung des Robinienmosaiks und über seine ursachlichen Viren. Arch. Forstw. 17:621–628.

3453. Schmelzer, K. 1969. Das Ulmenscheckungs-Virus. Phytopathol. Z. 64:39–67.

3454. Schmelzer, K. 1974. Untersuchungen an Viren der Zier- und Wildgehölze 8. Mitt.: Neue Befunde an *Forsythia*, *Hydrangea* und *Philadelphus* sowie Viren und Virosen an *Rhamnus*, *Centaurea*, *Galvezia*, *Cistus*, *Forestiera*, *Abeliophyllum*, *Celastrus*, *Staphylea* und *Crambe*. Zentralbl. Bakteriol. Parasitenk. Infektionskr. Hyg. II 129:139–168.

3455. Schmelzer, K., and Schmelzer, A. 1968. Virusbefall an Magnolien (*Magnolia* spp.). Acta Phytopathol. Acad. Sci. Hung. 3:411–413.

3456. Schmidt, R. A., and Fergus, C. L. 1965. Branch canker and dieback of *Quercus prinus* L. caused by a species of *Botryodiplodia*. Can. J. Bot. 43:731–737.

3457. Schmidt, R. A., Gramacho, K. P., Miller, T., and Young, C. H. 2000. Components of partial resistance in the slash pine–fusiform rust pathosystem. Phytopathology 90:1005–1010.

3458. Schmidt, R. A., Holley, R. C., Klapproth, M. C., and Miller, T. 1986. Temporal and spatial patterns of fusiform rust epidemics in young plantations of susceptible and resistant slash and loblolly pines. Plant Dis. 70:661–666.

3459. Schmidt, W. C., and McDonald, K. J., comp. 1995. Ecology and management of *Larix* forests: a look ahead. USDA For. Serv. Gen. Tech. Rep. INT-GTR-319. 521 pp.

3460. Schmieden, U., and Wild, A. 1995. The contribution of ozone to forest decline. Physiol. Plant. 94:371–378.

3461. Schmitt, C. L., Parmeter, J. L., and Kleijunas, J. T. 2000. Annosus root rot of western conifers. USDA For. Serv. For. Insect Dis. Leafl. 172. 10 pp. Internet pub.

3462. Schmitt, U., and Liese, W. 1991. Suberin in wound reaction parenchyma of birch xylem (*Betula pendula* Roth)—an electron microscopic study. Holzforschung 45:313–315.

3463. Schmitt, U., and Liese, W. 1993. Response of xylem parenchyma by suberization in some hardwoods after mechanical injury. Trees 8:23–30.

3464. Schmull, M., and Thomas, F. M. 2000. Morphological and physiological reactions of young deciduous trees (*Quercus robur* L., *Q. petraea* [Matt.] Liebl., *Fagus sylvatica* L.) to waterlogging. Plant Soil 225:227–242.

3465. Schnabel, G., Schnabel, E. L., and Jones, A. L. 1999. Characterization of ribosomal DNA from *Venturia inaequalis* and its phylogenetic relationship to rDNA from other tree-fruit *Venturia* species. Phytopathology 89:100–108.

3466. Schneider, A., and Dargent, R. 1977. Localisation et comportement du mycélium de *Taphrina deformans* dans le mésophyll et sous la cuticule des feuilles de pêcher (*Prunus persica*). Can. J. Bot. 55:2485–2495.

3467. Schneider, A., and Sutra, G. 1969. Les modalités de l'infection de *Populus nigra* L. par *Taphrina populina* Fr. C. R. Hebd. Séances Acad. Sci. (Paris) D 269:1056–1059.

3468. Schneider, R. 1961 Untersuchungen über das Auftreten der Guignardia-blattbräune der Rosskastanie (*Aesculus hippocastanum*) in Westdeutschland und ihren Erreger. Phytopathol. Z. 42:272–278.

3469. Schneider, R., and Arx, J. A. von. 1966. Zwei neue, als Erreger von Zweigsterben nachgewiesene Pilze: *Kabatina thujae* n.g., n.sp., und *K. juniperi* n.sp. Phytopathol. Z. 57:176–182.

3470. Schneider, R., and Sauthoff, W. 1972. Absterbeerscheinungen an *Carpinus betulus* L. (Erreger: *Monostichella robergei* [Desm.] v. Höhn.). Nachrichtenbl. Deutsch. Pflanzenschutzd. 24:117–119.

3471. Schnell, G. R. 1987. Investigations of plant diseases in reforestations of the subalpine region in the central Swiss Alps. Eur. J. For. Pathol. 17:19–33.

3472. Schnelle, M. A., Feucht, J. R., and Klett, J. E. 1989. Root systems of trees—facts and fallacies. J. Arboric. 15:201–205.

3473. Schoch, C. L., Crous, P. W., Wingfield, B. D., and Wingfield, M. J. 1999. The *Cylindrocladium candelabrum* species complex includes four distinct mating populations. Mycologia 91:286–298.

3474. Schoch, C. L., Crous, P. W., Wingfield, M. J., and Wingfield, B. D. 2000. Phylogeny of *Calonectria* and selected hypocrealean genera with cylindrical macroconidia. Stud. Mycol. 45:45–62.

3475. Schoeneweiss, D. F. 1967. Susceptibility of weakened cottonwood stems to fungi associated with blackstem. Plant Dis. Rep. 51:933–935.

3476. Schoeneweiss, D. F. 1969. Susceptibility of evergreen hosts to the juniper blight fungus, *Phomopsis juniperovora*, under epidemic conditions. J. Am. Soc. Hortic. Sci. 94:609–611.

3477. Schoeneweiss, D. F. 1974. *Tubercularia ulmea* canker of tallhedge: influence of freezing stress on disease susceptibility. Plant Dis. Rep. 58:937–941.

3478. Schoeneweiss, D. F. 1975. Predisposition, stress, and plant disease. Annu. Rev. Phytopathol. 13:193–211.

3479. Schoeneweiss, D. F. 1981. Infectious diseases of trees associated with water and freezing stress. J. Arboric. 7:13–18.

3480. Schoeneweiss, D. F. 1981. The role of environmental stress in diseases of woody plants. Plant Dis. 65:308–314.

3481. Schoeneweiss, D. F. 1983. Drought predisposition to Cytospora canker in blue spruce. Plant Dis. 67:383–385.

3482. Schoeneweiss, D. F., and Wene, E. G. 1977. Freezing stress predisposes *Euonymus alatus* stems to attack by *Nectria cinnabarina*. Plant Dis. Rep. 61:921–925.

3483. Scholian, U. 1996. The tinder fungus (*Fomes fomentarius*) and its use. Schweiz. Z. Forstwesen 147:647–665.

3484. Schoulties, C. L., Civerolo, E. L., Miller, J. W., Stall, R. E., Krass, C. J., Poe, S. R., and DuCharme, E. P. 1987. Citrus canker in Florida. Plant Dis. 71:388–395.

3485. Schouten, H. J. 1991. Simulation of pressure caused by multiplication and swelling of *Erwinia amylovora* in intercellular space of host tissue. Neth. J. Plant Pathol. 97:139–149.

3486. Schreiber, L. R., and Green, R. J. Jr. 1959. Die-back and root rot disease of *Taxus* spp. in Indiana. Plant Dis. Rep. 43:814–817.

3487. Schreiber, L. R., and Mayer, J. S. 1992. Seasonal variations in susceptibility and in internal inoculum densities in maple species inoculated with *Verticillium dahliae*. Plant Dis. 76:184–187.

3488. Schreiner, E. J. 1931. Two species of *Valsa* causing disease in *Populus*. Am. J. Bot. 18:1–29.

3489. Schrenk, H. von. 1901. A disease of the black locust (*Robinia pseudoacacia* L.). Mo. Bot. Gard. Annu. Rep. 12:21–31.

3490. Schrenk, H. von. 1903. A disease of white ash caused by *Polyporus fraxinophilus*. USDA Bur. Plant Indus. Bull. 32. 20 pp.

3491. Schrenk, H. von. 1914. A trunk disease of the lilac. Ann. Mo. Bot. Gard. 1:253–262.

3492. Schroeder, W. R., and Lindquist, C. H. 1989. Assiniboine poplar. Can. J. Plant Sci. 69:351–353.

3493. Schroers, H.-J., Geldenhuis, M. M., Wingfield, M. J., Schoeman, M. H., and Wingfield, B. D. 2005. Introduction of *Nalanthamala*, anamorphic *Rubrinectria*, for the guava wilt fungus, *N. guajavae* sp. nov., the palm-pathogenic *Gliocladium vermoesenii*, and the persimmon wilt fungus *Acremonium diospyri*. Mycologia (in press).

3494. Schroth, M. N., McCain, A. H., Foott, J. H., and Huisman, O. C. 1988. Reduction in yield and vigor of grapevine caused by crown gall disease. Plant Dis. 72:241–246.

3494a. Schubert, K., Ritschel, A., and Braun, U. 2003. A monograph of *Fusicladium s.lat.* (Hyphomycetes). Schlechtendalia 9:1–132.

3495. Schubert, R., Bahnweg, G., Nechwatal, J., Jung, T., Cooke, D. E. L., Duncan, J. M., Muller, S. G., Langebartels, C., Sandermann, H. Jr., and Osswald, W. 1999. Detection and quantification of *Phytophthora* species which are associated with root-rot diseases in European deciduous forests by species-specific polymerase chain reaction. Eur. J. For. Pathol. 29:169–188.

3496. Schubert, T. S., and El-Gholl, N. E. 1997. Annotated list of valid worldwide *Calonectria/Cylindrocladium* spp. Plant Diagnos. Q. 18:28–30.

3497. Schubert, T. S., Leahy, R. M., and El-Gholl, N. E. 1999. *Cylindrocladium perseae* sp. nov. Mycotaxon 73:465–475.

3498. Schuldt, P. H. 1955. Comparison of anthracnose fungi on oak, sycamore, and other trees. Contrib. Boyce Thompson Inst. 18:85–107.

3499. Schultz, T. P., Harms, W. B., Fisher, T. H., McMurtrey, K. D., Minn, J., and Nicholas, D. D. 1995. Durability of angiosperm heartwood: the importance of extractives. Holzforschung 49:29–34.

3500. Schultz, T. P., Hubbard, T. F., Jin, L. H., Fisher, T. H., and Nicholas, D. D. 1990. Role of stilbenes in the natural durability of wood—fungicidal structure activity relationships. Phytochemistry 29:1501–1507.

3501. Schulz, U. 1981. Histologische Untersuchungen der Eintrittspforten von *Cytospora*-Arten. Angew. Bot. 55:441–455.

3502. Schulz, U. 1981. Untersuchungen zur biologischen Bekämpfung von *Cytospora*-Arten. Z. Pflanzenkr. Pflanzenschutz 88:132–141.

3503. Schulze, E. D., Lange, O. L., Ziegler, H., and Gebauer, G. 1991. Carbon and nitrogen isotope ratios of mistletoes growing on nitrogen and non-nitrogen fixing hosts and on CAM plants in the Namib Desert confirm partial heterotrophy. Oecologia 88:457–462.

3504. Schulze, S. 1999. Rapid detection of European *Heterobasidion annosum* intersterility groups and intergroup gene flow using taxon-specific competitive-priming PCR (TSCP-PCR). J. Phytopathol. 147:125–127.

3505. Schulze, S., and Bahnweg, G. 1998. Critical review of identification techniques for *Armillaria* spp. and *Heterobasidion annosum* root and butt rot diseases. J. Phytopathol. 146:61–72.

3506. Schütt, P. 1972. Untersuchungen über den Einfluss von Cuticularwachsen auf die Infektionsfähigkeit pathogenen Pilze. 2. *Rhytisma acerinum*, *Microsphaera alphitoides*, und *Fusarium oxysporum*. Eur. J. For. Pathol. 2:43–59.

3507. Schwarze, F. 1993. *Piptoporus betulinus* (Bull.: Fr.) Karsten. Mycologist 7:122–123.

3508. Schwarze, F. 2001. Development and prognosis of decay in the sapwood of living trees. Arboric. J. 25:321–337.

3509. Schwarze, F. W. M. R., and Baum, S. 2000. Mechanisms of reaction zone penetration by decay fungi in wood of beech (*Fagus sylvatica*). New Phytol. 146:129–140.

3510. Schwarze, F. W. M. R., Engels, J., and Mattheck, C. 2000. Fungal strategies of wood decay in trees. English trans. W. Linnard, of Holzzersetzende Pilze in Bäumen: Strategien der Holzzersetzung. Springer, Berlin, New York. 185 pp.

3511. Schwarze, F. W. M. R., and Fink, S. 1997. Reaction zone penetration and prolonged persistence of xylem rays in London plane wood degraded by the basidiomycete *Inonotus hispidus*. Mycol. Res. 101:1207–1214.

3512. Schwarze, F. W. M. R., Lonsdale, D., and Fink, S. 1995. Soft rot and multiple T-branching by the basidiomycete *Inonotus hispidus* in ash and London plane. Mycol. Res. 99:813–820.

3513. Schwintzer, C. R., and Tjepkema, J. D. 1990. The biology of *Frankia* and actinorhizal plants. Academic Press, San Diego. 408 pp.

3514. Scortichini, M. 2002. Bacterial canker and decline of European hazelnut. Plant Dis. 86:704–709.

3515. Scortichini, M., Janse, J. D., Rossi, M. P., and Derks, J. H. J. 1996. Characterization of *Xanthomonas campestris* pv. *pruni* strains from different hosts by pathogenicity tests and analysis of whole-cell fatty acids and whole-cell proteins. J. Phytopathol. 144:69–74.

3516. Scortichini, M., Marchesi, U., Rossi, M. P., and Prospero, P. di. 2002. Bacteria associated with hazelnut (*Corylus avellana* L.) decline are of two groups: *Pseudomonas avellanae* and strains resembling *P. syringae* pv. *syringae*. Appl. Environ. Microbiol. 68:476–484.

3517. Scorza, R. 1992. Evaluation of foreign peach and nectarine introductions in the U.S. for resistance of leaf curl [*Taphrina deformans* (Berk.) Tul.]. Fruit Var. J. 46:141–145.

3518. Scott, E. S. 1984. Populations of bacteria in poplar stems. Eur. J. For. Pathol. 14:103–112.

3519. Scott, K. J., and Chakravorty, A. K., eds. 1982. The rust fungi. Academic Press, New York. 288 pp.

3520. Scott, S. W., and Barnett, O. W. 1984. Some properties of an isolate of broad bean wilt virus from dogwood (*Cornus florida*). Plant Dis. 68:983–985.

3521. Seaby, D. 1977. *Rhizina undulata* on *Picea abies* transplants. Eur. J. For. Pathol. 7:186–188.

3522. Seaver, F. J. 1922. Phyllostictales. N. Am. Flora 6 (Pt. 1):1–84.

3523. Seaver, F. J. 1951. The North American cup-fungi (inoperculates). Published by the author, New York. 428 pp.

3524. Secor, G. A., and Nyland, G. 1978. Rose ring pattern: a component of the rose-mosaic complex. Phytopathology 68:1005–1010.

3525. Seeler, E. V. Jr. 1940. A monographic study of the genus *Thyronectria*. J. Arnold Arbor. 21:429–460.

3526. Seeler, E. V. Jr. 1940. Two diseases of *Gleditsia* caused by a species of *Thyronectria*. J. Arnold Arbor. 21:405–427.

3527. Seemüller, E. 1992. Pear decline. Pages 308–334 in: Plant diseases of international importance. Vol. 3: Diseases of fruit crops. J. Kumar, H. S. Chaube, V. S. Singh, and A. N. Mukhopadhyay, eds. Prentice-Hall, Englewood Cliffs, NJ.

3528. Seemüller, E., Lorenz, K. H., and Lauer, U. 1998. Pear decline resistance in *Pyrus communis* rootstocks and progenies of wild and ornamental *Pyrus* taxa. Acta Hortic. No. 472:681–691.

3529. Seemüller, E., Marcone, C., Lauer, U., Ragozzino, A. and Göschl, M. 1998. Current status of molecular classification of the phytoplasmas. J. Plant Pathol. 80:3–26.

3530. Seemüller, E., Schaper, U., and Zimbelmann, F. 1984. Seasonal variation in the colonization patterns of mycoplasmalike organisms associated with apple proliferation and pear decline. Z. Pflanzenkr. Pflanzenschutz 91:371–382.

3531. Seemüller, E., and Schneider, B. 2004. 'Candidatus Phytoplasma mali,' 'Candidatus Phytoplasma pyri' and 'Candidatus Phytoplasma prunorum,' the causal agents of apple proliferation, pear decline and European stone fruit yellows, respectively. Int. J. Syst. Evol. Microbiol. 54:1217–1226.

3532. Seidel, M., Steffen, E., Seidel, D., and Walter, A. 1994. Survival of *Erwinia amylovora* (Burrill) Winslow et al. on bird feet. Arch. Phytopathol. Plant Prot. 29:25–27.

3533. Seifers, D., and Ammon, V. 1980. Mode of penetration of sycamore leaves by *Gloeosporium platani*. Phytopathology 70:1050–1055.

3534. Seifert, K. A., Gams, W., Crous, P. W., and Samuels, G. J., eds. 2000. Molecules, morphology and classification: towards monophyletic genera in the Ascomycetes. Stud. Mycol. No. 45. 230 pp.

3535. Seischab, F. K., Bernard, J. M., and Eberle, M. D. 1993. Glaze storm damage to western New York forest communities. Bull. Torrey Bot. Club 120:64–72.

3536. Seliskar, C. E. 1950. Some investigations on the wetwood diseases of American elm and Lombardy poplar. Ph.D. diss., Cornell Univ., Ithaca, NY. 148 pp.

3537. Seliskar, C. E. 1976. Mycoplasmalike organism found in the phloem of bunch-diseased walnuts. For. Sci. 22:144.

3538. Seliskar, C. E., KenKnight, G. E., and Bourne, C. E. 1974. Mycoplasmalike organism associated with pecan bunch disease. Phytopathology 64:1269–1272.

3539. Sendak, P. E., Bove, J. R., Bergdahl, D. R., Tobi, D. R., and Huyler, N. K. 1997. Effects of Eutypella canker and sugar maple borer in merchantable volume loss in sugar maple. North. J. Appl. For. 14:26–31.

3540. Setliff, E. C. 1988. Hyphal deterioration in *Ganoderma applanatum*. Mycologia 80:447–454.

3541. Setliff, E. C. 2002. The wound pathogen *Chondrostereum purpureum*, its history and incidence on trees in North America. Austral. J. Bot. 50:645–651.

3542. Setliff, E. C., Hoch, H. C., and Patton, R. F. 1974. Studies on nuclear division in basidia of *Poria latemarginata*. Can. J. Bot. 52:2323–2333.

3543. Sewell, G. W. F., and Wilson, J. F. 1964. Occurrence and dispersal of *Verticillium* conidia in xylem sap of the hop (*Humulus lupus* L.). Nature 204:901.

3544. Sewell, G. W. F., and Wilson, J. F. 1973. Phytophthora collar rot of apple: seasonal effects on infection and disease development. Ann. Appl. Biol. 74:149–158.

3545. Sewell, G. W. F., Wilson, J. F., and Dakwa, J. T. 1974. Seasonal variations in the activity in soil of *Phytophthora cactorum*, *P.*

syringae, and *P. citricola* in relation to collar rot disease of apple. Ann. Appl. Biol. 76:179–186.

3546. Seymour, C. P. 1969. Charcoal rot of nursery-grown pines in Florida. Phytopathology 59:89–92.

3547. Shabi, E., Rotem, J., and Lobenstein, G. 1973. Physiological races of *Venturia pyrina* on pear. Phytopathology 63:41–43.

3548. Shahin, E. A., and Claflin, L. E. 1978. The occurrence and distribution of Sirococcus shoot blight of spruce in Kansas. Plant Dis. Rep. 62:648–650.

3549. Shain, L. 1967. Resistance of sapwood in stems of loblolly pine to infection by *Fomes annosus*. Phytopathology 57:1034–1045.

3550. Shain, L. 1971. The response of sapwood of Norway spruce to infection by *Fomes annosus*. Phytopathology 61:301–307.

3551. Shain, L. 1988. Evidence for formae speciales in the poplar leaf rust fungus *Melampsora medusae*. Mycologia 80:729–732.

3552. Shain, L., and Franich, R. A. 1981. Induction of Dothistroma blight symptoms with dothistromin. Physiol. Plant Pathol. 19:49–55.

3553. Shain, L., and Jarlfors, U. 1987. Ultrastructure of eastern cottonwood clones susceptible or resistant to leaf rust. Can. J. Bot. 65:1586–1598.

3554. Shan, Y. 2000. Phenological disorder induced by atmospheric nitrogen deposition: original causes of pine forest decline over Japan. Part II. Relationship among earlier phenological development, extreme of minimum air temperature, and forest decline of pines over the Japan. Water Air Soil Pollut. 117:205–215.

3555. Shannon, M. C., Banuelos, G. S., Draper, J. H., Ajwa, H., Jordahl, J., and Licht, L. 1999. Tolerance of hybrid poplar (*Populus*) trees irrigated with varied levels of salt, selenium, and boron. Int. J. Phytoremed. 1:273–288.

3556. Sharland, P. R., Rayner, A. D. M., Ofong, A. U., and Barrett, D. K. 1988. Population structure of *Rosellinia desmazieresii* [sic] causing ring-dying of *Salix repens*. Trans. Br. Mycol. Soc. 90:654–656.

3557. Sharma, A. K., and Johri, B. N., eds. 2002. Arbuscular mycorrhizae: interactions in plants, rhizosphere, and soils. Science Pub., Enfield, NH. 311 pp.

3558. Shattock, R. C. 2003. Rust. Pages 165–168 in: Encyclopedia of rose science. Vol. 1. A. Roberts, T. Debener, and S. Gudin, eds. Elsevier Academic Press, Amsterdam.

3559. Shattock, R. C., and Bhatti, M. H. R. 1983. The effect of *Phragmidium mucronatum* on rose understocks and maiden rose bushes. Plant Pathol. 32:61–66.

3560. Shaw, B. D., Kuo, K., and Hoch, H. C. 1998. Germination and appressorium development of *Phyllosticta ampelicida* pycnidiospores. Mycologia 90:258–268.

3561. Shaw, C. G. 1949. Nomenclatorial problems in the Peronosporaceae. Mycologia 41:320–338.

3562. Shaw, C. G. 1951. New species of the Peronosporaceae. Mycologia 43:445–455.

3563. Shaw, C. G., and Leaphart, C. D. 1960. Two serious foliage diseases of western white pine in the Inland Empire. Plant Dis. Rep. 44:655–659.

3564. Shaw, C. G. III. 1991. Spread, intensification, and upward advances of dwarf mistletoe in thinned young stands of western hemlock in southeast Alaska. Plant Dis. 75:363–367.

3565. Shaw, C. G. III, and Kile, G. A. 1991. Armillaria root disease. USDA Agric. Handb. 691. 233 pp.

3566. Shaw, C. G. III, and Loopstra, E. M. 1991. Development of dwarf mistletoe infections on inoculated western hemlock trees in southeast Alaska. Northwest Sci. 65:48–52.

3567. Shaw, C. G. III, and Roth, L. F. 1976. Persistence and distribution of a clone of *Armillaria mellea* in a ponderosa pine forest. Phytopathology 66:1210–1213.

3568. Shaw, D. C., and Weiss, S. B. 2000. Canopy light and the distribution of hemlock dwarf mistletoe (*Arceuthobium tsugense* (Rosendahl) G. N. Jones subsp. *tsugense*) aerial shoots in an old-growth Douglas-fir/western hemlock forest. Northwest Sci. 74:306–315.

3569. Shaw, P. J. A., Holland, M. R., Darrall, N. M., and McLeod, A. R. 1993. The occurrence of SO_2-related foliar symptoms on Scots pine (*Pinus sylvestris* L.) in an open-air forest fumigation experiment. New Phytol. 123:143–152.

3570. Shear, C. L. 1902. Mycological notes and new species. Bull. Torrey Bot. Club 29:449–457.

3571. Shear, C. L., Stevens, N. E., and Tiller, R. J. 1917. *Endothia parasitica* and related species. USDA Bull. 380. 77 pp.

3572. Shear, C. L., Stevens, N. E., and Wilcox, M. S. 1925. *Botryosphaeria* and *Physalospora* in the eastern United States. Mycologia 17:98–107.

3573. Sheikh, A. H., and Ghaffar, A. 1987. Time-temperature relationships for the inactivation of sclerotia of *Macrophomina phaseolina*. Soil Biol. Biochem. 19:313–315.

3574. Shepard, D. P., and Zehr, E. I. 1994. Epiphytic persistence of *Xanthomonas campestris* pv. *pruni* on peach and plum. Plant Dis. 78:627–629.

3575. Sheppard, L. J., Leith, I. D., Murray, M. B., Cape, J. N., and Kennedy, V. H. 1998. The response of Norway spruce seedlings to simulated acid mist. New Phytol. 138:709–723.

3576. Sherald, J. L. 1993. Pathogenicity of *Xylella fastidiosa* in American elm and failure of reciprocal transmission between strains from elm and sycamore. Plant Dis. 77:190–193.

3577. Sherald, J. L., Hearon, S. S., Kostka, S. J., and Morgan, D. L. 1983. Sycamore leaf scorch: culture and pathogenicity of fastidious xylem-limited bacteria from scorch-affected trees. Plant Dis. 67:848–852.

3578. Sherald, J. L., and Kostka, S. J. 1992. Bacterial leaf scorch of landscape trees caused by *Xylella fastidiosa*. J. Arboric. 18:57–63.

3579. Sherald, J. L., and Lei, J. D. 1991. Evaluation of a rapid ELISA test kit for detection of *Xylella fastidiosa* in landscape trees. Plant Dis. 75:200–203.

3580. Sherald, J. L., Patton, E. N., Stidham, T. M., and Favre, C. L. 1994. Incidence and development of bacterial leaf scorch of elm on the National Mall. J. Arboric. 20:18–23.

3581. Sherald, J. L., Santamour, F. S. Jr., Hajela, R. K., Hajela, N., and Sticklen, M. B. 1994. A Dutch elm disease resistant triploid elm. Can. J. For. Res. 24:647–653.

3582. Sherald, J. L., Stidham, T. M., Hadidian, J. M., and Hoeldtke, J. E. 1996. Progression of the dogwood anthracnose epidemic and the status of flowering dogwood in Catoctin Mountain Park. Plant Dis. 80:310–312.

3583. Sherald, J. L., Wells, J. M., Hurtt, S. S., and Kostka, S. J. 1987. Association of fastidious, xylem-inhabiting bacteria with leaf scorch in red maple. Plant Dis. 71:930–933.

3584. Sherwood, C. H., Weigle, J. L., and Denisen, E. L. 1970. 2,4-D as an air pollutant: effects on growth of representative horticultural plants. HortScience 5:211–213.

3585. Shew, H. D., and Benson, D. M. 1981. Fraser fir root rot induced by *Phytophthora citricola*. Plant Dis. 65:688–689.

3586. Shew, H. D., and Benson, D. M. 1983. Influence of soil temperature and inoculum density of *Phytophthora cinnamomi* on root rot of Fraser fir. Plant Dis. 67:522–524.

3587. Shiel, P. J., and Castello, J. D. 1985. Detection of tobacco mosaic and tobacco ringspot viruses in herbaceous and woody plants near virus-infected white ash trees in central New York. Plant Dis. 69:791–795.

3588. Shigo, A. L. 1963. Fungi associated with the discolorations around rot columns caused by *Fomes igniarius*. Plant Dis. Rep. 47:820–823.

3589. Shigo, A. L. 1967. Successions of organisms in discoloration and decay of wood. Int. Rev. For. Res. 2:237–299.

3590. Shigo, A. L. 1969. How *Poria obliqua* and *Polyporus glomeratus* incite cankers. Phytopathology 59:1164–1165.

3591. Shigo, A. L. 1974. Relative abilities of *Phialophora melinii*, *Fomes connatus*, and *F. igniarius* to invade freshly wounded tissues of *Acer rubrum*. Phytopathology 64:708–710.

3592. Shigo, A. L. 1979. Compartmentalization of decay associated with *Heterobasidion annosum* in roots of *Pinus resinosa*. Eur. J. For. Pathol. 9:341–347.

3593. Shigo, A. L. 1983. Tree defects: a photo guide. USDA For. Serv. Gen. Tech. Rep. NE-82. 167 pp.

3594. Shigo, A. L. 1984. Compartmentalization: a conceptual framework for understanding how trees grow and defend themselves. Annu. Rev. Phytopathol. 22:189–214.

3595. Shigo, A. L. 1989. A new tree biology. 2nd ed. Shigo and Trees, Assoc., Durham, NH. 619 pp.

3596. Shigo, A. L. 1991. Modern arboriculture. Shigo and Trees, Assoc., Durham, NH. 424 pp.

3597. Shigo, A. L. 1994. Tree anatomy. Shigo and Trees, Assoc., Durham, NH. 104 pp.

3598. Shigo, A. L., and Hillis, W. E. 1973. Heartwood, discolored wood, and microorganisms in living trees. Annu. Rev. Phytopathol. 11:197–222.

3599. Shigo, A. L., and Marx, H. G. 1977. Compartmentalization of decay in trees. USDA Agric. Inf. Bull. 405. 73 pp.

3600. Shigo, A. L., and Marx, H. G. 1979. Tree decay. An expanded concept. USDA Agric. Inf. Bull. 419. 73 pp.

3601. Shigo, A. L., and Sharon, E. M. 1970. Mapping columns of discolored and decayed tissues in sugar maple, *Acer saccharum*. Phytopathology 60:232–237.

3602. Shigo, A. L., and Shortle, W. C. 1979. Compartmentalization of discolored wood in heartwood of red oak. Phytopathology 69:710–711.

3603. Shigo, A. L., and Tippett, J. T. 1981. Compartmentalization of American elm tissues infected by *Ceratocystis ulmi*. Plant Dis. 65:715–718.

3604. Shigo, A. L., and Tippett, J. T. 1981. Compartmentalization of decayed wood associated with *Armillaria mellea* in several tree species. USDA For. Serv. Res. Pap. NE-488. 20 pp.

3605. Shin, H.-D. and Lee, H.-T. 1999. A new species of *Marssonina* on *Celastrus orbiculatus*. Mycotaxon 72:199–203.

3606. Shoemaker, R. A. 1964. Conidial states of some *Botryosphaeria* species on *Vitis* and *Quercus*. Can. J. Bot. 42:1297–1301.

3607. Shortle, W. C. 1979. Compartmentalization of decay in red maple and hybrid poplar trees. Phytopathology 69:410–413.

3608. Showalter, T. D., and Filip, G. M., eds. 1993. Beetle-pathogen interactions in conifer forests. Academic Press, San Diego. 292 pp.

3609. Shukla, A. N., Schmidt, R. A., and Miller, T. 2001. Symptoms in slash pine seedlings following inoculation with the cone rust fungus *Cronartium strobilinum*. For. Pathol. 31:345–352.

3610. Shurtleff, M. C., and Averre, C. W. III. 2000. Diagnosing plant diseases caused by nematodes. APS Press, St. Paul, MN. 187 pp.

3611. Shurtleff, W., and Aoyagi, A. 1977. The book of kudzu. Autumn Press, Brookline, MA. 102 pp.

3612. Siccama, T. G., Weir, G., and Wallace, K. 1976. Ice damage in a mixed hardwood forest in Connecticut in relation to *Vitis* infestation. Bull. Torrey Bot. Club 103:180–183.

3613. Sickle, G. A. Van. 1969. Occurrence of *Cronartium comptoniae* in the Maritime Provinces. Plant Dis. Rep. 53:369–371.

3614. Sickle, G. A. Van. 1973. A quantitative survey for needle rust of balsam fir. Plant Dis. Rep. 57:765–766.

3615. Sickle, G. A. Van. 1973. A survey of production losses due to witches' broom of blueberry in the Maritime Provinces. Plant Dis. Rep. 57:608–611.

3616. Sickle, G. A. Van. 1974. Growth loss caused by a needle rust (*Pucciniastrum goeppertianum*) of balsam fir. Can. J. For. Res. 4:138–140.

3617. Sickle, G. A. Van. 1974. Nectria canker: a problem on black locust in New Brunswick. Plant Dis. Rep. 58:872–874.

3618. Sickle, G. A. Van. 1975. Basidiospore production and infection of balsam fir by a needle rust, *Pucciniastrum goeppertianum*. Can. J. Bot. 53:8–17.

3619. Sickle, G. A. Van. 1977. Seasonal periodicity in the discharge of *Pucciniastrum goeppertianum* basidiospores. Can. J. Bot. 55:745–751.

3620. Sicoli, G., Gioia, T. de, Luisi, N., and Lerario, P. 1998. Multiple factors associated with oak decline in southern Italy. Phytopathol. Medit. 37:1–8.

3621. Siddiqi, M. R. 2000. *Tylenchida*: parasites of plants and insects. 2nd ed. CABI Pub., Wallingford, UK; New York. 833 pp.

3622. Sieber, T. N. 1989. Endophytic fungi in twigs of healthy and diseased Norway spruce and white fir. Mycol. Res. 92:322–326.

3623. Sieber, T. N., Rys, J., and Holdenrieder, O. 1999. Mycobiota in symptomless needles of *Pinus mugo* ssp. *uncinata*. Mycol. Res. 103:306–310.

3624. Sieber, T. N., Sieber, C. F., Petrini, O., Ekramoddoullah, A. K. M., and Dorworth, C. E. 1991. Characterization of Canadian and European *Melanconium* from some *Alnus* species by morphological, cultural, and biochemical studies. Can. J. Bot. 69:2170–2176.

3625. Sierota, Z. 1998. *Rhizina undulata* on stem and roots of 84-year-old Scots pine trees. Acta Mycol. 33:69–76.

3626. Sierotzki, H., and Gessler, C. 1998. Inheritance of virulence of *Venturia inaequalis* toward *Malus × domestica* cultivars. J. Phytopathol. 146:509–514.

3627. Sierra, A. P., Whitehead, D. S., and Whitehead, M. P. 1999. Investigation of a PCR-based method for the routine identification of British *Armillaria* species. Mycol. Res. 103:1631–1636.

3628. Sigee, D. C. 1993. Bacterial plant pathology. Cambridge Univ. Press, Cambridge, UK. 300 pp.

3629. Siggers, P. V. 1944. The brown spot needle blight of pine seedlings. USDA Tech. Bull. 870. 36 pp.

3630. Sikora, E. J., and Malek, R. B. 1988. Transmission of the pinewood nematode (*Bursaphelenchus xylophilus*) to six pine species by *Monochamus carolinensis*. Plant Dis. 72:734.

3631. Sikorowski, P. P., and Roth, L. F. 1962. *Elytroderma* mycelium in the phloem of ponderosa pine. Phytopathology 52:332–336.

3632. Silva, F. R. da, Vettore, A. L., Kemper, E. L., Leite, A., and Arruda, P. 2001. Fastidian gum: the *Xylella fastidiosa* exopolysaccharide possibly involved in bacterial pathogenicity. FEMS Microbiol. Lett. 203:165–171.

3633. Silveira, R. L. V. A., Krugner, T. L., Silveira, R. I., and Gonçalves, A. N. 1996. Effect of boron on the susceptibility of *Eucalyptus citriodora* to *Botryosphaeria ribis* and *Lasiodiplodia theobromae*. Fitopatol. Brasil. 21:482–485.

3634. Simard, M., Rioux, D., and Laflamme, G. 2001. Formation of ligno-suberized tissues in jack pine resistant to the European race of *Gremmeniella abietina*. Phytopathology 91:1128–1140.

3635. Simay, E. I. 1987. Stigmina leaf spot of *Yucca filamentosa* in Hungary. Novenyvedelem 23:413–417.

3636. Simmons, E. G. 1995. *Alternaria* themes and variations (112–144); (150). Mycotaxon 55:55–163; 59:319–335.

3637. Simms, H. R. 1967. On the ecology of *Herpotrichia nigra*. Mycologia 59:902–909.

3638. Simons, M. D., Rothman, P. J., and Michel, L. J. 1979. Pathogenicity of *Puccinia coronata* from buckthorn and from oats adjacent to and distant from buckthorn. Phytopathology 69:156–158.

3639. Simpson, A. J. G., Reinach, F. C., and 117 authors. 2000. The genome sequence of the plant pathogen *Xylella fastidiosa*. Nature 406:151–157.

3640. Sinclair, W. A., and Campana, R. J., eds. 1978. Dutch elm disease. Perspectives after 60 years. Cornell University Agric. Exp. Stn. Search Agric. 8, No. 5. 52 pp.

3641. Sinclair, W. A., Cowles, D. P., and Hee, S. M. 1975. Fusarium root rot of Douglas-fir seedlings: suppression by soil fumigation, fertility management, and inoculation with spores of the fungal symbiont *Laccaria laccata*. For. Sci. 21:390–399.

3642. Sinclair, W. A., Gleason, M. L., Griffiths, H. M., Iles, J. K., Zriba, N., Charlson, D. V., Batzer, J. C., and Whitlow, T. H. 2000. Responses of 11 *Fraxinus* cultivars to ash yellows phytoplasma strains of differing aggressiveness. Plant Dis. 84:725–730.

3643. Sinclair, W. A., and Griffiths, H. M. 1994. Ash yellows and its relationship to dieback and decline of ash. Annu. Rev. Phytopathol. 32:49–60.

3644. Sinclair, W. A., and Griffiths, H. M. 1995. Epidemiology of a slow-decline phytoplasmal disease: ash yellows on old-field sites in New York State. Phytopathology 85:123–128.

3645. Sinclair, W. A., and Griffiths, H. M. 2000. Variation in aggressiveness of strains of ash yellows phytoplasmas. Plant Dis. 84:282–288.

3646. Sinclair, W. A., Griffiths, H. M., and Davis, R. E. 1996. Ash yellows and lilac witches'-broom: phytoplasmal diseases of concern in forestry and horticulture. Plant Disease 80:468–475.

3647. Sinclair, W. A., Griffiths, H. M., and Treshow, M. 1994. Ash yellows in velvet ash in Zion National Park, Utah: high incidence but low impact. Plant Dis. 78:486–490.

3648. Sinclair, W. A., Griffiths, H. M., and Whitlow, T. H. 1997. Comparisons of tolerance of ash yellows phytoplasmas in *Fraxinus* species and rootstock-scion combinations. Plant Dis. 81:395–398.

3649. Sinclair, W. A., and Hudler, G. W. 1980. Tree and shrub pathogens new or noteworthy in New York State. Plant Dis. 64:590–592.

3650. Sinclair, W. A., and Hudler, G. W. 1988. Tree declines: four concepts of causality. J. Arboric. 14:29–35.

3651. Sinclair, W. A., Smith, K. L., and Larsen, A. O. 1981. Verticillium wilt of maples: symptoms related to movement of the pathogen in stems. Phytopathology 71:340–345.

3652. Sinclair, W. A., and Stone, E. L. 1974. Boron toxicity to pines subject to home laundry waste water. Arborist's News 39:71–72.

3653. Sinclair, W. A., Stone, E. L., and Scheer, C. F. Jr. 1975. Toxicity to hemlocks grown in arsenic-contaminated soil previously used for potato production. HortScience 10:35–36.

3654. Sinclair, W. A., Townsend, A. M., Griffiths, H. M., and Whitlow, T. H. 2000. Responses of six Eurasian *Ulmus* cultivars to a North American elm yellows phytoplasma. Plant Dis. 84:1266–1270.

3655. Sinclair, W. A., Townsend, A. M., and Sherald, J. L. 2001. Elm yellows phytoplasma lethal to Dutch elm disease-resistant *Ulmus americana* cultivars. Plant Dis. 85:560.

3656. Sinclair, W. A., Whitlow, T. H., and Griffiths, H. M. 1997. Heritable tolerance of phytoplasmal infection in green ash. Can. J. For. Res. 27:1928–1935.

3657. Singh, P. 1978. Broom rusts of balsam fir and black spruce in Newfoundland. Eur. J. For. Pathol. 8:25–36.

3658. Singh, P., and Carew, G. C. 1989. Impact of eastern dwarf mistletoe in black spruce forests of Newfoundland. Eur. J. For. Pathol. 19:305–322.

3659. Singh, P., and Carew, G. C. 1990. Inland spruce cone rust of black spruce: effect on cone and seed yield, and seed quality. Eur. J. For. Pathol. 20:397–404.

3660. Singh, R. K., and Dwivedi, R. S. 1991. Ecology and biology of *Sclerotium rolfsii* Sacc. Int. J. Trop. Plant Dis. 9:161–171.

3661. Singh, S. J., and Heather, W. A. 1982. Temperature sensitivity of qualitative race-cultivar interactions in *Melampsora medusae* Thüm. and *Populus* species. Eur. J. For. Pathol. 12:123–127.

3662. Singh, S. J., and Heather, W. A. 1982. Temperature-light sensitivity of infection types expressed by cultivars of *Populus deltoides* Marsh. to races of *Melampsora medusae* Thüm. Eur. J. For. Pathol. 12:327–331.

3663. Singh, S. N. 1980. Effect of rootstock on growth, flowering, and disease resistance of hybrid tea roses. Prog. Hortic. 12(3):5–14.

3664. Sinha, M. K., Singh, R., and Jeyarajan, R. 1970. Graphiola leaf spot on date palm (*Phoenix dactylifera*): susceptibility of date varieties and effect on chlorophyll content. Plant Dis. Rep. 54:617–619.

3665. Sinha, R. C., and Chiykowski, L. N. 1980. Transmission and morphological features of mycoplasmalike bodies associated with peach X-disease. Can. J. Plant Pathol. 2:119–124.

3666. Sisinni, S. M., Zipperer, W. C., and Pleninger, A. G. 1995. Impacts from a major ice storm: street-tree damage in Rochester, New York. J. Arboric. 21:156–167.

3667. Sivanesan, A. 1970. *Cumminsiella mirabilissima*. C.M.I. Descr. Pathogenic Fungi Bact. No. 261. 2 pp.

3668. Sivanesan, A. 1977. The taxonomy and pathology of *Venturia* species. Bibl. Mycol. 59. 138 pp.

3669. Sivanesan, A. 1984. The bitunicate ascomycetes and their anamorphs. Cramer, Vaduz, Austria. 701 pp.

3670. Sivanesan, A. 1990. *Mycosphaerella populorum*. C.M.I. Descr. Pathogenic Fungi Bact. No. 988. 2 pp.

3671. Sivanesan, A., and Gibson, I. A. S. 1972. *Herpotrichia juniperi*. C.M.I. Descr. Pathogenic Fungi Bact. No. 328. 2 pp.

3672. Sivanesan, A., and Holliday, P. 1976. *Mycosphaerella citri*. C.M.I. Descr. Pathogenic Fungi Bact. No. 510. 2 pp.

3673. Sivapalan, A. 1993. Effects of water on germination of powdery mildew conidia. Mycol. Res. 97:71–76.

3674. Siwecki, R., and Liese, W., eds. 1991. Oak decline in Europe: proceedings of an international symposium, Kornik, Poland, 1990. Polish Acad. Sci., Kornik. 360 pp.

3675. Siwecki, R., and Ufnalski, K. 1998. Review of oak stand decline with special reference to the role of drought in Poland. Eur. J. For. Pathol. 28:99–112.

3676. Sjamsuridzal, W., Nishida, H., Ogawa, H., Kakishima, M., and Sugiyama, J. 1999. Phylogenetic positions of rust fungi parasitic on ferns: evidence from 18S rDNA sequence analysis. Mycoscience 40:21–27.

3677. Skarby, L., Ro, P. H., Wellburn, F. A. M., and Sheppard, L. J. 1998. Impacts of ozone on forests: a European perspective. New Phytol. 139:109–122.

3678. Skarmoutsos, G., and Michalopoulos, S. H. 2000. Pathogenicity of *Bursaphelenchus sexdentati*, *Bursaphelenchus leoni* and *Bursaphelenchus hellenicus* on European pine seedlings. For. Pathol. 30:149–156.

3679. Skarmoutsou, H., and Skarmoutsos, G. 1999. First report of Fusarium wilt disease of mimosa in Greece. Plant Dis. 83:590.

3680. Skelly, J. M. 2000. Tropospheric ozone and its importance to forests and natural plant communities of the northeastern United States. Northeast. Nat. 7:221–236.

3681. Skelly, J. M., Davis, D. D., Merrill, W., Cameron, E. A., Brown, H. D., Drummond, D. B., Dochinger, L. S., and Hellmann, R. 1988. Shadendiagnose an Waldbäumen im Osten der USA: ein Hanbuch zur Identifizierung von Shäden versursacht durch Luftschadstoffe, Pathogene, Insekten, und abiotischen Stress. USDA For, Serv. For. Pest Manag. and Penn. State Univ., University Park. 122 pp.

3682. Skelly, J. M., and Wood, F. A. 1966. The occurrence and etiology of an annual canker of sugar maple in Pennsylvania. Can. J. Bot. 44:1401–1411.

3683. Skerman, V. D. B., McGowan, V., and Sneath, P. H. A., eds. 1989. Approved lists of bacterial names. Am. Soc. Microbiol., Washington, DC. 188 pp.

3684. Skilling, D. D., and Nicholls, T. H. 1974. Brown spot needle disease—biology and control in Scotch pine plantations. USDA For. Serv. Res. Pap. NC-109. 19 pp.

3685. Skilling, D. D., Schneider, B., and Fasking, D. 1986. Biology and control of Scleroderris canker in North America. USDA For. Serv. Res. Pap. NC-275. 18 pp.

3686. Skroch, W. A., and Catanzaro, C. J. 1994. Variation in sensitivity of azaleas to herbicides. Int. Plant Prop. Soc. Comb. Proc. 43:395–396.

3687. Slagg, C. M., and Wright, E. 1943. Diplodia blight in coniferous seedbeds. Phytopathology 33:390–393.

3688. Sleeth, B., and Bidwell, C. B. 1937. *Polyporus hispidus* and a canker of oaks. J. For. 35:778–785.

3689. Slippers, B., Crous, P. W., Denman, S., Coutinho, T. A., Wingfield, B. D., and Wingfield, M. J. 2004. Combined multiple gene gen-ealogy and phenotypic characters differentiate several species previously identified as *Botryosphaeria dothidea*. Mycologia 96:83–101.

3690. Slippers, B., Johnson, G. I., Crous, P. W., Coutinho, T. A., Wingfield, B. D., and Wingfield, M. J. 2005. Phylogenetic and morphological re-evaluation of the *Botryosphaeria* species causing diseases of *Mangifera indica*. Mycologia (in press).

3691. Slovik, S. 1996. Early needle senescence and thinning of the crown structure of *Picea abies* as induced by chronic SO_2 pollution. II. Field data basis, model results and tolerance limits. Global Change Biol. 2:459–477.

3692. Smalley, E. B., and Guries, R. P. 1993. Breeding elms for resistance to Dutch elm disease. Annu. Rev. Phytopathol. 31:325–352.

3693. Smallidge, P. J., Leopold, D. J., and Castello, D. 1991. Structure and composition of forest stands affected and unaffected by ash yellows. Plant Dis. 75:13–18.

3694. Smart, C. D., Schneider, B., Blomquist, C. L., Guerra, L. J., Harrison, N. A., Ahrens, U., Lorenz, K. H., Seemüller, E., and Kirkpatrick, B. C. 1996. Phytoplasma-specific PCR primers based on sequences of the 16S–23S rRNA spacer region. Appl. Environ. Microbiol. 62:2988–2993.

3695. Smeltzer, D. L. K., and French, D. W. 1981. Factors affecting spread of *Cronartium comptoniae* on the sweetfern host. Can. J. For. Res. 11:400–408.

3696. Smerlis, E. 1962. Taxonomy and morphology of *Potebniamyces balsamicola* sp. nov. associated with a twig and branch blight of balsam fir in Quebec. Can. J. Bot. 40:351–359.

3697. Smerlis, E. 1967. Pathogenicity of *Phacidium abietis*. Plant Dis. Rep. 51:678–679.

3698. Smerlis, E. 1968. Pathogenicity of *Phacidium taxicolum* on Canadian yew. Plant Dis. Rep. 52:403–404.

3699. Smerlis, E. 1970. Pathogenicity of *Aleurodiscus amorphus*. Can. For. Serv. Bi-mon. Res. Notes 26:18.

3700. Smerlis, E. 1971. Pathogenicity tests of *Valsa* species occurring on conifers in Québec. Phytoprotection 52:28–31.

3701. Smerlis, E. 1973. Pathogenicity tests of some discomycetes occurring on conifers. Can. J. For. Res. 3:7–16.

3702. Smerlis, E., and Saint-Laurent, M. 1966. Pathogenicity of *Lophophacidium hyperboreum* Lagerberg. Plant Dis. Rep. 50:356–357.

3703. Smilansky, Z., and Umiel, N. 1984. Mutations and breeding in roses. I. Spontaneous mutations for flower colours in the Mercedes group of cultivars (Mercedes, Gabriela, Jaguar). Hassadeh 64:952–955.

3704. Smiley, E. T., Kielbaso, J. J., and Proffer, T. J. 1986. Maple disease epidemic in southeastern Michigan. J. Arboric. 12:126–128.

3705. Smit, W. A., Viljoen, C. D., Wingfield, B. D., Wingfield, M. J., and Calitz, F. J. 1996. A new canker disease of apple, pear, and plum rootstocks caused by *Diaporthe ambigua* in South Africa. Plant Dis. 80:1331–1335.

3706. Smit, W. A., Wingfield, B. D., and Wingfield, M. J. 1997. Vegetative incompatibility in *Diaporthe ambigua*. Plant Pathol. 46:366–372.

3707. Smith, B. J., Shearer, B. L., and Sivasithamparam, K. 1997. Compartmentalization of *Phytophthora cinnamomi* in stems of highly susceptible *Banksia brownii* treated with phosphonate. Mycol. Res. 101:1101–1107.

3708. Smith, C. O. 1928. Oleander bacteriosis in California. Phytopathology 18:503–518.

3709. Smith, D. G., and Pearce, C. M. 2000. River ice and its role in limiting woodland development on a sandy braid-plain, Milk River, Montana. Wetlands 20:232–250.

3710. Smith, D. H., Lewis, F. H., and Wainwright, S. H. 1970. Epidemiology of the black knot disease of plums. Phytopathology 60:1441–1444.

3711. Smith, D. R., Bronson, J. J., and Stanosz, G. R. 2003. Host-related variation among isolates of the Sirococcus shoot blight pathogen from conifers. For. Pathol. 33:141–156.

3712. Smith, D. R., and Stanosz, G. R. 2001. Molecular and morphological differentiation of *Botryosphaeria dothidea* (anamorph *Fuisicoccum aesculi*) from some other fungi with *Fusicoccum* anamorphs. Mycologia 93:505–515.

3713. Smith, E. M. 1975. Tree stress from salts and herbicides. J. Arboric. 1:201–205.

3714. Smith, E. M., and Treaster, S. A. 1981. Preventing habitual iron chlorosis of woody landscape plants. Ohio Agric. Res. Devel. Cent. Res. Circ. 263:26–29.

3715. Smith, F. A., and Smith, S. E. 1997. Structural diversity in (vesicular)-arbuscular mycorrhizal symbioses. New Phytol. 137:373–388.

3716. Smith, G. C., and Brennan, E. G. 1984. Response of honeylocust cultivars to air pollution stress in an urban environment. J. Arboric. 10:289–293.

3717. Smith, H., Crous, P. W., Wingfield, M. J., Coutinho, T., and Wingfield, B. D. 2001. *Botryosphaeria eucalyptorum* sp. nov., a new species in the *B. dothidea*–complex on *Eucalyptus* in South Africa. Mycologia 93:277–285.

3718. Smith, H., Wingfied, M. J., and Coutinho, T. A. 2002. The role of latent *Sphaeropsis sapinea* infections in post-hail associated dieback of *Pinus patula*. For. Ecol. Manag. 164:177–184.

3719. Smith, H., Wingfield, M. J., Crous, P. W., and Coutinho, T. A. 1996. *Sphaeropsis sapinea* and *Botryosphaeria dothidea* endophytic in *Pinus* spp. and *Eucalyptus* spp. in South Africa. S.Afr. J. Bot. 62:86–88.

3720. Smith, H., Wingfield, M. J., and Petrini, O. 1996. *Botryosphaeria dothidea* endophytic in *Eucalyptus grandis* and *Eucalyptus nitens* in South Africa. For. Ecol. Manag. 89:189–195.

3721. Smith, H. C., and MacDonald, W. L., eds. 1982. Proceedings of the USDA Forest Service American chestnut cooperators' meeting. West Va. Univ. Books, Morgantown. 229 pp.

3722. Smith, I. M., Dunez, J., Phillips, D. H., Lelliott, R. A., and Archer, S. A., eds. 1988. European handbook of plant diseases. Blackwell Scientific, Boston. 583 pp.

3723. Smith, J. A., Blanchette, R. A., Ostry, M. E., and Anderson, N. A. 2002. Etiology of bronze leaf disease of *Populus*. Plant Dis. 86:462–469.

3724. Smith, K. T., and Houston, D. R. 1994. Metal concentrations in wood of sugar maple infected with sapstreak disease. Can. J. For. Res. 24:185–188.

3725. Smith, K. T., and Sutherland, E. K. 1999. Fire-scar formation and compartmentalization in oak. Can. J. For. Res. 29:166–171.

3726. Smith, L. D. 1983. Major nutrient influence on *Verticillium dahliae* infections of *Acer saccharum*. J. Arboric. 9:277–281.

3727. Smith, L. D., and Neely, D. 1979. Relative susceptibility of tree species to *Verticillium dahliae*. Plant Dis. Rep. 63:328–332.

3728. Smith, L. J., and Skroch, W. A. 1995. Turf herbicide injury to landscape trees as influenced by mulch. J. Environ. Hortic. 13:60–63.

3729. Smith, M. L., Bruhn, J. N., and Anderson, J. B. 1992. The fungus *Armillaria bulbosa* is among the largest and oldest living organisms. Nature 356:428–431.

3730. Smith, R. B. 1973. Factors affecting dispersal of dwarf mistletoe seeds from an overstory western hemlock tree. Northwest Sci. 47:9–19.

3731. Smith, R. B., and Wass, E. F. 1979. Infection trials with three dwarf mistletoe species within and beyond their known ranges in British Columbia. Can. J. Plant Pathol. 1:47–57.

3732. Smith, R. B., Wass, E. F., and Meagher, M. D. 1993. Evidence of resistance to hemlock dwarf mistletoe (*Arceuthobium tsugense*) in western hemlock (*Tsuga heterophylla*) clones. Eur. J. For. Pathol. 23:163–170.

3733. Smith, R. S. Jr. 1966. Effect of diurnal temperature fluctuations on the charcoal root disease of *Pinus lambertiana*. Phytopathology 56:61–64.

3734. Smith, R. S. Jr. 1967. Verticicladiella root disease of pines. Phytopathology 57:935–938.

3735. Smith, R. S. Jr. 1973. Sirococcus tip dieback of *Pinus* spp. in California forest nurseries. Plant Dis. Rep. 57:69–73.

3736. Smith, S. E., Dickson, S., and Smith, F. A. 2001. Nutrient transfer in arbuscular mycorrhizas: how are fungal and plant processes integrated? Austral. J. Plant Physiol. 28:683–694.

3737. Smith, S. E., and Read, D. J. 1997. Mycorrhizal symbiosis. 2nd ed. Academic Press, San Diego. 605 pp.

3738. Smith, V. L. 1993. Canker of Japanese maple caused by *Colletotrichum acutatum*. Plant Dis. 77:197–198.

3739. Smith, V. L. 1994. First report of anthracnose caused by *Discula destructiva* on *Cornus florida × kousa* hybrid dogwood. Plant Dis. 78:100.

3740. Smith, V. L., Punja, Z. K., and Jenkins, S. F. 1986. A histological study of infection of host tissue by *Sclerotium rolfsii*. Phytopathology 76:755–759.

3741. Smith, W. H. 1981. Air pollution and forests. Interactions between air contaminants and forest ecosystems. Springer-Verlag, New York. 379 pp.

3742. Smith, W. H. 2000. Ice and forest health. North. J. Appl. For. 17:16–19.

3743. Smith-Fiola, D. 1995. Pest resistant ornamental plants. Rutgers Coop. Exten., Toms River, NJ. 29 pp.

3744. Smits, B. G., and Noguera, R. 1988. The ontogeny and morphogenesis of sclerotia and pycnidia of *Macrophomina phaseolina*. Agron. Trop. 1988(4–6):69–78.

3745. Sneh, B., Burpee, L., and Ogoshi, A. 1991. Identification of *Rhizoctonia* species. APS Press, St. Paul, MN. 133 pp.

3746. Sneh, B., Jabaji-Hare, S., Neate, S., and Dijst, G., eds. 1996. *Rhizoctonia* species: taxonomy, molecular biology, ecology, pathology, and disease control. Kluwer Academic, Dordrecht, Boston. 578 pp.

3747. Snetsinger, R., and Himelick, E. B. 1957. Observations on witches'-broom of hackberry. Plant Dis. Rep. 41:541–544.

3748. Snow, G. A., Beland, J. W., and Czabator, F. J. 1974. Formosan sweetgum susceptible to North American *Endothia gyrosa*. Phytopathology 64:602–605.

3749. Snyder, B. A., Adams, G. C., and Fulbright, D. W. 1989. Association of a virus-like particle with a diseased isolate of *Leucostoma persoonii*. Mycologia 81:241–247.

3750. Snyder, E. B., and Derr, H. J. 1972. Breeding longleaf pines for resistance to brown spot needle blight. Phytopathology 62:325–329.

3751. Snyder, M. A., Fineschi, B., Linhart, Y. B., and Smith, R. H. 1996. Multivariate discrimination of host use by dwarf mistletoe *Arceuthobium vaginatum* subsp. *cryptopodum*: inter- and intraspecific comparisons. J. Chem. Ecol. 22:295–305.

3752. Snydor, T. D., and Kuhns, L. 1976. Visual symptoms of copper toxicity on woody ornamentals. Weeds Trees Turf 15(7):58–59.

3753. Sobers, E. K. 1964. Cercospora diseases of *Ligustrum*. Proc. Fla. State Hortic. Soc. 77:486–489.

3754. Sobers, E. K. 1967. The perfect stage of *Coniothyrium concentricum* on leaves of *Yucca aloifolia*. Phytopathology 57:234–235.

3755. Sobers, E. K., and Seymour, C. P. 1967. *Cylindrocladium floridanum* sp. n. associated with decline of peach trees in Florida. Phytopathology 57:389–393.

3756. Solel, Z., and Bruck, R. I. 1989. Effect of nitrogen fertilization and growth suppression on pitch canker development on loblolly pine seedlings. J. Phytopathol. 125:327–335.

3757. Solel, Z., and Kimchi, M. 1997. Susceptibility and resistance of citrus genotypes to *Alternaria alternata* pv. *citri*. J. Phytopathol. 145:389–391.

3758. Solel, Z., and Kimchi, M. 1998. Histopathology of infection of Minneola tangelo by *Alternaria alternata* pv. *citri* and the effect of host and environmental factors on lesion development. J. Phytopathol. 146:557–561.

3759. Solel, Z., Madar, Z., Kimchi, M., and Golan, Y. 1987. Diplodia canker of cypress. Can. J. Plant Pathol. 9:115–118.

3760. Solla, A., Tomlinson, F., and Woodward, S. 2002. Penetration of *Picea sitchensis* root bark by *Armillaria mellea*, *Armillaria ostoyae* and *Heterobasidion annosum*. For. Pathol. 32:55–70.

3761. Sombrero, C. 1996. Enigma of variegations. New Plantsman 3:158–169.

3762. Sommer, N. F. 1955. Sunburn predisposes walnut trees to branch wilt. Phytopathology 45:607–613.

3763. Sorbo, G. del, Scala, F., Parrella, G., Lorito, M., Comparini, C., Ruocco, M., and Scala, A. 2000. Functional expression of the gene *cu*, encoding the phytotoxic hydrophobin cerato-ulmin, enables *Ophiostoma quercus*, a nonpathogen on elm, to cause symptoms of Dutch elm disease. Mol. Plant-Microbe Interact. 13:43–53.

3764. Sousa, A. J. T. de, 1985. Control of *Rosellinia necatrix* (Hartig) Berlese, causal agent of white root rot: susceptibility of several plant species and chemical control. Eur. J. For. Pathol. 15:323–332.

3765. Southey, J. F., ed. 1978. Plant nematology. 3rd ed. HMSO, London. 440 pp.

3766. Spaine, P. C., and Kaneko, S. 1996. Nuclear behavior during basidiospore germination in *Cronartium quercuum* f.sp. *fusiforme*. Mycologia 88:892–896.

3767. Spaink, H. P., Kondorosi, A., and Hooykaas, P. J. J. 1998. The Rhizobiaceae: molecular biology of model plant-associated bacteria. Kluwer Academic, Dordrecht, Boston. 566 pp.

3768. Spanos, K. A., Pirrie, A., and Woodward, S. 2001. Screening for resistance to *Seiridium cardinale*, *S. cupressi*, and *S. unicorne* isolates in glasshouse-grown seedlings of Cupressaceae. Silvae Genet. 50:258–264.

3769. Spanos, Y. A., and Woodward, S. 1994. The effects of *Taphrina betulina* infection on growth of *Betula pubescens*. Eur. J. For. Pathol. 24:277–286.

3770. Spaulding, P. 1922. Investigations of the white-pine blister rust. USDA Bull. 957. 100 pp.

3771. Spaulding, P., and Bratton, A. W. 1946. Decay following glaze storm damage in woodlands of central New York. J. For. 44:515–519.

3772. Spaulding, P., and Hansbrough, J. R. 1932. *Cronartium comptoniae*, the sweetfern blister rust of pitch pines. USDA Tech. Bull. 217. 21 pp.

3773. Spaulding, P., and MacAloney, H. J. 1931. A study of organic factors concerned in the decadence of birch on cut-over lands in northern New England. J. For. 29:1134–1149.

3774. Spencer, D. M., ed. 1978. The powdery mildews. Academic Press, New York. 565 pp.

3775. Spencer, D. M., ed. 1981. The downy mildews. Academic Press, New York. 636 pp.

3776. Spencer, J. A., and Wood, O. W. 1992. Resistance of selected rose cultivars to variants of *Marssonina rosae* in Mississippi. J. Environ. Hortic. 10:235–238.

3777. Spielman, L. J. 1983. Taxonomy and biology of *Valsa* species on hardwoods in North America, with special reference to species on maples. Ph.D. diss. Cornell Univ., Ithaca, NY. 175 pp.

3778. Spielman, L. J. 1985. A monograph of *Valsa* on hardwoods in North America. Can. J. Bot. 63:1355–1378.

3779. Spiers, A. G. 1983. Host range and pathogenicity studies of *Marssonina brunnea* to poplars. Eur. J. For. Pathol. 13:181–196.

3780. Spiers, A. G. 1983. Host range and pathogenicity studies of *Marssonina castagnei* to poplars. Eur. J. For. Pathol. 13:218–227.

3781. Spiers, A. G. 1984. Comparative studies of host specificity and symptoms exhibited by poplars infected with *Marssonina brunnea, Marssonina castagnei*, and *Marssonina populi*. Eur. J. For. Pathol. 14:202–218.

3782. Spiers, A. G. 1985. Factors affecting basidiospore release by *Chondrostereum purpureum* in New Zealand. Eur. J. For. Pathol. 15:111–126.

3783. Spiers, A. G. 1988. Comparative studies of type and herbarium specimens of *Marssonina* species pathogenic to poplars. Eur. J. For. Pathol. 18:140–156.

3784. Spiers, A. G. 1990. Influence of environmental, host and cultural factors on conidium morphology of *Marssonina* species pathogenic to poplars. Eur. J. For. Pathol. 20:154–166.

3785. Spiers, A. G., Brewster, D. T., Bus, V. G., and Hopcroft, D. H. 1998. Seasonal variation in susceptibility of xylem tissue of *Malus, Pyrus, Prunus*, and *Salix* species to *Chondrostereum purpureum* in New Zealand. Mycol. Res. 102:881–890.

3786. Spiers, A. G., Edwards, W. R. N., and Hopcroft, D. H. 1987. Effects of silverleaf infection on ultrastructure of foliage of *Prunus, Rosa* and *Populus*. N.Z. J. Bot. 25:411–423.

3787. Spiers, A. G., and Hopcroft, D. H. 1983. Ultrastructural study of the pathogenesis of *Marssonina* species to poplars. Eur. J. For. Pathol. 13:414–427.

3788. Spiers, A. G., and Hopcroft, D. H. 1985. Ultrastructural studies of the spermatial and aecial stages of *Melampsora larici-populina* and *Melampsora epitea* on *Larix decidua*. N.Z. J. Bot. 23:101–116.

3789. Spiers, A. G., and Hopcroft, D. H. 1985. Ultrastructural studies of pathogenesis and uredinial development of *Melampsora larici-populina* and *M. medusae* on poplar and *M. coleosporioides* and *M. epitea* on willow. N.Z. J. Bot. 23:117–133.

3790. Spiers, A. G., and Hopcroft, D. H. 1988. Factors affecting *Chondrostereum purpureum* infection of *Salix*. Eur. J. For. Pathol. 18:257–278.

3791. Spiers, A. G., and Hopcroft, D. H. 1988. Penetration and infection of poplar leaves by urediniospores of *Melampsora larici-populina* and *Melampsora medusae*. N.Z. J. Bot. 26:101–111.

3792. Spiers, A. G., and Hopcroft, D. H. 1988. Ultrastructural studies of basidial and basidiospore development and basidiospore release in *Chondrostereum purpureum*. Eur. J. For. Pathol. 18:367–381.

3793. Spiers, A. G., and Hopcroft, D. H. 1988. Ultrastructural studies of the telial, basidial, and spermatial stages of the willow rust fungus *Melampsora coleosporioides* in New Zealand. N.Z. J. Bot. 26:423–430.

3794. Spiers, A. G., and Hopcroft, D. H. 1990. Ultrastructural studies of interactions between resistant and susceptible poplar cultivars and the rusts, *Melampsora medusae* and *Melampsora larici-populina*. N.Z. J. Bot. 28:307–322.

3795. Spiers, A. G., and Hopcroft, D. H. 1992. Some electron microscope observations of conidium ontogeny of *Sphaceloma murrayae* on *Salix*. N.Z. J. Bot. 30:353–358.

3796. Spiers, A. G., and Hopcroft, D. H. 1993. Black canker and leaf spot of *Salix* in New Zealand caused by *Glomerella miyabeana* (*Colletotrichum gloeosporioides*). Eur. J. For. Pathol. 23:92–102.

3797. Spiers, A. G., and Hopcroft, D. H. 1994. Comparative studies of the poplar rusts *Melampsora medusae, M. larici-populina* and their interspecific hybrid *M. medusae-populina*. Mycol. Res. 98:889–903.

3798. Spiers, A. G., and Hopcroft, D. H. 1998. Morphology of *Drepanopeziza* species pathogenic to poplars. Mycol. Res. 102:1025–1037.

3799. Spiers, A. G., and Wenham, H. T. 1983. Poplar seed-transmission of *Marssonina brunnea*. Eur. J. For. Pathol. 13:305–314.

3800. Spooner, D. M. 1983. The northern range of eastern mistletoe, *Phoradendron serotinum* (Viscaceae), and its status in Ohio. Bull. Torrey Bot. Club 110:489–493.

3801. Spotts, R. A. 1980. Infection of grape by *Guignardia bidwellii*—factors affecting lesion development, conidial dispersal, and conidial populations on leaves. Phytopathology 70:252–255.

3802. Spotts, R. A., and Cervantes, L. A. 1994. Factors affecting maturation and release of ascospores of *Venturia pyrina* in Oregon. Phytopathology 84:260–264.

3803. Spotts, R. A., and Cervantes, L. A. 1995. Factors affecting the severity of bacterial canker of pear caused by *Pseudomonas syringae* pv. *syringae*. Plant Pathol. 44:325–331.

3804. Spotts, R. A., Cervantes, L. A., and Niederholzer, F. J. A. 1997. Effect of dolomitic lime on production of asci and pseudothecia of *Venturia inaequalis* and *V. pirina*. Plant Dis. 81:96–98.

3805. Spotts, R. A., Covey, R. P., and Chen, P. M. 1981. Effect of low temperature on survival of apple buds infected with the powdery mildew fungus. HortScience 16:781–783.

3806. Sprague, R., and Heald, F. D. 1927. A witches' broom of the service berry. Trans. Am. Microscop. Soc. 46:219–247.

3807. Sproston, T. Jr., and Scott, W. W. 1954. *Valsa leucostomoides*, the cause of decay and discoloration in tapped sugar maples. Phytopathology 44:12–13.

3808. Sproule, A. 1996. Branch age in jack pine at the time of dwarf mistletoe infection. For. Chron. 72:307.

3809. Sproule, A. 1996. Impact of dwarf mistletoe on some aspects of the reproductive biology of jack pine. For. Chron. 72:303–306.

3810. Squillace, A. E., Dinus, R. J., Hollis, C. A., and Schmidt, R. A. 1978. Relation of oak abundance, seed source, and temperature to geographic patterns of fusiform rust incidence. USDA For. Serv. Res. Pap. SE-186. 20 pp.

3811. Sreenivasaprasad, S., Mills, P. R., Meehan, B. M., and Brown, A. E. 1996. Phylogeny and systematics of 18 *Colletotrichum* species based on ribosomal DNA spacer sequences. Genome 39:499–512.

3812. Srinivasan, K. G., Narendrakumar, R., and Wong, S. M. 2002. Hibiscus virus S is a new subgroup II tobamovirus: evidence from its unique coat protein and movement protein sequences. Arch. Virol. 147:1585–1598.

3813. Srivastava, L. M., and Esau, K. 1961. Relation of dwarfmistletoe (*Arceuthobium*) to the xylem tissue of conifers. I. Anatomy of parasite sinkers and their connection with host xylem. Am. J. Bot. 48:159–167.

3814. Srivastava, P. S., and Glock, H. 1987. FPA-induced fasciation of shoots of birch in vitro. Phytomorphology 37:395–399.

3815. Stacey, G., Burris, R. H., and Evans, H. J. 1992. Biological nitrogen fixation. Chapman & Hall, London. 943 pp.

3816. Stack, R. W., and Ash, C. L. 1979. Botryodiplodia canker of American elm. (Abstr.) J. Arboric. 5:163–164.

3817. Stackebrandt, E. 1997. Identifying culturable and uncultured prokaryotes. Symbiosis 22:47–58.

3818. Stafford, H. A. 2000. Crown gall disease and *Agrobacterium tumefaciens*: a study of the history, present knowledge, missing information, and impact on molecular genetics. Bot. Rev. 66:99–118.

3819. Stahl, S. A., Rogers, J. D., and Adams, M. J. 1988. Observations on *Hendersonia pinicola* and the needle blight of *Pinus contorta*. Mycotaxon 31:323–338.

3820. Staley, J. M. 1964. A new *Lophodermium* on ponderosa pine. Mycologia 56:757–762.

3821. Staley, J. M. 1965. Decline and mortality of red and scarlet oaks. For. Sci. 11:2–17.

3822. Staley, J. M. 1979. *Lophodermella cerina*, a pathogen of pine foliage. Phytopathology 69:1045–1046.

3823. Staley, J. M., and Hawksworth, F. G. 1967. *Bifusella crepidiformis* on Englemann spruce. Plant Dis. Rep. 51:791–792.

3824. Stalpers, J. A. 1996. The Aphyllophoraceous fungi—II. Keys to the species of the Hericiales. Stud. Mycol. No. 40. 185 pp.

3825. Stamps, D. J., Waterhouse, G. M., Newhook, F. J., and Hall, G. S. 1990. Revised tabular key to the species of *Phytophthora*. CAB Int. Mycol. Pap. No. 162. 28 pp.

3826. Standiford, R., McCreary, D., and Purcell, K. L., tech coords. 2003. Proceedings of the Fifth Symposium on Oak Woodlands: Oaks in California's Changing Landscape. USDA For. Serv. Gen. Tech. Rep. PSW-GTR-184. 846 pp. Released also as Internet pub.

3827. Standish, E. D., MacDonald, J. D., and Humphrey, W. A. 1982. Phytophthora root and crown rot of junipers in California. Plant Dis. 66:925–928.

3828. Stanek, W., Hopkins, J. C., and Simmons, C. S. 1986. Effects of spacing in lodgepole pine stands on incidence of Atropellis canker. For. Chron. 62:91–95.

3829. Stanghellini, M. E., Rasmussen, S. L., and Kim, D. H. 1999. Aerial transmission of *Thielaviopsis basicola*, a pathogen of corn-salad, by adult shore flies. Phytopathology 89:476–479.

3830. Stanosz, G. R. 1992. Effect of cherry leaf spot on nursery black cherry seedlings and potential benefits from control. Plant Dis. 76:602–604.

3831. Stanosz, G. R. 1993. Symptoms, association and pathogenicity of *Discula campestris*, a cause of sugar maple seedling anthracnose. Plant Dis. 77:1022–1026.

3832. Stanosz, G. R. 1994. Benomyl and acephate applications increase survival of sugar maple seedlings during their first growing season in northern Pennsylvania. Can. J. For. Res. 24:1107–1111.

3833. Stanosz, G. R., Blodgett, J. T., Smith, D. R., and Kruger, E. L. 2001. Water stress and *Sphaeropsis sapinea* as a latent pathogen of red pine seedlings. New Phytol. 149:531–538.

3834. Stanosz, G. R., and Carlson, J. C. 1996. Association of mortality of recently planted seedlings and established saplings in red pine plantations with Sphaeropsis collar rot. Plant Dis. 80:750–753.

3835. Stanosz, G. R., and Moorman, G. W. 1997. Branch dieback of Savin juniper in Pennsylvania caused by *Diplodia mutila*. Plant Dis. 81:111.

3836. Stanosz, G. R., and Patton, R. F. 1990. Stump colonization by *Armillaria* in Wisconsin aspen stands following clearcutting. Eur. J. For. Pathol. 20:339–346.

3837. Stanosz, G. R., Smith, D. R., Guthmiller, M. A., and Stanosz, J. C. 1997. Persistence of *Sphaeropsis sapinea* on or in asymptomatic shoots of red and jack pines. Mycologia 89:525–530.

3838. Stanosz, G. R., Swart, W. J., and Smith, D. R. 1998. Similarity between fungi identified as *Diplodia pinea* f.sp. *cupressi* in Israel and *Botryosphaeria stevensii* or *Diplodia mutila* on *Juniperus* in the United States. Eur. J. For. Pathol. 28:33–42.

3839. Stanosz, G. R., Swart, W. J., and Smith, D. R. 1999. RAPD marker and isozyme characterization of *Sphaeropsis sapinea* from diverse coniferous hosts and locations. Mycol. Res. 103:1193–1202.

3840. Stapleton, J. J., Ferguson, L., McKenry, M. V., Dougherty, D. S., and Stapleton, S. C. 1999. Using solarization to disinfest soil for olive nursery production. Acta Hortic. No. 474:589–591.

3841. Starkey, D. A., Oak, S. W., Ryan, G. W., Tainter, F. H., Redmond, C., and Brown, H. D. 1989. Evaluation of oak decline areas in the South. USDA For. Serv. Prot. Rep. R8-PR-17. 36 pp.

3842. Stasiak, M. A., Hofstra, G., Payne, N. J., Prasad, R., and Fletcher, R. A. 1991. Alterations of growth and shikimic acid levels by sublethal glyphosate applications on pin cherry and trembling aspen. Can. J. For. Res. 21:1086–1090.

3843. Stathis, P. D., and Plakidas, A. G. 1958. Anthracnose of azaleas. Phytopathology 48:256–260.

3844. Stathis, P. D., and Plakidas, A. G. 1959. Entomosporium leaf spot of *Photinia glabra* and *Photinia serrulata*. Phytopathology 49:361–365.

3845. Steinmetz, H. F., and Hilborn, M. T. 1937. A histological evaluation of low temperature injury to apple trees. Maine Agric. Exp. Stn. Bull. 388. 32 pp.

3846. Stelzer, H. E., Doudrick, R. L., Kubisiak, T. L., and Nelson, C. D. 1999. Prescreening slash pine and *Cronartium* pedigrees for evaluation of complementary gene action in fusiform rust disease. Plant Dis. 83:385–389.

3847. Stenlid, J., and Johansson, M. 1987. Infection of roots of Norway spruce (*Picea abies*) by *Heterobasidion annosum*. II. Early changes in phenolic content and toxicity. Eur. J. For. Pathol. 17:217–226.

3848. Stensvand, A., Amundsen, T., Semb, L., Gadoury, D. M., and Seem, R. C. 1998. Discharge and dissemination of ascospores by *Venturia inaequalis* during dew. Plant Dis. 82:761–764.

3849. Stephan, B. R. 1980. Prüfung von Douglasien-Herkünften auf Resistenz gegen *Rhabdocline pseudotsugae* in Infektionsversuchen. Eur. J. For. Pathol. 10:152–161.

3850. Stephan, B. R., and Hyun, S. K. 1983. Studies on the specialization of *Cronartium ribicola* and its differentiation on the alternate hosts *Ribes* and *Pedicularis*. Z. Pflanzenkr. Pflanzenschutz 90:670–678.

3851. Stephan, B. R., and Millar, C. S., comps. 1975. *Lophodermium* on pines. Proceedings of the 5th European colloquium on forest pathology, Schmalenbeck, 1975. Mitt. Bundes. Forst- u. Holzw. 108:1–201.

3852. Stermer, B. A., Scheffer, R. P., and Hart, J. H. 1984. Isolation of toxins of *Hypoxylon mammatum* and demonstration of some toxin effects on selected clones of *Populus tremuloides*. Phytopathology 74:654–658.

3853. Stern, A. C., ed. 1976. Air pollution. 3rd ed. Vol. 2: The effects of air pollution. Academic Press, New York. 684 pp.

3854. Stevens, N. E. 1926. Two species of *Physalospora* on citrus and other hosts. Mycologia 18:206–217.

3855. Stevens, N. E. 1933. Life history and synonomy of *Physalospora glandicola*. Mycologia 25:504–508.

3856. Stevens, N. E. 1933. Two apple black rot fungi in the United States. Mycologia 25:536–548.

3857. Stevens, N. E. 1936. Two species of *Physalospora* in England. Mycologia 28:331–336.

3858. Stewart, E. L., Liu, Z., Crous, P. W., and Szabo, L. J. 1999. Phylogenetic relationships among some cercosporoid anamorphs of *Mycosphaerella* based on rDNA sequence analysis. Mycol. Res. 103:1491–1499.

3859. Stewart, F. C. 1917. Witches-brooms on hickory trees. Phytopathology 7:185–187.

3860. Stewart, J. E., Halik, S., and Bergdahl, D. R. 2004. Viability of *Sirococcus clavigignenti-juglandacearum* conidia on exoskeletons of three coleopteran species. Plant Dis. 88:1085–1091.

3861. Stewart, V. B. 1916. The leaf blotch disease of horse-chestnut. Phytopathology 6:5–19.

3862. Steyaert, R. 1967. Considérations générales sur le genre *Ganoderma* et plus spécialement sur les espèces européennes. Bull. Soc. R. Bot. Belg. 100:189–211.

3863. Steyaert, R. 1967. Les *Ganoderma* palmicoles. Bull. Jard. Bot. Natl. Belg. 37:465–492.

3864. Steyaert, R. 1975. The concept and circumscription of *Ganoderma tornatum*. Trans. Br. Mycol. Soc. 65:451–467.

3865. Steyaert, R. L. 1980. Study of some *Ganoderma* species. Bull. Jard. Bot. Natl. Belg. 50:135–186.

3866. Sticklen, M. B., and Sherald, J. M., eds. 1993. Dutch elm disease research: cellular and molecular approaches. Springer-Verlag, New York. 344 pp.

3867. Stiles, K. C., and Martin, C. E. 1996. Effects of drought stress on CO_2 exchange and water relations in the CAM epiphyte *Tillandsia utriculata* (Bromeliaceae). J. Plant Physiol. 149:721–728.

3868. Stillwell, M. A. 1964. The fungus associated with woodwasps occurring in beech in New Brunswick. Can. J. Bot. 42:495–496.

3869. Stimm, B., and Dong, P. H. 2001. The Kaiserslautern Douglas fir provenance trial after nine decades of observation. Forstwiss. Centralbl. 120:173–186.

3870. Stipes, R. J. Jr. 1972. Comparative susceptibility of eight species of *Albizia* to fusarium wilt caused by *Fusarium oxysporum* f.sp. *perniciosum*. Va. J. Sci. 35(2):72.

3871. Stipes, R. J. [Jr.], and Campana, R. J., eds. 1981. Compendium of elm diseases. Am. Phytopathol. Soc., St. Paul, MN. 96 pp.

3872. Stipes, R. J. [Jr.], and Phipps, P. M. 1971. A species of *Endothia* associated with a canker disease of pin oak (*Quercus palustris*) in Virginia. Plant Dis. Rep. 55:467–469.

3873. Stoessl, A., Abramowski, Z., Lester, H. H., Rock, G. L., and Towers, G. H. N. 1990. Further toxic properties of the fungal metabolite dothistromin. Mycopathologia 112:179–186.

3874. Stone, E. L. Jr. 1952. An unusual type of frost injury in pine. J. For. 50:560–561.

3875. Stone, E. L. Jr. 1953. Magnesium deficiency of some northeastern pines. Proc. Soil Sci. Soc. Am. 17:297–300.

3876. Stone, E. L. [Jr.] 1968. Microelement nutrition of forest trees: a review. Pages 132–175 in: Forest fertilization: theory and practice. Tenn. Valley Authority, Muscle Shoals, AL. 306 pp.

3877. Stone, E. L. [Jr.] 1990. Boron deficiency and excess in forest trees: a review. For. Ecol. Manag. 37:49–75.

3878. Stone, E. L. [Jr.], and Baird, G. 1956. Boron level and boron toxicity in red and white pine. J. For. 54:11–12.

3879. Stone, E. L. [Jr.], and Cornwell, S. 1968. Basal bud burls in *Betula populifolia*. For. Sci. 14:64–65.

3880. Stone, E. L. [Jr.], and Greweling, T. 1971. Arsenic toxicity in red pine and the persistence of arsenic in nursery soils. Tree Planters' Notes 22(1):5–7.

3881. Stone, E. L. [Jr.], Morrow, R. R., and Welch, D. S. 1954. A malady of red pine on poorly drained sites. J. For. 52:104–114.

3882. Stone, J. K. 1988. Fine structure of latent infections by *Rhabdocline parkeri* on Douglas-fir, with observations on uninfected epidermal cells. Can. J. Bot. 66:45–54.

3883. Storer, A. J., Gordon, T. R., and Clark, S. L. 1998. Association of the pitch canker fungus, *Fusarium subglutinans* f.sp. *pini*, with Monterey pine seeds and seedlings in California. Plant Pathol. 47:649–656.

3884. Storey, R., and Walker, R. R. 1999. Citrus and salinity. Sci. Hortic. 78:39–81.

3885. Stowell, E. A., and Backus, M. P. 1966, 1967. Morphology and cytology of *Diplocarpon maculatum* on *Crataegus*. I. The *Entomosporium* stage. II. Initiation and development of the apothecium. Mycologia 58:949–960; 59:623–636.

3886. Straker, C. J. 1996. Ericoid mycorrhiza: ecological and host specificity. Mycorrhiza 6:215–225.

3887. Strandberg, J. O. 2001. A new disease of flowering dogwood caused by *Colletotrichum acutatum*. Plant Dis. 85:229.

3888. Streets, R. B., and Bloss, H. E. 1973. Phymatotrichum root rot. Monogr. 8. Am. Phytopathol. Soc., St. Paul, MN. 38 pp.

3889. Strider, D. L., and Jones, R. K. 1978. Rust of fuchsia in North Carolina. Plant Dis. Rep. 62:745–746.

3890. Strobel, N. E., Hussey, R. S., and Roncadori, R. W. 1982. Interaction of vesicular-arbuscular mycorrhizal fungi, *Meloidogyne incognita*, and soil fertility on peach. Phytopathology 72:690–694.

3891. Strobel, N. E., and Sinclair, W. A. 1991. Role of flavanolic wall infusions in the resistance induced by *Laccaria bicolor* to *Fusarium oxysporum* in primary roots of Douglas-fir. Phytopathology 81:420–425.

3892. Struckmeyer, B. E., Kuntz, J. E., and Riker, A. J. 1958. Histology of certain oaks infected with the oak wilt fungus. Phytopathology 48:556–561.

3893. Struckmeyer, B. E., and Riker, A. J. 1951. Wound-periderm formation in white-pine trees resistant to blister rust. Phytopathology 41:276–281.

3894. Stuntz, D. E., and Seliskar, C. E. 1943. A stem canker of dogwood and madrona. Mycologia 35:207–221.

3895. Subirats, F. J., and Self, R. L. 1971. A new petal blight of camellias incited by *Pestalotia* sp. Plant Dis. Rep. 55:697–700.

3896. Sucoff, E. 1975. Effect of deicing salts on woody vegetation along Minnesota roads. Minn. Agric. Exp. Stn. Tech. Bull. 303. 49 pp.

3897. Sugui, J. A., Pascholati, S. F., Kunoh, H., Howard, R. J., and Nicholson, R. L. 1998. Association of *Pestalotia malicola* with the plant cuticle: visualization of the pathogen and detection of cutinase and non-specific esterase. Physiol. Mol. Plant Pathol. 52:213–221.

3898. Suleman, P., and Steiner, P. W. 1994. Relationship between sorbitol and solute potential in apple shoots relative to fire blight symptom development after infection by *Erwinia amylovora*. Phytopathology 84:1244–1250.

3899. Summerell, B. A., Leslie, J. F., Backhouse, D., Bryden, W. L., and Burgess, L. W., eds. 2001. Fusarium: Paul E. Nelson memorial symposium. APS Press, St. Paul, MN. 392 pp.

3900. Sun, L. J. 1986. Identification of two viruses isolated from *Paulownia tomentosa*. Sci. Silvae Sinicae 22:142–146.

3901. Sun-Li, J., and Punja, Z. K. 1999. Morphological and molecular characterization of *Chalara elegans* (*Thielaviopsis basicola*), cause of black root rot on diverse plant species. Can. J. Bot. 77:1801–1812.

3902. Surico, G. 1993. Symptom development in olive and oleander leaves inoculated with *Pseudomonas syringae* subsp. *savastanoi* and scanning electron microscopy of infections. Petria 3:117–127.

3903. Surico, G., Comai, L., and Kosuge, T. 1984. Pathogenicity of strains of *Pseudomonas syringae* pv. *savastanoi* and their indoleacetic acid–deficient mutants on olive and oleander. Phytopathology 74:490–493.

3904. Surico, G., Iacobellis, N. S., and Sisto, A. 1985. Studies on the role of indole-3-acetic acid and cytokinins in the formation of knots on olive and oleander plants by *Pseudomonas syringae* pv. *savastanoi*. Physiol. Plant Pathol. 26:309–320.

3905. Sury, R. von, and Flückiger, W. 1991. Effects of air pollution and water stress on leaf blight and twig cankers of London planes (*Platanus acerifolia* (Ait.) Willd.) caused by *Apiognomonia veneta* (Sacc. & Speg.) Höhn. New Phytol. 118:397–405.

3906. Sutherland, J. R. 1977. Corky root disease of Douglas-fir seedlings: pathogenicity of the nematode *Xiphinema bakeri* alone and in combination with the fungus *Cylindrocarpon destructans*. Can. J. For. Res. 7:41–46.

3907. Sutherland, J. R., Hopkinson, S. J., and Farris, S. H. 1984. Inland spruce cone rust, *Chrysomyxa pirolata*, in *Pyrola asarifolia* and cones of *Picea glauca* and morphology of the spore stages. Can. J. Bot. 62:2441–2447.

3908. Sutherland, J. R., Lock, W., and Farris, S. H. 1981. Sirococcus blight: a seed-borne disease of container-grown spruce seedlings in coastal British Columbia forest nurseries. Can. J. Bot. 59:559–562.

3909. Sutherland, J. R., and Eerden, E. Van. 1980. Diseases and insect pests in British Columbia forest nurseries. B.C. Minist. For., Can. For. Serv. Joint Rep. 12. 55 pp.

3910. Suto, Y., and Kobayashi, T. 1993. Taxonomic studies on the species of *Pestalotiopsis*, parasitic on conifers in Japan. Trans. Mycol. Soc. Jap. 34:323–344.

3911. Sutton, B. C. 1969. Forest microfungi. III. The heterogeneity of *Pestalotia* de Not. Section *Sexloculatae* Klebahn *sensu* Guba. Can. J. Bot. 47:2083–2094.

3912. Sutton, B. C. 1973. *Tubakia* nom. nov. Trans. Br. Mycol. Soc. 60:164.

3913. Sutton, B. C. 1980. The Coelomycetes: Fungi Imperfecti with pycnidia, acervuli, and stromata. Commonw. Mycol. Inst., Kew, Surrey, UK. 696 pp.

3914. Sutton, B. C., and Dyko, B. J. 1989. Revision of *Hendersonula*. Mycol. Res. 93:466–488.

3915. Sutton, B. C., and Hennebert, G. L. 1994. Interconnection amongst anamorphs and their possible contribution to ascomycete systematics. Ascomycete systematics. Problems and perspectives in the nineties. D. L. Hawksworth, ed. NATO ASI Series 269:77–100.

3916. Sutton, B. C., and Hodges, C. S. 1990. Revision of *Cercospora*-like fungi on *Juniperus* and related conifers. Mycologia 82:313–325.

3917. Sutton, B. C., and Lawrence, J. J. 1969. Black rib of willows in Manitoba and Saskatchewan. Plant Dis. Rep. 53:101–102.

3918. Sutton, B. C., and Pascoe, I. G. 1989. Reassessment of *Peltosoma*, *Stigmina* and *Batcheloromyces* and description of *Hyphothyrium* gen. nov. Mycol. Res. 92:210–222.

3919. Sutton, B. C., and Pollack, F. G. 1973. *Gloeosporium cercocarpi* and *Sphaceloma cercocarpi*. Mycologia 65:1125–1134.

3920. Sutton, T. B. 1981. Production and dispersal of ascospores and conidia by *Physalospora obtusa* and *Botryosphaeria dothidea* in apple orchards. Phytopathology 71:584–589.

3921. Sutton, T. B., Brown, E. M., and Hawthorne, D. J. 1987. Biology and epidemiology of *Mycosphaerella pomi*, cause of Brooks fruit spot of apple. Phytopathology 77:431–437.

3922. Sutton, T. B., and Shane, W. W. 1983. Epidemiology of the perfect stage of *Glomerella cingulata* on apples. Phytopathology 73:1179–1183.

3923. Suzuki, K., and Kiyohara, T. 1978. Influence of water stress on development of pine wilting disease caused by *Bursaphelenchus lignicolus*. Eur. J. For. Pathol. 8:97–107.

3924. Svejda, F., and Bolton, A. T. 1980. Resistance of rose hybrids to three races of *Diplocarpon rosae* blackspot. Can. J. Plant Pathol. 2:23–25.

3925. Svihra, P., and McCain, A. H. 1992. Susceptibility of plane trees to anthracnose and powdery mildew in California. J. Arboric. 18:161–163.

3926. Svircev, A. M., Biggs, A. R., and Miles, N. W. 1991. Isolation and partial purification of phytotoxins from liquid cultures of *Leucostoma cincta* and *Leucostoma persoonii*. Can. J. Bot. 69:1998–2003.

3927. Swai, I. S., and Hindal, D. F. 1981. Selective medium for recovering *Verticicladiella procera* from soils and symptomatic white pines. Plant Dis. 65:963–965.

3928. Swart, H. J. 1973. The fungus causing cypress canker. Trans. Br. Mycol. Soc. 61:71–82.

3929. Swart, H. J., and Walker, J. 1988. Australian leaf-inhabiting fungi. XXVIII. *Hendersonia* on *Eucalyptus*. Trans. Br. Mycol. Soc. 90:633–641.

3930. Swart, L., Crous, P. W., Kang, J., McHau, G. R. A., Pascoe, I., Palm, M. E., and Kang, J. C. 2001. Differentiation of species of *Elsinoë* associated with scab disease of Proteaceae based on morphology, symptomatology, and ITS sequence phylogeny. Mycologia 93:366–379.

3931. Swart, W. J., Wingfield, M. J., and Grant, W. S. 1993. Comparison of *Sphaeropsis sapinea* and *Sphaeropsis sapinea* f.sp. *cupressi*. Mycol. Res. 97:1253–1260.

3932. Swart, W. J., Wingfield, M. J., and Wyk, P. van 1993. Variation in conidial morphology among geographic isolates of *Sphaeropsis sapinea*. Mycol. Res. 97:832–838.

3933. Swedjemark, G., Johannesson, H., and Stenlid, J. 1999. Intraspecific variation in *Heterobasidion annosum* for growth in sapwood of *Picea abies* and *Pinus sylvestris*. Eur. J. For. Pathol. 29:249–258.

3934. Sweet, J. B. 1980. Fruit tree virus infections of woody exotic and indigenous plants in Britain. Acta Phytopathol. Acad. Sci. Hung. 15:231–238.

3935. Sweet, J. B., and Barbara, D. J. 1979. A yellow mosaic disease of horse chestnut (*Aesculus* spp.) caused by apple mosaic virus. Ann. Appl. Biol. 92:335–341.

3936. Sweet, J. B., and Campbell, A. I. 1976. Pome fruit virus infections of some woody ornamental and indigenous species of Rosaceae. J. Hortic. Sci. 51:91–97.

3937. Swensen, S. M., and Mullin, B. C. 1997. The impact of molecular systematics on hypotheses for the evolution of root nodule symbioses and implications for expanding symbioses to new host plant genera. Plant Soil 194:185–192.

3938. Swingle, R. U. 1938. A phloem necrosis of elm. Phytopathology 28:757–759.

3939. Swingle, R. U., and Bretz, T. W. 1950. Zonate canker, a virus disease of American elm. Phytopathology 40:1018–1022.

3940. Swings, J. G., and Civerolo, E. L., eds. 1993. *Xanthomonas*. Chapman & Hall, New York. 399 pp.

3941. Sydnor, T. D., Chatfield, J. A., Zondang, R. H., Bennett, P. J., Boggs, J. F., and Cochran, K. D. 1995. Winter injury in woody ornamental plants in Ohio: the winter of 1993–1994. Ohio Agric. Res. Devel. Cent. Spec. Circ. 142:18–23.

3942. Sylvestre-Guinot, G. 1981. Étude de l'émission des ascospores du *Lachnellula willkommii* (Hartig) Dennis dans l'Est de la France. Eur. J. For. Pathol. 11:275–283.

3943. Sylvestre-Guinot, G. 1986. Study of the infection sites of *Lachnellula willkommii* on *Larix decidua*. Ann. Sci. For. 43:199–206.

3944. Sylvestre-Guinot, G., Paques, L., and Delatour, C. 1994. An inoculation method for early assessment of the behaviour of larch towards *Lachnellula willkommii*. Eur. J. For. Pathol. 24:160–170.

3945. Sylvestre-Guinot, G., Paques, L. E., and Delatour, C. 1999. Hybrid larch resistance to *Lachnellula willkommii*. Ann. Sci. For. 56:485–492.

3946. Szirmai, J. 1972. An *Acer* virus disease in maple trees planted in avenues. Acta Phytopathol. Acad. Sci. Hung. 7:197–207.

3947. Sztejnberg, A., and Madar, Z. 1980. Host range of *Dematophora necatrix*, the cause of white root rot disease in fruit trees. Plant Dis. 64:662–664.

3948. Sztejnberg, A., Madar, Z., and Chet, I. 1980. Induction and quantification of microsclerotia in *Rosellinia necatrix*. Phytopathology 70:525–527.

3949. Tabata, M. 1991. Distribution and host range of *Seiridium unicorne* in Japan. Trans. Mycol. Soc. Jap. 32:259–264.

3950. Tainter, F. H. 1973. Development of *Cronartium comandrae* in *Comandra umbellata*. Can. J. Bot. 51:1369–1372.

3951. Tainter, F. H., and Anderson, R. L. 1993. Twenty-six new pine hosts of fusiform rust. Plant Dis. 77:17–20.

3952. Tainter, F. H., and Baker, F. A. 1996. Principles of forest pathology. Wiley, New York. 805 pp.

3953. Tainter, F. H., and Fraedrich, S. W. 1986. Compartmentalization of *Ceratocystis fagacearum* in turkey oak in South Carolina. Phytopathology 76:698–701.

3954. Tainter, F. H., and French, D. W. 1971. The role of wound periderm in the resistance of eastern larch and jack pine to dwarf mistletoe. Can. J. Bot. 49:501–504.

3955. Tainter, F. H., and Gubler, W. D. 1973. Natural biological control of oak wilt in Arkansas. Phytopathology 63:1027–1034.

3956. Tainter, F. H., and Gubler, W. D. 1974. Effect of invasion by *Hypoxylon* and other microorganisms on carbohydrate reserves of oak-wilted trees. For. Sci. 20:337–342.

3957. Tainter, F. H., and Ham, D. L. 1983. The survival of *Ceratocystis fagacearum* in South Carolina. Eur. J. For. Pathol. 13:102–109.

3958. Tainter, F. H., O'Brien, J. G., Hernandez, A., Orozco, F., and Rebolledo, O. 2000. *Phytophthora cinnamomi* as a cause of oak mortality in the state of Colima, Mexico. Plant Dis. 84:394–398.

3959. Tainter, F. H., Retzlaff, W. A., Starkey, D. A., and Oak, S. W. 1990. Decline of radial growth in red oaks is associated with short-term changes in climate. Eur. J. For. Pathol. 20:95–105.

3960. Tainter, F. H., Williams, T. M., and Cody, J. B. 1983. Drought as a cause of oak decline and death on the South Carolina coast. Plant Dis. 67:195–197.

3961. Takai, S. 1980. Relationship of the production of the toxin, cerato-ulmin, to synnemata formation, pathogenicity, mycelial habit, and growth of *Ceratocystis ulmi* isolates. Can. J. Bot. 58:658–662.

3962. Takai, S., Richards, W. C., and Stevenson, K. J. 1983. Evidence for the involvement of cerato-ulmin, the *Ceratocystis ulmi* toxin, in the development of Dutch elm disease. Physiol. Plant Pathol. 23:275–280.

3963. Takayama, T., Yoshida, H., Araki, K., and Nakayama, K. 1983. Microbial production of abscisic acid with *Cercospora rosicola*. I. Stimulation of abscisic acid accumulation by plant extracts. Biotech. Lett. 5:55–58.

3964. Tan, P., and Whitlow, T. 2001. Physiological responses of *Catharanthus roseus* (periwinkle) to ash yellows phytoplasmal infection. New Phytol. 150:757–769.

3965. Taris, B., and Avenard, J.-C. 1969. Comparaison de l'évolution au printemps des attaques de deux Taphrinales, *Taphrina aurea* (Pers.) Fr. et *Taphrina deformans* (Berk.) Tul., se développant sur les jeunes bouquets foliaires de *Populus* et de *Prunus persica*. C. R. Hebd. Séances Acad. Sci., Paris D 268:3078–3081.

3966. Tarjan, A. C., and Howard, F. L. 1954. Detrimental effects of copper sprays to Norway maple in Rhode Island. Plant Dis. Rep. 38:58.

3967. Tate, R. L. 1980. Detection, description, and treatment of girdling roots on urban Norway maple trees. J. Arboric. 6:168.

3968. Tate, R. L. 1981. Characteristics of girdling roots on urban Norway maples. J. Arboric. 7:268–270.

3969. Taubenhaus, J. J., and Ezekiel, W. N. 1936. A rating of plants with reference to their relative resistance or susceptibility to Phymatotrichum root rot. Tex. Agric. Exp. Stn. Bull. 527. 52 pp.

3970. Tauer, C. G. 1978. Sweetfern rust resistance in jack pine seedlings: geographic variation. Can. J. For. Res. 8:416–423.

3971. Tayal, M. S., Sharma, S. M., and Agarwal, M. L. 1981. Studies on the polyphenols, proteins, chlorophylls, IAA-oxidase, and amylases of normal and false smut infected leaves of *Phoenix sylvestris*. Indian Phytopathol. 34:337–339.

3972. Taylor, C. H., and Long, P. G. 2000. Review of literature on camellia flower blight caused by *Ciborinia camelliae*. N.J. J. Crop Hortic. Sci. 28:123–138.

3973. Taylor, G. S. 1983. Cryptosporiopsis canker of *Acer rubrum*: some relationships among host, pathogen, and vector. Plant Dis. 67:984–986.

3974. Taylor, G. S., and Moore, R. E. B. 1979. A canker of red maples associated with oviposition by the narrow-winged tree cricket. Phytopathology 69:236–239.

3975. Taylor, J. 2001. Pycnothyrium ultrastructure in *Tubakia dryina*. Mycol. Res. 105:119–121.

3976. Taylor, J., and Birdwell, D. O. 2000. A scanning electron microscopic study of the infection of water oak (*Quercus nigra*) by *Taphrina caerulescens*. Mycologia 92:309–311.

3977. Taylor, J., and Clark, S. 1996. Infection and fungal development of *Tubakia dryina* on sweet gum (*Liquidambar styraciflua*). Mycologia 88:613–618.

3978. Taylor, J. E., and Mathiasen, R. L. 1999. Limber pine dwarf mistletoe. USDA For. Serv. For. Insect Dis. Leafl. 171. Internet pub.

3979. Taylor, K. C., Albrigo, L. G., and Chase, C. D. 1996. Purification of a Zn-binding phloem protein with sequence identity to chitin-binding proteins. Plant Physiol. 110:657–664.

3980. Tehon, L. R., and Daniels, E. 1925. Notes on the parasitic fungi of Illinois—II. Mycologia 17:240–249.

3981. Tehon, L. R., and Jacks, W. R. 1933. Smooth patch, a bark lesion of white oak. J. For. 31:430–433.

3982. Tekauz, A., and Patrick, Z. A. 1974. The role of twig infections on the incidence of perennial canker of peach. Phytopathology 64:683–688.

3983. Téliz, D., Grogan, R. G., and Lownsbery, B. F. 1966. Transmission of tomato ringspot, peach yellow bud mosaic, and grape yellow vein viruses by *Xiphinema americanum*. Phytopathology 56:658–663.

3984. Temple, B., and Horgen, P. A. 2000. Biological roles for cerato-ulmin, a hydrophobin secreted by the elm pathogens, *Ophiostoma ulmi* and *O. novo-ulmi*. Mycologia 92:1–9.

3985. Tennakoon, J. U., and Pate, J. S. 1996. Effect of parasitism by a mistletoe on the structure and functioning of branches of its host. Plant Cell Environ. 19:517–528.

3986. Terashima, K., Kawashima, Y., Cha, J., and Miura, K. 1998. Identification of *Armillaria* species from Hokkaido by analysis of the intergenic spacer (IGS) region of ribosomal DNA using PCR-RFLP. Mycoscience 39:179–183.

3987. Terashita, T. 1963. Studies on the diseases of *Acacia dealbata*. III. Taxonomic opinion on the anthracnose fungus. Bull. Gov. For. Exp. Stn., Tokyo, No. 155. 22 pp.

3988. Terashita, T. 1973. Studies of an anthracnose fungus on broad leaved trees in Japan, with special reference to the latency of the fungus. Bull. Gov. For. Exp. Stn., Tokyo, No. 252. 85 pp.

3989. Terho, M., and Uotila, A. 1999. Virulence of two Finnish *Gremmeniella abietina* types (A and B). Eur. J. For. Pathol. 29:143–152.

3990. Têtu-Bernier, P., Allen, E., and Hiratsuka, Y. 1983. Bibliography of western gall rust. Can. For. Serv. Inf. Rep. NOR-X-250. 10 pp.

3991. Teviotdale, B. L., and Harper, D. H. 1991. Infection of pruning and small bark wounds in almond by *Ceratocystis fimbriata*. Plant Dis. 75:1026–1030.

3992. Teviotdale, B. L., Michailides, T. J., and Pscheidt, J. W., eds. 2002. Compendium of nut crop diseases in temperate zones. APS Press, St. Paul, MN. 89 pp.

3993. Teviotdale, B. L., Sibbett, G. S., Fitch, L., and Harper, D. H. 1991. Budwood transmission of *Erwinia rubrifaciens*, causal agent of deep bark canker disease of English walnut. Plant Dis. 75:360–363.

3994. Thielges, B. A., and Adams, J. C. 1975. Genetic variation and heritability of *Melampsora* leaf rust resistance in eastern cottonwood. For. Sci. 21:278–282.

3995. Thies, W. G. 1983. Determination of growth reduction in Douglas-fir infected by *Phellinus weirii*. For. Sci. 29:305–315.

3996. Thies, W. G., and Patton, R. F. 1970. The biology and control of *Cylindrocladium scoparium* in Wisconsin forest tree nurseries. Phytopathology 60:1662–1668.

3997. Thies, W. G., and Sturrock, R. N. 1995. Laminated root rot in western North America. USDA For. Serv. Gen. Tech. Rep. PNW-GTR-349. 32 pp.

3998. Thirumalachar, M. J., Rao, D. V. S., and Ravindranath, V. 1950. Telia of the rust on cultivated figs. Curr. Sci. 19:27–28.

3999. Thomas, B. J. 1981. Studies on rose mosaic disease in field-grown roses produced in the United Kingdom. Ann. Appl. Biol. 98:419–429.

4000. Thomas, B. J. 1982. The effect of Prunus necrotic ringspot virus on field-grown roses. Ann. Appl. Biol. 100:129–134.

4001. Thomas, B. J. 1984. Rose mosaic disease: symptoms induced in roses by graft inoculation with both Prunus necrotic ringspot and apple mosaic viruses. Plant Pathol. 33:155–160.

4002. Thomas, C. S., and Hart, J. H. 1986. Site factors associated with Nectria canker on black walnut in Michigan. Plant Dis. 70:1117–1121.

4003. Thomas, D. L., and Donselman, H. M. 1979. Mycoplasmalike bodies and phloem degeneration associated with declining Pandanus in Florida. Plant Dis. Rep. 63:911–916.

4004. Thomas, F. M. 2000. Growth and water relations of four deciduous tree species (Fagus sylvatica L., Quercus petraea [Matt.] Liebl., Q. pubescens Willd., Sorbus aria [L.] Cr.) occurring at Central-European tree-line sites on shallow calcareous soils: physiological reactions of seedlings to severe drought. Flora 195:104–115.

4005. Thomas, F. M., Blank, R., and Hartmann, G. 1996. The effects of stem exposure, nitrogen status and insect defoliation on the frost hardiness of bark tissue of mature oaks. Verhandl. Gesellsch. Okol. 26:153–160.

4006. Thomas, F. M., Blank, R., and Hartmann, G. 2002. Abiotic and biotic factors and their interactions as causes of oak decline in Central Europe. For. Pathol. 32:277–307.

4007. Thomas, F. M., and Hartmann, G. 1998. Tree rooting patterns and soil water relations of healthy and damaged stands of mature oak (Quercus robur L. and Quercus petraea [Matt.] Liebl.). Plant Soil 203:145–158.

4008. Thomas, G. P. 1958. Studies in forest pathology. XVIII. The occurrence of the Indian paint fungus, Echinodontium tinctorium E. & E., in British Columbia. Can. Dep. Agric. Pub. 1041. 30 pp.

4009. Thomas, H. E. 1933. The quince-rust disease caused by Gymnosporangium germinale. Phytopathology 23:546–553.

4010. Thomas, H. E. 1934. Studies on Armillaria mellea (Vahl) Quel., infection, parasitism, and host resistance. J. Agric. Res. 48:187–218.

4011. Thomas, H. E., and Mills, W. D. 1929. Three rust diseases of the apple. Cornell Univ. Agric. Exp. Stn. Mem. 123. 21 pp.

4012. Thomas, H. E., and Scott, C. E. 1953. Rosette of rose. Phytopathology 43:218–219.

4013. Thomas, L. K. Jr. 1980. The impact of three exotic plant species on a Potomac island. U.S. Natl. Park Serv. Sci. Monogr. Ser. No. 13. 179 pp.

4014. Thomas, V., Premakumari, D., Reghu, C. P., Panikkar, A. O. N., and Amma, C. K. S. 1995. Anatomical and histochemical aspects of bark regeneration in Hevea brasiliensis. Ann. Bot. 75:421–426.

4015. Thompson, E. O. 1939. Morphological differences in Taphrina caerulescens upon different species of Quercus. Univ. Kans. Sci. Bull. 26:357–366.

4016. Thompson, G. E. 1939. A leaf blight of Populus tacamahaca Mill. caused by an undescribed species of Linospora. Can. J. Res. C17:232–238.

4017. Thompson, G. E. 1941. Leaf-spot diseases of poplars caused by Septoria musiva and S. populicola. Phytopathology 31:241–254.

4018. Thompson, G. E. 1963. Decay of oaks caused by Hypoxylon atropunctatum. Plant Dis. Rep. 47:202–205.

4019. Thompson, J. H., and Tattar, T. A. 1973. Rhizina undulata associated with disease of 80-year-old red spruce in Vermont. Plant Dis. Rep. 57:394–396.

4020. Thomson, A. J., Alfaro, R. I., Bloomberg, W. J., and Smith, R. B. 1985. Impact of dwarf mistletoe on the growth of western hemlock trees having different patterns of suppression and release. Can. J. For. Res. 15:665–668.

4021. Thomson, A. J., Muir, J. A., and Lewis, K. J. 1997. Variability in subregional impact of dwarf mistletoe on mature lodgepole pine. For. Chron. 73:371–375.

4022. Thomson, V. E., and Mahall, B. E. 1983. Host specificity by a mistletoe, Phoradendron villosum (Nutt.) Nutt. subsp. villosum, on three oak species in California. Bot. Gaz. 144:124–131.

4023. Thornton, F. C., Joslin, J. D., Pier, P. A., Neufeld, H., Seiler, J. R., and Hutcherson, J. D. 1994. Cloudwater and ozone effects upon high elevation red spruce: a summary of study results from Whitetop Mountain, Virginia. J. Environ. Qual. 23:1158–1167.

4024. Tiangco, E. S., and Varney, E. H. 1970. A vein chlorosis or yellow-net disease of Forsythia caused by tobacco ringspot virus. (Abstr.) Phytopathology 60:579.

4025. Tibbetts, T. J., and Ewers, F. W. 2000. Root pressure and specific conductivity in temperate lianas: exotic Celastrus orbiculatus (Celastraceae) vs. native Vitis riparia (Vitaceae). Am. J. Bot. 87:1272–1278.

4026. Tiedemann, A. R., Clary, W. P., and Barbour, R. J. 1987. Underground systems of Gambel oak (Quercus gambelii) in central Utah. Am. J. Bot. 74:1065–1071.

4027. Tilney-Bassett, R. A. E. 1986. Plant chimeras. Arnold, Baltimore, London. 199 pp.

4028. Timmer, L. W., Agostini, J. P., Zitko, S. E., and Zulfiqar, M. 1994. Postbloom fruit drop, an increasingly prevalent disease of citrus in the Americas. Plant Dis. 78:329–334.

4029. Timmer, L. W., Garnsey, S. M., Grimm, G. R., El-Gholl, N. E., and Schoulties, C. L. 1979. Wilt and dieback of Mexican lime caused by Fusarium oxysporum. Phytopathology 69:730–734.

4030. Timmer, L. W., Gottwald, T. R., and Zitko, S. E. 1991. Bacterial exudation from lesions of Asiatic citrus canker and citrus bacterial spot. Plant Dis. 75:192–195.

4031. Timmer, L. W., Roberts, P. D., Darhower, H. M., Bushong, P. M., Stover, E. W., Peever, T. L., and Ibáñez, A. M. 2000. Epidemiology and control of citrus greasy spot in different citrus-growing areas in Florida. Plant Dis. 84:1294–1298.

4032. Timmer, L. W., Solel, Z., Gottwald, T. R., Ibáñez, A. M., and Zitko, S. E. 1998. Environmental factors affecting production, release, and field populations of conidia of Alternaria alternata, the cause of brown spot of citrus. Phytopathology 88:1218–1223.

4033. Timonin, M. I., and Self, R. L. 1955. Cylindrocladium scoparium Morgan on azaleas and other ornamentals. Plant Dis. Rep. 39:860–863.

4034. Tinnin, R. O., and Knutson, D. M. 1980. Growth characteristics of the brooms on Douglas-fir caused by Arceuthobium douglasii. For. Sci. 26:149–158.

4035. Tinnin, R. O., Parks, C. G., and Knutson, D. M. 1999. Effects of Douglas-fir dwarf mistletoe on trees in thinned stands in the Pacific Northwest. For. Sci. 45:359–365.

4036. Tinus, R. W. 1981. Root system configuration is important to long tree life. Plants Landscape 4(1):1–5.

4037. Tippett, J. T., and Shigo, A. L. 1981. Barrier zone formation: a mechanism of tree defense against vascular pathogens. IAWA Bull., n.s., 2:163–168.

4038. Tippett, J. T., and Shigo, A. L. 1981. Barriers to decay in conifer roots. Eur. J. For. Pathol. 11:51–59.

4039. Tisserat, N. 1987. Stem canker of black walnut caused by Fusarium solani in Kansas. Plant Dis. 71:557.

4040. Tisserat, N., and Kuntz, J. E. 1983. Dispersal gradients of conidia of the butternut canker fungus in a forest during rain. Can. J. For. Res. 13:1139–1144.

4041. Tisserat, N., and Kuntz, J. E. 1983. Longevity of conidia of Sirococcus clavigignenti-juglandacearum in a simulated airborne state. Phytopathology 73:1628–1631.

4042. Tisserat, N., and Kuntz, J. E. 1984. Butternut canker: development on individual trees and increase within a plantation. Plant Dis. 68:613–616.

4043. Tisserat, N. A., and Pair, J. C. 1997. Susceptibility of selected juniper cultivars to cedar-apple rust, Kabatina tip blight, Cercospora needle blight and Botryosphaeria canker. J. Environ. Hortic. 15:160–163.

4044. Tisserat, N. A., Nus, A., and Barnes, L. W. 1991. A canker disease of the Cupressaceae in Kansas and Texas caused by Seiridium unicorne. Plant Dis. 75:138–140.

4045. Tisserat, N. A., Rossman, A. Y., and Nus, A. 1988. A canker disease of Rocky Mountain juniper caused by Botryosphaeria stevensii. Plant Dis. 72:699–701.

4046. Tjamos, E. C., and Beckman, C. H., eds. 1988. Vascular wilt diseases of plants. NATO ASI Series H. Cell Biol. Vol. 28. Springer-Verlag, New York, Berlin, Heidelberg. 590 pp.

4047. Tjamos, E. C., Biris, D. A., and Paplomatas, E. J. 1991. Recovery of olive trees with Verticillium wilt after individual application of soil solarization in established olive orchards. Plant Dis. 75:557–562.

4048. Tjamos, E. C., Rowe, R. C., Heale, J. B., and Fravel, D. R., eds. 2000. Advances in Verticillium: research and disease management. APS Press, St. Paul, MN. 357 pp.

4049. Tkacz, B. M., and Baker, F. A. 1991. Survival of Inonotus tomentosus in spruce stumps after logging. Plant Dis. 75:788–790.

4050. Tobi, D. R., Wargo, P. M., and Bergdahl, D. R. 1995. Growth response of red spruce after known periods of winter injury. Can. J. For. Res. 25:669–681.

4051. Tobiessen, P., and Buchsbaum, S. 1976. Ash dieback and drought. Can. J. Bot. 54:543–545.

4052. Toda, T., and Kurinobu, S. 2002. Realized genetic gains observed in progeny tolerance of selected red pine (Pinus densiflora) and black pine (P. thunbergii) to pine wilt disease. Silvae Genet. 51:42–44.

4053. Todhunter, M. N., and Beineke, W. F. 1984. Effect of anthracnose on growth of grafted black walnut. Plant Dis. 68:203–204.

4054. Tomasino, S. F., and Conway, K. E. 1987. Spatial pattern, inoculum density–disease incidence relationship and population dynamics of *Sclerotium rolfsii* on apple rootstock. Plant Dis. 71:719–724.

4055. Toole, E. R. 1949. Fusarium wilt of staghorn sumac. Phytopathology 39:754–759.

4056. Toole, E. R. 1952. Two races of *Fusarium oxysporum* f. *perniciosum* causing wilt of *Albizia* spp. Phytopathology 42:694.

4057. Toole, E. R. 1954. Rot and cankers on oak and honeylocust caused by *Poria spiculosa*. J. For. 52:941–942.

4058. Toole, E. R. 1955. *Polyporus hispidus* on southern bottomland oaks. Phytopathology 45:177–180.

4059. Toole, E. R. 1959. Decay after fire injury to southern bottom-land hardwoods. USDA Tech. Bull. 1189. 25 pp.

4060. Toole, E. R. 1963. Cottonwood canker caused by *Fusarium solani*. Plant Dis. Rep. 47:1032–1035.

4061. Toole, E. R. 1966. Root rot caused by *Polyporus lucidus*. Plant Dis. Rep. 50:945–946.

4062. Toole, E. R. 1966. Stem canker of red oaks caused by *Fusarium solani*. Plant Dis. Rep. 50:160–161.

4063. Topa, M. A., and McLeod, K. W. 1988. Promotion of aerenchyma formation in *Pinus serotina* seedlings by ethylene. Can. J. For. Res. 18:276–280.

4064. Torkelsen, A.-E., and Eckblad, F.-E. 1977. Encoelioidae (Ascomycetes) of Norway. Norweg. J. Bot. 24:133–149.

4065. Torrey, J. G. 1988. Cellular interactions between host and endosymbiont in dinitrogen-fixing root nodules of woody plants. NATO ASI Ser. H. Cell Biol. 17:1–25.

4066. Torrey, J. G. 1992. Can plant productivity be increased by inoculation of tree roots with soil microorganisms? Can. J. For. Res. 22:1815–1823.

4067. Torsello, M. L., Davis, D. D., and Nash, B. L. 1994. Incidence of *Cryphonectria parasitica* cankers on scarlet oak (*Quercus coccinea*) in Pennsylvania. Plant Dis. 78:313–315.

4068. Toti, L., Chapela, I. H., and Petrini, O. 1992. Morphometric evidence for host-specific strain formation in *Discula umbrinella*. Mycol. Res. 96:420–424.

4069. Touliatos, P., and Roth, E. 1971. Hurricanes and trees: ten lessons from Camille. J. For. 69:285–289.

4070. Townsend, A. M., and Douglass, L. W. 2001. Variation among American elm clones in long-term dieback, growth, and survival following *Ophiostoma* inoculation. J. Environ. Hortic. 19:100–103.

4071. Townsend, A. M., and Douglass, L. W. 2004. Evaluation of elm clones for tolerance to Dutch elm disease. J. Arboric. 30:179–183.

4072. Townsend, A. M., Schreiber, L. R., Hall, T. J., and Bentz, S. E. 1990. Variation in response of Norway maple cultivars to *Verticillium dahliae*. Plant Dis. 74:44–46.

4073. Trappe, J. M., and Fogel, R. D. 1977. Ecosystematic functions of ectomycorrhizae. Pages 205–214 in: The belowground ecosystem: a synthesis of plant-associated processes. J. K. Marshall, ed. Colo. State Univ., Range Sci. Dep. Sci. Ser. No. 26.

4074. Trappe, J. M., Stahly, E. A., Benson, N. R., and Duff, D. M. 1973. Mycorrhizal deficiency of apple trees in high arsenic soils. HortScience 8:52–53.

4075. Traquair, J. A. 1988. Oxalic acid and calcium oxalate produced by *Leucostoma cincta* and *L. persoonii* in culture and in peach bark tissues. Can. J. Bot. 65:1952–1956.

4076. Tredici, P. del. 1999. Redwood burls: immortality underground. Arnoldia 59:14–22.

4077. Tremblay, L., Levasseur, C., and Tremblay, F. M. 1999. Frequency of somaclonal variation in plants of black spruce (*Picea mariana*, Pinaceae) and white spruce (*P. glauca*, Pinaceae) derived from somatic embryogenesis and identification of some factors involved in genetic instability. Am. J. Bot. 86:1373–1381.

4078. Treshow, M. 1971. Fluorides as air pollutants affecting plants. Annu. Rev. Phytopathol. 9:21–44.

4079. Treshow, M., Anderson, F. K., and Harner, F. 1967. Responses of Douglas-fir to elevated atmospheric fluorides. For. Sci. 13:114–120.

4080. Trigiano, R. N., Caetano-Anollés, G., Bassam, B. J., and Windham, M. T. 1995. DNA amplification fingerprinting provides evidence that *Discula destructiva*, the cause of dogwood anthracnose in North America, is an introduced pathogen. Mycologia 87:490–500.

4081. Trimble, G. R. Jr., and Tryon, E. H. 1974. Grapevines a serious obstacle to timber production on good hardwood sites in Appalachia. North. Logger 23(5):22, 23, 44.

4082. Triolo, E., Materazzi, A., and Toni, S. 1996. An isolate of tobacco mosaic tobamovirus from *Olea europaea* L. Adv. Hortic. Sci. 10:39–45.

4083. Trione, E. J. 1959. The pathology of *Phytophthora lateralis* on native *Chamaecyparis lawsoniana*. Phytopathology 49:306–310.

4084. Tripepi, R. R., and Mitchell, C. A. 1984. Metabolic response of river birch and European birch roots to hypoxia. Plant Physiol. 76:31–35.

4085. Triplett, E. W., ed. 2000. Prokaryotic nitrogen fixation: a model system for the analysis of a biological process. Horizon Scientific, Wymondham, UK. 800 pp.

4086. Trockenbrodt, M. 1994. Light and electron microscopic investigations on wound reactions in the bark of *Salix caprea* L. and *Tilia tomentosa*. Flora 189:131–140.

4087. Trolinger, J. C., Elliott, E. S., and Young, R. J. 1978. Host range of *Cristulariella pyramidalis*. Plant Dis. Rep. 62:710–714.

4088. Trotter, D., Shrimpton, G., and Kope, H. 1994. The effects of Keithia blight on outplanting performance of western redcedar container seedlings at two reforestation sites in British Columbia—preliminary results. USDA For. Serv. Gen. Tech. Rep. RM-257:196–202.

4089. True, R. P. 1938. Gall development on *Pinus sylvestris* attacked by the Woodgate *Peridermium* and morphology of the parasite. Phytopathology 28:24–49.

4090. True, R. P., Barnett, H. L., Dorsey, C. K., and Leach, J. G. 1960. Oak wilt in West Virginia. West Va. Agric. Exp. Stn. Bull. 448T. 119 pp.

4091. True, R. P., and Tryon, E. H. 1956. Oak stem cankers initiated in the drought year 1953. Phytopathology 46:617–622.

4092. True, R. P., Tryon, E. H., and King, J. F. 1955. Cankers and decays of birch associated with two *Poria* species. J. For. 53:412–415.

4093. Tryon, E. H. 1971. Frost damage to tree species related to time of budbreak. Proc. West Va. Acad. Sci. 43:1–8.

4094. Tryon, E. H., and True, R. P. 1952. Blister-shake of yellow poplar. West Va. Agric. Exp. Stn. Bull. 350T. 15 pp.

4095. Tsahouridou, P. C., and Thanassoulopoulos, C. C. 2000. First report of *Hendersonula toruloidea* as a foliar pathogen of strawberry-tree (*Arbutus unedo*) in Europe. Plant Dis. 84:487.

4096. Tsoumis, G. 1965. Structural deformities in an epidemic tumor of white spruce, *Picea glauca*. Can. J. Bot. 43:176–181.

4097. Tsuneda, A., Hiratsuka, Y., and Maruyama, P. J. 1980. Hyperparasitism of *Scytalidium uredinicola* on western gall rust, *Endocronartium harknessii*. Can. J. Bot. 58:1154–1159.

4098. Tsuneda, I., and Kennedy, L. L. 1980. Basidiospore germination and substrate preference in *Fomes fomentarius* and *Fomitopsis cajaderi*. Mycologia 72:204–208.

4099. Tu, C. C., Hsieh, T. F., Tsai, W. H., and Kimbrough, J. W. 1992. Induction of basidia and morphological comparison among isolates of *Athelia* (*Sclerotium*) *rolfsii*. Mycologia 84:695–704.

4100. Tucker, C. M., and Milbrath, J. A. 1942. Root rot of *Chamaecyparis* caused by a species of *Phytophthora*. Mycologia 34:94–103.

4101. Tucker, D. P. H., Lee, R. F., Timmer, L. W., Albrigo, L. G., and Brlansky, R. H. 1984. Experimental transmission of citrus blight. Plant Dis. 68:979–980.

4102. Tuno, N. 1999. Insect feeding on spores of a bracket fungus, *Elfvingia applanata* (Pers.) Karst. (Ganodermataceae, Aphyllophorales). Ecol. Res. 14:97–103.

4103. Tuset, J. J., and Portilla, M. T. 1989. Taxonomic status of *Fusicoccum amygdali* and *Phomopsis amygdalina*. Can. J. Bot. 67:1275–1280.

4104. Tuttle, M. A., and Gotlieb, A. R. 1985. Graft union histology and distribution of tomato ringspot virus in infected McIntosh/Malling Merton 106 apple trees. Phytopathology 75:347–351.

4105. Tuttle, M. A., and Gotlieb, A. R. 1985. Histology of Delicious/Malling Merton 106 apple trees affected by apple union necrosis and decline. Phytopathology 75:342–347.

4106. Tymon, A. M., Jones, P., and Harrison, N. A. 1998. Phylogenetic relationships of coconut phytoplasmas and the development of specific oligonucleotide PCR primers. Ann. Appl. Biol. 132:437–452.

4107. Tyree, M. T., and Sperry, J. S. 1989. Vulnerability of xylem to cavitation and embolism. Annu. Rev. Plant Physiol. 40:19–38.

4108. U.S. Agricultural Research Service, Systematic Botany and Mycology Laboratory. Fungal Databases. Beltsville, MD. Internet resource.

4109. U.S. Agricultural Research Service, Systematic Botany and Mycology Laboratory. Germ Plasm Resources Information Network. GRIN Taxonomy. Vascular Plant Database. Beltsville, MD. Internet resource.

4110. U.S. Natural Resources Conservation Service. National Plant Data Center. The PLANTS Database. Baton Rouge, LA. Internet resource.

4111. Uchida, J. Y., Aragaki, M., and Yoshimura, M. A. 1984. Alternaria leaf spots of *Brassaia actinophylla*, *Dizygotheca elegantissima*, and *Tupidanthus calyptratus*. Plant Dis. 68:447–449.

4112. Uddin, W., and McCarter, S. M. 1996. First report of Rhododendron leaf spot caused by *Pseudomonas cichorii*. Plant Dis. 80:960.

4113. Uddin, W., McCarter, S. M., and Gitaitis, R. D. 1996. First report of oakleaf hydrangea bacterial leaf spot caused by a pathovar of *Xanthomonas campestris*. Plant Dis. 80:599.

4114. Uddin, W., Stevenson, K. L., and Pardo, S. R. A. 1997. Pathogenicity of a species of *Phomopsis* causing a shoot blight on peach in Georgia and evaluation of possible infection courts. Plant Dis. 81:983–989.

4115. Uecker, F. A. 1988. A world list of *Phomopsis* names with notes on nomenclature, morphology and biology. Cramer, Berlin. 231 pp.

4116. Uemura, M., and Steponkus, P. L. 1999. Cold acclimation in plants: relationship between the lipid composition and the cryostability of the plasma membrane. J. Plant Res. 112:245–254.

4117. Ulloa, M., and Hanlin, R. T. 2000. Illustrated dictionary of mycology. APS Press, St. Paul, MN. 448 pp.

4118. Ulrich, B. 1995. The history and possible causes of forest decline in central Europe, with particular attention to the German situation. Environ. Rev. 3:262–276.

4119. Umbach, D. M., and Davis, D. D. 1984. Severity and frequency of SO_2-induced leaf necrosis on seedlings of 57 tree species. For. Sci. 30:587–596.

4120. Uotila, A. 1990. Infection of pruning wounds in Scots pine by *Phacidium coniferarum* and selection of pruning season. Acta Forestal. Fenn. No. 215. 36 pp.

4121. Uotila, A. 1992. Mating system and apothecia production in *Gremmeniella abietina*. Eur. J. For. Pathol. 22:410–417.

4122. Uotila, A., Hantula, J., Vaatanen, A. K., and Hamelin, R. C. 2000. Hybridization between two biotypes of *Gremmeniella abietina* var. *abietina* in artificial pairings. For. Pathol. 30:211–219.

4123. Upadhyay, H. P. 1981. A monograph of *Ceratocystis* and *Ceratocystiopsis*. Univ. Ga. Press, Athens. 176 pp.

4124. Uphof, J. C. T. 1931. *Tillandsia usneoides* als Pflanzenschädling. Z. Pflanzenkr. 41:593–607.

4125. Utomo, C., and Niepold, F. 2000. Development of diagnostic methods for detecting *Ganoderma*-infected oil palms. J. Phytopathol. 148:507–514.

4126. Uyemoto, J. K., Kirkpatrick, B. C., and Cummins, J. N. 1991. Susceptibility of selected cherry clones and related species to western X-disease. HortScience 26:1510–1511.

4127. Vajna, L. 1986. Branch canker and dieback of sessile oak (*Quercus petraea*) in Hungary caused by *Diplodia mutila*. I. Identification of the pathogen. Eur. J. For. Pathol. 16:223–229.

4128. Vajna, L. 1999. Pathogenic *Botryosphaeria* species in Hungary. II. *Botryosphaeria melanops* (Tul. et C. Tul.) G. Winter. Novenyvedelem 35:195–198.

4129. Vajna, L. 2002. *Diaporthe oncostoma* causing stem canker of black locust in Hungary. Plant Pathol. 51:393.

4130. Vancon, S. 1993. Fertilization affects growth and incidence of grey mold on container-grown giant sequoia. Tree Planters' Notes 44:68–72.

4131. Vanev, S. G. 1995. *Discosia alboferruginea,* comb. nov. Mycotaxon 56:85–87.

4132. Vanev, S. G., and Aa, H. A. van der. 1998. An annotated list of the published names in *Asteromella*. Persoonia 17:47–67.

4133. Vann, D. R., Johnson, A. H., and Casper, B. B. 1994. Effect of elevated temperatures on carbon dioxide exchange in *Picea rubens*. Tree Physiol. 14:1339–1349.

4134. Vann, D. R., Strimbeck, G. R., and Johnson, A. H. 1992. Effects of ambient levels of airborne chemicals on freezing resistance of red spruce foliage. For. Ecol. Manag. 51:69–79.

4135. Vanneste, J. L., ed. 2000. Fire blight: the disease and its causative agent, *Erwinia amylovora*. CABI Pub., Wallingford, UK. 370 pp.

4136. Vannini, A., Biocca, M., and Paparatti, B. 1996. Contribution to the knowledge of the biological cycle of *Hypoxylon mediterraneum* on *Quercus cerris*. Inf. Fitopatol. 46:53–55.

4137. Vannini, A., Caruso, C., Leonardi, L., Rugini, E., Chiarot, E., Caporale, C., and Buonocore, V. 1999. Antifungal properties of chitinases from *Castanea sativa* against hypovirulent and virulent strains of the chestnut blight fungus *Cryphonectria parasitica*. Physiol. Mol. Plant Pathol. 55:29–35.

4138. Vannini, A., Paganini, R., and Anselmi, N. 1996. Factors affecting discharge and germination of ascospores of *Hypoxylon mediterraneum* (De Not.) Mill. Eur. J. For. Pathol. 26:12–24.

4139. Vannini, A., and Valentini, R. 1994. Influence of water relations on *Quercus cerris–Hypoxylon mediterraneum* interaction: a model of drought-induced susceptibility to a weakness parasite. Tree Physiol. 14:129–139.

4140. Vannini, A., Valentini, R., and Luisi, N. 1996. Impact of drought and *Hypoxylon mediterraneum* on oak decline in the Mediterranean region. Ann. Sci. For. 53:753–760.

4141. Varma, A., and Hock, B., eds. 1999. Mycorrhiza: structure, function, molecular biology, and biotechnology. 2nd ed. Springer, Berlin, New York. 704 pp.

4142. Vasiliauskas, R., and Stenlid, J. 2001. Homothallism in the postfire ascomycete *Rhizina undulata*. Mycologia 93:447–452.

4143. Vassey, W. E., Gould, C. J., and Ryan, G. F. 1977. Disease resistant pyracantha for the Pacific Northwest. Ornamentals Northwest 1977–78 (Dec–Jan):4–6.

4144. Vauterin, L., Hoste, B., Kersters, K., and Swings, J. 1995. Reclassification of *Xanthomonas*. Int. J. Syst. Bacteriol. 45:472–489.

4145. Vauterin, L., Rademaker, J., and Swings, J. 2000. Synopsis on the taxonomy of the genus *Xanthomonas*. Phytopathology 90:677–682.

4146. Vauterin, L., Yang, P., Hoste, B., Vancanneyt, M., Civerolo, E. L., Swings, J., and Kersters, K. 1991. Differentiation of *Xanthomonas campestris* pv. *citri* strains by sodium dodecyl sulfate–polyacrylamide gel electrophoresis of proteins, fatty acid analysis, and DNA-DNA hybridization. Int. J. Syst. Bacteriol. 41:535–542.

4147. Venkatasubbaiah, P., and Chilton, W. S. 1992. Phytotoxins produced by *Tubakia dryina*. Mycopathologia 120:33–37.

4148. Venkatasubbaiah, P., Sutton, T. B., and Chilton, W. S. 1991. Effect of phytotoxins produced by *Botryosphaeria obtusa,* the cause of black rot of apple fruit and frogeye leaf spot. Phytopathology 81:243–247.

4149. Venter, M., Myburg, H., Wingfield, B. D., Coutinho, A., and Wingfield, M. J. 2002. A new species of *Cryphonectria* from South Africa and Australia, pathogenic to *Eucalyptus*. Sydowia 54:98–117.

4150. Venter, M., Wingfield, M. J., Coutinho, T. A., and Wingfield, B. D. 2001. Molecular characterization of *Endothia gyrosa* isolates from *Eucalyptus* in South Africa and Australia. Plant Pathol. 50:211–217.

4151. Verhoeff, K., Malathrakis, N. E., and Williamson, B., eds. 1992. Recent advances in *Botrytis* research. Proceedings of the 10th international *Botrytis* symposium, Heraklion, Crete, Greece. Pudoc Scientific, Wageningen. 294 pp.

4152. Verkley, G. J. M. 1999. A monograph of the genus *Pezicula* and its anamorphs. Stud. Mycol. No. 44. 180 pp.

4153. Verma, H. N., Vivek, P., Prasad, V., Heywood, W. H., and McNeil, J. 1986. Virus diseases in pawpaw (papaya). Rev. Trop. Plant Pathol. 2:311–327.

4154. Verrall, A. F., and May, C. 1937. A new species of *Dothiorella* causing die-back of elm. Mycologia 29:321–324.

4155. Vettraino, A. M., Barzanti, G. P., Bianco, M. C., Ragazzi, A., Capretti, P., Paoletti, E., Luisi, N., Anselmi, N., and Vannini, A. 2002. Occurrence of *Phytophthora* species in oak stands in Italy and their association with declining oak trees. For. Pathol. 32:19–28.

4156. Victor, D., Crous, P. W., Janse, B. J. H., and Wingfield, M. J. 1997. Genetic variation in *Cylindrocladium floridanum* and other morphologically similar *Cylindrocladium* species. Syst. Appl. Microbiol. 20:268–285.

4157. Viennot-Bourgin, G. 1941. Un *Fusicladium* sur *Crataegus pyracantha* Medik. Rev. Mycol., n.s., 6:147–155.

4158. Vike, E., and Habjorg, A. 1995. Variation in fluoride content and leaf injury on plants associated with three aluminium smelters in Norway. Sci. Total Environ. 163:25–34.

4159. Vilgalys, R., and Cubeta, M. A. 1994. Molecular systematics and population biology of *Rhizoctonia*. Annu. Rev. Phytopathol. 32:135–155.

4160. Viljoen, A., Marasas, W. F. O., Wingfield, M. J., and Viljoen, C. D. 1997. Characterization of *Fusarium subglutinans* f.sp. *pini* causing root disease of *Pinus patula* seedlings in South Africa. Mycol. Res. 101:437–445.

4161. Viljoen, C. D., Wingfield, B. D., and Wingfield, M. J. 1993. Comparison of *Seiridium* isolates associated with cypress canker using sequence data. Exp. Mycol. 17:323–328.

4162. Vineet, K., Babu, A. M., Sharma, D. D., and Datta, R. K. 1998. Penetration and infection process of *Phyllactinia corylea* on mulberry leaf causing powdery mildew—I. Asexual stage. J. Phytopathol. 146:469–472.

4163. Vineet, K., Gupta, V. P., Babu, A. M., Mishra, R. K., Thiagarajan, V., Datta, R. K., and Kumar, V. 2001. Surface ultrastructural studies on penetration and infection process of *Colletotrichum gloeosporioides* on mulberry leaf causing black spot disease. J. Phytopathol. 149:629–633.

4164. Vingnana-Singam, V., Long, P. G., Rowland, R. E., and Zydenbos, S. M. 2000. Infection processes of *Ciborinia camelliae* on camellia flower tissue. N.Z. Plant Prot. Conf. Proc. 53:151–156.

4165. Viskari, E. L., and Karenlampi, L. 2000. Roadside Scots pine as an indicator of deicing salt use—a comparative study from two consecutive winters. Water Air Soil Pollut. 122:405–419.

4166. Vlamis, J., and Raabe, R. D. 1985. Copper deficiency of manzanita grown in a bark-sand mixture. HortScience 20:61–62.

4167. Vogler, D. R., and Bruns, T. D. 1998. Phylogenetic relationships among the pine stem rust fungi (*Cronartium* and *Peridermium* spp.). Mycologia 90:244–257.

4168. Vogler, D. R., Cobb, F. W. Jr., Geils, B. W., and Nelson, D. L. 1996. Isozyme diversity among hard pine stem rust fungi in the western United States. Can. J. Bot. 74:1058–1080.

4169. Vogler, D. R., Epstein, L., and Cobb, F. W. Jr. 1997. Nuclear behaviour and evolution of two populations of the western gall rust fungus. Mycol. Res. 101:791–797.

4170. Volk, T. J., and Burdsall, H. H. 1995. A nomenclatural study of *Armillaria* and *Armillariella* species: Basidiomycotina, Tricholomataceae. Fungiflora, Oslo. 121 pp.

4171. Voorhees, R. K. 1942. Life history and taxonomy of the fungus *Physalospora rhodina*. Fla. Agric. Exp. Stn. Bull. 371. 91 pp.

4172. Vrain, T. C., and Copeman, R. J. 1987. Interactions between *Agrobacterium tumefaciens* and *Pratylenchus penetrans* in the roots of two red raspberry cultivars. Can. J. Plant Pathol. 9:236–240.

4173. Vujanovic, V., Cogliastro, A., St. Arnaud, M., Neumann, P., and Gagnon, D. 1999. First report of *Fusarium solani* canker and wilt symptoms on red oak (*Quercus rubra*) in Quebec, Canada. Plant Dis. 83:78.

4174. Wade, G. C., and Cruickshank, R. H. 1992. The establishment and structure of latent infections with *Monilinia fructicola* on apricots. J. Phytopathol. 136:95–106.

4175. Wadsworth, F. H. 1943. Lightning damage in ponderosa pine stands of northern Arizona. J. For. 41:684–685.

4176. Wagener, W. W. 1925. Mistletoe in the lower bole of incense cedar. Phytopathology 15:614–616.

4177. Wagener, W. W. 1939. The canker of *Cupressus* induced by *Coryneum cardinale* n.sp. J. Agric. Res. 58:1–46.

4178. Wagener, W. W. 1949. Top dying of conifers from sudden cold. J. For. 47:49–53.

4179. Wagener, W. W. 1957. The limitation of two leafy mistletoes of the genus *Phoradendron* by low temperatures. Ecology 38:142–145.

4180. Wagener, W. W. 1958. Infection tests with two rusts of Jeffrey pine. Plant Dis. Rep. 42:888–892.

4181. Wagener, W. W., and Davidson, R. W. 1954. Heart rots in living trees. Bot. Rev. 20:61–134.

4182. Wagener, W. W., and Mielke, J. L. 1961. A staining-fungus root disease of ponderosa, Jeffrey, and pinyon pines. Plant Dis. Rep. 45:831–835.

4183. Wagner, T., and Fischer, M. 2001. Natural groups and a revised system for the European poroid Hymenochaetales (Basidiomycota) supported by nLSU rDNA sequence data. Mycol. Res. 105:773–782.

4184. Wagnon, H. K., Traylor, J. A., Williams, H. E., and Weiner, A. C. 1968. Investigations of cherry rasp leaf disease in California. Plant Dis. Rep. 52:618–622.

4185. Wahlstrom, K. T., and Johansson, M. 1992. Structural responses in bark to mechanical wounding and *Armillaria ostoyae* infection in seedlings of *Pinus sylvestris*. Eur. J. For. Pathol. 22:65–76.

4186. Wahyuno, D., Kakishima, M., and Ono, Y. 2001. Morphological analyses of urediniospores and teliospores in seven *Phragmidium* species parasitic on ornamental roses. Mycoscience 42:519–533.

4187. Wahyuno, D., Kakishima, M., and Ono, Y. 2002. Aeciospore-surface structures of *Phragmidium* species parasitic on roses. Mycoscience 43:159–167.

4188. Wainhouse, D. 1980. Dispersal of first instar larvae of the felted beech scale, *Cryptococcus fagisuga*. J. Appl. Ecol. 17:523–532.

4189. Wainwright, S. H., and Lewis, F. H. 1970. Developmental morphology of the black knot pathogen on plum. Phytopathology 60:1238–1244.

4190. Walker, L. C. 1956. Foliage symptoms as indicators of potassium-deficient soils. For. Sci. 2:113–120.

4191. Wall, R. E. 1986. Effects of black knot disease on pin cherry. Can. J. Plant. Pathol. 8:71–77.

4192. Wall, R. E. 1991. Pathological effects of *Chondrostereum purpureum* in inoculated yellow birch and beech. Can. J. Plant Pathol. 13:81–87.

4193. Wall, R. E., and Kuntz, J. E. 1964. Water-soluble substances in dead branches of aspen (*Populus tremuloides* Michx.) and their effects on *Fomes igniarius*. Can. J. Bot. 42:969–977.

4194. Wall, R. E., Macey, D. E., and Sela, E. 1996. Virulence and interfertility of *Chondrostereum purpureum* isolates. Biol. Control 7:205–211.

4195. Wall, R. E., and Magasi, L. P. 1976. Environmental factors affecting Sirococcus shoot blight of black spruce. Can. J. For. Res. 6:448–452.

4196. Walla, J. A. 1984. Incidence of *Phellinus punctatus* on living woody plants in North Dakota. Plant Dis. 68:252–253.

4197. Walla, J. A., Jacobi, W. R., Tisserat, N. A., Harrell, M. O., Ball, J. J., Neill, G. B., Reynard, D. A., Guo, Y. H., and Spiegel, L. 2000. Condition of green ash, incidence of ash yellows phytoplasmas, and their association in the Great Plains and Rocky Mountain regions of North America. Plant Dis. 84:268–274.

4198. Walla, J. A., and Riffle, J. W. 1981. *Fomes fraxinophilus* on green ash in North Dakota windbreaks. Plant Dis. 65:669–670.

4199. Walla, J. A., and Stack, R. W. 1980. Dip treatment for control of blackstem on *Populus* cuttings. Plant Dis. 64:1092–1095.

4200. Walla, J. A., and Stack, R. W. 1988. Tubercularia canker of honeylocust in North Dakota. Plant Dis. 72:734.

4201. Walla, J. A., Tuskan, G. A., and Lundquist, J. E. 1991. Comparison of white- and orange-spored *Peridermium harknessii* in native North Dakota *Pinus ponderosa*. Mycologia 83:251–256.

4202. Walla, J. A., Tuskan, G. A., Lundquist, J. E., and Wang, C. 1997. Expression of inoculum and family specific responses in the ponderosa pine–western gall rust pathosystem. Plant Dis. 81:57–62.

4203. Wallis, G. W. 1976. Growth characteristics of *Phellinus* (*Poria*) *weirii* in soil and on root and other surfaces. Can. J. For. Res. 6:229–232.

4204. Wallis, G. W., and Bloomberg, W. J. 1981. Estimating the total extent of *Phellinus weirii* root rot centers using above- and below-ground disease indicators. Can. J. For. Res. 11:827–830.

4205. Wallis, G. W., and Reynolds, G. 1965. The initiation and spread of *Poria weirii* root rot of Douglas-fir. Can. J. Bot. 43:1–9.

4206. Walter, J. M. 1946. Canker stain of planetrees. USDA Circ. 742. 12 pp.

4207. Walter, J. M., Rex, E. G., and Schreiber, R. 1952. The rate of progress and destructiveness of canker stain of planetrees. Phytopathology 42:236–239.

4208. Walter, M. 1993. The pH value and the occurrence of low molecular weight fatty acids in wetwood of beech (*Fagus sylvatica* L.). Eur. J. For. Pathol. 23:1–10.

4209. Walters, J. W., Hinds, T. E., Johnson, D. W., and Beatty, J. 1982. Effects of partial cutting on diseases, mortality, and regeneration of Rocky Mountain aspen stands. USDA For. Serv. Res. Pap. RM-240. 12 pp.

4210. Wang, D., Iezzoni, A., and Adams, G. 1998. Genetic heterogeneity of *Leucostoma* species in Michigan peach orchards. Phytopathology 88:376–381.

4211. Wang, J., and Kamp, B. J. van der. 1992. Resistance, tolerance, and yield of western black cottonwood infected by *Melampsora* rust. Can. J. For. Res. 22:183–192.

4212. Wang, S. H., Gergerich, R. C., Wickizer, S. L., and Kim, K. S. 2002. Localization of transmissible and nontransmissible viruses in the vector nematode *Xiphinema americanum*. Phytopathology 92:646–653.

4213. Wang, Y. C., and Guo, L. 1985. Taxonomic studies on *Gymnosporangium* in China. Acta Mycol. Sinica 4:24–34.

4214. Wanner, J. L., and Tinnin, R. O. 1986. Respiration in lodgepole pine parasitized by American dwarf mistletoe. Can. J. For. Res. 16:1375–1378.

4215. Wanner, J. L., and Tinnin, R. O. 1989. Some effects of infection by *Arceuthobium americanum* on the population dynamics of *Pinus contorta* in Oregon. Can. J. For. Res. 19:736–742.

4216. Ward, J. C. 1986. The effect of wetwood on lumber drying times and rates: an exploratory evaluation with longitudinal gas permeability. Wood Fiber Sci. 18:288–307.

4217. Ward, J. C., and Pong, W. Y. 1980. Wetwood in trees: a timber resource problem. USDA For. Serv. Gen. Tech. Rep. PNW-112. 56 pp.

4218. Wardlaw, T. J. 1999. *Endothia gyrosa* associated with severe stem cankers on plantation grown *Eucalyptus nitens* in Tasmania, Australia. Eur. J. For. Pathol. 29:199–208.

4219. Ware, G. H. 1982. Decline in oaks associated with urbanization. Pages 61–64 in: Urban and suburban trees: pest problems, needs, prospects, and solutions. Conference proceedings, Mich. State Univ. April 1982. B. O. Parks, F. A. Fear, M. T. Lambur, and G. A. Simmons, eds. Dep. Resour. Dev. and Dep. Entomol., Mich. State Univ. East Lansing. 253 pp.

4220. Wargo, P. M. 1977. *Armillariella mellea* and *Agrilus bilineatus* and mortality of defoliated oak trees. For. Sci. 23:485–492.

4221. Wargo, P. M. 1981. Defoliation and secondary-action organism attack: with emphasis on *Armillaria mellea*. J. Arboric. 7:64–69.

4222. Wargo, P. M. 1988. Amino nitrogen and phenolic constituents of bark of American beech, *Fagus grandifolia*, and infestation by beech scale, *Cryptococcus fagisuga*. Eur. J. For. Pathol. 18:279–290.

4223. Wargo, P. M. 1996. Consequences of environmental stress on oak: predisposition to pathogens. Ann. Sci. For. 53:359–368.

4224. Wargo, P. M., and Houston, D. R. 1974. Infection of defoliated sugar maple trees by *Armillaria mellea*. Phytopathology 64:817–822.

4225. Wargo, P. M., Houston, D. R., and La Madeleine, L. A. 1983. Oak decline. USDA For. Serv. For. Insect Dis. Leafl. 165. Internet pub.

4226. Wargo, P. M., Parker, J., and Houston, D. R. 1972. Starch content in roots of defoliated sugar maple. For. Sci. 18:203–204.

4227. Warmelo, K. T. Van, and Sutton, B. C. 1981. Coelomycetes VII. *Stegonsporium.* Commonw. Mycol. Inst. Mycol. Pap. No. 145. 45 pp.

4228. Warner, J. 1990. Field susceptibility of scab-resistant apple cultivars and selections to cedar apple rust, quince rust and hawthorn rust. Fruit Var. J. 44:216–224.

4229. Warner, J. 1992. Field susceptibility of 68 apple cultivars to cedar apple rust, quince rust and hawthorn rust. Fruit Var. J. 46:6–10.

4230. Warokka, J. S. 1999. The impact and etiology of coconut wilt disease in India. Pages 27–35 in: Proceedings of the international workshop on lethal diseases of coconut by phytoplasmas in Southeast Asia. D. Allorerung, H. C. Harries, P. Jones, and J. S. Warokka, eds. APCC Pub., Jakarta.

4231. Warren, J. E., and Linit, M. J. 1994. Effect of *Monochamus carolinensis* on the life history of the pinewood nematode, *Bursaphelenchus xylophilus.* J. Nematol. 25:703–709.

4232. Wass, E., and Mathiasen, R. 2003. A new subspecies of *Arceuthobium tsugense* (Viscaceae) from British Columbia and Washington. Novon 13:268–276.

4233. Waterhouse, G. M. 1970. The genus *Phytophthora* de Bary. 2nd ed. Commonw. Mycol. Inst. Mycol. Pap. No. 122. 59 pp.

4234. Waterman, A. M. 1943. *Diplodia pinea,* the cause of a disease of hard pines. Phytopathology 33:1018–1031.

4235. Waterman, A. M. 1947. *Rhizosphaera kalkhoffii* associated with a needle cast of *Picea pungens.* Phytopathology 37:507–511.

4236. Waterman, A. M. 1954. Septoria canker of poplars in the United States. USDA Circ. 947. 24 pp.

4237. Waterman, A. M. 1955. The relation of *Valsa kunzei* to cankers on conifers. Phytopathology 45:686–692.

4238. Waterman, A. M. 1957. Canker and dieback of poplars caused by *Dothichiza populea.* For. Sci. 3:175–183.

4239. Waterman, A. M., and Marshall, R. P. 1947. A new species of *Cristulariella* associated with a leaf spot of maple. Mycologia 39:690–698.

4240. Waters, C. W. 1962. Significance of life history studies of *Elytroderma deformans.* For. Sci. 8:250–254.

4241. Waterworth, H. E. 1972. Purification, serology, and properties of a virus from lilac, *Syringa oblata affinis.* Plant Dis. Rep. 56:923–926.

4242. Waterworth, H. E., and Lawson, R. H. 1973. Purification, electron microscopy, and serology of the dogwood ringspot strain of cherry leafroll virus. Phytopathology 63:141–146.

4243. Waterworth, H. E., Lawson, R. H., and Monroe, R. L. 1976. Purification and properties of Hibiscus chlorotic ringspot virus. Phytopathology 66:570–575.

4244. Waterworth, H. E., and Povish, W. R. 1972. Tobacco ringspot virus from naturally infected dogwood, autumn crocus, and forsythia. Plant Dis. Rep. 56:336–337.

4245. Waterworth, H. E., and Povish, W. R. 1977. A yellow leafspot disease of *Ilex crenata* caused by tobacco ringspot virus. Plant Dis. Rep. 61:104–105.

4246. Watling, R., Kile, G. A., and Gregory, N. M. 1982. The genus *Armillaria*—nomenclature, typification, the identity of *Armillaria mellea,* and species differentiation. Trans. Br. Mycol. Soc. 78:271–285.

4247. Watmough, S., Brydges, T., and Hutchinson, T. 1999. The tree-ring chemistry of declining sugar maple in central Ontario, Canada. Ambio 28:613–618.

4248. Watson, A. J. 1941. Studies of *Botryosphaeria ribis* on *Cercis* and *Benzoin.* Plant Dis. Rep. 25:29–31.

4249. Watson, G. W., and Clark, S. 1993. Regeneration of girdling roots after removal. J. Arboric. 19:278–280.

4250. Watson, G. W., and Clark, S. 1996. When the roots go round and round. Arnoldia 56:15–21.

4251. Watson, G. W., Clark, S., and Johnson, K. 1990. Formation of girdling roots. J. Arboric. 16:197–202.

4252. Watson, W. A., Zehr, E. I., and Grimes, L. W. 2002. Influence of temperature and wetting period on inoculum production by *Monilinia fructicola* in peach twig cankers. Plant Dis. 86:666–668.

4253. Watson, W. T., Kenerley, C. M., and Appel, D. N. 2000. Visual and infrared assessment of root colonization of apple trees by *Phymatotrichopsis omnivora.* Plant Dis. 84:539–543.

4254. Wattad, C., Kobiler, D., Dinoor, A., and Prusky, D. 1997. Pectate lyase of *Colletotrichum gloeosporioides* attacking avocado fruits: cDNA cloning and involvement in pathogenicity. Physiol. Mol. Plant Pathol. 50:197–212.

4255. Waxman, S. 1975. Witches'-brooms sources of new and interesting dwarf forms of *Picea, Pinus,* and *Tsuga* species. Acta Hortic. No. 54:25–32.

4256. Waxman, S. 1989. Variability among witches'-broom seedlings. Comb. Proc. Int. Plant Prop. Soc. 39:433–436.

4257. Wean, R. E. 1937. The parasitism of *Polyporus schweinitzii* on seedling *Pinus strobus.* Phytopathology 27:1124–1142.

4258. Weaver, D. J. 1974. Effect of root injury on the invasion of peach roots by *Clitocybe tabescens.* Mycopathol. Mycol. Appl. 52:313–317.

4259. Weaver, D. J. 1979. Role of conidia of *Botryosphaeria dothidea* in the natural spread of peach tree gummosis. Phytopathology 69:330–334.

4260. Weaver, M. J., and Stipes, R. J. 1988. White pine decline: a case study from Virginia landscapes. J. Arboric. 14:109–120.

4261. Webber, J. F., and Hansen, E. M. 1990. Susceptibility of European and north-west American conifers to the North American vascular pathogen *Leptographium wageneri.* Eur. J. For. Pathol. 20:347–354.

4262. Weber, D. J., Gang, D., Halls, S., and Nelson, D. L. 1995. Juniper decline in Natural Bridges National Monument and Canyonlands National Park. USDA For. Serv. Gen. Tech. Rep. INT-GTR-315:258–262.

4263. Weber, G. F. 1957. Cold injury to young pine trees. Plant Dis. Rep. 41:494–495.

4264. Weber, G. F., and Roberts, D. A. 1951. Silky threadblight of *Elaeagnus pungens* caused by *Rhizoctonia ramicola* n.sp. Phytopathology 41:615–621.

4265. Weber, H. 1993. Allgemeine Mykologie. Gustav Fischer, Jena. 541 pp.

4266. Weber, H. C, and Forstreuter, W., eds. 1987. Proceedings of the fourth international symposium on parasitic flowering plants. Philipps Univ., Marburg, Germany. 844 pp.

4267. Webster, R. K., and Butler, E. E. 1967. A morphological and biological concept of the species *Ceratocystis fimbriata.* Can. J. Bot. 45:1457–1468.

4268. Wegwitz, E. 1993. Needle and broom rusts of true firs. Can. For. Serv. Pac. For. Cent. For. Pest Leafl. 45. 7 pp.

4269. Wehlburg, C., and Cox, R. S. 1966. Rhizoctonia leaf blight of azalea. Plant Dis. Rep. 50:354–355.

4270. Wehmeyer, L. E. 1933. The genus *Diaporthe* Nitschke and its segregates. Univ. Mich. Press, Ann Arbor. 349 pp.

4271. Wehmeyer, L. E. 1941. A revision of *Melanconis, Pseudovalsa, Prosthecium,* and *Titania.* Univ. Mich. Press, Ann Arbor. 161 pp.

4272. Weidensaul, T. C., Leben, C., and Ellett, C. W. 1977. Reducing decay losses in hardwood forests and farm woodlots. Ohio State Univ. Coop. Exten. Serv. Bull. 629. 14 pp.

4273. Weidensaul, T. C., and Wood, F. A. 1973. Sources of species of *Fusarium* in northern hardwood forests. Phytopathology 63:367–371.

4274. Weimer, J. L. 1917. The origin and development of the galls produced by two cedar rust fungi. Am. J. Bot. 4:241–251.

4275. Weimer, J. L. 1917. Three cedar rust fungi, their life histories, and the diseases they produce. Cornell Univ. Agric. Exp. Stn. Bull. 390:509–549.

4276. Weingartner, D. P., and Klos, E. J. 1975. Etiology and symptomatology of canker and dieback diseases on highbush blueberries caused by *Godronia* (*Fusicoccum*) *cassandrae* and *Diaporthe* (*Phomopsis*) *vaccinii.* Phytopathology 65:105–110.

4277. Weir, J. R. 1916. *Keithia thujina,* the cause of a serious leaf disease of the western red cedar. Phytopathology 6:360–363.

4278. Weir, J. R. 1921. *Cenangium piniphilum* n.sp., an undescribed canker-forming fungus on *Pinus ponderosa* and *P. contorta.* Phytopathology 11:294–296.

4279. Weir, J. R. 1921. Note on *Cenangium abietis* (Pers.) Rehm on *Pinus ponderosa* Laws. Phytopathology 11:166–170.

4280. Weir, J. R. 1925. Notes on the parasitism of *Endothia gyrosa* (Schw.) Fr. Phytopathology 15:489–491.

4281. Weir, J. R., and Hubert, E. E. 1918. Notes on the overwintering of forest tree rusts. Phytopathology 8:55–59.

4282. Weiser, R. L., and Wallner, S. J. 1988. Freezing woody plant stems produces acoustic emissions. J. Am. Soc. Hortic. Sci. 113:636–639.

4283. Weiss, F., and Smith, F. F. 1940. A flower-spot disease of cultivated azaleas. USDA Circ. 556. 28 pp.

4284. Welch, B. L., and Martin, N. E. 1974. Invasion mechanisms of *Cronartium ribicola* in *Pinus monticola* bark. Phytopathology 64:1541–1546.

4285. Welch, D. S. 1934. The range and importance of Nectria canker on hardwoods in the Northeast. J. For. 32:997–1002.

4286. Wells, B. W. 1939. A new forest climax: the salt spray climax of Smith Island, N.C. Bull. Torrey Bot. Club 66:629–634.

4287. Wells, J. M., Raju, B. C., Hung, H. Y., Weisburg, W. G., Mandelco, P. L., and Brenner, D. J. 1987. *Xylella fastidiosa* gen. nov., sp. nov.:

Gram-negative, xylem-limited, fastidious plant bacteria related to *Xanthomonas* spp. Int. J. Syst. Bacteriol. 37:136–143.

4288. Wene, E. G., and Schoeneweiss, D. F. 1980. Localized freezing predisposition to Botryosphaeria canker in differentially frozen woody stems. Can. J. Bot. 58:1455–1458.

4289. Wenefrida, I., and Spencer, J. A. 1993. *Marssonia rosae* variants in Mississippi and their virulence on selected rose cultivars. Plant Dis. 77:246–248.

4290. Wenner, J. J. 1914. A contribution to the morphology and life history of *Pestalozzia funerea* Desm. Phytopathology 4:375–383.

4291. Wenner, N. G., and Merrill, W. 1998. Pathological anatomy of needles of *Pinus strobus* exposed to carbon-filtered air or to three times ambient ozone concentrations, or infected by *Canavirgella banfieldii*. Can. J. Bot. 76:1331–1339.

4292. Wensley, R. N. 1964. Occurrence and pathogenicity of *Valsa* (*Cytospora*) species and other fungi associated with peach canker in southern Ontario. Can. J. Bot. 42:841–857.

4293. Werner, A., and Siwecki, R. 1978. Histological studies of infection processes by *Dothichiza populea* Sacc. et Briard in susceptible and resistant poplar clones. Eur. J. For. Pathol. 8:217–226.

4294. Werner, D. 1992. Symbiosis of plants and microbes. Chapman & Hall, New York. 389 pp.

4295. Werner, D. J., Ritchie, D. F., Cain, D. W., and Zehr, E. I. 1986. Susceptibility of peaches and nectarines, plant introductions, and other *Prunus* species to bacterial spot. HortScience 21:127–130.

4296. Werres, S., Marwitz, R., Man In't Veld, W. A., Cock, A. W. A. M. de, Bonants, P. J. M., Weerdt, M. de, Themann, K., Ilieva, E., and Baayen, R. P. 2001. *Phytophthora ramorum* sp. nov., a new pathogen on *Rhododendron* and *Viburnum*. Mycol. Res. 105:1155–1165.

4297. Werther, F., and Havranek, W. M. 2000. Effects of nutrient-deficient soil on gas-exchange, chlorophyll fluorescence and C-allocation in young *Picea abies* (L.) Karst. Phyton 40:179–184.

4298. Wessels, J. G. H. 2000. Hydrophobins, unique fungal proteins. Mycologist 14:153–159.

4299. West, E. 1933. Powdery mildew of crape myrtle caused by *Erysiphe lagerstroemiae* n.sp. Phytopathology 23:814–819.

4300. Weste, G. 1994. Impact of *Phytophthora* species on native vegetation of Australia and Papua New Guinea. Australas. Plant Pathol. 23:190–209.

4301. Weste, G. 1997. The changing status of disease caused by *Phytophthora cinnamomi* in Victorian open forests, woodlands and heathlands. Australas. Plant Pathol. 26:1–9.

4302. Weste, G., and Marks, G. C. 1987. The biology of *Phytophthora cinnamomi* in Australasian forests. Annu. Rev. Phytopathol. 25:207–229.

4303. Wester, H. V., Davidson, R. W., and Fowler, M. E. 1950. Cankers of linden and redbud. Plant Dis. Rep. 34:219–223.

4304. Wester, H. V., and Jylkka, E. W. 1959. Elm scorch, graft transmissible virus of American elm. Plant Dis. Rep. 43:519.

4305. Wester, H. V., and Jylkka, E. W. 1963. High incidence of Dutch elm disease in American elms weakened by elm scorch associated with breeding attacks of *Scolytus multistriatus*. Plant Dis. Rep. 47:545–547.

4306. Westerman, R. L., ed. 1990. Soil testing and plant analysis. 3rd ed. Soil Sci. Soc. Am., Madison, WI. 784 pp.

4307. Westing, A. H. 1966. Sugar maple decline: an evaluation. J. Econ. Bot. 20:196–212.

4308. Westing, A. H. 1969. Plants and salt in the roadside environment. Phytopathology 59:1174–1179.

4309. Westwood, M. N., and Cameron, H. R. 1978. Environment-induced remission of pear decline symptoms. Plant Dis. Rep. 62:176–179.

4310. Wet, J. de, Burgess, T., Slippers, B., Preisig, O., Wingfield, B. D., and Wingfield, M. J. 2003. Multiple gene genealogies and microsatellite markers reflect relationships between morphotypes of *Sphaeropsis sapinea* and distinguish a new species of *Diplodia*. Mycol. Res. 107:557–566.

4311. Wet, J. de, Wingfield, M. J., Coutinho, T., and Wingfield, B. D. 2002. Characterisation of the 'C' morphotype of the pine pathogen *Sphaeropsis sapinea*. For. Ecol. Manag. 161:181–188.

4312. Whalley, A. J. S. 1996. The xylariaceous way of life. Mycol. Res. 100:897–922.

4313. Whalley, A. J. S., and Edwards, R. L. 1985. *Nummulariella marginata*: its conidial state, secondary metabolites and taxonomic relationships. Trans. Br. Mycol. Soc. 85:385–390.

4314. Whigham, D. 1984. The influence of vines on the growth of *Liquidambar styraciflua* L. (sweetgum). Can. J. For. Res. 14:37–39.

4315. Whitcomb, C. E. 1986. Solving the iron chlorosis problem. J. Arboric. 12:44–48.

4316. Whitcomb, C. E., and Hickman, G. G. 1986. 'Prairie Shade' elm. HortScience 21:162–163.

4317. Whitcomb, R. F., and Tully, J. G., eds. 1979. The mycoplasmas. Vol. 3: Plant and insect mycoplasmas. Academic Press, New York. 351 pp.

4318. Whitcomb, R. F., and Tully, J. G., eds. 1989. The mycoplasmas. Vol. 5: Spiroplasmas, acholeplasmas, and mycoplasmas of plants and arthropods. Academic Press, San Diego. 653 pp.

4319. White, B. L., and Merrill, W. 1969. Pathological anatomy of *Abies balsamea* infected with *Melampsorella caryophyllacearum*. Phytopathology 59:1238–1242.

4320. White, E. E., Allen, E. A., Ying, C. C., and Foord, B. M. 2000. Seedling inoculation distinguishes lodgepole pine families most and least susceptible to gall rust. Can. J. For. Res. 30:841–843.

4321. White, E. E., Dubetz, C. P., Cruickshank, M. G., and Morrison, D. J. 1998. DNA diagnostic for *Armillaria* species in British Columbia: within and between species variation in the IGS-1 and IGS-2 regions. Mycologia 90:125–131.

4322. White, J. H. 1919. On the biology of *Fomes applanatus* (Pers.) Wallr. Trans. R. Can. Inst. 12:133–174.

4323. White, O. E. 1948. Fasciation. Bot. Rev. 14:319–358.

4324. White, P. M. 1972. Plant tolerance for standing water: an assessment. Cornell Plantations 28:50–52.

4325. White, P. R., and Millington, W. F. 1954. The structure and development of a woody tumor affecting *Picea glauca*. Am. J. Bot. 41:353–361.

4326. White, P. R., Tsoumis, G., and Hyland, F. 1967. Some seasonal aspects of the growth of tumors on the white spruce, *Picea glauca*. Can. J. Bot. 45:2229–2232.

4327. White, R. P. 1930. Pathogenicity of *Pestalotia* spp. N.J. Agric. Exp. Stn. Annu. Rep. 43 (1929–30):264–268.

4328. White, R. P. 1930. Pathogenicity of *Pestalotia* spp. on rhododendron. Phytopathology 20:85–91.

4329. White, R. P. 1937. Rhododendron wilt and root rot. N.J. Agric. Exp. Stn. Bull. 615. 32 pp.

4330. White, R. P., and McCulloch, L. 1934. A bacterial disease of *Hedera helix*. J. Agric. Res. 48:807–815.

4331. Whitehead, A. G. 1998. Plant nematode control. CAB International, Wallingford, UK. 384 pp.

4332. Whiteside, J. O. 1970. Etiology and epidemiology of citrus greasy spot. Phytopathology 60:1409–1414.

4333. Whiteside, J. O. 1971. Some factors affecting the occurrence and development of foot rot on citrus trees. Phytopathology 61:1233–1238.

4334. Whiteside, J. O. 1972. Histopathology of citrus greasy spot and identification of the causal fungus. Phytopathology 62:260–263.

4335. Whiteside, J. O. 1974. Environmental factors affecting infection of citrus leaves by *Mycosphaerella citri*. Phytopathology 64:115–120.

4336. Whiteside, J. O. 1981. Aberrant behavior of *Mycosphaerella citri* on freeze-killed citrus leaf tissue and its taxonomic and epidemiologic implications. Phytopathology 71:1108–1110.

4337. Whiteside, J. O. 1988. The history and rediscovery of citrus canker in Florida. Citrograph 73:197–206.

4338. Whiteside, J. O., Garnsey, S. M., and Timmer, L. W., eds. 1988. Compendium of citrus diseases. APS Press, St. Paul, MN. 80 pp.

4339. Whiting, E. C., Roncadori, R. W., Bongarten, B. C., and Reynolds, K. L. 1997. Relationship between environmental factors, effects of fungicide applications, and premature defoliation of black locust trees (*Robinia pseudoacacia* L.). Int. J. Pest Manag. 43:213–219.

4340. Whitney, H. E., and Johnson, W. C. 1984. Ice storms and forest succession in southwestern Virginia. Bull. Torrey Bot. Club 111:429–437.

4341. Whitney, R. D. 1962. Studies in forest pathology. XXIV. *Polyporus tomentosus* Fr. as a major factor in stand-opening disease of white spruce. Can. J. Bot. 40:1631–1658.

4342. Whitney, R. D. 1966. Germination and inoculation tests with basidiospores of *Polyporus tomentosus*. Can. J. Bot. 44:1333–1343.

4343. Whitney, R. D. 1977. *Polyporus tomentosus* root rot of conifers. Can. For. Serv. For. Tech. Rep. 18. 11 pp.

4344. Whitney, R. D. 1993. Damage by tomentosus root rot in white spruce plantations in Ontario, and the effects of thinning on the disease. For. Chron. 69:445–449.

4345. Whitney, R. D. 1995. Root-rotting fungi in white spruce, black spruce, and balsam fir in northern Ontario. Can. J. For. Res. 25:1209–1230.

4346. Whitney, R. D., and Bohaychuk, W. P. 1976. Pathogenicity of *Polyporus tomentosus* and *P. tomentosus* var. *circinatus* on seedlings of 11 conifer species. Can. J. For. Res. 6:129–131.

4347. Wichman, R. L., and Hopkins, D. L. 2002. Differentiation of pathogenic groups of *Xylella fastidiosa* strains with whole-cell protein profiles. Plant Dis. 86:875–879.

4348. Wichman, R. L., Hopkins, D. L., and Wichman, T. A. 2000. First report of oleander leaf scorch caused by *Xylella fastidiosa* in Florida. Plant Dis. 84:198.

4349. Wichmann, C., Wulf, A., and Kehr, R. 1991. Evaluation of tree damage after injection treatment. Nachrichtenbl. Deut. Pflanzenschutzd. 43:176–183.

4350. Wick, R. L., and Moore, L. D. 1983. Histopathology of root disease incited by *Thielaviopsis basicola* in *Ilex crenata*. Phytopathology 73:561–564.

4351. Wicker, E. F. 1965. A Phomopsis canker on western larch. Plant Dis. Rep. 49:102–105.

4352. Wicker, E. F. 1981. Natural control of white pine blister rust by *Tuberculina maxima*. Phytopathology 71:997–1000.

4353. Wicker, E. F., and Hawksworth, F. G. 1991. Upward advance, intensification, and spread of dwarf mistletoe in a thinned stand of western larch. USDA For. Serv. Res. Note RM-504. 4 pp.

4354. Wicker, E. F., Laurent, T. H., and Israelson, S. 1978. Sirococcus shoot blight damage to western hemlock regeneration at Thomas Bay, Alaska. USDA For. Serv. Res. Pap. INT-198. 11 pp.

4355. Wicker, E. F., and Wells, J. M. 1983. Intensification and lateral spread of *Arceuthobium laricis* in a young stand of western larch with stocking control. Can. J. For. Res. 13:314–319.

4356. Widin, K. D., and Schipper, A. L. Jr. 1980. Epidemiology of *Melampsora medusae* leaf rust of poplars in the north central United States. Can. J. For. Res. 10:257–263.

4357. Widin, K. D., and Schipper, A. L. Jr. 1981. Effect of *Melampsora medusae* leaf rust infections on yield of hybrid poplars in the north-central United States. Eur. J. For. Pathol. 11:438–448.

4358. Wiersum, L. K., and Harmanny, K. 1983. Changes in the water-permeability of roots of some trees during drought stress and recovery, as related to problems of growth in urban environment. Plant Soil 75:443–448.

4359. Wiggers, R. J., West, J. G., and Taylor, J. 1997. Conidial germination and infection by *Diplocarpon rosae* on susceptible and resistant rose species. Mycologia 89:103–108.

4360. Wikström, C. 1976. Occurrence of *Phellinus tremulae* (Bond.) Bond. and Borisov as a primary parasite in *Populus tremula* L. Eur. J. For. Pathol. 6:321–328.

4361. Wikström, C., and Unestam, T. 1976. Decay pattern of *Phellinus tremulae* (Bond.) Bond. et Borisov in *Populus tremula* L. Eur. J. For. Pathol. 6:291–301.

4362. Wilcox, P. L., Amerson, H. V., Kuhlman, E. G., Liu, B., O'Malley, D. M., and Sederoff, R. R. 1996. Detection of a major gene for resistance to fusiform rust disease in loblolly pine by genomic mapping. Proc. Natl. Acad. Sci. USA 93:3859–3864.

4363. Wilcox, W. F., and Ellis, M. A. 1989. Phytophthora root and crown rots of peach trees in the eastern Great Lakes region. Plant Dis. 73:794–798.

4364. Wilding, N., Collins, N. M., Hammond, P., M., and Webber, J. F., eds. 1989. Insect-fungus interactions. Academic Press, London. 344 pp.

4365. Wilhelm, S., and Taylor, J. B. 1965. Control of Verticillium wilt of olive through natural recovery and resistance. Phytopathology 55:310–316.

4366. Wilkins, W. H. 1933–1939. Studies in the genus *Ustulina* with special reference to parasitism. Trans. Br. Mycol. Soc. 18:320–346; 20:133–156; 22:47–93; 23:65–85, 171–185.

4367. Wilkinson, R. W. 1953. *Berberis thunbergii*, a host of cucumber mosaic virus (*Marmor cucumeris*). (Abstr.) Phytopathology 43:489.

4368. Wilks, D. S., Gersper, P. L., and Cobb, F. W. Jr. 1985. Association of soil moisture with spread of *Ceratocystis wageneri* in ponderosa pine disease centers. Plant Dis. 69:206–208.

4369. Williams, E. N. D., Todd, N. K., and Rayner, A. D. M. 1981. Spatial development of populations of *Coriolus versicolor*. New Phytol. 89:307–319.

4370. Williamson, B., Duncan, G. H., Harrison, J. G., Harding, L. A., Elad, Y., and Zimand, G. 1995. Effect of humidity on infection of rose petals by dry-inoculated conidia of *Botrytis cinerea*. Mycol. Res. 99:1303–1310.

4371. Willig, J., and Schlechte, G. B. 1995. Fungus succession on wood after windthrow in a natural beech forest reserve. Allg. Forst Z. 50:814–818.

4372. Wills, W. H., and Lambe, R. C. 1978. Pathogenicity of *Thielaviopsis basicola* from Japanese holly (*Ilex crenata*) to some other host plants. Plant Dis. Rep. 62:1102–1106.

4373. Wilson, A. D. 1990. The genetics of sexual incompatibility in the Indian paint fungus, *Echinodontium tinctorium*. Mycologia 82:332–341.

4374. Wilson, A. D. 1991. Somatic incompatibility in dikaryotic-monokaryotic and dikaryotic pairings of *Echinodontium tinctorium*. Can. J. Bot. 69:2716–2723.

4375. Wilson, A. D., and Lester, D. G. 2002. Trench inserts as long-term barriers to root transmission for control of oak wilt. Plant Dis. 86:1067–1074.

4376. Wilson, C. L. 1962. Brooming and galling of shagbark hickory in Arkansas. Plant Dis. Rep. 46:448–450.

4377. Wilson, C. L. 1963. Wilting of persimmon caused by *Cephalosporium diospyri*. Phytopathology 53:1402–1406.

4378. Wilson, C. L. 1965. Consideration of the use of persimmon wilt as a silvicide for weed persimmons. Plant Dis. Rep. 49:789–791.

4379. Wilson, D. 1995. Fungal endophytes which invade insect galls: insect pathogens, benign saprophytes, or fungal inquilines? Oecologia 103:255–260.

4380. Wilson, D., and Carroll, G. C. 1994. Infection studies of *Discula quercina*, an endophyte of *Quercus garryana*. Mycologia 86:635–647.

4381. Wilson, E. E. 1935. The olive knot disease: its inception, development, and control. Hilgardia 9:233–264.

4382. Wilson, E. E. 1947. The branch wilt of Persian walnut trees and its cause. Hilgardia 17:413–436.

4383. Wilson, E. E. 1965. Pathological histogenesis in oleander tumors induced by *Pseudomonas savastanoi*. Phytopathology 55:1244–1249.

4384. Wilson, E. E., and Magie, A. R. 1964. Systemic invasion of the host plant by the tumor-inducing bacterium *Pseudomonas savastanoi*. Phytopathology 54:576–579.

4385. Wilson, E. E., Starr, M. P., and Berger, J. A. 1957. Bark canker, a bacterial disease of the Persian walnut tree. Phytopathology 47:669–673.

4386. Wilson, E. E., Zeitoun, F. M., and Fredrickson, D. L. 1967. Bacterial phloem canker, a new disease of Persian walnut trees. Phytopathology 57:618–621.

4387. Wilson, M., and Henderson, D. M. 1966. British rust fungi. Cambridge Univ. Press, Cambridge, UK. 324 pp.

4388. Wilson, M., Sigee, D. C., and Epton, H. A. S. 1989, 1990. *Erwinia amylovora* infection of hawthorn blossom. I. The anther. II. The stigma. III. The nectary. J. Phytopathol. 127:1–14, 15–28; 128:62–74.

4389. Windham, M. T., Witte, W. T., and Trigiano, R. N. 2003. Three white-bracted cultivars of *Cornus florida* resistant to powdery mildew. HortScience 38:1253–1255.

4390. Wingfield, M. J. 1983. Association of *Verticicladiella procera* and *Leptographium terebrantis* with insects in the Lake States. Can. J. For. Res. 13:1238–1245.

4391. Wingfield, M. J. 1985. Reclassification of *Verticicladiella* based on conidial development. Trans. Br. Mycol. Soc. 85:81–93.

4392. Wingfield, M. J. 1986. Pathogenicity of *Leptographium procerum* and *L. terebrantis* on *Pinus strobus* seedlings and established trees. Eur. J. For. Pathol. 16:299–308.

4393. Wingfield, M. J., ed. 1987. Pathogenicity of the pine wood nematode. APS Press, St. Paul, MN. 122 pp.

4394. Wingfield, M. J., Blanchette, A., and Kondo, E. 1983. Comparison of the pine wood nematode, *Bursaphelenchus xylophilus*, from pine and balsam fir. Eur. J. For. Pathol. 13:360–370.

4395. Wingfield, M. J., and Knox-Davies, P. S. 1980. Association of *Diplodia pinea* with a root disease of pines in South Africa. Plant Dis. 64:221–223.

4396. Wingfield, M. J., Seifert, K. A., and Webber, J. F., eds. 1993. *Ceratocystis* and *Ophiostoma*. Taxonomy, ecology, and pathogenicity. APS Press, St. Paul, MN. 293 pp.

4397. Wingfield, M. J., Zyl, L. M. van, Heerden, S. van, Myburg, H., and Wingfield, B. D. 1999. Virulence and the genetic composition of the *Cryphonectria cubensis* Bruner population in South Africa. Pages 163–172 in: Physiology and genetics of tree phytophage interactions. F. Leutier, W. J. Mattson, and M. R. Wagner, eds. Inst. Natl. Recher. Agron., Paris.

4398. Winkler, A. J., ed. 1949. Pierce's disease investigations. Hilgardia 19:207–264.

4399. Winter, S., and Nienhaus, F. 1989. Identification of viruses from European beech (*Fagus sylvatica* L.) of declining forests in Northrhine-Westfalia (FRG). Eur. J. For. Pathol. 19:111–118.

4400. Winton, L. M., and Hansen, E. M. 2001. Molecular diagnosis of *Phytophthora lateralis* in trees, water, and foliage baits using multiplex polymerase chain reaction. For. Pathol. 31:275–283.

4401. Winton, L. M., Manter, D. K., Stone, J. K., and Hansen, E. M. 2003. Comparison of biochemical, molecular, and visual methods to quantify *Phaeocryptopus gaeumannii* in Douglas-fir foliage. Phytopathology 93:121–126.

4402. Wise, F. C., McCulloch, S. M., Britt, J. L., Brand, M. H., Kiyomoto, R., Mudge, K. W., Lardner, J. P., Mahoney, H. K., Good, G. L., Rowland, L. J., Levi, A., Zimmerman, R. H., LaMondia, J. A., Smith, V. L., and Rathier, T. M. 1997. Tissue proliferation in micropropagated rhododendrons. HortScience 32:985–1003.

4403. Wisniewski, M., Bogle, A. L., and Wilson, C. L. 1984. Histopathology of canker development on peach trees after inoculation with *Cytospora leucostoma*. Can. J. Bot. 62:2804–2813.

4404. Witcosky, J. J., Schowalter, T. D., and Hansen, E. M. 1986. *Hylastes nigrinus* (Coleoptera: Scolytidae), *Pissodes fasciatus,* and *Steremnius carinatus* (Coleoptera: Curculionidae) as vectors of black-stain root disease of Douglas-fir. Environ. Entomol. 15:1090–1095.

4405. Witcosky, J. J., Schowalter, T. D., and Hansen, E. M. 1987. Host-derived attractants for the beetles *Hylastes nigrinus* (Coleoptera: Scolytidae) and *Steremnius carinatus* (Coleoptera: Curculionidae). Environ. Entomol. 16:1310–1313.

4406. Withers, P. C., Cowling, W. A., and Wills, R. T., eds. 1994. Plant diseases in ecosystems: threats and impacts in south-western Australia. J. R. Soc. W. Austral. 77:97–186.

4407. Witthuhn, R. C., Harrington, T. C., Steimel, J. P., Wingfield, B. D., and Wingfield, M. J. 2000. Comparison of isozymes, rDNA spacer regions and MAT-2 DNA sequences as phylogenetic characters in the analysis of the *Ceratocystis coerulescens* complex. Mycologia 92:447–452.

4408. Witzell, J., and Karlsson, A. 2002. *Anisogramma virgultorum* on saplings of *Betula pendula* and *Betula pubescens* in a district of northern Sweden. For. Pathol. 32:207–212.

4409. Wohlgemuth, H., Mittelstrass, K., Kschieschan, S., Bender, J., Weigel, H. J., Overmyer, K., Kangasjarvi, J., Sandermann, H., and Langebartels, C. 2002. Activation of an oxidative burst is a general feature of sensitive plants exposed to the air pollutant ozone. Plant Cell Environ. 25:717–726.

4410. Wolf, F. A. 1912. A new *Gnomonia* on hickory leaves. Ann. Mycol. 10:488–491.

4411. Wolf, F. A. 1930. A parasitic alga, *Cephaleuros virescens* Kunze, on citrus and certain other plants. J. Elisha Mitchell Sci. Soc. 45:187–205.

4412. Wolf, F. A. 1939. Leafspot of ash and *Phyllosticta viridis*. Mycologia 31:258–266.

4413. Wolf, F. A. 1958. Mechanism of apothecial opening and ascospore expulsion by the cup-fungus *Urnula craterium*. Mycologia 50:837–843.

4414. Wolf, F. A., and Barbour, W. J. 1941. Brown-spot needle disease of pines. Phytopathology 31:61–74.

4415. Wolf, F. A., and Davidson, R. W. 1941. Life cycle of *Piggotia fraxini,* causing leaf disease of ash. Mycologia 33:526–539.

4416. Wolf, F. T., and Wolf, F. A. 1939. A study of *Botryosphaeria ribis* on willow. Mycologia 31:217–227.

4417. Wolf, F. T., and Wolf, F. A. 1952. Pathology of *Camellia* leaves infected by *Exobasidium camelliae* var. *gracilis* Shirai. Phytopathology 42:147–149.

4418. Wolfe, L. M., and Rissler, L. J. 1999. Reproductive consequences of a gall-inducing fungal pathogen (*Exobasidium vaccinii*) on *Rhododendron calendulaceum* (Ericaceae). Can. J. Bot. 77:1454–1459.

4419. Wolswinkel, P., Ammerlaan, A., and Peters, H. F. C. 1984. Phloem unloading of amino acids at the site of attachment of *Cuscuta europaea*. Plant Physiol. 75:13–20.

4420. Woo, J. Y., and Martin, N. E. 1981. Scanning electron microscopy of *Cronartium ribicola* infecting *Ribes* leaves. Eur. J. For. Pathol. 11:7–15.

4421. Woo, J. Y., and Partridge, A. D. 1969. The life history and cytology of *Rhytisma punctatum* on bigleaf maple. Mycologia 61:1085–1095.

4422. Woo, K. S., Fins, L., McDonald, G. I., and Wiese, M. V. 2001. Differences in needle morphology between blister rust resistant and susceptible western white pine stocks. Can. J. For. Res. 31:1880–1886.

4423. Wood, A. K., and Tainter, F. H. 2002. First report of *Phytophthora cinnamomi* on *Quercus laurifolia*. Plant Dis. 86:441.

4424. Wood, B. W., Tedders, W. L., and Reilly, C. C. 1988. Sooty mold fungus on pecan foliage suppresses light penetration and net photosynthesis. HortScience 23:851–853.

4425. Wood, F. A., and French, D. W. 1963. *Ceratocystis fimbriata,* the cause of a stem canker of quaking aspen. For. Sci. 9:232–235.

4426. Wood, F. A., and French, D. W. 1965. *Hypoxylon* canker of aspen: seasonal development of cankers and ascospore ejection in winter. Phytopathology 55:771–774.

4427. Wood, F. A., and Skelly, J. M. 1969. Relation of the time of year to canker initiation by *Fusarium solani* in sugar maple. Plant Dis. Rep. 53:753–755.

4428. Woodcock, H., Davies, K. M. Jr., and Patterson, W. A. III. 1997. White ash decline hazard assessment and management strategies in Massachusetts stands. North. J. Appl. For. 14:10–15.

4429. Woodcock, H., Patterson, W. A. III, and Davies, K. M. Jr. 1993. The relationship between site factors and white ash (*Fraxinus americana* L.) decline in Massachusetts. For. Ecol. Manag. 60:271–290.

4430. Woods, A. J. 2003. Species diversity and forest health in northwest British Columbia. For. Chron. 79:892–897.

4431. Woodward, S., and Pearce, R. B. 1988. The role of stilbenes in resistance of Sitka spruce (*Picea sitchensis* (Bong.) Carr.) to entry of fungal pathogens. Physiol. Mol. Plant Pathol. 33:127–149.

4432. Woodward, S., and Pocock, S. 1996. Formation of the ligno-suberized barrier zone and wound periderm in four species of European broad-leaved trees. Eur. J. For. Pathol. 26:97–105.

4433. Woodward, S., Stenlid, J., Karjalainen, R., and Hutterman, A. 1998. *Heterobasidion annosum*: biology, ecology, impact and control. CAB International, Wallingford, UK. 589 pp.

4434. Woolhouse, H. W. 1983. Toxicity and tolerance in the responses of plants to metals. Pages 245–300 in: Physiological plant ecology. III. Responses to the chemical and biological environment. O. L. Lange, P. S. Nobel, C. B. Osmond, and H. Ziegler, eds. Springer-Verlag, Berlin.

4435. Working Group on International Cooperation in Forest Disease Research, IUFRO Sect. 24, Forest Protection. 1963. Internationally dangerous forest tree diseases. USDA Misc. Pub. 939. 122 pp.

4436. Worley, R. E., Littrell, R. L., and Dutcher, J. D. 1980. A comparison of tree trunk injection and implantation of zinc capsules for correction of zinc deficiency. J. Arboric. 6:253–257.

4437. Worrall, J. J. 1994. Population structure of *Armillaria* species in several forest types. Mycologia 86:401–407.

4438. Worrall, J. J., Correll, J. C., and McCain, A. H. 1986. Pathogenicity and teleomorph-anamorph connection of *Botryosphaeria dothidea* on *Sequoiadendron giganteum* and *Sequoia sempervirens*. Plant Dis. 70:757–759.

4439. Worrall, J. J., and Parmeter, J. R. Jr. 1982. Formation and properties of wetwood in white fir. Phytopathology 72:1209–1212.

4440. Worrall, J. J., and Parmeter, J. R. Jr. 1983. Inhibition of wood-decay fungi by wetwood of white fir. Phytopathology 73:1140–1145.

4441. Worrall, J. J., and Sullivan, K. 2002. Discoloration of ponderosa pine on the San Juan National Forest, 1999–2001. USDA For. Serv. Rocky Mtn. Region, For. Health Manag. Biol. Eval. R2–02–06. 20 pp.

4442. Wright, E. 1942. *Cytospora abietis,* the cause of a canker of true firs in California and Nevada. J. Agric. Res. 65:143–153.

4443. Wright, E. 1957. Cytospora canker of Rocky Mountain Douglas-fir. Plant Dis. Rep. 41:811–813.

4444. Wright, E., and Wells, H. R. 1948. Tests on the adaptability of trees and shrubs to shelterbelt planting on certain Phymatotrichum root rot infested soils of Oklahoma and Texas. J. For. 46:256–262.

4445. Wright, J. G., Johnson, G. I., and Hyde, K. D. 1998. Studies on the endophytic mycota of *Citrus* spp. ACIAR (Austral. Cent. Int. Agric. Res.) Proc. Ser. 80:167–173.

4446. Wright, S. F., and Upadhyaya, A. 1998. A survey of soils for aggregate stability and glomalin, a glycoprotein produced by hyphae of arbuscular mycorrhizal fungi. Plant Soil 198:97–107.

4447. Wu, H. X., and Ying, C. C. 1997. Genetic parameters and selection efficiencies in resistance to western gall rust, stalactiform blister rust, needle cast, and sequoia pitch moth in lodgepole pine. For. Sci. 43:571–581.

4448. Wu, H. X., and Ying, C. C. 1998. Stability of resistance to western gall rust and needle cast in lodgepole pine provenances. Can. J. For. Res. 28:439–449.

4449. Wu, W. D., Jeng, R. S., and Hubbes, M. 1989. Toxic effects of elm phytoalexin mansonones on *Ophiostoma ulmi,* the causal agent of Dutch elm disease. Eur. J. For. Pathol. 19:343–357.

4450. Wu, W., Sutton, B. C., and Gange, A. C. 1996. Revision of *Septoria* species on *Hebe* and *Veronica* and description of *Kirramyces hebes* sp. nov. Mycol. Res. 100:1207–1217.

4451. Wutscher, H. K., and Derrick, K. S. 1995. Development of citrus blight symptoms and recovery of 'Valencia' orange trees in solution culture. Proc. Fla. State Hortic. Soc. 108:106–109.

4452. Wutscher, H. K., Cohen, M., and Young, R. H. 1977. Zinc and water-soluble phenolics levels in the wood for the diagnosis of citrus blight. Plant Dis. Rep. 61:572–576.

4453. Wutscher, H. K., Gottwald, T. R., Berger, R. D., Parys-Naday, L. van, and Parys-de Wit, C. van. 1992. Progression curves of citrus blight in Brazil and Florida. Proc. Interamer. Soc. Trop. Hortic. 36:20–25.

4454. Wyman, D. 1939. Salt water injury of woody plants resulting from the hurricane of September 21, 1938. Arnold Arbor. Bull. Pop. Inf. Ser. 4, Vol. 7:45–51.

4455. Wyss, U. 1982. Virus-transmitting nematodes: feeding behaviour and effect on root cells. Plant Dis. 66:639–644.

4456. Xenopoulos, S. G. 1991. Pathogenic variability of various isolates of Seiridium cardinale, S. cupressi and S. unicorne inoculated on selected Cupressus clones and seedlings. Eur. J. For. Pathol. 21:129–135.

4457. Xu, M.-O., Walla, J. A., and Zhao, W.-X., eds. 2003. Proceedings of the second IUFRO rusts of forest trees working party conference, Yangling, China, 2002. For. Res. 16 (Suppl.):1–228.

4458. Xu, X., Butt, D. J., and Ridout, M. S. 1998. The effects of inoculum dose, duration of wet period, temperature and wound age on infection by Nectria galligena of pruning wounds on apple. Eur. J. Plant Pathol. 104:511–519.

4459. Xu, X., Butt, D. J., and Xu, X. M. 1998. Effects of temperature and atmospheric moisture on the early growth of apple powdery mildew (Podosphaera leucotricha) colonies. Eur. J. Plant Pathol. 104:133–140.

4460. Xu, X. M., and Robinson, J. D. 2000. Effects of temperature on the incubation and latent periods of hawthorn powdery mildew (Podosphaera clandestina). Plant Pathol. 49:791–797.

4461. Xu, Y., Rohrig, E., and Folster, H. 1997. Reaction of root systems of grand fir (Abies grandis Lindl.) and Norway spruce (Picea abies Karst.) to seasonal waterlogging. For. Ecol. Manag. 93:9–19.

4462. Xu, Z., Leininger, T. D., Lee, A. W. C., and Tainter, F. H. 2001. Chemical properties associated with bacterial wetwood in red oaks. Wood Fiber Sci. 33:76–83.

4463. Xu, Z., Leininger, T. D., Lee, A. W. C., and Tainter, F. H. 2001. Physical, mechanical, and drying properties associated with bacterial wetwood in red oaks. For. Prod. J. 51:79–84.

4464. Xue, A. G., and Davidson, C. G. 1998. Components of partial resistance to black spot disease (Diplocarpon rosae Wolf) in garden roses. HortScience 33:96–99.

4465. Yakota, S. 1975. Scleroderris canker of Todo-fir in Hokkaido, northern Japan. III. Dormant infection of the causal fungus. IV. An analysis of climatic data associated with the outbreak. Eur. J. For. Pathol. 5:7–12, 13–21.

4466. Yakota, S., Uozumi, T., and Matsuzaki, S. 1974. Scleroderris canker of Todo-fir in Hokkaido, northern Japan. I. Present status of damage and features of infected plantations. II. Physiological and pathological characteristics of the causal fungus. Eur. J. For. Pathol. 4:65–74, 155–166.

4467. Yamamoto, F., Sakata, T., and Terazawa, K. 1995. Physiological, morphological and anatomical responses of Fraxinus mandshurica seedlings to flooding. Tree Physiol. 15:713–719.

4468. Yamaoka, Y., and Katsuya, K. 1985. Evaluation of morphological characteristics in the rust taxonomy with special reference to the cultural state of Melampsora and Melampsoridium. Trans. Mycol. Soc. Jap. 26:231–246.

4469. Yan, Z. 1993. Germination and seedling development of two mistletoes, Amyema preissii and Lysiana exocarpi: host specificity and mistletoe-host compatibility. Austral. J. Ecol. 18:419–429.

4470. Yan, Z. 1993. Resistance to haustorial development of two mistletoes, Amyema preissii (Miq.) Tieghem and Lysiana exocarpi (Behr.) Tieghem ssp. exocarpi (Loranthaceae), on host and nonhost species. Int. J. Plant Sci. 154:386–394.

4471. Yang, D., Bernier, L., and Dessureault, M. 1995. Phaeotheca dimorphospora increases Trichoderma harzianum density in soil and suppresses red pine damping-off caused by Cylindrocladium scoparium. Can. J. Bot. 73:693–700.

4472. Yang, Y.-S., Skelly, J. M., and Chevone, B. I. 1982. Clonal response of eastern white pine to low doses of O_3, SO_2, and NO_3, singly and in combination. Can. J. For. Res. 12:803–808.

4473. Yao, Y. J., Pegler, D. N., and Young, T. W. K. 1996. Genera of Endogonales. R. Bot. Gard., Kew, UK. 229 pp.

4474. Yarwood, C. E. 1957. Powdery mildews. Bot. Rev. 23:235–300.

4475. Yarwood, C. E. 1981. The occurrence of Chalara elegans. Mycologia 73:524–530.

4476. Yarwood, C. E., and Gardner, M. W. 1972. Powdery mildews favored by man. Plant Dis. Rep. 56:852–855.

4477. Yde-Anderson, A. 1979. Disease symptoms, taxonomy, and morphology of Lachnellula willkommii. A literature review. Eur. J. For. Pathol. 9:220–228.

4478. Yde-Andersen, A. 1979. Host spectrum, host morphology, and geographic distribution of larch canker, Lachnellula willkommii. A literature review. Eur. J. For. Pathol. 9:211–219.

4479. Yde-Andersen, A. 1979. Lachnellula willkommii—canker formation and the role of microflora. A literature review. Eur. J. For. Pathol. 9:347–355.

4480. Yde-Andersen, A. 1980. Infection process and the influence of frost damage in Lachnellula willkommii. A literature review. Eur. J. For. Pathol. 10:28–36.

4481. Ye, J., Qi, G., Bao, H., and Feng, W. 2000. Studies on the mechanisms of the brown spot needle blight fungus toxin in making host cell damage. Sci. Silvae Sinicae 36:82–86.

4482. Yeh, Z., Chen, Z., and Kimbrough, J. W. 2000. Ganoderma australe from Florida. Mycotaxon 75:233–240.

4483. Yeoman, M. M., Kilpatrick, D. C., Miedzybrodzka, M. B., and Gould, A. R. 1978. Cellular interactions during graft formation in plants: a recognition phenomenon? Symp. Soc. Exp. Biol. 32:139–160.

4484. Ying, C. C., and Hunt, R. S. 1987. Stability of resistance among Pinus contorta provenances to Lophodermella concolor needle cast. Can. J. For. Res. 17:1596–1601.

4485. Ylimartimo, A., Laflamme, G., Simard, M., and Rioux, D. 1997. Ultrastructure and cytochemistry of early stages of colonization by Gremmeniella abietina in Pinus resinosa seedlings. Can. J. Bot. 75:1119–1132.

4486. Yokoya, K., Kandasamy, K. I., Walker, S., Mandegaran, Z., and Roberts, A. V. 2000. Resistance of roses to pathotypes of Diplocarpon rosae. Ann. Appl. Biol. 136:15–20.

4487. York, H. H., Wean, R. E., and Childs, T. W. 1936. Some results of investigations on Polyporus schweinitzii Fr. Science 84:160–161.

4488. Young, J. M., Bull, C. T., Boer, S. H. De, Firrao, G., Gardan, L., Saddler, G. E., Stead, D. E., and Takikawa, Y. 2001. Classification, nomenclature, and plant pathogenic bacteria—a clarification. Phytopathology 91:617–620.

4489. Young, J. M., Kuykendall, L. D., Martinez-Romero, E., Kerr, A., and Sawada, H. 2001. A revision of Rhizobium Frank 1889, with an emended description of the genus, and the inclusion of all species of Agrobacterium Conn 1942 and Allorhizobium Undicola de Lajudie et al. 1998 as new combinations: Rhizobium radiobacter, R. rhizogenes, R. rubi, R. undicola and R. vitis. Int. J. Syst. Evol. Microbiol. 51:89–103.

4490. Young, J. M., Saddler, G. S., Takikawa, Y., Boer, S. H. De, Vauterin, L., Gardan, L., Gvozdyak, R. I., and Stead, D. E. 1996. Names of plant pathogenic bacteria 1864–1995. Rev. Plant Pathol. 75:721–763.

4491. Yuan, Z., and Mohammed, C. 1998. Infection of wounds in young Eucalyptus nitens by ascospores and conidia of Endothia gyrosa. N.Z. J. For. Sci. 28:316–324.

4492. Zabel, R. A. 1976. Basidiocarp development in Inonotus obliquus and its inhibition by stem treatments. For. Sci. 22:431–437.

4493. Zabel, R. A., and Morrell, J. J. 1992. Wood microbiology: decay and its prevention. Academic Press, San Diego. 476 pp.

4494. Zagula, K. R., Aref, N. M., and Ramsdell, D. C. 1989. Purification, serology, and some properties of a mechanically transmissible virus associated with green ring mottle disease in peach and cherry. Phytopathology 79:451–456.

4495. Zalasky, H. 1964. Nomenclature and description of Diplodia tumefaciens (Shear) Zalasky [Macrophoma tumefaciens Shear apud Hubert]. Can. J. Bot. 42:1049–1055.

4496. Zalasky, H. 1964. The histopathology of Macrophoma tumefaciens infections in black poplar. Can. J. Bot. 42:385–391.

4497. Zalasky, H. 1965. Process of Ceratocystis fimbriata infection in aspen. Can. J. Bot. 43:1157–1162.

4498. Zalasky, H. 1968. Penetration and initial establishment of Nectria galligena in aspen and peachleaf willow. Can. J. Bot. 46:57–60.

4499. Zalasky, H. 1974. Keissleriella emergens, a perfect state of Diplodia tumefaciens in roots of poplar. Can. J. Bot. 52:11–13.

4500. Zalasky, H. 1975. Chimeras, hyperplasia, and hypoplasia in frost burls induced by low temperature. Can. J. Bot. 53:1888–1898.

4501. Zalasky, H. 1975. Low-temperature-induced cankers and burls in test conifers and hardwoods. Can. J. Bot. 53:2526–2535.

4502. Zalasky, H. 1976. Xylem in galls of lodgepole pine caused by western gall rust, Endocronartium harknessii. Can. J. Bot. 54:1586–1590.

4503. Zalasky, H. 1978. Stem and leaf spot infections caused by Septoria musiva and S. populicola on poplar seedlings. Phytoprotection 59:43–50.

4504. Zalasky, H. 1980. Lodgepole pine (Pinus contorta Dougl. var. latifolia Engelm.) shoot abnormalities from frost injury. Can. For. Serv. Bi-mon. Res. Notes 36:21–22.

4505. Zambino, P. J. 2000. Evaluating white pine blister rust resistance in *Ribes* after artificial inoculation. HortTechnology 10:544–545.

4506. Zambino, P. J., and Harrington, T. C. 1990. Heterokaryosis and vegetative compatibility in *Leptographium wageneri*. Phytopathology 80:1460–1469.

4507. Zarnovican, R. 2001. Glaze damage in a young yellow birch stand in southern Quebec, Canada. North. J. Appl. For. 18:14–18.

4508. Zehr, E. I., Shepard, D. P., and Bridges, W. C. Jr. 1996. Bacterial spot of peach as influenced by water congestion, leaf wetness duration, and temperature. Plant Dis. 80:339–341.

4509. Zeikus, J. G., and Ward, J. C. 1974. Methane formation in living trees: a microbial origin. Science 184:1181–1183.

4510. Zentmyer, G. A. 1980. *Phytophthora cinnamomi* and the diseases it causes. Monogr. 10. Am. Phytopathol. Soc., St. Paul, MN. 96 pp.

4511. Zhang, J. W., Klopfenstein, N. B., and Peterson, G. W. 1997. Genetic variation in disease resistance of *Juniperus virginiana* and *J. scopulorum* grown in eastern Nebraska. Silvae Genet. 46:11–16.

4512. Zhang, N., and Blackwell, M. 2001. Molecular phylogeny of dogwood anthracnose fungus (*Discula destructiva*) and the Diaporthales. Mycologia 93:355–365.

4513. Zhang, P. G., and Sutton, J. C. 1994. High temperature, darkness, and drought predispose black spruce seedlings to gray mold. Can. J. Bot. 72:135–142.

4514. Zhang, Y. C., and Walker, J. T. 1995. Factors affecting infection of water oak, *Quercus nigra*, by *Tubakia dryina*. Plant Dis. 79:568–571.

4515. Zhang, Y. P., Kirkpatrick, B. C., Smart, C. D., and Uyemoto, J. K. 1998. cDNA cloning and molecular characterization of cherry green ring mottle virus. J. Gen. Virol. 79:2275–2281.

4516. Zhang, Y. P., Kirkpatrick, B. C., Terlizzi, B. D., and Uyemoto, J. K. 2000. Comparison of cherry green ring mottle virus strains using RT-PCR and coat protein sequence phylogeny. J. Plant Pathol. 82:49–53.

4517. Zhao, J.-D., and Zhang, X.-Q. 1992. The polypores of China. Bibl. Mycol. No. 145. 524 pp.

4518. Zhao, W. X., and Guo, Y. L. 1993. Studies on Hyphomycetes of Zhangjiajie in Hunan. II. *Cercospora* and *Pseudocercospora*. Acta Mycol. Sinica 12:193–199.

4519. Zhou, S., Smith, D. R., and Stanosz, G. R. 2001. Differentiation of *Botryosphaeria* species and related anamorphic fungi using Inter Simple or Short Sequence Repeat (ISSR) fingerprinting. Mycol. Res. 105:919–926.

4520. Zhou, S., and Stanosz, G. R. 2001. Relationships among *Botryosphaeria* species and associated anamorphic fungi inferred from the analyses of ITS and 5.8S rDNA sequences. Mycologia 93:515–527.

4521. Zhu, J., Oger, P. M., Schrammeijer, B., Hooykaas, P. J. J., Farrand, S. K., and Winans, S. C. 2000. The bases of crown gall tumorigenesis. J. Bacteriol. 182:3885–3895.

4522. Zhu, X. B., Cox, R. M., and Arp, P. A. 2000. Effects of xylem cavitation and freezing injury on dieback of yellow birch (*Betula alleghaniensis*) in relation to a simulated winter thaw. Tree Physiol. 20:541–547.

4523. Zhu, X. B., Cox, R. M., Meng, F.-R., and Arp, P. A. 2001. Responses of xylem cavitation, freezing injury and shoot dieback to a simulated winter thaw in yellow birch seedlings growing in different nursery culture regimes. For. Ecol. Manag. 145:243–253.

4524. Ziller, W. G. 1955–1970. Studies of western tree rusts. II. *Melampsora occidentalis* and *M. albertensis*, two needle rusts of Douglas-fir. V. The rusts of hemlock and fir caused by *Melampsora epitea*. VI. The aecial host ranges of *Melampsora albertensis, M. medusae*, and *M. occidentalis*. VII. Inoculation experiments with pine stem rusts (*Cronartium* and *Endocronartium*). VIII. Inoculation experiments with conifer needle rusts (Melampsoraceae). Can. J. Bot. 33:177–188; 37:109–119; 43:217–230; 48:1313–1319, 1471–1476.

4525. Ziller, W. G. 1968. Studies of hypodermataceous needle diseases. I. *Isthmiella quadrispora* sp. nov. causing needle blight of alpine fir. Can. J. Bot. 46:1377–1381.

4526. Ziller, W. G. 1974. The tree rusts of western Canada. Can. For. Serv. Pub. 1329. 272 pp.

4527. Ziller, W. G., and Funk, A. 1973. Studies of hypodermataceous needle diseases. III. The association of *Sarcotrochila macrospora* n.sp. and *Hemiphacidium longisporum* n.sp. with pine needle cast caused by *Davisomycella ampla* and *Lophodermella concolor*. Can. J. Bot. 51:1959–1963.

4528. Zoeten, G. A. de, Lauritis, J. A., and Mircetich, S. M. 1982. Cytopathology and properties of cherry leaf roll virus associated with walnut blackline disease. Phytopathology 72:1261–1265.

4529. Zoina, A., Raio, A., Peluso, R., and Spasiano, A. 2001. Characterization of agrobacteria from weeping fig (*Ficus benjamina*). Plant Pathol. 50:620–627.

4530. Zotz, G., and Andrade, J. L. 1998. Water relations of two co-occurring epiphytic bromeliads. J. Plant Physiol. 152:545–554.

4531. Zulfiqar, M., Brlansky, R. H., and Timmer, L. W. 1996. Infection of flower and vegetative tissues of citrus by *Colletotrichum acutatum* and *C. gloeosporioides*. Mycologia 88:121–128.

4532. Zweifel, R., and Hasler, R. 2000. Frost-induced reversible shrinkage of bark of mature subalpine conifers. Agric. For. Meteorol. 102:213–222.

4533. Zwet, T. van der, and Beer, S. V. 1991. Fire blight—its nature, prevention, and control: a practical guide to integrated disease management. USDA Agric. Inf. Bull. 631. 83 pp.

4534. Zwet, T. van der, and Keil, H. L. 1979. Fire blight. A bacterial disease of rosaceous plants. USDA Agric. Handb. 510. 200 pp.

4535. Zwolinski, J. B., Swart, W. J., and Wingfield, M. J. 1995. Association of *Sphaeropsis sapinea* with insect infestation following hail damage of *Pinus radiata*. For. Ecol. Manag. 72:293–298.

4536. Zyl, L. M. van, and Wingfield, M. J. 1999. Wound response of *Eucalyptus* clones after inoculation with *Cryphonectria cubensis*. Eur. J. For. Pathol. 29:161–167.

4537. Zyl, L. M. van, Wingfield, M. J., Alfenas, A. C., and Crous, P. W. 1999. Hypovirulence detected in Brazilian isolates of *Cryphonectria cubensis*. Plant Pathol. 48:267–272.

Diseases, other disorders, pathogens, and some symptoms are indexed both directly and under the names of plant genera and species. Common names of plants and some pathogens are cross-indexed to Latin names. To locate information about a disease or disorder when searching under plant names, check under both genus and species names. If multiple species of a plant genus are indicated to be affected by a particular pathogen or disorder, then the pathogen or disorder is indexed for the plant genus and the illustrated plant species but is not usually indexed for all relevant plant species. If multiple species of a pathogen genus are mentioned on a page, then only the genus name and any illustrated species are usually indexed for that page. A page number in **bold type** indicates presence of a relevant illustration on the page or the facing page.

Space limitations prevented the presentation of lists of susceptible plant species or genera for some diseases. If no list appears for a particular disease, woody genera commonly affected by it may be indexed nonetheless. For diseases treated in this manner, readers who need lists of plants affected are directed to the indices by Conners (779), Ginns (1298), and Farr et al. (1099) and to the fungal database maintained by the U.S. Agricultural Research Service, Systematic Botany and Mycology Laboratory (ref. 4108).

242, 460, Verticillium wilt **242**, 468, virus 422, water stress 492, wood discoloration and decay **514**, *Xylella fastidiosa* 388, zone lines **306**

Acer spicatum (mountain maple): leaf spot 42, 92, *Phyllosticta minima* 42, *Rhytisma* tar spots 66, *Venturia acerina* 92

Acer tataricum (tatarian maple): leaf spot 42, *Phyllosticta minima* 42

Acer truncatum (Shantung maple): drought tolerance 494

Aceria celtis **14**

acidic fog 464

Acerviclypeatus poriformans **44**

acervulus 1, **18**, **82**, **96–114**

anthracnose **82**, **96–114**

actinorhizae **524**

Acoelorraphe wrightii (Everglades palm, saw cabbage palm): *Ganoderma zonatum* 346, leaf spot 30, lethal yellowing avoidance 402, root and butt rot 346, *Stigmina palmivora* 30

Acremonium diospyri **246**

Acrocomia: *Ganoderma zonatum* 346, root and butt rot 346

Actinidia: anthracnose 114, Armillaria root disease 326, *Botryosphaeria* 120–122, *Colletotrichum* 114, dieback 122, Verticillium wilt 242

Actinidia arguta (tara vine): leaf spot 34, *Pseudocercospora handelii* 34

Actinidia deliciosa (kiwifruit): bacterial disease 368–370, blossom blight 368, *Botrytis cinerea* 72, bud rot 368, charcoal root rot 218, gray mold rot 72, leaf blight 190, leaf spot 368, *Macrophomina phaseolina* 218, *Pestalotiopsis menezesiana* 190, *Phytophthora lateralis* 362, *Pseudomonas* 368, *P. syringae* 370, root rot 218, 362, *Sclerotium rolfsii* 226, southern blight 226

Adelges abietis (eastern spruce gall adelgid) 144

Adelges cooleyi (Cooley spruce gall adelgid) 58

Adonidia: *Ganoderma zonatum* 346, root and butt rot 346

Adonidia merrillii (Christmas or Manila palm): leaf spot 30, lethal yellowing **402**, phytoplasma **402**, *Stigmina palmivora* 30

adventitious: roots **506**, 510, shoots **504**, 510

aecium, see rusts

Aeglopsis chevalieri: greasy spot 22, *Mycosphaerella citri* 22

Aesculus (buckeye, horse-chestnut): anthracnose 114, Armillaria root disease 326, basal canker 206, bleeding canker 358, *Botryosphaeria* 120, 126, canker 120, 142, 176, 182, 206, 318, 358–362, canker-rot 318, *Cerrena unicolor* 318, *Chondrostereum purpureum* 320, collar rot 362, *Colletotrichum* 114, *Diaporthe ambigua* 142, dieback 120, 126,

142, 176, 320, *Erysiphe flexuosa* 8, *Ganoderma applanatum* 350, *Guignardia aesculi* **40**, herbicide tolerance 474–476, *Inonotus dryadeus* 344, *Kretzschmaria deusta* 206, leaf blotch **40**, lightning injury 506, mistletoe 442, Nectria canker 182, *Nectria cinnabarina* 176, *Neonectria galligena* 182, *Oxyporus populinus* 310, ozone tolerance 484, *Phoradendron serotinum* 442, *Phyllactinia guttata* 10, *Phytophthora* 358–362, Phytophthora canker 358–362, *Phytophthora ramorum* 360, *Polyporus squamosus* 310, powdery mildew 8–10, root and butt rot 206, 344, 350, root rot 326, salt tolerance 470, sapwood decay 318–320, *Schizophyllum commune* 318, SO_2 tolerance 484, trunk decay 310, 350, Verticillium wilt 242

Aesculus californica (California buckeye): bacterial scorch **386–388**, *Guignardia aesculi* 40, leaf blister 4, leaf blotch 40, mistletoe 442, *Phoradendron serotinum* 442, *Phytophthora ramorum* 360, *Taphrina aesculi* 4, *Xylella fastidiosa* **386**–388

Aesculus ×carnea (red horse-chestnut): *Apple mosaic virus* 416, *Guignardia aesculi* 40, leaf blotch 40, mosaic 416, virus 416

Aesculus flava (yellow buckeye): flood tolerance 502, *Inonotus dryadeus* 344, root and butt rot 344

Aesculus glabra (Ohio buckeye): *Guignardia aesculi* 40, leaf blotch 40, ozone sensitivity 482

Aesculus hippocastanum (horse-chestnut, European horse-chestnut): *Apple mosaic virus* 416, bacterial canker **378**, bleeding canker **378**, *Guignardia aesculi* **40**, leaf blotch **40**, leaf scorch **492**, mosaic 416, salt spray tolerance 468, virus 416

Aesculus parviflora (bottlebrush buckeye): *Guignardia aesculi* 40, leaf blotch 40, ozone sensitivity 482

Aesculus pavia (red buckeye): *Guignardia aesculi* 40, leaf blotch 40

Aesculus turbinata: (Japanese horse-chestnut): *Guignardia aesculi* 40, leaf blotch 40

African plume, see *Bauhinia galpinii*

Agathis (kauri): canker 176, dieback 176, *Nectria cinnabarina* 176, *Phaeolus schweinitzii* 340, root and butt rot 340

Agave: leaf spot 36, *Microsphaeropsis concentrica* 36

Agave americana (century plant): *Phymatotrichopsis* resistance 222

Agrilus 124, *A. anxius* 464, *A. bilineatus* 500

Agrobacterium 382, 510, *A. pseudotsugae* 510, *A. rhizogenes* 382, 510, *A. tumefaciens* **382**

Ailanthus: anthracnose 114, Armillaria root disease 326, *Botryosphaeria* 120, 126–128, butt rot 350, canker 120, 176, 186, canker-rot 318, *Cerrena unicolor* 318, *Colletotrichum* 114, *Cristulariella moricola* 74, dieback 120, 126–128, 176, 186, *Fusarium lateritium* 186, *Ganoderma applanatum* 350, herbicide tolerance 474, leaf spot 74, *Nectria cinnabarina* 176, *Phymatotrichopsis omnivora* 222, root rot 222, 326, sapwood decay 318, *Schizophyllum commune* 318, trunk decay 318, 350

Ailanthus altissima (tree-of-heaven): fluoride tolerance 486, *Fusarium oxysporum* f.sp. *perniciosum* **248**, Fusarium wilt **248**, ice glaze damage 506, ozone tolerance 482, salt tolerance 470, SO_2 tolerance 484, *Verticillium dahliae* 242, Verticillium wilt 242

Aiphanes: *Ganoderma zonatum* 346, root and butt rot 346

Aiphanes lindeniana: lethal yellowing 402, phytoplasma 402

air pollutants: injury by **478–486**, landscape damage 478, **484**, overview 478, pollution stress 334, predisposing to disease 334

airplant **526**

Ajuga reptans (bugleweed): *Sclerotium rolfsii* **226**, southern blight **226**

Albizia (mimosa, silktree): Armillaria root disease 326, *Botryosphaeria* 120, 128, canker 120, 176, 186, canker-rot 318, dieback 120, 128, 176, 186, *Fusarium lateritium* 186, *F. oxysporum* f.sp. *perniciosum* **248**, Fusarium wilt **248**, gall 150, *Ganoderma* 348, herbicide tolerance 474, *Nectria cinnabarina* 176, Nectriella gall 150, *Nectriella pironii* 150, *Oxyporus latemarginatus* 344, ozone tolerance 482, root and butt rot 348, root rot 326, salt tolerance 470, sapwood decay 318, *Schizophyllum commune* 318, trunk decay 344, Verticillium wilt 242

Albizia julibrissin (mimosa, silktree): alcoholic flux 248, **384**, dieback 180, *Fusarium oxysporum* f.sp. *perniciosum* **248**, Fusarium wilt **248**, *Nectria austroamericana* 180, striped chlorosis **428**, viral disease **428**, wilt 248

Albizia lebbek (lebbek): gall 134, *Sphaeropsis tumefaciens* 134

alcoholic flux 378, **384**

alder, see *Alnus*

black or European: *Alnus glutinosa*

green: *A. viridis*

Italian: *A. cordata*

mountain: *A. incana* subsp. *tenuifolia*

red: *A. rubra*

speckled: *A. incana*

Sierra: *A. rhombifolia*

thinleaf: *A. incana* subsp. *tenuifolia*

white: *A. rhombifolia*

Aleurites fordii, see *Vernica*

Aleurodiscus 520, *A. oakesii* **520**

Alfalfa mosaic virus 426

Alfalfamovirus 426

alkaline soil 466

Allagoptera arenaria (seashore palm): lethal yellowing 402, phytoplasma 402

Allantophomopsis pseudotsugae 144

almond, see *Prunus dulcis*; *Terminalia*

desert: *P. fasciculata*

dwarf Russian: *P. tenella*

flowering: *P. triloba*

tropical: *Terminalia*

Alnus (alder): alder yellows 390, *Apiosporopsis carpinea* 112, Armillaria root disease 326, bacterial disease 370, *Biscogniauxia mediterranea* 200, *Botryosphaeria* 120, 126, *Botrytis cinerea* 72, butt rot 350, canker 120, 176, 182, 358, 362, canker-rot 318, catkin hypertrophy 4, *Cerrena unicolor* 318, *Chondrostereum purpureum* 320, collar rot 362, *Cristulariella moricola* 74, *Cryptodiaporthe salicella* 156, dieback 120, 126, 176, 200, 320, *Entoleuca mammata* 198, *Erysiphe* 8, flood tolerance 502, *Fomes fomentarius* 308, *Fomitopsis pinicola* 322, *Ganoderma* 348–350, gray mold blight 72, herbicide tolerance 474–476, *Kretzschmaria deusta* 206, leaf blister 4, leaf spot 74, *Melampsoridium* 274, mistletoe 442, Nectria canker 182, *Nectria cinnabarina* 176, *Neonectria galligena* 182, *Onnia tomentosa* 336, *Oxyporus latemarginatus* 344, ozone tolerance 482, *Phellinus laevigatus* 306, *Phoradendron serotinum* 442, *Phyllactinia guttata* 10, *Phytophthora* 358, 362, Phytophthora canker 358, 362, phytoplasma 390, *Polyporus squamosus* 310, powdery mildew 8–10, *Pseudomonas syringae* 370, *Rhytisma salicinum* 66, root and butt rot 206, 348, root rot 326, rust 274, salt tolerance 470, sapwood decay 318–320, *Schizophyllum commune* 318, *Scorias spongiosa* **16**, sooty mold **16**, *Taphrina* 4, tar spot 66, trunk decay 306–310, 322, 344, 350

Alnus cordata (Italian alder): bacterial spot 372, leaf blight 372, *Xanthomonas* 372

Alnus glutinosa (black or European alder): fluoride tolerance 486, SO_2 tolerance 484

Alnus incana (mountain, speckled, or thinleaf alder): SO_2 tolerance 484

Alnus rhombifolia (Sierra or white alder): bacterial spot **372**, leaf blight **372**, *Xanthomonas* **372**

Carpentaria (carpentaria palm): *Ganoderma zonatum* 346, root and butt rot 346

Carpinus (hornbeam): *Aleurodiscus oakesii* 520, anthracnose 112, *Apiosporopsis carpinea* 112, Armillaria root disease 326, *Biscogniauxia* 200–202, *Botryosphaeria obtusa* 126, butt rot 350, *Camillea tinctor* 204, canker 182, 186, canker-rot 318, *Cerrena unicolor* 318, *Chondrostereum purpureum* 320, *Cristulariella moricola* 74, dieback 126, 186, 200–202, 320, *Entoleuca mammata* 198, *Erysiphe ellisii* 8, *Fusarium lateritium* 186, *Ganoderma* 348–350, *Kretzschmaria deusta* 206, leaf blight 112, leaf blister 4, leaf spot 74, *Melampsoridium carpini* 274, Nectria canker 182, *Neonectria galligena* 182, ozone tolerance 482, *Phyllactinia guttata* 10, powdery mildew 8–10, root and butt rot 206, 326, 348, root rot 326, rust 274, salt sensitivity 470, sapwood decay 318–320, *Schizophyllum commune* 318, smooth patch 520, *Taphrina australis* 4, trunk decay 350

Carpinus betulus (European hornbeam): *Apioplagiostoma carpinicolum* 94, Apple mosaic virus 416, fluoride tolerance 486, leaf blight 94, virus 416

Carpinus caroliniana (American hornbeam, blue-beech, ironwood): anthracnose 112, *Apiosporopsis carpinea* 112, flood tolerance 502, leaf blight 112, *Monostichella robergei* 112, SO₂ tolerance 484

carrotwood, see *Cupaniopsis*

Carya (hickory, pecan): *Aleurodiscus oakesii* 520, anthracnose 114, Armillaria root disease 326, *Asteroma caryae* 104, *Biscogniauxia atropunctata* 200, *Botryosphaeria* 120, 124–128, bunch disease **392–394**, butt rot 206, 350, *Calonectria* 220, *Camillea punctulata* 204, canker 120, 176, 182, canker-rot 312, 316–318, *Cerrena unicolor* 318, *Chondrostereum purpureum* 320, *Climacodon septentrionalis* 308, cold acclimation 498, *Colletotrichum* 114, *Conoplea globosa* 216, *Cristulariella moricola* 74, *Cylindrocladium* 220, decline 224, *Dendrothele* 520, dieback 120, 124–128, 176, 200, 320, downy spot **252**, drought stress 492, *Erysiphe caryae* 8, flood tolerance 502, *Fomes fomentarius* 308, *Fomitopsis pinicola* 322, gall 148, *Ganoderma* 348–350, *Globifomes graveolens* 308, *Gnomonia caryae* **104**, *Hendersonia* 36, herbicide tolerance 474–476, *Hericium erinaceus* 306, ice glaze damage 506, *Inonotus dryadeus* 344, *I. hispidus* 316, *Kretzschmaria deusta* 206, leaf spot 36, 44, 74, **104**, leaf yellowing 492, *Microstroma juglandis* **252**, mistletoe 442, Nectria canker 182, *Nectria cinnabarina* 176, *Neonectria galligena* 182, *Oxyporus populinus* 310, *O. latemarginatus* 344, *Phellinus everhartii* 308, *P. spiculosus* 312, Phomopsis gall 148, *Phoradendron serotinum* 442, *Phyllactinia guttata* 10, *Phymatotrichopsis omnivora* 222, Phymatotrichum root rot 222, phytoplasma **392–394**, powdery mildew 8–10, root and butt rot 206, 348, root rot 220, 224, 326, 344, salt tolerance 470, sapwood rot 318–320, *Schizophyllum commune* 318, smooth patch 520, Strumella canker 216, supercooling 498, trunk decay 306–312, 316, 322, 344, 350, *Tubakia dryina* 44, twig canker 124, 176, *Urnula craterium* 216, Verticillium resistance 242, white leaf mold **252**, witches'-broom 252, **392–394**, *Xylaria polymorpha* 224

Carya aquatica (water hickory): *Hericium erinaceus* 306, trunk decay 306

Carya cordiformis (bitternut hickory): Phomopsis gall **148**

Carya glabra (pignut hickory): dieback 136, *Melanconis juglandis* 136

Carya illinoinensis (pecan): anthracnose 18, *Articularia quercina* var. *minor* 252, bacterial scorch 386, bunch disease 390, **392–394**, burrowing nematode 432, catkin blight 252, downy spot 20, 252, *Elsinoë randii* 18, fruit rot 358, *Gnomonia* 104, leaf blotch 20, leaf spot 104, liver spot 104, *Microstroma juglandis* 252, mouse ear 490, *Mycosphaerella* 20, nematode 432, nickel deficiency 490, *Phytophthora cactorum* 358, phytoplasma 390, **392–394**, *Radopholus similis* 432, *Rosellinia necatrix* resistance 224, vein spot 104, white leaf mold 252, witches'-broom 390, **392–394**, *Xylella fastidiosa* 386, zinc deficiency 490

Carya laciniosa (shellback hickory): SO₂ tolerance 484

Carya ovata (shagbark hickory): *Asteroma caryae* 104, *Biscogniauxia marginata* 202, bunch disease 390, **392–394**, decline **224**, dieback 202, **224**, *Cristulariella moricola* **74**, downy spot **252**, *Gnomonia caryae* **104**, leaf spot **74**, **104**, *Microstroma juglandis* **252**, mistletoe 442, *Phoradendron serotinum* 442, phytoplasma 390, **392–394**, root rot **224**, white leaf mold **252**, witches'-broom 252, **392–394**, *Xylaria polymorpha* **224**, zonate leaf spot **74**

Carya pallida (sand hickory): downy spot 252, *Microstroma juglandis* 252

Carya texana (black or Texas hickory): downy spot 252, *Microstroma juglandis* 252, witches'-broom 252

Carya tomentosa (mockernut hickory): *Biscogniauxia marginata* 202, canker-rot 316, dieback 136, 202, *Inonotus andersonii* 316, *Melanconis juglandis* 136, trunk decay 316

Caryopteris: Arabis mosaic virus 412

Caryota (fishtail palm): anthracnose 114, *Botryosphaeria* 120, *Colletotrichum* 114, *Ganoderma zonatum* 346, leaf spot 190, *Pestalotiopsis palmarum* 190, root and butt rot 346

Caryota mitis (Burmese fishtail palm): leaf spot 30, lethal yellowing 402, phytoplasma 402, *Stigmina palmivora* 30

Caryota rumphiana (Albert or giant fishtail palm): lethal yellowing 402, phytoplasma 402

Caryota urens (wine palm): leaf spot 30, *Stigmina palmivora* 30

cashew, see *Anacardium*

cassandra, see *Chamaedaphne calyculata*

Cassia (cassia, senna, see also *Senna*): *Botryosphaeria* 120, 128, *Calonectria* 220, canker 120, *Cylindrocladium* 220, dieback 120, 128, *Erysiphe diffusa* 8, *Ganoderma* 348, powdery mildew 8, root rot 220, 348, *Sclerotium rolfsii* 226, southern blight 226

cassina, see *Ilex vomitoria*

Cassiope (mountain heather): *Exobasidium* 250, red shoot syndrome 250, shoot deformity 250

Cassytha filiformis (dodder laurel, devil's gut) **456**

Castanea (chestnut, chinkapin): anthracnose 114, *Apiosporopsis carpinea* 112, Armillaria root disease 326, bacterial disease 370, bark rot 520, *Biscogniauxia mediterranea* 200, bleeding canker 358, *Botryosphaeria* 120, 126–128, brown cubical rot 310, butt rot 350, *Calonectria* 220, *Camillea punctulata* 204, canker 120, **160**, canker-rot 316–318, *Cerrena unicolor* 318, chestnut blight **160**, *Chondrostereum purpureum* 320, *Ciborinia foliicola* 68, collar rot 356, 362, *Colletotrichum* 114, *Conoplea globosa* 216, *Cryphonectria parasitica* **160**, *Cylindrocladium* 220, *Dendrothele* 520, dieback 120, 126–128, **160**, 200, 320, *Erysiphe* 10, *Fomitopsis pinicola* 322, *Ganoderma applanatum* 350, *Hericium erinaceus* 306, *Inonotus dryadeus* 344, *I. glomeratus* 316, *Kretzschmaria deusta* 206, *Laetiporus sulphureus* 310, leaf blight 68, leaf spot 44, mistletoe 442, *Perenniporia medulla-panis* 520, *Phellinus everhartii* 308, *Phoradendron serotinum* 442, *Phyllactinia* 10, *Phytophthora* 356–358, 362–364, *P. cambivora* 356, *P. cinnamomi* 364, Phytophthora cankers and collar rots 358–360, phytoplasma 390, powdery mildew 10, *Pseudomonas syringae* 370, root and butt rot 206, 344, root rot 220, 224, 326, 364, *Rosellinia necatrix* 224, sapwood rot 318–320, smooth patch 520, *Stereum gausapatum* 306, Strumella canker 216, *Trametes versicolor* 310, trunk decay 306–310, 316, 322, 350, *Tubakia dryina* 44, *Urnula craterium* 216, Verticillium resistance 242, white rot 310, witches'-broom 390

Castanea crenata (Japanese chestnut): chestnut blight 160, *Cryphonectria parasitica* 160, leaf spot 44, phytoplasma 390, *Tubakia dryina* 44, witches'-broom 390

Castanea dentata (American chestnut): blight **160**, canker **160**, 164, *Cryphonectria parasitica* **160**, *Endothia gyrosa* 164, *Inonotus dryadeus* 344, root and butt rot 344

Castanea mollissima (Chinese chestnut): blight **160**, *Cryphonectria parasitica* **160**

Castanea sativa (European chestnut): chlorotic spots 422, dieback 136, *Fusicoccum mangiferum* 136, leaf deformity 422, *Nattrassia mangiferae* 136, *Scytalidium dimidiatum* 136, sooty canker 136, tobamovirus 422

Castanopsis (chinkapin): Armillaria root disease 326, *Ganoderma* 348, leaf blister 4, root- and butt rot 348, root rot 326, *Taphrina castanopsidis* 4

Castilleja (paintbrush): *Cronartium* 292–294, rust 292–294, 300–304

Casuarina (sheoak, Australian pine): *Armillaria tabescens* 330, bark crack **516**, *Frankia* 524, freeze damage **516**, *Ganoderma* 346–348, *Kretzschmaria deusta* 206, nitrogen-fixing nodules 524, root and butt rot 206, 330, 346–348, root rot 330

Casuarina equisetifolia (beach sheoak, horsetail casuarina): flood tolerance 502, slime flux **384**, small ballmoss **526**, *Tillandsia recurvata* **526**, wetwood 384

Catalpa (catalpa): *Alternaria* 84, anthracnose 114, Armillaria root disease 326, *Botryosphaeria* 120, 126–128, canker 120, 176, canker-rot 318, *Colletotrichum* 114, *Cristulariella moricola* 74, dieback 120, 126–128, 176, *Erysiphe elevata* 8, herbicide tolerance 474–476, ice glaze damage 506, leaf spot 74, 84, *Nectria*

318, *Sclerotium rolfsii* 226, *Scytalidium dimidiatum* 136, seedling tip necrosis 84, SO$_2$ tolerance 484, sooty canker 136, sooty mold **16**, southern blight 226, Sphaeropsis knot 134, *Sphaeropsis tumefaciens* 134, *Spiroplasma citri* **390**, 394, spreading decline 432, stubborn disease **390**, 394, stunting **390**, *Thielaviopsis basicola* 222, thread blight 256, trunk decay 316, 350, Verticillium resistance 242, virus 414, 418, 424, wilt 232, witches'-broom 134, *Xanthomonas axonopodis* pv. *citri* **372**, *Xylaria polymorpha* 224, *Xylella fastidiosa* 466, xylem plugging 466, zinc deficiency 466

Citrus aurantifolia (Mexican lime): *Fusarium oxysporum* f.sp. *citri* 248, Fusarium wilt 248

Citrus aurantium (sour orange): *Phymatotrichopsis omnivora* **222**, Phymatotrichum root rot **222**

Citrus jambhiri (rough lemon): gall 134, *Sphaeropsis tumefaciens* 134

Citrus leprosis virus 424

Citrus limon (lemon): anthracnose **114**, *Armillaria tabescens* 330, Asiatic citrus canker **372**, *Colletotrichum* **114**, *Glomerella cingulata* 114, leaf blotch **114**, leaf spot **372**, mushroom root rot 330, *Spiroplasma citri* **390**, stubborn disease **390**, *Xanthomonas axonopodis* pv. *citri* **372**

Citrus madurensis (calamondin): *Elsinoë fawcettii* 18, scab 18

Citrus medica (citron): gall 134, *Sphaeropsis tumefaciens* 134

Citrus ×paradisi (grapefruit): Asiatic citrus canker **372**, bacterial fruit spot **372**, blight **466**, burrowing nematode 432, decline **360**, **466**, dieback **360**, **466**, epicormic sprouts **466**, foot rot **360**, greasy spot **22**, leaf spot **372**, *Mycosphaerella citri* **22**, *Phytophthora nicotianae* **360**, *Radopholus similis* 432, spreading decline 432, *Xanthomonas axonopodis* pv. *citri* 372

Citrus reticulata (Mandarin orange): dieback **360**, fluoride sensitivity 486, foot rot **360**, *Phytophthora nicotianae* **360**

Citrus sinensis (sweet orange): Asiatic citrus canker **372**, burrowing nematode **432**, greasy spot **22**, leaf spot **372**, *Mycosphaerella citri* **22**, spreading decline **432**, *Radopholus similis* **432**, *Xanthomonas axonopodis* pv. *citri* **372**

Citrus variegation virus 414

Cladosporium 16, 20, 30, 92

Cladrastis (yellowwood): basal canker **224**, *Botryosphaeria* 120, canker 120, dieback 120, herbicide tolerance 476, *Oxyporus latemarginatus* 344, *Phyllactinia*

guttata 10, powdery mildew 10, root rot **224**, trunk decay 344, Verticillium wilt 242, *Xylaria* **224**

Cladrastis lutea (yellowwood): *Bean yellow mosaic virus* 420, flood tolerance 502, leaf deformity and mottle 420, veinal chlorosis 420, virus 420

Clarkia (farewell-to-spring): *Pucciniastrum epilobii* 278, rust 278

Clausena lansium (wampi): *Elsinoë fawcettii* 18, scab 18

cleistothecium **8–12**

Clematis (clematis, virgin's-bower): anthracnose 114, *Botryosphaeria obtusa* 126, chlorotic flecks 418, *Colletotrichum* 114, *Cucumber mosaic virus* 418, dieback 126, *Phymatotrichopsis* resistance 222, vein banding 418, virus 418

Clerodendrum (glory-bower): gall 150, *Nectriella pironii* 150

Cliftonia (buckwheat tree): leaf spot 190, *Pestalotiopsis versicolor* 190

Climacodon septentrionalis **308**

climate change 464

Clitocybe root rot 330

Clostridium: in wetwood 384

clove, see *Syzygium aromaticum*

clover proliferation phytoplasma 396

cobalt as nutrient 490

Coccoloba (pigeon-plum, seagrape): algal leaf spot 438, Armillaria root disease 326, *Alternaria* 84, anthracnose 114, canker 186, *Cassytha filiformis* 456, *Cephaleuros virescens* 438, *Colletotrichum* 114, dieback 186, dodder laurel 456, *Fusarium solani* 186, leaf spot 84, root rot 326

Coccoloba diversifolia (pigeon-plum, tietongue): *Ganoderma australe* 350, wood decay 350

Coccoloba uvifera (sea-grape): leaf spot 190, *Pestalotiopsis versicolor* 190, salt tolerance 470

Coccothrinax argentata (silver palm): false smut 254, *Graphiola phoenicis* 254, *Ganoderma zonatum* 346, root rot 346

Coccothrinax crinita (thatch palm): false smut 254, *Graphiola phoenicis* 254

Cocculus (coralbead): *Botryosphaeria* 120, 126, dieback 120, 126

coconut, see *Cocos*

Cocos nucifera (coconut palm): *Botryosphaeria* 120, 128, bud rot 358, burrowing nematode 432, *Calonectria* 220, *Cylindrocladium* 218, decline 432, false smut, *Ganoderma* 346, *Graphiola phoenicis* 254, leaf rot 220, leaf spot 66, 190, lethal decline 390, lethal yellowing 390, **402**, nematode 432, nut drop 358, *Pestalotiopsis palmarum* 190, *Phyllachora* 66, *Phytophthora heveae* 358, phytoplasma 390, **402**, phytoplasma resistance 392, *Radopholus similis* 432, root and butt rot 346, root rot 220, tar spot 66

Codiaeum variegatum (croton): anthracnose 114, canker 186, *Colletotrichum* 114, dieback 186, *Fusarium solani* 186, gall **150**, *Kutilakesa pironii* **150**, leaf deformity 424, leaf spot **370–372**, *Nectriella pironii* **150**, rhabdovirus 424, *Sclerotium rolfsii* 226, southern blight 226, virus 424, *Xanthomonas codiaei* **370–372**

coelomycete 1, 20, 24–28, 36–44, 56, 78, 82, 96–150, 154–156, 174, 190–196, 218, 460

Coffea (coffee): bacterial disease 368–370, 386, bacterial scorch 386, *Ceratocystis fimbriata* 232, canker 232, leaf spot 368, *Pseudomonas cichorii* 368, *P. syringae* 370, sapwood lesions 232, sooty mold 16, *Xylella fastidiosa* 386

coffee, wild-, see *Psychotria*

coffeeberry, see *Frangula californica*

coffeetree, see *Gymnocladus*; *Polyscias guilfoylei*

Kentucky: *G. dioica*

cold hardiness 498

Coleosporium 290, *C. asterum* 290, *C. pacificum* **290**

collar rot 128, 190, 344, **354–360**, **362–364**, 460

Colletogloeum 20

Colletotrichum 92, 96, 110, **114**, *C. acutatum* 92, **114**, *C. dematium* 114, *C. gloeosporioides* 92, 110, **114**, 512, *C. neriicolum* 114, *C. trichellum* 372

colored canker of *Platanus* **234**

comandra, see *Comandra*; *Geocaulon*

northern: *G. lividum*

Comandra (comandra, bastard-toadflax): *Cronartium comandrae* 292, rust 292

comandra blister rust 292, **300**

Comandra umbellata: *Cronartium comandrae* 300, comandra blister rust 300

compartmentalization of wound-associated defects in wood **512–514**

compartmentalization responses to cankers and canker-rot pathogens **186**, **312–314**, **318**

Comptonia peregrina (sweetfern): *Botryosphaeria* 120, canker 120, *Cronartium comptoniae* 292, 298, dieback 120, *Gymnosporangium ellisii* 260, 268, rust 260, 268, sweetfern blister rust **292**, 298

Condalia (snakewood): mistletoe 444, *Phoradendron californicum* 444

cone bug 192

cone moth 302

cone rusts **286–288**, 292, **302**

conidiomata, types 1

conidium 1

conifer-aspen rust 284

conifers: annosum root rot **332–334**, Armillaria root rot 326, **328–330**, bark beetle attack re-

lated to root disease 328, *Botrytis cinerea* 72, brown cubical butt rot 340, brown root and butt rot 340, canker 144, damping-off 72, dieback 144, drought crack 494, 516, foliar diseases **16**, **32**, **42–64**, Fusarium root diseases **218**, gray mold blight 72, *Heterobasidion annosum* 332, *Inonotus dryadeus* 344, needle rusts **276–290**, *Onnia circinata* 336, *O. tomentosa* 338, parch blight 494, *Phaeolus schweinitzii* 340, red belt injury 494, red root and butt rot 336, resinosis 336, root and butt rot **338**, 344, root rot **218–220**, **328–344**, rusts **264–268**, **276–304**, snow blight **62**, 144, soils predisposing to root disease 502, trunk decay **322–324**, **340**, viruses *Apple chlorotic leafspot virus* **416**, 422, 428

Coniothyrium 36, *C. fuckelii* **36**

conk 1, **306–334**

Conocarpus (buttonwood): *Calonectria* 220, *Cylindrocladium* 220, root rot 220, *Sphaceloma* 18

Conophthorus radiatae 188

Conoplea globosa **216**

conserved names 2

Copernicia: *Ganoderma zonatum* 346, root rot 346

copper deficiency **490**

copper fungicide injury 476

copperleaf, see *Acalypha*

coralbead, see *Cocculus*

coralberry, see *Ardisia*; *Symphoricarpos orbiculatus*

coral-spot canker and dieback **176–178**

coral tree, see *Erythrina*

cordgrass, see *Spartina*

Coreopsis (coreopsis, tickseed): *Coleosporium inconspicuum* 290, rust 290

cork, see periderm

cork cambium 246, 266, 508, **518**, 526

cork tree, see *Phellodendron*

cornel, see *Cornus*

small rough-leaf: *C. asperifolia*

Cornus (dogwood): anthracnose 18, **106**, 114, Armillaria root disease 326, bacterial diseases 370, **394**, *Biscogniauxia mediterranea* 200, *Botryosphaeria* 120–**122**, **126**, *Calonectria* 220, canker **106**, 120–**122**, 176, 182, 186, 362, canker-rot 318, *Ceratobasidium stevensii* 256, *Cerrena unicolor* 318, *Cherry leafroll virus* 410, *Chondrostereum purpureum* 320, collar rot 362, *Colletotrichum* 114, *Cristulariella* 74, crown canker **354**, *Cylindrocladium* 220, dieback **106**, 120, 126, 176, 186, 200, 320, *Discula destructiva* **106**, *Elsinoë corni* 18, *Erysiphe* 10, flood tolerance 502, fluoride tolerance 486, *Fusarium solani* 186, herbicide tolerance 476, leaf blight **106**, leaf spot 24, 40–42, 74, **106**, *Meloidogyne*

627

Euterpe (Assai palm): *Ganoderma zonatum* 346, root rot 346

Euthamia graminifolia (goldentop): *Coleosporium delicatulum* 290, rust 290

Eutypella 196, *E. parasitica* **196**

Eutypella canker **196**

Exobasidium **250**, *E. camelliae* var. *gracile* **250**, *E. japonicum* **250**, *E. vaccinii* **250**

Fagaceae as rust hosts 292

Fagus (beech): Armillaria root disease 326, bark rot 520, basal canker 206, *Biscogniauxia* 200, bleeding canker 358, *Botryosphaeria* 120, *B. quercuum* 124, *Botrytis cinerea* 72, brown cubical rot 310, *Camillea punctulata* 204, canker 120, 124, 176, 182, 360–362, canker-rot **316–318**, *Ceratocystis virescens* 236, *Cerrena unicolor* 318, *Chondrostereum purpureum* 320, *Climacodon septentrionalis* 308, collar rot 356, 362, *Conoplea globosa* 216, decline 224, *Dendrothele* 520, dieback 120, 124, 176, 182, 200, 320, *Entoleuca mammata* 198, *Erysiphe erineophila* 8, flood tolerance 502, *Fomes fomentarius* 308, *Fomitopsis pinicola* 322, *Ganoderma* 348, *Globifomes graveolens* 308, gray mold blight 72, herbicide tolerance 474–476, *Hericium erinaceus* 306, *Inonotus* 314–318, *I. dryadeus* 344, *Kretzschmaria deusta* 206, *Laetiporus sulphureus* 310, lightning injury 506, mistletoe 442, *Nectria cinnabarina* 176, *Neonectria galligena* 182, *Oxyporus latemarginatus* 344, ozone tolerance 482, *Perenniporia fraxinophila* 310, *P. medulla-panis* 520, *Phellinus everhartii* 308, *P. laevigatus* 306, *Phoradendron serotinum* 442, *Phyllactinia* 10, *Phytophthora* 356–362, *P. cambivora* 356, *P. inflata* 358, *P. ramorum* 360, *Polyporus squamosus* 310, powdery mildew 8–10, root and butt rot 206, 326, 344, 348, root rot 224, 326, 348, salt tolerance 470, sapwood decay 318–320, sapwood stain 236, *Schizophyllum commune* 318, *Scorias spongiosa* 16, smooth patch 520, SO_2 tolerance 484, sooty mold 16, Strumella canker 216, sunscald 498, *Tomato black ring virus* 414, trunk decay 306–310, **316**, 322, 344, *Urnula craterium* 216, Verticillium resistance 242, virus 414, *Xylaria polymorpha* 224

Fagus grandifolia (American beech): basal canker **206**, beech bark disease **184**, beech scale **184**, butt rot **206**, **350**, canker 164, **184**, **204**, canker-rot 314–**316**, cavity formation **512**, *Cryptococcus fagisuga* **184**, dieback 204, *Endothia gyrosa* 164, frost sensitivity 496, *Ganoderma applan-*

atum **350**, heart rot **306**, *Hypoxylon cohaerens* **204**, ice glaze damage 506, *Inonotus glomeratus* **316**, *I. obliquus* 314, *Kretzschmaria deusta* **206**, *Neonectria coccinea* 184, *N. coccinea* var. *faginata* **184**, *N. galligena* 184, *Phellinus igniarius* **306**, pseudosclerotial plates **306**, trunk decay **306**, 314–**316**, **350**, **512**, *Xylococcus betulae* **184**, zone lines **306**

Fagus sylvatica (European beech): air-pollution-associated growth loss 482, *Apiognomonia errabunda* 100, beech bark disease 184, beech scale 184, bleeding canker **356**, canker **164**, 184, 204, **356**, canker-rot 314–318, *Cerrena unicolor* 318, *Cherry leafroll virus* 410, chlorotic mottle 410, *Cryptococcus fagisuga* 184, dieback 204, *Discula umbrinella* 100, *Endothia gyrosa* **164**, fluoride tolerance 486, *Hypoxylon cohaerens* 204, *Inonotus glomeratus* 316, *I. obliquus* 314, leaf roll 410, *Neonectria coccinea* 184, *Phytophthora* **356–358**, sapwood decay 318, trunk decay 314–318, virus 410

false smut **254**

false velvet-top fungus **338**

farkleberry, see *Vaccinium arboreum*

fasciation **508**

Fatsia japonica (Japanese fatsia, paper-plant): *Alternaria panax* 84, bacterial spot 372, gall 150, leaf spot 84, *Nectriella pironii* 150, Verticillium wilt 242, *Xanthomonas hortorum* pv. *hederae* 372

Feijoa, see *Acca*

Fendlera (fendlerbush): *Gymnosporangium speciosum* 260, 268, rust 260, 268

fermentation 384

fern, see below and Pteridophyta
bladder: *Cystopteris*
bracken: *Pteridium*
chain: *Woodwardia*
flowering: *Osmunda*
lady: *Athyrium*
oak: *Gymnocarpium*
ostrich: *Matteuccia*
polypody: *Polypodium*
sensitive: *Onoclea*
wood: *Dryopteris*

fernbush, see *Chamaebatiaria millefolium*

ferns: fir-fern rusts 280, *Hyalopsora aspidotus* 280, *Milesina* 280, *Uredinopsis* 280

fetterbrush, see *Lyonia*

fetterbush, see *Leucothoe*; *Lyonia*

Ficus (fig): *Agrobacterium fici* 382, *Alternaria* 84, Armillaria root disease 326, black leaf spot 44, *Botryosphaeria* 120–122, 126–128, *Botrytis cinerea* 72, *Calonectria* 220, canker 120, 176, 186, canker-rot 318, *Cerotelium fici* **274**, *Cristulariella moricola* 74, *Cylin-*

drocladium 220, dieback 120–122, 126–128, 176, 186, *Fusarium lateritium* 186, gall 150, 382, gray mold blight 72, leaf spot 74, 84, *Meloidogyne* 430, *Nectria cinnabarina* 176, *Nectriella pironii* 150, nematodes 430–432, *Ophiodothella* 44, *Phymatotrichopsis omnivora* 222, Phymatotrichum root rot 222, *Pratylenchus* 432, root knot 430, root rot 220–224, 326, *Rosellinia necatrix* 224, rust 274, sapwood decay 318, *Schizophyllum commune* 318, *Sclerotium rolfsii* 226, southern blight 226, tobamovirus 422, tar spot 44, Verticillium wilt 242

Ficus aurea (strangler or Florida strangler fig): *Cerotelium fici* 274, *Ophiodothella* 44, rust 274, tar spot 44

Ficus benjamina (weeping fig): canker 142, *Diaporthe cinerascens* **142**, dieback **142**, gall 134, *Phomopsis cinerascens* **142**, *Sphaeropsis tumefaciens* 134

Ficus carica (fig, common fig): canker 142, *Cerotelium fici* **274**, *Diaporthe cinerascens* 142, dieback 136, 142, fig mosaic **428**, *Fusicoccum mangiferum* 136, *Nattrassia mangiferae* 136, nematode 432, rust **274**, *Scytalidium dimidiatum* 136, sooty canker 136, viral disease **428**, *Xiphinema index* 432

Ficus citrifolia (short-leaf fig): *Cerotelium fici* 274, rust 274

Ficus elastica (Indian rubber-tree, rubberplant): anthracnose **114**, *Armillaria tabescens* 330, canker **114**, *Cerotelium fici* 274, *Colletotrichum gloeosporioides* **114**, *Glomerella cingulata* **114**, leaf blotch **114**, root rot 330, rust 274

Ficus erecta: leaf spot 368, *Pseudomonas ficuserectae* 368, shoot blight 368

Ficus lyrata (fiddle-leaf fig): *Cerotelium fici* 274, leaf spot 368, *Pseudomonas cichorii* 368, rust 274

Ficus pumila (climbing or creeping fig): *Cerotelium fici* 274, rust 274

fig, see *Ficus*
climbing or creeping: *F. pumila*
common: *F. carica*
fiddle-leaf: *F. lyrata*
short-leaf: *F. citrifolia*
strangler, Florida strangler: *F. aurea*
weeping: *F. benjamina*

filbert, see *Corylus*
American: *C. americana*
beaked: *C. cornuta*
European: *C. avellana*
giant: *C. maxima*
Turkish: *C. colurna*

fir, see *Abies*; *Cunninghamia*; *Pseudotsuga*
alpine: *A. lasiocarpa*
balsam: *A. balsamea*

China: *C. lanceolata*
Douglas-: *P. menziesii*
Fraser: *A. fraseri*
grand: *A. grandis*
Greek: *A. cephalonica*
lowland white: *A. grandis*
noble: *A. procera*
Pacific silver: *A. amabilis*
red or California red: *A. magnifica*
Rocky Mountain subalpine: *A. bifolia*
sacred: *A. religiosa*
Sakhalin: *A. sachalinensis*
Shasta red: *Abies ×shastensis*
Siberian: *A. sibirica*
silver: *A. alba*
subalpine: *A. lasiocarpa*
Todo: *A. sachalinensis*
white or Colorado white: *A. concolor*

fir-blueberry rust **278**

fir broom rust **282**

fir-fern rusts **280**

fir-fireweed rust 276–278

fire blight **376**

fire damage 496, **512**

fire scar **512**

firethorn, see *Pyracantha*
Formosan: *P. koidzumii*
scarlet: *P. coccinea*

firetree, see *Morella*

fireweed, see *Chamerion*

Firmiana (parasol tree, phoenix-tree): *Botryosphaeria* 120, canker 176, dieback 176, *Nectria cinnabarina* 176

flame-tree, see *Brachychiton*

flannelbush, see *Fremontodendron*

fleabane, daisy, see *Erigeron*

flood damage **502**

flood tolerance 502

Florida hibiscus virus 422

flower blight **70–72**

flower spot **72**

fluoride injury 478, **486**

foliar senescence and shedding, normal **520**

Fomes fomentarius **308**

Fomitiporia dryadea: see *Inonotus*

Fomitopsis pinicola **322**

forest decline 478

Forestiera: bacterial knot 380, *Erysiphe syringae* 10, powdery mildew 10, *Pseudomonas savastanoi* pv. *savastanoi* 380, *Tomato black ring virus* 414

Forestiera acuminata (swamp privet): *Coleosporium minutum* 290, flood tolerance 502, *Puccinia sparganioides* 270, rust 270, 290

Forestiera pubescens (desert olive): *Erysiphe neomexicana* 8, powdery mildew 8

Forestiera segregata (Florida privet): *Puccinia sparganioides* 270, rust 270, salt tolerance 470, *Sphaceloma* 18, spot anthracnose 18

forsythia, see *Abeliophyllum*; *Forsythia*
white-: *A. distichum*

Forsythia (forsythia, goldenbells): *Alternaria* 84, *Arabis mosaic virus* 404, 410–412, bacterial disease

Hydrangea macrophylla (French hydrangea): chlorosis 408, green blotches 408, leaf crinkle 408, leaf deformity 408, *Oidium hortensiae* 10, powdery mildew 10, stunting 408, *Tomato ringspot virus* 408, *Tobacco ringspot virus* 408

Hydrangea mosaic virus 414

Hydrangea paniculata (panicled hydrangea): *Pucciniastrum hydrangeae* 276, rust 276, SO$_2$ tolerance 484, *Thekopsora hydrangeae* 276, *Tobacco ringspot virus* 410

Hydrangea quercifolia (oak-leaf hydrangea): bacterial spot 372, leaf spot **370**, *Xanthomonas* **370**, *X. campestris* 372

hydrogen fluoride 486

Hylastes 230

Hylobius 228, 338, *H. pales* (pales weevil) 228, *H. radicis* (pine root-collar weevil) 228

Hylurgopinus *rufipes* 240

Hyophorbe (pignut palm, spindle palm): *Ganoderma zonatum* 346, root rot 346

Hyophorbe *verschaffeltii* (spindle palm): lethal yellowing 402, phytoplasma 402

hyperparasite, hyperparasitism 304, 442

hyphae, hyphal karyotypes 1

hyphal pillar **116**, **198**

Hyphoderma 520

hyphomycetes 1, **16**, 20, **30–34**, **72–74**, **84–92**, **150–152**, **176–178**, **186–188**, **216–222**, **228–**230, 240–248, 252, 464

Hypodermella *laricis* 60

hypovirulence in fungi 160–162

Hypovirus 162

Hypoxylon 198, **204**–206, 360 (see also *Biscogniauxia*; *Camillea*; *Kretzschmaria*), *H. cohaerens* **204**, *H. deustum* **206**, *H. mammatum* **198**, *H. punctulatum* **204**, *H. thouarsianum* 360, *H. tinctor* **204**

Hypoxylon canker **198**

hysterothecium **46–52**

ice damage **504**–506

ice nucleation 368, 496–498

Ilarvirus **412**, 414–418

Ilex (holly): algal leaf spot 438, *Alternaria alternata* 84, anthracnose 114, Armillaria root disease 326, *Botryosphaeria* 120, 126–128, *Botrytis cinerea* 72, *Calonectria avesiculata* 220, canker 120, 182, 186, canker-rot 318, *Cephaleuros virescens* 438, *Ceratobasidium stevensii* 256, *Cerrena unicolor* 318, *Chondrostereum purpureum* 320, *Colletotrichum* 114, *Cylindrocladium* 220, dieback 120, 126–128, 186, *Erysiphe nemopanthi* 8, flood tolerance 502, fluoride tolerance 486, fungicide injury 476, *Fusarium solani* 186, gall **134**, gray mold blight 72, *Kretzschmaria deusta* 206, *Meloido-*

gyne 430, leaf spot 84, 220, nematodes 430, *Neonectria galligena* 182, ozone tolerance 482, *Phyllactinia guttata* 10, *Phymatotrichopsis* resistance 222, powdery mildew 8, Rhizoctonia blight 256, *Rhytisma* 66, root and butt rot 206, root knot 430, root rot 220–222, 326, salt tolerance 470, sapwood decay 318, *Schizophyllum commune* 318, *Sphaeropsis tumefaciens* **134**, tar spot 66, Thielaviopsis root rot **222**, thread blight 256, Verticillium resistance 242, web blight 256, witches'-broom **134**

Ilex aquifolium (English holly): canker 142, *Diaporthe eres* 142, dieback 142, freeze damage **498**, fruit rot 358, *Phytophthora ilicis* 358, SO$_2$ tolerance 484

Ilex cassine (dahoon holly): gall **134**, *Sphaeropsis tumefaciens* **134**, witches'-broom **134**

Ilex cornuta (Chinese holly): *Elsinoë ilicis* 18, spot anthracnose 18

Ilex crenata (box-leaved or Japanese holly): *Chalara elegans* 222, decline 222, root rot **222**, *Thielaviopsis basicola* **222**, *Tobacco ringspot virus* 408, yellow spots 408

Ilex glabra (inkberry): root rot 222, *Thielaviopsis basicola* 222

Ilex ×meserveae (blue or Meserve hybrid holly): root rot 222, *Thielaviopsis basicola* 222

Ilex opaca (American holly): canker 164, *Chrysomyxa ilicina* 288, decline 222, *Endothia gyrosa* 164, root rot 222, rust 288, *Thielaviopsis basicola* 222

Ilex vomitoria (cassina, yaupon): salt-induced hypertrophy 468, salt spray tolerance 468

Illicium (anise-tree): *Botryosphaeria rhodina* 128, dieback 128

impervious tissue **518**

incense-cedar, see *Calocedrus decurrens*

incubation: in disease cycle 2

Indian paint fungus **322**

indigo, false, see *Amorpha*

infection: in disease cycle 2

Inga: anthracnose 114, *Colletotrichum* 114

Inkberry, see *Ilex glabra*

ink-spot leaf blight 68

inoculation: in disease cycle 2

Inonotus 312, **316**, 336–338, **344**; *I. andersonii* 312, **316;** *I. circinatus*, see *Onnia circinata* 330, **336**–338; *I. dryadeus* **344;** *I. glomeratus* **316;** *I. hispidus* 312, **316;** *I. sulphurascens*, see *Phellinus weirii* **342**, *I. tomentosus*, see *Onnia tomentosa* 336–**338**

insect defoliators 466

insect-pathogen relationships 100, 144, 174, 186–188, 192, 198, 210, 228–230, 318, 324, 338

insect-vectored pathogens 234–236, 376, 386–402

insect vectors 188, 192, 228–232, 238–240, 376, 386–402

insecticide phytotoxicity **476**

International Code of Botanical Nomenclature 2

intersterility groups 332

Ipomoea (morning-glory): *Coleosporium ipomoeae* 290, rust 290

Ips 236

Irenopsis 16

iron deficiency **488–490**

ironweed, see *Vernonia*

ironwood, see *Carpinus*; *Ostrya*

Irpex *lacteus* 318

Isthmiella 56, *I. crepidiformis* **56**

Itea (sweetspire): *Botryosphaeria* 120, dieback 120

Iva *axillaris* (povertyweed): *Coleosporium helianthi* 290, rust 290

ivy, see *Cissus*; *Hedera*; *Parthenocissus*

ivy, aralia-: ×*Fatshedera*

ivy, Boston: *P. tricuspidata*

ivy, English: *H. helix*

ivy, grape: *Cissus*

Ixora: *Alternaria* 84, Armillaria root disease 326, gall 134, leaf blotch **190**, leaf spot 84, *Pestalotiopsis* **190**, root rot 326, *Sphaeropsis tumefaciens* 134

Ixora coccinea (scarlet ixora, scarlet jungleflame): leaf scorch **492**, sooty mold **16**

Jacaranda: Armillaria root disease 326, canker-rot 318, sapwood decay 318, *Schizophyllum commune* 318

jambolan, see *Syzygium cumini*

jasmine, see *Gardenia*; *Gelsemium*; *Jasminum*; *Murraya*
   cape: *Gardenia jasminoides*
   Carolina: *Gelsemium*
   Japanese: *J. mesnyi*
   orange: *M. paniculata*
   primrose: *J. mesnyi*
   winter: *J. nudiflorum*

Jasminum (jasmine): *Alternaria alternata* 84, anthracnose 114, *Arabis mosaic virus* 412, Armillaria root disease 326, bacterial disease 370, *Botryosphaeria obtusa* 126, *Calonectria* 220, carlavirus 414, chlorotic variegation 414, *Colletotrichum* 114, *Cucumber mosaic virus* 418, *Cylindrocladium* 220, dieback 126, *Elsinoë jasmini* 18, gall 150, leaf spot 84, *Nectriella pironii* 150, *Pseudomonas syringae* 370, root rot 220, 326, scab 18, *Sclerotium rolfsii* 226, southern blight 226, *Tobacco ringspot virus* 408, Verticillium wilt 242, virus 410–412, 418

Jasminum mesnyi (primrose jasmine): bacterial knot 380, *Pseudomonas savastanoi* pv. *savastanoi* 380

Jasminum nudiflorum (winter jasmine): bacterial knot 380, Phomopsis gall 148, *Pseudomonas savastanoi* pv. *savastanoi* 380

Jasminum officinale (poet's jasmine): dieback 408, *Tobacco*

ringspot virus 408, yellow leaf blotch 408

Jatropha (nettlespurge): Armillaria root disease 326, gall 134, *Sphaeropsis tumefaciens* 134

Jerusalem-artichoke, see *Helianthus*

Jerusalem thorn, see *Parkinsonia*

jessamine, yellow, see *Gelsemium*

jet-bead, see *Rhodotypos*

Judas-tree, see *Cercis*

Juglans (butternut, walnut): anthracnose **104**, 114, Armillaria root disease 326, bacterial blight 374, bacterial disease 370, 374, *Biscogniauxia mediterranea* 200, blackline disease **408**–412, 508, *Botryosphaeria* 120–122, 126, brown cubical rot 310, bunch disease 390, **392**–394, butt rot 350, *Calonectria* 220, canker 182, 186, 358, canker-rot 316–318, *Cherry leafroll virus* 410–412, collar rot 360, *Colletotrichum* 114, *Cristulariella moricola* 74, *Cylindrocladium* 220, decline 224, *Diaporthe ambigua* 142, dieback 120–122, 126, 136, 176, 186, 200, downy spot **252**, flatlimb 428, flood tolerance 502, fluoride tolerance 486, Fusarium canker 186, *Fusarium* 186, *Ganoderma applanatum* 350, *Gnomonia leptostyla* **104**, herbicide injury 472, herbicide tolerance 476, ice glaze damage 506, *Inonotus hispidus* 316, *Laetiporus sulphureus* 310, leaf blight 374, leaf spot **20**, 22, 74, **104**, *Marssoniella juglandis* **104**, *Marssonina juglandis* 104, *Melanconis juglandis* 136, *Microstroma juglandis* **252**, mistletoe 442–444, *Mycosphaerella juglandis* **20**–22, Nectria canker 182, *Nectria cinnabarina* 176, nematodes 432, *Neonectria galligena* 182, *Oxyporus latemarginatus* 344, ozone tolerance 482, *Pestalotiopsis maculans* 190, *Phellinus everhartii* 308, *Phoradendron* 442–444, *Phyllactinia guttata* 10, *Phytophthora* 358–360, 364, *P. citricola* **364**, Phytophthora canker 358, phytoplasma 390, **392**–394, powdery mildew 10, *Pratylenchus* 432, *Pseudomonas syringae* 370, root rot 220, 224, 326, **364**, *Rosellinia necatrix* 224, salt tolerance 470, sapwood decay 318, *Schizophyllum commune* 318, *Sclerotium rolfsii* 226, SO$_2$ tolerance 484, southern blight 226, trunk decay 308–310, 316, 344, 350, twig blight 190, twig dieback 142, Verticillium resistance 242, virus **408**–412, 508, white leaf mold **252**, witches'-broom 390, **392**–394, *Xanthomonas arboricola* pv. *juglandis* 374, *Xylaria polymorpha* 224

Juglans ailanthifolia (Japanese walnut): anthracnose 104, canker 116, dieback 136, *Gnomonia leptostyla* 104, leaf spot 22, *Mel-*

Lonicera oblongifolia (swamp fly honeysuckle): *Insolibasidium deformans* 254, leaf blight 254

Lonicera periclymenum: Honeysuckle latent virus 414, leaf mottle 414

Lonicera prolifera (grape honeysuckle): *Insolibasidium deformans* 254, leaf blight 254

Lonicera sempervirens (trumpet honeysuckle): *Insolibasidium deformans* 254, leaf blight 254

Lonicera tatarica (Tatarian honeysuckle): *Insolibasidium deformans* **254**, leaf blight **254**

Lonicera xylosteum (dwarf or European fly honeysuckle): salt tolerance 470

loosestrife, see *Lysimachia*

Lophodermella 48, 54, *L. cerina* 48, 54, *L. concolor* **54**

Lophodermium 36, **46**, *L. juniperi* **46**, *L. pinastri* **46**, *L. seditiosum* **46–48**

Lophophacidium *hyperboreum* 62

loquat, see *Eriobotrya japonica*

Loranthaceae 440

lousewort, see *Pedicularis*

luolu, see *Pritchardia affinis*

Lycium: powdery mildew 10, *Podosphaera pannosa* 10

Lycium barbarum (matrimony-vine): chlorotic mottle 418, *Cucumber mosaic virus* 418

Lymantria *dispar* 466

Lyonia (fetterbrush, fetterbush, huckleberry, staggerbush): *Botryosphaeria* 120, 126, dieback 120, 126, *Erysiphe vaccinii* 10, *Exobasidium* 250, leaf rust 276, powdery mildew 10, *Pucciniastrum minimum* 276, *Rhytisma* 66, tar spot 66, *Thekopsora minima* 276

Lyonia ferruginea (staggerbush): leaf spot 44, *Ophiodothella* 44

Lyonia ligustrina (he-huckleberry, maleberry): rust 276, *Naohidemyces vaccinii* 276, *Pucciniastrum vaccinii* 276

Lysiloma: gall 150, *Ganoderma* 348, *Nectriella pironii* 150, root rot 348

Lysimachia (loosestrife): *Coleosporium tussilaginis* 290, rust 290

Macadamia: anthracnose 114, *Botryosphaeria* 120, 128, *Botrytis cinerea* 72, canker 120, *Colletotrichum* 114, dieback 120, 128, gray mold blight 72

Macadamia integrifolia (macadamia): root rot 206, decline 206, *Kretzschmaria clavus* 206

Machaeranthera (tansyaster): *Coleosporium asterum* 290, rust 290

Maclura (osage-orange): *Botryosphaeria* 126–128, *Botrytis cinerea* 72, canker 176, canker-rot 318, dieback 126–128, 176, gray mold blight 72, mistletoe 442, *Nectria cinnabarina* 176, *Phoradendron* 442, *Phymatotrichopsis* resistance 222, sapwood decay

318, *Schizophyllum commune* 318, Verticillium wilt 242

Maclura pomifera (osage-orange): *Cerotelium fici* 274, cucumovirus 418, flood tolerance 502, line pattern 418, rust 274, salt tolerance 470, virus 418

Macrodiplodiopsis 36

Macrophomina *phaseolina* **218**

Madia (tarweed): *Coleosporium* 290, rust 290

madricacao, see *Gliricidia sepium*

madrone, see *Arbutus menziesii*

magnesium deficiency **488**

magnolia, see *Magnolia*
   bigleaf: *M. macrophylla*
   Fraser: *M. fraseri*
   saucer: *M. ×soulangeana*
   southern: *M. grandiflora*
   star: *M. stellata*
   yulan: *M. denudata*

Magnolia (bay, magnolia): *Alternaria alternata* 84, anthracnose 114, Armillaria root disease 326, *Botryosphaeria* 120, 126–128, butt rot 350, *Calonectria* 220, *Camillea tinctor* 204, canker 120, 176, 182, canker-rot 318, *Cercosporidium magnoliae* 20, *Cerrena unicolor* 318, chlorotic markings 418, *Chondrostereum purpureum* 320, *Ciborinia gracilipes* **68**, *Colletotrichum* 114, *Coniothyrium fuckelii* 36, *Cristulariella moricola* 74, *Cucumber mosaic virus* 418, *Cylindrocladium* 220, dieback 120, 126–128, 176, *Erysiphe magnifica* 8, flood tolerance 502, fluoride tolerance 486, *Fomitopsis pinicola* 322, *Ganoderma* 348–350, *Hendersonia* 36, ice glaze damage 506, leaf spot 20, 36, 74, 84, 220, **370**, line pattern 418, mosaic 418, *Mycosphaerella milleri* 20, Nectria canker 182, *Nectria cinnabarina* 176, nematodes 432, *Neonectria galligena* 182, *Oxyporus latemarginatus* 344, ozone tolerance 482, *Pestalotiopsis maculans* 190, *Phyllactinia guttata* 10, powdery mildew 8–10, *Pratylenchus* 432, *Pseudomonas* **370**, root and butt rot 326, 348, root rot 220, salt tolerance 470, sapwood decay 318, *Schizophyllum commune* 318, trunk decay 318, 322, 344, 350, *Tylenchorhynchus* 432, Verticillium wilt 242, virus 418

Magnolia acuminata (cucumber tree): *Cucumber mosaic virus* 418, 428, mosaic 418, 428

Magnolia denudata (yulan magnolia): *Cucumber mosaic virus* 428, oak leaf pattern 428

Magnolia fraseri (Fraser magnolia): canker 182, frost sensitivity 496, *Nectria magnoliae* 182

Magnolia grandiflora (bull bay, southern magnolia): algal leaf spot **438**, *Cephaleuros virescens* **438**, *C. parasiticus* 438, *Conio-

thyrium fuckelii* 36, *Elsinoë magnoliae* 18, leaf spot 18, **36**, 368–**370**, *Pseudomonas cichorii* 368–**370**, scab 18

Magnolia macrophylla (bigleaf magnolia): mosaic **426–428**, ringspot **426**, virus **426**–428

Magnolia ×soulangeana (saucer magnolia): bacterial leaf spot 370, *Cucumber mosaic virus* 428, leaf spot 370, oak leaf pattern 428, *Pseudomonas cichorii* 370, virus 428

Magnolia stellata (star magnolia): blossom blight **72**, *Botrytis cinerea* **72**

Magnolia virginiana (sweet bay): *Biscogniauxia marginata* 202, *Ciborinia gracilipes* 68, dieback 202, petal blight 68

mahogany, see *Cercocarpus*; *Persea*; *Swietenia*
   birch-leaf mountain-: *C. montanus* var. *glaber*
   Florida: *P. borbonia*

mahonia, see *Berberis*; *Mahonia*
   Beale: *M. bealei*
   Cascades: *M. nervosa*
   cluster: *M. pinnata*
   creeping: *M. repens*

Mahonia: *Calonectria* 220, *Cumminsiella mirabilissima* **272**, *Cylindrocladium* 220, mistletoe 444, *Phoradendron bolleanum* 444, *Puccinia* 270–272, *P. graminis* 270, root rot 220, rust 270–**272**, salt tolerance 470, winter injury **492**

Mahonia aquifolium (Oregon grape): *Cumminsiella mirabilissima* **272**, fluoride tolerance 486, rust **272**, SO_2 tolerance 484

Mahonia bealei (Beale mahonia): *Cumminsiella mirabilissima* 272, rust 272

Mahonia dictyota (shining netvein barberry): *Cumminsiella mirabilissima* 272, rust 272

Mahonia nervosa (Cascades mahonia): *Cumminsiella mirabilissima* 272, rust 272

Mahonia pinnata (cluster mahonia): *Cumminsiella mirabilissima* 272, rust 272

Mahonia repens (creeping mahonia): *Cumminsiella mirabilissima* 272, rust 272

maleberry, see *Lyonia ligustrina*

mallow, see *Hibiscus*; *Malvaviscus*
   wax: *M. arboreus*

Malpighia *coccigera* (Singapore holly): Cassytha 456, dodder laurel 456

Malus (apple, crabapple): *Agrobacterium tumefaciens* 382, *Alternaria* 84, anthracnose 114, apple proliferation 390, *Apple mosaic virus* 414, apple scab **86**, *Apple stem pitting virus* 420, Armillaria root disease 326–330, bacterial disease 370, **376**, 382, *Biscogniauxia* 200–202, blister canker 202, black rot 126, *Botryosphaeria* 120–128, *B. obtusa* **126**, *Bo-*

trytis cinerea 72, brown cubical rot 310, brown line disease 408, brown rot of fruit 76, butt rot 350, *Calonectria* 220, canker 36, 120–**126**, 142, 170–176, 182, 186, canker-rot 310, **318**, cedar-apple rust 260–262, *Ceratobasidium stevensii* 256, *Cerrena unicolor* 318, *Chondrostereum purpureum* **320**, *Climacodon septentrionalis* 308, cold acclimation 498, *Colletotrichum* 114, *Coniothyrium fuckelii* 36, *Cristulariella moricola* 74, crown gall 382, *Cryptosporiopsis curvispora* 174, *Cylindrocladium* 220, *Cytospora* 170, 174, decline 224, 408, *Diaporthe* 142, dieback 120–**126**, 128, 142, 170, 174–176, 186, 200, 320, *Diplocarpon mespili* 78, disease-resistant cultivars 86, *Elsinoë piri* 18, *Entoleuca mammata* 198, Entomosporium leaf spot 78, *Erwinia amylovora* **376**, fire blight **376**, flood tolerance 502, *Fomes fomentarius* 308, *Fomitopsis pinicola* 322, frogeye leaf spot **126**, fruit rot 84, 120, 126, 142, 190, fruit spot 20, *Fusarium lateritium* 186, *Ganoderma* 348–350, graft union necrosis **404**, 408, gray mold blight 72, *Gymnosporangium* **260–268**, *G. clavipes* 264, *G. globosum* 262, *G. juniperi-virginianae* **260**–262, hail damage **504**, hawthorn rust 262, *Hendersonia* 36, herbicide tolerance 476, juniper broom rust 266, *Laetiporus sulphureus* 310, leaf blister 4, leaf fleck 20, leaf spot 18, 36, 44, 74, 78, 84, *Leucostoma persoonii* 170, *Monilinia* 76, *Mycosphaerella* 20, nailhead canker 202, Nectria canker 176, 180, *Nectria cinnabarina* 176, *Neofabraea* 174, *Neonectria galligena* 180–182, ozone tolerance 482, *Pestalotiopsis malicola* 190, *Pezicula* 174, *Phellinus everhartii* 308, *Phyllactinia guttata* 10, *Phymatotrichopsis omnivora* 222, Phymatotrichum root rot 222, *Phytophthora* resistance 362, phytoplasma 390–392, *Podosphaera clandestina* 10, *P. leucotricha* 10–12, powdery mildew 10–12, *Pseudomonas syringae* 370, quince rust 264, root and butt rot 326, 348, root rot 220–224, 326–330, 348, *Rosellinia necatrix* 224, rust **260**–268, rust resistant cultivars 262, salt-induced loss of cold hardiness 468, salt tolerance 470, sapwood decay 310, **318**–320, *Schizophyllum commune* **318**, *Sclerotium rolfsii* 226, silverleaf **320**, SO_2 tolerance 484, southern blight 226, spot anthracnose 18, stem pitting 420, *Taphrina bullata* 4, thread blight 256, tobamovirus 422, *Tobacco* ringspot virus 408, *Tomato ringspot virus* 404, 408, *Trametes versicolor* 310, trunk decay 308–310, **318–322**, *Tubakia dryina* 44, *Valsa* 172–174, *Venturia inaequalis* 86, Verticillium wilt and resistance 242, virus 404, 408, 414, 420, white rot 120, witches'-broom 390, *Xylaria mali* 224

Malus baccata (Siberian crabapple): *Erwinia amylovora* **376**, fire blight **376**

Malus coronaria (American or sweet crabapple): *Biscogniauxia marginata* 202, blister canker 202, dieback 202

Malus ×domestica, see *M. pumila*

Malus floribunda (Japanese flowering or showy crabapple): *Apple mosaic virus* 416, mosaic 416

Malus hilleri: fire blight resistance 376

Malus ioensis (prairie crabapple): *Apple mosaic virus* 416, mosaic 416

Malus pumila: (apple): *Apple mosaic virus* **412**, 416, apple scab **86**, *Armillaria tabescens* 330, *Biscogniauxia marginata* 202, blister canker 202, *Botryosphaeria dothidea* 120, *B. obtusa* 126, brown line disease **404**, 408, canker 180, 202, canker-rot **318**, cedar-apple rust **260**, *Cerato-basidium stevensii* **256**, *Cherry leafroll virus* 412, *Cherry rasp leaf virus* 412, *Chondrostereum purpureum* 320, crown rot **360–362**, decline 224, **360**, **404**, 432, *Dematophora necatrix* **224**, dieback 202, 320, disease-resistant cultivars 86, enations 412, *Erwinia amylovora* **376**, fire blight **376**, flatlimb **426–428**, fluoride tolerance 486, freeze damage 12, graft union necrosis **404**, 408, 432, *Gymnosporangium juniperi-virginianae* **260**, herbicide injury **476**, ice glaze damage 506, leaf blight 256, leaf spot **260**, mistletoe **442–444**, mosaic **412**, 416, nematodes 432, *Oxyporus latemarginatus* 344, ozone injury **482**, *Phytophthora* 362, *Podosphaera leucotricha* **12**, powdery mildew **12**, *Pratylenchus penetrans* **432**, root lesions **432**, root rot 224, 330, root stunting **432**, *Rosellinia necatrix* 224, rust **260**, sapwood decay 320, *Schizophyllum commune* **318**, *Sclerotium rolfsii* **226**, silverleaf **320**, southern blight **226**, stem flattening 428, thread blight **256**, *Tobacco necrosis virus* 424, *Tobacco ringspot virus* 408, tobamovirus 422, *Tomato ringspot virus* **404**, 408, 432, trunk and limb decay **318–320**, 344, vein clearing **412**, *Venturia inaequalis* 86, virus 404, 408, **412**, 416, 422, **426–428**, 432, *Viscum album* **442–444**,

Xiphinema americanum **432**, *Xylaria* 224

Malus sargentii (Sargent apple): fire blight resistance 376

Malvaviscus arboreus (Turk's cap, wax mallow): Armillaria root disease 326, *Phymatotrichopsis* resistance 222

manganese deficiency **488–490**

Mangifera indica (mango): *Alternaria alternata* 84, anthracnose 114, *Botryosphaeria* 120, 128, *Botrytis cinerea* 72, canker 120, *Ceratocystis fimbriata* 232, *Colletotrichum* 114, decline 84, dieback 120, 128, *Elsinoë mangiferae* 18, fruit rot 84, gray mold rot 72, leaf spot 190, nematode 432, *Pestalotiopsis mangiferae* 190, *Radopholus similis* 432, scab 18, sooty mold 16, Verticillium wilt 242

mango, see *Mangifera indica*

mangrove, see *Avicennia; Laguncularia; Rhizophora*
    black: *A. germinans*
    red: *R. mangle*
    white: *L. racemosa*

Manilkara zapota (sapodilla): Armillaria root disease 326, *Elsinoë lepagei* 18, scab 18

manzanita, see *Arctostaphylos*
    greenleaf: *A. patula*
    pine-mat: *A. nevadensis*

maple, see *Acer*
    Amur: *A. ginnala*
    bigleaf: *A. macrophyllum*
    black: *A. nigrum*
    hedge: *A. campestre*
    Japanese: *A. palmatum*
    Manitoba: *A. negundo*
    mono: *A. mono*
    mountain: *A. spicatum*
    Norway: *A. platanoides*
    red: *A. rubrum*
    Rocky Mountain: *A. glabrum*
    Shantung: *A. truncatum*
    silver: *A. saccharinum*
    striped: *A. pensylvanicum*
    sugar: *A. saccharum*
    sycamore: *A. pseudoplatanus*
    Tatarian: *A. tataricum*
    vine: *A. circinatum*

marigold, see *Tagetes*

marlberry, see *Ardisia*

Marssoniella juglandis **104**

Marssonina 20, 22, **80–82**, 104, *M. betulae* 82, *M. brunnea* 82, *M. castagnei* 82, *M. celastri* 82, *M. fraxini* 22, *M. juglandis* 104, *M. populi* 82, *M. rosae* 80, *M. thomasiana* 82

matrimony-vine, see *Lycium*

Matsucoccus 210

Matteuccia (ostrich fern): fir-fern rust 280, *Uredinopsis struthiopteridis* 280

medlar, see *Mespilus*
    Japanese: *Eriobotrya japonica*

Melaleuca: Armillaria root disease 326, *Botryosphaeria* 120, *Calonectria* 220, canker 120, *Cylindrocladium* 220, dieback 120, root rot 220, 326, shoot blight

220

Melampsora 154, 280, **284–286**, *M. abietis-canadensis* 284, *M. albertensis* 284, *M. epitea* 286, *M. larici-populina* 286, *M. medusae* **284–286**, *M. occidentalis* 286, *M. populnea* 286

Melampsora rusts **284–286**

Melampsorella 280–**282**, *M. caryophyllacearum* 282

Melampsoridium 274, *M. betulinum* **274**

Melampyrum lineare (narrowleaf cow-wheat): *Cronartium coleosporioides* 292, 300, rust 292, 300

Melanconis juglandis 116, **136**, *M. stilbostoma* **136**

Melanconium bicolor **136**, *M. oblongum* **136**

Melia (bead tree, chinaberry): anthracnose 114, Armillaria root disease 326, *Botryosphaeria* 120, 126–128, canker 120, 176, 186, canker-rot 318, *Colletotrichum* 114, *Cristulariella moricola* 74, dieback 120, 126–128, 176, 186, *Fusarium lateritium* 186, gall 150, leaf spot 74, mistletoe 442, *Nectria cinnabarina* 176, *Nectriella pironii* 150, *Phoradendron serotinum* 442, *Phyllactinia guttata* 10, *Oxyporus latemarginatus* 344, *Phymatotrichopsis omnivora* 222, Phymatotrichum root rot 222, powdery mildew 10, root rot 326, sapwood decay 318, *Schizophyllum commune* 318, trunk decay 318, 344

Melia azedarach (chinaberry): bacterial gall 368, flood tolerance 502, *Pseudomonas meliae* 368

Meliola 16, *M. floridensis* **16**, *M. palmicola* **16**

Meloidogyne 244, 248, **430**, **434**, interaction with crown gall bacteria 382, *M. hapla* **430**

Menziesia (mock-azalea): *Erysiphe vaccinii* 10, *Exobasidium* 250, powdery mildew 10, *Rhytisma arbuti* 66, tar spot 66

Menziesia ferruginea (rusty menziesia): leaf rust 276, *Naohidemyces vaccinii* 276, *Pucciniastrum vaccinii* 276

Meria parkeri 58, *M. laricis* **58–60**

Mespilus: Diplocarpon mespili 78, Entomosporium leaf spot 78, fruit rot 76, leaf spot 78, *Monilinia* 76, *Podosphaera clandestina* 12, powdery mildew 12

Mespilus germanica (medlar): *Gymnosporangium* 262–264, 268, *G. globosum* 262, rust 262–264, 268

mesquite, see *Prosopis*

Metacapnodium spongiosum **16**

Metasequoia glyptostroboides (dawn redwood): *Botryosphaeria dothidea* 120, canker 120, *Diaporthe lokoyae* 144, dieback 120, 144, leaf spot 190, ozone tolerance 482, *Pestalotiopsis foedans* 190, *Phomopsis lokoyae*

144, salt sensitivity 470, shoot blight 190, SO_2 tolerance 484, twig canker 144

methane 384

Methanobrevibacter: in wetwood 384

Metrosideros (ohia): *Botryosphaeria rhodina* 128, dieback 128

Michelia figo (banana shrub): algal leaf spot 438, *Cephaleuros virescens* 438

Miconia: Botryosphaeria 120, dieback 120

microbial colonization of wounds 512–514, succession in wounded xylem 514

microsclerotia 1, **218**, 224, **244**

Microsphaera, see *Erysiphe*

Microsphaeropsis 36, *M. olivacea* 36

Microstroma album 252, *M. juglandis* 252

mildew, downy 352

mildew, powdery **8–14**

Milesina 280

Millettia (pongam): *Cristulariella moricola* 74, leaf spot 74

mimosa, see *Albizia julibrissin*

mimosa wilt **248**

mistletoe **440–454**, American or leafy (*Phoradendron*) **440–444**, autoparasitism and hyperparasitism 442, Christmas (*P. serotinum*) 440, Colorado desert (*P. serotinum* subsp. *macrophyllum*) 442, dwarf (*Arceuthobium*) 440, 444, **446–454**, eastern or oak (*P. serotinum* **440–442**, European (*Viscum album*) 440–**442**, hairy (*P. serotinum* subsp. *tomentosum*) 440–**442**, host resistance 442, mahogany (*P. rubrum*) 442, mesquite or desert (*P. californicum*) **442–444**, Pacific (*P. serotinum* subsp. tomentosum) 442

mites as virus vectors **418**, 422–428

Miuraea 20

mock-azalea, see *Menziesia*

mock-orange, see *Philadelphus*
    Lewis': *P. lewisii*
    littleleaf: *P. microphyllus*
    scentless: *P. inodorus*
    sweet: *P. coronarius*

Mollicutes 390

molybdenum deficiency 490

mombin, purple, see *Spondias purpurea*

Moneses uniflora (one-flowered shinleaf, single-delight): *Chrysomyxa monesis* 288, rust 288

Monilia 76

Monilinia 76, *M. fructicola* **76**, *M. laxa* **76**

monk's-hood vine, see *Ampelopsis acontifolia*

monkey-puzzle, see *Araucaria*

Monochamus as nematode vectors 436, *M. carolinensis* **436**

Monostichella 96, **108**, **112**, *M. coryli* **108**, *M. robergei* **112**

Montagnula 36

Morella (bayberry, firetree): Armillaria root disease 326, *Botryo-*

palm *(continued)*
  jelly: *Butia*
  kentia: *Howeia forsteriana*
  key: *Thrinax morrisii*
  Kona: *Pritchardia affinis*
  lady: *Rhapis excelsa*
  latan: *Latania*
  llume: *Gaussia attenuata*
  Loureir date: *Phoenix loureiri*
  Luolu: *Pritchardia affinis*
  MacArthur: *Ptychosperma macarthurii*
  Madagascar: *Dypsis lutescens*
  Manila: *Adonidia merrillii*
  Mazari: *Nannorrhops ritchiana*
  Mexican blue: *Brahea armata*
  Mexican fan: *Washingtonia robusta*
  miniature sugar: *Arenga engleri*
  Montgomery: *Veitchia montgomeryana*
  oil: *Elaeis guineensis*
  pignut: *Hyophorbe*
  pindo: *Butia capitata*
  princess: *Dictyosperma album*
  Puerto Rican gaussia: *Gaussia attenuata*
  pygmy date: *Phoenix roebelenii*
  queen: *Syagrus romanzoffiana*
  royal: *Roystonea*
  sago-: *Cycas revoluta*
  saw-cabbage: *Acoelorraphe wrightii*
  seashore: *Allagoptera arenaria*
  Senegal date: *Phoenix reclinata*
  silver: *Coccothrinax argentata*
  solitaire: *Ptychosperma elegans*
  South American jelly: *Butia capitata*
  spindle: *Hyophorbe verschaffeltii*
  spiny-club: *Bactris*
  sugar: *Arenga, A. pinnata*
  thatch: *Coccothrinax crinita*
  Thurston: *Pritchardia thurstonii*
  toddy: *Borassus flabellifer*
  triangle: *Dypsis decaryi*
  Washington: *Washingtonia*
  wild date: *Phoenix sylvestris*
  windmill: *Trachycarpus fortunei*
  yellow: *Dypsis*
palm: *Alternaria citri* 84, anthracnose 114, Armillaria root disease 326, *Botryosphaeria* 120–122, 128, *Colletotrichum* 114, false smut **254**, *Fusarium oxysporum* 248, Fusarium wilt 248, *Ganoderma* **346**–348, *Graphiola phoenicis* **254**, leaf spot 30, 84, 190, lethal decline 402, lethal yellowing **402**, *Oxyporus latemarginatus* 344, *Palm mosaic virus* 420, *Pestalotiopsis palmarum* 190, phytoplasma **402**, root and butt rot **346**, *Stigmina palmivora* 30, trunk decay 344, virus 420
Palm Beach bells, see *Kalanchoe*
Palm mosaic virus 420
palma jipijapa: see *Carludovica palmata*
palmetto, see *Sabal; Serenoa*
  cabbage: *Sabal palmetto*
  dwarf or scrub: *Sabal minor*
  saw or scrub: *Serenoa repens*

paloverde, see *Parkinsonia*
Pandanus (screw-pine): *Alternaria* 84, anthracnose 114, *Botryosphaeria* 120, 128, *Colletotrichum* 114, leaf spot 84, 190, *Pestalotiopsis palmarum* 190
Pandanus utilis (screw-pine): lethal decline 402, phytoplasma 402
Pantoea: in wetwood 384
papaya, see *Carica papaya*
Papaya ringspot virus 420
paper-mulberry, see *Broussonetia*
paper-plant, see *Fatsia*
Paralongidorus 408
Paraphaeosphaeria 36
parasitic plants **438–456**
parasol tree, see *Firmiana*
parch blight 494
Parkinsonia (Jerusalem thorn, paloverde): Armillaria root disease 326, *Botryosphaeria rhodina* 128, dieback 128, gall 150, *Ganoderma* 348, mistletoe 444, *Nectriella pironii* 150, *Phoradendron californicum* 444, root rot 326, 348
Parkinsonia aculeata (Jerusalem thorn): gall 134, *Sphaeropsis tumefaciens* 134
Parthenium argentatum (guayule): Verticillium wilt 242
Parthenocissus (Boston ivy, Virginia creeper): *Botryosphaeria obtusa* 126, canker 36, *Coniothyrium fuckelii* 36, *Cristulariella moricola* 74, dieback 126, downy mildew 352, *Elsinoë parthenocissi* 18, *Erysiphe necator* 8, *Guignardia bidwellii* 36, leaf spot 18, 36, 74, *Plasmopara viburni* 352, powdery mildew 8, spot anthracnose 18, Verticillium wilt 242
Parthenocissus quinquefolia (Virginia creeper, woodbine): fluoride tolerance 486, *Guignardia bidwellii* 36, leaf spot 36, ozone tolerance 482, as pest 458, SO_2 tolerance 484
Parthenocissus tricuspidata (Boston ivy): *Guignardia bidwellii* 36, leaf spot 36
Passalora 20, 30–32, *P. rosicola* 20–22, *P. sequoiae* 32
pathovar (pv.) 368
Paulownia: *Biscogniauxia mediterranea* 200, *Botryosphaeria* 120, 126, canker 120, 142, canker-rot 318, *Diaporthe eres* 142, dieback 126, 142, 200, phytoplasma 390, sapwood decay 318, *Schizophyllum commune* 318, witches'-broom 390
Paulownia tomentosa (empress tree, karri tree, princess tree): *Cucumber mosaic virus* 418, fluoride tolerance 486, interveinal yellowing 422, mosaic 418, ozone tolerance 482, *Phymatotrichopsis* resistance 222, *Tobacco mosaic virus* 422, virus 418, 422
pawpaw, see *Asimina; Carica*
pea tree, see *Caragana; Sesbania*
  pea tree or pea shrub, Siberian: *C. arborescens*

peach, see *Prunus persica*
  Chinese wild: *P. davidiana*
peach stunt 414
peach X-disease **400**
peachbush, see *Prunus texana*
Peanut stunt virus 418
pear, see *Pyrus*
  Asian: *P. pyrifolia*
  'Bradford': *P. calleryana* cv.
  callery: *P. calleryana*
  Chinese: *P. pyrifolia, P. ussuriensis*
  common or European: *P. communis*
  Nashi: *P. pyrifolia*
  Oriental: *P. betulaefolia, P. pyrifolia*
  sand: *P. pyrifolia, P. ussuriensis*
  ussury: *P. ussuriensis*
pear psylla 392
pear scab 88
pecan, see *Carya illinoinensis*
Pedicularis (lousewort): *Cronartium* 292–294, rust 292–294
Pentas: *Cristulariella moricola* 74, leaf spot 74
peony, tree, see *Paeonia suffruticosa*
Peperomia: Rhizoctonia blight **256**, web blight **256**
pepper, black, see *Piper nigrum*
pepper, kava, see *Piper methysticum*
pepper-tree, see *Schinus*
peppervine, see *Ampelopsis*
Peraphyllum (squaw-apple): *Gymnosporangium nelsonii* 268, rust 268
Perenniporia fraxinophila **310**, *P. medulla-panis* 520, *P. phloiophila* **520**
Periconiella 200
periderm formation 86, 172, 376–378, 442, 456, 496–498, 508, **518**
Peridermium 292, **294–304**, *P. bethelii* 292, *P. filamentosum* 292, 302, *P. harknessii* 292, **304**
perithecium 1, **16**, **64**, **94–98**, **108**, **150–152**, **156–160**, **164–168**, **176**, **180**, **196**, **202**, **232**
periwinkle, see *Catharanthus roseus*
Peronospora **352**, *P. grisea* 352, *P. rubi* 352, *P. sparsa* **352**
Persea (avocado, bay): Armillaria root disease 326, bacterial disease 370, brown cubical rot 310, *Calonectria* 220, canker 182, 358, canker-rot 318, *Cylindrocladium* 220, *Laetiporus sulphureus* 310, leaf blotch 220, Nectria canker 182, *Neonectria galligena* 182, *Phytophthora* 358–360, 364, *P. cinnamomi* 364, Phytophthora canker 358, *Pseudomonas syringae* 370, root rot 220, 326, 364, salt tolerance 470, sapwood decay 318, *Schizophyllum commune* 318, *Sclerotium rolfsii* 226, southern blight 226, trunk decay 310, 318
Persea americana (avocado): *Alternaria* 84, anthracnose 114, *Botryosphaeria* 120, 126–128, canker

120, collar rot 362, *Colletotrichum* 114, *Cristulariella moricola* 74, dieback 120, 126–128, flood tolerance 502, fruit rot 84, gall 134, *Ganoderma* 348, leaf spot 74, 372, nematodes 432, ozone tolerance 482, *Phytophthora* 358, 362, 364, *P. cinnamomi* 364, *Pseudomonas* sp. 372, root and butt rot 348, root rot 224, 364, *Rosellinia necatrix* 224, scab 18, *Sphaceloma perseae* 18, *Sphaeropsis tumefaciens* 134, tobamovirus 422, Verticillium 242
Persea borbonia (Florida mahogany, red bay): algal leaf spot **438**, *Cephaleuros virescens* **438**
persimmon, see *Diospyros*
persimmon wilt **246**
Pestalosphaeria 190
Pestalotia 190
Pestalotiopsis 190, *P. funerea* **190**, *P. maculans* **190**
pesticide phytotoxicity **472–476**
petal blight **70**
Pezicula 174
Phacidiopycnis balsamicola 144
Phacidium 62, 144, *P. coniferarum* 144, *P. abietis* **62**, *P. infestans* **62**
Phaeocryptopus 58, *P. gaeumannii* **58**
Phaeoisariopsis 20, 30
Phaeolus schweinitzii 338, **340**
Phaeophleospora 20, 32, 36, *P. epicoccoides* 32
Phaeoramularia 30
Phellinus 306–308, **312**, 314–316, **324**, 332, 336, **342**, *P. arctostaphyli* 306, *P. cancriformans* **324**, *P. chrysoloma* 324, *P. everhartii* **308**, *P. igniarius* 306–308, *P. laevigatus* 306, 314, *P. pini* **324**, *P. pini* var. *cancriformans* 324, *P. robiniae* **308**, *P. spiculosus* **312**, 316, *P. tremulae* 306–308, *P. weirianus* **308**, *P. weirii* 332, 336, **342**, *P. yamanoi* 324
Phellodendron (cork tree): *Calonectria* 220, *Cylindrocladium* 220, herbicide tolerance 476, leaf blight 220, root rot 220, Verticillium wilt 242
phellogen, see cork cambium
phenolic compounds in plant defense 2, 162, 172, 198, 234, 236, 328–330, 442, 480, 512, 518
Phialophora parasitica 246
Philadelphus (mock-orange): bacterial disease 370, *Botrytis cinerea* 72, canker 176, chlorotic mottle 418, *Cucumber mosaic virus* 418, dieback 176, *Elm mottle virus* 418, gray mold blight 72, *Gymnosporangium speciosum* 260, 268, leaf mottle 418, mosaic 418, *Nectria cinnabarina* 176, *Phyllactinia guttata* 10, *Phymatotrichopsis* resistance 222, powdery mildew 10, *Pseudomonas syringae* 370, rust 260, 268, salt tolerance 470, virus 418

Platanus (continued)
poria fraxinophila 310, *Phoraden-dron* 444, *Phyllactinia guttata* 10, *Phytophthora* 362, powdery mildew 10, root and butt rot 206, 348, root rot 220, 224, 326, 348, salt tolerance 470, sapwood decay 310, 318–320, sapwood lesions 232, *Schizophyllum commune* 318, shoot blight **102**, smooth patch 520, SO$_2$ tolerance 486, *Trametes versicolor* 310, trunk decay 306, 310, 320, 350, twig dieback 36, Verticillium resistance 242, wetwood 384, 500, wilt 232, *Xylaria polymorpha* 224, *Xylella fastidiosa* **386**
Platanus ×*acerifolia* (London plane): anthracnose **102**, *Apiognomonia veneta* **102**, bacterial scorch 388, bark formation **518**, *Camillea tinctor* 204, canker **102**, 204, canker-stain 234, *Ceratocystis fimbriata* f. *platani* 234, colored canker 234, decline 234, dieback 204, 234, **492**, **516**, *Discula platani* **102**, drought stress 516, herbicide injury **476**, leaf blight **102**, shoot blight **102**, SO$_2$ tolerance 484, trunk cracks **516**, winter injury **492**, wood discoloration and decay *Xylella fastidiosa* 388
Platanus occidentalis (sycamore, eastern sycamore): air-pollution-associated growth loss 482, anthracnose 100–**102**, *Apiognomonia quercina* 100, *A. veneta* **102**, *Asteromella* **22**, bacterial scorch **388**, *Botryosphaeria rhodina* **128**, 234, 388, *Camillea tinctor* **204**, canker **102**, **128**, **142**, cankerstain **234**, *Ceratocystis fimbriata* f. *platani* **234**, colored canker **234**, decline 128, **234**, *Diaporthe arctii* 142, dieback **128**, **204**, 234, 388, *Discula platani* **102**, flood tolerance 502, ice glaze damage 506, leaf blight **102**, leaf spot 22, 44, leaf scorch **388**, *Mycosphaerella* 22, *Ophiodothella leucospila* 44, *Oxyporus latemarginatus* 344, *Phomopsis* 142, *P. scabra* **128**, **142**, *Pseudocercospora* 22, root and butt rot 344, sapwood stain **234**, shoot blight **102**, strip canker **204**, twig canker **142**, *Xenostigmina* 22, *Xylella fastidiosa* 128, **388**
Platanus orientalis: bacterial scorch 388, canker-stain 234, *Ceratocystis fimbriata* f. *platani* 234, decline 234, dieback 234, *Xylella fastidiosa* 388
Platanus wrightii (Arizona sycamore): *Inonotus dryadeus* 344, *Phymatotrichopsis* resistance 222, root rot 344
Platycladus orientalis (Oriental arborvitae, Oriental cedar): *Calonectria* 220, canker 166, *Cercospora* blight 32, *Cylindrocladium* 220, *Didymascella thujina* 60, dieback 166, leaf blight 32, 60,

Passalora sequoiae 32, root rot 220, *Seiridium* canker **192**, *Seiridium unicorne* **192**, *Valsa abietis* 166
Pleochaeta polychaeta 10
Ploioderma 48, *P. hedgcockii* 48, *P. lethale* **48**
plum, see *Carissa*; *Coccoloba*; *Oemleria*; *Prunus*
American: *P. americana*
beach: *P. maritima*
bullace: *P. domestica* subsp. *insititia*
Canadian: *P. nigra*
cherry: *P. cerasifera*
chickasaw: *P. angustifolia*
common or European: *P. domestica*
Damson: *P. domestica* subsp. *insititia*
flatwoods or hog: *P. umbellata*
flowering: *P.* ×*blireiana*, *P. triloba*
Indian: *O. cerasiformis*
Japanese: *P. salicina*
Klamath: *P. subcordata*
myrobalan: *P. cerasifera*
Natal: *Carissa macrocarpa*
pigeon-: *Coccoloba diversifolia*
sand: *P. angustifolia*
sierra: *P. subcordata*
wild: *P. americana*
wild-goose: *P. hortulana*, *P. munsoniana*
Plum pox virus 420
plum-yew, see *Cephalotaxus*
Plumeria (plumeria, frangipani, temple tree): anthracnose 114, *Botryosphaeria* 120, canker 120, dieback 120, *Frangipani mosaic virus* 422
Podocarpus: anthracnose 114, Armillaria root disease 326, *Colletotrichum* 114
Podocarpus macrophyllus (yew-plum pine): salt tolerance **470**
Podosphaera 8, **10–12**, *P. aphanis* var. *physocarpi* 10–**12**, *P. clandestina* 8–**10**, 12, *P. leucotricha* **12**, *P. mors-uvae* 12, *P. pannosa* **8–10**, 12
poinsettia, see *Euphorbia pulcherrima*
Pollaccia 86, **90–92**, *P. elegans* **90**, *P. letifera* **90**, *P. saliciperda* **92**
polymorphism in fungi 1
Polypodium (polypody): fir-fern rusts 280, *Milesina* 280
Polyporus squamosus **310**
Polyscias (aralia): *Botryosphaeria rhodina* 128, dieback 128
Polyscias fruticosa: bacterial spot 372, *Xanthomonas hortorum* pv. *hederae* 372
Polyscias guilfoylei (aralia, coffeetree, geranium-leaf aralia): *Alternaria panax* **84**, anthracnose 114, *Colletotrichum* 114, *Elsinoë mulleri* 18, gall 150, leaf spot **84**, *Nectriella pironii* 150, spot anthracnose 18
pomegranate, see *Punica granatum*
Poncirus trifoliata (citrumelo, trifoliate orange): *Alternaria citri* 84, anthracnose 114, Armillaria root

disease 326, Asiatic citrus canker 372, bacterial spot 372, *Botryosphaeria* 120, 126–128, canker 120, canker-rot 318, *Colletotrichum* 114, decline 234, dieback 120, 126–128, *Elsinoë fawcettii* 18, fruit spot 372, gall 134, *Ganoderma* 348, greasy spot 22, *Mycosphaerella citri* 22, nematode 432, *Radopholus similis* 432, root and butt rot 348, root rot 224, 326, sapwood decay 318, scab 18, *Schizophyllum commune* 318, seedling blight 84, *Sphaeropsis tumefaciens* 134, *Xanthomonas axonopodis* pv. *citri* 372, *Xylaria polymorpha* 224
pongam, see *Millettia*
poplar, see *Liriodendron*; *Populus*
balsam: *P. balsamifera*
Berlin: *P.* ×*berolinensis*
black: *P. nigra*
Carolina: *P.* ×*canadensis*
Chinese white: *P. tomentosa*
gray: *P.* ×*canescens*
Japanese: *P. maximowiczii*
Lombardy: *P. nigra* 'Italica'
'Robusta': *P.* ×'Robusta'
Simon: *P. simonii*
white: *P. alba*
yellow-: *L. tulipifera*
poplar leaf rusts 284–286
Poplar mosaic virus 404, **410**, 414
poplar potyvirus 420–422
Populus (aspen, cottonwood, poplar): adaptation to flooding 502, *Agrobacterium tumefaciens* 382, air-pollution-associated growth loss 482, *Alternaria* 84, *A. raphani* 84, anthracnose 114, *Apioplagiostoma populi* **94**, *Arabis mosaic virus* 412, 426, Armillaria root disease 326, bacterial disease 370–372, 384, bark rot 520, *Biscogniauxia mediterranea* 200, *Botryosphaeria* 120–122, 126, bronze leaf disease **94**, 420, brown cubical rot 310, brown spot **82**, *Calonectria* 220, canker **24**, 120, 142, **154**, **172**, 176, 182, 186, **194**, **198**, 208, 232, 372, canker-rot 318, *Ceratocystis fimbriata* 232, *Cerrena unicolor* 318, chlorotic spots 414, **422**, *Chondrostereum purpureum* **320**, *Ciboriania whetzelii* 68, *Cladosporium humile* 92, *Climacodon septentrionalis* 308, collar rot 362, *Colletotrichum* 114, crown gall 382, *Cryptodiaporthe populea* 154, *C. salicella* 156, *Cryptosphaeria* canker **194**, *Cryptosphaeria lignyota* **194**, *Cylindrocladium* 220, Cytospora canker **172**, *Diaporthe santonensis* 142, dieback 120–122, 126, 142, **154**, **172**, 176, 186, 200, 320, 384, *Discosporium populeum* **154**, *Discula* **94**, Dothichiza canker **154**, *Drepanopeziza* **82**, drought stress 492, *Encoelia pruinosa* **208**, *Entoleuca mammata* **198**, *Erysiphe adunca* 8, flood

tolerance 502, fluoride tolerance 486, *Fomes fomentarius* 308, *Fomitopsis pinicola* 322, *Fusarium* 186, *Ganoderma* 348–**350**, growth suppression **422**, heart rot 306, herbicide tolerance 474–476, hypertrophy 4, Hypoxylon canker **198**, *Hypoxylon mammatum* 198, ice glaze damage 506, ink spot leaf blight 68, *Inonotus andersonii* 316, *I. dryadeus* 344, *Kretzschmaria deusta* 206, *Laetiporus sulphureus* 310, leaf and twig blight **90**, **94**, leaf blight 68, **82**, leaf blister 4, leaf distortion **422**, leaf spot 24, **82–84**, **90**–92, leaf yellowing 492, *Leucostoma niveum* **172**, lightning injury 506, *Marssonina* **82**, *Melampsora* 284–**286**, *M. medusae* **286**, *M. occidentalis* **286**, mistletoe 444, mosaic 414, mycorrhizae 522, *Mycosphaerella populicola* 24, *M. populorum* 24, Nectria canker 182, *Nectria cinnabarina* 176, nematodes 432, *Neonectria galligena* 182, *Oxyporus populinus* 310, *O. latemarginatus* 344, ozone tolerance 482, *Perenniporia medulla-panis* 520, *Phellinus everhartii* 308, *P. tremulae* **306**–308, *Phyllactinia guttata* 10, *Pollaccia* **90**, *Phomopsis macrospora* 142, *Phoradendron serotinum* 442, *Phymatotrichopsis omnivora* 222, *Phymatotrichum* root rot 222, *Phytophthora* 362, phytoplasma 390, *Polyporus squamosus* 310, *Poplar mosaic virus* **410**, 414, poplar potyvirus 420, powdery mildew 8–10, *Pseudomonas syringae* 370, punk knot 306, *Raspberry ringspot virus* 426, root and butt rot 206, 344, 348–**350**, root rot 220–224, 326, 344, 348, *Rosellinia necatrix* 224, rust 284–**286**, salt tolerance 470, sapwood decay 318–**320**, *Schizophyllum commune* 318, *Sclerotium rolfsii* 226, Septoria canker **24**, *Septoria musiva* 24, SO$_2$ tolerance 484, sooty-bark canker **208**, southern blight 226, *Taphrina* 4–**6**, *Tobacco necrosis virus* 424, *Tomato black ring virus* 414, trunk decay **194**, **306**–310, 316, 322, *Valsa sordida* **168**, **172**, vein banding **422**, vein yellowing **410**, *Venturia* **90**, Verticillium wilt and resistance 242, virus 412–414, 420–**422**, 426, wetwood 384, witches'-broom 390, wood stain **194**, *Xanthomonas populi* 372, *Xiphinema* 432
Populus ×*acuminata* (lanceleaf cottonwood): canker 154, *Cryptodiaporthe populea* 154, *Melampsora occidentalis* 286, rust 286
Populus adenopoda (Chinese aspen): *Entoleuca mammata* 198, Hypoxylon canker 198
Populus alba (white poplar): *Apioplagiostoma populi* 94, Cyto-

Prunus *(continued)*

80, *Phomopsis amygdali* **140**, *Phoradendron serotinum* 442, *Phyllactinia guttata* 10, *Phymatotrichopsis omnivora* 222, Phymatotrichum root rot 222, *Phytophthora* 358, 362, Phytophthora canker 358, 362, Phytophthora root and crown rot 344, phytoplasma **390**–392, **400**, plum pocket 4–6, *Podosphaera* 10–12, powdery mildew 10–12, *Pratylenchus* 432, *Prune dwarf virus* 414–416, *Prunus necrotic ringspot virus* 416, Prunus yellow leafroll 392, *Pseudocercospora circumscissa* 20, *Pseudomonas syringae* **368**–370, root and butt rot 206, 348–350, root rot 220–224, 326, 348, *Rosellinia necatrix* 224, salt tolerance 470, sapwood decay 310, **318**–320, sapwood lesions 232, *Schizophyllum commune* 318, *Sclerotium rolfsii* 226, shoot blight **76**, SO₂ tolerance 484, sour cherry yellows 414–416, southern blight 226, *Sphaeropsis tumefaciens* 134, stem pitting 408, 432, stone fruit yellows **390**, *Taphrina* **4**–6, thread blight 256, tobamovirus 422, *Tomato ringspot virus* 408, virus 408, 412–414, 420–422, 432, *Trametes versicolor* 310, trunk decay 308–310, 322, 350, twig blight **76**, 142, *Valsa sordida* 172, Verticillium wilt **244**, virus **404**, 410, **412**–**414**, 416, 420, witches'-broom 4–6, *Xanthomonas arboricola* pv. *pruni* **374**, X-disease 390–392, **400**, *Xiphinema* 432, *Xylaria mali* 224, *Xylella fastidiosa* 386

Prunus americana (American or wild plum): charcoal root rot 218, *Macrophomina phaseolina* 218, plum pockets, *Taphrina communis* 6

Prunus angustifolia (chickasaw or sand plum): *Apiosporina morbosa* 152, bacterial spot 374, black knot 152, bladder plum 6, *Blumeriella jaapii* 80, canker 170, leaf spot 80, *Leucostoma persoonii* 170, *Taphrina communis* 6, *Xanthomonas arboricola* pv. *pruni* 374

Prunus armeniaca (apricot): alcoholic flux 384, bacterial canker 368, blossom blight **76**, brown rot **76**, canker 368–370, fruit rot 76, leaf curl 4, *Monilinia laxa* **76**, phytoplasma 394, *Pseudomonas* 368, *P. syringae* 370, shoot blight **76**, twig canker **76**, witches'-broom 394

Prunus avium (sweet or mazzard cherry): *Agrobacterium tumefaciens* **382**, bacterial canker **368**, blossom blight, brown rot **76**, bud blight 368–370, canker **368**–370, *Cherry leafroll virus* **408**–410, cherry necrotic rusty mottle virus **414**, 420, cherry raspleaf

408, crown gall **382**, decline **400**, 420, enations **408**, 412, fruit rot **76**, leaf roll 410, leaf spot 370, 420, leaf tatter **414**, *Monilinia* 76, *M. laxa* **76**, mosaic 410, necrotic rusty mottle **414**, phytoplasma **400**, *Prune dwarf virus* 416, *Prunus necrotic ringspot virus* **412**–416, *Pseudomonas syringae* pv. *syringae* **368**–370, ringspot **412**–416, rugose mosaic 416, shoot blight 76, stunting 410, *Taphrina* 4, *Tobacco ringspot virus* 410, twig canker 76, virus **408**–410, **412**–**414**, 416, 420, X-disease **400**

Prunus besseyi, see *P. pumila*

Prunus ×*blireiana* (flowering plum): bacterial spot 374, salt tolerance 470, *Xanthomonas arboricola* pv. *pruni* 374

Prunus caroliniana (cherry-laurel): bacterial spot 374, *Blumeriella jaapii* 80, canker 170, leaf spot 80, *Leucostoma persoonii* 170, *Xanthomonas arboricola* pv. *pruni* 374

Prunus cerasifera (cherry plum, myrobalan plum): *Apiosporina morbosa* 152, black knot 152, fluoride tolerance 486, salt tolerance 470, SO₂ tolerance 484

Prunus cerasus (sour or tart cherry): *Blumeriella jaapii* **80**, canker-rot **318**, *Cerrena unicolor* **318**, *Cherry green ring mottle virus* **414**, 420, cherry leaf spot **80**, decline 400, phytoplasma **400**, *Prune dwarf virus* **412**–**414**, *Prunus necrotic ringspot virus* 416, sour cherry yellows **412**–416, trunk and limb decay **318**, *Verticillium dahliae* **244**, Verticillium wilt **244**, virus 412–414, 420, X-disease **400**

Prunus davidiana (Chinese wild peach): bacterial spot 374, canker 170, *Leucostoma persoonii* 170, *Xanthomonas arboricola* pv. *pruni* 374

Prunus domestica (plum, prune): *American plum line pattern virus* 414, *Apple mosaic virus* 414–416, bacterial spot 374, blossom blight 76, bud blight 370, canker 370, *Cucumber mosaic virus* 418, decline **390**, dieback 420, fruit deformity 420, fruit rot **76**, fruit spot 374, leaf mottle 420, leaf spot 370, line pattern 416, *Monilinia fructicola* **76**, phytoplasma **390**–392, *Plum pox virus* 420, *Prune dwarf virus* 413, *Prunus necrotic ringspot virus* 416, Prunus yellow leafroll 392, *Pseudomonas syringae* 370, shoot blight **76**, stone fruit yellows 390, twig canker 374, twig dieback **76**, 374, virus 414, 418, *Xanthomonas arboricola* pv. *pruni* 374

Prunus dulcis (almond): alcoholic flux 384, *Apple mosaic virus* 416, bacterial scorch **386**–388,

Brenneria nigrifluens 378, bud blight 368–370, calico disease 416, canker 136, **232**, 358, 368–370, 378, *Ceratocystis fimbriata* **232**, decline 386, 390, dieback 136, *Fusicoccum mangiferum* 136, gummosis **232**, *Inonotus dryadeus* 344, leaf blight 36, leaf curl 4, leaf scorch 386, **470**, leaf spot 368–370, *Nattrassia mangiferae* 136, *Phytophthora syringae* 358, phytoplasma 390–392, *Prunus necrotic ringspot virus* 416, Prunus yellow leafroll 392, *Pseudomonas* 368, *P. syringae* 370, root and butt rot 344, salt toxicity **470**, *Scytalidium dimidiatum* 136, *Seimatosporium lichenicola* 36, sooty canker 136, stone fruit yellows 390, *Taphrina deformans* 4, twig canker 378, virus 416, witches'-broom 390, *Xylella fastidiosa* **386**–388

Prunus emarginata (bitter or Oregon cherry): SO₂ tolerance 484

Prunus fasciculata (desert almond): *Prunus necrotic ringspot virus* 416

Prunus fruticosa (European dwarf or European ground cherry): *Blumeriella jaapii* 80, leaf curl 6, leaf spot 80, *Taphrina wiesneri* 6, witches'-broom 6

Prunus hortulana (wild-goose plum): bacterial spot 374, bladder plum 6, *Taphrina communis* 6, *Xanthomonas arboricola* pv. *pruni* 374

Prunus ilicifolia (holly-leaf cherry): *Podosphaera clandestina* **10**, powdery mildew 10, *Sphaceloma* 18, spot anthracnose 18

Prunus laurocerasus (cherry-laurel): bacterial spot **374**, canker 164, *Endothia gyrosa* 164, leaf spot 370, **374**, *Pseudomonas syringae* **370**, shothole **370**, **374**, *Xanthomonas arboricola* pv. *pruni* 374

Prunus lusitanica (Portugal laurel): dieback **492**, winter injury **492**

Prunus mahaleb (Mahaleb cherry): *Cherry rasp leaf virus* 412, decline 224, enations 412, *Inonotus dryadeus* 344, root and butt rot 344, root rot 224, virus 412, *Xylaria mali* 224

Prunus maritima (beach plum): *Apiosporina morbosa* 152, bacterial spot 374, black knot 152, bladder plum 6, *Taphrina communis* 6, *Xanthomonas arboricola* pv. *pruni* 374

Prunus munsoniana (wild-goose plum): *Apiosporina morbosa* 152, black knot 152, canker 170, *Leucostoma persoonii* 170, plum pocket 6, *Taphrina communis* 6

Prunus necrotic ringspot virus 416–418

Prunus nigra (Canadian plum): bladder plum **6**, plum pocket **4**–6, *Taphrina communis* **4**–6, *T. pruni* 6

Prunus padus (bird or European

bird cherry): *Apiosporina morbosa* 152, black knot 152, *Blumeriella jaapii* 80, canker 170, leaf spot 80, *Leucostoma* 170

Prunus pensylvanica (pin cherry): *Eutypella parasitica* 196, leaf curl **6**, lichens **526**, phytoplasma 390, 394, *Prunus necrotic ringspot virus* 416, *Taphrina wiesneri* **6**, virus 416, witches'-broom **6**, 390, 394

Prunus persica (nectarine, peach): alcoholic flux 384, *Apple mosaic virus* 416, *Armillaria tabescens* 330, bacterial spot 374, blossom blight 76, *Botryosphaeria* 122–124, **126**–**128**, brown rot **76**, bud blight 370, burrowing nematode 432, *Calonectria* 220, 'Candidatus Phytoplasma prunorum' 390–392, canker 122, **126**–**128**, **170**, 370, *Cherry rasp leaf virus* 412, constriction canker **140**, *Cylindrocladium scoparium* 220, *Cytospora* **170**, decline 220, 390, **400**, 404, dieback 122, 126–128, 140, **170**, enations 412, freeze damage **498**, frosty mildew 20, fruit lesions 368, **374**, fruit rot 76, 358, fruit spot **374**, Fusicoccum canker **140**, gummosis **126**, **170**, *Inonotus dryadeus* 344, leaf spot 370, *Leucostoma cincta* 170, *L. persoonii* **170**, line pattern 416, *Meloidogyne* 430, *Monilinia* 76, *M. laxa* **76**, mushroom root rot 330, *Mycosphaerella prunipersicae* 20, nematodes 430–432, *Oxyporus latemarginatus* **344**, peach leaf curl 4, peach yellow leafroll 392, perennial canker **170**, *Phomopsis amygdali* **140**, *Phytophthora cactorum* 358, phytoplasma **390**–394, **400**, powdery mildew 12, *Podosphaera pannosa* 12, *Prune dwarf virus* 414, *Pseudomonas cichorii* 368, *P. syringae* 370, *Radopholus similis* 432, root knot 430, root rot 220, **344**, shoot blight 76, sour cherry yellows 414, stem pitting **404**, stone fruit yellows **390**, *Taphrina deformans* 4, *Tomato ringspot virus* **404**, 408, trunk decay 344, twig canker **374**, twig dieback 374, *Valsa canker* **170**, virus **404**, 408, 412–414, witches'-broom 394, *Xanthomonas arboricola* pv. *pruni* **374**, X-disease **400**, yellow bud mosaic 408

Prunus pumila (sand cherry): *Apiosporina morbosa* 152, bacterial spot 374, black knot 152, *Blumeriella jaapii* 80, leaf spot 80, *Xanthomonas arboricola* pv. *pruni* 374

Prunus salicina (Japanese plum): *Apiosporina morbosa* 152, black knot 152, bladder plum 6, *Blumeriella jaapii* 80, leaf spot 80, phytoplasma 400, salt tolerance 470, *Taphrina* 6, X-disease 400